Teacher's Wraparound Edition

Mathematics with Business Applications

Fourth Edition

Teacher's Wraparound Edition

Mathematics with Business Applications

Fourth Edition

Walter H. Lange
The University of Toledo

Temoleon G. Rousos
The University of Toledo

Robert D. Mason
Professor Emeritus
The University of Toledo

New York, New York Columbus, Ohio Woodland Hills, California Peoria, Illinois

A Division of The McGraw-Hill Companies

Glencoe/McGraw-Hill
A Division of The McGraw-Hill Companies

Formerly published under the title *Business Mathematics.*

Send all inquiries to:

Glencoe/McGraw-Hill
21600 Oxnard Street, Suite 500
Woodland Hills, CA 91367

ISBN 0-02-814730-8 (Student Edition)
ISBN 0-02-814731-6 (Teacher's Wraparound Edition)

Printed in the United States of America

4 5 6 7 8 9 027 04 03 02 01

Teaching Mathematics with Business Applications

Introduction

A key element in motivating students to learn mathematics is relating the content of the course to the real world. This is done by using applications that students find interesting and meaningful in their lives. The refrain, "Why do I have to learn this?" is familiar to all teachers. But few students using this textbook ask this question because the answer is so obvious. The content of *MATHEMATICS WITH BUSINESS APPLICATIONS* teaches students how to get along in the world and they know it.

At the heart of this course is the real-world application of computational skills to solve business and consumer problems. Of course, many students still have difficulty with computational skills. The text recognizes this fact and provides ample opportunities for review and reteaching.

The Basic Lesson

A textbook is a learning tool for students and a teaching tool for teachers. It must be written and organized in a way that effectively serves the purposes of both groups. The most fundamental part of a textbook is the basic lesson and how it is organized. It is through the structure of each lesson that the subject matter of the course is presented. Thus, each lesson provides a blueprint for teachers and students to follow as they work their way through the content of the course.

Each lesson in this textbook is easily identified by a number such as 1-1, meaning Unit 1, Lesson 1. Thus 5-3 means Unit 5, Lesson 3. The lessons in each unit are numbered consecutively. Each lesson also follows the same plan throughout the book. First the objective of the lesson is stated. New terms are then defined and facts important to the lesson are explained. These terms and facts are now presented in the form of an equation and used in an Example showing a completely worked out solution. The instructional part of the lesson ends at this point. However, one or two Self-Check problems follow the Example-Solution and can be used to ascertain if there are any difficulties that need to be addressed before moving on to the Problems.

The Problems provide practice on the objective of the lesson. They are organized in an increasing level of difficulty to help students succeed. It is by working the problems that students will arrive at a mastery of the objective of the lesson. At the end of each lesson are computational skill exercises to help students maintain their level of skill proficiency.

A Method of Instruction

The basic lesson suggests a method of instruction that is commonly used in many mathematics classes. New terms are introduced and explained by the teacher. Then the Example is worked at the chalkboard by the teacher. Some practice problems may now be assigned for classwork, or answers to homework problems are reviewed. Most of the time, the teacher is doing all the work, while students listen passively in their seats. There is nothing wrong with this method of instruction except that its *continuous use* may result in a loss of motivation and achievement. A motivated student is a successful student. Motivation can be sustained by actively involving students in the content of the course.

Motivation: The Key to Learning

A powerful technique for keeping students motivated and participating in class is to vary your means of instruction. Variety is a key element in creating a stimulating classroom environment. The traditional pattern of lecture followed by assignment of exercises can be modified in many ways so that your students will enter the classroom in a mood of expectancy. One of the best ways to stimulate and maintain students' interest is to vary the means of instruction by increasing their participation in class.

Varying the Methods of Instruction

One way to introduce variety in class is to use cooperative learning groups. These are small groups of four to six students that work on common tasks together. The introduction and use of small groups help to involve students actively in the content of the course.

From time to time, you may want to individualize instruction. Although this technique is often associated with remediation, it also can be used as enrichment by capitalizing on a student's interest. For instance, a student who knows a great deal about cars might want to prepare guidelines for the class on what to look for when buying a used car. You can individualize instruction by having students write reports, make surveys, do outside readings, or give presentations on topics of special interest to them.

Other variations in the pattern of classroom instruction include field trips, guest speakers, films, television, and software programs. Using a variety of methods challenges both you and your students. As a result, everyone will find the course more stimulating and, most important, students will learn faster, retain more, and have fun as they learn.

Enriching the Content of the Course

The content of *MATHEMATICS WITH BUSINESS APPLICATIONS* can be broadened or enriched by incorporating the techniques discussed in this section of the TWE. Throughout each unit of this TWE, specific suggestions are given for incorporating the ideas in the articles into your instruction. Additional notes within various lessons present more background information on business topics. Varying the means of instruction and embedding the content of the course within a larger framework of concepts and skills will reward your students and yourself with an enriching experience.

Developing Thinking Skills

Introduction

The word *think* is used all the time in schools, at home, and on the job. Teachers often ask students, “What do you think?” Parents respond to their childrens’ requests for more spending money or some new clothing by saying, “I’ll think about it.” A boss may say to an employee working on a new project, “Let’s think about that for awhile before making a decision.”

What is it that people do when they think? This is not an easy question to answer. Many books have been written on the subject, and it is an area in which intensive research is being done today at universities throughout the world.

The goal of this article is to help you, the teacher, use the content of this book to help your students understand, develop, and strengthen a few fundamental thinking skills that are intrinsic to the study of mathematics.

What Thinking Skills Should Students Develop?

The answer to this question is that students should develop deductive and inductive thinking skills in this course. In mathematics, deductive thinking involves the application of general ideas, whose truth has already been established, to the solution of specific problems. This is the type of thinking that is used to organize and formalize mathematics into a coherent body of knowledge.

Inductive thinking involves the construction or creation of new ideas from prior experience. Inductive thinking is rooted in both physical experience and abstract knowledge, and is used to generate new ideas. Both types of thinking are indispensable to learning and using mathematics in the real world.

Can Students Learn These Thinking Skills in This Course?

The answer to this question is a resounding *yes.* Deductive and inductive thinking skills can be taught in any secondary mathematics course, not only in algebra or geometry as some may think. The content of *Mathematics with Business Applications* contains essentially two major components, namely, arithmetic and applications of arithmetic to business and consumer topics. The treatment of algebraic and geometric topics in this book is limited but appropriate for the kinds of applications presented. To teach students to think deductively and inductively, you can use both the mathematical content of the course, as well as its real-world applications.

Prerequisites for Learning

Let's think for a moment about the idea of *learning* itself. What does it mean *to learn* something? What do students have to do to learn arithmetic, algebraic, or geometric skills and concepts?

First, it is important to recognize and to stress to students that learning is goal oriented. Just as learning to swim or ride a bicycle is goal oriented, so is learning to find the percent of a number. Students need to know what the goals of each lesson are in order to learn them. These goals are clearly stated in both this Teacher's Edition and in the student's textbook.

A second very important aspect of learning is that it requires the use of one's memory. Students need to memorize how to perform computational skills and the meanings of business terms. This is extremely important because learning also involves linking new information to prior knowledge. Skills and concepts that have been memorized can now be connected to the new information.

One final word about learning is appropriate before we discuss the notion of learning to think. Learning also involves organizing information; for students, this means keeping a notebook with questions, answers, and solutions to problems. Many students have difficulty learning because they are very disorganized. The organization of information physically is a basic step to the organization of information mentally.

Teaching Deductive and Inductive Thinking Skills

Students will not learn to use deductive or inductive thinking skills unless they are given explicit instruction and opportunities to use them throughout the school year. An example of deductive thinking is presented in every lesson of this textbook. In Lesson 1-1 on hourly pay, for example, the ideas of hourly rate and straight-time pay are linked together by a formula: **Straight-time Pay = Hourly Rate × Hours Worked.** This formula can be used to solve many specific problems involving the computation of straight-time pay. Students should be told that each application of a formula to specific data in a problem is a process called deductive thinking.

Other techniques also can be used to teach deductive thinking skills. They can be taught by asking students to explain and justify their answers to problems, or by asking them to explain the logic of their arguments. If a student makes an assertion about something, ask other students if they can think of examples (called counterexamples) for which the assertion is not true.

As you work through the content of this book with students, continue to reinforce the use of deductive thinking skills by using examples and problems from the real world as well as those in the book itself.

Inductive thinking is a process in which people arrive at a general understanding of a situation or problem by examining many specific examples. The use of inductive thinking can be a valuable teaching tool to help students make sense of discrete but related data. For example, in teaching some lessons, a valid approach would be to start by discussing many specific examples. Then see if students can construct the formula that would tie together the solutions of the different examples. Such an approach illustrates the use of inductive thinking skills.

Inductive skills also can be reinforced by using students' experiences outside of school as a source of data. Helping students to convert their practical experience to more general ideas would teach them to think inductively. The generalization of specific experiences to form new and broader ideas can help students form a different perspective of their experiences and thus better prepare them to solve problems in the real world.

Problem Solving and Applications

Introduction

The ability to solve problems is a crucial survival skill in today's world. Most people are employees and consumers, and as such they need to be able to solve problems on the job and in the home. Individuals without problem-solving skills generally hold the lowest paying jobs in our society, if they can find any work at all. In businesses around the world, machines perform jobs that require automatic responses. People work at jobs that require flexibility and problem-solving skills.

First and foremost, *MATHEMATICS WITH BUSINESS APPLICATIONS* is a course in problem solving. This textbook offers a comprehensive plan for strengthening students' skills in problem solving by focusing on two key elements—business applications and problem-solving strategies.

Business Applications

A quick glance through this book will confirm that problem solving is the thrust of the course. Every lesson begins with a brief expository paragraph followed by a highlighted Example. The Example consists of a representative problem with its solution presented in detail to show the sequence of thinking and the computation involved.

The problems are structured into three groups. In most cases, the first group appears in table form and provides practice in the computation required to solve the problems that follow. Students then proceed to the second group of "mini problems" that provide a bridge from computation to the full-fledged word problems that comprise the third group. This structure adheres to one of the basic tenets of learning theory—it enables students to take the familiar and apply it to a new situation.

Problem solving is a skill that is developed primarily through experience with a variety of problems and possible strategies. The structure of this book and the format of each lesson enable you to focus on the problem-solving aspects of business topics.

Problem-Solving Strategies

Students should think of problem-solving strategies as resources they can draw upon to solve complex problems. This book presents twelve strategies for their use. The strategies appear in Workshops 19-30 at the beginning of the book and range from the straightforward to the sophisticated. These strategies are helpful in solving the more complex problems that appear at the end of the problem sets.

The workshops on problem-solving strategies can be used in a variety of ways. You may want to work through them at the beginning of the year, before starting any of the units, in order to prepare students for the problem-solving nature of the course. Another approach is to teach a particular strategy as the need arises. This may be done with the entire class or with those individuals who are unfamiliar with a particular strategy. Whichever approach you take, you should probably teach the workshop on the four-step method (Workshop 19) to the entire class at the start of the school year. Most students will find this method helpful in solving all types of problems.

Problem Solving in Groups

Many of the Cooperative Learning activities in the side-column notes of this TWE present opportunities for small-group problem solving. Having students solve problems by working together in groups is important for a variety of reasons. First, students can learn from one another, and the more capable students in a group can help their less capable classmates. Also, in a small group, some students feel more free to try to solve problems than they would in a large group. A small-group approach also simulates those conditions that most of us face on the job or in the community when we work with others to solve problems. Some problems in the TWE are challenging enough to require that group members be assigned to work on various aspects of the problem. Each individual must then accept responsibility for fulfilling his or her assignment!

No textbook can ever include or simulate every type of problem that students will encounter in life. However, by the end of this course, students will have a thorough background in everyday consumer and business topics. In addition to this experience, they will be armed with the computational skills and the problem-solving tools needed to solve a wide range of problems. This combination will make them knowledgeable and tough-minded consumers.

Making Estimates

Estimation is an essential part of the problem-solving process. Students need to know when an answer is reasonable or unreasonable, and an estimate helps them to make this determination. Estimates can be used not only to check the reasonableness of an answer, but they also can be used to reveal how to solve a problem. In other words, making an estimate of the answer prior to solving a problem may reveal the solution process. Also, estimates can be helpful while actually solving a problem because they may help students to avoid making computational errors

Although estimating answers is important in solving problems, students must understand also that most situations require exact answers. For example, while grocery shopping, you might keep a running estimate of your total cost. You could not leave the store with your purchases, however, without paying the cashier the *exact* total cost.

Problem Solving: A Unifying Theme

In any mathematics course, there are many important objectives that have to be met. In this particular course, in addition to the lesson objectives, you have been encouraged to pursue other, broader objectives such as developing communication skills, critical thinking skills, calculator skills, estimation skills, and cooperative learning skills. All of these skill areas can be strengthened in students by concentrating on solving problems. This can be done by using the four-step problem-solving method. Let's examine each step of that method to see exactly what it entails.

> Step 1: **Understand.** What is the problem? What information is given? What are you asked to do?

The essential aspect of this step is based upon the *communication skills:* reading, writing, and speaking. Students must first read a problem to understand it; they need to write out the given information and the question asked.

> Step 2: **Plan.** What do you need to do to solve the problem? Choose a problem-solving strategy.

The key skills in this step are *thinking skills.* Making a plan means that students have to think about how to solve the problem. What operations are involved? Would a picture be helpful? Can the problem be solved by solving a simpler problem first? These are some questions that students can ask themselves in trying to formulate a plan .

> Step 3: **Work.** Carry out the plan. Do any necessary calculations.

Calculations can be done using paper and pencil or by *using a calculator.* It is important that students be able to do paper and pencil calculations, but being able to use a calculator also is essential in this course and in many jobs today.

> Step 4: **Answer.** Is your answer reasonable? Did you answer the question?

Estimation skills can provide the answers to these questions.

Finally, *cooperative learning skills* can be developed by having students solve problems in small groups.

Reteaching Concepts and Skills

Introduction

The mathematical content of this textbook has been taught to students in their first eight years of school. Operations with whole numbers, fractions and decimals, percent problems, some basic geometric concepts, and the solution of simple linear equations is content that is covered exhaustively in grades K–8. However, many students in high school still have serious difficulties with these topics, and their difficulties can be major stumbling blocks to mastering the objectives of the course.

The first step in providing help to those students who need it is to find out not only who they are, but also what kinds of deficiencies they have. This information will become apparent as you work with students on the Guided Practice and Independent Practice assignments in each lesson. An alternative approach, however, would be to assign the skill exercises on the Reviewing the Basics page at the end of each unit before teaching the unit. An analysis of students' performance on these skills will alert you to the kinds of difficulties some students may be having. The Basic Math Skills Workshops 1–30 are provided to help students improve their skills. They can be used individually by students, or you can use them for instructional purposes with the entire class. Specific reteaching suggestions are given in this TWE for each workshop.

Is Reteaching Really Necessary?

For some students with serious skill and conceptual deficiencies, reteaching is absolutely necessary. Reteaching these students means going back to basic ideas and using techniques and methods that teach again what the student does not know. It does not mean simply assigning students more practice exercises. If students have not already mastered the skills presented in all of the workshops in this book, then it is very likely that (1) their understanding of underlying concepts is faulty; (2) they have forgotten certain basic facts; (3) they are applying an algorithm or procedure incorrectly; or (4) they are organizing their work poorly, resulting in careless mistakes.

Any effective reteaching plan has to be based upon an understanding of the kinds of errors students make and why they make them. This means that the teacher needs to examine carefully students' errors in order to provide the right remedy.

Error Analysis

In recent years, much attention has centered on the importance of analyzing students' errors. For many math teachers, however, identifying and remediating areas of difficulty has always been a natural part of teaching mathematics. Most teachers would agree that it is impossible to check answers without examining students' work to see where they are encountering stumbling blocks.

The first step in analyzing an error is to decide whether it reflects a lack of conceptual understanding or a failure to master the procedure involved. Consider the following example, which involves incorrectly finding the elapsed time between 7:45 PM and 10:00 PM.

$$\begin{array}{rcrr} {}^{9} & & {}^{9} & {}^{10} \\ 10 & : & 0 & 0 \\ -7 & : & 4 & 5 \\ \hline 2 & : & 5 & 5 \end{array}$$

Here is a conceptual error at work. The student ignored or did not recognize the need to rename 1 hour as 60 minutes in order to subtract. This is quite different from the following computational error, in which the student has not yet mastered the algorithm for the division of fractions.

$$\frac{3}{5} \div \frac{1}{2} = \frac{3}{5} \times \frac{1}{2} = \frac{3}{10}$$

With computational errors, it is not clear from one example whether the error is due simply to carelessness or to an incomplete understanding of the algorithm. A more thorough examination of the student's work is needed to pinpoint the source of the difficulty. At times, it may be necessary to interview the student concerning the thought processes that were used, since the written work may not reveal fully the misunderstanding.

An important part of any plan for analyzing and remediating errors should be to encourage students to watch for and correct their own errors. Encourage students to become more critical by having them double-check their work on all quizzes and tests. Also, you may want to have students practice estimating answers before doing the actual computations. With such practice, students will acquire a better awareness of reasonable answers and will be able to recognize unreasonable ones. This approach is particularly helpful when using a calculator.

When Is More Practice Appropriate?

Very often, the approach to students who make computational errors, or who cannot solve problems, is to assign more practice. This approach has no validity whatsoever because, without some reteaching effort being made, these students can simply continue to make the same types of mistakes. Once the source of an error has been identified, the next steps are first to reteach the concept or skill, and then to provide practice so the student can master the skill. Files at the back of this book contain abundant materials for providing additional practice after reteaching has taken place.

The SQ3R Study Method

Introduction

The SQRRR study method can be used by students to increase their reading comprehension of the textbook. In particular, the method can be applied to the study of each lesson in the book and thereby improve the learning efficiency of students using it. The letters SQRRR have the following meanings: Survey, Question, Read, Recite, and Review. Let's look now at each of these five steps in detail as they apply to the lessons of *MATHEMATICS WITH BUSINESS APPLICATIONS.*

Step 1: Survey

As students begin each new unit, they should survey it by reading the introduction to the unit and by looking over the lesson titles. They also should look over the pages at the end of each unit, noting in particular the Reviewing the Basics and Unit Test pages.

After surveying the unit, students should then survey the first lesson in the unit. As they do this, they should read the objective at the top of the page and the formula(s) that precedes the Example. This will give them a general idea of what the lesson is about. Since each lesson in a unit has a consistent format, students can follow this survey procedure throughout the course.

Step 2: Question

The essence of this step is to turn the objective into a question. For example, in Unit 1, Lesson 1, the objective states: Compute the straight-time pay. By turning this objective into a question, students ask themselves: How can straight-time pay be computed? Then, by looking at the formula above the Example, they have the answer to their question: Straight-time pay $=$ Hourly rate $\times$ Hours Worked.

By beginning their study of a lesson with a question, students are preparing themselves to learn the material by seeking an answer to the question. Thus, they are now active participants in the search for answers rather than passive observers being given information. This is a very significant difference in how one approaches a new learning task.

Step 3: Read

At this step, students read the introductory material of the lesson and the Example and Solution that follows. The opening paragraph of the lesson introduces and defines new terms. These terms are always printed in color to make them stand out. As students read, they should keep their question in mind and relate the terms to those used in the formula. This method of reading will give more meaning to the formula and help students understand the answer to their question.

Business terms and mathematical formulas need to be read slowly and more than once. Understanding the meanings of the terms and how the formula ties them together is the foundation of the lesson. If these basic ideas are not understood, students will have difficulty with the Example and the Problems that follow.

In mathematics, there is an application step that nonscientific subjects do not have. In addition to understanding what the new terms mean and how their use in a formula answers the question of Step 2, students must now be able to apply this knowledge to specific situations. The first part of Step 3 involves learning new terms and their relationships. The second part of Step 3 involves applying the knowledge from part one to solving problems.

At the application step, students must use problem-solving skills and computational or calculator skills. Problem-solving skills include reading skills because a problem must first be read and understood before it can be solved. Then the data given in the problem are analyzed and a plan is formulated to solve the problem. Very often the plan is a decision to apply a particular formula. This type of thinking about a problem—understanding it and planning how to solve it—is at a conceptual level. The next step, solving the problem by actually carrying out the plan (using either paper and pencil calculations or a calculator) is essentially a skill step.

In the lesson, the application part of Step 3 involves reading and understanding the Example and its Solution. As students work through the Solution, their understanding of the terms and formulas is given a deeper meaning. They can now see exactly how the concepts of the lesson are applied to specific situations. In summary, Step 3 has two parts: (1) read the terms and formula and relate their meanings to the objective of the lesson in question form, and (2) apply the knowledge from (1) to solving a problem as given in the Example.

Step 4: Recite

In this step, students do a *self-check* of what they have learned. They go back and restate the objective of the lesson as a question and ask themselves if they can answer it. They should recite the question and how they answered it in their own words. This procedure helps students to remember what information was important. Using their books, students should do the Self-Check problems as a key part of the recite step.

Step 5: Review

After students do the Self-Check problems, they then go back and review the entire lesson up to the start of the problems. They should reread the terms and formula as well as the Example and its Solution; they should again work through all the steps of the Solution. In class, you can help students review by having a discussion of the material covered up to this point in the lesson.

As students move ahead in the lesson and work on the problems, they have many opportunities for review. As each problem is read and solved, students are applying what they have learned. An inability to solve the problems is a sure indicator that something is wrong. If this is the case with some students, they may need to go back to Step 1.

Summary

The SQ3R method will work if students use it. Therefore, discuss the method with students and illustrate its use with two or three lessons. Give students a list of the steps with their key words and examples of how the steps apply to the lessons. Also, continually remind students of SQ3R as you progress through the course and check to see if they are using it.

As the teacher, you also can use the method as a technique for instruction. In other words, you can work with the students in class, using the method as a vehicle for teaching a lesson. This would keep the method before students' eyes as a means of improving their study habits and increasing their learning efficiency.

Communication Skills

Introduction

In the classroom, students need to communicate with one another in order to learn effectively. The fundamental communication skills are reading, writing, speaking, and listening. As you work through the subject matter of this course with students, use every opportunity to engage students in all four communication skills. In too many mathematics classrooms, students spend too much of their time listening to the teacher. This does not have to be the situation in your classroom. In an active and stimulating learning environment, students are the major participants in the learning process. They can do this if encouraged to communicate by reading, writing, and speaking, in addition to listening. The following discussion presents some of the key aspects of the four communication skills as they relate to the content of *MATHEMATICS WITH BUSINESS APPLICATIONS.*

Reading Effectively

In order to learn the content of this course, students must read the textbook. You can encourage them to do so by making explicit reading assignments, or by having students read occasionally in class, both silently and aloud. Question students on their reading and provide assistance as needed.

To help students read this book effectively, focus their attention on the three key elements of each lesson: (l) the explanatory text at the beginning of each lesson; (2) the Examples; and (3) the Problems. Stress to students that they should read slowly to comprehend the content of these three parts of a lesson. To check for comprehension of the new terms introduced at the beginning of each lesson, ask students to explain their meanings in their own words. Students also should be encouraged to read the Examples with paper and pencil in hand so that the Solutions can be worked through. If a student has trouble understanding an Example and its Solution, a careful rereading of the problem may help.

Special emphasis should be given to the reading of directions. A student may be tempted to let his or her eyes fly down the page in search of operation signs and numbers, completely bypassing the directions. It is difficult to arrive at a correct answer without knowing the question. Directions may call for estimates rather than exact answers, or for answers rounded to a given decimal place.

Word problems can be intimidating to many students. Again, the key is careful and slow reading, beginning with the directions. Paper and pencil are essential also. Writing down a series of questions and answering them can help students organize the material: what information does the problem give; what kind of answer is expected; will drawing a diagram or making a table help? A major step in solving problems is to read and understand the problem. Stress the importance of reading for understanding throughout the course.

Writing Mathematics

In this course, writing mathematics means solving problems and doing computations. It also can mean writing brief reports on business topics you assign for research projects. In solving problems and doing computations, legibility and neatness are essential. Mistakes are often the result of the careless placement of letters, numbers, or symbols. It is impossible for students to find the correct solution to a problem if the computation is illegible.

Help students to realize the importance of correct alignment in computations. In addition and subtraction exercises, the correct place-value alignment of digits is essential, both in copying an exercise and in writing the answer. In multiplication, the focus is on the correct alignment of the partial products. In division, correct alignment of the first digit in the quotient is important. Also, stress the importance of the alignment of decimals when applicable.

Writing brief research reports gives student an opportunity to formulate, organize, internalize, evaluate, and share their work and ideas. Expect to find a wide range of responses from your students in their written reports, and be prepared to take the time that will be necessary for students to improve their writing skills. Reading students' work will provide you with a valuable method for evaluating their progress.

Speaking in Class

Outside of school, the most commonly used communication skill is speaking. Students speak to one another all the time, yet in mathematics class, they tend to avoid sharing their work or ideas. The reason for this, generally, is that they are fearful of making a mistake. This pattern must be broken if students are to enjoy mathematics and learn it.

As the teacher, you can set the tone of the class and control the means of communication with students. Once students feel comfortable that they can make an error and not be ernbarrassed in front of their peers, they will begin to participate more frequently. In so doing, they can enhance their own education and contribute to their classmates' knowledge as well.

The notes in the side-columns of this TWE encourage students to participate in class. Many specific suggestions are given to engage students in class discussions so that they can bring their own experiences into the course. Also, through the use of cooperative learning groups, students have many opportunities to speak with each other about their work.

Point out to students the importance of verbal communications in the business world, and tell them that they can learn to improve their speaking skills in this course. In each class, ask students to explain the meanings of new terms and formulas, describe how they solved a problem, or volunteer personal experiences related to the topic being studied. Call upon different students to take part so the same people do not dominate the discussions. When students ask questions, have other students try to give the answers.

It is not necessary for the teacher to be the center of every discussion in order that a lesson be a worthwhile learning experience for students. If a positive and encouraging climate exists in the classroom, some student leaders might emerge to help other students. One very important outcome in a classroom in which there is a great deal of student participation is the opportunity for the teacher to make informal assessments of what the students have learned.

Is Listening a Skill?

The answer to this question is a resounding *yes!* The best learners are very often the best listeners. They listen attentively and take notes. They think about what they are hearing and ask questions. They pay attention to a speaker, teacher or student, and do not interrupt. Very few students are listeners in the ways just described. But you can help students to improve their listening skills. First, just talking about listening as a communication skill will make students aware of this fact. Ask students to suggest ways to improve listening skills. You might make a list of characteristics that describe good listeners and post it on the bulletin board. Since most of the time a student spends in school is spent listening to others, any improvements made in listening skills can benefit students greatly.

Using Cooperative Learning Groups

Introduction

Life in the United States today is fast moving and ever changing. High school students are exposed to many activities in their lives. They go to school, have part-time jobs, take part in social and athletic activities, and spend time with family and friends. Students' lives seem to be full of action, except perhaps in the classroom where they are often passive observers, watching and listening to a teacher lecture in front of the class. Students need to be more involved with one another in class to get the most out of being in school.

There are many ways that instruction can be varied to increase student participation. One very effective way is the use of cooperative learning groups.

What is a Cooperative Learning Group?

Cooperative learning groups have been around for a long time. Have you ever taught a small class with perhaps six to ten students in it? Wasn't it enjoyable? Why? There are probably many reasons *why*. You, as the teacher, could relate better to each student—you got to know that person as an individual. The students themselves generally are more relaxed. They talk to one another and are not afraid to ask questions or volunteer answers. You probably asked more students to explain their work to the class. In a small group, students and teacher can work better together to enhance learning.

Most classes, however, have many more than six to ten students and thus teachers tend to rely on the traditional pattern of lecturing. This approach is effective from the point of view of covering the objectives of the course, and managing the behavior of students; however, over time, it is boring to students and places them in a passive role.

A cooperative learning group is a small group of four to six students working together within a traditional classroom setting that may contain 20 to 30 students. The class is not organized every day in this manner, but the approach is used often enough to keep students motivated and involved with the content of the course.

Are Small Groups Really Effective?

Research studies in psychology and education have verified that students learn faster and retain more when they are actively involved in the learning process. One method of accomplishing this goal is the use of cooperative learning groups.

Let's look now, very specifically, at the purposes small groups serve. The use of cooperative learning groups accomplishes the following purposes.

1. Students take a more active role by working in small groups. Many students will participate by asking questions or making suggestion in a small group that they would not do in a larger class.
2. Participation in a small group improves students' attitudes and motivation toward learning the subject matter of the course.
3. In a small group, capable students have an opportunity to share their ideas with less capable students. This can improve the overall competency level of the class.
4. Students learn valuable social and problem-solving skills by working together with others on common problems. These kinds of skills are useful in many real-world occupations.
5. Students can improve their communication skills by more actively taking part in the learning process.
6. Students can be involved in the solution of problems that they might not have been able to solve on their own.

Techniques for Organizing Groups

The following techniques have proved themselves useful in working with small groups within a traditional classroom environment.

1. Explain to students that you will be organizing them into small groups to work on certain topics together throughout the year.
2. Decide on the size of the groups for your class, using four students in a group as the ideal.
3. Plan how to arrange the room physically so students can work in small groups.
4. Assign students to groups to maximize their effectiveness. Try to balance each group in terms of the capabilities of the students in it. Students learn well from one another and most often groups should be composed of both capable and less capable students.
5. Change the composition of the groups occasionally so different students have an opportunity to work together.
6. Explain the tasks that the groups will be doing.
7. Require groups to show evidence of their progress as they plan and execute their work—for example, a solution to a problem, a list of references to be obtained, or a graphic display of their findings.

What is the Teacher's Role?

Clearly, your role as the teacher changes in terms of the methods used for instruction. Instead of lecturing and working with one large group, you now need to interact with and guide the work of many groups. Physically, you should move about the room, listening to the groups as they work, and giving guidance, assistance or encouragement when needed. A good technique to use with groups is to ask thought-provoking questions rather than giving answers to problems. Try to get students to think about and discuss what they are doing and thus arrive at their own answers.

Another valuable technique with small groups is to use the team approach. Prepare the groups to think of themselves as teams. Each member of a team has a responsibility to contribute to the team's work, and the teams can compete against one another on certain assignments to see which one reaches its goals first, or does the best job. You and the students can have some fun using the team approach while accomplishing your objectives in a positive and constructive learning environment.

When Should Groups Be Used and What Do They Do?

Specific suggestions for using cooperative learning groups are given at least once in each unit of this Teacher's Wraparound Edition (TWE). Certainly, more frequent use of small groups is entirely your decision. The work of the groups can be focused around the major topics discussed in the articles of this book. For example, groups can work together to *solve problems* or to enhance *communication skills* by discussing a particular topic or by writing a brief report. Cooperative learning groups also can be used for *reteaching* purposes by using student tutors, or for *alternative assesssment* purposes. As students work together in groups, the teacher has a perfect opportunity to interact by asking questions or viewing their written work.

Group assignments can be given for using *calculators,* on working through the *simulations,* and on *job opportunities* in the community and the kinds of skills they require. All of these activities can make this course more stimulating and rewarding for your students.

Multicultural Education

Introduction

The teaching of topics in business mathematics provides a wonderful opportunity to give students a multicultural perspective of business. At the most basic level, business is conducted by individuals in their personal lives. They have jobs and earn income. Money is deposited in checking and savings accounts and then spent for consumer goods, using cash or credit cards. People borrow money to buy automobiles or homes, which they insure against loss. Savings can be invested in certificates of deposits, stocks, bonds, and other financial instruments. All of these activities are done by people of various cultures and ethnic groups.

Business is also conducted in a formal and organized way by groups of people working in small businesses of their own, operating at the local level, or by corporations that operate at a state or national level. Many large corporations are multinational; that is, they not only sell their products around the world, but they also manufacture their products in foreign countries. Thus, business today is truly a multicultural activity, and it is the purpose of this article to offer some ideas and suggestions for incorporating a multicultural perspective into this course.

Local Resources

The first place to look to broaden students' perspectives about the contributions made to business by peoples of different cultural and ethnic backgrounds is within your own community. Your students may know successful business people who they can identify and explain what they do. Perhaps some of these people would be willing to visit your classroom to discuss their accomplishments in business. Founders of small businesses generally have many stories to tell about the effort and desire that goes into making a business successful. They can also discuss other important aspects of their businesses such as product development, quality control, customer relations, pricing, inventory, budgets, personnel, and so on. All of this information can help to round out students' understanding of many of the topics presented in this course.

In addition to local business people, other resources exist at the community level. Newspapers are an excellent source of information and should be made an integral part of your instruction. News stories and advertising can be used effectively to illustrate multicultural aspects of business as well as to connect the subject matter of the textbook to the real world. Magazines can play the same role.

Students need to be made aware of the fact that running a small local business is a challenging job, one that requires many skills and many hours of work. Small business people make a substantial contribution to the economic well-being of their communities as well as providing essential services. Their involvement in your course might even encourage some students to be young entrepreneurs themselves.

Beyond the Community

Business today is conducted on a worldwide level by both large and small corporations. One of the most effective ways to give students a multicultural perspective of business is to make them aware of the international aspect of trade. During the past 40 years, the United States has gradually shifted from a manufacturing economy to a more service-oriented economy. Many products that were once made in U.S. factories are now made in foreign nations and imported to the United States for sale. One of the best examples is the automobile. Japanese and European cars are now a common sight on our roads. Students can learn a great deal about business by exploring why this is so. They can also learn something about the people and countries in which imported cars are made. This can be done by having students work on special projects related to automobiles. Some basic questions to consider are: From what countries does the U.S. import cars? Are the manufacturing procedures and techniques the same or different in foreign automobile plants than in U.S. plants? How are cars shipped to the U.S.? How do pay rates in foreign auto plants compare to those in the U.S.? These are only a few questions that students can consider. You might wish to have students explore some questions of their own regarding the manufacturing of imported cars.

Automobiles are only one product that can be used as a means to explore the multicultural aspects of business. Engage students in a discussion of other products that are imported into the United States. Some that they might mention are clothing, watches, TV sets, oil to make gasoline, art and craft objects, food products, and so on. Use students' suggestions to have them learn about the countries from which the products come and also what products these countries buy from U.S. companies.

Techniques for Introducing Multicultural Topics

In each unit of this TWE, there is a multicultural note that can be used to introduce multicultural topics into the course. In addition to using these notes, you might wish to take a more in-depth approach to multicultural education. One way to do this would be to assign each student a different country in the world to research. Students can deliver written or oral reports to the class about the country and its business activities. Be sure to select a representative set of countries from each continent, and ask students to relate part of their reports, if possible, to some aspect of what they have learned in this course. This type of activity can start early in the school year and be continued on an on-going basis throughout the year.

Another technique would be to look at the different ethnic and racial groups that make up the population of the United States. With a little research, students can report on the business, cultural, political, or athletic contributions these people have made to American society during this century. Reports also can be prepared on famous individuals and their achievements.

As mentioned above, the use of printed materials, such as magazines and newspapers, can also help to bring a multicultural perspective into the classroom. By watching for opportunities to share multicultural topics that appear in the press with students, you can broaden their horizons and understanding of the contributions made by various groups to our culture.

Students in school today will be living their adult lives in the next century. The world today is a much smaller place than it was only 50 years ago, and it is growing still smaller. Economically, all nations are dependent to some degree upon one another. Cooperation, understanding, and mutual respect are essential for all people living in an interdependent and internationalized world. The subject matter of this course can be used as a vehicle to start students thinking about these ideas.

Alternative Assessment

Introduction

Evaluation of students' performance is fundamental to the teaching and learning process. Typically, this is done by using written tests. In this textbook, Unit Tests are provided in order to evaluate students' mastery of the concepts, skills, and problems presented in the unit. However, written tests do not evaluate students' communication skills or critical thinking skills. Nor do they always evaluate students' problem-solving abilities or their understanding of the mathematical and business ideas presented in the course. Other means of assessment should be used to form a complete picture of a student's performance.

Assessment Methods

In addition to written tests, the following methods also can be employed for assessment purposes.

1. Questioning students in class
2. Asking students to show their work at the chalkboard
3. Class discussions and oral presentations
4. Analyzing students' questions in class
5. Homework
6. Group work
7. Take-home tests
8. Written reports
9. Special business projects
10. Student interviews
11. Oral tests

Two Types of Assessments

Essentially there are two types of assessments that a teacher makes when teaching a mathematics course, namely, formal and informal assessments. Formal assessments, such as written tests, are used to grade students. Informal assessments, such as asking questions in class, usually are made during the instruction process and are used by the teacher to adjust the instruction to the class. Through questioning, a teacher may decide to review a particular topic, discuss a certain problem, or go over another example. With informal assessments, the unit of assessment is generally the class as a whole. Formal assessments are directed at individual students.

Class Assessment

The following methods can be grouped under class assessment: (1) questioning students in class; (2) asking students to show their work at the chalkboard; (3) class discussions;

(4) analyzing students' questions in class; and (5) group work. All of these methods are used by mathematics teachers to modify and pace their instruction.

As students take an active role in class, the teacher is in a position to assess and evaluate their performance. Some students like to participate in class. They answer questions and ask them. Their arms always seem to be raised in an attempt to contribute. Teachers quickly recognize these students as their best achievers since they often do well on the written tests also.

Some students may find participation in small learning groups more comfortable than in a large class, and their demonstrated abilities in these groups also should be taken into account when formal evaluations are made. Thus, class assessments can and should have a bearing on evaluating a student's overall performance. In particular, they can be used to evaluate students' verbal and thinking skills. Problem-solving skills can be evaluated by having students explain how they solved a problem. On a written test, a student may get a wrong answer to a problem, and the teacher may have no way of knowing why the answer is wrong. Has a computational error been made? Or perhaps the student is having difficulty reading or using the most appropriate problem-solving strategy. These kinds of difficulties show up when students have to explain how to solve a problem.

Evaluating Individual Students

To achieve the broader goals of this course, a variety of teaching methods and assessment techniques should be used. Specific suggestions in the side-column notes of this book provide many interesting and worthwhile activities that students can pursue. There are also critical thinking questions and problem-solving activities that can be used to gauge students' abilities to reason and to solve challenging problems.

Using different methods of assessing students' performance is necessary in order to make an accurate and fair judgment of what students know. Homework papers need to be reviewed and feedback provided to the student. On occasion, you might give students take-home tests. These tests can be used to evaluate students' abilities to solve challenging problems or to provide essay type answers to questions. Many students like these kinds of tests because they feel less pressure to perform quickly. They can work at their own pace and, in so doing, they may reveal understandings and skills that a classroom test would not.

Two infrequently used but valuable assessment techniques are oral tests and interviews. It can be very enlightening to talk with students about their work. This personal touch to evaluation may have a powerful motivational effect on students. It is an opportunity for the teacher to express directly to a student his or her strengths and weaknesses in a non-threatening way, and to provide encouragement.

Using Calculators

Introduction

Calculators are used extensively by people today in their homes, when working on personal business tasks, and on the job, when doing business calculations. Students need to become proficient in using calculators if they are to develop their problem-solving skills and make a successful transition from students to productive members of society.

Although this course makes ample provision for the reteaching and maintenance of computations skills, the use of calculators in the classroom and at home is recommended.

Calculators and Estimation Skills

When using calculators, students can get incorrect answers by using the wrong sequence of keys or by inadvertently pressing the wrong key. This possibility makes it important for students to have an idea of what the answer to a problem may be by making an estimate first.

In this textbook, estimation is viewed as a natural part of the problem-solving process. Students are encouraged to use a variety of estimation techniques to simplify computations or to judge the appropriateness of a numerical result. This, in turn, makes finding the solution to a problem easier and faster. Also, students should be encouraged to estimate before and after solving a problem to make sure the answer is reasonable. This use of estimation is always appropriate when using a calculator.

Problems Associated With Using Calculators in Class

The first problem associated with using calculators in the classroom is that not all students may own one. If only a few students in your class own a calculator or have access to one at home, then you may decide not to allow their use in class at all. Clearly, when working problems in class, those students using a calculator would have an unfair advantage over those who do not have one. However, if most students have a calculator, then you probably should allow their use in class. Students with calculators may be willing to share them with students who do not have one. Students sharing calculators can work in pairs or in small groups of three or four.

Ideally, each student in your class would have a calculator so you could proceed to integrate their use into the course work in a meaningful way.

Another problem with using calculators in the classroom is that even though students may own one, they may not know how to use it. Therefore, instruction in using a calculator may be necessary for some students. The following section of this article gives a suggested sequence of instruction for familiarizing students with their calculators. Since there are many different kinds of calculators there are also differences in the way they work. Encourage students to use their own instruction booklets to learn how to operate their calculators.

Providing Instruction in Using a Calculator

In this brief article, it would be impossible to go into detail on how to teach students to use a calculator. What is possible, however, is to outline a sequence of objectives necessary to achieve proficiency with a calculator in this course. Using a calculator, students should be able to:

1. Enter a number and clear it.
2. Clear an entire computational problem.
3. Add and subtract whole numbers.
4. Multiply and divide whole numbers.
5. Add and subtract decimals.
6. Multiply and divide decimals.
7. Change a fraction to a decimal by dividing the numerator by the denominator.
8. Multiply and divide fractions and mixed numbers by first converting them to decimals.
9. Use the percent key to find the percent of a number.
10. Use the memory keys.

One approach to implementing the objectives would be to devote one or two class periods to using a calculator at the beginning of the school year. The advantage of this approach is that you can introduce the objectives in the order listed, that is, from simple objectives to more advanced ones.

Another approach is to teach the lessons in the textbook and provide instruction on using the calculator for each lesson. If students are familiar with calculators and can use them somewhat effectively, this approach might be the more desirable one. As you progress through the textbook and students become accustomed to using calculators, you should still check their computational skills occasionally.

Calculators and Problem Solving

A major goal of this textbook is to make your students good problem solvers. This means that they must focus their attention on solving problems and not get bogged down in doing numerical computations. In many lessons of the book, the subject matter is such that it requires detailed calculations to arrive at the solutions to problems. To do these calculations using paper and pencil implies that students would have time to work on only a few problems. Thus, not using a calculator would shift the focus of the course to doing computations rather than to solving problems. This is why the use of a calculator is so important. It allows students to solve problems more quickly and, in so doing provides much more time for them to work on additional problems. And good problem solvers are people who have worked on and solved many problems.

Assignment Guide for
Mathematics with Business Applications

Unit 1 Gross Income

LESSON	PROGRAM Basic	PROGRAM Average
1-1	GP: 1–8 IP: 9–15 odd HW: 9–18, 20–36	GP: 1–8 IP: 9–15 odd HW: 10–16 even, 17–19, 20–36 even
1-2	GP: 1–8 IP: 10–14 even HW: 9–14, 16–25	GP: 1–8 IP: 10–14 even HW: 9–15 odd, 16–24 even
1-3	GP: 1–8 IP: 9–15 HW: 9–19, 22–36	GP: 1–8 IP: 9–15 HW: 16–21, 22–36 even
1-4	GP: 1–10 IP: 11–17 odd HW: 11–18, 22–34	GP: 1–10 IP: 11–17 odd HW: 12–16 even, 18–21, 22–34 even
1-5	GP: 1–8 IP: 9–15 odd HW: 9–17, 20–30	GP: 1–8 IP: 9–15 odd HW: 10–16 even, 17–19, 20–30 even
1-6	GP: 1–8, 12–15 IP: 9–11, 15–17, 19–23 odd HW: 9–11, 15–17, 18–26, 29–45 odd	GP: 1–8, 12–15 IP: 9–11, 15–17, 19–23 odd HW: 18–24 even, 25–28, 30–46 even
1-7	GP: 1–6 IP: 7–11 odd HW: 7–12, 14–25	GP: 1–6 IP: 7–11 odd HW: 8–12 even, 13, 14–24 even
	Reviewing the Basics	Reviewing the Basics
	Unit Test	Unit Test

Unit 2 Net Income

LESSON	PROGRAM Basic	PROGRAM Average
2-1	GP: 1–8 IP: 9–15 odd HW: 9–19, 21–33	GP: 1–8 IP: 9–15 odd HW: 10–16 even, 17–20, 21–33 odd
2-2	GP: 1–6 IP: 7–11 odd HW: 7–13, 15–22	GP: 1–6 IP: 7–11 odd HW: 8–12 even, 13, 14, 15–21 odd
2-3	GP: 1–3, 5 IP: 4, 6, 7 HW: 4–9, 11–21	GP: 1–3, 5 IP 4, 6, 7 HW: 8–10, 11–21
2-4	GP: 1–7 IP: 8, 9, 11 HW: 8–12, 16–27	GP: 1–7 IP: 8, 9, 11 HW: 10, 12–15, 17–27 odd
2-5	GP: 1–5 IP: 6–8 HW: 6–10, 12–19	GP: 1–5 IP: 6–8 HW: 9–11, 13–19 odd
2-6	GP: 1–3 IP: 4–6 HW: 4–8, 10–21	GP: 1–3 IP: 4–6 HW: 7–9, 11–21 odd
	Reviewing the Basics	Reviewing the Basics
	Unit Test	Unit Test

KEY: **GP = Guided Practice** **IP = Independent Practice** **HW = Homework**

Assignment Guide for Mathematics with Business Applications

Unit 3 Checking Accounts

LESSON	PROGRAM Basic	PROGRAM Average
3-1	GP: 1–8 IP: 9, 10, 13, 15 HW: 9–17, 19–28	GP: 1–8 IP: 9, 10, 13, 15 HW: 11, 12, 14, 16–18, 20–28 even
3-2	GP: 1, 2, 3–13 odd IP: 4–14 even, 15, 16 HW: 4–14 even, 15–18, 21–29	GP: 1, 2, 3–13 odd IP: 4–14 even, 15, 16 HW: 17–20, 21–29 odd
3-3	GP: 1–8 IP: 9, 11 HW: 9–14, 16–27	GP: 1–8 IP: 9, 11 HW: 10, 12–15, 16–26 even
3-4	GP: 1–8 IP: 9, 11 HW: 9–12, 14–25	GP: 1–8 IP: 9, 11 HW: 10, 12, 13, 14–22 even
3-5	GP: 1–6 IP: 7, 9 HW: 7–9, 11–19	GP: 1–6 IP: 7, 9 HW: 8, 10, 11–19 odd
	Reviewing the Basics	Reviewing the Basics
	Unit Test	Unit Test

Unit 4 Savings Accounts

LESSON	PROGRAM Basic	PROGRAM Average
4-1	GP: 1–7 IP: 8, 11, 15 HW: 8–16, 18–27	GP: 1–7 IP: 8, 11, 15 HW: 9, 10, 12–14, 17, 18–26 even
4-2	GP: 1, 2, 3–9 odd IP: 4–10 even, 11, 13 HW: 4–15, 17–25	GP: 1, 2, 3–9 odd IP: 4–10 even, 11, 13 HW: 12, 14–16, 17–25 odd
4-3	GP: 1–7 IP: 8 HW: 8, 9, 11–20	GP: 1–7 IP: 8 HW: 9, 10, 11–19 odd
4-4	GP: 1–5 IP: 6–10 even HW: 6–12, 14–24	GP: 1–5 IP: 6–10 even HW: 7–11 odd, 12, 13, 14–24 even
4-5	GP: 1–7 IP: 9–13 odd HW: 8–15, 17–28	GP: 1–7 IP: 9–13 odd HW: 8–14 even, 15, 16, 17–27 odd
4-6	GP: 1–6 IP: 8–12 even HW: 7–13, 15–29	GP: 1–6 IP: 8–12 even HW: 7–13 odd, 14, 15–29 odd
4-7	GP: 1–7 IP: 9, 11, 14 HW: 8–14, 17–28	GP: 1–7 IP: 9, 11, 14 HW: 8–14, 17–27 odd
4-8	GP: 1–7 IP: 9–13 odd HW: 8–13, 15–26	GP: 1–7 IP: 9–13 odd HW: 8–14 even, 15–25 odd
	Reviewing the Basics	Reviewing the Basics
	Unit Test	Unit Test

Unit 5 Cash Purchases

LESSON	PROGRAM Basic	Average
5-1	GP: 1–8 IP: 9, 10, 12 HW: 9–16, 18–37	GP: 1–8 IP: 9, 10, 12 HW: 11, 13–17, 18–36 even
5-2	GP: 1–4 IP: 5–13 odd HW: 5–17, 20–33	GP: 1–4 IP: 5–13 odd HW: 6–14 even, 15–18, 21–33 odd
5-3	GP: 1–6 IP: 8–12 even HW: 8–16, 18–33	GP: 1–6 IP: 8–12 even HW: 7–13 odd, 14–17, 19–33 odd
5-4	GP: 1–5 IP: 7–11 odd HW: 6–15, 17–22	GP: 1–5 IP: 7–11 odd HW: 6–12 even, 13–16, 17–23 odd
5-5	GP: 1–6 IP: 7–11 HW: 7–15, 17–21	GP: 1–6 IP: 7–11 HW: 12–16, 17–21 odd
5-6	GP: 1–9 IP: 10–16 even HW: 10–21, 24–31	GP: 1–9 IP: 10–16 even HW: 11–17 odd, 19–23, 25–31 odd
5-7	GP: 1–7 IP: 8–10 HW: 8–12, 15–23	GP: 1–7 IP: 8–10 HW: 11–14, 16–22 even
	Reviewing the Basics	Reviewing the Basics
	Unit Test	Unit Test
	Cumulative Review	Cumulative Review
	Cumulative Review Test	Cumulative Review Test

Unit 6 Charge Accounts and Credit Cards

LESSON	PROGRAM Basic	Average
6-1	GP: 1–9 IP: 10–13, 14–18 even HW: 10–21, 24, 25–40	GP: 1–9 IP: 10–13, 14–18 even HW: 15–19 odd, 20–24, 25–39 odd
6-2	GP: 1–6 IP: 7, 8, 9, 11 HW: 7–14, 16–25	GP: 1–6 IP: 7, 8, 9, 11 HW: 10, 12–15, 16–24 even
6-3	GP: 1–8 IP: 9–14 HW: 9–17, 19–32	GP: 1–8 IP: 9–14 HW: 15–18, 19–31 odd
6-4	GP: 1–7, 11 IP: 8–10, 13, 15 HW: 8–17, 19–33	GP: 1–7, 11 IP: 8–10, 13, 15 HW: 11, 12, 14, 16–18, 20–32 even
6-5	GP: 1–12 IP: 13, 16 HW: 13–18, 20–35	GP: 1–12 IP: 13, 16 HW: 14, 15, 17–19, 21–35 odd
6-6	GP: 1–11 IP: 12, 13 HW: 12–14, 15–30	GP: 1–11 IP: 12,13 HW: 14, 16–30 even
	Reviewing the Basics	Reviewing the Basics
	Unit Test	Unit Test

Assignment Guide for Mathematics with Business Applications

Unit 7 Loans

LESSON	PROGRAM Basic	PROGRAM Average
7-1	GP: 1–8 IP: 9–11 HW: 9–14, 16–30	GP: 1–8 IP: 9–11 HW: 12–15, 16–30 even
7-2	GP: 1–6 IP: 7–10, 13 HW: 7–16, 19–31	GP: 1–6 IP: 7–10, 13 HW: 11, 12, 14–18, 19–31 odd
7-3	GP: 1–7 IP: 8–10, 13 HW: 8–18, 21–30	GP: 1–7 IP: 8–10, 13 HW: 11, 12, 14–20, 22–30 even
7-4	GP: 1–10 IP: 11–15 odd HW: 11–18, 27–35	GP: 1–10 IP: 11–15 odd HW: 12–16 even, 17–26, 28–34 even
7-5	GP: 1–7 IP: 8–10 HW: 8–13, 16–24	GP: 1–7 IP: 8–10 HW: 11–15, 17–23 odd
7-6	GP: 1–7 IP: 8–14 even HW: 8–22 even, 25–40	GP: 1–7 IP: 8–14 even HW: 9–23 odd, 24, 26–40 even
7-7	GP: 1–8 IP: 9–11 HW: 9–14, 16–23	GP: 1–8 IP: 9–11 HW: 12–15, 17–23 odd
	Reviewing the Basics	Reviewing the Basics
	Unit Test	Unit Test

Unit 8 Automobile Transportation

LESSON	PROGRAM Basic	PROGRAM Average
8-1	GP: 1–5 IP: 6–9 HW: 6–10, 12–17	GP: 1–5 IP: 6–9 HW: 10, 11, 13–17 odd
8-2	GP: 1–5 IP: 6–10 even HW: 6–11, 13–23	GP: 1–5 IP: 6–10 even HW: 7–11 odd, 12, 14–22 even
8-3	GP: 1–4 IP: 5–7 HW: 5–9, 11–22	GP: 1–4 IP: 5–7 HW: 8–10, 11–21 odd
8-4	GP: 1–4 IP: 5–9 odd, 10 HW: 5–10, 11–17 odd, 20–34	GP: 1–4 IP: 5–9 odd, 10 HW: 4–10 even, 11–17 odd, 18, 19, 21–33 odd
8-5	GP: 1–8 IP: 9–11, 16 HW: 9–11, 12–18 even, 19, 20, 23–38	GP: 1–8 IP: 9–11, 16 HW: 12–15, 17–22, 24–38 even
8-6	GP: 1–8 IP: 9–13 odd HW: 9–16, 18–21	GP: 1–8 IP: 9–13 odd HW: 10–14 even, 15–17, 18–21
8-7	GP: 1–4, 7–9, 12 IP: 5, 6, 10–15 HW: 5, 6, 10–22, 25–34	GP: 1–4, 7–9, 12 IP: 5, 6, 10–15 HW: 16–24, 26–34 even
	Reviewing the Basics	Reviewing the Basics
	Unit Test	Unit Test

Unit 9 Housing Costs

LESSON	PROGRAM Basic	Average
9-1	GP: 1–6 IP: 7–11 odd HW: 7–14, 16–26	GP: 1–6 IP: 7–11 odd HW: 8–12 even, 13–15, 16–26 even
9-2	GP: 1–4 IP: 5, 6, 8, 9 HW: 5–12, 14–20	GP: 1–4 IP: 5, 6, 8, 9 HW: 7, 10–13, 14–20 even
9-3	GP: 1–3 IP: 4–8 even HW: 4–8, 10–21	GP: 1–3 IP: 4–8 even HW: 5–9 odd, 10–20 even
9-4	GP: 1–9 IP: 10–14 even, 17 HW: 10–19, 22–27	GP: 1–9 IP: 10–14 even, 17 HW: 11–15 odd, 16, 18–21, 22–26 even
9-5	GP: 1–3, 5, 7 IP: 4–8 even HW: 4–12, 14–25	GP: 1–3, 5, 7 IP: 4–8 even HW: 9–13, 14–24 even
9-6	GP: 1–3, 5, 7 IP: 4–8 even HW: 4–10, 12–17	GP: 1–3, 5, 7 IP: 4–8 even HW: 5–9 odd, 10, 11, 13–17 odd
9-7	GP: 1, 2, 4 IP: 3, 5, 6 HW: 3, 5–8, 10–18	GP: 1, 2, 4 IP: 3, 5, 6 HW: 7–9, 10–18 even
9-8	GP: 1–9 IP: 10–14 even HW: 10–16, 18–36 even	GP: 1–9 IP: 10–14 even HW: 11–15 odd, 16, 17, 18–36 even
	Reviewing the Basics	Reviewing the Basics
	Unit Test	Unit Test

Unit 10 Insurance and Investments

LESSON	PROGRAM Basic	Average
10-1	GP: 1–5 IP: 6–8, 10 HW: 6–11, 13–20	GP: 1–5 IP: 6–8, 10 HW: 9, 11, 12, 14–20 even
10-2	GP: 1–5 IP: 6–8 HW: 6–9, 11–19	GP: 1–5 IP: 6–8 HW: 9, 10, 12–18 even
10-3	GP: 1–4 IP: 5–8 HW: 5–10, 12–29	GP: 1–4 IP: 5–8 HW: 9–11, 13–29 odd
10-4	GP: 1–4 IP: 5, 7 HW: 5–8, 10–25	GP: 1–4 IP: 5, 7 HW: 5–8, 11–25 odd
10-5	GP: 1–5 IP: 6–8 HW: 6–10, 12–25	GP: 1–5 IP: 6–8 HW: 9–11, 12–24 even
10-6	GP: 1–7 IP: 8–10 HW: 8–12, 14–22	GP: 1–7 IP: 8–10 HW: 11–13, 14–22 even
10-7	GP: 1–6 IP: 7–11 odd HW: 7–12, 14–22	GP: 1–6 IP: 7–11 odd HW: 8–12 even, 13, 14–22 even
10-8	GP: 1–6, 8 IP: 7–11 odd HW: 7, 9–13, 15–23	GP: 1–6, 8 IP: 7–11 odd HW: 10, 12–14, 16–22 even
10-9	GP: 1–7 IP: 9–13 odd HW: 8–13, 15–23	GP: 1–7 IP: 9–13 odd HW: 8–14 even, 15–23 odd
10-10	GP: 1–7 IP: 8, 9 HW: 8–10, 12–23	GP: 1–7 IP: 8, 9 HW: 10, 11, 13–23 odd
	Reviewing the Basics	Reviewing the Basics
	Unit Test	Unit Test

Assignment Guide for Mathematics with Business Applications

Unit 11 Recordkeeping		
	PROGRAM	
LESSON	Basic	Average
11-1	GP: 1–5, 8 IP: 9–12 HW: 6, 7, 9–14, 16–19	GP: 1–5, 8 IP: 9–12 HW: 6, 7, 13–15, 16–19
11-2	GP: 1–4 IP: 5–9 HW: 5–15, 17–30	GP: 1–4 IP: 5–9 HW: 10–16, 18–30 even
11-3	GP: 1–7 IP: 8–15 HW: 8–15, 17–24, 25–33	GP: 1–7 IP: 8–15 HW: 16–24, 25–33 odd
	Reviewing the Basics	Reviewing the Basics
	Unit Test	Unit Test
	Cumulative Review	Cumulative Review
	Cumulative Review Test	Cumulative Review Test

Unit 12 Personnel		
	PROGRAM	
LESSON	Basic	Average
12-1	GP: 1–5 IP: 6–9 HW: 6–10, 12–19	GP: 1–5 IP: 6–9 HW: 10, 11, 13–19 odd
12-2	GP: 1–4 IP: 5–9 HW: 5–12, 14–22	GP: 1–4 IP: 5–9 HW: 10–13, 14–22 even
12-3	GP: 1–3 IP: 4–8 HW: 4–9, 11–19	GP: 1–3 IP: 4–8 HW: 9, 10, 12–18 even
12-4	GP: 1–5, 14–18 IP: 6–11, 19–23 HW: 6–12, 19–24, 27–39	GP: 1–5, 14–18 IP: 6–11, 19–23 HW: 12, 13, 24–26, 27–39 odd
12-5	GP: 1–4 IP: 5–9 HW: 5–10, 12–27	GP: 1–4 IP: 5–9 HW: 10, 11, 12–26 even
12-6	GP: 1–4 IP: 5–8, 10 HW: 5–11, 13–24	GP: 1–4 IP: 5–8, 10 HW: 9, 11, 12, 14–24 even
	Reviewing the Basics	Reviewing the Basics
	Unit Test	Unit Test

Unit 13 Production

LESSON	PROGRAM: Basic	Average
13-1	GP: 1–5 IP: 6–8 HW: 6–11, 13–30	GP: 1–5 IP: 6–8 HW: 9–12, 14–30 even
13-2	GP: 1–6 IP: 7–11 HW: 7–13, 15–23	GP: 1–6 IP: 7–11 HW: 12–14, 16–22 even
13-3	GP: 1–9 IP: 10, 11, 13 HW: 10–15, 18–32	GP: 1–9 IP: 10, 11, 13 HW: 12, 14–17, 18–32 even
13-4	GP: 1–6 IP: 7, 9 HW: 7–9, 11–18	GP: 1–6 IP: 7, 9 HW: 8, 10, 11–18
13-5	GP: 1–7 IP: 8–11 HW: 8–12, 14–21	GP: 1–7 IP: 8–11 HW: 12, 13, 15–21 odd
13-6	GP: 1–14 IP: 15, 17 HW: 15–18, 20–42 even	GP: 1–14 IP: 15, 17 HW: 16, 18, 19, 21–43 odd
	Reviewing the Basics	Reviewing the Basics
	Unit Test	Unit Test

Unit 14 Purchasing

LESSON	PROGRAM: Basic	Average
14-1	GP: 1–6 IP: 7–9, 11–15 odd HW: 7–15, 17–25	GP: 1–6 IP: 7–9, 11–15 odd HW: 10–16 even, 17–25 odd
14-2	GP: 1–6 IP: 7–9, 11, 13 HW: 7–15, 17–30	GP: 1–6 IP: 7–9, 11, 13 HW: 10–14 even, 15, 16, 18–30 even
14-3	GP: 1–6 IP: 7–9, 11, 12 HW: 7–13, 15–26	GP: 1–6 IP: 7–9, 11, 12 HW: 10, 13, 14, 16–26 even
14-4	GP: 1–5 IP: 6, 7, 9, 11 HW: 6–12, 14–31	GP: 1–5 IP: 6, 7, 9, 11 HW: 8–12 even, 13, 14–30 even
14-5	GP: 1–5 IP: 6–8, 10 HW: 6–12, 14–28	GP: 1–5 IP: 6–8, 10 HW: 9, 11–13, 14–28 even
14-6	GP: 1–5 IP: 6–9 HW: 6–10, 12–21	GP: 1–5 IP: 6–9 HW: 10, 11, 13–21 odd
14-7	GP: 1–4 IP: 5–9 HW: 5–10, 12–20	GP: 1–4 IP: 5–9 HW: 10, 11, 12–20 even
	Reviewing the Basics	Reviewing the Basics
	Unit Test	Unit Test

Assignment Guide for Mathematics with Business Applications

Unit 15 Sales

LESSON	PROGRAM Basic	PROGRAM Average
15-1	GP: 1–6 IP: 7–10, 12–16 even HW: 7–18, 21–29	GP: 1–6 IP: 7–10, 12–16 even HW: 11–17 even, 18–20, 22–26 even
15-2	GP: 1–4 IP: 5–8 HW: 5–13, 15–26	GP: 1–4 IP: 5–8 HW: 9–14, 15–25 odd
15-3	GP: 1–6 IP: 7–11 odd HW: 7–12, 15–29	GP: 1–6 IP: 7–11 odd HW: 8–12 even, 13, 14, 15–29 odd
15-4	GP: 1–8 IP: 9–13 odd HW: 9–15, 18–26	GP: 1–8 IP: 9–13 odd HW: 10–14 even, 15–17, 18–26 even
15-5	GP: 1–6 IP: 7–9, 11 HW: 7–13, 15–29	GP: 1–6 IP: 7–9, 11 HW: 10, 12–14, 15–29 odd
15-6	GP: 1–8 IP: 9–13 odd HW: 9–14, 16–24	GP: 1–8 IP: 9–13 odd HW: 10–14 even, 15, 16–24 even
15-7	GP: 1–9 IP: 10–14 even HW: 10–16, 19–30	GP: 1–9 IP: 10–14 even HW: 11–15 odd, 16–18, 19–29 odd
15-8	GP: 1–7 IP: 8–12 even HW: 8–13, 15–29 odd	GP: 1–7 IP: 8–12 even HW: 9–13 odd, 14, 16–28 even
	Reviewing the Basics	Reviewing the Basics
	Unit Test	Unit Test

Unit 16 Marketing

LESSON	PROGRAM Basic	PROGRAM Average
16-1	GP: 1–5 IP: 6–9 HW: 6–10, 12–18	GP: 1–5 IP: 6–9 HW: 10, 11, 12–18
16-2	GP: 1–4 IP: 5–8 HW: 5–10, 12–15	GP: 1–4 IP: 5–8 HW: 9–11, 12–15
16-3	GP: 1–6 IP: 7–11 odd HW: 7–17, 19–26	GP: 1–6 IP: 7–11 odd HW: 8–12 even, 13–18, 19–25 odd
16-4	GP: 1–7 IP: 8 HW: 8, 9, 11–20	GP: 1–7 IP: 8 HW: 9, 10, 11–19 odd
16-5	GP: 1–7 IP: 8–10 HW: 8–13, 15–23	GP: 1–7 IP: 8–10 HW: 11–14, 15–23 odd
16-6	GP: 1–7 IP: 8–10 HW: 8–11, 13–20	GP: 1–7 IP: 8–10 HW: 11, 12, 14–20 even
16-7	GP: 1–6 IP: 7–9 HW: 7–10, 12–16	GP: 1–6 IP: 7–9 HW: 10, 11, 12–16
16-8	GP: 1–4 IP: 5, 6 HW: 5–7, 9–12	GP: 1–4 IP: 5, 6 HW: 7, 8, 9–12
	Reviewing the Basics	Reviewing the Basics
	Unit Test	Unit Test
	Cumulative Review	Cumulative Review
	Cumulative Review Test	Cumulative Review Test

Unit 17 Warehousing and Distribution

LESSON	PROGRAM Basic	PROGRAM Average
17-1	GP: 1–9 IP: 10–14 even HW: 10–16, 19–27	GP: 1–9 IP: 10–14 even HW: 11–15 odd, 16–18, 19–27 odd
17-2	GP: 1–7 IP: 8, 9, 11 HW: 8–16, 19–28	GP: 1–7 IP: 8, 9, 11 HW: 10, 12–18, 20–28 even
17-3	GP: 1–3 IP: 4–7 HW: 4–8, 14–29	GP: 1–3 IP: 4–7 HW: 8–13, 14–28 even
17-4	GP: 1–6 IP: 7, 8, 10, 11 HW: 7–13, 15–25 odd	GP: 1–6 IP: 7, 8, 10, 11 HW: 9, 12–14, 16–26 even
17-5	GP: 1–6 IP: 7, 9 HW: 7–10, 12–19	GP: 1–6 IP: 7, 9 HW: 8, 10, 11, 13–19 odd
17-6	GP: 1–8 IP: 9–11 HW: 9–14, 16–30 even	GP: 1–8 IP: 9–11 HW: 12–15, 17–33 odd
	Reviewing the Basics	Reviewing the Basics
	Unit Test	Unit Test

Unit 18 Services

LESSON	PROGRAM Basic	PROGRAM Average
18-1	GP: 1–5 IP: 6, 7, 9 HW: 6–12, 14–19	GP: 1–5 IP: 6, 7, 9 HW: 8, 10–13, 14–19
18-2	GP: 1–5 IP: 6–12 even HW: 6–12, 14–19	GP: 1–5 IP: 6–12 even HW: 7–13 odd, 14–19
18-3	GP: 1–5 IP: 6–8 HW: 6–11, 13–16	GP: 1–5 IP: 6–8 HW: 9–12, 13–16
18-4	GP: 1–4 IP: 5–7, 9 HW: 5–10, 12–15	GP: 1–4 IP: 5–7, 9 HW: 8, 10, 11, 12–15
18-5	GP: 1–3 IP: 4, 5, 7 HW: 4–8, 10–13	GP: 1–3 IP: 4, 5, 7 HW: 6, 8, 9, 10–13
18-6	GP: 1–6 IP: 7–10 HW: 7–13, 15–20	GP: 1–6 IP: 7–10 HW: 11–14, 15–20
	Reviewing the Basics	Reviewing the Basic
	Unit Test	Unit Test

Assignment Guide for Mathematics with Business Applications

Unit 19 Accounting

LESSON	PROGRAM Basic	PROGRAM Average
19-1	GP: 1–7 IP: 8–10 HW: 8–12, 14–19	GP: 1–7 IP: 8–10 HW: 11–13, 15–19 odd
19-2	GP: 1–4 IP: 5–8 HW: 5–9, 11–19	GP: 1–4 IP: 5–8 HW: 9, 10, 12–18 even
19-3	GP: 1–5 IP: 6–8 HW: 6–10, 12–19	GP: 1–5 IP: 6–8 HW: 9–11, 12–18 even
19-4	GP: 1–6 IP: 7–9 HW: 7–12, 15–26	GP: 1–6 IP: 7–9 HW: 10–13, 15–25 odd
19-5	GP: 1–5 IP: 6–8 HW: 6–10, 12–20	GP: 1–5 IP: 6–8 HW: 9–11, 12–20 even
19-6	GP: 1–6 IP: 7–9 HW: 7–10, 12–18	GP: 1–6 IP: 7–9 HW: 10, 11, 12–18 even
19-7	GP: 1–4 IP: 5–8 HW: 5–9, 11–18	GP: 1–4 IP: 5–8 HW: 9, 10, 12–18 even
19-8	GP: 1–4 IP: 5–7 HW: 5–9, 11–14	GP: 1–4 IP: 5–7 HW: 8–10, 11–14
	Reviewing the Basics	Reviewing the Basics
	Unit Test	Unit Test

Unit 20 Accounting Records

LESSON	PROGRAM Basic	PROGRAM Average
20-1	GP: 1–8 IP: 9, 10 HW: 9–11, 13–20	GP: 1–8 IP: 9, 10 HW: 11, 12, 14–20 even
20-2	GP: 1–6 IP: 7–9 HW: 7–11, 13–17	GP: 1–6 IP: 7–9 HW: 10–12, 13–17 odd
20-3	GP: 1–6 IP: 7–9 HW: 7–13, 15–24	GP: 1–6 IP: 7–9 HW: 10–14, 15–23 odd
20-4	GP: 1–5 IP: 6, 7, 9 HW: 6–12, 15–22	GP: 1–5 IP: 6, 7, 9 HW: 8, 10–14, 15–21 odd
20-5	GP: 1–6, 8, 9 IP: 7, 10, 11 HW: 7, 10–13, 15–20	GP: 1–6, 8, 9 IP: 7, 10, 11 HW: 12–14, 15–20
20-6	GP: 1–5 IP: 6–8 HW: 6–11, 13–17	GP: 1–5 IP: 6–8 HW: 9–12, 13–17 odd
	Reviewing the Basics	Reviewing the Basics
	Unit Test	Unit Test

Unit 21 Financial Management

LESSON	PROGRAM Basic	Average
21-1	GP: 1–5 IP: 6–9 HW: 6–10, 12–22	GP: 1–5 IP: 6–9 HW: 10, 11, 12–22 even
21-2	GP: 1–5 IP: 6, 7, 9 HW: 6–11, 13–16	GP: 1–5 IP: 6, 7, 9 HW: 8, 10–12, 13–16
21-3	GP: 1–5 IP: 6–8 HW: 6–10, 12–16	GP: 1–5 IP: 6–8 HW: 9–11, 12–16
21-4	GP: 1–6 IP: 6–8 HW: 6–10, 12–19	GP: 1–6 IP: 6–8 HW: 9–11, 12–18 even
21-5	GP: 1–4 IP: 5–7 HW: 5–8, 10–13	GP: 1–4 IP: 5–7 HW: 8, 9, 10–13
21-6	GP: 1–4 IP: 5, 6 HW: 5–8, 10–14	GP: 1–4 IP: 5–6 HW: 7–9, 10–14 even
	Reviewing the Basics	Reviewing the Basics
	Unit Test	Unit Test

Unit 22 Corporate Planning

LESSON	PROGRAM Basic	Average
22-1	GP: 1–9 IP: 10–12, 14 HW: 10–18, 20–25	GP: 1–9 IP: 10–12, 14 HW: 13, 15–19, 21–25 odd
22-2	GP: 1–6 IP: 7–11 HW: 7–18, 21–26	GP: 1–6 IP: 7–11 HW: 12–20, 21–25 odd
22-3	GP: 1–9 IP: 10–13 HW: 10–23, 26–36 even	GP: 1–9 IP: 10–13 HW: 14–25, 27–37 odd
22-4	GP: 1–8 IP: 9–13 odd HW: 9–16, 18–23	GP: 1–8 IP: 9–13 odd HW: 10–14 even, 15–17, 18–22 even
	Reviewing the Basics	Reviewing the Basics
	Unit Test	Unit Test
	Cumulative Review	Cumulative Review
	Cumulative Review Test	Cumulative Review Test

Mathematics with Business Applications

Fourth Edition

Walter H. Lange
The University of Toledo

Temoleon G. Rousos
The University of Toledo

Robert D. Mason
Professor Emeritus
The University of Toledo

New York, New York   Columbus, Ohio   Woodland Hills, California   Peoria, Illinois

A Division of The McGraw-Hill Companies

Glencoe/McGraw-Hill

A Division of The ***McGraw-Hill*** *Companies*

Formerly published under the title *Business Mathematics.*

Send all inquiries to:

Glencoe/McGraw-Hill
21600 Oxnard Street, Suite 500
Woodland Hills, CA 91367

ISBN 0-02-814730-8 (Student Edition)
ISBN 0-02-814731-6 (Teacher's Wraparound Edition)

Printed in the United States of America

3 4 5 6 7 8 9 027 04 03 02 01 00 99

TABLE OF CONTENTS

PART 1 BASIC MATH SKILLS WORKSHOP: A REVIEW OF FUNDAMENTALS

TABLE OF CONTENTS

PART 2 PERSONAL BUSINESS MATHEMATICS 

TABLE OF CONTENTS

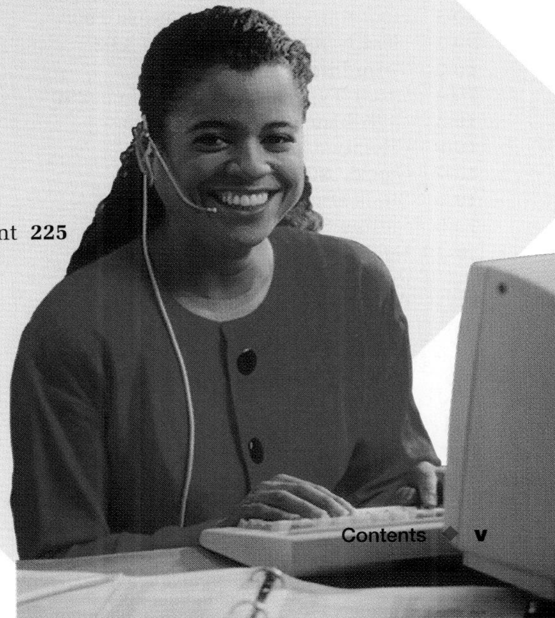

TABLE OF CONTENTS

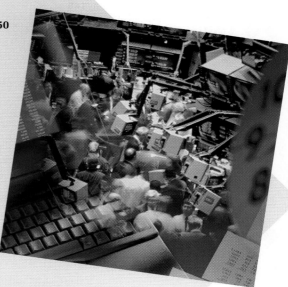

TABLE OF CONTENTS

TABLE OF CONTENTS

TABLE OF CONTENTS

TABLE OF CONTENTS

Unit 22 Corporate Planning 553

TABLE OF CONTENTS

TABLE OF CONTENTS

TABLE OF CONTENTS

SPREADSHEET APPLICATIONS

MEMO TO STUDENTS

To: The Student
From: The Authors

This is your textbook. It is specially designed to help you become a knowledgeable consumer and business person. This text contains information and resources to help you achieve these goals.

Each lesson begins with a brief description of the topic and highlights the key formula for you.

*The **Example** shows the topic in an applied situation. It provides you with a step-by-step method to solve for the answer. If you have any trouble or need to review, there is helpful information in the Reference Files at the back of the book. The references at the top of the example tell you just which items to study.*

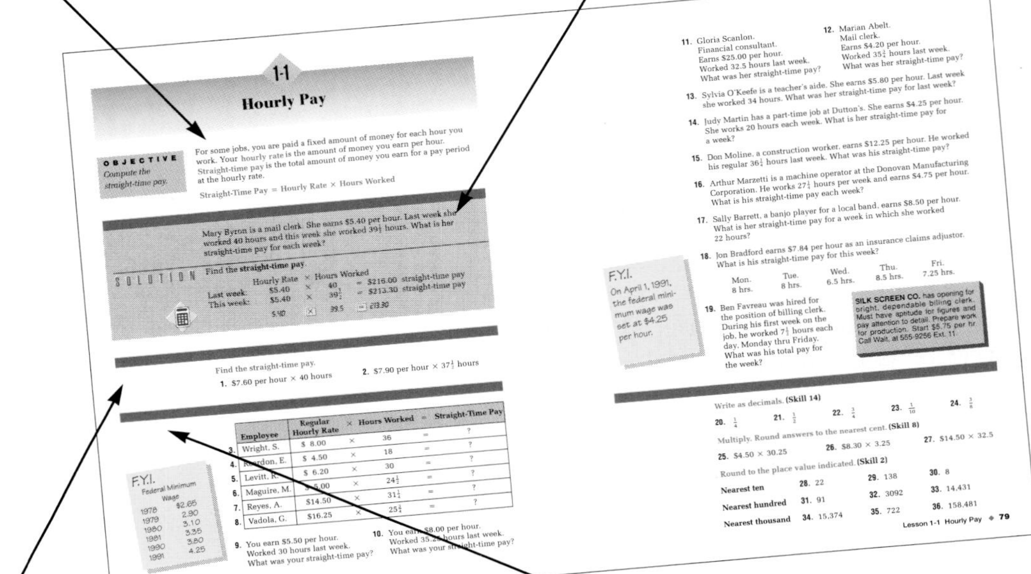

1-1

Hourly Pay

OBJECTIVE
Compute the straight-time pay.

For some jobs, you are paid a fixed amount of money for each hour you work. Your hourly rate is the amount of money you earn per hour. Straight-time pay is the total amount of money you earn for a pay period at the hourly rate.

Straight-Time Pay = Hourly Rate × Hours Worked

Mary Byron is a mail clerk. She earns $5.40 per hour. Last week she worked 40 hours and this week she worked $39\frac{1}{2}$ hours. What is her straight-time pay for each week?

SOLUTION Find the **straight-time pay**.

	Hourly Rate	×	Hours Worked		
Last week:	$5.40	×	40	=	$216.00 straight-time pay
This week:	$5.40	×	$39\frac{1}{2}$	=	$213.30 straight-time pay
	5.40	×	39.5	=	213.30

Find the straight-time pay.

1. $7.60 per hour × 40 hours
2. $7.90 per hour × $37\frac{1}{2}$ hours

	Employee	Regular Hourly Rate	×	Hours Worked	=	Straight-Time Pay
3.	Wright, S.	$ 8.00	×	36	=	?
4.	Reardon, E.	$ 4.50	×	18	=	?
5.	Levitt, R.	$ 6.20	×	30	=	?
6.	Maguire, M.	$ 5.00	×	$24\frac{1}{2}$	=	?
7.	Reyes, A.	$14.50	×	$31\frac{1}{4}$	=	?
8.	Vadola, G.	$16.25	×	$25\frac{3}{4}$	=	?

F.Y.I.
Federal Minimum Wage

1978	$2.65
1979	2.90
1980	3.10
1981	3.35
1990	3.80
1991	4.25

9. You earn $5.50 per hour. Worked 30 hours last week. What was your straight-time pay?
10. You earn $8.00 per hour. Worked 35.25 hours last week. What was your straight-time pay?

78 ◆ Unit 1 Gross Income

11. Gloria Scanlon. Financial consultant. Earns $25.00 per hour. Worked 32.5 hours last week. What was her straight-time pay?
12. Marian Abelt. Mail clerk. Earns $4.20 per hour. Worked $35\frac{3}{4}$ hours last week. What was her straight-time pay?
13. Sylvia O'Keefe is a teacher's aide. She earns $5.80 per hour. Last week she worked 34 hours. What was her straight-time pay for last week?
14. Judy Martin has a part-time job at Dutton's. She earns $4.25 per hour. She works 20 hours each week. What is her straight-time pay for a week?
15. Don Moline, a construction worker, earns $12.25 per hour. He worked his regular $36\frac{1}{4}$ hours last week. What was his straight-time pay?
16. Arthur Marzetti is a machine operator at the Donovan Manufacturing Corporation. He works $27\frac{1}{4}$ hours per week and earns $4.75 per hour. What is his straight-time pay each week?
17. Sally Barrett, a banjo player for a local band, earns $8.50 per hour. What is her straight-time pay for a week in which she worked 22 hours?
18. Jon Bradford earns $7.84 per hour as an insurance claims adjustor. What is his straight-time pay for this week?

Mon.	Tue.	Wed.	Thu.	Fri.
8 hrs.	8 hrs.	6.5 hrs.	8.5 hrs.	7.25 hrs.

F.Y.I.
On April 1, 1991, the federal minimum wage was set at $4.25 per hour.

19. Ben Favreau was hired for the position of billing clerk. During his first week on the job, he worked $7\frac{1}{2}$ hours each day, Monday thru Friday. What was his total pay for the week?

SILK SCREEN CO. has opening for bright, dependable billing clerk. Must have aptitude for figures and pay attention to detail. Prepare work for production. Start $5.75 per hr. Call Walt, at 555-9256 Ext. 11

Write as decimals. **(Skill 14)**

20. $\frac{1}{4}$ 21. $\frac{1}{2}$ 22. $\frac{3}{4}$ 23. $\frac{1}{10}$ 24. $\frac{3}{8}$

Multiply. Round answers to the nearest cent. **(Skill 8)**

25. $4.50 × 30.25 26. $8.30 × 3.25 27. $14.50 × 32.5

Round to the place value indicated. **(Skill 2)**

Nearest ten	28. 22	29. 138	30. 8
Nearest hundred	31. 91	32. 3092	33. 14,431
Nearest thousand	34. 15,374	35. 722	36. 158,481

Lesson 1-1 Hourly Pay ◆ 79

*The **Self-Check** lets you practice the concept and use the formula on your own before you begin the lesson problems.*

After you work through the Example and the Self-Check, complete the problems assigned by your instructor. These problems are arranged systematically to apply gradually what you have just studied. Answers to selected problems are found at the back of your book. Use the answers to check your work.

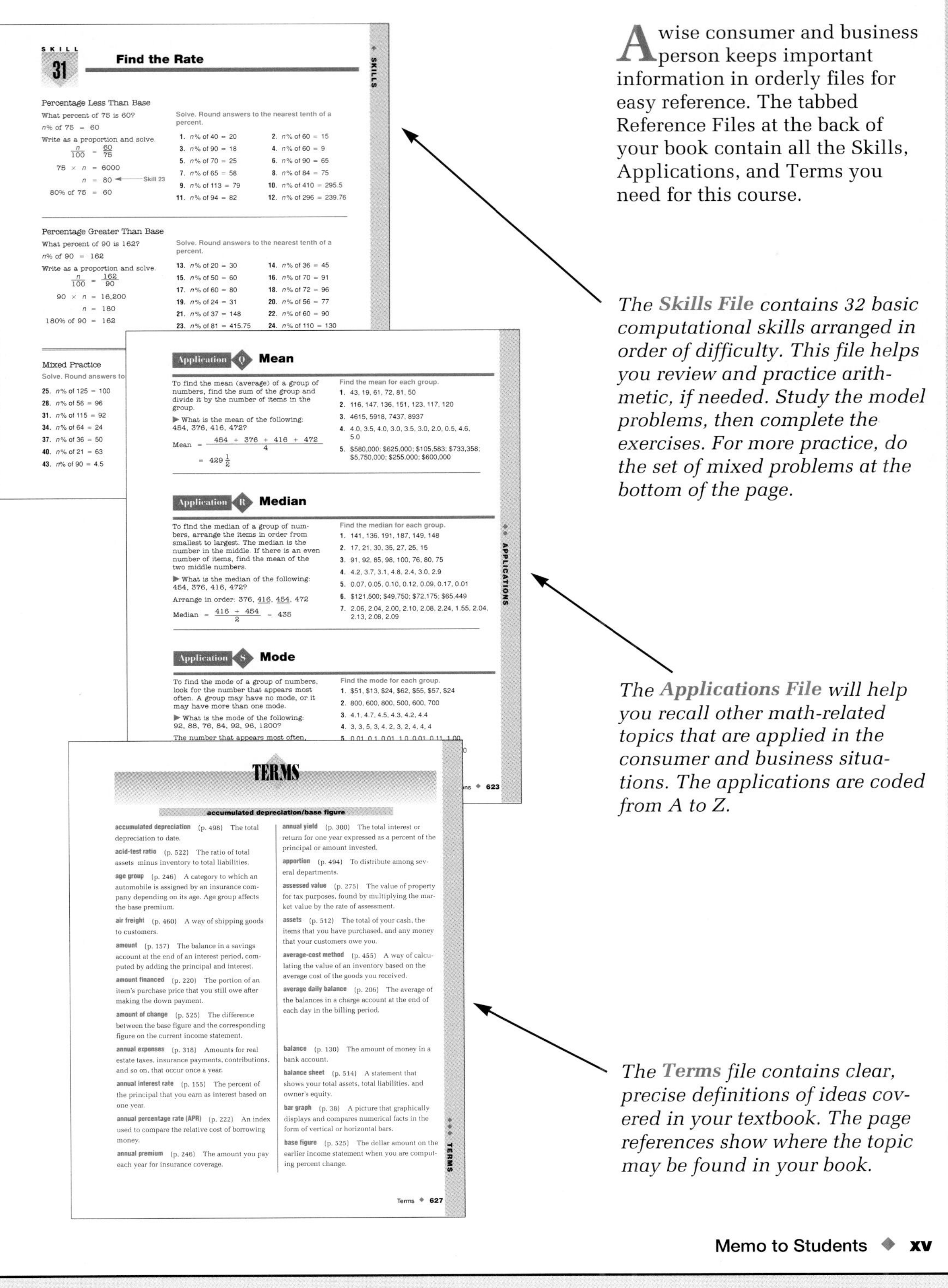

SKILL 31

Find the Rate

Percentage Less Than Base

What percent of 75 is 60?

n% of 75 = 60

Write as a proportion and solve.

$\frac{n}{100} = \frac{60}{75}$

75 × n = 6000

n = 80 ← Skill 23

80% of 75 = 60

Solve. Round answers to the nearest tenth of a percent.

1. n% of 40 = 20
2. n% of 60 = 15
3. n% of 90 = 18
4. n% of 60 = 9
5. n% of 70 = 25
6. n% of 90 = 65
7. n% of 65 = 58
8. n% of 84 = 75
9. n% of 113 = 79
10. n% of 410 = 295.5
11. n% of 94 = 82
12. n% of 296 = 239.76

Percentage Greater Than Base

What percent of 90 is 162?

n% of 90 = 162

Write as a proportion and solve.

$\frac{n}{100} = \frac{162}{90}$

90 × n = 16,200

n = 180

180% of 90 = 162

Solve. Round answers to the nearest tenth of a percent.

13. n% of 20 = 30
14. n% of 36 = 45
15. n% of 50 = 60
16. n% of 70 = 91
17. n% of 60 = 80
18. n% of 72 = 96
19. n% of 24 = 31
20. n% of 56 = 77
21. n% of 37 = 148
22. n% of 60 = 90
23. n% of 81 = 415.75
24. n% of 110 = 130

Mixed Practice

Solve. Round answers to

25. n% of 125 = 100
28. n% of 56 = 96
31. n% of 115 = 92
34. n% of 64 = 24
37. n% of 36 = 50
40. n% of 21 = 63
43. n% of 90 = 4.5

SKILLS

Application Q **Mean**

To find the mean (average) of a group of numbers, find the sum of the group and divide it by the number of items in the group.

▶ What is the mean of the following: 454, 376, 416, 472?

Mean = $\frac{454 + 376 + 416 + 472}{4}$

= $429\frac{1}{2}$

Find the mean for each group.

1. 43, 19, 61, 72, 81, 50
2. 116, 147, 136, 151, 123, 117, 120
3. 4615, 5918, 7437, 8937
4. 4.0, 3.5, 4.0, 3.0, 3.5, 3.0, 2.0, 0.5, 4.6, 5.0
5. $580,000; $625,000; $105,583; $733,358; $5,750,000; $255,000; $600,000

Application R **Median**

To find the median of a group of numbers, arrange the items in order from smallest to largest. The median is the number in the middle. If there is an even number of items, find the mean of the two middle numbers.

▶ What is the median of the following: 454, 376, 416, 472?

Arrange in order: 376, 416, 454, 472

Median = $\frac{416 + 454}{2}$ = 435

Find the median for each group.

1. 141, 136, 191, 187, 149, 148
2. 17, 21, 30, 35, 27, 25, 15
3. 91, 92, 85, 98, 100, 76, 80, 75
4. 4.2, 3.7, 3.1, 4.8, 2.4, 3.0, 2.9
5. 0.07, 0.05, 0.10, 0.12, 0.09, 0.17, 0.01
6. $121,500; $49,750; $72,175; $65,449
7. 2.06, 2.04, 2.00, 2.10, 2.08, 2.24, 1.55, 2.04, 2.13, 2.08, 2.09

Application S **Mode**

To find the mode of a group of numbers, look for the number that appears most often. A group may have no mode, or it may have more than one mode.

▶ What is the mode of the following: 92, 88, 76, 84, 92, 96, 120?

The number that appears most often,

Find the mode for each group.

1. $51, $13, $24, $62, $55, $57, $24
2. 800, 600, 800, 500, 600, 700
3. 4.1, 4.7, 4.5, 4.3, 4.2, 4.4
4. 3, 3, 5, 3, 4, 2, 3, 2, 4, 4, 4

APPLICATIONS

...ns ◆ 623

TERMS

accumulated depreciation/base figure

accumulated depreciation (p. 498) The total depreciation to date.

acid-test ratio (p. 522) The ratio of total assets minus inventory to total liabilities.

age group (p. 246) A category to which an automobile is assigned by an insurance company depending on its age. Age group affects the base premium.

air freight (p. 460) A way of shipping goods to customers.

amount (p. 157) The balance in a savings account at the end of an interest period, computed by adding the principal and interest.

amount financed (p. 220) The portion of an item's purchase price that you still owe after making the down payment.

amount of change (p. 525) The difference between the base figure and the corresponding figure on the current income statement.

annual expenses (p. 318) Amounts for real estate taxes, insurance payments, contributions, and so on, that occur once a year.

annual interest rate (p. 155) The percent of the principal that you earn as interest based on one year.

annual percentage rate (APR) (p. 222) An index used to compare the relative cost of borrowing money.

annual premium (p. 246) The amount you pay each year for insurance coverage.

annual yield (p. 300) The total interest or return for one year expressed as a percent of the principal or amount invested.

apportion (p. 494) To distribute among several departments.

assessed value (p. 275) The value of property for tax purposes, found by multiplying the market value by the rate of assessment.

assets (p. 512) The total of your cash, the items that you have purchased, and any money that your customers owe you.

average-cost method (p. 455) A way of calculating the value of an inventory based on the average cost of the goods you received.

average daily balance (p. 206) The average of the balances in a charge account at the end of each day in the billing period.

balance (p. 130) The amount of money in a bank account.

balance sheet (p. 514) A statement that shows your total assets, total liabilities, and owner's equity.

bar graph (p. 38) A picture that graphically displays and compares numerical facts in the form of vertical or horizontal bars.

base figure (p. 525) The dollar amount on the earlier income statement when you are computing percent change.

TERMS

Terms ◆ 627

A wise consumer and business person keeps important information in orderly files for easy reference. The tabbed Reference Files at the back of your book contain all the Skills, Applications, and Terms you need for this course.

The ***Skills File*** *contains 32 basic computational skills arranged in order of difficulty. This file helps you review and practice arithmetic, if needed. Study the model problems, then complete the exercises. For more practice, do the set of mixed problems at the bottom of the page.*

The ***Applications File*** *will help you recall other math-related topics that are applied in the consumer and business situations. The applications are coded from A to Z.*

The ***Terms*** *file contains clear, precise definitions of ideas covered in your textbook. The page references show where the topic may be found in your book.*

PREFACE

A knowledge of basic mathematics is absolutely essential for survival in today's world. Mathematics as a survival tool is implemented in almost every phase of personal and business life. How effectively one is able to use that important tool may determine how well one is able to prepare for the future, manage one's personal and business resources, and benefit from one's efforts as a consumer, worker, or business person.

The main purpose of *Mathematics with Business Applications,* Third Edition, is to assist students in learning to use mathematics effectively as a tool in their personal and business lives. After students have completed their study of the textbook and related ancillaries, they will be able to

1. Understand terminology relating to personal and business mathematics applications.
2. Apply basic math skills to the solution of both personal and business applications.
3. Use common mathematics formulas to solve a variety of personal and business mathematics problems.

Organization of the Textbook

While retaining the sound organization and approach of the first and second editions *Mathematics with Business Applications,* Third Edition focuses even more strongly on the review, remediation, and reinforcement of basic math skills. An expanded workshop section offers methods students can use to approach the solving of problems in a more logical manner. Computer activities, accompanied by an optional template diskette, provide the opportunity for students to use the computer in solving selected problems. Additional mini-simulations in the workbook add a touch of realism and heighten student interest.

The textbook is organized into four sections: Part One, Basic Math Skills Workshop: A Review of Fundamentals; Part Two, Personal Application Mathematics; Part Three, Business Applications; and Reference Files.

Basic Math Skills Workshop: A Review of Fundamentals

Basic math skills are an essential element of functional literacy. As an aid to students who need to review basic math skills or who need reinforcement of those skills, we have expanded the basic skills workshops.

Sixteen new workshops have been added to the third edition for a new total of thirty workshops. Each workshop includes a description of the covered skill, a variety of examples, and numerous applications and problems that allow students to apply each specific skill. These workshops provide students with a foundation of basic skills they will need during their study of Parts Two and Three of the textbook.

PREFACE

Personal Application Mathematics

In Part Two of the textbook, Personal Application Mathematics, students examine the personal applications of basic math skills used by workers and consumers. These applications include the calculation of gross income and net income, the use of both checking and savings accounts, cash purchases, charge purchases, loans, the cost of owning and operating an automobile, housing costs, insurance and investments, and budgets.

Business Applications

In Part Three, Business Applications, students apply mathematics fundamentals to realistic business situations as they pertain to eleven different departments of a large business. Representative business departments covered include personnel, production, purchasing, sales, marketing, warehousing and distribution, services, accounting, accounting records, financial, and corporate planning.

Parts Two and Three of the textbook are organized into units covering a specific type of application or a business department. Each unit is composed of numbered lessons. Each lesson begins with a clear, concise statement of the objective to be met followed by a brief explanation of the personal or business application to be examined in the lesson. The basic mathematical formula to be used in the lesson is followed by a worked-out, sample problem that illustrates in detail how to solve the application. Table, phrase, and word problems—all graded in order of difficulty—give the students a variety of opportunities to apply the lesson formula. Many lessons contain "A Brief Case" problem, which challenges students to use critical thinking skills as they apply the lesson formula to a real world situation. Each lesson concludes with a set of basic math review problems that help students maintain or improve their basic math skills.

Each unit is followed by review activities; a feature on spreadsheet applications; and a one-page unit test. A career feature is included at the end of each odd-number unit. Also, five mini-simulations are spaced throughout the textbook to provide interesting and realistic opportunities for students to apply their math skills.

Reference Files

Extensive reference files are included at the end of the student textbook to help students who may need to review basic skills or may want to use the files as study aids in completing individual lessons in the textbook. The reference files contain a Skills File of thirty-two numbered computational math skills with practice problems; an Applications File of twenty-six common practice applications; and a Terms File of key terms used in business and consumer situations. These skills, applications, and terms

xvii

PREFACE

are keyed to the examples within the textbook lessons. Students who need further explanation or practice can turn to the reference files and complete the problems found there to reinforce lesson content or basic math skills.

In addition to the preceding files, the textbook also contains several tables that can be used in solving related problems, as well as the answers to the Self-Check and odd-number problems in the text lessons.

Supporting Materials

In addition to the student textbook, the program includes the following: a new teacher's wraparound edition of the textbook; a problems and simulations workbook with additional practice, review, and reinforcement; a teacher's edition of the workbook; a teacher's resource book that contains additional test material, a solution key to all the problems in the textbook, and answers to the Spreadsheet Application problems as well as instructions for using the template diskette to solve the spreadsheet activities in the textbook.

Acknowledgments

We wish to acknowledge the valuable contributions made to this textbook by teachers, students, colleagues, friends, and collaborators throughout its various stages of revision. In particular, we would like to express our appreciation to Professors James Arbaugh, Clayton L. Ziegler, and Robert A. Siddens of the University of Toledo for their expert advice in accounting, engineering, and transportation. We would also like to thank Carla Christy and Danuta T. Lange, who checked solutions and helped prepare the manuscript.

We would like to acknowledge the significant contributions of Robert D. Mason, Professor Emeritus, The University of Toledo—College of Business, with whom we co-authored previous editions of this textbook.

Walter H. Lange

Temoleon G. Rousos

PART 1

Basic Math Skills Workshop: A Review of Fundamentals

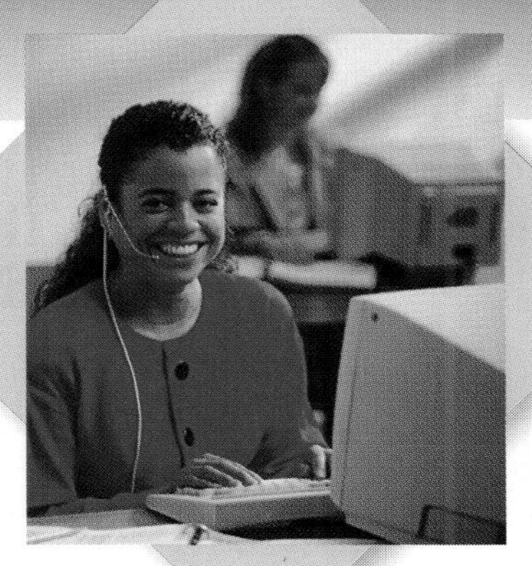

As you use this book, you will need to know certain mathematical fundamentals. To be sure you are prepared, here are lessons for review and practice. Each workshop uses realistic business problems to show how math is used in different job settings.

LESSON PLAN Workshop 1

OBJECTIVES

1. To write numbers as words and words as numbers.
2. To round whole numbers and decimals.
3. To round dollar and cent amounts.

RETEACHING

The key concept underlying the skills of this workshop is place value. Work with students on identifying correctly the place value of each digit in a number. As an example, use the number 444 to illustrate that each 4 has a different value. Extend this example by using 44,444 and then 44,444.44.

Stress the importance of memorizing the names of each place in a number, using the example at the top of this page in the student's book.

WORKSHOPS

WORKSHOP 1

Writing Numbers as Words and Rounding Numbers

Look up Skills 1 and 2 on pages 582–584 for more practice.

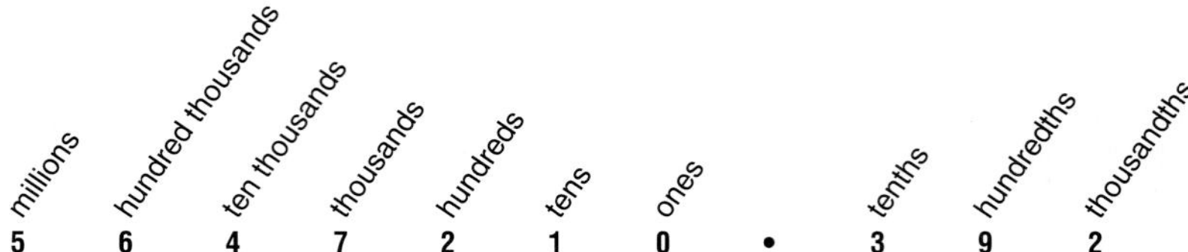

The place-value chart shows the value of each digit in the number 5,647,210.392. For example, the digit 7 is in the thousands place and has a value of 7000. The digit 9 is in the hundredths place and has a value of $\frac{9}{100}$, or 0.09. The place-value chart can help you write numbers.

EXAMPLE	SOLUTION (read as or written as)
536	five hundred thirty-six
72,051	seventy-two thousand fifty-one
9.227	nine and two hundred twenty-seven thousandths
$45.63	forty-five and sixty-three one hundredths dollars
	or forty-five and $\frac{63}{100}$ dollars
	or forty-five dollars and sixty-three cents

✓ SELF-CHECK Complete the problems, then check your answers in the back of the book.

Write in words.

1. 241 **two hundred forty-one**
2. 7.317 **seven and three hundred seventeen thousandths**
3. $761.13 **seven hundred sixty-one and thirteen one hundredths dollars**
   or seven hundred sixty-one and $\frac{13}{100}$ dollars,
   or seven hundred sixty-one dollars and thirteen cents

Place value is also used in rounding numbers. If the digit to the right of the place value you want to round is 5 or more, round up by adding 1 to the number in the place value and then change all the digits to the right of the place value to zeros. If the number is 4 or less, round down by changing all the numbers to the right of the place value to zeros.

EXAMPLE Round 7862 to the nearest hundred.

SOLUTION

7**8**62 — **A.** Find the digit in the hundreds place. It is 8.

78**6**2 — **B.** Is the digit to the right 5 or more? Yes.

7**9**00 — **C.** Add 1 to the hundreds place. Change the digits to the right to zeros.

EXAMPLE Round 0.637 to the nearest tenth.

SOLUTION

0.**6**37 — **A.** Find the digit in the tenths place. It is 6.

0.6**3**7 — **B.** Is the digit to the right 5 or more? No.

0.**6** — **C.** Do not change the tenths digit. Drop the digits to the right.

Dollar and cents amounts are often rounded to the nearest cent, or the hundredths place. Begin with the digit to the right of the hundredths place, or the thousandths place.

EXAMPLE	SOLUTION
$13.6341	$13.63
$112.4171	$112.42
$7918.325	$7918.33

✓ SELF-CHECK Complete the problems, then check your answers in the back of the book.

Round to the nearest cent.

4. $21.277 **$21.28**

5. $967.461 **$967.46**

6. $138.7836 **$138.78**

7. $647.555 **$647.56**

PROBLEMS

Write as numbers.

8. five thousand six hundred seventy-two **5672**

9. sixty-five and $\frac{75}{100}$ dollars **$65.75**

10. one hundred four and forty-nine thousandths **104.049**

11. ninety-five and $\frac{7}{100}$ dollars **$95.07**

Write in word form.

12. 216,798 **13.** $15.43 **14.** 214.093 **15.** 300.5 **16.** $9.81

12. two hundred sixteen thousand seven hundred ninety-eight

13. fifteen and forty three one hundredths of a dollar or fifteen and $\frac{43}{100}$ dollars or fifteen dollars and 43 cents

14. two hundred fourteen and ninety-three one thousandths

15. three hundred and five tenths

16. nine and eighty-one hundredths of a dollar or nine and $\frac{81}{100}$ dollars or nine dollars and 81 cents

ERROR ANALYSIS

In any reteaching situation, it is important to examine the kinds of errors students make. If students have not already mastered the skills presented in all of the workshops in this book, then it is very likely that (1) their understanding of underlying concepts is faulty; (2) they have forgotten certain basic facts; (3) they are applying an algorithm or procedure incorrectly; or (4) they are organizing their work poorly, resulting in careless mistakes.

A typical error that students make when writing numbers as words, or simply reading a number, is to use the word *and* in the wrong place. For example, they may write 647 as "six hundred and forty-seven" instead of "six hundred forty-seven" Stress that *and* is used to replace the decimal point only.

Stress also that all numbers between twenty-one and ninety-nine, other than multiples of ten, should be hyphenated.

PRACTICE AND APPLY

The following problems can be assigned for classwork and the answers checked in class to help students master the objectives of the workshop.

- Guided Practice: 10–14, 19–29 odd
- Independent Practice: 47–59 odd

Round to the nearest place value shown.

Ten thousand	**17.** 416,719 **420,000**	**18.** 514,917 **510,000**	**19.** 960,817 **960,000**	**20.** 5,149,601 **5,150,000**
Thousand	**21.** 312,410 **312,000**	**22.** 42,614 **43,000**	**23.** 867,501 **868,000**	**24.** 7,523,614 **7,524,000**
Hundred	**25.** 41,515 **41,500**	**26.** 3156 **3200**	**27.** 17,771 **17,800**	**28.** 21,910 **21,900**
Ten	**29.** 7291 **7290**	**30.** 694 **690**	**31.** 21,945 **21,950**	**32.** 50,740 **50,740**
One	**33.** 3.341 **3**	**34.** 516.17 **516**	**35.** 42.716 **43**	**36.** 147.88 **148**
Tenth	**37.** 21.217 **21.2**	**38.** 86.554 **86.6**	**39.** 4718.24 **4718.2**	**40.** 717.97178 **718.0**
Hundredth	**41.** 5.2167 **5.22**	**42.** 15.518 **15.52**	**43.** 31.79861 **31.80**	**44.** 14.52499 **14.52**

Round 15,748,516 to the place value given.

45. millions **16,000,000** **46.** ten millions **20,000,000** **47.** thousands **15,749,000**

48. hundreds **15,748,500** **49.** ten thousands **15,750,000** **50.** hundred thousands **15,700,000**

Round to the nearest place value shown.

Cent	**51.** $14.7151 **$14.72**	**52.** $49.592 **$49.59**	**53.** $72.195 **$72.20**	**54.** $94.1249 **$94.12**
Ten cents	**55.** $119.864 **$119.90**	**56.** $8.7199 **$8.70**	**57.** $7.0451 **$7**	**58.** $89.975 **$90**
One dollar	**59.** $151.912 **$152**	**60.** $510.47 **$510**	**61.** $623.97 **$624**	**62.** $53.099 **$53**
Ten dollars	**63.** $3145 **$3150**	**64.** $1789.75 **$1790**	**65.** $515.11 **$520**	**66.** $710.00 **$710**
Hundred dollars	**67.** $19,876 **$19,900**	**68.** $21,050 **$21,100**	**69.** $4132 **$4100**	**70.** $74,656 **$74,700**

APPLICATIONS

71. In his job as an accountant for the Tarwood Manufacturing Co., Steve Alexander needs to write numbers as words. Write each check amount in words.

a. $7.17 **b.** $9.75 **c.** $42.91 **d.** $51.80

e. $217.23 **f.** $531.49 **g.** $9751.63 **h.** $1270.82

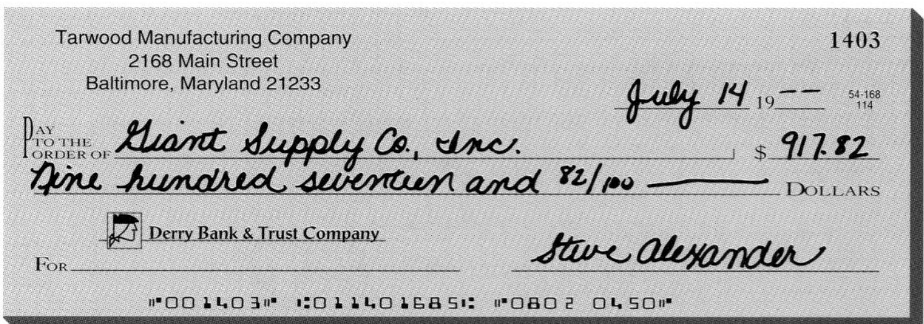

Tarwood Manufacturing Company
2168 Main Street
Baltimore, Maryland 21233

1403

July 14 19 — —

54-168
114

PAY TO THE ORDER OF Giant Supply Co. Inc. $ 917.82

Nine hundred seventeen and 82/100 ——— DOLLARS

Derry Bank & Trust Company

FOR

Steve Alexander

⑈001403⑈ ⑆011401685⑆ ⑈0802 0450⑈

a. Seven and $\frac{17}{100}$ dollars

b. Nine and $\frac{75}{100}$ dollars

c. Forty-two and $\frac{91}{100}$ dollars

d. Fifty-one and $\frac{80}{100}$ dollars

e. Two hundred seventeen and $\frac{23}{100}$ dollars

f. Five hundred thirty-one and $\frac{49}{100}$ dollars

g. Nine thousand seven hundred fifty-one and $\frac{63}{100}$ dollars

h. One thousand two hundred seventy and $\frac{82}{100}$ dollars

72. Amy Cole records the attendance data for the sporting events at the Glass Bowl. Round each number to the amount indicated.

	Football Game	Number	Nearest Thousand	Nearest Hundred	Nearest Ten
a.	Austin Peay	18,971	? **19,000**	? **19,000**	? **18,970**
b.	Bowling Green	25,687	? **26,000**	? **25,700**	? **25,690**
c.	Ohio University	20,119	? **20,000**	? **20,100**	? **20,120**
d.	Western Michigan	24,567	? **25,000**	? **24,600**	? **24,570**
e.	Central Michigan	19,424	? **19,000**	? **19,400**	? **19,420**
f.	Ball State	21,972	? **22,000**	? **22,000**	? **21,970**
g.	Kent State	19,876	? **20,000**	? **19,900**	? **19,880**
h.	Miami	23,947	? **24,000**	? **23,900**	? **23,950**

73. To get an overall sales picture of first-quarter sales, the marketing executive wants the sales figures in round numbers. Round the sales of each company to the nearest thousand dollars.

a. $618,000
b. $719,000
c. $272,000
d. $915,000
e. $535,000
f. $420,000
g. $178,000
h. $572,000
i. $423,000

FIRST-QUARTER SALES, 19--

	Company	Sales
a.	Able Computers	$617,571
b.	Crawford Electronics	719,417
c.	Diamond Research, Inc.	271,710
d.	Laser Systems, Inc.	914,620
e.	Meltex Graphics	535,498
f.	Perfect Software	419,501
g.	Quality Communications	178,071
h.	Superior Limited	571,719
i.	Total Surveys, Inc.	423,412

74. John Brown, an inventory clerk, often rounds inventory figures for easier handling. Round the numbers from the inventory list to the nearest ten.

a. 120
b. 220
c. 480
d. 220
e. 430
f. 580
g. 360
h. 600

CORN INVENTORY

	Description	Quantity	Rounded
a.	Golden Cross	119 lb	?
b.	Golden Beauty	221 lb	?
c.	Early Glo	476 lb	?
d.	X-tra Sweet	219 lb	?
e.	White Sunglo	431 lb	?
f.	Silver Queen	578 lb	?
g.	Honey-Cream	364 lb	?
h.	Country Gtlm	597 lb	?

WRAP-UP

Write the number 6,709,431.528 on the chalkboard and call upon individual students to give the name of each place you select randomly.

Assignment Guide

- Basic: 41–65 odd, 66–74
- Average: 24–38, 41–65 odd, 69, 72

WORKSHOPS

LESSON PLAN Workshop 2

OBJECTIVES

1. To compare whole numbers to see which one is greater.
2. To compare decimals to see which one is greater.

RETEACHING

The use of a number line can help those students having difficulty comparing numbers. Start with some simple examples using whole numbers and make sure students understand that the larger number is always to the right.

Ask students to compare 9.84 and 9.67 by using a number line. Have them start with 9.00 and arrange 9.10, 9.20, 9.30, . . . , 9.90, 10.00 on their number lines.

Before discussing the Examples in the book, review the concept of place value. Without an understanding of place value the comparison of numbers, particularly decimals, becomes a rote and meaningless process.

WORKSHOP 2

Comparing Numbers

Look up Skill 1 on pages 582–583 for more practice.

To compare numbers to see which one is greater, first find out if one of the numbers has more digits to the left of the decimal point. For example, 100 is greater than 99.62 because 100 has three digits to the left of the decimal point and 99.62 has only two. If both numbers have the same number of digits to the left of the decimal point, compare each digit, beginning at the left. One way is to write the value of each digit.

EXAMPLE Compare 5972 and 5983. Which number is greater?

SOLUTION

5972 = 5000 + 900 + 70 + 2
5983 = 5000 + 900 + 80 + 3

5000: same; 900: same; 80 is greater than 70, so 5983 is greater than 5972.

You may not have to write the value of each digit. Just compare each digit, starting at the left.

EXAMPLE Compare 237.51 and 237.29. Which number is greater?

SOLUTION

2	3	7	• 5	1
2	3	7	• 2	9
same	same	same	5 is greater than 2, so 237.51 is greater than 237.29.	

✓ SELF-CHECK Complete the problems, then check your answers in the back of the book.

Which number is greater?

1. 8891 and 8889 **8891**
2. 759.39 and 759.80 **759.80**
3. 3064 and 3079 **3079**
4. 56.633 and 56.629 **56.633**

When you are comparing decimals, you may have to write zeros to the right of the decimal point so that the numbers being compared have the same number of decimal places.

EXAMPLE Compare 5.17 and 5.149 (without writing the value of each digit). Which number is greater?

SOLUTION

5.170 ← Write in the zero.
5.149

7 is greater than 4, so 5.17 is greater than 5.149.

✓ SELF-CHECK Complete the problems, then check your answers in the back of the book.

Which number is greater?

5. 284.4 and 284.396 **284.4**
6. 0.06 and 0.006 **0.06**
7. 62.3 and 62.313 **62.313**

PROBLEMS

Which number is greater?

8. 26 or 29
29

9. 545 or 554
554

10. 4201 or 4210
4210

11. 3943 or 3934
3943

12. 2.65 or 2.56
2.65

13. 81.2 or 80.9
81.2

14. 0.696 or 0.695
0.696

15. 0.4 or 0.04
0.4

16. 3.54 or 4
4

17. 0.1 or 0.9
0.9

18. 0.03 or 0.003
0.03

19. 2.234 or 2.244
2.244

20. $8.79 or $8.97
$8.97

21. $329.02 or $329.20
$329.20

22. $13.11 or $11.13
$13.11

23. "29 miles to the gallon in highway driving. 24.9 miles to the gallon in city driving." Which mileage is greater? **29 miles to the gallon**

24. "Still available for the play-offs: 3428 box seats, 3128 grandstand seats." Which number is greater? **3428 box seats**

Write the numbers in order from lowest to highest.

25. 1.37, 1.36, 1.39
1.36, 1.37, 1.39

26. 5.11, 5.09, 5.10
5.09, 5.10, 5.11

27. 7.18, 7.38, 7.58
7.18, 7.38, 7.58

28. 5.86, 5.95, 5.81
5.81, 5.86, 5.95

29. 15.321, 15.231, 15.270
15.231, 15.270, 15.321

30. 40.004, 40.04, 40.4
40.004, 40.04, 40.4

31. 121.012, 121.021, 121.210
121.012, 121.021, 121.210

32. 365.15, 365.51, 365.490
365.15, 365.490, 365.51

33. 0.1234, 0.1342, 0.1423
0.1234, 0.1342, 0.1423

34. 0.2539, 0.4139, 0.3239
0.2539, 0.3239, 0.4139

35. 10.100, 11.11, 11.01
10.100, 11.01, 11.11

36. 57.57, 57.571, 57.75
57.57, 57.571, 57.75

37. 3.1, 3.062, 3.09
3.062, 3.09, 3.1

38. 8.001, 8.02, 8.0135
8.001, 8.0135, 8.02

APPLICATIONS

39. Janet Swick works part-time shelving books in the school library. Arrange the following library call numbers from lowest to highest.

a. Science: 513.12, 519.03, 532.626, 571.113, 587.41
513.12, 519.03, 532.626, 571.113, 587.41

b. Literature: 94.79, 32.615, 11.7, 67.192, 34.9
11.7, 32.615, 34.9, 67.192, 94.79

c. Religion: 46.94, 18.7, 15.04, 71.21, 26.311
15.04, 18.7, 26.311, 46.94, 71.21

d. Language: 22.5, 67.21, 48.275, 38.9, 93.047
22.5, 38.9, 48.275, 67.21, 93.047

40. Fourteen books were returned in the last hour. To help Janet shelve them faster, put them in numerical order from lowest to highest.

824.62, 392.4, 329.213, 419.52, 723.6, 122.51, 918.7, 749.02, 728.31, 927.263, 918.712, 826.401, 615.02, 131.06 **122.51, 131.06, 329.213, 392.4, 419.52, 615.02, 723.6, 728.31, 749.02, 824.62, 826.401, 918.7, 918.712, 927.263**

41. Janet also files the cards from checked-out books, according to the card numbers. Between which two cards would she file the checked-out books?

a. Card number of the checked-out book is 874.192. **873.15 and 877.142**
Filed cards are: 872.41, 873.15, 877.142, 879.190

b. Card number of the checked-out book is 332.75. **332.749 and 333.54**
Filed cards are: 309.8, 311.75, 332.075, 332.749, 333.54

WORKSHOPS

ERROR ANALYSIS

Ask those students who continue to get wrong answers to explain what they are doing. Their explanations may provide a clue to why they cannot compare numbers successfully.

Also observe students' written work. Carelessness in arranging numbers vertically can result in errors. Tell students that the decimal points must be aligned vertically to compare decimals. Then all they have to do is look for the place where the digits are different to find the greater number. Using the Examples in the text, ask students *why* 237.51 is greater than 237.29. Try to elicit an answer based upon an understanding of place value.

PRACTICE AND APPLY

The following problems can be assigned for classwork and the answers checked in class to help students master the objective of the lesson.

- Guided Practice: 8–20
- Independent Practice: 21–43 odd

WRAP-UP

Call upon four volunteers to write the answers to Problem 32 on the chalkboard. Ask other students if they see any errors and, if so, to correct them.

Assignment Guide

- Basic: 21–45 odd
- Average: 27–45 odd

WORKSHOPS

42. Use the tax table at the right to complete the following problem.

	Amount	At least	But less than
	$197.50	$195	$200
a.	$193.40	? **$190**	? **$195**
b.	$178.30	? **$175**	? **$180**
c.	$180.00	? **$180**	? **$185**
d.	$181.50	? **$180**	? **$185**
e.	$195.01	? **$195**	? **$200**

FEDERAL WITHHOLDING TAX
WEEKLY Payroll Period

And the wages are:

At least	But less than
$170	$175
175	180
180	185
185	190
190	195
195	200

43. Shipping charges often depend on the dollar amount of the order, as shown in the table. Find the cost of shipping each order.

a. $2.45 **b.** $2.95 **c.** $2.45 **d.** $2.95 **e.** $2.95 **f.** $3.95 **g.** $4.45 **h.** $1.95 **i.** $2.95 **j.** $3.45

a. $9.90 **b.** $14.35
c. $8.54 **d.** $15.00
e. $12.44 **f.** $21.22
g. $45.00 **h.** $6.54
i. $11.70 **j.** $16.15

SHIPPING AND HANDLING

Up to $8.00	Add $1.95
$8.01 to $11.00	Add $2.45
$11.01 to $15.00	Add $2.95
$15.01 to $20.00	Add $3.45
$20.01 to $30.00	Add $3.95
Over $30.00	Add $4.45

44. The minimum payment required on a charge account can be determined from a minimum payment schedule. Find the minimum payment for each balance.

a. $25 **b.** $30 **c.** $40 **d.** $50 **e.** $9.30 **f.** $35 **g.** $55 **h.** $25 **i.** $30 **j.** $45

a. $75.90 **b.** $322.82
c. $520.00 **d.** $782.29
e. $9.30 **f.** $427.20
g. $800.01 **h.** $26.13
i. $330.85 **j.** $700.00

MINIMUM PAYMENT SCHEDULE

New Balance	Minimum Monthly Payment
$ .01 to $ 300.00	$25.00
$300.01 to $ 400.00	$30.00
$400.01 to $ 500.00	$35.00
$500.01 to $ 600.00	$40.00
$600.01 to $ 700.00	$45.00
$700.01 to $ 800.00	$50.00
$800.01 to $ 900.00	$55.00
$900.01 to $1000.00	$60.00

45. Olympic scores are often given in decimals. In racing events, the lowest time wins. List the order of finish for each event.

a. Men's 1500 m

2	Baumann	3:15.52
1	Ngugi	3:11.70
3	Kunze	3:15.73

b. Women's 1500 m

3	Samoienko	4:00.30
1	Ivan	3:53.96
2	Baikauskaite	4:00.24

c. Men's 400 m relay

1	Spain	38.19 s
3	France	38.40 s
2	Britain	38.28 s

In some events, the highest number wins. Determine the winner for each of following events.

d. Men's discus

2	Oubartas	221 ft $4\frac{1}{2}$ in
1	Schult	225 ft $9\frac{1}{4}$ in
3	Danneberg	221 ft $0\frac{1}{2}$ in

e. Women's shot put

3	Meisu	69 ft 1 in
1	Lisovskaya	72 ft $11\frac{1}{2}$ in
2	Neimke	69 ft $1\frac{1}{2}$ in

f. Men's archery

1	S. Korea	986 pts.
3	Britain	968 pts.
2	USA	972 pts.

Adding Decimals

Look up Skill 5 on page 587 for more practice.

When adding decimals, write the addition problem in vertical form. Be sure to line up the decimal points. Write a decimal point in the answer directly below the decimal points in the problem. Then add as you would whole numbers.

EXAMPLE 26.94 + 31.71 + 19.47

SOLUTION Line up decimals.

```
26.94
31.71
19.47
```

Add.

```
  26.94
  31.71
+ 19.47
  78.12
```

✓ SELF-CHECK Complete the problems, then check your answers in the back of the book.

1. 17.56 + 11.73 + 12.47 = ? 41.76
2. 31.71 + 22.46 + 41.97 = ? 96.14
3. 53.15 + 41.71 + 79.84 = ? 174.70
4. 41.98 + 74.71 + 81.96 = ? 198.65

When adding amounts with different numbers of decimal places, you may want to write zeros in the empty decimal places. Put a decimal point in any whole number included in the problem.

EXAMPLE 41.51 + 117.483 + 7 + 19.6

SOLUTION

Line up decimals.

```
 41.51
117.483
  7.
 19.6
```

Write zeros.

```
 41.510
117.483
  7.000
 19.600
```

Add.

```
   41.510
  117.483
    7.000
+  19.600
  185.593
```

Adding amounts of money is just like adding decimals. The decimal point separates the dollars and cents. Remember to put a dollar sign in the total.

EXAMPLE \$43.78 + \$6.93

SOLUTION Line up decimals.

```
$43.78
  6.93
```

Add.

```
  $43.78
+   6.93
  $50.71
```

✓ SELF-CHECK Complete the problems, then check your answers in the back of the book.

5. 114.7 + 31.83 + 51.678 = ? 198.208
6. 13.7 + 6.012 + 17.412 + 6 = ? 43.124
7. \$14.71 + \$21.84 + \$7.51 = ? \$44.06
8. \$75 + \$14.40 + \$21.72 = ? \$111.12

PROBLEMS

9.
```
  36.74
+ 51.27
  88.01
```

10.
```
  78.91
+  4.17
  83.08
```

11.
```
  715.17
+ 612.94
 1328.11
```

12.
```
  513.88
+ 721.93
 1235.81
```


WORKSHOPS

LESSON PLAN
Workshop 3

OBJECTIVES

1. To add decimals with the same number of decimal places.
2. To add decimals with a different number of decimal places
3. To add amounts of money.

RETEACHING

Work a few problems with whole numbers before having students add decimals. Start with the following example: 1473 + 2609 = 4082. Then change the problem to adding 14.73 and 40.82. Ask students to read the decimals 14.73 and 40.82 and have them name each place in both decimals.

Write the following decimals on the chalkboard and ask students if they are equal: 42.76 and 42.760. Use similar examples and ask students why the zero does not change the value of the decimal. Review place value concepts from Workshop 1 if necessary.

ERROR ANALYSIS

Write the following decimals on the chalkboard and ask students to write them in vertical form so they can be added: 107.42, 63.971, 4.063. Stress that the decimal points must be aligned vertically when adding decimals. Ask students to write 9 as a decimal.

Accept any of the following answers: 9.0, 9.00, 9.000 and so on. Then have students add the following numbers: 12.42, 7, 13.06, 8.4, and 159. Review students' written work if they did not get the correct answer (199.88) to see what kinds of errors are being made.

13. 71.172 + 14.835 **86.007**	**14.** 64.574 + 76.915 **141.489**	**15.** 3.171 + 7.847 **11.018**	**16.** 7.974 + 4.157 **12.131**
17. 71.91 84.76 + 48.53 **205.2**	**18.** 44.76 53.12 + 87.94 **185.82**	**19.** 7.4189 4.5567 + 3.7124 **15.688**	**20.** 17.471 35.683 + 71.457 **124.611**
21. 417.9 2.171 + 43.42 **463.491**	**22.** 119.71 14.861 + 9.5 **144.071**	**23.** 2.178 47.5 + 917. **966.678**	**24.** 61.178 191.41 + 7.786 **260.374**
25. 7.456 517.1 + 120.8 **645.356**	**26.** 7180.9 15.754 + 617.9 **7814.554**	**27.** 14.79 3.1451 + 976.8 **994.7351**	**28.** 0.42 817. + 41.7181 **859.1381**
29. $17.98 14.31 + 19.47 **$51.76**	**30.** $161.91 72.87 + 54.36 **$289.14**	**31.** $17.56 71. + 41.84 **$130.40**	**32.** $517. 14.97 + 416.34 **$948.31**

33. 17.47 + 34.71 + 56.78 + 15.07 **124.03**

34. 0.17 + 17.94 + 13 + 147.7 **178.81**

35. 14.6 + 19.314 + 4.71 + 264.5176 **303.1416**

36. 4.917 + 6 + 4.37 + 15.971 **31.258**

37. $1.98 + $71.49 + $0.49 + $50 **$123.96**

38. $7.79 + $0.89 + $412.37 + $7 **$428.05**

39. $71.84 + $2.79 + $143.54 + $71 **$289.17**

40. $7.98 + $4.14 + $71.84 + $0.47 **$84.43**

APPLICATIONS Find the net deposit for each bank deposit slip by adding the deposits.

41.

		DOLLARS	CENTS
CASH	CURRENCY	17	00
	COINS		
CHECKS	LIST SEPARATELY 89-1	114	79
	89-2	17	08
	SUBTOTAL		
	LESS CASH RECEIVED		
	TOTAL DEPOSIT	?	

148.87

42.

		DOLLARS	CENTS
CASH	CURRENCY		
	COINS	4	75
CHECKS	LIST SEPARATELY 85-12	17	85
	85-10	71	94
	85-11	114	47
	SUBTOTAL		
	LESS CASH RECEIVED		
	TOTAL DEPOSIT	?	

209.01

43.

		DOLLARS	CENTS
CASH	CURRENCY	7	00
	COINS		
CHECKS	LIST SEPARATELY 95-76	314	78
	98-11	37	14
	95-13	212	71
	SUBTOTAL		
	LESS CASH RECEIVED		
	TOTAL DEPOSIT	?	

571.63

Complete the sales receipts by finding the subtotal and the total.

44.

DATE	AUTH. NO	IDENTIFACTION	CLERK	REG/DEPT.	☑TAKE ☐SEND
3/14/-	42		JR		

QTY	CLASS	DESCRIPTION	PRICE	AMOUNT
1		dress shirt		24 95
1		sweater		49 98
a.			SUBTOTAL	?
CUSTOMER SIGNATURE X Betty Clark			TAX	4 50
b.		SALES SLIP	TOTAL	?

a. $74.93
b. $79.43

45.

DATE	AUTH. NO	IDENTIFACTION	CLERK	REG/DEPT.	☑TAKE ☐SEND
6/1/-	86430		DL		

QTY	CLASS	DESCRIPTION	PRICE	AMOUNT
1		Knit shirt		32 98
2		Pairs Socks	3.19 ea.	6 38
1		Can Tennis Balls		2 19
a.			SUBTOTAL	?
CUSTOMER SIGNATURE X Ed Hack			TAX	3 32
b.		SALES SLIP	TOTAL	?

a. $41.55
b. $44.87

46. The personnel office of Color Sales, Inc., received this partially completed travel expense report for Beatrice Casey. Find the missing totals. First add the total expenses for each category horizontally. Then add all expenses for each day vertically. Add the Daily Totals horizontally. The sum of the vertical totals should equal the sum of the horizontal totals.

Color Sales, Inc.
419 N. Second St.
Allen, NY 01212

Travel Expense Report

Name *Beatrice Grey* Position *Sales Representative*
Destination *Philadelphia, PA*
Purpose of Trip *Regional Sales Meeting*
Date(s) of Trip *12/5-9*

DATES

Answer		Expense Item	Sun.	Mon. 12-5	Tue. 12-6	Wed. 12-7	Thur. 12-8	Fri. 12-9	Sat.	TOTALS
$418.96	a.	Hotel		104.74	104.74	104.74	104.74			?
37.15	b.	Breakfast			8.14	9.16	8.71	11.14		?
31.14	c.	Lunch			9.17	8.24	6.54	7.19		?
89.21	d.	Dinner		21.74	19.91	22.41	25.15			?
25.00	e.	Tips		5.00	6.00	6.00	6.00	2.00		?
		Entertainment								
22.00	f.	Taxi		10.00				12.00		?
		Public trans.								
		Parking								
		Tolls								
		Car rental								
240.00	g.	Airfare		240.00						?
27.85	h.	Telephone		6.50	7.15		9.50	4.70		?
		DAILY TOTALS		?	?	?	?	?		? Total Before Mileage
				i. 387.98	j. 155.11	k. 150.55	l. 160.64	m. 37.03		n. $891.31

47. These 5 contestants at the gymnastic tryouts received the scores listed below. Find the total for each contestant.

Answer		Judge:	A	B	C	D	E	F	G	H	Total
		Contestant									
a. 39.4	a.	Mark	4.7	4.9	5.1	5.0	4.9	5.3	4.7	4.8	?
b. 48.8	b.	Steve	6.1	6.3	6.0	5.9	6.4	6.3	6.0	5.8	?
c. 77.6	c.	Tom	9.8	9.6	9.5	10.0	9.7	9.6	10.0	9.4	?
d. 75.5	d.	John	9.5	9.4	9.6	9.3	9.2	9.6	9.5	9.4	?
e. 70.5	e.	Brad	8.7	8.8	9.1	8.5	8.7	8.8	8.9	9.0	?

WORKSHOPS

PRACTICE AND APPLY

The following problems can be assigned for classwork and the answers checked in class to help students master the objective of the lesson.

- Guided Practice: 11–22
- Independent Practice: 23–41 odd

WRAP-UP

Call upon individual students to read their answers to Problems 21–39 to the class. Go over any problem for which a wrong answer is given.

Assignment Guide

- Basic: 25–43 odd
- Average: 41–47

LESSON PLAN
Workshop 4

OBJECTIVES

1. To subtract decimals with the same number of decimal places.
2. To subtract decimals with a different number of decimal places.
3. To subtract amounts of money.

RETEACHING

Use the Examples in the text to reteach the subtraction of decimals. Ask for a few volunteers to work some problems at the chalkboard. Then read the following decimals to students and tell them to arrange the decimals in vertical form and subtract:

(1) 72.69 − 48.32 24.37
(2) 104.7 − 39.46 65.24
(3) 15 − 8.98 6.02
(4) 8.901 − 3.79 5.111

Ask students to read aloud their answers to check for misunderstandings in reading decimals.

WORKSHOPS

WORKSHOP 4

Subtracting Decimals

Look up Skill 6 on page 588 for more practice.

When subtracting decimals, write the subtraction problem in vertical form. Be sure to line up the decimal points. Write a decimal point in the answer directly below the decimal points in the problem. Then subtract as you would with whole numbers.

EXAMPLE

85.29 − 34.72 ➡

SOLUTION

Line up decimals.
85.29
34.72 ➡

Subtract.
85.29
− 34.72
50.57

When subtracting amounts with different numbers of decimal places, you may want to write zeros in the empty decimal places. Put a decimal point in any whole number included in the problem.

EXAMPLE

838.5
− 39.248 ➡

SOLUTION

Put in zeros and subtract.
838.500
− 39.248
799.252

EXAMPLE

47.49
− 9 ➡

SOLUTION

Put in zeros and subtract.
47.49
− 9.00
38.49

✓ SELF-CHECK Complete the problems, then check your answers in the back of the book.

1. 87.86 − 34.25 = ? **53.61**
2. 125.9 − 87.6 = ? **38.3**
3. 675.4 − 65.32 = ? **610.08**
4. 76.76 − 8 = ? **68.76**

Subtracting amounts of money is just like subtracting decimals. The decimal point separates the dollars and cents. Remember to put a dollar sign in the answer.

EXAMPLE

$89.59
− 56.37 ➡

SOLUTION

$89.59
− 56.37
$33.22

EXAMPLE

$572.
− 89.64 ➡

SOLUTION

$572.00
− 89.64 ➡

$572.00
− 89.64
$482.36

✓ SELF-CHECK Complete the problems, then check your answers in the back of the book.

5. $47.85 − 31.07 **$16.78**	6. $998.41 − 86.35 **$912.06**	7. $679.13 − 346.98 **$332.15**	8. $114. − 74.95 **$39.05**

PROBLEMS

9. 94.7 − 31.4 **63.3**	10. 98.6 − 88.5 **10.1**	11. 19.87 − 8.54 **11.33**	12. 7.93 − 2.03 **5.90**
13. 49.64 − 10.34 **39.30**	14. 96.13 − 12.37 **83.76**	15. 38.065 − 33.426 **4.639**	16. 68.111 − 9.648 **58.463**

17. 28.37 − 3.067 **25.303**

18. 28.36 − 4.239 **24.121**

19. 6.1536 − .1985 **5.9551**

20. 6.9909 − 3.9999 **2.9910**

21. 87.492 − 31.875 **55.617**

22. 43.686 − 20.320 **23.366**

23. 754.8 − 34.632 **720.168**

24. 895.1 − 33.47 **861.63**

25. 420. − 12.78 **407.22**

26. 201. − 9.764 **191.236**

27. 3.2 − .459 **2.741**

28. 36. − .357 **35.643**

29. $99.85 − 32.16 **$67.69**

30. $75.67 − 28.30 **$47.37**

31. $953.22 − 287.32 **$665.90**

32. $506.37 − 243.70 **$262.67**

33. $347. − 82.97 **$264.03**

34. $275. − 85.12 **$189.88**

35. $55,553.65 − 38,872.58 **$16,681.07**

36. $47,005.45 − 3257.64 **$43,747.81**

37. 335.4 − 217.9 **117.5**

38. 148.1 − 132.5 **15.6**

39. 5.21 − 0.71 **4.5**

40. 348.1 − 25.38 **322.72**

41. 23.57 − 0.6421 **22.9279**

42. 537 − 46.11 **490.89**

43. $61.92 − $33.01 **$28.91**

44. $75.99 − $31.49 **$44.50**

45. $8 − $2.63 **$5.37**

46. $434.66 − $51.43 **$383.23**

47. $3479.31 − $2616.16 **$863.15**

48. $6000 − $4333.83 **$1666.17**

APPLICATIONS

49. You work as a cashier in a restaurant. Compute the correct change for each of the following orders.

	Customer's Order	Customer Gives You	Change	
a.	$ 6.94	$ 7.00	?	**$0.06**
b.	$ 9.12	$10.00	?	**$0.88**
c.	$16.97	$17.00	?	**$0.03**
d.	$ 3.42	$ 5.42	?	**$2.00**
e.	$ 5.01	$ 5.01	?	**$0.00**
f.	$23.11	$25.00	?	**$1.89**
g.	$41.97	$42.07	?	**$0.10**
h.	$27.42	$30.00	?	**$2.58**
i.	$20.13	$21.13	?	**$1.00**
j.	$150.84	$151.00	?	**$0.16**

Complete the computational part of the following bank deposit tickets.

50. Subtotal **$52.56**; Total deposit **$42.56**

			DOLLARS	CENTS
CASH	CURRENCY		2	00
CASH	COINS			
CHECKS	LIST SEPARATELY	1-43	22	76
CHECKS		1-44	27	80
CHECKS				
a.	SUBTOTAL		?	
	LESS CASH RECEIVED		10	00
b.	TOTAL DEPOSIT		?	

51. Subtotal **$132.55**; Total deposit **$112.55**

			DOLLARS	CENTS
CASH	CURRENCY			
CASH	COINS		3	50
CHECKS	LIST SEPARATELY	57-12	42	50
CHECKS		57-10	86	55
CHECKS				
a.	SUBTOTAL		?	
	LESS CASH RECEIVED		20	00
b.	TOTAL DEPOSIT		?	

52. Subtotal **$215.91**; Total deposit **$170.91**

			DOLLARS	CENTS
CASH	CURRENCY		13	00
CASH	COINS		6	41
CHECKS	LIST SEPARATELY	85-76	155	40
CHECKS		88-11	25	35
CHECKS			15	75
a.	SUBTOTAL		?	
	LESS CASH RECEIVED		45	00
b.	TOTAL DEPOSIT		?	

ERROR ANALYSIS

Many errors students make when subtracting decimals are computational errors in subtraction. If this is the case with your students, you will need to review the subtraction of whole numbers. A thorough understanding of place value concepts can help students having difficulty subtracting whole numbers and decimals. Also, a discussion of examples involving dollar and cent amounts can make the subtraction of decimals more understandable to many students.

PRACTICE AND APPLY

The following problems can be assigned for classwork and the answers checked in class to help students master the objective of the lesson.

- Guided Practice: 9–20
- Independent Practice: 21–48

WRAP-UP

Work Problem 49 with the class by calling upon individual students to do each part.

Assignment Guide

- Basic: 22–56 even
- Average: 37–57 odd

53.

		DOLLARS	CENTS
CASH	CURRENCY		
	COINS	17	25
CHECKS	LIST SEPARATELY 50-50	16	40
	50-75	35	83
	45-30	27	11
a.	SUBTOTAL	?	
	LESS CASH RECEIVED	40	00
b.	TOTAL DEPOSIT	?	

Subtotal $96.59
Total deposit $56.59

54.

		DOLLARS	CENTS
CASH	CURRENCY	137	00
	COINS	12	35
CHECKS	LIST SEPARATELY 66-40	15	27
	60-50	19	38
	65-70	214	51
a.	SUBTOTAL	?	
	LESS CASH RECEIVED	75	00
b.	TOTAL DEPOSIT	?	

Subtotal $398.51
Total deposit $323.51

55.

		DOLLARS	CENTS
CASH	CURRENCY	21	00
	COINS	37	75
CHECKS	LIST SEPARATELY 40-10	43	33
	45-50	37	40
	42-13	437	65
a.	SUBTOTAL	?	
	LESS CASH RECEIVED	75	00
b.	TOTAL DEPOSIT	?	

Subtotal $577.13
Total deposit $502.13

56. A partially completed budget comparison statement for the Roadway Bicycle Club is shown below.

ROADWAY BICYCLE CLUB **BUDGET COMPARISON STATEMENT**
For the year 19--

ITEM	ESTIMATED	ACTUAL	DIFFERENCE	OVER/UNDER?
Receipts				
Dues	360 00	520 00	?	?
Bike-a-thon	400 00	324 00	?	?
Raffle	1000 00	1142 00	?	?
T-shirt sales	275 00	231 60	?	?
Total Receipts	?	?		

160.00 over
76.00 under
142.00 over
43.40 under

2035 **2217.60**

a. Find the club's total estimated and total actual receipts.

b. Complete the difference column by subtracting the Actual column from the Estimated column. Label the difference as being "over" or "under" the estimate.

57. The marketing department for Jackson Sporting Goods Company prepared a comparison sheet that shows sales projections and actual sales for ten products. Find the difference between the projected sales and the actual sales by subtracting actual sales from projected sales.

	Jackson Sporting Goods Company			
	Product	Sales Projections	Actual Sales	Difference
a.	Electric fishing motors	$ 3000.00	$ 2749.67	**$250.33**
b.	Fishing tackle	$32,000.00	$31,897.40	**$102.60**
c.	Wilderness boots	$ 6500.00	$ 5607.15	**$892.85**
d.	Boat moccasin	$ 950.00	$ 571.28	**$378.72**
e.	Enamel campware sets	$ 400.00	$ 339.67	**$60.33**
f.	Geodesic dome tent	$ 6400.00	$ 5379.58	**$1020.42**
g.	Backpacking tent	$ 2500.00	$ 1999.40	**$500.60**
h.	Sleeping bags	$ 6300.00	$ 5919.75	**$380.25**
i.	Sleeping bag pads	$ 800.00	$ 589.20	**$210.80**
j.	Folding cot	$ 740.00	$ 731.52	**$8.48**

Multiplying Decimals

Look up Skill 8 on page 590 for more practice.

When multiplying decimals, multiply as if the decimal numbers were whole numbers. Then count the total number of decimal places to the right of the decimals in the factors. This number will be the number of decimal places in the product.

EXAMPLE

```
  18.1  ← factor
× 0.35  ← factor
   905
  543
  6335  ← product
```

SOLUTION

```
  18.1  ←    1 decimal place
× 0.35  ←  + 2 decimal places
   905
  543
 6.335  ←    3 decimal places
```

If the product does not have enough digits to place the decimal in the correct position, you will need to write zeros. Start at the right of the product in counting the decimal places and write zeros at the left.

EXAMPLE

```
  0.72
× 0.03
   216
```

SOLUTION

```
  0.72  ←    2 decimal places
× 0.03  ←  + 2 decimal places
0.0216  ←    4 decimal places
```

✓ SELF-CHECK Complete the problems, then check your answers in the back of the book.

1. 24.7 × 0.33 = 8.151 **2.** 41.8 × 2.14 = 89.452 **3.** 0.78 × 0.11 = 0.0858 **4.** 0.74 × 0.08 = 0.0592

When multiplying amounts of money, round the answer off to the nearest cent. Remember to put a dollar sign in the answer.

EXAMPLE

```
  $3.35
×   4.5
  15075
```

SOLUTION

```
  $3.35  ←    2 places    $3.35 × 4.5 = $15.075
×   4.5  ←  + 1 place                 = $15.08
$15.075  ←    3 places    rounded to nearest cent
```

When multiplying by 10, 100, or 1000, count the number of zeros and then move the decimal point to the right the same number of spaces.

EXAMPLE 1

6.7 × 10

SOLUTION

6.7 × 10 = 6.7 = 67. — 10 has 1 zero; move decimal 1 place.

EXAMPLE 2

5.24 × 100

SOLUTION

5.24 × 100 = 5.24 = 524. — 100 has 2 zeros; move decimal 2 places.

EXAMPLE 3

9.4718 × 1000

SOLUTION

9.4718 × 1000 = 9.4718 = 9471.8 — 1000 has 3 zeros; move decimal 3 places.

LESSON PLAN Workshop 5

OBJECTIVES

1. To multiply decimals.
2. To multiply amounts of money.
3. To multiply decimals by 10, 100, or 1000.

RETEACHING

Remind students that multiplication of decimals is similar to multiplying whole numbers. Emphasize the need for setting the problems up neatly and properly aligning the numbers in columns. Neatness is very important. Discuss the rule for finding the number of decimal places in the answer.

In the second example, make sure the students start at the right of the product in computing decimal places and place any needed zeros at the left of the product.

Stress that multiplication by 10, 100, 1000, or 10,000 is simply a matter of counting zeros and moving the decimal point that many places to the right.

ERROR ANALYSIS

Review students' written work to identify the kinds of errors they may be making. Very often, errors in the basic multiplication facts are common. If this is the case, ask students to write out the multiplication tables, starting with the two times table and going through the nine times table. Review their tables to find and correct the errors in them.

✓ SELF-CHECK Complete the problems, then check your answers in the back of the book.

5. $4.15 × 8.5 **$35.275 = $35.28**
6. 71.4 × 10 **714**
7. 41.861 × 100 **4186.1**
8. 3.1794 × 1000 **3179.4**

PROBLEMS

9. 41.3 × 0.2 **8.26**
10. 78.4 × 0.3 **23.52**
11. 84.8 × 0.25 **21.2**
12. 74.5 × 0.15 **11.175**
13. 4.31 × 0.71 **3.0601**
14. 51.7 × 0.72 **37.224**
15. 97.8 × 0.31 **30.318**
16. 51.7 × 0.67 **34.639**
17. 0.62 × 0.06 **0.0372**
18. 0.86 × 0.08 **0.0688**
19. 0.41 × 0.02 **0.0082**
20. 0.74 × 0.08 **0.0592**
21. 0.51 × 0.06 **0.0306**
22. 0.93 × 0.07 **0.0651**
23. 0.635 × 0.005 **0.003175**
24. 0.051 × 0.003 **0.000153**
25. 4.171 × 0.04 **0.16684**
26. 7.823 × 0.09 **0.70407**
27. 3.471 × 0.63 **2.18673**
28. 9.875 × 0.42 **4.1475**
29. 0.071 × 0.002 **0.000142**
30. 0.207 × 0.034 **0.007038**
31. 0.973 × 1.918 **1.866214**
32. 0.573 × 6.423 **3.680379**
33. $2.25 × 20 **$45**
34. $5.15 × 85 **$437.75**
35. $5.85 × 3.5 **$20.475 = $20.48**
36. $3.35 × 7.6 **$25.46**
37. $18.74 × 2.3 **$43.102 = $43.10**
38. $87.98 × 8.7 **$765.426 = $765.43**

Multiply by 10 **39.** 31.7 **317** **40.** 5.71 **57.1** **41.** 6.517 **65.17** **42.** 0.9186 **9.186**

Multiply by 100 **43.** 32.85 **3285** **44.** 41.786 **4178.6** **45.** 3.987 **398.7** **46.** 412.42 **41,242**

Multiply by 1000 **47.** 72.716 **72,716** **48.** 7.1956 **7195.6** **49.** 0.415 **415** **50.** 916.518 **916,518**

Multiply by 10,000 **51.** 6.9178 **69,178** **52.** 3.42876 **34,287.6** **53.** 94.7189 **947,189** **54.** 723.471 **7,234,710**

APPLICATIONS

55. Find the total price for each item and the total order amount.

	Quantity	Description	Price Each	Total Price
$3.87 a.	3	*Notebooks*	*$1.29*	?
$7.96 b.	2	*Automatic pencils*	*3.98*	?
$2.98 c.	2	*Pads of 3 × 5 cards*	*1.49*	?
$9.98 d.	1	*Pen*	*9.98*	?
$8.76 e.	4	*Packs of paper*	*2.19*	?
$9.27 f.	3	*Boxes of paper clips*	*3.09*	?
$7.49 g.	1	*Stapler with staples*	*7.49*	?
$50.31 h.		*Total*		?

56. Find the total price of each item, the total list price, the 10% trade discount, the net price, the 2% cash discount, the cash price, the 6% sales tax, and the final price.

	Quantity	Item Number	Unit Price	Total Price	
a.	32	2Z-143	$47.85	?	$1531.20
b.	15	5R-719	$14.73	?	$220.95
c.	75	6S-427	$ 3.87	?	$290.25
d.		Total List Price		?	$2042.40
e.		− Trade Discount (× 0.1)		?	$204.24
f.		Net Price		?	$1838.16
g.		− Cash Discount (× 0.02)		?	$36.76 ($36.7632 rounded)
h.		Cash Price		?	$1801.40
i.		+ Sales Tax (× 0.06)		?	$108.08 ($108.084 rounded)
j.		Final Price		?	1909.48

57. Find the gross pay for each employee by multiplying the hourly rate times the number of hours. Find social security by multiplying the gross pay by 0.062, medicare by multiplying by 0.0145, FIT by multiplying by 0.25, SIT by multiplying by 0.045. Round each deduction to the nearest cent. Find the total deductions and subtract from the gross pay to find the net pay.

	Employee	Hourly Rate	Number of Hours	Gross Pay	Deductions				Total Deductions	Net Pay
					Soc. Sec.	Medicare	FIT	SIT		
a.	Faye White	$7.85	40	? $314.00	? $19.47	? $4.55	? $78.50	? $14.13	? $116.65	? $197.35
b.	John Zielinski	$8.50	36	? $306.00	? $18.97	? $4.44	? $76.50	? $13.77	? $113.68	? $192.32
c.	Bill Adams	$8.75	40	? $350.00	? $21.70	? $5.08	? $87.50	? $15.75	? $130.03	? $219.97
d.	Jane Boyce	$9.70	35	? $339.50	? $21.05	? $4.92	? $84.88	? $15.28	? $126.13	? $213.37
e.	Al Tone	$9.70	40	? $388.00	? $24.06	? $5.63	? $97.00	? $17.46	? $144.15	? $243.85
f.			Total	? $1697.50	? $105.25	? $24.62	? $424.38	? $76.39	? $630.64	? $1066.86

WORKSHOPS

PRACTICE AND APPLY

The following problems can be assigned for classwork and the answers checked in class to help students master the objective of the lesson.

- Guided Practice: 9–23
- Independent Practice: 24–38

WRAP-UP

Ask five students to show their work for Problems 28–32 at the chalkboard.

Assignment Guide

- Basic: 12–44
- Average: 39–57

LESSON PLAN Workshop 6

OBJECTIVES

1. To divide decimals.
2. To divide an amount of money by a whole number.
3. To divide by 10, 100, or 1000.

RETEACHING

Dividing decimals is difficult for many students because a distinct series of steps must be carried out to get the correct answer. Work through each Example very carefully. Then review students' work on the Self-Check. Ask students why the decimal point is moved the same number of places in both the divisor and the dividend.

To help students remember how to divide decimals, you might wish to type the steps given in the first paragraph of this Workshop in list form and reproduce copies for all students. For example:

1. First check if there is a decimal point in the divisor.
2. If there is, move the decimal point to the right to make the divisor a whole number.

Continue the list until you have five steps.

WORKSHOPS

WORKSHOP 6

Dividing Decimals

Look up Skills 9, 10, and 11 on pages 591, 592, and 593 for more practice.

When dividing decimals, first check if there is a decimal point in the divisor. If there is, move the decimal point to the right to make the divisor a whole number. Then move the decimal point in the dividend to the right the same number of places you moved the decimal point in the divisor. Write the decimal point in the quotient directly above the decimal point in the dividend. Then divide as with whole numbers.

```
               231  ← quotient
divisor → 3 ) 693  ← dividend
```

EXAMPLE

23.78 ÷ 5.8

or

5.8) 23.78

SOLUTION

```
                          4.1
5.8 ) 23.78   ➡   58 ) 237.8
                      − 232
                          5 8
                        − 5 8
```

Add zeros to the right of the decimal point in the dividend if needed.

EXAMPLE

0.147 ÷ 0.42

or

0.42) 0.147

SOLUTION

```
                            0.35
0.42 ) 0.147   ➡   42 ) 14.70  ← zero added
                       − 12 6
                          2 10
                        − 2 10
```

✓ SELF-CHECK Complete the problems, then check your answers in the back of the book.

1. 35.96 ÷ 5.8 **6.2**
2. 12.9) 55.341 **4.29**
3. 1.47 ÷ 0.05 **29.4**
4. 0.052) 1.872 **36**

When the dividend is an amount of money, remember to place the dollar sign in the quotient and round the answer to the nearest cent.

EXAMPLE

24) $47.56

SOLUTION

```
     $1.981
24 ) $47.560      $47.56 ÷ 24 = $ 1.98  ← rounded to
                                            nearest cent
```

When dividing by 10, 100, or 1000, count the number of zeros in 10, 100, or 1000 and move the decimal point to the left the same number of places.

EXAMPLE 1 9.3 ÷ 10

SOLUTION 9.3 ÷ 10 = 09.3 = 0.93 — 10 has 1 zero; move decimal 1 place.

EXAMPLE 2 742.64 ÷ 100

SOLUTION 742.64 ÷ 100 = 742.64 = 7.4264 — 100 has 2 zeros; move decimal 2 places.

EXAMPLE 3 13,436.1 ÷ 1000

SOLUTION 13,436.1 ÷ 1000 = 13436.1 = 13.4361 — 1000 has 3 zeros; move decimal 3 places.

✓ SELF-CHECK Complete the problems, then check your answers in the back of the book.

5. $16.32 ÷ 12 **$1.36**
6. 7.9 ÷ 10 **0.79**
7. 138.9 ÷ 100 **1.389**
8. 9862.8 ÷ 1000 **9.8628**

PROBLEMS

9. 5.3)9.54 **1.8**
10. 3.2)11.2 **3.5**
11. 2.6)24.18 **9.3**
12. 3.2)40.32 **12.6**
13. 4.8)35.52 **7.4**
14. 2.2)70.84 **32.2**
15. 46)354.2 **7.7**
16. 9.9)564.3 **57**
17. 6.02)51.17 **8.5**
18. 3.12)39.936 **12.8**
19. 3.87)10.836 **2.8**
20. 3.04)29.336 **9.65**
21. 0.004)0.2928 **73.2**
22. 0.007)0.2415 **34.5**
23. 0.108)0.5616 **5.2**
24. 0.101)0.70498 **6.98**

Divide. Round answers to the nearest hundredth or to the nearest cent.

25. 4.3)7.871 **1.83**
26. 5.9)5.343 **0.91**
27. 7.36)88.34 **12.00**
28. 2.95)177.9 **60.31**
29. 0.88)46.98 **53.39**
30. 0.97)602.7 **621.34**
31. 14.32)8.027 **0.56**
32. 31.44)9.109 **0.29**
33. 31)$82.42 **$2.66**
34. 28)$23.98 **$0.86**
35. 78)$678.40 **$8.70**
36. 43)$177.79 **$4.13**
37. 62)$267.32 **$4.31**
38. 57)$793.28 **$13.92**
39. 961)$9803.88 **$10.20**
40. 817)$4635.79 **$5.67**

Divide by 10	41. 9.3 **0.93**	42. 14.42 **1.442**	43. 726.81 **72.681**	44. 0.034 **0.0034**
Divide by 100	45. 429.8 **4.298**	46. 133.39 **1.3339**	47. 8462.65 **84.6265**	48. 0.746 **0.00746**
Divide by 1000	49. 5896.9 **5.8969**	50. 321.29 **0.32129**	51. 22.098 **0.022098**	52. 5.432 **0.005432**
Divide by 10,000	53. 63,652.18 **6.365218**	54. 3879.19 **0.387919**	55. 415.49 **0.041549**	56. 64.858 **0.0064858**

57. Ken Kowlinski drove 495 miles on 15.0 gallons of gas. How many miles per gallon did he get? **33**

58. A 64-ounce bottle of detergent costs $3.49. What is the cost per ounce? **$0.05**

59. A 4.25-kilogram roast costs $17.39. What is the cost per kilogram? **$4.09**

WORKSHOPS

ERROR ANALYSIS

Students can make a variety of errors when dividing decimals. Those errors must be analyzed individually and corrected. Some types of errors are the following: (1) moving the decimal points incorrectly; (2) not moving the decimal point in the dividend; (3) errors in basic multiplication facts; (4) subtracting incorrectly; (5) forgetting to put a zero in the quotient; (6) not completing the division to arrive at the correct quotient.

PRACTICE AND APPLY

The following problems can be assigned for classwork and the answers checked in class to help students master the objective of the lesson.

- Guided Practice: 9–24
- Independent Practice: 25–39

WRAP-UP

Work out the complete solutions to Problems 39 and 40 at the chalkboard.

Assignment Guide

- Basic: 25–59
- Average: 41–62

WORKSHOPS

APPLICATIONS

60. As a clerk and stockperson at the Quick Stop Carryout, you are required to compute the cost per item (to the nearest cent) for the following items.

	Item	Cost per Case	Number per Case	Cost per Item
a.	Orange juice, 64 oz	$ 17.94	6	? **$2.99**
b.	Wild rice	$ 40.56	24	? **$1.69**
c.	Cake mix	$ 30.96	24	? **$1.29**
d.	Mushrooms, 4 oz	$ 6.09	12	? **$0.51**
e.	Cat food, 6 oz	$ 19.20	48	? **$0.40**
f.	Tuna fish, 6.5 oz	$ 25.36	36	? **$0.70**
g.	Batteries, 9 V	$108.00	60	? **$1.80**
h.	Cream cheese, 8 oz	$ 39.60	40	? **$0.99**
i.	Raisins, 15 oz	$ 31.20	24	? **$1.30**

61. Compute the miles per gallon (to the nearest tenth) for the types of vehicles leased by the Pilloid Plastics Company the first week in May.

	Type of Vehicle	Miles	Gallons of Fuel	Miles per Gallon
a.	Standard	665.4	40.3	? **16.5**
b.	Intermediate	371.3	19.9	? **18.7**
c.	Compact	407.0	16.8	? **24.2**
d.	Subcompact	511.6	17.3	? **29.6**
e.	Passenger van	291.5	22.1	? **13.2**
f.	Motor home	423.7	64.2	? **6.6**

62. The unit price of an item can be found by dividing the selling price by the weight. Determine the unit price of each item to the nearest thousandth.

	Item	Selling Price	Weight	Unit Price (3 decimal places)
a.	Vegetable oil	$1.79	32 oz	? **$0.056**
b.	Orange juice	$1.29	12 oz	? **$0.108**
c.	Olives	$1.39	7 oz	? **$0.199**
d.	Crackers	$1.59	16 oz	? **$0.099**
e.	Tea bags	$2.39	100 ctns	? **$0.024**
f.	Onion soup	$0.99	1.25 oz	? **$0.792**
g.	Dog food	$8.49	25 lb	? **$0.340**
h.	Pie spice	$1.69	1.125 oz	? **$1.502**

Fraction to Decimal, Decimal to Fraction

Look up Skill 14 on page 596 for more practice.

Any fraction can be renamed as a decimal and any decimal can be renamed as a fraction.

To rename a fraction as a decimal, use division. Think of the fraction bar in the fraction as meaning "divide by." For example, $\frac{5}{8}$ means "5 divided by 8." After the 5, write a decimal point and as many zeros as are needed. Then divide by 8.

EXAMPLE

Change $\frac{5}{8}$ to a decimal.

SOLUTION

$\frac{5}{8}$ ➡ 8)5.000 = 0.625

```
   0.625
8)5.000
 - 4 8
    20
  - 16
    40
  - 40
```

EXAMPLE

Change $\frac{2}{5}$ to a decimal.

SOLUTION

$\frac{2}{5}$ ➡

```
   0.4
5)2.0
 - 2 0
```

If a fraction does not divide out evenly, divide to one more decimal place than you are rounding to.

EXAMPLE

Change $\frac{3}{7}$ to a decimal rounded to the nearest hundredth. (Divide to the thousandths place.)

SOLUTION

$\frac{3}{7}$ ➡

```
   0.428 = 0.43
7)3.000
 - 2 8
    20
  - 14
     60
   - 56
      4
```

$\frac{3}{7} = 0.43$ (rounded)

EXAMPLE

Change $\frac{5}{6}$ to a decimal rounded to the nearest thousandth. (Divide to the ten thousandths place.)

SOLUTION

$\frac{5}{6}$ ➡

```
   0.8333 = 0.833
6)5.0000
 - 4 8
    20
  - 18
     20
   - 18
      20
    - 18
       2
```

$\frac{5}{6} = 0.833$ (rounded)

✓ SELF-CHECK Complete the problems, then check your answers in the back of the book.

Change the fractions to decimals. Round to the nearest thousandth. (Divide to the ten thousandths place.)

1. $\frac{7}{8}$ **0.875** **2.** $\frac{3}{5}$ **0.6** **3.** $\frac{7}{9}$ **0.778** **4.** $\frac{2}{3}$ **0.667**

WORKSHOPS

LESSON PLAN Workshop 7

OBJECTIVES

1. To rename a fraction as a decimal.
2. To rename a decimal as a fraction.

RETEACHING

Remind students that fractions and decimals are names for numbers. For example, the fraction $\frac{1}{2}$ and the decimal 0.5 are just two different ways to name the number one-half. Names for numbers are called *numerals.* The objectives of this workshop are concerned with ways to convert from one numeral form (fraction) to another numeral form (decimal). Confusion arises because students think that the numeral or symbol $\frac{1}{2}$ is actually the number, but it is only one way to name the number. Point out that $\frac{2}{4}$, $\frac{3}{6}$, $\frac{4}{8}$, and so on all are names for one-half.

ERROR ANALYSIS

When renaming a decimal as a fraction, students make errors by not using the correct denominator. Review place value concepts if these kinds of errors occur. You will probably also have to review how to write a fraction in lowest terms by dividing out common factors in the numerator and denominator.

WORKSHOPS

To rename a decimal as a fraction, name the place value of the digit at the far right. This is the denominator of the fraction.

$0.79 = \frac{79}{100}$	$0.003 = \frac{3}{1000}$
9 is in the hundredths place, so the denominator is 100.	3 is in the thousandths place, so the denominator is 1000.

Note that the number of zeros in the denominator is the same as the number of places to the right of the decimal point. The fraction should always be written in lowest terms.

$0.05 = \frac{5}{100} = \frac{1}{20}$ $\qquad$ $4.625 = 4\frac{625}{1000} = 4\frac{5}{8}$

5. $\frac{4}{10} = \frac{2}{5}$
6. $\frac{12}{100} = \frac{3}{25}$
7. $1\frac{125}{1000} = 1\frac{1}{8}$
8. $7\frac{82}{100} = 7\frac{41}{50}$

✓ SELF-CHECK Complete the problems, then check your answers in the back of the book.

Change the decimals to fractions reduced to lowest terms.

5. 0.4 **6.** 0.12 **7.** 1.125 **8.** 7.82

PROBLEMS

Change the fractions to decimals. Round to the nearest thousandth.

9. $\frac{1}{5}$ 0.2	**10.** $\frac{3}{4}$ 0.75	**11.** $\frac{5}{7}$ 0.714	**12.** $\frac{7}{15}$ 0.467
13. $\frac{9}{20}$ 0.45	**14.** $\frac{1}{6}$ 0.167	**15.** $\frac{4}{9}$ 0.444	**16.** $\frac{17}{120}$ 0.142
17. $\frac{17}{40}$ 0.425	**18.** $\frac{3}{8}$ 0.375	**19.** $\frac{7}{10}$ 0.7	**20.** $\frac{13}{50}$ 0.26
21. $\frac{9}{25}$ 0.36	**22.** $\frac{1}{2}$ 0.5	**23.** $\frac{5}{12}$ 0.417	**24.** $\frac{117}{200}$ 0.585

Change the fractions to decimals. Round to the nearest hundredth.

25. $\frac{1}{4}$ 0.25	**26.** $\frac{4}{5}$ 0.8	**27.** $\frac{1}{8}$ 0.13	**28.** $\frac{5}{9}$ 0.56
29. $\frac{6}{7}$ 0.86	**30.** $\frac{1}{3}$ 0.33	**31.** $\frac{3}{10}$ 0.3	**32.** $\frac{7}{16}$ 0.44
33. $\frac{3}{100}$ 0.03	**34.** $\frac{17}{25}$ 0.68	**35.** $\frac{13}{14}$ 0.93	**36.** $\frac{71}{75}$ 0.95

Change the fractions to decimals. Round to the nearest thousandth.

37. $\frac{1}{7}$ 0.143	**38.** $\frac{8}{9}$ 0.889	**39.** $\frac{7}{12}$ 0.583	**40.** $\frac{11}{13}$ 0.846
41. $\frac{4}{15}$ 0.267	**42.** $\frac{11}{16}$ 0.688	**43.** $\frac{17}{30}$ 0.567	**44.** $\frac{13}{40}$ 0.325
45. $\frac{27}{50}$ 0.54	**46.** $\frac{173}{200}$ 0.865	**47.** $\frac{491}{500}$ 0.982	**48.** $\frac{171}{1000}$ 0.171

Change the decimals to fractions reduced to lowest terms.

49. 0.25 $\frac{1}{4}$	**50.** 0.6 $\frac{3}{8}$	**51.** 0.5 $\frac{1}{2}$	**52.** 0.375 $\frac{3}{8}$
53. 0.3 $\frac{3}{10}$	**54.** 0.17 $\frac{17}{100}$	**55.** 0.45 $\frac{9}{20}$	**56.** 0.75 $\frac{3}{4}$
57. 0.125 $\frac{1}{8}$	**58.** 0.48 $\frac{12}{25}$	**59.** 0.755 $\frac{151}{200}$	**60.** 0.05 $\frac{1}{20}$
61. 0.1875 $\frac{3}{16}$	**62.** 0.65 $\frac{13}{20}$	**63.** 0.325 $\frac{13}{40}$	**64.** 0.68 $\frac{17}{25}$
65. 0.34 $\frac{17}{50}$	**66.** 0.2125 $\frac{17}{80}$	**67.** 0.9375 $\frac{15}{16}$	**68.** 0.95 $\frac{19}{20}$
69. 1.1 $1\frac{1}{10}$	**70.** 14.35 $14\frac{7}{20}$	**71.** 7.08 $7\frac{2}{25}$	**72.** 4.5 $4\frac{1}{2}$
73. 5.117 $5\frac{117}{1000}$	**74.** 10.085 $10\frac{17}{200}$	**75.** 3.4375 $3\frac{7}{16}$	**76.** 2.85 $2\frac{17}{20}$
77. 21.975 $21\frac{39}{40}$	**78.** 57.44 $57\frac{11}{25}$	**79.** 43.34375 $43\frac{11}{32}$	**80.** 76.313 $76\frac{313}{1000}$

81. $30\frac{1}{4}$ yards equal one square rod. Write $30\frac{1}{4}$ as a decimal. **30.25**

82. A basketball player made 163 free throws in 200 attempts. Write $\frac{163}{200}$ as a decimal. What was the player's free throw average? **0.815 81.5%**

83. Three months is $\frac{1}{4}$ of a year. Write $\frac{1}{4}$ as a decimal. **0.25**

84. The price of a stock is $\$16\frac{1}{4}$. Write $\$16\frac{1}{4}$ as a decimal. **$16.25**

85. One cubic foot is about 0.8 bushel. Write 0.8 as a fraction in lowest terms. $\frac{4}{5}$

WORKSHOPS

APPLICATIONS

86. The Major Indoor Lacrosse League standings are given. Convert the decimals to fractions reduced to lowest terms.

		MAJOR INDOOR LACROSSE LEAGUE				
		National Division	**W**	**L**	**Pct.**	**GB**
1	a.	Detroit	3	0	1.000	—
$\frac{1}{2}$	b.	Pittsburgh	1	1	.500	$1\frac{1}{2}$
about $\frac{1}{3}$	c.	New England	1	2	.333	2
		American Division	**W**	**L**	**Pct.**	**GB**
$\frac{1}{2}$	d.	New York	1	1	.500	—
about $\frac{1}{3}$	e.	Philadelphia	1	2	.333	$\frac{1}{2}$
about $\frac{1}{3}$	f.	Baltimore	1	2	.333	$\frac{1}{2}$

87. Stock prices are quoted as dollars and fractions of a dollar. Change the stock prices to dollars and cents. Round to the nearest cent.

	Stock	Price	
a.	Alcoa	$52\frac{1}{2}$	**$52.50**
b.	BnkAm	$22\frac{7}{8}$	**$22.88**
c.	Chrysler	$12\frac{5}{8}$	**$12.63**
d.	E Kodak	$45\frac{1}{4}$	**$45.25**
e.	JP Ind	$8\frac{3}{8}$	**$8.38**
f.	Ford	$26\frac{3}{8}$	**$26.38**
g.	GenMtrs	$39\frac{1}{2}$	**$39.50**
h.	Wendys	$5\frac{5}{8}$	**$5.63**

88. The stock market summary lists certain key measures in decimal form. Change the decimals to fractions reduced to lowest terms.

		STOCK MARKET SUMMARY	
		Dow Jones Average	**Decimal**
2643 7/100	a.	30 Industrials	2643.07
1036 8/25	b.	20 Transportation	1036.32
206 49/100	c.	15 Utilities	206.49
954 21/100	d.	65 Stocks	954.21
92 9/50	e.	20 Bonds	92.18
93 17/20	f.	10 Pub util bonds	93.85
90 2/25	g.	10 Ind bonds	90.08
124 69/100	h.	Commodity futures	124.69

89. Individual bowling averages in The Blade Classic League are carried to the nearest hundredth. Convert the decimals to fractions reduced to lowest terms.

	Name	Average	
a.	Kevin Taber	222.08	$222\frac{2}{25}$
b.	Dan Koles	216.32	$216\frac{8}{25}$
c.	Eric Roberts	216.01	$216\frac{1}{100}$
d.	Terry Jacobs	212.29	$212\frac{29}{100}$
e.	Jim Ferguson	210.05	$210\frac{1}{20}$
f.	Steve Hold	208.42	$208\frac{21}{50}$
g.	Paul Tyler	208.25	$208\frac{1}{4}$

PRACTICE AND APPLY

The following problems can be assigned for classwork and the answers checked in class to help students master the objective of the lesson.

- Guided Practice: 9–12, 25–28, 37–39, 49–52
- Independent Practice: 13–21, 29–32, 41–44, 53–56

WRAP-UP

Use the following Problems to review the skills of this workshop: 21, 33, 45, 65, and 69.

Assignment Guide

- Basic: 17–20, 31–36, 45–48, 51–61
- Average: 21–24, 31–36, 45–48, 61–64, 81–89

LESSON PLAN
Workshop 8

OBJECTIVES

1. To write a percent as a decimal.
2. To write a decimal as a percent.

RETEACHING

Review the meaning of percent by using this example: If there are 100 people in a movie theatre and 47 of them are females, then we can say that 47 percent are females. Ask students what percent are males. (53%)

Stress that percent means so many parts in one hundred. Make sure students understand that when using percent the reference is always to one hundred.

WORKSHOPS

WORKSHOP 8

Percent to Decimal, Decimal to Percent

Look up Skills 26 and 28 on pages 608 and 610 for more practice.

Percent is an abbreviation of the Latin words *per centum*, meaning "by the hundred." So percent means "divide by 100." A percent can be written as a decimal. To change a percent to a decimal, first write the percent as a fraction with a denominator of 100, then divide by 100.

EXAMPLE Change 42% to a decimal.

SOLUTION

$42\% = \frac{42}{100} = 0.42$

EXAMPLE Change 19.4% to a decimal.

SOLUTION

$19.4\% = \frac{19.4}{100} = 0.194$

When dividing by 100, you can just move the decimal point two places to the left. So when you write a percent as a decimal, you are moving the decimal point two places to the left and dropping the percent sign (%). If necessary, use zero as a placeholder.

EXAMPLE	SOLUTION
A. 42%	42% = 42. = 0.42 ← Drop % sign.
B. 37.2%	37.2% = 37.2 = 0.372 ← Move decimal 2 places.
C. 435%	435% = 435. = 4.35
D. 5%	5% = 05. = 0.05

Insert a zero as a placeholder.

✓ SELF-CHECK Complete the problems, then check your answers in the back of the book.

Change the percents to decimals.

1. 25% 0.25 **2.** 37.5% 0.375 **3.** 142% 1.42 **4.** 9% 0.09

To write a decimal as a percent, move the decimal point two places to the right and add a percent sign (%).

EXAMPLE	SOLUTION
A. 0.42	0.42 = 0.42 = 42% ← Add % sign.
B. 0.05	0.05 = 0.05 = 5% ← Move decimal 2 places.
C. 0.5	0.5 = 0.50 = 50%
D. 7.5	7.5 = 7.50 = 750%
E. 0.005	0.005 = 0.005 = 0.5%

✓ SELF-CHECK Complete the problems, then check your answers in the back of the book.

Change the decimals to percents.

5. 0.85 **85%** **6.** 0.07 **7%** **7.** 0.3 **30%** **8.** 1.55 **155%** **9.** 0.004 **0.4%**

PROBLEMS

Write as decimals.

10. 54% **0.54** **11.** 34% **0.34** **12.** 49% **0.49** **13.** 60% **0.60**
14. 23% **0.23** **15.** 87% **0.87** **16.** 71% **0.71** **17.** 99% **0.99**
18. 52.1% **0.521** **19.** 97.6% **0.976** **20.** 47.3% **0.473** **21.** 81.8% **0.818**
22. 20.6% **0.206** **23.** 34.4% **0.344** **24.** 60.9% **0.609** **25.** 73.6% **0.736**
26. 252% **2.52** **27.** 817% **8.17** **28.** 798% **7.98** **29.** 376% **3.76**
30. 485% **4.85** **31.** 508% **5.08** **32.** 2578% **25.78** **33.** 1783% **17.83**
34. 8% **0.08** **35.** 3% **0.03** **36.** 2% **0.02** **37.** 9% **0.09**
38. 1.1% **0.011** **39.** 4.7% **0.047** **40.** 5.5% **0.055** **41.** 7.2% **0.072**
42. 3.05% **0.0305** **43.** 8.03% **0.0803** **44.** 8.0765% **0.080765** **45.** 5.0921% **0.050921**

Write as percents.

46. 0.84 **84%** **47.** 0.13 **13%** **48.** 0.67 **67%** **49.** 0.35 **35%**
50. 0.75 **75%** **51.** 0.24 **24%** **52.** 0.15 **15%** **53.** 0.97 **97%**
54. 0.03 **3%** **55.** 0.05 **5%** **56.** 0.08 **8%** **57.** 0.01 **1%**
58. 0.016 **1.6%** **59.** 0.042 **4.2%** **60.** 0.0251 **2.51%** **61.** 0.0965 **9.65%**
62. 0.001 **0.1%** **63.** 0.005 **0.5%** **64.** 0.0032 **0.32%** **65.** 0.0061 **0.61%**
66. 4.125 **412.5%** **67.** 9.371 **937.1%** **68.** 4.5 **450%** **69.** 7.2 **720%**
70. 1.11 **111%** **71.** 2.98 **298%** **72.** 3.004 **300.4%** **73.** 7.000 **700%**
74. 0.004 **0.4%** **75.** 0.009 **0.9%** **76.** 9 **900%** **77.** 46 **4600%**
78. 0.5 **50%** **79.** 0.1 **10%** **80.** 750.4 **75,040%** **81.** 1083 **108,300%**

82. "Coats on sale, 33% off." Write 33% as a decimal. **0.33**

83. "Cost of food up 112.3%." Write 112.3% as a decimal. **1.123**

84. "42 out of 100 attempts." Write $\frac{42}{100}$ as a percent. **42%**

85. "All furniture 30%–50% off." Write 30% and 50% as decimals. **0.30, 0.50**

86. "Batteries on sale, 25% off." Write 25% as a decimal. **0.25**

WORKSHOPS

ERROR ANALYSIS

Students make errors with percent because they do not understand the concept itself. After students work the Self-Checks, ask them to explain how they got their answers. If a response is based only on a mechanical procedure of moving a decimal point or adding a percent sign, then ask why the answer is correct. Mechanical rules are forgotten easily, but concepts are not.

PRACTICE AND APPLY

The following problems can be assigned for classwork and the answers checked in class to help students master the objective of the lesson.

- Guided Practice: 10–17, 46–53
- Independent Practice: 18–25, 54–61

WRAP-UP

Ask a student to explain the meaning of percent. Ask a few others to give some examples of changing a percent to a decimal and a decimal to a percent.

Assignment Guide

- Basic: 18–34 even, 54–74 even
- Average: 27–45 odd, 64–90 even

WORKSHOPS

APPLICATIONS

87. The percent changes in retail sales were reported as a decimal in the October issue of *Modern Merchandise* magazine. Change the decimals to percents.

a. 1.6%
b. 0.2%
c. 1.5%
d. 4.5%
e. 4.2%
f. 1.3%
g. 1.9%

	RETAIL SALES	
	Month	**Change**
a.	February	0.016
b.	March	0.002
c.	April	0.015
d.	May	0.045
e.	June	0.042
f.	July	0.013
g.	August	0.019

88. Erica Stubenhofer is an investment counselor for Foxboro Investment Company. When working with clients, she uses this chart. Change the percents to decimals.

a. 0.01
b. 0.02
c. 0.04
d. 0.06
e. 0.08
f. 0.093

	Taxable Income	Marginal Rate
a.	$0–7,300	1.0%
b.	7,301–17,300	2.0%
c.	17,301–27,300	4.0%
d.	27,301–37,900	6.0%
e.	37,901–47,900	8.0%
f.	47,901 and up	9.3%

89. The commission rate schedule for a stockbroker is shown. Change the percents to decimals.

a. 0.017
b. 0.013
c. 0.011
d. 0.009
e. 0.007
f. 0.005

	COMMISSION RATE SCHEDULE	
	Dollar Amount	**% of Dollar Amount**
	Stocks:	
a.	$0 – $2,499	1.7%, minimum $30
b.	$2,500 – $4,999	1.3%, minimum $42
c.	$5,000 – $9,999	1.1%, minimum $65
d.	$10,000 – $14,999	0.9%, minimum $110
e.	$15,000 – $24,999	0.7%, minimum $135
f.	$25,000 – $49,999	0.5%, minimum $175
	$50,000 and above	negotiated

90. During the National Basketball Association season, the teams had these won–lost records. The Pct. column shows the percent of games won, expressed as a decimal. Change the decimals to percents.

EASTERN CONFERENCE

Atlantic Division

Answer		Team	W	L	Pct.	GB
76.3%	**a.**	Boston	29	9	.763	—
56.4%	**b.**	Philadelphia	22	17	.564	$7\frac{1}{2}$
44.7%	**c.**	New York	17	21	.447	12
44.7%	**d.**	Washington	17	21	.447	12
31.6%	**e.**	New Jersey	12	26	.316	18
27.5%	**f.**	Miami	11	29	.275	19

Central Division

Answer		Team	W	L	Pct.	GB
71.8%	**g.**	Chicago	28	11	.718	—
70.0%	**h.**	Detroit	28	12	.700	
67.5%	**i.**	Milwaukee	27	13	.675	$1\frac{1}{2}$
61.5%	**j.**	Atlanta	24	15	.615	4
38.5%	**k.**	Indiana	15	24	.385	13
31.6%	**l.**	Charlotte	12	26	.316	$15\frac{1}{2}$
31.6%	**m.**	Cleveland	12	26	.316	$15\frac{1}{2}$

WESTERN CONFERENCE

Midwest Division

Team	W	L	Pct.	GB		Answer
San Antonio	27	10	.730	—	**n.**	73.0%
Utah	26	13	.667	2	**o.**	66.7%
Houston	20	19	.513	8	**p.**	51.3%
Dallas	13	24	.351	14	**q.**	35.1%
Minnesota	13	24	.351	14	**r.**	35.1%
Orlando	10	31	.244	19	**s.**	24.4%
Denver	9	30	.231	19	**t.**	23.1%

Pacific Division

Team	W	L	Pct.	GB		Answer
Portland	34	7	.829	—	**u.**	82.9%
L.A. Lakers	27	11	.711	$5\frac{1}{2}$	**v.**	71.1%
Phoenix	25	12	.676	7	**w.**	67.6%
Golden State	21	17	.553	$11\frac{1}{2}$	**x.**	55.3%
Seattle	17	19	.472	$14\frac{1}{2}$	**y.**	47.2%
L.A. Clippers	14	27	.341	20	**z.**	34.1%
Sacramento	10	26	.278	$21\frac{1}{2}$	**aa.**	27.8%

Finding a Percentage

Look up Skill 30 on page 612 for more practice.

Finding a percentage means finding a percent of a number. To find a percent of a number, you change the percent to a decimal, then multiply it by the number.

EXAMPLE 1 20% of 95 is what number?

SOLUTION

20% × 95 = n In mathematics, *of* means "times" and *is* means "equals." Let n stand for the unknown number.

0.20 × 95 = n Change the percent to a decimal.

19 = n Multiply.

20% of 95 = 19 Write the answer.

EXAMPLE 2 The delivery charge is 7% of the selling price of $140.00. Find the delivery charge.

SOLUTION

7% × $140.00 = n

0.07 × $140.00 = n

$9.80 = n

7% × $140.00 = $9.80 delivery charge

EXAMPLE 3 The student had 85% correct out of 60 questions. How many answers were correct?

SOLUTION

85% × 60 = n

0.85 × 60 = n

51 = n

85% × 60 = 51 correct

✓ SELF-CHECK Complete the problems, then check your answers in the back of the book.

1. Find 40% of 70. **28**
2. Find 25% of 120. **30**
3. Find 5% of 30. **1.5**
4. Find 145% of 200. **290**

PROBLEMS

Find the percentage.

5. 35% of 70 **24.5**
6. 60% of 95 **57**
7. 20% of 54 **10.8**
8. 40% of 216 **86.4**
9. 42% of 335 **140.7**
10. 75% of 815 **611.25**
11. 32% of 315 **100.8**
12. 64% of 320 **204.8**
13. 50% of 419 **209.5**
14. 8% of 50 **4**
15. 4% of 95 **3.8**
16. 6% of 48 **2.88**
17. 3% of 217 **6.51**
18. 2% of 371 **7.42**
19. 9% of 912 **82.08**
20. 5% of 41.6 **2.08**
21. 1% of 76.4 **0.764**
22. 8% of 64 **5.12**
23. 130% of 600 **780**
24. 175% of 615 **1076.25**
25. 195% of 860 **1677**
26. 420% of 470 **1974**
27. 425% of 746 **3170.5**
28. 650% of 416 **2704**
29. 315% of 2170 **6835.5**
30. 515% of 3350 **17,252.5**
31. 715% of 5218 **37,308.7**

LESSON PLAN Workshop 9

OBJECTIVE

To find a percent of a number.

RETEACHING

The words *percent* and *percentage* are often used by students interchangeably, but they do have different meanings. A percent is a ratio of two numbers while a percentage is a single number. A percentage is the result of finding a percent of a number. Using Example 1, point out to students that since 20% of 95 is 19, 19 has the same relationship to 95 that 20 has to 100; that is, $\frac{19}{95} = \frac{20}{100}$. This equation can be checked by cross multiplying: 19 × 100 = 20 × 95, or 1900 = 1900. Stress also that 20% means $\frac{20}{100}$ and that 19 (20% of 95) is called the percentage.

ERROR ANALYSIS

The types of errors students make when finding the percent of a number are either computational, that is they multiply incorrectly, or are errors resulting from putting the decimal point in the wrong place in the answer. Both types of errors are easy to identify and correct.

PRACTICE AND APPLY

The following problems can be assigned for classwork and the answers checked in class to help students master the objective of the lesson.

- Guided Practice: 5–10, 45–47
- Independent Practice: 12–17, 48–50

32. 6.5% of 30 **1.95**
33. 8.25% of 75 **6.1875**
34. 1.67% of 64 **1.0688**
35. 5.5% of 136 **7.48**
36. 7.45% of 234 **17.433**
37. 4.5% of 584 **26.28**
38. 24.5% of 419 **102.655**
39. 32.5% of 591 **192.075**
40. 56.25% of 691 **388.6875**
41. 71.2% of 875 **623**
42. 84.6% of 340 **287.64**
43. 99.9% of 742 **741.258**

Round answers to the nearest cent.

44. 5.5% of $60 **$3.30**
45. 6% of $70 **$4.20**
46. 7.75% of $30 **$2.33**
47. 6.5% of $420 **$27.30**
48. 7.5% of $160 **$12**
49. 4.25% of $470 **$19.98**
50. 8.25% of $76 **$6.27**
51. 4.5% of $36 **$1.62**
52. 7.455% of $246 **$18.34**
53. 1.75% of $71.80 **$1.26**
54. 1.25% of $117.45 **$1.47**
55. 1.85% of $48.79 **$0.90**
56. 0.8% of $24,000 **$192**
57. 0.25% of $174,000 **$435**
58. 2.0% of 96,500 **$1930**

APPLICATIONS

59. Allison Welles works for Klein Department Store as a salesclerk. When items are on sale, she computes the amount saved. She also subtracts the amount saved from the regular price to find the sale price. These insulated guide boots are marked down 30%. Complete the computations for her.

	Height	Color	Order No.	Was	Amount Saved	Sale Price
a.	8"	Brown	699 B 7011	$29.99	?	?
		Black	699 B 7012		**$9**	**$20.99**
b.	10"	Brown	699 B 7013	$33.99	?	?
		Black	699 B 7014		**$10.20**	**$23.79**
c.	12"	Brown	699 B 7015	$39.99	?	?
		Black	699 B 7016		**$12**	**$27.99**

60. Use the chart at the right to find the postage and handling charges for the amounts listed below.

TO FIGURE POSTAGE AND HANDLING
. . . add to order

Sale items, total order:
$3.00 or less, add 75¢
$3.01–$10, add 20% of total
$10.01–$25, add 15% of total
Over $25, add 10% of total

a. $3.00 **b.** $6.00
c. $20.00 **d.** $40.00
e. $4.60 **f.** $9.80
g. $18.50 **h.** $196.40
i. $2.89 **j.** $7.49
k. $19.94 **l.** $94.49

a. **75¢** b. **$1.20**
c. **$3** d. **$4**
e. **$0.92** f. **$1.96**
g. **$2.78** h. **$19.64**
i. **75¢** j. **$1.50**
k. **$2.99** l. **$9.45**

61. Student Bert Trace received these test scores. How many answers were correct on each test?

Answer		Subject	Test Score	Number of Items
54	**a.**	Math	90%	60
40	**b.**	English	80%	50
33	**c.**	Science	75%	44
34	**d.**	Spanish	85%	40
108	**e.**	Government	90%	120

62. Schedule X of the tax rate schedule is used if the filing status is single. Find the tax for each of the incomes listed.

$2475 **a.** $16,500

$4578.70 **b.** $24,640

$20,244.28 **c.** $74,916

$9761.50 **d.** $43,150

SCHEDULE X Taxable Income			
Over	But not over	Your tax is	of amount over
$0	$17,850	15%	$0
17,850	43,150	$2677.50 + 28%	17,850
43,150	89,560	9761.50 + 33%	43,150
89,560	See further instructions.		

63. Betty Einstein is a stockbroker. She charges a percent of the dollar amount of stocks sold. Use the rate schedule and find how much she charges on these sale amounts. Round to the nearest cent.

COMMISSION RATE SCHEDULE Dollar Amount	% of Dollar Amount
Stocks:	
$0–$2499	2%, minimum $30
$2500–$4999	1.7%, minimum $42
$5000–$9999	1.4%, minimum $65
$10,000–$14,999	1%, minimum $110
$15,000–$24,999	0.8%, minimum $135
$25,000–$49,999	0.6%, minimum $175
$50,000 and above	negotiated

a. $1400 $30.00 **b.** $3000 $51.00

c. $8400 $117.60 **d.** $11,700 $117.00

e. $3640 $61.88 **f.** $18,670 $149.36

g. $41,948 $251.69 **h.** $25,148 $175.00

i. $12,147.75 $121.48 **j.** $56,148.95 Negotiated

64. Sales taxes are found by multiplying the tax rate times the selling price of the item. The total purchase price is the selling price plus the sales tax. Find the sales tax and total purchase price for each selling price. Round to the nearest cent.

	Selling Price	Tax Rate	Sales Tax	Total Purchase Price
a.	$ 12.40	4%	? $0.50	? $12.90
b.	19.49	5%	? $0.97	? $20.46
c.	2.19	6%	? $0.13	? $2.32
d.	74.79	6.5%	? $4.86	? $79.65
e.	119.49	7.25%	? $8.66	? $128.15
f.	96.79	8.25%	? $7.99	? $104.78
g.	44.99	7.455%	? $3.35	? $48.34
h.	21.19	4.625%	? $0.98	? $22.17

WRAP-UP

Ask students to find 15% of 250 and have one student show the solution at the chalkboard. Ask another student to name the percentage. (37.5)

Assignment Guide

- Basic: 17–31 odd, 47, 49
- Average: 18–44 even, 50–64 even

LESSON PLAN Workshop 10

OBJECTIVE

To find the average, or mean, of a group of numbers.

RETEACHING

Work through the four Examples with students so that they understand how to find the average of a group of numbers. Then ask students to think about what an average represents. In other words, for Example 1, what meaning can be given to the statement that the average of 7, 9, 4, 6, and 4 is 6. Point out that 7 and 9 are greater than 6, and 4 is less than 6. Thus, 6 lies somewhere in the middle of the group. Tell students that the average, or mean, is a number that represents the middle, or center, of a group of numbers. Use Examples 2, 3, and 4 to show that the average may not always be one of the numbers in the group, as it is in Example 1.

WORKSHOP 10

Average (Mean)

Look up Application Q on page 625 for more practice.

The average, or mean, is a single number used to represent a group of numbers. The average, or mean, of two or more numbers is the sum of the numbers divided by the number of items added.

EXAMPLE 1 Find the average of 7, 9, 4, 6, and 4.

SOLUTION $\frac{7 + 9 + 4 + 6 + 4}{5} = \frac{30}{5} = 6$

Add to find the total.

Divide by the number of items.

EXAMPLE 2 Find the average of 693, 367, 528, and 626.

SOLUTION $\frac{693 + 367 + 528 + 626}{4} = \frac{2214}{4} = 553.5$

EXAMPLE 3 Find the average of 5.7, 6.3, 4.2, 5.8, and 3.4. Round to the nearest tenth.

SOLUTION $\frac{5.7 + 6.3 + 4.2 + 5.8 + 3.4}{5} = \frac{25.4}{5} = 5.08 = 5.1$

EXAMPLE 4 Find the average of \$17, \$24, \$38, \$23, \$19, and \$26. Round to the nearest dollar.

SOLUTION $\frac{\$17 + \$24 + \$38 + \$23 + \$19 + \$26}{6} = \frac{\$147}{6} = \$24.50 = \$25$

✔ SELF-CHECK Complete the problems, then check your answers in the back of the book.

Find the average for each set of numbers. Round to the nearest cent.

1. 3, 6, 2, 5, 8, 6 **5**

2. 3.2, 1.8, 6.5, 8.1, 5.9 **5.1**

3. 134, 126, 130 **130**

4. \$25, \$37, \$49, \$53, \$42, \$42 **\$41.33**

PROBLEMS

Find the average for each group.

5. 7, 8, 9, 12, 14 **10**

6. 4, 6, 10, 12, 8 **8**

7. 70, 85, 90, 75 **80**

8. 44, 86, 35, 95 **65**

9. 197, 108, 116 **140.3**

10. 225, 432, 321 **326**

11. 776, 709, 754, 733 **743**

12. 526, 387, 431, 388 **433**

13. 2.4, 3.5, 4.7, 2.9, 8.4 **4.38**

14. 6.8, 5.6, 3.4, 2.5, 4.7 **4.6**

15. 1.4, 2.5, 4.8, 3.7, 2.0, 3.9 **3.05**

16. 8.1, 5.3, 3.8, 7.9, 4.6, 3.2 **5.48**

17. $14, $12, $16, $14, $15 **$14.20**

18. $64, $38, $92, $51, $65 **$62**

19. $86, $75, $82, $87, $64, $70 **$77.33**

20. $44, $44, $40, $42, $42, $40 **$42**

Find the average for each group. Round to the amount indicated.

21. 7.8, 6.3, 8.3, 4.9, 7.7, 6.9, 5.1 (nearest tenth) **6.7**

22. 9.2, 7.6, 8.2, 5.9, 9.5, 7.8 (nearest tenth) **8.0**

23. 31.7, 33.9, 36.1, 33.8 (nearest tenth) **33.9**

24. 4.37, 3.74, 4.90, 5.74 (nearest hundredth) **4.69**

25. 34.87, 42.90, 46.21, 36.34, 39.89 (nearest hundredth) **40.04**

26. $37.50, $44.50, $39.65, $34.25, $15.61, $11.22 (nearest cent) **$30.46**

27. $450,000; $386,000; $425,000; $372,000 (nearest thousand dollars) **$408,000**

28. Ben Agars had bowling scores of 175, 132, and 142. What was his average? **150**

29. Marcia McComber recorded her pulse rate on 4 occasions as follows: 68, 85, 77, and 82. What was her average pulse rate? **78**

30. Rachel Kelley's tips from being a bellhop were $4.00, $2.00, $3.50, $1.00, $4.00, $2.00, $1.00, and $3.00. What was her average tip? **$2.56**

31. During a 6-day period in June, you earned an average of $25 a day for mowing lawns. What were your total earnings? **$150.00**

32. Last year, Andre Barsotti's telephone bills averaged $35.45 a month. What was his total bill for the year? **$425.40**

33. Hung Lee had an average grade of 92 on his first 4 business math tests. He had a 95 and a 98 on the next 2 tests. What is his average grade for the 6 tests? **93.5**

34. Christy Houck recorded her math test scores this quarter. What is her average? **85**

Test Number	1	2	3	4	5	6	7	8
Score	87	75	98	95	82	77	78	88

35. What does she need on the next test to have an average of 86? **94**

36. If there are a total of 10, 100-point tests for the quarter, is it possible for her to raise her average to 90? **No**

WORKSHOPS

ERROR ANALYSIS

Few students should make errors with computing averages after working through the Examples. Any errors that are made can usually be traced to carelessness. For example, students may miscount the number of numbers in the group and thus divide by the wrong number. Or they make careless errors in addition. Emphasize the importance of being organized and orderly when doing mathematics.

PRACTICE AND APPLY

The following problems can be assigned for classwork and the answers checked in class to help students master the objective of the lesson.

- Guided Practice: 5–12
- Independent Practice: 13–20

WRAP-UP

Use Problem 21 to summarize the concept of an average and how it is computed.

Assignment Guide

- Basic: 13–29 odd
- Average: 22–40 even

WORKSHOPS

APPLICATIONS

37. Erik Pitts earns extra money each spring by rototilling gardens. His brother, John, went along to record the time spent doing each part of the job. Complete the chart in order to find the average time for each part of the job and the average time for each job. Round to the nearest minute.

	Task	Job 1	Job 2	Job 3	Job 4	Job 5	Average
a.	Unload tiller	3	2	2	3	3	**3** ?
b.	Till garden	55	40	65	45	50	**51**?
c.	Clean up tiller	6	10	11	8	6	**8** ?
d.	Load tiller	5	4	3	5	6	**5** ?
e.	Travel time	16	20	25	12	8	**16**?
f.	Average	**17** ?	**15** ?	**21** ?	**15** ?	**15** ?	

38. As captain of his bowling team, Burton McKaig has to complete this form after each match. Help him by computing the total and the average for each bowler. He also computes the total and the team average for each game. Round to the nearest whole number.

	Bowler	Game 1	Game 2	Game 3	Total	Average
a.	McKaig	145	171	163	**479**?	**160**?
b.	Pegorsch	220	215	203	**638**?	**213**?
c.	Hoskinson	175	160	172	**507**?	**169**?
d.	Goldberg	214	210	190	**614**?	**205**?
e.	Hallauer	136	181	172	**489**?	**163**?
f.	Total	**890**?	**937**?	**900**?		
g.	Team average	**178**?	**187**?	**180**?		

39. During a recent vacation trip, Denise and Dennis Hogan purchased 10 gallons of gasoline for \$12.79, 11 gallons for \$14.29, 8 gallons for \$10.63, and 7 gallons for \$9.51. What was the average cost per gallon of gasoline? **\$1.31**

40. While on a business trip, you purchased 11.2 gallons of gasoline at \$1.299 per gallon, 8.6 gallons at \$1.319 per gallon, 7.3 gallons at \$1.369 per gallon, and 10.3 gallons at \$1.429 per gallon. What was the average cost per gallon of gasoline? **\$1.35**

41. Jeri Keefer wanted to estimate the number of words in her 12-page term paper for her history class. She picked out 8 lines and counted the number of words per line: 12, 9, 7, 11, 13, 8, 10, and 12. Next she counted the lines on four pages: 28, 26, 27, and 27.

a. What was the average number of words per line? **10.25**

b. What was the average number of lines per page? **27**

c. Use the averages to estimate the number of words in the term paper. **3321**

Elapsed Time

Look up Application F on page 619 for more practice.

To find elapsed time, subtract the earlier time from the later time.

EXAMPLE Find the elapsed time for Shirley Miller who worked from:

A. 2:15 p.m. to 10:30 p.m. **B.** 6:45 a.m. to 12:56 p.m.

SOLUTIONS

A. 10:30 − 2:15 = 8:15 = 8 hours 15 minutes written as 8 h:15 min

B. 12:56 − 6:45 = 6:11 = 6 h:11 min

You cannot subtract 45 minutes from 30 minutes unless you borrow an hour and add it to the 30 minutes. Remember that 1 hour = 60 minutes.

EXAMPLE Find the elapsed time from 1:45 p.m. to 8:30 p.m.

SOLUTION

8:30	=	7:30 + :60	=	7:90 borrowed 1 hour
− 1:45	=	− 1:45	=	− 1:45
				6:45 = 6 h:45 min

✓ SELF-CHECK Complete the problems, then check your answers in the back of the book.

Find the elapsed time for a person who worked from:

1. 4:30 p.m. to 11:45 p.m. **7 h:15 min**
2. 6:30 a.m. to 11:45 a.m. **5 h:15 min**
3. 4:15 p.m. to 10:10 p.m. **5 h:55 min**
4. 7:43 a.m. to 10:40 a.m. **2 h:57 min**

To find elapsed time when the time period spans 1 o'clock, add 12 hours to the later time before subtracting.

EXAMPLE 1 Find the elapsed time from 8:30 a.m. to 4:40 p.m.

SOLUTION

4:40	=	4:40 + 12:00	=	16:40
− 8:30	=	− 8:30	=	− 8:30
				8:10 = 8 h:10 min

EXAMPLE 2 Find the elapsed time from 7:50 p.m. to 3:34 a.m.

SOLUTION

3:34	=	15:34	=	14:34 + :60	=	14:94
− 7:50	=	− 7:50	=	− 7:50	=	− 7:50
						7:44 = 7 h:44 min

✓ SELF-CHECK Complete the problems, then check your answers in the back of the book.

WORKSHOPS

LESSON PLAN Workshop 11

OBJECTIVE

To find elapsed time.

RETEACHING

Use a clock face to review the twelve hours shown in one complete revolution of the hour hand from 12 to 12. Then discuss the 24-hour clock by using examples: 1 p.m. is 12 + 1 or 13, and is called 13 hundred hours. The 24-hour clock will help students understand how to find elapsed time when the time period spans 1 o'clock. Ask students to give the 24-hour clock time for 2 a.m.; 5 a.m.; 7 p.m; 11 a.m.

ERROR ANALYSIS

The computations in this lesson are straightforward. Students will make the usual errors in subtraction, but the major problems with finding elapsed time are conceptual. Students may not understand the need to borrow one hour (60 minutes) in certain problems, or they may not use the 24-hour clock correctly. Both problems can be overcome with practice.

PRACTICE AND APPLY

The following problems can be assigned for classwork and the answers checked in class to help students master the objective of the lesson.

- Guided Practice: 9–18
- Independent Practice: 19–24

WRAP-UP

Work Problems 25–28 to summarize the objective of this workshop.

Assignment Guide

- Basic: 17–39
- Average: 25–41 odd

WORKSHOPS

Find the elapsed time for a person who worked from:

5. 8:00 a.m. to 5:10 p.m. **9 h:10 min** **6.** 11:15 p.m. to 7:30 a.m. **8 h:15 min**

7. 7:37 a.m. to 3:20 p.m. **7 h:43 min** **8.** 10:07 p.m. to 6:00 a.m. **7 h:53 min**

PROBLEMS

Find the elapsed time.

4 h:15 min **9.** From 1:15 p.m. to 5:30 p.m. **10.** From 2:12 p.m. to 10:25 p.m. **8 h:13 min**

6 h:15 min **11.** From 5:40 a.m. to 11:55 a.m. **12.** From 4:25 a.m. to 11:50 a.m. **7 h:25 min**

6 h:50 min **13.** From 3:30 a.m. to 10:20 a.m. **14.** From 1:30 p.m. to 10:15 p.m. **8 h:45 min**

3 h:37 min **15.** From 6:35 p.m. to 10:12 p.m. **16.** From 5:40 a.m. to 11:14 a.m. **5 h:34 min**

8 h:30 min **17.** From 8:00 a.m. to 4:30 p.m. **18.** From 9:15 a.m. to 6:25 p.m. **9 h:10 min**

3 h:53 min **19.** From 7:35 a.m. to 11:28 a.m. **20.** From 2:50 a.m. to 11:05 a.m. **8 h:15 min**

4 h:45 min **21.** From 6:20 p.m. to 11:05 p.m. **22.** From 1:37 a.m. to 9:28 a.m. **7 h:51 min**

7 h:45 min **23.** From 2:50 p.m. to 10:35 p.m. **24.** From 1:48 p.m. to 9:33 p.m. **7 h:45 min**

7 h:15 min **25.** From 12:15 a.m. to 7:30 a.m. **26.** From 12:30 p.m. to 8:45 p.m. **8 h:15 min**

8 h:35 min **27.** From 8:15 a.m. to 4:50 p.m. **28.** From 8:45 a.m. to 5:10 p.m. **8 h:25 min**

8 h **29.** From 9:00 a.m. to 5:00 p.m. **30.** From 9:30 a.m. to 5:00 p.m. **7 h:30 min**

8 h:45 min **31.** From 8:30 a.m. to 5:15 p.m. **32.** From 8:17 a.m. to 5:13 p.m. **8 h:56 min**

7 h:55 min **33.** From 6:30 a.m. to 2:25 p.m. **34.** From 6:47 a.m. to 3:28 p.m. **8 h:41 min**

6 h:37 min **35.** From 6:45 p.m. to 1:22 a.m. **36.** From 10:57 a.m. to 6:12 p.m. **7 h:15 min**

6 h:34 min **37.** From 9:37 p.m. to 4:11 a.m. **38.** From 6:46 p.m. to 2:14 a.m. **7 h:28 min**

APPLICATIONS

39. Sam Watson worked from 9:45 a.m. to 6:12 p.m. How long did he work? **8 h:27 min**

40. Helen Angell took a bus that left Detroit at 9:25 a.m. and arrived in Chicago at 3:20 p.m. How long was the trip? Disregard time zones. **5 h:55 min**

41. Determine the elapsed time for each flight. Disregard time zones.

AIRLINE SCHEDULE

	From	To	Departure Time	Arrival Time	Elapsed Time
a.	Toledo	Detroit	7:30 a.m.	8:12 a.m.	? **0 h:42 min**
b.	Chicago	Houston	11:30 a.m.	2:45 p.m.	? **3 h:15 min**
c.	Los Angeles	Detroit	8:35 a.m.	1:07 p.m.	? **4 h:32 min**
d.	New York	Cleveland	12:30 p.m.	2:56 p.m.	? **2 h:26 min**
e.	Boston	Detroit	11:15 p.m.	1:10 a.m.	? **1 h:55 min**
f.	Atlanta	Miami	12:47 p.m.	2:15 p.m.	? **1 h:28 min**
g.	Pittsburgh	St. Louis	12:15 a.m.	3:10 a.m.	? **2 h:55 min**

Reading Tables and Charts

Look up Application C on page 617 for more practice.

To read a table or chart, find the *column* containing one of the pieces of information you have. Look across the *row* containing the other piece of information. Read down the column and across the row. Read the information you need where the column and row cross.

ANY FRACTION OF A POUND OVER THE WEIGHT SHOWN TAKES THE NEXT HIGHER RATE

WEIGHT NOT TO EXCEED	RATE CHART TO GROUND ZONES						
	2	3	4	5	6	7	8
1 lb	$1.25	$1.28	$1.32	$1.36	$1.42	$1.48	$1.55
2 lb	1.34	1.40	1.47	1.55	1.67	1.79	1.93
3 lb	1.43	1.52	1.63	1.75	1.92	2.11	2.32
4 lb	1.52	1.64	1.78	1.94	2.18	2.42	2.70
5 lb	1.61	1.75	1.93	2.14	2.43	2.74	3.09
6 lb	1.70	1.87	2.09	2.33	2.68	3.05	3.47
7 lb	1.79	1.99	2.24	2.53	2.94	3.37	3.86
8 lb	1.88	2.11	2.40	2.72	3.19	3.68	4.24
9 lb	1.97	2.23	2.55	2.92	3.44	4.00	4.63
10 lb	2.05	2.34	2.70	3.11	3.69	4.31	5.01
11 lb	2.14	2.46	2.86	3.31	3.95	4.63	5.40
12 lb	2.23	2.58	3.01	3.50	4.20	4.94	5.78
13 lb	2.32	2.70	3.17	3.70	4.45	5.26	6.17
14 lb	2.41	2.82	3.32	3.89	4.71	5.57	6.55
15 lb	2.50	2.93	3.47	4.09	4.96	5.89	6.94

EXAMPLE What is the cost to ship a 10-lb package to Zone 4?

SOLUTION **a.** Find the Zone 4 column. **b.** Find the 10-lb row.
c. Read across the 10-lb row to the Zone 4 column.
The cost is $2.70.

✓ SELF-CHECK Complete the problems, then check your answers in the back of the book.

1. $1.25
2. $1.55
3. $2.34
4. $3.32

Find the cost to ship each package to the indicated zone.

1. 1 lb, Zone 2 **2.** 1 lb, Zone 8 **3.** 10 lb, Zone 3 **4.** 13.2 lb, Zone 4

To classify an item, find the row that contains the known data. Then read the classification from the head of the column.

SIZE CHART—MEN'S SIZES

Suits and Sportcoats	Order size	36	37	38	39	40	42	44	46
Sizes 36 to 46 Order by chest size	If chest is (inches)	35–36	36–37	37–38	38–39	39–40	41–42	43–44	45–46
Be sure waist will fit comfortably.	And waist is (inches)	28–31	29–32	30–33	31–34	32–35	34–37	36–39	38–41

Jackets	Order size	36	38	40	42	44	46
Sizes 36 to 46 Order by chest size	If chest is (inches)	34½–36	36½–38	38½–40	40½–42	42½–44	44½–46

EXAMPLE What size suit should a man with a 41-inch chest order?

SOLUTION **a.** Find the Suits and Sportcoats section.
b. Find the row If chest is (inches). **c.** Read across the row to 41–42.
e. Read the number at the head of the column (42).
A man with a 41-inch chest should order a size 42 suit.

LESSON PLAN Workshop 12

OBJECTIVES

1. To read a table or chart.
2. To classify an item in a table or chart.

RETEACHING

Discuss the meanings of the terms *column* and *row.* Stress that a column goes down a page and a row goes across a page. You might want to tell students that a column runs "North to South" on a page and a row runs "East to West". These directional ideas can help students remember the difference between a column and a row.

ERROR ANALYSIS

Typical problems involve reading information from the wrong column or row or reading the wrong crossing point. Emphasize the need to be careful when reading data in a table or chart.

WORKSHOPS

✓ SELF-CHECK Complete the problems, then check your answers in the back of the book.

Determine what size garment should be ordered.

5. Suit, chest size 38 inches. **Size 38 or 39**

6. Jacket, chest size 41 inches. **Size 42**

PROBLEMS

Use the shipping chart on page 35 to find the cost to ship each package to the indicated zone.

7. 6 lb, Zone 3 **$1.87**
8. 2 lb, Zone 8 **$1.93**
9. 11 lb, Zone 3 **$2.46**
10. 9 lb, Zone 5 **$2.92**
11. 12 lb, Zone 4 **$3.01**
12. 9 lb, Zone 2 **$1.97**
13. 6.5 lb, Zone 2 **$1.79**
14. 9.75 lb, Zone 4 **$2.70**
15. 11.2 lb, Zone 3 **$2.58**
16. 2.6 lb, Zone 5 **$1.75**
17. 1.35 lb, Zone 3 **$1.40**
18. 0.75 lb, Zone 7 **$1.48**

APPLICATIONS

19. Using the same shipping chart, determine the maximum weight a package can weigh.

	a.	b.	c.	d.	e.	f.
Shipping to Zone	2	4	5	8	7	3
Shipping Cost	$1.34	$2.86	$3.70	$5.01	$5.57	$2.11
Maximum Weight	2 ?lb	11?lb	13?lb	10?lb	14?lb	8 ?lb

20. Use the size chart on page 35 to determine what size garment should be ordered.

	a.	b.	c.	d.	e.	f.
Chest Size (inches)	37	42	45	36	40	38
Waist Size (inches)	30	35	39	32	35	34
Suit Order Size	37 or 38	42	46	37	40	39
Jacket Order Size	38	42	46	36	40	38

21. What size suit should a man with a 36-inch chest and a 31-inch waist order? **36 or 37**

22. What size suit should a man with a 39-inch chest and a 33-inch waist order? **39 or 40**

Use the New England Weather chart to answer the following.

NEW ENGLAND WEATHER

Station	Today	High	Low	Tomorrow	High
Bangor	Cloudy	36	24	Snow	32
Bedford	Snow	31	27	"	31
Boston	"	33	29	"	33
Brockton	"	33	28	"	32
Framingham	"	31	26	"	29
Gloucester	"	33	30	"	33

23. What was the high temperature in Boston today? **33**

24. What was the low temperature in Bedford today? **27**

25. What will be the high temperature in Framingham tomorrow? **29**

26. In which city was a low temperature of 28 degrees recorded? **Brockton**

27. In which city was the highest temperature recorded for today? **Bangor**

Use the federal income tax table to find the following.

FEDERAL WITHHOLDING TAX—WEEKLY Payroll Period, Employee MARRIED									
And the wages are—		And the number of withholding allowances claimed is—							
At least	But less than	0	1	2	3	4	5	6	7
		The amount of income tax to be withheld shall be—							
$480	$490	$63	$56	$50	$44	$38	$32	$25	$19
490	500	64	58	52	45	39	33	27	21
500	510	66	59	53	47	41	35	28	22
510	520	67	61	55	48	42	36	30	24
520	530	69	62	56	50	44	38	31	25
530	540	70	64	58	51	45	39	33	27
540	550	72	65	59	53	47	41	34	28
550	560	73	67	61	54	48	42	36	30
560	570	75	68	62	56	50	44	37	31
570	580	76	70	64	57	51	45	39	33
580	590	78	71	65	59	53	47	40	34
590	600	79	73	67	60	54	48	42	36
600	610	81	74	68	62	56	50	43	37
610	620	82	76	70	63	57	51	45	39
620	630	84	77	71	65	59	53	46	40

	28.	29.	30.	31.	32.	33.
Income	$481.50	$558.75	$612.31	$485.03	$522.00	$520.00
Allowances	1	3	0	2	4	2
Amount Withheld	? $56.00	? $54.00	? $82.00	? $56.00	? $44.00	? $56.00

	Number of Allowances	Tax Withheld	Income At least	Income But less than
34.	2	$56	? $520	? $530
35.	3	$63	? $610	? $620
36.	0	$79	? $590	? $600
37.	1	$61	? $510	? $520
38.	6	$30	? $510	? $520
39.	4	$45	? $530	? $540

Use the tax table above to find the amount of tax withheld.

40. Ralph Adams earns $483 and claims 2 allowances. **$50**

41. Barbara Knighten earns $562.29 and claims 1 allowance. **$68**

42. Luther Whittier earns $498.60 and claims 2 allowances. **$52**

43. Stacey Hagley earns $571.30 and claims 1 allowance. **$70**

44. Verna Denbridge earns $602.19 and claims 3 allowances. **$62**

45. Shirley Sagamore earns $596.78 and claims 0 allowances. **$79**

WORKSHOPS

PRACTICE AND APPLY

The following problems can be assigned for classwork and the answers checked in class to help students master the objective of the lesson.

- Guided Practice: 7–13
- Independent Practice: 14–19

WRAP-UP

Use problem 7 to illustrate and review the objective of the workshop.

Assignment Guide

- Basic: 15–35
- Average: 17–45 odd

LESSON PLAN
Workshop 13

OBJECTIVES

1. To construct a vertical bar graph.
2. To construct a line graph.

RETEACHING

Use the Example to review the procedure for constructing a vertical bar graph. Then have students work the Self-Check Problem independently. Ask a few students to reproduce their graphs on the chalkboard. Discuss what the graphs show and their value in comparing data. (It's easy to see the comparisons in the data when the information is presented in graph form.) Repeat the above procedure for constructing a line graph.

WORKSHOPS

Constructing Graphs

Look up Applications M and N on pages 622–623 for more practice.

A **bar graph** is a picture that displays and compares numerical facts in the form of vertical or horizontal bars. To construct a vertical bar graph, follow these steps:

a. Draw the vertical and horizontal axes.
b. Scale the vertical axis to correspond to the given data.
c. Draw one bar to represent each quantity.
d. Label each bar and the vertical and horizontal axes.
e. Title the graph.

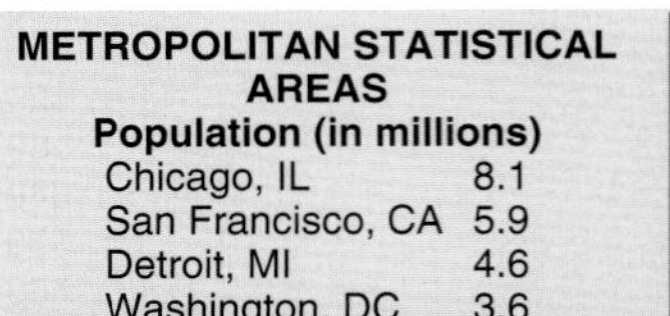

METROPOLITAN STATISTICAL AREAS Population (in millions)	
Chicago, IL	8.1
San Francisco, CA	5.9
Detroit, MI	4.6
Washington, DC	3.6

EXAMPLE Construct a vertical bar graph of the given data.

SOLUTION

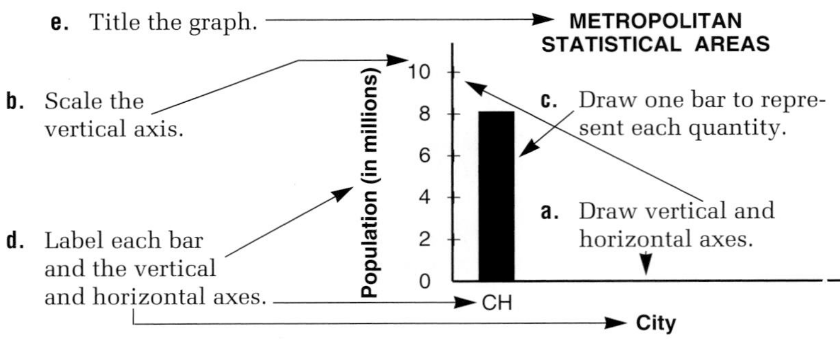

A **line graph** is a picture used to compare data over a period of time. It is an excellent way to show trends (increases or decreases). To construct a line graph, follow these steps:

a. Draw the vertical and horizontal axes.
b. Scale the vertical axis to correspond to the given data.
c. Label the axes.
d. Place a point on the graph to correspond to each item of data.
e. Connect the points from left to right.
f. Title the graph.

COMPUTER CLASSES Enrollment	
1987	15
1988	20
1989	30
1990	45
1991	50

EXAMPLE Construct a line graph of the given data.

SOLUTION

COMPUTER CLASSES

f. Title the graph.

a. Draw the vertical and horizontal axes.

b. Scale the vertical axis.

c. Label the axes.

d. Place a point to correspond to each item of data.

e. Connect the points from left to right.

Enrollment: 0, 10, 20, 30, 40, 50

Year: 87, 88, 89, 90, 91, 92

✓ SELF-CHECK Complete the problems, then check your answers in the back of the book.

1. Complete construction of the vertical bar graph started in the Example.
2. Complete construction of the line graph started in the Example.

PROBLEMS Construct vertical bar graphs of the given data.

3.

CITY GOVERNMENT EMPLOYMENT National Summary (in thousands)	
Schools	355
Hospitals	133
Highways	131
Police	427
Fire	236
Parks & Rec.	152

4.

HEBAN'S DEPARTMENT STORE Total Sales by Department (in thousands)	
Sports	145
Housewares	82
Men's Clothing	120
Women's Clothing	112
Appliances	75
Electronics	130

5. Read the vertical bar graph.

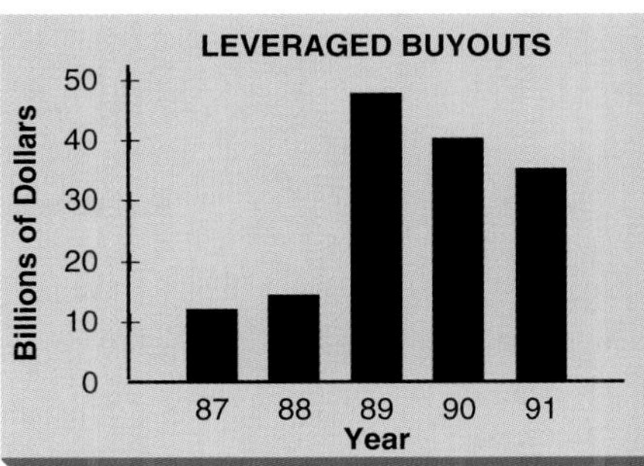

1989; About $48 billion — **a.** In what year was the most money spent on leveraged buyouts? How much?

1982; About $12 billion — **b.** In what year was the least money spent on leveraged buyouts? How much?

About $35 billion — **c.** What was the total value in 1991?

WORKSHOPS

ERROR ANALYSIS

Students enjoy drawing graphs, but they tend to have difficulty deciding upon an appropriate scale to use. The choice of a scale depends upon the given data. Point out that the scale chosen must allow all the data to be represented on the graph, and that it should be convenient to use. By having students draw their graphs on the chalkboard, you can analyze the scales used and guide students in understanding how to select a scale that works well.

Remind students to title their graphs. Titles should be short but accurate in describing the data graphed.

PRACTICE AND APPLY

The following problems can be assigned for classwork and the answers checked in class to help students master the objectives of the lesson.

- Guided Practice: 3
- Independent Practice: 4

WRAP-UP

Discuss the steps involved in constructing bar graphs and line graphs.

Assignment Guide

Basic: 5–8
Average: 5–8

WORKSHOPS

Construct line graphs for the given data.

6.

ACME INDUSTRIES STOCK	
Monthly	**Average**
Jan.	$2.00
Feb.	$2.50
Mar.	$2.75
Apr.	$3.25
May	$4.00
June	$4.50

7.

TAX RETURNS FILED (in millions)	
1980	90
1981	93
1982	94
1983	95
1984	96
1985	96
1986	99
1987	100
1988	101
1989	102
1990	102

8. Read the line graph.

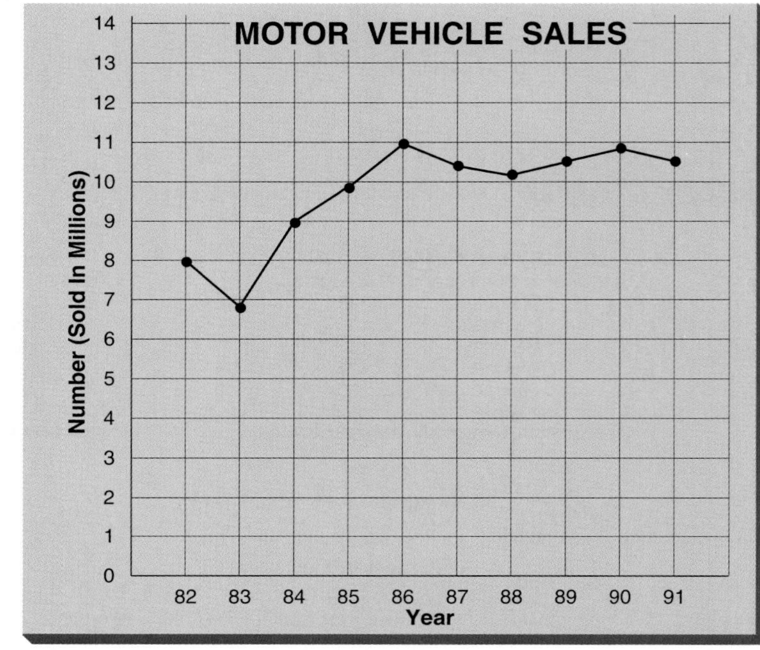

a. Which year shows the highest motor vehicle sales? **1986**

b. Which year shows the lowest motor vehicle sales? **1983**

c. Indicate the time span that showed an increase 3 years in a row. How much was the increase? **1983–1986; Increased from 7 million to 11 million**

d. Compare the decrease from 1981 to 1982 with the decrease from 1985 to 1986. **1982 to 1983 decrease of about one million while the decrease from 1986 to 1987 is about half a million.**

e. Estimate the sales for 1992. **11 million based on the downward trend of the graph from 1990 to 1991.**

WORKSHOP 14

Units of Measure

Here are the abbreviations and conversions for units of measure in the U.S. Customary System.

Length	Volume	Weight
12 inches (in) = 1 foot (ft)	2 cups (c) = 1 pint (pt)	16 ounces (oz) = 1 pound (lb)
3 ft = 1 yard (yd)	2 pt = 1 quart (qt)	2000 lb = 1 ton (t)
5280 ft = 1 mile (mi)	4 qt = 1 gallon (gal)	

Here are the symbols and conversions for units of measure in the metric system.

Length	Volume
1000 millimeters (mm) = 1 meter (m)	1000 milliliters (mL) = 1 liter (L)
100 centimeters (cm) = 1 m	**Mass**
1000 m = 1 kilometer (km)	1000 grams (g) = 1 kilogram (kg)

To convert from one unit of measure to another, use the conversion lists above.

When converting to a smaller unit, multiply.

EXAMPLE Convert 6 ft to inches.

SOLUTION Use 12 in = 1 ft
6 ft: 6 × 12 = 72
6 ft = 72 in

EXAMPLE Convert 2 m to centimeters.

SOLUTION Use 100 cm = 1 m
2 m: 2 × 100 = 200
2 m = 200 cm

When converting to a larger unit, divide.

EXAMPLE Convert 12 pt to quarts.

SOLUTION Use 2 pt = 1 qt
12 pt: 12 ÷ 2 = 6
12 pt = 6 qt

EXAMPLE Convert 8400 g to kg.

SOLUTION Use 1000 g = 1 kg
8400 g: 8400 ÷ 1000 = 8.4
8400 g = 8.4 kg

✓ SELF-CHECK Complete the problems, then check your answers in the back of the book.

Convert:

1. 9 ft to inches **108 in**
2. 0.15 L to milliliters **150 mL**
3. 24 ft to yards **8 yd**
4. 350 cm to meters **3.5 m**

PROBLEMS Do the following conversions.

5. 9 yd to feet **27 ft**
6. 14 gal to quarts **56 qt**
7. 7 lb to ounces **112 oz**
8. 3 ft to inches **36 in**
9. 6 lb to ounces **96 oz**
10. 4 L to milliliters **4000 mL**
11. 3.8 km to meters **3800 m**
12. 16 pt to cups **32 c**
13. 3.2 kg to grams **3200 g**
14. 96 in to yards **2.7 yd**
15. 9 qt to gallons **2.25 gal**
16. 42 oz to pounds **2.6 lb**

LESSON PLAN Workshop 14

OBJECTIVE

To convert from one unit of measure to another.

RETEACHING

Discuss the units of measure in both the U. S. Customary System and the metric system. Be sure to point out that none of the symbols for the units of measure use periods. However, the symbol for inch can use a period *(in.)* to avoid confusion in certain situations. After working through the Examples, ask students why it is necessary to multiply when converting to a smaller unit. (The number of smaller units is greater than the given larger unit.) Ask why it is necessary to divide when converting to a larger unit. (The number of larger units is less than the given smaller unit.)

ERROR ANALYSIS

Typical errors are made by using the wrong units of measure for conversions, for example, 2 cups = 1 quart. Allow students to use the textbook for reference until they have memorized the conversions. Students also multiply when they should have divided and vice versa. If they understand why it is necessary to multiply or divide, these kinds of errors can be avoided.

PRACTICE AND APPLY

The following problems can be assigned for classwork and the answers checked in class to help students master the objective of the lesson.

- Guided Practice: 5–18
- Independent Practice: 19–27

WRAP-UP

Use Problems 28 and 29 to illustrate the objective of the workshop.

Assignment Guide

- Basic: 15–31 odd
- Average: 28–46 even, 47

WORKSHOPS

17. 14 qt to gallons **3.5 gal**
18. 33 oz to pounds **2.1 lb**
19. 2000 g to kilograms **2 kg**
20. 90 cm to meters **0.9 m**
21. 3300 mL to liters **3.3 L**
22. 450 cm to meters **4.5 m**
23. 72.1 kg to grams **72 100 g**
24. 3.4 L to milliliters **3400 mL**
25. 723 g to kilograms **0.723 kg**
26. 11.316 mL to liters **0.011316 L**
27. 18 cm to millimeters **180 mm**
28. 383.2 cm to meters **3.832 m**
29. 1 yd 4 in to inches **40 in**
30. 4 ft 2 in to inches **50 in**
31. 5 qt 1 pt to pints **11 pt**
32. 3 lb 7 oz to ounces **55 oz**
33. 2 gal 2 qt to quarts **10 qt**
34. 2 yd 2 ft 2 in to inches **98 in**
35. 3 gal 2 qt 1 pt to pints **29 pt**
36. 2 m 142 cm 45 mm to millimeters **3465 mm**

APPLICATIONS

37. How many quarts will a 5-gallon plastic bag hold? **20 qt**
38. How many milliliters will a one-liter bottle hold? **1000 mL**
39. How many cups of coffee does a 2-quart coffeepot hold? **8 c**
40. How many cups of hot chocolate will a 2-gallon thermos jug hold? **32 c**
41. How many inches long is a $6\frac{2}{3}$-yard roll of aluminum foil? **240 in**
42. Cottage cheese is sold in 1-pint containers. How many pints must be bought to have enough for a recipe that calls for 5 cups? **3**
43. James Hartman knows that his jogging stride is about 1 meter long. The jogging track he uses is 3.9 kilometers long. How many strides does it take him to go around the track once? **3900**
44. The cafeteria receives 32 cases of milk each day. Each case contains 24 half-pint cartons. How many gallons of milk are received each day? **48 gal**
45. A soft drink is sold in 355 mL cans. How many liters are in a six-pack? **2.13 L**
46. John Patroulus, a pastry chef, baked a walnut cake weighing 2.4 kilograms. How many 75-gram servings can be cut from the cake? **32**
47. Joan Wagoner ordered baseboard molding for the rooms of a new house. Joan needs to complete this chart to determine the total number of feet of molding needed. How much molding is needed?

	Length	Width	2 Lengths	+	2 Widths	=	Perimeter
	12 ft	9 ft	24 ft	+	18 ft	=	42 ft
a.	10 ft	8 ft	20 ft	+	16 ft	=	**36 ft?**
b.	18 ft	24 ft	**36 ft?**	+	**48 ft?**	=	**84 ft?**
c.	10 ft 6 in	9 ft 4 in	**21 ft?**	+	**18 ft?8 in**	=	**39 ft?8 in**
d.	11 ft 3 in	7 ft 8 in	**22 ft?6 in**	+	**15 ft?4 in**	=	**37 ft?10 in**
e.	11 ft 4 in	13 ft 2 in	**22 ft?8 in**	+	**26 ft?4 in**	=	**49 ft?**
f.	17 ft 8 in	12 ft 9 in	**35 ft?4 in**	+	**25 ft?6 in**	=	**60 ft?10 in**
g.	8 ft 5 in	7 ft 9 in	**16 ft?10 in**	+	**15 ft?6 in**	=	**32 ft?4 in**
					Total	=	**381 ft 8 in**

WORKSHOP 15

Estimation: Rounding

Rounding is often used to estimate an answer. First, round the numbers to the highest place value. Then perform the indicated computation. If all the numbers do not have the same number of digits, round the numbers to the highest place value of the smaller number.

ESTIMATE

$$\begin{array}{r} 76.761 \\ -\ 39.302 \\ \hline \end{array} \qquad \begin{array}{r} 80 \\ -\ 40 \\ \hline 40 \end{array}$$

The answer is about 40.
By computation, it is 37.459.

ESTIMATE

$$6.1\overline{)23.79} \qquad \begin{array}{r} 4 \\ 6\overline{)24} \end{array}$$

The answer is about 4.
By computation, it is 3.9.

ESTIMATE

$$\begin{array}{r} 5.65 \\ \times\ 4.45 \\ \hline \end{array} \qquad \begin{array}{r} 6 \\ \times\ 4 \\ \hline 24 \end{array}$$

The answer is about 24.
By computation, it is 25.1425.

ESTIMATE

$$\begin{array}{r} \$265.88 \\ +\ 32.47 \\ \hline \end{array} \qquad \begin{array}{r} \$270 \\ +\ 30 \\ \hline \$300 \end{array}$$

The answer is about \$300.
By computation, it is \$298.35.

ESTIMATE

$2\frac{3}{4} \times 8\frac{1}{4} \qquad 3 \times 8 = 24$

The answer is about 24.
By computation, it is $22\frac{11}{16}$.

ESTIMATE

33% of \$62 $\quad \frac{1}{3} \times \$60 = \$20$

The answer is about \$20.
By computation, it is \$20.46.

✓ SELF-CHECK Complete the problems, then check your answers in the back of the book.

First round the numbers to the highest place value, estimate the answer, then perform the computation.

1. $43.986 - 27.491$
2. $7.1\overline{)44.872}$
3. 63.91×9.43
4. $5\frac{1}{8} \times 4\frac{2}{3}$

Self-Check Solutions: 1. 40 − 30 = 10; 16.495 2. 7)50 = 7; 6.32 3. 60 × 10 = 600; 602.6713 4. 5 × 5 = 25; $23\frac{11}{12}$

PROBLEMS

First estimate, then perform the indicated computation.

5. $5965 + 1824$ **8000; 7789**
6. $7.791 + 2.151$ **10; 9.942**
7. $15.86 - 13.72$ **2; 2.14**
8. $\$72.75 - 6.47$ **\$67.00 \$66.28**
9. 9.34×7.92 **72; 73.9728**
10. $\$28.40 \times 5.20$ **\$150.00; \$147.68**
11. $6.8\overline{)48.52}$ **7; 7.1**
12. $21.3\overline{)57.723}$ **3; 2.71**
13. $2\frac{3}{4} \times 1\frac{1}{2}$ **6; $4\frac{1}{8}$**
14. $12\frac{1}{2} \times 2\frac{3}{5}$ **36; $32\frac{1}{2}$**
15. $\frac{1}{4} \times 12\frac{1}{2}$ **3; $3\frac{1}{8}$**
16. 31% of 18 **6; 5.58**
17. 70% of \$49.95 **\$35; \$34.97**
18. 27% of \$12 **\$3; \$3.24**
19. 52% of 160 **80; 83.2**
20. You purchase items for \$39.45, \$17.55, and \$32.53. Estimate the total, then calculate. **\$88; \$89.53**

WORKSHOPS

LESSON PLAN Workshop 15

OBJECTIVE

To estimate an answer by rounding.

RETEACHING

Using the numbers 76.761 and 39.302, ask students to name the highest place value (tens place). Then have them round the numbers to 80 and 40 respectively. Ask students to round 75.761 to the highest place value (80). Then have them round 74.761 (70). Using the numbers 23.79 and 6.1 in the text, ask students why 23.79 was rounded to 24 rather than 20. (The highest place value of 6.1 is the ones place so 23.79 is also rounded to the ones place.) Ask similar types of questions for the other examples given in the text.

ERROR ANALYSIS

Students do not really make errors in rounding in the same sense that errors are made in computations. If different students round the same numbers differently, then different estimates may be given. The result is that some estimates will be closer to the actual computational answer than others, and in that sense they are better estimates. The key to making a good estimate is to follow the rule of rounding the numbers to the highest place value of the smaller number. This is why an understanding of place value is important when making estimates by rounding.

PRACTICE AND APPLY

The following problems can be assigned for classwork and the answers checked in class to help students master the objective of the lesson.

- Guided Practice: 5–9
- Independent Practice: 10–13

WRAP-UP

Use Problems 15–19 to make sure that students understand how to make estimates by rounding numbers.

Assignment Guide

- Basic: 10–26 even
- Average: 21–27 odd

WORKSHOPS

21. Forty-two people charter a bus for $388. Estimate the amount each person pays, then calculate. **$10; $9.24**

22. A living room measures $16\frac{1}{4}$ ft by $12\frac{1}{2}$ ft. Estimate the area in square feet, then calculate. **192 ft; 203.125 ft**

23. Blue jeans are on sale for 33% off the regular price of $26.95. Estimate the savings, then calculate. **$9; $8.89**

APPLICATIONS First estimate, then find the total for these grocery store receipts.

24.

SUPER VALUES MARKET	
FACE TISSUE	1.15
COFFEE	3.95
FACE TISSUE	1.15
FACE TISSUE	1.15
CORN CHIPS	2.19
BACON	1.58
RYE CRACKERS	1.39
SMALL LINKS	1.98
SMALL LINKS	1.98
SMALL LINKS	1.98
CAT FOOD	3.60
DETERGENT	1.88
DETERGENT	1.08
TOTAL	?

$25; $25.06

25.

SUPER VALUES MARKET	
CORN CHIPS	2.19
VANILLA	4.19
BNL SHK HAM	25.33
ORANGE JUICE	5.48
CASING LINKS	3.65
KIELBASA	3.18
KIELBASA	2.43
PICKLES	1.89
CEREAL	1.73
MAYO	1.69
ALUMINUM FOIL	1.09
CHEDDAR	2.99
MACARONI	.65
STEAK	6.23
RIPE OLIVES	1.29
SPAN OLIVES	.99
SPAN OLIVES	.99
SPAN OLIVES	.99
VINEGAR	1.49
TOTAL	?

$67; $68.47

26.

SUPER VALUES MARKET	
BABY CLAMS	1.49
RICE	1.03
CHESTNUT	.97
CHEESE	3.65
SPROUTS	.95
CRANAPLE	1.23
SALAD DRSG	.89
HERB DRSG	.89
HERB DRSG	.89
SALAD DRSG	.89
HALF HALF PT	.85
WAX PAPER	.89
ITAL SAUS	2.37
ITAL SAUS	2.98
MINI CHP	1.95
SAUSAGE 2 LB	4.35
MINTS	.35
GUM	.45
CRANPLE	1.23
TOTAL	?

$27; $28.30

Use the menu from the Columbian House to estimate the bill for the following orders.

27. Two persons **$26**

1 Soup	1 Chocolate Torte
1 Fruit Cup	1 Apple Strudel
1 Baked Ham	1 Milk
1 Red Snapper	1 Herbal Tea

28. Party of six **$105**

4 Soup	3 Chocolate Torte
2 Fruit Cup	3 Walnut Dream Cake
1 Baked Ham	2 Milk
3 Prime Rib	2 Soft Drinks
1 Red Snapper	2 Coffee
1 Filet Mignon	

COLUMBIAN HOUSE

Appetizers	
Soup Romaine	$1.25
Fruit Cup	1.55
Shrimp Cocktail	5.95
Entrees	
Vegetarian Plate	5.95
Baked Ham	6.50
Red Snapper	7.95
Roast Prime Rib	13.95
Filet Mignon	12.75
Desserts	
Chocolate Torte	3.25
Apple Strudel	2.75
Walnut Dream Cake	3.50
Beverages	
Coffee	.95
Milk	.75
Herbal Tea	.75
Soft Drinks	1.00

WORKSHOP 16

Estimation: Front End

Estimation is a very valuable tool in business mathematics. It can be used as a quick method of checking the reasonableness of a calculation, or when an exact answer is not needed. It is important to know how to estimate. One way to estimate a sum is to add the front-end digits.

ESTIMATE

```
  6477       6
  2142       2
+ 1321     + 1
           ---
             9
```

The answer is about 9000.
By computation, it is 9940.

ESTIMATE

```
  $8.60      9  (by rounding)
   3.19      3
+  0.65    + 0
           ---
            12
```

The answer is about $12.00.
By computation, it is $12.44.

Each estimate in the examples above is less than the correct sum. A closer estimate can be found by adjusting the sum of the front-end digits.

ESTIMATE

```
   5|67|   about 100
   3|35|
   1|24|   about 100
 + |84|
about 900    +   200 = 1100
```

The answer is about 1100.
By computation, it is 1110.

ESTIMATE

```
  $3.85 ← about $1.00
   4.44
   1.13    about $1.00
 + 0.65
about $8.00   +   $2.00 = $10.00
```

The answer is about $10.00.
By computation, it is $10.07.

✓ SELF-CHECK Complete the problems, then check your answers in the back of the book.

Estimate by first adding the front-end digits and then by adjusting the sum of the front-end digits. Perform the computations.

1. 389; 467; 15; + 240 — **900; 1100; 1111**

2. 5623; 221; 9879; + 1061 — **15,000; 17,000; 16,784**

3. $ 3.45; 1.49; 9.72; + 0.35 — **$13.00; $15.00; $15.01**

PROBLEMS

Estimate by adding the front-end digits. Perform the computations.

4. 4427; 3274; + 1245 — **8000; 8946**

5. 7178; 4298; + 5370 — **16,000; 16,846**

6. 112; 448; 515; + 324 — **1300; 1399**

7. 25.5; 86.6; 7.7; + 0.6 — **120; 120.4**

8. $64.50; 43.60; 6.90; + 53.20 — **$150; $168.20**

Estimate by adjusting the sum of the front-end digits. Perform the computations.

9. 335; 660; 74; + 126 — **1200; 1195**

10. 285; 315; 544; + 361 — **1500; 1505**

11. $39.37; 7.49; 23.75; + 4.35 — **$73; $74.96**

12. $15.95; 54.20; 12.57; + 21.54 — **$102; $104.26**

13. $340.05; 21.65; 47.32; + 53.79 — **$400; $462.81**

WORKSHOPS

LESSON PLAN
Workshop 16

OBJECTIVES

1. To estimate an answer by adding the front-end digits.
2. To estimate an answer by adjusting the sum of the front-end digits.

RETEACHING

Point out to students that the front-end digits are the digits having the highest place value. In the first example, the 6, 2, and 1 are in the thousands place, therefore the estimate is 9000. Tell students that estimates made by adding the front-end digits are rough estimates that can be improved by considering the other digits in each number. Then work the other examples to illustrate how to do this.

ERROR ANALYSIS

If the estimates of some students are not close to the answers given in the text, ask them to explain what they are doing. Then review the examples in the text and work some of the Problems with them. When adjusting the sum of the front-end digits, some students may make errors in mental arithmetic. They may also have some trouble in seeing how to pair the non-front-end digits to refine the front-end estimate. Continue working closely with these students until they understand the process and can give good estimates.

PRACTICE AND APPLY

The following problems can be assigned for classwork and the answers checked in class to help students master the objectives of the lesson.

- Guided Practice: 4–6, 9–11
- Independent Practice: 7–8, 12–14

WRAP-UP

Use Problem 15 as a means of reviewing the objectives of the workshop.

Assignment Guide

- Basic: 7–8, 12–14, 16–23
- Average: 16–23

WORKSHOPS

14. 21,000; 21,647

15. $23.00; $23.39

16. $1200; $1181.05

14. Estimate first, then calculate the total attendance for six home soccer games. September 12, 3187; September 19, 2234; October 5, 2108; October 17, 3421; October 24, 6790; and November 2, 3907.

15. Estimate first, then calculate the cost of these grocery items: tissues $1.29, coffee $6.99, orange juice $1.45, T-bone steak $6.79, oatmeal $1.99, cocoa mix $2.89, and a jar of salad dressing $1.99.

16. Estimate first, then calculate the cost of these options on a new car: two-tone paint $147.30, air-conditioning $578.50, oil gauge $39.50, tinted glass $75.75, automatic transmission $340.00.

APPLICATIONS

First estimate, then find the total for each receipt.

17.

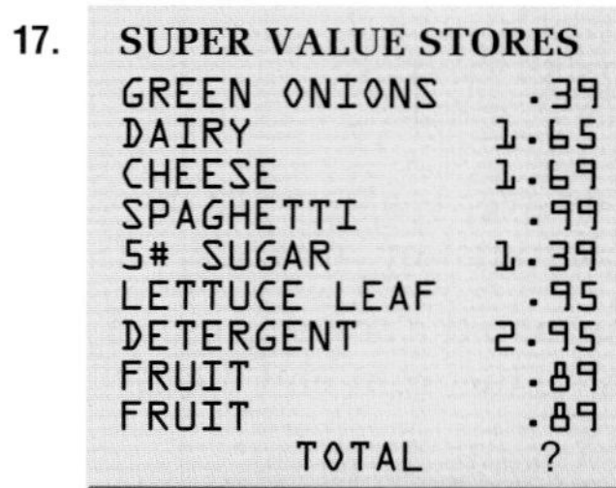

SUPER VALUE STORES

GREEN ONIONS	.39
DAIRY	1.65
CHEESE	1.69
SPAGHETTI	.99
5# SUGAR	1.39
LETTUCE LEAF	.95
DETERGENT	2.95
FRUIT	.89
FRUIT	.89
TOTAL	?

$12.00; $11.79

18.

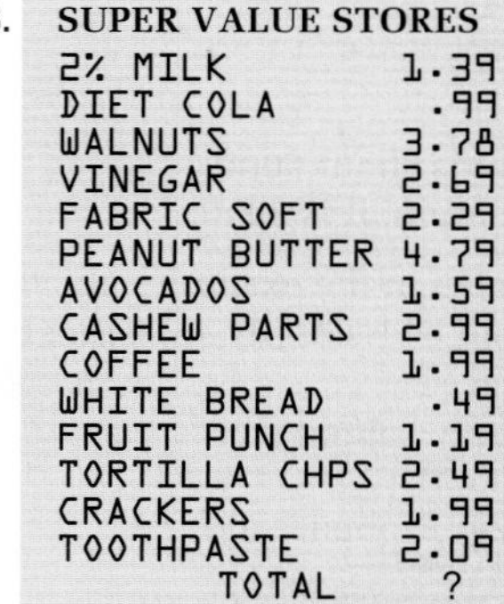

SUPER VALUE STORES

2% MILK	1.39
DIET COLA	.99
WALNUTS	3.78
VINEGAR	2.69
FABRIC SOFT	2.29
PEANUT BUTTER	4.79
AVOCADOS	1.59
CASHEW PARTS	2.99
COFFEE	1.99
WHITE BREAD	.49
FRUIT PUNCH	1.19
TORTILLA CHPS	2.49
CRACKERS	1.99
TOOTHPASTE	2.09
TOTAL	?

$30.00; $30.75

19.

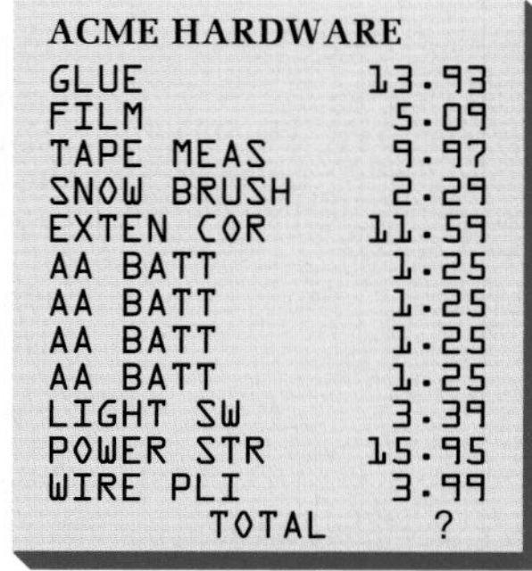

ACME HARDWARE

GLUE	13.93
FILM	5.09
TAPE MEAS	9.97
SNOW BRUSH	2.29
EXTEN COR	11.59
AA BATT	1.25
AA BATT	1.25
AA BATT	1.25
AA BATT	1.25
LIGHT SW	3.39
POWER STR	15.95
WIRE PLI	3.99
TOTAL	?

$70.00; $71.20

First estimate, then find the total deposit.

20.

CASH	CURRENCY	74	00
	COINS	322	05
CHECKS		191	80
		201	20
		121	00
	Total From Other Side		
	SUBTOTAL		
	LESS CASH RECEIVED		
	TOTAL DEPOSIT		?

900; 910.05

21.

CASH	CURRENCY	40	00
	COINS	121	43
CHECKS		37	20
		73	40
	Total From Other Side	158	25
	SUBTOTAL		
	LESS CASH RECEIVED		
	TOTAL DEPOSIT		?

450; 430.28

22.

CASH	CURRENCY	2317	00
	COINS	819	95
CHECKS		501	21
		9213	76
		8324	05
	Total From Other Side	913	17
	SUBTOTAL		
	LESS CASH RECEIVED		
	TOTAL DEPOSIT		?

22,00; 22,089.14

23. Helen Hazelton used a calculator to solve the following problems. If she entered each number properly into the calculator, her answers should be correct. Estimate each answer and decide if Helen's answers are correct.

	Problem	Her Answer on Calculator	Your Estimate	Is She Right? Yes/No
a.	577 + 321 + 225 + 70	1193	**1200** ?	**Yes** ?
b.	$8.60 + $4.90 + $0.50	$13.50	**$14** ?	**No, $14.00** ?
c.	12.3 + 11.2 + 9.8 + 7.6 + 8.4	42.3	**48** ?	**No, 49.3** ?
d.	32 + 53 + 41 + 77 + 58 + 90	351	**360** ?	**Yes** ?
e.	31.26 + 44.51 + 31.07 + 28.46	189.3	**130** ?	**No, 135.3** ?
f.	34.671 + 9.902 + 1.009 + 0.103	45.685	**45** ?	**Yes** ?
g.	1.19 + 0.9 + 1.39 + 0.08 + 1.09	12.75	**4** ?	**No, 4.65** ?
h.	$345.70 + $12.35 + $75.95 + $5.45	$489.45	**$450** ?	**No, $439.45** ?

Estimation: Compatible Numbers

Sometimes a reasonable estimate can be arrived at by changing the numbers in the problem to numbers that can be computed easily. These are called compatible numbers.

DIVIDE	ESTIMATE
36,414 ÷ 9	$9\overline{)36{,}000}$ = 4000

The estimate is easy since 36,000 is close to 36,414 and is compatible with 9. The answer is about 4000. By computation, it is 4046.

DIVIDE	ESTIMATE
798 ÷ 42	$40\overline{)800}$ = 20

The estimate is easy since 800 is close to 798 and 40 is close to 42. 800 and 40 are compatible. The answer is about 20. By computation, it is 19.

FIND	ESTIMATE
$\frac{1}{3} \times 8\frac{3}{4}$	$\frac{1}{3} \times 9 = 3$

The estimate is easy since 9 is close to $8\frac{3}{4}$ and is compatible with $\frac{1}{3}$. The answer is about 3. By computation, it is $2\frac{11}{12}$.

FIND	ESTIMATE
25.5% of $420	$\frac{1}{4} \times 400 = 100$

The estimate is easy since 25.5% is about $\frac{1}{4}$ and $420 is about 400. 400 and $\frac{1}{4}$ are compatible. The answer is about $100. By computation, it is $107.10.

✓ SELF-CHECK Complete the problems, then check your answers in the back of the book.

Estimate using compatible numbers, then perform the computations.

1. 661.74 ÷ 82 — 640 ÷ 80 = 8; 8.07
2. 8763.3 ÷ 321 — 9000 ÷ 300 = 30; 27.3
3. 35% × 926 — $\frac{1}{3} \times 900$ = 300; 324.1

PROBLEMS

For problems 4–21, estimate using compatible numbers. Perform the computations.

4. 8824 ÷ 8 — 1000; 1103
5. 4879 ÷ 7 — 700; 697
6. 6095 ÷ 5 — 1200; 1219
7. 642 ÷ 6 — 110; 107
8. 896 ÷ 32 — 30; 28
9. 24,564 ÷ 575 — 40; 42.72
10. 8766.63 ÷ 81 — 100; 108.23
11. $6447.6 ÷ 7.2 — $900; $895.50
12. $\frac{1}{3} \times 8\frac{1}{2}$ — 3; $2\frac{5}{6}$
13. $\frac{5}{7} \times 44\frac{1}{2}$ — 30; $31\frac{11}{14}$
14. 50% of $430 — $215; $215
15. 65% of $75 — $50; $48.75
16. Delbert Rowell drove his tractor trailer rig 84,572.5 miles in 6 months. Estimate how many miles he drove each month. 14,000; 14,095.42
17. Anna Nethery earned $47,500 this year as a stock analyst. Estimate how much she earns each month. $4000; $3958.33
18. A walking path is $20\frac{1}{2}$ miles long. You walk $\frac{1}{3}$ of the path by noon. Estimate the distance walked. 7; $6\frac{5}{6}$
19. Harriet Murdock saves 33% of her paycheck each week. Last week her check was for $247.95. Estimate the amount saved. $80; $81.82
20. During the high school state basketball finals, 75% of the 22,000 tickets were sold. Estimate how many tickets were not sold. 5,000; 5,500

LESSON PLAN Workshop 17

OBJECTIVES

To make an estimate by using compatible numbers.

RETEACHING

Discuss the idea of compatible numbers by using the examples in the text. Point out that compatible numbers are found by rounding the original numbers. Generally, numbers can be rounded to the nearest 10, 100, or 1000 to make them compatible. In the first example, 36,414 is rounded to the nearest thousand 36,000. When working with percents or fractions, compatible numbers usually can be found by rounding to fractions such as $\frac{1}{2}$, $\frac{1}{4}$, $\frac{1}{3}$, $\frac{1}{5}$ and so on. However, remind students that there are no hard and fast rules for finding compatible numbers. The idea is simply to use rounded numbers that can be easily computed with to arrive at an estimate.

ERROR ANALYSIS

In using compatible numbers, there is a chance that an estimate may be too far off from the actual answer and thus be misleading. In the example 36,314 ÷ 9, if a student rounded the numbers to 36,000 ÷ 10, then the estimate would be 3600 rather than 4000. The number 3600 is not a wrong estimate, it is just not as good an estimate as 4000. Caution students to try to use one of the original numbers, if possible, or to find two numbers as close to the original numbers as possible.

PRACTICE AND APPLY

The following problems can be assigned for classwork and the answers checked in class to help students master the objective of the lesson.

- Guided Practice: 4–11
- Independent Practice: 12–18

WRAP-UP

To demonstrate the use of compatible numbers to make estimates use Problems 19–21.

Assignment Guide

- Basic: 16–30
- Average: 16–30 even

WORKSHOPS

21. The hotel-motel tax in Pittsburgh is 9.5% of the price of a room. If a room cost $112.50 per night, estimate the tax for 3 nights.
21. $33.00; $32.06

APPLICATIONS

22. Louis Watts was hired as the accounting clerk/bookkeeper at an annual salary of $12,500. Estimate his monthly salary. **$1000; $1041.67**

ACCOUNTING CLERK/ BOOKKEEPER $12–16K Salary. Knowledge of accounts payable/receivable. Career Position!

ACCOUNT EXECUTIVE/ MARKETING: $23–26K Salary (Fee paid). Degreed! 2 yrs. Marketing/ promotional experience! Kathleen 555-2222.

23. Ed Cooper learned at the job interview that he would be paid an annual salary of $24,600 as account executive/marketing. Estimate his monthly salary. **$2000; $2050**

BOOKKEEPER-FEE PAID. $15,000. IBM computerized bookkeeping to trial balance. Prefer complete charge small company experience. Free parking. BOARDROOM PERSONNEL, 626 Madison, 555-4271

24. Betty Borman took a job as a bookkeeper at an annual salary of $15,000. Estimate her weekly salary. **$300; $288.46**

Manufacturing Outlet
No Experience
Start pay for new employees is up to
$350 PER WEEK
To apply call Mon. or Tues. 9 am to 4 pm
555-3580

25. Linda Jackson took a job at a manufacturing outlet. She earns $350 a week. Estimate her daily income if she works 5 days a week. Estimate her annual salary. **$70; $18,200**

DATA TYPIST: $700–$800/ mo. Entry level! Type 45 wpm, phones. Call Polly 555-2284.

26. Jianguo Wang is a data typist earning $725 per month. Estimate his annual salary. **$8400; 8700**

OFFICE CLERK: $8–9K/ year. Great benefits! Filing, light typing. Call Sue 555-7670.

27. Suppose you take a job as an office clerk earning $8,000 to $9,000 a year. Estimate your weekly salary range. **$160 to $180; $153.85 to $173.08**

30% OFF MFG. LIST 3.30
LADIES' ANKLETS
Cotton/nylon blend. White and fashion colors.

28. Amy DiGrazia saw the sale ad for ladies' anklets. Estimate the savings. **$1.10; $0.99**

29. Gonzalo Carlos purchased a cordless telephone that had a regular price of $71.50. If he received 25% off, estimate how much he saved. **$18; $17.88**

25% OFF OUR REG. 59.99–119.99
ALL CORDLESS TELEPHONES
SALE 44.99 TO 89.99 Choose from a large selection of quality brand names.

30. Lynn Hatch purchased some jewelry that had a regular price of $19.75. If she received 40% off the regular price, estimate how much she paid for the jewelry. **$12; $11.85**

40% OFF OUR REG. 5.99–75.99
ALL 14K GOLD FILLED JEWELRY
SALE 3.59 TO 45.59 Earrings, lockets, pendants, more. Men's and ladies' styles.

WORKSHOP 18

Estimation: Clustering

Estimation by clustering is another way of projecting what an answer will be. When the numbers to be added are close to the same quantity, the sum can be found by clustering.

ESTIMATE

$4.85	All of the
5.15	numbers cluster
4.89	around $5.00, so
+ 5.17	$5 × 4 = $20.

The answer is about $20.00.
By computation, it is $20.06.

ESTIMATE

$15.95		
16.50	about	$48.00
15.75		
+ 7.95		+ 8.00
	about	$56.00

The answer is about $56.00.
By computation, it is $56.15.

Clustering cannot always be used, but when it can be used, it is usually a fast method since it can be done mentally.

✓ SELF-CHECK Complete the problems, then check your answers in the back of the book.

Estimate the sums by clustering, then compute the sums.

1.		2.		3.	
	563		$27.95		$1.59
	598		31.42		1.79
	559		30.25		2.21
	+ 612		+ 29.47		+ 0.75

4 × 600 = 2400; 2332 **4 × $30 = $120; $119.09** **(3 × $2) + $1 = $6 + $1 = $7; $6.34**

PROBLEMS

Estimate the sums by clustering. Compute the actual amount.

4.		5.		6.		7.		8.	
	763		525		$53.39		$36.20		88,026
	781		496		55.24		39.67		91,521
	773		512		49.26		41.78		87,842
	+ 810		+ 530		+ 49.97		+ 43.59		+ 94,819
	3200; 3127		**2000; 2063**		**$200; $207.86**		**$160; $161.24**		

8. 360,000; 362,208

9.		10.		11.		12.		13.	
	7.95		16.05		$2.29		$25.95		547.89
	7.87		15.95		2.39		26.30		546.90
	8.13		15.50		2.25		25.70		789.32
	8.25		15.50		2.40		37.95		804.90
	+ 3.67		+ 5.50		+ 8.95		+ 38.50		+ 811.66
	36; 35.87		**70; 68.50**		**$17; $18.28**		**$158; $154.40**		

13. 3600; 3500.67

14. School supplies costing: $6.95, $7.25, $7.45, $6.65, and $6.79. **$35.00; $35.09**
15. Groceries costing: $2.89, $3.15, $3.29, $2.67, $3.25, and $3.35. **$18; $18.60**
16. Work clothes costing: $25.95, $23.95, $26.50, $39.95, and $41.25. **$158; $157.60**
17. School supplies costing: $1.09, $0.99, $0.89, $2.29, $1.99, and $2.15. **$9; $9.40**
18. Clothes costing: shirt $17.95, shoes $39.95, six pairs of socks $18.25, tie $16.95, and belt $16.95. **$112; $110.05**
19. Sylvanus Schaibley drove the miles indicated: Mon. 367, Tues. 390, Wed. 405, Thurs. 386, Fri. 402, and Sat. 396. Estimate first, then calculate the total miles for the week. **2400; 2346**

LESSON PLAN Workshop 18

OBJECTIVES

To make an estimate by clustering.

RETEACHING

Clustering can be used only when the numbers fall within a narrow range of one another. If the numbers are dispersed over a broad range of values, then they cannot be clustered. However, it may be possible, in the latter case, to cluster some of the numbers in the group. This is done in one of the examples at the top of the page. In this workshop, students again use their rounding skills to find estimates by clustering.

ERROR ANALYSIS

If a group of numbers has one or more numbers that do not cluster, then students should treat the number or numbers separately. See Problems 9 and 12 for examples. In Problem 9, 3.67 is rounded to 4 and then added to 8 × 4, or 32, to arrive at the estimate of 36. In Problem 12, two clusters can be used: (26 × 3) + (40 × 2) = 158. Poor estimates can result if students do not cluster the given numbers in the right way.

PRACTICE AND APPLY

The following problems can be assigned for classwork and the answers checked in class to help students master the objective of the lesson.

- Guided Practice: 4–8
- Independent Practice: 9–13

WRAP-UP

Ask three students to work Problems 16, 17, and 18 and to explain how they arrived at their estimates.

Assignment Guide

- Basic: 10–20
- Average: 17–24

WORKSHOPS

20. Lincoln McClure had the following sales: First Qtr. $52,900, Second Qtr. $88,900, Third Qtr. $91,980, and Fourth Qtr. $89,830. Estimate first, then calculate the total sales for the year. **$320,000; $323,610**

21. Esther Radner had the following long-distance phone charges: $6.15, $2.15, $5.98, $1.81, $5.87, $1.89, and $11.71. Estimate first, then calculate the total long-distance charges. **$36; $35.56**

APPLICATIONS

22. $81; $76.25
- a. $16.00
- b. $6.00
- c. $4.50
- d. $22.50
- e. $7.75
- f. $9.00
- g. $4.00
- h. $6.50
- i. $76.25

23. $92; $88.46

22. During an extended stay at the Union Square Hotel, Mary Anderson sent the items indicated to be dry-cleaned. Estimate first, then calculate the total.

23. Mary's total dry-cleaning bill is subject to a 6% sales tax, and Mary gave the bellhop a tip equal to 10% of the dry-cleaning charges. Estimate the bill rounded to the nearest one dollar. What is the total bill?

HOTEL VALET
GUEST DRY CLEANING

	No.	Ladies'	Clean & Press	Press Only	Charges
a.	2	Suits	(8.00)	5.00	
b.	2	Capris & Slacks	4.00	(3.00)	
c.	1	Blazers	(4.50)	3.00	
d.	5	Blouses	(4.50)	3.00	
e.	1	Dresses	(7.75)	5.00	
f.	2	Skirts	(4.50)	3.00	
		Sweaters	4.50	3.00	
		Coats	8.00up	5.00	
g.	1	Coats-Car	7.50up	(4.00)	
h.	1	Jumpsuits	8.50up	(6.50)	
		Scarves	2.50up	1.50	
		Belts			
i.		Rm. No.	TOTAL		?

24. Greg Leininger used a calculator to solve the following problems. If he entered each number properly into the calculator, his answers should be correct. Estimate each answer and decide if Greg's answers are correct.

24.
- a. 2200 Yes
- b. $40 Yes
- c. 27 No, 26.98
- d. 330 No, 331
- e. 150 Yes
- f. 1260 Yes
- g. 29 No, 28.13
- h. 63,000 No, 63795

	Problem	His Answer on Calculator	Your Estimate	Is He Right? Yes/No
a.	672 + 703 + 725 + 130	2230	?	?
b.	$9.80 + $9.95 + $10.35 + $10.01	$40.11	?	?
c.	1.23 + 1.95 + 7.8 + 7.6 + 8.4	24.98	?	?
d.	82 + 75 + 79 + 34 + 29 + 32	299	?	?
e.	44.20 + 44.51 + 30.07 + 29.35	148.13	?	?
f.	625.1 + 615.2 + 12.35 + 10.75	1263.4	?	?
g.	9.89 + 2.9 + 9.49 + 3.1 + 2.75	24.73	?	?
h.	12,341 + 25,452 + 13,021 + 12,981	61,231	?	?

Problem Solving: Using the Four-Step Method

The problem-solving process consists of several interrelated actions. The solutions to some problems are obvious and require very little effort. Others require a step-by-step procedure. Using a procedure such as the four-step method should help you to solve word problems.

The Four-Step Method

Step 1:	Understand	What is the problem? What is given? What are you asked to do?
Step 2:	Plan	What do you need to do to solve the problem? Choose a problem-solving strategy.
Step 3:	Work	Carry out the plan. Do any necessary calculations.
Step 4:	Answer	Is your answer reasonable? Did you answer the question?

EXAMPLE A small office building is being remodeled. It will take 3 plumbers 7 days to install all the pipes. Each plumber works 8 hours a day at $24 per hour. How much will it cost for the plumbers?

SOLUTION

Step 1: Given — 3 plumbers, 7 days, 8 hours, $24 per hour

Find — The cost per day for 1 plumber.
The cost per day for 3 plumbers.
The cost of 3 plumbers for 7 days.

Step 2: Plan — Find the cost per day for 1 plumber, then multiply by the number of plumbers, and then multiply by the number of days.

Step 3: Work — 8 hrs per day × $24 per hr = $192 per day for 1 plumber
3 plumbers × $192 per day for 1 plumber = $576 per day for 3 plumbers
7 days × $576 per day for 3 plumbers = $4032 for 3 plumbers for 7 days

Step 4: Answer — It will cost $4032 for 3 plumbers for 7 days.

✓ SELF-CHECK Complete the problem, then check your answer in the back of the book.

1. It takes 2 finish carpenters 8 days to do the work. Each finish carpenter earns $26.50 per hour and works $7\frac{1}{2}$ hours per day. How much will it cost for the finish carpenters?
$3180

LESSON PLAN Workshop 19

OBJECTIVES

To use the four-step problem solving method to solve word problems.

RETEACHING

Write the problem given in the Example on the chalkboard. Then lead students in a discussion of the four steps of the problem solving method, writing the key words on the chalkboard: *Understand, Plan, Work, Answer.* Discuss each step thoroughly. Then illustrate each step using the problem, as shown in the textbook. In order to answer the question in Step 4, "Is your answer reasonable?", an estimate can be made.

ERROR ANALYSIS

Many students have a great deal of difficulty solving word problems. One of the major difficulties is just knowing where to begin. The four-step method will help students overcome their difficulty by giving them specific questions to answer in trying to understand the problem. Once a problem is understood, a plan can be formulated to solve it.

Of course, students can make errors in choosing a plan or in carrying out the plan. Very often the choice of a plan rests upon an understanding of *when* numbers need to be added or subtracted, or multiplied or divided. These are conceptual errors based upon a lack of understanding of the operations of arithmetic. Errors made in carrying out the plan usually are computational errors. Only by working closely with students and observing their work, or by asking questions and analyzing students' answers, will you be able to isolate the kinds of errors students make and thus be able to help them correct the errors.

PRACTICE AND APPLY

The following problem can be assigned for classwork and the answers checked in class to help students master the objective of the lesson.

- Guided Practice: 2–3
- Independent Practice: 4–5

WRAP-UP

Work Problem 6 together with the class as a means to review the four-step problem solving method.

Assignment Guide

- Basic: 7–13
- Average: 7–13

PROBLEMS

Identify the plan, work, and answer for each problem.

2. Henry Pitulski makes a car payment of $214.50 every month. His car loan is for 5 years. How much will he pay in 5 years? **$12,870**
3. Barb Allen and Joe Casey spent a total of $115.75 on their prom date. Dinner cost $45.87. How much did everything else cost Barb and Joe? **$69.88**
4. Elaine Wong purchased 2 sweaters at $24.99 each, a belt for $14.49, slacks for $19.79, shoes for $54.49, and 5 pairs of socks at $3.99 a pair. How much did Elaine spend? **$158.70**
5. A builder is building 5 new homes. It will take 4 electricians 2 days to wire each home. The electricians work 8 hours per day and earn $27.45 per hour. How much will it cost for the electricians? **$8784**
6. A builder is building 5 new homes. Each home has a foyer measuring 9 feet by 18 feet. Wood parquet floors for each foyer cost $46.80 per square yard. What is the cost of the wood parquet floors for the foyers in all 5 homes? **$4212**
7. Nadine Huffman charges $1.75 per page for typing rough drafts and an additional 50¢ per page for changes and deletions. A manuscript had 212 pages, of which 147 pages had changes and deletions. What was the total cost of typing the manuscript? **$444.50**
8. Nick Elsass is paying $10.50 per week for a compact disc player. The total cost of the CD player was $504. How long will it take Nick to pay for the CD player? **48 weeks**
9. Karen Johnson rode her 27" bicycle to the store and back. The store is 1 mile from Karen's home. Approximately how many rotations did Karen's bicycle wheels make in going to the store and back? (Hint: The circumference of a circle is approximately 3.14 times the diameter.) **Approximately 1495 rotations**
10. Ben Cornell and Tom Ingulli drove to Chicago, a distance of 510 miles. Their car gets 17 miles per gallon of gasoline. Gasoline costs them 96¢ per gallon. How much did Ben and Tom spend for gasoline on their trip? **$28.80**
11. Nancy Greer drove due north for 3 hours at 46 miles per hour. From the same spot, Ellen Young drove due south for 2 hours at 43 miles per hour. How far apart were they after their trip? **224 miles**
12. Peter Norris bought 3 boxes of cereal at $1.79 each, a roll of paper towels for 78¢, and 2 pounds of margarine at 74¢ a pound. How much change would Peter get back from $10? **$2.37**
13. Bread is sliced from a 30 cm loaf into equal slices 2 cm thick. All the slices are then toasted for a large breakfast requiring 180 pieces of toast. However, 1 out of every 5 pieces of toast is too well done for use. How many loaves of bread are needed? **15 loaves**

WORKSHOP 20

Problem Solving: Identifying Information

Before you begin to solve a word problem, first read the problem carefully and answer these questions:

- What are you asked to find?
- What facts are given?
- Are enough facts given? Do you need more information than the problem provides?

Some word problems provide more information than is needed to solve the problem. Others cannot be solved without additional information. Identifying what is wanted, what is given, and what is needed allows you to organize the information and plan your solution.

EXAMPLE 1 Mary Ising is a systems analyst for EDP Consultants, Inc. She earns $17.25 per hour. She is married and claims 2 withholding allowances. Last week she worked 40 hours at the regular rate and 4 hours at the weekend rate. She is 28 years old. Find her gross pay last week.

SOLUTION

A. Wanted: Mary Ising's gross pay last week

B. Facts given: $17.25 hourly rate
40 hours worked at regular rate
4 hours worked at weekend rate

C. Facts needed: Weekend rate

This problem cannot be solved.

EXAMPLE 2 John Skaggs, age 28, runs 6 miles every day. How many miles does John run in a week?

SOLUTION

A. Wanted: Number of miles run in 1 week

B. Facts given: Runs 6 miles every day

C. Facts needed: None

This problem can be solved. Multiply the number of miles run per day (6) by the number of days in 1 week (7). The answer is 42 miles.

✓ SELF-CHECK Complete the problem, then check your answer in the back of the book.

1. Tonia Walsh bought a new car with a $2000 down payment and monthly payments of $274.50. How much did Tonia pay, in total, for her new car?
Problem cannot be solved. Need number of payments.

WORKSHOPS

LESSON PLAN Workshop 20

OBJECTIVES

1. To solve word problems that may provide more information than is needed to solve the problem.
2. To identify word problems for which additional information is needed and thus cannot be solved.

RETEACHING

Go over the information given in the textbook prior to Example 1. Then work through the Examples with students. Point out that in the real world, people encounter many problems that often have too much or too little information. Thus, they first have to identify pertinent information before attempting to solve the problem..

ERROR ANALYSIS

Sometimes students will say that a problem cannot be solved when, in fact, it can be solved. This is simply a situation in which the student cannot identify *how* to solve the problem. Have these students list the four problem solving steps from Workshop 19 as an aid in organizing their work when attempting to solve problems. This should help them to identify the given information and make a plan to solve the problem.

PRACTICE AND APPLY

The following problems can be assigned for classwork and the answers checked in class to help students master the objectives of the lesson.

- Guided Practice: 2–3
- Independent Practice: 4–5

WRAP-UP

Work Problems 8 and 9 to wrap up the workshop.

Assignment Guide

- Basic: 4–15
- Average: 4–15

PROBLEMS

Identify the wanted, given, and needed information. If enough information is given, solve the problem.

2. The Camp Store is having a sale on camping equipment. It has 2-person tents for $79.49, cookstoves for $27.45, and cooking sets for $24.79. How much does a lantern and a tent cost? **Cannot be solved. Need cost of a lantern.**
3. The D & J Fruit Farm pays pickers 45¢ per pound to pick blueberries. The berries are packed in pint baskets and sold to grocery stores for 95¢ per pint. How many pint baskets are needed for 300 pounds of berries? **Cannot be solved. Need relationship between pints and pounds.**
4. Don Diamond paid $84 each way to fly round-trip from Detroit to Pittsburgh. Bob Tucker paid $139 for the round-trip fare. Who paid more? How much more? **Don paid $29 more for the round-trip.**
5. Edith Fairmont paid $126 for 3 tickets to a stage play. She paid for the tickets with three $50 bills. How much change did she receive? **$24**
6. Find the cost of 3 tablecloths, each 68 inches long and 52 inches wide. Each tablecloth costs $21.95. **$65.85**
7. Adam Larson paid for 2 watermelons with a $10 bill. He received $2.10 in change. What did the watermelons cost per pound? **Cannot be solved. Need weight of watermelons.**
8. Food for the party cost $44.95. Party supplies cost $17.48. Tula Drake and her friends have agreed to share the total cost of food and supplies equally. How much will each pay? **Cannot be solved. Need to know how many friends.**
9. Lou Hart is 6 feet 2 inches tall and weighs 195 pounds. He grew 3 inches in the past year. How tall was Lou 1 year ago? **5' 11"**
10. Alice Petroll has finished 25 of the 30 mathematics problems on her test. It is now 11:50 a.m. The 1-hour test started at 11:00 a.m. What is the average number of minutes she can spend on each of the remaining problems? **2 minutes**
11. A tennis racket and a can of balls cost a total of $59.89. What is the cost of the tennis racket? **Cannot be solved. Need cost of tennis balls.**
12. A bottle of cider costs 95¢. The cider costs 65¢ more than the bottle. How much does the cider cost? **80¢**
13. Richard Anderson sells magazine subscriptions and receives a weekly salary of $145. He also receives a $3 bonus for each subscription that he sells. Last week his gross pay was $202. How many subscriptions did Richard sell last week? **19**
14. In shopping for the latest recording of her favorite artist, Cheryl Nowokoski found that the cost of the compact disc was $9.20 more than the cost of the record album, and the cost of the cassette was $1.84 less than the cost of the record album. How much more than the cost of the cassette was the compact disc? **$11.04**
15. Assume you are driving on a 2-mile circular racetrack. For the first half of the track, you average 30 miles per hour. What speed must you maintain for the second half of the racetrack to average 60 miles per hour? **Impossible to average 60 mph.**

Problem Solving: Using More Than One Operation

Some problems require several operations to solve. After deciding which operations to use, you must decide the correct order in which to perform them.

EXAMPLE 1 The cash price of a new car is $18,750. Arthur Dennis cannot pay cash, so he is making a down payment of $2750 and 60 monthly payments of $325 each. How much more does it cost to buy the car this way?

SOLUTION

A. Given: Cash price of $18,750
$2750 down + 60 payments of $325 each

B. Multiply: To get total of payments
60 × $325 = $19,500

Add: $2750 to total of payments
$2750 + $19,500 = $22,250

Subtract: Total payments from cash price of car
$22,250 − $18,750 = $3500

It costs $3500 more to buy the car this way. In this example, the order of operations is very important; that is, to first multiply, then add, then subtract.

EXAMPLE 2 Nancy Paris bought 3 notebooks costing $3.98 each. She gave the cashier a $20 bill. How much change did she receive if there was no sales tax?

SOLUTION

A. Given: Bought 3 notebooks at $3.98 each, no sales tax
Gave cashier $20.00

B. Multiply: To get total cost
3 × $3.98 = $11.94

Subtract: To find change
$20.00 − $11.94 = $8.06

Nancy received $8.06 in change.

✓ SELF-CHECK Complete the problem, then check your answer in the back of the book.

1. Spent $7.95, $15.20, and $12.47 on entertainment. Entertainment budget is $50. How much is left in the entertainment budget? **$14.38**

PROBLEMS Give the sequence of operations needed to solve the problems, then solve.

2. Donna Preski works 8 hours a day, 5 days a week. So far this year, she has worked 680 hours. How many weeks has she worked? ×, ÷, **17**

LESSON PLAN Workshop 21

OBJECTIVES

To solve problems that require using several operations.

RETEACHING

It is important to work through both Examples very carefully with students. Ask students *why questions* as you do the Examples. In Example 1, ask: Why do you multiply 60 × $325 to get the total of payments? Why is $2,750 added to $19,500? Why is $18,750 subtracted from $22,250? These types of questions will force students to think about the operations involved and the order in which they are used.

ERROR ANALYSIS

Some students try to formalize rules for solving problems. For example, they will think along these lines: Most problems require addition to solve. So, when in doubt, add.

Or they think that when two numbers are given, one very large and the other very small, then divide the numbers. Of course, these types of approaches are completely wrong, and demonstrate that the student has no understanding of arithmetic operations and when they should be used. Questioning students as to what they are doing and why can help to reveal the kinds of errors students are making. Then the correct reteaching methods can be used.

PRACTICE AND APPLY

The following problems can be assigned for classwork and the answers checked in class to help students master the objective of the lesson.

- Guided Practice: 2–4
- Independent Practice: 5–6

WRAP-UP

Problem 7 can be used to review the objective of this workshop.

Assignment Guide

- Basic: 6–16
- Average: 6–16

WORKSHOPS

3. The band boosters sell cider and doughnuts at home football games. Last week they sold 318 cups of cider at 50¢ per cup and 12 dozen doughnuts at 40¢ per doughnut. What were the total sales? ×, ×, +, **$216.60**

4. Patsy Cole paid monthly electric bills of $51.72, $47.75, and $53.21. Her electric budget is $150 for 3 months. Is she over or under her budget? By how much? +, **compare,** −, **$2.68 over**

5. Victor Haddad sold 8 pumpkins for $2.75 each, 9 for $2 each, 24 for $1.50 each, and 15 for $1 each. He receives 35¢ for each pumpkin sold plus a $10 bonus if his sales total $75 or more. How much did he receive? ×, +, +, ×, **compare,** +, **$29.60**

6. Debra Smith worked through 174 pages of a 408-page computer training manual. It took her 2 days to work through the remaining pages. If she worked through the same number of pages each day, how many pages did she work each day? −, ÷, **117**

7. Mark Quincy worked 40 hours for $3.95 per hour. He worked 5 hours for $5.93 an hour. How much money did Mark earn? ×, ×, +, **$187.65**

8. The Parkers spent $42.78, $45.91, and $41.15 in 3 visits to the grocery store. Their food budget is $150. How much money do they have left to spend for food? +, −, **$20.16**

9. The temperature in the production department is 21 degrees Celsius at 12 noon. If the temperature increases 1.5 degrees Celsius every hour, what will the temperature be at 5 p.m.? −, ×, +, **28.5 degrees C**

10. Flora Sturgeon walks 4 miles round-trip 3 times a week to work. How far will she walk in one year? ×, ×, **624**

11. John Piotrowski assembled a total of 642 circuit boards in 3 days of work. During the first 2 days, he assembled 211 and 208 circuit boards, respectively. How many did he assemble the last day? +, −, **223**

12. In a one-month sales contest, Ernie Johnkovich earned 4 two-point certificates, 2 three-point certificates, and 6 one-point certificates. How many points did he earn for the month? ×, +, **20**

13. Seventeen hundred tickets costing $5 each were sold for a scholarship fund-raiser. One prize of $2000, three prizes of $1000, and five prizes of $250 were given away. How much money did the scholarship fund-raiser make? ×, +, −, **$2250**

14. Tom Kramer, sales leader for the past month, earned 47 one-point certificates. If he earned a total of 68 points, how many three-point certificates did he earn? −, ÷, **7**

15. Paula Cohn bought 3 T-shirts for $7.50 each and a sweatshirt for $22.95. How much change did she receive from a $50 bill? ×, +, −, **$4.55**

16. Rich Spangler saved $212. After he earned an additional $124, he spent $149 for a small color TV, $35 for a rugby shirt, and $79 for a pair of sneakers. How much money did Rich have left? +, +, −, **$73**

WORKSHOP 22

Problem Solving: Using Estimation

An important part of problem solving is determining the reasonableness of an answer. Checking an answer doesn't mean that you must recalculate it. Quite often, it is sufficient simply to determine if your answer makes sense. Estimation can be used to check the reasonableness of an answer. Some problems may ask for just an estimate.

EXAMPLE 1 Three cans of juice cost $1.19, six cans of soda cost $1.49, and six peaches cost $0.95. About how much will it cost for one of each item?

SOLUTION
$1.19 ÷ 3 is about $0.40
$1.49 ÷ 6 is about $0.25
$0.95 ÷ 6 is about $0.15
Total is about $0.80

EXAMPLE 2 Sandra Kaselman used her calculator to find 25% of $198.50. The result is shown at the right. Is her answer reasonable? Why or why not?

$4962.50

SOLUTION Her answer is not reasonable.
25% is equal to $\frac{1}{4}$, and $\frac{1}{4}$ of $198.50 is about $50.
It looks as if she multiplied by 25, not 25%.

✓ SELF-CHECK Complete the problem, then check your answer in the back of the book.

No, he used 93 instead of $0.93. Answer is $9.30.

1. Tom Lucas estimated the cost of 10 gallons of gas at 93¢ a gallon to be $930. Is his estimate reasonable? What error did he make?

PROBLEMS

In problems 2–9, determine the reasonableness of the estimate. If it is not reasonable, state what error was made.

	Problem	Estimate	Reasonable	Error, If Any
2.	20% of $496.98	$100	?	?
3.	5 gallons of gas at 89¢ a gallon	$450	?	?
4.	$14,203.00 ÷ 200	$700	?	?
5.	98.7 × 516	50,000	?	?
6.	Tip of 15% on $19.47	$3	?	?
7.	Sales tax of 5% on $1014.74	$5	?	?
8.	$15.00 per sq yd carpeting for 2' × 5' area	$150	?	?
9.	$49.79 − $19.49	$70	?	?

2. Yes; No error
3. No; Used 89¢ as $89; estimate: $4.50
4. No; Decimal point; estimate: $70
5. Yes; No error
6. Yes; No error
7. No; Decimal point; estimate: $50
8. No; Used 10 sq ft instead of 1 sq yd; estimate: $15
9. No; Added instead of subtracted; estimate: $30

WORKSHOPS

LESSON PLAN Workshop 22

OBJECTIVES

1. To estimate an answer.
2. To use estimation to check the reasonableness of an answer.

RETEACHING

Before working through the Examples with students, review the various ways of finding an estimate; rounding, front-end, compatible numbers, and clustering. Ask students which of these methods has been used in Example 1. (Compatible numbers). In discussing Example 2, point out that $198.50 is about $200 and $\frac{1}{4}$ of $200 is $50.

Discuss with students the notion that to say an answer *makes sense* or that it is *reasonable* means that it is likely the answer is correct. An estimate can show whether or not the calculated answer is likely to be the correct answer. If the estimate is close to the calculated answer, then the latter is probably correct. An estimate does not show that the answer, in fact, is correct. This can only be done by recalculating the answer.

ERROR ANALYSIS

When using estimates to check the reasonableness of an answer, students may make a poor estimate and conclude the answer is not reasonable. However, this is not very likely to happen as most estimates can readily establish the degree of magnitude of the answer; that is to say, the correct answer is in the hundreds, or thousands, or ten thousands, or millions, for example.

Example 2 illustrates well the fact that the answer of $4962.50 cannot possibly be correct because Sandra is taking 25% of a number much smaller than $4962.50, namely $198.50, and therefore the answer must be less than $198.50. Finding the estimate shows the answer to be about $50.

PRACTICE AND APPLY

The following problems can be assigned for classwork and the answers checked in class to help students master the objective of the lesson.

- Guided Practice: 2–5
- Independent Practice: 6–9

WRAP-UP

In Example 2, change the percent to 125% and ask students to explain why the answer of $4962.50 is much too large. (125% of $198.50 is about $250. By mental math, 100% of a number is the number itself, and 25% of $198.50 is about $50. Therefore, $198.50 + $50 is about $250.)

Assignment Guide

- Basic: 10–19
- Average: 10–19

WORKSHOPS

10. A picture frame requires 36 inches of frame molding at $5.99 a foot. About how much will the molding cost? **About $18**

11. Ursula VanMeer bought 100 shares of stock at 18\frac{1}{2}$ per share. One year later, she sold all her shares at 23\frac{1}{4}$ per share. About how much did she make on her stock? **About $500**

12. A new Vacation Plus mini van costs $24,897. About how much would 4 of these mini vans cost? **About $100,000**

13. If the sales tax rate is 5%, about how much sales tax will be due on a $24,897 mini van? **About $1250**

14. If a 10% down payment is required to finance the purchase of a $24,897 mini van, about how much money would you need for a down payment? **About $2500**

15. Ben Grasser is buying a new car with a total purchase price including interest of $14,987. He plans to make a $3000 down payment and to finance the rest for 5 years. About how much will his monthly payments be? **About $200**

16. Molly Noonan purchased 6 dozen cupcakes for the 25 children attending a birthday party. All the cupcakes were eaten. About how many cupcakes did each child eat? **About 3**

17. Peter Cummings purchased the following school supplies on sale: a $2.79 notebook for $1.99, a $1.49 ballpoint pen for 99¢, a $1.99 pack of notebook paper for 99¢, and a $3.49 automatic pencil for $2.79. Estimate the total savings. **About $3**

18. It is about 1300 feet around the bicycle test track. How many times must you ride around the track to ride about 1 mile? **4 times**

19. Fruit baskets containing 6 apples, 4 oranges, and 2 grapefruits are on sale for $2.98. If you have 148 apples, 121 oranges, and 64 grapefruits, about how many fruit baskets can be made? **About 25**

WORKSHOP 23

Problem Solving: Constructing a Table

Constructing a table can be a good way of solving some problems. By organizing the data into a table, it is easier to identify the information that you need. A table is useful in classifying information.

EXAMPLE 1 Meredith McCall is a car salesperson. For each new car she sells, she earns 10 bonus points; for each used car she sells, she earns 5 bonus points. She earned 125 bonus points by selling 16 cars last week. How many of each type of car did she sell?

SOLUTION Construct a table in order to evaluate the possibilities.

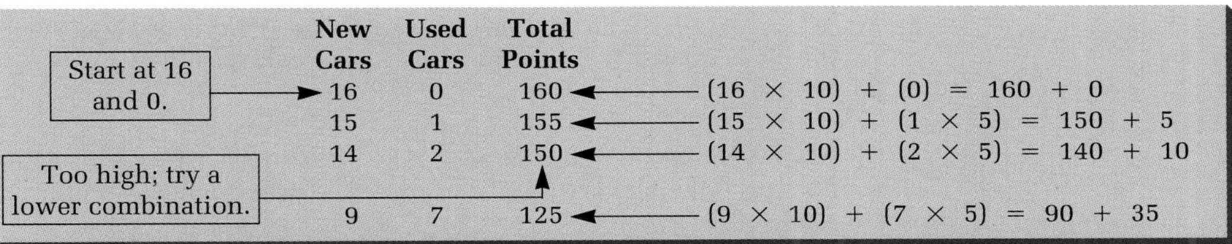

New Cars	Used Cars	Total Points	
16	0	160	$(16 \times 10) + (0) = 160 + 0$
15	1	155	$(15 \times 10) + (1 \times 5) = 150 + 5$
14	2	150	$(14 \times 10) + (2 \times 5) = 140 + 10$
9	7	125	$(9 \times 10) + (7 \times 5) = 90 + 35$

Start at 16 and 0.
Too high; try a lower combination.

Meredith sold 9 new cars and 7 used cars.

EXAMPLE 2 Roger Bitter has a total of 20 coins consisting of dimes and quarters. The total value of the coins is $4.70. How many of each coin does he have?

SOLUTION Construct a table in order to evaluate the possibilities.

Number of		Value of		
Quarters	Dimes	Quarters	Dimes	Total Value
20	0	$5.00	$ 0	$5.00
19	1	4.75	0.10	4.85
18	2	4.50	0.20	4.70

Roger has 18 quarters and 2 dimes.

✓ SELF-CHECK Complete the problem, then check your answer in the back of the book.

1. A total of 11 vehicles consisting of unicycles (1 wheel) and bicycles (2 wheels) went by. Eighteen wheels were counted. How many of each were there? **7 bicycles; 4 unicycles**

WORKSHOPS

LESSON PLAN Workshop 23

OBJECTIVES

To solve problems by constructing a table.

RETEACHING

Use the Example to reteach solving problems by constructing a table. A table can be used to list all the possible solutions and, in so doing, the actual solution can be found. In the solution to Example 1, 16 new cars and 0 used cars are used to start the list because a total of 16 cars were sold. After three possibilities are listed, it is clear that 14 and 2 is still too high a combination. Instead of continuing to list all combinations, have students choose a lower combination. In the Example, 9 and 7 is the correct combination. Point out to students that the power of using a table is that the solution will eventually be found as the possibilities are listed. So the method of constructing a table provides a somewhat mechanical way to solve problems.

ERROR ANALYSIS

The key to avoiding errors in constructing a table is to label the headings correctly. Then the data will fall into place naturally. Stress the importance of reading the problem carefully and identifying the information given.

Also, the question asked in the problem must be understood. Then a start can be made on constructing the table. Encourage students to check their first few entries in the table to see that they meet the conditions of the problem.

PRACTICE AND APPLY

The following problems can be assigned for classwork and the answers checked in class to help students master the objective of the lesson.

- Guided Practice: 2
- Independent Practice: 3

WRAP-UP

The following problem can be used to summarize the objective of this workshop. Solve by constructing a table. Chuck, Amy, Jim, and Sue each subscribe to one of the following magazines: *Sports Today, Teen Times, Car Care,* and *Newsworld.* Use the following clues to determine which person subscribes to which magazine.

- Chuck's magazine has a one-word title.
- Amy hates sports.
- Jim changed the type of car wax he uses after reading an article in his magazine.

Chuck—*Newsworld,*
Jim—*Car Care,*
Amy—*Teen Times,*
Sue—*Sports Today*

Assignment Guide

- Basic: 3–11 odd
- Average: 4–12 even

WORKSHOPS

PROBLEMS

2. Lieutenant Sampson is a recruitment officer for the Marines. For each high school graduate he recruits, he gets 5 points; for each college graduate, 11 points. He earned 100 points last week by recruiting 14 high school and college graduates. How many of each did he recruit? **9 high shool; 5 college**

3. Nine cycles were produced using 21 wheels. How many bicycles and how many tricycles were produced? **6 bicycles; 3 tricycles**

4. Third National Bank charges a monthly service fee of $5 plus 25¢ per check. The bank is changing its policy to a $6 service charge and 15¢ per check. A bank officer said you will save money with the new system. How many checks must you write each month in order to save money? **11**

5. Wanda Cross has a total of 40 coins consisting of nickels and quarters with a total value of $6. How many of each coin does she have? **20 nickels; 20 quarters**

6. The Theatre Club sold a total of 415 tickets. The adult tickets cost $5 and the children's tickets cost $3. If $1615 was collected, how many adult tickets were sold? **185**

7. There are 56 stools in the storeroom. Some stools have 3 legs and some have 4 legs. If there are 193 legs, how many 4-legged stools are in the storeroom? **25**

8. Steve Swartz has 4 dimes and 3 nickels. List the amounts of all the exact-change telephone calls Steve could make using 1 or more of these coins. **11**

9. How many different ways can you make change for a quarter? **12**

10. Pam Young has exactly 20 dimes, 20 nickels, and 20 pennies. Find all the ways Pam can choose 22 coins whose total value is $1 if she must use at least 1 coin of each type. **2 (5 pennies, 15 nickels, 2 dimes or 10 pennies, 6 nickels, 6 dimes)**

11. Frank DeGeorge has 69¢ in coins. Bob White asked Frank for change for a half-dollar. Frank tried to make change but found that he didn't have the coins to do so. What coins did Frank have if each coin was less than a half-dollar? **4 pennies; 4 dimes; 1 quarter**

12. Carol Meeks was playing darts. She threw 6 darts, all of which hit the target shown. Which of the following scores could be hers? 2, 19, 58, 28, 33, 37 **28**

Problem Solving: Looking for a Pattern

Some problems can be solved more easily if the information is first organized into a list or table. Then the list or table can be examined to see if a pattern exists. A pattern may not "jump out" at you, but you may be able to discover a pattern after manipulating the information.

EXAMPLE Joe Cobb has 15 coins consisting of dimes and quarters. The total value of the coins is $2.55. How many of each coin does he have?

SOLUTION Given: 15 coins consisting of dimes and quarters

The table displays different combinations of dimes and quarters with the total number of coins equaling 15.

Number of Dimes	15	14	13	12	11
Number of Quarters	0	1	2	3	4
Total Value	$1.50	$1.65	$1.80	$1.95	$2.10

← Look for a pattern.

You could continue the table, but it is easier if you see the pattern. Each time you take away a dime and add a quarter, the total value increases by $0.15. The difference between $2.55 and $1.50 is $1.05, and $1.05 ÷ $0.15 = 7. Therefore, subtract 7 from 15 and conclude that there are 8 dimes and 7 quarters.

Check: The total value of 7 quarters and 8 dimes is:

(7 × $0.25) + (8 × $0.10) = $1.75 + $0.80 = $2.55

✓ SELF-CHECK Complete the problems, then check your answers in the back of the book.

Write the next 3 numbers for the established pattern.

1. 1, 3, 9, 27, . . . **81, 243, 729**

2. 2, 4, 7, 11, . . . **16, 22, 29**

PROBLEMS

In problems 3–10, look for a pattern and then write the next 3 numbers.

3. 10, 16, 22, 28, . . . **34, 40, 46**

4. 2, 4, 8, 16, . . . **32, 64, 128**

5. 30, 27, 24, 21, . . . **18, 15, 12**

6. 8, 4, 2, 1, . . . $\frac{1}{2}$, $\frac{1}{4}$, $\frac{1}{8}$

7. 1, 4, 9, 16, . . . **25, 36, 49**

WORKSHOPS

LESSON PLAN Workshop 24

OBJECTIVES

To solve a problem by using a pattern.

RETEACHING

Discuss the Example. Ask students what purpose the list or table serves (It helps to organize the data so that a pattern can be identified.) Point out that the difficult part about using a pattern to solve a problem usually is to find the pattern.

Ask students why $1.50 is subtracted from $2.55. (To get the total difference between the starting point in the pattern, $1.50, and the end point, $2.55. Dividing this difference, $1.05, by each increase in the pattern, $0.15, yields the number of quarters in the pattern, 7. Therefore, there are 8 dimes, since the total number of coins is 15.)

ERROR ANALYSIS

When working with patterns, students may construct a pattern that does not solve the problem. A check of the result will show that the pattern did not work. Thus, they need to start over.

In the Example, if students have difficulty understanding the solution, point out that a continuation of the table from 11 dimes to 10, 9, and 8 dimes reveals the final result: 8 dimes, 7 quarters, totaling $2.55. Point out however, that if this problem required 50, or 100, or 1000 steps to extend the table, then finding the pattern would be a much more practical way to solve the problem.

PRACTICE AND APPLY

The following problems can be assigned for classwork and the answers checked in class to help students master the objective of the lesson.

- Guided Practice: 3–6
- Independent Practice: 7–12

WRAP-UP

Work Problem 13 with the students as a cooperative activity to illustrate the objective of this workshop.

Assignment Guide

- Basic: 10–20
- Average: 10–20

WORKSHOPS

8. 2, 3, 5, 9, 17, 33, . . . **65, 129, 257**

9. 1, 4, 13, 40, 121, 364, . . . **1093, 3280, 9841**

10. 1, 1, 2, 3, 5, 8, 13, . . . **21, 34, 55**

11. Sy Mah has 20 coins consisting of dimes and quarters. Their total value is $3.80. How many of each coin does Sy have? **8 dimes; 12 quarters**

12. Peggy Mays had 185 tickets to the school play. Adult tickets sold for $4 each and children's tickets sold for $2 each. The total value was $594. How many of each ticket did she sell? **112 adult; 73 children**

13. Three-sided numbers, such as 3, are so named because dots can be used to form a triangle with an equal number of dots, such as 3, on each side. What three-sided number has 12 dots on a side? **33**

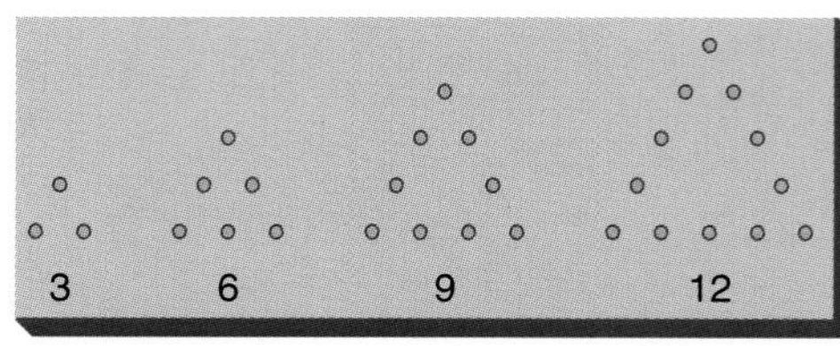

14. On the first day of school, your teacher agrees to allow 1 minute of "fun" at the end of the first day, 2 minutes on the second day, 4 minutes on the third day, 8 minutes on the fourth day, and so on. How much time will you have for "fun" at the end of 10 days? **1024**

15. At 9 a.m., there were 7 students in the computer room. At 9:30 a.m., 2 students left, and at 10 a.m., 1 student arrived. At 10:30 a.m., 2 students left, and at 11 a.m., 1 student arrived. This pattern continued with 2 students leaving at half past the hour and 1 student arriving on each hour. At what time did the computer room first become empty? **2:30 p.m.**

16. Bob and Ray Hunt were responsible for total lawn care of the factory grounds. The first week, Bob mowed half the lawn. The next week, he mowed two-thirds as much as he had the first week. The third week, he mowed three-fourths as much as he had the second week, and so on. The tenth week, he mowed ten-elevenths as much as he mowed the ninth week. How much of the lawn did Bob mow the tenth week? **1/11**

17. There are 2 rectangular storage rooms whose sides are whole numbers and whose area and perimeter are the same number. What are their dimensions? **4 × 4; 6 × 3**

18. A 5-pound bag of lawn food sells for $3.25 and a 3-pound bag sells for $2.29. You need 17 pounds of lawn food. What is the least amount you can pay and buy at least 17 pounds? **$12.04**

19. When it is 12 o'clock, the hands on the face of the clock overlap. How many more times will the hands overlap as the clock runs until it is again 12 o'clock? (Count the second time it is 12 o'clock but not the first.) **11**

20. Each year, before the U.S. Supreme Court opens, each of the 9 justices shakes hands with every other justice. How many handshakes are involved in this ceremony? **36**

WORKSHOP 25

Problem Solving: Using Guess and Check

One way to solve a problem is by using guess and check, or trial and error. Guessing at a solution doesn't mean making a blind guess in the hopes that it is correct. It means making an informed guess and then checking it against the conditions stated in the problem to determine how to make a better guess. The process is repeated until the answer is found.

EXAMPLE 1 You need 80 sandwich buns. You can buy 8 for $1.19 or 12 for $1.69. What do you buy to obtain at least 80 sandwich buns at the lowest cost?

SOLUTION Guess and check until you find the lowest cost. Keep your information organized by using a table.

Number		Total Buns	Cost		Total Cost
12/pkg.	8/pkg.		12/pkg.	8/pkg.	
7	0	84	(7 × $1.69) +	0	$11.83
6	1	80	(6 × $1.69) +	(1 × $1.19)	$11.33
5	3	84	(5 × $1.69) +	(3 × $1.19)	$12.02
4	4	80	(4 × $1.69) +	(4 × $1.19)	$11.52
3	6	84	(3 × $1.69) +	(6 × $1.19)	$12.21
2	7	80	(2 × $1.69) +	(7 × $1.19)	$11.71
1	9	84	(1 × $1.69) +	(9 × $1.19)	$12.40
0	10	80	0	+ (10 × $1.19)	$11.90

Six packages of 12 sandwich buns and one package of 8 sandwich buns cost $11.33, the least amount, and result in 80 sandwich buns.

EXAMPLE 2 The factory building is 4 times as old as the equipment. Three years from now, the factory building will be 3 times as old. How old is the equipment?

SOLUTION

Current Age		Three Years From Now		Check
Equipment	Building	Equipment	Building	
7	28	10	31	No
8	32	11	35	No, wrong direction
6	24	9	27	Yes

The equipment is now 6 years old.

✓ SELF-CHECK Complete the problem, then check your answer in the back of the book.

1. The product of 3 consecutive whole numbers is 504. What are the numbers? 7, 8, 9

WORKSHOPS

LESSON PLAN Workshop 25

OBJECTIVES

To solve problems using the guess and check or trial and error method.

RETEACHING

Work through Example 1 with the students showing all the information in the table. Then have students look at the total cost figures and ask them if they see a pattern. (Pairs of values are increasing: $11.83 and $11.33 to $12.02 and $11.52. With this observation, there is no need to continue the table; $11.33 is the lowest cost.) Thus, emphasize the fact that problem-solving strategies can be combined to solve problems. Point out also that the guess and check method employs a good deal of common sense.

ERROR ANALYSIS

Encourage students not to make wild guesses. The successful use of this strategy rests upon understanding the problem first and them making a fairly accurate guess. The result of the guess should be examined and then a second, improved guess made. This procedure is followed until the answer is found. When using guess and check, students need to think about the possibility of there being more than one answer. In Example 1, the possibility of more than one answer is eliminated by using a pattern. In Problem 2, however, if students stop after finding one answer, six other answers would be missed.

PRACTICE AND APPLY

The following problems can be assigned for classwork and the answers checked in class to help students master the objective of the lesson.

- Guided Practice: 2–3
- Independent Practice: 4–5

WRAP-UP

Use the following problem to summarize the discussion of the guess and check method.

Roy O'Neal gave the clerk a twenty-dollar bill. His change was $1.00 more than he paid for a cassette tape. How much did he pay for the tape? ($9.50)

Assignment Guide

- Basic: 4–11
- Average: 4–11

PROBLEMS

2.

Triads:	Tetrads:
27	0
23	3
19	6
15	9
11	12
7	15
3	18

5. All sums are 12.

8. 3 ways; two 3's and six 4's would probably be the cheapest

9. Ford = 10; Chevy = 8; Dodge = 4

2. The planet Tetriad only has creatures with 3 legs (triads) or 4 legs (tetrads). Astronauts Peter North and Sally Clark could not bear to look at these ugly creatures, so they kept their eyes on the ground. On their first day on Tetriad, they counted 81 legs. How many Triads and Tetrads did they meet?

3. It costs 19¢ to mail a postcard and 29¢ to mail a letter. Sam Checkers wrote to 15 friends and spent $3.55 for postage. How many letters and how many postcards did he write? **7 letters; 8 postcards**

4. Paula McDale earns $8 per hour Monday through Friday and $12 per hour on weekends. One week, she worked 49 hours and earned $404. How many hours did she work on the weekend? **3 hours**

5. Arrange the 4 dominoes below into a domino donut so that all sides equal the same sum (not necessarily the same sum as that of the example at the right).

Example

6. Ti Sun wants to fence off a storage area for surplus lumber. She has 96 meters of new fencing to put along an existing fence. What are the dimensions that will give Ti the largest storage area? **48 m × 24 m**

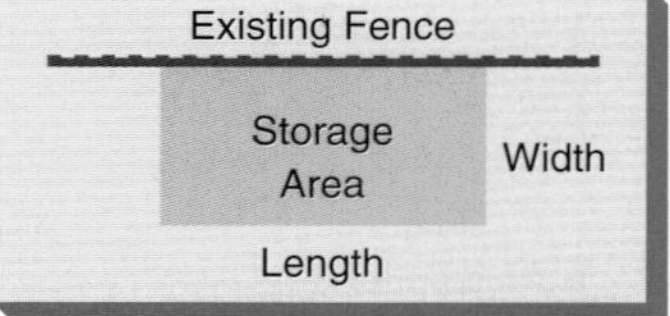

7. In the square at the right, a rule applies from top to bottom and from left to right. Find the rule and figure out the missing number. **2**

8. Batteries come in packs of 3 or 4. If your class needs 30 batteries, how many different ways are there of buying exactly 30 batteries? Which combination do you think would be the cheapest?

9. The ages of 3 delivery vans total 22 years. The Ford is the oldest and is 10 years old. The Dodge is 6 years younger than the Ford. The third delivery van is a Chevy. What are the ages of the vans?

10. The product of 3 consecutive whole numbers is 120. What are the numbers? **4, 5, 6**

11. Arrange the numbers 5 through 13 in a magic square whose rows, columns, and diagonals add up to the same number. **Sum is 27. Top: 8, 13, 6; Middle: 7, 9, 11; Bottom: 12, 5, 10**

Problem Solving: Working Backward

Problems that involve a sequence of events or actions can sometimes be solved by working backward. If the final result of the problem is given, start your solution with that result and work backward to arrive at the beginning conditions of the problem.

EXAMPLE 1 Each year, a delivery van is worth three fourths of its value from the previous year. A van is now worth $9000. What was its value last year?

SOLUTION

A. What is the van worth now? $9000

B. How does last year's value relate to this year's value?

$\frac{3}{4}$ of last year's value = this year's value

$\frac{3}{4}\ x = \$9000$

C. Solve: $x = \$9000 \div \frac{3}{4}$

$x = \$12{,}000$

The van was worth $12,000 last year.

EXAMPLE 2 Central Bakery baked some cookies and put one half of them away for the next day. Then Central Bakery divided the remaining cookies evenly among its 3 sales outlets so that each outlet received 40 dozen. How many cookies did Central Bakery bake?

SOLUTION Each of the 3 sales outlets received 40 dozen cookies. Thus, Central Bakery divided a total of 120 dozen cookies. The 120 dozen cookies represent one half of what was baked; therefore, they baked 240 dozen cookies.

✓ SELF-CHECK Complete the problem, then check your answer in the back of the book.

1. A water lily doubles itself in size each day. It takes 30 days from the time the original plant is placed in a pond until the surface of the pond is completely covered with lilies. How long does it take for the pond to be half covered? **29 days**

PROBLEMS

2. Checker Cab Co. charges a flat fee of $3.75 plus $0.30 for every $\frac{1}{4}$ mile driven. Dick Lewis paid a driver a total of $7.95 for a trip from the airport to his office. How many miles did he travel? $3\frac{1}{2}$

WORKSHOPS

LESSON PLAN
Workshop 26

OBJECTIVES

To solve problems by working backwards.

RETEACHING

After working the two Examples with students, emphasize that the following conditions necessitate working backward to find a solution. (1) A *series* of actions or events is described, and the end result is given; (2) the task is to determine the conditions at a previous action.

Now go back over the Examples and point out that the series of actions in Example 1 is the year-by-year value of the van. Ask students what the end result is. ($9000) What is the previous event? (value last year) Ask similar questions about Example 2.

ERROR ANALYSIS

When starting with the final result, students must then link it correctly to the previous event. This connection is crucial if previous actions or events are to be determined. Errors in working backwards generally occur at this step. In Example 1, the equation $\frac{3}{4}x = \$9000$ provides the key connection between the last event ($9000) and the previous event (van is worth $\frac{3}{4}$ of its value the previous year.)

PRACTICE AND APPLY

The following problems can be assigned for classwork and the answers checked in class to help students master the objective of the lesson.

- Guided Practice: 2
- Independent Practice: 3

WRAP-UP

Use Problem 5 to illustrate the objective of this workshop.

Assignment Guide

- Basic: 5–10 even
- Average: 4–11 odd

WORKSHOPS

3. A recipe for 24 medium-sized pancakes requires 2 eggs. Eggs are sold in cartons of 12 eggs. The band boosters plan to serve 240 people an average of 3 pancakes each. How many cartons of eggs are needed? **5**

4. Two barrels, A and B, contain unspecified amounts of cider, with A containing more than B. From A, pour into B as much cider as B already contains. Then from B, pour into A as much cider as A now contains. Finally, pour from A into B as much cider as B presently has. Both barrels now contain 80 liters of cider. How many liters of cider were in each barrel at the start of the process? **A:110 L; B: 50 L**

5. Your company automobile is now worth $12,000. You read an article that indicates that each year an automobile is worth 80% of what its value was the previous year. What was the value of the company automobile last year? **$15,000**

6. Tiffany Cole starts at point A and enters a fun house. She pays $2 to get in and loses half of the money in her possession while she is in the fun house. She then pays $1 when she exits at B. She goes to the next entrance (C), pays $2 to get in, loses half of her money, and pays $1 to exit at D. This is repeated until she exits at H and gives her last $1 to get out. How much money did she start with? **$60**

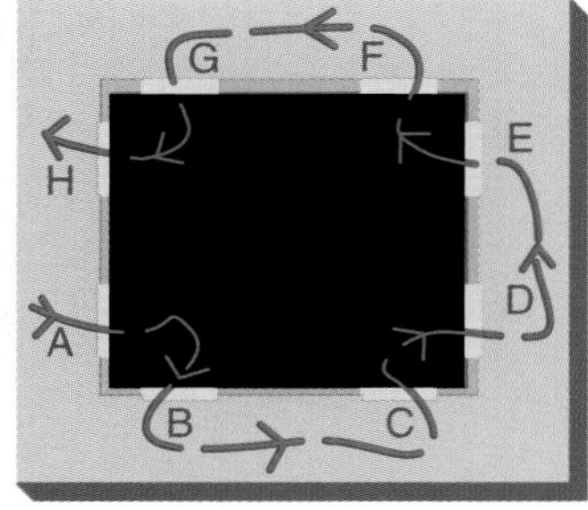

7. Merril Bakery baked some cookies in their new convection oven. Stores #1, #2, #3, and #4 sold a total of 4 dozen cookies each. The remainder were put in storage. Later, another dozen cookies were sold and one third of what was left was sent to stores #5 and #6. Those two stores sold 2 dozen cookies each and had 9 dozen left between them. How many cookies did Merril Bakery bake? **56 dozen**

8. Sixty-four players are entered in the company single elimination tennis tournament. How many matches must be played to determine the winner? **63**

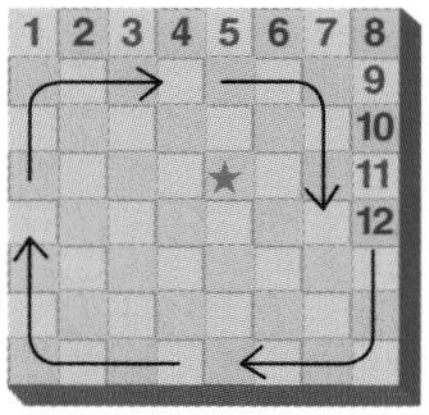

9. If you were to continue the number pattern until you got to the star, what number would you put in the star's square? **62**

10. Three girls played a game in which two of them won and one lost on each play. The girl who lost had to double the points that each winner had at that time by subtracting from her own points. The girls played the game 3 times, each winning twice and losing once. At the end of the 3 plays, each girl had 40 points. How many points did each girl have to start? **20, 35, and 65**

11. There were 8 women and 16 men at the last board of directors meeting. Every few minutes, 1 man and 1 woman (a couple) left the meeting together. How many couples must leave before there are exactly 5 times as many men as women left at the meeting? **6**

Problem Solving: Writing an Equation

A word problem can be translated into an equation that is solved by performing the same mathematical operation (adding, subtracting, multiplying, or dividing) to both sides. Solving the equation then leads to the solution of the problem.

To set up the equation, look for words in the problem that suggest which of the four mathematical operations to use.

Words	Symbol	Operation
The total, how many in all, the sum, plus	$+$	Addition
The difference, how much more, how much smaller, minus	$-$	Subtraction
The total for a number of equal items, the product	$\times$	Multiplication
The number left over, the quotient	$\div$	Division

EXAMPLE 1 In 40 hours at your regular rate of pay plus 5 hours of double time (twice your regular rate of pay), you earn $425. What is your regular rate of pay?

SOLUTION Use the letter x to stand for your regular rate of pay.

$$40x + 5(2x) = \$425.00$$

$$40x + 10x = \$425.00$$

$$50x = \$425.00$$

$$x = \$8.50 \text{ (Divide each side by 50.)}$$

EXAMPLE 2 A rectangle with a perimeter of 48 mm is 20 mm long. What is the width of the rectangle?

SOLUTION Let w equal the width of the rectangle.

$$w + 20 + w + 20 = 48$$

$$2w + 40 = 48 \text{ (Subtract 40 from both sides.)}$$

$$2w = 8 \text{ (Divide both sides by 2.)}$$

$$w = 4 \text{ mm wide}$$

✓ SELF-CHECK Complete the problem, then check your answer in the back of the book.

1. The sum of 2 consecutive numbers is 23. What is the smaller number? **11**

PROBLEMS

2. One brand of computer scanner can read 74 documents per hour while a second scanner can read 92 documents per hour. How many hours will it take to read 747 documents? **4.5 hours**

WORKSHOPS

LESSON PLAN Workshop 27

OBJECTIVES

To solve problems by writing an equation.

RETEACHING

Review the meaning of an equation by using some examples. Write $N + 7 = 10$ on the chalkboard. Point out that in an equation the quantities on both sides of the equals sign must be equal. Therefore, N must be equal to 3. Ask students to solve the following equations: $R - 20 = 50$ $(R = 70)$; $A + 4 = 13$ $(A = 9)$; $2W = 10$ $(W = 5)$; $100 \div B = 20$ $(B = 5)$.

Remind students that writing an equation to solve a word problem involves translating words into mathematical symbols. Go over the words, symbols, and operations in the book before working through the Examples.

ERROR ANALYSIS

There are many opportunities for students to make errors when solving problems by writing an equation. If some students continue to get wrong answers, you will need to examine their written work to see what errors they may be making. Errors occur when students use the incorrect operation in the translation process, or they may not use the correct coefficient with the variable. In other words, they do not write the correct equation as a model to solve the problem.

Other types of errors involve not knowing how to solve the equation. Students often forget the fundamental rule that an equation is solved by performing the *same* mathematical operations to both sides of the equation. Emphasize this point when working the Examples and going over the solutions to the Problems.

PRACTICE AND APPLY

The following problems can be assigned for classwork and the answers checked in class to help students master the objective of the lesson.

- Guided Practice: 2–3
- Independent Practice: 4–5

WRAP-UP

Write the following sentences on the chalkboard. Have students write an equation for each sentence and then solve it.

1. Fifty dollars minus the cost of a sweater equals twenty-two dollars.
   $50 − $C = $22;
   C = $28
2. 35 hours multiplied by an hourly rate equals a gross weekly pay of $287.
   35 × $R = $287;
   R = $8.20
3. The distance from Cobbs Landing to Green Hills is 520 miles. That equals twice the distance from Cobbs Landing to Glenview minus twenty miles.
   520 = 2D − 20;
   D = 270

Assignment Guide

- Basic: 3–15 odd
- Average: 4–14 even

WORKSHOPS

3. A bottle and a cork cost $1.10. The bottle costs $1.00 more than the cork. How much does each cost? **Cork: 5¢; Bottle: $1.05**

4. A robotic delivery unit travels 54 meters in traveling completely around the edge of a rectangular mail room. If the rectangle is twice as long as it is wide, how long is each side? **18 m × 9 m**

5. A football field is 100 yards long and has a distance around of 308 yards. How wide is it? **54 yards**

6. David and Liz Northrup make monthly payments of $727.20 on their $80,000 mortgage. They will have paid $138,160 in interest when their mortgage is paid off. For how many years is their mortgage? **25 years**

7. Edith Harris had gross pay of $464.86 last week. She earns $5.65 per hour plus a 3% commission on all sales. She knows she worked 40 hours last week but can't remember her total sales. What were her total sales? **$7962**

8. Steve Sutton earns $7.60 per hour plus double time for all hours over 40 per week. How much did Steve earn for working 46 hours last week? **$395.20**

9. Wilson Davis has 3 Guernsey cows and 2 Holstein cows that give as much milk in 4 days as 2 Guernsey and 4 Holstein cows give in 3 days. Which kind of cow is the better milk producer, the Guernsey or the Holstein? **Holstein**

10. The Karis Tool & Die Company building is 5 times as old as the equipment. The building was 24 years old when the equipment was purchased. How old is the equipment? **6 years**

11. The sum of 3 consecutive odd numbers is 27. What are the 3 numbers? **7, 9, 11**

12. Universal Inc. stock sells for 17\frac{1}{4}$ a share. Future Discount Brokers charges a flat fee of $40 for every transaction. How many shares could you buy for $730? **40**

13. Write an equation expressing the relationship between A and B given in the table at the right. What would B equal when A is 40?
    B = 3A − 2; 118

A	1	2	3	4	5	...
B	1	4	7	10	13	...

14. In problem 13, write an equation for A in terms of B. What would A equal when B is 253? **A = (B + 2) ÷ 3; 85**

15. Harry, Jerry, and Darrel have a combined weight of 599 pounds. Jerry weighs 13 pounds more than Harry, while Harry weighs 5 pounds more than Darrel. How much does each man weigh? **Harry = 197; Jerry = 210; Darrel = 192**

Problem Solving: Making a Venn Diagram

Some problems can be solved with a diagram. Making a diagram can be an effective way of showing the information in a problem. In this workshop, we will use diagrams called Venn diagrams. They can be used to show the relationship among several groups of people, animals, or objects.

EXAMPLE 1 Of 53 employees surveyed, 43 have degrees in management, 20 have degrees in marketing, and 10 in both. How many have degrees in marketing but not management?

SOLUTION Draw two intersecting circles, one for management and one for marketing. Work out from the middle region.

A. Fill in the number in both classes (10).
B. Fill in the remaining number in management: $43 - 10 = 33$
C. Fill in the remaining number in marketing: $20 - 10 = 10$
D. Add the numbers: $33 + 10 + 10 = 53$

Ten employees have degrees in marketing but not management.

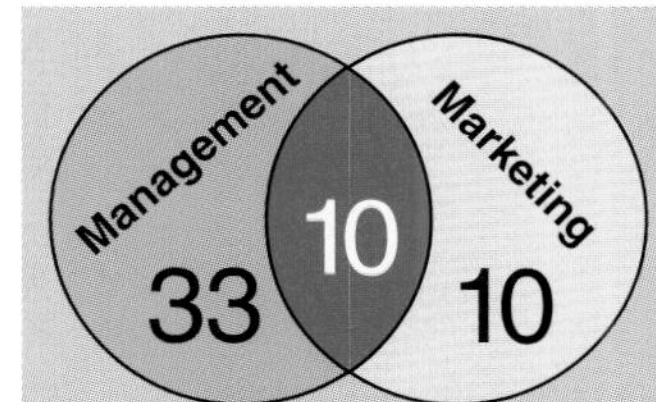

EXAMPLE 2 Two hundred people returning from a trip to Europe were asked which countries they had visited. One hundred forty-eight had been to England, 116 had been to France, and 96 had been to Spain. Eighty-two had been to England and France, 71 had been to France and Spain, 56 had been to England and Spain, and 44 had visited all 3 countries.

- How many had visited France but not England or Spain?
- How many had not visited any of these 3 countries?

SOLUTION

A. Seven people visited France only. This was found by the following procedure:

(1) Draw a Venn diagram as shown.

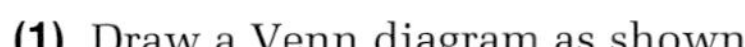

(2) Forty-four visited all 3 countries.
(3) Twelve visited England and Spain only. $(56 - 44)$
(4) Twenty-seven visited France and Spain only. $(71 - 44)$
(5) Thirty-eight visited England and France only. $(82 - 44)$
(6) Thirteen visited Spain only. $[96 - (12 + 44 + 27)]$
(7) Seven visited France only. $[116 - (38 + 44 + 27)]$
(8) Fifty-four visited England only. $[148 - (38 + 44 + 12)]$

B. Five people did not visit any of the 3 countries. $[200 - (54 + 7 + 13 + 38 + 27 + 12 + 44)]$

WORKSHOPS

LESSON PLAN Workshop 28

OBJECTIVE
To solve problems by using a Venn diagram.

RETEACHING
Use the Examples to reteach the objective of this workshop. A few key points to emphasize are the following: (1) The size of the circles is not relevant to the problem; (2) intersections of circles contain data from more than one group; (3) work out from intersecting circles first.

Remind students that a Venn diagram is a visual model of a problem that can clarify the relationships among the various elements of the problem.

ERROR ANALYSIS
Most students probably will not have any difficulty with Example 1. Example 2, however, is another matter. The relationships in this Example are more complex, and you will need to help students not only draw the diagram correctly, but you will also need to help them find the correct number in each group. Again, stress that the place to begin is with those people who had visited all three countries. This group of 44 is represented in the diagram by the intersection of all three circles. Secondly, the groups represented by the intersections of two circles should be determined. Thirdly, those people who have visited only one country are found. Knowing all of the above groups allows students to find the number of people who had not visited any of the three countries.

If students can understand and solve Example 2 successfully, they should not have any difficulty with the Problems. Generally, students make errors in finding the numbers in the intersection groups because they tend to get overwhelmed by all the data. The way to avoid this problem is to break down the original problem into a series of simpler problems and to solve the simpler problems one at a time.

PRACTICE AND APPLY

The following problems can be assigned for classwork and the answers checked in class to help students master the objective of the lesson.

- Guided Practice: 2
- Independent Practice: 3

WRAP-UP

Have students draw a Venn diagram to represent the following.

A survey of 100 teenagers found that

- 59 have seen the movie *Teen Angel*
- 51 have seen *Racetrack*
- 48 have seen *February 29th*
- 32 have seen *Teen Angel* and *Racetrack*
- 30 have seen *Racetrack* and *February 29th*
- 27 have seen *Teen Angel* and *February 29th*
- 20 have seen all three movies

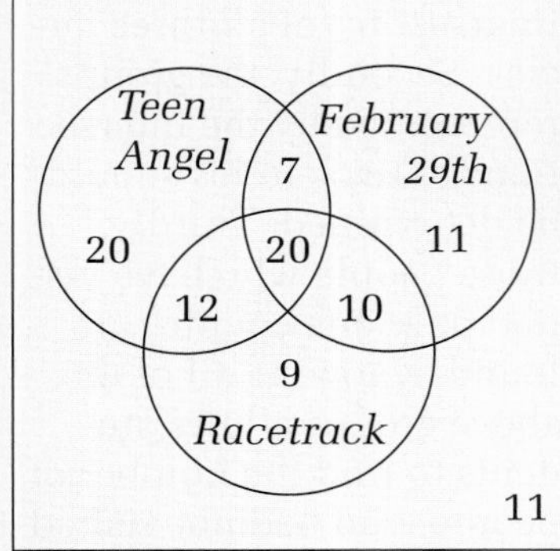

Assignment Guide

- Basic: 3–6
- Average: 3–6

WORKSHOPS

✓ SELF-CHECK Complete the problem, then check your answer in the back of the book.

1. Of 12 classmates, 6 went to the game, 7 went to the dance, and 4 went to both. How many went to neither? **3**

PROBLEMS

2. There are 197 delegates at the International Trade Association. Eighty-five of them speak Spanish, 74 speak French, and 15 speak both French and Spanish.

a. How many speak Spanish but not French? **70**

b. How many speak French but not Spanish? **59**

c. How many speak neither French nor Spanish? **53**

3. A survey of 150 people revealed that 121 people watch the early evening TV news, 64 watch the noon news, and 47 watch both. How many do not watch either one? **12**

4. A survey of 120 college junior business students produced these results: 40 read *Business Journal*, 48 read the local paper, 70 read the campus paper, 25 read *Business Journal* and the local paper, 28 read the local paper and the campus paper, 21 read the campus paper and *Business Journal*, and 18 read all three papers.

a. How many do not read any of the papers? **18**

b. How many read *Business Journal* and the local paper but not the campus paper? **7**

5. Of 20 students eating at a restaurant, 14 ordered salad, 10 ordered soup, and 4 ordered both salad and soup. How many did not order either? **none**

6. One hundred employees were asked which sports they play. Fifty play football, 48 play basketball, 54 play baseball; 24 play both football and basketball, 22 play both basketball and baseball, 25 play both football and baseball; and 14 play all 3 sports.

a. How many play basketball only? **16**

b. How many play football and baseball but not basketball? **11**

c. How many play none of the 3 sports? **5**

7. In a marketing survey, 500 people were asked their 2 favorite colors. Red was chosen by 380, blue by 292, and 10 chose neither red nor blue.

a. How many like both colors? **182**

b. How many like red but not blue? **198**

Problem Solving: Drawing a Sketch

Some word problems, particularly those that involve lengths, widths, and dimensions, can be simplified if you draw a sketch. Sketches and diagrams can also help you keep track of information in multistep problems.

EXAMPLE 1 A 30-cm piece of pipe is cut into 3 pieces. The second piece is 2 cm longer than the first piece, and the third piece is 2 cm shorter than the first piece. How long is each piece?

SOLUTION

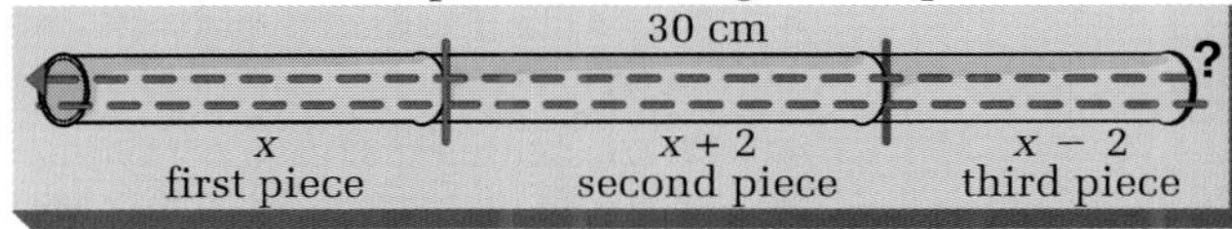

$x + (x + 2) + (x - 2) = 30$

$3x = 30$

$x = 10$ cm length of first piece

$x + 2 = 12$ cm length of second piece

$x - 2 = 8$ cm length of third piece

EXAMPLE 2 A 2-volume set of classics is bound in 1/4-inch covers. The text in each volume is 3 inches thick. The 2 volumes are side by side on the shelf. A bookworm travels from inside the front cover of Volume I to the inside back cover of Volume II. How far does the bookworm travel?

SOLUTION

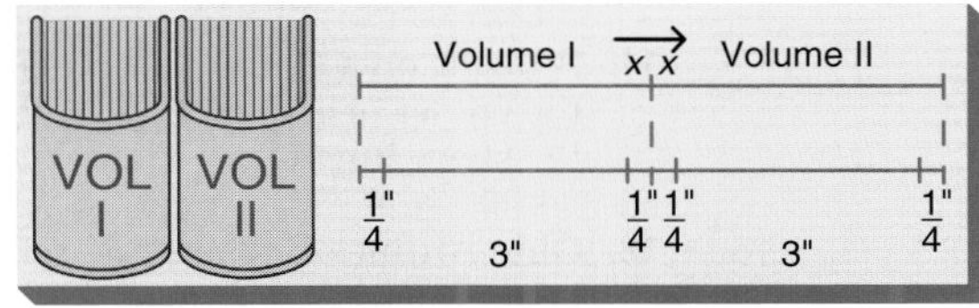

Bookworm travels: $\frac{1}{4}'' + \frac{1}{4}'' = \frac{1}{2}''$

The bookworm travels $\frac{1}{2}$ inch from inside the front cover of Volume I to inside the back cover of Volume II.

✓ SELF-CHECK Complete the problem, then check your answer in the back of the book.

1. Use an 18-inch, 10-inch, and 7-inch length of board to mark off a length of 15 inches. **(18" + 7") − 10"**

PROBLEMS

2. Tom Brown leaves his house and jogs 8 blocks west, 5 blocks south, 3 blocks east, 9 blocks north, 6 blocks east, 12 blocks south, 3 blocks west, and then stops to rest. Where is he in relation to his house? **2 blocks west and 8 blocks south**

WORKSHOPS

LESSON PLAN Workshop 29

OBJECTIVES
To draw a sketch as a means to solving a word problem.

RETEACHING
Example 1 illustrates nicely that drawing a sketch helps to organize the conditions of the problem. The sketch is used as a means to write the equation that yields the solution to the problem. In Example 2, the use of a sketch helps students to avoid the common error made with this problem of having the worm travel through the entire two books. You might suggest that students use graph paper to work the Problems.

ERROR ANALYSIS
The purpose of drawing a sketch to help solve a problem is to avoid making errors. In Example 2, the correct sketch shows the solution immediately. Of course, an incorrect sketch, or an inaccurate sketch, can lead to an incorrect answer. Point out to students that a sketch is a visual model of the problem and that it must satisfy the data given in the problem. Otherwise, it is worthless. Check students' sketches to see that they are drawn carefully and accurately.

PRACTICE AND APPLY

The following problems can be assigned for classwork and the answers checked in class to help students master the objective of the lesson.

- Guided Practice: 2–3
- Independent Practice: 4–5

WRAP-UP

Have students use graph paper to draw sketches of the following:

1. a square with an area of 4 square units
2. a rectangle with a width of 3 units and a length that is 2 times the width
3. a path that starts at 0, goes up 3 units, left 2 units, down 4 units, and right 5 units
4. a path that starts at 0, goes south 2 units, east 3 units, north 4 units, west 5 units, and south 5 units

Assignment Guide

- Basic: 5–11 odd
- Average: 6–12 even

WORKSHOPS

3. Allison Kelly bicycles to work. She travels 3 miles north of her house, turns right and bikes 2 miles, turns left and bikes 3 miles, and turns left and bikes 5 miles. At this point, where is she in relation to her home? **3 miles west and 6 miles north**
4. The rectangular area allotted to the shoe department of a department store has a perimeter of 96 feet. The length of the rectangle is 8 feet longer than the width. What is the width of the rectangle? **20 feet**
5. A barn has dimensions of 50 feet by 60 feet. A cow is tethered to 1 corner of the barn with a 60-foot rope. The cow always stays outside the barn. How many square feet of grazing land can the cow reach? **8556.5 sq ft**
6. A 60-foot piece of fencing is cut into 3 pieces. The second piece is 3 feet longer than the first piece, and the third piece is 9 feet longer than the first piece. Find the lengths of the pieces. **16 feet; 19 feet; 25 feet**
7. Assuming that each corner must be tacked, what is the least number of tacks that you need to display eight 8-inch by 10-inch photographs? **15**
8. Making identical cuts, a lumberjack can saw a log into 4 pieces in 12 minutes. How long would it take to cut a log of the same size and shape into 8 pieces? **28 minutes**
9. Master Chemical Company has containers with capacities of 4 L, 7 L, and 10 L. How could you use these containers to measure exactly 1 L? **(4 + 7) − 10**
10. Three book volumes are arranged as shown. The thickness of each cover is 0.2 cm. The text in each volume is 3 cm thick. What is the distance from the first page of Volume I to the last page of Volume III? **3.8 cm**

VOL I VOL II VOL III

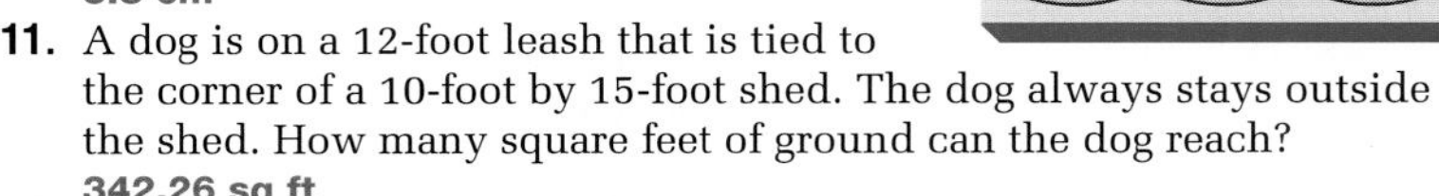

11. A dog is on a 12-foot leash that is tied to the corner of a 10-foot by 15-foot shed. The dog always stays outside the shed. How many square feet of ground can the dog reach? **342.26 sq ft**
12. Ohio Airlines is to provide service between cities as shown on the map. The airline employed 5 new people to sit in the control tower at each of the 5 cities. The people are Carol, Connie, Clare, Charles, and Cedric. The 2 people in the cities with connecting routes will be talking to each other a great deal, so it would be helpful if these people were friends. The pairs of friends are: Charles-Connie, Carol-Cedric, Charles-Carol, Clare-Cedric, and Carol-Clare. Place the 5 people in the 5 cities so that the ones in connecting cities are friends.

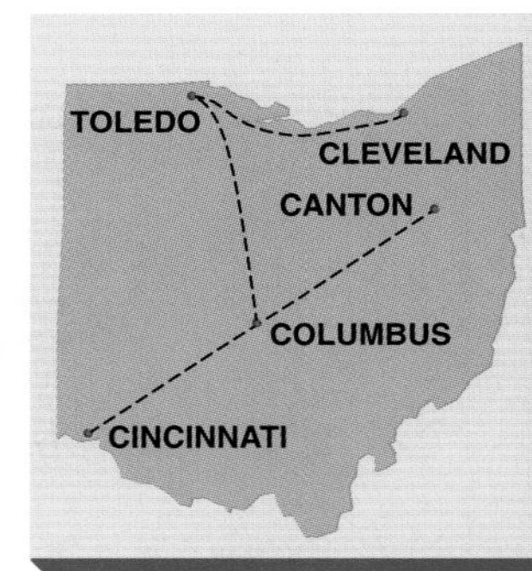

Cedric or Clare in Canton or Cincinnati; Connie–Cleveland; Charles–Toledo; Carol–Columbus

WORKSHOP 30

Problem Solving: Restating the Problem

To solve a problem, you may find it helpful to restate the problem in a different way. A difficult problem can be restated as a simpler problem and become easier to solve. Or, a problem may involve a series of simpler problems that will lead to the solution of the original problem.

EXAMPLE 1 A single elimination tournament is to be held among 124 contestants. Single elimination means that you are out of the tournament if you have 1 loss. How many matches will have to be played to determine a winner? (Hint: A similar problem was encountered in the "Working Backward" workshop.)

SOLUTION Given that it is a single elimination tournament, every player (but the winner) will have to lose 1 match. Since 123 contestants will have to lose 1 match, 123 matches will have to be held.

EXAMPLE 2 A 27-inch bicycle makes 100 revolutions per minute. A stone stuck in the treads will travel how many feet in 1 hour?

SOLUTION Restate the problem to follow a series of simpler problems.

A. How many inches will the stone travel in 1 revolution?
$C = \pi d$ or 3.14×27 inches $= 84.78$ inches

B. How many inches will the stone travel in 1 minute?
100 rpm $\times$ 84.78 inches per revolution $=$ 8478 inches

C. How many inches will the stone travel in 1 hour?
8478 inches per minute $\times$ 60 minutes per hour $=$ 508,680 inches

D. How many feet will the stone travel in 1 hour?
508,680 inches $\div$ 12 inches per foot $=$ 42,390 feet

✔ SELF-CHECK Complete the problem; then check your answer in the back of the book.

1. A 27-inch bicycle wheel makes 100 revolutions per minute. A stone stuck in the treads will travel how many miles in 1 hour? (Round answer to the nearest mile.) **42,390 $\div$ 5280 $=$ 8**

PROBLEMS

2. Your automobile's gas mileage is 27.5 miles per gallon. Last month your automobile was driven 1595 miles, and a gallon of gasoline cost 96¢. What did it cost to buy gasoline for your automobile last month? **$55.68**

WORKSHOPS

LESSON PLAN Workshop 30

OBJECTIVES

To solve a problem by restating it as a simpler problem or as a series of simpler problems.

RETEACHING

Very often, restating a problem as a series of simpler problems is the key to solving a difficult problem. Work through both Examples with students to illustrate this point. In Example 1, the problem becomes simple by asking the right questions: How many matches does a player have to lose to be out of the tournament? (1) Then who is left? (the winner)

Example 2 shows clearly how the solution of a series of simple problems leads to the solution of the original problem. Go over each step carefully so students see how they all relate to one another to yield the final answer.

ERROR ANALYSIS

Finding the simple problem or simpler problems embedded in a more complicated problem is not an easy task. Some insight into the problem may be required in order to do this. Also, some approaches may lead into a "blind alley." These approaches should not be considered errors. Rather, students can learn from them by finding out what does not work and then trying a different approach. As students work the Problems, ask questions and provide assistance as needed.

PRACTICE AND APPLY

The following problems can be assigned for classwork and the answers checked in class to help students master the objective of the lesson.

- Guided Practice: 2–3
- Independent Practice: 4–5

WRAP-UP

Write the following problems on the chalkboard. Have students solve each problem.

1. Carol earns $8.00 an hour as a word processor. She earns time-and-a-half for any hours over a 35-hour work week. How much does she earn if she works a 55-hour week? $520
2. A rotating sprinkler can water a lawn 40 ft in every direction. How many square feet of lawn can the sprinkler water? (Use $\pi \approx 3.14$.) 5024 ft^2
3. A bicyclist travels at an average rate of 15 mph. If she rides for 6 hours a day, 6 days a week, how many weeks will it take her to travel 21,060 miles? 39 weeks

Assignment Guide

- Basic: 5–11
- Average: 5–11

WORKSHOPS

3. Vanna Burzinski works in the stockroom at Keynote Auto Supply. Vanna always stacks crates using the triangular pattern shown. The top row always has 1 crate. Each row always has 2 fewer crates than the row before it.
  a. How many crates are in a stack that has 20 rows? **400**
  b. How many crates are in a stack if the bottom row has 29 crates? **225**

4. Rachel Murphy is having a party. The first time the doorbell rings, 3 guests arrive. Each time the doorbell rings after that, a group arrives that has 2 more guests than the preceding group. **a. 15; b. 120**
  a. How many guests arrive on the seventh ring?
  b. If the doorbell rings 10 times, how many guests came to the party?

5. Note that $1 + 3 = 4$, $1 + 3 + 5 = 9$, and $1 + 3 + 5 + 7 = 16$. What is the sum of the first 25 odd numbers? **625**

6a. Either 94 or 95
6b. 94 if winner is undefeated, 95 if winner has 1 loss

6. Forty-eight teams are entered in a double elimination tournament. Double elimination means that a team is out of the tournament if it loses two games.
  a. How many games will have to be played to determine a winner?
  b. Why are there 2 possible answers?

7. A train traveling at 60 miles per hour takes 4 seconds to enter a tunnel and another 50 seconds to pass completely through the tunnel.
  a. What is the length of the train? **352 ft**
  b. What is the length of the tunnel? **4400 ft**

8. There are 26 teams in the National Football League. To conduct their annual draft, teams in each city must have a direct telephone line to each of the other cities.
  a. How many direct telephone lines must be installed to accomplish this? **325**
  b. How many direct telephone lines must be installed if the league expands to 30 teams? **435**

9. Borrow 1 horse from neighbor, then Rick = 9, Mike = 6, and Peter = 2.

9. Grandpa Moyer wanted to leave his 17 horses to his 3 grandsons. Rick was to get $\frac{1}{2}$ the horses, Mike was to get $\frac{1}{3}$, and Peter was to get $\frac{1}{9}$. How could he accomplish this?

10. Two cyclists start toward each other from points 25 miles apart. One cyclist travels 15 mph, while the other cyclist travels 10 mph. A trained bird leaves the shoulder of 1 cyclist and travels to the shoulder of the other cyclist and back and forth until the cyclists meet. If the bird flies at 40 mph, how far will the bird fly? **40 miles**

11. Three men agree to share a hotel room and pay a total of $60 when they check in. Later, the clerk finds that he should have charged them only $55. He sends the bellhop up with a $5 refund for the room, but because the bellhop cannot divide $5 evenly by 3, he returns only $3 to the men and keeps $2. Therefore, each man paid only $19 as his share of the room for a total of $57. The bellhop kept the other $2 for a total of $59. What happened to the extra $1? **There is no extra dollar.**

PART 2

Personal Application Mathematics

Part 2. Personal Application Mathematics, covers the personal applications of basic math skills that you will use in your roles as worker and knowledgeable consumer. By learning how to handle your money wisely, you can use your income to meet day-to-day living expenses and still save and invest part of your income for use in the future. Eleven consumer topics are covered in Part 2.

Terms to Know

The following definitions provide an overview of the important areas that affect how you earn, spend, and save your money.

Gross Income The total amount of money you earn for work done during a pay period, before any deductions are taken.

Net Income Your take-home pay after all tax withholdings and personal deductions have been taken from your gross income.

Checking Accounts Funds deposited in a bank from which you can write checks to pay for goods and services instead of paying by cash.

Savings Accounts Funds deposited in a bank for which you earn interest.

Cash Purchases Items or services that you buy at regular or sales prices.

Charge Accounts and Credit Cards Systems of buying in which you receive goods and services now and spread the payments, along with finance charges, if any, over a period of time.

Loans Money that is borrowed and must be repaid along with interest owed.

Automobile Transportation 1. The costs of buying, insuring, and maintaining your own automobile. 2. The costs of leasing an automobile or renting from a rental agency.

Housing Costs The costs of purchasing, owning, and maintaining your own home.

Insurance and Investments 1. Financial protection purchased for your dependents.
2. Sums of money paid out to gain future income for yourself.

Planning Activities related to planning for business decisions; includes inflation, GNP, consumer price index, and budget analysis.

Recordkeeping Maintaining accurate accounts of your expenses as tools for managing your money.

UNIT

1

Gross Income

Your *income,* the money that you earn on a job, may be determined in one of several ways. Income may be computed on hours worked, or *hourly rate,* the number of items produced, or *piecework,* or *commission.* Commission may be a specified amount of money you are paid for selling a product, or it may be a percent of the selling price. You may also be paid on a *graduated commission,* which increases as your sales increase. Many workers are paid a fixed annual *salary,* which they receive *weekly, biweekly, semimonthly,* or *monthly.*

OUTLINE

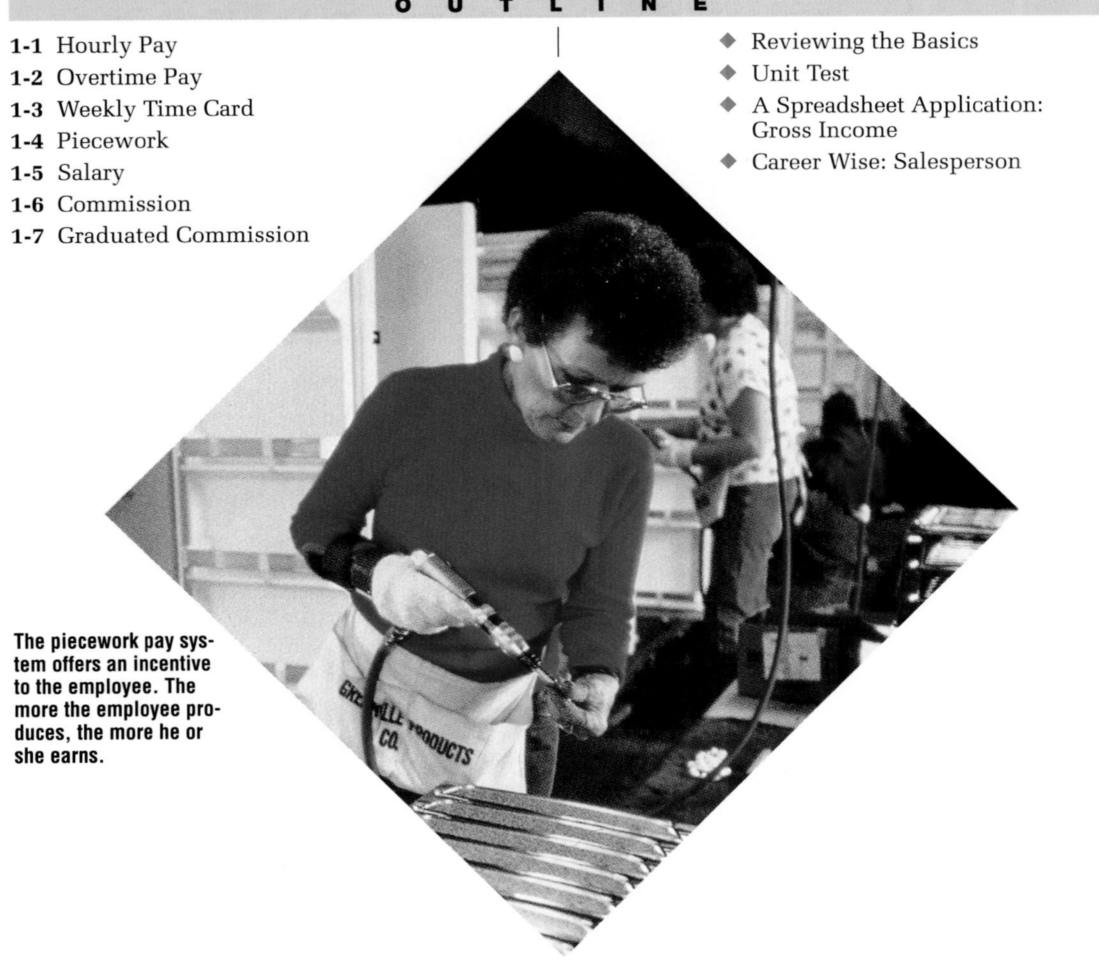

The piecework pay system offers an incentive to the employee. The more the employee produces, the more he or she earns.

UNIT 1
GROSS INCOME

INTRODUCING THE UNIT

Introduce the unit, Gross Income, by discussing the concept of the "average work week."

In 1938, Congress passed the Fair Labor Standards Act which defined the normal working week as 40 hours, required that an employee receive time-and-a-half pay for all hours worked over 40, and also established a federal minimum wage.

LESSON PLAN
1-1 Hourly Pay

FOCUS
You can motivate the discussion of straight-time pay by asking those students who have part-time jobs to raise their hands. Ask a few of these students to explain how they are paid and how many hours per week each works. Then compute the straight-time pay for each student at the chalkboard.

TEACH
The key terms *hourly rate* and *straight-time pay* need to be explained. Relate these terms to the examples worked above with the students' data. Then work through the Example in the text. Use additional examples by reading some want-ads from your local newspaper and have students do the computations. Ask volunteers to give their answers.

Warm-Up Exercises
These exercises can be used to review prerequisite skills for this lesson and to ascertain any arithmetic deficiencies that students may have. Work through these exercises with students before having them do the Self-Check.

Write as a decimal.

1. $5\frac{1}{2}$ 5.5
2. $17\frac{3}{4}$ 17.75
3. $42\frac{1}{4}$ 42.25
4. $5.00 × 4 $20.00
5. $2.00 × $7\frac{1}{2}$ $15.00
6. $3.60 × $6\frac{1}{2}$ $23.40
7. $3.87 × $1\frac{1}{4}$ $4.84

1-1 Hourly Pay

OBJECTIVE
Compute the straight-time pay.

For some jobs, you are paid a fixed amount of money for each hour you work. Your **hourly rate** is the amount of money you earn per hour. **Straight-time pay** is the total amount of money you earn for a pay period at the hourly rate.

Straight-Time Pay = Hourly Rate × Hours Worked

EXAMPLE *Skills* 14, 8, 2 *Application* A *Term* Straight-time pay

Mary Byron is a mail clerk. She earns $5.40 per hour. Last week she worked 40 hours and this week she worked $39\frac{1}{2}$ hours. What is her straight-time pay for each week?

SOLUTION Find the **straight-time pay**.

	Hourly Rate	×	Hours Worked		
Last week:	$5.40	×	40	=	$216.00 straight-time pay
This week:	$5.40	×	$39\frac{1}{2}$	=	$213.30 straight-time pay

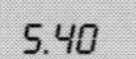  5.40 × 39.5 = 213.30

✓ SELF-CHECK Complete the problems, then check your answers in the back of the book.

Find the straight-time pay.

1. $7.60 per hour × 40 hours **$304**
2. $7.90 per hour × $37\frac{1}{2}$ hours **$296.25**

PROBLEMS

	Employee	Regular Hourly Rate	×	Hours Worked	=	Straight-Time Pay
3.	Wright, S.	$ 8.00	×	36	=	**$288**
4.	Reardon, E.	$ 4.50	×	18	=	**$81**
5.	Levitt, R.	$ 6.20	×	30	=	**$186**
6.	Maguire, M.	$ 5.00	×	$24\frac{1}{2}$	=	**$122.50**
7.	Reyes, A.	$14.50	×	$31\frac{1}{4}$	=	**$453.13**
8.	Vadola, G.	$16.25	×	$25\frac{3}{4}$	=	**$418.44**

F.Y.I.
Federal Minimum Wage

1978	$2.65
1979	2.90
1980	3.10
1981	3.35
1990	3.80
1991	4.25

9. You earn $5.50 per hour. Worked 30 hours last week. What was your straight-time pay? **$165**
10. You earn $8.00 per hour. Worked 35.25 hours last week. What was your straight-time pay? **$282**

COMMUNICATION SKILLS
During the first few weeks of school, call upon different students to read aloud the Examples or Problems that are being worked on in class. Listen for and correct any reading errors and stress the importance of reading and writing money amounts correctly.

11. Gloria Scanlon.
Financial consultant.
Earns $25.00 per hour.
Worked 32.5 hours last week.
What was her straight-time pay?
$812.50

12. Marian Abelt.
Mail clerk.
Earns $4.20 per hour.
Worked $35\frac{3}{4}$ hours last week.
What was her straight-time pay?
$150.15

13. Sylvia O'Keefe is a teacher's aide. She earns $5.80 per hour. Last week she worked 34 hours. What was her straight-time pay for last week? **$197.20**

14. Judy Martin has a part-time job at Dutton's. She earns $4.25 per hour. She works 20 hours each week. What is her straight-time pay for a week? **$85**

15. Don Moline, a construction worker, earns $12.25 per hour. He worked his regular $36\frac{1}{4}$ hours last week. What was his straight-time pay? **$444.06**

16. Arthur Marzetti is a machine operator at the Donovan Manufacturing Corporation. He works $27\frac{1}{4}$ hours per week and earns $4.75 per hour. What is his straight-time pay each week? **$129.44**

17. Sally Barrett, a banjo player for a local band, earns $8.50 per hour. What is her straight-time pay for a week in which she worked 22 hours? **$187**

F.Y.I.
On April 1, 1991, the federal minimum wage was set at $4.25 per hour.

18. Jon Bradford earns $7.84 per hour as an insurance claims adjustor. What is his straight-time pay for this week? **$299.88**

Mon.	Tue.	Wed.	Thu.	Fri.
8 hrs.	8 hrs.	6.5 hrs.	8.5 hrs.	7.25 hrs.

19. Ben Favreau was hired for the position of billing clerk. During his first week on the job, he worked $7\frac{1}{2}$ hours each day, Monday thru Friday. What was his total pay for the week? **$215.63**

SILK SCREEN CO. has opening for bright, dependable billing clerk. Must have aptitude for figures and pay attention to detail. Prepare work for production. Start $5.75 per hr. Call Walt, at 555-9256 Ext. 11.

MAINTAINING YOUR SKILLS Look up the skills in parentheses if you need help or more practice.

Write as decimals. **(Skill 14)**

20. $\frac{1}{4}$ **0.25** 21. $\frac{1}{2}$ **0.50** 22. $\frac{3}{4}$ **0.75** 23. $\frac{1}{10}$ **0.10** 24. $\frac{3}{8}$ **0.375**

Multiply. Round answers to the nearest cent. **(Skill 8)**

25. $4.50 × 30.25 **$136.13** 26. $8.30 × 3.25 **$26.98** 27. $14.50 × 32.5 **$471.25**

Round to the place value indicated. **(Skill 2)**

Nearest ten	28. 22 **20**	29. 138 **140**	30. 8 **10**
Nearest hundred	31. 91 **100**	32. 3092 **3100**	33. 14,431 **14,400**
Nearest thousand	34. 15,374 **15,000**	35. 722 **1000**	36. 158,481 **158,000**

PRACTICE AND APPLY

The following problems can be assigned for classwork and the answers checked in class to help students master the objective of the lesson.

- Guided Practice: 1–8
- Independent Practice: 9–15 odd

WRAP-UP

Read Problem 15 to the class. Write the formula for computing straight-time pay on the chalkboard and use it to solve the problem. Ask a student to explain the meanings of hourly rate and straight-time pay.

Assignment Guide

- Basic: 9–18, 20–36
- Average: 10–16 even, 17–19, 20–36 even

ALTERNATIVE STRATEGIES: Enrichment

Make this lesson a real world experience by having students read the **Help Wanted Ads** in the newspaper. These ads are an excellent source of up-to-date wage and salary data. Have them compute the gross pay using the wages suggested in the ads for various hours worked.

LESSON PLAN
1-2 Overtime Pay

FOCUS

To compute the straight-time, overtime, and total pay, students must first understand the idea of *regular hours.* Ask students how many hours they think most adults work each week on their jobs. Point out that overtime pay is not usually given to a person who works less than 35 or 40 hours a week.

TEACH

Explain that *time and a half* means the regular hourly rate of pay plus one-half of the regular hourly rate. Use a few simple examples to illustrate this point. For example, if the regular rate is $8.00 per hour, then the time-and-a-half rate is $8.00 + $4.00, or $12.00. Now show students that this problem can be worked by multiplying $8.00 by 1.5: $8.00 × 1.5 = $12.00

Warm-Up Exercises

These exercises can be used to review prerequisite skills for this lesson and to ascertain any arithmetic deficiencies that students may have. Work through these exercises with students before having them do the Self-Check.

Write as a decimal.

1. $48\frac{1}{4}$ 48.25
2. $54\frac{3}{4}$ 54.75
3. $42\frac{1}{2}$ 42.5
4. $8.00 × 40 $320.00
5. $12.00 × $5\frac{1}{2}$ $66.00
6. $9.42 × 40 $376.80
7. ($1\frac{1}{2}$ × $9.00) × 10 $135.00
8. ($1\frac{1}{2}$ × $10.00) × $3\frac{1}{4}$ $48.75
9. ($1\frac{1}{2}$ × $6.45) × 9 $87.08

1-2 Overtime Pay

OBJECTIVE
Compute the straight-time, overtime, and total pay.

When you work more than your regular hours, you may receive **overtime pay.** The overtime rate may be $1\frac{1}{2}$ times your regular hourly rate. This is called **time and a half.** You may receive **double time,** 2 times your regular hourly rate, for overtime on Sundays or holidays.

Overtime Pay = Overtime Rate × Overtime Hours Worked

Total Pay = Straight-Time Pay + Overtime Pay

EXAMPLE *Skill* 5 *Application* A *Term* Overtime pay

Bill Learner is paid $8.20 an hour for a regular 40-hour week. His overtime rate is $1\frac{1}{2}$ times his regular hourly rate. This week Bill worked his regular 40 hours plus 10 hours of overtime. What is his total pay?

SOLUTION

A. Find the **straight-time pay.**
Hourly Rate × Regular Hours Worked
$8.20 × 40 = $328.00 straight-time pay

B. Find the **overtime pay.**
Overtime Rate × Overtime Hours Worked
(1.5 × $8.20) × 10 = $123.00 overtime pay

C. Find the **total pay.**
Straight-Time Pay + Overtime Pay
$328.00 + $123.00 = $451.00 total pay

8.20 [×] 40 [=] 328 [M+] 1.5 [×] 8.20 [×] 10 [=] 123 [M+] [RM] 451

✓ SELF-CHECK Complete the problems, then check your answers in the back of the book.

Find the total pay.

1. $9.00 per hour for 40 hours. Time and a half for 6 hours. **$441**
2. $11.50 per hour for 40 hours. Time and a half for 7 hours. **$580.75**

PROBLEMS

	3.	4.	5.	6.	7.	8.
Hourly Pay (40 hrs.)	$6.00	$10.60	$4.80	$5.55	$17.00	$19.25
Straight-Time Pay	? $240	? $424	? $192	? $222	? $680	? $770
Overtime Rate	$1\frac{1}{2}$	$1\frac{1}{2}$	$1\frac{1}{2}$	2	2	$1\frac{1}{2}$
Overtime Hours	8	5	4	8	13	6
Overtime Pay	? $72	? $79.50	? $28.80	? $88.80	? $442	? $173.25
Total Pay	? $312	? $503.50	? $220.80	? $310.80	? $1,122	? $943.25

USE CALCULATORS

Calculators are an essential and integral part of the business world, and students should be encouraged to use them in this course. In this lesson, students should realize that they can use their memory keys to find the sum of the straight-time pay and overtime pay.

9. Steven Kellogg, machine operator. Regular hourly rate of $9.60. Time and a half for overtime. Worked 37 regular hours. Worked 8 hours overtime. What is his straight-time pay? What is his overtime pay? What is his total pay?

10. Dorothy Kaatz, programmer. Regular hourly rate of $9.15. Double time for overtime. Worked 40 regular hours. Worked 12 hours overtime. What is her straight-time pay? What is her overtime pay? What is her total pay?

9.	10.
$355.20	$366
$115.20	$219.60
$470.40	$585.60

11. Cindy Haskins is paid $7.00 an hour for a regular 35-hour week. Her overtime rate is $1\frac{1}{2}$ times her regular hourly rate. This week Cindy worked her regular 35 hours plus 8 hours of overtime. What is her total pay? **$329**

F.Y.I.
In 1994, the average hourly rate of production workers in the food manufacturing industries was $10.67.

12. As a maintenance engineer, Bonnie Zoltowski earns $7.85 an hour plus time and a half for weekend work. Last week she worked her regular 40 hours plus 16 hours of overtime on the weekend. What was her total pay for the week? **$502.40**

13. Judy Sweeney is a carryout cashier. She earns $5.15 an hour for a regular $36\frac{1}{2}$-hour week. She earns double time for work on Sundays. Last week Judy worked her regular hours plus $7\frac{1}{4}$ hours on Sunday. What was her total pay for the week? **$262.66**

CARRYOUT CASHIER
$5.15/ hr.
Double time on weekends

14. Bill McCrate is a cable TV installer. He is paid $7.57 an hour for a 40-hour week and time and a half for overtime. What is Bill's total pay for a week in which he worked 61 hours? **$541.26**

15. George Keller earns a regular hourly rate of $8.64. He earns time and a half for overtime work on Saturdays and double time for overtime on Sundays. This week he worked 40 hours from Monday through Friday, 8 hours on Saturday, and 7 hours on Sunday. What is his total pay for the week? **$570.24**

MAINTAINING YOUR SKILLS Look up the skills in parentheses if you need help or more practice.

Add. **(Skill 5)**

16. $42.50 + 24.76 = **$67.26**

17. $160.45 + 86.90 = **$247.35**

18. $10.62 + 9.34 + 0.45 = **$20.41**

19. 56.7 + 0.007 + 121.33 + 1568.125 = **1746.162**

Multiply. **(Skills 8, 14)**

20. $(1\frac{1}{2} \times \$8.40) \times 7$ **$88.20**

21. $(2 \times \$18.40) \times 12$ **$441.60**

22. $(1\frac{1}{4} \times \$6.00) \times 9$ **$67.50**

23. $(1.5 \times \$12.50) \times 8$ **$150**

24. $(2 \times \$4.50) \times 6$ **$54**

25. $(1.25 \times \$10.00) \times 10$ **$125**

PRACTICE AND APPLY

The following problems can be assigned for classwork and the answers checked in class to help students master the objective of the lesson.

- Guided Practice: 1–8
- Independent Practice: 10–14 even

WRAP-UP

Ask a student to name the new terms introduced in this lesson. Ask other students to explain the meanings of the terms.

Assignment Guide

- Basic: 9–14, 16–25
- Average: 9–15 odd, 16–24 even

ALTERNATIVE STRATEGIES: Reteaching

Relate this lesson to the real world by using a newspaper **Help Wanted Ad** for an hourly rate. Suggest that an employee worked 46 hours at $5.50 an hour. Calculate the gross pay.

STEP 1: Regular hours worked (40) × Rate per hour ($5.50) = $220.00
STEP 2: Overtime hours × 1.5 × Rate per hour
(46 − 40) × 1.5 × 5.50 = +49.50
STEP 3: Add to get NET PAY $269.50

LESSON PLAN 1-3 Weekly Time Card

FOCUS

Ask those students who have part-time jobs how they keep track of the number of hours they work each week. Use their answers to lead into a discussion of the Example, showing how to compute the total hours on a weekly time card.

TEACH

After discussing the Example, ask students to suggest some examples of businesses that use time cards. Most department stores, restaurants, banks, and manufacturers, for example, require hourly employees to keep a weekly time card. A time clock automatically records the time employees come to work, when they break for lunch and return, and when they leave for the day. Total daily and weekly hours are determined either by a computer or by hand. Many businesses round to the nearest quarter hour; however, computerized timekeeping has made it possible for some businesses to pay employees to the nearest minute.

Some students may need help in computing elapsed time when the minutes of the minuend are less than the minutes of the subtrahend. For example, when subtracting time 1:45 from 6:34, tell students to rewrite 6:34 as 5:94.

1-3 Weekly Time Card

OBJECTIVE
Compute the total hours on a weekly time card.

When you work for a business that pays you on an hourly basis, you are usually required to keep a time card. A **weekly time card** shows the time you reported for work and the time you departed each day of the week. The hours worked are computed for each day. The daily hours are added to arrive at the total hours for the week. The hours worked each day must be rounded to the nearest quarter hour.

Total Hours = Sum of Daily Hours

EXAMPLE *Skills* 2, 16 *Applications* E, F *Term* Weekly time card

Gail Stough is required to keep a weekly time card. What are her daily hours for 9/18? What are her total hours for the week?

EMPLOYEE TIME CARD	DATE	IN	OUT	IN	OUT	HOURS
	9/15	8:15	12:15	1:00	5:00	8
NAME: GAIL STOUGH	9/16	8:30	12:00	12:35	5:20	8 1/4
DEPT.: CREDIT	9/17	8:28	12:05	12:30	4:30	7 1/2
Gail Stough	9/18	8:15	1:45	2:15	4:59	?
	9/19	8:30	12:20	12:50	5:00	8
EMPLOYEE SIGNATURE					TOTAL HOURS	

SOLUTION

A. Find the hours worked on 9/18.

(1) Time between 8:15 a.m. and 1:45 p.m.
(1:45 + 12:00) − 8:15
13:45 − 8:15 .... 5h:30min

(2) Time between 2:15 p.m. and 4:59 p.m.
4:59 − 2:15 .... 2h:44min
5h:30 min + 2 h:44min 7h:74min
Rounded to the nearest quarter hour = $8\frac{1}{4}$ hours

B. Find the **total hours.**
Sum of daily hours

$8 + 8\frac{1}{4} + 7\frac{1}{2} + 8\frac{1}{4} + 8$

$8 + 8\frac{1}{4} + 7\frac{2}{4} + 8\frac{1}{4} + 8 = 39\frac{4}{4} = 40$ total hours

✓ SELF-CHECK Complete the problems, then check your answers in the back of the book.

1. Karl West worked from 8:00 to 11:45 and from 12:30 to 4:15. Find the total hours. $7\frac{1}{2}$
2. Rod Abet worked from 7:30 to 11:55 and from 1:00 to 4:50. Find the total hours. $8\frac{1}{4}$

BUSINESS PROJECT

1. Ask those students who work part time to bring to class one of the time cards used by their employers. Show the time cards to all the students in the class.
2. Ask some students to get a copy of an airline, train, or bus schedule. The schedules will show departure and arrival times of various flights, trains, or buses. Have them calculate elapsed time. (Be careful with time zones.)

PROBLEMS

Eddie Irwin works as a computer operator in the data processing department at The Needham Medical Center. He is required to keep a weekly time card. What are his hours for each day? What are his total hours for the week?

3. $7\frac{3}{4}$ 4. 8
5. $7\frac{3}{4}$ 6. $7\frac{1}{2}$
7. $8\frac{1}{2}$ 8. $39\frac{1}{2}$

	EMPLOYEE TIME CARD NEEDHAM MEDICAL CENTER			NAME: Eddie Irwin DEPARTMENT: Data Processing		
	Date	**In**	**Out**	**In**	**Out**	**Total**
3.	12/13	8:15	12:00	12:30	4:30	?
4.	12/14	8:30	12:30	1:10	5:10	?
5.	12/15	8:35	12:50	1:30	5:00	?
6.	12/16	8:15	12:10	12:45	4:25	?
7.	12/17	8:25	1:15	1:50	5:30	?
8.					Total Hours	?

Lucy Wallace works for Appliance Parts, Inc. Use her time card below to find her total hours worked each day and her total hours for the week.

9. $8\frac{1}{4}$ 10. $9\frac{1}{2}$
11. $8\frac{3}{4}$ 12. $8\frac{1}{4}$
13. 9 14. $5\frac{3}{4}$
15. $49\frac{1}{2}$

	EMPLOYEE TIME CARD APPLIANCE PARTS, INC.			NAME: Lucy Wallace DEPARTMENT: Parts			
	Day	**Date**	**In**	**Out**	**In**	**Out**	**Total**
9.	Mon.	6/7	7:00	11:00	12:00	4:20	?
10.	Tue.	6/8	7:10	11:30	12:00	5:15	?
11.	Wed.	6/9	6:50	11:15	11:55	4:20	?
12.	Thu.	6/10	6:58	10:45	11:25	3:50	?
13.	Fri.	6/11	7:15	11:49	12:30	5:02	?
14.	Sat.	6/12	7:05	12:45	—	—	?
15.						Total Hours	?

16. Gail Stough earns \$6.40 per hour. Use her time card on page 82 to find her total pay for the week. **\$256**

17. Eddie Irwin is paid \$7.25 for a regular 37-hour week. His overtime rate is $1\frac{1}{2}$ times his regular hourly rate. Use his time card above to find his total pay for the week. **\$295.44**

18. Lucy Wallace is paid \$7.80 for a regular 40-hour week. Her overtime rate is $1\frac{1}{2}$ her regular hourly rate. Use her time card above to find her total pay for the week. **\$423.15**

Warm-Up Exercises

These exercises can be used to review prerequisite skills for this lesson and to ascertain any arithmetic deficiencies that students may have. Work through the exercises with students before having them do the Self-Check.

Find the total time.

1. 8:10 to 12:15
   4 h: 5 min
2. 1:45 to 5:34
   3 h: 49 min
3. 8:07 to 1:15
   5 h: 8 min

Round to the nearest quarter hour.

4. 4 hours 40 minutes
   $4\frac{3}{4}$ hours
5. 6 hours 21 minutes
   $6\frac{1}{4}$ hours
6. 9 hours 52 minutes
   $9\frac{3}{4}$ hours
7. 4 hours 30 minutes
   $4\frac{1}{2}$ hours

ALTERNATIVE STRATEGIES: Reteaching

Refer to the **Problems and Simulations Manual**. In Lesson 1-3, Problem 2, what are the hours worked on 4/7? Show this procedure.

12:05	=	11h 65m	5:30 + 12:00	= 17:30	=	16h 80m
−7:55	=	− 7h 55m	−12:50		=	−12h 50m
		4h 10m				4h 30m = 8h 40m = 8.75h

PRACTICE AND APPLY

The following problems can be assigned for classwork and the answers checked in class to help students master the objective of the lesson.

- Guided Practice: 1–8
- Independent Practice: 9–15

WRAP-UP

Remind students of what they have learned by checking the answers to Problems 9–15. Use an actual time card, if you can locate one, to emphasize the objective of the lesson.

Assignment Guide

- Basic: 9–19, 22–36
- Average: 16–21, 22–36 even

19. You are scheduled to work from 7:00 a.m. to 11:00 a.m. and from 12:00 noon to 4:00 p.m. You are not allowed to work overtime. Your hourly rate is $7.50. Find the hours worked and the total pay for the week.

Day	In	Out	In	Out	Hrs.	
Sun.						
Mon.	7:00	11:00	12:00	4:04	8	
Tue.	7:05	11:10	11:55	3:59	?	$8\frac{1}{4}$
Wed.	7:01	10:59	12:05	3:57	?	$7\frac{3}{4}$
Thu.	6:59	10:55	12:01	4:03	8	
Fri.	6:55	11:01	12:02	3:49	8	
Sat.						

40 hours worked; $300 total pay

20. A portion of Kim Fong's time card is shown below. Find the number of hours he worked each day.

Day	In	Out	In	Out	Hrs.	
Sun.	1:00	6:15			?	$5\frac{1}{4}$
Mon.	1:00	5:00	6:00	9:10	?	$7\frac{1}{4}$
Tue.	10:02	3:04	4:00	8:15	?	$9\frac{1}{4}$
Wed.						
Thu.	11:05	4:04	5:10	9:25	?	$9\frac{1}{4}$
Fri.	1:30	5:25	6:15	10:08	?	$7\frac{3}{4}$
Sat.	9:50	2:05	3:00	6:25	?	$7\frac{3}{4}$

21. In problem 20, how many hours did Kim work on Saturday and Sunday? How many hours did he work Monday through Friday? If he earns $7.40 per hour plus time and a half for work on Saturday and Sunday, what is his total pay for the week?

S + S = 13 hrs.; M − F = $33\frac{1}{2}$ hrs.; $392.20

MAINTAINING YOUR SKILLS Look up the skills in parentheses if you need help or more practice.

Add. **(Skill 16)**

22. $\frac{1}{4} + \frac{1}{4} + \frac{1}{2}$ **1** **23.** $\frac{3}{4} + \frac{1}{8}$ $\frac{7}{8}$ **24.** $\frac{3}{8} + \frac{1}{4} + \frac{1}{4}$ $\frac{7}{8}$ **25.** $\frac{5}{8} + \frac{1}{3}$ $\frac{23}{24}$

Round to the nearest quarter hour. **(Application E)**

26. 8:09 $8\frac{1}{4}$ **27.** 7:47 $7\frac{3}{4}$ **28.** 11:55 **12** **29.** 3:39 $3\frac{3}{4}$ **30.** 5:23 $5\frac{1}{2}$

Round to the place value indicated. **(Skill 2)**

Nearest thousand	**31.** 39,972 **40,000**	**32.** 21,944 **22,000**	**33.** 68,498 **68,000**
Nearest hundred	**34.** 842 **800**	**35.** 257 **300**	**36.** 3580 **3600**

PROBLEM SOLVING

At this point in the unit, you may wish to discuss with students some important problem-solving skills such as being sure to *read* each problem carefully. They should then identify the *key question* in the problem and note the *given facts.* The problem usually can be solved by applying the appropriate formula.

1-4 Piecework

OBJECTIVE

Compute the total pay on a piecework basis.

Some jobs pay on a **piecework** basis. You receive a specified amount of money for each item of work that you complete.

Total Pay = Rate per Item × Number Produced

EXAMPLE *Skill* 8 *Application* A *Term* Piecework

Ramon Hernandez works for National Cabinet Company. He is paid $8.00 for each cabinet he assembles. Last week he assembled 45 cabinets. This week he assembled 42. What is his pay for each week?

SOLUTION Find the **total pay.**

	Rate per Item	×	Number Produced		
Last week:	$8.00	×	45	=	$360.00 total pay
This week:	$8.00	×	42	=	$336.00 total pay

✓ SELF-CHECK Complete the problems, then check your answers in the back of the book.

Find the total pay.

1. $3.20 per item × 140 items produced **$448**
2. 15¢ per item × 1494 items produced **$224.10**

PROBLEMS

	Employee	Rate per Item	×	Number Produced	=	Total Pay
3.	Wo, E.	$2.05	×	180	=	**$369**
4.	Washington, G.	$0.63	×	360	=	**$226.80**
5.	Forest, J.	$1.55	×	174	=	**$269.70**
6.	Gibley, T.	$0.95	×	610	=	**$579.50**
7.	Jackman, L.	$9.50	×	87	=	**$826.50**
8.	Mohammed, H.	$0.28	×	831	=	**$232.68**
9.	Lewis, L.	$3.34	×	79	=	**$263.86**
10.	Thompson, R.	$0.97	×	818	=	**$793.46**

ALTERNATIVE STRATEGIES: Enrichment

Discuss the fact that many consumer products, such as clothing, are made in many countries around the world and sold in the United States. In South America or Asia, for example, labor costs are less than in the U.S. and therefore products can be made for less money and sold at lower prices.

LESSON PLAN 1-4 Piecework

FOCUS

Ask students how they would be paid to wash a car, deliver newspapers, paint a house, or cut a lawn. Discuss why some jobs pay on a piecework basis rather than an hourly rate basis. Ask students to calculate the total amount of money raised at a school fund-raising activity if 74 cars were washed at $3.50 per car. ($259)

TEACH

Point out that for some manufacturing jobs, employers use a piecework method of payment to encourage employees to produce more. However, as the United States has moved from a manufacturing economy to a more service-oriented economy during the latter half of this century, piecework jobs have declined substantially. Ask students if they know any adults who are paid on a piecework basis and what these people actually do.

Warm-Up Exercises

These exercises can be used to review prerequisite skills for this lesson and to ascertain any arithmetic deficiencies that students may have. Work through these exercises with students before having them do the Self-Check.

1. $0.20 × 100 $20.00
2. $0.07 × 235 $16.45
3. $1.50 × 240 $360.00
4. $0.02 × 1458 $29.16
5. ($0.10 × 250) + ($0.17 × 125) + ($21 × 23) $529.25

PRACTICE AND APPLY

The following problems can be assigned for classwork and the answers checked in class to help students master the objective of the lesson.

- Guided Practice: 1–10
- Independent Practice: 11–17 odd

WRAP-UP

Ask students if they would like to work on a piecework basis and to explain why or why not.

Assignment Guide

- Basic: 11–18, 22–34
- Average: 12–16 even, 18–21, 22–34 even

11. Hanley Fellhour.
Beauty shop operator.
Rate per haircut is $10.50.
Gave 60 haircuts.
Find his total pay. **$630**

12. Ellen Kolazinski.
Strawberry picker.
Rate per quart is $0.25.
Picked 1053 quarts.
Find her total pay. **$263.25**

13. Paul Aymes.
Chrome plater.
Rate per item is $1.25.
Plated 321 items.
What is his total pay? **$401.25**

14. Jules Gartner.
Shirt silk screener.
Rate per shirt is $0.45.
Silk screened 388 shirts.
What is his total pay? **$174.60**

15. Carol Ying assembles calculators at Central Electronics. She is paid $0.45 for each calculator she assembles. What is her pay for a day in which she assembles 134 calculators? **$60.30**

16. Leah Elliot runs a carpet cleaning service. She charges $15.95 per room. On Monday, she did 3 rooms in one house, 2 in another, and 4 in a third. Find her total cleaning charges. **$143.55**

17. Jeremy Sullivan delivers newspapers for the Dispatch. He receives 4.2¢ per paper, 6 days a week, for the daily paper, and 25¢ for the Sunday paper. He delivers 124 daily papers each day and 151 Sunday papers each week. What is his total pay for the week? **$69**

Use the table to find the charges for the typing services in problems 18–20.

Spacing	Rate per Page
Single-space	$2.25
Double-space	1.25
Space and a half	1.75
Duplicate copies: $0.15 per page	

$30.60 **18.** A 12-page, single-spaced report plus 2 copies of each page.

$49 **19.** A 35-page, double-spaced term paper plus 1 copy of each page.

$126.65 **20.** A financial report with 14 single-spaced pages, 31 double-spaced pages, and 6 space-and-a-half pages. Plus 6 copies of each page.

21. Rex Moore operates a Quick Stop oil change and tune-up service. He charges $25.95 per oil change, $52.50 to tune a 4-cylinder engine, $58.50 to tune a 6-cylinder engine, and $64.50 to tune an 8-cylinder engine. What are the charges for a week in which he did 35 oil changes, tuned five 4-cylinder engines, seven 6-cylinder engines, and two 8-cylinder engines? **$1709.25**

MAINTAINING YOUR SKILLS Look up the skills in parentheses if you need help or more practice.

Multiply. Round answers to the nearest cent. **(Skill 8)**

22. $0.23 × 89 **$20.47**
23. $1.10 × 240 **$264.00**
24. $0.06 × 4192 **$251.52**
25. 6¢ × 906 **$54.36**
26. $5.20 × 23 **$119.60**
27. $0.04 × 3200 **$128.00**
28. $0.66 × 350 **$231.00**
29. 4.7¢ × 731 **$34.357 = $34.36**

Multiply. **(Skill 8)**

30. 1.17 × 100 = **117**
31. 15.876 × 100 = **1587.8**
32. 242 × 0.04 = **9.68**
33. 456 × 1000 = **456,000**
34. 693.25 × 482 = **334,146.5**

CRITICAL THINKING

In Problem 15, ask students if they see any problems that might arise in paying a person on a piecework basis to assemble calculators.

(In trying to earn more money, working faster would increase the risk of making errors and producing faulty calculators.)

1-5 Salary

OBJECTIVE

Compute the salary per pay period.

A **salary** is a fixed amount of money that you earn on a regular basis. Your salary may be paid weekly, biweekly, semimonthly, or monthly. Your annual salary is the total salary you earn during a year. There are 52 weekly, 26 biweekly, 24 semimonthly, and 12 monthly pay periods per year.

$$\text{Salary per Pay Period} = \frac{\text{Annual Salary}}{\text{Number of Pay Periods per Year}}$$

EXAMPLE *Skill* 11 *Application* K *Term* Salary

Beth Huggins is a computer programmer. Her annual salary is $22,560. What is her monthly salary? What is her weekly salary?

SOLUTION

Find the **salary per pay period.**

	Annual Salary	÷	Number of Pay Periods per Year		
Monthly:	$22,560.00	÷	12	=	$1880.00 monthly salary
Weekly:	$22,560.00	÷	52	=	$433.85 weekly salary
	22560	÷	52	=	433.84615

✓ SELF-CHECK Complete the problems, then check your answers in the back of the book.

1. Find the biweekly salary.
   $42,900 ÷ 26 = ?
   $1650
2. Find the semimonthly salary.
   $18,200 ÷ 24 = ?
   $758.33

PROBLEMS

	Employee	Pay Period	Annual Salary	÷	Pay Periods per Year	=	Salary per Pay Period
3.	K. O. Sanchez	Semimonthly	$ 15,600	÷	24	=	$650
4.	B. J. Molley	Weekly	$ 16,900	÷	52	=	$325
5.	O. L. Farrell	Monthly	$ 16,500	÷	12	=	$1,375
6.	T. B. Reston	Semimonthly	$ 34,650	÷	? 24	=	$1,443.75
7.	R. R. Halston	Weekly	$ 66,598	÷	? 52	=	$1,280.73
8.	K. C. Ying	Biweekly	$132,475	÷	? 26	=	$5,095.19

9. You are a legal clerk.
   Annual salary is $12,500.
   What is your weekly salary?
   $240.38
10. You are a store manager.
    Annual salary is $28,320.
    What is your semimonthly salary?
    $1180

LESSON PLAN 1-5 Salary

FOCUS

Read a few newspaper ads for jobs that are paid on a salary basis and discuss the meaning of *salary.* Ask students to suggest some jobs that would earn a salary. Have students find the monthly salary for a person earning $30,000 a year. ($2500)

TEACH

Discuss the different meanings of the terms *weekly, biweekly, semimonthly,* and *monthly.* Point out that in a biweekly schedule, typically there is a payday every other Friday, but twice during the year there will be three biweekly paydays in a month. Typically, a semimonthly pay period means that the employee is paid on the 1st and 15th of each month.

Point out also that a salary, like an hourly rate, is also based upon time, but the unit of time is a week, a month, or a year. Ask students if they think a salaried employee is paid for overtime work.

Warm-Up Exercises

These exercises can be used to review prerequisite skills for this lesson and to ascertain any arithmetic deficiencies that students may have. Work through these exercises with students before having them do the Self-Check.

1. $12,000 ÷ 12 $1000
2. $54,000 ÷ 24 $2250
3. $37,700 ÷ 26 $1450
4. $325,000 ÷ 52 $6250
5. $624 × 52 $32,448
6. $550 × 26 $14,300

BUSINESS PROJECT

Have students read the employment ads in their local newspapers and make a list of ten different jobs and how they pay. Ask students to try to find examples of jobs having different pay plans. Brief verbal reports can be given to the class in future

COOPERATIVE LEARNING

Have students discuss the advantages and disadvantages of the "13-month" year for salaried employees. Payments under the "13-month" year plan occur on the same day of the week every fourth week of the calendar year.

PRACTICE AND APPLY

The following problems can be assigned for classwork and the answers checked in class to help students master the objective of the lesson.

- Guided Practice: 1–8
- Independent Practice: 9–15 odd

WRAP-UP

Review the solutions to Problems 9, 11, and 13. Ask students how many pay periods in a year there are under weekly, semimonthly, and monthly pay schedules.

Assignment Guide

- Basic: 9–17, 20–30
- Average: 10–16 even, 17–19, 20–30 even

11. Geraldine Piela is a sales clerk. Her annual salary is $17,040. What is her semimonthly salary? **$710**

12. Gary Wilder is a claims adjustor. His annual salary is $21,840. What is his biweekly salary? **$840**

13. Susan Snell was just hired as a mechanical engineer for the Howwel Company. Her starting salary is $32,400 per year. What is her monthly salary? What is her weekly salary? **$2700; $623.08**

14. Morgan Behnke works as a court reporter. Her annual salary is $19,380. What is her weekly salary? What is her biweekly salary? **$372.69; $745.38**

15. Ben Rodebaugh is a medical lab assistant. His annual salary is $17,400. Ben is paid on a monthly basis. What is his monthly salary? **$1450**

TECHNICAL SERVICES CLERK
$15,600 annual salary. Basic Math background helpful. Call Creative Research, 555-1212 or come to 3540 Lincoln Rd., Suite 3100.

16. Juan Rodriguez qualifies for the technical services position described in the advertisement. If he is hired, what will his weekly salary be? **$300**

F.Y.I.

Average Starting Salaries

Engineering	$32,304
Accounting	27,408
Sales/marketing	27,828
Business administration	26,496
Mathematics	26,712
Computer science	29,100

17. Louis Rahn is currently earning an annual salary of $15,090 at Budgett Electronics. He has been offered a job at Delta Tech at an annual salary of $16,660. How much more would Louis earn per week at Delta Tech than at Budgett Electronics? **$320.38 − $290.19 = $30.19**

18. When Paul Sellers first started working at Custom Computers, he was paid a biweekly salary of $615. Custom Computers is now converting to a new payroll system and will be paying its employees on a semimonthly basis. What will Paul's semimonthly salary be? **$666.25**

19. Sune Fung earns a weekly salary of $350 at Howard's Department Store. Next month she will be promoted from assistant buyer to buyer. In her new position, she will be paid $895 semimonthly. How much more per year will Sune earn as a buyer than as assistant buyer? **$3280**

Critical Thinking . . .

20. Assume that your present job pays a monthly gross salary of $1560. You are offered a new position that pays $8.60 per hour with $1\frac{1}{2}$ for all hours over 40 per week. How many hours of overtime per week would you need to earn the same amount per week as your present job? **1.24 hours**

MAINTAINING YOUR SKILLS Look up the skills in parentheses if you need help or more practice.

Divide. Round answers to the nearest hundredth. (Skill 11)

21. 50)$14,290 **$285.80**
22. 8.6)41.62 **4.84**
23. 14.7)191.3 **13.01**
24. 12)5473.2 **456.1**
25. 0.032)17.9 **559.38**
26. 26)21,860 **840.77**
27. 52)69,146 **1329.73**
28. 24)10,549.9 **439.58**

Find the number of pay periods. (Application K)

29. Weekly for 2 years **104**
30. Semimonthly for 3 years **72**
31. Biweekly for 4 years **104**

ALTERNATIVE STRATEGIES: Reteaching

Ask students which is the higher salary:

1. $1200 biweekly or $1300 semimonthly? (it is the same) Why?
   $1200 × 26 = $31,200
   $1300 × 24 = $31,200
2. $350 a week or $760 semimonthly? ($760)
   $350 × 52 = $18,200
   $760 × 24 = $18,240

1-6
Commission

OBJECTIVE
Compute the straight commission and determine the gross pay.

A **commission** is an amount of money that you are paid for selling a product or service. Your **commission rate** may be a specified amount of money for each sale or it may be a percent of the total value of your sales. If the commission is the only pay you receive, you work on **straight commission.**

Commission = Total Sales × Commission Rate

EXAMPLE *Skills* 30,2 *Application* A *Term* Commission

Milton Arps sells real estate at a $7\frac{1}{2}$% straight commission. Last week his sales totaled $90,000. What was his commission?

SOLUTION Find the **commission.**

Total Sales	×	Commission Rate	
$90,000.00	×	$7\frac{1}{2}$%	
$90,000.00	×	0.075	= $6750.00 straight commission

90000 [×] .075 [=] 6750 or 90000 [×] 7.5 [%] 6750

✓ SELF-CHECK Complete the problems, then check your answers in the back of the book.

Find the commission.

1. $9400 × 8% commission rate
2. $143,400 × $5\frac{1}{2}$ % commission rate

1. $752
2. $7887

Instead of working only on commission, you may be guaranteed a minimum weekly or monthly salary. The commission you earn during a week or month is compared with your minimum salary. Your gross pay is the greater of the two amounts.

Gross Pay = Salary or Commission

EXAMPLE *Skill* 30 *Application* A *Term* Commission

Owen Theil is guaranteed a minimum salary of $250 a week or 7% of his total sales, whichever is greater. What is his gross pay for a week in which his total sales were $3614?

SOLUTION

A. Find the **commission.**

Total Sales	×	Commission Rate	
$3614.00	×	7%	= $252.98 commission

B. Find the **gross pay.**
Salary or Commission
$250.00 or $252.98 Gross pay is $252.98 (greater amount).

LESSON PLAN
1-6 Commission

FOCUS
The focus of this lesson is on computing straight commission and gross pay. Engage students in a brief discussion of why many businesses pay sales employees on a commission basis.

TEACH
After working through the Example with students, point out that the concept of a commission is similar to that of piecework. Commission pay is based upon the number of items sold while piecework pay depends on the number of items produced.

A commission is a way to provide an employee with an incentive to sell more items while piecework provides an incentive to make more items.

ALTERNATIVE STRATEGIES: Enrichment

1. Have Students read the **Help Wanted Ads** in the newspaper. Some ads will give the rate of commission or indicate conditions for a bonus, etc.
2. Have students talk to a real estate salesperson or an automobile salesperson to find out about rate of commission.

Warm-Up Exercises

These exercises can be used to review prerequisite skills for this lesson and to ascertain any arithmetic deficiencies that students may have. Work through these exercises with students before having them do the Self-Check.

1. Write $8\frac{1}{2}\%$ as a decimal. 0.085
2. Write $5\frac{3}{4}\%$ as a decimal. 0.0575
3. $4\frac{1}{2}\%$ of $500 $22.50
4. 3% of $1545 $46.35
5. $11\frac{1}{2}\%$ of $1250 $143.75
6. 8% of $590.42 $47.23

✓ SELF-CHECK Complete the problems, then check your answers in the back of the book.

3. Minimum salary: $160.
   Commission: $5\frac{1}{2}\%$ on $2900.
   $160

4. Minimum salary: $2000.
   Commission: $6\frac{1}{4}\%$ on $34,000.
   $2125

PROBLEMS

	Position	Total Sales	×	Commission Rate	=	Commission
5.	Real estate sales	$98,000	×	8%	=	? **$7840**
6.	Computer sales	$18,100	×	12%	=	? **$2172**
7.	Major appliance sales	$ 9,598	×	16%	=	? **$1535.68**
8.	Dress sales	$ 1,311	×	9%	=	? **$117.99**
9.	Computer supplies sales	$ 929	×	15%	=	? **$139.35**
10.	Siding contract sales	$ 754	×	$12\frac{1}{2}\%$	=	? **$94.25**
11.	Auto sales	$68,417	×	$3\frac{1}{2}\%$	=	? **$2394.60**

	Salesperson	Minimum Monthly Salary	Total Monthly Sales	×	Commission Rate	=	Monthly Commission	Gross Pay
12.	Anne Moser	$2,100	$28,000	×	8.0%	=	? **$2,240**	? **$2,240**
13.	Peter Zinn	$1,600	$23,000	×	6.5%	=	? **$1495**	? **$1,600**
14.	Vern Yoder	$3,140	$31,000	×	9.25%	=	? **$2,867.50**	? **$3,140**
15.	Rene Vershum	$ 850	$10,400	×	3.5%	=	? **$364**	? **$850**
16.	Virgil Ulrich	$1,200	$45,000	×	5.5%	=	? **$2475**	? **$2,475**
17.	Robin Coy	$3,410	$29,100	×	12.1%	=	? **$3,521.10**	? **$3,521.10**

18. Roger Tussing.
    3% commission on sales.
    $9500 in sales for a week.
    What is his commission? **$285**

19. Sam Taylor.
    $6\frac{1}{2}\%$ commission on sales.
    $4226 in sales for a week.
    What is his commission?
    $274.69

20. Theresa Britsch.
    Minimum weekly salary is $225.
    Rate of commission is 6.5%.
    Weekly sales are $3420.
    What is the commission? **$222.30**
    What is the gross pay? **$225**

21. Clair Ming.
    Minimum weekly salary is $160.
    Rate of commission is 7.5%.
    Weekly sales are $2420.
    What is the commission? **$181.50**
    What is the gross pay? **$181.50**

BUSINESS NOTES

One of the most important decisions a business makes is how to pay its employees. A proper compensation plan for employees is a crucial factor in motivating them to perform. Most large businesses have more than one way to pay employees who usually are involved in many different kinds of jobs.

22. Marie Busack sells used cars for Bonded Auto. She receives a straight commission of 4% of the selling price of each car. What commission will she receive for selling a $19,420 van? **776.80**

BONDED AUTO SALES
We are looking for someone to sell used cars. Earn 4% commission on every car you sell. Call 555-8081.

23. Bruce Clay sells real estate. His commission is $5\frac{3}{4}$% of the price of every house he sells. What is his commission when he sells a $145,450 house? **$8363.38**

24. Auto mechanics at Cline's Garage receive a straight commission of 22% of the service income they bring in each week. What is the commission for a mechanic who brings in $1465 service income in a week? **$322.30**

25. John Navarro is a salesperson in the appliance department at Morris Appliance, Inc. He is guaranteed a minimum salary of $185 per week or 5.5% of his total sales, whichever is greater. What is his gross pay for a week in which his total sales were $3422? **$188.21**

26. Maude Eggert sells cosmetics for Soft Touch, Inc. She is guaranteed a salary of $790 a month or $7\frac{1}{4}$% of her total sales, whichever is greater. What is her gross pay for a month in which her total sales were $10,984? **$796.34**

27. Irma DeWolfe is paid either a commission or $4.75 per hour plus time and a half overtime for all hours over 8 per day, whichever is greater. Her commission consists of $5\frac{1}{2}$% of sales. Find her gross pay for a week in which she worked 6 hours on Monday, 8 hours on Tuesday, 10 hours on Wednesday, 6 hours on Thursday, 10 hours on Friday, and 9 hours on Saturday. Her total sales for the week were $4100. **$244.63**

28. Some jobs pay a commission plus a bonus at the end of the year. The bonus may be a percent of the salesperson's total commission for the year.

   a. Madelyn Carr is a sales representative. She receives a 7% commission on all sales. At the end of the year, she receives a bonus of 5% of her commission. What is her total pay for a year in which she had sales totaling $412,454? **$30,315.37**

   b. What would Madelyn's total pay be if her sales were $316,250? **$23,244.38**

MAINTAINING YOUR SKILLS Look up the skills in parentheses if you need help or more practice.

Write as a decimal. Round answers to the nearest hundredth. **(Skill 14)**

29. $2\frac{1}{8}$ **2.13** **30.** $5\frac{1}{2}$ **5.50** **31.** $6\frac{3}{4}$ **6.75** **32.** $4\frac{1}{3}$ **4.33** **33.** $9\frac{1}{4}$ **9.25**

Find the percentage. Round to nearest cent or tenth. **(Skill 30)**

34. 4% of $1250 **$50** **35.** $8\frac{1}{2}$ % of $4300 **$365.50** **36.** $7\frac{1}{4}$ % of $8200 **$594.50** **37.** 9.2% of $3600 **$331.20**

38. 7.3% of $120 **$8.76** **39.** 8.92% of 1380 **123.1** **40.** 39% of 281 **109.6** **41.** 5.7% of 9140 **521**

Round answers to the nearest tenth. **(Skill 2)**

42. 0.081 **0.1** **43.** 0.608 **0.6** **44.** 0.92 **0.9** **45.** 0.821 **0.8** **46.** 0.22 **0.2**

PRACTICE AND APPLY

The following problems can be assigned for classwork and the answers checked in class to help students master the objective of the lesson.

- Guided Practice: 1–8, 12–15
- Independent Practice: 9–11, 15–17, 19–23 odd

WRAP-UP

Ask students to explain the difference between a commission and a commission rate.

Assignment Guide

- Basic: 9–11, 15–17, 18–26, 29–45 odd
- Average: 18–24 even, 25–28, 30–46 even

LESSON PLAN
1-7 Graduated Commission

FOCUS
Point out to students that sometimes a business may set different commission rates for different levels of sales to motivate employees to sell more. Thus, as sales increase so do the rates of commission. In this lesson, students compute the total graduated commission.

TEACH
Explain the meaning of the term *graduated*. Point out that schools are arranged in a series of levels, called grades, and thus schools are graduated. Ask students to give some other examples for which the term *graduated* could apply. Suggest that they think of things that involve heights, depth, or difficulty.

Warm-Up Exercises
These exercises can be used to review prerequisite skills for this lesson and to ascertain any arithmetic deficiencies that students may have. Work through these exercises with students before having them do the Self-Check.

1. 8% of $800 $64
2. $12\frac{1}{2}$% of $920 $115
3. 18% × $3920 $705.60
4. 22% × $459.50 $101.09

1-7 Graduated Commission

OBJECTIVE
Compute the total graduated commission.

Your commission rate may increase as your sales increase. A **graduated commission** offers a different rate of commission for each of several levels of sales. It provides an extra incentive to sell more.

Total Graduated Commission = Sum of Commissions for All Levels of Sales

EXAMPLE *Skills* 5, 28, 30 *Application* A *Term* Graduated Commission

Irene Tomas sells appliances at Twin City Sales. She receives a graduated commission of 4% on her first $1000 of sales, 6% on the next $2000, and 8% on sales over $3000. Irene's sales for the past month totaled $9840. What is her commission for the month?

SOLUTION Find the **sum of commissions for all levels of sales.**

	Sales	× Commission Rate	
A. First $1000:	$1000 ×	4%	= $ 40.00
B. Next $2000:	$2000 ×	6%	= $120.00
C. Over $3000:	($9840 − $3000) ×	8%	
	$6840 ×	8%	= $547.20
	Total Graduated Commission		$707.20

1000 [×] 4 [%] 40 [M+] 2000 [×] 6 [%] 120 [M+] 9840 [−] 3000 [=] 6840 [×] 8 [%] 547.2 [M+] [RM] 707.2

✓ SELF-CHECK Complete the problems, then check your answers in the back of the book.

1. Commission: 10% on first $5000; 15% over $5000. Find the total graduated commission on $22,000. **$3050**
2. Commission: 5% first $2000; 8% over $2000. Find the total graduated commission on $7740. **$559.20**

PROBLEMS

	3.	4.
Amount of Sales	$3,400	$2,500
Commission		
First $1000: 10%	$100	$100
Over $1000: 15%	$360	$225
Total Commission	$460	$325

	5.	6.
Amount of Sales	$3,900	$8,250
Commission		
First $1500: 6%	$90	$90
Next $2000: 8%	$160	$160
Over $3500: 10%	$40	$475
Total Commission	$290	$725

MATHEMATICAL NOTES
Be sure to stress that if the graduated-commission structure has three levels, then: (a) the lowest rate, say 8%, applies only to the lowest level of sales, say the first $800; (b) the next higher rate, say 10%, applies to the next higher level of sales, say the next $2000; (c) the highest rate, say 14%, applies to the highest level of sales, say over $2800.

7. Bill Weston, part-time sales.
20% commission on first $500.
25% on next $1000.
30% on sales over $1500.
$1940 total sales. **$482**
What is his total commission?

8. Mary Draper, sporting goods sales.
5% commission on first $6000.
$7\frac{1}{2}$% on next $6000.
10% on amount over $12,000.
$14,640 total sales. **$1014**
What is her total commission?

9. Jean Gray sells office supplies. She receives a graduated commission of 4% on her first $2000 of sales and $8\frac{1}{2}$% on all sales over $2000. Jean's sales for the past week totaled $3925. What is her commission for the week? **$243.63**

10. Charles Beaudry sells computer hardware for a computer firm. He is paid a 4% commission on the first $6000 of sales, 6% on the next $10,000, and 8% on sales over $16,000. What is his commission on $24,550 in sales? **$1524**

11. Ralph King demonstrates cookware at the National Food Fair. He is paid $6.00 each for the first 10 demonstrations in one day and $7.50 for each demonstration over 10. What is Ralph's commission for a day in which he makes 21 demonstrations? **$142.50**

12. Odessa Dilulo demonstrates home fire alarm systems. She is paid $10.00 per demonstration for the first 10 demonstrations given during a week and $15.50 for each demonstration over 10. Also, for every sale, she gets a bonus of $15.00. What is her commission for a week in which she gives 18 demonstrations and makes 6 sales? **$314**

13. Some sales positions pay a commission only if sales exceed a **sales quota.** The salesperson is rewarded only for having sales beyond an expected amount.
Joyce Doyle is a sales trainee. She is paid $550 per month plus a commission of 7.5% on all sales over a quota of $9000 per month. What is her total pay for a month in which she has sales totaling $10,650? **$673.75**

MAINTAINING YOUR SKILLS Look up the skills in parentheses if you need help or more practice.

Add. **(Skill 5)**

14. $123.72 + 112.69 + 45.23 = **$281.64**

15. $567.89 + 34.69 + 431.73 = **$1034.31**

16. $1350.23 + 946.00 + 18.11 = **$2314.34**

17. 0.007 + 1.384 + 0.569 = **1.960**

Write as a decimal. **(Skill 28)**

18. $5\frac{1}{2}$% **0.055**

19. $10\frac{1}{4}$% **0.1025**

20. 15% **0.15**

21. 97% **0.97**

22. 342% **3.42**

Find the percentage. **(Skill 30)**

23. $5\frac{1}{4}$% of $2000 **$105.00**

24. 7% of $4560 **$319.20**

25. $15\frac{1}{2}$% of $3500 **$542.50**

PRACTICE AND APPLY

The following problems can be assigned for classwork and the answers checked in class to help students master the objective of the lesson.

- Guided Practice: 1–6
- Independent Practice: 7–11 odd

WRAP-UP

Remind students of the objective of the lesson and work the following problem as a summary.
Total Sales: $35,000
1. First $5000: 2% of $5000
2. Next $10,000: 4% of $10,000
3. Over $15,000: 6% of $20,000 ($1700)

Assignment Guide

- Basic: 7–12, 14–25
- Average: 8–12 even, 13, 14–24 even

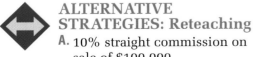

ALTERNATIVE STRATEGIES: Reteaching

A. 10% straight commission on sale of $100,000
Commission = 10% × $100,000
= $10,000

B. 10% commission on first $60,000 of sale
12% commission on sales over $60,000
Commission = 10% × $60,000 = $6,000
= 12% × $40,000 = 4,800
$10,800

LESSON PLAN
Reviewing the Basics

The exercises on this page review skills, applications, and terms used in the unit. You can use the exercises to assess informally students' proficiency with this material.

The page can be used for guided practice and independent practice. You can work through a selection of the exercises together with students, and thus see immediately if they know how to do them, and you can then assign some of the exercises for independent practice. Be sure to go over the answers to all assigned exercises.

Reviewing the Basics

Skills

Round to the nearest cent.

(Skill 2)
1. \$45.672 **\$45.67** **2.** \$469.3591 **\$469.36** **3.** \$0.045 **\$0.05** **4.** \$53.744 **\$53.74** **5.** \$172.255 **\$172.26**

Solve.

(Skill 5)
6. \$672.01 + \$45.32 **\$717.33** **7.** 0.004 + 1.5 + 19.03 **20.534** **8.** \$32.51 + \$95.98 **\$128.49**

(Skill 8)
9. \$45.00 × 0.055 **\$2.48** **10.** 11.563 × 100 **1156.3** **11.** 0.007 × 0.03 **0.00021**

(Skill 11)
12. \$23,660 ÷ 26 **\$910** **13.** \$55,920 ÷ 12 **\$4660** **14.** 27.3 ÷ 0.05 **546**

(Skill 16)
15. $5\frac{1}{4} + 7\frac{1}{2} + 6\frac{3}{4}$ **$19\frac{1}{2}$** **16.** $5\frac{1}{2} + 2\frac{2}{3} + 7\frac{5}{6}$ **16** **17.** $8\frac{1}{2} + 2\frac{1}{4}$ **$10\frac{3}{4}$**

(Skill 30)
18. $15\frac{1}{2}$% of \$650 **\$100.75** **19.** 110% of \$310 **\$341.00** **20.** $\frac{1}{4}$% of \$24 **\$0.06**

Write as a decimal.

(Skill 14)
21. $\frac{1}{2}$ **0.5** **22.** $\frac{1}{4}$ **0.25** **23.** $\frac{3}{4}$ **0.75** **24.** $\frac{8}{10}$ **0.8** **25.** $\frac{1}{8}$ **0.125**

(Skill 28)
26. 11.5% **0.115** **27.** 1.25% **0.0125** **28.** 0.125% **0.00125** **29.** 385% **3.85** **30.** $10\frac{1}{4}$% **0.1025**

Applications

Round to the nearest quarter hour.

(Application E)
31. 8 hours 35 minutes **$8\frac{1}{2}$ hrs.** **32.** 7 hours 50 minutes **$7\frac{3}{4}$ hrs.**

Find the number of occurrences.

(Application K)
33. Weekly for 2 years **104** **34.** Monthly for 3 years **36**

35. Quarterly for 4 years **16** **36.** Semimonthly for 2 years **48**

Terms

Write a correct definition for each term. Refer to the terms file. **Answers will vary.**

37. Straight-time pay **38.** Piecework pay

39. Salary **40.** Commission

41. Graduated commission **42.** Overtime pay

Refer to your reference files at the back of the book if you need help.

Unit Test

LESSON PLAN
Unit Test

Students should do the Unit Test on their own. Each problem on the test is keyed to a lesson in the unit. Students having difficulty with any particular problem should review the Example in the appropriate lesson and be assigned some of the Independent Practice problems for additional practice.

Lesson 1-1

1. Judy Enright, a painter, earns $8.60 an hour during a regular workweek. Last week she worked her regular 40 hours. This week she worked 32.5 hours. What is her straight-time pay for each week? **$344.00; $279.50**

Lesson 1-2

2. Resmund Downs works as a hotel clerk. He earns $5.60 per hour for a 40-hour week. His overtime rate is $1\frac{1}{2}$ times his regular hourly rate. This week he worked his regular 40 hours plus $8\frac{3}{4}$ hours of overtime. What is his total pay? **$297.50**

Lesson 1-3

3. Stella Gordon fills out a weekly time card. What are her total hours for the week? (Round to the nearest quarter hour.)

Day	In	Out	In	Out	Hrs.	
Sun.						
Mon.	8:30	12:00	12:35	5:35	?	$8\frac{1}{2}$
Tue.	8:25	12:15	1:00	5:30	?	$8\frac{1}{4}$
Wed.	8:28	12:45	1:15	4:25	?	$7\frac{1}{2}$
Thu.	8:15	1:08	1:45	5:32	?	$8\frac{3}{4}$
Fri.	8:50	12:30	1:00	5:08	?	$7\frac{3}{4}$
Sat.						
				Total Hours	?	$40\frac{3}{4}$

Lesson 1-4

4. Duane Smith is paid on a piecework basis. He is paid $14.50 for each chair he assembles. This week he assembled 32 chairs. What is his pay for the week? **$464.00**

Lesson 1-5

5. Frank DuByne is a production manager for National Plastics. He earns an annual salary of $32,420. What is his semimonthly pay? What is his weekly pay? **$1350.83; $623.46**

Lesson 1-6

6. Bettie Ray sells real estate for a 6% straight commission. Her sales totaled $392,400 in September. What is her pay for the month? **$23,544**

Lesson 1-7

7. Eric Erford is an appliance salesman. He earns a graduated commission on sales as shown. His sales for 1 week totaled $5240. What was his total commission? **$508.80**

Commission	Level of Sales
8%	First $1500
10%	Next $3000
12%	Over $4500

LESSON PLAN
A Spreadsheet Application

USING TECHNOLOGY

In this application, students use a computer to find the straight-time pay, overtime pay, and total or gross pay. You might want to use the application as a basis for a *cooperative learning activity,* with small groups of students working together at the computer.

As a *business note,* point out to students that computers are used throughout the business world to compute employees' pay. The power of a computer enables it to keep track of large amounts of data accurately. Thus, very large businesses, which have thousands of employees, can process efficiently their pay records. However, even very small businesses consisting of only a few employees, use computers for pay purposes.

A SPREADSHEET APPLICATION

Gross Income

To complete this spreadsheet application, you will need the template diskette for *Mathematics with Business Applications.* Follow the directions in the User's Guide portion of the TRB to complete this activity.

Input the information in the following problems to find the straight-time pay, the overtime pay, and the total or gross pay.

1. Employee: D. Wyse.
Rate per hour: $9.60.
Regular hours: 40.
Overtime hours: 6.
What is the straight-time pay?
What is the overtime pay?
What is the total pay?

2. Employee: K. Yackee.
Rate per hour: $12.50.
Regular hours: 36.
Overtime hours: 8.
What is the straight-time pay?
What is the overtime pay?
What is the total pay?

3. Employee B. Sullivan.
Rate per hour: $4.50.
Regular hours: 25.
Overtime hours: 4.
What is the straight-time pay?
What is the overtime pay?
What is the total pay?

4. Employee: G. Stambaugh.
Rate per hour: $22.00.
Regular hours: 40.
Overtime hours: 12.
What is the straight-time pay?
What is the overtime pay?
What is the total pay?

5. Employee: A. Roth.
Rate per hour: $27.22.
Regular hours: 26.
Overtime hours: 0.
What is the straight-time pay?
What is the overtime pay?
What is the total pay?

6. Employee: W. Pawlowicz.
Rate per hour: $8.55.
Regular hours: 36.
Overtime hours: 9.
What is the straight-time pay?
What is the overtime pay?
What is the total pay?

7. Marc Pempertolie earns $5.50 per hour plus time and a half for overtime. Last week he worked his regular 40 hours plus 6 hours overtime. What is his straight-time pay? What is his overtime pay? What is his total pay?

8. Carey Nofziger earns $6.95 per hour plus time and a half for overtime. Last week she worked her regular 40 hours plus 12 hours overtime. What is her straight-time pay? What is her overtime pay? What is her total pay?

A SPREADSHEET APPLICATION (CONTINUED)

9. Hilarie Mapes earns $16.50 per hour plus time and a half for overtime. Last week she worked her regular 36 hours plus 7.5 hours overtime. What is her straight-time pay? What is her overtime pay? What is her total pay?

10. Judy Dohm earns $19.25 per hour plus time and a half for overtime. She worked her regular 42 hours plus 9.25 hours overtime. What is her straight time pay? What is her overtime pay? What is her total pay?

11. Vernon Cymbola earns $8.63 per hour plus time and a half for overtime on weekends. Last week he worked 35 hours Monday through Friday, 8 hours on Saturday, and 8 hours on Sunday. What is his total pay for the week?

12. Sherri Cordesa earns $18.875 per hour plus double time for overtime on weekends. Last week she worked 28 hours Monday through Friday, 7.5 hours on Saturday, and 9.5 hours on Sunday. What is her total pay for the week?

13. Angela Perez earns $11.36 an hour plus time and a half for overtime. Last week she worked her regular 39 hours and 3 hours overtime. What is her total pay for the week?

14. Robert Lewis earns $12.42 an hour plus double time for overtime. Last week he worked his regular 40 hours plus 6.5 hours overtime. What is his total pay for the week?

15. Stephanie Martin earns $8.50 an hour plus time and a half for overtime. Last week she worked her regular 37 hours plus 10 hours overtime. What is Stephanie's total pay for the week?

16. Dwight Madsen earns $7.25 an hour plus time and a half for overtime. Last week he worked his regular 39 hours plus 7.25 hours on Saturday. What is his total pay for the week?

LESSON PLAN
Career Wise

APPLICATIONS ON THE JOB

Students can work on these Career Wise pages independently in school or at home. In either case, it will be worthwhile to discuss them in class and go over the answers to the questions.

When discussing this page, point out to students that graphs are an excellent way to communicate information to other people. Making and using a graph can be thought of as a *communication skill* because the graph helps to explain data in a visual way. Mention that many people in business use graphs when making presentations. Ask students if Mountaintop Ski Shop could present its entire yearly sales by using a bar graph. (Yes)

CAREER WISE

Salesperson

Denise Hobson is a salesperson in the Mountaintop Ski Shop. She enjoys meeting new people and helping them purchase ski equipment.

When Denise began work at the shop about four years ago, she entered a training program. In it, she learned how to relate to customers and how to advise them, and she learned the company's product line inside and out. Due to her enthusiasm and hard work, Denise now earns $850 a month as a salary and 6% of each sale she makes in commission.

Last year, Denise decided to make a chart of her sales performance. She constructed a bar graph in which her sales total for each month of the year was represented by a bar or column. The bar graph below shows, for example, that in January (the first J on the horizontal axis) of last year, she sold $12,500 worth of ski equipment.

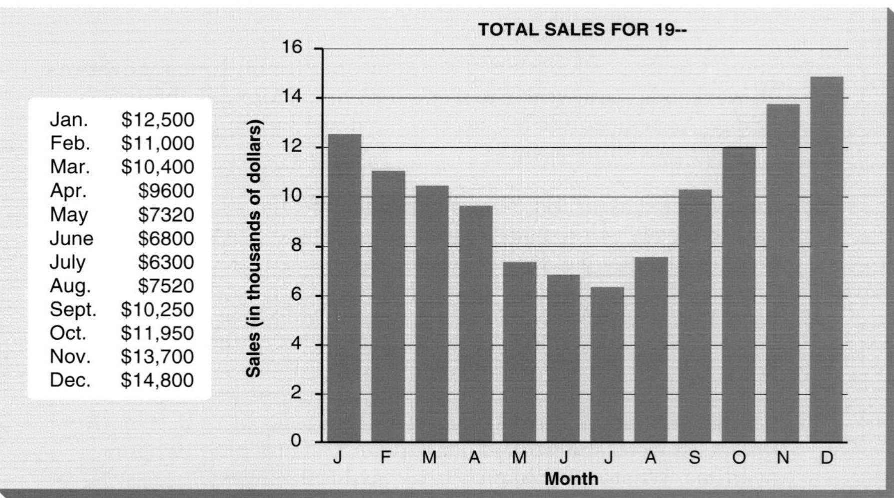

Check Your Understanding

1. From the graph, in which month was Denise's income the least? In which month was her income the greatest? **Least: July; Greatest: December**
2. Compute Denise's income for last February. **$850 + (0.06 × $11,000) = $1510**
3. Compute Denise's income for last July. **$850 + (0.06 × $6300) = $1228**
4. To offset the sag in summer sales, Denise might suggest to the owner that Mountaintop offer a product line in addition to its ski line. What product or products might she suggest? Discuss your answers.
Answers will vary: swimwear, beach umbrellas, coolers, etc.

U N I T

2

Net Income

Out of your *gross pay*, the total amount of money you earn, employers withhold money for federal, state, social security, and medicare taxes. These *deductions* from your gross pay result in your *net pay*, the money you actually receive each pay period. *Federal income taxes* depend on your salary and the number of people you support, or your *dependents* (also called withholding allowances or personal exemptions). Deductions from your gross pay may include the cost of medical insurance, unemployment insurance, disability insurance, savings, stocks, union dues, and charitable contributions. These deductions are shown on your *earnings statement*, which is attached to your paycheck.

O U T L I N E

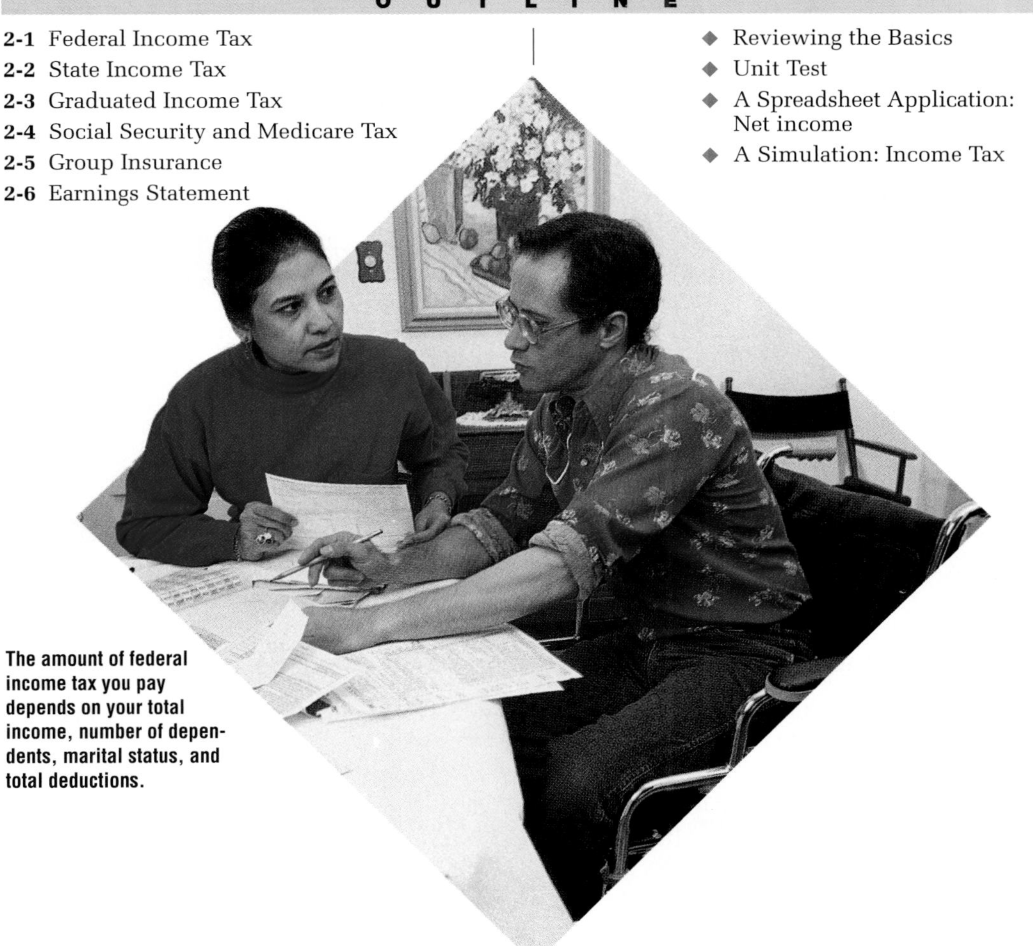

The amount of federal income tax you pay depends on your total income, number of dependents, marital status, and total deductions.

UNIT 2
NET INCOME

INTRODUCING THE UNIT

Explain that when a person begins a job, the employer will ask the new employee to complete an "Employee's Withholding Allowance Certificate (Form W-4)." This form requires you to list your name, address, social security number, marital status, and number of allowances. From this information and the employee's gross salary, the employer will use the appropriate tax table to determine the amount of tax to withhold.

LESSON PLAN 2-1 Federal Income Tax

FOCUS

To motivate the lesson and unit, ask students if they have ever heard someone complaining about how much money is taken out of their paycheck for taxes. Did they ever wonder how these tax amounts are calculated? In this lesson, students will use tax tables to find the amount withheld for federal income tax.

TEACH

The thrust of this lesson is reading income tax tables. To help students familiarize themselves with the tax table on this page, point out the format of the determining factors: marital status, wages, allowances, and payroll schedule. Regarding wages, be sure and emphasize the column headings "At least" and "But less than." Mention that there are also tax tables for other payroll schedules, such as monthly and biweekly.

2-1 Federal Income Tax

OBJECTIVE
Use tables to find the amount withheld for FIT.

Employers are required by law to withhold a certain amount of your pay for **federal income tax** (FIT). The Internal Revenue Service provides employers with tables that show how much money to withhold. The amount withheld depends on your income, marital status, and **withholding allowances.** You may claim one allowance for yourself and one allowance for your spouse if you are married. You may claim additional allowances for any others you support.

WEEKLY Payroll MARRIED

Wages		Allowances		
At least	But less than	0	1	2
280	290	33	26	20
290	300	34	28	22
300	310	36	29	23
310	320	37	31	25
320	330	39	32	26
330	340	40	34	28
340	350	42	35	29
350	360	43	37	31
360	370	45	38	32
370	380	46	40	34

EXAMPLE *Skill* 1 *Application* C *Term* Income tax

Velma Miller's gross pay for this week is $321.85. She is married and claims 2 allowances, herself and her husband. What amount will be withheld from Velma's pay for federal income tax?

SOLUTION

A. Find the **income range** from the table.

At least	But less than
320	330

B. Find the column for 2 **allowances.**

C. Amount of income to be withheld is $26.00.

✓ SELF-CHECK Complete the problems, then check your answers in the back of the book.

1. Weekly income: $285.40. 1 allowance. **$26.00**

2. Weekly income: $333.83. 1 allowance. **$34.00**

PROBLEMS

Use the MARRIED Persons—WEEKLY Payroll Period table above to find the amount withheld for federal income tax.

	3.	**4.**	**5.**	**6.**	**7.**
Employee	G. Chesson	B. Brown	J. R. Foos	J. Isch	F. Restivo
Weekly Income	$283.00	$348.00	$289.50	$334.20	$320.00
Allowances	2	1	0	2	0
Amount Withheld	? $20	? $35	? $33	? $28	? $39

SINGLE Persons—WEEKLY Payroll Period												
And the wages are—		And the number of withholding allowances claimed is—										
Wages		0	1	2	3	4	5	6	7	8	9	10
At least	But less than	The amount of income tax to be withheld shall be—										
170	175	22	16	10	4	0	0	0	0	0	0	0
175	180	23	17	11	4	0	0	0	0	0	0	0
180	185	24	18	11	5	0	0	0	0	0	0	0
185	190	25	18	12	6	0	0	0	0	0	0	0
190	195	25	19	13	7	0	0	0	0	0	0	0
195	200	26	20	14	7	1	0	0	0	0	0	0
200	210	27	21	15	9	2	0	0	0	0	0	0
210	220	29	22	16	10	4	0	0	0	0	0	0
220	230	30	24	18	12	5	0	0	0	0	0	0
230	240	32	25	19	13	7	1	0	0	0	0	0
240	250	33	27	21	15	8	2	0	0	0	0	0
250	260	35	28	22	16	10	4	0	0	0	0	0
260	270	36	30	24	18	11	5	0	0	0	0	0
270	280	38	31	25	19	13	7	0	0	0	0	0
280	290	39	33	27	21	14	8	2	0	0	0	0
290	300	41	34	28	22	16	10	3	0	0	0	0
300	310	42	36	30	24	17	11	5	0	0	0	0
310	320	44	37	31	25	19	13	6	0	0	0	0
320	330	45	39	33	27	20	14	8	2	0	0	0
330	340	47	40	34	28	22	16	9	3	0	0	0

F.Y.I.
In 1988, the federal income taxes collected from individuals by the IRS amounted to $586 billion.

Use the table above to find the amount withheld.

8. Jeannie Cooley, single.
Earns $330.15 weekly.
Claims 2 allowances.
What is the FIT withheld? **$34**

9. Matt Osborn, single.
Earns $170 weekly.
Claims 3 allowances.
What is the FIT withheld? **$4**

10. Hubert Pacetta, single.
Earns $232.40 weekly.
Claims 3 allowances.
What is the FIT withheld? **$13**

11. Emily Zasada, single.
Earns $300 weekly.
Claims 2 allowances.
What is the FIT withheld? **$30**

12. Shirley Delaney earns $325 a week. She is single and claims 2 allowances. What amount is withheld weekly for federal income tax? **$33**

13. Hank Michalat earns $321 a week. He is single and claims no allowances. What amount is withheld weekly for federal income tax? **$45**

14. Harold Leu is single and earns $300 a week. He claims 1 allowance. What amount is withheld weekly for federal income tax? **$36**

15. Tina Ford is single, earns $252 a week, and claims 1 allowance. What amount is withheld weekly for federal income tax? **$28**

Use the MARRIED Persons—WEEKLY Payroll Period table on page 643 to find the amount withheld for federal income tax.

16. Carol Meinka is married, earns $615 a week, and claims 3 allowances. What amount is withheld weekly for federal income tax? **$63**

Discuss the headings "At least" and "But less than" when finding the correct range of a person's income.

You might wish to explain that by law a person cannot claim more allowances than those to which he or she is entitled. However, a person can claim fewer allowances.

Warm-Up Exercises

Work through these exercises with students before having them do the Self-Check.

1. How much is withheld from the weekly wage of a married person earning $282 with 1 withholding allowance? $26
2. How much is withheld from the weekly wage of a married person earning $330 with 2 allowances? $28

ALTERNATIVE STRATEGIES: Enrichment

Although some citizens of the United States feel they are taxed too heavily, the U.S. tax on individuals currently remains lower than taxes on citizens of Japan, and much lower than those of most European countries. Ask students to find data on the tax rate of various countries for discussion in class.

PRACTICE AND APPLY

The following problems can be assigned for classwork and the answers checked in class to help students master the objective of the lesson.

- Guided Practice: 1–8
- Independent Practice: 9–15 odd

WRAP-UP

Ask students to list the four factors that are used to determine the amount to withhold for federal income tax. Ask a volunteer to create a situation as an employee using the four factors. Select another student to find the tax amount that would be withheld for that employee.

Assignment Guide

- Basic: 9–19 odd, 21–33
- Average: 10–16 even, 17–20, 21–33 odd

17. Luther Myers earns $455 a week. He is married and claims 2 allowances. What amount is withheld weekly for federal income tax? **$46**

18. Georgette Young is paid a weekly wage of $491.60. She is married and claims 2 allowances. Beginning next year, she will claim another allowance for supporting her mother. How much less will be withheld for federal income tax for the year? **$364**

19. Lamoin Kelb earns $272.44 a week. He is married and claimed 2 allowances last year. He hopes to receive a refund on his next tax return by claiming no allowances this year. How much more in withholdings will be deducted weekly if he claims no allowances? **$12**

Critical Thinking . . .

20. Some companies use a percentage method instead of the tax tables to compute the income tax withheld. Use the table at the right to find the amount withheld for the SINGLE employees below. Each weekly allowance is $41.35.

WEEKLY PAYROLL PERIOD, SINGLE PERSON

If the amount of wages (after subtracting withholding allowances) is: — *The amount of income tax to be withheld shall be:*

Not over $24 0

Over—	*But not over—*		*of excess over—*
$24	—$415	. . . .15%	—$24
$415	—$972	. . . .$58.65 plus 28%	—$415
$972		. . . .$214.61 plus 31%	—$972

a. Find the **allowance amount.**
Number of Allowances × $41.35

b. Find the **taxable wages.**
Weekly Wage − Allowance Amount

c. Find the **amount withheld** for the employees below.

Employee	Weekly Wage	Number of Allowances	Allowance Amount	Taxable Wages	Amount Withheld
D. Faust	$ 184.00	1	$ 41.35	$142.65	$17.80
J. Wray	$ 490.00	3	$124.05	$365.95	**$51.29**
E. Kohn	$ 690.00	2	**$82.70**	**$607.30**	**$112.49**
K. Bird	$1150.00	4	**$165.40**	**$984.60**	**$218.52**

MAINTAINING YOUR SKILLS Look up the skills in parentheses if you need help or more practice.

Give the place and value of the underlined digit. **(Skill 1)**

21. 84 **ones, 4** **22.** 965 **tens, 60** **23.** 6532 **24.** 36.94 **25.** 672 **26.** 14.783

23. thousands, 6000
24. hundredths, $\frac{4}{100}$
25. hundreds, 600
26. thousandths, $\frac{3}{1000}$

Find the percentage. **(Skill 30)**

27. 22.5% × $120 **$27** **28.** 17.50% × $430 **$75.25**

29. $\frac{1}{2}$% × $298 **$1.49** **30.** 128% × $82 **$104.96**

Subtract. **(Skill 6)**

31. $344.50 − $68.40 **$276.10** **32.** $432.60 − $83.70 **$348.90** **33.** $1298.65 − $423.76 **$874.89**

CRITICAL THINKING

Problems 18 and 19 provide examples of what happens when a person raises or lowers his or her allowances. The greater the number of allowances, the lower the withholding tax; the lower the allowance, the greater the withholding tax. Ask students if there are any disadvantages to claiming fewer allowances than one is entitled to? (The individual could put the exta money into a savings account; the individual is giving the government an interest-free loan.)

2-2 State Income Tax

OBJECTIVE
Compute the state tax on a straight percent basis.

Most states require employers to withhold a certain amount of your pay for state income tax. In some states, the tax withheld is a percent of your taxable wages. Your taxable wages depend on **personal exemptions,** or withholding allowances, allowed for supporting yourself and others in your family.

Taxable Wages = Annual Gross Pay − Personal Exemptions

Annual Tax Withheld = Taxable Wages × Tax Rate

EXAMPLE *Skills* 6, 30 *Application* A *Term* Personal exemptions

James Bowler's gross pay is $18,800 a year. The state income tax rate is 3% of taxable wages. James takes a married exemption for himself and his wife and 2 personal exemptions for his 2 children. How much is withheld a year from his gross earnings for state income tax?

PERSONAL EXEMPTIONS
Single—$1500
Married—$3000
Each Dependent—$700

SOLUTION

A. Find the **taxable wages.**
Annual Gross Pay − Personal Exemptions
$18,800.00 − ($3000.00 + $700.00 + $700.00) = $14,400.00

B. Find the **annual tax withheld.**
Taxable Wages × Tax Rate
$14,400.00 × 3% = $432.00 annual tax withheld

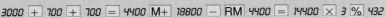

3000 [+] 700 [+] 700 [=] 4400 [M+] 18800 [−] [RM] 4400 [=] 14400 [×] 3 [%] 432

✓ SELF-CHECK Complete the problems, then check your answers in the back of the book.

Find the annual tax withheld.

1. Gross pay: $18,900.
Single, 1 dependent.
State income tax rate: 4%. **$668**

2. Gross pay: $34,000.
Married, 3 dependents.
State income tax rate: 5%. **$1445**

PROBLEMS

	(Annual Gross Pay	−	Personal Exemptions	=	Taxable Wages)	×	State Tax Rate	=	Tax Withheld
3.	($17,000	−	$3000	=	$14,000)	×	2%	=	**$280**
4.	($24,000	−	$2200	=	**$21,800**)	×	5%	=	**$1090**
5.	($ 9,500	−	$1500	=	**$8000**)	×	3.5%	=	**$280**

LESSON PLAN 2-2 State Income Tax

FOCUS

In this lesson, students will compute a state tax on a straight percent basis, which is one method to generate money for a state's expenses. The method used and the tax rate vary from state to state. To motivate this lesson, ask students why the federal government needs to collect money from taxes (examples: to pay for human services, national defense, government employees' salaries and expenses) Ask why states need to collect taxes (examples: education, highways, police protection)

TEACH

Before working the Example with students, write the formulas for taxable wages and annual tax withheld on the chalkboard. Stress that the tax rate is used with taxable wages, not on gross pay. Therefore, students must first find the taxable wages.

Warm-Up Exercises

Work through these exercises with students before having them do the Self-Check.

Write Exercises 1–4 as decimals.

1. 9% 0.09
2. $4\frac{1}{2}$% 0.045
3. $2\frac{1}{4}$% 0.0225
4. 7.24% 0.0724
5. 2.5% × $560 $14
6. $3\frac{1}{4}$% × $754.50 $24.52
7. $\frac{1}{4}$% of $800 $2.00
8. 0.9% of $600 $5.40
9. $10,500 ÷ 12 $875
10. $59,800 ÷ 26 $2300

ALTERNATIVE STRATEGIES: Reteaching

Suppose you have an annual wage of $15,720 and claim two exemptions. The state allows a $1500 exemption and uses a tax rate of 5%. What is the weekly tax? Follow this procedure.

STEP 1. Find total exemptions:
2 × $1500 = $3000
STEP 2. Find total taxable wage:
$15,720 − $3000 = $12,720
STEP 3. Find annual state tax:
$12,720 × 5% = $636
STEP 4. Find weekly tax:
$636 ÷ 52 = $12.23

PRACTICE AND APPLY

The following problems can be assigned for classwork and the answers checked in class to help students master the objective of the lesson.

- Guided Practice: 1–6
- Independent Practice: 7–11 odd

Problems 13 and 14 require that students use the tax tables from the previous lesson. Thus, students will have to first convert the annual salary to a weekly salary. Remind students to read the problems carefully.

WRAP-UP

Write the terms displayed in the formulas of the lesson on the chalkboard, as well as the symbols −, × and =, vertically. They should not be listed in the order they appear. Ask volunteers to come to the chalkboard to write the terms and symbols in the correct order to find the annual tax withheld. (Annual Gross Pay − Personal Exemptions = Taxable Wages × Tax Rate = Annual Tax Withheld)

Assignment Guide

- Basic: 7–13, 15–22
- Average: 8–12 even, 13, 14, 15–21 odd

Use the table on page 103 for personal exemptions and find the amount withheld.

F.Y.I.
Alaska, Florida, Nevada, South Dakota, Texas, Washington, and Wyoming do not have a state income tax.

6. Linda Raabe.
Earns $15,900 annually.
Single, no dependents.
What are her personal exemptions? **$1500**

7. Harold Gibbons.
Earns $13,840 annually.
Married, no dependents.
What are his personal exemptions? **$3000**

8. Roger Hoblet.
Earns $79,500 annually.
Married, 3 dependents.
State tax rate is 4.6%.
What are his personal exemptions? **$5100**
What is withheld for state tax? **$3422.40**

9. Sarah Krouse.
Earns $17,300 annually.
Single, 1 dependent.
State tax rate is 2.5%.
What are her personal exemptions? **$2200**
What is withheld for state tax? **$377.50**

10. Henry Altman earns $24,200 annually as a traffic analyst. He is married and supports 2 children. The state tax rate in his state is 2% of taxable income. What amount is withheld yearly for state income tax? **$396**

11. Kristi Maher earns $39,940 per year. Her personal exemptions include herself and her husband. The state tax rate in her state is 4.5% of taxable income. What amount is withheld yearly for state income tax? **$1662.30**

12. Heidi Harse is a registered nurse. She earns $29,830 a year and is single. The state income tax rate is 5% of taxable income. What amount is withheld yearly for state income tax? **$1416.50**

Use the tables on pages 642–643 for federal withholdings.

13. Joseph Ryczke earns $32,000 a year as a city planner. He is paid on a weekly basis. He is married, has no dependents, and claims 2 withholding allowances for federal income tax purposes. The state tax rate is 2% of taxable income. How much is withheld annually from Joseph's gross pay for state and federal income taxes? **$4220**

14. Wendy Chou earns $26,320 a year as a nurse's aide. She is paid on a weekly basis. She is single, has 1 dependent, and claims 2 withholding allowances. The state tax rate is 2.5% of taxable income. How much is withheld each year from her gross pay for state and federal income taxes? **$3775**

MAINTAINING YOUR SKILLS Look up the skills in parentheses if you need help or more practice.

Subtract. **(Skill 6)**

15. 82.19 − 16.32 **65.87**
16. 39 − 16.2 **22.8**
17. 46.2 − 14.297 **31.903**
18. 900.12 − 612.89 **287.23**

Find the percentage. **(Skill 30)**

19. 120% of 160 **192**
20. 8% of 122 **9.76**
21. $2\frac{1}{2}$% of 500 **12.5**
22. $14\frac{3}{5}$% of 200 **29.2**

USE OF CALCULATORS

If students have a percent key on their calculators, you may want to remind them that it is not necessary to convert the percent to a decimal when using the percent key.

2-3

Graduated State Income Tax

OBJECTIVE

Compute the state tax on a graduated income basis.

Some states have a **graduated income tax.** Graduated income tax involves a different tax rate for each of several levels of income. The tax rate increases as income increases. The tax rate on low incomes is usually 1% to 3%. The tax rate on high incomes may be as much as 20%.

$$\text{Tax Withheld per Pay Period} = \frac{\text{Annual Tax Withheld}}{\text{Number of Pay Periods per Year}}$$

EXAMPLE *Skills* 5, 11 *Application* K *Term* Graduated income tax

Lois Ryan's annual salary is $24,800. She is paid semimonthly. Her personal exemptions total $1500. How much does her employer deduct from each of Lois's semimonthly paychecks for state income tax?

STATE TAX

Annual Gross Pay	Tax Rate
First $1000	1.5%
Next $2000	3.0%
Next $2000	4.5%
Over $5000	5.0%

SOLUTION

A. Find the **taxable wages.**

Annual Gross Pay − Personal Exemptions
$24,800.00 − $1500.00 = $23,300.00

B. Find the **annual tax withheld.**

(1) First $1000: 1.5% of $1000.00 .... $ 15.00
(2) Next $2000: 3.0% of $2000.00 .... $ 60.00
(3) Next $2000: 4.5% of $2000.00 .... $ 90.00
(4) Over $5000: 5.0% of ($23,300.00 − $5000.00)
5.0% of $18,300.00 = $ 915.00
Total $1080.00

C. Find the **tax withheld per pay period.**

Annual Tax Withheld ÷ Number of Pay Periods per Year
$1080.00 ÷ 24 = $45.00 tax withheld semimonthly

24800 [−] 1500 [=] 23300 1000 [×] 1.5 [%] 15 [M+] 2000 [×] 3 [%] 60 [M+] 2000 [×] 4.5 [%] 90 [M+] 23300 [−] 5000 [=] 18300 × 5 [%] 915 [M+] [RM] 1080 [÷] 24 [=] 45

✓ SELF-CHECK Complete the problems, then check your answers in the back of the book.

Find the tax withheld per pay period.

1. Annual salary: $19,400.
Personal exemptions: $1500.
24 pay periods. **$33.75**

2. Annual salary: $21,350.
Personal exemptions: $3000.
26 pay periods. **$32.02**

LESSON PLAN 2-3 Graduated State Income Tax

FOCUS

To motivate this lesson, ask students to give as many definitions of the word *graduate* as they can. Elicit the definitions "to change gradually;" "to divide into grades;" "to mark with degrees of measurement." This lesson has students compute state taxes using the graduated income tax method.

TEACH

Work carefully through each step of the Solution to the Example with students. Remind students that personal exemptions are first subtracted from gross pay to find the amount of income to be taxed.

In the second step of the Solution, point out that the $5000 subtracted from the taxable wage is the total of the amount that has already been taxed at the lower rates.

Review with students the number of pay periods in a year for weekly, monthly and semimonthly pay periods.

Warm-Up Exercises

Work through these exercises with students before having them do the Self-Check.

Write as decimals.

1. 2.2% 0.022
2. 16.5% 0.165
3. 0.9% 0.009

Add.

4. $22.25 + $54.60 + $81.54 = $158.39
5. $68.00 + $127.86 + $191.41 = $387.27

ALTERNATIVE STRATEGIES: Enrichment

Have students investigate and report on which states do not have a state income tax. Do those states have a state sales tax? Does your city have a city income tax? What is the rate?

PRACTICE AND APPLY

The following problems can be assigned for classwork and the answers checked in class to help students master the objective of the lesson.

- Guided Practice: 1–3, 5
- Independent Practice: 4, 6, 7

WRAP-UP

Ask students to explain the difference between a state tax based on the straight percent method and one that is based on a graduated income method. Select two students to explain how each is calculated.

Assignment Guide

- Basic: 4–9, 11–21
- Average: 8–10, 11–21

PROBLEMS

3. Anna Vail. **$3200**
   Annual gross pay of $6200.
   Personal exemptions of $3000.
   1.5% state tax on first $2000.
   3% tax on amount over $2000.
   What is her taxable income?

4. Clyde Browning. **$5500**
   Annual gross pay of $7500.
   Personal exemptions of $2000.
   1.5% state tax on first $2000.
   3% tax on amount over $2000.
   What is his taxable income?

5. Melissa Brossia.
   Annual gross pay of $22,000.
   Personal exemptions of $4400.
   1% state tax on first $2000.
   3% tax on next $3000.
   4.5% tax on amount over $5000.
   How much state tax is withheld? **$677**

6. Curtis Spiess.
   Annual gross pay of $12,350.
   Personal exemptions of $1000.
   1% state tax on first $2000.
   3% tax on next $3000.
   4.5% tax on amount over $5000.
   How much state tax is withheld? **$395.75**

Use the tables below to compute the state tax for problems 7–10.

PERSONAL EXEMPTIONS	
Single	$1500
Married	3000
Each dependent	700

STATE TAX	
Annual Gross Pay	**Tax Rate**
First $3500	3%
Next $3500	4.5%
Over $7000	7%

7. Milton Chandler's annual gross pay is $14,400. He is single and is paid on a monthly basis. How much is withheld monthly for state tax? **$56.29**

8. Nancy Palino's annual gross pay is $34,460. She is married with no dependents. How much is withheld from her biweekly paycheck for state income tax? **$75.95**

9. August Daily earns $24,600 a year as a parks manager. He is married and has 2 children. How much is withheld from his weekly paycheck for state income tax? **$22.82**

10. Suzanne Pollitz earns $18,235 a year as a legal secretary. She is single and is paid on a weekly basis. How much is withheld each week for state income tax? **$18.15**

MAINTAINING YOUR SKILLS Look up the skills in parentheses if you need help or more practice.

Add. **(Skill 5)**

11. 0.005 + 0.319 + 0.223 = **0.547**
12. 465.6 + 1.627 + 15.11 = **482.337**
13. $ 42.60 + 187.90 + 5.42 = **$235.92**
14. $2041.42 + 106.90 + 22.84 = **$2171.16**

Divide. Round answers to the nearest hundredth. **(Skill 11)**

15. 397 ÷ 24 **16.54**
16. 28.618 ÷ 1.3 **22.01**
17. 0.00891 ÷ 0.31 **0.03**
18. 27 ÷ 15 **1.80**

Find the number of occurrences. **(Application K)**

19. Quarterly for 3 years **12**
20. Monthly for $1\frac{1}{2}$ years **18**
21. Weekly for 4 years **208**

CRITICAL THINKING

After working the Example and the Self-Check exercises, ask students what are the tax amounts for the first $5000 of taxable wages. (The amount of tax is always the same.)

2-4

Social Security and Medicare Taxes

OBJECTIVE
Compute the amount of income withheld for social security and medicare taxes.

The Federal Insurance Contributions Act (FICA) requires employers to deduct 7.65% of your income for social security and medicare taxes. Social security (6.2%) is deducted on the first $62,700 of income, but medicare (1.45%) is paid on all your earnings. The employer must contribute an amount that equals your contribution. The federal government uses social security to pay for retirement and disability benefits and medicare to provide hospitalization insurance.

Tax Withheld = Gross Pay × Tax Rate

EXAMPLE *Skills* 30, 2 *Application* A *Terms* Social security, medicare

Carl Nichol's gross weekly pay is $232.00. His earnings to date for the year total $11,136. What amount is deducted from his pay this week for social security taxes? For medicare taxes?

SOLUTION

His earnings to date are less than $62,700.

A. Find the **social security tax withheld.**
Gross Pay × Tax Rate
$232.00 × 6.2% = $14.384 = $14.38 social security tax

B. Find the **medicare tax withheld.**
Gross Pay × Tax Rate
$232.00 × 1.45% = $3.364 = $3.36 medicare tax

✓ SELF-CHECK Complete the problems, then check your answers in the back of the book.

Find the social security and medicare taxes withheld for this pay period.

1. Monthly salary: $2400.
   Earnings to date: $26,400.
   social security $148.80, medicare $34.80
2. Weekly salary: $350.
   Earnings to date: $16,800.
   social security $21.70, medicare $5.08

PROBLEMS

	Gross Pay	Soc. Sec. Tax Rate	Medicare Tax Rate	Soc. Sec. Tax Withheld	Med. Tax Withheld
3.	$ 67.00	6.2%	1.45%	? **$4.15**	? **$0.97**
4.	$ 345.00	6.2%	1.45%	? **$21.39**	? **$5.00**
5.	$ 139.50	6.2%	? **1.45%**	? **$8.65**	? **$2.02**
6.	$1500.00	? **6.2%**	? **1.45%**	? **$93.00**	? **$21.75**
7.	$4820.00	? **6.2%**	? **1.45%**	? **$298.84**	? **$69.89**

LESSON PLAN
2-4 Social Security and Medicare Tax

FOCUS

The discussion of social security and medicare tax can be motivated by asking how students think the elderly pay for living expenses when they are no longer working. Do students know they will be contributing to their own retirement income and health insurance when they become employed? This is done through the social security and medicare taxes that are deducted from their pay.

TEACH

Stress that you and your employer pay taxes for social security and medicare. In 1996 you each paid 7.65% of your gross salary, up to a limit of $62,700. A self-employed person paid 15.3%. The limit normally rises yearly based on increases in average wages.

You pay medicare on all your earnings. This portion of the tax is 1.45% for both employers and employees and 2.9% for self-employed persons. In 1937, when social security first began, the tax rate was 1% and the maximum was $3000. In 1976 the tax rate was 5.85% and the maximum was $15,300.

Warm-Up Exercises

Work through these exercises with students before having them do the Self-Check.

Write the smaller amount.

1. $55,980 or $62,700 $55,980
2. $64,100 or $62,700 $62,700
3. 7.65% of $600 $45.90
4. 6.2% of $11,000 $682.00
5. 1.45% of $27,500 $398.75

ALTERNATIVE STRATEGIES: Reteaching

Point out that in 1989 the combined social security and medicare tax was 7.51%, while in 1986 it was 7.15%. Have them compute the combined social security and medicare tax on $24,000 for each year.

A. 1986 @ 7.15% = $1716.00
B. 1989 @ 7.51% = $1802.40
C. 1996 @ 7.65% = $1836.00

PRACTICE AND APPLY

The following problems can be assigned for classwork and the answers checked in class to help students master the objective of the lesson.

- Guided Practice: 1–7
- Independent Practice: 8, 9, 11

WRAP-UP

To ensure students understand how year-to-date earnings affect social security deductions, present the following problem. As a manager, Lisa Turner earns $6300 a month. Will she have social security deducted from her paycheck at the end of November? (No, October YTD earnings were $63,000; $63,000 is greater than $62,700.)

Assignment Guide

- Basic: 8–12, 14–25
- Average: 10, 12–15. 17–27 odd

Use the social security tax rate of 6.2% of the first $62,700 and medicare tax rate of 1.45% on all income in solving.

8. Matt Chester, firefighter. $23,560 earned this year to date. $980 gross pay this week. How much deducted this week for social security and for medicare? **$60.76; $14.21**

9. Alicia Tsongas, pilot. $9,750 gross pay this check. $63,000 earned this year to date. How much deducted this pay period for social security and for medicare? **$0, over $62,700; $141.38**

F.Y.I.
Maximum Annual Wage Subject to:

Social Security Tax	
1992	$55,500
1993	57,600
1994	60,600
1995	61,200
1996	62,700
Medicare Tax	
1992	$130,200
1993	135,000
1994	Unlimited
1995	Unlimited
1996	Unlimited

10. Cassius Mortier is paid monthly. His gross pay this month is $2174. His earnings to date for this year are $15,218. How much is deducted from his paycheck this month for social security? for medicare? **$134.79; $31.52**

11. Ann Brewer earns $32,400 a year, paid on a semimonthly. How much is deducted per pay period for social security tax? for medicare tax? **$83.70; $19.58**

12. Beth Deiger was hired for the position of receptionist. How much will be deducted from each of her biweekly paychecks for social security? for medicare? **$40.30; $9.43**

RECEPTIONIST
$16,900 salary. Meet, greet clients. Park free. Great company. Call Charles Guidry 555-1325 for appointment.

Use the tables on pages 642–643 for federal withholding taxes.

13. Halle Poole is a designer for Madison Toys. She is married, earns $472 weekly, and claims no allowances. Her gross pay to date this year is $9912. How much is deducted from her paycheck this week for federal income, social security, and medicare taxes? **$61; $29.26; $6.84**

14. Jasin Pawloski is married and claims 2 allowances. How much is withheld from his weekly paycheck of $550 for the last week of December for federal income, social security, and medicare taxes? **$61; $34.10; $7.98**

15. Nick Peralta was hired on January 2 for the supervisory position shown. He's paid monthly. How much is withheld in December for social security and medicare? **$60,500 YTD; $136.40; $79.75**

MECHANICAL ENGINEER
Supervise engineers in machine design. Good communicator. Downtown. $66,000 salary.

MAINTAINING YOUR SKILLS Look up the skills in parentheses if you need help or more practice.

Find the percentage. **(Skill 30)**

16. 6.2% of $480 **$29.76** **17.** 1.45% of 500 **$7.25** **18.** 7.65% of 80 **$6.12** **19.** $\frac{3}{4}$% of 20 **$0.15**

Round to the place value indicated. **(Skill 2)**

Nearest ten **20.** 219 **220** **21.** 322.46 **320** **22.** 3954.97 **3950** **23.** 23,981.98 **23,980**

Nearest hundred **24.** 7928 **7900** **25.** 3764 **3800** **26.** 63,782 **63,800** **27.** 147,366.20 **147,400**

COMMUNICATION SKILLS

Have students research the Federal Insurance Contributions Act and write a brief report on the history of social security, medicare, some of the current benefits it provides, the maximum wage subject to tax, the law with respect to self-employed persons, possible future of social security, or any other aspect of the system they choose. If time allows, ask a few volunteers to read their reports to the class.

2-5
Group Insurance

OBJECTIVE
Compute the deduction for group insurance.

Many businesses offer **group insurance** plans to their employees. You can purchase group insurance for a lower cost than individual insurance. Businesses often pay part of the cost of the insurance. The remaining cost is deducted from your pay.

$$\text{Deduction per Pay Period} = \frac{\text{Total Annual Amount Paid by Employee}}{\text{Number of Pay Periods per Year}}$$

EXAMPLE *Skills* 6, 11, 30 *Application* A *Term* Group insurance

Nikki Bort is a carpenter for Houck Construction Co. She has family medical coverage through the group medical plan that Houck provides for its employees. The annual cost of Nikki's family membership is $4200. The company pays 80% of the cost. How much is deducted from her weekly paycheck for medical insurance?

SOLUTION

A. Find the **percent paid by employee.**
100% − 80% = 20%

B. Find the **total amount paid by employee.**
$4200.00 × 20% = $840.00

C. Find the **deduction per pay period.**
Total Amount Paid by Employee ÷ Number of Pay Periods per Year
$840.00 ÷ 52 = $16.153 = $16.15 deducted per pay period

100 [−] 80 [=] 20 4200 [×] 20 [%] 840 [÷] 52 [=] 16.1538

✓ SELF-CHECK Complete the problems, then check your answers in the back of the book.

Find the deduction per pay period.

1. Annual cost of insurance: $2400.
   Employer pays 80%.
   12 pay periods. **$40.00**

2. Annual cost of insurance: $3272.
   Employer pays 70%.
   52 pay periods. **$18.88**

PROBLEMS

3. Rebbekah Roots, geologist.
   Annual group insurance costs $3800.
   Company pays 40% of the cost.
   How much does Rebbekah pay semimonthly? **$95**

4. Ron Stover, hydrologist.
   Annual group insurance costs $3400.
   Employer pays 75% of the cost.
   How much does Ron pay yearly? **$850**

LESSON PLAN 2-5 Group Insurance

FOCUS

To motivate this lesson, list the following items and approximate costs on the chalkboard.

Emergency Room: $100
X-Rays: $50
Surgery: $3000
Medication & Supplies: $100

Explain that these costs are what could be incurred should you break an arm or a leg. A portion of these costs will be paid for by your medical insurance if you are enrolled in a group medical plan through your company. Your cost for the plan is deducted from your paycheck.

TEACH

Point out that group insurance is an employee benefit that many companies offer. Usually, dependents also are covered under a family medical plan.

After working through the Example, remind students to use estimation to check that the answer is reasonable. If the amount an employee pays appears to be too high, the error could be caused by not subtracting the percent the employer pays. Ask: Is $800 about 20% of $4000.

Warm-Up Exercises

Work through these exercises with students before having them do the Self-Check.

Write as decimals.

1. 80% 0.80.
2. 75% 0.75
3. $78\frac{1}{2}$% 0.785
4. $1100 × 70% $770
5. $780 × 25% $195
6. $580 ÷ 12 $48.33
7. $1196 ÷ 52 $23

ALTERNATIVE STRATEGIES: Enrichment

BUSINESS NOTES–ENRICHMENT

Using the example, have the students calculate how much the employer pays for the employee's insurance ($3360). Suppose the person in the example earned $22,400 per year. What percent of the annual salary is the employer's insurance? $3360 ÷ $22,400 = .15 or 15%. How much would the employer pay if there were 1000 employees? ($3,360,000)

PRACTICE AND APPLY

The following problems can be assigned for classwork and the answers checked in class to help students master the objective of the lesson.

- Guided Practice: 1–5
- Independent Practice: 6–8

WRAP-UP

Present the Example situation to students, but change the company to Goodson Builders and the amount the company pays for the same medical plan to 40%. Ask: If Nikki left Houck Construction to work for Goodson Builders at the same salary, how much less would Nikki have in her weekly paycheck? ($32.31)

Assignment Guide

- Basic: 6–10, 12–19
- Average: 9–11, 13–19 odd

5. Janet Ingel's group medical insurance coverage costs $3580 a year. The company pays 85% of the cost. How much is deducted each month from her paycheck for medical insurance? **$44.75**

6. Bev Katz, a clerk at La Mirage Motel, earns $245.22 weekly. Her group medical insurance costs $2350 a year, of which the company pays 60% of the costs. How much is deducted weekly from her paycheck for medical insurance? **$18.08**

7. Veronica McMame earns $312.48 a week as a security guard for the city. Because she is an employee of the city, the city pays 90% of the cost of any insurance coverage. Her family medical insurance costs $3228 a year. How much is deducted each week for medical insurance? **$6.21**

8. Joel LaVita is a painter for the city. He earns an annual salary of $26,000. His annual medical insurance costs $4312 a year. The city pays 90% of the costs. How much is deducted each week from his paycheck for medical insurance? **$8.29**

9. Lee Munger is employed at McDermott International as a TV technician. Her annual salary is $28,600 and she is paid on a semimonthly basis. Her annual group medical coverage costs $3360, of which the company pays 75%. How much is deducted each pay period from Lee's paycheck for medical coverage? **$35.00**

10. Phil Akers is a surveyor for Globe Land Co. He is covered by travel insurance in addition to the basic medical coverage. Medical insurance costs $3844 a year and travel insurance costs another $347 a year. The company pays 80% of all insurance costs. How much is deducted monthly for travel insurance? How much is deducted monthly for basic medical coverage? **$5.78; $64.07**

11. Helen Poling has medical, dental, and term life insurance coverages through the company for which she works. Medical coverage costs $2444 a year, dental coverage costs $298 a year, and term life insurance costs $96.72 a year. The company pays 75% of the medical and 50% of the dental coverage. Term life insurance is entirely paid for by the company. What is the total amount deducted weekly for these coverages? **$14.62**

MAINTAINING YOUR SKILLS Look up the skills in parentheses if you need help or more practice.

Divide. Round answers to the nearest hundredth. **(Skill 11)**

12. 600.48 ÷ 24 **25.02** **13.** 16.97 ÷ 1.3 **13.05**

14. 0.00392 ÷ 0.30 **0.01** **15.** 0.692 ÷ 0.008 **86.50**

Find the percentage. **(Skill 30)**

16. 8% of 240 **19.2** **17.** 9.6% of 80 **7.68** **18.** $5\frac{1}{4}$% of 1600 **84** **19.** $8\frac{3}{10}$% of 95 **7.89**

BUSINESS PROJECT

Have students talk with parents, relatives, and employers to find out what other deductions besides taxes and medical insurance are taken from paychecks. Students should make a list and share it with the class.

2-6

Earnings Statement

OBJECTIVE
Compute the net pay per pay period.

You may have additional deductions taken from your gross pay for union dues, contributions to community funds, savings plans, and so on. The earnings statement attached to your paycheck lists all your deductions, your gross pay, and your **net pay** for the pay period.

Net Pay = Gross Pay − Total Deductions

EXAMPLE *Skills* 5, 6 *Application* A *Term* Net pay

Juan Teijeiro's gross weekly salary is $400. He is married and claims 2 allowances. The social security tax is 6.2% of the first $62,700. The medicare tax is 1.45% of gross pay. The state tax is 1.5% of gross pay. Each week he pays $10.40 for medical insurance and $2.50 for charity. Is Juan's earnings statement correct?

DEPT.	EMPLOYEE		CHECK #	WEEK ENDING		GROSS PAY	NET PAY
04	Teijeiro, J.		20566	9/17/—		400.00	312.50
TAX DEDUCTIONS				PERSONAL DEDUCTIONS			
FIT	SOC. SEC.	MEDICARE	STATE	LOCAL	MEDICAL	UNION DUES	OTHERS
38.00	24.80	5.80	6.00	—	10.40	—	2.50

SOLUTION

A. Find the **total deductions.**

(1) Federal withholding:
(from table on page 643) $38.00
(2) Social security: 6.2% of $400.00 .... 24.80
(3) Medicare: 1.45% of $400.00 .... 5.80
(4) State tax: 1.5% of $400.00 .... 6.00
(5) Medical insurance 10.40
(6) Charity 2.50
Total $87.50

B. Find the **net pay.**

Gross Pay − Total Deductions
$400.00 − $87.50 = $312.50 net pay; his statement is correct.

38 [M+] 400 [×] 6.2 [%] 24.8 [M+] 400 [×] 1.45 [%] 5.8 [M+] 400 [×] 1.5 [%] 6 [M+] 10.40 [M+] 2.50 [M+] 400 [−] [RM] 87.5 [=] 312.5

✓ SELF-CHECK Complete the problem, then check your answer in the back of the book.

1. Ron Rice is single and claims 1 allowance. His gross weekly salary is $320. Each week he pays federal, social security, and medicare taxes, $16.20 for medical insurance, and $25 for the credit union. What is his net pay? **$215.32**

LESSON PLAN
2-6 Earnings Statement

FOCUS

This lesson on computing net pay can be motivated by asking students to recall all the deductions they have studied in the previous lessons. Write the gross salary of $170 on the chalkboard, and as students name the deductions, subtract the following amounts for each from the $170: federal tax: $16; state tax: $8; social security: $10.54; medicare: $2.46; and medical insurance: $18. Point out that the remaining $115 is about two-thirds of the gross salary this person earned.

TEACH

Allow students time to familiarize themselves with the earnings statement. Direct students to where each piece of information in the Example is shown on the statement.

Encourage students to check their addition of the deductions, since errors can occur when adding lists of numbers. Have students find subtotals for taxes and for personal deductions, then add the subtotals to check the final answer.

MATHEMATICAL NOTES

The computations in this lesson include skills that were practiced in the preceding lessons of this unit; namely, reading a tax table and finding the percentage of state tax, social security, medicare, and medical insurance deductions. Use Problem 8 to check that students have mastered these skills.

Warm-Up Exercises

Work through these exercises with students before having them do the Self-Check.

Add.

1. $28.91
   17.79
   9.18
   $55.88
2. $10.31
   11.20
   20.00
   5.00
   $46.51
3. $456.78 − $56.19
   $400.59
4. $45.23 + $41.50
   $86.73

PROBLEMS

Find the deductions and the net pay. Social security is 6.2% of the first $62,700. Medicare is 1.45% of all income. Use the tax tables on pages 642–643 for federal tax. For problems 2–4, the state tax is 2% of gross pay and the local tax is 1.5% of gross pay.

2. Terry Medley is single and claims 1 allowance.

DEPT.	EMPLOYEE	CHECK #	WEEK ENDING	GROSS PAY	NET PAY
07	MEDLEY, T.	54601	8/24/—	250.50	? (181.82)

TAX DEDUCTIONS					PERSONAL DEDUCTIONS		
FIT	SOC. SEC.	MEDICARE	STATE	LOCAL	MEDICAL	UNION DUES	OTHERS
?	?	?	?	?	12.75	—	—
28.00	15.53	3.63	5.01	3.76			

3. Brad Herzig is married and claims 3 allowances.

DEPT.	EMPLOYEE	CHECK #	WEEK ENDING	GROSS PAY	NET PAY
12	HERZIG, B.	11352	5/03/—	304.20	? (224.69)

TAX DEDUCTIONS					PERSONAL DEDUCTIONS		
FIT	SOC. SEC.	MEDICARE	STATE	LOCAL	MEDICAL	UNION DUES	OTHERS
?	?	?	?	?	16.60	10.00	2.00
17.00	18.86	4.41	6.08	4.56			

4. Sue Shore is married and claims 1 allowance.

DEPT.	EMPLOYEE	CHECK #	WEEK ENDING	GROSS PAY	NET PAY
A	SHORE, S.	8002349	1/16/—	455.00	? (323.01)

TAX DEDUCTIONS					PERSONAL DEDUCTIONS		
FIT	SOC. SEC.	MEDICARE	STATE	LOCAL	MEDICAL	UNION DUES	OTHERS
?	?	?	?	?	17.25	12.00	—
52.00	28.21	6.60	9.10	6.83			

ALTERNATIVE ASSESSMENT

To assess that students understand how each deduction is found, change the data in the Example. Call on individual students to explain the deductions, and other students to explain how to compute the deduction.

COOPERATIVE LEARNING

You might want to have students work in small groups to solve Problem 5. When completed, groups should share their answers with the other groups.

5. Veronica Bell, an interior decorator for Crown Interior Design, is married and claims 1 allowance. Her state personal exemption is $35.46 a week. The state tax rate is 2.5% of taxable wages. Local tax is 1.75% of gross pay.

DEPT.	EMPLOYEE	CHECK #	WEEK ENDING	GROSS PAY	NET PAY
M	BELL, V.	347528	11/02/—	421.25	?
					303.01

TAX DEDUCTIONS					PERSONAL DEDUCTIONS		
FIT	SOC. SEC.	MEDICARE	STATE	LOCAL	MEDICAL	UNION DUES	OTHERS
?	?	?	?	?	22.00	—	—
47.00	26.12	6.11	9.64	7.37			

6. Glen Graybar is a painter and earns $24,700 a year. He is single and claims 3 allowances. Weekly deductions include FIT, social security, medicare, state tax of 3.0% and local tax of 1.75% on gross earnings, $14.50 for medical insurance, and union dues of $12 a week.

DEPT.	EMPLOYEE	CHECK #	WEEK ENDING	GROSS PAY	NET PAY
15	GRAYBAR, G.	91666	3/31/—	?	?
				475.00	340.60

TAX DEDUCTIONS					PERSONAL DEDUCTIONS		
FIT	SOC. SEC.	MEDICARE	STATE	LOCAL	MEDICAL	UNION DUES	OTHERS
?	?	?	?	?	?	?	—
49.00	29.45	6.89	14.25	8.31	14.50	12.00	

7. Donna Ovsky is employed by Hiram Associates and works in the marketing department. She is single and claims 2 allowances. Weekly deductions include FIT, social security, medicare, state tax of 2.5% on gross earnings, local tax of 1.75% on gross earnings, and medical insurance. The company pays 70% of the $3448 annual medical cost.

DEPT.	EMPLOYEE	CHECK #	WEEK ENDING	GROSS PAY	NET PAY
14	OVSKY, D.	10-142	10/31/—	460.50	?
					331.81

TAX DEDUCTIONS					PERSONAL DEDUCTIONS		
FIT	SOC. SEC.	MEDICARE	STATE	LOCAL	MEDICAL	UNION DUES	OTHERS
?	?	?	?	?	?	—	—
54.00	28.55	6.68	11.51	8.06	19.89		

PRACTICE AND APPLY

The following problems can be assigned for classwork and the answers checked in class to help students master the objective of the lesson.

- Guided Practice: 1–3
- Independent Practice: 4–6

ALTERNATIVE STRATEGIES: Reteaching

Have students find the following.

1. The FIT on $224 for a single person claiming one exemption ($24).
2. The SIT if rate is 4.5% ($10.08).
3. The social security if rate is 6.2% ($13.89)
4. The medicare if rate is 1.45% ($3.25).
5. The insurance deduction is 25% of $54 ($13.50)
6. The net pay for the above ($224 − $64.72 = $159.28).

WRAP-UP

The terms *gross pay* and *net pay* are key words students should know. Ask a student to give the definition of gross pay. Ask another student to give the definition of net pay.

Assignment Guide

- Basic: 4–8, 10–21
- Average: 7–9, 11–21 odd

8. Todd Damask is employed as a payroll supervisor and earns $9.60 per hour for a 40-hour week. He is married and claims 6 allowances. The state tax is 3.5% of gross earnings and the local tax is 0.5% of gross earnings. He pays $27.40 for medical insurance and $15 in union dues. During the week ending 4/12, he worked 40 hours.

DEPT.	EMPLOYEE	CHECK #	WEEK ENDING	GROSS PAY	NET PAY
PY	DAMASK, T.	10-942	4/12/—	? **384.00**	? **286.86**

TAX DEDUCTIONS					PERSONAL DEDUCTIONS		
FIT	SOC. SEC.	MEDICARE	STATE	LOCAL	MEDICAL	UNION DUES	OTHERS
? **10.00**	? **23.81**	? **5.57**	? **13.44**	? **1.92**	? **27.40**	? **15.00**	—

9. John Clark, a purchasing agent at the Barnett Co., earns $8.00 an hour with time and a half for working over 40 hours a week. He is married and claims 4 allowances. The state tax rate is 25% of the federal tax. The local tax is 2.5% of gross earnings. Medical insurance costs $2432 a year, of which the company pays 80% of the cost. Credit union deductions are $35 a week. What is his net pay for a week in which he worked 48 hours? **$295.68**

MAINTAINING YOUR SKILLS Look up the skills in parentheses if you need help or more practice.

Add. **(Skill 5)**

10. 16.002 + 161.320 + 342.117 = **519.439**

11. 16 + 9.3 + 42.016 = **67.316**

12. 429.9 + 76.107 + 3.05 = **509.057**

13. 62.90 + 3.52 + 186.81 = **253.23**

14. 0.005 + 0.319 + 0.223 = **0.547**

15. 465.6 + 1.627 + 15.11 = **482.337**

16. 42.60 + 187.90 + 5.42 = **235.92**

17. 2041.42 + 106.90 + 22.84 = **2171.16**

Subtract. **(Skill 6)**

18. 571.21 − 16.96 = **554.25**

19. 69.3 − 6.24 = **63.06**

20. 62.39 − 86.14 = **−23.75**

21. 88.19 − 22.81 = **65.38**

Reviewing the Basics

LESSON PLAN Reviewing the Basics

The exercises on this page review skills, applications, and terms used in the unit. You can use the exercises to assess informally students' proficiency with this material.

The page can be used for guided practice and independent practice. You can work through a selection of the exercises together with students, and thus see immediately if they know how to do them, and you can then assign some of the exercises for independent practice. Be sure to go over the answers to all assigned exercises.

Skills

Write the number that is greater.

(Skill 1)
1. $239 or $230.98 **$239**
2. $194.45 or $200.45 **$200.45**

Round to the nearest cent.

(Skill 2)
3. $154.807 **$154.81**
4. $25.209 **$25.21**
5. $40.0953 **$40.10**
6. $1.523 **$1.52**

Solve. Round answers to the nearest cent.

(Skill 5)
7. $48.72 + $54.59 **$103.31**
8. $10.41 + $11.03 + $6.88 **$28.32**

(Skill 6)
9. $84.72 − $60.71 **$24.01**
10. $8.89 − $0.62 **$8.27**
11. $143.32 − $7.89 **$135.43**

(Skill 11)
12. $892.50 ÷ 12 **$74.38**
13. $988 ÷ 52 **$19.00**
14. $1147.27 ÷ 24 **$47.80**

(Skill 30)
15. 4.5% of $105.45 **$4.75**
16. 3.5% of $404.10 **$14.14**
17. 3.2% of $148.89 **$4.76**

Applications

Use the following formula to solve problems 18–19.

Tax Withheld = Tax Rate × Gross Pay

(Application A)
18. Tax rate: 5%. Weekly wage: $320.19. Find the tax withheld. **$16.01**
19. Tax rate: 6.5%. Monthly wage: $979.67. Find the tax withheld. **$63.68**

(Application C)
20. Marge Fuller is single, earns $253.86 weekly, and claims 2 allowances. Use the table to find the weekly federal income tax withheld. **$22**

WEEKLY Payroll SINGLE Persons—

Wages		Allowances		
		Tax Withheld		
At least	But less than	0	1	2
240	250	33	27	21
250	260	35	28	22
260	270	36	30	24
270	280	38	31	25
280	290	39	33	27

Find the number of pay periods.

(Application K)
21. Semimonthly for 1 year **24**
22. Biweekly for 2 years **52**
23. Monthly for 6 years **72**

Terms

Write your own definition for each item. **Answers will vary.**

24. Group insurance
25. Income tax
26. Net pay
27. Social security
28. Graduated income tax
29. Personal exemptions
30. Medicare

Refer to your reference files at the back of the book if you need help.

LESSON PLAN
Unit Test

Students should do the Unit Test on their own. Each problem on the test is keyed to a lesson in the unit. Students having difficulty with any particular problem should review the Example in the appropriate lesson and be assigned some of the Independent Practice problems for additional practice.

Unit Test

Lesson 2-1

1. Julie Miles, a project manager, earns $445.20 a week. She is married and claims 2 allowances. How much is withheld from her weekly paycheck for federal income tax? **$44**

WEEKLY Payroll MARRIED Persons—

Wages		Allowances		
		Tax Withheld		
At least	But less than	0	1	2
430	440	55	49	43
440	450	57	50	44
450	460	58	52	46
460	470	60	53	47
470	480	61	55	49

Lesson 2-2

2. Melvia Hoskins earns $18,000 a year as a librarian. The state income tax rate is 3.6% of taxable income. Her personal exemptions total $3700. How much is withheld each week from Melvia's gross pay for state income tax? **$9.90**

Lesson 2-3

3. Waylon Lewis, a meteorologist, earns an annual salary of $37,420. He is paid biweekly. His personal exemptions total $3000. How much is deducted each pay period from his paycheck for state income tax? **$69.43**

STATE TAX

Taxable Wages	Tax Rate
First $2000	2%
Next $4000	3%
Next $4000	4.5%
Over $10,000	6%

Lesson 2-4

4. Chris Huen, an oceanographer, is paid $576.20 a week. His earnings to date this year total $21,895.60. The social security tax rate is 6.2% of the first $62,700 earned. How much is deducted from his paycheck this week for social security tax? **$35.72**

Lesson 2-5

5. Sherry Reese, a technical writer for Ace Electronics, earns $423.08 a week. Her medical insurance costs $3219 a year, of which her company pays 75% of the costs. How much is deducted each week from her paycheck for medical insurance? **$15.48**

Lesson 2-6

6. John Jacobson, a title insurance officer, is married and claims 2 allowances. He earns $432.75 a week. The social security tax rate is 6.2% of the first $62,700 earned. The medicare tax rate is 1.45% of gross. The state tax is $6.95 a week. He has weekly deductions of $21 for medical insurance and $30 for payroll savings. Use the FIT table above to find John's federal tax withheld. What is his net pay for a week? **$298.70**

Lesson 2-6

7. You are a travel agent earning $27,400 annually, single, and claim 1 allowance. The social security rate is 6.2%. The medicare tax rate is 1.45% of gross. The state tax is 1% of your gross pay and the local tax is 1.25% of gross pay. You pay $25 a week to the credit union and $10.50 a week for medical insurance. What is your net pay for a week? You will need to use the table on page 642. **$361.25**

A SPREADSHEET APPLICATION

Net Income

To complete this spreadsheet application, you will need the template diskette for *Mathematics with Business Applications.* Follow the directions in the *User's Guide* to complete this activity.

The Kreo Ice Cream Store employs high school students after school and in the summer. Kreo pays a standard hourly rate of $5.15. Deductions are taken for federal withholding (FIT), social security, medicare, and city income tax (CIT). Input the information in the following problems to determine the net income.

1. Week of: June 15

	Employee	Employee Number	Hours Worked	Income Tax Information
a.	Cole, Dean	1001	32	Single, 1 allowance
b.	Drake, Ann	1002	36	Single, 0 allowances
c.	Lusetti, Marie	1003	25	Single, 1 allowance
d.	Pappas, Mike	1004	30	Single, 0 allowances
e.	Smith, Luellen	1005	32	Single, 1 allowance
f.	Trotter, Robert	1006	38	Single, 1 allowance
g.	Young, Ann	1007	40	Single, 0 allowances

2. Week of: June 22

	Employee	Employee Number	Hours Worked	Income Tax Information
a.	Cole, Dean	1001	34	Single, 1 allowance
b.	Drake, Ann	1002	35	Single, 0 allowances
c.	Lusetti, Marie	1003	27	Single, 1 allowance
d.	Pappas, Mike	1004	30	Single, 0 allowances
e.	Smith, Luellen	1005	20	Single, 1 allowance
f.	Trotter, Robert	1006	28	Single, 1 allowance
g.	Young, Ann	1007	39	Single, 0 allowances

3. Week of: June 29

	Employee	Employee Number	Hours Worked	Income Tax Information
a.	Cole, Dean	1001	39	Single, 1 allowance
b.	Drake, Ann	1002	40	Single, 0 allowances
c.	Lusetti, Marie	1003	33	Single, 1 allowance
d.	Pappas, Mike	1004	38	Single, 0 allowances
e.	Smith, Luellen	1005	29	Single, 1 allowance
f.	Trotter, Robert	1006	27	Single, 1 allowance
g.	Young, Ann	1007	38	Single, 0 allowances

LESSON PLAN
A Spreadsheet Application

USING TECHNOLOGY

Before students work at their computers with this application, review the idea of net income. In addition to the deductions taken here, ask students to name some other common deductions; for example, medical insurance, unemployment taxes, union dues, or savings.

To enhance students' *communication skills,* encourage them to use the acronyms FIT, FICA, and CIT. Make sure they also know the meanings of the acronyms.

Discuss with students some advantages of using a computer in a business to compute net income. (accuracy of records and results, speed of computation)

LESSON PLAN
A Simulation

USING THE SIMULATION

Students can work on these simulations individually, or you may wish to use them with *cooperative learning groups.* Each student should have copies of the necessary forms. You can begin by reading with students the introductory material at the top of the page. Then discuss briefly the W-2 form.

Check students' work as they complete the forms and provide help as needed. After students have finished the simulation, provide them with completed forms so they can check their work.

A SIMULATION

If you are employed, some money is probably withheld by your employer from each of your paychecks for federal income tax. You must, by law, prepare an income tax return by April 15 of each year and send it to the Internal Revenue Service (IRS).

On an income tax return, you report your adjusted gross income, which is the total of your wages, salaries, tips, interest, and other income. Based on your adjusted gross income, you use tax tables to figure out your tax liability, the amount of income tax you must pay. Some of your tax liability is already paid by your withholdings. If your tax liability is greater than your withholdings, you must pay the IRS an amount called the "amount you owe." If your tax liability is less than your withholdings, the IRS will return the extra money to you as a tax refund.

Each year your employer must send you a Wage and Tax Statement form, called a W-2 form. This form tells how much money you earned, how much was withheld for FICA (social security tax and medicare tax), and how much was withheld for federal, state, and local income taxes. You will receive copies of your W-2 form to send with your federal, state, and local income tax returns, as well as a copy to keep for your records. If you have earned interest on a bank account, the bank will send you a form showing the amount of interest.

Suppose that last year you had a part-time job as a clerk at a sporting goods store. Your W-2 form might look like this.

1 Control number	OMB No. 1545-0008		
2 Employer's name, address, and ZIP code		6 Statutory employee / Deceased / Pension plan / Legal rep. / 942 emp. / Subtotal / Deferred compensation / Void	
		7 Allocated tips	8 Advance EIC payment
		9 Federal income tax withheld $396	10 Wages, tips, other compensation $5775
3 Employer's identification number	4 Employer's state I.D. number	11 Social security tax withheld	12 Social security wages
5 Employee's social security number		13 Social security tips	14 Nonqualified plans
19 Employee's name, address and ZIP code		15 Dependent care benefits	16 Fringe benefits incl. in Box 10
		17	18 Other
20	21	22	23

24 State income tax	25 State wages, tips, etc.	26 Name of state	27 Local income tax	28 Local wages, tips, etc.	29 Name of locality

Copy B To be filed with employee's FEDERAL tax return — Dept. of the Treasury—Internal Revenue Service

Form W-2 Wage and Tax Statement 1991

A SIMULATION

Preparing a 1040EZ Income Tax Return

You are allowed to use the 1040EZ form in preparing your income tax because (1) your income was all from wages, salaries, tips, and taxable scholarships or fellowships; (2) you did not have more than $400 in taxable interest and dividends; (3) your filing status is single; (4) you do not claim any dependents; (5) you are under 65 and not blind and; (6) your income is less $50,000. This return is shown on page 120.

Use the information in the W-2 form on page 118 and assume that you received $10.66 in interest on your savings account.

1. What amount would you write on line 1? **$5775**
2. What amount would you write on line 2? **$10.66**
3. Add the amounts on lines 1 and 2 to find your adjusted gross income. Write this sum on line 3. **$5785.66**
4. Assume you will be claimed as a dependent on another person's return. Then check "Yes" on line 4 and complete the following:

Standard deduction worksheet for dependents

A. Enter the amount from line 1. A. $5775

B. Minimum amount. B. 550

C. **Compare** the amounts on lines A and B above. Enter the larger of the two amounts here. C. $5775

D. Maximum amount. D. 3,400.00

E. **Compare** the amounts on lines C and D above. Enter the smaller of the two amounts here and on line 4. E. $3400

5. Subtract line 4 from line 3 and write this difference on line 5. This is your taxable income. **$2386.66**
6. What amount would you write on line 6? (See your W-2 form.) **$396**
7. Refer to the tax table on page 122. What amount would you write as your tax on line 7? **$358**
8. Is line 6 larger than line 7? If yes, subtract line 7 from line 6. This is your refund. **$396 – $358 = $38**
9. Is line 7 larger than line 6? If yes, subtract line 6 from line 7. This is the amount you owe. **No**

To complete the tax return, you would sign your name, write the date, and attach a copy of your W-2 form. If you owe a balance due, you would also attach a check or money order for the amount you owe. If you owe less than $1, you do not have to pay. You would make a copy of your completed return for your records. Then you would send the return to the address listed in the instruction booklet that came with the return.

A SIMULATION (CONTINUED)

Form **1040EZ**

Department of the Treasury—Internal Revenue Service
Income Tax Return for Single Filers With No Dependents (T) **1991**

OMB No. 1545-0675

Name & address

Use the IRS label (see page 10). If you don't have one, please print.

LABEL HERE

Print your name (first, initial, last)

Home address (number and street). (If you have a P.O. box, see page 11.) | Apt. no.

City, town or post office, state, and ZIP code. (If you have a foreign address, see page 11.)

Please print your numbers like this: 9876543210

Your social security number

Please see instructions on the back. Also, see the Form 1040EZ booklet.

Presidential Election Campaign (see page 11) Do you want $1 to go to this fund? *Note: Checking "Yes" will not change your tax or reduce your refund.* ▶ Yes No

Dollars Cents

Report your income

Attach Copy B of Form(s) W-2 here. Attach tax payment on top of Form(s) W-2.

1 Total wages, salaries, and tips. This should be shown in Box 10 of your W-2 form(s). (Attach your W-2 form(s).) 1

2 Taxable interest income of $400 or less. If the total is more than $400, you cannot use Form 1040EZ. 2

3 Add line 1 and line 2. This is your **adjusted gross income.** 3

Note: *You **must** check Yes or No.*

4 Can your parents (or someone else) claim you on their return?
☐ **Yes.** Do worksheet on back; enter amount from line E here.
☐ **No.** Enter 5,550.00. This is the total of your standard deduction and personal exemption. 4

5 Subtract line 4 from line 3. If line 4 is larger than line 3, enter 0. This is your **taxable income.** 5

Figure your tax

6 Enter your Federal income tax withheld from Box 9 of your W-2 form(s). 6

7 **Tax.** Use the amount on **line 5** to find your tax in the tax table on pages 16-18 of the booklet. Enter the tax from the table on this line. 7

Refund or amount you owe

8 If line 6 is larger than line 7, subtract line 7 from line 6. This is your **refund.** 8

9 If line 7 is larger than line 6, subtract line 6 from line 7. This is the **amount you owe.** Attach your payment for full amount payable to the "Internal Revenue Service." Write your name, address, social security number, daytime phone number, and "1991 Form 1040EZ" on it. 9

Sign your return

Keep a copy of this form for your records.

I have read this return. Under penalties of perjury, I declare that to the best of my knowledge and belief, the return is true, correct, and complete.

Your signature X | Date | Your occupation

For IRS Use Only — Please do not write in boxes below.

For Privacy Act and Paperwork Reduction Act Notice, see page 4 in the booklet. Cat. No. 11329W Form 1040EZ (1991)

1991 **Instructions for Form 1040EZ**

Use this form if

- Your filing status is single.
- You were under 65 and not blind at the end of 1991.
- You do not claim any dependents.
- Your taxable income (line 5) is less than $50,000.
- You had **only** wages, salaries, tips, and taxable scholarship or fellowship grants, and your taxable interest income was $400 or less. **Caution:** *If you earned tips (including allocated tips) that are not included in Box 13 and Box 14 of your W-2, you may not be able to use Form 1040EZ. See page 12 in the booklet.*
- You did not receive any advance earned income credit payments.

If you are not sure about your filing status, see page 6 in the booklet. If you have questions about dependents, see Tele-Tax (topic no. 155) on page 25 in the booklet.

If you can't use this form, see Tele-Tax (topic no. 152) on page 25 in the booklet.

Completing your return

Please print your numbers inside the boxes. Do not type your numbers. Do not use dollar signs.

Most people can fill out the form by following the instructions on the front. But you will have to use the booklet if you received a scholarship or fellowship grant or tax-exempt interest income (such as on municipal bonds). Also use the booklet if you received a 1099-INT showing income tax withheld (backup withholding) or if you had two or more employers and your total wages were more than $53,400.

Remember, you must report your wages, salaries, and tips even if you don't get a W-2 form from your employer. You must also report all your taxable interest income, including interest from savings accounts at banks, savings and loans, credit unions, etc., even if you don't get a Form 1099-INT.

If you paid someone to prepare your return, that person must also sign it and show other information. See page 15 in the booklet.

Standard deduction worksheet for dependents who checked "Yes" on line 4

Fill in this worksheet to figure the amount to enter on line 4 if someone can claim you as a dependent (even if that person chooses not to claim you).

A. Enter the amount from line 1 on front.	**A.**	
B. Minimum amount.	**B.**	550.00
C. Compare the amounts on lines A and B above. Enter the LARGER of the two amounts here.	**C.**	
D. Maximum amount.	**D.**	3,400.00
E. Compare the amounts on lines C and D above. Enter the SMALLER of the two amounts here and on line 4 on front.	**E.**	

If you checked "No" because no one can claim you as a dependent, enter 5,550.00 on line 4. This is the total of your standard deduction (3,400.00) and personal exemption (2,150.00).

Avoid common mistakes

This checklist is to help you make sure that your form is filled out correctly.

1. Are your name, address, and social security number on the label correct? If not, did you correct the label?
2. If you didn't get a label, did you enter your name, address (including ZIP code), and social security number in the spaces provided on page 1 of Form 1040EZ?
3. Did you check the "Yes" box on line 4 if your parents (or someone else) can claim you as a dependent on their 1991 return (even if they choose not to claim you)? If no one can claim you as a dependent, did you check the "No" box?
4. Did you enter an amount on line 4? If you checked the "Yes" box on line 4, did you fill out the worksheet above to figure the amount to enter? If you checked the "No" box, did you enter 5,550.00?
5. Did you check your computations (additions, subtractions, etc.) especially when figuring your taxable income, Federal income tax withheld, and your refund or amount you owe?
6. Did you use the amount from **line 5** to find your tax in the tax table? Did you enter the correct tax on line 7?
7. Did you attach your W-2 form(s) to the left margin of your return? And, did you sign and date Form 1040EZ and enter your occupation?

Mailing your return

Mail your return by **April 15, 1992**. Use the envelope that came with your booklet. If you don't have that envelope, see page 19 in the booklet for the address to use.

A SIMULATION
(CONTINUED)

Tax Tables

For line 7 of a 1040EZ return, you need to find your tax liability. To do this, you use tax tables like the one below. Tax tables are included in the instruction booklet that comes with the return. Your tax liability depends on your filing status, the number of exemptions you claim, and your taxable income (line 5). For example, suppose your taxable income is $2012, you are single, and a 1040EZ filer. Notice that your taxable income is at least $2000 but less than $2025, so your tax liability is $302.

Use the tax table to find the tax liabilities for taxpayers with the following taxable incomes. Assume that all are 1040EZ filers.

	10.	11.	12.	13.
Taxable income	$2310	$11,180	$21,390	$2380
Tax	? $347	? $1706	? $3569	? $358

1991 1040EZ Tax Table

If line 5 is at least—	But less than—	Your tax is—	If line 5 is at least—	But less than—	Your tax is—	If line 5 is at least—	But less than—	Your tax is—
2,000			**11,000**			**21,000**		
2,000	2,025	302	11,000	11,025	1,654	21,000	21,025	3,359
2,025	2,050	306	11,025	11,050	1,661	21,025	21,050	3,373
2,050	2,075	309	11,050	11,075	1,669	21,050	21,075	3,387
2,075	2,100	313	11,075	11,100	1,676	21,075	21,100	3,401
2,100	2,125	317	11,100	11,125	1,684	21,100	21,125	3,415
2,125	2,150	321	11,125	11,150	1,691	21,125	21,150	3,429
2,150	2,175	324	11,150	11,175	1,699	21,150	21,175	3,443
2,175	2,200	328	11,175	11,200	1,706	21,175	21,200	3,457
2,200	2,225	332	11,200	11,225	1,714	21,200	21,225	3,471
2,225	2,250	336	11,225	11,250	1,721	21,225	21,250	3,485
2,250	2,275	339	11,250	11,275	1,729	21,250	21,275	3,499
2,275	2,300	343	11,275	11,300	1,736	21,275	21,300	3,513
2,300	2,325	347	11,300	11,325	1,744	21,300	21,325	3,527
2,325	2,350	351	11,325	11,350	1,751	21,325	21,350	3,541
2,350	2,375	354	11,350	11,375	1,759	21,350	21,375	3,555
2,375	2,400	358	11,375	11,400	1,766	21,375	21,400	3,569
2,400	2,425	362	11,400	11,425	1,774	21,400	21,425	3,583
2,425	2,450	366	11,425	11,450	1,781	21,425	21,450	3,597
2,450	2,475	369	11,450	11,475	1,789	21,450	21,475	3,611
2,475	2,500	373	11,475	11,500	1,796	21,475	21,500	3,625

A SIMULATION (CONTINUED)

Calculating Your Tax

A few years later, you are working full-time. You are single and have no dependents. Your only sources of income are your salary and the interest on your savings account. Here are your W-2 form and your statement of earnings from your savings bank. You are not claimed as a dependent on another's return.

1 Control number		OMB No. 1545-0008	
2 Employer's name, address, and ZIP code		6 Statutory employee ☐ Deceased ☐ Pension plan ☐ Legal rep. ☐ 942 emp. ☐ Subtotal ☐ Deferred compensation ☐ Void ☐	
		7 Allocated tips	8 Advance EIC payment
		9 Federal income tax withheld $3900	10 Wages, tips, other compensation $26,800
3 Employer's identification number	4 Employer's state I.D. number	11 Social security tax withheld	12 Social security wages
5 Employee's social security number		13 Social security tips	14 Medicare wages and tips
19 Employee's name, address, and ZIP code		15 Medicare tax withheld	16 Nonqualified plans
		17 See Instrs. for Box 17	18 Other
20	21	22 Dependent care benefits	23 Benefits included in Box 10

24 State income tax	25 State wages, tips, etc.	26 Name of state	27 Local income tax	28 Local wages, tips, etc.	29 Name of locality

Copy B To Be Filed With Employee's FEDERAL Tax Return — Department of the Treasury—Internal Revenue Service

Form **W-2 Wage and Tax Statement 1991**

Use the formulas above, the return on page 120, and the tax table on page 122 to answer these questions. **$26,800; $41.25; $26841.25**

14. What is your salary? your interest income? your taxable income?

15. How much was withheld for federal income tax? What is your tax liability? **$3900; $3513**

16. Will you have a tax refund or a balance due? How much? **Refund; $387**

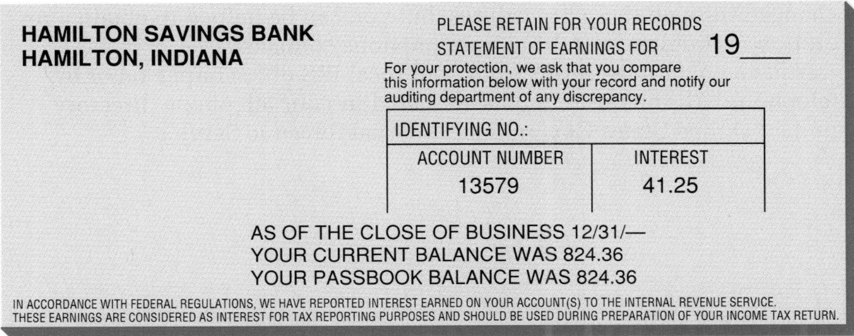

HAMILTON SAVINGS BANK
HAMILTON, INDIANA

PLEASE RETAIN FOR YOUR RECORDS
STATEMENT OF EARNINGS FOR 19___

For your protection, we ask that you compare this information below with your record and notify our auditing department of any discrepancy.

IDENTIFYING NO.:	
ACCOUNT NUMBER	INTEREST
13579	41.25

AS OF THE CLOSE OF BUSINESS 12/31/—
YOUR CURRENT BALANCE WAS 824.36
YOUR PASSBOOK BALANCE WAS 824.36

IN ACCORDANCE WITH FEDERAL REGULATIONS, WE HAVE REPORTED INTEREST EARNED ON YOUR ACCOUNT(S) TO THE INTERNAL REVENUE SERVICE. THESE EARNINGS ARE CONSIDERED AS INTEREST FOR TAX REPORTING PURPOSES AND SHOULD BE USED DURING PREPARATION OF YOUR INCOME TAX RETURN.

A SIMULATION (CONTINUED)

Other Forms and Services

There are other tax return forms besides the 1040EZ form. In some cases, you should use a 1040A or 1040 form, called the long form. For example, you must use a 1040 form if you itemize deductions. By itemizing deductions, people who had large medical expenses, certain interest payments, or certain other expenses may be able to reduce their tax liability. If you itemize deductions, you must file another form with the 1040 Form that lists the expenses you are deducting.

You must file additional forms if, for example:

- You had more than $400 in interest or dividends.
- You had business expenses that were not paid by your employer.
- You had capital gains or losses.
- You had supplemental income from rents, royalties, etc.
- You had farm income and expenses.
- You claim credit for elderly or permanently disabled people.

Check your return carefully after you complete it. The sooner you send it in, the more quickly you will receive your refund.

It is important to be aware that income tax rules change and to consult the Internal Revenue Service when you feel it is necessary. The IRS offers free tax help in preparing tax returns in most areas to older, handicapped, and non-English-speaking individuals. Some institutions and public-spirited groups often make themselves available to the community prior to April 15 to offer help and advice in preparing returns. Federal tax information is available toll free by telephone throughout the United States. The government also offers a telephone service that provides tax refund informtion. See the "Tele-Tax" listing under "IRS."

The special toll-free telephone listing "Problem Resolution" is for those taxpayers who have been unable to resolve their problems with the IRS. Of course, it is advisable to write directly to the IRS District Director with any tax problem that you cannot resolve in the ordinary manner. It is also a good idea to contact your local IRS office and ask for Problem Resolution assistance. Tax laws or technical decisions cannot be changed through these procedures, but you can be helped to obtain a resolution to problems resulting from previous contacts.

You can obtain tax forms from your local IRS office in person or by telephone. All the IRS numbers are listed in your telephone directory under "United States Government, Internal Revenue Service."

UNIT

3

Checking Accounts

A bank *checking account* allows you to pay for goods and services by *check* instead of cash. A check is a form of payment to another person. The bank deducts the sum from your account and pays it to the person named on the check. You keep a record of your *deposits* and the checks you have written in a *check register*. A monthly *bank statement* sent by the bank also records your deposits, withdrawals, and checks written. The amount of money in your check register and bank statement should agree, or *balance*. Many checking accounts earn interest.

OUTLINE

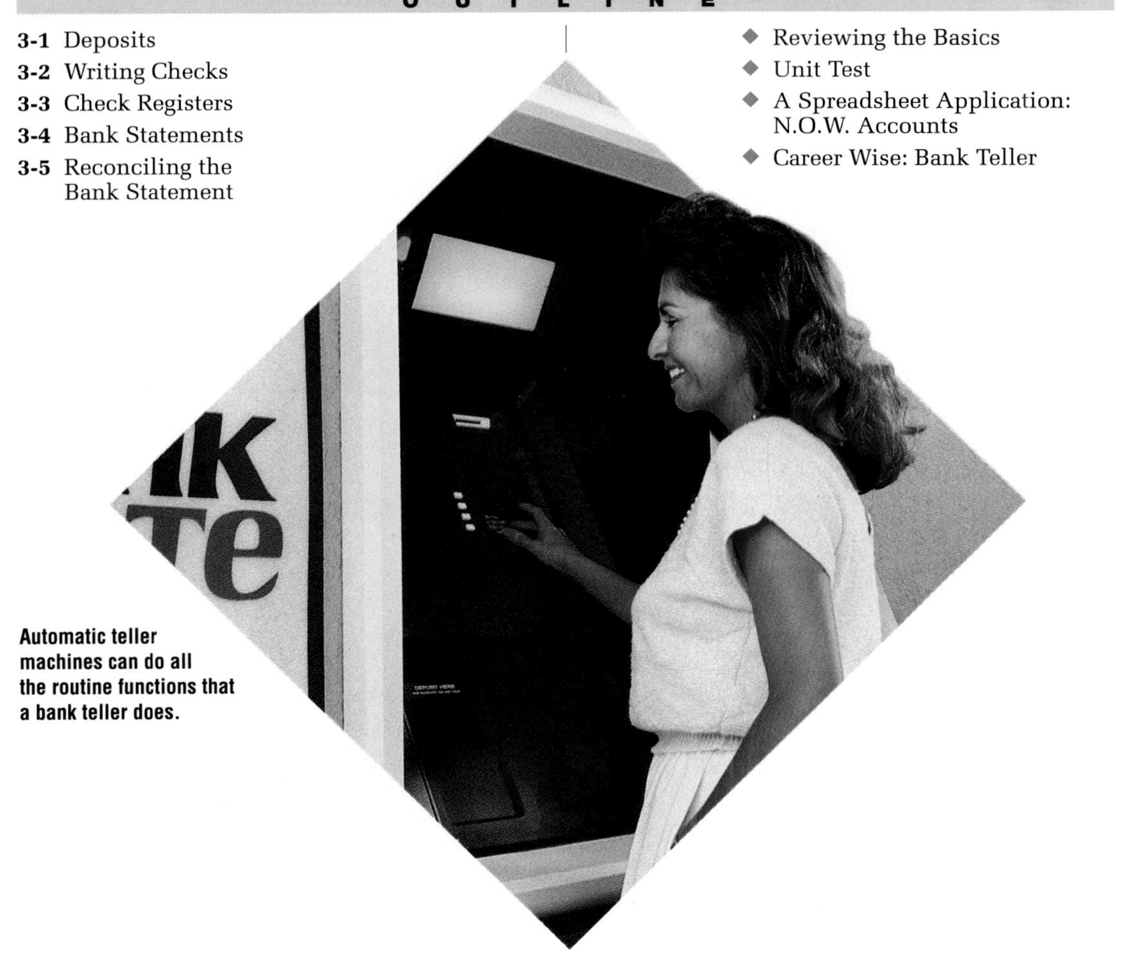

Automatic teller machines can do all the routine functions that a bank teller does.

UNIT 3
CHECKING ACCOUNTS

INTRODUCING THE UNIT

Have students discuss the advantages and disadvantages of (1) having employees send paychecks directly to banks for deposits; (2) having certain bills paid automatically each month as a part of the checking account services.

LESSON PLAN
3-1 Deposits

FOCUS

To motivate the lesson and unit, ask students if any of them have a checking account: Ask these students if they deposit their checks themselves, or if their employer arranges for the check to be sent to the student's account.

TEACH

Have students discuss the benefits of having a checking account. You may also wish to discuss services offered with a checking account. Discuss the steps in opening a checking account. Point out that many checking accounts are interest-bearing accounts.

Warm-Up Exercises

Work through these exercises with students before having them do the Self-Check.

1. $6.82 ` $18.00 ` $14.79 ` $212.50 $252.11
2. $41.92 ` $171.80 1 $75.00 $138.72
3. Find the total deposit: Currency: 7 one-dollar bills, 3 five-dollar bills, and 6 ten-dollar bills. Coins: 22 dimes, 41 quarters, and 5 half-dollars. Checks of $17.89 and $31.16. $146.00

3-1 Deposits

OBJECTIVE
Compute the total checking account deposit.

A **deposit** is an amount of money that you put into a bank account. You use a deposit slip to record the amounts of currency, coins, and checks you deposit. To open a checking account, you must make a deposit.

Total Deposit = (Currency + Coins + Checks) − Cash Received

EXAMPLE *Skills* 5, 6 *Application* A *Term* Deposit

Margaret Miller has a check for $235.42 and a check for $55.47. She would like to receive $40 in cash and deposit the rest of the money in her checking account. What is Margaret's total deposit?

SOLUTION

In the account of
Margaret C. Miller
DATE March 11 19—
FIRST CITY BANK
Check's bank number
Currency + Coins + Checks
531⑈87652 ← Margaret's account number

CHECKING ACCOUNT DEPOSIT

		DOLLARS	CENTS
CASH	CURRENCY		
	COINS		
CHECKS	LIST SEPARATELY 76-39	235	42
	117-52	55	47
	SUBTOTAL	290	89
	LESS CASH RECEIVED	40	00
	TOTAL DEPOSIT	250	89

✔ SELF-CHECK Complete the problems, then check your answers in the back of the book.

1. ($60.00 + $0.90 + $14.00) − $0.00 = total deposit of how much? $74.90
2. ($45.00 + $80.00) − $20.00 = total deposit of how much? $105.00

PROBLEMS

Find the subtotal and total deposit.

	3.	4.	5.	6.	7.	8.
Currency	$30.00	$74.00	—	—	$400.00	$975.00
Coins	$11.80	$ 9.65	—	—	—	$ 40.00
Checks	—	—	$84.50	$124.26	$734.40	$986.53
	—	—	$93.70	$ 48.79	$141.55	$ 91.11
Subtotal	$41.80	$83.65	$178.20	$173.05	$1275.95	$2092.64
Less Cash Received	—	—	$10.00	$ 20.00	$ 35.00	$ 80.00
Total Deposit	$41.80	$83.65	$168.20	$153.05	$1240.95	$2012.64

BUSINESS NOTES

The Federal Deposit Insurance Corporation (FDIC) was created in 1933. If a bank is a member of FDIC, the money that a person deposits in that bank is protected, up to $100,000, in case the bank fails and cannot return the money deposited.

Find the subtotal and total deposit.

		DOLLARS	CENTS
CASH	CURRENCY		
	COINS	5	85
CHECKS	LIST SEPARATELY 14-2	87	18
	7-43	342	41
9.	SUBTOTAL	$435	44
	LESS CASH RECEIVED	75	00
10.	TOTAL DEPOSIT	$360	44

		DOLLARS	CENTS
CASH	CURRENCY	74	00
	COINS	3	89
CHECKS	LIST SEPARATELY	121	74
	120-18	399	59
11.	SUBTOTAL	$599	22
	LESS CASH RECEIVED		
12.	TOTAL DEPOSIT	$599	22

13. Olive Baker has a paycheck for $173.45 and a refund check for $3. She would like to receive $25 in cash and deposit the remaining amount. What is her total deposit? **$151.45**

14. Jess Norton deposited a check for $474.85 and a check for $321.15. He received $50 in cash. What was his total deposit? **$746.00**

15. Bill and Mary Randall deposited their paychecks for $611.33 and $701.45 and a check from their insurance company for $75.25. They received $150 in cash. What was their total deposit? **$1238.03**

16. Morgan Meers deposited his paycheck for $201.20, a refund check from a store for $19.78, and $34.23 in cash. What was his total deposit? **$255.21**

17. Carole Winer deposits the following in her checking account: 7 five-dollar bills, 3 two-dollar bills, 18 one-dollar bills, 9 half dollars, 15 quarters, 99 dimes, 48 nickels, 16 pennies, and a check for $28.32. What is her total deposit? **$108.03**

18. Duane Coldren has a check for $343 and a check for $88.91. He would like to deposit the checks and receive 7 ten-dollar bills, 4 one-dollar bills, 9 quarters, and 15 dimes. What is his total deposit? **$354.16**

MAINTAINING YOUR SKILLS Look up the skills in parentheses if you need help or more practice.

Add. **(Skill 5)**

19. $321.00 + $9.30 + $65.75 **$396.05**

20. $400.00 + $0.95 + $374.65 **$775.60**

21. $350.00 + $8.00 + $15.93 **$373.93**

22. $694.75 + $321.57 **$1016.32**

Subtract. **(Skill 6)**

23. $540.00 − $33.00 **$507.00**

24. $920.00 − $630.00 **$290.00**

25. $734.40 − $75.00 **$659.40**

26. $619.20 − $40.00 **$579.20**

27. $718.32 − $65.00 **$653.32**

28. $316.37 − $55.00 **$261.37**

PRACTICE AND APPLY

The following problems can be assigned for classwork and the answers checked in class to help students master the objective of the lesson.

- Guided Practice: 1–8
- Independent Practice: 9, 10, 13, 15

WRAP-UP

Ask students what is meant by a deposit to a checking account. Ask students to figure the deposit if Crystal has a check for $82.50 but wants to keep $25 in cash.

Assignment Guide

- Basic: 9–17, 19–28
- Average: 11, 12, 14, 16–18, 20–28 even

ALTERNATIVE STRATEGIES: Reteaching

Using the Deposit Slip found in the TRB, walk through the following transactions with the class:

1) Deposit a check for $213.82 and currency of $75.00.
2) Deposit a check for $126.84 and a check for $79.50 and currency of $40.00 and coins of $1.45.
3) Deposit a check for $2478.20 and withdraw $200.00 in cash.

LESSON PLAN
3-2 Writing Checks

FOCUS

Ask students if any of them have ever written a check, or if they have ever watched someone write a check. Ask students the reasons the checks were written.

TEACH

When filling out the date, emphasize the importance of using the correct date. Checks cannot be post-dated nor cashed after so many months have passed.

Point out that others can fill in the line *Pay to the Order of,* but only the person whose account it is can sign the check. The signature can be compared to the one on the signature card if forgery is suspected.

Warm-Up Exercises

Work through these exercises with students before having them do the Self-Check.

Write in words:

1. $12.32 twelve and $\frac{32}{100}$
2. $105.08 one hundred five and $\frac{08}{100}$
3. $24.75 twenty-four and $\frac{75}{100}$
4. $550.63 five hundred fifty and $\frac{63}{100}$

3-2 Writing Checks

OBJECTIVE
Write a check.

After you have opened a checking account and made a deposit, you can write **checks.** A check directs a bank to deduct money from your checking account to make a payment. Your account must contain as much money as the amount of the check you are writing so that you do not **overdraw** your account.

EXAMPLE *Skill* 1 *Term* Check

Margaret Miller is buying a gift at Hud's Department Store. The cost of the gift is $45.78. Margaret is paying by check. How should Margaret write the check?

SOLUTION

A. Write the date.
B. Write the name of the person or organization to whom payment will be made.
C. Write the amount of the check as a numeral.
D. Write the amount of the check in words with cents expressed as a fraction of a dollar.
E. Make a notation on the check to indicate its purpose.
F. Sign the check.

Check's bank number

Margaret C. Miller 202
VOID
March 23, 19__ 6-39/1
PAY TO THE ORDER OF Hud's Department Store $ 45.78
Forty-five and 78/100 DOLLARS
First City Bank
MEMO Jullen's birthday Margaret C Miller
531⑈87652

✓ SELF-CHECK Complete the problems, then check your answers in the back of the book.

1. Write two hundred fifteen and $\frac{32}{100}$ dollars as a numeral. **$215.32**
2. Write $143.32 in words with cents expressed as a fraction of a dollar. **One hundred forty-three and $\frac{32}{100}$**

PROBLEMS

Write each amount in words as it would appear on a check.

3. $40.40	**4.** $703.00	**5.** $63.74	**6.** $7.94
7. $34.06	**8.** $66.00	**9.** $1917.00	**10.** $17,200.00
11. $201.09	**12.** $172.61	**13.** $5327.17	**14.** $47,983.39

3.–14. Refer to odd answers in the back of the book.

CULTURAL ANGLES

When traveling in a foreign country, a person may encounter difficulty in cashing a personal check but no difficulty in cashing a Traveler's Check. Point out that a Traveler's Check has been guaranteed by the bank to be good, while the personal check has not.

15. Alice Chino wrote check number 311 to Pugh Health Clinic for $98.72. Did she write the amount in words correctly? If not, write the amount in words correctly.

No; Ninety-eight and $\frac{72}{100}$

Alice Chino 311
8-26 19—
PAY TO THE ORDER OF Pugh Health Clinic $ 98.72
Ninety-eight and 72 DOLLARS
Bank of Canfield
MEMO July bill Alice Chino

16. For what purpose did Alice Chino write the check to the Pugh Health Clinic? **July bill**

Jim Liebert 073
3/2 19—
PAY TO THE ORDER OF Allied Auto Body $ 247.25
Two hundred forty-seven and 25/100 DOLLARS
The Farmers Bank
MEMO Door and truck repairs Jim Liebert

> **F.Y.I.**
> The amount written in words takes precedence over the amount written as a numeral.

17. Jim Liebert is paying for an auto repair bill with a check in the amount of $247.25. Did Jim write the amount of the check correctly, both as a numeral and in words? **Yes**

Amy Lepon 055
7-10 19—
PAY TO THE ORDER OF Downtown Company $ 29.78
Twenty-nine and 78/100 DOLLARS
THIRD STREET BANK
MEMO June electric bill Amy Lepon

18. What is the number of the check that Jim Liebert used to pay for his auto repair bill? **073**

19. Amy Lepon wrote check number 055 to Downtown Electric Company to pay her bill for the month of June. The bill totaled $29.78. Is the check made out in the correct amount? **Yes**

20. Is the check Amy Lepon wrote made out to the correct company? If not, write the correct company name. **No; Downtown Electric Company**

MAINTAINING YOUR SKILLS Look up the skill in parentheses if you need help or more practice.

Write the number. **(Skill 1)**

21. Thirty-five and $\frac{15}{100}$ dollars **$35.15**

22. Seventy-four and $\frac{31}{100}$ dollars **$74.31**

Write in word form with cents expressed as a fraction of a dollar. **(Skill 1)**

23. $19.25
24. $50.32
25. $435.00
26. $345.42
27. $5274.19
28. $11,871.63
29. $6110.50
30. $39,974.12

23.–30. Refer to odd answers in the back of the book.

PRACTICE AND APPLY

The following problems can be assigned for classwork and the answers checked in class to help students master the objective of the lesson.

- Guided Practice: 1, 2, 3–13 odd
- Independent Practice: 4–14 even, 15, 16

WRAP-UP

Ask students to name the six items that should be filled out on a check. As a student gives one of the items, have him or her fill in that line for the following: A check written on Jan. 16, 19__, to School House Supplies for $24.39, by Sara Milo, for purchase of a book.

Assignment Guide

- Basic: 4–14 even, 15–18, 21–29
- Average: 17–20, 21–29 odd

ALTERNATIVE STRATEGIES: Reteaching

Using the blank check found in the TRB, walk through the writing of the following check with the class:

Check number 27, dated today, payable to Corner Loan Company for $212.45 for this month's car payment.

Stress that the amount written in words takes precedence over the amount written as a numeral.

LESSON PLAN
3-3 Check Registers

FOCUS

You can motivate this lesson by asking students if they have ever seen a sign by a cash register stating that there will be a certain dollar charge if a check is returned for insufficient funds. Point out that one way to make sure a checking account has enough money to cover a check written is by means of a written record.

TEACH

Point out that the balance in a check register changes with every transaction. It is possible to both make a deposit and write a check on the same day.

Emphasize the use of two columns in the register — one for checks and one for deposits.

3-3 Check Registers

OBJECTIVE
Compute the balance in a check register.

You use a **check register** to keep a record of your automatic teller deposits, regular deposits, electronic transfers, and the checks you have written. The **balance** is the amount of money in your account. When you make a deposit, add the amount of the deposit to the balance. When you write a check, subtract the amount of the check from the balance.

New Balance = Previous Balance − Check Amount

New Balance = Previous Balance + Deposit Amount

EXAMPLE *Skills* 5, 6 *Application* A *Term* Balance

Margaret Miller's checking account had a balance of $313.54. She wrote a check for $45.78 on March 23. On March 25, she made a deposit of $240.32. What is the new balance in Margaret's account?

SOLUTION

CHECK NO.	DATE	CHECKS ISSUED TO OR DESCRIPTION OF DEPOSIT	AMOUNT OF CHECK		✓	AMOUNT OF DEPOSIT		BALANCE	
		Previous Balance − Check Amount	BALANCE BROUGHT FORWARD →					313	54
202	3/23	Hud's Dept. Store	45	78				267	76
	3/25	Deposit				240	32	508	08
		Previous Balance + Deposit Amount							

✓ SELF-CHECK Complete the problems, then check your answers in the back of the book.

1. Balance $1236.29
Amount of check − 35.28
Balance ?
$1201.01

2. Balance $992.71
Deposit + 138.84
Balance ?
$1131.55

PROBLEMS

3. **$401.43**
4. **$618.27**
5. **$366.72**
6. **$399.20**
7. **$367.33**
8. **$475.72**

Find the new balance after each check or deposit.

	AMOUNT OF CHECK		✓	AMOUNT OF DEPOSIT		BALANCE	
	BALANCE BROUGHT FORWARD →					448	35
3.	46	92				?	
4.				216	84	?	
5.	251	55				?	

	AMOUNT OF CHECK		✓	AMOUNT OF DEPOSIT		BALANCE	
	BALANCE BROUGHT FORWARD →					475	19
6.	75	99				?	
7.	31	87				?	
8.				108	39	?	

COMMUNICATIONS SKILLS

Have students contact different banks in your area to see what kinds of checking accounts are available. Students should find some that offer interest, some that require minimum balances, and some accounts offered especially to students. Have students report their findings to the class.

9. Your balance is $89.75 on May 23. **$211.86**
Deposit $156.90 on May 30.
Write a $34.79 check on April 2.
What is your new balance?

10. Your balance is $131.02 on April 4. **$53.82**
Write a $31.28 check on April 9.
Write a $45.92 check on April 14.
What is your new balance?

11. Mac Valent opened a new checking account by depositing his paycheck for $209.81. The check register shows his transactions since opening his account. What is his balance after each transaction?

	CHECK NO.	DATE	CHECKS ISSUED TO OR DESCRIPTION OF DEPOSIT	AMOUNT OF CHECK		✓	AMOUNT OF DEPOSIT		BALANCE	
							BALANCE BROUGHT FORWARD→		209	81
162.66	101	8/12	Acme Light & Power	47	15					?
109.63	102	8/18	Foodtown	53	03					?
2.33	103	8/20	Shireen Sportswear	107	30					?

12. Paula Melinte's checkbook balance was $149.21 on October 5. Her check register shows her transactions since. What is Paula's balance after each transaction?

	CHECK NO.	DATE	CHECKS ISSUED TO OR DESCRIPTION OF DEPOSIT	AMOUNT OF CHECK		✓	AMOUNT OF DEPOSIT		BALANCE	
							BALANCE BROUGHT FORWARD→		149	21
103.42	571	10/6	Pettisville Flowers	45	79					?
317.22		10/9	Deposit				213	80		?
300.28	572	10/10	Grisier's Music Inc.	16	94					?
225.03	573	10/19	Cellular Phone Co.	75	25					?

13. Ralph Snow's latest transactions are shown on the check register. Find his balance after each transaction.

	CHECK NO.	DATE	CHECKS ISSUED TO OR DESCRIPTION OF DEPOSIT	AMOUNT OF CHECK		✓	AMOUNT OF DEPOSIT		BALANCE	
							BALANCE BROUGHT FORWARD→		314	30
268.30	472	11/17	Oak Park Garden Ctr.	46	00					?
218.30	473	11/18	Cash	50	00					?
505.00		11/20	Deposit				286	70		?
378.65	474	11/21	Swanton Health Care	126	35					?
669.65		12/4	Deposit				291	00		?

Warm-Up Exercises

Work through these exercises with students before having them do the Self-Check.

1. $719.63 − $124.91
   $594.72
2. $112.61 + $75.49
   $188.10
3. Balance: $96.25
   Deposit: $71.93
   New balance? $168.18
4. Balance: $271.80
   Check: $86.95
   New balance? $184.85
5. Balance: $423.67
   Check: $45.24
   Deposit: $68.09
   Check: $13.54
   Deposit: $11.90
   New balance? $444.88

PRACTICE AND APPLY

The following problems can be assigned for classwork and the answers checked in class to help students master the objective of the lesson.

- Guided Practice: 1–8
- Independent Practice: 9, 11

ALTERNATIVE STRATEGIES: Reteaching

Using the blank checkbook register in the TRB, walk through entering these transactions in the register with the class:

Balance brought forward $350; Check #27, dated today, issued to Corner Loan Co. in the amount of $212.45; Also dated today, a deposit of $378.90 is made; Check #28, dated tomorrow, issued to Gentry Food Store in the amount of $50.

WRAP-UP

Draw a portion of a check register on the chalkboard. Have one student at a time come to the board, write the figure in the appropriate place, and find the new balance.

Balance: $800.00
Deposit: $50.34 (850.34)
Check: $324.60 (525.74)
Deposit: $85.42 (611.16)
Check: $32.10 (579.06)

Assignment Guide

- Basic: 9–14, 16–27
- Average: 10, 12–15, 16–26 even

14. Brad Adams opened a new checking account by depositing his IRS tax refund of $145.75. The check register shows his transactions to date. What is his balance after each transaction?

	CHECK NO.	DATE	CHECKS ISSUED TO OR DESCRIPTION OF DEPOSIT	AMOUNT OF CHECK		✓	AMOUNT OF DEPOSIT		BALANCE	
						BALANCE BROUGHT FORWARD→			145	75
133.75	1	4/20	Delta High School	12	00					?
102.32	2	4/25	Ames Department Store	31	43					?
102.32	3	4/26	VOID		0					?
222.32		5/1	Deposit				120	00		?
199.73	4	5/3	Wyse Office Supply	22	59					?
165.06	5	5/3	Kayling Fabric Store	34	67					?
66.56	6	5/4	Georgio's Restaurant	98	50					?
9.24	7	5/15	Crown Bookstore	57	32					?

15. Leah Ahren's latest transactions are shown on the check register. Find her balance after each transaction.

	CHECK NO.	DATE	CHECKS ISSUED TO OR DESCRIPTION OF DEPOSIT	AMOUNT OF CHECK		✓	AMOUNT OF DEPOSIT		BALANCE	
						BALANCE BROUGHT FORWARD→			397	16
313.88	916	9/20	Baggett Auto Store	83	28					?
358.98		9/22	Deposit				45	10		?
224.32	917	10/3	Rolland Buchele	134	67					?
22.32	918	10/4	First Federal S&L	201	99					?
161.72		10/5	Deposit				139	40		?
50.57	919	10/19	Vollmars Ceramics	111	15					?
20.57	920	10/20	Cash	30	00					?
−28.35; overdrawn	921	10/22	Rapids Pharmacy	48	92					?

MAINTAINING YOUR SKILLS Look up the skills in parentheses if you need help or more practice.

Add. **(Skill 5)**

16. $414.85 + $265.50 **$680.35** **17.** $845.96 + $400.00 **$1245.96**

18. $192.78 + $112.50 **$305.28** **19.** $72.85 + $393.36 **$466.21**

20. $2371.81 + $491.48 **$2863.29** **21.** $371.80 + $444.75 **$816.55**

Subtract. **(Skill 6)**

22. $579.23 − $212.60 **$366.63** **23.** $347.89 − $99.92 **$247.97**

24. $261.85 − $8.47 **$253.38** **25.** $141.82 − $64.73 **$77.09**

26. $3427.80 − $635.60 **$2792.20** **27.** $671.82 − $314.91 **$356.91**

3-4 Bank Statements

OBJECTIVE
Compute the present balance on a checking account bank statement.

When you have a checking account, you receive a **statement** and **canceled checks** from the bank each month. Canceled checks are the checks that the bank has paid by deducting money from your account. Your statement lists all your checks that the bank has paid and your deposits that the bank has recorded since your last statement. The statement may include a **service charge** for handling the account, and it may also show an interest credit if you have an interest bearing account.

Present Balance = Previous Balance + Deposits Recorded − Checks Paid − Service Charge + Interest

EXAMPLE *Skills* 5, 6 *Application* A *Term* Statement

Margaret Miller received her bank statement and canceled checks for March. She checks the statement. What is her present balance?

SOLUTION

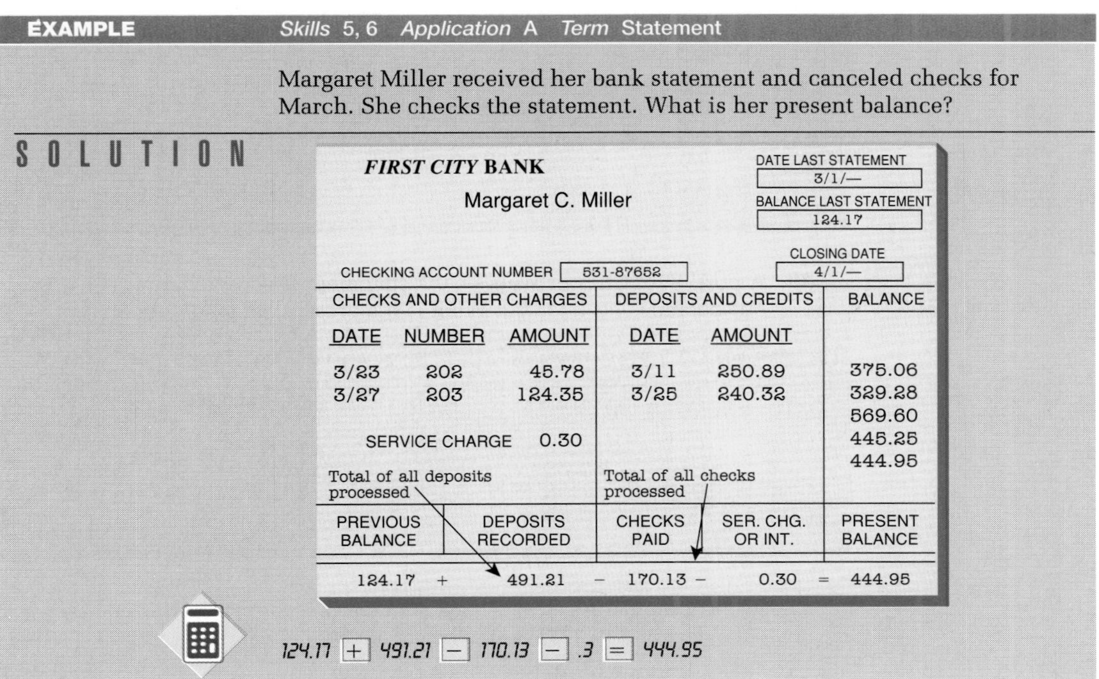

FIRST CITY BANK
Margaret C. Miller

DATE LAST STATEMENT: 3/1/—
BALANCE LAST STATEMENT: 124.17
CLOSING DATE: 4/1/—
CHECKING ACCOUNT NUMBER: 531-87652

CHECKS AND OTHER CHARGES			DEPOSITS AND CREDITS		BALANCE
DATE	NUMBER	AMOUNT	DATE	AMOUNT	
3/23	202	45.78	3/11	250.89	375.06
3/27	203	124.35	3/25	240.32	329.28
					569.60
SERVICE CHARGE		0.30			445.25
					444.95

Total of all deposits processed — Total of all checks processed

PREVIOUS BALANCE		DEPOSITS RECORDED		CHECKS PAID		SER. CHG. OR INT.		PRESENT BALANCE
124.17	+	491.21	−	170.13	−	0.30	=	444.95

124.17 [+] 491.21 [−] 170.13 [−] .3 [=] 444.95

✓ SELF-CHECK Complete the problems, then check your answers in the back of the book.

	Previous Balance	+	Deposits Recorded	−	Checks Paid	−	Service Charge	+	Interest	=	Present Balance
1.	$280.00	+	$120.00	−	$140.00	−	$2.50	+	$1.20	=	**$258.70**
2.	$275.50	+	$105.00	−	$312.60	−	$4.00	+	0	=	**$63.90**

LESSON PLAN 3-4 Bank Statements

FOCUS

Just as anyone with a checking account must keep a written record of deposits made and checks written, so must a bank. The bank periodically sends each person with an account a record of the transactions as the bank has recorded them.

TEACH

Students may not be familiar with a service charge. You may wish to mention that the service charge helps to defray the bank's cost of handling a checking account. Some banks may charge per check written; some use a flat rate; and some may charge based on a combination of the two.

Banks that charge a service fee may also offer interest on some accounts that maintain a minimum balance.

ALTERNATIVE STRATEGIES: Enrichment

Have your students investigate and report on:

1) Various types of checking accounts available in your area. These should include NOW accounts, share drafts, cash management accounts.
2) What happens when a checking account is overdrawn; that is, there are not sufficient funds (NSF) to cover the check?

Warm-Up Exercises

Work through these exercises with students before having them do the Self-Check.

1. $161.75 + $421.60 − $390.57 − $2.00 $190.78
2. $315.61 + $171.52 − $87.63 $399.50
3. Previous balance: $93.51; Deposit: $50.00; Check: $46.75; Service Charge: $1.23. Present balance? $95.53
4. Previous balance: $216.91; Checks: $116.91 and $23.56; Deposit: $267.21; Service charge: $3.55. Present balance? $340.10

PROBLEMS

	3.	4.	5.	6.	7.	8.
Previous Balance	$ 40.10	$487.67	$949.07	$500.00	$9421.99	$5513.11
Total Deposits	$200.00	$430.75	$401.00	$373.96	$7509.23	$1412.46
Total Checks	$190.10	$598.17	$319.80	$289.34	$1397.86	$1209.32
Service Charge	$ 2.34	0	$ 4.15	0	0	$ 5.60
Interest	0	$ 1.38	0	$ 2.49	0	$ 27.41
Present Balance	?	?	?	?	?	?
	$47.66	**$321.63**	**$1026.12**	**$587.11**	**$15,533.36**	**$5738.06**

9. A portion of Susan Dixon's bank statement is shown. Her previous balance was $271.31. What is her present balance? **$709.66**

CHECKS AND OTHER CHARGES			DEPOSITS AND CREDITS		BALANCE
DATE	NUMBER	AMOUNT	DATE	AMOUNT	
6/11	304	19.45	6/12	115.90	
6/15	305	21.02	6/19	115.90	
6/30	307	95.98	6/26	345.85	
SERVICE CHARGE		2.85			

10. A portion of Chester Weis's bank statement is shown below. His previous balance was $341.72. What is his present balance? **$358.94**

CHECKS AND OTHER CHARGES			DEPOSITS AND CREDITS		BALANCE
DATE	NUMBER	AMOUNT	DATE	AMOUNT	
11/2	1031	312.00	11/9	215.55	
11/5	1033	52.38	11/19	112.48	
11/20	1032	47.98	11/29	106.80	
SERVICE CHARGE		5.25			

11. A portion of Kim Hudik's bank statement is shown. Her previous balance was $39.37. What is her present balance? **$401.24**

CHECKS AND OTHER CHARGES			DEPOSITS AND CREDITS		BALANCE
DATE	NUMBER	AMOUNT	DATE	AMOUNT	
1/18	386	27.50	1/20	504.60	
1/22	389	35.00	1/31	121.37	
1/30	387	317.78	2/12	164.22	
2/5	AUTO TELLER	50.00	INTEREST	2.46	
SERVICE CHARGE		.50			

CRITICAL THINKING

One type of service charge a bank may offer is: no charge if (1) a minimum balance is kept; and (2) no more than 50 checks a year are written. When would this type of service charge be an advantage to the customer?

12. Dale Holt received his checking account statement and canceled checks for September. His previous balance was $372.48. What is his present balance?

CHECKS AND OTHER CHARGES			DEPOSITS AND CREDITS		BALANCE
DATE	NUMBER	AMOUNT	DATE	AMOUNT	
9/3	451	37.15	9/7	100.00	335.33
9/9	452	268.75	9/21	200.00	435.33
					166.58
9/21	453	41.89			366.58
9/21	454	45.00			324.69
SERVICE CHARGE		3.50			279.69
					276.19

PREVIOUS BALANCE	DEPOSITS RECORDED	CHECKS PAID	SER. CHG. OR INT.	PRESENT BALANCE
?	?	?	?	?
372.48	300.00	392.79	3.50	276.19

13. Kay Hoover received her checking account statement and canceled checks for June. Her previous balance was $581.63. What is her present balance?

CHECKS AND OTHER CHARGES			DEPOSITS AND CREDITS		BALANCE
DATE	NUMBER	AMOUNT	DATE	AMOUNT	
6/4	916	515.25	6/10	1314.50	66.38
6/12	917	185.00			1380.88
					1195.88
6/20	918	217.85			978.03
6/21	919	112.35	6/24	1015.40	865.68
			INTEREST	7.17	1881.08
					1888.25

PREVIOUS BALANCE	DEPOSITS RECORDED	CHECKS PAID	SER. CHG. OR INT.	PRESENT BALANCE
?	?	?	?	?
581.63	2329.90	1030.45	7.17	1888.25

MAINTAINING YOUR SKILLS Look up the skills in parentheses if you need help or more practice.

Add. **(Skill 5)**

14. $346.50 + $215.50 + $35.97 **$597.97**

15. $543.07 + $172.40 + $351.23 **$1066.70**

16. $917.35 + $448.55 + $627.25 **$1993.15**

17. $43,906.54 + $3172 + $4.20 **$47,082.74**

Subtract. **(Skill 6)**

18. $915.87 − 748.42 = **$167.45**

19. $684.31 − 183.49 = **$500.82**

20. $342.18 − 191.84 = **$150.34**

21. $2346.39 − 983.42 = **$1362.97**

22. 462.28 − 128.59 = **333.69**

23. 1329.84 − 483.49 = **846.35**

24. 363.790 − 44.461 = **319.329**

25. 1191.863 − 326.399 = **865.464**

PRACTICE AND APPLY

The following problems can be assigned for classwork and the answers checked in class to help students master the objective of the lesson.

- Guided Practice: 1–8
- Independent Practice: 9, 11

WRAP-UP

Ask a student to name the new terms introduced in this lesson. Ask other students to explain the meanings of the terms.

Assignment Guide

- Basic: 9–12, 14–25
- Average: 10, 12, 13, 14–22 even

BUSINESS PROJECT

Have students investigate what a bank does when someone's checking account is overdrawn. Have students choose 2 or 3 merchants and ask each what happens when a check is returned unpaid.

LESSON PLAN
3-5 Reconciling the Bank Statement

FOCUS

Ask students for a definition of *reconcile* (to agree, to make consistent). Explain that in this lesson, they will learn how to make sure that the bank's records agree with the records of a person with a checking account.

TEACH

Make sure students understand what is meant by an *outstanding deposit.*

Many people do not subtract their monthly service charge until the bank statement arrives. Interest, also, may not be known until this time. Both need to be entered into the register to complete the records.

Point out that services are usually available at a bank to help someone having difficulty in reconciling his or her statement.

3-5 Reconciling the Bank Statement

OBJECTIVE
Reconcile a check register and a bank statement.

When you receive your bank statement, you compare the canceled checks, the bank statement, and your check register to be sure they agree. You may find some **outstanding checks** and **deposits** that appear in your register but did not reach the bank in time to be processed and listed on your statement. You **reconcile** the statement to make sure that it agrees with your check register.

$$\text{Adjusted Balance} = \text{Statement Balance} - \text{Outstanding Checks} + \text{Outstanding Deposits}$$

EXAMPLE *Skills* 5, 6 *Application* A *Term* Reconcile

Margaret Miller's check register balance is $447.38. She compared her statement, canceled checks, and check register. For each check and deposit listed on her statement, she placed a check mark next to the information in her register. Margaret found these outstanding checks and deposits:

check #201 $81.32 check #204 $39.00 deposit $122.45

How does Margaret reconcile her statement?

SOLUTION

RECONCILIATION STATEMENT

CHECK REGISTER BALANCE	$ 447.38	STATEMENT BALANCE	$ 444.95
INTEREST	+ 0	OUTSTANDING CHECKS #201 $81.32, #204 $39.00	− 120.32
SERVICE CHARGE	− .30		
	$ 447.08		324.63
		OUTSTANDING DEPOSITS $122.45	+ 122.45
		ADJUSTED BALANCE	$ 447.08

Service charge must be entered in the register and subtracted from the balance. Interest must be added to the balance.

NEW BALANCE

If these balances agree, the records are correct.

PROBLEM SOLVING

If a bank statement does not reconcile, a problem is presented in why the two balances did not agree. Ask students what they can look at in order to solve the problem of making the two statements agree.

Answers should include making sure numbers were entered correctly, and checking that the addition and subtraction were computed correctly.

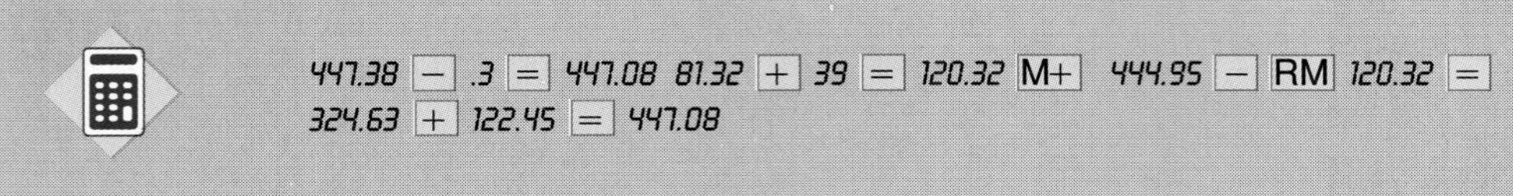

447.38 − .3 = 447.08 81.32 + 39 = 120.32 M+ 444.95 − RM 120.32 =
324.63 + 122.45 = 447.08

✓ SELF-CHECK

Complete the problems, then check your answers in the back of the book.

1. Statement balance $374.47
   Outstanding checks − 238.98
   Outstanding deposits + 140.00
   Adjusted balance ? **$275.49**

2. Statement balance $772.33
   Outstanding checks − 283.75
   Outstanding deposits + 427.75
   Adjusted balance ? **$916.33**

PROBLEMS

Complete the table.

Do the register and statement balances agree?

	3.	4.	5.	6.
Check Register Balance	$147.60	$505.85	$1439.76	$4581.62
Interest	0	0	$ 7.13	$ 22.91
Service Charge	$ 4.70	$ 9.80	0	0
NEW BALANCE	? **$142.90**	? **$496.05**	? **$1446.89**	? **$4604.53**
Statement Balance	$388.29	$507.21	$1360.44	$1328.66
Outstanding Checks	$345.39	$132.90	$ 432.81	$ 421.99
Outstanding Deposits	$100.00	$121.74	$ 519.26	$3697.86
ADJUSTED BALANCE	? **$142.90**	? **$496.05**	? **$1446.89**	? **$4604.53**

Warm-Up Exercises

Work through these exercises with students before having them do the Self-Check.

1. $241.50 − $121.60 + $175.42 $295.32
2. $34.80 − $19.32 + $172.90 + $45.81 $234.19
3. $20.00 + $25.60 − $25.26 + $3.20 $23.54
4. $78.45 + $67.31 − $13.29 − $56.32 + $181.32 $257.47
5. $1398.65 − $234.11 + $56.85 − $11.00 − $93.68 $1116.71

ALTERNATIVE STRATEGIES: Reteaching

Using the checkbook reconciliation form in the TRB, walk through the reconciliation of the following checking account:

Check register balance $148.95; Statement balance $137.34; Outstanding checks $23.86; Outstanding deposits $35.47.

PRACTICE AND APPLY

The following problems can be assigned for classwork and the answers checked in class to help students master the objective of the lesson.

- Guided Practice: 1–6
- Independent Practice: 7, 9

7. After comparing her bank statement, canceled checks, and checkbook register, Jane Scott completes the reconciliation statement shown below. What is the adjusted balance? Do the register and statement balances agree? **$316.54; Yes**

RECONCILIATION STATEMENT

CHECK REGISTER BALANCE	$ 321.04	STATEMENT BALANCE		$ 398.70
INTEREST	+ 0	OUTSTANDING CHECKS		
SERVICE CHARGE	− 4.50	#071 $87.00	−	
NEW BALANCE	$ 316.54	#080 $91.44	$	
		OUTSTANDING DEPOSITS		
		$96.28	+	
** RECONCILIATION STATEMENT FOR YOUR CONVENIENCE		ADJUSTED BALANCE	$	

8. Charlie Tanner completes the reconciliation statement shown. What are the new and adjusted balances? Do they agree? **$796.14; $796.14; Yes**

RECONCILIATION STATEMENT

CHECK REGISTER BALANCE	$ 792.51	STATEMENT BALANCE		$ 621.89
INTEREST	+ 3.63	OUTSTANDING CHECKS		
SERVICE CHARGE	− 0	#609 $123.96	−	
NEW BALANCE	$	#610 $36.87		
		#611 $159.28		
		OUTSTANDING DEPOSITS		
		$294.36	+	
		$200.00		
** RECONCILIATION STATEMENT FOR YOUR CONVENIENCE		ADJUSTED BALANCE	$	

9. Pattie Millberg received her bank statement and canceled checks for the period ending November 20. She compared the check register with the canceled checks and deposits listed on the statement, then placed a check mark next to the items processed. She reconciles the bank statement.

   a. What total amount does she have in outstanding checks? **$44.89**

   b. What total amount does she have in outstanding deposits? **$191.37**

MATHEMATICAL NOTES

The mathematics in this lesson involves adding and subtracting decimal numbers. Remind students that they need to subtract the service charge and add the interest to the check register before comparing the two balances, since these numbers are usually not known before the bank statement arrives.

c. What is her adjusted balance? **$794.83**

d. What is her new check register balance? **$794.83**

e. Do the register and adjusted balances agree? **Yes**

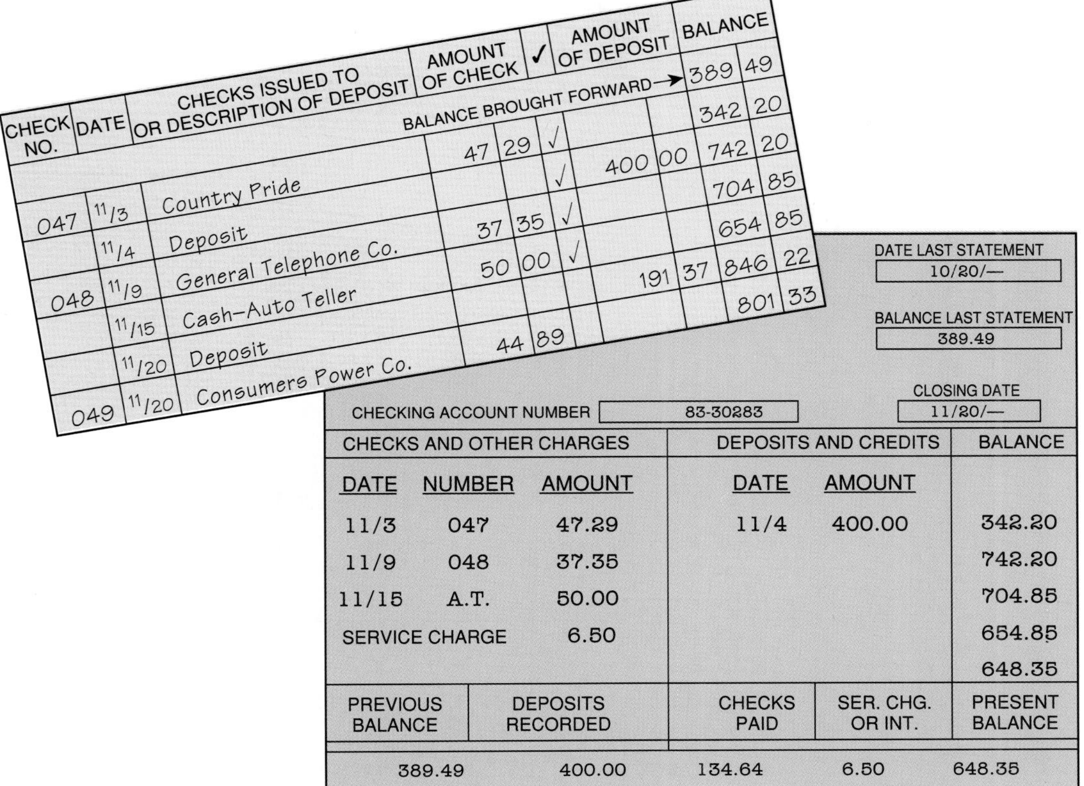

CHECK NO.	DATE	CHECKS ISSUED TO OR DESCRIPTION OF DEPOSIT	AMOUNT OF CHECK	✓	AMOUNT OF DEPOSIT	BALANCE
		BALANCE BROUGHT FORWARD →				389 49
047	11/3	Country Pride	47 29	✓		342 20
	11/4	Deposit		✓	400 00	742 20
048	11/9	General Telephone Co.	37 35	✓		704 85
	11/15	Cash–Auto Teller	50 00	✓		654 85
	11/20	Deposit			191 37	846 22
049	11/20	Consumers Power Co.	44 89			801 33

DATE LAST STATEMENT 10/20/—

BALANCE LAST STATEMENT 389.49

CLOSING DATE 11/20/—

CHECKING ACCOUNT NUMBER 83-30283

CHECKS AND OTHER CHARGES			DEPOSITS AND CREDITS		BALANCE
DATE	NUMBER	AMOUNT	DATE	AMOUNT	
11/3	047	47.29	11/4	400.00	342.20
11/9	048	37.35			742.20
11/15	A.T.	50.00			704.85
SERVICE CHARGE		6.50			654.85
					648.35

PREVIOUS BALANCE	DEPOSITS RECORDED	CHECKS PAID	SER. CHG. OR INT.	PRESENT BALANCE
389.49	400.00	134.64	6.50	648.35

CHECK REGISTER BALANCE	$ 803.33	STATEMENT BALANCE		$ 648.35
INTEREST	+ 0	OUTSTANDING CHECKS		
SERVICE CHARGE	− 6.50	#49 $44.89	−	44.89
NEW BALANCE	$ 794.83		$	
		OUTSTANDING DEPOSITS		
		$191.37	+	191.37
** RECONCILIATION STATEMENT FOR YOUR CONVENIENCE		ADJUSTED BALANCE		$ 794.83

WRAP-UP

Review the solutions to Problem 9. Ask students why it is important to check off items in the check register as they are found on the bank statement.

10. Jasper Henning received his bank statement and canceled checks for the month of May. He compared them against the items listed in his check register. He found the items shown below outstanding. Reconcile his bank statement.

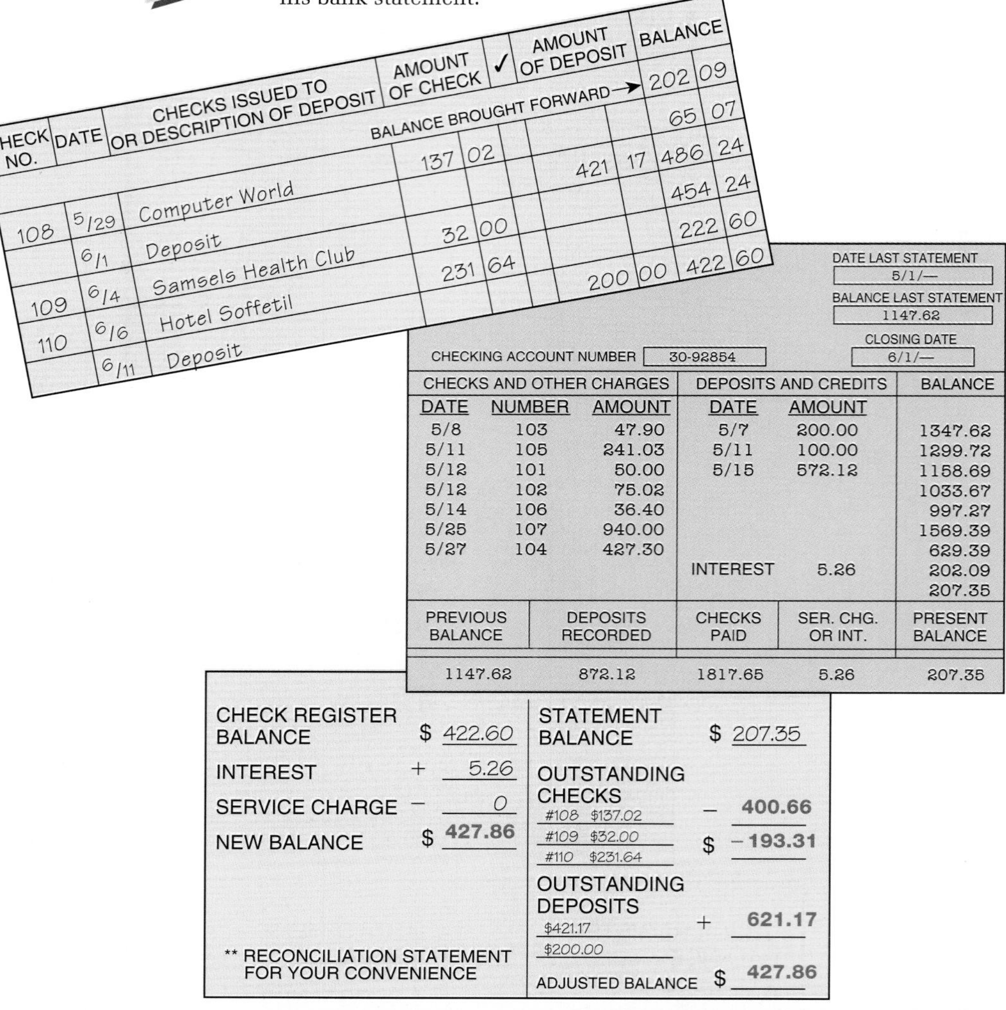

CHECK NO.	DATE	CHECKS ISSUED TO OR DESCRIPTION OF DEPOSIT	AMOUNT OF CHECK	✓	AMOUNT OF DEPOSIT	BALANCE
		BALANCE BROUGHT FORWARD →				202.09
108	5/29	Computer World	137.02			65.07
	6/1	Deposit			421.17	486.24
109	6/4	Samsels Health Club	32.00			454.24
110	6/6	Hotel Soffetil	231.64			222.60
	6/11	Deposit			200.00	422.60

DATE LAST STATEMENT 5/1/—
BALANCE LAST STATEMENT 1147.62
CLOSING DATE 6/1/—
CHECKING ACCOUNT NUMBER 30-92854

CHECKS AND OTHER CHARGES			DEPOSITS AND CREDITS		BALANCE
DATE	NUMBER	AMOUNT	DATE	AMOUNT	
5/8	103	47.90	5/7	200.00	1347.62
5/11	105	241.03	5/11	100.00	1299.72
5/12	101	50.00	5/15	572.12	1158.69
5/12	102	75.02			1033.67
5/14	106	36.40			997.27
5/25	107	940.00			1569.39
5/27	104	427.30			629.39
			INTEREST	5.26	202.09
					207.35

PREVIOUS BALANCE	DEPOSITS RECORDED	CHECKS PAID	SER. CHG. OR INT.	PRESENT BALANCE
1147.62	872.12	1817.65	5.26	207.35

CHECK REGISTER BALANCE	$ 422.60	STATEMENT BALANCE	$ 207.35
INTEREST	+ 5.26	OUTSTANDING CHECKS #108 $137.02, #109 $32.00, #110 $231.64	− 400.66
SERVICE CHARGE	− 0		$ −193.31
NEW BALANCE	$ 427.86	OUTSTANDING DEPOSITS $421.17, $200.00	+ 621.17
** RECONCILIATION STATEMENT FOR YOUR CONVENIENCE		ADJUSTED BALANCE	$ 427.86

MAINTAINING YOUR SKILLS Look up the skill in parentheses if you need help or more practice.

Subtract. **(Skill 6)**

11. \$412.30 − \$1.25 **\$411.05**
12. \$219.63 − \$2.50 **\$217.13**
13. \$96.78 − \$3.00 **\$93.78**
14. \$349.82 − \$116.78 **\$233.04**
15. \$491.60 − \$247.83 **\$243.77**
16. \$127.80 − \$64.92 **\$62.88**
17. \$421.92 − \$250.00 **\$171.92**
18. \$216.91 − \$150.00 **\$66.91**
19. \$97.83 − \$51.47 **\$46.36**

Reviewing the Basics

Skills

Write the amount in words with cents expressed as a fraction of a dollar.

(Skill 1)

1. $25.79 **2.** $299.46

3. $1372.35 **4.** $14,551.00

1. Twenty-five and $\frac{79}{100}$
2. Two hundred ninety-nine and $\frac{46}{100}$
3. One thousand three hundred seventy-two and $\frac{35}{100}$
4. Fourteen thousand five hundred fifty-one and $\frac{00}{100}$

Solve.

(Skill 5)

5. $81.27 + $99.64 + $81.27 **$262.18**

6. $1408.34 + $57.84 + $2.57 **$1468.75**

7. $473.93 + $147.85 **$621.78**

8. $8402.03 + $998.57 **$9400.60**

(Skill 6)

9. $416.37 − $57.02 **$359.35**

10. $783.19 − $531.90 **$251.29**

11. $527.05 − $315.37 **$211.68**

12. $172.21 − $132.41 **$39.80**

Applications

Use the following formula to solve problems 13–14.

Total Deposit = (Currency + Coins + Checks) − Cash Received

(Application A)

13. Deposit slip for Jill Dohr.
Check for $701.32.
Coins totaling $15.97.
Currency totaling $108.
What is the total deposit? **$825.29**

14. Deposit slip for Bill Shockey.
Check for $1321.17.
Currency totaling $3217.
Received a $50 bill.
What is the total deposit? **$4488.17**

Terms

Match each term with its definition on the right.

c **15.** balance

a **16.** reconcile

b **17.** deposit

f **18.** statement

d **19.** check

a. to obtain agreement between two financial records by accounting for outstanding items

b. an amount of money that you put into an account

c. the amount of money in your account

d. a written order directing the bank to deduct money from your checking account to make a payment

e. items that did not reach the bank in time to be processed and listed in your statement

f. a record prepared by the bank listing all transactions the bank has recorded

Refer to your reference files at the back of the book if you need help.

LESSON PLAN Reviewing the Basics

The exercises on this page review skills, applications, and terms used in the unit. You can use the exercises to assess informally students' proficiency with this material.

The page can be used for guided practice and independent practice. You can work through a selection of the exercises together with students, and thus see immediately if they know how to do them, and you can then assign some of the exercises for independent practice. Be sure to go over the answers to all assigned exercises.

LESSON PLAN
Unit Test

Students should do the Unit Test on their own. Each problem on the test is keyed to a lesson in the unit. Students having difficulty with any particular problem should review the Example in the appropriate lesson and be assigned some of the Independent Practice problems for additional practice.

Unit Test

Lesson 3-1

1. A portion of Cheryl Dodge's deposit slip is shown. What is her total deposit? **$539.81**

		DOLLARS	CENTS
CASH	CURRENCY		
	COINS	79	25
CHECKS (LIST SEPARATELY)	9-14	143	85
	18-4	396	71
SUBTOTAL			
LESS CASH RECEIVED		80	00
TOTAL DEPOSIT			

Lesson 3-2

2. One thousand three hundred ninety-four and $\frac{45}{100}$

2. Herman Stone wrote a check to Acton Electronics in the amount of $1394.45 for the purchase of a microcomputer. The check number was 363. Write the amount of the check in words with cents expressed as a fraction of a dollar.

Lesson 3-3

3. A portion of Ralph Ray's check register is shown listing his transactions since June 5. What is his new balance? **$1117.28**

CHECK NO.	DATE	CHECKS ISSUED TO OR DESCRIPTION OF DEPOSIT	AMOUNT OF CHECK		✓	AMOUNT OF DEPOSIT		BALANCE	
		BALANCE BROUGHT FORWARD →						827	31
	6/5	Deposit				418	90		
082	6/9	Miller's Hardware	83	97					
083	6/18	Corner Drug	44	96					

Lesson 3-4

4. A portion of Marie Dunn's bank statement is shown. Her balance from last month's statement was $491.34. What is her present balance? **$850.79**

CHECKS AND OTHER CHARGES			DEPOSITS AND CREDITS		BALANCE
DATE	NUMBER	AMOUNT	DATE	AMOUNT	
10/4	902	42.95	10/8	335.35	
10/6	904	31.21	10/22	391.61	
10/15	903	285.00			
SERVICE CHARGE		8.35			

Lesson 3-5

5. Marie Dunn's statement balance is $952.64. She finds that she has check numbers 901 and 905 outstanding in the amounts of $47.27 and $59.23. She also has an outstanding deposit of $200. She reconciles the statement. What should her adjusted balance be? **$1046.14**

A SPREADSHEET APPLICATION

N.O.W. Accounts

To complete this spreadsheet application, you will need the template diskette for *Mathematics with Businss Applications.* Follow the directions in the User's Guide to complete this activity.

Input the transactions for a N.O.W. account in the following problems to find the interest earned and the new balance. Assume the statement period begins on the first day and ends on the last day of the month.

1. Here is a portion of Alan Cook's bank statement. The previous balance is $827.80.

CHECKS AND OTHER CHARGES			DEPOSITS AND CREDITS		BALANCE
DATE	NUMBER	AMOUNT	DATE	AMOUNT	
6/9	301	24.50	6/10	150.00	
6/11	302	145.67	6/15	45.00	
6/15	303	9.86			

2. Here is a portion of Kelly Beck's bank statement. The previous balance is $248.68.

CHECKS AND OTHER CHARGES			DEPOSITS AND CREDITS		BALANCE
DATE	NUMBER	AMOUNT	DATE	AMOUNT	
7/12	101	54.50	7/15	250.00	
7/17	103	284.90	7/29	65.00	
7/24	105	38.78			

3. Here is a portion of Diana Roger's bank statement. The previous balance is $949.99.

CHECKS AND OTHER CHARGES			DEPOSITS AND CREDITS		BALANCE
DATE	NUMBER	AMOUNT	DATE	AMOUNT	
11/5	819	983.78	11/1	450.00	
11/23	820	421.99	11/19	1095.00	
11/28	823	18.23	11/28	120.50	

4. Here is a portion of R. J. Perez's bank statement. The previous balance is $1124.53.

CHECKS AND OTHER CHARGES			DEPOSITS AND CREDITS		BALANCE
DATE	NUMBER	AMOUNT	DATE	AMOUNT	
3/1	235	763.91	3/5	321.98	
3/6	238	21.80			
3/15	239	108.77			

5. Here is a portion of Fong Lo's bank statement. The previous balance is $596.89.

CHECKS AND OTHER CHARGES			DEPOSITS AND CREDITS		BALANCE
DATE	NUMBER	AMOUNT	DATE	AMOUNT	
2/7	522	224.50	2/10	355.16	
2/11	524	391.16	2/20	445.00	
2/22	521	31.68	2/25	40.00	
2/27	527	15.19			

LESSON PLAN
A Spreadsheet Application

USING TECHNOLOGY

In order to use this application, you should first discuss the concept of *interest* earned on money deposited in an account. Interest is money earned on the funds in the account. A N.O.W. account is a savings account against which a *withdrawal order,* which is just like a check, can be written. N.O.W. stands for *negotiable orders of withdrawal.* These checking account substitutes have been created by banks and savings and loan associations because the law, at one time, prevented them from paying interest on checking accounts. Today, banks do have interest-bearing checking accounts.

LESSON PLAN
Career Wise

APPLICATIONS ON THE JOB

You might wish to discuss what traveler's checks are and why people use them when traveling in foreign countries. Also, ask students what they think Colleen does when she *reconciles* the day's transactions at closing time. Point out that the type of graph used on this page is called a *pictograph*.

Ask students if they know the names of any other foreign currencies. Also, as a *multicultural* activity, you might wish to bring a map of the world to class and locate the countries in the graph on the map. Then ask students to research one cultural or business contribution each country has made to the world and report their findings to the class.

CAREER WISE

Bank Teller

Colleen Miller is a bank teller at a local savings bank. At the bank, she handles customers' deposits and withdrawals. Some people bring their checks for their mortgage payments to her for processing. Many people come to her to purchase money orders and traveler's checks. Using the bank's computer, Colleen can easily provide customers with accurate statements about their current balances.

Colleen is comfortable with numbers, and so she finds it easy to reconcile the day's transactions at closing time.

Hired for her dependability, knowledge of banking, and courteous attitude, she hopes to become a head teller some day.

Many travelers purchase traveler's checks and convert some cash into foreign currency before going on a trip. The graphic below shows how nine different foreign currencies compared recently to one American dollar.

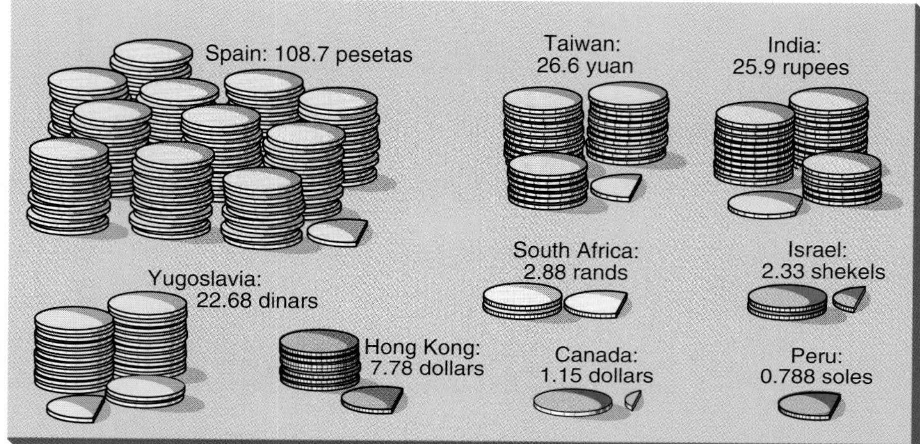

Check Your Understanding

1. Complete: 108.7 pesetas has the same value as ______ shekels. **2.33**
2. Of the monetary units above, which is the smallest relative to the American dollar? **Peseta. It takes about 109 pesetas to equal the American dollar.**
3. Which monetary unit above is the closest in value to the American dollar? **Canadian dollar**
4. How many dinars are equivalent to \$150 in American money? **$150(22.68) = 3402$**
5. How much of an American dollar is 1 rupee? $\frac{1}{25.9} \approx$ **0.038, about 4 cents**

UNIT

4

Savings Accounts

A *savings account* is used for savings only. You cannot write checks on this account. Instead of a check register, you keep track of your deposits, withdrawals, and *interest* with a *passbook* or an *account statement.* Each time you deposit or withdraw money, your bank records the transaction in your passbook. The bank also records the interest earned on the money deposited in your savings account. Interest is money earned on the funds, or *principal,* in your account. The more often the interest is computed, or *compounded,* the more money your account will earn.

OUTLINE

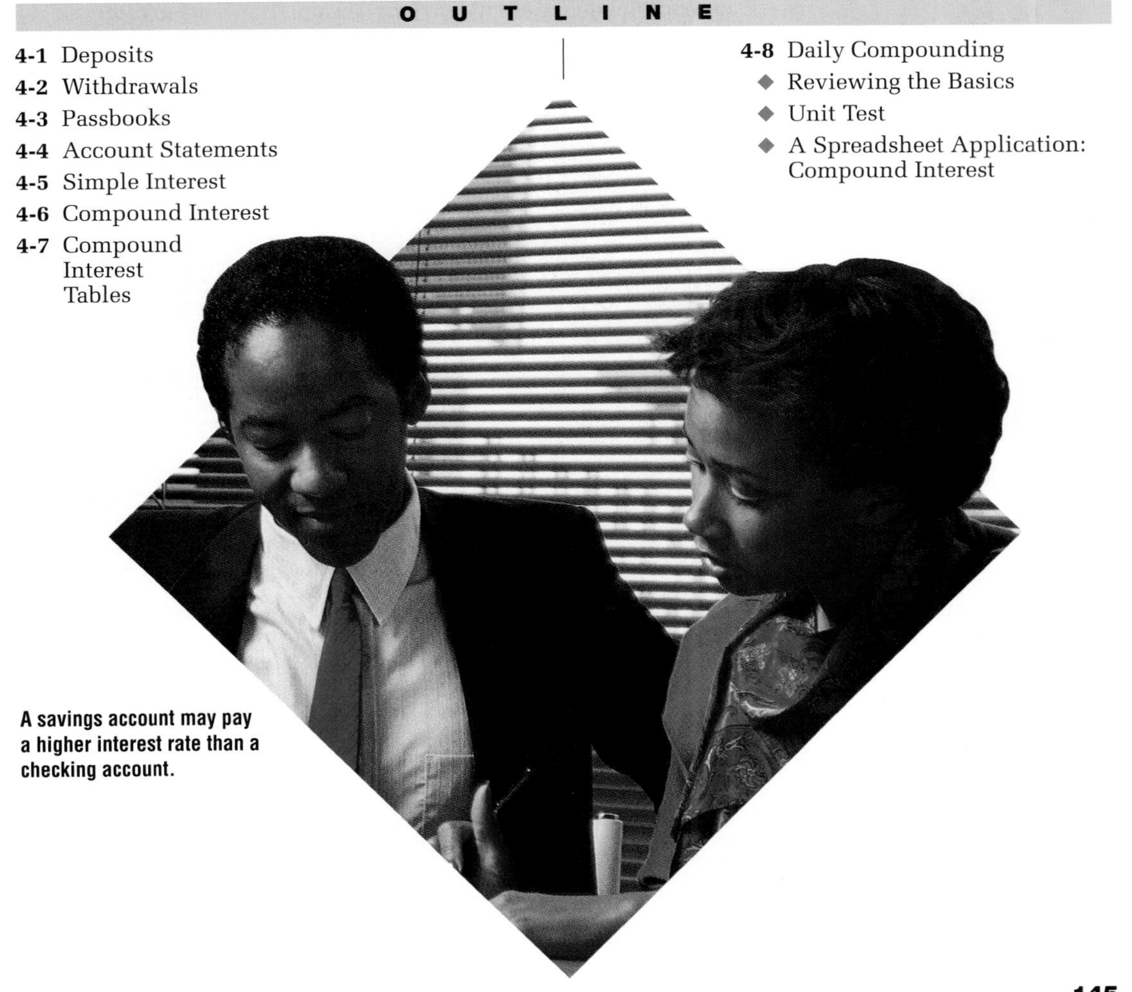

A savings account may pay a higher interest rate than a checking account.

UNIT 4
SAVINGS
ACCOUNTS

INTRODUCING THE UNIT

Introduce the unit by asking students to discuss the different reasons they might want to establish a savings account. Why do people keep money in a savings account instead of a checking account?

LESSON PLAN
4-1 Deposits

FOCUS
To motivate the lesson and unit, ask students if any of them have a savings account. For those who do, ask them if they have a plan for saving—say 10% of any money received, $5 every time money is received, and so on. This lesson deals with how to make a deposit to a savings account.

TEACH
Many banks require a minimum initial deposit in order to open a savings account. After that, any amount may be deposited. There is usually a minimum balance that is required in order to earn interest.

Discuss how to fill out a deposit slip. Mention that if cash is to be received at the time of making the deposit, most banks require the depositor to sign the deposit slip.

Warm-Up Exercises
1. $97.40 ` $22.35
   $119.75
2. $71.50 ` $142.93
   $214.43
3. $356.90 ` $146.20
   1 $50 $453.10
4. $161.81 ` $39.47
   1 $40 $161.28

4-1 Deposits

OBJECTIVE
Complete a savings account deposit slip and compute the total deposit.

To open a savings account, you must make a **deposit**. Each time you make a deposit, it is added to the balance of your account. You fill out on savings account deposit slips the cash and checks that you deposit. If you want to receive cash, you subtract the amount from the subtotal to find the total deposit amount.

Total Deposit = (Currency + Coins + Checks) − Cash Received

EXAMPLE *Skills* 5, 6 *Application* A *Term* Deposit

Robert Cassidy has 28 one-dollar bills in currency, $7.27 in coins, and a check for $29.34 to deposit in his savings account. He wants to receive a twenty-dollar bill in cash. How much will he deposit?

SOLUTION

BRIDGETOWN BANK & TRUST
DATE *November 12, 19—*
Currency + Coins + Checks
ACCOUNT NUMBER
0 1 1 3 0 1 4
NAME *Robert Cassidy*
ADDRESS *18 Laurel Lane*
Bridgetown, CT 05120

		DOLLARS	CENTS
CASH	CURRENCY	28	00
	COINS	7	27
CHECKS	LIST SEPARATELY 14-6	29	34
	SUBTOTAL	64	61
	LESS CASH RECEIVED	20	00
	TOTAL DEPOSIT	44	61

SAVINGS ACCOUNT DEPOSIT

28 + 7.27 + 29.34 = 64.61 − 20 = 44.61

✓ SELF-CHECK Complete the problems, then check your answers in the back of the book.

Find the total deposit.

1. Deposited currency of $37. Deposited checks for $40 and $86. **$163**
2. Deposited checks for $32 and $94. Received $25 in cash. **$101**

PROBLEMS

Find the subtotal and total deposit.

	3.	4.	5.	6.	7.
Deposits	$44.00 8.35 26.80	$76.00 9.27 44.38	$52.96 39.75	$180.81 115.35	$ 64.89 39.57 928.12
Subtotal	79.15	129.65	92.71	296.16	1032.58
Less Cash Received	0	0	$30.00	$150.00	$ 20.00
Total Deposit	79.15	129.65	62.71	146.16	1012.58

COOPERATIVE LEARNING
Have groups of students work together on Problem 16. Have one student find the value of the twenty-dollar bills, one find the value of the one-dollar bills, and one find the value of the quarters. Have the last student find the total cash received and the deposit.

Then have students make up similar types of problems and exchange them with other groups to solve.

8. Sandi Spencer.
Deposits $74 in cash.
Deposits $3.95 in coins.
What is the total deposit?
$77.95

9. Ernest McMahon.
Deposits $73.23 in currency.
Deposits a check for $124.17.
What is the total deposit?
$197.40

10. Kenneth Hall.
Deposits a check for $335.28.
Deposits a check for $29.50.
Deposits $90 in cash.
What is the total deposit
$454.78

11. Susan Allen.
Deposits a check for $823.40.
Deposits a check for $61.88.
Deposits $50 in cash.
What is the total deposit?
$935.28

12. Hazel Bruot fills out the savings deposit form shown. What is her total deposit? **$226.83**

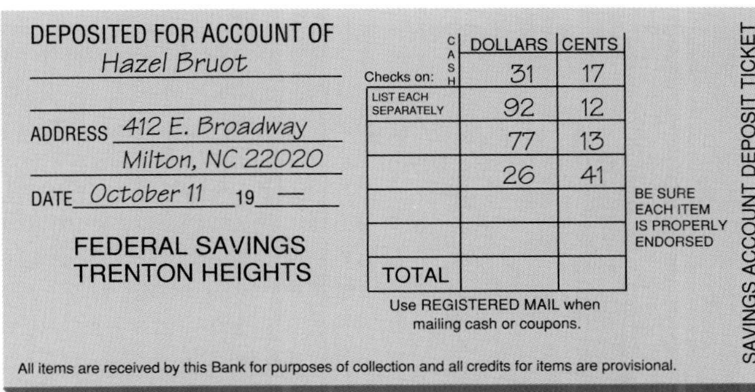

DEPOSITED FOR ACCOUNT OF *Hazel Bruot*

ADDRESS *412 E. Broadway*
Milton, NC 22020

DATE *October 11* 19 —

FEDERAL SAVINGS
TRENTON HEIGHTS

CASH	DOLLARS	CENTS
Checks on:	31	17
LIST EACH SEPARATELY	92	12
	77	13
	26	41
TOTAL		

BE SURE EACH ITEM IS PROPERLY ENDORSED

Use REGISTERED MAIL when mailing cash or coupons.

All items are received by this Bank for purposes of collection and all credits for items are provisional.

SAVINGS ACCOUNT DEPOSIT TICKET

13. Joe Gryster deposited a check for $475.77 and another check for $94.26 in his savings account. He received $70 in cash. What was his total deposit? **$500.03**

14. Norman Frances completed a savings account deposit slip on which he recorded checks for $327.19 and $52.88 for deposit. He received $38 in cash. What was his total deposit? **$342.07**

15. Barbara Hagedorn deposited 4 twenty-dollar bills, 9 ten-dollar bills, 35 quarters, 8 dimes, 97 pennies, and a check for $75.96 in her savings account. What was her total deposit? **$256.48**

16. David Rodero had 2 checks for $478.80 and $54.29. He would like to deposit the 2 checks and receive 4 twenty-dollar bills, 1 one-dollar bill, and 25 quarters. What would be his total deposit? **$445.84**

17. Loretta Miller would like to deposit 2 checks for $136.40 and $889.11. She would like to receive $75 in cash. What would be her total deposit? **$950.51**

MAINTAINING YOUR SKILLS Look up the skills in parentheses if you need help or more practice.

Add. **(Skill 5)**

18. $31.50 + 42.45 = **$73.95**

19. $40.46 + 18.32 = **$58.78**

20. $173.79 + 45.93 = **$219.72**

21. $551.16 + 146.81 = **$697.97**

22. $89.70 + 4.32 + 26.92 = **$120.94**

Subtract. **(Skill 6)**

23. $98.93 − 20.00 = **$78.93**

24. $692.57 − 35.40 = **$657.17**

25. $103.33 − 60.00 = **$43.33**

26. $687.28 − 75.00 = **$612.28**

27. $68.25 − 17.65 = **$50.60**

PRACTICE AND APPLY

The following problems can be assigned for classwork and the answers checked in class to help students master the objective of the lesson.

- Guided Practice: 1–7
- Independent Practice: 8, 11, 15

WRAP-UP

Read Problem 15 to the class. Have volunteers explain how they would find the total value of the bills, the total value of the coins, and the total deposit.

Assignment Guide

- Basic: 8–16, 18–27
- Average: 9, 10, 12–14, 17, 18–26 even

ALTERNATIVE STRATEGIES: Reteaching

Using the Deposit Slip for savings accounts in the TRB, complete the slip for the following deposits:

1) Deposit a check for $68.50 and currency of $80.00. $148.50
2) Deposit a check for $310.75 and a check for $82.50 and currency of $50.00 and coins of $4.75. $448.00
3) Deposit a check for $1327.68 and withdraw $150.00 in cash. $1177.68

LESSON PLAN
4-2 Withdrawals

FOCUS
The focus of this lesson is on completing a savings account withdrawal slip. Point out that this form must be used when no money is being deposited, only being taken out of a savings account.

TEACH
Point out the similarities between filling out the withdrawal slip and writing a check. If there is a discrepancy between the numeral amount and the amount in words of the withdrawal, the amount in words takes precedence.

Warm-Up Exercises
Write in words.

1. $47.83
   forty-seven and $\frac{83}{100}$ dollars
2. $261.87
   two hundred sixty-one and $\frac{87}{100}$ dollars
3. $1391.46
   one thousand, three hundred ninety-one and $\frac{46}{100}$ dollars

Write as numerals.

4. fifty-six and $\frac{71}{100}$ dollars
   $56.71
5. five hundred seventeen and $\frac{00}{100}$ dollars
   $517.00
6. four thousand, six hundred ten and $\frac{07}{100}$ dollars
   $4610.07

4-2 Withdrawals

OBJECTIVE
Complete a savings account withdrawal slip.

A **withdrawal** is a sum of money that you take out of your savings account. Your withdrawal is subtracted from the balance of your account. You fill out on withdrawal slips the money you are taking from your account.

EXAMPLE *Skills* 1 *Term* Withdrawal

Robert Cassidy would like to withdraw $45 from his savings account. His account number is 0113014. How should he fill out the withdrawal slip?

SOLUTION

A. Write the date of withdrawal.
B. Write the savings account number.
C. Write the amount withdrawn in words with cents expressed as a fraction of a dollar.
D. Write the amount withdrawn as a numeral.
E. Sign the withdrawal slip.

BRIDGETOWN BANK & TRUST
A. DATE *November 14* 19– –
B. 0 1 1 3 0 1 4 SAVINGS ACCOUNT NUMBER
C. *Forty-five and 00/100* DOLLARS
D. $ 45 00 dollars cents
E. *Robert Cassidy* SIGN HERE
WITHDRAWAL

✓ SELF-CHECK Complete the problems, then check your answers in the back of the book.

1. Write $45.76 in words with cents expressed as a fraction of a dollar.
   Forty-five and $\frac{76}{100}$
2. Write two hundred ninety-one and $\frac{42}{100}$ dollars as a numeral.
   $291.42

PROBLEMS

Write each amount in words as it would appear on a withdrawal slip.

3. $17.35
4. $21.25
5. $44.93
6. $68.74
7. $406.00
8. $137.51
9. $7852.03
10. $9182.14

3.–10. Refer to selected answers in the back of the book.

BUSINESS NOTES
Money that is withdrawn can be in the form of cash or a certified check. Any money that is to be mailed should not be cash. A certified check can be used for this purpose, and does not require a person to have a checking account.

For each of the following exercises, write (a) the account number, (b) the amount as a numeral, and (c) the amount in words.

11.–16. Refer to selected answers in the back of the book.

11.

FARMER'S MERCANTILE | SAVINGS ACCOUNT NUMBER 1 7 5 9 4 1 7 9

DATE 5/19/— | $ 831 | 95

PAY TO MYSELF

OR TO Home Finance Co.

DOLLARS

AND CHARGE TO THE ABOVE NUMBERED ACCOUNT

SIGN HERE Calvin Gordon

WITHDRAWAL

12.

FARMER'S MERCANTILE | SAVINGS ACCOUNT NUMBER 1 6 0 1 0 3 6 8

DATE Jan. 12, 19— | $ 374 | 28

PAY TO MYSELF

OR TO Odessa French

DOLLARS

AND CHARGE TO THE ABOVE NUMBERED ACCOUNT

SIGN HERE Odessa French

WITHDRAWAL

13. Avis Bogart's savings account number is 81-0-174927. He fills out a savings withdrawal slip for $318.29 to purchase a gift.

14. Mona Hornack has been saving for a trip abroad. Her travel agent has arranged a trip to Europe that will cost $2460. Mona withdraws the amount from her savings account. Her account number is 13-122-541.

15. Mike Reed has been saving to buy a commemorative stamp for his stamp collection. He fills out the savings withdrawal slip shown.

NAME Mike Reed DATE 11/8/19—

ADDRESS 9 Bellvue Drive

Mission Hill, KA 58050

0 6 0 2 9 1 7 5 $ 76.60

SAVINGS ACCOUNT NUMBER AMOUNT

RECEIVED THE SUM OF

CITIZEN SAVINGS AND TRUST

Mike Reed

SIGNATURE

⑆041206261⑆ SAVINGS WITHDRAWAL

16. Clyde Murphy has finally saved enough money to buy a video camera. The total purchase price is $947.31. He fills out a withdrawal slip for the amount. His savings account number is 7086-5.

MAINTAINING YOUR SKILLS Look up the skills in parentheses if you need help or more practice.

Write in words with cents expressed as a fraction of a dollar. **(Skill 1)**

17.–21. Refer to selected answers in the back of the book.

17. $94.78 **18.** $219.34 **19.** $162.05 **20.** $15.71 **21.** $4211.15

Write as a numeral. **(Skill 1)**

22. Thirty-nine and $\frac{41}{100}$ dollars **$39.41** **23.** Two hundred fifty-one and $\frac{27}{100}$ dollars **$251.27**

24. Six thousand three hundred forty and $\frac{22}{100}$ dollars. **$6340.22**

25. Twenty-five thousand six hundred ninety-six and $\frac{29}{100}$ dollars **$25,696.29**

PRACTICE AND APPLY

The following problems can be assigned for classwork and the answers checked in class to help students master the objective of the lesson:

- Guided Practice: 1, 2, 3–9 odd
- Independent Practice: 4–10 even, 11, 13

WRAP-UP

Ask students to explain when they need a withdrawal slip. Then have them tell what information is required on the withdrawal slip.

Assignment Guide

- Basic: 4–15, 17–25
- Average: 12, 14–16, 17–25 odd

ALTERNATIVE STRATEGIES: Reteaching

Using the blank Savings Account Withdrawal Slip in the TRB, slowly go through with your class the filling out of the slip for the following withdrawals:

1) Today's date; Savings account number 1234567; the amount of $425.00.
2) Today's date; Savings account number 7396815; the amount of three hundred fifty-two dollars and thirty-five cents.

LESSON PLAN
4-3 Passbooks

FOCUS
Money cannot be withdrawn if there is not enough money in a savings account. A written record is kept in a savings account passbook. The focus of this lesson is to compute the new balance in a savings account passbook after a transaction has been made.

TEACH
Go over the formula for computing the new balance from the old balance. Point out that interest is only entered in the passbook on a periodic basis, such as quarterly. The interest is given in this lesson; however, students will learn to compute the interest at a later time.

Some banks and savings and loan associations use savings registers rather than savings account passbooks. This register is similar to a checkbook register.

Warm-Up Exercises
1. $691.75 − $124.90 + $215.16 + $12.35
   $794.36
2. $41.50 + $116.71 + 90.73 + $8.42
   $257.36
3. Find the new balance if the previous balance is $314.53, deposits are $17.50 and $276.93, a withdrawal of $50.00 is made, and interest of $12.38 is earned.
   $571.34

4-3 Passbooks

OBJECTIVE
Compute the new balance in a savings account passbook.

Your bank may provide you with a savings account **passbook.** When you make a deposit or withdrawal, a bank teller records in your passbook the transaction, any interest earned, and the new balance.

New Balance = Previous Balance + Interest + Deposits − Withdrawals

EXAMPLE *Skills* 5, 6 *Application* A *Term* Passbook

Ann Hardy has a passbook savings account. On April 1, she deposited $173.12 in her account. The bank teller recorded the transaction and the interest earned during the past savings period. What is the new balance in her account?

DEPOSITOR: Ann Hardy
ACCOUNT NO.: 08-021235-4

	DATE	WITH-DRAWAL	DEPOSIT	INTEREST	BALANCE
18	01-01			9.75	793.90
19	01-12		250.00		1043.90
20	02-11		35.50		1079.40
21	03-03	30.00			1049.40
22	04-01			13.72	?
23	04-01		173.12		?
24					

SOLUTION Find the **new balance.**

Previous Balance + Interest + Deposits − Withdrawals
$1049.40 + $13.72 + $173.12 − 0 = $1236.24 new balance

1049.4 [+] 13.72 [+] 173.12 [−] 0 [=] 1236.24

✓ SELF-CHECK Complete the problems, then check your answers in the back of the book.

Find the new balance.

1. Previous balance, $900; interest, $11; deposit, 0; withdrawal, $60. **$851.00**
2. Previous balance, $3074; interest, $42.30; deposit, $75; withdrawal, 0. **$3191.30**

PROBLEMS

	3.	4.	5.	6.	7.
Previous Balance	$481.32	$737.02	$81.31	$3019.20	$478.97
Interest	$ 6.80	$ 20.83	$ 1.77	$ 170.71	$ 27.08
Deposit	$ 80.00	—	$40.00	—	$110.00
Withdrawal	—	$145.00	$10.00	$ 921.00	—
New Balance	**$568.12**	**$612.85**	**$113.08**	**$2268.91**	**$616.05**

ALTERNATIVE ASSESSMENT
Have one student act as a bank teller. Have another student act as a customer. Using play money, coins, and fake checks, have the customer make a deposit. Have the bank teller verify the passbook figures. Students then switch roles.

8. On November 20, Consuela Mendez withdrew $430 from her savings account. What is the new balance in her account?
$2810.26

DEPOSITOR: Consuela Mendez				
DATE	WITHDRAWAL	DEPOSIT	INTEREST	BALANCE
10/2				3151.70
11/15			88.56	3240.26
11/20	430.00			?

9. On January 16, Keith Fair deposited $98.25 in his savings account. The teller also recorded $18.60 in interest to his account. What is the new balance in his account?
$1357.10

DEPOSITOR: Keith Fair				
DATE	WITHDRAWAL	DEPOSIT	INTEREST	BALANCE
1/01				1000.00
1/05		240.25		1240.25
1/16		98.25	18.60	?

10. Find the balance for each date on this savings account passbook.

	PASSBOOK ACCOUNT NUMBER: 07-17615				
	DATE	WITHDRAWAL	DEPOSIT	INTEREST	BALANCE
	2/06				847.51
1088.51	3/15		241.00		?
1096.49	4/01			7.98	?
996.49	5/15	100.00			?
1297.92	6/01		301.43		?
1308.04	7/01			10.12	?
1522.80	8/20		214.76		?
1422.80	9/15	100.00			?
1438.24	10/01			15.44	?
1313.24	10/05	125.00			?
1213.24	10/12	100.00			?
1340.69	10/20		127.45		?
1265.69	11/1	75.00			?

MAINTAINING YOUR SKILLS Look up the skills in parentheses if you need help or more practice.

Add. **(Skill 5)**

11. $350.00 + $88.00
$438.00

12. $325.17 + $11.05
$336.22

13. $858.60 + $314.58
$1173.18

14. $921.73 + $9.15 + $370.00
$1300.88

Subtract. **(Skill 6)**

15. $250 − $75
$175

16. $66.97 − $12.50
$54.47

17. $517.54 − $120.00
$397.54

18. $815.35 − $140
$675.35

19. $6114.98 − $39.73
$6075.25

20. $84,317.21 − $5829.37
$78,487.84

PRACTICE AND APPLY

The following problems can be assigned for classwork and the answers checked in class to help students master the objective of the lesson.

- Guided Practice: 1–7
- Independent Practice: 8

WRAP-UP

Ask students what is meant by the old balance and the new balance. Ask one student for an old balance, another student for a transaction (type and amount), and a third student for the new balance.

Assignment Guide

- Basic: 8, 9, 11–20
- Average: 9, 10, 11–19 odd

ALTERNATIVE STRATEGIES: Enrichment

Some individuals have labeled this society as the "cashless society" because so many financial transactions are taken care of through electronic transfer of funds. Many financial institutions are no longer using passbooks in their passbook savings accounts. Have your students find out how the financial institutions in your area notify their savings account customers of the status of their accounts.

LESSON PLAN
4-4 Account Statements

FOCUS

Ask students if they know anyone who has automatic transfer of funds—that is, money from savings can automatically be transferred to checking if needed to cover a check. This is one of many reasons why the reconciliation of a savings passbook or register and the bank's statement is so important.

TEACH

Point out that reconciliation is extremely important for anyone who uses a savings register, as they are usually expected to update these registers themselves.

If any transactions were made after the statement was sent, they should be ignored in the reconciliation. The passbook or register balance should be compared to the balance of the date on the statement. Persons who use registers may have to add the amount of interest earned to their register at this time.

4-4 Account Statements

OBJECTIVE

Compute the new balance on a savings account statement.

When you have a savings account, your bank may mail you a monthly or quarterly account **statement**. The account statement shows all deposits, withdrawals, and interest credited to your account.

New Balance = Previous Balance + Interest + Deposits − Withdrawals

EXAMPLE *Skills* 5, 6 *Application* A *Term* Statement

Mary Witkew received this savings account statement. After checking her passbook and transactions to be sure all items have been recorded correctly, she checks the calculations. What is the balance in her account on July 1?

DEPOSITOR: Mary Witkew

DATE	WITHDRAWAL	DEPOSIT	INTEREST	BALANCE
4-15		250.00		524.50
5-11		125.00		649.50
6-10	100.00			549.50
7-01			6.77	556.27
7-01		80.00		

PREVIOUS STATEMENT		THIS STATEMENT	
DATE	BALANCE	DATE	BALANCE
04-01	$274.50	07-01	?

SOLUTION

Find the **new balance.**

Previous Balance + Interest + Deposits − Withdrawals

$274.50 + $6.77 + $455.00 − $100.00 = $636.27 new balance

274.5 [+] 250 [=] 524.5 [+] 125 [=] 649.5 [−] 100 [=] 549.5 [+] 6.77 [=] 556.27 [+] 80 [=] 636.27

✓ SELF-CHECK Complete the problem, then check your answers in the back of the book.

1. Previous balance, $700; interest, $9.50; deposits of $100 and $250; withdrawals of $80 and $110. What is the new balance? **$869.50**

PROBLEMS

	Previous Balance	+	Interest	+	Deposits	−	Withdrawals	=	New Balance
2.	$ 400.00	+	$19.00	+	$ 50.00	−	$150.00	=	**$319.00**
3.	$ 485.00	+	$19.50	+	$125.00	−	$200.00	=	**$429.50**
4.	$ 674.00	+	$12.19	+	$160.00	−	$190.00	=	**$656.19**
5.	$7381.19	+	$96.41	+	$231.43	−	$180.00	=	**$7529.03**

6. Judi Imhoff. **$1033.83**
Previous balance: $717.52.
Interest: $4.36.
Deposits: $125 and $276.95.
Withdrawal: $90.
What is her new balance?

7. Scott Roan. **$1947.25**
Previous balance: $2161.41.
Interest: $20.04.
Deposits: $345 and $575.80.
Withdrawals: $210 and $945.
What is his new balance?

8. Ruth Schuler. **$62,091.53**
Previous balance: $74,561.49.
Interest: $1017.98.
Deposits: $918.37 and $944.56.
Withdrawals: $959.40 and $14,391.47.
What is her new balance?

9. Louis Robinson. **$3832.44**
Previous balance: $6817.86.
Interest: $60.41.
Deposits: $439.35 and $865.82.
Withdrawals: $235 and $4116.
What is his new balance?

10. Harry Karras checks the calculations on his savings account statement. What is the balance in his account on September 30? **$334.94**

DEPOSITOR: Harry Karras			ACC. NO. 020-79-98	
DATE	WITHDRAWAL	DEPOSIT	INTEREST	BALANCE
7/09	314.00			421.34
8/30		217.50		638.84
9/29	310.00			328.84
9/30			6.10	

PREVIOUS STATEMENT		THIS STATEMENT	
DATE	BALANCE	DATE	BALANCE
06/30	$735.34	9/30	?

11. Ed Bogin received his savings account statement. What is the balance in his account on July 15? **$579.86**

NAME: Edward Bogin		ACCOUNT NUMBER: 12-36-5000			
DATE	WITHDRAWAL	DEPOSIT	INTEREST	BALANCE	
01-28		45.00		548.27	
02-03		80.40		628.67	
02-15			2.85	631.52	
03-15			2.86	634.38	
04-10	400.00			234.38	
04-15			2.37	?	**236.75**
05-01		335.60		?	**572.35**
05-15			2.28	?	**574.63**
06-15			2.61	?	**577.24**
07-15			2.62	?	**579.86**

PREVIOUS STATEMENT		THIS STATEMENT		
DATE	BALANCE	DATE	BALANCE	
12-15	$503.27	07-15	?	**$579.86**

Warm-Up Exercises

1. $420 + $15 + $25 − $75 $385
2. $210.78 + $8.47 + $75.80 + $146.90 − $120 $321.95
3. $976.75 + $9.77 + ($134.90 + $14.76 + $475.83) − $419.20 $1192.81
4. Find the new balance if the previous balance was $417, interest earned was $16.68, deposits were $117.80 and $235.93, and a withdrawal of $217.25 was made. $570.16

PRACTICE AND APPLY

The following problems can be assigned for classwork and the answers checked in class to help students master the objective of the lesson.

- Guided Practice: 1–5
- Independent Practice: 6–10 even

ALTERNATIVE STRATEGIES: Reteaching

Review the finding of the new balance for a savings account. If we think of withdrawals as checks being written these transactions are very similar to those of a checking account. Find the balance after each date below:

1/1/—Prev. bal.	$375.80	
1/6/—Withdrawal	$168.50	$207.30
1/17/—Deposit	$ 75.00	$283.30
1/25/—Interest	$ 3.45	$286.75

WRAP-UP

Write the terms *new balance, previous balance, interest, deposits,* and *withdrawals* on the chalkboard in any order. Ask students which one they would start with to check an account statement and why, the next term they would use and how, and so on.

Assignment Guide

- Basic: 6–12, 14–24
- Average: 7–11 odd, 12, 13, 14–24 even

12. Sylvia Road received her savings account statement. Find the balance for each date on her statement.

1037.10
857.10
1077.10
1027.10
1030.55

$1030.55

DEPOSITOR: Sylvia Road			ACC. NO. 10-257-190	
DATE	WITHDRAWAL	DEPOSIT	INTEREST	BALANCE
7/01		21.43		953.00
7/07		123.88		1076.88
7/10	300.00			776.88
7/15		220.00	5.07	1001.95
7/22		35.15		?
8/01	180.00			?
8/07		220.00		?
8/10	50.00			?
8/15			3.45	?

PREVIOUS STATEMENT		THIS STATEMENT	
DATE	BALANCE	DATE	BALANCE
6/30	$931.57	8/15	?

13. Leona Biniakiewicz received her savings account statement. Find the balance for each date on her statement.

	DEPOSITOR: Leona Biniakiewicz			ACC. NO. 16-86-1453	
	DATE	WITHDRAWAL	DEPOSIT	INTEREST	BALANCE
	9/30		500.00	118.08	18,519.75
	10/5		100.00		18,619.75
18,219.75	10/15	400.00			?
18,339.93	10/30			120.18	?
17,939.93	11/15	400.00			?
18,559.39	11/30		500.00	119.46	?
18,159.39	12/15	400.00			?
18.059.39	12/20	100.00			?
18,179.24	12/30			119.85	?

	PREVIOUS STATEMENT		THIS STATEMENT	
	DATE	BALANCE	DATE	BALANCE
$18,179.24	9/30	$17,901.67	12/31	?

MAINTAINING YOUR SKILLS Look up the skills in parentheses if you need help or more practice.

Add. **(Skill 5)**

14. $450.00 + $9.50 + $40.00 **$499.50**
15. $385 + $7.52 + $875 **$1267.52**
16. $793.60 + $2.38 + $5.00 **$800.98**
17. $426.30 + $278.41 + $342.91 **$1047.62**
18. $7812.39 + $37.90 + $527.84 + $477.91 **$8856.04**

Subtract. **(Skill 6)**

19. $7942.70 − $3453.80 **$4488.90**
20. $16,865.95 − $14,991.39 **$1874.56**
21. $338.49 − $299.39 **$39.10**
22. $41,215.24 − $11,645.91 **$29,569.33**
23. $3647.98 − $1647.08 **$2000.90**
24. $10,451.47 − $9.99 **$10,441.48**

PROBLEM SOLVING

See Problem 12. If Sylvia accidentally recorded the amount of $220.00 as a deposit on 8/7, when it should have been a withdrawal, what would be her correct balance? ($590.55)

4-5 Simple Interest

OBJECTIVE
Compute the simple interest.

When you deposit money in a savings account, you are permitting the bank to use the money. The amount you earn for permitting the bank to use your money is called **interest**. The **principal** is the amount of money earning interest. The **annual interest rate** is the percent of the principal that you earn as interest based on one year. **Simple interest** is interest paid on the original principal.

Interest = Principal × Rate × Time

EXAMPLE *Skills* 28, 14, 30 *Application* J *Term* Simple interest

Joyce Tyler deposited $900 in a savings account at Hamler State Bank. The account pays an annual interest rate of $5\frac{1}{2}$%. She made no other deposits or withdrawals. After 3 months, the interest was calculated.

A. How much simple interest did her money earn?
B. Suppose the account paid 6%. How much would she have earned?

SOLUTION

A. Find the **interest** at $5\frac{1}{2}$%.
Principal × Rate × Time
$900.00 × $5\frac{1}{2}$% × $\frac{3}{12}$
$900.00 × 0.055 × 0.25 =
$12.375 = $12.38 interest

B. Find the **interest** at 6%.
Principal × Rate × Time
$900.00 × 6% × $\frac{3}{12}$
$900.00 × 0.06 × 0.25 =
$13.50 interest

900 [×] 5.5 [%] 49.5 [×] 3 [÷] 12 [=] 12.375 or 900 [×] .055 [×] .25 [=] 12.375

✓ SELF-CHECK Complete the problems, then check your answers in the back of the book.

1. Principal: $400.
Annual interest rate: $6\frac{1}{2}$%.
What is the interest after 3 months? **$6.50**

2. Principal: $1500.
Annual interest rate: $7\frac{1}{2}$%.
What is the interest after 6 months? **$56.25**

PROBLEMS

	3.	4.	5.	6.	7.
Principal	$720.00	$960.00	$327.00	$4842.00	$3945.37
Rate	× 0.06	× 0.06	× 0.12	× 0.10	× 0.065
	? $43.20	? $57.60	? $39.24	? $484.20	? $256.45
Time (Months)	× $\frac{3}{12}$	× $\frac{9}{12}$	× $\frac{6}{12}$	× $\frac{12}{12}$	× $\frac{3}{12}$
Interest	? $10.80	? $43.20	? $19.62	? $484.20	? $64.11

LESSON PLAN 4-5 Simple Interest

FOCUS
You can motivate this lesson by asking students if any of them have accounts that earn interest. Point out that students will learn how to compute simple interest in this lesson.

TEACH
Go over the terms *annual interest rate, interest, principal, rate,* and *time.* Emphasize that time must be expressed as a part of a year. For example, 3 months is written as $\frac{1}{4}$ year, and 7 months is written as $\frac{7}{12}$ year. Point out that the interest rate is written as a decimal when computing interest.

Banks and savings and loan associations often use simple interest for loans of less than one year's duration and for home mortgages.

Warm-Up Exercises
1. Write as a fraction of a year and then as a decimal.
   a. 6 months $\frac{1}{2}$; 0.5
   b. 9 months $\frac{3}{4}$; 0.75
   c. 8 months $\frac{2}{3}$; 0.667
2. Write as a decimal
   a. 6% 0.06
   b. 12% 0.12
3. $250 × 0.12 × 0.75 $22.50

ALTERNATIVE STRATEGIES: Reteaching
There are many interest rates to be found in the newspaper: rates for money markets, certificates of deposit (CD) of varying length, NOW accounts, passbook savings accounts, and so on. Have your students bring in some of these rates and tie them in with the problems in the book.

PRACTICE AND APPLY

The following problems can be assigned for classwork and the answers checked in class to help students master the objective of the lesson.

- Guided Practice: 1–7
- Independent Practice: 9–13 odd

WRAP-UP

Read Problem 13 to the class. Ask students which figure is the principal, which is the interest rate, and which is the time. Then ask them what three numbers, in decimal form, they would multiply to find the simple interest.

Assignment Guide

- Basic: 8–15, 17–28
- Average: 8–14 even, 15, 16, 17–27 odd

8. Lillian Hanby's savings account.
$975 on deposit.
$5\frac{1}{2}$% annual interest rate.
Time is 3 months.
Find the interest. **$13.41**

9. Trevor Tyler's savings account.
$1342 on deposit.
$6\frac{1}{2}$% annual interest rate.
Time is 6 months.
Find the interest. **$43.62**

10. Ameil Trost's savings account.
$8460 on deposit.
$5\frac{3}{4}$% annual interest rate.
Time is 9 months.
Find the interest.
$364.84

11. Norman Robert's savings account.
$927.15 on deposit.
$8\frac{1}{2}$% annual interest rate.
Time is 5 months.
Find the interest. **$32.84**

12. Perry Wells deposited $760 in a new savings account at Maumee Savings and Loan Association. No other deposits or withdrawals were made. After 3 months, the interest was computed at an annual interest rate of $5\frac{1}{2}$%. How much simple interest did his money earn? **$10.45**

13. Loretta Patterson deposited $429 in a new savings account at Henry County Bank and Trust. She made no other deposits or withdrawals. After 6 months, the interest was computed at an annual rate of $6\frac{3}{4}$%. How much simple interest did her money earn? **$14.48**

14. On May 1, Harold Bothan opened a savings account at Fulton Savings Bank with a deposit of $1250. No other deposits or withdrawals were made. How much simple interest did his money earn by August 1? **$17.19**

$ FULTON SAVINGS BANK $

GOOD NEWS to our regular savings account customers!

$5\frac{1}{2}$ %

****Interest computed quarterly and paid quarterly.***

****Effective May 1.***

15. Vernon Taber deposited his $2000 scholarship money in a savings account at State Home Savings Bank on June 1. At the end of July, interest was computed at an annual interest rate of $6\frac{1}{2}$%. How much simple interest did his money earn? **$21.67**

16. On March 1, Tessa Obee deposited a refund check for $9364.85 in a savings account at Napoleon Trust Co. At the end of August, interest was computed at an annual interest rate of $7\frac{1}{4}$%. How much simple interest did her money earn? **$339.48**

MAINTAINING YOUR SKILLS Look up the skills in parentheses if you need help or more practice.

Change the fractions and mixed numbers to decimals. (Skill 14)

17. $\frac{1}{2}$ **0.5** **18.** $\frac{3}{4}$ **0.75** **19.** $\frac{1}{4}$ **0.25** **20.** $5\frac{1}{2}$ **5.5** **21.** $6\frac{1}{4}$ **6.25** **22.** $7\frac{3}{4}$ **7.75**

Write the percents as decimals. (Skill 28)

23. $5\frac{1}{2}$% **0.055** **24.** $6\frac{1}{4}$% **0.0625** **25.** $9\frac{1}{2}$% **0.095** **26.** $7\frac{3}{4}$% **0.0775** **27.** $10\frac{5}{8}$% **0.10625** **28.** $5\frac{3}{8}$% **0.05375**

COMMUNICATION SKILLS

Banks are required by law to pay simple interest on six-month money market certificates. Have students write a brief report on money market certificates, how they are used, and the interest rates typically offered.

4-6

Compound Interest

OBJECTIVE

Compute the compound interest and the amount.

Interest that you earn in a savings account during an interest period is added to your account. The new balance is used to calculate the interest for the next interest period. **Compound interest** is interest earned not only on the original principal but also on the interest earned during previous interest periods. The **amount** is the balance in your account at the end of an interest period.

Amount = Principal + Interest

EXAMPLE *Skills* 28, 14, 30 *Application* K *Term* Compound interest

John Heldt deposited $1800 in a savings account that earns 6% interest compounded quarterly. He made no deposits or withdrawals. What was the amount in the account at the end of the second quarter?

SOLUTION Find the **amount** for each quarter.

First Quarter: Interest = Principal × Rate × Time
Interest = $1800.00 × 6% × $\frac{1}{4}$ = $27.00
Amount = Principal + Interest
Amount = $1800.00 + $27.00 = $1827.00

Second Quarter: Interest = $1827.00 × 6% × $\frac{1}{4}$ = $27.41
Amount = $1827.00 + $27.41 = $1854.41

1800 [M+] [×] 6 [%] 108 [×] .25 [=] 27 [+] [RM] 1800 [=] 1827 [CM] [M+] [×] 6 [%] 109.62 [×] .25 [=] 27.405 [+] [RM] 1827 [=] 1854.405

✓ SELF-CHECK Complete the problems, then check your answers in the back of the book.

1. $2000 is deposited at 8% compounded semiannually for 1 year. Find the amount after 1 year. **$2163.20**

PROBLEMS

	2.	3.	4.	5.	6.
Principal	$900.00	$400.00	$2,360.00	$18,260.00	$27,721.00
Annual Interest Rate	6%	6%	$8\frac{1}{2}$%	$7\frac{1}{2}$%	9.513%
Interest Period	quarterly	monthly	semiannually	quarterly	annually
First Period Interest	$ 13.50	$ 2.00	$ 100.30	**$342.38**	**$2637.10**
Amount	$913.50	$402.00	**$2,460.30**	**$18,602.38**	**$30,358.10**
Second Period Interest	**$13.70**	**$2.01**	**$104.56**	**$348.79**	**$2,887.97**
Amount	**$927.20**	**$404.01**	**$2,564.86**	**$18,951.17**	**$33,246.07**

LESSON PLAN 4-6 Compound Interest

FOCUS

The focus of this lesson is on computing the compound interest and the amount in a savings account. Locate some advertisements in newspapers that deal with compound interest. Ask students if any of them have accounts paying compound interest.

TEACH

Emphasize the fact that the interest must be added to the principal at the end of each interest period. This figure becomes the principal for the new interest period.

Make sure students understand the difference between compounding interest annually, semiannually, quarterly, and monthly. Write the following example on the chalkboard to illustrate how the same principal and interest rate can result in different amounts of money at the end of the year.

original deposit: $10,000
interest rate: 6%

compounded annually: $10,600.00
compounded semiannually: $10,609.00
compounded quarterly: $10,613.64
compounded monthly: $10,616.78

Warm-Up Exercises

1. $1319.50 + $18.41
   $1337.91
2. $1951.21 + $16.43
   $1967.64
3. $1000 × 12% × $\frac{1}{4}$
   $30
4. $6000 × 6% × $\frac{1}{12}$
   $30

ALTERNATIVE STRATEGIES: Enrichment

Emphasize the power of compound interest by using this example:

The Anchorage, Alaska Bicentennial Commission had $1,000 left after the celebration in 1976. The money was not to be touched until the tricentennial celebration in 2076. At 6% interest compounded monthly the $1,000 will be worth $397,442.32 in 2076! At 6% simple interest the $1,000 will be worth $7000 in 2076. A difference of $390,442.32!!

PRACTICE AND APPLY

The following problems can be assigned for classwork and the answers checked in class to help students master the objective of the lesson.

- Guided Practice: 1–6
- Independent Practice: 8–12 even

WRAP-UP

Have students explain how many payment periods there are if interest is compounded monthly, semiannually, quarterly, or annually. Read Problem 12 to the class. Ask students how many interest periods there are from July 1 to April 1 if the interest is compounded quarterly.

Assignment Guide

- Basic: 7–13, 15–29
- Average: 7–13 odd, 14, 15–29 odd

7. Alicia Martin's savings account. Principal is $800. 6% interest compounded quarterly. What is the amount in the account at the end of the second quarter? **$824.18**

8. Angelo Larragu's savings account. Principal is $1200. 6% interest compounded quarterly. What is the amount in the account at the end of the third quarter? **$1254.81**

9. Elmer Pasture deposited $860 in a new regular savings account that earns 5.5% interest compounded semiannually. He made no other deposits or withdrawals. What was the amount in the account at the end of 1 year? **$907.95**

10. Jana Dejute deposited $1860 in a new credit union savings account on the first of a quarter. The principal earns 8% interest compounded quarterly. She made no other deposits or withdrawals. What was the amount in her account at the end of 6 months? **$1935.14**

11. Betty and Sam Eassavore's savings account had a balance of $9544 on May 1. The account earns interest at a rate of 5.25% compounded monthly. What is the amount in their account at the end of August if no deposits or withdrawals were made during the period? **$9712.13**

12. Ernie Boddy had $3620 on deposit at Suburban Savings Bank on July 1. The money earns interest at a rate of 6.5% compounded quarterly. What is the amount in the account on April 1 of the following year if no deposits or withdrawals were made? **$3799.36**

13. The Vassillis opened a savings account with a deposit of $1800 on July 1. The account pays interest at 9% compounded semiannually. What amount will they have in their account on July 1 one year later? **$1965.65**

14. Elaine Deck had $10,000 put into a trust fund for her on her 16th birthday. The trust fund pays $9\frac{1}{4}$% interest compounded annually. What will her trust fund be worth on her 21st birthday? **$15,563.50**

MAINTAINING YOUR SKILLS Look up the skills in parentheses if you need help or more practice.

Write the percents as decimals. **(Skill 28)**

15. $5\frac{1}{4}$% **0.0525** **16.** $8\frac{1}{2}$% **0.085** **17.** $5\frac{3}{4}$% **0.0575** **18.** 6.9% **0.069** **19.** 7.25% **0.0725**

Find the percentage. **(Skill 30)**

20. $950 × 8% **$76**
21. $760 × $6\frac{1}{2}$% **$49.40**
22. $3620 × 9% **$325.80**
23. $12,380 × $9\frac{1}{2}$% **$1176.10**
24. $23,260 × $4\frac{1}{2}$% **$1046.70**
25. $36,200 × 1% **$362**

Write the fractions as decimals. Round answers to the nearest hundredth. **(Skill 14)**

26. $\frac{3}{10}$ **0.30** **27.** $\frac{4}{12}$ **0.33** **28.** $\frac{5}{8}$ **0.63** **29.** $\frac{3}{5}$ **0.60**

USE OF CALCULATORS

Have students work the following on a calculator.

1800 M+ 1800 × 6 % × 0.25 = M+ RM × 6 % × 0.25 = M+ RM

1854.405 display

Ask students what this calculation is doing. (finding the compound interest for two quarters; same as in the Example)

4-7 Compound Interest Tables

OBJECTIVE
Find compound interest with tables.

To compute compound interest quickly, you can use a compound interest table, which shows the amount of $1.00 for many interest rates and interest periods. To use the table, you must know the total number of interest periods and the interest rate per period.

AMOUNT OF $1.00			
Total Interest Periods	Interest Rate per Period		
	1.250%	1.375%	1.500%
1	1.01250	1.01375	1.01500
2	1.02515	1.02768	1.03022
3	10.3797	1.04182	1.04567
4	1.05094	1.05614	1.06136
5	1.06408	1.07066	1.07728
6	1.07738	1.08538	1.09344
7	1.09085	1.10031	1.10984
8	1.10448	1.11544	1.12649

Amount = Original Principal × Amount of $1.00

Compound Interest = Amount − Original Principal

EXAMPLE *Skills* 11, 8 *Application* C *Term* Compound interest

Farmers and Merchants Bank pays 6% interest compounded quarterly on regular savings accounts. Carol Whiteside deposited $3000 for 2 years. She made no other deposits or withdrawals. How much interest did Carol earn during the 2 years?

SOLUTION

A. Find the **total interest periods.**
2 years × 4 quarters per year = 8 total interest periods

B. Find the **interest rate per period.**
6% ÷ 4 quarters = 1.5% interest rate per period

C. Find the **amount of $1.00** from the table for 8 periods (2 years).
1.12649

D. Find the **amount.**
Original Principal × Amount of $1.00
$3000.00 × 1.12649 = $3379.47

E. Find the **compound interest.**
Amount − Original Principal
$3379.47 − $3000 = $379.47 compound interest

✓ SELF-CHECK Complete the problems, then check your answers in the back of the book.

1. $2000 invested at 5.5% compounded quarterly for 2 years. Find the amount. **$2230.88**
2. $4500 invested at 3% compounded semiannually for 2 years. Find the amount. **$4776.12**

LESSON PLAN 4-7 Compound Interest Tables

FOCUS
Ask students how many times they would have to compute the compound interest on an original amount if it is left in the account for two years and is compounded monthly. (24) The focus of this lesson is finding compound interest with tables. Tables can be useful in situations where many calculations would be necessary.

TEACH
Point out that the table is based on $1 so that the entries in the table can be applied to any dollar amount.

In the Example, there are 8 total interest periods. Point out that an interest rate is an *annual* interest rate. This is why, in step 2, 6% is divided by 4 quarters per year, not by the 8 quarters total for the problem. To find the Amount of $1.00, students must understand that they choose the column for 1.500% and the row for 8 total interest periods. When they multiply this figure times the original principal, they are finding the *Amount,* or both the principal and the compound interest.

CULTURAL NOTES
The Algonquins (Native Americans) sold Manhattan to Captain Stuyvesant for $24 in 1624. In 1986, the $24 was worth: $545.28 at 6% simple interest, and $34,748,372,000.00 at 6% compounded annually.

CRITICAL THINKING
See Problem 3. Suppose there were 5 interest periods per year. What interest rate per period would you need available on the table in order to find the interest? (1.1%)

Warm-Up Exercises

1. $100 × 1.025 $102.50
2. $700 × 1.772 $1240.40
3. $2000 × 1.1154 $2230.80
4. 5.75% ÷ 4   1.4375%
5. 6.5% ÷ 12   0.5417%
6. $754.68 − $700.00 $54.68

PRACTICE AND APPLY

The following problems can be assigned for classwork and the answers checked in class to help students master the objective of the lesson.

- Guided Practice: 1–7
- Independent Practice: 9, 11, 14

AMOUNT OF $1.00

Total Interest Periods	Interest Rate per Period						
	1.250%	1.375%	1.500%	2.750%	2.875%	3.000%	3.125%
1	1.01250	1.01375	1.01500	1.02749	1.02875	1.03000	1.03125
2	1.02515	1.02768	1.03022	1.05575	1.05832	1.06090	1.06347
3	1.03797	1.04182	1.04567	1.08478	1.08875	1.09272	1.09671
4	1.05094	1.05614	1.06136	1.11462	1.12005	1.12550	1.13098
5	1.06408	1.07066	1.07728	1.14527	1.15225	1.15927	1.16632
6	1.07738	1.08538	1.09344	1.17676	1.18538	1.19405	1.20277
7	1.09085	1.10031	1.10984	1.20912	1.21946	1.22987	1.24036
8	1.10448	1.11544	1.12649	1.24237	1.25452	1.26677	1.27912
9	1.11829	1.13078	1.14339	1.27654	1.29059	1.30477	1.31909
10	1.13227	1.14632	1.16054	1.31165	1.32769	1.34391	1.36031
11	1.14642	1.16209	1.17795	1.34772	1.36586	1.38423	1.40282
12	1.16075	1.17806	1.19562	1.38478	1.40513	1.42576	1.44666
13	1.17526	1.19426	1.21355	1.42286	1.44553	1.46853	1.49187
14	1.18995	1.21068	1.23175	1.46199	1.48709	1.51258	1.53849
15	1.20482	1.22733	1.25023	1.50219	1.52984	1.55796	1.58657
16	1.21988	1.24421	1.26898	1.54350	1.57382	1.60470	1.63615
17	1.23513	1.26132	1.28802	1.58595	1.61907	1.65284	1.68728
18	1.25057	1.27866	1.30734	1.62956	1.66562	1.70243	1.74000
19	1.26620	1.29624	1.32695	1.67438	1.71351	1.75350	1.79438
20	1.28203	1.31406	1.34685	1.72042	1.76277	1.80611	1.85045
21	1.29806	1.33213	1.36706	1.76773	1.81345	1.86029	1.90828
22	1.31428	1.35045	1.38756	1.81635	1.86559	1.91610	1.96791
23	1.33071	1.36902	1.40838	1.86630	1.91922	1.97358	2.02941
24	1.34735	1.38784	1.42950	1.91762	1.97440	2.03279	2.09283

PROBLEMS

Use the compound interest table above to solve. Round answers to the nearest cent.

	3.	4.	5.	6.	7.
Principal	$900.00	$640.00	$1340.00	$6231.40	$3871.67
Annual Interest Rate	5.5%	6%	5%	5.75%	18%
Interest Periods per Year	4	2	4	2	12
Total Time	2 years	1 year	3 years	5 years	6 months
Amount	? **$1003.90**	? **$678.98**	? **$1555.41**	? **$8273.37**	? **$4233.44**
Compound Interest	? **$103.90**	? **$38.98**	? **$215.41**	? **$2041.97**	? **$361.77**

8. Principal is $936.
   5.5% annual interest rate.
   Compounded quarterly for 1 year.
   What is the compound interest? **$52.55**

9. Principal is $1236.35.
   6% annual interest rate.
   Compounded quarterly for 2 years.
   What is the compound interest? **$156.39**

ALTERNATIVE STRATEGIES: Reteaching

Help students find the "interest rate per period" and the "total interest periods" in these situations:

10. National Savings and Trust pays 5.5% interest compounded semiannually on regular savings accounts. Iva Howe deposited $2800 in a regular savings account for 2 years. She made no other deposits or withdrawals during the period. How much interest did her money earn? **$320.94**

11. Alistair Russel deposited $900 in a savings plan with the employees' credit union where she works. The credit union savings plan pays 6% interest compounded quarterly. If she makes no other deposits or withdrawals, how much interest will her money earn in 1 year? **$55.22**

F.Y.I.
The compound interest formula is:
$A = P(1 + i)^n$
where P is the principal, i is the rate per period, n is the number of periods, and A is the amount.

12. Dime Bank pays 6.25% interest compounded semiannually on special notice savings accounts. Jessie McKenzie deposited $3438.70 in a special notice savings account for $2\frac{1}{2}$ years. No other deposits or withdrawals were made during the period. How much interest did his money earn? **$571.92**

13. Union Savings and Loan pays 5% interest compounded quarterly on regular savings accounts. Rose and Bob Yung had $4000 on deposit for 1 year. At the end of the year, they withdrew all their money and deposited $4000 at Volunteer Co-Operative Bank. How much more did the $4000 earn at Volunteer Co-Operative for 1 year? **$29.52**

VOLUNTEER CO-OPERATIVE BANK
announces new interest rate!
5.75%
on regular savings accounts.
INTEREST COMPOUNDED SEMIANNUALLY.
MINIMUM OF $10 DEPOSIT REQUIRED.

14. Wilma Bracken opened a savings account at Dallas Trust Bank on March 1. Dallas Trust pays 15% compounded monthly on money left on deposit for at least 6 months. Wilma opened her account with an initial deposit of $10,000. What is the $10,000 worth on October 1? What did the $10,000 earn by October 1? **$10,908.50; $908.50**

15. Maryann Radner opened a savings account at New England Savings and Loan on January 1 with a deposit of $800. New England Savings and Loan pays 11.5% compounded quarterly. What will the $800 be worth on January 1 of the following year? What will the $800 have earned by January 1 of the following year? **$896.04; $96.04**

16. Martha and Mike Aieta deposited $4000 in a savings account in their child's name when their child was born. The $4000 earns interest at $6\frac{1}{4}$ % compounded semiannually. What will the $4000 be worth on the child's 12th birthday? **$8371.32**

MAINTAINING YOUR SKILLS Look up the skill in parentheses if you need help or more practice.

Divide. **(Skill 11)**

17. 8% ÷ 4 **2%**	18. 4% ÷ 4 **1%**	19. 6% ÷ 12 **0.5%**	20. 7% ÷ 4 **1.75%**
21. 8.5% ÷ 2 **4.25%**	22. 3.5% ÷ 4 **0.875%**	23. 5.25% ÷ 4 **1.3125%**	24. 16.5% ÷ 2 **8.25%**
25. $7\frac{1}{2}$% ÷ 4 **1.875%**	26. $6\frac{1}{4}$% ÷ 2 **3.125%**	27. $19\frac{1}{2}$% ÷ 12 **1.625%**	28. $11\frac{1}{2}$% ÷ 4 **2.875%**

	Annual Rate of Interest	Compounded	Number of Years	Interest rate per Period	Total Interest Periods
1)	6%	Monthly	5	(0.5%)	(60)
2)	8%	Quarterly	7	(2%)	(28)
3)	5%	Annually	15	(5%)	(15)
4)	5.5%	Quarterly	10	(1.375%)	(40)
5)	5%	Semiannually	8	(2.5%)	(16)

WRAP-UP

Read Problem 11 to the class. Ask one student to explain how to find the total interest periods. Ask another student how to find the interest rate per period. Ask a third student how to find the entry in the table. Ask a fourth student how to find the compound interest earned.

Assignment Guide

- Basic: 8–14, 17–28
- Average: 8, 10, 13, 14–16, 17–27 odd

LESSON PLAN
4-8 Daily Compounding

FOCUS

You can motivate this lesson by pointing out that students have found that compounding semiannually earns more interest than compounding annually; compounding quarterly earns more than semiannually; and monthly earns more than quarterly. The most competitive method of computing interest is compounding daily. That is the focus of this lesson.

TEACH

In order to compound interest on a daily basis, students have to be able to figure out how many days the original principal was in the account. Have students practice using the table on page 644.

Point out that a separate table would be needed for each annual interest rate. Daily compounding of interest relies on the use of computers.

4-8 Daily Compounding

OBJECTIVE
Find compound interest using tables.

Usually the more frequently interest is compounded, the more interest you will earn. Many banks offer savings accounts with daily compounding. When interest is compounded daily, it is computed each day and added to the account balance. The account will earn interest from the day of deposit to the day of withdrawal. A table can be used to calculate the amount and interest for daily compounding.

AMOUNT OF $1.00 AT 5.5% COMPOUNDED DAILY, 365-DAY YEAR			
Day	Amount	Day	Amount
21	1.00316	31	1.00468
22	1.00331	32	1.00483
23	1.00347	33	1.00498
24	1.00362	34	1.00513
25	1.00377	35	1.00528

Amount = Original Principal × Amount of $1.00

Compound Interest = Amount − Original Principal

EXAMPLE *Skill* 8 *Application* G *Term* Compound interest

On May 30, Deloris Zelms deposited $1000 in a savings account that pays $5\frac{1}{2}$% interest compounded daily. On July 1, how much interest had been earned on the principal in her account?

SOLUTION

A. Find the **number of days** from May 30 to July 1. Use the table on page 645.
July 1 is day 182. May 30 is day 151.
182 − 151 = 31 days

B. Find the **amount.** Use the table on page 644.
Original Principal × Amount of $1.00
$1000.00 × 1.00468 = $1004.68

C. Find the **compound interest.**
Amount − Original Principal
$1004.68 − $1000.00 = $4.68 compound interest

✓ SELF-CHECK Complete the problems, then check your answers in the back of the book.

$6000 deposited at 5.5% compounded daily for 25 days.

1. Find the amount. **$6022.62** **2.** Find the compound interest. **$22.62**

ALTERNATIVE STRATEGIES: Enrichment

Peter Stuyvesent (last Dutch Governor of New Amsterdam) purchased the island of Manhattan from the Algonquin Indians for $24 in the year 1624.

Find the value of the $24 in the year 1995 if it had been invested at:

1) 7% interest compounded daily (let's not concern ourselves with leap years)

PROBLEMS

Use the daily compound interest table on page 644 to solve. Round answers to the nearest cent. Interest is 5.5% compounded daily.

	3.	4.	5.	6.	7.
Principal	\$80,000	\$900	\$6,500	\$3,800	\$15,321
Number of Days	25	31	50	90	120
Amount	**\$80,301.60**	**\$904.21**	**\$6549.08**	**\$3851.83**	**\$15,600.30**
Compound Interest	**\$301.60**	**\$4.21**	**\$49.08**	**\$51.83**	**\$279.30**

8. Welton Bower's savings account. \$980 on deposit for 90 days. $5\frac{1}{2}$% annual interest rate. Interest compounded daily. What is the compound interest? **13.37**

9. Donna Wayne's savings account. \$4160 on deposit for 50 days. $5\frac{1}{2}$% annual interest rate. Interest compounded daily. What is the compound interest? **\$31.41**

10. On June 10, Bertha Scisloski deposited \$8241.78 in a savings account that pays $5\frac{1}{2}$% interest compounded daily. How much interest will the money earn in 31 days? **\$38.57**

11. Marylin Robertson has a savings account that earns $5\frac{1}{2}$% interest compounded daily. On May 5, the amount in the account was \$28,214.35. How much interest will the money earn in 90 days? **\$384.84**

12. On April 11, Rob Walthall had \$6521.37 in his savings account. The account pays $5\frac{1}{2}$% interest compounded daily. How much interest will the money earn by June 30? **\$79.04**

13. On August 23, Richard McCormick had \$1432.19 in his savings account at Camden Savings and Trust. The account earns $5\frac{1}{2}$% interest compounded daily. What will be the amount in his savings account when he closes his account on October 1? **\$1440.63**

14. Sandy Hane's savings account shows a balance of \$904.31 on March 1. The same day, she made a deposit of \$375 to the account. The bank pays interest at a rate of $5\frac{1}{2}$% compounded daily. What will be the amount in her account on May 10? on June 29? **\$1292.86; \$1302.62**

MAINTAINING YOUR SKILLS

Look up the skills in parentheses if you need help or more practice.

Multiply. **(Skill 8)**

15. \$4000 × 1.02131 **\$4085.24**
16. \$9000 × 1.00135 **\$9012.15**
17. \$550 × 1.00907 **\$554.99**
18. \$1437 × 1.00392 **\$1442.63**
19. \$950 × 1.00392 **\$953.72**
20. \$1416 × 1.01059 **\$1431.00**
21. \$7370 × 1.00347 **\$7395.57**
22. \$41,520 × 1.00407 **\$41,688.99**
23. \$94 × 1.00196 **\$94.18**
24. \$389 × 1.00301 **\$390.17**
25. \$7925.14 × 1.01670 **\$8057.49**
26. \$327.78 × 1.00015 **\$327.83**

Warm-Up Exercises

1. \$500 × 1.01670
   \$508.35
2. \$5147.85 × 1.00452
   \$5171.12
3. \$5617.21 − \$5500
   \$117.21
4. \$4288.83 − \$4268.85
   \$19.98

Round to the nearest cent.

5. \$5417.205 \$5417.21
6. \$6429.4012 \$6429.40

PRACTICE AND APPLY

The following problems can be assigned for classwork and the answers checked in class to help students master the objective of the lesson.

- Guided Practice: 1–7
- Independent Practice: 9–13 odd

WRAP-UP

Have a student explain how to use the table on page 645 to compute the number of days between two dates. Have a second student explain how to use this answer to find the factor (amount) used in finding interest earned at 5.5% for this number of days.

Assignment Guide

- Basic: 8–13, 15–26
- Average: 8–14 even, 15–25 odd

P = \$24, i = 0.0001918, and n = 135,415. The \$24 would be worth \$4,559,163,942,000

2) 7% simple interest. The \$24 would be worth \$647.28. WOW!! A difference of \$4,559,163,941,352.72!!!

LESSON PLAN Reviewing the Basics

The exercises on this page review skills, applications, and terms used in the unit. You can use the exercises to assess informally students' proficiency with this material.

The page can be used for guided practice and independent practice. You can work through a selection of the exercises together with students, and thus see immediately if they know how to do them, and you can then assign some of the exercises for independent practice. Be sure to go over the answers to all assigned exercises.

Reviewing the Basics

Skills

Write in words.

(Skill 1)
1. \$479.00 **2.** \$37.74 **3.** \$3091.47 **4.** \$15,263.59

Solve. Round answers to the nearest cent.

(Skill 5)
5. \$193.74 + \$18.31 **\$212.05** **6.** \$42,481.91 + \$53.40 + \$6.19 **\$42,541.50**

(Skill 6)
7. \$871.95 − \$136.99 **\$734.96** **8.** \$1400.00 − \$519.87 **\$880.13**

(Skill 8)
9. 1.12051 × \$6000 **\$6723.06** **10.** 1.104361 × \$961.82 **\$1062.20**

(Skill 10)
11. 8.5 ÷ 4 **2.125** **12.** 7.5 ÷ 2 **3.75** **13.** 6.74 ÷ 4 **1.685** **14.** 9.5 ÷ 12 **0.7917**

Write as a decimal.

(Skill 14)
15. $8\frac{1}{4}$ **8.25** **16.** $4\frac{1}{2}$ **4.5** **17.** $21\frac{3}{4}$ **21.75** **18.** $7\frac{5}{8}$ **7.625**

(Skill 28)
19. $6\frac{3}{4}$% **0.0675** **20.** $4\frac{7}{8}$% **0.04875** **21.** 19.95% **0.1995** **22.** 12.351% **0.12351**

Applications

Use the table on page 644 to find the amount of \$1.00.

(Application C)
23. 9 interest periods. Rate per period 1.5%. **1.14339** **24.** 4 interest periods. Rate per period 1.25%. **1.05094**

(Application G)
Find the elapsed time in days. Use the table on page 645.

25. From March 4 to May 31 **88 days** **26.** From June 10 to July 30 **50 days**

Write as a fraction of a year.

(Application J)
27. 3 months $\frac{3}{12} = \frac{1}{4}$ **28.** 6 months $\frac{6}{12} = \frac{1}{2}$ **29.** 9 months $\frac{9}{12} = \frac{3}{4}$ **30.** 18 months $\frac{18}{12} = 1\frac{1}{2}$

(Application K)
Find the total number of interest periods.

31. Quarterly for 4 years **16 periods** **32.** Semiannually for $2\frac{1}{2}$ years **5 periods**

Terms

Use each term in a sentence that explains the term. **Answers will vary.**

33. Passbook **34.** Withdrawal **35.** Compound interest

36. Simple interest **37.** Deposit **38.** Statement

Refer to your reference files in the back of the book if you need help.

Unit Test

Lesson 4-1

1. George Pinewood deposited $54.21 in currency and a check for $384.19 in his savings account. What was his total deposit? **$438.40**

Lesson 4-2

2. Lynn Moyer has to withdraw $218.31 from her savings account. Write the amount as a numeral and in words. **$218.31; Two hundred eighteen and $\frac{31}{100}$**

Lesson 4-3

3. On September 13, Calvin Muir deposited $341.75 in his savings account. What is the new balance? **$757.55**

DEPOSITOR: Calvin Muir				
DATE	WITHDRAWAL	DEPOSIT	INTEREST	BALANCE
8-24				$415.80
9-13		341.75		?

Lesson 4-4

4. Susan Allard's savings account statement shows the following: a previous balance of $387.21, a withdrawal of $218.30, total deposits of $522.61, and interest of $1.23. What is the new balance? **$692.75**

Lesson 4-5

5. Rhea Franken deposited $419.31 in a savings account that pays an annual interest rate of $6\frac{3}{4}$ %. She made no other deposits or withdrawals. How much simple interest did her money earn in 3 months? **$7.08**

Lesson 4-6

6. Dan Dorsey deposited $981.40 in a regular savings account that earns 5.5% interest compounded quarterly. He made no other deposits or withdrawals. What is the amount at the end of the second quarter? **$1008.57**

Lesson 4-7

7. Boston Savings Bank pays 6% interest compounded quarterly on regular savings accounts. Fred Pilliod deposited $1800 in a regular savings account for $1\frac{1}{2}$ years. He made no other deposits or withdrawals. Use the table to find the interest earned. **$168.19**

AMOUNT OF $1.00		
Total Interest Periods	Rate Per Period	
	1.375%	1.500%
4	1.05614	1.06136
5	1.07067	1.07728
6	1.08539	1.09344

Lesson 4-8

8. Irene Myers deposited $571.31 in a savings account that pays 5.5% interest compounded daily. How much interest will her money earn in 31 days? **$2.67**

AMOUNT OF $1.00 AT 5.5% COMPOUNDED DAILY, 365-DAY YEAR			
Day	Amount	Day	Amount
11	1.00165	31	1.00468
12	1.00180	32	1.00483
13	1.00196	33	1.00498

LESSON PLAN
Unit Test

Students should do the Unit Test on their own. Each problem on the test is keyed to a lesson in the unit. Students having difficulty with any particular problem should review the Example in the appropriate lesson and be assigned some of the Independent Practice problems for additional practice.

LESSON PLAN
A Spreadsheet Application

USING TECHNOLOGY

You can broaden students' knowledge and appreciation of the power of computers by asking them how banks compute compound interest for millions of accounts daily. (by using computers) Point out to students that large amounts of money move through banks around the world on a daily basis. Keeping track of these transactions and calculating interest on them would be impossible in today's business world without computers.

A SPREADSHEET APPLICATION

Compound Interest

To complete this spreadsheet applicaton, you will need the template diskette for *Mathematics with Business Applications.* Follow the directions in the *User's Guide* to complete this activity. (Then pick up copy pertinent to each unit's problem.)

1. Mike Oyer.
Principal: $1000.
Interest rate: 6%
Periods per year: 4.
Total time: 2 years.
What is the amount?
What is the interest?

2. Max Markley
Principal: $10,000.
Interest rate: 6.5%.
Periods per year: 12.
Total time: 1 year.
What is the amount?
What is the interest?

3. Helen Hallett.
Principal: $5000.
Interest rate: 7.5%.
Periods per year: 4.
Total time: 0.5 years.
What is the amount?
What is the interest?

4. Peggy Delaney
Principal: $6540.
Interest rate: 8%
Periods per year: 2.
Total time: 4 years.
What is the amount?
What is the interest?

5. Jean Deilman
Principal: $568.65.
Interest rate: 5.5%.
Periods per year: 365.
Total time: 2 years.
What is the amount?
What is the interest?

6. Angelo Diaz.
Principal: $6987.25.
Interest rate: 9.5%.
Periods per year: 365.
Total time: 0.25 years.
What is the amount?
What is the interest?

7. Elmer O'Neal deposited $40,000 in a savings plan paying 9.85% interest compounded monthly for 5 years. What is the amount? What is the interest?

8. Cheryle Contikos opened a $40,000 savings account at 9.85% interest compounded quarterly for 5 years. What is the amount? What is the interest?

9. Juan Carrero deposited $2000 in a passbook savings account paying 5.5% interest compounded quarterly for 2 years. What is the amount? What is the interest?

10. Elva and Paul Schrerez deposited $5000 in a 5-year certificate of deposit paying 15% interest compounded daily (365). How much interest did they earn?

11. Laticia Olrich deposited $1308.25 in a passbook savings account earning 5.5% interest compounded quarterly for 1 year. How much interest did she earn?

12. Maria and Randall Zapata deposited $50,000 in a 20-year certificate of deposit paying 6.5% interest compounded monthly. How much interest did they earn in 20 years?

UNIT

5

Cash Purchases

A *sales receipt* is a proof of purchase showing the cost of your purchase and the *sales tax,* where applicable. To find the best buy based on price, you can use the *unit price,* or the cost per unit of an item. Many stores offer *coupons* and *rebates* as an incentive for consumers to purchase an item. Stores may also sell products at *sale prices,* which are lower than the regular selling price. The *markdown* is the difference between the regular selling price and the sale price.

OUTLINE

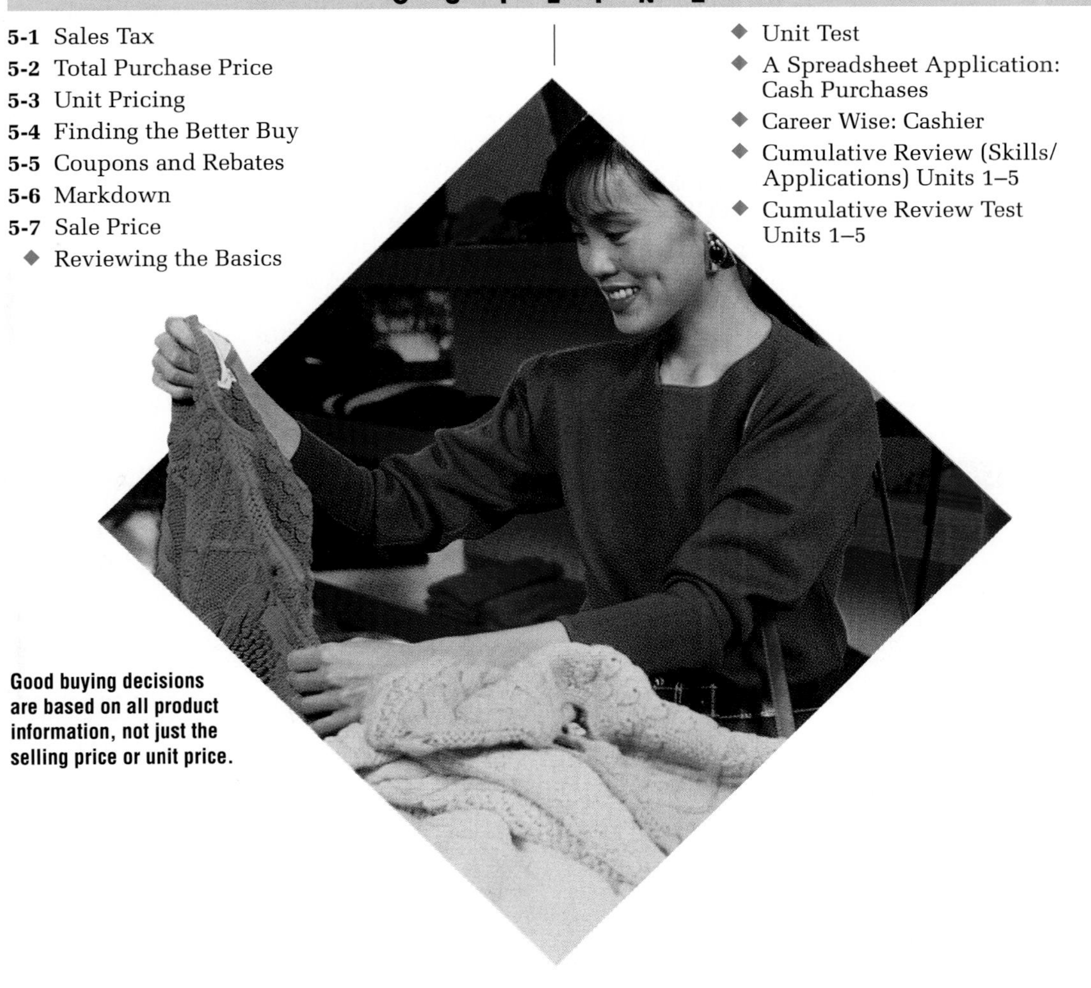

Good buying decisions are based on all product information, not just the selling price or unit price.

INTRODUCING THE UNIT

Point out that all but five states have a sales tax. Also mention that sales taxes range from 2% to 7.5% and that some states allow counties and cities to add on an additional tax. Mention the types of items and services on which tax is imposed differ among states. Emphasize, however, that tax is not imposed on every item or service. In most states food, medicine, and medical services are nontaxable.

LESSON PLAN
5-1 Sales Tax

FOCUS

The focus of this lesson is on computing the sales tax. If your state has a sales tax, ask students if they know what the rate is.

TEACH

Some of your students who have part-time jobs may have worked with a sales tax already. Ask them how they find the sales tax on the total selling price. Some stores use a table showing the tax, while others may have a sales tax key on the cash register that automatically calculates the tax. Or the tax may be found by doing the actual computation, as shown in the Example.

Remind students that percent means per hundred. Thus, a tax of 5% means 5 cents per 100 cents, or 5 cents per dollar.

Warm-Up Exercises

1. 5% of $19.00 $.95
2. 4.5% of $7.50 $.34
3. $6\frac{3}{4}$% of $76.90 $5.19
4. $3\frac{1}{4}$% of $1350.60 $43.89

5-1 Sales Tax

OBJECTIVE
Compute the sales tax.

Many state, county, and city governments charge a **sales tax** on certain items or services that you buy. The sales tax rate is usually expressed as a percent.

Sales Tax = Total Selling Price × Sales Tax Rate

EXAMPLE *Skills* 30, 2 *Application* A *Term* Sales tax

Ron Engle bought a camera that had a selling price of $129.99 and 3 rolls of film that had a selling price of $18.95. What is the sales tax on his purchases if the tax rate is: (a) 4%? (b) 5.5%?

SOLUTION

A. Find the **total selling price.**
$129.99 + $18.95 = $148.94

B. Find the **sales tax.**
Total Selling Price × Sales Tax Rate

(1) $148.94 × 4%
$148.94 × 0.04 = $5.9576 = $5.96 sales tax
(2) $148.94 × 5.5%
$148.94 × 0.055 = $8.1917 = $8.19 sales tax

129.99 [+] 18.95 [=] 148.94 [×] 4 [%] 5.9576

✓ SELF-CHECK Complete the problems, then check your answers in the back of the book.

Find the sales tax.

1. Total selling price: $420. Sales tax rate: 5%. **$21**
2. Total selling price: $11,400. Sales tax rate: 4.5%. **$513**

PROBLEMS

	Item	Total Selling Price	×	Sales Tax Rate	=	Sales Tax
3.	Clothing	$ 81.39	×	3%	=	**$2.44**?
4.	Pillowcases	$ 22.99	×	4%	=	**$0.92**?
5.	Calculator	$ 21.75	×	5%	=	**$1.09**?
6.	Compact disc	$ 12.95	×	4.5%	=	**$0.58**?
7.	Computer	$995.00	×	6%	=	**$59.70**?
8.	Breakfast	$ 7.95	×	6.25%	=	**$0.50**?

BUSINESS PROJECT

Have the students investigate and report on:

a. items and services that are taxed in their community.
b. items and services that are not taxed.
c. the tax rates, both state and local.
d. a city or county add-on tax.

Find the sales tax.

9.

PNAPL	.79
DAIRY	.89
SUGAR	1.39
CHEESE	1.69
SPAGHETTI	1.19

Tax rate: 5%
$0.30

10.

13.95	ATLAS
7.95	MAPS
8.49	TOTE BG
29.95	BOOK
15.95	BOOK
19.95	BOOK

Tax rate: 4%
$3.85

11.

PICTURE FRAME	$49.50
HOOKS	1.29
MATTING	11.45
MATTING	14.55

Tax rate: 6%
$4.61

F.Y.I.
Alaska, Delaware, Montana, New Hampshire, and Oregon have no statewide sales tax.

12. Lloyd DeVoe purchases a $29.95 blanket, a $14.95 mini blind, an adjustable screen for $15.95, and a 12-foot extension cord for $9.49. The sales tax rate is 6%. What is the total selling price? What is the sales tax? **$70.34; $4.22**

13. Wilber Haddad purchases a weed trimmer for $79, potting soil for $4.95, a 50-foot garden hose for $9.44, and a chaise lounge for $17.95. The sales tax rate is 5.5%. What is the sales tax? **$6.12**

14. Sally Hahn purchases the 13" color television shown. The sales tax rate is 4.5%. What is the sales tax? **$9.00**

13" Color Television
$199.99

15. At Alvin's Book Shop Rick Reimer purchases a record for $9.95, a dictionary for $21.50, a paperback book for $4.25, and a magazine for $4.00. The sales tax rate is 4.5%. What is the sales tax on his purchases? **$1.79**

16. Paula Ducat purchases a $109.95 phone answering machine, a $169.95 cordless phone, and a phone cord for $9.45. The sales tax rate is 6.5%. What is the sales tax on her purchases? **$18.81**

17. Fran Oates purchases a $25.95 leather handbag, a pair of running shoes for $37.95, and four polyester knit tops at $8.75 each. The sales tax rate is 3.75%. What is the sales tax on her purchases? **$3.71**

MAINTAINING YOUR SKILLS Look up the skills in parentheses if you need help or more practice.

Find the percentage. **(Skill 30)**

18. 8% of $29.9 **$2.39** **19.** $4\frac{1}{2}$% of $13,840 **$622.80** **20.** 120% of $90 **$108** **21.** $\frac{1}{2}$% of $240 **$1.20**

Round to the place value indicated. **(Skill 2)**

Nearest thousand	**22.** 85,713 **86,000**	**23.** 4139 **4000**	**24.** $31,234 **$31,000**	**25.** 2,436,541 **2,437,000**
Nearest ten	**26.** 435 **440**	**27.** 35,229 **35,230**	**28.** 94 **90**	**29.** 1131 **1130**
Nearest cent	**30.** $34.125 **$34.13**	**31.** $279.5429 **$279.54**	**32.** $357.346 **$357.35**	**33.** $2.8556 **$2.86**
Nearest tenth	**34.** 37.614 **37.6**	**35.** 357.982 **358.0**	**36.** 131.025 **131.0**	**37.** 521.681 **521.7**

PRACTICE AND APPLY

The following problems can be assigned for classwork and the answers checked in class to help students master the objective of the lesson.

- Guided Practice: 1–8
- Independent Practice: 9, 10, 12

WRAP-UP

Ask students to find the sales tax on the total of items costing $27.50, $13.99, $17.05, and $42.70. Use a rate of 4%. ($4.05)

Assignment Guide

- Basic: 9–16, 18–37
- Average: 11, 13–17, 18–36 even

ALTERNATIVE STRATEGIES: Reteaching

Emphasize that calculating sales tax is an application of Skill 30, finding a percentage, and Skill 2, rounding numbers. Review changing a percent to a decimal.

LESSON PLAN
5-2 Total Purchase Price

FOCUS
To compute the total purchase price, the total selling price and the sales tax are added. Ask students if they ever check their total purchase price by adding the cost of each purchase, finding the sales tax, and then adding the two figures together.

TEACH
Discuss the importance of the sales receipt as a record of purchase when manufacturers' rebates and special offers are involved. Review the important parts of a sales receipt, namely, (a) the quantity and selling price of each item or service purchased, (b) the total selling price of the items, (c) the sales tax, if any, and (d) the total purchase price.

Emphasize the importance of checking the sales receipt.

5-2 Total Purchase Price

OBJECTIVE
Compute the total purchase price.

Most stores give you a **sales receipt** as proof of purchase. The sales receipt may be a handwritten sales slip or a cash register tape. The receipt shows the selling price of each item or service you purchased, the total selling price, any sales tax, and the total purchase price.

Total Purchase Price = Total Selling Price + Sales Tax

EXAMPLE *Skills* 8, 5 *Application* A *Term* Sales recepit

Mike Gustweller purchases the golf items listed on the sales receipt. He checks the information on the receipt. The sales tax rate is 5%. What is the total purchase?

SOLUTION

WILEY'S PRO SHOP: 5th and Main Street

QUANTITY	DESCRIPTION	PRICE	AMOUNT
1	ProFlite 8-iron set	139.95	139.95
2	3-ball packages-golf balls	2.89	5.78
1	Deluxe golf cart cover	9.68	9.68
	SALES SLIP Customer's copy	Subtotal	155.41
		Sales tax	7.77
		Total	163.18

Labels: NUMBER PURCHASED; PRICE PER ITEM; TOTAL PRICE; TOTAL SELLING PRICE; TOTAL PURCHASE PRICE

F.Y.I.
Note that $163.18 is 105% of $155.41.

139.95 M+ 2 × 2.89 = 5.78 M+ 9.68 M+ RM 155.41 × 5 % 7.7705 + RM 155.41 = 163.1805

✓ SELF-CHECK Complete the problem, then check your answer in the back of the book.

1. Kate Farison bought a dome tent priced at $179, 2 coolers priced at $18 each, and a propane stove priced at $40. The sales tax rate is 2%. Find the total purchase price. **$260.10**

PROBLEMS

	Item	Total Selling Price	×	Sales Tax Rate	=	Sales Tax	Total Purchase Price
2.	Rubber cement	$ 1.65	×	5%	=	$0.08	? **$1.73**
3.	Note pad	$ 1.15	×	6%	=	$0.07	? **$1.22**
4.	Pencil/pen set	$23.95	×	4%	=	? **$0.96**	? **$24.91**

CULTURAL ANGLES
Assign each student the task of listing the names of the currencies in six foreign countries around the world. Ask one student to name all the countries that use the *dollar* as a basic unit of currency. (U.S., Australia, Canada, Hong Kong, New Zealand, Singapore, Taiwan) Ask students if the dollars used in foreign countries are the same as U.S. dollars. (No)

The sales tax rate is 6%. Find the total purchase price.

5.

2.49	GRAP
1.69	MILK
2.75	PROD
1.49	PROD
.65	GROC
2.29	DELI
.40	ONIO

$12.47

6.

.95	LETT
.79	FLOU
2.15	BRED
.52	MUSH
.99	CRAN
1.15	RICE
2.29	ORAN

$9.37

7.

BIRDSEED	9.94
THISTLES	8.95
FEEDER	27.45
SUET	1.39
SUNFLOWER	12.79

$64.15

8. Sales slip for Anita Ewing.
Shirt for $11.99.
Tie for $9.95.
Sales tax of $1.21. **$23.15**
What is the total purchase price?

9. Sales slip for Carl Gonzales.
Briefcase for $59.99.
2 notepads at $0.59 each.
Sales tax of $2.45. **$63.62**
What is the total purchase price?

10. Sales slip for Aurelio Esparza.
Computer disk file for $14.95.
Surge protector for $19.95.
2 boxes of labels at $19.99 each.
Sales tax rate of 4%. **$77.88**
What is the total purchase price?

11. Sales slip for Norma Engel.
Microwave for $149.99.
Microwave cart for $119.95.
Microwave cookware for $19.95
Sales tax rate of 5.5%. **$305.83**
What is the total purchase price?

12. Ralph McDounagh purchases a $45.85 power saw and a $12.95 extension cord. The sales tax rate is 6%. What is the total purchase price? **$62.33**

13. Gloria LeVect purchases the items shown on the sales slip. The sales tax rate is 2.5%. What is the total purchase price? **$132.23**

2 SWEATSHIRTS	29.50 EA
2 SWEATPANTS	24.50 EA
3 TURTLENECKS	7.00 EA

14. Frank Norris purchases a 35-mm camera body for $299.97, a 50-mm lens for $69.97, a tripod for $49.97, and 3 rolls of film at $2.95 each. The sales tax rate is 5.4%. What is the total purchase price? **$451.91**

15. Tanya Burley purchases 6 cans of motor oil at $1.09 a can, 8 spark plugs at $2.25 per plug, and an oil filter for $4.99. The sales tax rate is 5.5%. What is the total purchase price? **$31.15**

16. At Big Discount Store, Liz Doyle purchases a 22-lb turkey at 99¢ per pound, 6 oranges at 4 for $1.00, a $3.99-box of laundry detergent, a gallon of milk for $1.69, a 99¢ frozen pie, and 6 cans of condensed soup at 2 cans for 74¢. The sales tax rate is 3%. There is no sales tax charged on food items. What is the total purchase price? **$32.29**

Warm-Up Exercises

1. Add: $12.49, $3.16, $0.07, $10.40 $26.12
2. $129.55 + $37.50 + $1.22 + $0.67 $168.94
3. 5% of $136.75 $6.84
4. $8\frac{1}{2}\%$ × $67.20 $5.71

PRACTICE AND APPLY

The following problems can be assigned for classwork and the answers checked in class to help students master the objective of the lesson.

- Guided Practice: 1–4
- Independent Practice: 5–13 odd

ALTERNATIVE STRATEGIES: Enrichment

In the example, point out that the $155.41 is the total selling price. Since the total purchase price is the selling price plus the tax, it follows that:

Total selling price	+	*Sales tax*	=	*Total purchase price*
100% of the total selling price	+	*5% of the total selling price*	=	*105% of the total selling price*
100% of $155.41	+	*5% of $155.41*	=	*105% of $155.41*

WRAP-UP

Use Problems 9 and 11 to review the objective of the lesson. You might wish to ask for two volunteers to show their work at the chalkboard.

Assignment Guide

- Basic: 5–17, 20–33
- Average: 6–14 even, 15–18, 21–33 odd

17. Luther Strik purchases the items shown on the sales slip. The sales tax rate is 5.225%. There is no sales tax charged on items of clothing. What is the total purchase price?
$139.92

1	Calfskin Golves	$36.95
1	Sunglasses	34.95
2	Pair Socks	7.95 EA
2	Flashlights	8.95 EA
2	Lawnchairs	14.95 EA

18. Nancy Topok purchases a suit for $129.75, a blouse for $29, a pair of shoes for $46, and a matching bracelet and necklace for $33.95 each. She pays a state tax of 4% and a local tax of 2%. What is the total sales tax on her purchases? What is the total purchase price?
$16.36 **$289.01**

19. Jodi Lodyard purchased the camping items listed on the following sales receipt. The sales tax is 6%. Complete the sales receipt.

	SPORTLAND			
	QUANTITY	**DESCRIPTION**	**PRICE**	**AMOUNT**
$29.98	1	Propane cylinder	$ 29.98	?
$119.95	1	Propane cooker	119.95	?
$23.96	2	Pie cookers	11.98	?
$26.94	3	Folding camp stools	8.98	?
$7.72	2	Insect repellent	3.86	?
$208.55			Subtotal	?
$12.51			Sales tax	?
$221.06			Total	?

MAINTAINING YOUR SKILLS Look up the skills in parentheses if you need help or more practice.

Multiply. **(Skill 8)**

20. 7.4 × 9.2 **68.08**
21. 15.6 × 9.142 **142.6152**
22. 0.91 × 0.004 **0.00364**
23. 0.03 × 1.07 **0.0321**

Add. **(Skill 5)**

24. 322.29 + 174.00 + 4.52 = **500.81**
25. 19.38 + 7.8 + 0.007 = **27.187**
26. 9.65 + 0.0024 + 342.1 = **351.7524**
27. 708.326 + 57.942 + 9.888 = **776.156**
28. 542.36 + 400.50 + 0.724 = **943.584**
29. 414.6 + 8.05 + 960 = **1382.65**
30. 37.3 + 84.9 + 260.05 = **382.25**
31. 33 + 2.45 + 90.8 = **126.25**
32. 475.86 + 861.2 + 994.9 = **2331.96**
33. 416.015 + 513.02 + 322.811 = **1251.846**

5-3

Unit Pricing

OBJECTIVE
Compute the unit price.

Grocery stores often give unit price information for their products. Shoppers can use this information to determine which size of a product is the better buy based solely on price. The unit price of an item is its cost per unit of measure or count, such as dollars per pound or cents per dozen.

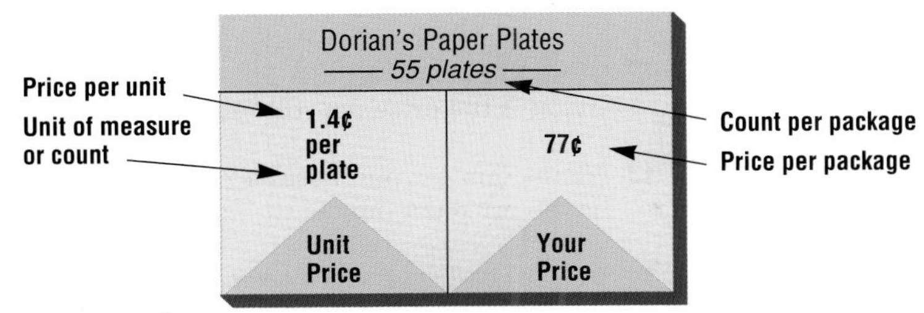

$$\text{Unit Price} = \frac{\text{Price per Item}}{\text{Measure or Count}}$$

EXAMPLE *Skills* 11, 2 *Application* A *Term* Unit price

Shane Burns purchased a 6.5-oz can of tuna for 89¢ and a dozen oranges for $3.96. What is the unit price of the items to the nearest tenth of a cent?

SOLUTION Find the **unit price.**

Price per Item ÷ Measure or Count

Tuna: $0.89 ÷ 6.5 = $0.1369 = 13.69¢ = 13.7¢ unit price per ounce

Oranges: $3.96 ÷ 12 = $0.33 = 33¢ unit price per orange

✓ SELF-CHECK Complete the problems, then check your answers in the back of the book.

1. A 32-oz can of spaghetti sauce costs 99¢. Find the unit price per ounce to the nearest tenth of a cent. **3.09¢ = 3.1¢**
2. A package of 8 hot dog buns costs 99¢. Find the price per bun to the nearest hundredth of a cent. **12.375¢ = 12.38¢**

PROBLEMS

Round to the nearest tenth of a cent.

Answer		Price per Item	÷	Measure or Count	=	Unit Price
3. **8.3¢**	**3.**	99¢	÷	12 oz	=	? per oz
4. **20.0¢**	**4.**	$ 7.99	÷	40 oz	=	? per oz
5. **65.9¢**	**5.**	$10.54	÷	16 m	=	? per m
6. **44.5¢**	**6.**	89¢	÷	2 L	=	? per L

LESSON PLAN
5-3 Unit Pricing

FOCUS

The use of some advertisements for a product based upon common units of measure can be used to motivate the topic of computing the unit price.

TEACH

Discuss the idea of a unit of measure and a count, using the illustration in the textbook. Ask students to name some other units of measure. (pound, gallon, yard, ounce, and so on) The computation in this lesson can be rather tedious unless calculators are used. Point out that the amount $0.1369 can be written in another way as 13.7 cents, or 13.7¢. This is the form required when the direction line states "to the nearest tenth of a cent."

Warm-Up Exercises

1. $.96 ÷ 30 $.032
2. $2.75 ÷ 25 $.11
3. $1.57 ÷ 115 $.014
4. $0.57 ÷ 85 $.007
5. $4.50 ÷ 8 $.563
6. $54.50 ÷ 18 $3.028

ALTERNATIVE STRATEGIES: Reteaching

Relate this lesson to the real world by using a newspaper advertisement or sale flyers for a grocery store or supermarket. The ads are a good source of current prices.

PRACTICE AND APPLY

The following problems can be assigned for classwork and the answers checked in class to help students master the objective of the lesson.

- Guided Practice: 1–6
- Independent Practice: 8–12 even

WRAP-UP

Ask a student to come to the chalkboard and write the formula for computing the unit price. Then have students use the data in Problem 10 to find the unit price.

Assignment Guide

- Basic: 8–16, 18–33
- Average: 7–13 odd, 14–17, 19–33 odd

Round to the nearest cent.

7. Roasted peanuts.
10 ounces for $1.89.
What is the price per ounce?
$0.19

8. Soft drinks.
12 cans for $2.69.
What is the price per can?
$0.22

9. 9-volt batteries.
2 per pack.
Sells for $1.99.
What is the price per battery?
$1.00

10. Mustard.
10.5-oz jar.
Sells for $1.29.
What is the price per ounce?
$0.12

11. Laura Zeedy recently purchased 4 tires for her automobile. The total purchase price was $178.44. What was the price per tire? **$44.61**

12. Alan Kinsman sees that Polaroid film is on sale for 2 rolls for $15.99. What is the price per roll? **$8.00**

13. Tania Ramirez purchases a 7.25-oz package of egg rolls for $1.09. What is the price per ounce? **$0.15**

14. A jar of barbecue sauce costs $1.29. There are 346 milliliters in the jar. What is the unit price to the nearest tenth of a cent? **$0.004**

15. Tracy Haserman purchases a package containing 3 video cassettes for $11.25. What is the unit price? **$3.75**

16. A 4-roll pack of bath tissue is priced at 2 packs for $1.75. What is the price per pack? What is the unit price? **$0.88; $0.22**

17. What is the unit price for each item in the advertisement?
a. Paper plates **$0.03**
b. Paper towels **$0.75**
c. Two cans of peaches **$0.55**
d. Potatoes **$0.20**

Shopper's Specials

PAPER PLATES, 50 ct.	$1.49
PAPER TOWELS, 2 rolls	$1.49
PEACHES, 59¢ ea.	2 / $1.10
POTATOES, 5-lb bag	$0.98

MAINTAINING YOUR SKILLS Look up the skills in parentheses if you need help or more practice.

Divide. Round answers to the nearest hundredth. **(Skill 11)**

18. 190.50 ÷ 76 **2.51** **19.** 13.4 ÷ 2.9 **4.62** **20.** 27.218 ÷ 7.6 **3.58** **21.** 0.91 ÷ 40 **0.02**

Round to the place value indicated. (Skill 2)

Nearest ten	**22.** 62.9 **63**	**23.** 19.8 **20**	**24.** 411.61 **410**	**25.** 1347.9 **1350**
Nearest tenth	**26.** 21.78 **21.8**	**27.** 741.92 **741.9**	**28.** 2.348 **2.3**	**29.** 1132.562 **1132.6**
Nearest hundredth	**30.** 3.218 **3.22**	**31.** 87.161 **87.16**	**32.** 430.144 **430.14**	**33.** 999.999 **1000.00**

COMMUNICATION SKILLS

Lead students in a discussion of the following statement: " 'Cheaper is better' is not always true." Guide the discussion to include: quality of goods and services, convenience of location, quantity, and availability. Ask if it makes sense to purchase a gallon of salad dressing for a family of 3. Have a student list the key points of the discussion on the chalkboard.

5-4

Finding the Better Buy

OBJECTIVE
Find the better buy based on unit price.

Compare the unit prices of products when you're deciding which size to buy. The size with the lower unit price is the better buy. If the larger size of a product is the better buy, consider whether you will be able to use up the product before it spoils.

EXAMPLE *Skills* 11, 1 *Application* A *Term* Unit price

Delman's raisins are sold in two sizes. The price of a 12-oz package is $1.29. The price of a 24-oz package is $1.99. Based on price alone, which package is the better buy?

SOLUTION

A. Find the **unit price** for each item.
Price per Item ÷ Measure or Count

12-oz package: $1.29 ÷ 12 = $0.1075 = 10.75¢ unit price per ounce

24-oz package: $1.99 ÷ 24 = $0.0829 = 8.29¢ unit price per ounce

B. Find the **better buy.**
Compare: 10.75¢ and 8.29¢ ← Better Buy

✓ SELF-CHECK Complete the problems, then check your answers in the back of the book.

1. 20 oz at 3.1¢ an ounce
2. 100 tea bags at 2.39¢ each

Find the better buy.

1. A 20-oz package costs 62¢. A 15-oz package costs 54¢.

2. A box of 48 tea bags costs $1.99. A box of 100 tea bags costs $2.39.

PROBLEMS

Find the unit price to the nearest tenth of a cent. Then determine the better buy based on price alone.

	Small Size	Large Size	Better Buy
3.	59¢ ÷ 16 oz = ? 3.7¢	89¢ ÷ 22 oz = ? 4.0¢	? small
4.	65¢ ÷ 10 = ? 6.5¢	$6.70 ÷ 100 = ? 6.7¢	? small
5.	$9.25 ÷ 14 oz = ? 66.1¢	$19.95 ÷ 30 oz = ? 66.5¢	? small

6. Scouring pads.
a. Box of 18 for $1.69.
b. Box of 10 for $1.05.
Which is the better buy? a.

7. Tortilla chips.
a. 16-oz bag for $2.49.
b. 12-oz bag for $1.75.
Which is the better buy? b.

LESSON PLAN
5-4 Finding the Better Buy

FOCUS
In choosing a product having different sizes, a decision can be made based upon the unit price. Ask students how they would do this. (Choose the size with the lower unit price.)

TEACH
After working through the Example, discuss some aspects of shopping by using unit prices. For example, one gallon of milk might not be the better buy for a family of two if some of the milk spoils before it is used up. Also, a person must be very careful when comparing the unit prices of two different brands of a product because they may not be of equal quality.

Warm-Up Exercises
Which number is greater?
1. 0.0264 or 0.0239 0.0264
2. 23.621 or 23.692 23.692
3. 18.296 or 17.999 18.296
4. 1.9 or 9.1 9.1

Divide. Do not round.
5. $0.22 ÷ 8 $0.0275
6. 1.44 ÷ 12 0.12
7. 0.50 ÷ 3.2 0.15625
8. 19.20 ÷ 32 0.6

ALTERNATIVE STRATEGIES: Enrichment
Give the following situations for purchasing liquid detergent and point out the inability to distinguish if the answers are rounded too much.
$1.78 ÷ 32 oz = $0.0556
$1.29 ÷ 22 oz = $0.0586
$2.89 ÷ 48 oz = $0.0602
If all three are rounded to the nearest cent $0.06, you cannot distinguish the best buy.

PRACTICE AND APPLY

The following problems can be assigned for classwork and the answers checked in class to help students master the objective of the lesson.

- Guided Practice: 1–5
- Independent Practice: 7–11 odd

WRAP-UP

Ask a student to explain how to find the better buy of two products based on unit price.

Assignment Guide

- Basic: 6–15, 17–22
- Average: 6–12 even, 13–16, 17–23 odd

8. Alicia Murphy has a choice of two sizes of spray paint. The 16-oz can sells for $3.29 while a 13-oz can sells for $2.69. Which size is the better buy? **16-oz can**

9. Jeff Evans wants to purchase some insect repellent. A 6-oz can sells for $3.86, while a 9-oz can sells for $5.59. Which size can is the better buy? **9-oz can**

10. Kathy Kruse is shopping in the meat department of the grocery store and sees the turkey advertised. Which turkey is the better buy? **18–20-lb turkey**

HOLIDAY SPECIAL *This week only* TURKEY
10 – 12lb $14.25
18 – 20lb $23.75

11. Nathan Ayers reads in the newspaper that his favorite cola is on sale at two stores in his neighborhood. The Country Store is selling a 24-can box for $5.99, while Food Mart is selling a 15-can box for $3.49. Which size box is the better buy? **15-can box**

12. Guiseppe Gerken wants to purchase some paint thinner. A 64-oz can costs $2.78, while a 16-oz can costs $1.09. Which is the better buy? **64-oz can**

13. Kenny Chape's favorite peanut butter is available in the following sizes: a 12-oz jar for $1.69, an 18-oz jar for $2.19, a 28-oz jar for $3.49, and a 40-oz jar for $4.79. Which size jar is the best buy? **40-oz jar**

14. Wu Chong is buying mint tea bags. The brand he prefers is sold in boxes of 16 bags for 89¢, 48 bags for $1.99, 100 bags for $2.49, and 150 bags for $3.39. Which size box is the best buy? **150 bags**

15. The Co-Op Food Chain sells 4 sizes of a generic laundry detergent. The family size contains 25 pounds and costs $10.99. The giant size contains 14 pounds and costs $5.89. The regular size contains 6 pounds 15 ounces and costs $3.09. The small size contains 4 pounds and costs $1.99.What is the unit price of each? Which size is the better buy? **2.7¢; 2.6¢; 2.8¢; 3.1¢ giant size 14-lb box**

16. Heather Baker is shopping for the best buy of a brand of dog food. The dog food comes in the following sizes: a 2.27-kg bag for $2.29, a 4.54-kg bag for $3.19, and an 11.35-kg bag for $5.89. What is the unit price of each? Which size bag is the best buy? **$1.01; 70.3¢; 51.9¢; 11.35-kg bag**

MAINTAINING YOUR SKILLS Look up the skills in parentheses if you need help or more practice.

Divide. Round answers to the nearest hundredth. **(Skill 11)**

17. $2.3\overline{)49.16}$ **21.37**
18. $0.250\overline{)18.5}$ **74.00**
19. $0.006\overline{)0.8703}$ **145.05**
20. $24\overline{)328.20}$ **13.68**

Which number is greater? **(Skill 1)**

21. $1568 or $1586 **$1586**
22. $0.2326 or $0.2349 **$0.2349**
23. 5.1 or 5.01 **5.1**
24. $96 or $106 **$106**

CRITICAL THINKING

Ask students to determine the best buy on a soft drink that is sold in a 16-oz bottle, a 12-oz can, and a 2-liter bottle. They should choose a product in the supermarket and report their findings to the class. (1 liter = 33.6 oz)

5-5 Coupons and Rebates

OBJECTIVE
Compute the final price after using a coupon or rebate.

Many manufacturers, stores, and service establishments offer customers discounts through **coupons** and **refunds** or **rebates.** These special discounts are an incentive to purchase a particular item. Manufacturer's or store coupons are redeemed at the time of purchase. In order to obtain a manufacturer's rebate, the consumer must mail in a rebate coupon along with the sales slip and the UPC (Universal Product Code) label from the items.

Final Price = Total Selling Price − Total Savings

EXAMPLE *Skill* 6 *Application* A *Terms* Coupon, rebate

Gus Senton purchased a 32-ounce bottle of liquid detergent for $2.19. He had a coupon for $1.00 off. What is the final price of the detergent?

$100
MANUFACTURER'S COUPON
NO EXPIRATION DATE
$100
SAVE
$100
on SAVON Liquid Detergent
Any Size
$100
$100

SOLUTION Find the **final price.**

Total Selling Price − Total Savings
$2.19 − $1.00 = $1.19 final price

✓ SELF-CHECK Complete the problems, then check your answers in the back of the book.

1. Cheese slices cost $2.09. Coupon for $1.00. Find the final price. **$1.09**
2. Trash bags cost $4.67. Coupon for 40¢. Find the final price. **$4.27**

PROBLEMS

Find the final price of each item.

	Item	Total Selling Price	Coupon	Final Price
3.	Fabric softener	$2.39	25¢	**$2.14** ?
4.	Deodorant	$1.98	35¢	**$1.63** ?
5.	Shampoo	$2.89	50¢	**$2.39** ?
6.	Crackers	$1.89	50¢	**$1.39** ?
7.	Freezer bags	$1.83	25¢	**$1.58** ?
8.	Cleanser	$2.79	20¢	**$2.59** ?

LESSON PLAN
5-5 Coupons and Rebates

FOCUS

Ask students if they have ever received coupons in the mail at home. Where else have they seen coupons? (newspapers, magazines) Discuss the value of coupons; that is, their value to a consumer.

TEACH

If you can bring some coupons to class, they would serve as good examples for this lesson. Use them to compute the final price by making up the selling price, if necessary. Then work through the Example.

Point out that coupons are usually found in local newspapers, or advertising literature, on the product container, or inside a container, such as in a box of cereal. Coupons are redeemed at the time of purchase. To obtain a manufacturer's refund or rebate, you must fill out a form, enclose a sales slip and some proof of purchase, such as the universal product code label. Also mention the inconvenience of having to include the sales slip or cash register receipt (you lose your own record of purchase). Don't forget the cost of the envelope and a stamp.

Warm-Up Exercises

1. $1.79 − $0.25 $1.54
2. $2.59 − $0.50 $2.09
3. $12.50 − $1.00 $11.50
4. $479.00 − $50.00 $429.00

ALTERNATIVE STRATEGIES: Enrichment

Have students bring coupons or rebate information to class. Ask them to also bring in newspaper advertisements that indicate the selling price of items for which they have a coupon. What effect do "double" or "triple" coupon offers have on the selling price of an item?

PRACTICE AND APPLY

The following problems can be assigned for classwork and the answers checked in class to help students master the objective of the lesson.

- Guided Practice: 1–6
- Independent Practice: 7–11

WRAP-UP

Ask students to explain the meaning of the term *rebate.* Ask another student how to compute the final price using a coupon or rebate.

Assignment Guide

- Basic: 7–15, 17–21
- Average: 12–16, 17–21 odd

F.Y.I.
Some stores will double and sometimes triple the value of a coupon.

9. Audrey Bailey purchased a 22-ounce can of cherry pie filling for $1.79. She had a store coupon for 35¢. What is the final price of the pie filling? **$1.44**

10. Jonnie Marker purchased 3 boxes of crackers for $1.79 per box. He had two coupons, one for 30¢ and one for 25¢. What is the final price of the crackers? **$4.82**

11. Shawn Derricotte purchased an AM/FM headset radio (7-1625S) for $28.95. What is the price after the rebate? **$25.95**

12. Tammy Duffey purchased a personal stereo (3-5415) for $127.45. What is the price after the rebate? **$124.45**

13. Joel Jurcevich purchased a clock radio (7-4646) for $57.67. What is the price after the rebate? **$57.67**

CASH BACK BONUS REBATES

Mail-in certificate (Not payable at the retail store)

Clock Radio/Telephone	
7-4712	$5.00
7-4712WH	$5.00
AM/FM Headset Radios	
7-1625BLS	$3.00
7-1625S	$3.00
7-1625	$3.00
Personal Stereo	
3-5415	$3.00
3-5415S	$3.00
3-5432	$3.00
3-5432S	$3.00

14. An AM/FM headset radio (7-1625BLS) sells for $48.95. What is the price after the rebate? **$45.95**

15. A personal stereo (3-5432) sells for $84.60. What is the final price after the rebate if an envelope costs $0.15 and a stamp costs $0.29? **$82.04**

16. Tom Keener does his own service work on his car. He made use of the manufacturer's rebates. He purchased these items: 2 battery cables for $7.95 each, 6 spark plugs at $1.29 each, 1 can of carburetor cleaner for $1.19, 1 can of engine cleaner for $1.29, an air filter for $8.97, an oil filter for $3.99, a spark plug wire set for $14.58, and 6 quarts of oil at $0.99 a quart. Determine the total cost of the service work on the car. How much will Tom have paid after he receives the rebates?
$59.60; $52.10

BONUS REBATES

25¢ off on Spark Plugs
50¢ off on Automotive Chemicals:
- Gas Treatment
- Oil Treatment
- Carburetor Cleaner
- Power Steering Fluid
- Engine Cleaner
- Brake Fluid

$1.00 off on Filters
(Oil Filter or Air Filter)
$3.00 off on Spark Plug Wire Sets

MAINTAINING YOUR SKILLS Look up the skill in parentheses if you need help or more practice.

Subtract. **(Skill 6)**

17.	**18.**	**19.**	**20.**	**21.**
$1.65	$11.65	$1.79	$127.98	$1.01
− .35	− 3.00	− .25	− 5.00	− .15
$1.30	**$8.65**	**$1.54**	**$122.98**	**$0.86**

MATHEMATICAL NOTES

Discuss the effect that using a coupon has on the unit price of an item, such as the cherry pie filling in Problem 9. (It lowers the unit price.) Have students do the calculations and compare the unit prices. (6.5 cents per oz verses 8.1 cents per oz)

5-6 Markdown

OBJECTIVE
Compute the dollar amount of the markdown.

Stores often sell products at **sale prices,** which are lower than their regular selling prices. The **markdown,** or **discount,** is the amount of money that you save by purchasing a product at the sale price.

The **markdown rate,** or **discount rate,** of an item is its markdown expressed as a percent of its regular selling price. Businesses frequently advertise the markdown rate rather than the sale price.

Markdown = Regular Selling Price − Sale Price

Markdown = Regular Selling Price × Markdown Rate

EXAMPLE 1 — *Skill* 6 *Application* A *Term* Markdown

Nora Maag purchased a VCR at a sale price of $399.99. The regular selling price is $549.99. What was the markdown?

SOLUTION

Find the **markdown.**

Regular Selling Price − Sale Price
$549.99 − $399.99 = $150.00 markdown

EXAMPLE 2 — *Skills* 30, 2 *Application* A *Term* Markdown rate

Furniture World is selling all summer furniture for 35% off the regular selling price. Chad Wolfrum purchased furniture there that regularly sold for $799. How much did he save by buying the furniture during the sale?

END OF SEASON SALE!
This month only
Save 35% on all
Summer Furniture
FURNITURE WORLD

SOLUTION

Find the **markdown.**

Regular Selling Price × Markdown Rate
$799.00 × 35% = $279.65 markdown

✔ SELF-CHECK Complete the problems, then check your answers in the back of the book.

1. The regular selling price of a camcorder is $999.95. The sale price is $419.95. Find the markdown. **$580**
2. Shoes are on sale for 40% off the regular selling price of $89.50. Find the markdown. **$35.80**

LESSON PLAN 5-6 Markdown

FOCUS

To introduce the ideas of this lesson, ask students if they know the meaning of the term *discount.* Then ask them to give some examples of when discounts are used.

TEACH

Use your students' examples from the Focus to introduce the other terms used in this lesson. Point out that markdowns on sale items are often advertised either by giving the dollar amount or the markdown rate. The rate could be a percent off, such as "30% off" or it could be a fraction off, such as "$\frac{1}{3}$ off."

Stress to students that in order to find the amount of savings on a sale item, it is necessary to know the regular selling price. The regular selling price serves as the base upon which the markdown is computed.

You may wish to bring to class advertisements for sales from such resources as newspapers, magazines, and advertisements. Depending on the information given in an advertisement, have students compute the markdown.

COOPERATIVE LEARNING

You can organize students into small groups to work on their assignment in class. Have one student from each group give the answer to at least one problem.

Warm-Up Exercises

1. $8.99 − $4.50 $4.49
2. $16.00 − $12.88 $3.12
3. 20% of $85 $17
4. $12\frac{1}{2}$% of $96 $12
5. $39.95 × 40% $15.98
6. $1390 × 15% $208.50

PRACTICE AND APPLY

The following problems can be assigned for classwork and the answers checked in class to help students master the objective of the lesson.

- Guided Practice: 1–9
- Independent Practice: 10–16 even

PROBLEMS

	Item	Regular Selling Price	−	Sale Price	=	Markdown
3.	Table	$169.99	−	$99.99	=	$70 ?
4.	Shirt	$ 16.99	−	$11.99	=	$5 ?
5.	Handbag	$ 25.99	−	$15.50	=	$10.49 ?

	Item	Regular Selling Price	−	Sale Price	=	Markdown
6.	Cookware	$19.99	×	40%	=	$8 ?
7.	Toaster	$12.99	×	23%	=	$2.99 ?
8.	Lamp	$80.00	×	35%	=	$28 ?
9.	Humidifier	$49.99	×	$\frac{1}{3}$ off	=	$16.66 ?

10. Computer software.
Regularly sells for $74.95.
Sale price is $59.95.
What is the markdown? **$15**

11. Mini van.
Regularly sells for $14,495.
Sale price is $11,595.
What is the markdown? **$2900**

12. A pair of dress pants.
Regularly sells for $87.99.
Markdown rate is 45%.
What is the markdown? **$39.60**

13. Plastic cups.
Regularly sell for $2.29.
Markdown rate is 25%.
What is the markdown? **$0.57**

14. Bargains on games usually occur in January. Angie Sigler purchases a backgammon set that regularly sells for $39.95. The set is on sale for $24.95. What is the markdown? **$15**

15. April is usually a good month to purchase home improvement supplies. A sheet of cedar paneling at Wetzler Lumber Yard regularly sells for $38.50. Jud Stassi purchases the paneling on sale for $24.95 a sheet. What is the markdown? **$13.55**

16. A 27″ color television that regularly sells for $899.95 is marked down 33% during McSurley's May Sale. What is the markdown? **$296.98**

17. Tori Topps sees a winter coat that regularly sells for $174.95 on sale at 50% off. How much can she save by purchasing the coat during the sale? **$87.48**

18. TBA Auto Parts holds a sale on car care products. All polishing cloths and cellulose sponges are marked down 35%. What is the markdown on polishing cloths that regularly sell for 90¢ each? What is the markdown on cellulose sponges that regularly sell for $1.35 each? **$0.32; $0.47**

19. Tires are generally on sale during the month of September. Melanie Peters purchased 4 of the glass-belted tires shown on sale. What was the total markdown on her purchase? **$64**

Ojai Tire Center

A78-13 Glass-belted
****** RADIALS ******

reg. $49.95 per tire

SALE PRICE $33.95 ea.

20. Typewriter sales often occur in the month of June. Pitre Office Supplies has 3 electronic typewriters Model 118E remaining in stock. The Model 118E regularly sells for $835 but is on sale for $585. What is the markdown? **$250**

$ 7.00
$13.75
$14.75
$ 6.25
$19.50

21. Gift items, such as the ones shown, are frequently discounted after the holidays in December and January. What is the markdown for each item advertised?

CRYSTAL SERVING ACCESSORIES
50% off regular prices

Salt and pepper set, reg. $14.00
Sugar and creamer, reg. $27.50
Handled mugs, set of 4, reg. $29.50
Jam jar with spoon, reg. $12.50
3-pc. salad set, reg. $38.99

22. Jeremy Berkowitz purchases a down jacket during an April coat sale at Westside Ski Shop. All coats are discounted 35%. The down jacket regularly sells for $74.50. What is the markdown? **$26.08**

White Sale	30% off
$2.11	$2.10
$3.99	$3.90
$5.99	$6.00
$1.99	$1.80
$5.19	$2.40
$19.27	$16.20

23. January, May, July, and August are months in which many department stores hold "white sales." Use the advertisement to find the markdown on each item. Then compare the results with a markdown rate of 30% off the regular selling price on each item. Which results in a larger markdown? **White sale**

August White Sale

Full flat or fitted sheets
Reg. $6.99 . . . . 2 / $9.75
Queen flat or fitted sheets
Reg. $12.99 . . . . 2 / $18.00
King flat or fitted sheets
Reg. $19.99 . . . . 2 / $28.00
Standard cases,
Reg. $5.99 . . . . $7.99 pr.
King cases
Reg. $7.99 . . . . $5.59 pr.

MAINTAINING YOUR SKILLS Look up the skills in parentheses if you need help or more practice.

Subtract. **(Skill 6)**

24. 401.44 − 98.25 **303.19**

25. 79.7 − 9.924 **69.776**

26. 5.6 − 4.6301 **0.9699**

27. 73.291 − 62.824 **10.467**

Find the percentage. **(Skill 30)**

28. 30% of $120 **$36**

29. 40% of $280 **$112**

30. $\frac{1}{4}$ % of 850 **2.125**

31. 230% of 550 **1265**

WRAP-UP

Call upon individual students to explain the difference between the markdown and the markdown rate and between the discount and discount rate.

Assignment Guide

- Basic: 10–21, 24–31
- Average: 11–17 odd, 19–23, 25–31 odd

ALTERNATIVE STRATEGIES: Reteaching

Relate this lesson to the real world by using newspaper advertisements as a source of problems. Point out different types of markdown ads.

A. Reg. price $229
Sale price − 199
Markdown $ 30

B. Reg. price $44
Markdown: 25%
Markdown = 25% of $44 = $11

LESSON PLAN
5-7 Sale Price

FOCUS
In this lesson students compute the sale price when the markdown rate is known. Ask students what they must multiply the markdown rate by in order to find the amount of the markdown.

TEACH
Discuss the Example. See the calculator note below for showing students how to work the Example by using a calculator.

Have students make a list of the different ways stores usually give the markdown on sale items. For example, $10 off regular selling price; $\frac{1}{3}$ off regular selling price; 20% off regular selling price. Emphasize that the sale price of an item can be determined if the regular selling price and the markdown are given. Remind students that the discount can be given as a dollar amount or as a percent.

Warm-Up Exercises
1. $235.00 − $23.50 $211.50
2. $7.88 − $1.56 $6.32
3. $899.95 × 25% $224.99
4. $565 × 50% $282.50

5-7 Sale Price

OBJECTIVE
Compute the sale price when markdown rate is known.

For items that are on sale, some stores advertise the amount of markdown and the regular selling price. If you know this information, you can calculate the sale price.

Sale Price = Regular Selling Price − Markdown

EXAMPLE *Skills* 30, 6 *Application* A *Term* Sale price

F.Y.I.
Note that 38% off implies that you pay 62%. Thus, 62% of $47.95 is $29.73.

Debralee Medford is purchasing this clock radio at Highland's Hi-fi. What will she pay for the radio?

Dual-Alarm Clock Radio
Cut 38% Reg. 47.95

SOLUTION

A. Find the **markdown.**
Regular Selling Price × Markdown Rate
$47.95 × 38% = $18.221 = $18.22 markdown

B. Find the **sale price.**
Regular Selling Price − Markdown
$47.95 − $18.22 = $29.73 sale price

47.95 M+ × 38 % 18.221 M− RM 29.729

✔ SELF-CHECK Complete the problems, then check your answers in the back of the book.

1. The regular selling price of a skirt is $45. The markdown rate is 30%. Find the sale price. $31.50
2. The regular selling price of a bicycle is $219. The markdown rate is 20%. Find the sale price. $175.20

PROBLEMS

	Regular Selling Price	×	Markdown Rate	=	Markdown	Sale Price
3.	$120.00	×	40%	=	$48	$72
4.	$ 79.90	×	30%	=	$23.97	$55.93
5.	$ 10.99	×	25%	=	$2.75	$8.24
6.	$175.00	×	15%	=	$26.25	$148.75
7.	$450.00	×	65%	=	$292.50	$157.50

ALTERNATIVE ASSESSMENT
Have students make a list of all the terms introduced in this lesson. Then call upon one student to read a term aloud and another student to give its meaning.

USE OF CALCULATORS
Show students the following sequence of keys to work the Example:
47.95 M+ × 38 % 18.221 M− RM 29.729.

8. Rechargeable lantern.
Regularly sells for $21.95.
Markdown rate is 30%.
What is the sale price? **$15.36**

9. Compact stereo.
Regularly sells for $129.95.
Markdown rate is 53%.
What is the sale price? **$61.08**

10. Amity's Shoe Store has marked down men's work and outdoor boots 25% during its spring sale. What is the sale price of a pair of boots with a regular price of $49.99? **$37.49**

25% OFF
ALL MEN'S WORK and OUTDOOR BOOTS
Regularly
14.99 – 49.99

11. During an August pre-school sale, Casual Apparel, Inc. has marked down its line of dress shirts 30%. These shirts are regularly priced at $17.95 each. What is the sale price of these shirts? **$12.56**

12. The Wyoming Pizza Shack discounts 15% off the regular selling price on the purchase of 3 or more pizzas. What is the sale price on the purchase of 3 pizzas that regularly sell for $5.75, $7.75, and $9.95?
$4.89; $6.59; $8.46

13. Owen's Discount Store placed this advertisement in the local papers. Verify that you save 29%.
$27.88 − $19.79 = $8.09 = 29% of $27.88

14. Ruiz's Furniture Market carried this ad for recliner rockers. Determine the regular selling price. Determine the percent markdown. **$449.99; 44.4%**

MAINTAINING YOUR SKILLS Look up the skills in parentheses if you need help or more practice.

Find the percentage. Round answers to the nearest hundredth. **(Skill 30)**

15. 25% of $74.80 **$18.70**
16. 7.5% of 425 **31.88**
17. 8% of 322 **25.76**
18. 35% of 82.5 **28.88**

Subtract. **(Skill 6)**

19. 471.814 − 389.008 = **82.806**
20. 37 − 26.321 = **10.679**
21. 0.016 − 0.0019 = **0.0141**
22. $138.26 − 125.18 = **$13.08**
23. $22.85 − 18.36 = **$4.49**

PRACTICE AND APPLY

The following problems can be assigned for classwork and the answers checked in class to help students master the objective of the lesson.

- Guided Practice: 1–7
- Independent Practice: 8–10

WRAP-UP

Using the Example, have a student change the markdown rate and the regular selling price. Have the other students in the class compute the sale price.

Assignment Guide

- Basic: 8–12, 15–23
- Average: 11–14, 16–22 even

ALTERNATIVE STRATEGIES: Enrichment

Have students look at the example and determine what percent they pay. 38% off implies 62% is paid. Have them complete this statement: The sales price is _?_% of the regular price (62%). Have them solve problems 1 through 7 using this concept.

LESSON PLAN
Reviewing the Basics

The exercises on this page review skills, applications, and terms used in the unit. You can use the exercises to assess informally students' proficiency with this material.

The page can be used for guided practice and independent practice. You can work through a selection of the exercises together with students, and thus see immediately if they know how to do them, and you can then assign some of the exercises for independent practice. Be sure to go over the answers to all assigned exercises.

Reviewing the Basics

Skills

Write the number that is smaller.

(Skill 1)
1. 14.4¢ or 15.2¢ **14.4¢**
2. 50.01¢ or 50.1¢ **50.01¢**
3. 17.25¢ or 17.3¢ **17.25¢**

(Skill 2) Round to the nearest cent.

4. $12.2806 **$12.28**
5. $1.723 **$1.72**
6. 49.36¢ **49¢**
7. 421.71¢ **422¢**

Solve.

(Skill 5)
8. $4.49 + $2.57 + $26.08 **$33.14**
9. $139.15 + $12.20 + $127.85 **$279.20**

(Skill 6)
10. $86.50 − $18.59 **$67.91**
11. $103.71 − $32.51 **$71.20**
12. $75 − $1.07 **$73.93**

(Skill 8)
13. $4.69 × 8 **$37.52**
14. $45.60 × 24 **$1094.40**
15. $74.27 × 1.04 **$77.24**

(Skill 11)
16. 95¢ ÷ 16 **5.94¢**
17. 85¢ ÷ 12 **7.08¢**
18. $4.32 ÷ 10 **$0.432**

(Skill 30)
19. 18% of $18.95 **$3.41**
20. 3.2% of $60 **$1.92**
21. 8.5% of $372.80 **$31.69**

Applications

Use the following formula to find the markdown.

Markdown = Regular Selling Price × Markdown Rate

(Application A)
22. Sewing machine. Regular price $479.95. Markdown rate of 30%. **$143.99**
23. TV set. Regular price $799.95. Markdown rate of 35%. **$279.98**

Terms

Match each term with its definition on the right.

g **24.** Sales tax
e **25.** Sales receipt
f **26.** Unit price
a **27.** Markdown
d **28.** Markdown rate
b **29.** Sale price

a. the amount of money that is saved by buying an item at the sale price
b. the purchase price of an item after markdown
c. the total purchase price including sales tax
d. the markdown of an item expressed as a percent of the regular selling price
e. a cash register tape or sales slip showing the selling price of items purchased, any sales tax, and the total purchase price
f. the cost per unit measure or count of an item
g. a tax charged on the selling price of an item or service provided

Refer to your reference files at the back of the book if you need help.

Unit Test

Lesson 5-1

1. Debra Rothchild purchases a garden hose for $12.95, 5 packs of seeds at $0.59 each, and a flat of flowers for $7.25. The sales tax rate is 6%. What is the sales tax on her purchases? **$1.39**

Lesson 5-2

2. Jack Roth purchases 6 cans of oil at 99¢ a can, 8 spark plugs at $2.25 each, a $7.49 air filter, and an oil filter for $4.89. The sales tax rate is 4%. What is the total purchase price of his purchases? **$37.77**

Lesson 5-3

3. A brand of salad dressing that Sandy Beckman prefers is on sale at Foodway. What is the unit price of each package? **16.1¢; 15.8¢**

> *Shop and Save at*
> **FOODWAY**
> This week's specials:
> **Creamy Dressing**
> 8-oz bottle for $1.29
> 12-oz bottle for $1.89

Lesson 5-4

4. Martin Gillis compares the prices for paper towels. A 3-roll package is priced at $2.89. A single roll is priced at $0.89. Based on price alone, which is the better buy? **$0.89**

Lesson 5-5

5. Jim Masters purchased 2 boxes of crackers for $1.99 per box. He had two coupons, one for 30¢ and one for 25¢. What is the final price of the crackers? **$3.43**

Lesson 5-5

6. A roll of film sells for $3.97. What is the price after a $1.00 rebate if an envelope costs 15¢ and a stamp costs 29¢? **$3.41**

Lesson 5-6

7. Mary Patterson purchases a CB radio on sale for $79.95. The regular selling price was $139.95. What is the markdown on the radio? **$60**

Lesson 5-6

8. Fuller's Bedland regularly sells a posture classic king-sized mattress for $309.95. For an August clearance sale, the mattress is discounted 30% off the regular selling price. How much is saved by purchasing the mattress during the sale? **$92.99**

Lesson 5-7

9. Bob Kenton purchases a jacket that regularly sells for $39.95. The jacket has been discounted 33%. What is the sale price of the jacket? **$26.77**

Lesson 5-7

10. A rechargeable flashlight regularly sells for $15.97. It is on sale for $11.97 and has a rebate of $2.00. What is the final price after the rebate if an envelope costs $0.15 and a stamp costs $0.29? How much is saved altogether? **$10.41; $5.56**

LESSON PLAN
Unit Test

Students should do the Unit Test on their own. Each problem on the test is keyed to a lesson in the unit. Students having difficulty with any particular problem should review the Example in the appropriate lesson and be assigned some of the Independent Practice problems for additional practice.

LESSON PLAN
A Spreadsheet Application

USING TECHNOLOGY

Take a few minutes to review with students the meanings of the terms used in this application: *selling price, percent markdown, sale price, sales tax rate,* and *total purchase price.* After students have worked the problems using their computers, work through Problem 1 using a calculator. Write each step on the chalkboard to illustrate the sequence of steps necessary to arrive at the total purchase price. Tell students that this sequence of steps can be called an *algorithm*. Ask students to research the meaning of this term and report their findings to the class.

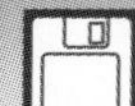

A SPREADSHEET APPLICATION

Cash Purchases

To complete this spreadsheet application, you will need the template diskette for *Mathematics with Business Applications*. Follow the directions in the User's Guide to complete this activity.

Input the information in the following problems to find the sale price and total purchase price.

1. Item: chair.
Selling price: $340.
Percent markdown: 25%.
What is the sale price?
Sales tax rate: 6.5%.
What is the total purchase price?

2. Item: shirt.
Selling price: $15.25.
Percent markdown: 10%.
What is the sale price?
Sales tax rate: 4%.
What is the total purchase price?

3. Item: couch.
Selling price: $568.99.
Percent markdown: 20%.
What is the sale price?
Sales tax rate: 7.55%.
What is the total purchase price?

4. Item: socks.
Selling price: $3.65.
Percent markdown: 33%.
What is the sale price?
Sales tax rate: 7.25%.
What is the total purchase price?

5. Item: wallet.
Selling price: $15.99.
Percent markdown: 25%.
What is the sale price?
Sales tax rate: 6.50%.
What is the total purchase price?

6. Item: jeans.
Selling price: $24.99.
Percent markdown: 15%.
What is the sale price?
Sales tax rate: 7%.
What is the total purchase price?

7. Mary Zimmerman purchased a microwave oven that was marked down 20%. The regular price is $156 and the sales tax rate is 8.5%. What is the sale price? What is the total purchase price?

8. Cleo Voss purchased a computer that was marked down 30%. The regular price is $2345 and the sales tax rate is 6.5%. What is the sale price? What is the total purchase price?

A SPREADSHEET APPLICATION
(CONTINUED)

9. A lawn mower regularly sells for $239.99. If it is currently marked down 17%, what is the sale price? If the sales tax rate is 6%, what is the total purchase price?

10. A washing machine regularly sells for $519.99. If it is currently marked down 23%, what is the sale price? If the sales tax rate is 6.50%, what is the total purchase price?

11. Carter Auto Sales advertised 10% off the list price of all automobiles. A station wagon has a list price of $15,550. The sales tax rate is 8%. What is the total purchase price?

12. Perkins Clothing Store advertised 45% off all winter clothing. Diane Peebles picked out several items with a total selling price of $435.42. The sales tax rate is 5.25%. What is the total purchase price?

13. An oak desk regularly sells for $589.99. If it is marked down 24%, what is the sale price? If the sales tax rate is 6.5%, what is the total price?

14. A curio cabinet regularly sells for $319.99. If it is marked down 14%, what is the sale price? The sales tax rate is 5.75%. What is the total purchase price?

15. Everhard's Appliance Store advertised 25% off the list price of all washing machines in stock.The most popular model has a list price of $639.99. The sales tax rate is 7%. What is the total purchase price?

16. Cassie's Health Foods advertised 10% off the list price of Choice Herbs. Sue Trundell selected items that totaled $122.43. The sales tax rate is 6.25%. What is the total selling price?

LESSON PLAN
Career Wise

APPLICATIONS ON THE JOB

Ask students if any of them has worked as a cashier after school. Do they use a computerized cash register? How does it work? Then turn their attention to the pie chart. Ask why they think it is called a pie chart. How are the sizes of each piece determined? (by the percents represented)

Go over the four questions at the bottom of the page. Extend Question 4 by finding the number of other types of books identified in the pie chart.

CAREER WISE

Cashier

In 1991 there were 47,489 new books and new editions published in the United States. This is no surprise to Tomas Ortiz who works as a cashier at a downtown bookstore. In a typical day, he will sell books for students, children's books, computer and technology books, cookbooks, craft books, and many other types of books.

Recently, he set aside a collection of books that had been marked down for clearance. Markdowns ranged from 25% to 40%. The paperback biographies were marked down the most and sold quickly.

As a cashier, Tomas uses a computerized cash register that automatically records the sale price, computes the tax, and gives the total purchase. The register also indicates the correct change. When the day is done, he reconciles the day's transactions with the starting and ending cash amounts.

The pie chart at the left below shows a percentage breakdown by subject area of the 47,489 new books and new editions for 1991. The chart at the right below shows how the academic books comprise the "academic slice" of the pie.

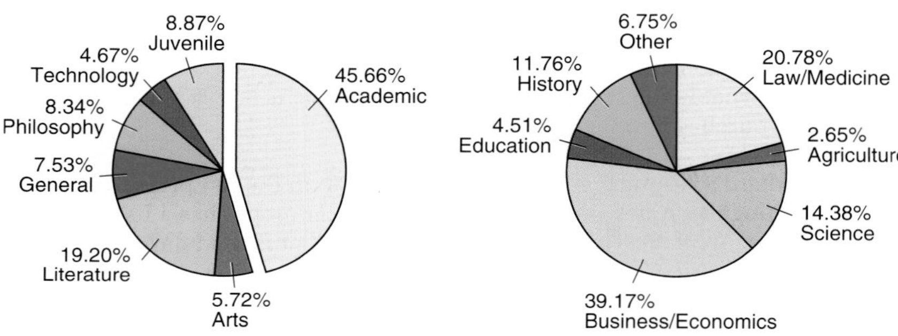

Check Your Understanding

1. Refer to the pie chart at the left above. In which category might you find "how-to" books? **General**
2. Which are the two largest slices of the pie chart for academic books? **Business/Economics, Law/Medicine**
3. How do the two answers to question 3 compare? **Business/Economics is about twice as big as Law/Medicine.**
4. Of the 47,489 new books and new editions in 1991, how many were academic? How many were science books?
45.66% of 47,489 books = 21,683 books;
14.38% of 45.66% of 47,489 books = 3118 books

Cumulative Review

Skills

Write the amount in words with cents expressed as a fraction of a dollar.

(Skill 1)

1. $47.59
2. $385.63
3. $2035.26

1. Forty-seven and $\frac{59}{100}$ dollars
2. Three hundred eighty-five and $\frac{63}{100}$ dollars
3. Two thousand thirty-five and $\frac{26}{100}$ dollars

Write the number that is less.

4. 7.3¢ or 6.9¢ **6.9¢**
5. $321 or $298 **$298**
6. $0.013 or $0.0129 **$0.0129**

Round to the nearest cent.

(Skill 2)

7. $33.175 **$33.18**
8. $0.1456 **$0.15**
9. $137.997 **$138.00**
10. $4716.3042 **$4716.30**

Round to the nearest tenth of a cent.

11. 15.415¢ **15.4¢**
12. 8.573¢ **8.6¢**
13. 33.0192¢ **33.0¢**
14. 95.1775¢ **95.2¢**

Solve.

(Skill 4)

15. $96 − $74 **$22**
16. $619 − $479 **$140**
17. $8914 − $7105 **$1809**
18. $3734 − $1573 **$2161**
19. $10,913 − $315 **$10,598**
20. $1117 − $927 **$190**

(Skill 5)

21. $421.00 + $785.00 **$1206**
22. $47.69 + $13.87 **$61.56**
23. $629.78 + $26.05 + $9.00 **$664.83**
24. $3.67 + $12.19 + $0.89 **$16.75**

(Skill 6)

25. $900.24 − $300.29 **$559.95**
26. $321.83 − $200.44 **$121.39**
27. $136.75 − $23.84 **$112.91**
28. $461.90 − $322.21 **$139.69**
29. $4325.60 − $190.25 **$4135.35**

(Skill 8)

30. $7.25 × 40 **$290**
31. $9.50 × 36 **$342**
32. $4.25 × 27.5 **$116.875**
33. $0.99 × 0.207 **$0.20493**
34. $0.25 × 0.004 **$0.001**
35. $3.70 × 0.06 **$0.222**

(Skill 11)

36. $9.66 ÷ 24 **$0.4025**
37. $4.56 ÷ 12 **$0.38**
38. $352.96 ÷ 52 **$6.79**
39. $43.02 ÷ 0.32 **$134.44**
40. $9.75 ÷ 0.25 **$39**
41. $149.28 ÷ 0.75 **$199.04**

(Skill 30)

42. 35% of $375.95 **$131.58**
43. 3.5% of $9290 **$325.15**
44. 33.5% of $2240 **$750.40**
45. 8.13% of $11,800.50 **$959.38**
46. $7\frac{3}{4}$% of $9600 **$744**
47. $8\frac{1}{2}$% of $4968.15 **$422.29**

(Skill 16)

48. $\frac{1}{2} + \frac{3}{4}$ **$1\frac{1}{4}$**
49. $3\frac{1}{2} + 2\frac{1}{2}$ **6**
50. $8\frac{3}{4} + 6\frac{1}{2}$ **$15\frac{1}{4}$**

Write as a decimal.

(Skill 14)

51. $\frac{1}{2}$ **0.5**
52. $\frac{1}{10}$ **0.1**
53. $25\frac{1}{4}$ **25.25**
54. $42\frac{3}{8}$ **42.375**

(Skill 28)

55. 7.25% **0.0725**
56. 37.30% **0.3730**
57. $12\frac{1}{2}$% **0.125**
58. $45\frac{3}{4}$% **0.4575**

LESSON PLAN
Cumulative Review (Skills/Applications) Units 1-5

The exercises on this page review skills, applications, and terms used in Units 1–5. You can use the exercises to assess informally students' proficiency with this material.

These two pages can be used for guided practice and independent practice. You can work through a selection of the exercises together with students, and thus see immediately if they know how to do them, and you can then assign some of the exercises for independent practice. Be sure to go over the answers to all assigned exercises.

Applications

(Application A)

Use the following formula to find the markdown.

Markdown = Regular Selling Price × Markdown Rate

59. Luggage bag.
Regular price is $39.95.
Markdown rate is 25%. **$9.99**

60. Barbecue grill.
Regular price is $79.95.
Markdown rate is 40%. **$31.98**

61. Sports coat.
Regular price is $125.00.
Markdown rate is 20%. **$25**

62. Lawn mower.
Regular price is $219.95.
Markdown rate is 33%. **$72.58**

Use the table on page 644 to find the amount of $1.

63. 4 interest periods.
Rate per period is 2.875%.
1.12005

64. 9 interest periods.
Rate per period is 3.000%.
1.30477

65. 7 interest periods.
Rate per period is 2.875%.
1.21946

66. 20 interest periods.
Rate per period is 1.500%.
1.34685

(Application F)

Find the elapsed time.

67. From 9:15 a.m. to 11:55 a.m.
2hrs. 40 min.

68. From 1:10 p.m. to 6:30 p.m.
5 hrs. 20 min.

69. From 7:52 a.m. to 5:20 p.m.
9 hrs. 28 min.

70. From 8:17 a.m. to 5:30 p.m.
9 hrs. 13 min.

(Application G)

Use the table on page 645 to find the elapsed time in days.

71. From May 7 to May 26
19

72. From June 2 to June 30
28

73. From April 15 to July 23
99

74. From March 10 to May 21
72

Write as a fraction of a year.

75. 3 months $\frac{1}{4}$

76. 6 months $\frac{1}{2}$

77. 10 months $\frac{5}{6}$

78. 16 months $1\frac{1}{3}$

Find the number of occurrences.

79. Weekly for 3 years **156**

80. Biweekly for 2 years **52**

81. Monthly for $1\frac{1}{2}$ years **18**

82. Semiannually for 4 years **8**

Terms

Write you own definition for each term. **Answers wll vary.**

83. Straight-time pay
84. Overtime pay
85. Commission
86. Income tax
87. Social security
88. Group insurance
89. Sales tax
90. Unit price
91. Markdown
92. Sale price
93. Deposit
94. Reconcile
95. Withdrawal
96. Simple interest
97. Compound interest
98. Time card
99. Passbook
100. Piecework

Refer to your reference files in the back of the book if you need help.

Cumulative Review Test

Units 1–5

Lesson 1-2

1. Julius Harrison works as a truck driver and earns $9.40 an hour for a regular 40-hour week. His overtime rate is $1\frac{1}{2}$ times his regular hourly rate. This week he worked his regular 40 hours plus $7\frac{3}{4}$ hours of overtime. What is his total pay? **$485.28**

Lesson 1-5

2. Bonnie Webster is an accountant for Radio News Corp. She earns an annual salary of $42,312 paid on a semimonthly basis. What is her semimonthly pay? **$1763**

Lesson 1-6

3. R. T. Sloan sells hospital equipment for a 9% straight commission. Her sales totaled $94,321 in September. What is her pay for the month? **$8488.89**

Lesson 2-1

4. Royce Adill, a shipping supervisor, earns $448.50 a week. He is married and claims 2 allowances. How much is withheld from his weekly paycheck for federal income tax? **$44**

WEEKLY Payroll Married

Wages		Allowances			
At least	But less than	0	1	2	3
420	430	54	47	41	35
430	440	55	49	43	36
440	450	57	50	44	38
450	460	58	52	46	39
460	470	60	53	47	41
470	480	61	55	49	42
480	490	63	56	50	44
490	500	64	58	52	45
500	510	66	59	53	47
510	520	67	61	55	48

Lesson 2-2

5. Julia Capper earns $32,564 a year as a systems analyst. The state income tax rate is 3.6% of taxable income. Her personal exemptions total $4692. How much is withheld each week from her gross pay for state income tax? **$19.30**

Lesson 2-4

6. Wilbur Clark, a medical technician, is paid $418.25 a week. His earnings to date this year total $15,057. The medicare tax rate is 1.45% of all income. How much is deducted from Wilbur's paycheck this week for the medicare tax? **$6.06**

Lesson 2-6

7. Cletus Carr, a research assistant for Winzler Chemicals, is married and claims no allowances. He earns $505.20 a week. The social security tax rate is 6.2% of the first $62,700 earned and the medicare tax rate is 1.45% of all income. The state tax is $10.15 a week. He has weekly deductions of $12 for medical insurance and $60 for payroll savings. Use the table above to find his federal tax withheld. What is his net pay for a week? **$318.40**

Lesson 3-2

8. Gary Webster wrote a check to Seaton Oil Co. in the amount of $337.19. The check number was 363. Write the amount of the check in words with cents expressed as a fraction of a dollar.
Three hundred thirty-seven and $\frac{19}{100}$

LESSON PLAN
Cumulative Review Test
Units 1-5

Students should do the Cumulative Review Test on their own. Each problem on the test is keyed to a lesson in Units 1–5. Students having difficulty with any particular problem should review the Example in the appropriate lesson and be assigned some of the Independent Practice problems for additional practice.

Lesson 3-4

9. A portion of Betty Bigg's bank statement is shown. Her balance from last month's statement was $829.32. What is her present balance? **$1135.55**

CHECKS AND OTHER CHARGES			DEPOSITS AND CREDITS		BALANCE
DATE	NUMBER	AMOUNT	DATE	AMOUNT	
10/4	902	41.94	10/8	347.90	
10/4	904	29.63	10/22	271.40	
10/15	903	237.00			
SERVICE CHARGE		4.50			

Lesson 3-5

10. Betty Biggs compares her checkbook register, canceled checks, and bank statement. She finds that she has check numbers 901 and 905 outstanding in the amounts of $41.83 and $95.92. She also has an outstanding deposit of $400. She reconciles the statement. What should her adjusted balance be? **$1397.80**

Lesson 4-4

11. Viola Eyster's savings account statement shows the following: a previous balance of $913.82, a withdrawal of $241.93, total deposits of $518.25, and interest of $5.12. What is the new balance? **$1195.26**

Lesson 4-5

12. Esther Marshal deposited $3472.50 in an account that pays an annual interest rate of $6\frac{1}{2}$%. She made no other deposits or withdrawals. How much simple interest did Beverly's money earn in 6 months? **$112.86**

Lesson 4-7

13. Swanton Savings Bank pays $5\frac{1}{2}$% interest compounded quarterly on regular savings accounts. Willard Boehler deposited $4400 in a regular savings account for $1\frac{1}{2}$ years. He made no other deposits or withdrawals. Use the table to find the interest earned. **$375.72**

Total Interest Periods	AMOUNT OF $1.00 Rate Per Period	
	1.375%	1.500%
4	1.05614	1.06136
5	1.07067	1.07728
6	1.08539	1.09344

Lesson 5-2

14. Michael Noll purchases 6 cans of soup at 77¢ a can, 3 pounds of green beans at 89¢ a pound, a $3.49-box of laundry detergent, and a package of chicken for $6.74. The sales tax rate is 6.5%. What is the total purchase price? **$18.66**

Lesson 5-4

15. Sue Duffey compares the prices for fabric softener dryer sheets. A box of 20 is priced at $0.89. A box of 40 is priced at $1.59 and a box of 60 is priced at $2.29. Based on price alone, which is the better buy? **Box of 60**

Lesson 5-6

16. Conner Appliance regularly sells a Model K2504E 25″ console color television for $849.99. For an August clearance sale, the Model K2504E is discounted 40% off the regular selling price. How much is saved by purchasing the television during the sale? **$340**

U N I T

6

Charge Accounts and Credit Cards

A *charge account* allows you to "buy now and pay later." You pay for goods or services with a *credit card.* At the end of a month, the credit company bills you for all the items you purchased on credit. This account *statement* shows any *finance charges* you must pay if your previous bill was not paid in full. A finance charge is interest charged on the amount you owe. Finance charges are figured by the *previous-balance method,* the *unpaid-balance method,* or the *average-daily-balance method.*

O U T L I N E

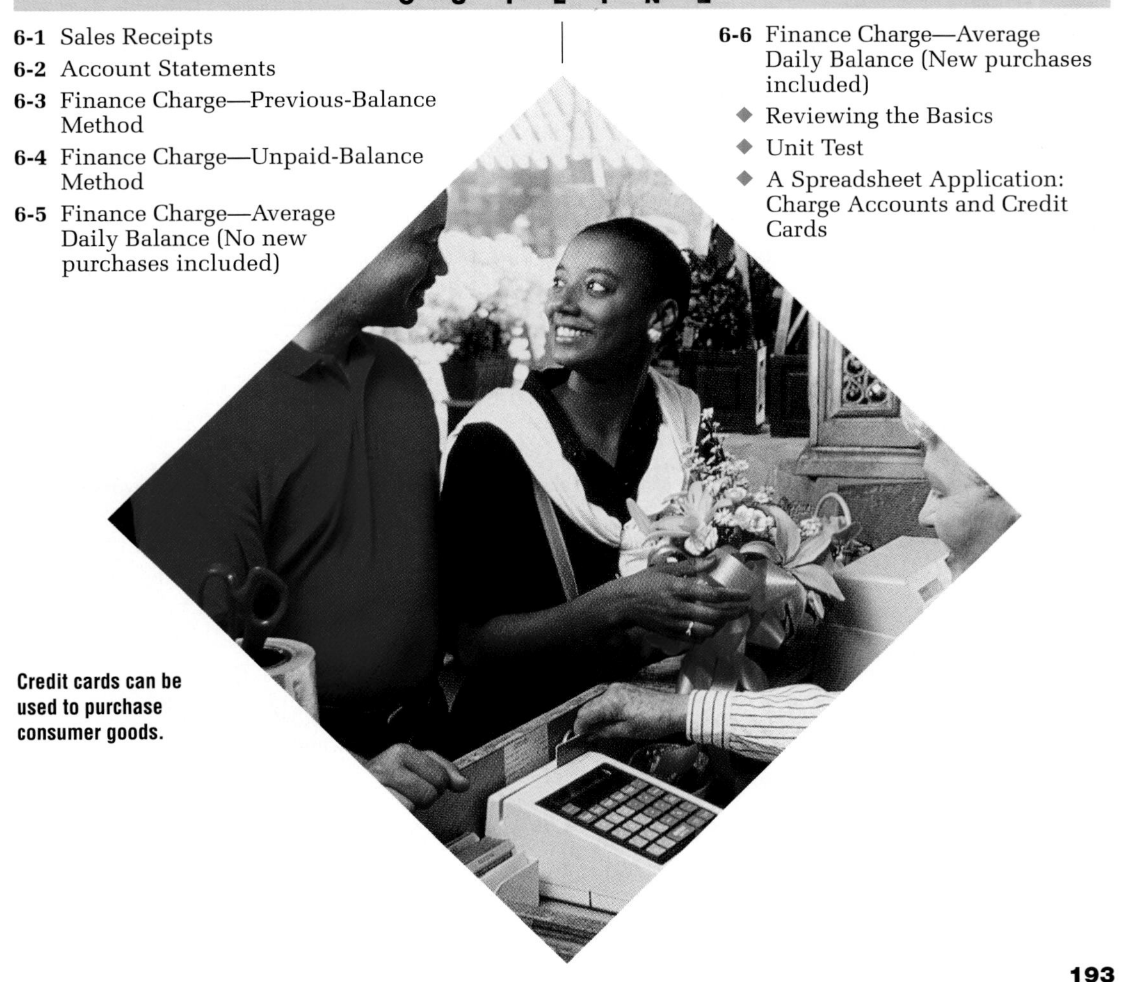

Credit cards can be used to purchase consumer goods.

UNIT 6
CHARGE ACCOUNTS AND CREDIT CARDS

INTRODUCING THE UNIT

Discuss with students the responsibilities that accompany owning a credit card. In many stores, charge receipts and charge slips are used interchangeably with sales receipts. Errors can be made either in writing a sales receipt or in entering an amount in the cash register. Stress that all amounts, including the total and sales tax, should be checked by the customer. Discuss the implications of signing an inaccurate sales receipt. Point out there is usually a membership fee, which is paid on an annual basis.

193

LESSON PLAN
6-1 Sales Receipts

FOCUS
Ask students to suggest an alternative to paying cash when purchasing an item in a store, buying gasoline, or for paying for a dinner at a restaurant. (Charge it!) Have students name various credit cards with which they are familiar. In this lesson, the total purchase price is computed when using a credit or charge card.

TEACH
Point out the various parts of the sales receipt in the Example. Explain that when signing a sales receipt, you are agreeing to pay the amount shown. Therefore, it is important to check the total. Errors can be made either in writing a sales receipt or in entering an amount in the cash register.

Discuss with students the reasons why people use charge cards. (You don't have to carry a lot of cash; it is easier to return merchandise rather than asking for cash back; you can purchase by telephone.) Ask students to give one disadvantage. (Many people buy more than they can afford.)

6-1 Sales Receipts

OBJECTIVE
Compute the total purchase price.

When you make a purchase with a credit or charge card, the salesclerk prepares a **sales receipt**. The receipt shows your name and account number, the price of each item you purchased, the sales tax, and the total purchase price. You sign the sales receipt and receive a copy for your records.

ECONO-CHARGE
1212 430 235 170 19
GOOD THRU
LAST DAY OF ▶ 10/93
X002250
MARY M. KEY

Total Purchase Price = Total Selling Price + Sales Tax

EXAMPLE *Skill* 30 *Application* A *Term* Sales receipt

Martha Palmer purchased the items listed on the sales receipt. The sales tax rate is 4%. Is the total purchase price correct?

SOLUTION

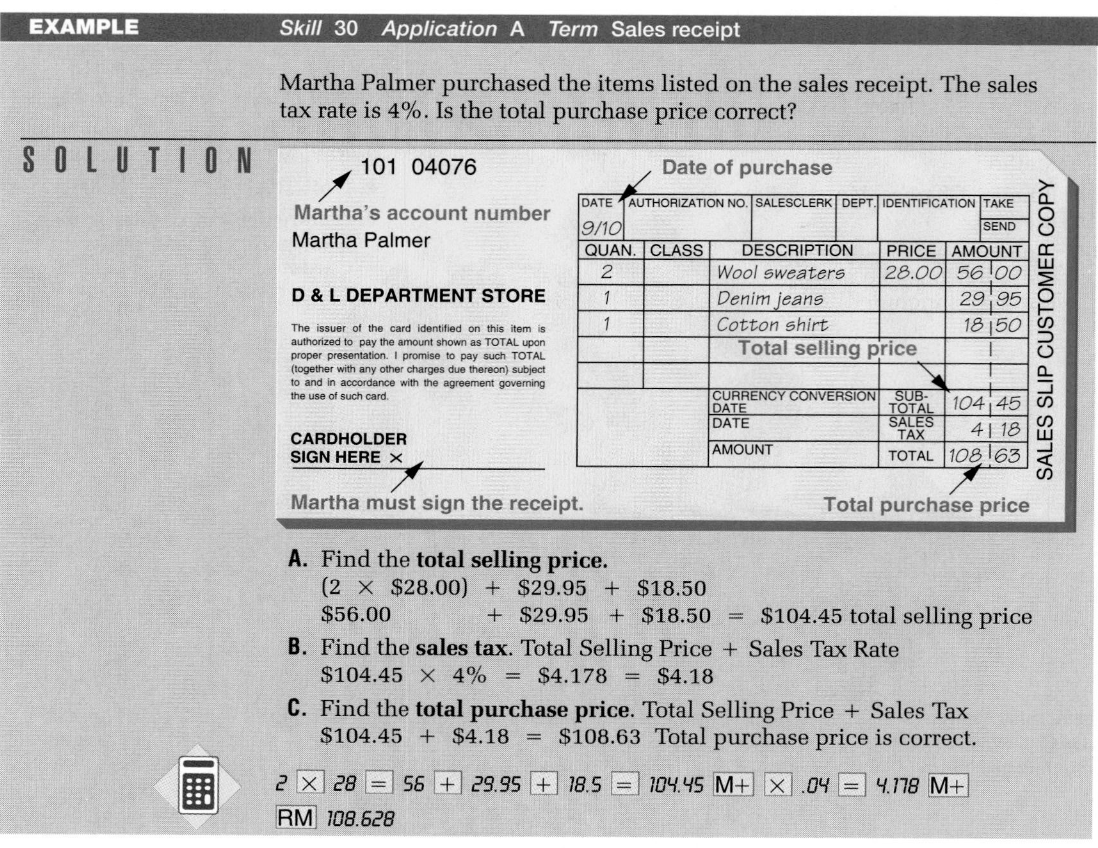

101 04076
Martha's account number
Martha Palmer
Date of purchase
D & L DEPARTMENT STORE
The issuer of the card identified on this item is authorized to pay the amount shown as TOTAL upon proper presentation. I promise to pay such TOTAL (together with any other charges due thereon) subject to and in accordance with the agreement governing the use of such card.
CARDHOLDER SIGN HERE ×
Martha must sign the receipt.
Total selling price
Total purchase price
SALES SLIP CUSTOMER COPY

DATE	AUTHORIZATION NO.	SALESCLERK	DEPT.	IDENTIFICATION	TAKE / SEND
9/10					

QUAN.	CLASS	DESCRIPTION	PRICE	AMOUNT
2		Wool sweaters	28.00	56 00
1		Denim jeans		29 95
1		Cotton shirt		18 50
		CURRENCY CONVERSION DATE	SUB-TOTAL	104 45
		DATE	SALES TAX	4 18
		AMOUNT	TOTAL	108 63

A. Find the **total selling price.**
(2 × $28.00) + $29.95 + $18.50
$56.00 + $29.95 + $18.50 = $104.45 total selling price

B. Find the **sales tax.** Total Selling Price + Sales Tax Rate
$104.45 × 4% = $4.178 = $4.18

C. Find the **total purchase price.** Total Selling Price + Sales Tax
$104.45 + $4.18 = $108.63 Total purchase price is correct.

2 [×] 28 [=] 56 [+] 29.95 [+] 18.5 [=] 104.45 [M+] [×] .04 [=] 4.178 [M+] [RM] 108.628

✓ SELF-CHECK Complete the problems, then check your answers in the back of the book.

Find the sales tax and the total purchase price.

1. Cost of coat: $116.
   Sales tax rate: 6%.
   $6.96; $122.96
2. Price of boots: $99.95.
   Sales tax rate: 5%.
   $5.00; $104.95

PROBLEMS

	Total Selling Price	×	Sales Tax Rate	=	Sales Tax	Total Purchase Price
3.	$144.00	×	7%	=	**$10.08**	**$154.08**
4.	$ 63.00	×	4%	=	**$2.52**	**$65.52**
5.	$129.55	×	6.5%	=	**$8.42**	**$137.97**

Complete the sales receipt.

DATE 3/14/– | AUTH NO 42 | IDENTIFICATION | CLERK J.R. | REG/DEPT | ☑TAKE ☐SEND

QTY	CLASS	DESCRIPTION	PRICE	AMOUNT
1		Lamp		49 95
1		Lock set		29 95
			SUBTOTAL	**6.** $79.90
CUSTOMER SIGNATURE × Dacia Adams			TAX	4 19
SALES SLIP			TOTAL	**7.** $84.09

DATE 6/1/– | AUTH NO 86430 | IDENTIFICATION | CLERK D.L. | REG/DEPT | ☑TAKE ☐SEND

QTY	CLASS	DESCRIPTION	PRICE	AMOUNT
1		Knit Shirt		15 98
2		Pairs Socks	2.59 ea	5 18
1		Golves		5 75
			SUBTOTAL	**8.** $26.91
CUSTOMER SIGNATURE × Steve Riggo			TAX	1 35
SALES SLIP			TOTAL	**9.** $28.26

50057 394856 009 DATE: 10 / 6 / —
LEWIS PETERRO
BARNEY'S SERVICE

PRODUCTS AND SERVICES	QTY	Price	Amount
Supreme / Regular / ~~Unleaded~~	12.8	1 15	14 72
Motor oil	1 Qt	2 25	2 25
Windshield washer fluid			1 35
	SALES TAX		19
SIGN HERE × Lewis Peterro	TOTAL		**10.** $18.51

DATE 9/2 | AUTH NO — | IDENTIFICATION | CLERK R.H. | REG/DEPT A | ☑TAKE ☐SEND

QTY	CLASS	DESCRIPTION	PRICE	AMOUNT
1		Fan		39 95
3		Drill Bits	2.09	**11.** $6.27
1		Storage Box		13 39
			SUBTOTAL	**12.** $59.61
CUSTOMER SIGNATURE × Janet Hye			TAX	3 13
SALES SLIP			TOTAL	**13.** $62.74

14. Sales slip for Jim Baker.
Tune-up for $44.75.
2 tires at $45.99 each.
Headlight for $8.79.
4 quarts of oil at $1.85 each.
Sales tax is $11.09.
What is the total purchase price? **$164.01**

15. Sales slip for Lisa Chong.
2 bags of lawn seed at $29.95 each.
1 bag of potting soil for $7.95.
12 packs of seeds at $1.09 each.
Lawn sprinkler for $24.95.
Sales tax is $7.41.
What is the total purchase price? **$113.29**

F.Y.I.
In 1989, there were 109.5 million credit cardholders with a total of 956.9 million cards.

16. Darlene Zink charges 12.5 gallons of gasoline. The gas costs $1.299 per gallon. What is the total purchase price? **$16.24**

17. Allen Zintec purchases a ring at Tillman's Jewelers. The ring costs $42.95 plus a 4% sales tax. He charges the purchase on his bank charge card. What is the total purchase price? **$44.67**

18. Roy Horst purchases a typewriter at Hall Office Supplies. The typewriter costs $199.99 plus a sales tax of 5.5%. What is the total purchase price? **$210.99**

Explain to students that the charge slip in restaurants has another space which is labeled "tip." Draw on the chalkboard the amount column of a sales receipt showing spaces for subtotal, tax, tip, and total. Present the following information: Lunch cost $20; tip given was 15% of the cost; tax was 6%. Note that the sales tax and tip each are computed on the lunch cost only. Fill in the spaces as students compute each amount. Explain that 15% is the standard tip for service in a restaurant. You might want to show students how to find 15% mentally: 10% of $20 is $2; 5% is one-half of 10% or $1; thus, $2 + $1 = $3, which is 15% of $20.

Warm-Up Exercises

1. $149 + $2.74 + $1.50 $153.24
2. $1.49 × 3 $4.47
3. $22.03 × 4.5% $0.99
4. $46.38 × $6\frac{1}{2}$% $3.01

ALTERNATIVE STRATEGIES: Reteaching

Refer to the *Problems and Simulations Manual.* In Lesson 6-1, use the form from problem 5. Suggest that a small storage shed is being built and the following materials are being purchased:

4 lb of nails at $1.89 per pound	=	$ 7.56
12 sheets of plywood at $15.45 ea.	=	185.40
40 2×4 wall studs at $1.59 each	=	63.60
2 windows at $59.90 each	=	119.80
Subtotal	=	$376.36
7% sales tax	=	26.35
Total	=	$402.71

PRACTICE AND APPLY

The following problems can be assigned for classwork and the answers checked in class to help students master the objective of the lesson.

- Guided Practice: 1–9
- Independent Practice: 10–13, 14–18 even

WRAP-UP

Ask students what information is shown on a sales receipt when charging an item. (date of purchase, account number, total selling price, tax, total purchase price, and signature)

Assignment Guide

- Basic: 10–21, 24, 25–40
- Average: 15–19 odd, 20–24, 25–39 odd

19. Dorothy Mathias purchases a stereo at Electronics, Inc. The stereo costs $379.99 plus a sales tax of 6%. What is the total purchase price? **$402.79**

20. Quentin Cassidy charges the following items on his bank charge card: a suit for $259.99, 3 shirts at $24.99 each, 2 ties at $18.50 each, and a pair of shoes for $89.99. There is a sales tax of 5%. What is the total purchase price? **$485.05**

21. Mary Jane Drolshagen charges the following items on her department store charge card: a suit for $179.99, 2 skirts for $65.99 each, 2 sweaters at $25.99 each, and 2 pair of shoes for $49.99 each. There is a sales tax of 6%. What is the total purchase price? **$491.77**

22. At Bargain City, Emily Williams purchases a gallon of paint thinner for $3.98, 3 gallons of paint at $17.95 a gallon, 2 tubes of caulking at $1.89 each, 2 paint rollers at $2.79 each, and 2 paintbrushes at $5.99 each. There is a sales tax of 6%. What is the total purchase price? **$83.92**

23. At Budget Mart, Leonard Milum purchases a baby stroller for $34.97, 3 plush animals at $9.79 each, a wastebasket for $9.95, and a box of detergent for $4.99. There is a sales tax of 3.5%. He charges the purchases on his bank charge card. What is the total purchase price? **$82.05**

24. At the Harvest Inn, Theora and Frankie Watson dined on clams and lobster. A portion of the check for the dinner is shown. There is an 8% meal tax. Frankie charged the dinner on his bank charge card and added a tip of 15% on the cost of the meal before tax. What was the total cost of dining out? **$62.52**

HARVEST INN

1	Steamed clams	4	95
2	Lobster dinners	43	90
	@ $21.95 each		
2	Cups coffee	1	80

MAINTAINING YOUR SKILLS Look up the skills in parentheses if you need help or more practice.

Write the percents as decimals. **(Skill 28)**

25. 8% **0.08** **26.** 7% **0.07** **27.** 4.5% **0.045** **28.** 8.75% **0.0875** **29.** 9.2% **0.092**

Add. **(Skill 5)**

30. $34.65 + $22.55 + $45.88 **$103.08**
31. $223.43 + $64.89 + $7.95 **$296.27**
32. $46.75 + $8.65 + $324.45 **$379.85**
33. $8.25 + $35.76 + $42.03 + $1.23 **$87.27**

Find the percentage. **(Skill 30)**

34. 9% of 120 **10.8** **35.** 7% of 42 **2.94** **36.** 6.5% of 20 **1.3** **37.** 8.2% of 86 **7.052**
38. 9.2% of $120 **$11.04** **39.** 8.4% of $425 **$35.70** **40.** 6.75% of $93.60 **$6.32**

BUSINESS NOTES

Businesses usually have to pay credit card companies between 2% and 6% of each purchase charged. Discuss with students the advantages and disadvantages to a merchant of offering credit services to its customers or accepting cash only.

6-2 Account Statements

OBJECTIVE
Compute the new balance in a charge account.

With a credit card or charge account, you receive a monthly **statement** listing all transactions processed by the closing date for that month. If your previous bill was not paid in full, a **finance charge** is added. A finance charge is interest that is charged for delaying payment.

New Balance = Previous Balance + Finance Charge + New Purchases − (Payments + Credits)

EXAMPLE *Skills* 6, 5 *Application* A *Term* Finance charge

Martha Palmer received this charge account statement on October 2. What is her new balance?

MAIL THIS PORTION WITH PAYMENT TO P.O. BOX 1027

D & L DEPARTMENT STORE
☐ IF ADDRESS IS INCORRECT, PLEASE CHECK BOX AND CORRECT ON REVERSE SIDE

Martha Palmer
1234 MAIN ST.
TOLEDO, OH 43605

Martha's account number → ACCOUNT NUMBER 101-04076
NEW BALANCE ?
TOTAL MINIMUM AMOUNT NOW DUE 43.00
SEND PAYMENT TO REACH US BY 10/30/--

DATE	DEPT. NO.	DESCRIPTION	PURCHASES	CREDITS	PAYMENTS	ITEM NO.
09/07	162	SPORTSWEAR		23.42		0-06-18005
09/10	363	SPORTSWEAR	108.63			0-07-20015
09/17		PAYMENT-THANK YOU			50.00	0-06-47023
09/19	214	JR SPORTSWEAR		8.52		0-07-45005
09/19	53	HOUSEWARES	19.53			0-07-45006
09/27	34	COSMETICS	25.62			0-07-64004
09/29	1	LINENS-TOWELS	19.25			0-07-70012

Last date transactions were processed — Amount Martha owed last month — $50 payment toward last month's bill — Amount Martha now owes — Credit for items returned

BILL CLOSING DATE	PREVIOUS BALANCE	FINANCE CHARGE	TOTAL PURCHASES THIS MONTH	CREDITS	PAYMENTS	NEW BALANCE
9/30	332.64	5.64	173.03	31.94	50.00	?

Pay new balance in full within 25 days of bill closing date to avoid finance charge next month. Finance charge, if any, is computed on the average daily balance of $375.72 by applying monthly periodic rates as follows:

MONTHLY PERIODIC RATE	BALANCE	ANNUAL PERCENTAGE RATE
1.50 %	FIRST $600	18%
1.00 %	OVER $600	12%

SUBJECT TO A MINIMUM FINANCE CHARGE OF $.50.
NOTICE: SEE REVERSE SIDE FOR IMPORTANT INFORMATION

ACCOUNT NUMBER 101-04076
NEW BALANCE
TOTAL MINIMUM AMOUNT NOW DUE
SEND PAYMENT TO REACH US BY

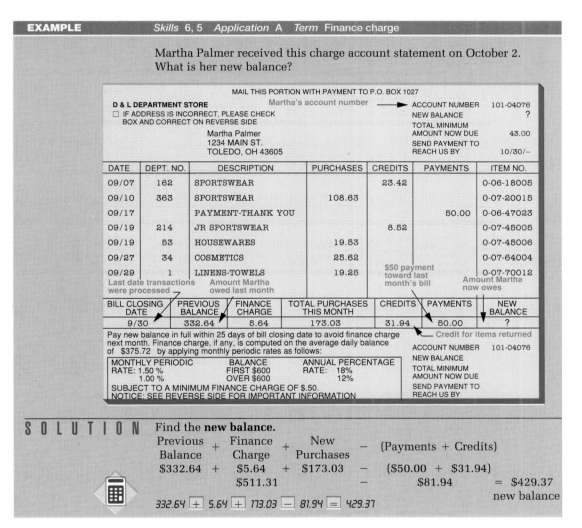

SOLUTION Find the **new balance.**

Previous Balance + Finance Charge + New Purchases − (Payments + Credits)
$332.64 + $5.64 + $173.03 − ($50.00 + $31.94)
$511.31 − $81.94 = $429.37 new balance

332.64 [+] 5.64 [+] 173.03 [−] 81.94 [=] 429.37

LESSON PLAN 6-2 Account Statements

FOCUS

To motivate this lesson, ask students if they know how a person actually pays for the purchases made with a credit card. Explain that a credit card company sends a monthly statement to the customer. In this lesson, students will compute the new balance on a charge account statement.

TEACH

Explain that the format of monthly statements varies from company to company, but that the information given is the same: previous balance; an itemized list of purchases which includes the date of purchase, payments, credits, new balance; and so on.

Stress that if the balance of the previous month had been paid in full, there is no finance charge. Thus, you essentially can borrow the money for purchases without paying interest.

Mention that although a statement shows the minimum amount due, there is no penalty for paying more than this amount. The more you pay, the lower the finance charge.

ALTERNATIVE STRATEGIES: Reteaching

Relate this lesson to the real world by sharing one of your own, or a friend's, account statement. Go over the various computations with students.

Warm-Up Exercises

1. $170 + $3.40 + $50 − $70 $153.40
2. $250 + 0 + $125.60 + $249.95 $625.55
3. $315.67 + $3.15 + $75.33 − $50 − $12.50 $331.65
4. $451.70 + $6.78 + $119.99 − ($70.00 + $11.92) $496.55

PRACTICE AND APPLY

The following problems can be assigned for classwork and the answers checked in class to help students master the objective of the lesson.

- Guided Practice: 1–6
- Independent Practice: 7, 8, 9, 11

✓ SELF-CHECK Complete the problems, then check your answers in the back of the book.

Find the new balance.

	Previous Balance	+	Finance Charge	+	New Purchases	−	Payments & Credits	New Balance
1.	$600.00	+	$7.50	+	$90.00	−	$100.00	= $597.50
2.	$278.75	+	$4.18	+	$35.85	−	$48.00	= $270.78

PROBLEMS

What is the new balance for the credit statements shown?

	Answer	BILLING DATE	PREVIOUS BALANCE	FINANCE CHARGE	NEW PURCHASES	PAYMENTS & CREDITS	NEW BALANCE
3.	$649.00	8/15/--	$600.00	$9.00	$140.00	$100.00	?
4.	$442.95	1/22/--	$410.75	$7.20	$175.00	$150.00	?
5.	$337.66	3/1/--	$450.95	$6.76	$39.95	$160.00	?
6.	$142.69	9/4/--	$233.23	$2.33	$40.36	$133.23	?
7.	$416.34	6/15/--	$675.19	0	$416.34	$675.19	?
8.	$2225.11	8/1/--	$2494.21	$43.65	$137.25	$450.00	?

9. What is the new balance for the credit statement shown?

PAYMENTS (P) & CREDIT (C) DEDUCTED-DATES
20.00 P 04-07
20.00 P 03-28

PREVIOUS BALANCE	PURCHASES ADDED
265.69	145.99
FINANCE CHARGE ADDED 3.98	NEW BALANCE ? $375.66
MINIMUM PAYMENT 25.00	MUST BE RECEIVED BY 05-15 TO AVOID ADDITIONAL FINANCE CHARGE

ANNUAL PERCENTAGE RATE			PERIODIC MONTHLY RATE
TO	$500	18 %	1-1/2 %
OVER	$500	12 %	1 %

10. What is the new balance for the credit statement shown?

ACCOUNT NUMBER	BILLING DATE
064-173-388-62	03-10

PREVIOUS BALANCE	− PAYMENTS	RECEIVED ON
366 72	366 72	02-26
− CREDITS	+ FINANCE CHARGE	+ PURCHASES
24 59	5 50	45 23
+ INS. PREMIUMS	= NEW BALANCE	AMOUNT DUE
	$26.14 ?	40 00
	NEXT BILLING DATE 04-12	

		PERIODIC RATE(S)	ANNUAL PERCENTAGE RATE(S)
UP TO	$500	1.50%	18.0%
OVER	$500	1.00%	12.0%
MINIMUM CHARGE		.50	

CULTURAL ANGLES

When you purchase items in other countries with a credit card, the credit card company will figure the foreign exchange rate on the day you made the purchase or the day your account was billed. Have students investigate what the current value is of a foreign currency in relation to one U.S. dollar. *The Wall Street Journal* is an excellent source to help in assigning countries to students. You might want to extend this activity and have students find, for example, the cost of an item in U.S. dollars purchased in Italy for 6140 lira. (6140 ÷ current exchange rate)

11. Monthly statement for Mae Keyes.
Credit Master bankcard.
Previous balance of $307.85.
Payment of $40.
New purchases: $9.50, $41.75.
Finance charge of $4.62.
What is the new balance? **$323.72**

12. Monthly statement for Bob Ross.
J-MART Store charge card.
Previous balance of $144.79.
Payment and credit totaling $144.79.
Finance charge of $2.53.
No new purchases.
What is the new balance? **$2.53**

F.Y.I.
Americans charged $242.1 billion on Visa and MasterCard in 1990.

13. Monthly statement for Walt Klin.
Department store charge card.
Previous balance of $787.29.
Payment of $100.
New purchases: $47.97, $49.28.
Finance charge of $11.81.
What is the new balance? **$796.35**

14. Monthly statement for Jane Cook.
All-Charge bankcard.
Previous balance of $529.78.
Payment of $85.
New purchases: $277.32, $38.20.
Finance charge of $7.95.
What is the new balance? **$768.25**

15. Marci Cassidy received this charge account statement. Find her new balance.

DEPT.	DESCRIPTION	CHARGES	PAYMENT/CREDIT	DATE	REF. #
109	Garden Shop	42.75		1/25	6004
85	Menswear	145.98		1/25	7018
	PAYMENT		74.40	2/1	8014
71	Appliances		35.50	2/2	3113

BILLING DATE: 2/16

PREVIOUS BALANCE	PAYMENTS & CREDITS	UNPAID BALANCE	FINANCE CHARGE	NEW PURCHASES	NEW BALANCE
$285.92	?	——	$4.29	?	?
	$109.90			**$188.73**	**$369.04**

MAINTAINING YOUR SKILLS Look up the skills in parentheses if you need help or more practice.

Add. **(Skill 5)**

16. $532.75 + $45.90 + $38.90 + $16.55 **$634.10**

17. $44.29 + $324.60 + $8.65 + $27.50 **$405.04**

18. $44.52 + $923.17 + $9.20 + $337.63 **$1314.52**

19. $8.16 + $24.50 + $39.25 + $234.78 **$306.69**

Subtract. **(Skill 6)**

20. $41.50 − $9.50 **$32.00**
21. $321.65 − $12.35 **$309.30**
22. $427.16 − $45.12 **$382.04**
23. $337.42 − $49.58 **$287.84**
24. $798.87 − $389.68 **$409.19**
25. $4296.83 − $999.49 **$3297.34**

WRAP-UP

Review with students the items that are added to and subtracted from the previous balance when finding the new balance. Give students a previous balance, say $150, and then other dollar amounts for some items and have students find the new balance. Then ask, if the new balance was paid in full, would there be a finance charge?

Assignment Guide

- Basic: 7–14, 16–25
- Average: 10, 12–15, 16–24 even

LESSON PLAN
6-3 Finance Charge— Previous-Balance Method

FOCUS
Remind students that money in a savings account earns interest. Ask why. (The bank has your money to invest.) In the previous lesson, it was stated that credit card companies have finance charges, which is interest you pay for delaying payment. Why? (They are lending you money.) You might want to show a comparison of current rates offered for savings accounts and rates that are used for finance charges. (The national average in late 1991 was 18.78%.)

TEACH
Explain to students that annual interest rates for finance charges can vary, and so can the method used to calculate the finance charge.

Refer students to Lesson 6-2 and to the bottom portion of the account statement, which shows an annual rate of 18% and a monthly periodic rate of 1.5%. Ask how the periodic rate is found. Demonstrate how to find one or the other using different annual rates and periodic rates.

You might want to point out the minimum finance charge of $0.50 on the account statement of Lesson 6-2. This means that the store's policy is to charge $0.50 even though the interest computed is less than $0.50.

6-3 Finance Charge— Previous-Balance Method

OBJECTIVE
Compute the finance charge by the previous-balance method.

Some credit card companies use the **previous-balance** method to compute finance charges. They compute the finance charge based on the amount you owed on the closing date of your last statement. The **periodic rate** is the monthly finance charge rate.

Finance Charge = Previous Balance × Periodic Rate

New Balance = Previous Balance + Finance Charge + New Purchases − (Payments + Credits)

EXAMPLE *Skills* 30, 2 *Application* A *Term* Previous balance

Carl Byers has a charge account with a store that charges 1.5% of the previous balance for the finance charge. A portion of Carl's statement is shown. He checks the computations. What is the new balance of his account on October 5?

PREVIOUS BALANCE	CLOSING DATE THIS MONTH		CLOSING DATE LAST MONTH	
$125.60	October 5, 19--		September 6, 19--	
TOTAL PURCHASES	PAYMENTS & CREDITS	FINANCE CHARGE	NEW BALANCE	MINIMUM PAYMENT
$122.15	$48.75	?	?	$20.00

SOLUTION

A. Find the **finance charge.**
Previous Balance × Periodic Rate
$125.60 × 1.5% = $1.884 = $1.88 finance charge

B. Find the **new balance.**
Previous Balance + Finance Charge + New Purchases − (Payments + Credits)
$125.60 + $1.88 + $122.15 − $48.75
$249.63 − $48.75 = $200.88 new balance

✓ SELF-CHECK Complete the problems, then check your answers in the back of the book.

The periodic rate is 1.5%. Find the finance charge.

1. Previous balance of $180. **$2.70**
2. Previous balance of $87.90. **$1.318 = $1.32**

CRITICAL THINKING
When buying on credit, you are spending more for the item than its original price. Have students compute the total cost of a $400 VCR charged on a credit card. The annual interest rate is 18%, and the buyer pays $50 a month. (Finance charges for 9 months would be $29.39; thus, the buyer paid $429.39 for the VCR.) Ask how much interest the buyer would have earned each month by putting the $50 in a savings account paying 5.5% interest for 9 months. ($10.33) How much did the buyer lose by using credit? ($29.39 + $10.33 = $39.72)

PROBLEMS

Use the previous-balance method to find the finance charge. Round answers to the nearest cent.

	3.	4.	5.	6	7.
Previous Balance	$60.00	$150.00	$148.00	$197.30	$287.42
Periodic Rate	× 1.5%	× 1.5%	× 1.6%	× 1.25%	× 1.75%
Finance Charge	**$0.90**	**$2.25**	**$2.37**	**$2.47**	**$5.03**

The finance charge is 1.5% of the previous balance. Find the finance charge and the new balance for each statement shown.

	BILLING DATE	PREVIOUS BALANCE	FINANCE CHARGE	NEW PURCHASES	PAYMENTS & CREDITS	NEW BALANCE
8.	4/11/--	$400.00	**$6.00**	$200.00	$160.00	**$446.00**
9.	5/7/--	$170.00	**$2.55**	$64.00	$45.00	**$191.55**
10.	3/1/--	$32.50	**$0.49**	$15.50	$20.00	**$28.49**
11.	9/15/--	$497.00	**$7.46**	$35.95	$80.00	**$460.41**
12.	12/12/--	$564.28	**$8.46**	$221.82	$125.00	**$669.56**
13.	10/30/--	$1282.29	**$19.23**	0	$225.00	**$1076.52**

Use the previous-balance method of computing finance charges to solve.

14. Rick Demski's charge account.
Previous balance of $188.
Periodic rate is 2.0%.
What is the finance charge? **$3.76**
New purchases totaling $42.50.
Payment and credit totaling $25.
What is the new balance? **$209.26**

15. Marie Burch's charge card.
Previous balance of $144.30.
Periodic rate is 1.25%.
What is the finance charge? **$1.80**
New purchases totaling $97.32.
Payments totaling $120.
What is the new balance? **$123.42**

Warm-Up Exercises

1. 1.5% of $400 $6
2. 1.25% of $300 $3.75
3. 1.65% of $500 $8.25
4. 2.5% of $450 $11.25
5. 1.75% of $319.60 $5.59
6. 2.0% of $19.38 $0.39

PRACTICE AND APPLY

The following problems can be assigned for classwork and the answers checked in class to help students master the objective of the lesson.

- Guided Practice: 1–8
- Independent Practice: 9–14

ALTERNATIVE STRATEGIES: Enrichment

Have students bring in charge account applications from various stores, banks, gas stations, etc. These application forms are an excellent source of data on current interest rates and methods of computing the finance charge. Students may need to read the "fine print" to ascertain some of the information.

WRAP-UP
Review the solutions to Problem 9 and ask students to compute the finance charge on the next month's statement ($2.87)

Assignment Guide
- Basic: 9–17, 19–32
- Average: 15–18, 19–31 odd

16. Kay Maxwell has a charge account at Simon's Department Store where the finance charge is 1.5% of the previous balance. A portion of her account statement is shown. Find the finance charge and the new balance.

PREVIOUS BALANCE	CLOSING DATE THIS MONTH		CLOSING DATE LAST MONTH	
$157.53	July 20, 19--		June 21, 19--	
TOTAL PURCHASES	PAYMENTS & CREDITS	FINANCE CHARGE	NEW BALANCE	MINIMUM PAYMENT
$42.91	$27.18	? $2.36	? $175.62	$30.00

17. Dale Watson has a charge account where the finance charge is 1.25% of the previous balance. A portion of his statement for May is shown. What is the total of his payments and credits? What is the finance charge? What is the new balance of his account?

DATE	REFERENCE #	DEPT.	DESCRIPTION	PURCHASES	PAYMENTS	CREDITS
5-04	34029	03	Sport coat	135.29		
5-12	40085	06	Robe			49.79
5-23	29450	10	Payment		50.00	
5-23	37047	03	Men's slacks	47.35		
5-25	77460	04	Hardware	27.15		

PREVIOUS BALANCE	CLOSING DATE THIS MONTH		CLOSING DATE LAST MONTH	
$141.58	May 31, 19--		April 30, 19--	
TOTAL PURCHASES	PAYMENTS & CREDITS	FINANCE CHARGE	NEW BALANCE	MINIMUM PAYMENT
$209.79	? $99.79	? $1.77	? $253.35	$20.00

18. Many credit account companies charge periodic rates that vary depending on the amount of the previous balance. Use the periodic rates on the statement shown to calculate the finance charge and the new balance according to the previous-balance method.

PREVIOUS BALANCE	– PAYMENTS	RECEIVED ON
849 74	120 00	12-09
– CREDITS	+ FINANCE CHARGE	+ PURCHASES
35 93	? $11.09	221 75
+ INS. PREMIUMS	= NEW BALANCE	AMOUNT DUE
	? $926.59	90 00
PAY NEW BALANCE BY TO AVOID FINANCE CHARGE ▶	NEXT BILLING DATE	
	01-27	

BALANCES		PERIODIC RATE(S)	ANNUAL PERCENTAGE RATE(S)
UP TO	$500	1.50%	18.0%
OVER	$500	1.00%	12.0%
MINIMUM CHARGE		.50	

MAINTAINING YOUR SKILLS Look up the skills in parentheses if you need help or more practice.

Round answers to the nearest hundredth. **(Skill 2)**

19. 43.155 **43.16** **20.** 8.241 **8.24** **21.** 9.5371 **9.54** **22.** 14.8981 **14.90** **23.** 2.9971 **3.00**

Round answers to the nearest cent. **(Skill 2)**

24. $2.385 **$2.39** **25.** $3.7521 **$3.75** **26.** $9.5145 **$9.51** **27.** $0.4354 **$0.44** **28.** $18.015 **$18.02**

Find the percentage. **(Skill 30)**

29. 8% of 438 **35.04** **30.** 3.75% of 988 **37.05** **31.** 1.25% of 42.24 **0.528** **32.** 1.75% of 365.5 **6.39625**

COOPERATIVE LEARNING
You may want to have students work in small groups to solve Problem 18. Have one group show its computations at the chalkboard and ask the other groups if they agree with the answers.

6-4

Finance Charge—Unpaid-Balance Method

OBJECTIVE
Compute the finance charge by the unpaid-balance method.

Some companies use the **unpaid-balance** method of computing finance charges. They compute the finance charge based on that portion of the previous balance that you have not paid.

Unpaid Balance = Previous Balance − (Payments + Credits)

Finance Charge = Unpaid Balance × Periodic Rate

New Balance = Unpaid Balance + Finance Charge + New Purchases

EXAMPLE *Skill* 30 *Application* A *Term* Unpaid balance

A portion of Lucille Sherman's charge account statement is shown. The monthly finance charge is 1.5% of the unpaid balance. What is the new balance of her account?

88	PAYMENT/Thank you		40.00		
BILLING DATE: 2/16					
PREVIOUS BALANCE	PAYMENTS & CREDITS	UNPAID BALANCE	FINANCE CHARGE	NEW PURCHASES	NEW BALANCE
$132.40	$40.00	?	?	$79.55	?

SOLUTION

A. Find the **unpaid balance.**
Previous Balance − (Payments + Credits)
$132.40 − $40.00 = $92.40 unpaid balance

B. Find the **finance charge.**
Unpaid Balance × Periodic Rate
$92.40 × 1.5% = $1.386 = $1.39 finance charge

C. Find the **new balance.**
Unpaid Balance + Finance Charge + New Purchases
$92.40 + $1.39 + $79.55 = $173.34 new balance

✓ SELF-CHECK Complete the problems, then check your answers in the back of the book.

The periodic rate is 1.5%. Find the unpaid balance, the finance charge, and the new balance.

	Previous Balance	Payments and Credits	New Purchases
1.	$400.00	$100.00	$70.00
2.	$220.00	$150.00	$95.00

1. $300; $4.50; $374.50
2. $70; $1.05; $166.05

LESSON PLAN
6-4 Finance Charge—Unpaid-Balance Method

FOCUS

To motivate this lesson, write on the chalkboard the formula from the previous lesson: Finance Charge = Previous Balance × Periodic Rate. Have a student give the definition of the term *previous balance.* (the amount you owe on the closing date of your last statement) Ask students this question: If you received a statement and made a payment toward the new balance, would your payment be considered in the finance charge using the previous-balance method? (no) Explain that there is another method used to compute the finance charge, called the unpaid-balance method, in which the payment is considered.

TEACH

To ascertain that students understand the difference between the previous-balance method and the unpaid-balance method, work the Example using both methods. Students should see that the unpaid-balance method yields the smaller finance charge. Ask why a credit card company would choose to use the unpaid-balance method, which results in less interest being received. (It is an incentive for customers to pay more on their new balance.)

PROBLEM SOLVING

You might wish to have small groups of students work the following problem. Michael Taylor misplaced the bottom portion of his account statement, which showed the purchases made during the last month. He knows that in the previous month his balance was $122, and he made a payment of $55. He also knows that his new balance is $257. If the credit card company uses the unpaid-balance method for finance charges, and the periodic rate is 1.5%, how much did Michael spend for purchases? ($122 − $55 = $67; $67 × 1.5% = $1.01; $257 − $67 − $1.01 = $188.99)

Warm-Up Exercises

1. $71.40 − $21.40 $50
2. $415.20 − $200 $215.20
3. ($171.50 − $71.50) × 1.5% $1.50
4. ($78.50 − $24.90) × 2.0% $1.07

PRACTICE AND APPLY

The following problems can be assigned for classwork and the answers checked in class to help students master the objective of the lesson.

- Guided Practice: 1–7, 11
- Independent Practice: 8–10, 13, 15

PROBLEMS

For problems 3 through 10, use a periodic rate of 1.5% and the unpaid-balance method of computing the finance charge.

	(Previous Balance	−	Payments + Credits	=	Unpaid Balance)	+	Finance Charge	+	New Purchases	=	New Balance
3.	($500.00	−	$100.00	=	$400	)+	$6	+	$ 80.00	=	$486
4.	($300.00	−	$150.00	=	$150	)+	$2.25	+	$ 45.00	=	$197.25
5.	($350.00	−	$ 75.00	=	$275	)+	$4.13	+	$ 90.00	=	$369.13
6.	($125.50	−	$ 45.50	=	$80	)+	$1.20	+	$ 42.50	=	$123.70
7.	($473.50	−	$ 57.50	=	$416	)+	$6.24	+	$222.50	=	$644.74
8.	($173.43	−	$100.00	=	$73.43	)+	$1.10	+	$127.91	=	$202.44
9.	($491.87	−	$119.00	=	$372.87	)+	$5.59	+	$147.94	=	$526.40
10.	($738.27	−	$145.00	=	$593.27	)+	$8.90	+	$199.95	=	$802.12

11. Ruth Kean's account statement.
Unpaid balance of $88.
Periodic rate is 1.5%.
What is the finance charge? **$1.32**
New purchases of $40.
What is the new balance? **$129.32**

12. Don Vester's account statement.
Unpaid balance of $19.70.
Periodic rate is 1.25%. **$0.25**
What is the finance charge?
New purchases of $431.85.
What is the new balance? **$451.80**

13. Midge Duez's account statement.
Unpaid balance of $121.60.
Periodic rate is 1.6%.
What is the finance charge? **$1.95**
New purchases of $72.19.
What is the new balance? **$195.74**

14. Liz Cole's account statement.
Unpaid balance of $921.35.
Periodic rate is 1.75%. **$16.12**
What is the finance charge?
New purchases of $75.43.
What is the new balance? **$1012.90**

15. A portion of Alvin Sujkowski's charge account statement is shown. The finance charge is 2% of the unpaid balance. What is the new balance?

PREVIOUS BALANCE	PAYMENTS & CREDITS	UNPAID BALANCE	FINANCE CHARGE	NEW PURCHASES	NEW BALANCE
$419.29	$45.00	?	?	$79.31	?
		$374.29	**$7.49**		**$461.09**

16. A portion of Verda Buell's charge account statement is shown. The finance charge is 1.25% of the unpaid balance. What is the new balance?

PREVIOUS BALANCE	PAYMENTS & CREDITS	UNPAID BALANCE	FINANCE CHARGE	NEW PURCHASES	NEW BALANCE
$556.71	$147.55	?	?	$21.64	?
		$409.16	**$5.11**		**$435.91**

17. June Ray has a charge account at Parker's Discount Store where the finance charge is 2% of the unpaid balance. Find the indicated amounts on the statement shown.

DEPT.	DESCRIPTION	CHARGES	PAYMENT/CREDIT	DATE	REF. #
55	Battery	65.67		3/2	6982
44	Smoke alarm	9.85		3/2	8640
32	Muffler	54.96		3/2	9000
98	Daywear		21.45	3/5	7640
88	PAYMENT		50.00	3/9	500

BILLING DATE: 3/10

PREVIOUS BALANCE	PAYMENTS & CREDITS	UNPAID BALANCE	FINANCE CHARGE	NEW PURCHASES	NEW BALANCE
$379.13	$71.45	$307.68	$6.15	$130.48	$444.31

18. Gary Green has a charge account at Knapp's Department Store where the finance charge is 1.75% of the unpaid balance. Find the indicated amounts on the statement shown.

DEPT.	DESCRIPTION	CHARGES	PAYMENT/CREDIT	DATE	REF. #
09	Linens	84.39		9/12	9064
14	PAYMENT		50.00	9/13	A345
03	Sporting goods	27.83		9/26	11309
08	Sporting goods		27.83	10/02	3960
15	Electronics	239.95		10/02	11714

BILLING DATE: 10/10

PREVIOUS BALANCE	PAYMENTS & CREDITS	UNPAID BALANCE	FINANCE CHARGE	NEW PURCHASES	NEW BALANCE
$338.65	$77.83	$260.82	$4.56	$352.17	$617.55

MAINTAINING YOUR SKILLS Look up the skills in parentheses if you need help or more practice.

Add. **(Skill 5)**

19. $425.10 + $38.75 + $29.51 + $4.22 **$497.58**

20. $5.95 + $38.75 + $71.19 + $314.75 **$430.64**

Subtract. **(Skill 6)**

21. $499.24 − $88.31 **$410.93**
22. $391.37 − $79.43 **$311.94**
23. $523.89 − $154.79 **$369.10**
24. $310.01 − $58.75 **$251.26**
25. $87.01 − $9.54 **$77.47**
26. $808.76 − $39.41 **$769.35**

Find the percentage. Round answers to the nearest hundredth. (Skill 30)

27. 4% of 220 **8.8**
28. 8% of 60 **4.8**
29. 3.9% of 500 **19.5**
30. 2.8% of 460 **12.88**
31. 0.7% of 500 **3.5**
32. 3.15% of 181.2 **5.71**
33. 1.75% of 851 **14.89**

WRAP-UP

Have students close their textbooks, and ask volunteers to write the three formulas of this lesson on the chalkboard. Ask students to explain the difference between the unpaid-balance method and the previous-balance method for computing the finance charge.

Assignment Guide

- Basic: 8–17, 19–33
- Average: 11, 12, 14, 16–18, 20–32 even

ALTERNATIVE STRATEGIES: Reteaching

Refer to the example in the previous lesson (6-3). Rework the example using the unpaid balance method.

Thus: $125.60 − $48.75 = $76.85
F.C. = 1.5% × $76.85 = $1.15

Compare finance charges:
$1.10 using the unpaid-balance method
$1.88 using the previous-balance method

LESSON PLAN
6-5 Finance Charge—Average Daily Balance (No new purchases included)

FOCUS

Lead students in a discussion of the meaning of *average daily balance.* Elicit from students the fact that the balance each day is the same until the day a transaction is made by the credit card company. After that date, the balance again remains the same. Ask students how to find the average of the balances in a 30-day period. (Find the sum of the daily balances and divide by 30.) Explain that some credit card companies compute the finance charge by using the daily balances. In this lesson, the only transactions considered are payments.

6-5

Finance Charge—Average Daily Balance (No new purchases included)

OBJECTIVE

Compute the finance charge based on the average daily balance, no new purchases included.

Many companies calculate the finance charge using the **average-daily-balance** method where no new purchases are included. The average daily balance is the average of the account balance at the end of each day of the billing period. For this method of computing finance charges, new purchases posted during the billing period are not included when figuring the balance at the end of the day.

$$\text{Average Daily Balance} = \frac{\text{Sum of Daily Balances}}{\text{Number of Days}}$$

EXAMPLE *Skills* 8, 11 *Application* G *Term* Average daily balance

A portion of Dewey Napp's credit card statement is shown.

REFERENCE	POSTING DATE	TRANSACTION DATE	DESCRIPTION	PURCHASES & ADVANCES	PAYMENTS & CREDITS
131809	9/05	8/24	Health Club	48.75	
265118	9/18		PAYMENT		44.85
407372	9/20	9/01	Wilson's	37.85	
329416	10/01	8/30	Ed's Discount	20.99	

BILLING PERIOD	PREVIOUS BALANCE	PERIODIC RATE	AVERAGE DAILY BALANCE	FINANCE CHARGE
9/04 - 10/03	$194.85	2%	?	?
PAYMENTS & CREDITS	**PURCHASES & ADVANCES**	**NEW BALANCE**	**MINIMUM PAYMENT**	**PAYMENT DUE**
$44.85	$107.59	?	$20.00	10/25

A finance charge was added to Dewey's account balance because he did not pay his last bill in full. The finance charge was computed using the average daily balance where new purchases were not included. Only the payment of $44.85 affected the average daily balance. What is the average daily balance?

ALTERNATIVE ASSESSMENT

Use the Example in this lesson. Have students work in cooperative groups to find the average daily balance, the finance charge, and the new balance when no new purchases are included in finding the average daily balance. While the groups are working, ask questions to ascertain that students understand the computations required.

SOLUTION

A. Find the **sum of daily balances.**

Dates	Payment	End-of-Day Balance		Number of Days	Sum of Balances
9/4–9/17		$194.85	×	14	$2727.90
9/18	$44.85	150.00	×	1	150.00
9/19–10/3		150.00	×	15	2250.00
			Total	30	$5127.90

B. Find the **average daily balance.**
Sum of Daily Balances ÷ Number of Days
$5127.90 ÷ 30 = $170.93 average daily balance

194.85 [×] 14 [=] 2727.9 [M+] 194.85 [−] 44.85 [=] 150 [M+] 150 [×] 15 [=] 2250 [M+] [RM] 5127.9 [÷] 30 [=] 170.9

✓ SELF-CHECK Complete the problems, then check your answers in the back of the book.

Find the average daily balance, excluding new purchases.

Dates	Payment	End-of-Day Balance	×	Number of Days	=	Sum of Balances
9/9–9/18		$500.00	×	10	=	**1.** ? $5000
9/19	$100.00	$400.00	×	1	=	**2.** ? $400
9/20–10/8		**3.** ? $400	×	**4.** ? 19	=	**5.** ? $7600
			Total	**6.** ? 30		**7.** ? $13,000

and **8.** ? ÷ **9.** ? = **10.** average daily balance
$13,000 30 $433.33

The finance charge is calculated by multiplying the average daily balance times the periodic rate.

Finance Charge = Average Daily Balance × Periodic Rate

New Balance = Unpaid Balance + Finance Charge + New Purchases

EXAMPLE *Skills* 8, 11 *Application* G *Term* Average daily balance

Dewey checks the finance charge and the new balance. The finance charge is 2% of the average daily balance. What is the new balance?

SOLUTION

A. Find the **unpaid balance.**
Previous Balance − (Payments + Credits)
$194.85 − $44.85 = $150.00 unpaid balance

B. Find the **finance charge.**
Average Daily Balance × Periodic Rate
$170.93 × 2% = $3.418 = $3.42 finance charge

C. Find the **new purchases.** $107.59

D. Find the **new balance.**
Unpaid Balance + Finance Charge + New Purchases
$150.00 + $3.42 + $107.59 = $261.01 new balance

TEACH

Work through the first Example of the lesson carefully with students. Finding the sum of daily balances may be confusing for some students. Emphasize the importance of setting up the computations in an orderly manner, such as the method shown in the solution.

Point out that the first date to use in finding the daily balances is the beginning date of the billing period. Explain also that the *posting* date is the date the payment is entered into the accounting records of the company.

Ask students why the average-daily balance method for computing finance charges provides an incentive for the customer. (Customers will want to make monthly payments as early in the billing cycle as possible in order to reduce the amount of the finance charge.)

You might want to discuss the term *advances* shown on the statement in the Example. Explain that a cash advance is considered a loan and is subject to a finance charge. Some companies use a higher percentage rate for advances than for purchases.

ALTERNATIVE STRATEGIES: Reteaching

Students can use the number-of-days table on page 645 to find the number of days between two dates. For example, from 9/4 (day 247) to 9/17 (day 260) is 13 days. As an additional teaching tool to show the balance of each day of the business cycle, you may want to make a copy of the calendar worksheet in the Resource Book that accompanies this program.

Warm-Up Exercises

1. (12 × $400) + (19 × $300) $10,500
2. (15 × $600) + (11 × $570) $15,270
3. $2754.85 ÷ 30 $91.83
4. $140 × 1.75% $2.45

✓ SELF-CHECK Complete the problems, then check your answers in the back of the book.

The finance charge is 1.5% of the daily balance.

11. Find the finance charge. **$2.48**

12. Find the new balance. **$192.48**

PREVIOUS BALANCE	TOTAL CHARGES	PAYMENTS AND CREDITS	AVERAGE DAILY BALANCE	FINANCE CHARGE	NEW BALANCE
$180.00	$40.00	$30.00	$165.00	?	?

PROBLEMS

F.Y.I.
On December 1, 1991, the average monthly credit card annual percentage rate was 18.87%.

	Billing Periods	Payment	End-of-Day Balance	Number of Days	Sum of Balances	
13.	6/01–6/15		$75.00	15	$1125.00	What is the average daily balance?
	6/16	$50.00	25.00	1	25.00	
	6/17–6/30		25.00	14	? $350.00	
			TOTALS	? 30	? $1500.00	$50
14.	7/15–8/2		$400.00	? 19	? $7600	What is the average daily balance?
	8/3	$100.00	300.00	? 1	? $300	
	8/4–8/14		300.00	? 11	? $3300	
			TOTALS	? 31	? $11,200	$361.29

15. Lee Hoshino has a bank charge card. Use the portion of the account statement shown to find the average daily balance, excluding new purchases, and finance charge.

REFERENCE	POSTING DATE	TRANSACTION DATE	DESCRIPTION	PURCHASES & ADVANCES	PAYMENTS & CREDITS
6646598	8/25	8/20	Al's Hardware	44.80	
7000507	8/28		PAYMENT RECEIVED		50.00

BILLING PERIOD	PREVIOUS BALANCE	PERIODIC RATE	AVERAGE DAILY BALANCE	FINANCE CHARGE
8/01 - 8/31	$250.00	1.5%	?	——
			$243.55	**$3.65**

The finance charge is computed using the average-daily-balance method where no new purchases are included. Find the finance charge and the new balance for the following statements.

16.

BILLING PERIOD	PREVIOUS BALANCE	PERIODIC RATE	AVERAGE DAILY BALANCE	FINANCE CHARGE
2/3 - 3/2	$196.00	2%	$156.00	? $3.12

PAYMENTS & CREDITS	PURCHASES & ADVANCES	NEW BALANCE	MINIMUM PAYMENT	PAYMEMT DUE
$60.00	0	? $139.12	$15.00	3/20

17.

BILLING PERIOD	PREVIOUS BALANCE	PERIODIC RATE	AVERAGE DAILY BALANCE	FINANCE CHARGE
11/15 - 12/14	$322.49	1.75%	$277.21	? $4.85
PAYMENTS & CREDITS	**PURCHASES & ADVANCES**	**NEW BALANCE**	**MINIMUM PAYMENT**	**PAYMEMT DUE**
$123.49	$49.51	? $253.36	$20.00	1/2

18.

BILLING PERIOD	PREVIOUS BALANCE	PERIODIC RATE	AVERAGE DAILY BALANCE	FINANCE CHARGE
7/1 - 8/1	$74.50	1.25%	$45.66	? $0.57
PAYMENTS & CREDITS	**PURCHASES & ADVANCES**	**NEW BALANCE**	**MINIMUM PAYMENT**	**PAYMEMT DUE**
$44.50	$23.95	? $54.52	$10.00	8/20

19. A portion of Helena Strege's account statement for March from Inbank Charge Company is shown. The finance charge is computed using the average-daily-balance method where new purchases are excluded. Find the average daily balance, the finance charge, and the new balance.

REFERENCE	POSTING DATE	TRANSACTION DATE	DESCRIPTION	PURCHASES & ADVANCES	PAYMENTS & CREDITS
450345	3/20		PAYMENT		24.66
458343	3/27	3/14	Aston Oil Co.	81.30	

BILLING PERIOD	PREVIOUS BALANCE	PERIODIC RATE	AVERAGE DAILY BALANCE	FINANCE CHARGE
3/4 - 4/3	$94.66	2%	?	?
PAYMENTS & CREDITS	**PURCHASES & ADVANCES**	**NEW BALANCE**	**MINIMUM PAYMENT**	**PAYMENT DUE**
$24.66	$81.30	?	$10.00	4/21
		$152.95	$82.73	$1.65

20. Edith Bertelli received this statement from Bertrand's. Find the average daily balance, the finance charge, and the new balance.

REFERENCE	POSTING DATE	TRANSACTION DATE	DESCRIPTION	PURCHASES & ADVANCES	PAYMENTS & CREDITS
1027485	4/11		PAYMENT		40.00
4500298	4/15	4/01	Menswear	39.95	
5473390	4/23	4/21	Housewares	15.99	
1374655	4/25		PAYMENT		50.00

BILLING PERIOD	PREVIOUS BALANCE	PERIODIC RATE	AVERAGE DAILY BALANCE	FINANCE CHARGE
4/1 - 5/1	$175.00	1.2%	? $136.61	? $1.64
PAYMENTS & CREDITS	**PURCHASES & ADVANCES**	**NEW BALANCE**	**MINIMUM PAYMENT**	**PAYMENT DUE**
$90.00	$55.94	? $142.58	$25.00	5/25

PRACTICE AND APPLY

The following problems can be assigned for classwork and the answers checked in class to help students master the objective of the lesson.

- Guided Practice: 1–12
- Independent Practice: 13, 16

COMMUNICATION SKILLS

The terms presented in this unit are important in the business world and necessary as a consumer when communicating with a credit card company. Encourage students to use the correct terminology when discussing the concepts of the lesson, working the Examples, or reviewing solutions to problems.

WRAP-UP

Ask students if the payment in the Example were posted on 9/10, would the average daily balance be more or less than the amount in the Example. (less) Rework the Example using 9/10 as the posting date for the payment.

Assignment Guide

Basic: 13–18, 20–38
Average: 14, 15, 17–19, 21–37 odd

21. Louis Minier received this statement from Crowley Bank. Find the average daily balance, the finance charge, and the new balance.

REFERENCE	POSTING DATE	TRANSACTION DATE	DESCRIPTION	PURCHASES & ADVANCES	PAYMENTS & CREDITS
1616787	4/25	4/19	A-1 Plumbing	61.45	
3945557	4/30		PAYMENT		79.60
3957475	5/5	5/1	Ace Hardware	32.45	
4000076	5/20		PAYMENT		50.00

BILLING PERIOD	PREVIOUS BALANCE	PERIODIC RATE	AVERAGE DAILY BALANCE	FINANCE CHARGE
4/21 - 5/20	$179.60	2%	$122.21	$2.44
PAYMENTS & CREDITS	**PURCHASES & ADVANCES**	**NEW BALANCE**	**MINIMUM PAYMENT**	**PAYMENT DUE**
$129.60	$93.90	$146.34	$30.00	6/12

Critical Thinking . . .

22. Alice Kruse received this statement from Garrison's. Find the indicated amounts on the statement shown.

REFERENCE	POSTING DATE	TRANSACTION DATE	DESCRIPTION	PURCHASES & ADVANCES	PAYMENTS & CREDITS
31784	7/25		PAYMENT		30.00
103645	7/25	7/12	Men's shoes	79.48	
116748	7/30	7/28	Electronics		19.48
345803	8/8		PAYMENT		40.00
57845	8/8	8/7	Bakery	12.45	

BILLING PERIOD	PREVIOUS BALANCE	PERIODIC RATE	AVERAGE DAILY BALANCE	FINANCE CHARGE
7/22 - 8/21	$379.46	1.75%	$319.85	$5.60
PAYMENTS & CREDITS	**PURCHASES & ADVANCES**	**NEW BALANCE**	**MINIMUM PAYMENT**	**PAYMENT DUE**
$89.48	$91.93	$387.51	$30.00	9/11

MAINTAINING YOUR SKILLS Look up the skills in parentheses if you need help or more practice.

Multiply. **(Skill 8)**

23. 12 × 200 **2400**
24. 18 × 150 **2700**
25. 22 × 37.5 **825**
26. 9 × 34.56 **311.04**
27. 7 × 225.5 **1578.5**
28. 14 × 42.50 **595**
29. 35 × 61.8 **2163**
30. 37 × 341.5 **12,635.5**

Divide. Round answers to the nearest hundredth. **(Skill 10)**

31. 3750 ÷ 30 **125**
32. 3360 ÷ 28 **120**
33. 1129.95 ÷ 31 **36.45**
34. 10,852.5 ÷ 30 **361.75**

Find the average. **(Application Q)**

35. 44, 73, 92, 88, 63 **72**
36. 324, 406, 958, 285, 785, 374 **522**
37. 135, 415, 364, 419, 84 **283.4**
38. 52.8, 63.41, 14.66, 34.89, 9.35 **35.022**

6-6

Finance Charge—Average Daily Balance (New purchases included)

OBJECTIVE
Compute the finance charge based on the average daily balance, new purchases included.

Some companies compute the finance charge using the average-daily-balance method where new purchases are included when figuring the daily balances during the posting period.

$$\text{Average Daily Balance} = \frac{\text{Sum of Daily Balances}}{\text{Number of Days}}$$

EXAMPLE *Skills* 8, 11 *Application* G *Term* Average daily balance

Scott McTique has a charge account where the finance charge is computed using the average-daily-balance method that includes new purchases. He checks to be sure the average daily balance is correct.

REFERENCE	POSTING DATE	TRANSACTION DATE	DESCRIPTION	PURCHASES & ADVANCES	PAYMENTS & CREDITS
1-32734	12/10	12/8	Housewares	25.85	
2-44998	12/20		PAYMENT		70.00

BILLING PERIOD	PREVIOUS BALANCE	PERIODIC RATE	AVERAGE DAILY BALANCE	FINANCE CHARGE
12/1 - 12/31	$125.80	2%	?	?
PAYMENTS & CREDITS	**PURCHASES & ADVANCES**	**NEW BALANCE**	**MINIMUM PAYMENT**	**PAYMENT DUE**
$70.00	$25.85	?	$20.00	1/21

SOLUTION

A. Find the **sum of daily balances.**

Dates	Payment	Purchase	End-of-Day Balance		Number of Days	Sum of Balances
12/1–12/9			$125.80	×	9	$1132.20
12/10		$25.85	151.65	×	1	151.65
12/11–12/19			151.65	×	9	1364.85
12/20	$70.00		81.65	×	1	81.65
12/21–12/31			81.65	×	11	898.15
				Total	31	$3628.50

B. Find the **average daily balance.**

Sum of Daily Balances ÷ Number of Days
$3628.50 ÷ 31 = $117.048 = $117.05 average daily balance

125.8 [×] 9 [=] 1132.2 [M+] 125.8 [+] 25.85 [=] 151.65 [M+] [×] 9 [=] 1364.85 [M+]
151.65 [−] 70 [=] 81.65 [M+] [×] 11 [=] 898.15 [M+] [MR] 3628.5 [÷] 31 [=] 117.048

LESSON PLAN
6-6 Finance Charge—Average Daily Balance (New purchases included)

FOCUS
Ask students to suggest a way that a credit card company can compute the sum of daily balances to yield a higher finance charge. (include new purchases made during the billing, or posting, period) Explain that many charge accounts do use new purchases in computing the finance charge.

TEACH
Encourage students to organize their computations in an orderly way, as is done in the Example. This will help to avoid errors when many transactions are added or subtracted.

BUSINESS PROJECT
Have students investigate and report on charge accounts offered at local stores as well as credit cards offered by the banks in the area. Examples of the information students should collect include: how the finance charge is computed, what percentage rate is used, and how the minimum payment due is calculated. If students have family members who have credit cards, this information usually can be found on the back of the account statement.

Warm-Up Exercises

1. (10 × $350) + (20 × $425) $12,000
2. (18 × $246) + (13 × $186) $6846
3. $8275.60 ÷ 30 $275.85
4. $2892.50 ÷ 31 $93.31
5. $91.55 × 1.25% $1.14

PRACTICE AND APPLY

The following problems can be assigned for classwork and the answers checked in class to help students master the objective of the lesson.

- Guided Practice: 1–11
- Independent Practice: 12, 13

✓ SELF-CHECK Complete the problems, then check your answers in the back of the book.

1. $3500
2. $600
3. $600
4. $3000
5. $450
6. $450
7. $450
8. $7200
9. 30
10. $14,750; average daily balance = $491.67

Find the average daily balance, with new purchases.

Dates	Payment	Purchase	End-of-Day Balance	×	Number of Days	=	Sum of Balances
9/9–9/15			$500.00	×	7	=	**1.** ?
9/16		$100.00	$600.00	×	1	=	**2.** ?
9/17–9/21			**3.** ?	×	5	=	**4.** ?
9/22	$150.00		**5.** ?	×	1	=	**6.** ?
9/23–10/8			**7.** ?	×	16	=	**8.** ?
			Total		**9.** ?		**10.** ?

The finance charge is calculated by multiplying the average daily balance times the periodic rate.

Finance Charge = Average Daily Balance × Periodic Rate
New Balance = Unpaid Balance + Finance Charge + New Purchases

EXAMPLE *Skills* 8, 11 *Application* G *Term* Average daily balance

Scott checks the finance charge and the new balance. The finance charge is 2% of the average daily balance. What is the new balance?

SOLUTION

A. Find the **unpaid balance.**
Previous Balance − (Payments + Credits)
$125.80 − $70.00 = $55.80 unpaid balance

B. Find the **finance charge.**
Average Daily Balance × Periodic Rate
$117.05 × 2% = $2.3418 = $2.34 finance charge

C. Find the **new purchases.** $25.85

D. Find the **new balance.**
Unpaid Balance + Finance Charge + New Purchases
$55.80 + $2.34 + $25.85 = $83.99 new balance

PROBLEMS

Find the average daily balance, new purchases included.

11.

Dates	Payment	Purchase	End-of-Day Balance	Number of Days	Sum of Balances	
9/6–9/17			$600.00	12	$7200	What is the average daily balance? **$640.00**
9/18		$140.00	740	1	740	
9/19–9/24		?	740	6	4440	
9/25	$120.00	?	620	1	620	
9/26–10/5		?	620	10	6200	
			TOTAL	30	19,200	

ALTERNATIVE STRATEGIES: RETEACHING

Use the calendar worksheet in the TRB in order to show the daily balances for the Example.

1	2	3	4	5	6	7
125.80	125.80	125.80	125.80	125.80	125.80	125.80
8	**9**	**10**	**11**	**12**	**13**	**14**
125.80	125.80	125.80 + 25.85 = 151.65	151.65	151.65	151.65	151.65

12. Edith Bertelli received this statement from Bertrand's. Find the average daily balance (new purchases included), the finance charge, and the new balance.

REFERENCE	POSTING DATE	TRANSACTION DATE	DESCRIPTION	PURCHASES & ADVANCES	PAYMENTS & CREDITS
1027485	4/11		PAYMENT		40.00
4500298	4/15	4/01	Menswear	39.95	
5473390	4/23	4/21	Housewares	15.99	
1374655	4/25		PAYMENT		50.00

BILLING PERIOD	PREVIOUS BALANCE	PERIODIC RATE	AVERAGE DAILY BALANCE	FINANCE CHARGE
4/1 - 5/1	$175.00	1.2%	$163.90	$1.97
PAYMENTS & CREDITS	**PURCHASES & ADVANCES**	**NEW BALANCE**	**MINIMUM PAYMENT**	**PAYMENT DUE**
$90.00	$55.94	$142.91	$25.00	5/25

F.Y.I.
On December 1, 1991, the average monthly credit card balance was $1626.00.

13. Louis Minier received this statement from Crowley Bank. Find the average daily balance (new purchases included), the finance charge, and the new balance.

REFERENCE	POSTING DATE	TRANSACTION DATE	DESCRIPTION	PURCHASES & ADVANCES	PAYMENTS & CREDITS
1616787	4/25	4/19	A-1 Plumbing	61.45	
3945557	4/30		PAYMENT		79.60
3957475	5/5	5/1	Ace Hardware	32.45	
4000076	5/20		PAYMENT		50.00

BILLING PERIOD	PREVIOUS BALANCE	PERIODIC RATE	AVERAGE DAILY BALANCE	FINANCE CHARGE
4/21 - 5/20	$179.60	2%	$192.78	$3.86
PAYMENTS & CREDITS	**PURCHASES & ADVANCES**	**NEW BALANCE**	**MINIMUM PAYMENT**	**PAYMENT DUE**
$129.60	$93.90	$147.76	$30.00	6/12

MAINTAINING YOUR SKILLS Look up the skills in parentheses if you need help or more practice.

Multiply. **(Skills 7, 8)**

14. 7 × 145 **1015**
15. 5 × 360 **1800**
16. 31 × 56.23 **1743.13**
17. 8 × 385.21 **3081.68**
18. 12 × 14.5 **174**
19. 11 × 99.78 **1097.58**
20. 18 × 1455 **26,190**
21. 2 × 891.52 **1783.04**

Divide. Round answers to the nearest hundredth. **(Skill 10)**

22. 1608.75 ÷ 30 **53.63**
23. 1329.89 ÷ 31 **42.90**
24. 1029 ÷ 28 **36.75**
25. 9505.08 ÷ 37 **256.89**

Find the average. Round answers to the nearest hundredth. **(Application Q)**

26. 36, 45, 58, 62, 48 **49.80**
27. 456.2, 364.8, 471.5, 392.18 **421.17**
28. 36.25, 56.87, 33.10, 93.51 **54.93**
29. 365.23, 577.10, 782.20 **574.84**

WRAP-UP

Review with students the solution to Problem 12. Have volunteers work the different computations at the chalkboard. If a student has a question, ask another student to provide the answer.

Assignment Guide

- Basic: 12–14, 15–30
- Average: 14, 16–30 even

15	16	17	18	19	20	21
151.65	151.65	151.65	151.65	151.65	151.65 − 70.00 81.65	81.65
22	**23**	**24**	**25**	**26**	**27**	**28**
81.65	81.65	81.65	81.65	81.65	81.65	81.65
29	**30**	**31**				
81.65	81.65	81.65				

LESSON PLAN
Reviewing the Basics

The exercises on this page review skills, applications, and terms used in the unit. You can use the exercises to assess informally students' proficiency with this material.

The page can be used for guided practice and independent practice. You can work through a selection of the exercises together with students, and thus see immediately if they know how to do them, and you can then assign some of the exercises for independent practice. Be sure to go over the answers to all assigned exercises.

Reviewing the Basics

Skills

Round to the nearest cent.

(Skill 2)

1. $173.1426 **$173.14**
2. $172.183 **$172.18**
3. $37.998 **$38.00**
4. $5.692 **$5.69**

Solve.

(Skill 5)

5. $485.43 + $41.97 + $9.98 **$537.38**
6. $538.26 + $6.19 + $16.71 + $157.22 **$718.38**

(Skill 6)

7. $478.51 − $358.92 **$119.59**
8. $624.84 − $37.89 **$586.95**
9. $38.10 − $5.99 **$32.11**

(Skill 8)

10. $460 × 8 **$3680**
11. $520 × 16 **$8320**
12. $1420.85 × 30 **$42,625.50**
13. $22.15 × 9 **$199.35**
14. $35.75 × 15 **$536.25**
15. $66.95 × 19 **$1272.05**

(Skill 11)

16. $150.60 ÷ 30 **$5.02**
17. $1923.75 ÷ 28 **$68.71**
18. $7859.60 ÷ 30 **$261.99**
19. $582.86 ÷ 31 **$18.80**
20. $7245.79 ÷ 31 **$233.74**
21. $22,812.59 ÷ 31 **$735.89**

(Skill 30)

22. 1.5% of $720 **$10.8**
23. 1% of 543.21 **5.4321**
24. 2% of 267.15 **5.343**

Applications

Find the elapsed time in days.

(Application G)

25. From June 4 to June 30 **26 days**
26. From April 4 to April 25 **21 days**
27. From May 21 to June 9 **19 days**
28. From July 15 to August 28 **44 days**

Find the average.

(Application Q)

29. 48, 42, 74, 91, 51, 65 **61.83**
30. 12.8, 14.1, 21.2, 18.3, 20.6 **17.4**
31. $45, $94, $65, $83, $66 **$70.60**
32. 4.54, 8.23, 9.10, 7.54 **7.3525**

Terms

Match each term with its definition on the right.

c. 33. Sales receipt

d. 34. Finance charge

f. 35. Previous balance

b. 36. Unpaid balance

a. 37. Average daily balance

a. the average of the balances in a charge account at the end of each day in the billing period

b. the amount of money that remains unpaid in a charge account

c. a record of the total purchase that a charge card user signs each time the account is used

d. interest that a charge or credit card user pays for delaying payment

e. the monthly finance charge rate

f. the amount that a charge or credit card user owed at the close of the last billing period

Refer to your reference files in the back of the book if you need help.

Unit Test

Students should do the Unit Test on their own. Each problem on the test is keyed to a lesson in the unit. Students having difficulty with any particular problem should review the Example in the appropriate lesson and be assigned some of the Independent Practice problems for additional practice.

Lesson 6-1

1. Alma Ying used her bank charge card to purchase a stereo. The stereo cost $995.99 plus a 6% sales tax. What was the total purchase price on the sales receipt? **$1055.75**

Lesson 6-2

2. Find the new balance on the charge account statement.

PREVIOUS BALANCE	CLOSING DATE THIS MONTH		CLOSING DATE LAST MONTH	
$175.41	November 13, 19--		October 14, 19--	
TOTAL PURCHASES	PAYMENTS & CREDITS	FINANCE CHARGE	NEW BALANCE	MINIMUM PAYMENT
$72.59	$50.00	$1.90	**$199.90**	——

Lesson 6-3

3. Ray Etzel has a charge account that charges 1.5% of the previous balance for a finance charge. His statement shows purchases totaling $98.15, a previous balance of $670, and a payment of $95. What is the new balance? **$683.20**

Lesson 6-4

4. A portion of Ethel Lewis's account statement is shown. The finance charge is 1.5% of the unpaid balance. Find the unpaid balance, the finance charge, and the new balance.

BILLING DATE	PREVIOUS BALANCE	PAYMENTS & CREDITS	UNPAID BALANCE	FINANCE CHARGE	NEW PURCHASES	NEW BALANCE
10/13	$413.65	$137.35	**$276.30**	**$4.14**	$37.10	**$317.54**

Lesson 6-5

5. Find the average daily balance where new purchases are not included, the finance charge, and the new balance.

REFERENCE	POSTING DATE	TRANSACTION DATE	DESCRIPTION	PURCHASES & ADVANCES	PAYMENTS & CREDITS
13100	9/14	8/23	Westford Theatre	65.00	
20046	9/25	8/27	Record Mart	14.50	
54545	9/28		PAYMENT		40.00

BILLING PERIOD	PREVIOUS BALANCE	PERIODIC RATE	AVERAGE DAILY BALANCE	FINANCE CHARGE
9/1 - 9/30	$150.00	1.5%	**$146.00**	**$2.19**
PAYMENTS & CREDITS	PURCHASES & ADVANCES	NEW BALANCE	MINIMUM PAYMENT	PAYMENT DUE
$40.00	**$79.50**	**$191.69**	$20.00	10/21

Lesson 6-6

6. In problem 5, find the average daily balance, finance charge, and the new balance using the average-daily-balance method where new purchases are included. **ADB = $185.73; FC = $2.79; NB = $192.29**

LESSON PLAN
A Spreadsheet Application

USING TECHNOLOGY

Before students work this application, you may want to discuss negative numbers briefly. Remind students that the subtraction of a positive number yields the same result as the addition of a negative number. For example, in Problem 1, leaving off the dollar signs, 675 − 200 = 475; alternatively, 675 + (−200) = 475. Ask students why a payment can be considered a negative number. (The payment reduces the balance owed to the credit card company.)

As a *critical thinking* question, ask students if the average daily balance for Problems 1–6 is being calculated with no new purchases or with new purchases included. (with new purchases included)

A SPREADSHEET APPLICATION

Charge Accounts and Credit Cards

To complete this spreadsheet application, you will need the diskette *Spreadsheet Applications for Business Mathematics,* which accompanies this textbook. Follow the directions in the *User's Guide* to complete this activity.

Input the transactions indicated in the following problems to find the average daily balance including new purchases, finance charge, and new balance.

Note that for a payment you must enter a negative number. You do this by entering a negative or a minus sign in front of the number. Thus, a payment of $30.00 would be entered as −30 (do not use a dollar sign). Enter the periodic rate as a decimal. Thus, a periodic rate of 2% is entered as 0.02. The billing period coincides with the calendar month.

1. Balance on January 1: $675.00.
   Purchase on January 5: $100.00.
   Purchase on January 23: $25.00.
   Payment on January 28: $200.00.
   Periodic rate: 2%.
   What is the average daily balance?
   What is the finance charge?
   What is the new balance?

2. Balance on January 1: $233.55.
   Purchase on January 9: $23.55.
   Purchase on January 18: $45.25.
   Payment on January 25: $45.00.
   Periodic rate: 1.50%.
   What is the average daily balance?
   What is the finance charge?
   What is the new balance?

3. Balance on March 1: $379.46.
   Purchase on March 11: $59.97.
   Purchase on March 19: $23.56.
   Purchase on March 25: $156.20.
   Payment on March 30: $180.00.
   Period rate: 1.75%.
   What is the average daily balance?
   What is the finance charge?
   What is the new balance?

4. Balance on March 1: $785.23.
   Purchase on March 3: $23.21.
   Purchase on March 5: $156.32.
   Purchase on March 10: $97.58.
   Payment on March 20: $450.00.
   Periodic rate: 1.75%.
   What is the average daily balance?
   What is the finance charge?
   What is the new balance?

5. Balance on July 1: $563.25.
   Payment on July 5: $350.00.
   Purchase on July 12: $123.20.
   Purchase on July 19: $98.54.
   Purchase on July 26: $31.25.
   Periodic rate: 1.65%.
   What is the average daily balance?
   What is the finance charge?
   What is the new balance?

6. Balance on July 1: $563.25.
   Purchase on July 12: $123.20.
   Purchase on July 19: $98.54.
   Purchase on July 26: $31.25.
   Payment on July 30: $350.00.
   Periodic rate: 1.65%.
   What is the average daily balance?
   What is the finance charge?
   What is the new balance?

UNIT

Loans

A *loan* is money that you have borrowed and must repay with *interest.* Interest is the cost of borrowing money. A *single-payment loan* (sometimes called a promissory note) is repaid in one payment after a specified period of time. An *installment loan* is a loan for which you pay a portion of the loan and a portion of the interest in several installments. Because all banks do not charge the same amount of interest for a sum of money, you should shop around for the best buy. Using the *annual percentage rate* is one way to compare the cost of borrowing money. If you repay a loan early, you may receive a *refund* of part of the finance charge.

OUTLINE

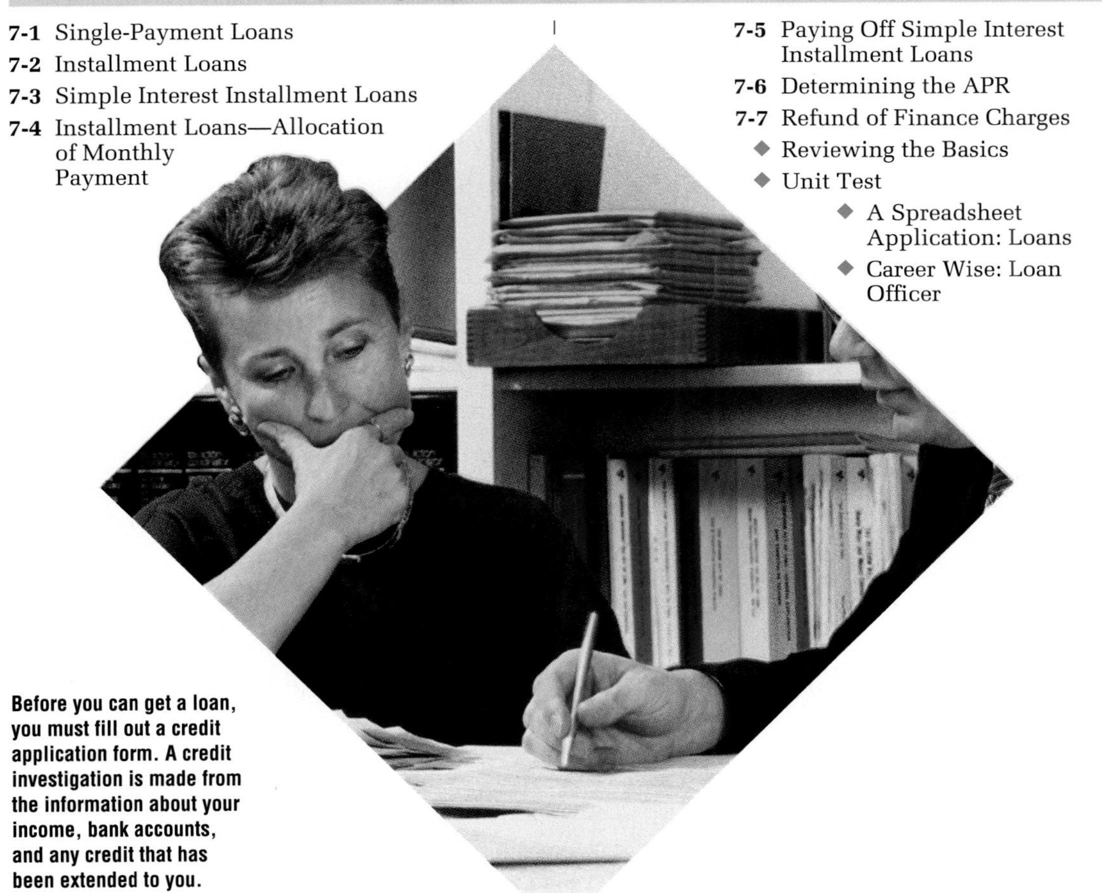

Before you can get a loan, you must fill out a credit application form. A credit investigation is made from the information about your income, bank accounts, and any credit that has been extended to you.

UNIT 7
LOANS

INTRODUCING THE UNIT

Lenders want to make sure a person borrowing money is a good "credit risk." Most rely upon information from credit bureaus, which keep data on how a person pays his or her bills. Another method is by "scoring" a person, where points are given in different categories. An example would be the following four categories: (1) **Credit history** — how fast do you pay your bills? (2) **Current debt** - can you pay all your current bills based upon your income? (3) **Employment stability** — how long have you been working at the same job, and what type of job do you have? (4) **Personal stability** — do you own a home or how long have you rented the same apartment? Do you have a checking and savings account? You might ask a bank official to speak with the class about how the bank's loan department qualifies a person for a loan.

LESSON PLAN 7-1 Single-Payment Loans

FOCUS

To motivate this lesson, ask students if they have ever loaned a person money. How long did it take to be paid back? Was it paid back in one payment or did the person repay a portion of the loan on different days? Explain that banks and other lending institutions grant loans to people; however, a bank charges interest. One type of loan is a single-payment loan, often called a promissory note, which is repaid in one payment.

TEACH

Spend ample time going over the terms presented in this lesson. Explain that single-payment loans are generally for short periods of time, usually less than one year.

Emphasize that when computing ordinary interest, the denominator in the fraction representing time is 360, which assumes that each month has 30 days. If a year is considered to have 365 days by a bank, as is the case for exact interest, the denominator is 365.

Warm-Up Exercises

1. $1000 × 12% × $\frac{30}{365}$
   $9.86
2. $1000 × 12% × $\frac{30}{360}$
   $10.00
3. $4800 × 15% × $\frac{73}{365}$
   $144
4. $2700 × 9% × $\frac{65}{360}$
   $43.88

7-1 Single-Payment Loans

OBJECTIVE

Compute the maturity value and interest rate of a single-payment loan.

A **single-payment loan** is a loan that you repay with one payment after a specified period of time. A **promissory note** is a type of single-payment loan. It is a written promise to pay a certain sum of money on a certain date in the future. The **maturity value** of the loan is the total amount you repay. It includes both the **principal** and the interest owed. The principal is the amount borrowed.

The **term** of a loan is the amount of time for which the loan is granted. A single-payment loan may be granted for a stated number of years, months, or days. When the term is a certain number of days, the lending agency may calculate interest in one of two ways. **Ordinary interest** is calculated by basing the time of the loan on a 360-day year. **Exact interest** is calculated by basing the time on a 365-day year.

Maturity Value = Principal + Interest Owed

EXAMPLE *Skills* 28, 14 *Application* J *Term* Single-payment loan

Anita Sloane's bank granted her a single-payment loan of $7200 for 91 days at an interest rate of 12%. What is the maturity value of the loan if: (1) her bank charges ordinary interest? (2) her bank charges exact interest?

SOLUTION

A. (1) Find the **ordinary interest owed.**
Principal × Rate × Time
$7200.00 × 12% × $\frac{91}{360}$ = $218.40 ordinary interest

(2) Find the **exact interest owed.**
Principal × Rate × Time
$7200.00 × 12% × $\frac{91}{365}$ = $215.408 = $215.41 exact interest

B. (1) Find the **maturity value.**
Principal + Interest Owed
$7200.00 + $218.40 = $7418.40 maturity value

(2) Find the **maturity value.**
Principal + Interest Owed
$7200.00 + $215.41 = $7415.41 maturity value

7200 M+ × 12 % × 91 ÷ 360 = 218.4 M+ RM 7418.4

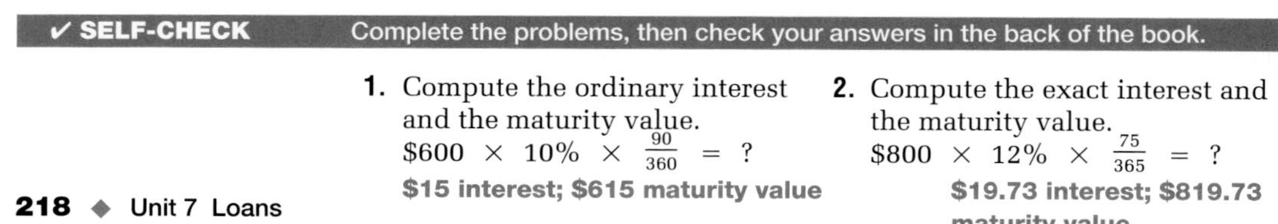

✓ SELF-CHECK Complete the problems, then check your answers in the back of the book.

1. Compute the ordinary interest and the maturity value.
   $600 × 10% × $\frac{90}{360}$ = ?
   $15 interest; $615 maturity value
2. Compute the exact interest and the maturity value.
   $800 × 12% × $\frac{75}{365}$ = ?
   $19.73 interest; $819.73 maturity value

COMMUNICATION SKILLS

You might want to have students keep a notebook of new terms and write the definitions in their own words as they are presented in this unit. This will help students remember what each term means and how it is used in the context of various loans.

PROBLEMS

	3.	4.	5.	6.	7.
Principal	$900	$9600	$850	$7500	$9435
Annual Interest Rate	12%	15%	11%	12.5%	18.72%
Term	45 days	30 days	72 days	123 days	275 days
Ordinary Interest	$13.50	**$120**	**$18.70**	**$320.31**	**$1349.21**
Maturity Value	**$913.50**	**$9720**	**$868.70**	**$7820.31**	**$10,784.21**

8. Maria Rodriquez.
Single-payment loan of $1000.
Interest rate of 15%.
108 days, ordinary interest.
What is the interest owed? **$45**
What is the maturity value? **$1045**

9. Manuel Bruins.
Single-payment loan of $8400.
Interest rate of 12%.
146 days, exact interest.
What is the interest owed? **$403.20**
What is the maturity value? **$8803.20**

10. Joseph Henning's bank granted him a single-payment loan of $2400 for 144 days at an interest rate of 18%. His bank charges ordinary interest. What is the maturity value of his loan? **$2572.80**

11. Vanessa Tackett's bank granted her a single-payment loan of $21,000 for 45 days at an interest rate of 15%. Her bank charges ordinary interest. What is the maturity value of her loan? **$21,393.75**

12. Jessie Ardella obtained a single-payment loan of $225 to pay a repair bill. He agreed to repay the loan in 31 days at an interest rate of 14.75% exact interest. What is the maturity value of his loan? **$227.82**

13. Gordon Hamen obtained a single-payment loan of $44,000 to pay for tree spray for his commercial orchard. He agreed to repay the loan in 270 days at an interest rate of 14.35% exact interest. What is the maturity value of his loan? **$48,670.63**

14. Mark Norwalk would like to borrow $8000 for 90 days. Advance Finance Company charges an interest rate of 18% at ordinary interest. AAA Finance Company also charges an interest rate of 18% but at exact interest. What would be the maturity value of the loan at Advance Finance Company? What would be the maturity value of the loan at AAA Finance Company? **$8360; $8355.07**

MAINTAINING YOUR SKILLS
Look up the skills in parentheses if you need help or more practice.

Write the percents as decimals. (Skill 28)

15. 40% **0.40**　**16.** 90.5% **0.905**　**17.** 7% **0.07**　**18.** 8.25% **0.0825**　**19.** 15.64% **0.1564**

Reduce the fractions to lowest terms. (Skill 12)

20. $\frac{40}{60}$ **$\frac{2}{3}$**　**21.** $\frac{180}{360}$ **$\frac{1}{2}$**　**22.** $\frac{90}{360}$ **$\frac{1}{4}$**　**23.** $\frac{120}{360}$ **$\frac{1}{3}$**　**24.** $\frac{73}{365}$ **$\frac{1}{5}$**

Change the fractions to decimals. (Skill 14)

25. $\frac{40}{60}$ **0.667**　**26.** $\frac{45}{360}$ **0.125**　**27.** $\frac{126}{360}$ **0.35**　**28.** $\frac{146}{365}$ **0.40**　**29.** $\frac{219}{365}$ **0.60**

PRACTICE AND APPLY
The following problems can be assigned for classwork and the answers checked in class to help students master the objective of the lesson.

- Guided Practice: 1–8
- Independent Practice: 9–11

WRAP-UP
Write $2121.67, $2000, and $121.67 on the chalkboard, and ask students to identify the principal, interest, and maturity value of the loan. Have a volunteer write the formula for finding the interest on the chalkboard. Have another volunteer do the computations for this loan, using an interest rate of 15% ordinary interest and the length of the loan as 144 days.

Assignment Guide
- Basic: 9–14, 16–29
- Average: 12–15, 16–28 even

ALTERNATIVE STRATEGIES: Enrichment
Have students do research and report on the "prime rate." They can read the *Wall Street Journal* or *USA Today* to find the current prime rate.

LESSON PLAN
7-2 Installment Loans

FOCUS

Have students suggest reasons for borrowing money. (to buy a car, furniture, appliance, or other expensive consumer items) Explain that when purchasing consumer items, it is more likely that the purchase is made through an installment loan, in which the amount borrowed is paid back in a series of equal payments over a specified time period. Ask students why they think people prefer this type of loan rather than a single-payment loan. (They can pay back the loan in smaller portions.) In this lesson, students will compute the amount financed on an installment loan.

TEACH

Discuss with students the similarities and differences between charge accounts and installment loans. Some points to bring out are that when a person uses a credit card, he or she can pay the account in full within a month or a finance charge is added to the unpaid balance. When purchasing an item through an installment loan, a person usually pays a small portion of the cash price as a down payment and then signs a contract to make equal weekly or monthly payments. The payments would include a finance charge, which is discussed in the next lesson. In both cases, the customer has immediate possession of the item.

7-2 Installment Loans

OBJECTIVE

Compute the amount financed on an installment loan.

An **installment loan** is a loan that you repay in several equal payments over a specified period of time. Usually when you purchase an item with an installment loan, you must make a **down payment**. The down payment is a portion of the cash price of the item you are purchasing. The **amount financed** is the portion of the cash price that you owe after making the down payment.

Amount Financed = Cash Price − Down Payment

EXAMPLE *Skills* 30, 28, 2 *Application* A *Term* Installment loan

Rebecca Clay purchased a washer and a dryer for $1140. She used the store's installment credit plan to pay for the items. She made a down payment and financed the remaining amount. What amount did Rebecca finance if she made: (1) a 20% down payment? (2) a 25% down payment?

SOLUTION

(1) Find the 20% **down payment.**
$1140.00 × 20% = $228.00

(2) Find the 25% **down payment.**
$1140.00 × 25% = $285.00

(1) Find the **amount financed.**
Cash Price − Down Payment
$1140.00 − $228.00
= $912.00 amount financed

(2) Find the **amount financed.**
Cash Price − Down Payment
$1140.00 − $285.00
= $855.00 amount financed

1140 M+ × 20 % 228 M− RM 912

✓ SELF-CHECK Complete the problems, then check your answers in the back of the book.

Find the down payment and the amount financed.

1. Waterbed.
Cash price of $1360.
20% down payment **$272; $1088**

2. Television set.
Cash price of $725.
30% down payment **$217.50; $507.50**

PROBLEMS

	3.	4.	5.	6.	7.	8.
Cash Price	$640	$4860	$1774	$3600	$9480	$5364
Percent Down Payment	20%	30%	25%	40%	15%	25%
Down Payment	$128	**$1458**	**$443.50**	**$1440**	**$1422**	**$1341**
Amount Financed	**$512**	**$3402**	**$1330.50**	**$2160**	**$8058**	**$4023**

COMMUNICATION SKILLS

Installment loan contracts can be very lengthy and the language difficult to understand. However, when signing a contract, it is very important to know what the contract actually states. If available, bring a typical installment contract from a bank or other lending source to class. Have students read the contract and discuss exactly what it states.

9. Cash price of $1265.
20% down payment.
What amount is financed?
$1012

10. Cash price of $14,470.
25% down payment.
What amount is financed?
$10,852.50

11. Cash price of $8371.39.
15% down payment.
What amount is financed?
$7115.68

12. Cash price of $18,936.50.
30% down payment.
What amount is financed?
$13,255.55

13. Milt Gibson purchased computer equipment for $4020. He used the store's credit plan. He made a 20% down payment. What amount did he finance? **$3216**

14. Linda Chevez purchased a stereo for her car. The stereo cost $279.50. Using the store's credit plan, she made a 30% down payment. What amount did she finance? **$195.65**

15. Ardella Haubert purchased living room furniture for $987.95. She made a down payment of 20% and financed the remaining amount using the store's installment plan. What amount did she finance? **$790.36**

16. Bev and Tom Hoffman went on a two-week vacation at a total cost of $1876. They financed the trip through Sentinel Bank. They made a 20% down payment and financed the remaining amount on the installment plan. What amount did they finance? **$1500.80**

SENTINEL BANK offers
TRAVEL and ADVENTURE
Go NOW .... Vacation loans
Pay ONLY
Later .... 20% down

17. Amy and Cliff Martin want to remodel their kitchen. They would like to finance part of the cost but do not want the amount financed to be more than $9000. The total cost of remodeling the kitchen is $12,000. What percent of the total cost should their down payment be? **25%**

18. Mack Casey wants to purchase a car costing $14,590. He will finance the car with an installment loan from the bank but would like to finance no more than $10,000. What percent of the total cost of the car should his down payment be? **31.5%**

MAINTAINING YOUR SKILLS Look up the skills in parentheses if you need help or more practice.

Write the percents as decimals. (Skill 28)

19. 32% **0.32** 20. 45% **0.45** 21. 25% **0.25** 22. 30% **0.30** 23. 20% **0.20**

Find the percentage. (Skill 30)

24. 440 × 30% **132** 25. 325 × 20% **65** 26. 1240 × 25% **310** 27. 950 × 15% **142.50**

Round to the nearest cent. (Skill 2)

28. $49.9638 **$49.96** 29. $178.3813 **$178.38** 30. $413.995 **$414.00** 31. $17.6309 **$17.63**

Point out that installment loans can be made through banks, finance companies, or credit unions. You might want to explain what credit unions are and how they operate.

Warm-Up Exercises

1. $800 − ($800 × 20%) $640
2. $8800 − ($8800 × 25%) $6600
3. $9172 − ($9172 × 15%) $7796.20
4. $17,360 − ($17,360 × 30%) $12,152

PRACTICE AND APPLY

The following problems can be assigned for classwork and the answers checked in class to help students master the objective of the lesson.

- Guided Practice: 1–6
- Independent Practice: 7–10, 13

WRAP-UP

Review the solutions to Problems 9 and 10. Ask students how much more the amount financed would be in Problem 9 if the down payment were 15%.

Assignment Guide

- Basic: 7–16, 19–31
- Average: 11, 12, 14–18, 19–31 odd

ALTERNATIVE STRATEGIES: Reteaching

Point out that another way to find the amount financed is to multiply the cash price by the complement of the percent down. Using the two examples:

A. 20% down implies 80% financed
80% × $1140 = $912
B. 25% down implies 75% financed
75% × $1140 = $855

LESSON PLAN
7-3 Simple Interest Installment Loans

FOCUS
Discuss the meaning of *interest* with students. Point out that a person can *earn* interest (on a savings account) or *pay* interest (on a loan). Point out that interest is always paid to the person who owns the money and is letting someone else use it. In this lesson, the finance charge is the interest paid on the installment loan.

TEACH
Discuss the loan table with students. Make sure they understand that the dollar amounts in the table are for a $100 loan. Therefore, when using the table, the total amount of a loan must first be divided by 100, and then this number is multiplied by the appropriate number from the table. Review the fact that dividing a number by 100 has the effect of moving the decimal point in the number two places to the left.

Discuss the formulas and ask students what the finance charge represents. (the interest charged on the amount financed) The annual percentage rate will be discussed in detail in Lesson 7-6. You might wish to point out that if the monthly interest rate for a loan is known, then the APR is 12 times the monthly rate.

7-3 Simple Interest Installment Loans

OBJECTIVE
Compute the monthly payment, total amount repaid, and finance charge on an installment loan.

When you obtain a **simple interest installment loan,** you must pay finance charges for the use of the money. Usually you repay the amount financed plus the finance charge in equal monthly payments. The amount of each monthly payment depends on the amount financed, the number of payments, and the **annual percentage rate** (APR). The annual percentage rate is an index showing the relative cost of borrowing money.

MONTHLY PAYMENT ON A $100 LOAN

Term in Months	Annual Percentage Rate			
	10%	12%	15%	18%
6	$17.16	$17.25	$17.40	$17.55
12	8.79	8.88	9.03	9.17
18	6.01	6.10	6.24	6.38
24	4.61	4.71	4.85	4.99
30	3.78	3.87	4.02	4.16
36	3.23	3.32	3.47	3.62
42	2.83	2.93	3.07	3.23
48	2.54	2.63	2.78	2.94

$$\text{Monthly Payment} = \frac{\text{Amount of Loan}}{\$100} \times \text{Monthly Payment for a \$100 loan}$$

$$\text{Total Amount Repaid} = \text{Number of Payments} \times \text{Monthly Payment}$$

$$\text{Finance Charge} = \text{Total Amount Repaid} - \text{Amount Financed}$$

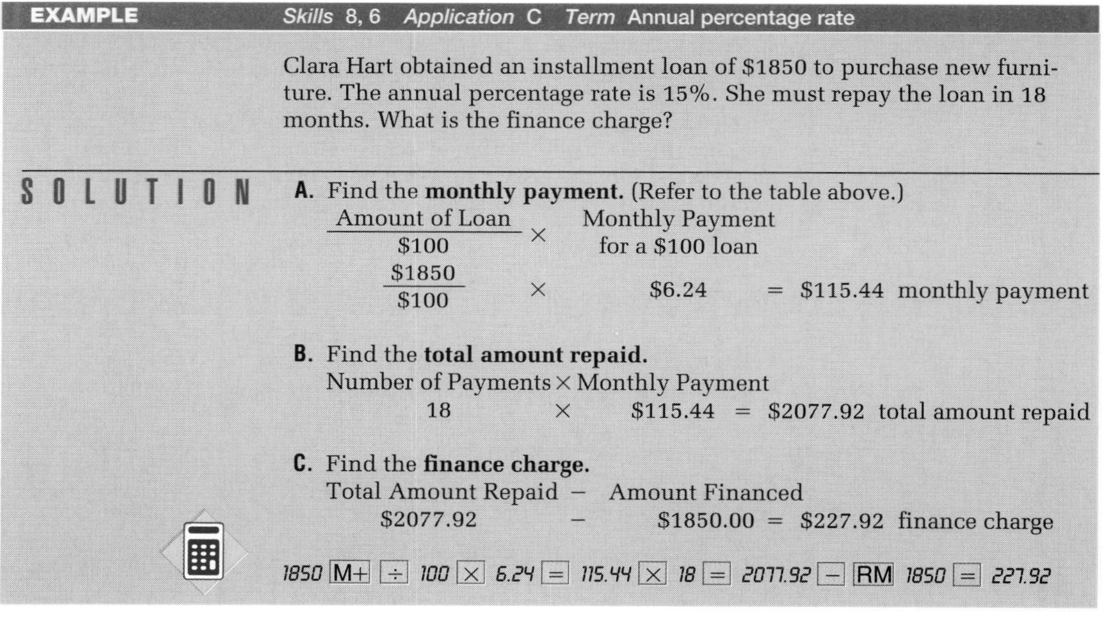

EXAMPLE *Skills* 8, 6 *Application* C *Term* Annual percentage rate

Clara Hart obtained an installment loan of $1850 to purchase new furniture. The annual percentage rate is 15%. She must repay the loan in 18 months. What is the finance charge?

SOLUTION

A. Find the **monthly payment.** (Refer to the table above.)

$$\frac{\text{Amount of Loan}}{\$100} \times \text{Monthly Payment for a \$100 loan}$$

$$\frac{\$1850}{\$100} \times \$6.24 = \$115.44 \text{ monthly payment}$$

B. Find the **total amount repaid.**

Number of Payments × Monthly Payment

18 × $115.44 = $2077.92 total amount repaid

C. Find the **finance charge.**

Total Amount Repaid − Amount Financed

$2077.92 − $1850.00 = $227.92 finance charge

1850 [M+] [÷] 100 [×] 6.24 [=] 115.44 [×] 18 [=] 2077.92 [−] [RM] 1850 [=] 227.92

USE OF CALCULATORS
To work the Solution to Problem 2 using a calculator, show students the following sequence of keys.
1000 [M+] [÷] 100 [×] 17.16 [=] 171.6 [×] 6 [=] 1029.6 [−] [RM] 1000 [=] 29.6.

✓ SELF-CHECK Complete the problem, then check your answer in the back of the book.

1. Find the monthly payment, total amount repaid, and the finance charge for a $1650 installment loan at 18% for 24 months. **$82.335 = $82.34; $1976.16; $326.16**

PROBLEMS

Use the table on page 222 to solve the following.

	2.	3.	4.	5.	6.
APR	10%	15%	18%	12%	15%
Term (Months)	6	18	24	6	36
Amount Financed	$1000	$1780	$600	$900	$4350
Monthly Payment	**$171.60**	**$111.07**	**$29.94**	**$155.25**	**$150.95**
Total Repaid	**$1029.60**	**$1999.26**	**$718.56**	**$931.50**	**$5434.20**
Finance Charge	**$29.60**	**$219.26**	**$118.56**	**$31.50**	**$1084.20**

7. Hazel Basnett.
   Installment loan of $2000.
   12 monthly payments.
   APR is 18%.
   What are the monthly payments?
   What is the finance charge?
   $183.40; $200.80

8. Brian Anderson.
   Installment loan of $750.
   24 monthly payments.
   APR is 10%.
   What are the monthly payments?
   What is the finance charge?
   $34.58; $79.92

9. Bob Wozniak obtained an installment loan of $2400 to put a roof on his house. The APR is 12%. The loan is to be repaid in 36 monthly payments. What is the finance charge? **$468.48**

10. Jim Wilson obtained an installment loan of $1450 to pay for some new furniture. He agreed to repay the loan in 18 monthly payments at an APR of 15%. What is the finance charge? **$178.64**

11. Mark and Pam Voss obtained an installment loan of $2460. They obtained the loan at an APR of 10% for 12 months. What is the finance charge? **$134.76**

12. Herb and Marci Rahla are purchasing a dishwasher with an installment loan that has an APR of 18%. The dishwasher sells for $699.95. They agree to make a down payment of 20% and to make 12 monthly payments. What is the finance charge? **$56.24**

13. Andrew and Ruth Bacon would like to obtain an installment loan of $1850 to repair the gutters on their home. They can get the loan at an APR of 15% for 24 months or at an APR of 18% for 18 months. Which loan costs less? How much do the Bacons save by taking the loan that costs less? **18%; $28.98**

14. Lola Samaria would like an installment loan of $950. Walton Savings and Loan will loan her the money at 15% for 12 months. Horton Finance Company will loan her the money at 18% for 24 months. Which loan costs less? How much will she save by taking the loan that costs less? **15%; $108.36**

Warm-Up Exercises

1. ($1240 ÷ 100) × $6 $74.40
2. ($6875 ÷ 100) × $5.50 $378.13
3. (6 × $146.63) − $850 $29.78
4. (24 × $79.85) − $1600 $316.40

PRACTICE AND APPLY

The following problems can be assigned for classwork and the answers checked in class to help students master the objective of the lesson.

- Guided Practice: 1–7
- Independent Practice: 8–10, 13

ALTERNATIVE STRATEGIES: Enrichment

Have students work these variations of the example by finding the finance charge if:

A. The rate is 12% for 18 months. ($181.30)
B. The rate is 18% for 18 months. ($274.54)
C. The time is 12 months at 15%. ($154.72)
D. The time is 24 months at 15%. ($303.52)

WRAP-UP

Change the data given in the Example to the following and ask students to solve the problem.

Installment Loan: $2450
APR: 18%
Time: 30 months
Finance charge: ?
($607.60)

Assignment Guide

- Basic: 8–18, 21–30
- Average: 11, 12, 14–20, 22–30 even

15. Adolfo Ramirez obtained an installment loan of $2800 for a used car. He financed the purchase with a finance company and agreed to repay the loan in 24 monthly payments at an APR of 22%. What is the finance charge? **$687.68**

MONTHLY PAYMENT ON A $100 LOAN				
Term in Months	Annual Percentage Rate			
	20%	22%	24%	26%
6	$17.65	$17.75	$17.85	$17.95
12	9.26	9.36	9.45	9.55
18	6.48	6.57	6.67	6.77
24	5.09	5.19	5.29	5.39
30	4.26	4.36	4.46	4.57
36	3.72	3.82	3.92	4.03
42	3.33	3.43	3.54	3.65
48	3.04	3.15	3.26	3.37

16. Aurora Kaylow obtained an installment loan of $6000 on a used sailboat. She financed the purchase through the boat dealer and agreed to repay the loan in 48 monthly payments at an APR of 24%. What is the finance charge? **$3388.80**

17. Pauline and Eldon Kharche would like to obtain an installment loan of $1800. They can get the loan at an APR of 22% for 24 months or at an APR of 20% for 18 months. Which loan costs less? How much do the Kharches save by taking the loan that costs less? **20%; $142.56**

18. Lucretia and Don Protsman would like an installment loan of $3280. City Loan will loan the money at 24% for 24 months. Economy Line Finance Company will loan the money at 22% for 30 months. Which loan costs less? How much will be saved by taking the loan that costs less? **City Loan; $126.06**

19. Sue and Tom Weber plan to buy a home computer that costs $2999.95. They have a down payment of $499.95 and will finance the remainder at an APR of 22% for 24 months. What is the finance charge? **$614**

20. Harold and Nora O'Desky obtained an installment loan to purchase the garage advertised for $3295. They made a down payment of $295. The bank has agreed to loan them the remainder of the money at an APR of 20% for 48 months. What is the finance charge? **$1377.60**

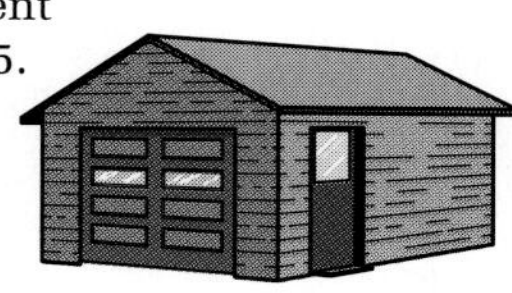

MAINTAINING YOUR SKILLS Look up the skills in parentheses if you need help or more practice.

Multiply. **(Skills 30, 20)**

21. $4000 \times 6\% \times \frac{1}{12}$ **20**

22. $1240 \times 12\% \times \frac{1}{12}$ **12.40**

23. $1800 \times 9\% \times \frac{1}{12}$ **13.50**

24. $\$1500 \times 12\% \times \frac{1}{12}$ **$15**

25. $\$4250 \times 15\% \times \frac{1}{12}$ **$53.13**

26. $\$3600 \times 15\% \times \frac{1}{12}$ **$45**

Subtract. **(Skill 6)**

27. 4200 − 42 = **4158**

28. 1240 − 15.50 = **1224.50**

29. 1224.50 − 15.31 = **1209.19**

30. 7321.65 − 29.86 = **7291.79**

BUSINESS NOTES

Point out to students that lenders are required by the Truth in Lending Law to inform borrowers of the finance charge and the APR.

7-4

Installment Loans—Allocation of Monthly Payment

OBJECTIVE
Compute the payment to interest, payment to principal, and the new balance.

A simple interest installment loan is repaid in equal monthly payments. Part of each payment is used to pay the interest on the unpaid balance of the loan, and the remaining part is used to reduce the balance. The interest is calculated each month using the simple interest formula. The amount of principal that you owe decreases with each monthly payment. A **repayment schedule** shows the distribution of interest and principal over the life of the loan. The repayment schedule here shows the interest and principal on an installment loan of $1800 for 6 months at 15%.

REPAYMENT SCHEDULE FOR AN $1800 LOAN AT 15% FOR 6 MONTHS

Payment Number	Monthly Payment	Amount for Interest	Amount for Principal	New Principal
1	$313.20	$22.50	$290.70	$1509.30
2	313.20	18.87	294.33	1214.97
3	313.20	15.19	298.01	916.96
4	313.20	11.46	301.74	615.22
5	313.20	7.69	305.51	309.71
6	313.20	3.87	309.33	0.38

Note that the last payment would be increased by $0.38 in order to zero out the loan.

Payment to Principal = Monthly Payment − Interest

New Principal = Previous Principal − Payment to Principal

EXAMPLE *Skills* 6, 8, 30 *Application* A *Term* Repayment schedule

Stephanie and Donald Cole obtained the loan of $1800 at 15% for 6 months shown in the repayment schedule. Show the calculation for the first payment. What is the interest? What is the payment to principal? What is the new principal?

SOLUTION

A. Find the **interest.**
Principal × Rate × Time
$1800.00 × 15% × $\frac{1}{12}$ = $22.50 interest

B. Find the **payment to principal.**
Monthly Payment − Interest
$313.20 − $22.50 = $290.70 payment to principal

C. Find the **new principal.**
Previous Principal − Payment to Principal
$1800.00 − $290.70 = $1509.30 new principal

✔ SELF-CHECK Complete the problems, then check your answers in the back of the book.

1. Interest the second month is: $1509.300 × 15% × 1/12 = ? **$18.87**
2. Payment to principal is: $313.20 − ? = ? **$18.87; $294.33**
3. The new balance is: $1509.30 − ? = ? **$294.33; $1214.97**

LESSON PLAN
7-4 Installment Loans—Allocation of Monthly Payment

FOCUS

Introduce the objective of this lesson by asking students to suppose they borrowed $500 from a bank. Then ask them if they would have to pay interest on the loan. (yes) Would they pay back more or less than $500? (more) What is the amount of $500 called? (the principal) How could they find the total amount of interest paid? (Subtract $500 from the total amount repaid on the loan.)

TEACH

Discuss the repayment schedule for the $1800 loan. Have students check the figures for payment number one. ($313.20 − $22.50 = $290.70; $1800 − $290.70 = $1509.30) Point out that the payment of $313.20 is found by using the table in Lesson 7-3. ($17.40 × 18) Review the formula for calculating simple interest and work through the solution to the Example. Point out that the concepts in this lesson will be used in Lesson 9-1 Mortgage Loans.

ALTERNATIVE STRATEGIES: Reteaching

Continue computations for the third month for the Example.

Step 1. Find the interest on the balance.
$1214.97 × 15% × $\frac{1}{12}$ = $15.19

Step 2. Subtract the interest from the payment.
$313.20 − $15.19 = $298.01

Step 3. Subtract the amount from Step 2 from the balance used in Step 1 to get a new balance.
$1214.97 − $298.01 = $916.96

Step 4. Start the process over again.

Warm-Up Exercises

1. $1575 − $12.55 $1562.45
2. $2000 − $16.67 $1983.33
3. $2000 × 10% $200
4. $4449.38 × 12% $533.93
5. $1417.23 × 15% $212.58
6. $212.58 × $\frac{1}{12}$ $17.72

PRACTICE AND APPLY

The following problems can be assigned for classwork and the answers checked in class to help students master the objective of the lesson.

- Guided Practice: 1–10
- Independent Practice: 11–15 odd

PROBLEMS

	Amount Financed	Interest Rate	Monthly Payment	Amount for Interest	Amount for Principal	New Principal
4.	$1200	12%	$106.56	$12.00	$94.56	**$1105.44**
5.	$2400	15%	$116.40	$30.00	**$86.40**	**$2313.60**
6.	$3460	10%	$207.95	**$28.83**	**$179.12**	**$3280.88**
7.	$1680	20%	$ 85.51	**$28.00**	**$57.51**	**$1622.49**
8.	$ 860	24%	$ 45.49	**$17.20**	**$28.29**	**$831.71**
9.	$ 975	26%	$ 66.01	**$21.13**	**$44.88**	**$930.12**

10. Joan and Bill Kelly.
Furniture loan of $2400.
Interest rate is 12%.
Monthly payment is $113.04.
How much of the first monthly payment is for interest? **$24**

11. Annie and Fred Petroff.
Appliance loan of $2000.
Interest rate of 18%.
Monthly payment is $99.80.
How much of the first monthly payment is for interest? **$30**

12. Wilma and Glen Barnes.
Used car loan of $3600.
Interest rate is 20%.
Monthly payment is $133.92.
How much of the first monthly payment is for interest? **$60**
How much of the first monthly payment is for principal? **$73.92**
What is the new balance? **$3526.08**

13. Cora and Jessie Vineyard.
Personal loan of $1500.
Interest rate of 24%.
Monthly payment is $79.35.
How much of the first monthly payment is for interest? **$30**
How much of the first monthly payment is for principal? **$49.35**
What is the new balance? **$1450.65**

14. Maxine Berlen obtained a 12-month loan of $1800 for tuition from First Federal Savings and Loan of Delta. The interest rate is 10%. Her monthly payment is $158.22. For the first payment, what is the interest? What is the payment to principal? What is the new principal? **$15; $143.22; $1656.78**

15. Arthur Greenler obtained a 24-month loan of $6000 for a used car from Economy Loan Inc. The interest rate is 20%. His monthly payment is $305.40. For the first payment, what is the interest? What is the payment to principal? What is the new principal? **$100; $205.40; $5794.60**

16. Orvil Marshall obtained a 42-month loan of $5550 for new windows for his home. The interest rate is 15%. His monthly payment is $170.39. For the first payment, what is the interest? What is the payment to principal? What is the new principal? **$69.38; $101.01; $5448.99**

17. Phyllis Peterson obtained an 18-month loan of $3200 for a used sailboat from the boat dealer. The interest rate is 22%. Her monthly payment is $213.44. For the first payment, what is the interest? What is the payment to principal? What is the new principal? **$58.67; $154.77; $3045.23**

BUSINESS PROJECT

Have students investigate current interest rates for new car loans at local banks, credit unions, car dealers, and finance companies.

Refer to the table of page 646–647 for problems 18 and 19.

18. Anna McGee obtained an 18-month loan of $3980 for a used car from the car dealer. The interest rate is 14%. What is the monthly payment? For the first payment, what is the interest? What is the payment to principal? What is the new principal? **$246.43; $46.43; $200; $3780**

19. Tyrone Murdoch obtained a 48-month loan of $8400 for a used truck from the dealer. The interest rate is 16%. What is the monthly payment? For the first payment, what is the interest? What is the payment to principal? What is the new principal? **$238.05; $112; $126.05; $8273.95**

For problems 20 to 26, complete the repayment schedule for a loan of $2400 at 12% for 12 months.

	REPAYMENT SCHEDULE FOR A $2400 LOAN AT 12% FOR 12 MONTHS				
	Payment Number	Monthly Payment	Amount for Interest	Amount for Principal	New Principal
	1	$213.12	$24.00	$189.12	$2210.88
	2	213.12	22.11	191.01	2019.87
	3	213.12	20.20	192.92	1826.95
	4	213.12	18.27	194.85	1632.10
	5	213.12	16.32	196.80	1435.30
20.	6	213.12	14.35	198.77	? **$1236.53**
21.	7	213.12	12.37	? **$200.75**	? **$1035.78**
22.	8	213.12	? **$10.36**	? **$202.76**	? **$833.02**
23.	9	213.12	? **$8.33**	? **$204.79**	? **$628.23**
24.	10	213.12	? **$6.28**	? **$206.84**	? **$421.39**
25.	11	213.12	? **$4.21**	? **$208.91**	? **$212.48**
26.	12	? **214.60**	? **$2.12**	? **$212.48**	? **$0**

MAINTAINING YOUR SKILLS Look up the skill in parentheses if you need help or more practice.

Find the percentage. **(Skill 30)**

27. 12% of $5000 **$600**

28. 15% of $6000 **$900**

29. 8% of $8400 **$672**

30. 22% of $1282.15 **$282.07**

31. 26% of $2348.90 **$610.71**

32. 20% of $456.21 **$91.24**

33. 6% of $340.80 **$20.45**

34. 15% of $9845.20 **$1476.78**

35. 7% of $12,346.97 **$864.29**

WRAP-UP

Go back to the Example and call upon a student to explain the steps of the solution.

Assignment Guide

- Basic: 11–18, 27–35
- Average: 12–16 even, 17–26, 28–34 even

LESSON PLAN
7-5 Paying Off Simple Interest Installment Loans

FOCUS

Ask students if they had a loan, would they like the option of paying it off early. (yes) Why? (save on interest) Tell them that in this lesson they will learn how to compute a final payment on an installment loan.

TEACH

Students should understand that the primary advantage of a simple interest installment loan is that you always know exactly how much you owe after each monthly payment. Thus, you can pay off the loan by simply paying the new principal after the last payment plus any interest due. In the Example, if the loan were paid off with the third payment, then the final payment would be the new balance after payment two, ($1212.08) plus the interest for payment three ($12.12). Thus, the final payment is $1212.08 + $12.12 or $1224.20.

Warm-Up Exercises

1. $\$4800 \times 12\% \times \frac{1}{12}$
   $48
2. $\$3200 \times 15\% \times \frac{1}{12}$
   $40
3. $\$6480 \times 18\% \times \frac{1}{12}$
   $97.20
4. $\$3256 \times 9\% \times \frac{1}{12}$
   $24.42

7-5 Paying Off Simple Interest Installment Loans

OBJECTIVE
To compute the final payment when paying off a simple interest installment loan.

The Truth-in-Lending Law specifies that if a loan is paid off early, the lender must disclose the method for paying off the loan. Because interest is always paid on the unpaid balance, you just pay the previous balance plus the current month's interest if you pay off a simple interest installment loan before the end of the term. The **final payment** is the previous balance plus the current month's interest.

Final Payment = Previous Balance + Current Month's Interest

EXAMPLE *Skills* 5, 6, 8, 30 *Application* A *Term* Final payment

The first three months of the repayment schedule for Doug and Donna Collins's loan of $1800 at 12% for 6 months is shown. What is the final payment if they pay the loan off with payment number 4?

REPAYMENT SCHEDULE FOR AN $1800 LOAN AT 12% FOR 6 MONTHS

Payment Number	Monthly Payment	Amount for Interest	Amount for Principal	New Principal
1	$310.50	$18.00	$292.50	$1507.50
2	310.50	15.08	295.42	1212.08
3	310.50	12.12	298.38	913.70

SOLUTION

A. Find the **previous balance.** It is $913.70.

B. Find the **interest** for the fourth month.
Principal × Rate × Time
$\$913.70 \times 12\% \times \frac{1}{12} = \$9.137 = \$9.14$ interest

C. Find the **final payment.**
Previous Balance + Current Month's Interest
$913.70 + $9.14 = $922.84 final payment

✔ SELF-CHECK Complete the problems, then check your answers in the back of the book.

Find the interest for a month and then the final payment.

1. Previous balance of $800 at 12%. $800 + $8 = $808
2. Previous balance of $1280 at 15%. $1280 + $16 = $1296

PROBLEMS

Find the interest and the final payment.

	3.	4.	5.	6.	7.
Interest Rate	12%	15%	10%	18%	22%
Previous Balance	$4800.00	$3000.00	$1460.80	$3987.60	$3265.87
Interest	$ 48.00	$37.50 ?	$12.17 ?	$59.81 ?	$59.87 ?
Final Payment	$4848 ?	$3037.50 ?	$1472.97 ?	$4047.41 ?	$3325.74 ?

CRITICAL THINKING

Ask students why the amount for interest goes down each month on an installment loan. (The principal is decreasing each month.) Ask also if they think many people pay off loans early and, if not, why not? (No; they have to come up with the full amount needed to pay off the loan.)

8. Chris Worthington.
Previous balance of $2460.
Interest rate is 20%.
What is the interest? **$41**
What is the final payment? **$2501**

9. Walter Tavinier.
Previous balance of $8258.
Interest rate is 15%.
What is the interest? **$103.23**
What is the final payment? **$8361.23**

10. Willard Hudson took out a simple interest loan of $6000 at 10% for 24 months. After 4 payments, the balance is $5081.23. He pays off the loan when the next payment is due. What is the interest? What is the final payment? **$42.34; $5123.57**

11. Lillian Hartwick took out a simple interest loan of $3600 at 18% for 12 months. After 6 payments, the balance is $1879.90. She pays off the loan when the next payment is due. What is the interest? What is the final payment? **$28.20; $1908.10**

12. Scott DuBois took out a simple interest loan of $1800 for home repairs. The loan is for 12 months at 10% interest. After 8 months, the balance is $620.26. He pays off the loan when the next payment is due. What is the final payment? **$625.43**

13. Carolyn Fricke took out a simple interest loan of $6000 for a used car. The loan is for 24 months at 20% interest. After 17 payments, the balance is $2001.46. She pays off the loan when the next payment is due. What is the final payment? **$2034.82**

14. Nicholas and Dorothea Schrodt were looking over the repayment schedule for their boat loan of $5550 at 15% for 42 months. They note the following:

a. Balance after payment 18 is $3525.03. **$3569.09**
b. Balance after payment 24 is $2742.99. **$2777.28**
c. Balance after payment 30 is $2045.25. **$2070.82**

Determine the final payment if they pay the loan off when the 19th payment is due. What if they wait until the 25th payment? The 31st payment?

15. Stanley Huston purchased a Rototiller for $1318.45. He made a down payment of 20%. The dealer financed the remainder at 12% simple interest for 1 year. Stan's custom gardening prospered, and he paid the loan off at the time the fifth payment was due. How much was his final payment? **$724.47**

MAINTAINING YOUR SKILLS Look up the skills in parentheses if you need help or more practice.

Multiply. **(Skill 8)**

16. 5489 × 0.15 **823.35**
17. 2729 × 0.22 **600.38**
18. 9032 × 0.18 **1625.76**
19. 983.54 × 0.12 **118.0248**
20. 657.80 × 0.06 **39.468**
21. 6118.53 × 0.09 **550.6677**

Find the percentage. **(Skill 30)**

22. 430 × 18% **77.40**
23. 3561.90 × 9% **320.571**
24. $10,907.45 × 15% **$1636.1175**

PRACTICE AND APPLY

The following problems can be assigned for classwork and the answers checked in class to help students master the objective of the lesson.

- Guided Practice: 1–7
- Independent Practice: 8–10

WRAP-UP

Call upon a student to work Problem 9 at the chalkboard and explain his or her solution.

Assignment Guide

- Basic: 8–13, 16–24
- Average: 11–15, 17–23 odd

ALTERNATIVE STRATEGIES: Enrichment

Point out that the main reason to pay off a loan early is to save paying interest. Have students determine how much interest is saved by paying off the loan in the Example with the fourth payment. ($8.66) Have them calculate the total interest ($63) and the interest if the loan were a single payment loan ($I = \$1800 \times 12\% \times \frac{6}{12} = \$108$).

LESSON PLAN 7-6 Determining the APR

FOCUS
Ask students if they have seen the APR of loans advertised in newspapers. Ask also which APR they would prefer for a loan of their own, 14% or 16%? Why? (14%, because interest on the loan would be less.) The focus of this lesson is to use a table to find the APR of a loan.

TEACH
Use the chart to show students how an APR can be found if the finance charge and term are known. For example, a finance charge of $9.77 for a term of 18 months yields an APR of 12%.

Ask students to think about the number that results from dividing the finance charge by the amount financed. What does it represent? (finance charge per one dollar of the loan) Why is the number multiplied by $100? (The APR chart gives the finance charge per $100 of amount financed.)

7-6

Determining the APR

OBJECTIVE
Use a table to find the annual percentage rate of a loan.

If you know the number of monthly payments and the finance charge per $100 of the amount financed, you can use a table to find the annual percentage rate of the loan. You can use the APR of loans to compare the relative cost of borrowing money.

APR	10.00%	10.25%	10.50%	10.75%	11.00%	11.25%	11.50%	11.75%	12.00%	12.25%	12.50%
Term	Finance Charge Per $100 of Amount Financed										
6	$ 2.94	$ 3.01	$ 3.08	$ 3.16	$ 3.23	$ 3.31	$ 3.38	$ 3.45	$ 3.53	$ 3.60	$ 3.68
12	5.50	5.64	5.78	5.92	6.06	6.20	6.34	6.48	6.62	6.76	6.90
18	8.10	8.31	8.52	8.73	8.93	9.14	9.35	9.56	9.77	9.98	10.19
24	10.75	11.02	11.30	11.58	11.86	12.14	12.42	12.70	12.98	13.26	13.54

$$\text{Finance Charge per \$100} = \$100 \times \frac{\text{Finance Charge}}{\text{Amount Financed}}$$

EXAMPLE *Skills* 11, 2 *Application* C *Term* Annual percentage rate

Paul Norris obtained an installment loan of $1500 to pay for ham radio equipment. The finance charge is $146.25. He agreed to repay the loan in 18 monthly payments. What is the annual percentage rate?

SOLUTION

A. Find the **finance charge per $100.**
$100 × (Finance Charge ÷ Amount Financed)
$100 × ($146.25 ÷ $1500.00)
$100 × 0.0975 = $9.75 finance charge per $100

B. Find the **APR.** (Refer to the table above.)
In the row for 18 payments, find the number closest to $9.75. It is $9.77. Read the APR at the top of the column. APR is 12.00%.

✓ SELF-CHECK Complete the problems, then check your answers in the back of the book.

1. 6-month loan.
$100 × ($24.64 ÷ $800) = ?
APR = ? **$3.08; 10.50%**

2. 24-month loan.
$100 × ($96.22 ÷ $850) = ?
APR = ? **$11.32; 10.50%**

PROBLEM SOLVING
Have students solve this problem. Suppose the APR of a $10,000 loan for 24 months is 12.5%. What is the finance charge for the loan? ($1354) Ask students to make up a similar problem.

PROBLEMS

Complete the table. Use the APR table on page 646–647.

	3.	4.	5.	6.	7.
Finance Charge	$33.10	$434.16	$421.50	$ 1652	$597.66
Amount Financed	$1,000	$ 2,400	$ 3,000	$5,400	$ 4,200
Finance Charge per $100	$3.31	? $18.09	? $14.05	? $30.59	? $14.23
Number of Payments	6	24	18	36	30
Annual Percentage Rate	? 11.25%	? 16.50%	? 17.00%	? 18.25%	? 10.50%

8. Ed Naiman.
Installment loan of $2500.
Finance charge of $430.50.
24 monthly payments.
What is the APR? **15.75%**

9. Betty Arca.
Installment loan of $800.
Finance charge of $170.40.
36 monthly payments.
What is the APR? **13.00%**

10. Webster Larkin.
Installment loan of $300.
Finance charge of $9.96.
6 monthly payments.
What is the APR? **11.25%**

11. Kenneth Bryant.
Installment loan of $9365.
Finance charge of $2823.
36 monthly payments.
What is the APR? **18.00%**

12. Marie Brenson obtained an installment loan of $460 to purchase computer software. The finance charge is $19.32. She agreed to repay the loan in 6 monthly payments. What is the annual percentage rate? **14.25%**

13. Herb Stanley obtained an installment loan of $6800 to pay his daughter's college tuition. The finance charge is $731. He agreed to repay the loan in 24 monthly payments. What is the APR? **10.00%**

14. Jeff Stapleton obtained an installment loan of $395 to pay for car repairs. The finance charge is $36.34. He agreed to repay the loan in 12 monthly payments. What is the APR? **16.50%**

F.Y.I.
The national average APR on simple interest installment loans for new cars was 10.95% on December 1, 1991.

15. Julia Bourne obtained an installment loan of $3800 to purchase a lawn and garden tractor. The finance charge is $722.38. She agreed to repay the loan in 24 monthly payments. What is the APR? **17.25%**

16. Oneta Correy wants to obtain an installment loan of $1960 to purchase the used car advertised. The bank has agreed to loan her the money. She must repay the loan in 24 months. The finance charge is $337.51. What is the APR of her loan? **15.75%**

SALE $1960.00
1983 4-dr sedan, a/c, power equipment, one owner. No rust. Easy financing.

17. Helen Olson needs an installment loan of $950 to purchase a VHS recorder. The store has agreed to loan her the money. She must repay the loan in 24 months. The finance charge is $150.10. What is the APR of her loan? **14.50%**

Warm-Up Exercises

1. (12 × $25) − $240
$60
2. (24 × $15.25) − $305
$61
3. $100 × ($40 ÷ $800)
$5
4. $100 × ($35 ÷ $75)
$46.67

PRACTICE AND APPLY

The following problems can be assigned for classwork and the answers checked in class to help students master the objective of the lesson.

- Guided Practice: 1–7
- Independent Practice: 8–14 even

ALTERNATIVE STRATEGIES: Enrichment

Have your students use this formula to find the APR for the Example.

M = Number of payments per year.
N = Number of scheduled payments.
F = Finance charge
A = Amount financed

$$APR = \frac{M \times F(95N + 9)}{12N(N + 1)(4A + F)}$$

$$= \frac{12 \times 146.25(95 \times 18 + 9)}{12 \times 18(18 + 1)(4 \times 1500 + 146.25)} = 11.96\%$$

WRAP-UP

Ask a student to explain how to use the APR chart. Ask another student to explain the meaning of the formula of finding the finance charge per $100.

Assignment Guide

- Basic: 8–22 even, 25–40
- Average: 9–23 odd, 24, 26–40 even

18. Brent and Lola Miller are buying a washer that costs $399.95 and a dryer that costs $249.99. To use the store's installment plan, they need a down payment of $49.94. They must make 18 monthly payments of $37.08 each. What is the APR on their installment loan? **13.75%**

19. Jorge Holland is buying a new furnace that costs $2600. The bank requires a down payment of 20% and 36 monthly payments of $66.95 each. What is the APR on his loan? **Less than 10%**

20. Ty Chin is buying a new color television set that costs $659.38. To use the installment plan available at the department store, he must make a down payment of 25% and make 30 monthly payments of $19.46 each. What is the APR on his loan? **13.25%**

21. Andrew Stachowick would like an installment loan of $5000 to be repaid in 36 months. Nathaniel Loan Company will grant the loan with a finance charge of $1215.60. City Finance Company will grant the loan with a finance charge of $1532.50. What is the APR on each loan? **14.75%; 18.25%**

22. Wayne Charles would like an installment loan of $8000 to be repaid in 24 months. ABC Finance Company will grant the loan with a finance charge of $1984. Atco Financial Service will grant the loan with a finance charge of $2010. What is the APR on each loan? **22.25%; 22.50%**

23. Eleanor Penny purchased a $1987 dining room set. She made a down payment of $87 and through the store's installment plan agreed to pay $169 a month for 12 months. What is the finance charge? What is the APR? What is the total amount paid for the set? **$128; 12.25%; $2115**

24. Kathleen Dunn purchased a home computer for $3249.29. She made a down payment of $649.29 and financed the remainder by agreeing to pay $129.23 per month for 24 months. What is the finance charge? What is the APR? What is the total amount paid for the computer? **$501.52; 17.50%; $3750.81**

MAINTAINING YOUR SKILLS Look up the skills in parentheses if you need help or more practice.

Round to the nearest cent. **(Skill 2)**

25. $19.439 **$19.44** **26.** $12.4162 **$12.42** **27.** $40.3072 **$40.31** **28.** $19.3019 **$19.30**

Divide. Round answers to the nearest hundredth. **(Skills 10, 11)**

29. 17)510 **30** **30.** 24)1060 **44.17** **31.** 47)2642 **56.21** **32.** 984)67,182 **68.27**

33. 1.4)98.22 **70.16** **34.** 8.2)219.6 **26.78** **35.** 76.1)94.88 **1.25** **36.** 64.2)104.76 **1.63**

Multiply. **(Skill 8)**

37. 100 × 4.21 **421** **38.** 10 × 0.3174 **3.174**

39. 100 × 0.0192 **1.92** **40.** 100 × 0.71442 **71.442**

CRITICAL THINKING

Ask students if they can compare the APR of a 6-month loan to the APR of a 12-month loan given the same principal. (No; the comparison can be made only if the terms of the loan are the same.)

7-7

Refund of Finance Charge

OBJECTIVE
Compute the finance charge refunded for early repayment of a loan.

If you repay an installment loan that is not a simple interest installment loan before the final due date, you may be entitled to a refund of part of the finance charge. The refund is usually stated as a percent of the finance charge. The lender may determine the percent refund using a **rebate schedule,** which is a table of percent refunds.

Refund = Finance Charge × Percent Refund

F.Y.I.
The rebate schedule is based on the "Rule-of-78."

Term of Loan	Number of Months Loan Has Run													
	1	2	3	4	5	6	7	8	9	10	11	12	13	14
3	50.00	16.67	0											
6	71.43	47.62	28.57	14.29	4.76	0								
9	80.00	62.22	46.67	33.33	22.22	13.33	6.67	2.22	0					
12	84.62	70.51	57.69	46.15	35.90	26.92	19.23	12.82	7.69	3.85	1.28	0		
15	87.50	75.83	65.00	55.00	45.83	37.50	30.00	23.33	17.50	12.50	8.33	5.00	2.50	0.83
18	89.47	79.53	70.18	61.40	53.22	45.61	38.60	32.16	26.32	21.05	16.37	12.28	8.77	5.85

EXAMPLE *Skills* 30, 2 *Application* C *Term* Rebate schedule

Viola Stambaugh had a 12-month installment loan. The total finance charge on the loan was $113.80. How much was Viola's refund if she paid off the loan with the:

A. 7th payment? **B.** 10th payment?

SOLUTION

A. (1) Find the **percent refund.**
Refer to the table above.
Term of loan in months 12
Number of months loan has run 7
Value from table is 19.23%

(2) Find the **refund.**
Finance Charge × Percent Refund
$113.80 × 19.23% = $21.883
= $21.88 refund

B. (1) Find the **percent refund.**
Refer to the table above.
Term of loan in months 12
Number of months loan has run 10
Value from table is 3.85%

(2) Find the **refund.**
Finance Charge × Percent Refund
$113.80 × 3.85% = $4.3813
= $4.38 refund

✓ SELF-CHECK Complete the problems, then check your answers in the back of the book.

A 15-month loan has run for 8 months. The finance charge was $252.80.

1. Find the percent refund.
23.33%

2. Find the refund.
$58.97824 = $58.98

LESSON PLAN 7-7 Refund of Finance Charge

FOCUS

In this lesson, students compute the amount of the finance charge refunded for early repayment of a loan. Ask students why a refund is given. (There is no finance charge for those months paid off early.)

TEACH

Stress to students that the table showing the number of months a loan has run gives percents. Thus, a loan having a term of 3 months that is paid off after 1 month would get a refund of 50% of the total finance charge. Point out that a refund of 50% means that 50 cents of every dollar of the finance charge is paid back to the borrower. Ask students what a rebate of 16.67% means. (About 17 cents of every dollar of the finance charge is refunded.)

Ask students if they think the terms *rebate* and *refund* have the same meaning. (yes)

Warm-Up Exercises

1. $460 × 80% $368
2. $400 × 33.33% $133.32
3. $300 × 70.51% $211.53
4. $840 × 7.69% $64.60

ALTERNATIVE STRATEGIES: Reteaching

Point out that the numbers in the table are percents and that the procedure is an application of finding a percentage **(Skill 30)** after correctly reading the table.

Use the data in the example and determine the refund if the loan is paid off with the third payment (57.69% of $113.80 = $65.65).

PRACTICE AND APPLY

The following problems can be assigned for classwork and the answers checked in class to help students master the objective of the lesson.

- Guided Practice: 1–8
- Independent Practice: 9–11

WRAP-UP

Use Problem 9 to wrap up the discussion of the lesson.

Assignment Guide

- Basic: 9–14, 16–23
- Average: 12–15, 17–23 odd

PROBLEMS

Use the table on page 233 to find the percent refund.

	3.	4.	5.	6.	7.	8.
Term of Loan (Months)	15	15	9	18	6	12
Finance Charge	$60.00	$140.00	$350.00	$417.75	$281.60	$727.12
Months Loan Has Run	7	4	5	13	4	6
Percent Refund	30.00%	55.00%	? 22.22%	? 8.77%	? 14.29%	? 26.92%
Amount of Refund	? $18	? $77	? $77.77	? $36.64	? $40.24	? $195.74

9. Chloe Carter.
15-month installment loan.
$224.50 total finance charge.
Repaid loan in 6 months.
What is the refund? **$84.19**

10. Kevin Taylor.
9-month installment loan.
$82.50 total finance charge.
Repaid loan in 6 months.
What is the refund? **$11.00**

11. Carlo Blanco had a 15-month installment loan that he repaid in 6 months. The total finance charge was $224. How much was his refund? **$84**

12. Ross Wyman had an 18-month installment loan that he repaid in 7 months. The total finance charge was $439.44. How much was his refund? **$169.62**

13. Ruth Royer took a 12-month installment loan to help pay for her new washer. The total finance charge was $49.96. Ruth repaid the loan in 4 months. How much was her refund? **$23.06**

14. Eileen Carl obtained an installment loan of $2400 from the bank. She agreed to pay $218.50 a month for 12 months. The finance charge totaled $222. How much of the finance charge can Eileen save by repaying the loan in 10 months rather than 12 months? **$8.55**

15. Teshona Ku had an installment loan of $8000 in which he agreed to pay $487.87 a month for 18 months. How much is the finance charge? How much can he save by repaying the loan in 12 months? **$781.66; $95.99**

MAINTAINING YOUR SKILLS

Look up the skills in parentheses if you need help or more practice.

Round to the nearest hundredth. **(Skill 2)**

16. 87.334 **87.33**
17. 421.065 **421.07**
18. 492.596 **492.60**
19. 339.996 **340.00**

Find the percentage. **(Skill 30)**

20. 80 × 18% **14.4**
21. 340 × 12% **40.8**
22. 68 × 7.82% **5.3176**
23. 39.4 × 17.50% **6.895**

BUSINESS NOTES

Emphasize that this refund procedure does not apply to simple interest installment loans. It is usually used for installment loans granted by some retailers, certain finance companies, and some auto dealers. This method is called The Rule-of-78. Have students find out why.

CULTURAL ANGLES

Have students research interest rates for loans in Canada and Mexico. How do they compare to rates in the U.S.? Do these countries have Truth-in-Lending Laws that stipulate the lender must disclose the method of computing any unearned finance charge?

Reviewing the Basics

Skills

Round to the nearest cent.

(Skill 2)
1. $732.513 **$732.51** 2. $48.266 **$48.27** 3. $84.599 **$84.60** 4. $24.683 **$24.68**

Solve.

(Skill 6)
5. $839.44 − $721.68 **$117.76** 6. $35.29 − $7.20 **$28.09** 7. $4113.97 − $873.98 **$3239.99**

(Skill 8)
8. $19 × 18 **$342** 9. $13.52 × 12 **$162.24** 10. $123.24 × 48 **$5915.52**

(Skill 11)
11. $48.00 ÷ $380 **$0.1263** 12. $32.25 ÷ $250 **$0.129** 13. $714.44 ÷ $1850 **0.3862**

(Skill 30)
14. 26% of $500 **$130** 15. 45% of $440 **$198** 16. 32.5% of $221.98 **$72.14**

Write as a decimal.

(Skill 14)
17. $\frac{3}{4}$ **0.75** 18. $\frac{36}{360}$ **0.1** 19. $\frac{7}{12}$ **0.583** 20. $\frac{250}{360}$ **0.694** 21. $\frac{73}{365}$ **0.2**

(Skill 28)
22. 73% **0.73** 23. 12% **0.12** 24. 9.25% **0.0925** 25. 15.3% **0.153** 26. 10.125% **0.10125**

Write as a percent.

(Skill 26)
27. 0.15 **15%** 28. 0.45 **45%** 29. 0.137 **13.7%** 30. 0.356 **35.6%** 31. 0.2235 **22.35%**
32. 0.125 **12.5%** 33. 0.454 **45.4%** 34. 1.571 **157.1%** 35. 20.512 **2051.2%** 36. 0.0051 **0.51%**

Applications

Use the table to find the monthly payment.

(Application C)
37. $700 financed for 12 months at 15% APR. **$63.21**

38. $3140 financed for 6 months at 12% APR. **$541.65**

MONTHLY PAYMENT ON A $100 LOAN

Term in Months	Annual Percentage Rate			
	10%	12%	15%	18%
6	$17.16	$17.25	$17.40	$17.55
12	8.79	8.88	9.03	9.17
18	6.01	6.10	6.24	6.38
24	4.61	4.71	4.85	4.99

Write as a fraction of a year.

(Application J)
39. 10 months $\frac{5}{6}$ 40. 6 months $\frac{1}{2}$ 41. 7 months $\frac{7}{12}$ 42. 3 months $\frac{1}{4}$

43. 185 days (exact year) $\frac{185}{365}$ 44. 170 days (ordinary year) $\frac{170}{360}$

45. 90 days (ordinary year) $\frac{90}{360}$ 46. 90 days (exact year) $\frac{90}{365}$

Terms

Use each term in a sentence that explains the term. **Answers will vary.**

47. Single-payment loan 48. Rebate schedule 49. Down payment

50. Annual percentage rate 51. Installment loan 52. Interest

Refer to your reference files in the back of the book if you need help.

LESSON PLAN
Reviewing the Basics

The exercises on this page review skills, applications, and terms used in the unit. You can use the exercises to assess informally students' proficiency with this material.

The page can be used for guided practice and independent practice. You can work through a selection of the exercises together with students, and thus see immediately if they know how to do them, and you can then assign some of the exercises for independent practice. Be sure to go over the answers to all assigned exercises.

LESSON PLAN
Unit Test

Students should do the Unit Test on their own. Each problem on the test is keyed to a lesson in the unit. Students having difficulty with any particular problem should review the Example in the appropriate lesson and be assigned some of the Independent Practice problems for additional practice.

Unit Test

Lesson 7-1

1. Charles Quick's bank granted him a single-payment loan of $3240 for 100 days at an annual interest rate of 14%. His bank charges ordinary interest. What is the maturity value of his loan? **$3366**

Lesson 7-2

2. Ed Wallace purchased a TV for $629 using the store's installment credit plan. Ed made a 25% down payment and financed the remaining amount. What amount did he finance? **$471.75**

Lesson 7-3

3. Lisa Snow obtained an installment loan of $2300. The annual percentage rate is 18%. She plans to repay the loan in 24 months. Use the table to find the finance charge. **$454.48**

MONTHLY PAYMENT ON A $100 LOAN				
Term in Months	Annual Percentage Rate			
	10%	12%	15%	18%
6	$17.16	$17.25	$17.40	$17.55
12	8.79	8.88	9.03	9.17
18	6.01	6.10	6.24	6.38
24	4.61	4.71	4.85	4.99

Lesson 7-4

4. Tom DuVall obtained a 36-month loan of $4350 for a used car. The interest rate is 15%. His monthly payment is $150.95. For the first payment, what is the interest? What is the payment to principal? What is the new principal? **$54.38; $96.57; $4253.43**

Lesson 7-5

5. Linda Hartman took out a simple interest loan of $3600 at 18% for 12 months. After 9 payments, the balance is $960.48. She pays off the loan when the next payment is due. What is the interest? What is the final payment? **$14.41; $974.89**

Lesson 7-6

6. Juan Corvez obtained an installment loan of $625 to pay for a new refrigerator. The finance charge is $102.44. He agreed to make 24 payments of $30.31 each. Use the table to find the annual percentage rate. **15.00%**

FINANCE CHARGE PER $100 FINANCED			
Number of Payments	APR		
	14.50%	14.75%	15.00%
6	$ 4.27	$ 4.35	$ 4.42
12	8.03	8.17	8.31
18	11.87	12.08	12.29
24	15.80	16.08	16.37

Lesson 7-7

7. Jane Tripp repaid a 12-month installment loan after 3 months. The total finance charge was $135.17. Use the rebate table to find Jane's refund. **$77.98**

Term of Loan	Number of Months Loan Has Run			
	1	2	3	4
3	50.00	16.67	0	0
6	71.43	47.62	28.57	14.29
9	80.00	62.22	46.67	33.33
12	84.62	70.51	57.69	46.15

A SPREADSHEET APPLICATION

Loans

To complete this spreadsheet application, you will need the template-diskette *Mathematics with Business Applications,* which accompanies this textbook. Follow the directions in the *User's Guide* to complete this activity.

Input the information in the following problems to determine the monthly payment, total interest, and the monthly payment allocation.

1. Auto loan of $4000.
12 monthly payments.
APR is 12%.
What is the payment?
What is the total interest?
What is the loan balance after payment 6?

2. Stereo loan of $1500.
12 monthly payments.
APR is 15%.
What is the payment?
What is the total interest?
What is the loan balance after payment 9?

3. Furniture loan of $2360.
24 monthly payments.
APR is 16.5%.
What is the payment?
What is the total interest?
What is the loan balance after payment 16?

4. Auto loan of $12,000.
48 monthly payments.
APR is 13.25%.
What is the payment?
What is the total interest?
What is the loan balance after payment 42?

5. Auto loan of $16,000.
60 monthly payments.
APR is 9%.
What is the payment?
What is the total interest?
What is the loan balance after payment 36?

6. Home repairs of $3425.
60 monthly payments.
APR is 16.25%.
What is the payment?
What is the total interest?
What is the loan balance after payment 30?

7. Boat loan of $19,960.
72 monthly payments.
APR is 14.5%.
What is the payment?
What is the total interest?
What is the loan balance after payment 71?

8. Personal loan of $8700.
48 monthly payments.
APR is 12.25%.
What is the payment?
What is the total interest?
What is the loan balance after payment 48?

LESSON PLAN
A Spreadsheet Application

USING TECHNOLOGY

Review the meaning of APR with students before they work on this application. If you wish, students can work in *cooperative groups* to do the problems. Ask students if the loan balance after a payment is made is the total amount paid off or the total amount still due on the loan. (still due)

APPLICATIONS ON THE JOB

Read the page with the class. Then help students interpret the pictograph. Make sure students understand that this type of graph yields estimates only and not exact numbers. Ask students why this is so. (The partial symbols cannot be interpreted exactly.) Ask students if they have had any experiences with credit cards and, if so, to share them with the class. Ask them also why it is important for people to establish a good credit rating.

CAREER WISE

Loan Officer

Keisha Themba handles applications for credit at a major credit card company. Typically, an application involves questions about the applicant, employment status, savings, other credit cards, and so on.

Keisha has met many college students through her job. Because many of them work part-time in the afternoons or evenings, they want to get their first credit card. After the information is verified, a credit card is issued. Their credit limit is usually lower than that of people with long and healthy credit histories. In recent years, she has helped thousands of people begin to establish credit.

Five stores that accept credit cards were surveyed. Sales figures for TVs and stereos are recorded in the pictograph below. A pictograph uses a picture to represent a quantity.

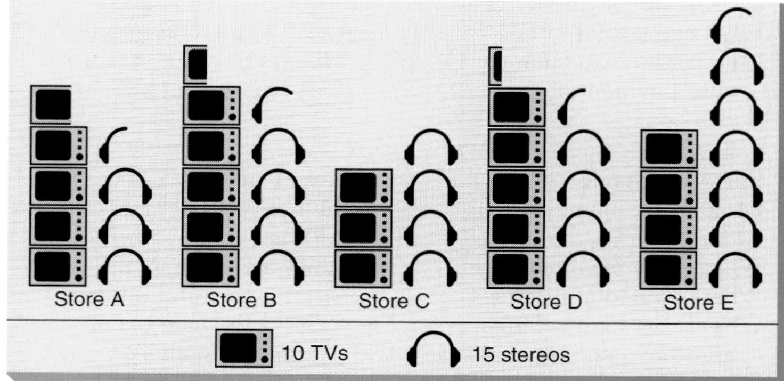

Check Your Understanding

1. How many TVs are represented by [TV symbol]? **10 TVs**
2. Estimate the number of stereos sold at Store E. **About 101 stereos**
3. If the average price of a TV at Store B is $249, estimate the sales from the TVs at Store B. **About 55 × $249 = $13,695**
4. The company's finance rate is 1.65% per month on any purchase not paid within the first 25 days after the purchase and any previous balance. What is the finance charge for a month on an outstanding balance of $875? **0.0165 × $875 = $14.44**
5. According to the survey, 82% of the stereos sold were purchased with credit cards. How many stereos is this? **About 0.82 × 351, or 288**

UNIT 8

Automobile Transportation

A *sticker price* on a new automobile shows the *base price,* the cost of *options,* and the *destination charge* for delivering the car from the factory to the dealer. The auto dealer may take less than the sticker price if you make an offer that is higher than the *dealer's cost.* If you buy a used car, there are *used-car guides* that show the approximate price you can expect to pay for a particular model. In addition to the cost of purchasing an automobile, your transportation costs include *insurance, maintenance,* fuel, and other operating costs.

OUTLINE

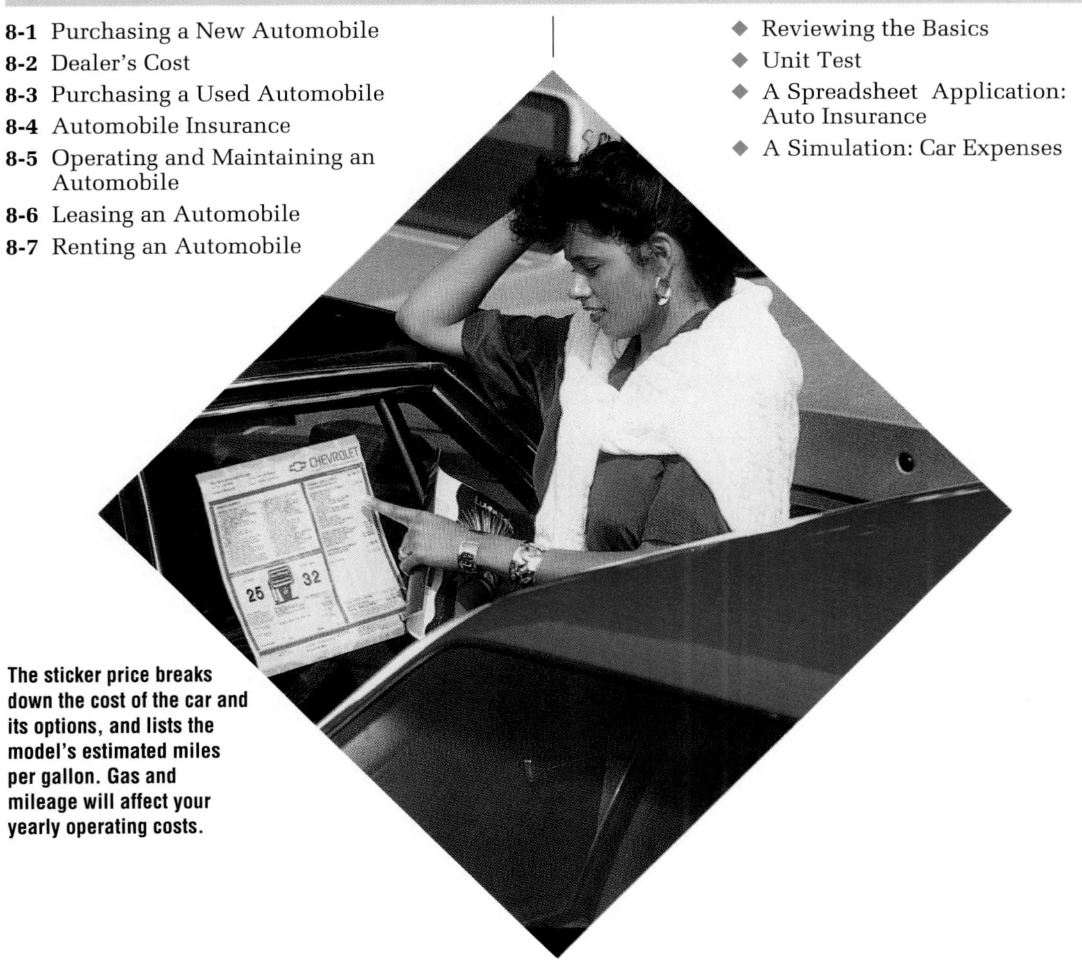

The sticker price breaks down the cost of the car and its options, and lists the model's estimated miles per gallon. Gas and mileage will affect your yearly operating costs.

INTRODUCING THE UNIT

Purchasing an automobile necessitates purchasing liability insurance. Obtain blank insurance forms and have students work in small groups to complete the forms. Encourage discussion of the issues raised by the forms, such as how an individual's answers to certain questions may affect the premium.

LESSON PLAN 8-1 Purchasing a New Automobile

FOCUS

Motivate the unit and this lesson by asking students if any of them have thought about buying a car. Ask how many have thought about buying a *new* car. The focus of this lesson is to compute the sticker price of a new automobile.

TEACH

Have students name other features that can be considered as standard equipment. Point out that some items of standard equipment may vary from model to model. Have students name other features that are considered optional. Advertisements found in newspapers or magazines may be helpful.

Ask your local auto dealer for a window sticker for classroom use. Point out that if a new car is ordered, you have a choice as to the optional equipment, not just what is listed on the sticker.

Point out that the destination charge is not taxed in some states.

Warm-Up Exercises

1. \$7400 + \$919 + \$80 + \$150 \$8549
2. \$8500 + \$3070 + \$440 + \$148 \$12,158
3. \$12,450 + \$2196 + \$319 + \$297 \$15,262
4. 5% × \$12,400 \$620
5. $6\frac{1}{2}$% × \$18,360 \$1193.40

8-1 Purchasing a New Automobile

OBJECTIVE
Compute the sticker price of a new automobile.

Automobile manufacturers are required by law to place a sticker on the window of each new car to show all charges for the car. The **base price** is the price of the engine, chassis, and any other piece of standard equipment for a particular model. **Options** are extras for convenience, safety, or appearance, such as a radio, air-conditioning, and tinted glass. The **destination charge** is the cost of shipping the car from the factory to the dealer. The **sticker price** is the total of the base price, options, and destination charge.

Sticker Price = Base Price + Options + Destination Charge

EXAMPLE *Skill* 5 *Application* A *Term* Sticker price

Scott Huber is shopping for a sports car. A portion of the sticker for one that he is interested in is shown below. What is the sticker price for this sports coupe?

LAMONT 2-DOOR SPORT COUPE		BASE PRICE \$11,495
C/C	OPTIONAL EQUIPMENT DESCRIPTION	LIST PRICE
WP1	Rear Window Defogger	145
AT0	Automatic Transmission W/overdrive	490
TG1	Tinted Glass	120
AC3	Air-Conditioning	610
PDL	Power Door Locks	195
D84	Custom Two-Tone Paint	93
	DESTINATION CHARGE	450

SOLUTION

A. Find the **options.**
\$145.00 + \$490.00 + \$120.00 + \$610.00 + \$195.00 + \$93.00 = \$1653.00

B. Find the **sticker price.**
Base Price + Options + Destination Charge
\$11,495.00 + \$1653.00 + \$450.00 = \$13,598.00 sticker price

145 [+] 490 [+] 120 [+] 610 [+] 195 [+] 93 [=] 1653 [M+] 11495 [+] [RM] 1653 [+] 450 [=] 13598

✓ SELF-CHECK Complete the problems, then check your answers in the back of the book.

Find the sticker price.

	Base Price	Options	Destination Charge	
1.	\$8975	\$450; \$610; \$126	\$390	**\$10,551**
2.	\$10,400	\$195; \$675; \$810; \$92	\$325	**\$12,497**

CULTURAL ANGLES

Many cars today are manufactured in foreign countries. Some are considered more economical due to higher gas mileage. Others are considered a status symbol. Consider the cost and availability of repair parts and mechanics, since this may be a problem of owning a foreign car.

PROBLEMS

Find the sticker price.

	3.	4.	5.	6.	7.
Base Price	$9000	$9900	$11,540	$12,654	$19,842
Options	$1600	$2400	$ 1260	$ 2865	$ 3861
Destination Charge	$ 400	$ 350	$ 345	$ 338	$ 425
Sticker Price	$11,000	$12,650	$13,145	$15,857	$24,128

8. Custom Ciera. **$12,271**
Base price is $9851.
Options: 2.8 Liter V6 for $660.
Auto. trans. for $495.
A/C for $725.
Tinted glass for $195.
Destination charge is $345.
What is the sticker price?

9. Salon Classic. **$18,335**
Base price is $15,495.
Options: GT package for $1295.
AM/FM stereo for $610.
Bucket seats for $345.
Power door locks for $190.
Destination charge is $400.
What is the sticker price?

F.Y.I.
In addition to the 185,000-mile national highway system, the United States has 3.9 million miles of other highways, roads, and city streets.

10. Marlo Barsotti is interested in a 4-door sedan that has a base price of $11,975. Factory-installed options total $1576. The destination charge from its assembly plant in St. Louis is $352. What is the sticker price for this car? **$13,903**

11. Andy Dorfman sees a Fleetwood mini van that he is interested in buying. There is a 5% state sales tax on the purchase of the van. If Andy pays the sticker price, how much tax must he pay? **$891.20**

C/C	OPTIONAL EQUIPMENT DESCRIPTION	LIST PRICE
UN2	FLEETWOOD MINI VAN	BASE PRICE 14,895
PS1	Six-Way Driver Power Seat	240
WRD	Electric Rear Window Defogger	145
WW4	Tinted Glass	310
AT2	Automatic Transmission	475
PDL	SB Radial Tires W/ws	88
ITC	Custom Interior	388
D84	Custom Two-Tone Paint	153
RA2	AM/FM Stereo W/disc player	690
	DESTINATION CHARGE	440

MAINTAINING YOUR SKILLS Look up the skill in parentheses if you need help or more practice.

Add. **(Skill 5)**

12. 8850 + 995 + 660 + 242 **10,747**

13. 6770 + 1217 + 648 + 344 + 85 **9064**

14. 19,453 + 2540 + 199 + 60 **22,252**

15. 9851 + 2889 + 401 + 75 + 90 **13,306**

16. 15,237 + 78.50 + 421.60 + 92 **15,829.10**

17. 7575.20 + 1243.26 + 791.24 **9609.70**

PRACTICE AND APPLY

The following problems can be assigned for classwork and the answers checked in class to help students master the objective of the lesson.

- Guided Practice: 1–5
- Independent Practice: 6–9

WRAP-UP

Ask students to name the three basic parts of a car's sticker price. Have them name features that are considered optional.

Assignment Guide

- Basic: 6–10, 12–17
- Average: 10, 11, 13–17 odd

ALTERNATIVE STRATEGIES: Enrichment

1. Have students research an auto issue of *Consumer Reports* or a magazine dealing with new cars. Students should list the type of information that is available on new cars, such as types of safety features and performance ratings.
2. Have students visit a "new car" dealership and report on the cost of various models, options, and destination charges.

LESSON PLAN
8-2 Dealer's Cost

FOCUS

Ask students if they have ever accompanied anyone to buy a new car? Did the person offer less than the sticker price? Dealers may sometimes be willing to accept an offer less than the sticker price because they have paid less to the manufacturer for the car.

TEACH

The amount of markup on a car will differ from dealer to dealer. The sticker price is a suggested starting price in negotiating a final selling price that is acceptable to both the dealer and the customer.

Point out that the percent used for options and the percent used for the base price can be different. These figures can be found in consumer magazines. Have students notice that there was no markup on the destination charge. This part of the price is not negotiable.

Warm-Up Exercises

1. 90% × $12,000 $10,800
2. 85% × $15,000 $12,750
3. 81% × $7780 $6301.80
4. 72% × $8900 $6408

8-2 Dealer's Cost

OBJECTIVE

Compute the dealer's cost of a new automobile.

Automobile dealers pay less than the prices on the sticker for both the basic automobile and the options. Consumer magazines often report the **dealer's cost** as a percent of the sticker price. You may save money when purchasing a new automobile by making an offer that is higher than the estimated dealer's cost but lower than the sticker price.

Dealer's Cost = Percent of Base Price + Percent of Options Price + Destination Charge

EXAMPLE *Skills* 30, 2 *Application* A *Term* Dealer's cost

Lisa and Tom Marker want to purchase a new Winsor Sedan. The car has a base price of $12,438, options totaling $2240, and a destination charge of $360. They read in a consumer magazine that the dealer's cost for a Winsor is about 80% of the base price and 77% of the options price. What should they estimate as the dealer's cost?

SOLUTION

A. Find the **percent of base price.**
$12,438.00 × 80% = $9950.40

B. Find the **percent of options price.**
$2240.00 × 77% = $1724.80

C. Find the **dealer's cost.**
Percent of Base Price + Percent of Options Price + Destination Charge
$9950.40 + $1724.80 + $360.00 = $12,035.20 dealer's cost

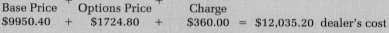

12438 × 80 % 9950.4 M+ 2240 × 77 % 1724.8 + RM 9950.4 + 360 = 12035.2

✓ SELF-CHECK Complete the problem, then check your answer in the back of the book.

1. The dealer's cost is 80% of the base price of $9000 and 75% of the options price of $3000 plus a destination charge of $340. Find the dealer's cost.
$7200 + $2250 + $340 = $9790

PROBLEMS

	(Base Price	× Dealer's Percent)	+ (Options Price	× Dealer's Percent)	+ Destination Charge	= Dealer's Cost
2.	($ 8000 ×	80%) +	($2200 ×	75%) +	$360	= $8410
3.	($17,000 ×	85%) +	($3400 ×	80%) +	$342	= $17,512
4.	($14,300 ×	82%) +	($1200 ×	78%) +	$465	= $13,127

MATHEMATICAL NOTES

The mathematics in this lesson involve adding decimals and finding a percentage. Answers can be rounded to the nearest dollar.

5. MX-7 Diesel pickup. **$10,018**
Base price is $10,800.
Options total $1260.
Destination charge is $370.
Dealer pays 80% of base price and 80% of options price.
What is the dealer's cost?

6. Rowley SXT. **$9167**
Base price is $8800.
Options total $1900.
Destination charge is $350.
Dealer pays 84% of base price and 75% of options price.
What is the dealer's cost?

7. Royal Sport, 2-door. **$12,460.75**
Base price is $12,795.
Options total $2845.
Destination charge is $290.
Dealer pays 78% of base price and 77% of options price.
What is the dealer's cost?

8. LeBou Series 720 SE. **$14,928.96**
Base price is $15,980.
Options total $1292.
Destination charge is $498.
Dealer pays 84% of base price and 78% of options price.
What is the dealer's cost?

9. Paul Dempsey is looking at a new Cameo GT-20 that has a base price of $8940, options totaling $1440, and a destination charge of $380. The dealer's cost is about 85% of the base price and 77% of the price of the options. What is the sticker price of the car? What should Paul estimate as the dealer's cost? **Sticker price: $10,760; est. dealer's cost: $9087.80**

10. Pauletta Kreger is considering the purchase of a recreational van. She sees one that has a base price of $16,950 and options totaling $2280. The destination charge is $695. She estimates that the dealer's cost is 79% of the base price and 80% of the price of the options. If Pauletta offers the dealer $250 over the estimated dealer's cost, what is her offer? **$16,159.50**

11. Julia Brown offered an automobile dealer $150 over the estimated dealer's cost on a car that has a base price of $12,900 and options totaling $2480. The dealer's cost is about 85% of the base price and 81% of the price of the options. The destination charge is $520. What was her offer? **$13,643.80**

12. Judith Estes is ordering a new car built to her specifications direct from the factory. The base price is $16,940 and options total $1180. There is a destination charge of $390. The dealer's cost is about 81% of the base price and 88% of the price of the options. The dealer will sell her the car for $200 over the estimated dealer's cost plus a 6% sales tax. What is the total cost of the car? **$16,270.79**

MAINTAINING YOUR SKILLS Look up the skill in parentheses if you need help or more practice.

Find the percentage. Round answers to the nearest hundredth. **(Skill 30)**

13. 8% of 700 **56**
14. 15% of 980 **147**
15. 22% of 756 **166.32**
16. 24% of 1520 **364.8**
17. 45% of 9800 **4410**
18. 78% of 3440 **2683.2**
19. 88% of 7780 **6846.4**
20. 92% of 2146 **1974.32**
21. 75% of 8971 **6728.25**
22. 30% of 1217 **365.1**
23. 87.25% of 6163 **5377.22**

PRACTICE AND APPLY

The following problems can be assigned for classwork and the answers checked in class to help students master the objective of the lesson.

- Guided Practice: 1–5
- Independent Practice: 6–10 even

WRAP-UP

Have students explain the term *dealer's cost.* Ask which of the three components of the sticker price does not have a dealer's cost.

Assignment Guide

- Basic: 6–11, 13–23
- Average: 7–11 odd, 12, 14–22 even

ALTERNATIVE STRATEGIES: Reteaching

Emphasize that calculating either percent of base price or percent of options is an application of Skill 30, finding a percentage. Review the use of the memory key on the calculator. Work these examples:

1. ($10,000 × 80%) + ($3000 × 75%) = $10,250
2. ($18,960 × 82%) + ($4880 × 70%) = $18,963.20

LESSON PLAN
8-3 Purchasing a Used Automobile

FOCUS
Many students may have had experiences shopping for, owning, or selling a used automobile. Ask students to share their experiences of trying to find a price agreeable to both the seller and the buyer.

TEACH
Make sure students understand the differences between *average trade-in*, *average loan*, and *average retail* in the table. The average trade-in is based on the trade-in value of hundreds of cars of that particular model. The average loan is what the bank can be expected to loan a customer. Point out that the average trade-in is usually less than the average retail, and that the average loan is less than this. Have students explain why this is so. Emphasize that having certain optional features adds to the value of the car, while not having others subtracts from the value.

Warm-Up Exercises
1. \$4000 + \$500 + \$50 − \$200 \$4350
2. \$8000 + \$250 + \$475 − \$125 \$8600
3. \$9500 + \$75 + \$25 − \$100 \$9500
4. \$1250 + \$125 + \$75 − \$75 − \$75 \$1300

8-3 Purchasing a Used Automobile

OBJECTIVE
Compute the average retail price of a used automobile.

Automobile dealers usually advertise used cars for prices that are higher than what they expect you to pay. **Used-car guides,** published monthly, give the average prices for cars that were purchased from dealers during the previous month. The information can help you make decisions about how much to pay for a used automobile.

Average Retail Price = Average Retail Value + Additional Options − Options Deductions − Mileage Deduction

EXAMPLE *Skills* 5, 6 *Application* C *Term* Used-car guide

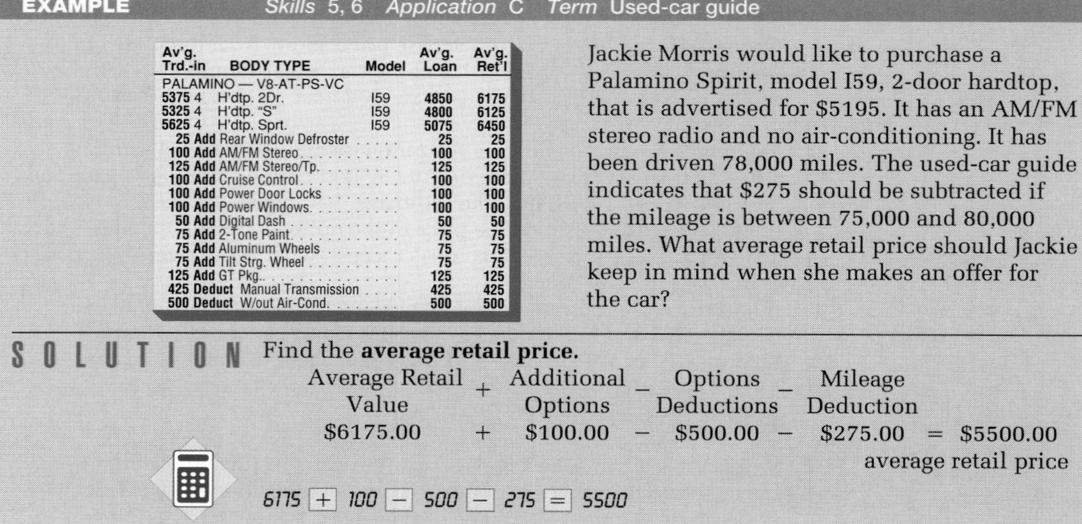

Av'g. Trd.-in	BODY TYPE	Model	Av'g. Loan	Av'g. Ret'l
	PALAMINO — V8-AT-PS-VC			
5375 4	H'dtp. 2Dr.	I59	4850	6175
5325 4	H'dtp. "S"	I59	4800	6125
5625 4	H'dtp. Sprt.	I59	5075	6450
25 Add	Rear Window Defroster		25	25
100 Add	AM/FM Stereo		100	100
125 Add	AM/FM Stereo/Tp.		125	125
100 Add	Cruise Control		100	100
100 Add	Power Door Locks		100	100
100 Add	Power Windows		100	100
50 Add	Digital Dash		50	50
75 Add	2-Tone Paint		75	75
75 Add	Aluminum Wheels		75	75
75 Add	Tilt Strg. Wheel		75	75
125 Add	GT Pkg.		125	125
425 Deduct	Manual Transmission		425	425
500 Deduct	W/out Air-Cond.		500	500

Jackie Morris would like to purchase a Palamino Spirit, model I59, 2-door hardtop, that is advertised for \$5195. It has an AM/FM stereo radio and no air-conditioning. It has been driven 78,000 miles. The used-car guide indicates that \$275 should be subtracted if the mileage is between 75,000 and 80,000 miles. What average retail price should Jackie keep in mind when she makes an offer for the car?

SOLUTION Find the **average retail price.**

Average Retail Value + Additional Options − Options Deductions − Mileage Deduction
\$6175.00 + \$100.00 − \$500.00 − \$275.00 = \$5500.00 average retail price

6175 + 100 − 500 − 275 = 5500

✓ SELF-CHECK Complete the problem, then check your answer in the back of the book.

	Retail Value	Additional Options	Options Deductions	Mileage Deduction	Average Retail Price
1.	\$4795	\$325	\$425	\$250	? \$4445

PROBLEMS

Use the table above to find the average retail price.

	Model	Retail Value	AM/FM Stereo	Power Locks	Power Windows	Digital Dash	Man. Trans.	A/C	Avg. Retail Price
2.	2 Dr.	\$6175	No	No	No	No	Yes	Yes	? \$5750
3.	H'dtp."S"	\$6125	Yes	No	No	Yes	Yes	No	? \$5350
4.	H'dtp. Sprt.	\$6450	Yes	No	Yes	No	No	Yes	? \$6650

BUSINESS NOTES
An odometer discloser statement is required by law to be provided to you when you buy a used car. Banks and other loan institutions use a "bluebook" as a price guide of how the odometer reading affects a car's value.

5. 5-year-old Elite compact.
Average retail value is $4725.
Add $50 for tilt steering.
Add $325 for air-conditioning.
Deduct $100 for manual transmission. $5000
What is the average retail price?

6. 2-year-old Cramer sedan.
Average retail value is $8650.
Add $175 for AM/FM stereo/tape.
Deduct $450 for no air-conditioning. $8050
Deduct $325 for excessive mileage.
What is the average retail price?

7. Sue Soto owns a 4-year-old Spector 4-door sedan that she wants to sell in order to purchase a new car. One used-car guide shows the average retail value of her car is $5225. She adds $50 for having a vinyl top, $125 for cassette player, $100 for power windows, and $75 for power locks. She deducts $475 for having no air conditioner. What is the average retail price she can ask as a selling price for her car? $5100

Use the used-car guide shown to solve problems 8–10.

Av'g. Trd.-in	BODY TYPE	Av'g. Loan	Av'g. Ret'l.
	MYSTIC V8-PS-PE		
8550	11 Spt. Cpe. 2D	7700	9575
9450	12 Berlinetta Cpe.	8525	10575
10350	Spt. Cpe. Z28	9325	11525
175	Add Sunroof	175	175
125	Add AM/FM Stereo	125	125
150	Add AM/FM Stereo/Tp.	150	150
125	Add Power Door Locks	125	125
125	Add Power Windows	125	125
125	Add Cruise Control	125	125
100	Add Tilt Strg. Wheel	100	100
75	Add Luggage Rack (S/W)	75	75
100	Add Aluminum Wheels	100	100
50	Add Rear Window Defroster	50	50
125	Add Digital Dash	125	125
250	Add 6 Cyl.	250	250
475	Deduct Manual Transmission	475	475
150	Deduct Convent Steer	150	150
575	Deduct W/out Air-Cond.	575	575

8. Jan Pohn owns a Mystic Sport Coupe 2D. The car has AM/FM stereo, air conditioning, power steering, power door locks, power windows, cruise control, and a manual transmission. What is the average retail price of the car? $9600

9. Ray Sanchez owns a used Mystic Sport Coupe Z28 with air-conditioning, an automatic transmission, AM/FM stereo/tape deck, rear window defroster, and tilt steering wheel. Recently he was asked if he wanted to sell his car. What is the average retail price he should keep in mind when pricing his car? $11,825

10. Joan Barryhill wishes to trade in her Mystic Berlinetta Coupe on the purchase of a new automobile. It has AM/FM stereo, air-conditioning, automatic transmission, power steering, power windows, and rear window defroster. She must deduct $375 from the trade-in value for high mileage. What is the average trade-in value of her car? $9,375

MAINTAINING YOUR SKILLS Look up the skills in parentheses if you need help or more practice.

Add. **(Skill 3)**

11. 4225 + 1200 + 375 + 245 6045

12. 4060 + 225 + 3950 + 325 + 75 8635

13. 4675 + 15 + 85 + 25 + 325 5125

14. 175 + 260 + 3470 + 25 + 25 + 10 3965

Subtract. **(Skill 4)**

15. 8450 − 475 7975
16. 3890 − 2530 1360
17. 2205 − 225 1980
18. 3625 − 425 3200

PRACTICE AND APPLY

The following problems can be assigned for classwork and the answers checked in class to help students master the objective of the lesson.

- Guided Practice: 1–4
- Independent Practice: 5–7

WRAP-UP

Write the terms *average retail price, average retail value, additional options, options deductions,* and *mileage deductions* in any order on the chalkboard. Have volunteers explain what each term means, then have a volunteer give the formula for finding the average retail price.

Assignment Guide

- Basic: 5–9, 11–18
- Average: 8–10, 11–17 odd

ALTERNATIVE STRATEGIES: Enrichment

A very practical out-of-class project is to have students investigate and report on the various used-car guides. Most bookstores will carry several different ones. Most auto dealers use the "bluebook" published by the National Automobile Dealers Association. Specifically, have them determine the value of various cars owned by family and friends.

LESSON PLAN
8-4 Automobile Insurance

FOCUS
The focus of this lesson is to use tables to compute the annual premium for automobile insurance. In many states, it is against the law to drive a car without auto insurance. This is an additional cost that needs to be considered when purchasing a car.

TEACH
Define the terms *liability, bodily injury, property damage, comprehensive,* and *collision insurance*. Point out that many states require liability insurance but make comprehensive insurance optional. An older car may not be worth enough to have comprehensive coverage. If you have a loan on a car, the bank often requires full coverage for the term of the loan to safeguard their money in case of damage or loss to the car.

8-4 Automobile Insurance

OBJECTIVE
Use tables to compute the annual premium for automobile insurance.

Liability insurance, which includes **bodily injury insurance** and **property damage insurance,** protects you against financial losses if your car is involved in an accident. Bodily injury limits of 25/100 mean that the insurance company will pay up to $25,000 to any one person injured and up to $100,000 if more than one person is injured. Property damage insurance protects you against financial loss if your automobile damages the property of others.

Comprehensive insurance on your automobile protects you from losses due to fire, vandalism, theft, and so on. **Collision insurance** will pay to repair the damage to your automobile if it is involved in an accident. Either policy may be written with a **deductible clause.** A $50-deductible clause means that you pay the first $50 of the repair bill. The annual **base premium** is determined by the amount of insurance you want, the **age group** of your car, and the **insurance-rating group.** The insurance-rating group depends on the size and value of your car.

The **annual premium** is the amount you pay each year for insurance coverage. Your annual premium depends on the base premium and your **driver-rating factor.** The base premium depends on the amount of coverage you want. The driver-rating factor depends on your age, marital status, the amount you drive each week, and so on. If several people drive your car, the highest driver-rating factor among those who drive your car is used to determine the annual premium.

Insurance companies use tables to determine your base premium.

BASE PREMIUM FOR A PRIVATE PASSENGER AUTOMOBILE

Property Damage Limits	Bodily Injury Limits 25/50	25/100	50/100	100/200	100/300	300/300
$ 25,000	$206.40	$218.80	$213.20	$252.00	$258.00	$286.80
50,000	212.40	224.80	237.20	258.00	264.00	293.20
100,000	220.80	233.20	245.60	266.40	272.40	301.20

PHYSICAL DAMAGE PREMIUM

Coverage	Age Group	Insurance-Rating Group 10	11	12	13	14	15
Comprehensive $50-Deductible	A	$76.80	$81.60	$95.20	$108.00	$122.00	$135.60
	B	65.20	77.60	90.40	102.40	115.60	128.40
	C	62.00	74.00	86.00	98.00	110.40	122.80
	D	59.20	70.40	82.00	93.20	105.20	116.80
Collision $50-Deductible	A	$225.60	$246.00	$266.80	$287.20	$307.60	$328.00
	B	214.00	233.20	253.20	272.40	291.60	311.20
	C	204.00	222.80	241.60	260.00	278.40	296.80
	D	194.40	212.00	230.00	247.60	265.20	282.80

$$\text{Annual Base Premium} = \text{Liability Premium} + \text{Comprehensive Premium} + \text{Collision Premium}$$

$$\text{Annual Premium} = \text{Annual Base Premium} \times \text{Driver-Rating Factor}$$

EXAMPLE *Skills* 5, 8 *Application* C *Term* Annual premium

Della Welch is the principal operator of her car. Her driver-rating factor is 2.20. Her insurance includes 50/100 bodily injury and $50,000 property damage. Her car is in age group A and insurance-rating group 13 (A, 13). She has $50-deductible comprehensive and $50-deductible collision insurance. What is her annual base premium? What is her annual premium?

SOLUTION

A. Find the **annual base premium.**

Liability Premium	+	Comprehensive Premium	+	Collision Premium		
$237.20	+	$108.00	+	$287.20	=	$632.40 annual base premium

B. Find the **annual premium.**

Annual Base Premium	×	Driver-Rating Factor		
$632.40	×	2.20	=	$1391.28 annual premium

237.2 [+] 108 [+] 287.2 [=] 632.4 [×] 2.2 [=] 1391.28

✓ SELF-CHECK Complete the problem, then check your answer in the back of the book.

$233.20
62.00
+ 204.00
$499.20 base premium
$748.80 annual premium

Find the annual base premium and the annual premium.

1. 25/100 bodily injury and $100,000 property damage.
Car is in age group C and insurance-rating group 10 (C, 10).
$50-deductible comprehensive and $50-deductible collision.
Driver-rating factor is 1.50.

PROBLEMS

2. $642.80; $835.64
3. $576.00; $921.60
4. $700.80; $805.92
5. $708.40; $2762.76

2. Driver-rating factor is 1.30.
Age, rating group is A, 14.
Coverage: 50/100 bodily injury.
$25,000 property damage.
$50-deductible comprehensive.
$50-deductible collision.
What is the annual base premium?
What is the annual premium?

3. Driver-rating factor is 1.60.
Age, rating group is D, 12.
Coverage: 100/300 bodily injury.
$50,000 property damage.
$50-deductible comprehensive.
$50-deductible collision.
What is the annual base premium?
What is the annual premium?

4. Driver-rating factor is 1.15.
Age, rating group is A, 15.
Coverage: 50/100 bodily injury.
$50,000 property damage.
$50-deductible comprehensive.
$50-deductible collision.
What is the annual base premium?
What is the annual premium?

5. Driver-rating factor 3.90.
Age, rating group is B, 14.
Coverage: 300/300 bodily injury.
$100,000 property damage.
$50-deductible comprehensive.
$50-deductible collision.
What is the annual base premium?
What is the annual premium?

Annual premiums also can be paid semiannually, quarterly, or monthly. There is sometimes a small handling fee charged if the payments are made quarterly or monthly. Make sure students know how to use the tables on page 246. Point out that factors such as prior driving records can also affect a person's insurance rating group.

Warm-Up Exercises

1. $180 × 1.20 $216
2. $70 × 3.10 $217
3. ($140 + $64) × 4.10 $836.40
4. ($180 + $90.20) × 2.90 $783.58

COOPERATIVE LEARNING
Have small groups of students look in the paper and find their "dream car." Then have them contact various insurance agents to determine the cost of insurance. Students should then report their findings to the class.

PRACTICE AND APPLY

The following problems can be assigned for classwork and the answers checked in class to help students master the objective of the lesson.

- Guided Practice: 1–4
- Independent Practice: 5–9 odd, 10

6. Paula Williams uses her car primarily to run errands. She has $50-deductible comprehensive, $50-deductible collision, 100/200 bodily injury, and $25,000 property damage coverage. Her driver-rating factor is 1.00 and her car is classified D, 14. What is her annual base premium? What is her annual premium? **$622.40; $622.40**

7. Scott Hanson uses his car primarily for business. He has $50-deductible comprehensive, $50-deductible collision, 100/300 bodily injury, and $100,000 property damage coverage. His driver-rating factor is 1.35 and his car is classified B, 15. What is his annual base premium? What is his annual premium? **$712.00; $961.20**

8. Cheryl Obritter uses her car mainly for pleasure. She has $50-deductible comprehensive, $50-deductible collision, 100/300 bodily injury, and $50,000 property damage coverage. Her driver-rating factor is 2.15 and her car is classified C, 12. What is her annual base premium? What is her annual premium? **$591.60; $1271.94**

9. Carl McPeak uses his car to deliver office supplies. He has $50-deductible comprehensive, $50-deductible collision, 100/200 bodily injury, and $50,000 property damage coverage. His driver-rating factor is 3.10 and his car is classified D, 15. What is his annual base premium? What is his annual premium? **$657.60; $2038.56**

10. Henry Klump delivers magazines to retail outlets. His driver-rating factor is 2.15. His insurance coverage includes 25/100 bodily injury and $25,000 property damage. He has $50-deductible comprehensive and $50-deductible collision. His car is in age group C and insurance-rating group 10. What is his annual base premium? What is his annual premium? **$484.80; $1042.32**

11. Esther Miller-Kralik drives to and from work. Her driver-rating factor is 1.85. Her insurance coverage includes 25/50 bodily injury and $25,000 property damage. She has $50-deductible comprehensive and $50-deductible collision. Her car is in age group B and insurance-rating group 11. What is her annual base premium? What is her annual premium? **$517.20; $956.82**

12. Elijah Kiboda owns and operates Kiboda's Imports, Inc. His driver-rating factor is 3.90. His insurance coverage includes 25/50 bodily injury and $25,000 property damage. He has $50-deductible comprehensive and $50-deductible collision. His convertible is in age group B and insurance-rating group 15. What is his annual base premium? What is his annual premium? **$646.00; $2519.40**

13. Tom Nome uses his car to drive to and from work. His driver-rating factor is 2.85. His insurance coverage includes 25/50 bodily injury and $50,000 property damage. He has $50-deductible comprehensive and $50-deductible collision. His car is in age group D and insurance-rating group 12. What is his annual base premium? What is his annual premium? **$524.40; $1494.54**

14. Lori Peterson uses her car for delivering newspapers. Her insurance coverage includes 25/50 bodily injury and $25,000 property damage. She has no comprehensive and no collision. Her rating factor is 1.90. What is her annual premium? **$392.16**

PROBLEM SOLVING

Refer to Problem 6. Have students call an insurance agent to see what price difference there is between a $50 deductible, a $100 deductible, and a $200 deductible. Have students rework this problem for the $100 and $200 deductibles.

15. Gordie Meeks drives a farm automobile. His rating factor is 0.90. His insurance coverage only includes 100/200 bodily injury and $100,000 property damage. What is his annual premium? **$239.76**

16. Nathaniel Steiner delivers eggs to retail stores. His 12-year-old car is not worth very much so he doesn't carry comprehensive or collision insurance. He carries 50/100 bodily injury and $50,000 property damage insurance. His rating factor is 3.10. What is his annual premium? **$735.32**

17. Glenn O'Malley owns a car that is classified D, 11. His driver-rating factor is 4.10. He has 25/100 bodily injury, $25,000 property damage, $50-deductible comprehensive, and $50-deductible collision coverage. What is his annual premium? How much more would he pay for 100/300 bodily injury? **$2054.92; $160.72**

18. Sandra Jabara owns and operates Jabara's Bicycle Shop. She purchased a van with which to make deliveries. The van is classified A, 15. She has the following coverage: $100,000 property damage, 100/300 bodily injury, $50-deductible comprehensive, and $50-deductible collision. Her driver-rating factor is 1.35. What is her annual premium? What would be her annual premium if the van were classified A, 12? **$993.60; $856.44**

19. Arnold Barbuto uses his car mainly for pleasure driving. His car is classified A, 12. His driver-rating factor is now 2.65. He learns that his rating factor will be 2.15 after he is married next year. How much less will his annual premium be if he has $25,000 property damage, 50/100 bodily injury, $50-deductible comprehensive, and $50-deductible collision? **$287.60**

MAINTAINING YOUR SKILLS Look up the skills in parentheses if you need help or more practice.

Add. **(Skill 5)**

20. 429.45 + 87.92 + 36.48 + 73.35 **627.2**

21. 49.55 + 2.82 + 34.59 + 733.20 **820.16**

22. 87.08 + 114.46 + 3.94 + 78.82 **284.3**

23. 480.88 + 65.44 + 749.45 + 9.80 **1305.57**

24. 2.50 + 0.89 + 31.30 + 7.97 **42.66**

25. 40.93 + 73.30 + 67.83 + 3.9 **185.96**

Multiply. **(Skill 8)**

26. 1.25 × 79.90 **99.875**
27. 2.40 × 360 **864**
28. 3.90 × 67.70 **264.03**
29. 34.42 × 73.20 **2519.544**
30. 654 × 2.6 **1700.4**
31. 2121 × 3.50 **7423.5**
32. 46.58 × 4.15 **193.307**
33. 371.19 × 3.20 **1187.808**
34. 431.50 × 2.25 **970.875**

WRAP-UP

Read Problem 10 to the class. Ask what is meant by 25/100 bodily injury insurance and $25,000 property damage insurance. Then ask what effect a deductible has on an insurance premium. Finally, ask students how to find an annual premium, given an annual premium base.

Assignment Guide

- Basic: 5–10, 11–17 odd, 20–34
- Average: 4–10 even, 11–17 odd, 18, 19, 21–33 odd

ALTERNATIVE STRATEGIES: Reteaching

Emphasize that the base premium is multiplied by the driver rating factor. Make these changes in the Example and solve.

1. Rating factor of 4.10 ($2592.84)
2. $100,000 property damage ($18.48 more)
3. Car is an A, 10 ($204.16 less)

LESSON PLAN
8-5 Operating and Maintaining An Automobile

FOCUS
Ask students to name the costs they or their families have encountered in operating and maintaining an automobile. List items on the chalkboard as they are named. Tell students that in today's lesson, they will learn to compute the cost per mile of operating and maintaining an automobile.

TEACH
Emphasize the difference between variable cost and fixed cost. Use the list on the chalkboard to have students categorize each item named earlier. Make sure that depreciation is included in the list, if it wasn't already named. Students may be surprised to find out how much a car depreciates the first year. Contact a car dealer and a bank in your area for more information. If payments are going to be made, the finance charge paid to the bank should be listed as one of the fixed costs.

8-5

Operating and Maintaining an Automobile

OBJECTIVE
Compute the total cost per mile of operating and maintaining an automobile.

Many costs are involved in operating and maintaining an automobile. **Variable costs,** like gasoline and tires, increase as the number of miles you drive increases. **Fixed costs,** like automobile insurance, registration fees, and **depreciation,** remain about the same regardless of how many miles you drive. Depreciation is a decrease in the value of your car because of its age and condition.

$$\text{Cost per Mile} = \frac{\text{Annual Variable Cost} + \text{Annual Fixed Cost}}{\text{Number of Miles Driven}}$$

EXAMPLE *Skills* 11, 2 *Application* A *Term* Depreciation

Ann Kory purchased a used automobile for $4000 one year ago. She drove 9000 miles during the year and kept a record of all her expenses. She estimates the car's present value at $3200. What was the cost per mile for Ann to operate her car last year?

Variable Costs		Fixed Costs	
Gasoline	$345.24	Insurance	$ 385.40
Oil changes	71.85	License/registration	76.25
Maintenance	114.36	Depreciation	
New tire	41.75	($4000 – $3200)	800.00
	$573.20		$1261.65

SOLUTION Find the **cost per mile.**

(Annual Variable Cost + Annual Fixed Cost) ÷ Number of Miles Driven
($573.20 + $1261.65) ÷ 9000
$1834.85 ÷ 9000 = $0.203 = $0.20 cost per mile

573.2 [+] 1261.65 [=] 1834.85 [÷] 9000 [=] 0.20387

✓ SELF-CHECK Complete the problems, then check your answers in the back of the book.

Find the total cost and the cost per mile.

	Variable Cost	Fixed Cost	Miles Driven	
1.	$900	$1700	10,000	$2600; $0.26/mile
2.	$1137.26	$2491.24	12,000	$3628.50; $0.30/mile

COMMUNICATION SKILLS
Have students visit a car dealer or use an automobile operation and maintenance manual to write a report on the following: (1) parts of an automobile covered under a typical warranty; (2) parts included under an extended warranty; (3) maintenance schedules that are recommended by manufacturers.

PROBLEMS

Solve. Round answers to the nearest cent.

	(Annual Variable Cost	+	Annual Fixed Cost	=	Total Annual Cost)	÷	Miles Driven	=	Cost per Mile
3.	($1000.00	+	$1250.00	=	**$2250** ?)	÷	9000	=	**$0.25**
4.	($1530.00	+	$1275.00	=	**$2805** ?)	÷	11,000	=	**$0.26**
5.	($2114.00	+	$3786.00	=	**$5900** ?)	÷	14,700	=	**$0.40**
6.	($1584.00	+	$934.35	=	**$2518.35**)	÷	6800	=	**$0.37**
7.	($2312.50	+	$4321.90	=	**$6634.40**)	÷	20,415	=	**$0.32**

8. $2200; $0.23
9. $6400; $0.26
10. $0.21
11. $0.17

8. Mabel Kite, student.
Drove 9500 miles last year.
Fixed costs totaled $1215.
Variable costs totaled $985.
What was the total annual cost?
What was the cost per mile?

9. Karim Yakobian, salesperson.
Drove 24,500 miles last year.
Fixed costs totaled $2460.
Variable costs totaled $3940.
What was the total annual cost?
What was the cost per mile?

10. John Davidson, accountant.
Drove 15,460 miles last year.
Fixed costs totaled $1127.40.
Variable costs totaled $2076.30.
What was the cost per mile?

11. Elska Rashidi, sales manager.
Drove 26,350 miles last year.
Fixed costs totaled $1527.32.
Variable costs totaled $2981.65.
What was the cost per mile?

12. Hope Kocinski drove her car 12,200 miles last year. Her variable costs totaled $780.35. Her fixed costs totaled $2439. What was the cost per mile for her to operate her car? **$0.26**

13. Alice Powers drove her car 13,550 miles last year. Her variable costs totaled $1776.90. Her fixed costs totaled $2457.15. What was the cost per mile for her to operate her car? **$0.31**

14. J. J. Olmstead drove his van 11,400 miles last year. His variable costs totaled $1965.89. His fixed costs totaled $1884.26. What was the cost per mile for him to operate his car? **$0.34**

15. Carl Collins drove his subcompact car 24,200 miles last year. His variable costs totaled $4059.33. His fixed costs totaled $1973.27. What was the cost per mile for him to operate his car? **$0.25**

16. Pinckney Keil purchased an automobile for $18,350 one year ago. He drove it 11,500 miles during the first year and kept a record of all his expenses. His variable costs were: gasoline, $533.60; oil changes, $95.84; parking, $115.71; and repairs, $91.35. His fixed costs were: insurance, $418; license, $76.75; and depreciation. He estimates the car's present value at $15,350. What is his cost per mile? **$0.38**

Point out that accurate records are important for many people who use their cars in their businesses. Part of these operating expenses can be deducted at tax time. For other people, the prior year's expenses of operating and maintaining an automobile can be used as a basis for preparing a budget for the new year. A large increase in the cost per mile driven is a signal to some individuals that they may need to consider trading for a car that is cheaper to own and operate.

Warm-Up Exercises

1. $67 + $33 + $24 + $136 $260
2. ($150 + $140) + ($35 + $75) $400
3. ($113 + $250 + $300 + $200) ÷ 4000 $0.22

PRACTICE AND APPLY

The following problems can be assigned for classwork and the answers checked in class to help students master the objective of the lesson.

- Guided Practice: 1–8
- Independent Practice: 9–11, 16

ALTERNATIVE STRATEGIES: Enrichment

1. Assign students to small groups of 3 or 4. Have them work through the Simulation starting on page 261 of the text.
2. Have students check with friends and relatives to see if anyone keeps track of their car expenses. Ask them to share the data with the class.

WRAP-UP

Read Problem 16 to the class. Have one student find the total variable cost, one find the total fixed costs, and one find the depreciation. Have another student explain how to use these figures to find the cost per mile.

Assignment Guide

- Basic: 9–11, 12–18 even, 19, 20, 23–38
- Average: 12–15, 17–22, 24–38 even

F.Y.I.
In 1991, the average cost of driving a four-door sedan 10,000 miles was 45¢ per mile.

17. Clara Nowiejski purchased an automobile for $16,940 one year ago. She drove it 14,500 miles during the first year and kept a record of all her expenses. Her variable costs were: gasoline, $1039; oil changes, $114.96; parking, $109.75; and repairs, $41.20. Her fixed costs were: insurance, $673; license, $46.75; and depreciation. She estimates the car's present value at $12,340. What was her cost per mile? **$0.46**

18. Nora Moskowitz and three friends agreed that they would share equally in the costs of operating and maintaining a car. Last year, the car was driven 11,342 miles. Fixed costs totaled $1399.56, while variable costs totaled $1626.39. How much did it cost per mile to operate the car? How much did it cost per person? **$0.27; $756.49**

19. Gary Halston and two neighbors carpool to and from work in Gary's subcompact. The three agreed that they would share equally in the cost of operating and maintaining the car. Last year, the car was driven 18,500 miles. Fixed costs totaled $2686.87, while variable costs totaled $3275.27. How much did it cost per mile to operate the car for the year? How much did it cost each person to carpool for the year? **$0.32; $1987.38**

V	Oil changes	$ 71.55
V	Tune - up	87.95
V	Alignment	27.95
F	Insurance	415.00
V	Parking	42.20
F	Registration	68.50
F	Loan interest	459.70
F	Depreciation	1520.00
V	Gasoline	366.24

20. Allison Pappas kept records on the operation and maintenance of her car for the previous year. In the summary shown, which of the items listed are fixed costs? Which are variable costs?

21. Last year, Allison Pappas drove her car 8400 miles. What was the total cost to operate and maintain her car for the year? What was her cost per mile? **$3059.09; $0.36**

22. Earl and Bella Ridder each own the same model car. Each kept records of the operating and maintenance costs. Last year, Earl drove his car 13,700 miles and Bella drove hers 9700 miles. Both had fixed costs totaling $1368.27. Earl's variable costs totaled $2056.95 and Bella's totaled $1597.30. Who spent more per mile to operate and maintain the car? How much more per mile? **Earl: $0.25; Bella: $0.31; Bella: $0.06**

MAINTAINING YOUR SKILLS Look up the skills in parentheses if you need help or more practice.

Round to the nearest hundredth. (Skill 2)

23. 21.751 **21.75** **24.** 15.352 **15.35** **25.** 4.3981 **4.40** **26.** 15.9061 **15.91**

27. 0.04126 **0.04** **28.** 0.3179 **0.32** **29.** 1.0711 **1.07** **30.** 0.0617 **0.06**

Divide. Round answers to the nearest hundredth. (Skill 10)

31. 641 ÷ 200 **3.21** **32.** 1500 ÷ 500 **3**

33. 850 ÷ 9000 **0.09** **34.** 3241 ÷ 15,000 **0.22**

35. 1875 ÷ 6800 **0.28** **36.** 3199 ÷ 23,400 **0.14**

37. 7876 ÷ 16,520 **0.48** **38.** 4135 ÷ 16,792 **0.25**

MATHEMATICAL NOTES

For Problem 19, point out that the sum of the fixed cost plus the total variable cost must be divided by 3 to arrive at the second part of the answer. If they worked 5 days a week what was the cost per person per day?

8-6 Leasing an Automobile

OBJECTIVE
Compute the total cost of leasing an automobile.

Rather than purchasing an automobile, you may want to lease one. When you lease an automobile, you make monthly payments to the leasing company, dealer, or bank for 2 to 5 years. At the end of the lease, you return the automobile to the dealer, or you may purchase it depending on the type of lease. The most common lease is a **closed-end lease.** With a closed-end lease, you make a specified number of payments, return the car, and owe nothing unless you damaged the car or exceeded the mileage limit. Another type of lease is an **open-end lease.** At the end of an open-end lease, you can buy the automobile for its **residual value.** The residual value is the expected value of the car at the end of the lease period. The residual value is established at the signing of the lease. With either lease, you must pay all the monthly payments, a security deposit, title fee, and license fee.

$$\text{Total Lease Cost} = \left(\text{Number of Payments} \times \text{Amount of Payment}\right) + \text{Deposit} + \text{Title Fee} + \text{License Fee}$$

EXAMPLE *Skills* 2, 5, 11 *Application* A *Term* Lease

Ralph Dunn leased an S-10 pickup truck for use in his lawn care business. He pays $168.97 per month for 48 months. His deposit was $200. He paid a $40 title fee and a $15 license fee. What is his total lease cost?

SOLUTION

Find the **total cost.**

Total of payments: 48 × $168.97 = ...	$8110.56
Deposit	200.00
Title fee	40.00
License fee..............................	+ 15.00
Total lease cost	$8365.56

48 [×] 168.97 [=] 8110.56 [+] 200 [+] 40 [+] 15 [=] 8365.56

✓ SELF-CHECK Complete the problems, then check your answers in the back of the book.

Find the total lease cost.

1. 54 payments of $139. **$8101**
   Deposit of $500.
   Title fee of $35.
   License fee of $60.

2. 60 payments of $249. **$16,215**
   Deposit of $1200.
   Title fee of $40.
   License fee of $35.

LESSON PLAN
8-6 Leasing an Automobile

FOCUS

Ask students what is meant by a lease in reference to an apartment. Point out that leasing an automobile has become a common business practice.

TEACH

For the closed-end lease, point out that the mileage limit is generally set at 12,000–14,000 miles. After that limit, there is an additional charge per mile. The car may not be purchased at the end of the closed-end lease.

The open-end lease may be preferable if the car is to be purchased. The lease payments are often less than the purchase payments when you finance all or most of the price of the car.

In addition to the security deposit, title fee, and license fee, you must also qualify for credit. Sales tax must sometimes be paid at the time of purchase. These problems assume the sales tax is part of the lease payment.

ALTERNATIVE STRATEGIES: Reteaching

Relate this lesson to the real world by using a newspaper advertisement for a car dealer. The ads are a good source of current prices. They usually give such pertinent information, as payment, number of payments, and down payment. Use a title fee and license fee typical to your area.

Warm-Up Exercises

1. 48 × $199 $9552
2. 36 × $175 $6300
3. 24 × $299.95 $7198.80
4. 9360 + 1200 + 125 + 60 10,745
5. 13,740 + 1500 + 400 + 75 15,715

PRACTICE AND APPLY

The following problems can be assigned for classwork and the answers checked in class to help students master the objective of the lesson.

- Guided Practice: 1–8
- Independent Practice: 9–13 odd

WRAP-UP

Change the number of months in the Example to 60 and the deposit to $500. Have students find the new total lease cost.

Assignment Guide

- Basic: 9–16, 18–21
- Average: 10–14 even, 15–17, 18–21

PROBLEMS

Find the total lease cost.

	(Number of Payments	× Amount of Payment	= Total of Payments)	+ Deposit	+ Title Fee	+ License Fee	= Total Lease Cost
3.	(24	× $219	= **$5256**	+ $419	+ $8	+ $36	= **$5719**
4.	(48	× $199	= **$9552**	+ $749	+ $15	+ $15	= **$10,331**
5.	(48	× $119	= **$5712**	+ $1200	+ $60	+ $75	= **$7047**
6.	(54	× $180	= **$9720**	+ $1200	+ $35	+ $96	= **$11,051**
7.	(60	× $374	= **$22,440**	+ $1500	+ $66	+ $55	= **$24,061**

8. Eagle Sport Coupe.
   48 payments of $199.
   Deposit of $1200.
   Title fee of $35.
   License fee of $25.
   What is the total lease cost? **$10,812**

9. Jeep Wrangler.
   48 payments of $149.
   Deposit of $200.
   Title fee of $60.
   License fee of $40.
   What is the total lease cost? **$7452**

10. Rhoda Solberg leased a Jeep Cherokee for personal use. She pays $239 a month for 48 months. She also paid a deposit of $1100, a title fee of $90, and a license fee of $125. What is the total lease cost? **$12,787**

11. Nadine Daniels leased an Accord LX Wagon for $249.50 a month for 60 months. She paid a deposit of $350, a title fee of $25, and a license fee of $215. What is the total lease cost? **$15,560**

12. Doris Boyer has an open-end lease for a Chevy Lumina for her fabric store. The lease costs $219 a month for 48 months. She paid a deposit of $250, a title fee of $25, and a license fee of $135. At the end of the lease, she can buy the car for its residual value of $5446. What is the total lease cost? What is the total cost if she buys the car? **$10,922; $16,368**

13. Homer Gilwee leased a Chevy Beretta for $199 a month for 48 months. He paid a deposit of $225, a title fee of $15, and a license fee of $60. The lease carried a stipulation that there would be a 10¢-per-mile charge for all miles over 60,000. He drove the car 68,515 miles. What is the total cost of leasing the automobile? **$9852.00 + $851.50 = $10,703.50**

14. Alicia Harper can lease a Ford Escort Sport Coupe for $254.95 a month for 48 months. She must pay a deposit of $250, a title fee of $75, and a license fee of $120. At the end of the 48 months, the car is expected to be worth $4117. Instead of leasing, she can purchase the car for $278.96 a month for 48 months plus a $978 down payment and the same title and license fees. Is it less expensive to lease or purchase the car? **Purchase: $16,799.60 to lease vs. $14,563.08 to purchase**

MAINTAINING YOUR SKILLS Look up the skill in parentheses if you need help or more practice.

Multiply. **(Skill 8)**

15. 33.90 × 5 **169.5**
16. 29.95 × 4 **119.8**
17. 7 × 54.65 **382.55**
18. 11 × 19.99 **219.89**

CRITICAL THINKING

Refer to Problem 14. What monthly fee would make it cheaper to lease the car before buying it?
$208.35

8-7 Renting an Automobile

OBJECTIVE
Compute the cost per mile of renting an automobile.

From time to time, you may need to **rent** a car. Some automobile rental agencies charge a daily rate plus a rate per mile, while others charge a daily rate with no mileage charge. In either case, you pay for the gasoline used. If you are under 21 years of age, your parents may be required to pay for the insurance on the automobile. If you are 21 or over, the rental agency usually pays for the insurance. The insurance generally has a collision deductible clause that states that you will pay for a portion of any damage to the car if it is in an accident. You can obtain complete insurance coverage with a **collision waiver** by paying an additional charge per day.

$$\text{Cost per Mile} = \frac{\text{Total Cost}}{\text{Number of Miles Driven}}$$

EXAMPLE *Skills* 11, 2, 5 *Application* A *Term* Rent

Joe Wozniak rented a compact car for 3 days at $27.95 per day plus 20¢ per mile. He purchased the collision waiver for $10.00 per day. Joe drove 468 miles and paid $21.70 for gasoline. What was the total cost of renting the car? What was the total cost per mile to rent the car?

SOLUTION

A. Find the **total cost.**

Daily cost: $27.95 × 3 = ...	$83.85
Mileage cost: $0.20 × 468 = ...	93.60
Gasoline cost:	21.70
Collision waiver: $10.00 × 3 = ...	$30.00
Total cost	$229.15

B. Find the **cost per mile.**

Total Cost ÷ Number of Miles Driven
$229.15 ÷ 468 = $0.489 = $0.49 cost per mile

27.95 [×] 3 [=] 83.85 [M+] .2 [×] 468 [=] 93.6 [M+] 21.7 [M+] 10 [×] 3 [=] 30 [M+]
[RM] 229.15 [÷] 468 [=] .4896

✓ SELF-CHECK Complete the problems, then check your answers in the back of the book.

Winona Simms rented a compact car for $30.00 for 4 days plus 22¢ per mile. Winona drove 430 miles and spent $18.90 on gasoline.

1. Find the total cost.
$120.00
94.60
+ 18.90
$233.50

2. Find the cost per mile.
$233.50 ÷ 430 = $0.543

LESSON PLAN
8-7 Renting An Automobile

FOCUS

Motivate this lesson by asking students if any of their families have ever rented a car either while on vacation or while the family car was being repaired. The focus of this lesson is to compute the cost per mile of renting an automobile.

TEACH

Go over each of the four components of the total cost of the Example. Emphasize that the insurance is charged per day. Some rental companies offer a flat fee, then charge an additional per-mile fee if you drive over a set limit of miles. The cost of renting an automobile varies from agency to agency. It also varies on the type of car, with subcompacts renting for the least and luxury cars for the most. Most agencies require either a deposit or a valid credit card.

COOPERATIVE LEARNING

Have students work in small groups to solve Problem 24. Discussion should include why the cost of gas can be $45 no matter which of the two trucks is used.

Warm-Up Exercises

1. $79 × 4 $316
2. $15 × 8 $120
3. 24¢ × 500 $120
4. $420 ÷ 8000 $0.0525

PRACTICE AND APPLY

The following problems can be assigned for classwork and the answers checked in class to help students master the objective of the lesson.

- Guided Practice: 1–4, 7–9, 12
- Independent Practice: 5, 6, 10–15

PROBLEMS

Solve. Find the cost per mile to the nearest cent.

	(Total Daily Cost	+	Total Mileage Cost	+	Gasoline Cost	=	Total Cost)	÷	Miles Driven	=	Cost per Mile
3.	($ 60.00	+	$45.00	+	$11.88	=	$116.88)	÷	300	=	$0.39
4.	($160.00	+	$94.00	+	$23.89	=	$277.89)	÷	500	=	$0.56
5.	($119.95	+	$74.40	+	$30.53	=	$224.88)	÷	620	=	$0.36
6.	($159.95	+	$63.55	+	$41.67	=	$265.17)	÷	420	=	$0.63

Solve. Find the cost per mile to the nearest cent. (No mileage charge)

	(Rental Cost	+	Gasoline Cost	=	Total Cost)	÷	Miles Driven	=	Cost per Mile
7.	($ 66.50	+	$10.39	=	$76.89)	÷	240	=	$0.32
8.	($ 96.00	+	$25.98	=	$121.98)	÷	300	=	$0.41
9.	($154.75	+	$28.27	=	$183.02)	÷	476	=	$0.38
10.	($ 89.95	+	$41.38	=	$131.33)	÷	518	=	$0.25
11.	($199.95	+	$74.39	=	$274.34)	÷	646	=	$0.42

12. Wo Chen rented a subcompact.
Cost $28 a day for 4 days.
Drove 520 miles at 33¢ a mile.
Gasoline cost $20.78.
What was the total cost? **$304.38**
What was the cost per mile? **$0.59**

13. Fred Bardi rented a sedan.
Cost $32 a day for 5 days.
Drove 1100 miles at 37¢ a mile.
Gasoline cost $54.95. **$621.95**
What was the total cost? **$0.57**
What was the cost per mile?

14. A. H. Wise rented a standard car.
Cost $159.95 for a week.
Gasoline cost $29.95.
Drove 415 miles.
What was the total cost? **$189.90**
What was the cost per mile? **$0.46**

15. Abrah Bahta rented a luxury car.
Cost $129.97 for a week.
Gasoline cost $17.90. **$147.87**
Drove 270 miles. **$0.55**
What was the total cost?
What was the cost per mile?

16. Sanchez Corado rented a sedan for 4 days at $38.00 a day plus 30¢ a mile. He purchased the collision waiver for $8.50 per day. He drove 420 miles and paid $19.64 for gasoline. What was the total cost of renting the car? What was the cost per mile? **$331.64; $0.79**

17. Freda Rochetti rented a midsize car for 2 days at $32.95 a day plus 33¢ a mile. She purchased the collision waiver for $7.50 per day. She drove 140 miles and paid $8.39 for gasoline. What was the total cost of renting the car? What was the cost per mile? **$135.49; $0.97**

18. Ace Car Rentals has a weekly rate of $129.95 for economy cars and no mileage charge. An economy car is driven 450 miles and uses $22.90 in gasoline. What is the cost per mile? **$0.34**

ALTERNATIVE ASSESSMENT

Have students call the nearest rental agency to find out the cost involved for renting a standard car for 3 weekdays. Ask for the estimated gas mileage for the car. Using the current price for gas, have students find the cost of renting this car. Do not include a collision waiver.

19. Riser Car Rentals has a weekly rate of $159.95 for station wagons and a mileage charge of 30¢ a mile with 100 free miles. A station wagon is driven 432 miles and uses $28.09 in gasoline. What is the cost per mile? **$0.67**

20. Jessica Vick is renting a subcompact car while her own car is being repaired. The subcompact rents for $99.95 a week with a mileage charge of 34¢ a mile with 75 free miles per week. She is going to rent the car for 2 weeks and expects to drive it about 460 miles. She estimates that the gasoline will cost about $20. What is the total cost of renting the car? What is the cost per mile? **$325.30; $0.71**

21. Wilma Wallace decides to rent a van for her vacation. The van rents for $62 a day or $249 a week with no mileage charge. She will use the van for 8 days and will drive it about 1500 miles. She estimates that gasoline will cost about $180. What will be the total cost to rent the van? How much will it cost per mile? **$491; $0.33**

22. Marty Collins and Paula Green plan to rent a luxury sedan for the prom. They will need the car for 1 day and will drive it about 60 miles. The car rents for $49.95 a day plus 30¢ a mile. They estimate that gasoline will cost about $10. If they share equally in the costs, how much will it cost each of them to rent the sedan? **$38.98**

23. The LaGuardias plan to fly to their vacation spot and then rent a car to tour the island. The vacation package they signed specifies that a sedan will be available for $49.90 a day with no charge for mileage. What will it cost the LaGuardias to rent the car for 5 days if they spend $28.35 for gasoline and $8.50 a day for the collision waiver? What will it cost per mile if they drive about 420 miles? **$320.35; $0.76**

24. Alice Coopersmith is moving to a new home. She can rent a 14-foot panel truck for $40 a day plus 26¢ a mile, or an 18-foot truck for $55 a day plus 29¢ a mile. It would take 4 trips to make the move in the 14-foot truck, but only 3 trips in the larger truck. She estimates that gasoline would cost about $45 for either truck. A round-trip to her new home is 60 miles. Regardless of the number of trips, Alice would need the truck for only 1 day. To save on expenses, which size truck should Alice rent? How much would it cost per mile to rent each truck? **14-foot truck; 14 ft = $0.61; 18 ft = $0.85**

MAINTAINING YOUR SKILLS Look up the skills in parentheses if you need help or more practice.

Add. **(Skill 5)**

25. 7.94 + 34.67 + 86.75 + 378.99 **508.35**

26. 86.03 + 8.75 + 94.01 + 378.19 **566.98**

27. 44.64 + 29.899 + 35.080 + 4.219 **113.838**

28. 39.37 + 58.35 + 19.70 + 88.01 **205.43**

Divide. Round answers to the nearest hundredth. **(Skill 11)**

29. 762.20 ÷ 32 **23.82**
30. 684.26 ÷ 42.2 **16.21**
31. 502.00 ÷ 361.9 **1.39**
32. 801.83 ÷ 15.5 **51.73**
33. 582.19 ÷ 395 **1.47**
34. 207.89 ÷ 385 **0.54**

WRAP-UP

Ask students to give the components that go into the total cost of renting an automobile. Ask what they would then do to find the cost per mile of renting an automobile.

Assignment Guide

- Basic: 5, 6, 10–22, 25–34
- Average: 16–24, 26–34 even

ALTERNATIVE STRATEGIES: Enrichment

Have students research travel magazines or the travel section of a newspaper to find advertisements for renting a car. Have them note the various rates for different parts of the country and report their findings to the class. They could also contact a travel agent for special deals.

LESSON PLAN
Reviewing the Basics

The exercises on this page review skills, applications, and terms used in the unit. You can use the exercises to assess informally students' proficiency with this material.

The page can be used for guided practice and independent practice. You can work through a selection of the exercises together with students, and thus see immediately if they know how to do them, and you can then assign some of the exercises for independent practice. Be sure to go over the answers to all assigned exercises.

Reviewing the Basics

Skills

Round to the nearest cent.

(Skill 2)

1. $312.7539 **$312.75**
2. $0.4462 **$0.45**
3. $223.1091 **$223.11**
4. $0.2981 **$0.30**

Solve. Round answers to the nearest cent.

(Skill 5)

5. $14.36 + $89.09 **$103.45**
6. $253.41 + $12.29 **$265.70**
7. $14 + $15.75 + $39.79 **$69.54**

(Skill 6)

8. $56.33 − $39.59 **$16.74**
9. $426.08 − $229.78 **$196.30**
10. $19 − $9.88 **$9.12**

(Skill 8)

11. $41.60 × 3.15 **$131.04**
12. $54.42 × 6.5 **$353.73**
13. $236.55 × 4.10 **$969.86**
14. $312.25 × 3.10 **$967.98**
15. $124.10 × 0.95 **$117.90**
16. $120.50 × 2.50 **$301.25**

(Skill 11)

17. $1280 ÷ 9500 **$0.13**
18. $3766 ÷ 7740 **$0.49**
19. $1532.35 ÷ 14,000 **$0.11**

(Skill 30)

20. 75% of $3600 **$2700**
21. 81% of $5180 **$4195.80**
22. 92% of $4575.24 **$4209.22**

Applications

Use the table to answer the following.

(Application C)

23. What is the average trade-in value of a G30 custom pickup, model A90, with a manual transmission? **$4200**
24. What is the average retail value of a G30 deluxe pickup, model A95, with power steering, air-conditioning, and 4-wheel drive? **$7050**

Av'g. Trd.-in	BODY TYPE Model	Av'g. Loan	Av'g. Ret'l.
	G30 Series E —V8 1-Tn AT		
4325	Customed Wdbed A90	3900	5300
5150	Deluxe Wdbed A95	4650	6175
500	Add 4-wheel drive	450	600
150	Add air-cond.	150	200
150	Add SS equip.	150	200
850	Add camper sp.	600	1000
50	Add power steering	50	75
125	Deduct man. trans.	125	150

Terms

Use each term in one of the sentences.

e 25. Sticker price

c 26. Rent

a 27. Used-car guide

d 28. Deductible clause

b 29. Depreciation

a. You can find price information on used cars that were purchased from dealers during the previous month in a ___?___.

b. A decrease in the value of your car because of its age or condition is called ___?___.

c. If you do not own an automobile but need to use one on occasion, you can ___?___. from a rental agency.

d. The ___?___ in your insurance policy states that you must pay a portion of any repair bill.

e. The total of the base price, options price, and destination charge is the ___?___.

Unit Test

Lesson 8-1

1. Duwayn Archer sees a mini van that has a base price of $13,785. Factory-installed options total $1732. There is a destination charge of $440. What is the sticker price of the mini van? **$15,957**

Lesson 8-2

2. Nancy Clark is interested in a 4-wheel drive pickup that has a base price of $11,050, options totaling $1349, and a destination charge of $390. She has read that the dealer's cost is about 82% of the base price and 80% of the price of the options. What is the estimated dealer's cost that Nancy should keep in mind when making an offer? **$10,530**

Lesson 8-3

3. August LaVoy owns a used Sport Van model WC21 with automatic transmission, a V8 engine, AM/FM stereo, and no air-conditioning. What is the average retail price that August should keep in mind when pricing her car for sale? **$13,475**

Av'g. Trd.-in	BODY TYPE Model	Av'g. Loan	Av'g. Ret'l.
	STARCEST VAN C10-V8-AT		
11025	Sport Van WC11	9925	12550
12000	Panel Van WC21	10800	13575
12200	Sport Van WC21	11000	13875
100	Add AM/FM Stereo	100	150
75	Add Cruise Control	75	100
550	Deduct W/o Air-Cond.	500	550

Lesson 8-4

4. Beth Houghton uses her car to drive to and from work. Her insurance coverage includes 50/100 bodily injury and $25,000 property damage. Her driver-rating factor is 1.15. She has $50-deductible comprehensive and $50-deductible collision insurance coverage on her car. Her car is in age group B and rating group 12. What is her annual base premium? What is her annual premium? **$556.80; $640.32**

Property Damage Limits	Bodily Injury Limits		
	25/50	25/100	50/100
$ 25,000	$206.40	$218.80	$213.20
50,000	212.40	224.80	237.20
100,000	220.80	233.20	245.60

Age Group	Insurance-Rating Group		
	12	13	14
	Comprehensive $50-Deductible		
A	$95.20	$108.00	$122.00
B	90.40	102.40	115.60
	Collision $50-Deductible		
A	$266.80	$287.20	$307.60
B	253.20	272.40	291.60

Lesson 8-5

5. Nora Hovey drove her car 13,220 miles last year. Her records show that fixed costs totaled $1290.60 and variable costs totaled $1940.40. How much did it cost per mile for Nora to operate her car last year? **$0.24**

Lesson 8-6

6. Nate Ruoff leased a GMC truck for his flower shop. He pays $299 per month for 48 months. His deposit was $300. He paid a $60 title fee and a $44 license fee. What is the total lease cost? **$14,756**

Lesson 8-7

7. Jen-Shiang Hong flew to Toronto on vacation. While there, he rented a car for 3 days at $35.95 a day plus 22¢ a kilometer. He drove the car 630 kilometers. Gasoline cost $32.60. How much did it cost per kilometer for him to rent the car? **$0.44**

LESSON PLAN
Unit Test

Students should do the Unit Test on their own. Each problem on the test is keyed to a lesson in the unit. Students having difficulty with any particular problem should review the Example in the appropriate lesson and be assigned some of the Independent Practice problems for additional practice.

LESSON PLAN
A Spreadsheet Application

USING TECHNOLOGY

Students can work in *cooperative groups* on this application. Before they start to work, you might wish to review the meanings of the different types of coverage. Ask students what the rating factor depends on. (age, marital status, the amount driven each week, and so on)

After all groups have finished the problems, go over the solutions and answer any questions students may have.

If possible, bring a copy of an actual computer printout of an auto insurance premium statement to class. Allow time for students to examine this document.

A SPREADSHEET APPLICATION

Auto Insurance

To complete this spreadsheet application, you will need the diskette *Spreadsheet Applications for Business Mathematics,* which accompanies this textbook. Follow the directions in the *User's Guide* to complete this activity.

Input the information in the following problems to find the insurance costs and to compare various amounts of coverage.

You will need to use the premium table on page 246 to determine the base premium. All problems are $50-deductible comprehensive and collision.

1. Driver-rating factor: 1.50.
Age, rating group: A, 12.
Property damage: $25,000.
Bodily injury: $25,000/$50,000.
What is the annual base premium?
What is the annual premium?

2. Driver-rating factor: 2.20.
Age, rating group: D, 12.
Property damage: $50,000.
Bodily injury: $25,000/$100,000.
What is the annual base premium?
What is the annual premium?

3. Driver-rating factor: 1.65.
Age, rating group: B, 15.
Property damage: $50,000.
Bodily injury: $100,000/$200,000.
What is the annual base premium?
What is the annual premium?

4. Driver-rating factor: 2.65.
Age, rating group: C, 15.
Property damage: $100,000.
Bodily injury: $100,000/$300,000.
What is the annual base premium?
What is the annual premium?

5. Michelle Ross is an occasional operator of the family car and drives mainly for pleasure. She has $25,000 property damage and 50/100 bodily injury coverage. Her driver-rating factor is 1.70, and the car is classified C, 14. What is her annual base premium? What is her annual premium?

6. Suppose Michelle in problem 5 becomes the principal operator and her rating factor is now 2.20. What is her annual premium? How much more does she have to pay?

7. Jeremy Gobans is an occasional operator of the family car and drives mainly for pleasure. He has $50,000 property damage and 100/200 bodily injury coverage. His driver-rating factor is 2.65, and the car is classified C, 10. What is his annual base premium? What is his annual premium?

8. Suppose Jeremy in problem 7 becomes the principal operator and his rating factor is now 4.10. What is his annual premium? How much more does he have to pay?

9. An unmarried female operator, Jenny Shockey, age 19, is the occasional operator of her father's farm Jeep. Her rating factor is 2.0. She has $25,000 property damage and 25/50 bodily injury coverage. The 1989 Jeep is classified C, 13. What is her annual base premium? What is her annual premium?

A SIMULATION

Car Expenses

According to the American Automobile Association, the average car owner spends approximately $3500 a year in operating and maintaining an automobile. The costs of owning a car include fixed costs, such as finance charges, depreciation, insurance, and license fees. You also have variable costs, such as gasoline, maintenance, repairs, parking fees, and tolls. As a car owner, you may want to calculate the cost per mile. If you drive to work, you may want to know the cost per day.

1. Suppose you drove your 4-year-old car 12,600 miles and the car gets 26 miles per gallon. How many gallons of gasoline did you use? At $1.229 per gallon, how much will it cost? **484.62 gal; $595.60**
2. In addition to gasoline, your variable costs include: 5 oil changes at $29.50 each, $89.00 for a tune-up, $225 for brake repairs, $185 for motor repairs, $49.40 for a tire, and $117 for parking and tolls. Your fixed costs include depreciation of $900 and an $85.50 registration fee. What is the total of these expenses? **$1798.40**
3. The base premium for your automobile insurance is $570. Assume you are unmarried, under 21, the principal operator, and drive for pleasure. Use the table on page 263 to find your driver-rating factor. What will your annual insurance premium be? **Female: $1254; Male: $2337**
4. What is the total annual cost for gasoline, maintenance, repairs, tires, parking, depreciation, registration, and insurance? **Female: $3648; Male: $4731**
5. What is your cost per mile? **Female: $0.29; Male: $0.38**
6. Suppose you drive 12 miles round-trip to work each day, 5 days a week, 50 weeks per year. How much does it cost you to drive to work each day? each week? each year? **Female: $3.48, $17.40, $870; Male: $4.56, $22.80, $1140**

Find the cost per mile and the cost to drive to work for each of these cases. Use your driver-rating factor from question 3.

	7.	8.	9.	10.
Annual Miles	12,150	11,950	13,700	21,300
Miles per Gallon	40	25	27	15
Cost per Gallon	$0.999	$1.059	$1.189	$1.299
Variable Cost	$690	$460	$330	$850
Fixed Cost	$1200	$600	$1400	$1200
Base Insurance Premium	$520	$660	$790	$672
Cost per Mile	F; M: 27¢; 36¢	25¢; 36¢	30¢; 41¢	25¢; 31¢
Daily Miles	45	30	26	63
Cost per Day to Go to Work	F: $12.15 M: $16.20	$7.50 $10.80	$7.80 $10.66	$15.75 $19.53

LESSON PLAN
Car Expenses: A Simulation

USING THE SIMULATION

Students can work on this simulation individually, or you may wish to assign it to *cooperative learning* groups. You can begin by reading with students the introductory material at the top of the page. Although many students want to own a car, few of them understand the costs involved, nor do they have the money to carry these costs.

After students have finished the simulation, provide them with the answers so they can check their work. Answer questions and provide assistance as required.

A SIMULATION (CONTINUED)

Purchase Price and Financing

You have decided to buy a car. One choice you have to make is whether to buy a new car or a used car. A new car has a higher purchase price and greater depreciation. Comprehensive and collision insurance are more expensive for a new car. A used car can be expected to have higher maintenance and repair costs.

There are two cars you are considering: a new Volney and a used Jess. To help choose which one to buy, you decide to calculate how much it would cost per month to run each car for the next three years.

CODE	DESCRIPTION	LIST PRICE
PH2	VOLNEY 2-DOOR HATCHBACK	8975.00
P36	SUNSET RED METALLIC	145.00
R58	AM/FM STEREO TAPE DECK	475.00
SA4	ADJUSTABLE SEATS	260.00
AT2	AUTOMATIC TRANSMISSION	490.00
V31	AIR-CONDITIONING	750.00
829	DESTINATION CHARGE	425.00

11. A portion of the sticker for the Volney is shown. What is the total sticker price? **$11,520**

12. The dealer's cost for the Volney is about 90% of the base price and 85% of the price of the options. If the dealer accepts your offer of $200 over the dealer's cost, what would the Volney cost you, including destination charge and 6% sales tax? **$11,134.77**

13. You would make a $2000 down payment for the Volney and finance the rest of the cost with a simple interest installment loan at 12% for 48 months.

a. What is the amount financed? **$9134.77**

b. Use the table on page 222 to find the monthly payment. What is the total amount repaid? What is the finance charge? **$240.24; $11,531.32; $2396.75**

14. The used-car guide entry for the Jess is shown. You are interested in a 2-door hardtop with power windows, AM/FM stereo with tape deck, and cruise control. If you bought the Jess for its average retail value, what would it cost, including 6% sales tax? **$5406**

Av'g. Trd.-in	BODY TYPE Model	Av'g. Loan	Av'g. Ret'l.
	JESS — Series C JB LT		
3875	H'dtp 2 Dr	3500	4800
4275	H'dtp "S"	3850	5225
125	Add AM/FM Stereo/Tp	125	125
75	Add Power Door Locks	75	75
100	Add Power Windows	100	100
100	Add Power Seats	100	100
75	Add Cruise Control	75	75

15. You would make a $2000 down payment for the Jess and finance the rest of the cost at 18% for 36 months. Using the table on page 222, find the monthly payment and the finance charge. **$123.30; $1032.80**

A SIMULATION
(CONTINUED)

Insurance

In order to choose an automobile insurance company, you have read articles in consumer and automobile magazines, talked with relatives and friends, and asked insurance companies about their coverage and premiums. The tables below show the base premiums of the company you have chosen.

Property Damage Limits	Bodily Injury Limits		
	25/100	50/100	100/200
$ 25,000	$218.80	$213.20	$252.00
50,000	224.80	237.20	258.00
100,000	233.20	245.60	266.40

Age Group	Insurance-Rating Group 10	11	12
	Comprehensive $50-Deductible		
A	$76.80	$81.60	$95.20
B	65.20	77.60	90.40
	Collision $50-Deductible		
A	$225.60	$246.00	$266.80
B	214.00	233.20	253.20

16. For either car you choose, you plan to have 50/100 bodily injury and $50,000 property damage coverage. What is the annual base premium? **$237.20**

17. You plan to have $50-deductible comprehensive and $50-deductible collision insurance coverage. The Volney is in age group A and rating group 12. What is the annual base premium? **$362.00**

18. The Jess is in age group B and rating group 10. What is the annual base premium for comprehensive and collision insurance? **$279.20**

19. Use the table below to find your driver-rating factor. Assume you are unmarried, under 21, the principal operator, and drive for pleasure. For each car, what is the total cost per month for insurance?
Female: Volney $109.85 Jess $94.67; Male: Volney $204.73, Jess $176.44

DRIVER-RATING FACTORS FOR AUTOMOBILE INSURANCE

Driver Classification	Multiple of Base Premium				
	Under 21 Years Occasional Operator	Under 21 Years Principal Operator	21 – 24 Years Occasional Operator	21 – 24 Years Principal Operator	25 – 29 Years Principal Operator
Unmarried youthful operators					
Females					
Pleasure	1.70	2.20	1.30	1.65	Classify
Farm auto	1.30	1.65	0.95	1.25	as Adult
Males					
Pleasure	2.65	4.10	1.80	2.75	1.65
Farm auto	2.00	3.05	1.35	2.05	1.30

A SIMULATION
(CONTINUED)

Maintenance, Depreciation, and Other Costs

A car depreciates in value most rapidly when it is new. You can estimate how a car will depreciate by looking at the values of older cars of the same model. Depreciation is often expressed as a percent of the purchase price.

Maintenance and repair costs tend to increase as a car ages. You can estimate these costs by talking with mechanics and with people who own the same model. Consumer and automobile magazines sometimes have articles comparing these costs for various models.

20. The Volney should lose 18% of its purchase price (less tax) the first year, 15% of its purchase price (less tax) the second year, and 12% the third year. What is the total percent? How much less will the Volney be worth in three years? What is the average depreciation per month? **45%; $4727.025; $131.31**

21. The Jess can be expected to have a value of $2094 in three years. What is the depreciation per month? **$83.50**

22. For the Volney, you expect to spend $100 the first year, $160 the second year, and $480 the third year for maintenance and repairs. What is the average cost per month? **$20.56**

23. You expect the Jess to have maintenance and repair costs of $440 the first year, $560 the second year, and $750 the third year. It will also need tires for $250 plus 6% sales tax. What is the average cost per month? **$55.97**

24. You plan to drive 12,000 miles per year. Gasoline costs $1.259 per gallon. The Volney gets 30 miles per gallon. What is the cost of gasoline per month? **$41.97**

25. The Jess gets 25 miles per gallon. What is the cost of gasoline per month? **$50.36**

26. Make a table like this one to compare the monthly costs of the cars.

	Monthly cost		Volney	Jess
Fixed Costs	Finance charge		$49.93	$28.69
	Insurance	F M	$109.85 $204.73	$94.67 $176.44
	Depreciation		$131.31	$83.50
	License and registration		$3.90	$3.90
Variable Costs	Gasoline		$41.97	$50.36
	Maintenance, repairs, tires		$20.56	$55.97
	TOTAL	F M	$357.52 $452.40	$317.09 $398.86
	Cost per Mile	F M	$0.36 $0.45	$0.32 $0.40

U N I T

9

Housing Costs

Most people finance the cost of a home by taking out a *mortgage* loan. You pay a portion of the selling price and then finance the remaining amount. Most mortgages must be repaid in equal monthly installments that include the amount financed (principal) and the interest. Some monthly payments include a monthly payment for *real estate taxes* and *homeowner's insurance.* Real estate taxes are paid to the city or county for schools, roads, and other expenses. Homeowner's insurance pays for loss by fire, theft of contents, and property damage. Other housing costs include costs for *utilities,* such as heating fuel, water, and electricity.

O U T L I N E

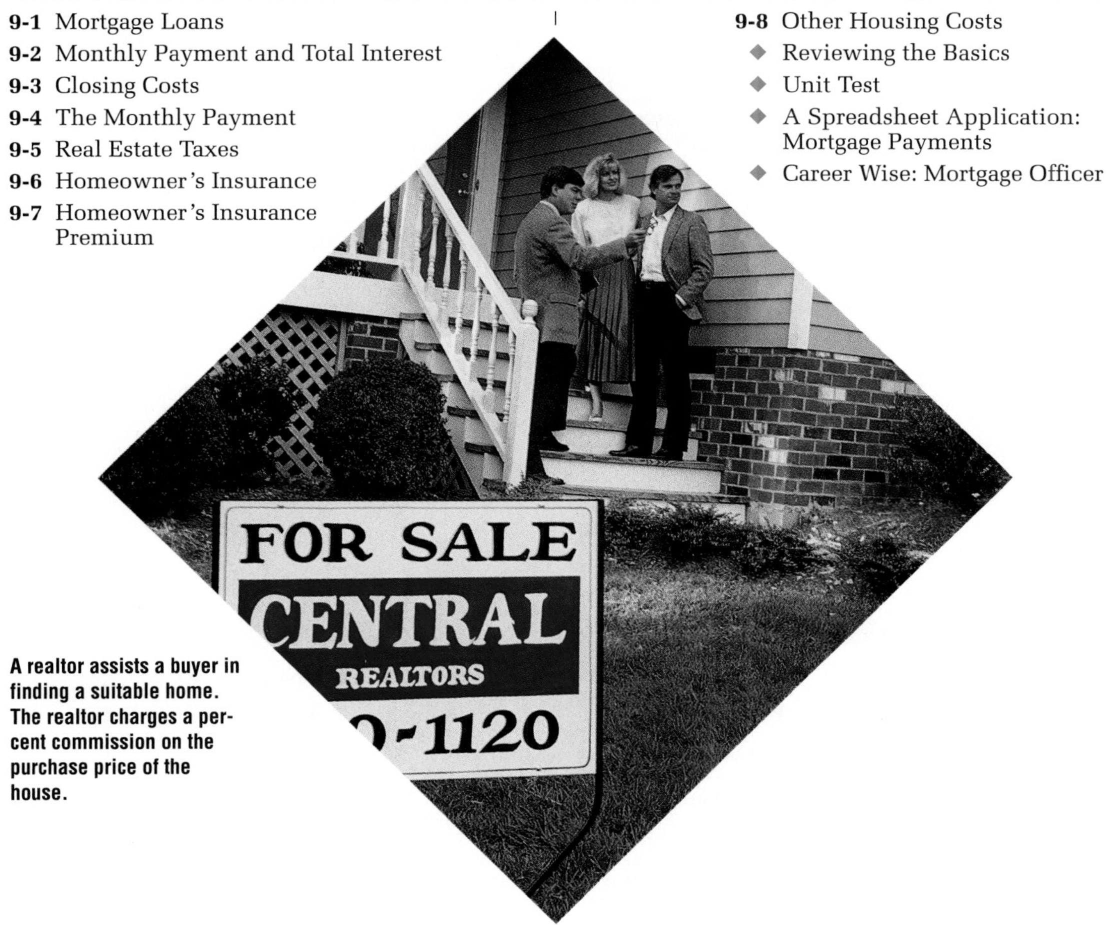

A realtor assists a buyer in finding a suitable home. The realtor charges a percent commission on the purchase price of the house.

UNIT 9
HOUSING COSTS

INTRODUCING THE UNIT

To introduce this unit, ask students to name the different costs that accompany buying and maintaining a home. Relate the ways in which we depend on our knowledge of mathematics to make important decisions about where we live and how to allocate our monthly pay to support our homes.

LESSON PLAN
9-1 Mortgage Loans

FOCUS

Bring to class the real estate section of your local newspaper. Read a variety of advertisements of houses for sale. Select one home in the $100,000 range, and ask students if they know anyone who could pay for the house in cash. Explain that very few homes today are purchased outright. Instead, most people need to borrow money. In this lesson, students will compute the amount that is borrowed, which is called a mortgage loan.

TEACH

When discussing the Example, point out to students that the complement of the down payment of 15% is 85%, which can be used to find the mortgage amount. Write on the chalkboard: 85% of $140,000 is __?__. Ask a volunteer to work the problem at the chalkboard. Suggest that students check their answers to mortgage amounts by using the complement method.

Mention that there are a variety of mortgage loans available (such as VA and FHA), and that lending institutions vary in their requirements for a mortgage.

Warm-Up Exercises

1. 20% of $40,000 $8000
2. 25% of $30,000 $7500
3. $60,000 − $12,000 $48,000
4. $32,500 − $8125 $24,375

9-1 Mortgage Loans

OBJECTIVE
Compute the mortgage loan amount.

When you purchase a home, you will probably make a down payment and finance the remaining portion of the selling price with a **mortgage loan** from a bank or savings and loan association. Generally, the down payment is between 10% and 40% of the selling price. The mortgage loan is usually repaid with interest in equal monthly payments. The mortgage gives the bank the right to sell the property if you fail to make the payments.

Mortgage Loan Amount = Selling Price − Down Payment

EXAMPLE *Skills* 30, 4, 6 *Application* A *Term* Mortgage loan

Jessica and Kirk Cramer are considering the purchase of a new home for $140,000. A 15% down payment is required. What is the amount of the mortgage loan needed to finance the purchase?

SOLUTION

A. Find the **down payment.**
$140,000 × 15% = $21,000 down payment

B. Find the **mortgage loan amount.**
Selling Price − Down Payment
$140,000 − $21,000 = $119,000 mortgage loan amount

140000 [×] 15 [%] 21000 140000 [−] 21000 [=] 119000

✓ SELF-CHECK Complete the problems, then check your answers in the back of the book.

Find the down payment and the amount of the mortgage.

1. $80,000 selling price, 25% down. **$20,000; $60,000**
2. $200,000 selling price, 30% down. **$60,000; $140,000**

PROBLEMS

	Buyer	Selling Price	Percent Down Payment	Down Payment	Mortgage Loan Amount
3.	Jeff Wilson	$ 87,000	20%	$17,400	**$69,600**
4.	Beth Ivanoski	$ 62,500	15%	**$9,375**	**$53,125**
5.	David Ellison	$ 98,800	25%	**$24,700**	**$74,100**
6.	Sherrie DeQuine	$156,000	18%	**$28,080**	**$127,920**

CRITICAL THINKING

Explain to students that the first step in buying a home is to determine whether or not you can afford it. The rule of thumb is that you should not buy a home that is more than $2\frac{1}{2}$ times your annual income. Extend the Example of the lesson by including the fact that the Cramers combined annual income is $46,500. Using the rule of thumb, can the Cramers afford the home they want to purchase? (No) What is the maximum price they should pay for a home? ($116,250)

F.Y.I.
In 1992, the average price of a new home was $107,000.

7. Marleen Belton.
Home priced at $48,500.
20% down payment required.
What is the down payment?
What is the mortgage loan amount? **$9700; $38,800**

8. Cliff and Megan Sammour.
Condominium priced at $75,000.
30% down payment required.
What is the down payment?
What is the mortgage loan amount? **$22,500; $52,500**

9. David and Peggy Chin.
Mobile home priced at $27,400.
Made a 20% down payment.
What is the mortgage loan amount? **$21,920**

10. Alvira and Berry Fukunaga.
Home priced at $280,000.
Made a 40% down payment.
What is the mortgage loan amount? **$168,000**

11. John and Maria Leivowitz decide that the home advertised is ideally suited to the needs of their family. They wish to make a 15% down payment and finance the remaining amount through their bank. What is the amount of their mortgage loan? **$195,415**

A home with character, charm, and personality, 4 bedrooms, 3 full baths, beautiful vaulted living room ceiling, deluxe kitchen, large master bedroom suite, big family room.

At $229,900—it's a steal!

12. Rita and Alfred Johnson offered $85,000 for a home that had been priced at $90,000. After some negotiating, the Johnsons and the seller agreed on a selling price of $86,500. What is the amount of the mortgage loan if they make a 20% down payment? **$69,200**

13. Richard Darman offered $112,500 for a home that had been priced at $125,000. The seller agreed to the offer. A bank is willing to finance the purchase if he can make a down payment of 25%. What is the amount of his mortgage loan? **$84,375**

14. The Hasbros have decided to purchase a duplex. They would use the rental income from one part of the duplex to help meet the mortgage payments. A selling price of $100,800 was agreed upon. What is the amount of the mortgage loan if a down payment of $22\frac{1}{2}$% is required? **$78,120**

15. Dan and Sue Willingham have saved $14,000 for a down payment on their future home. Their bank has informed them that the minimum down payment required to obtain a mortgage loan is 20%. What is the most that they can spend for a home and expect to receive bank approval for their loan? **$70,000**

MAINTAINING YOUR SKILLS Look up the skills in parentheses if you need help or more practice.

Find the percentage. **(Skill 30)**

16. 15% of 40,000 **6000**
17. 25% of 84,000 **21,000**
18. 20% of 90,000 **18,000**
19. 40% of 175,000 **70,000**
20. 22% of 190,000 **41,800**
21. 5% of 425,500 **21,275**

Subtract. **(Skill 4)**

22. 40,000 − 8400 **31,600**
23. 94,000 − 18,000 **76,000**
24. 180,000 − 36,000 **144,000**
25. 98,500 − 12,250 **86,250**
26. 31,000 − 3100 **27,900**
27. 125,500 − 25,100 **100,400**

PRACTICE AND APPLY
The following problems can be assigned for classwork and the answers checked in class to help students master the objective of the lesson.

- Guided Practice: 1–6
- Independent Practice: 7–11 odd

WRAP-UP
Read one of the real estate advertisements from the newspaper. Have students compute the down payment required and the mortgage amount using different percentages for the down payment.

Assignment Guide
- Basic: 7–14, 16–27
- Average: 8–12 even, 13–15, 16–26 even

ALTERNATIVE STRATEGIES: Reteaching
Finding the dollar amount of the down payment when given the percent is simply an application of finding the percentage. Review these exercises with your students:

	Cash Price	Percent Down	Amount Down	Mortgage Amount
1)	$ 80,000	20%	($16,000)	($ 64,000)
2)	50,000	10%	($ 5,000)	($ 45,000)
3)	160,000	25%	($40,000)	($120,000)
4)	200,000	15%	($30,000)	($170,000)

LESSON PLAN
9-2 Monthly Payment and Total Interest

FOCUS

When applying for a mortgage loan, it is smart to shop around for the lowest rate available. Ask students why people shop for a low rate. In this lesson, the answer to this question will be clear after students learn how to compute the total payment, amount paid, and total interest charged.

TEACH

After working through the Example, students will be surprised at the total amount paid for the loan. Point out that the total amount paid is over $3\frac{1}{2}$ times the amount of the original loan of $80,000.

Although the difference in a monthly payment for a loan having an interest rate of 11% and one having a 12% rate may seem relatively small, when the amount and time period of a loan are great, the difference can be substantial. Have students compute the loan of the Example using an 11% interest rate. Ask them also to find the amount saved if the interest rate were 11% rather than 12%. ($22,000)

Warm-Up Exercises

1. $10.67 × 42 $448.14
2. $448.14 × 20 × 12 $107,553.60
3. 300 × $384.84 $115,452
4. (120 × $450) − $30,000 $24,000

9-2 Monthly Payment and Total Interest

OBJECTIVE

Compute the monthly payment, total amount paid, and total interest charged.

Banks that make mortgage loans charge **interest** for the use of their money. The interest rate will vary from bank to bank, so it pays to shop around. If you know the annual interest rate, the amount of the loan, and the length of the loan, you can use a table to find the monthly payment, the total amount paid, and the total interest charged.

MONTHLY PAYMENT FOR A $1000 LOAN

Annual Interest Rate	Length of Loan (Years) 20	25	30
10.00%	$ 9.66	$ 9.09	$ 8.78
10.50%	9.99	9.45	9.15
11.00%	10.33	9.81	9.53
11.50%	10.67	10.17	9.91
12.00%	11.02	10.54	10.29
12.50%	11.37	10.91	10.68
13.00%	11.72	11.28	11.07
13.50%	12.08	11.66	11.46

$$\text{Monthly Payment} = \frac{\text{Amount of Mortgage}}{\$1000} \times \text{Monthly Payment for a \$1000 Loan}$$

Amount Paid = Monthly Payment × Number of Payments

Total Interest Charged = Amount Paid − Amount of Mortgage

EXAMPLE *Skills* 8, 6 *Application* C *Term* Interest

Carol and Adam Burke have applied for an $80,000 mortgage loan at an annual interest rate of 12.00%. The loan is for a period of 30 years and will be paid in equal monthly payments that include interest. What is the total amount of interest charged?

SOLUTION

A. Find the **monthly payment.** (Refer to the table above.)

$$\frac{\text{Amount of Mortgage}}{\$1000} \times \text{Monthly Payment for a \$1000 Loan}$$

$$\frac{\$80{,}000.00}{\$1000.00} \times \$10.29 = \$823.20 \text{ monthly payment}$$

B. Find the **amount paid.**

Monthly payment × Number of Months
$823.20 × (12 months × 30 years)
$823.20 × 360 = $296,352.00 amount paid

C. Find the **total interest charged.**

Amount Paid − Amount of Mortgage
$296,352.00 − $80,000.00 = $216,352.00 total interest

80000 ÷ 1000 × 10.29 = 823.2 × 12 × 30 = 296352 − 80000 = 216352

✓ SELF-CHECK Complete the problem, then check your answer in the back of the book.

$869.40; $208,656; $118,656

1. Find the monthly payment, amount paid, and interest charged for a $90,000 mortgage loan at an annual interest rate of 10% for 20 years.

CULTURAL ANGLES

As a research project, assign students one of the following countries and ask them to find what percent of the total population lives in different kinds of housing, such as private homes, apartment buildings, mobile homes, or condominiums: Argentina, Brazil, Canada, Egypt, France, Greece, Japan, Korea, Mexico, Poland, Sweden, United States. When all the data have been compiled, prepare a chart summarizing the results.

PROBLEMS

Use the table on page 268 to solve.

	2.	3.	4.	5.	6.
Mortgage	$50,000	$70,000	$95,000	$225,000	$395,000
Years	20	25	30	20	30
Rate	10.00%	12.00%	11.50%	13.00%	13.50%
Payment	**$483**	**$737.80**	**$941.45**	**$2637**	**$4526.70**
Amount Paid	**$115,920**	**$221,340**	**$338,922**	**$632,880**	**$1,629,612**
Total Interest	**$65,920**	**$151,340**	**$243,922**	**$407,880**	**$1,234,612**

Use the table on page 645 to solve.

7. $784.80; $235,440; $155,440
8. $870.80; $208,922; $138,992

7. Charles and Sandy Compton.
$80,000 mortgage.
Terms: 11% for 25 years.
What is the monthly payment?
What is the total amount paid?
What is the total interest charged?

8. Abigail and Karlis Krisjanis.
$70,000 mortgage.
Terms: 14% for 20 years.
What is the monthly payment?
What is the total amount paid?
What is the total interest charged?

9. Julie Hardy. **$209,640**
$150,000 mortgage.
Terms: 10.5% for 20 years.
What is the total interest charged?

10. Diane Novak. **$108,040**
$50,000 mortgage.
Terms: 10% for 30 years.
What is the total interest charged?

> F.Y.I.
> The national average mortgage rate dropped to a 15-year low of 9% in December of 1991. It was 10.18% in December of 1990.

11. Ivan and Vicki Egan have obtained a $60,000 mortgage loan at an annual interest rate of 12% for 15 years. What is the monthly payment? What is the total amount to be paid? **$720.60; $129,708**

12. Ellen and Clyde Perez reached an agreed upon price of $124,000 with the owner for the purchase of a house. They made a down payment of $14,000 and could finance the remaining amount in one of two ways: at 11.5% for 25 years or at 12% for 20 years. Which mortgage results in a larger amount of interest paid? How much greater?
11.5% for 25 years; $44,682

13. How much can be saved in total interest by financing $90,000 at 10% for 15 years rather than 20 years? **$34,506**

MAINTAINING YOUR SKILLS
Look up the skills in parentheses if you need help or more practice.

Multiply. **(Skill 8)**

14. 24 × 120.50 **2892** **15.** 36 × 431.2 **15,523.2** **16.** 12 × 832.40 **9988.8** **17.** 15 × 342.20 **5133**

Subtract. **(Skill 6)**

18. 75,500 − 22,200 **53,300** **19.** 92,461 − 12,420 **80,041** **20.** 453,821.50 − 100,000 **353,821.5**

PRACTICE AND APPLY
The following problems can be assigned for classwork and the answers checked in class to help students master the objective of the lesson.

- Guided Practice: 1–4
- Independent Practice: 5, 6, 8, 9

WRAP-UP
Ask students to name the three facts needed in order to find the amount of the monthly payment in the table. (mortgage amount, number of years of the mortgage, and the interest rate)

Assignment Guide
- Basic: 5–12, 14–20
- Average: 7, 10–13, 14–20 even

ALTERNATIVE STRATEGIES: Enrichment
In a given home purchase situation there are three ways of "saving" money on the total interest paid. These are 1) make a larger down payment; 2) find a lower interest rate; or 3) take the mortgage for a shorter number of years. Given: $120,000 selling price; 20% down payment; 10.50% annual interest rate for 30 years. How much total interest could be "saved" if you: (1) made a 25% down payment? (2) found a 10.00% mortgage? (3) took the mortgage for 25 years?

LESSON PLAN
9-3 Closing Costs

FOCUS
Present the following situation to the class. Last month, Marla and Glen Carleone found a $100,000 home they wanted to buy. They had the money for the 20% down payment and also found a bank to give them the mortgage loan. Today, they are going to the bank to actually get ownership of the house, which is called *closing day.* The Carleones brought their check for the down payment, as well as another check for $3135, which covers the bank's fees for the closing of the sale. This lesson will explain how their closing costs were computed.

TEACH
Explain to students that lenders, such as banks, can have different fees for closing costs. Just as it is important to shop around for a mortgage with low interest rates, it is also important to find a lender whose fees for closing costs are low.

Briefly explain some of the fees shown in the Example. Point out that the "Loan Origination Fee," the greatest amount, is paid to the bank to cover its administrative cost of processing the loan. Attorney fees, which cover the cost of the bank's lawyer for preparing or examining documents associated with the sale, can vary widely.

9-3
Closing Costs

OBJECTIVE
Compute the total closing cost.

At the time you sign the documents transferring ownership of your new home, you must pay any **closing costs** that the bank charges. The closing costs may include fees for lawyers, credit checks and title searches, taxes, and the preparation of the documents. Some banks charge a flat fee regardless of the amount of the loan. Some banks charge a percent of the amount of the loan. Other banks charge itemized fees at the closing.

Closing Costs = Sum of Bank Fees

EXAMPLE *Skills* 30, 3 *Application* A *Term* Closing costs

Marla and Glen Carleone have been granted an $80,000 mortgage loan. When they sign the papers to purchase their new home, they will have to pay the closing costs shown. What is the total of the closing costs?

Credit report:	$45
Loan origination fee:	2% of mortgage loan
Abstract of title:	$120
Attorney fee:	$250
Documentation stamp:	0.3% of mortgage
Processing fee:	1.10% of mortgage

SOLUTION

Find the **sum of the bank fees.**

a. Credit report	$ 45
b. Loan origination $80,000 x 2%	1600
c. Abstract of title	120
d. Attorney fee	250
e. Documentation stamp: $80,000 × 0.3%	240
f. Processing fee: $80,000 × 1.10%	880
Total closing cost	$3135

45 M+ *80000* × *2* % *1600* M+ *120* M+ *250* M+ *80000* × *.3* % *240* M+ *80000* × *1.1* % *880* M+ RM *3135*

✓ SELF-CHECK Complete the problem, then check your answer in the back of the book.

1. Use the list of closing costs in the example above to find the total closing cost on a $60,000 mortgage.
$45 + $1200 + $120 + $250 + $180 + $660 = $2455

PROBLEMS

Use the list of closing costs in the example above to solve.

2. Jeremy Roberts.
Mortgage loan of $50,000.
What is the total closing cost?
$2115

3. Vincent and Sue Hemsley.
Mortgage loan of $95,000.
What is the total closing cost?
$3645

BUSINESS PROJECT
Have small groups of students investigate and write a report on closing costs for a home. Some sources are banks, real estate offices, or friends or relatives in the real estate field. Suggested topics are: (1) listing of the various closing costs; (2) how a specific fee is computed; or (3) the purpose or description of a closing cost not discussed in class.

4. Ralph and Cristi Sheen. Mortgage loan of $271,000. What is the total closing cost? **$9629**

5. Jack and Dina King. Mortgage loan of $420,000. What is the total closing cost? **$14,695**

6. Jude and Rose McDermott are financing $39,700 at 11.5% for 30 years. They must pay these closing costs. What is the total of the closing costs? **$1520.50**

Appraisal fee	$250
Credit report	$45
Title search	$380
Service fee	1.5% of mortgage
Legal fees	$250

F.Y.I.
The percent of the mortgage paid to the bank up front averages 1.7% nationwide (USA TODAY, November 13, 1991).

7. Barry and Ella Ellerbee have agreed to purchase a house for $96,500. Kenmore Savings and Loan Association is willing to lend the money at 12.25% for 25 years provided they can make a $10,000 down payment. The total closing cost is 3.25% of the amount of the mortgage loan. What is their closing cost? **$2811.25**

8. Rene and Jefferson Franklin are interested in purchasing a $60,000 home. They plan to make a 20% down payment and finance the remaining amount through Peabody Savings Association. Peabody Savings has these closing costs: credit report, $30; appraisal report, $255; title insurance, $190; survey and photographs, $225; recording fee, $45; legal fees, $280; and property taxes, $389. If the loan is approved, how much cash will the Franklins need to secure the loan, including the down payment? **$13,414**

Critical Thinking . . .

9. Many states have usury laws that set a legal maximum interest rate that lending agencies can charge. When the trend in interest rates is above the maximum allowed, many lending agencies will add points to their basic mortgage charges. A **point** is 1% of the mortgage loan and is included in the closing costs. Five points is a one-time charge of 5% of the mortgage loan.

Jean and Jim Couric have agreed to purchase a $92,800 home. They plan to make a 25% down payment. Beacon Hill Bank is willing to finance the mortgage at 10.25% for 30 years plus 2 points. How much cash will they need to secure the loan, including the down payment? **$26,407**

Title search:	$155
Appraisal fee:	$200
Credit report:	$55
Legal fee:	$375
Property tax:	$750
Fire insurance:	$280

MAINTAINING YOUR SKILLS Look up the skill in parentheses if you need help or more practice.

Find the percentage. **(Skill 30)**

10. 15% of 9000 **1350**
11. 4% of 86,000 **3440**
12. 7% of 252,000 **17,640**
13. 2.4% of 78,000 **1872**
14. 3.3% of 83,000 **2739**
15. 1.2% of 30,000 **360**
16. 0.3% of 92,000 **276**
17. 0.15% of 85,100 **127.65**
18. 0.04% of 22,300 **8.92**
19. 1% of 81,500 **815**
20. 1.4% of 150,000 **2100**
21. 0.61% of 283,300 **1728.13**

Warm-Up Exercises

1. $70 + $900 + $60.50 + $115.80 + $240 + $751 $2137.30
2. 2% of $20,000 $400
3. 0.5% of $40,000 $200
4. 3.5% of $30,000 $1050

PRACTICE AND APPLY

The following problems can be assigned for classwork and the answers checked in class to help students master the objective of the lesson.

- Guided Practice: 1–3
- Independent Practice: 4–8 even

WRAP-UP

Review the solution to Problem 8, having individual students give their answers to the three computations: amount of down payment, total closing cost, and the sum of the down payment and closing costs.

Assignment Guide

- Basic: 4–8, 10–21
- Average: 5–9 odd, 10–20 even

ALTERNATIVE STRATEGIES: Enrichment

Every other Sunday most local newspapers will publish what they call a "Mortgage Watch" or "Sampling the best mortgage rates". These tables will list the name of the lender, the size of the loan in percent (80% being most common), the interest rate, and the points required. Have your students find such a report and, using the information from Problem 9, do a few problems comparing a lower rate mortgage that requires a few more points.

LESSON PLAN 9-4 The Monthly Payment

FOCUS

To motivate this lesson on how a monthly mortgage payment is allocated toward the principal and the interest, refer back to the Example of Lesson 9-2. Remind students that the interest charged is included in the monthly payment. Ask if they think the interest portion will be greater than, equal to, or less than the portion toward the mortgage amount. Why? Recall the fact that the total interest charged in that example on the $80,000 mortgage was $216,352, a great deal more than the mortgage itself.

TEACH

As you go over the chart shown on this page, ask students what they notice about the "amount for principal" and "amount for interest" columns. (The amount for principal increases as the amount for interest decreases.)

Explain that during the early years of a mortgage, the payment toward the original mortgage amount is very small. If you should have to sell the house in one year, you will most likely lose money. When calculating the payments of the Example for 1 year, the new balance, which has to paid to the bank if the house is sold, is $79,452.12. Only $547.88 was paid toward the principal.

9-4 The Monthly Payment

OBJECTIVE

Compute the allocation of monthly payment toward principal, interest, and the new principal.

Most mortgage loans are repaid in equal monthly payments. Each payment includes an amount for payment of interest and an amount for payment of the principal of the loan. The amount of interest is calculated using the simple interest formula (I = P × R × T). The amount of principal that you owe decreases with each payment that you make. The chart shows the interest and principal paid in the first 4 months of an $80,000 mortgage loan.

$80,000 MORTGAGE LOAN AT 12.00% FOR 25 YEARS

Payment Number	Monthly Payment	Amount for Interest	Amount for Principal	New Balance
1	$843.20	$800.00	$43.20	$79,956.80
2	843.20	799.57	43.63	79,913.17
3	843.20	799.13	44.07	79,869.10
4	843.20	798.69	44.51	79,824.59

Payment to Principal = Monthly Payment − Interest

New Principal = Previous Balance − Payment to Principal

EXAMPLE *Skills* 6, 8, 30 *Application* A *Term* Principal

Rod and Carey Finn obtained a 30-year, $80,000 mortgage loan from First Bank and Trust. The interest rate is 12%. Their monthly payment is $823.20. For the first payment, what is the interest? What is the payment to principal? What is the new principal?

SOLUTION

A. Find the **interest.**

Principal × Rate × Time

$80,000.00 × 12% × $\frac{1}{12}$ = $800.00 interest

B. Find the **payment to principal.**

Monthly Payment − Interest

$823.20 − $800.00 = $23.20 payment to principal

C. Find the **new principal.**

Previous Balance − Payment to Principal

$80,000.00 − $23.20 = $79,976.80 new principal

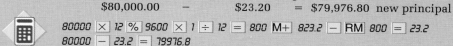

80000 × 12 % 9600 × 1 ÷ 12 = 800 M+ 823.2 − RM 800 = 23.2

80000 − 23.2 = 79976.8

✓ SELF-CHECK Complete the problems, then check your answers in the back of the book.

In the example above, the new principal is $79,976.80. For the second payment, find the:

1. Interest on $79,976.80. **$799.77**
2. Payment to principal. **$23.43**
3. New balance. **$79,953.37**

CRITICAL THINKING

Lenders usually will not allow monthly mortgage payments to exceed 20% to 25% of the borrower's monthly income. Ask students why they think this is a good policy. (A family will have other living costs, such as food, utilities, car payments, taxes, clothing, medical bills, and so on.)

PROBLEMS

	Mortgage Amount	Interest Rate	First Monthly Payment	Amount for Interest	Amount for Principal	New Principal
4.	$ 50,000	10%	$ 483.00	$416.67	$66.33	**$49,933.67**
5.	$ 70,000	12%	$ 737.80	**$700**	**$37.80**	**$69,962.20**
6.	$ 60,000	13%	$ 676.80	**$650**	**$26.80**	**$59,973.20**
7.	$120,000	14%	$1492.80	**$1400**	**$92.80**	**$119,907.20**
8.	$225,000	13%	$2637.00	**$2437.50**	**$199.50**	**$224,800.50**

9. Lois Larczyk.
Mortgage loan of $46,000.
Interest rate is 12%.
Monthly payment is $506.92.
How much of the first monthly payment is for interest? **$460**

10. Patrick Yunker.
Mortgage loan of $84,000.
Interest rate is 10%.
Monthly payment is $714.
How much of the first monthly payment is for interest? **$700**

11. Matthew Roberts.
Mortgage loan of $38,600.
Interest rate is 12.5%.
Monthly payment is $438.88.
How much of the first monthly payment is for interest? **$402.08**
How much of the first payment is for principal? **$36.80**
What is the new principal?
$38,563.20

12. Dee Pollom.
Mortgage loan of $98,000.
Interest rate is 14.5%.
Monthly payment is $1254.40.
How much of the first monthly payment is for interest? **$1184.17**
How much of the first payment is for principal? **$70.23**
What is the new principal?
$97,929.77

13. Jill Beyley obtained a 25-year, $60,000 mortgage loan from Peoples Savings and Loan Association. The interest rate is 12%. The monthly payment is $632.40. For the first payment, what is the interest? What is the payment to principal? What is the new balance?
$600; $32.40; $59,967.60

14. Norman Foster obtained a 30-year, $180,000 mortgage loan from First Federal Savings and Loan Association. The interest rate is 15%. His monthly payment is $2277. For the first payment, what is the interest? What is the payment to principal? What is the new balance?
$2250; $27; $179,973

15. Amelia McGuire obtained a 20-year, $36,000 mortgage loan from Society Trust Company. The interest rate is 11.5%. Her monthly payment is $384.12. For the first payment, what is the interest? What is the payment to principal? What is the new balance?
$345; $39.12; $35,960.88

16. Ken Burris obtained a 15-year, $88,500 mortgage loan from Swancreek Trust Company. The interest rate is 10.5%. His monthly payment is $978.81. For the first payment, what is the interest? What is the payment to principal? What is the new balance?
$774.38; $204.43; $88,295.57

During that same year, however, you have paid $9570.51 in interest, as well as additional monies for closing costs. Would it have been better to put the $20,000 used for the down payment in a savings account?

Warm-Up Exercises

1. $40,000 − $379.79
   $39,620.21
2. $12,000 − $360
   $11,640
3. $42,000 × 10%
   $4200
4. $21,000 × 15%
   $3150
5. $5875 ÷ 12 $489.58
6. $6768 ÷ 12 $564

PRACTICE AND APPLY

The following problems can be assigned for classwork and the answers checked in class to help students master the objective of the lesson.

- Guided Practice: 1–9
- Independent Practice: 10–14 even, 17

ALTERNATIVE STRATEGIES: Reteaching

Setting up an Amortization Table or Repayment Schedule is exactly like working with a Simple Interest Installment Loan (SIIL). Exactly! Compute the interest on the unpaid balance, subtract it from the payment and apply the rest to reduce the balance. Walk through the calculations necessary to set up the table for a $100,000 mortgage at 12% for 25 years.

WRAP-UP

Call on volunteers to write on the chalkboard the three formulas used in the lesson. Then compute the fifth payment of the Example using the formulas.

Assignment Guide

- Basic: 10–19, 22–27
- Average: 11–15 odd, 16, 18–21, 22–26 even

Refer to the table on page 645 for problems 17-21.

17. Emily Johnson obtained a 25-year, $82,000 mortgage loan from the Miners and Merchants Deposit Company. The interest rate is 11.5%. What is her monthly payment? For the first payment, what is the interest? What is the payment to principal? What is the new balance? **$833.94; $785.83; $48.11; $81,951.89**

18. Hollis McMahn obtained a 15-year, $90,500 mortgage loan from Pacific Trust Company. The interest rate is 10.5%. What is the monthly payment? For the first payment, what is the interest? What is the payment to principal? What is the new balance? **$1000.93; $791.88; $209.05; $90,290.95**

19. Bill and Julie Johnson purchased a home for $672,400. They made a 20% down payment and financed the remaining amount at 11% for 30 years. What is the monthly payment? How much of the first monthly payment is used to reduce the principal? **$5126.38; $195.45**

Rated X! X-tra nice, X-tra large master bedroom with dressing room, X-tra sized screen porch, X-tra space for a garden. Split plan, 3 bedroom, 2 bth, dbl. garage. X-cellent value too. $164,900.

20. Ed and Cathy Brehm purchased the home advertised. They made a down payment of $24,900 and financed the remaining amount at 12.5% for 30 years. What is the monthly payment? What is the new principal after the first monthly payment? **$1495.20; $139,963.13**

21. Jeff and Rita Contreras purchased a condominium for $67,600. They made a 15% down payment and financed the remaining amount at 11.5% for 20 years. They have made 167 payments to date.

a. Use this portion of the repayment schedule to find the remaining debt after the 170th payment.

Payment Number	Monthly Payment	Amount for Interest	Amount for Principal	New Balance
168	613.10	$306.04	$307.06	$31,627.99
169	613.10	**$303.10**	**$310.00**	**$31,317.99**
170	613.10	**$300.13**	**$312.97**	**$31,005.02**

F.Y.I.
It takes 22+ years for payment to principal to equal payment to interest on a 30-year mortgage.

b. Use this portion of the repayment schedule to find the last payment.

Payment Number	Monthly Payment	Amount for Interest	Amount for Principal	New Balance
238	613.10	$14.46	$593.64	$909.71
239	613.10	**$8.72**	**$604.38**	**$305.33**
240	**$308.26**	**$2.93**	**$305.33**	**0**

MAINTAINING YOUR SKILLS Look up the skills in parentheses if you need help or more practice.

Subtract. **(Skill 6)**

22. 48,000 − 29.46 **47,970.54**
23. 91,800 − 39.55 **91,760.45**
24. 24,400 − 23.12 **24,376.88**
25. 78,902 − 22.98 **78,879.02**
26. 18,185 − 45.11 **18,139.89**
27. 14,915 − 107.7 **14,807.3**

PROBLEM SOLVING

Before students work Problem 17, give them the data in the problem and refer them to the table on page 645. Ask: After the first payment on the mortgage, what is the new balance? Rather than solving the problem, have students work in cooperative groups to restate the problem to follow a series of simpler problems. For example, the first and second problems would be: (1) What is the monthly payment? (2) How much of the first monthly payment is for interest?

9-5

Real Estate Taxes

OBJECTIVE
Compute the assessed value and taxes of real estate.

When you own a home, you will have to pay city or county real estate taxes. The money collected is used to operate and maintain roads, parks, schools, government offices, and so on. The amount of real estate tax that you pay in one year depends on the assessed value of your property and the tax rate. The assessed value is found by multiplying the market value of your property by the rate of assessment. The market value is the price at which a house can be bought or sold. It is determined by an assessor hired by the municipality. The rate of assessment is a percent.

The tax rate is sometimes expressed in mills per dollar of valuation. A mill is $0.001. A tax rate of 80 mills is a tax rate of $80 per $1000 of assessed value. When working with mills, it is often convenient to express mills in dollars by dividing by 1000.

Assessed Value = Market Value × Rate of Assessment

Real Estate Tax = Tax Rate × Assessed Value

EXAMPLE *Skills* 8, 30 *Application* A, *Term* Real estate taxes

The Ottawa County tax assessor stated that the market value of the Courtland Farm is $340,000. The rate of assessment in Ottawa County is 40% of market value. The tax rate is 84.32 mills. What is the real estate tax on the Courtland Farm?

SOLUTION

A. Find the **assessed value.**
Market Value × Rate of Assessment
$340,000 × 40% = $136,000 assessed value

B. Express the **tax rate** as a decimal.
84.32 mills ÷ 1000 = 0.08432 tax rate

C. Find the **real estate tax.**
Tax Rate × Assessed Value
0.08432 × $136,000.00 = $11,467.52 real estate tax

340000 × 40 % 136000 M+ 84.32 ÷ 1000 = .08432 × RM 136000 = 11467.52

✓ SELF-CHECK Complete the problems, then check your answers in the back of the book.

The tax rate is 65.5 mills, market value is $70,000, and rate of assessment is 40%. Find the following:

1. The tax rate as a decimal. **0.0655**
2. The assessed value. **$28,000**
3. The real estate tax. **$1834**

LESSON PLAN 9-5 Real Estate Taxes

FOCUS

Ask students to name some different taxes that people have to pay. Discuss what these taxes are based on (income, purchases) and how they are computed (tax table, taxable wages, and tax rate) Explain that if you own a home, you also will pay another tax. In this lesson, students will learn how to compute real estate taxes, often referred to as property taxes, based on the assessed value of a home.

TEACH

To ascertain that students understand the terms of the lesson, have them give the definition of each term as you work through the Solution of the Example.

Point out that a tax rate of 84.32 mills means the same as a tax rate of $84.32 per $1000 of assessed value. Demonstrate the following alternative method of finding the real estate tax in the Example.
$136,000 ÷ $1000 = 136;
136 × $84.32 = $11,467.52

Students might be interested to know that both interest payments on a mortgage and real estate taxes can be deducted when filing income tax returns. That is one advantage of owning a home.

Warm-Up Exercises

1. 86.72 ÷ 1000 0.08672
2. 75.4 ÷ 1000 0.0754
3. $50,000 × 40% $20,000
4. $30,000 × 50% $15,000
5. 0.075 × $20,000 $1500
6. 0.05 × $15,000 $750

ALTERNATIVE STRATEGIES: Reteaching

Finding the assessed value is another application of finding the percentage, just like finding the dollar amount of the down payment. To find the real estate tax, multiply the assessed value by the tax rate expressed as a decimal. To convert the tax rate in mills to a decimal, divide by 1000, or just move the decimal point three places left.

PRACTICE AND APPLY

The following problems can be assigned for classwork and the answers checked in class to help students master the objective of the lesson.

- Guided Practice: 1–3, 5, 7
- Independent Practice: 4–8 even

WRAP-UP

Ask students how the following statement can be restated: The tax rate is 48.65 mills. (The tax rate is $48.65 per $1000 of assessed value.) If the market value of a home is $150,000 and the rate of assessment is 50%, have students explain how to find the real estate tax.

Assignment Guide

- Basic: 4–12, 14–25
- Average: 9–13, 14–24 even

PROBLEMS

4. The M. L. D'Limas' home.
Market value is $72,000.
Rate of assessment is 30%.
What is the assessed value? **$21,600**

5. The Simms' condominium.
Market value is $59,800.
Rate of assessment is 40%.
What is the assessed value? **$23,920**

6. The Posadny Farm.
Market value is $142,000.
Rate of assessment is 35%.
Tax rate is $86 per $1000.
What is the real estate tax? **$4274.20**

7. The Masons' home.
Market value is $36,000.
Rate of assessment is 50%.
Tax rate is 72.35 mills.
What is the real estate tax? **$1302.30**

8. Brenda Roth's home is located in a community where the rate of assessment is 50% of market value. The tax rate is $85 per $1000 of assessed value. Her home has a market value of $92,000. What is its assessed value? What is the property tax? **$46,000; $3910**

9. The rate of assessment in Fulton County is 35%. The tax rate is $72.25 per $1000 of assessed value. What is the real estate tax on a piece of property that has a market value of $236,000? **$5967.85**

10. Ali and Jackie Erwin live in a locality where the tax rate is 71.385 mills. The rate of assessment is 60%. The property that the Erwins own has a market value of $98,400. What is their real estate tax for a year? **$4214.57**

11. Jose and Trudy Engstrom own a home that has a market value of $675,000. They live in an area where the rate of assessment is 45% and the tax rate is 48.535 mills. What is the annual real estate tax? **$14,742.51**

12. Timothy Oakland owns a mobile home and the property on which it sits. His mobile home has a market value of $24,000 and his property has a market value of $13,000. Timothy lives in a county where the rate of assessment is 100% and the tax rate is $32.85 per $1000 of assessed value. What is his yearly real estate tax? **$1215.45**

13. Gina and Tony Jasinski received a tax statement showing that their land has an assessed value of $7500 and their buildings have an assessed value of $42,300. The rate of assessment in their locality is 40%. What is the market value of their property? **$124,500**

MAINTAINING YOUR SKILLS Look up the skills in parentheses if you need help or more practice.

Multiply. **(Skill 8)**

14. 37.3 × 78.4 **2924.32**
15. 13.18 × 9.42 **124.1556**
16. 13.3 × 4.37 **58.121**
17. 0.856 × 4.38 **3.74928**
18. 83.92 × 44.9 **3768.008**
19. 0.274 × 8356 **2289.544**

Find the percentage. **(Skill 30)**

20. 90,000 × 20% **18,000**
21. 140,000 × 45% **63,000**
22. 45,500 × 33% **15,015**
23. 328,800 × 92% **302,496**
24. 78,710 × 38% **29,909.8**
25. 49,500 × 70% **34,650**

MATHEMATICAL NOTES

Students may need to be reminded of how to divide decimals by 1000. For example, to express 84.32 mills as a decimal, move the decimal point to the left by the number of zeros in 1000. Since there are only two digits to the left of the decimal point in 84.32, a zero needs to be written in front of the 8 in the tenths place. Also, for a number less than 1, a zero is written in the ones place: 84.32 ÷ 1000 = 0.08432.

9-6

Homeowner's Insurance

OBJECTIVE
Compute the amount of coverage.

When you own a home, you will probably purchase **homeowner's insurance** to protect yourself and your home from losses due to fire, theft of contents, and personal liability. Homeowner's insurance also includes **loss-of-use coverage** to pay for some of the expenses of living away from home while damage to your home is being repaired. **Personal liability** and **medical coverage** protect you from financial losses if someone is injured on your property.

To receive full payment for any loss up to the amount of the policy, you must insure your home for at least 80% of its replacement value. Some companies may require you to insure your home for 90% to 100% of its replacement value. The **replacement value** is the amount required to reconstruct your home if it is destroyed. Insurance companies use the amount of coverage on your home to calculate the amount of coverage you receive on your garage, personal property, and for loss of use. Many companies use these percents of the amount of coverage on the home for each type of protection.

Coverage	Percent of Coverage
Personal property	50%
Loss of use	20%
Garage and other structures	10%

Amount of Coverage = Amount of Coverage on Home × Percent

EXAMPLE *Skill* 30 *Application* A *Term* Homeowner's insurance

The replacement value of Joy and Ron Amodeo's home is estimated at $94,000. They have insured their home for 80% of its replacement value. According to the guidelines above, what is the amount of coverage on the Amodeos' personal property?

SOLUTION

A. Find the **amount of coverage on home.**
$94,000 × 80% = $75,200 coverage on home

B. Find the **amount of coverage on personal property.**
Amount of Coverage on Home × Percent
$75,200 × 50% = $37,600 coverage on personal property

94000 [×] 80 [%] 75200 [×] 50 [%] 37600

✓ SELF-CHECK Complete the problems, then check your answers in the back of the book.

A home is insured for 90% of its replacement value of $120,000, or $108,000. Using the percents in the table above, find the coverage for:

1. Personal property. **$54,000**
2. Loss of use. **$21,600**
3. Garage. **$10,800**

LESSON PLAN
9-6 Homeowner's Insurance

FOCUS

Ask students if anyone has ever witnessed a house burning or seen the damage fire can do to the contents of a home. How can the owners of a home damaged by a fire replace the home and their personal belongings? In this lesson, students will learn about homeowner's insurance and how to compute the amount of coverage.

TEACH

Stress to students that the key computation in the Example is finding the amount of coverage on a home, which is a percent of the replacement value. The answer is the base for all other coverages.

Point out to students that if the home in the Example had damage by fire that cost $75,200 or less to repair, then the insurance company would pay the full amount. On the other hand, if the family spent $16,400 on expenses for living in an apartment while the home was being repaired, the insurance company would only pay 20% of $75,200, which is $15,040.

If you have a homeowner's insurance policy available, bring it to class. Students may be interested in learning what the policy covers and what it does not cover.

Warm-Up Exercises

1. 80% of $100,000 $80,000
2. 50% of $90,000 $45,000
3. 80% of $75,000 $60,000
4. 10% of $41,000 $4100

ALTERNATIVE STRATEGIES: Enrichment

There are many different types of homeowner's insurance as well as insurance for renters, condominium owners, and the like. Within homeowner's insurance there are basic or deluxe policies; policies which name the perils covered and those that name the perils that are excluded; policies that cover floods, earthquakes, hurricanes, and so on. Have your students examine their family insurance policy, or speak to an agent about various types of coverage, and report to the class.

PRACTICE AND APPLY

The following problems can be assigned for classwork and the answers checked in class to help students master the objective of the lesson.

- Guided Practice: 1–3, 5, 7
- Independent Practice: 4–8 even

WRAP-UP

Discuss with students why it would be more advantageous to insure a home for 90% or 100% of its replacement value rather than 80%.

Assignment Guide

- Basic: 4–10, 12–17
- Average: 5–9 odd, 10, 11, 13–17 odd

PROBLEMS

Use the table on page 277 to find the percent of coverage.

4. The Bahrs' home.
Replacement value is $70,000.
80% coverage on home. **$56,000**
What is the amount of insurance?

5. The Loos' duplex.
Replacement value is $95,000.
100% coverage on home. **$95,000**
What is the amount of insurance?

6. The Callahans' home.
Replacement value is $38,500.
90% coverage on home. **$34,650**
What is the amount of insurance?
What is the amount of coverage for personal property? **$17,325**

7. The Cloyds' home.
Replacement value is $144,000.
80% coverage on home. **$115,200**
What is the amount of insurance?
What is the amount of coverage for loss of use? **$23,040**

8. The Courtney family is purchasing a home that has a replacement value of $324,000. They wish to insure their new home for 90% of its replacement value. What is the amount of insurance on their home? What is the amount of coverage on their garage? **$291,600; $29,160**

9. The Frankels recently purchased a home that has a replacement value of $324,000. They insured their new home for 80% of its replacement value. What is the amount of insurance on their home? What is the amount of coverage for personal property? loss of use? garage? **$259,200; $129,600; $51,840; $25,920**

The Jenson Insurance Company uses these guidelines for homeowner's insurance coverages. The policies must be written for 80% of a home's replacement value. Use the guidelines to solve the following.

JENSON INSURANCE COMPANY	
Coverage—Home 80% Insured	
Garage/structures	10%
Personal property	50%
Loss of use	20%
Personal liability	50%
Medical payments	1%

10. The McCauleys own a home that has a replacement value of $62,000. They insure it for 80% of its replacement value. What is the amount of coverage on their personal property? **$24,800**

11. Hugh O'Neill has a wood-frame home that has a replacement value of $470,000. He has insured it for 80% of its replacement value. What is the amount of coverage on his home? What is the amount of coverage for each type of protection listed in the guidelines? **$376,000; 37,600; $188,000; $75,200; $188,000; $3760**

MAINTAINING YOUR SKILLS Look up the skill in parentheses if you need help or more practice.

Find the percentage. **(Skill 30)**

12. 10% of $90,000 **$9000**

13. 80% of $30,000 **$24,000**

14. 50% of 140,000 **$70,000**

15. 15% of 420,000 **63,000**

16. 15% of 95,500 **14,325**

17. 90% of 544,000 **489,600**

BUSINESS NOTES

Most insurance companies offer renter's insurance policies for those who live in apartments. The policy is similar to a homeowner's insurance because it provides coverage on the loss of personal property (such as furniture and clothing) due to fire, theft, smoke, and so on.

9-7

Homeowner's Insurance Premium

OBJECTIVE
Compute the annual homeowner's insurance premium.

The amount of your homeowner's policy **premium** depends on the amount of insurance, the location of your property, and the type of construction of your home. Your **fire protection class** is a number that reflects the quality of fire protection available in your area.

ANNUAL PREMIUMS FOR A TYPICAL HOMEOWNER'S POLICY

Amount of Insurance Coverage	Brick/Masonry Veneer — Fire Protection Class					Wood Frame — Fire Protection Class				
	1–6	7–8	9	10	11	1–6	7–8	9	10	11
$ 40,000	$ 128	$ 131	$ 173	$ 182	$ 208	$ 137	$ 141	$ 182	$ 191	$ 219
45,000	133	137	173	182	208	144	147	191	200	229
50,000	137	141	185	195	223	146	150	195	204	234
60,000	147	151	199	210	241	158	162	210	221	252
70,000	164	166	219	230	264	173	178	230	242	277
80,000	185	191	252	264	303	198	204	264	279	319
90,000	206	212	281	295	339	222	228	295	310	357
100,000	229	236	313	328	377	246	253	328	345	396
120,000	272	280	372	391	449	293	301	391	411	472
150,000	353	362	481	505	581	379	389	505	532	611
200,000	474	487	647	680	782	509	523	680	716	823
250,000	567	580	739	785	898	600	614	785	835	956
300,000	676	693	882	937	1072	716	733	937	996	1141
400,000	785	804	1024	1087	1244	831	850	1087	1157	1325
500,000	1007	1031	1313	1394	1595	1065	1091	1394	1484	1699

EXAMPLE *Skill* 30 *Application* C *Term* Premium

The replacement value of Marcia Syke's home is $150,000. She has insured her home for 80% of its replacement value. The home is of wood-frame construction and has been rated in fire protection class 4. What is the annual premium?

SOLUTION

A. Find the **amount of coverage.**
$150,000 × 80% = $120,000 coverage on home

B. Find the **annual premium.** (Refer to the table above.)
$120,000 coverage, wood frame,
fire protection class 4 . . . . $293 annual premium

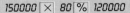

150000 [×] 80 [%] 120000

✓ SELF-CHECK Complete the problem, then check your answer in the back of the book.

1. A home is insured for $150,000. It has a brick/masonry veneer and is in fire protection class 9. Find the premium. **$481**

LESSON PLAN 9-7 Homeowner's Insurance Premiums

FOCUS

To motivate this lesson, recall with students the familiar childhood story of the three little pigs. The first pig built his house out of straw, the second pig built his out of sticks, and the third pig built his out of brick. When the wolf came to each house, he said, "I'll huff and I'll puff and I'll blow your house down." Ask students which of the three houses would they prefer to insure if employed by an insurance company? Point out that insurance companies consider the material used in the construction of a house to compute the premium.

TEACH

When reviewing the premium table, ask students to determine whether the lower numbers or higher numbers reflect a better fire protection class and how they came to their conclusion. (The lower numbers have lower premiums, which implies less risk for an insurance company.) See the Business Note below for descriptions of the different classes.

Students should recognize also that the premium is less for a brick house than for a wood frame house.

Warm-Up Exercises

1. 90% of $50,000
$45,000
2. 80% of $150,000
$120,000

ALTERNATIVE STRATEGIES: Reteaching

Walk through these problems with your students:

	Replacement Value	Percent of Coverage	Amount of Coverage	Protection Class	Frame/ Brick	Annual Premium
1)	$ 90,000	50%	($ 45,000)	7	Frame	($147)
2)	$150,000	80%	($120,000)	2	Brick	($272)
3)	80,000	75%	($ 60,000)	10	Frame	($221)

PRACTICE AND APPLY

The following problems can be assigned for classwork and the answers checked in class to help students master the objective of the lesson.

- Guided Practice: 1, 2, 4
- Independent Practice: 3, 5, 6

WRAP-UP

Ask individual students to name the different factors that one needs to know in order to insure a home. (replacement value, percent of coverage, amount of coverage, the type of construction of the home, and the fire protection class)

Assignment Guide

- Basic: 3, 5–8, 10–18
- Average: 7–9, 10–18 even

PROBLEMS

Use the table on page 279 to find the annual premium.

2. The Campbells' wood-frame house. $40,000 homeowner's policy. Fire protection class 8. **$141** What is the annual premium?

3. Kuen Yee Ngs's brick home. $80,000 homeowner's policy. Fire protection class 11. **$303** What is the annual premium?

4. The Norths' brick home. Replacement value of $100,000. Insured for 90%. Fire protection class 9. **$281** What is the annual premium?

5. The Wordeys' wood-frame home. Replacement value of $125,000. Insured for 96%. Fire protection class 5. **$293** What is the annual premium?

6. The Smiths own a wood-frame home in an area rated fire protection class 1. Their two-family home has a replacement value of $112,500 and is insured for 80%. What is their annual premium? **$222**

7. The Quicks own a brick home that has a replacement value of $375,000. They purchased a homeowner's policy for 80% of its replacement value. They live in an area rated fire protection class 9. What is their annual policy premium? **$882**

8. Carla Campodonico has insured her home for $300,000. The two-story, wood-frame house is located in an area rated fire protection class 11. What is the annual policy premium? **$1141**

9. When you obtain a mortgage loan, the bank may require you to include with your monthly principal and interest payment an amount equal to $\frac{1}{12}$ of the amounts needed to pay your real estate taxes and fire insurance premium. The bank places the money in an **escrow account.** When the real estate taxes and insurance premium are due, the bank uses the money from the escrow account to make the payments.

   Nelia and Gary Penn own a brick home with a market value and replacement value of $150,000. They insured their home for 100% of its replacement value. The Penns live in an area where the rate of assessment is 35%, the tax rate is 51.58 mills, and the fire protection is rated class 6. Their monthly principal and interest payment is $1581. How much is the monthly payment for principal, interest, real estate taxes, and insurance? **$1836.08**

MAINTAINING YOUR SKILLS Look up the skill in parentheses if you need help or more practice.

Find the percentage. **(Skill 30)**

10. 70% of 90,000 **63,000**
11. 80% of 140,000 **112,000**
12. 96% of 148,000 **142,080**
13. 50% of 70,000 **35,000**
14. 40% of 84,500 **33,800**
15. 15% of 71,400 **10,710**
16. 40% of 925,000 **370,000**
17. 20% of 345,500 **69,100**
18. 33.3% of 124,300 **41,391.9**

BUSINESS NOTE

The fire protection classes are usually defined as follows: Classes 1–6: Metropolitan area with good fire station; Classes 7–8: Rural area with fire hydrants; Class 9: Rural area without fire hydrants but with a good fire station within 6 miles; Class 10: rural area without fire hydrants and the fire station more than 6 miles away; Class 11: No fire department.

9-8 Other Housing Costs

OBJECTIVE
Compute the total housing cost and compare it with suggested guidelines.

In addition to your monthly mortgage payment, real estate taxes, and insurance payment, you will have expenses for utilities, maintenance, and home improvements. **Utilities costs** may include charges for electricity, gas, water, telephone, and heating fuel. The Federal Housing Administration (FHA) recommends that your total monthly housing cost be less than 35% of your monthly net pay.

EXAMPLE *Skills* 1, 5, 30 *Term* Utilities cost

Sue and Paul Kwan have a combined monthly take-home pay of $3320. They keep a record of their monthly housing expenses. The list of expenses for May is shown. Were their housing costs for May within the FHA guidelines?

Housing Expenses for May	
Mortgage payment	$698.24
Insurance ($303 ÷ 12)	25.25
Real estate taxes ($1885 ÷ 12)	157.08
Electricity	65.90
Heating fuel	54.20
Telephone	36.18
Water	26.20
Loan payment on oven	50.00
Repair storm door	38.68

SOLUTION

A. Find the **total monthly cost.**
Sum of expenses above = $1151.73

B. Find the **recommended maximum.**
$3320.00 × 35% = $1162.00

C. Compare. Is **total monthly cost** less than **recommended maximum**?
Is $1151.73 less than $1162.00? . . . . . Yes, the Kwans are within the guidelines.

698.24 [+] 25.25 [+] 157.08 [+] 65.9 [+] 54.2 [+] 36.18 [+] 26.2 [+] 50 [+] 38.68 [=] 1151.73 3320 [×] 35 [%] 1162

✓ SELF-CHECK Complete the problems, then check your answers in the back of the book.

Would the Kwans be within the FHA recommended guidelines if they had a combined monthly take-home pay of:

1. $3100 **No, $3100 × 35% = $1085** **2.** $3600 **Yes, $3600 × 35% = $1260**

PROBLEMS

	3.	4.	5.	6.	7.	8.
Monthly Net Pay	$1100	$3900	$880	$4284	$5439	$7942
Recommended FHA Maximum (round to $1)	**$385**	**$1365**	**$308**	**$1499**	**$1904**	**$2780**

LESSON PLAN 9-8 Other Housing Costs

FOCUS
Review with students the three costs of owning a home. (mortgage payments, real estate taxes, and homeowner's insurance) Ask students to name some other costs of owning a home (for example, utilities; repairs; purchases of large appliances, such as a refrigerator). Explain that usually all of these costs have to be paid each month. An important fact for a homeowner is to know that the total housing costs do not exceed the suggested guidelines based on monthly net income.

TEACH
Explain to students that monthly housing costs can be separated into two categories: (1) *fixed costs*, which are the same each month, and (2) *variable costs*, which can vary each month. Have students determine which of the costs in the Example are fixed (mortgage, taxes, and insurance) and which are variable (all others—the loan payment could also be considered fixed for the length of the loan). Ask why the utilities costs are variable. (They can change based on usage.)

USE OF CALCULATORS
When adding lists of numbers on a calculator, errors can be introduced by hitting the wrong key or forgetting to insert a decimal point. Suggest that students check their sums by entering the numbers again starting from the bottom.

Warm-Up Exercises

1. $500.00
   35.50
   26.50
   + 48.00
   $610.00
2. $480.00
   220.00
   65.45
   + 22.35
   $787.80
3. 35% of $600 $210
4. 35% of $1020 $357

PRACTICE AND APPLY

The following problems can be assigned for classwork and the answers checked in class to help students master the objective of the lesson.

- Guided Practice: 1–9
- Independent Practice: 10–14 even

9. Joshua and Peg Ryder.
Monthly net pay is $1980.
Housing expenses for June:

Mortgage payment	$328.65
Insurance	18.50
Real estate taxes	159.00
Electricity	55.44
Telephone service	44.98
Loan payment	82.35

Find the total housing cost.
Is it within the FHA recommendation? **$688.92; Yes, $693**

10. Frank and Yvette Shelby.
Monthly net pay is $2440.
Housing expenses for March:

Mortgage payment	$533.50
Insurance	19.75
Real estate taxes	132.40
Electricity	75.80
Telephone service	29.45
Water	22.00

Find the total housing cost.
Is it within the FHA recommendation? **$812.90; Yes, $854**

11. Manual and Irma Lopez.
Monthly net pay is $4300.
Housing expenses for July:

Mortgage payment	$932.80
Insurance	28.25
Real estate taxes	249.75
Utilities	232.40
Repairs	67.00
Loan payment	74.60

Find the total housing cost.
Is it within the FHA recommendation? **$1584.80; No, $1505**

12. Sally Pizzo.
Monthly net pay is $22,900.
Housing expenses for January:

Mortgage payment	$6174.00
Insurance	146.33
Real estate taxes	1100.00
Utilities	439.40
Repair driveway	240.00

Find the total housing cost.
Is it within the FHA recommendation? **$8099.73; No, $8015**

13. Fara Pinkston recorded her housing expenses for the month of August: mortgage payment, $347.90; insurance, $17; taxes, $84; electricity, $64.40; phone service, $33.50; fuel, $98.25; water, $17.44; and repairs, $79.87. Her monthly take-home pay is $1990. What is her total monthly cost? Is it within the FHA recommendation? **$742.36; No, $696.50**

14. Melvin Hayashi recorded his housing expenses for the month of December: mortgage payment of $548.36, $19.50 for insurance premium, $122.50 for real estate taxes, installment payment of $46.75 for refrigerator, $54.70 for electricity, $34.40 for telephone service, $86.70 for home heating oil, and $21.80 for water. His monthly take-home pay is $2500. What is his total monthly housing cost? Is it within the FHA recommendation? **$934.71; No, $875**

15. Eric Solomon has a monthly net income of $4400. He keeps a record of his monthly housing expenses. A list of expenses for January is shown. Were his housing costs for January within the FHA guidelines? **No, $1540; Total = $1605.73**

Housing Expenses for January	
Mortgage payment	$972.45
Insurance	30.00
Real estate taxes	216.00
Electricity	95.30
Heating fuel	122.50
Telephone	49.48
Water	35.00
Loan payment	85.00
Total	?

Water/sewer charges	$ 292.00
Electricity	940.00
Telephone service	345.30
Water heater	490.32
Repair storm windows	580.10
New air conditioner	1458.68
Replace gutters	760.00
New lawn mower	579.20
Total	?

16. David and Helen Voss have a combined monthly net income of $4750. Their records show that for last year they paid $10,789.20 in mortgage payments, $281 for insurance premiums, and $2085 in real estate taxes. In addition, they had the expenses shown. What was their average monthly housing cost for last year? Was it within the FHA recommendation? **$1550.07; Yes, $1662.50**

Use the table on page 645 to find the loan payment and the table on page 279 to find the insurance premium for the following.

Critical Thinking . . .

17. Molly and Chris Spaulding recently purchased a brick house for $150,000. They made a 20% down payment and financed the remaining amount at 12% for 30 years. The tax rate in their area is 71.57 mills and the rate of assessment is 40%. They purchased a homeowner's insurance policy for the purchase price of the house. The fire protection in the neighborhood is rated class 9. For the month of August, they recorded the following housing expenses: $69.20 for electricity, $44.85 for telephone service, $18.80 for water, and $74.65 to repair a door. They have a combined monthly net income of $5400. What is their monthly mortgage payment? What is the monthly insurance premium? What are their monthly taxes? What was their total monthly housing cost for August? Is it within the FHA recommendation?

Mort. = $1234.80;
Ins. = $40.08;
Taxes = $357.85;
Total = $1840.23;
FHA, Yes, $1890

WRAP-UP

Ask students to explain why it is a good idea to follow guidelines, such as FHA's on total housing costs and net income. (There are other living expenses.) Why is the guideline based on net income rather than gross income?

ALTERNATIVE STRATEGIES: Enrichment

Have your students identify all housing costs their family encounters. Ask them to further identify which costs remain fixed (F) and which vary from month to month (V). A partial list might include: mortgage P & I payment, (F); insurance payment to escrow, (F); real estate tax payment to escrow, (F); utility bills such as electricity, gas, water and sewer, telephone, heating oil, (V) unless on a budget plan in which case, (F); repairs and routine maintainance, (V); replacement of major appliances, (V); and lawn care, (V).

Assignment Guide

- Basic: 10–16, 18–36 even
- Average: 11–15 odd, 16, 17, 18–36 even

MAINTAINING YOUR SKILLS Look up the skills in parentheses if you need help or more practice.

Which number is greater? **(Skill 1)**

18. 2109.8 or 2107.9 **19.** 7484.08 or 74,846.50 **20.** 534.76 or 544.71

21. 823.40 or 826.70 **22.** 63,879 or 63,396 **23.** 268.74 or 378.47

Add. **(Skill 5)**

24. 85.89 + 74.84 + 35.30 + 306.24 **502.27**

25. 456.26 + 24.98 + 24.5 + 9.39 **515.13**

26. 974.79 + 997.32 + 9.81 + 35.8 **2017.72**

27. 527.86 + 368.2 + 41.88 + 9.2 **947.14**

Find the percentage. Round answers to the nearest hundredth. **(Skill 30)**

28. 35% of 8300 **2905** **29.** 40% of 4600 **1840** **30.** 42% of 7841 **3293.22**

31. 35% of 5089.7 **1781.395** **32.** 30% of 986.44 **295.932** **33.** 15% of 59,640 **8946**

34. 0.3% of 88,000 **264** **35.** 0.15% of 65,400 **98.1** **36.** 0.04% of 23,200 **9.28**

ALTERNATIVE ASSESSMENT

Problem 17 can be used to assess students understanding of how all housing costs are determined. Work the problem as a class activity, calling on selected students to perform the necessary computations to answer each question. Have other students verify that each answer is correct.

Reviewing the Basics

Skills

Solve. Round answers to the nearest hundredth.

(Skill 3)

1. \$744 + \$821 + \$50 **\$1615** **2.** \$278 + \$1435 + \$594 **\$2307**

3. \$28,040 + \$8152 **\$36,192**

(Skill 4)

4. \$795 − \$412 **\$383** **5.** \$56,000 − \$42,400 **\$13,600** **6.** \$93,000 − \$8794 **\$84,206**

(Skill 5)

7. \$2941.93 + \$2051.88 **\$4993.81** **8.** \$819.54 + \$583.32 **\$1402.86** **9.** \$41.71 + \$5.22 **\$46.93**

(Skill 6)

10. \$723.12 − \$90.07 **\$633.05** **11.** \$453.36 − \$79.64 **\$373.72** **12.** \$531.12 − \$379.84 **\$151.28**

(Skill 8)

13. 85.02 × 480 **40,809.6** **14.** 3017.45 × 0.324 **977.6538 = 977.65** **15.** \$4417 × 0.35 **\$1545.95**

(Skill 30)

16. \$44,000 × 25% **\$11,000** **17.** \$87,500 × 3.25% **\$2843.75** **18.** \$4192.50 × 35% **\$1467.375 = \$1467.38**

Applications

Use the table to find the monthly payment.

(Application C)

19. The Jericho National Bank is willing to loan Max Glenn \$80,000 at 12.50% for 20 years on the purchase of a home. What is the monthly payment? **\$909.60**

20. The Reynolds are purchasing a \$180,000 home. Everson Savings and Loan Association will finance the purchase at 13% for 25 years if the Reynolds can make a 20% down payment. What is the monthly payment? **\$1624.32**

MONTHLY PAYMENT FOR A \$1000 LOAN

Annual Interest Rate	Length of Loan (Years)		
	20	25	30
10.00%	\$ 9.66	\$ 9.09	\$ 8.78
10.50%	9.99	9.45	9.15
11.00%	10.33	9.81	9.53
11.50%	10.67	10.17	9.91
12.00%	11.02	10.54	10.29
12.50%	11.37	10.91	10.68
13.00%	11.72	11.28	11.07
13.50%	12.08	11.66	11.46

Terms

Match each term with its definition on the right.

21. Mortgage loan **b**

22. Closing costs **c**

23. Real estate taxes **f**

24. Premium **a**

25. Principal **e**

26. Utilities cost **d**

a. an amount paid for an insurance policy

b. a loan whereby the lender has the right to sell the property if payments are not made

c. fees paid at the time documents are signed transferring ownership of a home

d. charges for public services

e. an amount owed upon which interest charged is calculated

f. fees collected on the ownership of property used to support the operation of government

Refer to your reference files in the back of the book if you need help.

LESSON PLAN
Reviewing the Basics

The exercises on this page review skills, applications, and terms used in the unit. You can use the exercises to assess informally students' proficiency with this material.

The page can be used for guided practice and independent practice. You can work through a selection of the exercises together with students, and thus see immediately if they know how to do them, and you can then assign some of the exercises for independent practice. Be sure to go over the answers to all assigned exercises.

LESSON PLAN
Unit Test

Students should do the Unit Test on their own. Each problem on the test is keyed to a lesson in the unit. Students having difficulty with any particular problem should review the Example in the appropriate lesson and be assigned some of the Independent Practice problems for additional practice.

Unit Test

Lesson 9-1

1. The Dixons are purchasing a $95,000 home. What is the amount of the mortgage loan needed to finance the purchase if they wish to make a 15% down payment? **$80,750**

Lesson 9-2

2. The Mejias have a $60,000 mortgage loan from Home Loan Bank at an interest rate of 11% for 25 years. Use the table on page 645 to find the monthly payment and the total interest charged. **$588.60; $116,580**

Lesson 9-3

3. The McLaughlins were granted a $75,000 mortgage loan. At the time of the closing, they paid the closing costs shown. What is the total of the closing costs? **$2387.50**

Credit Report:	$50
Loan origination fee:	2.4% of loan
Abstract of title:	$215
Attorney fees:	$200
Property tax:	$122.50

Lesson 9-4

4. The Ramires obtained a 30-year, $85,000 mortgage loan at Swan Creek Bank at an interest rate of 10.5%. The monthly payment is $777.75. What is the new principal after the first monthly payment? **$84,966**

Lesson 9-5

5. The Hanlon County tax assessor stated that the market value of the Arjons' estate is $935,000. The rate of assessment in Dixon County is 30% of the market value. The tax rate is 67.23 mills. What is the annual real estate tax on the estate? **$18,858.02**

Lesson 9-6

6. The Rayans have insured their home for 90% of its replacement value of $85,000. The insurance company states that the amount of coverage on their personal property is 50% of the amount of coverage on their home. What is the amount of coverage on the Rayans' personal property? **$38,250**

Lesson 9-7

7. The replacement value of the Coreys' home is $60,000. They insured it for 100% of its replacement value. The townhouse is of masonry veneer and is rated in fire protection class 8. Use the table of annual premiums on page 279 to find the annual homeowner's policy premium. **$151**

Lesson 9-8

8. The Danelies have a combined monthly net income of $3155.00. Their housing expenses for June are listed below. Were their housing expenses for the month within the FHA recommendations?

Mortgage payment	$600.85	Telephone	$29.19
Insurance	34.35	Heating oil	84.21
Property taxes	176.67	Electricity	65.65
Water/sewer service	18.23	Furniture payment	49.95

Yes, $1104.25 is greater than $1059.10; Total = $1059.10

A SPREADSHEET APPLICATION

Mortgage Payments

To complete this spreadsheet application, you will need the diskette Spreadsheet *Applications for Business Mathematics*, which accompanies this textbook.

Select option 9, Mortgage Payments, from the menu. Input the information in the following problems to find the monthly payment and the total amount paid for each mortgage loan.

1. Susan and Willard Weber.
   Home selling for $86,400.
   Terms: $9400 down payment.
   a. Loan at 13% for 25 years.
   b. Loan at 13.25% for 25 years.
2. Rudolf and Charlotte Hess.
   Home selling for $376,000.
   Terms: $74,000 down payment.
   a. Loan at 14% for 25 years.
   b. Loan at 13.50% for 30 years.
3. Mobile home selling for $39,000.
   Terms: $9900 down payment.
   a. Loan at 14.25% for 15 years.
   b. Loan at 15% for 12 years.
4. Mobile home selling for $33,500.
   Terms: $3350 down payment.
   a. Loan at 15% for 10 years.
   b. Loan at 14.50% for 12 years.
5. Estate for $440,000.
   Terms: 10% down payment.
   a. Loan at 12.75% for 30 years.
   b. Loan at 12.75% for 35 years.
6. Farmhouse for $175,000.
   Terms: 30% down payment.
   a. Loan at 9.25% for 25 years.
   b. Loan at 9% for 23 years.
7. Walt Tyson plans to purchase a home for $106,900. He has an $8900 down payment. First Federal will give him a loan at 14.25% for 25 years. Baltimore Trust Company will give him a loan at 14.75% for 20 years. Find the monthly payment and total amount paid for each loan.
8. Marge McCarthy plans to purchase a commercial property for $539,000. She has $39,000 down payment. Hamilton Mortgage Company will give her a loan at 12.51% for 20 years. Grogan Investors will give her a loan at 12.92% for 18 years. Find the monthly payment and total amount paid for each loan.
9. Agnes Golinto plans to purchase a duplex for $125,400. After a 15% down payment, she plans to get a 30-year loan. Find the monthly payment and total amount paid for a loan at 14% and a loan at 14.50%.
10. Ernie Thomas plans to purchase a condominium for $86,000. After a 30% down payment, he plans to finance the remainder for 20 years. Find the monthly payment and total amount paid for a 10% loan and a 10.25% loan.

LESSON PLAN
A Spreadsheet Application

USING TECHNOLOGY
After students have completed the problems on this page, have them compare the total amounts paid for each mortgage loan in each problem. In Problems 1 and 5, discuss the effect of increasing the percent of a loan or increasing the time period of a loan on the total amount paid. Is the effect significant?

As a *critical thinking* question, ask students what their goals would be in shopping for a mortgage loan.

LESSON PLAN
Career Wise

APPLICATIONS ON THE JOB

After students have completed this page, point out that one of the major reasons people are turned down for a mortgage loan is a poor credit history. This is true even though they are employed and can afford to make the monthly payments. Students should be aware of the fact that a poor record on payments for a credit card, for example, may show up when a background check is made for a mortgage loan. This poor record might influence a loan officer to reject an application for a loan.

Ask students if they are aware of the fact that credit bureaus exist that track peoples' payments on credit cards and other loans, and that these "credit histories" are available to banks and other businesses if they wish to see them.

CAREER WISE

Mortgage Officer

Wu-Yi Wong is loan officer at a mortgage company, a company whose main goal is to make money available to people wishing to buy homes.

When someone has decided to purchase a particular house, he or she comes to Wu-Yi to apply for a long-term loan. Wu-Yi explains the terms of the loan, including the type of loan, rate, points involved, and closing costs. Wu-Yi also seeks information about the customer's credit history, employment status, and ability to make payments regularly.

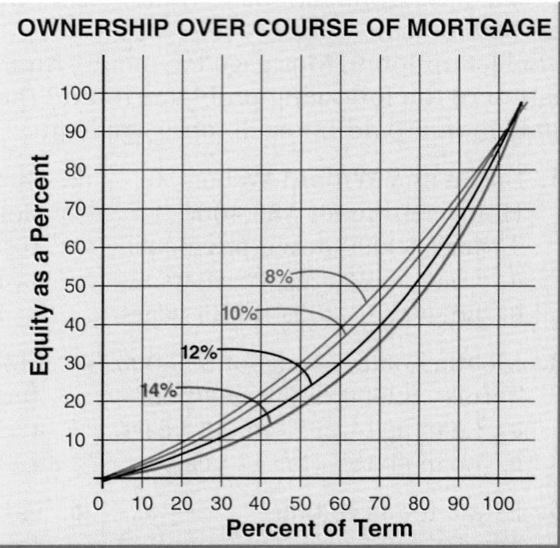

In effect, Wu-Yi's company and the prospective home buyer will own the home jointly. As payments are made to the mortgage company, the company's ownership decreases and the home buyer's ownership increases.

The graph at the right indicates how the share of ownership changes over the course of the mortgage. The graph is based on a 20-year mortgage with payments made monthly. Four different interest rates—8%, 10%, 12%, and 14%—are shown.

Interest rates change as the state of the economy changes. Wu-Yi has found that at some times of the year, many people seek loans, while at other times of the year, not many people can afford to apply for loans.

Check Your Understanding

1. What is the percentage of equity a homeowner will have after 50% of term if the interest rate is 8%? **about 30% equity**
2. After what percent of term will the equity be 30% if the interest rate is 12%? **about 60% of term**
3. Copy the axes and the curves in the graph above. Sketch the curve that you think corresponds to a rate of 9%.
4. Use the graph to explain this statement: At any given percent of term, equity at 14% is always less than equity at 10%. **The 14% graph is below the 10% graph for each percent of term.**

UNIT

10

Insurance and Investments

Health and *life insurance* protect you and your dependents against financial losses in case of illness or death. *Whole life* and *universal life insurance* offer a savings feature. *Certificates of deposit, stocks,* and *bonds* are other types of *investments,* which often earn more interest than money in a bank account. A certificate of deposit is purchased for specific amounts that remain on deposit in a bank for a specified period of time. A stock is a *share* in a company. The company may pay you a return on your investment in the form of *dividends.* A bond is money you loan to a company for a specified period. When the loan is due, you are paid the *face value* of the bond, which includes the interest on your original investment.

OUTLINE

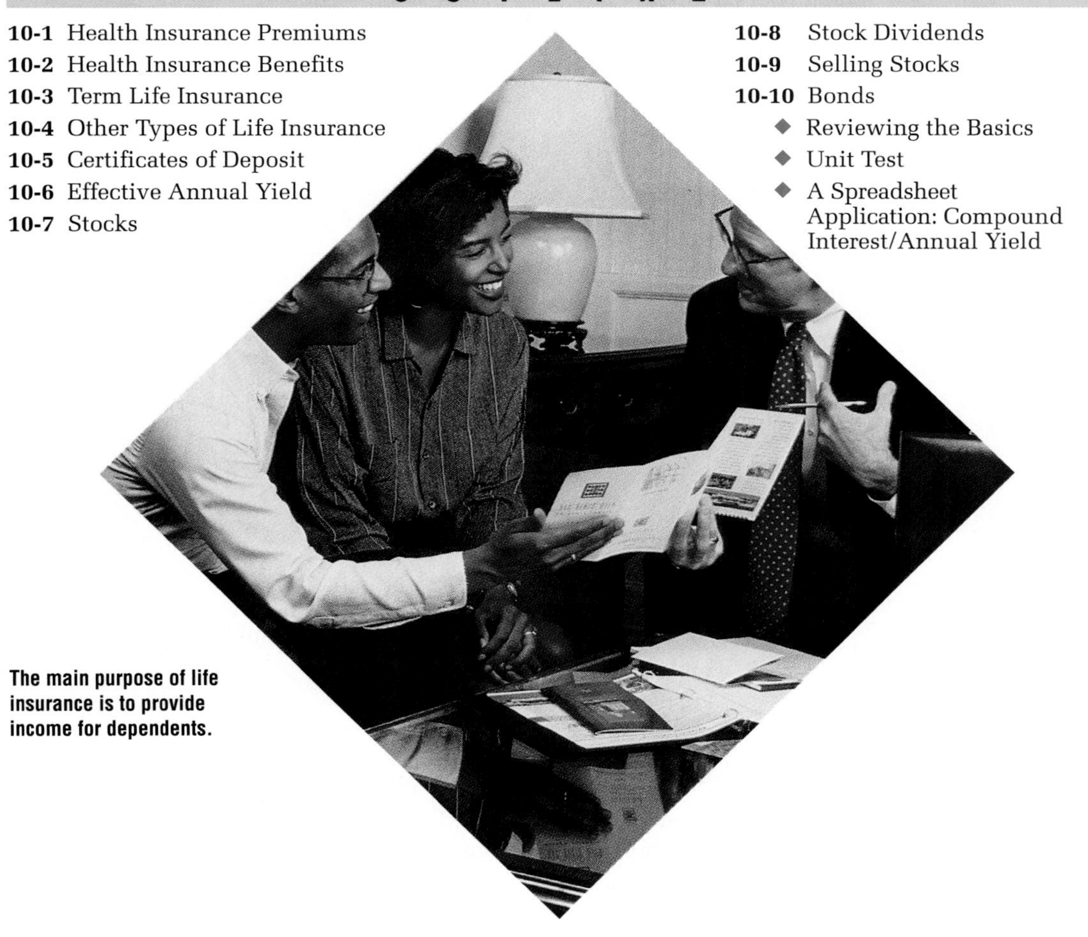

The main purpose of life insurance is to provide income for dependents.

UNIT 10 INSURANCE AND INVESTMENT

INTRODUCING THE UNIT

To introduce this unit, assign an outside project to be done while studying Unit 10.

Have students visit a local bank to find out the rates on the CD's for the time periods posted. The rates are usually posted in the lobby. Ask several students to share their data with the class during your study of Lesson 10-5.

Or, you may wish to assign library research for the class. Tell students to make a list of the major stock exchanges throughout the world. This information can be used in conjunction with Lesson 10-7.

LESSON PLAN 10-1 Health Insurance Premiums

FOCUS

Using a number cube or die, ask a student to choose a number he or she thinks the cube will land on when rolled. Then ask: If the cube doesn't land on the number, would you be willing to give up your most valued possession? Explain that not having some type of health insurance also is taking a big risk!

TEACH

Explain that most people are covered by some type of health insurance. Discuss the various health plan insurance companies offer, and mention that the less costly insurance usually is provided by an employer through a group plan.

Students have studied medical insurance deductions in Unit 2, and should be familiar with the formulas of this lesson. You might ask a student to read the Example; then, with their books closed, have students explain how to solve the problem.

You might want to review the term *deduction*, as well as different pay periods, such as biweekly and semimonthly.

Warm-Up Exercises

Divide. Round to the nearest hundredth.

1. 544 ÷ 12 45.33
2. 1195 ÷ 52 22.98

Add.

3. \$42.60 + 153.82 + 16.96 = \$213.38
4. \$1089.42 + 967.86 + 12.13 = \$2069.41
5. \$962 × 25% \$240.50
6. \$842 × 30% \$252.60

10-1 Health Insurance Premiums

OBJECTIVE
Compute health insurance premiums.

An accident or illness could cut off your income, wipe out your savings, and leave you in debt. To protect against overwhelming medical expenses, many people have **health insurance.** A **basic plan** includes hospital and surgical-medical insurance. A **comprehensive plan** includes hospital, surgical-medical, and major medical insurance.

One way to get health insurance is by joining a **group plan** where you work. Your employer may pay part or all of the premium. Health insurance companies also offer **nongroup plans** for people not enrolled in group plans. You can choose an individual plan that covers only yourself or, for a larger premium, a family plan that covers yourself, your spouse, and your children.

Employee's Percent = 100% − Employer's Percent

Employee's Contribution = Total Premium × Employee's Percent

EXAMPLE *Skills* 11, 30 *Application* K *Term* Health insurance

Sean Watts is employed by Schwyn Company. He has a family membership in the group comprehensive medical insurance program. The annual premium includes \$2523 for hospital insurance, \$795 for surgical-medical insurance, and \$138 for major medical insurance, for a total of \$3456. Sean's employer pays 70% of the total cost. His contribution is deducted monthly from his paycheck. What is Sean's annual contribution? What is his monthly deduction?

SOLUTION

A. Find the **employee's percent.**
100% − 70% = 30% employee's percent

B. Find the **employee's contribution.**
Total Premium × Employee's Percent
\$3456.00 × 30% = \$1036.80 employee's contribution

C. Find the **employee's monthly deduction.**
Employee's Contribution ÷ 12
\$1036.80 ÷ 12 = \$86.40 monthly deduction

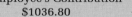

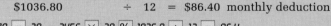

100 [−] 70 [=] 30   3456 [×] 30 [%] 1036.8 [÷] 12 [=] 86.4

✓ SELF-CHECK Complete the problems, then check your answers in the back of the book.

Annual premium includes \$2408 for hospital, \$804 for surgical-medical, and \$168 for major medical. The company pays 80% of the cost.

1. Find the employee's total annual contribution. **\$676**
2. Find the employee's monthly deduction. **\$56.33**

CRITICAL THINKING

Problem 12 deviates from the skills used to find the amount of the employee's contribution. Students are asked to find the percent the employer pays. You might want to discuss the problem in class, having students formulate questions that need to be answered: How much does the employee pay? \$48 is what percent of \$80? Skill 31 can help students who may have difficulty in finding the percent.

PROBLEMS

	Plan	Annual Premium for Hospital	+ Surgical-Medical	+ Major Medical	= Total Annual Premium	Employer's Percent	Employee's Annual Contribution
3.	Single	$1980 +	$ 568 +	$112	= $2660	60%	$1064
4.	Single	$1368 +	$ 365 +	$ 91	= $1824	80%	$364.80
5.	Family	$3066 +	$ 966 +	$168	= $4200	75%	$1050
6.	Family	$2540 +	$ 730 +	$210	= $3480	90%	$348
7.	Family	$3150 +	$1125 +	$225	= $4500	40%	$2700

8. Eric Ritter.
   Annual health insurance premium:
   Hospital $1984.
   Surgical-medical $568.
   Major medical $152.
   Employer pays 85% of cost.
   What is the total annual premium? **$2704**
   How much does Eric pay annually? **$405.60**
   How much is deducted from his weekly paycheck? **$7.80**

9. Kelli Lenz.
   Annual health insurance premium:
   Hospital $1145.
   Surgical-medical $350.
   Major medical $75.
   Employer pays 65% of cost.
   What is the total annual premium? **$1570**
   How much does Kelli pay annually? **$549.50**
   How much is deducted from her semimonthly paycheck? **$22.90**

10. Heather Farough has a family membership in the company's group medical insurance program. The total cost is $3817. Her employer pays 75% of the total cost. Her contribution is deducted biweekly from her paycheck. How much is her annual contribution? How much is her biweekly deduction? **$954.25; $36.70**

11. Rachel and Dustin Lutts are students at the state university. The comprehensive health insurance rate is $37.50 per month for each student. The premiums are paid quarterly (every three months). What does Rachel's and Dustin's health insurance cost per quarter? **$112.50 each**

12. Employees of the General Manufacturing Company may sign up for a group vision care health insurance at a cost of $4.00 per monthly pay period. The insurance actually costs the company $80 per employee per year. What percent of the cost of the insurance does the company pay? **40%**

MAINTAINING YOUR SKILLS Look up the skills in parentheses if you need help or more practice.

Find the rate. **(Skill 31)**

13. What percent of $60 is $3? **5%**
14. What percent of $150 is $60? **40%**
15. What percent of $475 is $95? **20%**
16. What percent of $210 is $73.50? **35%**

Find the percentage. **(Skill 30)**

17. 367 × 55% **201.85**
18. 940 × 14.2% **133.48**
19. 4200 × 13.65% **573.3**
20. 3500 × 4.81% **168.35**

PRACTICE AND APPLY

The following problems can be assigned for classwork and the answers checked in class to help students master the objective of the lesson.

- Guided Practice: 1–5
- Independent Practice: 6–8, 10

WRAP-UP

Review the solution to Problem 10. Ask students whether the amount deducted from Heather's paycheck would be more or less if the amount paid by the employer were 65%. (more)

Assignment Guide

- Basic: 6–11, 13–20
- Average: 9, 11, 12, 14–20 even

ALTERNATIVE STRATEGIES: Reteaching

First, review with your students that a monthly pay period is 12 times per year; semi-monthly is 24 times per year; bi-weekly is 26 times per year; and weekly is 52 times per year. Then, do the example if Sean is paid weekly ($19.94); semi-monthly ($43.20); or bi-weekly ($39.88). Ask these same types of questions for problems 2, 8, 9, 10, & 11.

LESSON PLAN 10-2 Health Insurance Benefits

FOCUS

If you have available a medical insurance claim form, pass it around for the class to see. Explain that when medical costs are incurred (such as doctor visits, hospital charges, or sometimes even medical prescriptions), you must complete a claims form and send it with the total bill to the insurance company. The insurance company will compute the amount they pay of the total cost and the amount you, the patient, must pay.

TEACH

Be sure students understand the meanings of the terms *deductible clause* and *coinsurance clause.*

Remind students to add the deductible amount to the coinsurance amount that the patient pays when finding the total amount paid by the patient. Although the deductible was subtracted in finding the amount subject to coinsurance, the actual dollars still have to be paid toward the medical bill itself.

You might want to mention that in most major medical plans, the deductible amount is used on an annual basis. That is, once a patient has paid the full deductible within one year, any other bills during that same year are not subject to the deductible; only the coinsurance clause applies.

10-2 Health Insurance Benefits

OBJECTIVE
Compute the amount paid by the patient for medical care.

A comprehensive health insurance plan includes three kinds of insurance. **Hospital insurance** pays most of the cost of hospitalization, including a semiprivate room and most laboratory tests. **Surgical-medical insurance** pays your doctor's fee, up to a certain amount, for surgery. **Major medical insurance** is designed to protect you against the other costs of a serious and expensive illness. Most major medical policies have a **deductible clause.** A $250 deductible clause means that you must pay the first $250 of the amount not covered by your hospital and surgical-medical insurance. Many major medical policies also have a **coinsurance clause.** A typical one states that you must pay 20% of the amount remaining after you pay the deductible. Your insurance company pays the other 80%.

Amount Paid by Patient = Deductible + Coinsurance Amount

EXAMPLE — *Skills* 4, 30 *Application* A *Term* Coinsurance clause

Brooke Kolodie's comprehensive medical insurance includes hospital, surgical-medical, and major medical insurance. The major medical policy has a $250 deductible and a 20% coinsurance clause. When she had surgery, her total bill for hospital care and physician's fee was $9350. The hospital and surgical-medical provisions of her insurance policy covered $6825. What amount did she pay?

SOLUTION

A. Find the **amount not paid by hospital and surgical-medical insurance.**
$9350 − $6825 = $2525 amount not paid

B. Find the **amount subject to coinsurance.**
$2525 − $250 deductible = $2275 subject to coinsurance

C. Find the **coinsurance amount paid by patient.**
$2275 × 20% = $455 coinsurance paid by patient

D. Find the **total amount paid by patient.**
Deductible + Coinsurance Amount
$250 + $455 = $705 total paid by patient

9350 [−] 6825 [=] 2525 [−] 250 [=] 2275 [×] 20 [%] 455 [+] 250 [=] 705

✓ SELF-CHECK — Complete the problems, then check your answers in the back of the book.

Krista Ridge had a total medical bill of $46,300. Her insurance covered $44,850. Her medical policy has a $200 deductible and a 20% coinsurance clause.

1. Find the amount subject to coinsurance. **$1250**
2. Find the total amount Krista had to pay. **$450**

COOPERATIVE LEARNING

You might wish to have students work in small groups to solve Problem 7. When completed, select one group to show at the chalkboard the complete solution. Individual members of the group could work each step.

PROBLEMS

	Total Medical Bill	Amount Not Paid by Hospital and Surgical-Medical Insurance	Major Medical Deductible	Amount Subject to Coinsurance	Coinsurance Rate	Amount Paid by Patient	
						Coinsurance	Total
3.	$ 44,500	$ 1,900	$250	$1650	20%	$330	**$ 580**
4.	$ 72,040	$ 2,970	$100	**$ 2870**	20%	**$ 574**	**$ 674**
5.	$148,215	$12,250	$250	**$12,000**	20%	**$2400**	**$2650**
6.	$140,000	$ 1,050	$500	**$ 550**	20%	**$ 110**	**$ 610**

F.Y.I.
In 1990, Americans spent $666 billion on health care.

7. Wanda Orsini.
   Total bill is $48,240.
   Amount paid by hospital and surgical-medical insurance is $44,790.
   Major medical has a $250 deductible and 20% coinsurance clause.
   How much is subject to coinsurance? **$3200**
   What is the total amount Wanda pays? **$890**

8. Kevin Bruce.
   Total bill is $92,980.
   Amount paid by hospital and surgical-medical insurance is $88,745.
   Major medical has a $500 deductible and 20% coinsurance clause.
   How much is subject to coinsurance? **$3735**
   What is the total amount Kevin pays? **$1247**

9. Mike Rodriguez has comprehensive medical insurance that includes hospital, surgical-medical, and major medical insurance. The major medical policy has a $250 deductible and a 20% coinsurance clause. When he had surgery, his total bill for hospital care and physician's fee was $26,454. The hospital and surgical-medical provisions of his insurance policy covered $25,322. What amount did Mike pay? **$426.40**

10. Sofia Carbondale's comprehensive medical insurance includes hospital, surgical-medical, and major medical insurance. The major medical policy has a $400 deductible and a 20% coinsurance clause. When she had surgery, her total bill for hospital care and physician's fee was $15,372. The hospital and surgical-medical provisions of her insurance policy covered $13,509. What amount did she pay? **$692.60**

MAINTAINING YOUR SKILLS Look up the skills in parentheses if you need help or more practice.

Find the percentage. Round answers to the nearest hundredth. **(Skill 30)**

11. 20% of $2490 **$498** **12.** 31.4% of 952 **298.93** **13.** $\frac{3}{4}$% of 120 **0.9**

Subtract. **(Skill 4)**

14. 978 − 865 **113**
15. 77,521 − 66,842 **10,679**
16. 997,341 − 942,876 **54,465**
17. 955 − 827 **128**
18. 82,321 − 32,966 **49,355**
19. 793,450 − 89,890 **703,560**

Warm-Up Exercises

1. 465.18 − 19.65 445.53
2. 1249 − 862 387
3. 14,963 − 8944 6019
4. $842.81 − $795.48 $47.33
5. $849 × 20% $169.80
6. $1260 × 15.5% $195.30

PRACTICE AND APPLY

The following problems can be assigned for classwork and the answers checked in class to help students master the objective of the lesson.

- Guided Practice: 1–5
- Independent Practice: 6–8

WRAP-UP

Write the following phrases on the chalkboard: (1) total bill, (2) cost not covered by insurance, (3) amount subject to coinsurance, and (4) payment by patient. Ask: Which of these uses the deductible in its computation? (2 and 4) Which uses the coinsurance rate? (3)

Assignment Guide

- Basic: 6–9, 11–19
- Average: 9, 10, 12–18 even

ALTERNATIVE STRATEGIES: Enrichment

Many varieties of health insurance plans are available. Have your students report on different plans either by talking to a family member or by talking to an insurance agent. In addition to *deductible* and *coinsurance,* your students might report on terms such as: first-dollar coverage; customary, usual, and reasonable (CUR); maximum out of pocket expense; and self-insure.

LESSON PLAN
10-3 Term Life Insurance

FOCUS

To motivate this lesson on computing the annual premiums for term life insurance, call on various students to explain what they think the purpose of life insurance is. Point out that, unlike other types of insurance, which provide financial protection *if* a certain event (such as an accident or illness, happens), life insurance provides financial protection *when* a person dies.

TEACH

Spend ample time discussing the basic concepts of term life insurance.

When discussing the insurance table on this page, ask students what two facts can be deduced from the table. (Rates for men are higher than for women and the amounts increase with age.) Elicit their reason as to why. (The risk to an insurance company is greater for men because women generally live longer.)

Stress the use of units when computing the annual premium. Remind students that when dividing by 1000, simply drop the same number of zeros in the dividend.

Warm-Up Exercises

1. 40,000 ÷ 1000 40
2. 65,000 ÷ 1000 65
3. 100,000 ÷ 1000 100
4. 150,000 ÷ 1000 150
5. 40 × $2.80 $112
6. 100 × $7.92 $792
7. 165 × $9.47 $1562.55
8. 87.5 × $4.98 $435.75

10-3 Term Life Insurance

OBJECTIVE
Use tables to compute the annual premium for term life insurance.

The main purpose of **life insurance** is to provide financial protection for your dependents in case of your death. **Term life insurance** is the least expensive form of life insurance that you can buy. You buy term insurance for a specified term, such as 5 years, or to a specified age. Unless you renew your policy at the end of each term, the insurance coverage ends.

If you should die during the term of the policy, your **beneficiary**, the person you name in the policy, will receive the **face value** of the policy. The face value is the amount of insurance coverage that you buy. The annual premium depends on your age at the time you buy the policy and the number of **units**. One unit of insurance has a face value of $1000. The annual premium for term life insurance usually increases with each new term.

ANNUAL PREMIUMS PER $1000 OF LIFE INSURANCE: 5-YR. TERM

AGE	MALE	FEMALE
18	$ 2.64	$ 1.85
20	2.70	1.91
25	3.01	2.05
30	3.20	2.26
35	3.83	2.99
45	8.86	6.36
50	14.10	9.26
55	23.45	13.87

Annual Premium = Number of Units Purchased × Premium per $1000

EXAMPLE *Skills* 11, 8 *Application* C *Term* Term life insurance

Doug Mason is 30 years old. He wants to purchase a $40,000, 5-year term life insurance policy. What is his annual premium?

SOLUTION

A. Find the **number of units purchased.**
$40,000 ÷ $1000 = 40 units purchased

B. Find the **premium per $1000.**
Refer to the table above. (male, age 30) $3.20

C. Find the **annual premium.**
Number of Units Purchased × Premium per $1000
40 × $3.20 = $128.00 annual premium

40000 ÷ 1000 = 40 × 3.2 = 128

✓ SELF-CHECK Complete the problems, then check your answers in the back of the book.

Use the table above to find the annual premium for a 5-year term policy.

1. $30,000, female, age 18. **$55.50**
2. $60,000, male, age 45. **$531.60**

BUSINESS NOTES

There are over 900 insurance companies in the U.S. These companies employ people in many different kinds of jobs, from sales to claims adjusters. Point out that the employees who help insurance companies set up their premium rates are called actuaries. An actuary, needs a very strong education in mathematics and statistics and must pass a series of examinations to qualify for the job.

PROBLEMS

Use the table on page 294 to answer the following.

	Insured	Age	Coverage	Number of Units	×	Annual Premium per $1000	=	Annual Premium
3.	Gloria Carver	20	$ 10,000	10	×	$1.91	=	**$ 19.10**
4.	John O'Neill	45	$ 20,000	**20**	×	**$8.86**	=	**$177.20**
5.	Kate Owens	35	$ 25,000	**25**	×	**$2.99**	=	**$ 74.75**
6.	Dale Payne	30	$200,000	**200**	×	**$3.20**	=	**$640.00**

7. Phil Davis's insurance policy.
5-year term life insurance.
$75,000 coverage.
He is 30 years old. **$240**
What is his annual premium?

8. Marica Deerfoot's insurance policy.
5-year term life insurance.
$120,000 coverage.
She is 45 years old. **$763.20**
What is her annual premium?

9. Joni Hauck wants to purchase a $15,000, 5-year term life insurance policy. She is 25 years old. What is her annual premium? **$30.75**

10. Peter and Edith Lichtner have 1 child. Peter is a career counselor and Edith is a child psychologist. Peter is 30 years old and Edith is 25 years old. Both want to purchase $40,000, 5-year term life insurance policies. What is Edith's annual premium? What is Peter's annual premium? **$82.00; $128.00**

Point out that the higher premium reflects a greater risk by the insurance company.

11. Sam and Kolleen Hastings have 2 children. Sam was 30 years old when he first purchased an $85,000, 5-year term life insurance policy. He will be 35 years old this year and his insurance agent has informed him that his policy comes up for renewal. What will be his annual premium if he renews his policy? What total amount did he pay during the previous 5-year term? What total amount will he pay for the next 5-year term? **$325.55; $1360; $1627.75**

MAINTAINING YOUR SKILLS
Look up the skills in parentheses if you need help or more practice.

Divide. Round answers to the nearest thousandth. **(Skill 11)**

12. 8.216 ÷ 6.12 **1.342**
13. 76.26 ÷ 0.14 **544.714**
14. 1.025 ÷ 0.05 **20.5**
15. 21,624 ÷ 1000 **21.624**
16. 93.40 ÷ 100 **0.934**
17. 18,400 ÷ 1000 **18.4**
18. 95,490 ÷ 10,000 **9.549**
19. 89.75 ÷ 10 **8.975**
20. 9.12 ÷ 100 **0.091**

Multiply. Round answers to the nearest hundredth. **(Skill 8)**

21. 34.362 × 100 **3436.2**
22. 0.95 × 0.16 **0.15**
23. 3.49 × 0.035 **0.12**
24. 42.6 × 32.914 **1402.14**
25. 0.052 × 1000 **52**
26. 3.481 × 1000 **3481**
27. 73.2 × 1000 **73,200**
28. 460 × 3500 **1,610,000**
29. 5.62 × 0.101 **0.57**

PRACTICE AND APPLY
The following problems can be assigned for classwork and the answers checked in class to help students master the objective of the lesson.

- Guided Practice: 1–4
- Independent Practice: 5–8

WRAP-UP
Call upon various students to give the definitions of the terms in the lesson. Change the numbers in the Example and have students work the new problem at their desks. Select volunteers to give their answers.

Assignment Guide
- Basic: 5–10, 12–29
- Average: 9–11, 13–29 odd

ALTERNATIVE STRATEGIES: Reteaching
Stress that the base premium has to be multiplied by the number of thousands.

	Insurance	Age	Gender	Units (000's)	Premium per 1000	Annual Premium
1)	$50,000	45	Male	(50)	($8.86)	($443.00)
2)	250,000	35	Female	(250)	($2.99)	($747.50)
3)	75,500	18	Male	(75.5)	($2.64)	(199.32)

LESSON PLAN
10-4 Other Types of Life Insurance

FOCUS
Review with students the fact that the purpose of term life insurance is to provide money to the beneficiary in case of a disaster. Explain that there are other types of life insurance coverage, which combine life insurance and a savings plan, but that the premiums are higher for the latter. The focus of this lesson is to compute the annual premiums for these types of insurance.

TEACH
Due to the complexity of the insurance business, this lesson only briefly discusses the other types of life insurance. The key ideas and terms presented are necessary to understand the basic concepts of life insurance. If time permits, you might ask a local insurance agent to address the class and answer any questions students may have.

Students should feel comfortable with the computations of this lesson as it is a continuation of the previous lesson. The key skill here is reading a more complicated insurance table.

Warm-Up Exercises
1. 80,000 ÷ 1000 80
2. 200,000 ÷ 1000 200
3. 25 × $38.14 $953.50
4. 125 × $87.52 $10,940
5. 80.5 × $62.90 $5063.45
6. 16.2 × $6.47 $104.81

10-4 Other Types of Life Insurance

OBJECTIVE
Use tables to compute the annual premiums for three types of life insurance.

Whole life insurance offers financial protection for your entire life. The premium usually remains the same each year. In addition to the face value that your beneficiary will receive, whole life insurance has a **cash value** and **loan value.** The cash value is the amount of money you will receive if you decide to cancel your policy. The loan value, usually the same amount as the cash value, is an amount that the insurance company will loan you if you request it. **Limited-payment life insurance** also offers lifetime protection. It has a cash value and a loan value. You pay premiums only for a specified number of years or until you reach a certain age. **Universal life insurance** is a combination of life insurance and a savings plan. The policy covers you for your entire life; however, you pay a minimum premium with anything over the minimum going into a savings account.

Age	ANNUAL PREMIUMS PER $1000 OF LIFE INSURANCE — Paid up at Age 65 Male	Paid up at Age 65 Female	Whole Life Male	Whole Life Female	MONTHLY PREMIUMS $50,000 UNIVERSAL LIFE Male or Female
20	$11.59	$ 9.61	$ 8.05	$ 6.29	$ 19.00
25	13.70	11.48	9.45	7.46	24.00
30	16.88	14.41	11.67	9.13	29.00
35	21.47	18.07	14.91	11.42	37.50
40	29.74	24.86	19.45	14.57	52.00
45	39.48	32.56	25.59	18.63	69.50
50	56.26	45.77	34.04	24.21	93.50
55	—	—	46.51	32.37	126.00

Annual Premium = Number of Units Purchased × Premium per $1000

EXAMPLE *Skills* 11, 8 *Application* C *Term* Whole life insurance

Phyllis Saul is 25 years old. She wants to purchase a whole life policy valued at $25,000. What is her annual premium?

SOLUTION

A. Find the **number of units purchased.**
$25,000 ÷ $1000 = 25 units purchased

B. Find the **premium per $1000.**
Refer to the table above. (female, age 25) $7.46

C. Find the **annual premium.**
Number of Units Purchased × Premium per $1000
25 × $7.46 = $186.50 annual premium

25000 ÷ 1000 = 25 × 7.46 = 186.5

✓ SELF-CHECK Complete the problems, then check your answers in the back of the book.

Find the annual premium.

1. 30-year old male. **$816.90** $70,000 whole life policy.
2. 40-year-old female. **$2237.40** Paid-up-at-age-65 policy of $90,000.

PROBLEMS

Use the table on page 296 to solve the following.

3. Ann Gosik's insurance policy. $50,000, whole life. She is 40 years old. **$728.50** What is her annual premium?

4. James Dolby's insurance policy. $35,000, paid up at age 65. He is 20 years old. **$405.65** What is his annual premium?

5. Terrance Gonzales is 30 years old and wants to purchase a $50,000 life insurance policy. He is considering a universal life insurance policy. What is his monthly premium? **$29.00**

6. Emerson Donohue is 40 years old. He wants to purchase a $50,000 universal life insurance policy. What is his monthly premium? How much will he pay in total premiums in a year? **$52; $624**

7. Robert and Lucy Dubbs each purchase a $60,000 whole life insurance policy. Both are 25 years of age. What is their total annual premium? How much more is Robert's annual premium than his wife's? **$1014.60; $119.40**

8. Leona Sowinski purchased a $50,000 universal life insurance policy at the age of 20. What is her annual premium? If Leona pays $50 a month, how much is she saving annually? **$228; $372**

9. Rather than make 1 annual payment, many insurance companies will allow you to make 2, 4, or 12 smaller payments. Due to the additional expense of collecting and handling the payments several times a year, a small fee is charged. Many companies use the guidelines shown below.

   Mabel Brossis is 20 years old and has purchased a $40,000 whole life insurance policy. She wants to make quarterly payments. What are her quarterly payments? How much can she save in 1 year by paying the annual premium? **$64.16; $5.04**

Optional Payment Plans Percent of Annual Premium
Semiannual Premiums = 50.5%
Quarterly Premiums = 25.5%
Monthly Premiums = 8.5%

MAINTAINING YOUR SKILLS Look up the skills in parentheses if you need help or more practice.

Divide. Round answers to the nearest hundredth. **(Skill 11)**

10. 18.4 ÷ 0.032 **575** **11.** 47.614 ÷ 15.62 **3.05**

12. 0.098 ÷ 1.9 **0.05** **13.** 344.62 ÷ 14.1 **24.44**

14. 285 ÷ 25 **11.4** **15.** 30,344 ÷ 110 **275.85**

16. 954 ÷ 9.25 **103.14** **17.** 9500 ÷ 6.50 **1461.54**

Multiply. Round answers to the nearest hundredth. **(Skill 8)**

18. 0.31 × 0.84 **0.26** **19.** 7.81 × 8.1 **63.26** **20.** 6.511 × 0.05 **0.33** **21.** 0.371 × 1000 **371**

22. 814 × 12 **9768** **23.** 159 × 11 **1749** **24.** 2431 × 75 **182,325** **25.** 95,175 × 0.30 **28,552.5**

PRACTICE AND APPLY

The following problems can be assigned for classwork and the answers checked in class to help students master the objective of the lesson.

- Guided Practice: 1–4
- Independent Practice: 5, 7

WRAP-UP

Go over the difference between term life insurance and the insurance coverages presented in this lesson. Focus on the cost of the premiums and the terms *cash value* and *loan value*. Ask which type of life insurance they would purchase and why.

Assignment Guide

- Basic: 5–8, 10–25
- Average: 6, 8, 9, 11–25 odd

ALTERNATIVE STRATEGIES: Enrichment

As mentioned under "TEACH" there are many different types of life insurance, and you may wish to invite an insurance agent to visit the class. Ask your students to investigate policies such as limited-pay, ordinary, whole, straight, endowment, variable, universal, and single premium. Ask your students the meaning of the phrase: "Term is best if you invest the rest."

LESSON PLAN 10-5 Certificates of Deposit

FOCUS

Banks will sometimes advertise their interest rates for certificates of deposit. Use such an advertisement, or copy the rates posted at a local bank, to motivate the computation of interest on certificates of deposit.

TEACH

Instruct students on how to read the interest table before discussing the Example. Point out that money invested in a CD earns a guaranteed rate of interest for the life of the CD. This is not the case with a regular savings account because a bank can set new rates anytime.

Tell students that interest rates for CD's can vary over time; that is, they can go up or down in any given time period. Then ask them if they see any advantages or disadvantages to being locked-in to a specific rate for a long period of time, say two to five years.

Warm-Up Exercises

1. 1.584017 × $5000 $7920.09
2. $7920.09 − $6000 $1920.09
3. 1.506731 × $7000 $10,547.12
4. (1.113025 × $2500) − $2500 $282.56

10-5 Certificates of Deposit

OBJECTIVE
Use tables to compute interest on certificates of deposit.

One way to invest your money is to purchase a **certificate of deposit** (CD), which earns interest at a higher rate than a regular savings account. You buy CDs for specific amounts, such as $500 or $1000, and must leave the money on deposit for a specified time, ranging up to 30 years. You are penalized for early withdrawal. Most certificates of deposit earn interest compounded daily, monthly, or quarterly. Although banks use computers to calculate interest earned, you can use a table to compute the interest earned.

AMOUNT PER $1.00 INVESTED, DAILY, MONTHLY, AND QUARTERLY COMPOUNDING

Annual Rate	Interest Period — 1 year			Interest Period — 4 years		
	Daily	Monthly	Quarterly	Daily	Monthly	Quarterly
7.00%	1.072500	1.072290	1.071859	1.323094	1.322053	1.319929
7. 25%	1.075185	1.074958	1.074495	1.336389	1.335261	1.332961
7.50%	1.077875	1.077632	1.077135	1.349817	1.348599	1.346114
7.75%	1.080573	1.080312	1.079781	1.363380	1.362066	1.359388
8.00%	1.083277	1.082999	1.082432	1.377079	1.375666	1.372785
8.25%	1.085988	1.085692	1.085087	1.390916	1.389398	1.386306
8.50%	1.088706	1.088390	1.087747	1.404891	1.403264	1.399951
8.75%	1.091430	1.091095	1.090413	1.419008	1.417266	1.413723
9.00%	1.094162	1.093806	1.093083	1.433265	1.431405	1.427621
9.25%	1.096900	1.096524	1.095758	1.447666	1.445682	1.441647

Amount = Original Principal × Amount per $1.00
Interest Earned = Amount − Original Principal

EXAMPLE *Skills* 2, 6, 8 *Application* C *Term* Certificate of deposit

Paul Crates invests $4000 in a 1-year certificate of deposit that earns interest at an annual rate of 8.00% compounded monthly. How much interest will he earn at the end of 1 year?

SOLUTION

A. Find the **amount.**
Original Principal × Amount per $1.00
$4000.00 × 1.082999 = $4331.996 = $4332.00

B. Find the **interest earned.**
Amount − Original Principal
$4332.00 − $4000.00 = $332.00 interest earned

4000 M+ × 1.082999 = 4331.996 − RM = 331.996

✓ SELF-CHECK Complete the problems, then check your answers in the back of the book.

Find the interest earned.

1. $8000 CD for 1 year. **$680.70**
   8.25% annual interest rate compounded quarterly.
2. $50,000 CD for 4 years. **$18,169**
   7.75% annual interest rate compounded daily.

USE OF CALCULATORS

Although the problems in this lesson can be done without using a calculator, the work is somewhat tedious and error prone. Encourage students to use a calculator to gain speed and accuracy. Point out that banks use computers to calculate interest on CD accounts and regular savings accounts.

Use the table on page 298 to find the amount per $1.00 invested.

	Annual Rate	Interest Period	Original Principal	×	Amount per $1.00	=	Amount	Interest Earned
3.	7.00%	1 year quarterly	$ 4500	×	1.071859	=	**$ 4,823.37**	**$ 323.37**
4.	9.25%	4 years daily	$ 18,000	×	**1.447666**	=	**$ 26,057.99**	**$ 8,057.99**
5.	8.00%	4 years monthly	$ 9000	×	**1.375666**	=	**$ 12,380.99**	**$ 3,380.99**
6.	9.25%	1 year quarterly	$140,000	×	**1.095758**	=	**$153,406.12**	**$13,406.12**

F.Y.I.
A 3-month certificate of deposit carried an interest rate of 3.65% on February 4, 1992. The rate was 7.98% on February 6, 1991.

7. $3000 certificate of deposit.
Interest period of 1 year.
8.75% annual interest rate.
Compounded quarterly.
What is the amount at maturity?
What is the interest earned?
$3271.24; $271.24

8. $3500 certificate of deposit.
Interest period of 4 years.
7.50% annual interest rate.
Compounded monthly.
What is the amount at maturity?
What is the interest earned?
$4720.10; $1220.10

9. Sophia Deles purchased a 1-year certificate of deposit for $500 that earns interest at a rate of 9.25% compounded daily. What is the amount of the CD at maturity? What is the interest earned?
$548.45; $48.45

10. Clifford and Hazel Ida purchased a 4-year certificate of deposit for $10,000. The CD earns interest at a rate of 8.75% compounded daily. What is the amount of the CD at maturity? What is the interest earned?
$14,190.08; $4190.08

F.Y.I.
The average rate paid for a 1-year certificate of deposit was 3.92% on February 4, 1992.

11. The LaGelleys have $9500 that they want to invest in a certificate of deposit. Granite Trust offers a 4-year certificate that earns interest at a rate of 8.50% compounded quarterly. Hancock Cooperative Bank offers a 4-year certificate of deposit that earns interest at a rate of 8.25% compounded daily. What CD earns the most interest? By how much?
Granite Trust; $85.83

MAINTAINING YOUR SKILLS Look up the skills in parentheses if you need help or more practice.

Round to the nearest ten. **(Skill 2)**

12. 78 **80** **13.** 144 **140** **14.** 3045 **3050** **15.** 337 **340** **16.** 3559 **3560** **17.** 798 **800** **18.** 5999 **6000**

Subtract. **(Skill 6)**

19. 0.099 − 0.032 **0.067** **20.** 921.7285 − 336.94 **584.7885** **21.** 5.435 − 0.794 **4.641**

Multiply. **(Skill 8)**

22. 0.073 × 8 **0.584** **23.** 1.42 × 1543 **2191.06** **24.** 26.49 × 85.76 **2271.7824** **25.** 7.61 × 1000 **7610**

PRACTICE AND APPLY

The following problems can be assigned for classwork and the answers checked in class to help students master the objective of the lesson:

- Guided Practice: 1–5
- Independent Practice: 6–8

WRAP-UP

Ask students to explain two key differences between a certificate of deposit and a regular savings account. (A CD has a fixed rate of interest for a fixed period of time. There is a penalty for early withdrawal of a CD.)

Assignment Guide

- Basic: 6–10, 12–25
- Average: 9–11, 12–24 even

ALTERNATIVE STRATEGIES: Reteaching

The table in Lesson 10-3 is "...per $1000"; the table in this lesson is "...per $1.00".

	Principal	Term	Rate	Compounded	Amt./$1.00	Amount	Interest
1)	$10,000	1 yr	7.50%	Daily	(1.077875)	($1077.88)	($77.88)
2)	2500	4 yr	7.00%	Quarterly	(1.319929)	(3299.82)	(799.82)
3)	90,000	4 yr	9.00%	Monthly	(1.431405)	(128,826.45)	(38,826.45)
4)	90,000	1 yr	7.00%	Quarterly	(1.071859)	(96,467.31)	(6,467.31)

LESSON PLAN 10-6 Effective Annual Yield

FOCUS

Ask students this question: If you have $1000 to invest for one year, would you invest it at 8% compounded daily or 8% compounded weekly? Explain your choice. In this lesson, students will learn how to compute the effective annual yields.

TEACH

Use the answers to the question above to discuss the concept of effective annual yield. Students need to understand that a valid comparison of the yields of two or more investments can be made only by using the effective annual yields of the investments.

The use of a calculator is highly recommended for working the problems in this lesson.

Warm-up Exercises

Multiply. Round answers to nearest hundredth.

1. 1.210620 × 4000
   4842.48
2. 1.009125 × 3000
   3027.38
3. $12,250.10 − $9067.80
   $3182.30
4. $9265.42 − $7770.00
   $1495.42

Divide. Do not round answers.

5. $859.75 ÷ 8000
   0.10746875
6. $19,259.67 ÷ $12,000.00 1.6049725

10-6 Effective Annual Yield

OBJECTIVE
Compute the effective annual yield.

Banks advertise not only the annual interest rates, but also the effective annual yields of their certificates of deposit. The **annual yield** is the rate at which your money earns simple interest in one year. The interest earned at the annual rate is affected by the frequency of compounding. The effective annual yield is not affected, so you can use it to compare investments.

$$\text{Effective Annual Yield} = \frac{\text{Interest for One Year}}{\text{Principal}}$$

EXAMPLE *Skills* 11, 26 *Application* A *Term* Annual yield

Randall Raye invested $5000 in a certificate of deposit for 3 years. The certificate earns interest at an annual rate of 9.25% compounded quarterly. What is the effective annual yield to the nearest hundredth of a percent?

SOLUTION

A. Find the **interest for one year.** (Refer to the table on page 298.)

Amount − Principal
($5000 × 1.095758) − $5000
$5478.79 − $5000 = $478.79 interest

B. Find the **effective annual yield.**

Interest for One Year ÷ Principal
$478.79 ÷ $5000.00 = 0.095758 = 9.58% effective annual yield

✓ SELF-CHECK Complete the problems, then check your answers in the back of the book.

Find the effective annual yield. (Refer to the table on page 298.)

1. $10,000 certificate of deposit. 8.75% annual interest rate compounded monthly. **9.11%**
2. $2500 certificate of deposit. 9% annual interest rate compounded quarterly. **9.31%**

PROBLEMS

Use the table on page 298 to find the amount per $1.00 invested.

	Annual Rate	Interest Period	Original Principal	×	Amount per $1.00	=	Amount	Interest Earned	Effective Annual Yield
3.	7.25%	1 year quarterly	$6000	×	1.074495	=	$6446.97	$446.97	7.45%
4.	8.00%	1 year quarterly	$8400	×	1.082432	=	$9092.43	$692.43	8.24%

	Annual Rate	Interest Period	Original Principal	×	Amount per $1.00	=	Amount	Interest Earned	Effective Annual Yield
5.	9.25%	1 year monthly	$5,000	×	**1.096524**	=	**$5482.62**	**$482.62**	**9.65%**
6.	7.50%	1 year daily	$1,800	×	**1.077875**	=	**$1940.18**	**$140.18**	**7.79%**

7. $9000 certificate of deposit.
9.25% annual interest rate.
Compounded quarterly.
4-year certificate. **9.58%**
What is the effective annual yield?

8. $4500 certificate of deposit.
8.75% annual interest rate.
Compounded daily.
$2\frac{1}{2}$-year certificate. **9.14%**
What is the effective annual yield?

9. Ollie Gibson invested $10,000 in a certificate of deposit of 4 years at an annual interest rate of 8.25% compounded daily. What is the interest earned for 1 year? What is the effective annual yield? **$859.88; 8.60%**

10. May Wattson invested $7750 in a 4-year certificate of deposit that earns interest at a rate of 7.75% compounded monthly. What is the interest earned for 1 year? What is the effective annual yield? **$622.42; 8.03%**

The interest in the Golden Dollar Accounts is compounded quarterly.

GOLDEN DOLLAR ACCOUNTS		
Terms Available	Interest Rate	Minimum Deposit
1 yr.	7.75%	$500
2 – 3 yrs.	8.00%	$1000
4 – 5 yrs.	8.75%	$1500
6 – 7 yrs.	9.00%	$2000
8 – 10 yrs.	9.25%	$2500

11. Paul Durant invested $4000 in a Golden Dollar Account that matures in 6 years. What is the interest rate that the bank will give? What is the interest earned for 1 year? What is the effective annual yield? **9.00%; $372.33; 9.31%**

12. Ben Garison invested $15,000 in a Golden Dollar Account that matures in 1 year. What is the interest rate that the bank will give? What is the interest earned at maturity? What is the effective annual yield? **7.75%; $1196.72; 7.98%**

13. Julie Whiteburn invested $2000 in a Golden Dollar Account for 4 years. What is the interest rate that the bank will give? What is the interest earned at maturity? What is the effective annual yield?
8.75%; $827.45; 9.04%

MAINTAINING YOUR SKILLS Look up the skills in parentheses if you need help or more practice.

Write as percents. Round answers to the nearest hundredth. **(Skill 26)**

14. 0.112741 **11.27%** **15.** 0.089146 **8.91%** **16.** 1.25642 **125.64%** **17.** 25.5 **2550.00%** **18.** 2.0 **200.00%**

Divide. **(Skill 11)**

19. 1500 ÷ 0.05 **30,000** **20.** 28.42 ÷ 1.2 **23.68** **21.** 679.2 ÷ 0.02 **33,960** **22.** 3.60 ÷ 7.2 **0.50**

PRACTICE AND APPLY

The following problems can be assigned for classwork and the answers checked in class to help students master the objective of the lesson.

- Guided Practice: 1–7
- Independent Practice: 8–10

WRAP-UP

Ask for volunteers to present their solutions to Problems 8 and 9 at the chalkboard. Call upon two students to explain the meaning of annual yield and effective annual yield, respectively.

Assignment Guide

- Basic: 8–12, 14–22
- Average: 11–13, 14–22 even

ALTERNATIVE STRATEGIES: Enrichment

The effective annual yield is the index to be used to compare various savings plans. Just as the annual percentage rate was to compare consumer credit plans, the effective annual yield plays a similar role with savings plans. Daily newspapers are filled with ads from financial institutions usually stating the interest rate paid and the effective annual yield. Have your students collect as many of these ads as they can and verify those that fit the rates and terms given in the tables in the text.

LESSON PLAN
10-7 Stocks

FOCUS

You can motivate the study of stocks by opening a discussion in starting a new business. Ask students to think about starting a business of their own. Who would invest in their business? If the business were successful, how could more money be raised to make it grow? Point out that both small and large businesses issue stocks to raise money. In this lesson, students compute the total cost of a stock investment.

TEACH

Ask students what they think a business does with money raised from the issue of new stock. Ask them also to think about the reasons a person would buy stocks. Students should understand that the sale of stocks can provide a business with new money for expansion into new markets, replace old equipment, hire people to do research and development, improve existing products, and so on. Individuals buy stocks in the hope that their value will increase and thus can be sold for a profit.

Review stock prices in eighths of a dollar and use a calculator to find their decimal equivalents. Write these numbers on the chalkboard for reference.

10-7 Stocks

OBJECTIVE

Compute the total cost of a stock investment.

You can invest your money in shares of **stocks.** Stock prices are published each day in the business sections of daily newspapers. Stock prices are usually quoted in eighths of a dollar. When you purchase a share of stock, you become a part owner of the corporation that issues the stock. You may receive a **stock certificate** as proof of ownership. The total amount you pay for the stock depends on the cost per share, the number of shares you purchase, and the **stockbroker's** commission.

Number of shares sold today, in hundreds → Sales 100s

Highest and lowest prices paid during the day → High, Low

Difference between last sale today and last sale yesterday → Net Chg.

Abbreviation of company name → Stock

Price paid for the last sale today → Last

52-week High	52-week Low	Stock	Div.	Yld.	PE	Sales 100s	High	Low	Last	Net Chg.
$39\frac{3}{4}$	$29\frac{1}{8}$	Bastek	2.24	6.9	6	68	$32\frac{3}{4}$	$32\frac{1}{4}$	$32\frac{1}{4}$	$-\frac{3}{4}$
$18\frac{3}{8}$	$14\frac{1}{2}$	BCA Co	1.24	8.5	6	187	15	$14\frac{5}{8}$	$14\frac{5}{8}$	$-\frac{3}{8}$
25	13	Beta	.28	2.1	12	432	$13\frac{3}{4}$	$13\frac{1}{4}$	$13\frac{1}{4}$	$-\frac{5}{8}$
$12\frac{1}{4}$	$9\frac{1}{8}$	BF Low	1	9.5	....	352	$10\frac{3}{4}$	$10\frac{1}{2}$	$10\frac{1}{2}$	$-\frac{1}{8}$
42	33	Bigley	1.82	5.5	6	845	$33\frac{1}{4}$	$32\frac{1}{2}$	$33\frac{1}{8}$	$+\frac{1}{8}$

Cost of Stock = Number of Shares × Cost per Share

Total Paid = Cost of Stock + Commission

EXAMPLE *Skill* 14 *Application* A *Term* Stocks

Melanie Lambert purchased 100 shares of stock at $\$37\frac{3}{8}$ per share. Her stockbroker charged her a $25 commission. What is the total amount that she paid for the stock?

SOLUTION

A. Find the **cost of stock.**

Number of Shares × Cost per Share

100 × $\$37\frac{3}{8}$

100 × $37.375 = $3737.50 cost of stock

B. Find the **total paid.**

Cost of stock + Commission

$3737.50 + $25.00 = $3762.50 total paid

✔ SELF-CHECK Complete the problems, then check your answers in the back of the book.

Find the total paid.

1. 150 shares of stock at $58 per share. Commission of $75. **$8775**

2. 400 shares of stock at $\$8\frac{1}{8}$ per share. Commission of $36.50. **$3286.50**

CULTURAL ANGLES

Students probably have heard of the New York Stock Exchange on Wall Street in New York City. Tell them that other large cities throughout the world, such as Tokyo, Hong Kong, London, and Paris also have stock exchanges where stocks are bought and sold. Each foreign exchange specializes in the stocks of its national companies.

	Company	(Number of Shares	×	Cost per Share	=	Cost of Stock)	+	Commission	=	Total Paid
3.	AvGlo	(100	×	$5\frac{1}{8}$	=	$512.50)	+	$ 24.00	=	$536.50
4.	Peaseley	(1,000	×	$40\frac{1}{2}$	=	$40,500)	+	$250.00	=	$40,750
5.	ICN Br	(500	×	$73\frac{1}{4}$	=	$36,625)	+	$318.00	=	$36,943
6.	HM Co	(250	×	$89\frac{5}{8}$	=	?)	+	$347.50	=	?
						$22,406.25				$22,753.75

7. Trudy Fahringer.
   500 shares of Air Waveson.
   Cost per share is $24.30.
   Commission is $122.
   What is the cost of the stock?
   What is the total paid?
   $12,150; $12,272

8. David Daly.
   215 shares of Atwood Tire.
   Cost per share is $50.125.
   Commission is $110.
   What is the cost of the stock?
   What is the total paid?
   $10,776.88; $10,886.88

9. Enice Brudley purchased 300 shares of GKI Petroleum at $40.75 per share. A broker's commission of $183.38 was charged. What was the cost of the stock? What was the total paid? **$12,225; $12,408.38**

10. Linda and Martin Sonoma purchased 300 shares of Hampton Publishing at $24\frac{3}{4}$ per share and 150 shares of A-1 Electronics at $28\frac{3}{4}$ per share. The broker charged a 1% commission. What was the total paid? **$11,854.88**

11. Dan Hostetler wants to purchase either 100 shares of Matell Scientific at $48\frac{1}{8}$ per share or 400 shares of Ballon Synergistics at $11\frac{7}{8}$ per share. Both companies show promising growth. The broker charges a 2% commission. Which purchase costs less? How much less? **Ballon; $63.75**

F.Y.I.
On February 5, 1992, there were 262,440,000 shares of stock involved in active trading on the New York Stock Exchange.

12. Hamilton Investment Group charges a fee plus commission for stock transactions. Cindy Siegel purchases 500 shares of stock at $13\frac{3}{4}$ per share through Hamilton Investment Group. What is the cost of the stock? What is the total paid? **$6687.50; $6821.07**

HAMILTON INVESTMENT GROUP

Sales	Commission Rate
Up to $4999	1%
$5000–$9999	1.25%
Over $10,000	1.5%

LOWEST FEES IN TOWN
$49.98 plus commission

13. Albert Nash purchases 700 shares of stock at $33\frac{7}{8}$ per share through Hamilton Investment Group. What is the total paid? **$24,118.17**

MAINTAINING YOUR SKILLS Look up the skills in parentheses if you need help or more practice.

Change the fractions to decimals. **(Skill 14)**

14. $15\frac{7}{8}$ **15.875** **15.** $3\frac{1}{8}$ **3.125** **16.** $9\frac{3}{4}$ **9.75** **17.** $14\frac{3}{8}$ **14.375** **18.** $22\frac{1}{2}$ **22.5**

Multiply. **(Skill 8)**

19. $45.50 × 400 **$18,200** **20.** $12.75 × 200 **$2550**

21. $57.375 × 400 **$22,950** **22.** $35.125 × 70 **$2458.75**

Warm-up Exercises

Change to decimals.
1. $\frac{5}{8}$ 0.625
2. $\frac{7}{8}$ 0.875
3. $25\frac{3}{8}$ 25.375

Calculate.
4. 100 × $49.625 $4962.50
5. $4.00 ÷ $40.00 $0.1
6. $2.50 ÷ $50.00 $0.05

PRACTICE AND APPLY

The following problems can be assigned for classwork and the answers checked in class to help students master the objective of the lesson.

- Guided Practice: 1–6
- Independent Practice: 7–11 odd

WRAP-UP

Change the numbers in the Example to 500 shares at $42\frac{1}{8}$ with a commission of $421.25. Have students work on the new problem independently and then check their answers. ($21,483.75)

Assignment Guide

- Basic: 7–12, 14–22
- Average: 8–12 even, 13, 14–22 even

ALTERNATIVE STRATEGIES: Reteaching

Review converting fractions to decimals.

	#/Shares	Cost/Share	Cost/Stock	Comm.	Total Paid
1)	100	$42.375	($4237.50)	2% ($84.75)	($4322.25)
			[$4237.50 + $84.75] or [102% × $4237.50]		
2)	500	18.50	($9250.00)	3% (103% × $9250 = $9527.50)	
3)	200	28.75	($5750.00)	1% (101% × $5750 = $5807.50)	

LESSON PLAN
10-8 Stock Dividends

FOCUS

Have students assume that each of them has $5000 to invest in stocks. Ask how they would decide which stocks to buy. Introduce the idea of a dividend and point out that many people choose to buy stocks of companies that pay high dividends. The focus of this lesson is on computing the annual dividend and annual yield of a stock investment.

TEACH

The annual yield of a stock can be used to compare its performance with other stocks or other investments, such as a CD. Students need to understand that there is a risk in investing in stocks. The price of a stock can go down and thus be worth less than an investor paid for it. Or, a company may have a poor year and decide not to pay a dividend. In such a case, the value of its stock would almost certainly go down.

Of course, money can also be made by buying and selling stocks. You might wish to have some students keep a daily record of the price fluctuations in the stocks of five or six different companies for about one month. Then the risks of investing in the stock market can be made real for students.

10-8 Stock Dividends

OBJECTIVE
Compute the annual yield and annual dividend of a stock investment.

When you buy stocks, you may receive **dividends.** Dividends are your return on your investment. You may receive an amount specified by the corporation for each share of stock that you own. The **annual yield** is the dividend expressed as a percent of the cost of the stock. The higher the annual yield, the greater the return on your investment.

Annual Dividend = Annual Dividend per Share × Number of Shares

$$\text{Annual Yield} = \frac{\text{Annual Dividend per Share}}{\text{Cost per Share}}$$

EXAMPLE *Skills* 11, 26 *Application* A *Term* Dividends, annual yield

Pam Schmidt bought 80 shares of stock at $42.50 per share. The company paid annual dividends of $2.75 per share. What is the annual dividend? What is the annual yield to the nearest hundredth of a percent?

SOLUTION

A. Find the **annual dividend.**
Annual Dividend per Share × Number of Shares
$2.75 × 80 = $220.00 annual dividend

B. Find the **annual yield.**
Annual Dividend per Share ÷ Cost per Share
$2.75 ÷ $42.50 = 0.06470 = 6.47% annual yield

2.75 [×] 80 [=] 220 2.75 [÷] 42.5 [=] .064705

✓ SELF-CHECK Complete the problem, then check your answer in the back of the book.

1. 180 shares at $4.70 per share. $0.25 annual dividend per share. Find the annual dividend and the annual yield. **$45; 5.32%**

PROBLEMS

	Annual Dividend per Share	÷	Cost per Share	=	Annual Yield
2.	$4.20	÷	$60.00	=	7.0%
3.	$0.34	÷	$ 6.50	=	5.23%
4.	$5.80	÷	$30.00	=	19.33%
5.	$6.10	÷	$80.00	=	7.63%

Round the annual yield to the nearest hundredth of a percent.

PROBLEM SOLVING

Problem 14 requires that students identify the appropriate information in order to compute the annual yield for each investment. Organizing the data in a table having two column headings, CD and Stock, may help students use the correct numbers to compute each annual yield.

6. Purchase price: $44.35 per share. Dividends were $2.50 per share. What is the annual yield? **5.64%**

7. Purchase price : $81.375 per share. Dividends were $3.70 per share. What is the annual yield? **4.55%**

8. Christine Gony.
Owns 400 shares of MPD Systems.
Purchase price: $\$19\frac{3}{4}$ a share.
Dividends were $1.10 per share.
What were her annual dividends?
What is the annual yield?
$440; 5.57%

9. Woody Davenport. **$807.50;**
Owns 850 shares of Moran Funds.
Purchase price: $\$17\frac{5}{8}$ a share.
Dividends were $0.95 per share.
What were his annual dividends?
What is the annual yield? **5.39%**

10. Joyce Kronecki bought 350 shares of Apex Coporation stock at $9.75 per share. Last year the company paid annual dividends of $1.00 per share. What were her annual dividends? What is the annual yield?
$350; 10.26%

11. Reed Hopkins purchased 400 shares of Computech stock at $\$30\frac{1}{8}$ per share. Recently the company paid annual dividends of $1.85 per share. What were his annual dividends? What is the annual yield?
$740; 6.14%

12. Duane Hartley owns 2000 shares of Baywater Energy stock, which he purchased at $\$8\frac{7}{8}$. Recently he read that the average selling price of his stock was $18. The company paid annual dividends of $0.90 per share last year. What is the annual yield on his stock? For an investor who purchased the stock at $18 per share, what is the annual yield?
10.14%; 5.00%

13. The Racitis own 300 shares of Balwin Aviation, which they purchased at $37.50 per share. What was the dividend per share paid by the company last year? What were the Racitis' annual dividends? What is the annual yield?
$0.32; $96; 0.85%

52-week High	52-week Low	Stock	Div.	High	Low	Close	Net Chg.
$12\frac{1}{4}$	$7\frac{7}{8}$	ASTEX	.50	$8\frac{3}{4}$	$8\frac{5}{8}$	$8\frac{3}{4}$	$+\frac{1}{4}$
$24\frac{1}{4}$	$5\frac{3}{8}$	BANLINE		$23\frac{1}{8}$	$22\frac{5}{8}$	$22\frac{7}{8}$	....
$42\frac{3}{4}$	23	BALW AVI	.32	$42\frac{3}{4}$	$42\frac{1}{4}$	$42\frac{3}{4}$	$+\frac{3}{4}$
$3\frac{1}{8}$	$1\frac{7}{8}$	BYTE TK		$2\frac{1}{4}$	$2\frac{1}{4}$	$2\frac{1}{4}$	....

Use the table on page 648 to find the amount per $1.00 of compound interest.

14. Rick LaMar purchased a 1-year, $5000 certificate of deposit that earns interest at a rate of 8.75% compounded daily. He also owns 185 shares of River Corporation, which he purchased at $\$28\frac{1}{8}$. Last year the company paid annual dividends of $1.65 per share. What is the interest earned on his certificate of deposit? What is the effective annual yield on the CD? What were his dividends on his stock? What is the annual yield? Which investment produced a greater annual yield? **$457.15; 9.14%; $305.25; 5.87%; CD**

MAINTAINING YOUR SKILLS Look up the skills in parentheses if you need help or more practice.

Divide. Round answers to the nearest hundredth. **(Skill 11)**

15. 24.0 ÷ 200 **0.12** **16.** 32.61 ÷ 35 **0.93** **17.** 0.587 ÷ 0.009 **65.22** **18.** 16.7 ÷ 1.1 **15.18**

Write the decimals as percents. **(Skill 26)**

19. 0.2583 **25.83%** **20.** 0.017 **1.7%** **21.** 0.03126 **3.126%** **22.** 1.125 **112.5%** **23.** 0.13145 **13.145%**

Warm-up Exercises

Divide. Round to the nearest hundredth.

1. 62.0 ÷ 400 0.16
2. 78.76 ÷ 20 3.94
3. 0.829 ÷ 0.006 138.17
4. 39.7 ÷ 1.2 33.08

Write the decimals as percents.

5. 0.357 35.7%
6. 2.15 215%
7. 0.053 5.3%
8. 1.03 103%

PRACTICE AND APPLY

The following problems can be assigned for classwork and the answers checked in class to help students master the objective of the lesson.

- Guided Practice: 1–6, 8
- Independent Practice: 7–11 odd

WRAP-UP

Ask students what they have learned in this lesson. Ask them to explain how to calculate an annual dividend and an annual yield.

Assignment Guide

- Basic: 7, 9–13, 15–23
- Average: 10, 12–14, 16–22 even

ALTERNATIVE STRATEGIES: Enrichment

Some stocks are classified as "growth and income" stocks. "Growth" anticipates the price of the stock increasing, while "income" would be represented by the dividend the stock pays.

Using the stock exchange reports from the local newspaper, find the dividend for various stocks and *verify* that the annual yield given is correct.

LESSON PLAN
10-9 Selling Stocks

FOCUS
Using the financial page of your newspaper, select a few stocks that had gone down the day before and, assuming a higher purchase price, calculate the loss for 100 shares. Then select a few stocks that had gone up and calculate the profit made. Use your real-world examples to lead into the topic of computing the profit or loss from a stock sale.

TEACH
Write the three formulas above the Example on the chalkboard for reference. Tell students that stocks are bought and sold through a person called a *broker.* A broker executes a buy or sell order through his or her brokerage firm. The brokerage firm must be a member of a stock exchange to buy or sell stocks. A stock exchange, such as the New York Stock Exchange, does not buy or sell stocks; it is simply a market place where the brokerage firms can buy and sell stocks.

Warm-up Exercises
1. $41.00 × 50 $2050
2. $42.50 × 120 $5100
3. $6550 − $5440 $1110
4. $8960 − $8470 $490

Write the amount that is greater.
5. $4080 or $6660 $6600
6. $3183.20 or $2815.20 $3183.20

10-9 Selling Stocks

OBJECTIVE
Compute the profit or loss from a stock sale.

When you sell your stocks, the sale can result in either a **profit** or a **loss.** If the amount you receive for the sale minus the sales commission is greater than the total amount you paid for the stocks, you have made a profit. If the amount you receive minus the sales commission is less than the total paid, your sale has resulted in a loss.

Net Sale = Amount of Sale − Commission

Profit = Net Sale − Total Paid   **Loss = Total Paid − Net Sale**

EXAMPLE   *Skills* 1, 6, 8   *Application* A   *Term* Profit or loss

Bill Tennyson paid a total of $3738.43 for 75 shares of stock. He sold the stock for $51.50 a share and paid a sales commission of $39.45. What is the profit or loss from the sale?

SOLUTION

A. Find the **net sale.**
Amount of Sale − Commission
($51.50 × 75) − $39.45
$3862.50 − $39.45 = $3823.05 net sale

B. Is **net sale** greater than **total paid?**
Is $3823.05 greater than $3738.43? Yes

C. Find the **profit.**
Net Sale − Total Paid
$3823.05 − $3738.43 = $84.62 profit

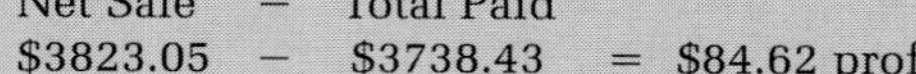

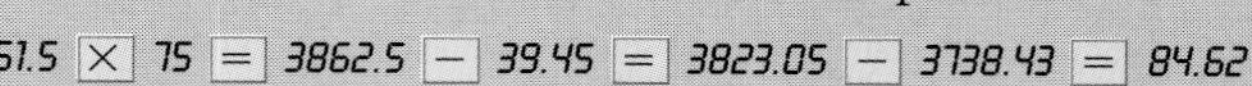

51.5 × 75 = 3862.5 − 39.45 = 3823.05 − 3738.43 = 84.62

✓ SELF-CHECK   Complete the problems, then check your answers in the back of the book.

Find the profit or loss.

1. Paid $1829 for 40 shares. Sold for $60 per share. Commission of $35.50. **$535.50 profit**
2. Paid $24,000 for 1000 shares. Sold for $22 per share. Commission of $98. **$2098 loss**

PROBLEMS

	Total Paid	(Selling Price per Share	×	Number of Shares	− Commission)	= Net Sale	Amount of Profit or Loss
3.	$4,000	($41	×	100)	− $32.75	= **$4067.25**	$67.25 profit
4.	$3,250	($18	×	200)	− $44.90	= **$3555.10**	**$305.10 profit**

CRITICAL THINKING
There is a great deal of information available to help investors make decisions about buying stocks. Books, financial magazines, company reports, and newspapers (*The Wall Street Journal*) all contain valuable financial data for making investment decisions. But there always are people who like a "hot tip." Ask students if they would use a "hot tip" to make an important investment decision. Have them give reasons for their answers.

	Total Paid	(Selling Price per Share	×	Number of Shares	−	Commission)	= Net Sale	Amount of Profit or Loss
5.	\$5925	(\$52$\frac{1}{4}$	×	100)	−	\$59.60	= **\$5165.40**	**\$759.60 loss**
6.	\$ 380	(\$2$\frac{1}{2}$	×	200)	−	\$26.45	= **\$473.55**	**\$93.55 profit**

7. Wendy and Bob Serta.
Bought 47 shares of ITA stock.
Paid a total of \$4512.85.
Sold at \$93$\frac{3}{4}$ per share.
Paid a \$95 sales commission.
What was the profit or loss?
\$201.60; loss

8. Sara and Eric Walton.
Bought 140 shares of Lap Top Stock.
Paid a total of \$9788.50.
Sold at \$81$\frac{7}{8}$ per share.
Paid a 2% sales commission.
What was the profit or loss?
\$1444.75; profit

9. Brian Rowell owned 200 shares of Big Q stock for which he paid a total of \$9187.50. He sold his stock at \$58.25 per share and paid a sales commission of \$150. What was the net amount of the sale? What was the profit or loss from the sale? **\$11,500; \$2312.50 profit**

10. Ira and Eve Marcucci sold 120 shares of United Tool stock at \$84$\frac{3}{4}$ per share and paid a \$110 sales commission. They had originally paid a total of \$6984.15 for the stock. What was the net amount of the sale? What was the profit or loss from the sale? **\$10,060; \$3075.85 profit**

11. Doris Hauler owned 300 shares of Watkins International stock for which she paid a total of \$2781. She sold the stock at \$3$\frac{1}{8}$ per share and paid a 1.5% sales commission. What was the net amount of the sale? What was the profit or loss from the sale? **\$923.44; \$1857.56 loss**

12. Bruce and Debi Maron sold 850 shares of Consumer Research stock at \$14$\frac{5}{8}$ per share. They paid a 2.5% sales commission. They had originally paid a total of \$14,320.45 for the stock. What was the profit or loss? **\$2199.98 loss**

13. Carl McCollun owned 80 shares of NewTech Computer stock for which he paid \$24$\frac{1}{2}$ per share plus a 2% commission. He sold at \$48$\frac{5}{8}$ and paid a 3% sales commission. What was the profit or loss from the sale? **\$1774.10 profit**

14. The Montvilles purchased 350 shares of Software Design Inc. stock at \$20$\frac{3}{4}$ and sold 3 years later at \$17$\frac{7}{8}$. They paid a 1.25% sales commission when they purchased the stock and paid a 3.25% sales commission when they sold. What was the profit or loss from the sale? **\$1300.36 loss**

MAINTAINING YOUR SKILLS Look up the skills in parentheses if you need help or more practice.

Which number is greater? **(Skill 1)**

15. 381 or 381.6 **381.6**

16. \$21.19 or \$21.91 **\$21.91**

17. \$219.84 or \$218.94 **\$219.84**

Subtract. **(Skill 6)**

18. 88.35 − 81.15 **7.2**

19. 53.83 − 38.91 **14.92**

20. \$8244.50 − \$3398.49 **\$4846.01**

Multiply. **(Skill 8)**

21. 471.2 × 0.035 **16.492**

22. 100 × 3.141 **314.1**

23. 15.058 × 7.2 **108.4176**

PRACTICE AND APPLY

The following problems can be assigned for classwork and the answers checked in class to help students master the objective of the lesson.

- Guided Practice: 1–7
- Independent Practice: 9–13 odd

WRAP-UP

Ask a student to explain the solution Problem 13. Select another student to identify the net sale for the transaction. (\$3773.30)

Assignment Guide

- Basic: 8–13, 15–23
- Average: 8–14 even, 15–23 odd

ALTERNATIVE STRATEGIES: Reteaching

Stress that the commission is *added* to the cost of stock to find the total paid when you *buy* stock and the commission is *subtracted* from the amount of sale to find the net sale when you *sell* stock. The difference is the profit or loss. Ask your students how they could lose money if they sold stock at the same price they paid for it? (Commission) Could they have made any money on this stock? (Yes, dividends) Could anyone make money on this transaction? (Yes, broker)

LESSON PLAN 10-10 Bonds

FOCUS
Some students may be familiar with U.S. government bonds, and you can use these types of bonds to initiate the discussion of this lesson. Students should understand that even the United States government borrows money to pay its bills by issuing bonds. In this lesson, the focus is on computing the annual interest and annual yield of a bond investment.

TEACH
Work through the Example step by step with students. Discuss the meaning of *face value* of a bond. Ask students why a person would want to buy bonds rather than putting the money in a bank savings account. (He or she could earn more interest.) Point out to students that cities can also sell bonds, called municipal bonds, which are used to finance city projects, including the building of new schools.

Warm-up Exercises
1. $\frac{1}{2}$ 0.5
2. $\frac{1}{4}$ 0.25
3. $\frac{3}{4}$ 0.75

Calculate.

4. 7% × $1000 $70
5. 85% × $1000 $850
6. $95\frac{1}{2}$% × $1000 $955
7. $65 ÷ $520 0.125
8. 100 ÷ 2000 0.05
9. 104 ÷ 2080 0.05

10-10 Bonds

OBJECTIVE
Compute the annual interest and annual yield of a bond investment.

Many corporations and governments raise money by issuing **bonds.** When you invest in a bond, you do not become a part owner of the corporation. Instead, you are lending money to the corporation or government. In return for the loan, you are paid interest. On the date the bond matures, you will receive the **face value** of the bond. The face value is the amount printed on the bond.

Corporations usually issue bonds that mature in 10 to 30 years with face values that are multiples of $1000. Bonds may sell at a **discount** for less than the face value, or at a **premium** for more than the face value. The cost of a bond is usually a percent of the face value and is referred to as the **quoted price.** A price of 90 on a $1000 bond means that the bond sells for 90% of $1000, or $900. The interest that you receive from a bond is calculated on its face value.

Annual Interest = Face Value × Interest Rate

Bond Cost = Face Value × Percent

$$\text{Annual Yield} = \frac{\text{Annual Interest}}{\text{Bond Cost}}$$

EXAMPLE *Skills* 30, 11 *Application* A *Term* Bond

George Vanderhill purchased a $1000 bond at the quoted price of $89\frac{1}{2}$. The bond paid interest at a rate of $6\frac{1}{2}$%. What is the annual yield, to the nearest hundredth of a percent?

SOLUTION

A. Find the **annual interest.**
Face Value × Interest Rate
$1000.00 × $6\frac{1}{2}$% = $65.00 annual interest

B. Find the **bond cost.**
Face Value × Percent
$1000.00 × $89\frac{1}{2}$% = $895.00 bond cost

C. Find the **annual yield.**
Annual Interest ÷ Bond Cost
$65.00 ÷ $895.00 = 0.07262 = 7.26% annual yield

1000 [×] 6.5 [%] 65 1000 [×] 89.5 [%] 895 65 [÷] 895 [=] .072625698

ALTERNATIVE ASSESSMENT
Unit 10 is a difficult unit for students because it covers many complex topics. Check students' understanding of the following basic concepts: health and life insurance, coinsurance clause, term life insurance, certificate of deposit, shares of stocks, dividends, and bonds.

✓ SELF-CHECK Complete the problems, then check your answers in the back of the book.

Find the interest, bond cost, and annual yield.

1. \$1000 bond at $80\frac{1}{2}$. Pays 6% interest. **\$60; \$805; 7.45%**
2. \$10,000 bond at 92. Pays $7\frac{1}{4}$% interest. **\$725; \$9200; 7.88%**

PROBLEMS

	Face Value of Bond	Quoted Price	Cost of Bond	Interest Rate	Annual Interest	Annual Yield
3.	\$10,000	95	\$9500	8%	\$800	8.42%
4.	\$ 1,000	$85\frac{1}{2}$	\$ 855	$7\frac{1}{4}$%	\$ 72.50	8.48%
5.	\$10,000	$92\frac{3}{8}$	\$9237.50	$8\frac{1}{2}$%	\$850	9.20%
6.	\$ 1,000	74.324	\$ 743.24	$9\frac{7}{8}$%	\$ 98.75	13.29%

7. Celinda Vasquez.
   Purchased a \$1000 bond at $87\frac{1}{2}$.
   Pays 4% annual interest.
   What is the annual interest?
   What was the cost of the bond?
   What is the annual yield?
   \$40; \$875; 4.57%

8. Giuseppe Caviness.
   Purchased a \$10,000 bond at $72\frac{5}{8}$.
   Pays $6\frac{1}{4}$% annual interest.
   What is the annual interest?
   What was the cost of the bond?
   What is the annual yield?
   \$625; \$7262.50; 8.61%

9. Sandy and Morry Doran purchased a \$10,000 bond at a quoted price of $94\frac{3}{8}$. The bond pays annual interest at a rate of $3\frac{3}{4}$%. What is the annual interest earned? What was the cost of the bond? What is the annual yield? **\$375; \$9437.50; 3.97%**

10. The Sonoma Housing Authority is offering \$5000 bonds that pay 7.4% annual interest. The quoted price of each bond is 92.128. Rollin Kowalski purchases 8 bonds through a broker who charges a 1.75% sales commission. What is the total cost of his purchase? What total interest will he earn yearly? What is the annual yield? **\$37,496.10; \$2960; 7.894%**

PRACTICE AND APPLY

The following problems can be assigned for classwork and the answers checked in class to help students master the objective of the lesson.

- Guided Practice: 1–7
- Independent Practice: 8, 9

ALTERNATIVE STRATEGIES: Reteaching

Emphasize the difference between stocks and bonds. With stocks you are a part owner of the company; with bonds you are lending the company money. Stock prices are per share; bond prices are given as a percent of the face value. What would a \$30,000 bond at 105% sell for? (\$31,500) Why would anyone pay a premium of \$31,500 (over 100%) for a bond with a face value of \$30,000? (High return on their money)

WRAP-UP

Called upon students to define the following terms: bond, face value, annual interest on a bond, bond cost, annual yield.

Assignment Guide

- Basic: 8–10, 12–23
- Average: 10, 11, 13–23 odd

11. Many newspapers give up-to-date bond information in the financial section. In the listing shown, the name of the corporation is given, then the annual interest rate the bond is paying, the year the bond matures, and so on. "08" means the bond matures in the year 2008.

Alex Wosick purchased six $1000 bonds of the Dyer Corporation at the closing price for the day. He had to pay a broker's commission of 2.25%. What was the total cost of his purchase? What will be his total yearly earnings? In what year will his bond mature?
$5046.04; $652.50; 2009

Bonds			Cur. Yld.	Vol.	High	Low	Close
DICLO	$9\frac{3}{8}$	08	10.5	2	72	72	72
DYER	$10\frac{7}{8}$	09	8.7	10	$82\frac{1}{4}$	$82\frac{1}{4}$	$82\frac{1}{4}$
EANLO	$8\frac{3}{4}$	12	13.9	28	$71\frac{1}{2}$	70	$71\frac{1}{2}$
EGG CO	$12\frac{1}{4}$	10	9.0	3	$93\frac{5}{8}$	$93\frac{5}{8}$	$93\frac{5}{8}$

MAINTAINING YOUR SKILLS Look up the skills in parentheses if you need help or more practice.

Find the percentage. **(Skill 30)**

12. $6\frac{1}{2}$% of $700 **$45.50** **13.** $93\frac{1}{4}$% of $15,000 **$13,987.50** **14.** $\frac{1}{2}$% of 40 **0.2** **15.** 120% of 900 **1080**

16. $15\frac{1}{4}$% of 350 **53.375** **17.** 110% of 940 **1034** **18.** $2\frac{1}{2}$% of 82 **2.05** **19.** 4.4% of 14,240 **626.56**

Divide. Round answers to the nearest hundredth. **(Skill 11)**

20. 2.22)56.80 **25.59** **21.** 19.4)608 **31.34** **22.** 0.004)96.28 **24,070** **23.** 14.10)48.215 **3.42**

COMMUNICATION SKILLS

Use the Alternative Assessment to also assess students' verbal skills as they attempt to define the basic concepts listed. You may want to have some students write the definitions on the chalkboard. Then allow all students to use their books to check the work on the board. Call upon some students to read aloud the definitions given in the textbook.

Reviewing the Basics

Skills

Round to the nearest cent.

(Skill 2)

1. $74.135 **$74.14** **2.** $1.374 **$1.37** **3.** $318.334 **$318.33** **4.** $6.997 **$7.00**

Solve. Round answers to the nearest cent.

(Skill 8)

5. $1411.25 × 30 **$42,337.50** **6.** $4.54 × 32 **$145.28** **7.** $838.06 × 205 **$171,802.30**

8. $7.35 × 85 **$624.75** **9.** $72.50 × 41 **$2972.50** **10.** $132.52 × 9 **$1192.68**

(Skill 11)

11. $908.14 ÷ 4 **$227.04** **12.** $43.20 ÷ 8 **$5.40** **13.** $983.62 ÷ 12 **$81.97**

(Skill 30)

14. 42.3% of $4914.23 **$2078.72** **15.** 6% of $180 **$10.80** **16.** 39.2% of $8974.17 **$3517.87**

Write as a decimal.

(Skill 14)

17. $\frac{5}{16}$ **0.3125** **18.** $\frac{7}{8}$ **0.875** **19.** $1\frac{5}{8}$ **1.625** **20.** $\frac{5}{32}$ **0.15625** **21.** $1\frac{1}{4}$ **1.25** **22.** $3\frac{5}{10}$ **3.5**

Write as a percent.

(Skill 26)

23. 0.09 **9%** **24.** 0.16 **16%** **25.** 0.188 **18.8%** **26.** 0.0875 **8.75%** **27.** 1.79 **179%**

Applications

Use the table to find the annual premium.

(Application C)

28. Tom Baker purchased a $25,000 whole life insurance policy at the age of 30. What is his annual premium? **$291.75**

29. Sandra Logan purchased a $40,000 paid-up-at-age-65 policy at the age of 30. What is her annual premium? **$576.40**

ANNUAL PREMIUMS PER $1000 OF LIFE INSURANCE

Age	Paid up at Age 65		Whole Life	
	Male	Female	Male	Female
20	$11.59	$ 9.61	$ 8.05	$6.28
25	13.70	11.48	9.45	7.46
30	16.88	14.41	11.67	9.13

Terms

Write your own definition for each term. **Answers will vary.**

30. Term life insurance **31.** Whole life insurance **32.** Dividends

33. Certificate of deposit **34.** Annual yield **35.** Stocks

36. Profit or loss **37.** Bonds **38.** Health insurance

Refer to your reference files in the back of the book if you need help.

LESSON PLAN
Reviewing the Basics

The exercises on this page review skills, applications, and terms used in the unit. You can use the exercises to assess informally students' proficiency with this material.

The page can be used for guided practice and independent practice. You can work through a selection of the exercises together with students, and thus see immediately if they know how to do them, and you can then assign some of the exercises for independent practice. Be sure to go over the answers to all assigned exercises.

LESSON PLAN
Unit Test

Students should do the Unit Test on their own. Each problem on the test is keyed to a lesson in the unit. Students having difficulty with any particular problem should review the Example in the appropriate lesson and be assigned some of the Independent Practice problems for additional practice.

Unit Test

Lesson 10-1

1. Russell Kirby is employed by Dental Associates. The annual premium for a comprehensive medical insurance program is $3160. His employer pays 70% of the cost. How much does he pay? **$948**

Lesson 10-2

2. Nancy King has a medical bill of $26,000. Her hospital and surgical-medical insurance did not pay for $2800 of the total bill. Her major medical insurance has a $250 deductible and a 20% coinsurance clause. How much did she pay? **$760**

Lesson 10-3

3. Monte Onstenk purchases a $60,000, 5-year term life insurance policy. The annual base premium is $14.10 per $1000. What is his annual premium? **$846**

Lesson 10-4

4. Paul Kenyon purchased a $40,000 whole life insurance policy. The annual base premium is $19.45 per $1000. What is his annual premium? **$778**

Lesson 10-5

5. Karolyn Cargill invested $8000 in a 1-year certificate of deposit. The CD earns interest at a rate of 10.00% compounded quarterly. How much interest will she earn on the date of maturity? **$830.50**

AMOUNT PER $1.00 INVESTED, COMPOUNDED QUARTERLY

Annual Rate	Interest Period 1 year	4 Years
10.00%	1.103813	1.484506
10.25%	1.106508	1.499055
10.50%	1.109207	1.513738

Lesson 10-6

6. Joe Ponderoza invested $4500 in a certificate of deposit that earns interest at a rate of 10.00% and matures after 4 years. Interest is compounded quarterly. What is the effective annual yield? **10.38%**

Lesson 10-7

7. Leah O'Grady purchased 300 shares of Penn Brothers stock at $\$44\frac{3}{8}$. She was charged a $30 commission. What was the total amount that she paid for the purchase? **$13,342.50**

Lesson 10-8

8. Linda Pitta owns 170 shares of Romez International for which she paid $\$28\frac{7}{8}$ per share. If the company pays dividends of $1.45 per share, what is the annual yield to the nearest hundredth of a percent? **5.02%**

Lesson 10-9

9. Rick Reilly owns 70 shares of DuPere stock for which he paid a total of $4472. He sold the stock for $84 per share and paid a sales commission of $50. What was his profit or loss on the investment? **$1358 profit**

Lesson 10-10

10. Denise Rutherford purchased a $1000 bond at the quoted price of $89\frac{3}{4}$. The bond pays interest at a rate of $4\frac{3}{4}$%. What is the annual yield to the nearest hundredth of a percent? **5.29%**

A SPREADSHEET APPLICATION

Compound Interest/ Annual Yield

To complete this spreadsheet application, you will need the template diskette for *Mathematics with Business Applications*. Follow the directions in the *User's Guide* to complete this activity.

Input the information in the following problems to find the amount, interest, and annual yield.

1. Investment: $5000.
Interest rate: 8%.
Compounded: quarterly.
Number of years: 10.
What is the amount?
What is the interest?
What is the annual yield?

2. Investment: $18,000.
Interest rate: 9%.
Compounded: monthly.
Number of years: 6.
What is the amount?
What is the interest?
What is the annual yield?

3. Investment: $1500.
Interest rate: 9.25%.
Compounded: quarterly.
Number of years: 4.
What is the amount?
What is the interest?
What is the annual yield?

4. Investment: $4000.
Interest rate: 8%.
Compounded: monthly.
Number of years: 12.
What is the amount?
What is the interest?
What is the annual yield?

5. Investment: $50,000.
Interest rate: 7.75%.
Compounded: daily.
Number of years: 4.
What is the amount?
What is the interest?
What is the annual yield?

6. Investment: $140,000.
Interest rate: 9.25%.
Compounded: quarterly.
Number of years: 1.
What is the amount?
What is the interest?
What is the annual yield?

7. Irving Pettit invested $5500 at 8.75% compounded daily for 6 years. What is the amount? What is the interest? What is the annual yield?

8. Kerri Keller invested $1000 at 6% compounded daily for 12 years. What is the amount? What is the interest? What is the annual yield?

LESSON PLAN A Spreadsheet Application

USING TECHNOLOGY

In this application, students use a computer to find the amount, interest, and annual yield for various investments. Point out to students that one of the most effective ways to make money grow is to invest it at compound interest rates. Ask them why this is so. (because interest is paid on interest over time)

In Problem 1, point out that an investment of $5000 at 8% compounded quarterly has more than doubled in 10 years. As a *critical thinking* question, ask students if an investment of any amount would double in less than 10 years at a rate less than 8% compounded quarterly. (yes) As a *problem solving* activity, have students change the rate in Problem 1 to find out what rate would double the $5000 in 10 years.

A SPREADSHEET APPLICATION
(CONTINUED)

9. Lyle Growden invested $5623.21 at 7.25% compounded monthly for 4 years. What is the amount? What is the interest? What is the annual yield?

10. Susie Bradford invested $6897 at 11.32% compounded monthly for 4 years. What is the amount? What is the interest? What is the annual yield?

11. At the end of the 4 years, Susie Bradford in problem 10 reinvested the amount at 10.72% compounded quarterly for 3 years. What is the amount? What is the interest? What is the annual yield?

12. Donald Huddleston has $10,000 to invest in a retirement account. He plans to leave the money on deposit for 30 years. Complete the chart below to help him decide which investment firm offers the best deal.

	Institution	Rate	Amount	Interest	Yield
a.	Farmers & Merchants	6.5%, annually	?	?	?
b.	First Federal S&L	6.5%, daily	?	?	?
c.	Metamora State Bank	7.5%, daily	?	?	?
d.	Century Insurance	8.5%, daily	?	?	?

UNIT

11

Recordkeeping

Recordkeeping is a way for you to manage your money. By keeping track of your monthly *expenditures,* you can find out how you have spent your money and how much money you need to live on. Your next step is to prepare a *budget sheet* showing your monthly expenses. The purpose of a budget is to allow you to compare how much money you are spending with how much money you are earning.

OUTLINE

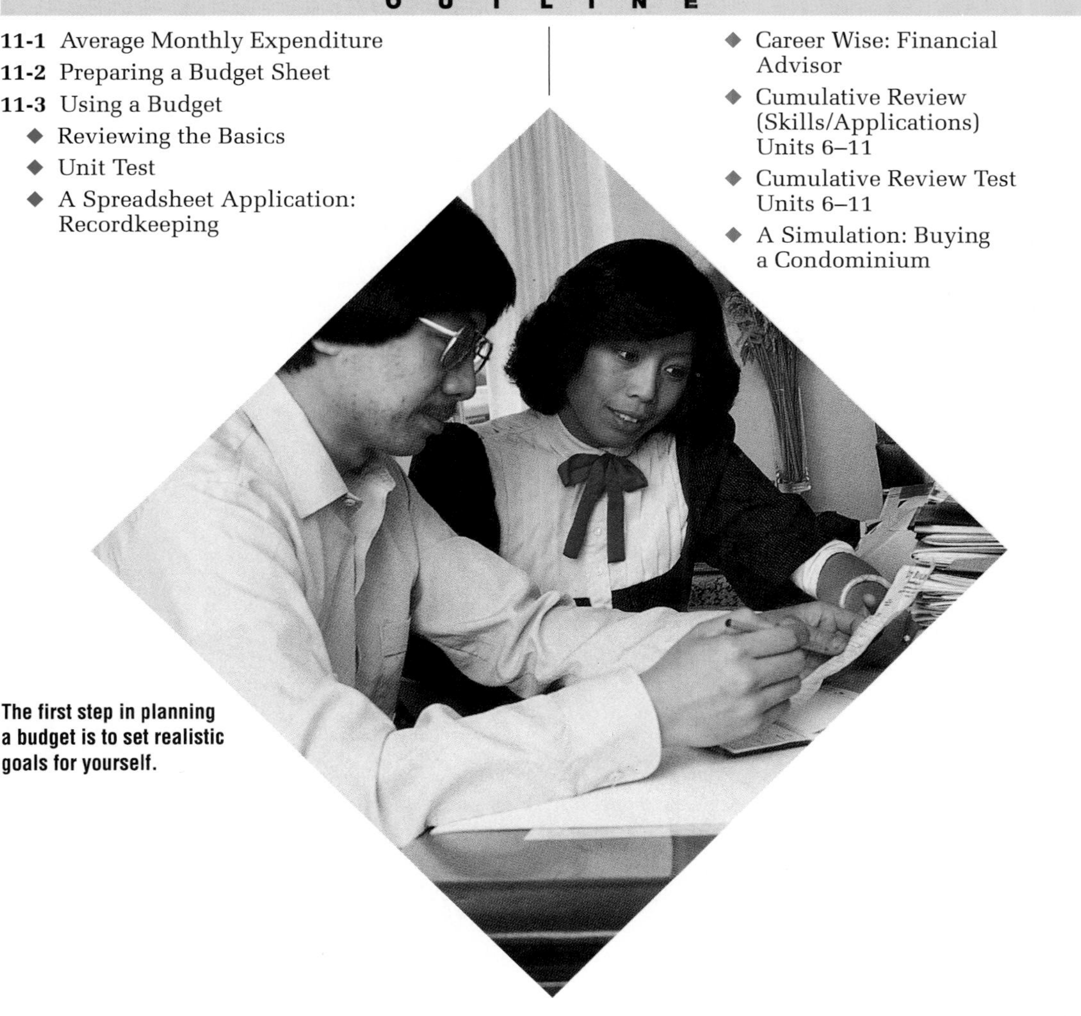

The first step in planning a budget is to set realistic goals for yourself.

UNIT 11 RECORDKEEPING

INTRODUCING THE UNIT

To introduce the lesson, discuss the importance of recordkeeping in our daily lives. Businesses and schools rely on recordkeeping to maintain accurate records. You may want to bring examples of some common recordkeeping forms (used in the school, perhaps) to class, and talk about when and why they are used.

In preparing budgets, recordkeeping is a critical writing skill. Discuss with students the importance of keeping an accurate record of expenditures. People are usually surprised by how much they spend in different ways when they examine their spending habits.

LESSON PLAN 11-1 Average Monthly Expenditure

FOCUS

You can motivate the unit and this lesson by asking students how many either earn money from a job or work for an allowance. Ask these students if they are expected to help supply their own needs from this money. Make a list of ways students spend the money they receive.

TEACH

Part of money management is seeing how much money is being spent and where. Point out that each monthly record has been written in the order in which the money was spent. This avoids missing any spending that was done. A simple notation will help keep track of how the money was spent. Point out that an average monthly expenditure is better than just looking at one month since many bills are seasonal or are not reoccurring.

Warm-Up Exercises

1. (\$48 + \$51 + \$42) ÷ 3 \$47
2. (\$17.49 + \$20.71 + \$17.45) ÷ 3 \$18.55
3. (\$212.50 + \$241.75 + \$227.80) ÷ 3 \$227.35
4. (\$51.60 + \$49.80 + \$52.30 + \$51.42) ÷ 4 \$51.28

11-1 Average Monthly Expenditure

OBJECTIVE
Compute the average monthly expenditure.

Your first step toward managing your money is to keep an accurate record of how you spend your money. Use a notepad to record any **expenditures** on the day you make them. At the end of the month, group them and total them. By keeping a record of your expenditures, you will be able to examine your spending habits.

$$\text{Average Monthly Expenditure} = \frac{\text{Sum of Monthly Expenditures}}{\text{Number of Months}}$$

EXAMPLE *Skills* 5, 11 *Applications* A, Q *Term* Expenditures

Sue and Bob Miller keep records of their expenditures. They want to know how much they spend each month, on the average. Here are their records for 3 months. What is their average monthly expenditure?

July		August		September	
mortgage payment	\$675.00	mortgage payment	\$675.00	mortgage payment	\$675.00
grocery bill	\$51.35	Beth's allowance	\$32.00	electric bill	\$51.42
Beth's allowance	\$32.00	electric bill	\$73.56	doctor bill	\$35.00
electric bill	\$71.47	restaurant	\$27.80	cleaners	\$17.65
dentist	\$43.50	movies	\$13.50	telephone bill	\$32.75
telephone bill	\$27.85	telephone bill	\$26.45	gasoline	\$16.75
gasoline	\$15.60	donation	\$25.00	grocery bill	\$59.74
water/sewer bill	\$31.45	grocery bill	\$62.35	football game	\$15.00
credit card payment	\$41.74	personal expenses	\$75.00	credit card payment	\$71.46
baseball game	\$19.50	credit card payment	\$54.92	gasoline	\$16.45
gift	\$45.00	gasoline	\$17.94	Beth's allowance	\$32.00
clothing	\$71.56	magazine subscription	\$31.50	grocery bill	\$56.74
car payment	\$178.50	car payment	\$178.50	car payment	\$178.50
grocery bill	\$63.70	grocery bill	\$71.48	fuel oil	\$78.75
TOTAL	\$1368.22	TOTAL	\$1365.00	TOTAL	\$1337.21

SOLUTION

Find the **average monthly expenditure.**

Sum of Monthly Expenditures ÷ Number of Months

\$1368.22 + \$1365.00 + \$1337.21 ÷ 3 = \$1356.81 average monthly expenditure

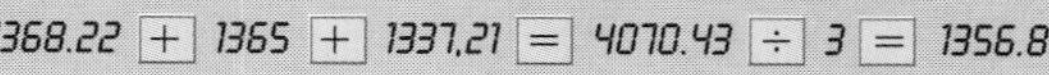

1368.22 [+] 1365 [+] 1337.21 [=] 4070.43 [÷] 3 [=] 1356.81

CULTURAL ANGLES

Thailand had a per-capita income in 1986 of \$771. South Korea's was \$2180. Sweden's, in 1989, was \$11,783. Point out to students that *per capita* means *per person,* and ask which of these countries has more money per person for its budget. (Sweden) Have students look in an almanac or other source for the current population to estimate the total income for each country based on the per capita income.

✔ SELF-CHECK Complete the problems, then check your answers in the back of the book.

Find the average monthly expenditure.

1. January, \$795; February, \$776; March, \$751. **\$774**

2. May, \$1571.83; June, \$1491.75; July, \$1543.85; August, \$1526.77. **\$1533.55**

PROBLEMS

Find the average monthly expenditure.

	3.	4.	5.	6.	7.
May	\$640.00	\$1178.50	\$1789.75	\$2311.75	\$112.11
June	710.00	1091.80	1741.36	2210.91	97.13
July	700.00	1207.70	1707.85	2371.85	106.45
August	685.00	1197.80	1751.63	2353.67	121.85
September	705.00	1245.90	1811.75	2412.91	107.91
Total	**\$3440.00**	**\$5921.70**	**\$8802.34**	**\$11,661.09**	**\$545.45**
Average	**\$688.00**	**\$1184.34**	**\$1760.47**	**\$ 2,332.22**	**\$109.09**

F.Y.I.
A typical family of four with a net income of \$20,000 spends 48% of their budget for food and housing.

Use the Millers' records of monthly expenditures on page 316 to answer the following.

8. What is the Millers' average monthly expenditure for groceries? **\$121.79**

9. Household costs include amounts for electric bills, telephone bills, water and sewer bills, home fuel oil bills, and so on. What is their average monthly expenditure for household expenses? **\$131.23**

10. Entertainment expenses include amounts for restaurants, movies, and recreation. What is their average monthly expenditure for entertainment? **\$25.27**

11. What do the Millers pay each month to repay their mortgage loan? **\$675.00**

12. Transportation costs include car payments and amounts for gasoline, oil, repairs, and so on. What is their average monthly expenditure for transportation costs? **\$200.75**

13. Can you determine how much the Millers save each month? Why or why not? **No; do not know their monthly net income.**

14. The Millers' records for October, November, and December show that their monthly expenditures totaled \$1375.80, \$1412.91, and \$1512.18, respectively. What is their average monthly expenditure for the past 6 months? **\$1395.22**

MAINTAINING YOUR SKILLS Look up the skills in parentheses if you need help or more practice.

Add. **(Skill 5)**

15. \$716.45 + \$820.97 **\$1537.42**

16. \$21.63 + \$22.71 + \$24.95 **\$69.29**

Find the average. **(Application Q)**

17. \$1170, \$1241, \$1193, \$1250 **\$1213.50**

18. \$17.91, \$18.43, \$16.25 **\$17.53**

PRACTICE AND APPLY

The following problems can be assigned for classwork and the answers checked in class to help students master the objective of the lesson.

- Guided Practice: 1–5, 8
- Independent Practice: 9–12

WRAP-UP

Have a student explain how to find an average monthly expenditure. Read Problem 10 to the class. Have students list, by month, which items should be included for entertainment.

Assignment Guide

- Basic: 6, 7, 9–14, 16–19
- Average: 6, 7, 13–15, 16–19

ALTERNATIVE STRATEGIES: Reteaching

Finding an average is the critical mathematical skill required in this lesson. Review with your students finding the average of these lists of expenditures:

1) \$470; \$486; \$490; \$450; (\$474)
2) \$1450; \$1470; \$1498; \$1500; \$1480; \$1410; (\$1468)
3) \$75.60; \$81.45; \$78.90; \$76.75; \$80.65; (\$78.67)
4) \$21,468.90; \$19,785.86; \$20,653.09; (\$20,635.95)

LESSON PLAN
11-2 Preparing a Budget Sheet

FOCUS
The focus of this lesson is to use records of past expenditures to prepare a monthly budget. Ask students if they have ever heard anyone talk about having to live on a budget. Ask students where else they have heard the term *budget.* Some may suggest a school budget, the federal budget, and so on.

TEACH
Go over the main categories on the budget sheet. Point out that the fixed expenses are ones which cannot be changed from month to month. Any adjustments in a budget are usually made in the living expenses. Have students discuss which living expenses would be easiest to change and which would be the most difficult to change. Point out that for items such as movies, the money can be carried from month to month, or used elsewhere, if it is not used in a particular month. Some annual expenses, such as real estate taxes or auto insurance, can sometimes be paid more often than annually. This would not change the monthly budgeted amount, however.

11-2 Preparing a Budget Sheet

OBJECTIVE
Use records of past expenditures to prepare a monthly budget.

If you have records of your past expenditures, you can use them to prepare a **budget sheet** outlining your total monthly expenses. **Living expenses** vary from month to month and include amounts for food, utility bills, pocket money, and so on. **Fixed expenses** are expenses that do not vary from one month to the next. **Annual expenses,** such as insurance premiums and real estate taxes, occur once a year.

Total Monthly Expenses = **Monthly Living Expenses** + **Monthly Fixed Expenses** + **Monthly Share of Annual Expenses**

EXAMPLE *Skills* 5, 11 *Application* A *Term* Budget sheet

The Millers use records of their past expenditures to complete the budget sheet below. What is the total of their monthly expenses?

A MONEY MANAGER FOR *Sue and Bob Miller* DATE *10/1/--*

MONTHLY LIVING EXPENSES		MONTHLY FIXED EXPENSES	
Food/Grocery Bill	$ *125.00*	Rent/Mortgage Payment	$ *675.00*
Household Expenses		Car Payment	$ *178.50*
Electricity	$ *70.00*	Other Installments	
Heating Fuel	$ *45.00*	Appliances	$
Telephone	$ *30.00*	Furniture	$
Water	$ *11.00*	Regular Savings	$ *75.00*
Garbage/Sewer Fee	$	Emergency Fund	$ *50.00*
Other	$	TOTAL	$ *978.50*
	$	**ANNUAL EXPENSES**	
Transportation		Life Insurance	$ *575.00*
Gasoline/Oil	$ *25.00*	Home Insurance	$ *240.00*
Parking	$	Car Insurance	$ *475.00*
Tolls	$	Real Estate Taxes	$ *1215.00*
Commuting	$	Car Registration	$ *26.50*
Other	$	Pledges/Contributions	$ *100.00*
Personal Spending		Other	$
Clothing	$ *30.00*	TOTAL	$ *2631.50*
Credit Payments	$ *60.00*	MONTHLY SHARE (Divide by 12)	$ *219.29*
Newspapers, Gifts, Etc.	$ *25.00*	**MONTHLY BALANCE SHEET**	
Pocket Money	$ *57.00*	Net Income (Total Budget)	$
Entertainment		Living Expenses:	$ *505.00*
Movies/Theater	$ *5.00*	Fixed Expenses:	$ *978.50*
Sporting Events	$ *12.00*	Annual Expenses:	$ *219.29*
Recreation	$	TOTAL MONTHLY EXPENSES	$
Dining Out	$ *10.00*		
TOTAL	$ *505.00*	BALANCE	$

COOPERATIVE LEARNING
Organize students into small groups and give each student a copy of the Money Manager from the Teacher's Resource Book. Have students write their names and today's date on the form. Tell them that expenditures for this month are the same as in the Example, except that you spend $221.50 for food/grocery, $520 for rent; thus, there is no home insurance nor real estate taxes, and $346 for car insurance. Your monthly net income is $1500. Have students complete the Money Manager and find the balance.

SOLUTION Find the **total monthly expenses.**

Monthly Living Expenses	+	Monthly Fixed Expenses	+	Monthly Share of Annual Expenses		
$505.00	+	$978.50	+	$219.29	=	$1702.79 total monthly expenses

505 [+] 978.5 [+] 219.29 [=] 1702.79

✔ SELF-CHECK Complete the problems, then check your answers in the back of the book.

Find the total monthly expenses.

1. Living, $670; fixed, $800; share of annual, $350. **$1820**
2. Living, $475.75; fixed, $679.65; share of annual, $291.17. **$1446.57**

PROBLEMS

Betty Kujawa is a landscaper. Walt Kujawa is a used car salesman. They complete the budget sheet shown using records of their past expenditures.

Use the budget sheet to answer the following.

3. What is the total of the Kujawas' monthly living expenses? **$751.50**
4. What is the total of their monthly fixed expenses? **$900**
5. What is the total of their annual expenses? **$1742**
6. What must be set aside each month for annual expenses? **$145.17**
7. What is their total monthly expenditure? **$1796.67**
8. Do the Kujawas live within their monthly net income? **Yes**
9. What individual expenses would be difficult for the Kujawas to cut back on? **Rent, life insurance, car insurance, car registration**

A MONEY MANAGER FOR *Walt and Betty Kujawa* DATE *4/10/--*

MONTHLY LIVING EXPENSES		MONTHLY FIXED EXPENSES	
Food/Grocery Bill	$ *160.00*	Rent/Mortgage Payment	$ *625.00*
Household Expenses		Car Payment	$
Electricity	$ *45.00*	Other Installments	
Heating Fuel	$ *50.00*	Appliances	$
Telephone	$ *35.00*	Furniture	$ *125.00*
Water	$ *24.50*	Regular Savings	$ *100.00*
Garbage/Sewer Fee	$	Emergency Fund	$ *50.00*
Other *Cable TV*	$ *25.00*	TOTAL	$
	$	**ANNUAL EXPENSES**	
Transportation		Life Insurance	$ *840.00*
Gasoline/Oil	$ *85.00*	Home Insurance	$
Parking	$ *5.00*	Car Insurance	$ *750.00*
Tolls	$ *10.00*	Real Estate Taxes	$
Commuting	$	Car Registration	$ *52.00*
Other	$	Pledges/Contributions	$ *100.00*
Personal Spending		Other	$
Clothing	$ *40.00*	TOTAL	$
Credit Payments	$ *50.00*	MONTHLY SHARE (Divide by 12)	$
Newspapers, Gifts, Etc.	$ *20.00*	**MONTHLY BALANCE SHEET**	
Pocket Money	$ *60.00*	Net Income (Total Budget)	$ *1800.00*
Entertainment		Living Expenses:	$
Movies/Theater	$ *10.00*	Fixed Expenses:	$
Sporting Events	$ *20.00*	Annual Expenses:	$
Recreation	$ *12.00*	TOTAL MONTHLY EXPENSES	$
Dining Out	$ *100.00*		
TOTAL	$	BALANCE	$

Warm-Up Exercises

1. $17.80 + 19.30 + 24.70 + 20.50 = $82.30
2. $74.60 + 81.90 + 71.60 + 78.70 = $306.80
3. $141.75 + 216.80 + 419.20 + 317.80 = $1095.55
4. $1791.86 + 1417.57 + 1921.43 + 2217.58 = $7348.44
5. $1590.72 ÷ 12 $132.56
6. $167.34 ÷ 6 $27.89

PRACTICE AND APPLY

The following problems can be assigned for classwork and the answers checked in class to help students master the objective of the lesson.

- Guided Practice: 1–4
- Independent Practice: 5–9

ALTERNATIVE STRATEGIES: Enrichment

Have your students complete a Money Manager form from the TRB using either themselves, their family, an older brother or sister, a friend, or a friend's family. It would be ideal if they could maintain this record over a period of months to get a feel for average monthly expenditures. If the income part of the form is getting too personal, omit that portion.

WRAP-UP

Ask different students to name each of the three components of the total monthly expenses. Then, use $600 for the monthly living expenses, $1500 for the monthly fixed expenses, and $2400 for the total annual expenses. Have students compute the total monthly expenses. ($2300)

Assignment Guide

- Basic: 5–15, 17–30
- Average: 10–16, 18–30 even

Nancy and Joe Thomas completed the budget sheet shown using records of their past expenditures.

Use the budget sheet to answer the following.

10. What is the total of their monthly living expenses? **$847.63**

11. What is the total of their monthly fixed expenses? **$1008.75**

12. What is the total of their annual expenses? **$2721.50**

13. What must be set aside each month for annual expenses? **$226.79**

14. What is their total monthly expenditure? **$2083.17**

15. Do the Thomases live within their monthly net income? **No**

16. What individual expenses would be difficult for the Thomases to cut back on? **Mortgage payment; life, home, car insurance; real estate taxes; car registration**

A MONEY MANAGER FOR *Nancy and Joe Thomas* DATE *7/20/--*

MONTHLY LIVING EXPENSES	
Food/Grocery Bill	$ *210.00*
Household Expenses	
Electricity	$ *55.65*
Heating Fuel	$ *63.75*
Telephone	$ *21.47*
Water	$ *31.80*
Garbage/Sewer Fee	$ *17.21*
Other *Security*	$ *25.00*
______	$ ______
Transportation	
Gasoline/Oil	$ *60.00*
Parking	$ *35.00*
Tolls	$ *12.00*
Commuting	$ *20.00*
Other *Misc.*	$ *35.00*
Personal Spending	
Clothing	$ *100.00*
Credit Payments	$ *25.00*
Newspapers, Gifts, Etc.	$ *16.75*
Pocket Money	$ *32.00*
Entertainment	
Movies/Theater	$ *20.00*
Sporting Events	$ *20.00*
Recreation	$ *15.00*
Dining Out	$ *32.00*
TOTAL	$ ______

MONTHLY FIXED EXPENSES	
Rent/Mortgage Payment (Mortgage circled)	$ *715.20*
Car Payment	$ ______
Other Installments	
Appliances	$ *57.75*
Furniture	$ *110.80*
Regular Savings	$ *75.00*
Emergency Fund	$ *50.00*
TOTAL	$ ______
ANNUAL EXPENSES	
Life Insurance	$ *480.00*
Home Insurance	$ *180.00*
Car Insurance	$ *475.00*
Real Estate Taxes	$ *1200.00*
Car Registration	$ *26.50*
Pledges/Contributions	$ *360.00*
Other ______	$ ______
TOTAL	$ ______
MONTHLY SHARE (Divide by 12)	$ ______
MONTHLY BALANCE SHEET	
Net Income (Total Budget)	$ *1800.00*
Living Expenses:	$ ______
Fixed Expenses:	$ ______
Annual Expenses:	$ ______
TOTAL MONTHLY EXPENSES	$ ______
BALANCE	$ ______

MAINTAINING YOUR SKILLS Look up the skills in parentheses if you need help or more practice.

Add. **(Skill 5)**

17. $75 + $45 + $53 + $68 **$241**

18. $475.80 + $519.20 + $647.80 **$1642.80**

19. $6.18 + $7.23 + $4.37 + $7.96 **$25.74**

20. $71.14 + $86.23 + $64.91 **$222.28**

21. $619.76 + $723.39 + $671.46 **$2014.61**

22. $1178.21 + $1371.89 + $1475.84 **$4025.94**

Divide. Round to the nearest cent. **(Skill 11)**

23. $241 ÷ 4 **$60.25**

24. $1642.80 ÷ 3 **$547.60**

25. $25.74 ÷ 4 **$6.44**

26. $222.28 ÷ 3 **$74.09**

BUSINESS PROJECT

Have students investigate and report on the percent of income that is recommended for various categories of expenditures. A possible source is the Bureau of Labor Statistics of the U.S. Department of Labor. Have students prepare a budget for a monthly net income of $2000.

11-3 Using a Budget

OBJECTIVE
Compare amount budgeted to actual expenditure and personal spending to "typical spending."

Once you have completed a budget sheet outlining your past expenditures, you can use it to plan for future spendings. You may want to prepare a monthly **expense summary** to compare the amounts that you spend to the amounts that you budgeted. When you draft a budget, you should include an **emergency fund** to provide for unpredictable expenses, such as medical bills and repair bills.

EXAMPLE *Skills* 1, 6 *Term* Emergency fund

The Zornows kept accurate records of their expenditures. At the end of March, they prepared an expense summary. They had planned to spend $220 on groceries. They actually spent $231.85. How much more or less did they spend on groceries than they had budgeted for?

SOLUTION

A. Compare.
Is **amount spent** more or less than **amount budgeted?**
Is $231.85 more or less than $220.00? More

B. Find the **difference.**
$231.85 − $220.00 = $11.85 more than amount budgeted

✓ SELF-CHECK Complete the problems, then check your answers in the back of the book.

Find how much more or less the amount spent is than the amount budgeted.

1. Budgeted $167.80, spent $158.90. **$8.90 less**
2. Budgeted $647.50, spent $671.92. **$24.42 more**

PROBLEMS

3. Mike Ogg's groceries for May.
Budgeted $176.80.
Actually spent $161.75.
Did he spend more or less than budgeted? **$15.05 less**
4. May Church's telephone bill for June.
Budgeted $25.
Actually spent $33.78.
Did she spend more or less than budgeted? **$8.78 more**
5. Mary Teal's water bill for June.
Budgeted $20. **$4.79 more**
Actually spent $24.79.
How much more or less did she spend than budgeted?
6. Burke Long's credit payment for May.
Budgeted $45.
Actually spent $43.45.
How much more or less did he spend than budgeted? **$1.55 less**

LESSON PLAN 11-3 Using a Budget

FOCUS
The focus of this lesson is to compare the amount budgeted to actual expenditure and personal spending to "typical spending." The overall purpose of a budget is for a person or family to check on how money is being spent, and to make adjustments where necessary. Point out that any budget must be somewhat flexible, since unbudgeted expenses are usually encountered at some time.

TEACH
Discuss the importance of the expense summary. In the example, emphasize that it is not necessarily careless spending that causes someone to spend more money than was budgeted originally. Rising costs of items such as food may require periodic adjustments to budgets. This may mean allowing less to be spent for entertainment, clothes, etc. Stress that it is not wise to spend more than you earn.

PROBLEM SOLVING
In a general sense, making and using a budget is a problem-solving activity. People often turn to using budgets when their expenses get out of control and they are forced to curtail expenditures. Ask students if they think the budget process is a problem-solving activity and why. Elicit specific examples of problem solving when using a budget.

Warm-Up Exercises

1. $245 − $240 $5
2. $79 − $67 $12
3. $1475 − $1279 $196
4. $71.74 − $65 $6.74
5. $247.86 − $225 $22.86
6. $75 − $68.71 $6.29
7. $1425 − $1397.63 $27.37
8. $2131.78 − $2125 $6.78
9. $3137.63 − $3078.74 $58.89

PRACTICE AND APPLY

The following problems can be assigned for classwork and the answers checked in class to help students master the objective of the lesson.

- Guided Practice: 1–17
- Independent Practice: 8–15

Here is the Kujawas' expense summary for the month of July. They want to compare what they had budgeted to what they actually spent.

EXPENDITURES FOR THE MONTH OF JULY

Expenses	Amount Budgeted	Actual Amount Spent
Food	$160.00	$175.70
Household		
Electric bill	45.00	44.35
Telephone bill	35.00	41.20
Heating fuel	50.00	15.00
Water bill	24.50	31.70
Cable TV bill	25.00	25.00
Transportation		
Gasoline purchases	85.00	101.70
Parking/tolls	15.00	15.00
Personal		
Clothing	40.00	31.75
Credit payments	50.00	41.74
Newspapers, gifts	20.00	11.65
Pocket money	60.00	72.00
Entertainment		
Movies	10.00	5.00
Sporting events/recreation	32.00	32.00
Dining out	100.00	63.80
Fixed		
Rent	625.00	625.00
Furniture	125.00	125.00
Savings	100.00	100.00
Emergency fund	50.00	50.00
Life/car insurance premiums	132.50	132.50
Car registration	4.33	0
Pledges, contributions	8.33	0

Use the Kujawas' expense summary to answer the following.

7. Which household expenses for the month were more than the amount budgeted? **Telephone bill, water bill**
8. Did they spend more or less than the amount budgeted for household expenses for the month? By how much? **Less; $22.25**
9. How much did they budget for transportation costs? Were the amounts spent for transportation during the month more or less than the amount budgeted? By how much? **$100.00; More; $16.70**
10. Which personal expenses did they spend more on than they had budgeted? **Pocket money**
11. Were their total personal expenditures for the month more or less than the amount budgeted? By how much? **Less; $12.86**
12. Were their total entertainment expenditures for the month more or less than the amount budgeted? By how much? **Less; $41.20**
13. Were there any monthly fixed expenses for which the Kujawas spent more than the amount budgeted? **No**
14. What annual expenses occurred during the month? Was the amount set aside for annual expenses more or less than the amount actually spent for annual expenses? **Life/car insurance; More**
15. What was the Kujawas' total expenditure for the month of July? Was this amount more or less than the amount they had originally budgeted? By how much? **$1740.09; less; $56.57**
16. Your total monthly expenditure will vary from month to month. During the winter months, your home heating bills may "push" your total monthly expenditure over the amount budgeted. In some months, spending will be less. Name some factors that might affect your spending for specific months. **Answers will vary.**

CRITICAL THINKING

Ask students to think about how making purchases using credit cards can affect a budget. Remind them of the finance charges that are incurred each month when the balance due is not paid in full. Should these charges be included in a budget? (yes) Discuss also the temptation to buy items that may not be in the budget. What effect would this have on a budget? (It could ruin the budget.)

Here is the Thomases' expense summary for December. They want to compare what they had budgeted to what they actually spent.

EXPENDITURES FOR THE MONTH OF DECEMBER		
Expenses	Amount Budgeted	Actual Amount Spent
Food	$210.00	$235.80
Household		
Electric	55.65	59.90
Telephone	21.47	17.95
Heating fuel	63.75	74.85
Water bill	31.80	35.50
Garbage/sewer fee	17.21	17.21
Security	25.00	25.00
Transportation		
Gasoline/oil	60.00	54.75
Parking/misc.	102.00	105.70
Personal		
Clothing	100.00	125.00
Credit payments	25.00	25.00
Newspapers, gifts	16.75	9.75
Pocket money	32.00	45.00
Entertainment		
Movies	20.00	32.00
Sports events	35.00	38.50
Dining out	32.00	27.50
Fixed		
Mortgage payments	715.20	715.20
Loan payments	168.55	168.55
Savings	125.00	125.00
Life/home/car insurance	94.58	94.58
Property taxes	100.00	100.00
Car registration	2.21	0
Pledges/contributions	30.00	0

Use the Thomases' expense summary to answer the following.

17. Which household expenses for the month were more than the amount budgeted? **Electric, heating fuel, water bill**

18. Did they spend more or less than the amount budgeted for household expenses for the month? By how much? **More; $15.53**

19. How much did they budget for transportation costs? Were the amounts spent for transportation more or less than the amount budgeted? By how much? **$162; less; $1.55**

20. Which personal expenses were more than the amount budgeted? **Clothing, pocket money**

21. Were their total personal expenditures for the month more or less than the amount budgeted? By how much? **More; $31**

22. Were their total entertainment expenditures for the month more or less than the amount budgeted? **$11 more**

23. Were there any monthly fixed expenses for which the Thomases spent more than the amount budgeted? **No**

24. What was the Thomases' total expenditure for the month of December? Was this amount more or less than the amount they had originally budgeted? By how much? **$2132.74; more; $49.57**

MAINTAINING YOUR SKILLS Look up the skills in parentheses if you need help or more practice.

Which number is greater? **(Skill 1)**

25. $174.85 or $159.94 **$174.85**
26. $35 or $37.19 **$37.19**
27. $2215.73 or $2231.61 **$2231.61**

Subtract. **(Skill 6)**

28. $47.50 − $43.86 **$3.64**
29. $171.84 − $165 **$6.84**
30. $19.47 − $15.50 **$3.97**
31. $1712.50 − $1697.43 **$15.07**
32. $2179.84 − $2050 **$129.84**
33. $3500 − $3147.81 **$352.19**

WRAP-UP

Ask a volunteer for the new terms of this lesson. Ask for definitions of these terms. Have a student explain why an emergency fund is important.

Assignment Guide

- Basic: 8–15, 17–24, 25–33
- Average: 16–24, 25–33 odd

ALTERNATIVE STRATEGIES: Enrichment

Have your students use their completed Money Manager form from Lesson 11-2, and maintain an actual list of expenditures and compare them to their budgeted expenditures. You could have them estimate their expenditures over the next week (or next day or next month) and then compare them to a record of their actual expenditures.

LESSON PLAN
Reviewing the Basics

The exercises on this page review skills, applications, and terms used in the unit. You can use the exercises to assess informally students' proficiency with this material.

The page can be used for guided practice and independent practice. You can work through a selection of the exercises together with students, and thus see immediately if they know how to do them, and you can then assign some of the exercises for independent practice. Be sure to go over the answers to all assigned exercises.

Reviewing the Basics

Skills

Which number is greater?

(Skill 1)

1. $56.91 or $54.92 **$56.91**
2. $124.61 or $122.93 **$124.61**
3. $111.11 or $99.99 **$111.11**

Solve.

(Skill 5)

4. $35.84 + $39.71 + $18.45 + $60.51 + $123.75 **$278.26**
5. $85 + $157 + $31.71 + $90.08 + $141.72 + $74.87 **$580.38**
6. $2.75 + $0.63 + $7 + $3.14 + $1.19 + $4.07 **$18.78**

(Skill 6)

7. $371.84 − $296.79 **$75.05**
8. $415.07 − $71.48 **$343.59**
9. $1243.71 − $906.74 **$336.97**

(Skill 11)

10. $634.29 ÷ 3 **$211.43**
11. $419.84 ÷ 4 **$104.96**
12. $894.84 ÷ 12 **$74.57**

Write as a percent. Round to the nearest tenth of a percent.

(Skill 26)

13. 0.24718 **24.7%**
14. 0.415612 **41.6%**
15. 0.096782 **9.7%**
16. 0.049951 **5.0%**

Applications

Find the average. Round answers to the nearest cent.

(Application Q)

17. $147.85, $216.47, $312 **$225.44**
18. $56.91, $62.54, $57.85, $60.19 **$59.37**
19. $375, $419.81, $407.63, $384.91, $397, $411.71 **$399.34**
20. $12.32, $14, $13.61, $10.71, $9.85, $10, $11.27, $12.50, $11.61 **$11.76**

Terms

Use each term correctly in the paragraphs below. Then rewrite the paragraphs.

21. Budget sheet
22. Annual expenses
23. Monthly fixed expenses
24. Emergency fund
25. Living expenses
26. Expenditures

A __21__ is used to summarize your spending habits and aid in planning for future spendings. Your budget is based on your take-home pay. To determine how you spend your money, keep records of your monthly __26__ for three months to a year.

Real estate taxes, insurance payments, pledges, and contributions are examples of __22__. Mortgage loan payments, rent, installment payments, and regular savings deposits are examples of __23__. Amounts for food, clothing, pocket money, and utility bills are examples of __25__. For your budget to succeed, an __24__ is necessary to provide for unpredictable expenses, such as medical bills and repair bills.

Refer to your reference files in the back of the book if you need help.

Unit Test

Lesson 11-1

1. Thomas O'Keefe keeps records of his family's expenses. Total monthly expenses for September, October, and November were $1787.43, $1891.74, and $1811.12, respectively. What was the average monthly expenditure? **$1830.10**

Lesson 11-2

2. Complete the budget sheet below. Find the total monthly living expenses, total monthly fixed expenses, total annual expenses, monthly share of annual expenses, and balance.

Use the budget sheet to answer the following.

3. What amount is set aside each month for annual expenses? What is the Bakers' total monthly expenditure? **$95.02; $1979.27**

Lesson 11-3

4. The Bakers had planned to spend $187.50 on transportation costs. They actually spent $212.45. How much more or less did they spend on transportation expenses than they had originally planned? **$24.95 more**

A MONEY MANAGER FOR *Pat and Charles Baker* DATE *1/15/--*

MONTHLY LIVING EXPENSES	
Food/Grocery Bill	$ *240.00*
Household Expenses	
Electricity	$ *61.50*
Heating Fuel	$
Telephone	$ *37.50*
Water	$
Garbage/Sewer Fee	$
Other *Child Care*	$ *60.00*
	$
Transportation	
Gasoline/Oil	$ *112.50*
Parking	$ *25.00*
Tolls	$
Commuting	$
Other *Repairs*	$ *50.00*
Personal Spending	
Clothing	$ *90.00*
Credit Payments	$ *50.00*
Newspapers, Gifts, Etc.	$ *15.00*
Pocket Money	$ *100.00*
Entertainment	
Movies/Theater	$ *20.00*
Sporting Events	$ *15.00*
Recreation	$ *20.00*
Dining Out	$ *50.00*
TOTAL	$ **946.50**

MONTHLY FIXED EXPENSES	
Rent/Mortgage Payment (Rent circled)	$ *575.00*
Car Payment	$ *167.75*
Other Installments	
Appliances	$
Furniture	$ *45.00*
Regular Savings	$ *100.00*
Emergency Fund	$ *50.00*
TOTAL	$ **937.75**
ANNUAL EXPENSES	
Life Insurance	$ *240.00*
Home Insurance	$
Car Insurance	$ *567.50*
Real Estate Taxes	$
Car Registration	$ *32.75*
Pledges/Contributions	$ *120.00*
Other *Renter's Ins.*	$ *180.00*
TOTAL	$ **1140.25**
MONTHLY SHARE (Divide by 12)	$ **95.02**
MONTHLY BALANCE SHEET	
Net Income (Total Budget)	$ *2000.00*
Living Expenses:	$ **946.50**
Fixed Expenses:	$ **937.75**
Annual Expenses:	$ **95.02**
TOTAL MONTHLY EXPENSES	$ **1979.27**
BALANCE	$ **20.73**

LESSON PLAN
Unit Test

Students should do the Unit Test on their own. Each problem on the test is keyed to a lesson in the unit. Students having difficulty with any particular problem should review the Example in the appropriate lesson and be assigned some of the Independent Practice problems for additional practice.

LESSON PLAN
A Spreadsheet Application

USING TECHNOLOGY

The use of computers by families to track monthly expenses against budgets is becoming more common each year. In addition to this specific application, computers can store all sorts of financial data for a family. Ask students to suggest the kinds of information they think a family could store with a computer.

When discussing this particular application involving the Kujkowskis, ask students if the computer established the amount budgeted. (no) Then who did? (the Kujkowskis) Ask students to explain what the computer actually is doing in this application. (storing data and doing computations)

A SPREADSHEET APPLICATION

Recordkeeping

To complete this spreadsheet application, you will need the diskette *Spreadsheet Applications for Business Mathematics,* which accompanies this textbook.

Input information in the following problem to find the total monthly budgeted amount, total monthly expenditures, and difference between amounts budgeted and spent for each month. Then answer the questions that follow.

MONTHLY BUDGET SHEET FOR JAMES AND SHIRLEY KUJKOWSKI

Budget Category	Amount Budgeted	April Actual Spent	Diff.	May Actual Spent	Diff.	June Actual Spent	Diff.	Three Months' Diff.
Groceries	$200.00	$195.00	?	$197.00	?	$206.00	?	?
Utilities	220.00	235.00	?	225.00	?	210.00	?	?
House payment	715.00	715.00	?	715.00	?	715.00	?	?
House insurance	30.00	30.00	?	30.00	?	35.00	?	?
Transportation	215.00	205.00	?	212.00	?	210.00	?	?
Clothing	40.00	25.50	?	45.00	?	48.00	?	?
Credit card payment	100.00	100.00	?	100.00	?	100.00	?	?
Entertainment	40.00	56.70	?	43.25	?	30.00	?	?
Dining out	50.00	42.85	?	61.95	?	51.85	?	?
Savings	200.00	200.00	?	180.00	?	200.00	?	?
Miscellaneous	100.00	89.80	?	107.90	?	93.45	?	?
TOTAL	?	?	?	?	?	?	?	?

A SPREADSHEET APPLICATION (CONTINUED)

1. What is the Kujkowskis' total amount budgeted?
2. What are the total expenditures for April?
3. What are the total expenditures for May?
4. What are the total expenditures for June?
5. Were they over or under their budget for April? By how much?
6. Were they over or under their budget for May? By how much?
7. Were they over or under their budget for June? By how much?
8. Were they over or under their budget for the three months combined? By how much?
9. For April, in which category did they differ most from their bud-geted amount? By how much?
10. For May, in which category did they differ most from their budgeted amount? By how much?
11. For June, in which category did they differ most from their budgeted amount? By how much?
12. Over the entire three months, in which category did they differ most from their budgeted amount? By how much?
13. Why did the actual amount of their house payment not change?
14. Explain why the actual amount of their house insurance payment went up.
15. Do you have any advice for the Kujowskis? Why or why not?

LESSON PLAN
Career Wise

APPLICATIONS ON THE JOB

Have students read the page and answer the questions. Point out that in any business it is important to control costs. In large businesses, *staff positions* are established, such as a financial advisor, to help control expenses. Ask students if they know what a staff position is. (an advisory position) Ask them also if they know what a *line* position is. (a position involved in actually running a business—the manager is in a line position)

You might wish to ask students what they think a spreadsheet program does. Discuss briefly the functions of production, advertising, marketing, facilities, and personnel.

CAREER WISE

Financial Advisor

Olga Chebyshev works as a financial advisor at a medium-sized mail-order firm. She spends a great deal of her time compiling the bookkeepers' financial records and making sense of the pages of data that result. Not only does she compare figures for the amount actually spent to figures for the projected expenditures, she also advises managers about ways that costs can be cut.

A computer network and an electronic spreadsheet program make much of the compilation work and analysis go more smoothly.

Of the actual and projected figures for production, advertising, marketing, facilities, and personnel, she decided to zero in on the costs of mailing the company's catalogs. The double bar graph below shows the actual cost figures (black bars) and projected cost figures (blue bars) for catalog mailings last year.

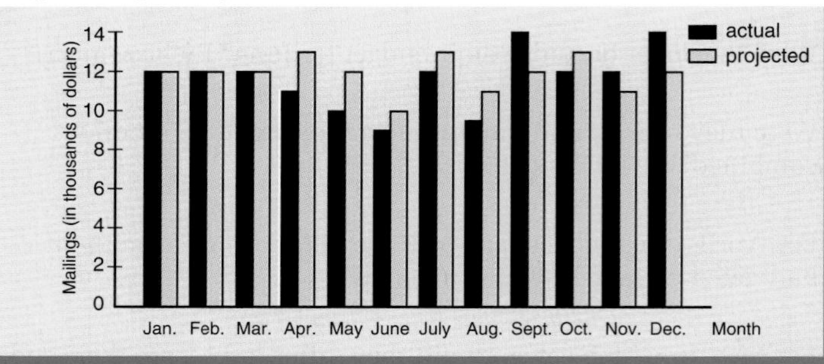

Check Your Understanding

1. For which months did the actual and projected mailings agree? **Jan., Feb., and Mar.**
2. For which months did actual mailings exceed projected mailings? **Sep., Nov., and Dec.**
3. Estimate the projected cost of mailing the catalogs in July. **$13,000**
4. Estimate how much the actual cost of mailings exceeded the projected cost in November. **$12,000 − $11,000 = $1000**
5. Estimate the total actual mailing costs for the year. **$139,500**
6. Estimate the total projected mailing costs for the year. **$143,000**

Cumulative Review

Skills

Find which number is greater.

(Skill 1)
1. 4.7 or 4.07 **4.7** **2.** 27.18 or 27.174 **27.18** **3.** 96.178 or 96.201 **96.201**

Round to the nearest cent.

(Skill 2)
4. $14.174 **$14.17** **5.** $219.676 **$219.68** **6.** $1.485 **$1.49** **7.** $74.997 **$75.00**

Solve. Round to the nearest cent.

(Skill 3)
8. $74 + $15 **$89** **9.** $71 + $39 **$110** **10.** $78 + $50 + $95 **$223**
11. $617 + $179 + $496 **$1292** **12.** $956 + $78 + $491 **$1525**

(Skill 4)
13. $78 − $43 **$35** **14.** $219 − $74 **$145** **15.** $640 − $571 **$69**
16. $7413 − $5527 **$1886** **17.** $9142 − $7291 **$1851** **18.** $8070 − $4793 **$3277**

(Skill 5)
19. $71.43 + $32.78 **$104.21** **20.** $49.56 + $17.72 **$67.28** **21.** $78.56 + $123.75 **$202.31**
22. $84.32 + $66.75 **$151.07** **23.** $516.45 + $491.84 **$1008.29** **24.** $718.45 + $709.91 **$1428.36**

(Skill 6)
25. $71.47 − $15.35 **$56.12** **26.** $65.86 − $41.93 **$23.93** **27.** $47.50 − $4.17 **$43.33**
28. $91.86 − $35.46 **$56.40** **29.** $607.08 − $49.78 **$557.30** **30.** $2174.81 − $471.55 **$1703.26**

(Skill 8)
31. $365.85 × 92 **$33,658.20** **32.** $2.17 × 35 **$75.95** **33.** $51.76 × 330 **$17,080.80**
34. $78.54 × 8.7 **$683.30** **35.** $8.75 × 6.8 **$59.50** **36.** $516.75 × 0.74 **$382.40**

(Skill 11)
37. $112.80 ÷ 26 **$4.34** **38.** $15.81 ÷ 24 **$0.66** **39.** $417.93 ÷ 52 **$8.04**
40. $956.74 ÷ 5.12 **$186.86** **41.** $75,750 ÷ 5.61 **$13,502.67** **42.** $2141.85 ÷ 74.6 **$28.71**

Write as a decimal.

(Skill 14)
43. $\frac{7}{8}$ **0.875** **44.** $\frac{9}{16}$ **0.5625** **45.** $\frac{21}{32}$ **0.65625** **46.** $\frac{145}{360}$ **$0.402\overline{7}$**

Write as a percent to the nearest tenth of a percent.

(Skill 26)
47. 0.748 **74.8%** **48.** 0.0971 **9.7%** **49.** 0.7137 **71.4%** **50.** 0.06449 **6.4%**

Write as a decimal.

(Skill 28)
51. 45% **0.45** **52.** 14.7% **0.147** **53.** 212% **2.12** **54.** 6.7% **0.067**

Solve. Round to the nearest cent.

(Skill 30)
55. 48% of $794 **$381.12** **56.** 53% of $145 **$76.85**
57. 8.7% of $217 **$18.88** **58.** 17.8% of $93.47 **$16.64**
59. 32% of $15.76 **$5.04** **60.** 6.5% of $119.79 **$7.79**
61. 4.05% of $19.47 **$0.79** **62.** 0.4% of $71.89 **$0.29**

LESSON PLAN
Cumulative Review (Skills/Applications) Units 6–11

The exercises on this page review skills, applications, and terms used in Units 6–11. You can use the exercises to assess informally students' proficiency with this material.

These two pages can be used for guided practice and independent practice. You can work through a selection of the exercises together with students, and thus see immediately if they know how to do them. You can then assign some of the exercises for independent practice. Be sure to go over the answers to all assigned exercises.

Applications

Use the table to answer the following.

Monthly Payments $1000 Loan at 10% APR	
Term in Months	Monthly Payment
6	$171.56
12	87.91
18	60.05
24	46.14

(Application C)

63. What is the monthly payment on a $1000 loan for a term of 18 months? **$60.05**

64. What is the total amount repaid if the loan is repaid in 12 months? **$1054.92**

65. For what term is the monthly payment lowest? **24 months**

66. For what term is the total amount repaid lowest? **6 months**

Use the table on page 645 to find the elapsed time in days.

(Application G)

67. From March 4 to March 30 **26**
68. From April 10 to July 20 **101**
69. From January 3 to February 20 **48**
70. April 15 to October 15 **183**
71. From October 16 to February 10 **117**
72. From July 20 to January 30 **194**

Write as a fraction of a year.

(Application J)

73. 4 months $\frac{1}{3}$
74. 3 months $\frac{1}{4}$
75. 90 days (ordinary year) $\frac{1}{4}$
76. 180 days (ordinary year) $\frac{1}{2}$
77. 182 days (exact year) $\frac{182}{365}$

Find the mean. Round answers to the nearest cent.

(Application Q)

78. $64.74, $71.83, $95.74, $67.85 **$75.04**
79. $7.15, $8, $6.45, $10, $7.15, $12.48 **$8.54**
80. $4175.85, $7181.74, $5671.24 **$5676.28**
81. $4.15, $5, $7.20, $6.15, $4, $3.98, $6.20, $7, $6.45, $8.10 **$5.82**

Terms

Write your own definition for each term. **Answers will vary.**

82. Sales receipt
83. Finance charge
84. Previous balance
85. Unpaid balance
86. Rebate schedule
87. Discount loan
88. Installment loan
89. Interest rate
90. Sticker price
91. Rent
92. Deductible clause
93. Depreciation
94. Mortgage loan
95. Closing costs
96. Premium
97. Principal
98. Term life insurance
99. Whole life insurance
100. Bonds
101. Stocks
102. Budget sheet
103. Fixed expenses
104. Emergency fund
105. Living expenses

Cumulative Review Test Units 6–11

Refer to your reference files in the back of the book if you need help.

Lesson 6-2

1. Ruth Warey has a charge account that charges 1.67% of the previous balance for a finance charge. Her statement shows purchases totaling $74.85, a previous balance of $294, and a payment of $100. What is the new balance? **$273.76**

Lesson 6-4

2. Adam North has a charge account that charges 1.75% of the unpaid balance. His statement shows purchases totaling $47.49, a previous balance of $374, and a payment of $70. Find the finance charge and the new balance. **$5.32; $356.81**

Lesson 6-6

3. Find the average daily balance where new purchases are included when figuring the daily balances. Then find the finance charge and the new balance. **$310.59; $4.66; $359.96**

REFERENCE	POSTING DATE	TRANSACTION DATE	DESCRIPTION	PURCHASES & ADVANCES	PAYMENTS & CREDITS
13100	9/14	8/23	Wesford Clothes	104.50	
20046	9/14	8/27	Record Mart	35.60	
54545	9/17		PAYMENT		30.00

BILLING PERIOD	PREVIOUS BALANCE	PERIODIC RATE	AVERAGE DAILY BALANCE	FINANCE CHARGE
9/1–9/30	$245.20	1.5%	?	?

PAYMENTS & CREDITS	PURCHASES & ADVANCES	NEW BALANCE	MINIMUM PAYMENT	PAYMENT DUE
$30.00	$140.10	?	$20.00	10/21

Lesson 7-1

4. Alice Russel's bank granted her a single-payment loan of $6500 for 90 days at an annual interest rate of 10%. Her bank charges ordinary interest. What is the maturity value of Alice's loan? **$6662.50**

Lesson 7-3

5. Tony Bonfiglio obtained an installment loan of $500 to purchase a stereo set. The annual percentage rate is 18%. He plans to repay the loan in 12 months. Use the table to find the finance charge. **$50.08**

APR	Term in Months	If You Finance . . . $200	$500	$1000	$1500
		Your Monthly Payments Are			
18%	6	35.10	87.76	175.52	263.28
	12	18.33	45.84	91.68	137.52
	18	12.76	31.90	63.80	95.70
	24	9.98	24.96	49.92	74.88

Lesson 8-2

6. Amy Green is interested in a luxury car that has a base price of $27,895, options totaling $6174.95, and a destination charge of $475. She has read that the dealer's cost is about 80% of the base price and 75% of the price of the options. What is the estimated dealer's cost? **$27,422.21**

LESSON PLAN Cumulative Review Test Units 6–11

Students should do the Cumulative Review Test on their own. Each problem on the test is keyed to a lesson in Units 6–11. Students having difficulty with any particular problem should review the Example in the appropriate lesson and be assigned some of the Independent Practice problems for additional practice.

Lesson 8-4

7. David Scott uses his car to drive to and from work. His insurance coverage includes 50/100 bodily injury and $100,000 property damage. His driver-rating factor is 4.10. He has $50-deductible comprehensive and $100-deductible collision insurance coverage on his car. His car is in age group A and rating group 8. What is his annual premium? **$1098.80**

Property Damage Limits	Bodily Injury Limits		
	25/50	25/100	50/100
$ 25,000	$103.20	$114.00	$119.20
50,000	107.20	118.00	123.20
100,000	110.00	121.20	126.00
Age Group	**Insurance-Rating Group 6**	**7**	**8**
	Comprehensive $50-ded.		
A	$32.80	$35.60	$39.60
B	30.80	33.20	36.80
	Collision $100-ded.		
A	$84.00	$83.20	$102.40
B	72.80	90.80	108.40

Lesson 9-4

8. The Sturgeons obtained a 30-year, $70,000 mortgage loan at First Federal Bank at an interest rate of 10.5%. Find the monthly payment. What is the new principal after the first monthly payment? Use the monthly payment table on page 645. **$640.50; $69,972**

Lesson 9-5

9. The Hocking County tax assessor stated that the market value of the King estate is $975,800. The rate of assessment in Hocking County is 35% of market value. The tax rate is 74.85 mills. What is the annual real estate tax on the King estate? **$25,563.52**

Lesson 10-1

10. Debbie Young is employed by A.I.M. Ind., Inc. The annual premium for comprehensive medical insurance is $2870. Her employer pays 50% of the cost. How much does Debbie pay? **$1435**

Lesson 10-4

11. Vicky Osinski purchaseed a whole life insurance policy with a face value of $75,000. The annual base premium is $22.68 per $1000. What is her annual premium? **$1701**

Lesson 10-6

12. Tom Green invested $8000 in a certificate of deposit that earns interest at a rate of 10.00% and matures after 4 years. Interest is compounded quarterly. What is the effective annual yield? **10.38%**

AMOUNT PER $1.00 INVESTED, COMPOUNDED QUARTERLY

Annual Rate	Interest Period 1 Year	4 Years
10.00%	1.103813	1.258577
10.25%	1.106508	1.271224
10.50%	1.109207	1.283998

Lesson 10-7

13. Manny Diaz purchased 150 shares of Petrie stock at $5\frac{7}{8}$. He was charged at $40 commission. What was the total amount that he paid for the purchase? **$921.25**

Lesson 11-1

14. Cynthia Pinziotti keeps records of family expenses. Total monthly expenses for September, October, and November were $2147.40, $2271.85, and $2141.47, respectively. What is the average monthly expenditure? **$2186.91**

A SIMULATION

Buying a Condominium

You have decided to purchase a condominium. You are chief systems programmer for Datacrunch, Inc., and your gross pay is $36,000 a year. You have $14,000 in a savings account.

Before you look at condominiums, you decide to figure out how much you can afford to spend.

Your bank suggests that you spend not more than 28% of your gross pay on principal, interest, and taxes.

Most condominium management companies in your area want a down payment of 20% of the selling price. You decide that you can use $12,000 of your savings account as a down payment.

Write your name on the Datacrunch badge.

Use this work sheet to figure out how much you can afford to spend for your condominium.

1.	Amount you can afford to spend per year on principal, interest, and taxes	? **$10,080**
2.	Amount you can afford to spend per month on principal, interest, and taxes	? **$840**
3.	Maximum selling price on which you can afford to make a 20% down payment	? **$60,000**

2 Employer's name, address, and ZIP code	DATACRUNCH, INC.
6 Statutory employee / Deceased / Pension plan / Legal rep. / 942 emp. / Subtotal / Deferred compensation / Void	☐ ☐ ☐ ☐ ☐ ☐ ☐ ☐
7 Allocated tips	
10 Advance EIC payment	
9 Federal income tax withheld	7124.00
10 Wages, tips, other compensation	36,000.00
3 Employer's identification number	
4 Employer's state I.D. number	
11 Social security tax withheld	2754.00
12 Social security wages	36,000.00
5 Employee's social security number	123-45-6789
13 Social security tips	
14 Nonqualified plans	
19 Employee's name, address, and ZIP code	
15 Dependant care benefits	
16 Fringe benefits incl. in Box 10	
17	
18 Other	
20	
21	
22	
23	
24 State income tax	1035.00
25 State wages, tips, etc.	36,000.00
26 Name of state	
27 Local income tax	900.00
28 Local wages, tips, etc.	36,000.00
29 Name of locality	

LESSON PLAN
A Simulation

USING THE SIMULATION

Students can work on this simulation individually, or you may wish to use it with *cooperative learning groups.* Each student should have copies of the necessary worksheets. You can begin by reading with students the introductory material at the top of the page.

Check students' worksheets as they complete them and provide help as needed. After students have finished the simulation, provide them with completed worksheets so they can check their work.

A SIMULATION
(CONTINUED)

Which Condominium?

After looking at several condominiums, you narrow your selection to two, one in the city and one in a suburb. The advertising booklet for each condominium gives the base price and the additional charges for extra features.

Country Squire Estates
One-Bedroom Unit
$56,000 850 sq ft
Fireplace: stone $3300
Refrigerator with ice maker 790
Microwave oven: built in 280
Plush carpeting: 80 sq yd @ $6.75 per sq yd extra cost

Townview
One-Bedroom Unit $52,000 790 sq ft
Fireplace: brick $3500
Refrigerator with ice maker 860
Microwave oven: built in 250
Plush carpeting: 72 sq yd @ $8.50 per sq yd extra cost

Choose one of these condominiums. You will keep the same condominium for the rest of this simulation. You decide to purchase all the extra features listed in the advertising booklet.

Use the prices in the booklet to find the total cost of your condominium.

			Townview	Country
4.	Condominium you have chosen	?		
5.	Base price	?	$52,000	$56,000
6.	Fireplace	?	$3,500	$3,300
7.	Refrigerator	?	$860	$790
8.	Microwave oven	?	$250	$280
9.	Carpeting: ____ sq yd × $ ____	?	$612	$540
10.	Total cost of condominium	?	$57,222	$60,910

A SIMULATION (CONTINUED)

Your Net Worth

After you choose the condominium you want, you fill out a mortgage application. The lender will check to make sure that you have as much money as you say you do, and that you have paid your debts on time.

The application also asks for a statement of assets and liabilities. You have $14,000 in a savings account and $950 in a checking account. You own a car worth $9000, on which you still owe $3000. Your life insurance has a net cash value of $3000. You have personal property, such as furniture and jewelry, with a total value of $6350. You owe $380 on your credit card.

Use this information to fill out the loan application and find your net worth.

RESIDENTIAL LOAN APPLICATION

Borrower	
Name	Student's name
Address	Student's address
Employer	Datacrunch, Inc.
Position or Title	Chief Systems Programmer
Gross Annual Income	$36,000

Statement of Assets and Liabilities			
Assets		**Liabilities**	
Description	**Cash or Market Value**	**Description**	**Unpaid Balance**
Cash		Installment debts	380 ?
Checking account	$ 950?	Real estate loans	
Savings account	14,000?	Automobile loans	3000 ?
Automobile	9000?	Other debt	
Life insurance net cash value	3000?		
Furniture and personal property	6350?		
Total assets	33,300?	Total liabilities	3380 ?
Net Worth (Total Assets Minus Total Liabilities)		$29,920 ?	

11. What are your total assets? $33,300

12. What are your total liabilities? $3380

13. What is your net worth? $29,920

A SIMULATION
(CONINTUED)

Bank Charges and Taxes

Now that you have selected the condominium you would like to buy, you will have to calculate what it will cost you each month.

14. Townview $45,777.60; Country $48,728

14. Look back at question 10 to find the total selling price of your condominium. If you make a down payment of 20% of the selling price, what is the mortgage loan amount?

Both development companies arrange financing. The mortgage for Townview is 11% for 25 years. The mortgage for Country Squires is 10.50% for 30 years.

15. Townview $449.08; Country $445.86

15. Use the mortgage table (below left) to find the monthly payment per $1000 financed. Then calculate your monthly payment for principal and interest.

MONTHLY MORTGAGE PAYMENTS PRINCIPAL AND INTEREST PER $1000 FINANCED

Interest Rate	15 Years	20 Years	25 Years	30 Years
10.50%	11.06	9.99	9.45	9.15
11.00%	11.37	10.33	9.81	9.53
11.50%	11.69	10.67	10.17	9.91
12.00%	12.01	11.02	10.54	10.29
12.50%	12.33	11.37	10.91	10.68

REAL ESTATE TAX RATES

Town	Rate of Assessment	Tax Rate per $1000
Sparta	63%	$109.40
Springfield	80%	96.20
Springton	75%	101.15
Stafford Valley	50%	136.50
Strawberry	100%	88.60

As a condominium owner, you will have to pay real estate taxes. Townview is in Springfield. Country Squires is in Stafford Valley.

Use the real estate tax table (above right) to find your tax.

			Townview	Country
16.	Value of your condominium	?	**$57,222**	**$60,910**
17.	Rate of assessment	?	**80%**	**50%**
18.	Assessed value: ____% × $____	?	**$45,777.60**	**$20,455**
19.	Tax rate per $1000	?	**$96.20**	**$136.50**
20.	Real estate taxes per year	?	**$4403.81**	**$4157.11**
21.	Real estate taxes per month	?	**$366.98**	**$346.43**

22. What is your total monthly cost for principal, interest, and taxes? Is this amount less than 28% of your gross income?
Townview $816.06; Country $792.29; yes, 28% = $10,080 ÷ 12 = $840

A SIMULATION
(CONTINUED)

Fees, Insurance, and Utilities.

Your condominium management company charges you an annual fee. This fee covers the cost of trash pickup, snow removal, landscaping, maintenance, management costs, and insurance on the buildings.

The Townview fee is $0.60 per square foot per year. The Country Squires fee is $0.75 per square foot per year.

23. Townview $474, $39.50; Country $637.50, $53.13

23. Look at the advertising booklets on page 334 to find the number of square feet in your condominium. What is the condominium fee per year? What is the fee per month?

The condominium fee includes insurance on the buildings. You have to get separate insurance on your furniture and other possessions. Your insurance company has a basic annual rate of $95. There are extra charges for insurance on certain valuable items.

Use these work sheets to calculate the cost of your insurance.

24. $6.60
25. $25.00
26. $5.60

	Item	Value	Annual Rate per $100	Annual Rate
24.	Camera	$400	$ 1.65	?
25.	Bicycle	250	10.00	?
26.	Jewelry	350	1.60	?

27. $95.00
28. $37.20
29. $132.20
30. $11.02

27.	Basic rate per year	?
28.	Extra charges per year	?
29.	Total rate per year	?
30.	Insurance cost per month	?

Townview uses electricity for heating, cooking, and lighting. It is estimated that the unit will use 9750 kW·h per year. The cost of electricity is $0.11 per kW·h.

Country Squires uses gas for heating and cooking. The estimated annual consumption of gas is 745 CCF at $0.65 per CCF. The unit would also use 2600 kW·h of electricity per year at $0.10 per kW·h.

Each unit will use about 8000 cubic feet of water per year. In Springfield, water costs $9 per 1000 cubic feet. Sewer charges, which are based on 70% of water usage or 5600 cubic feet, are $14 per 1000 cubic feet. In Stafford Valley, water and sewer charges are 50% more than in Springfield.

Find your utility costs.

			Townview	Country
31.	Annual electricity cost	?	$1072.50	$260.00
32.	Annual gas cost	?	0	$484.25
33.	Annual water and sewer cost	?	$150.40	$225.60
34.	Total annual utilities cost	?	$1222.90	$969.85
35.	Monthly utilities cost	?	$101.91	$80.82

36. What is your total monthly cost for fees, insurance, and utilities?
Townview $152.43; Country $144.97

A SIMULATION (CONTINUED)

Summary

Summarize your results so far.

			Townview	Country
37.	Your condominium	?	Townview	Country
38.	Total cost	?	$57,222	$60,910
39.	Your net worth	?	$29,920	$29,920
40.	Monthly cost for principal, interest, and taxes	?	$816.06	$792.29
41.	Monthly cost for fees, insurance, and utilities	?	$152.43	$144.97
42.	Total monthly cost	?	$968.49	$937.26

A common rule of thumb is that your monthly housing cost should be less than 35% of your monthly net pay. Your net pay is your gross pay minus deductions for taxes and FICA. Shown below is the stub attached to your monthly paycheck.

Use the stub to calculate your monthly net pay.

DATACRUNCH, INC.

EARNINGS	AMOUNT		DEDUCTIONS			YEAR TO DATE	
REGULAR	3000	00	FICA TAX	229	00	GROSS PAY	3000.00
			FED. TAX	596	50	FICA TAX	229.50
			STATE TAX	86	25	FED. TAX	596.00
			LOCAL TAX	75	00	STATE TAX	86.25
						LOCAL TAX	75.00
						FOR PERIOD ENDING	
						JANUARY 31, 19--	
						NET PAY	

43.	Monthly net pay	?	$2013.25
44.	35% of monthly net pay	?	$704.64

45. Is your total monthly housing cost less than 35% of your monthly net pay? **No**

46. Your monthly housing expenses are your mortgage, taxes, condominium fee, insurance, and utilities. Compare expenses with someone who chose the other condominium. For each expense, which condominium costs more?
Townview: mortgage, taxes, utilities; Country: fee.

47. If you had another chance to decide about buying a condominium, what might you do differently? Why? **Answers will vary.**

PART 3

Business Applications

In Part 3, you will learn how to apply mathematics to actual business practices. Different departments of one business are covered in Part 3. You will learn firsthand how mathematics is used in each department and how each department contributes to the entire operation of the business.

Terms to Know

The following definitions provide an overview of the activities involved in the operations of the departments within a business. The lessons within each unit explain in detail how business mathematics is used to carry out these activities.

Personnel The administrative activities of hiring staff, developing wage and salary scales, and providing benefits for employees within your business.

Production The activities involved in manufacturing and packaging the items that your business sells.

Purchasing The activities involved in buying goods at discount prices from your suppliers.

Sales The activities involved in determining the selling price of your product in the marketplace.

Marketing 1. The activities concerned with researching buyer preference. 2. The activities concerned with communicating information about your product to prospective buyers.

Warehousing and Distribution The activities of storing, shipping, and maintaining accurate inventory records of your products.

Services 1. The activities related to using buildings, equipment, or utilities from other businesses for a fee. 2. The activities related to hiring employees of another business to do work for your business.

Accounting The activities related to monitoring normal expenses of your business, such as payroll and depreciation.

Accounting Records The activities of maintaining and analyzing documents that show the income, expenses, and value of your business.

Financial The activities concerned with managing your business's money, including paying taxes, borrowing, and investing.

U N I T

12

Personnel

The personnel department of your business fills openings by *recruiting* new employees through advertising and interviews. The department also handles *wage and salary scales* for each position and *employee benefits,* such as disability, health, and life insurance and paid vacations and holidays.

O U T L I N E

Part of the cost of running a business goes to employee benefits. Employee benefits add real dollars to your paycheck.

UNIT 12
PERSONNEL

INTRODUCING THE UNIT

Introduce this unit by opening a discussion on the students' experiences of locating, applying, and interviewing for jobs. They have probably met with a member of the company's personnel department on at least one occasion. What information were they asked to give during that meeting? What information were they given?

LESSON PLAN
12-1 Hiring
New Employees

FOCUS
You can motivate this lesson by asking students what expenses they encountered when they looked for part-time jobs. The list may include cost of a newspaper, bus fare, or cost of a new outfit for the interview. Point out that it also costs an employer money to hire someone. The focus of this lesson is to compute the cost of recruiting new employees.

TEACH
Go over the formula for finding the total recruiting costs. Point out that advertising fees may include not only newspapers, but also trade journals for more highly skilled positions. If the company pays an employment agency for locating a new employee, this fee is included in the hiring expense. Point out that the company does not always pay the employment agency fee. Sometimes the person seeking employment must pay the fee.

Interviewing expenses may include travel expense, lodging, and meals. Hiring expenses may include moving expenses, real estate broker's commissions, and travel expense. It may sometimes include temporary lodging until permanent housing can be found.

12-1 Hiring New Employees

OBJECTIVE
Compute the cost of recruiting new employees.

To fill openings in your business, you may recruit new employees. The cost of recruiting includes advertising fees, interviewing expenses, such as travel expenses, and hiring expenses, such as moving expenses. In addition, you may pay an employment agency or your own employees to locate job candidates.

Total Recruiting Cost = Advertising Expense + Interviewing Expense + Hiring Expense

EXAMPLE *Skill* 5 *Application* A *Term* Recruiting costs

The Talbot Manufacturing Company is searching for a person to head the production department. The personnel department placed advertisements for a total cost of $2495 and employed the Empire Executive Search Company to locate candidates. Empire recommended Alice Welch, Thomas Snow, and Victor Grabowski. Talbot paid the candidates' travel expenses for interviews.

Alice Welch	Total travel expenses: $317
Thomas Snow	Total travel expenses: $435
Victor Grabowski	Total travel expenses: $474

After the interviews, Talbot hired Victor Grabowski at an annual salary of $54,900. Talbot paid these expenses to hire him:

Moving expenses	$1200
Sale of home (real estate broker's fee)	$9470
Empire's Fee (25% of Victor's first-year salary)	$13,725

What was the total expense of recruiting Victor Grabowski?

SOLUTION

Find the **total recruiting cost.**

Advertising Expense + Interviewing Expense + Hiring Expense

$2495 + ($317 + $435 + $474) + ($1200 + $9470 + $13,725)

$2495 + $1226 + $24,395 = $28,116 total recruiting cost

2495 M+ 317 + 435 + 474 = 1226 M+ 1200 + 9470 + 13725 = 24395 M+ RM 28116

✓ SELF-CHECK Complete the problems, then check your answers in the back of the book.

Find the total recruiting cost.

1. Advertising expenses, $2150; interviewing expenses, $245 and $315; hiring expenses, $1240 and $2170. **$6120**
2. Advertising expenses, $1975; interviewing expenses, $470 and $860; hiring expenses $475, $8600, and $11,415. **$23,795**

COMMUNICATION SKILLS
Have students choose a job they would like to have and write a want ad for it. The ad should include the necessary qualifications, pay range, and any other information students think important. Discuss the ads in class.

PROBLEMS

	Advertising Cost	+	Interviewing Cost	+	Hiring Cost	=	Total Recruiting Cost
3.	$ 420.70	+	$ 41.20	+	$ 517.20	=	**$979.10**
4.	$ 315.85	+	$ 79.80	+	$ 847.72	=	**$1243.37**
5.	$ 789.16	+	$415.25	+	$1213.49	=	**$2417.90**
6.	$1412.71	+	$614.91	+	$1971.44	=	**$3999.06**
7.	$8761.43	+	$971.84	+	$3147.43	=	**$12,880.70**

8. Wearite Tires recruits district sales manager. **$3803.66**
Advertising cost: $917.45.
Interviewing cost: $694.74.
Hiring cost: $2191.47.
What is the total recruiting cost?

9. Primo Company recruits data processing manager. **$6128.65**
Advertising cost: $1475.00.
Interviewing cost: $861.79.
Hiring cost: $3791.86.
What is the total recruiting cost?

10. The Mobile Communications Company hired Marilyn Curtiss as its new national distribution manager at an annual salary of $54,950.

Advertising costs: $2247.50
Interviewing expenses: Marilyn Curtiss, $147.43; Tom Hart, $216.94
Finder Agency fee: 20% of first year's salary **$13,601.87**

What was the total cost of hiring the national distribution manager?

11. Novi Discount Brokers hired the Wall Street Search Service to locate candidates for the position of manager, investment bonds. The agency's fee is 25% of the first year's salary if one of its candidates is hired. Novi also ran several advertisements at a total cost of $816.40. Novi interviewed three people:

David Gold	Nancy Cooper	Henry Little
Applied through agency	Answered advertisement	Applied through agency
Travel costs: $148.75	Travel costs: $216.40	Travel costs: $171.80

Novi hired Henry Little at an annual salary of $74,760. Novi paid his moving expenses of $419.20 and his real estate broker's fee of 7% to sell his $149,00 home. What was the recruiting cost? **$30,892.55**

MAINTAINING YOUR SKILLS
Look up the skill in parentheses if you need help or more practice.

Add. **(Skill 5)**

12. $148 + $74 + $865 **$1087**

13. $615 + $419 + $1291 **$2325**

14. $1484 + $815 + $11,650 **$13,949**

15. $2241 + $915 + $14,542 **$17,698**

16. $715.80 + $523.40 + $3120.50 **$4359.70**

17. $746.50 + $319.20 + $4314.70 **$5380.40**

18. $619.24 + $141.17 + $2512.64 **$3273.05**

19. $1171.89 + $471.75 + $12,147.94 **$13,791.58**

Warm-Up Exercises

1. $74.90 + $310.70 + $916.80 $1302.40
2. $1214.90 + $817.85 + $3165.96 $5198.71
3. $715 + (7.5% × $149,600) $11,935
4. $1247 + (20% × $53,157) $11,878.40

PRACTICE AND APPLY

The following problems can be assigned for classwork and the answers checked in class to help students master the objective of the lesson.

- Guided Practice: 1–5
- Independent Practice: 6–9

WRAP-UP

Have students name the three types of costs involved in hiring a new employee. Have them give an example of each type of cost.

Assignment Guide

- Basic: 6–10, 12–19
- Average: 10, 11, 13–19 odd

ALTERNATIVE STRATEGIES: Enrichment

Have students look at the Help Wanted section of the classified ads. Have them research and report on terms they find, such as "fee paid," "as a rep," "as a manager," "paid training," "equal opportunity employer," and "annual salary 22K"

LESSON PLAN
12-2 Administering Wages and Salaries

FOCUS
Ask students if they have ever received a raise in salary at the place where they work. Point out that, in this lesson, they will learn to compute the new salary after a merit increase or cost-of-living adjustment.

TEACH
Point out that a person's salary depends not only on the job title, but also on how much experience someone has and the degree of responsibility the job requires. In the Example, the level 2 operator's salary is probably different from the level 3 operator's. A level 2 with 4 years' experience is probably different from a level 2's with 2 years' experience.

Make sure students understand the difference between cost-of-living adjustments and merit increases. Mention that these increases do not have to happen together. Merit increases may be on a person's starting date anniversary, while cost-of-living adjustments may be on the first of the year.

Warm-Up Exercises
1. 4.1% of $17,680 $724.88
2. 7.2% of $28,474 $2050.13
3. $22,600 + (2.7% × $22,600) $23,210.20
4. $41,710 + (5.5% × $41,710) $44,004.05

12-2

Administering Wages and Salaries

OBJECTIVE
Compute the new salary after merit increase and cost-of-living adjustment.

Your business may have **wage and salary scales** for the positions in the company. You can use the information to compare various jobs or to estimate the cost of giving employees **cost-of-living adjustments** or **merit increases.** A cost-of-living adjustment is a raise in your salary to help you keep up with inflation. A merit increase is a raise in your salary to reward you for the quality of your work.

New Salary = Present Salary + Cost-of-Living Adjustment + Merit Increase

EXAMPLE *Skill* 30 *Application* C *Term* Salary scale

Elaine Taylor is a heavy equipment operator, level 2, for Construction Inc. The executive board of Construction Inc. voted to give all employees a cost-of-living adjustment of 4.1%. In addition, Elaine was awarded a merit increase of 2.8% for excellent work during the year. What will Elaine's salary be for the coming year?

Job Level	Equipment Maintenance	Lt. Equipment Operator	Hvy. Equipment Operator	Crew Supervisor
1	$18,000	$19,845	$22,975	$27,900
2	18,900	20,837	24,125	29,295
3	19,845	21,880	25,330	30,760
4	20,837	22,975	26,600	32,300

SOLUTION

A. Find the **cost-of-living adjustment.**
$24,125.00 × 4.1% = $989.125 = $989.13 cost-of-living adjustment

B. Find the **merit increase.**
$24,125.00 × 2.8% = $675.50 merit increase

C. Find the **new salary.**
Present Salary + Cost-of-Living Adjustment + Merit Increase
$24,125.00 + $989.13 + $675.50 = $25,789.63 new salary

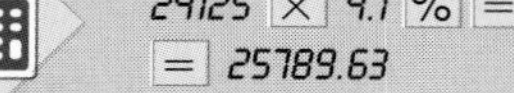
24125 × 4.1 % = 989.125 24125 × 2.8 % = 675.5 24125 + 989.13 + 675.5 = 25789.63

✓ SELF-CHECK Complete the problem, then check your answer in the back of the book.

1. A light equipment operator, level 1, receives a 4% cost-of-living adjustment and a 2% merit increase. Find the new salary. **$21,035.70**

MATHEMATICAL NOTES
You can suggest an alternate method for working the first two steps of the Example. Students can add the 4.1% + 2.8% = 6.9%, and then multiply Elaine's salary by 0.069 to find the total amount of her adjustment and merit increases. ($1664.63)

PROBLEMS

	Present Salary	+	Cost-of-Living Adjustment	+	Merit Increase	=	New Salary
2.	$13,400	+	$ 520	+	$ 780	=	**$14,700**
3.	$15,650	+	$ 620	+	$ 750	=	**$17,020**
4.	$19,860	+	$ 924	+	$1110	=	**$21,894**
5.	$27,847	+	$1365	+	$1087	=	**$30,299**
6.	$39,517	+	$1615	+	$1244	=	**$42,376**

Level	Systems Analyst	Senior Systems Analyst
1	$19,600	$27,600
2	22,100	32,100
3	25,100	37,600

7. Norm Young. **$22,984**
Systems analyst, level 2.
Cost-of-living adjustment: 4%.
What is his new salary?

8. May Song. **$29,973.60**
Senior systems analyst, level 1.
Cost-of-living adjustment: 5%.
Merit increase: 3.6%.
What is her new salary?

9. Beatrice Apptou. **$40,457.60**
Senior systems analyst, level 3.
Merit increase: 7.6%.
What is her new salary?

10. Robert Moore, systems analyst, level 1, received a 4.7% cost-of-living adjustment and a 4.5% merit increase. What is his new salary? **$21,403.20**

Refer to the table on page 344 for problems 11 and 12.

11. Ruth Tomasi is a crew supervisor, level 4. She receives a 4.8% cost-of-living adjustment and a 7.4% merit increase. What is her new salary? **$36,240.60**

12. Mike Rossi is in equipment maintenance, level 3. He receives a 3.7% cost-of-living adjustment and a 0.7% merit increase. What is his new salary? **$20,718.18**

13. Jim O'Reilly works 40 hours a week for the Metro Delivery Company. He earns $7.15 an hour.
    a. What is Jim's annual gross pay? **$14,872.00**
    b. If Jim receives a 4.9% merit increase, what will be his new annual gross pay? **$15,600.73**
    c. What would be his new hourly rate? **$7.50**

MAINTAINING YOUR SKILLS Look up the skills in parentheses if you need help or more practice.

Find the present salary by using the table on page 344. **(Application C)**

14. Heavy equipment operator, level 3 **$25,330**

15. Light equipment operator, level 4 **$22,975**

16. Crew supervisor, level 2 **$29,295**

Find the percentage. **(Skill 30)**

17. $17,740 × 4.5% **$798.30**

18. $21,510 × 8.6% **$1849.86**

19. $15,100 × 3.1% **$468.10**

20. $34,191 × 2.4% **$820.58**

21. $38,964 × 5.7% **$2220.95**

22. $64,315 × 1.8% **$1157.67**

PRACTICE AND APPLY

The following problems can be assigned for classwork and the answers checked in class to help students master the objective of the lesson.

- Guided Practice: 1–4
- Independent Practice: 5–9

WRAP-UP

Ask students for the new terms of the lesson. Ask for definitions of these terms. Have students compute the new hourly salary of an employee who was making $6.00 per hour and received a cost-of-living adjustment of 4% and a merit increase of 5%. ($6.54)

Assignment Guide

- Basic: 5–12, 14–22
- Average: 10–13, 14–22 even

ALTERNATIVE STRATEGIES: Reteaching

Point out that in the Example the new salary could have been found by multiplying the present salary by (100% + 4.1% + 2.8%) or 106.9%. As another example, go over the following with the students.

$40,000 + 5% of $40,000 + 3% of $40,000 = 108% of $40,000
$40,000 + $2000 + $1200 = 1.08 × $40,000
$43,200 = $43,200

LESSON PLAN
12-3 Employee Benefits

FOCUS

The focus of this lesson is on computing the rate of employee benefits based on annual gross pay. An employer considers benefits as part of the cost of employing a person, even though the employee does not receive the money in a paycheck.

TEACH

Since most students do not work full time, they probably are not eligible for any benefits. Benefits are usually reserved for full-time employees. If you can, bring in a booklet that describes employee benefits. Discussion can include types of health plans, dental plans, life insurance, and so on. Mention that some benefits are optional, and that the employee must pay for these. One type of optional benefit is life insurance. Also not all companies pay fully for medical insurance. Some businesses may pay a certain portion of the costs for life or medical insurance. Businesses may also pay unemployment insurance and retirement benefits.

Warm-Up Exercises

1. $13,520 ÷ 52 $260
2. 6.2% of $32,600 $2021.20
3. ($32,500 ÷ 52) × 2 $1250
4. 8 × $44.80 $358.40
5. $9730 is what percent of $27,800?   35%
6. $15,000 is what percent of $60,000?   25%

12-3 Employee Benefits

OBJECTIVE

Compute the rate of employee benefits based on annual gross pay.

Your business may offer several **employee benefits.** Employee benefits include health and dental insurance, life insurance, pensions, paid vacations and holidays, unemployment insurance, and sick leave. The total of the benefits may be figured as a percent of annual gross pay.

$$\text{Rate of Benefits} = \frac{\text{Total Benefits}}{\text{Annual Gross Pay}}$$

EXAMPLE *Skills* 30, 31 *Application* A *Term* Employee benefits

The personnel department of the Commercial Credit Company is preparing annual reports on employee benefits. Tamara Rey's total annual benefits are what percent of her annual salary?

Tamara Rey Annual Salary: $18,720.00	Weekly Salary: $360.00
Benefits	
Vacation: 2 weeks × $360.00	$ 720.00
Holidays: 8 days × ($360.00 ÷ 5)	576.00
Health insurance: 12 months × $147.80	1773.60
Dental insurance: 12 months × $18.74	224.88
Sick leave: 30 days × ($360.00 ÷ 5)	2160.00
Unemployment insurance: 4.6% of $18,720.00	861.12
Social security: 6.2% of $18,720.00	1160.64
Medicare: 1.45% of $18,720.00	271.44
Total	$7747.68

SOLUTION

Find the **rate of benefits.**

Total Benefits ÷ Annual Gross Pay

$7747.68 ÷ $18,720.00 = 0.413 = 41% rate of benefits

7747.68 ÷ 18720 = 0.4138

✓ SELF-CHECK Complete the problem, then check your answer in the back of the book.

1. Annual salary is $32,000. Total benefits are $9600. Find the rate of benefits. 30%

PROBLEMS

	Total Benefits	÷	Annual Gross Pay	=	Rate of Benefits
2.	$ 5937	÷	$23,748	=	? 25%
3.	$15,610	÷	$44,600	=	? 35%

CULTURAL ANGLES

Some countries, such as Canada, make medical programs available through the government. The federal government pays about half the cost of the program, while the provincial government pays the rest. This medical program is available to all. You may wish to have students research some European countries to see if any of them make medical programs available through the government.

	Total Benefits	÷	Annual Gross Pay	=	Rate of Benefits
4.	$ 2940	÷	$14,700	=	? **20%**
5.	$11,392	÷	$35,600	=	? **32%**
6.	$ 9860	÷	$24,650	=	? **40%**

7. Stock clerk's annual salary is $11,860.
Benefits:
Vacation and holidays $593.
Health insurance $1275.
Unemployment insurance $545.56.
Social security $735.32.
Medicare $171.97
Compensation insurance $486.26.
What are the total benefits? **$3807.11**
What is the rate of benefits? **32.1%**
(Round to nearest tenth percent.)

8. Teacher's annual salary is $19,740.
Benefits:
Retirement $1727.25.
Holidays $789.60.
Group life insurance $347.50.
Sick leave $2961.
Health insurance $271.80
What are the total benefits?
What is the rate of benefits?
(Round to nearest tenth percent.)
$6097.15; 30.9%

F.Y.I.
In 1993, a new college graduate in business administration was paid, on the average, $27,564 in salary and $33,685 in total compensation. (annual gross pay and total benefits)

9. a. Complete the benefits chart for the French Coffee Shoppe employees. (Round to the nearest cent.)
b. What is the rate of benefits for each employee? (Round to the nearest whole percent.) **All 19%**

Position	Annual Wage	2-Wk Vacation	3.6% Comp. Ins.	10-Day Sick Leave	6.2% Soc. Sec.	1.45% Medicare	Total Benefits
Manager	$30,680	$1180	$1104.48	$1180	$1902.16	$444.86	$5811.50
Service	$11,752	452	423.07	452	728.62	170.40	2226.09
Cook	$14,872	572	535.39	572	922.06	215.64	2817.09
Dishwasher	$ 7,488	288	269.57	288	464.26	135.72	1418.41

10. a. Complete the benefits chart for the Pathology Lab staff.
b. What is the rate of benefits for each employee? (Round to the nearest tenth percent.) **All 19.2%**
$10,209.29 **c.** How much does the lab pay in benefits for the 3 employees?
d. How much more would it cost the lab to give each employee a 3-week vacation? **$1024**

Position	Annual Wage	2-Wk Vacation	8 Holidays	4.6% Unempl. Insurance	6.2% Soc. Sec.	1.45% Medicare	Total Benefits
Lab technician	$20,488	$788	$630.40	$942.45	$1270.26	$297.08	$3928.19
Lab analyst	$23,400	900	720.00	1076.40	1450.80	339.30	4486.50
Receptionist	$ 9360	360	288.00	430.56	580.32	135.72	1794.60

MAINTAINING YOUR SKILLS Look up the skills in parentheses if you need help or more practice.

Find the percentage. **(Skill 30)**

11. $24,700 × 7.51% **$1854.97**
12. $18,960 × 4.6% **$872.16**
13. $26,418 × 3.6% **$951.05**

Find the rate. Round answers to the nearest tenth percent. **(Skill 31)**

14. $6500 ÷ $32,500 **20%**
15. $6720 ÷ $16,800 **40%**

PRACTICE AND APPLY

The following problems can be assigned for classwork and the answers checked in class to help students master the objective of the lesson.

- Guided Practice: 1–3
- Independent Practice: 4–8

WRAP-UP

Have students name different types of benefits. Use $5000 for total benefits and $20,000 for annual gross pay. Have students compute the rate of benefits. (25%)

Assignment Guide

- Basic: 4–9, 11–19
- Average: 9, 10. 12–18 even

BUSINESS NOTES

Point out to students that many small businesses cannot offer a wide range of employee benefits because of the costs involved. Ask students what benefit they would consider most essential if working for a small business.

ALTERNATIVE STRATEGIES: Enrichment

Have students determine the total compensation for the problems in the text. Total compensation is annual salary plus benefits.

LESSON PLAN
12-4 Disability Insurance

FOCUS

Disability benefits are often provided by an employer. Short-term disability usually lasts no longer than 26–52 weeks. The focus of this lesson is on computing the monthly benefit for long-term disability.

TEACH

There are many different long-term disability plans available. Some pay a flat monthly fee; some pay 50% of an employee's last salary; and some pay two thirds of that salary. Some assume retirement is at age 65; others, age 70. Point out that no matter what the plan, long-term disability pays less than the last salary amount. This is to serve as an incentive to try to get people back to work. Many plans require a doctor's approval to continue on disability each year. If social security or a pension is available, the amount of disability is often reduced so that a person does not receive more money by being off work than he or she would receive by working.

12-4 Disability Insurance

OBJECTIVE
Compute disability benefits under independent retirement systems and under social security.

Disability insurance pays benefits to individuals who must miss work due to an illness or injury. Short-term disability is covered by an employer or is obtained from private insurance companies. Long-term or permanent disability coverage is provided by social security or by an independent retirement system. Most independent retirement systems compute disability benefits based on a percent of the final average salary. To determine the annual disability benefit, multiply the sum of the years of service and the expected retirement age minus the present age times the rate of benefits times the final average salary. The monthly disability benefit is the annual benefit divided by 12.

$$\text{Annual Disability Benefit} = \left(\text{Years Worked} + \text{Expected Retirement Age} - \text{Present Age}\right) \times \text{Rate of Benefits} \times \text{Final Average Salary}$$

EXAMPLE *Skills* 5, 6, 30 *Application* A *Term* Disability insurance

Alicia Wormsley had worked at Northern State University for 21 years when she became permanently disabled and could not continue to work. Alicia was 52 years of age and had planned to retire in 13 years at Northern State's normal retirement age of 65. Her final average salary was $38,740. Northern State's rate of benefits is 2%. What is Alicia's monthly disability benefit?

SOLUTION

A. Find the **annual disability benefit.**

$$\left(\text{Years Worked} + \text{Expected Retirement Age} - \text{Present Age}\right) \times \text{Rate of Benefits} \times \text{Final Average Salary}$$

$$(21 + 65 - 52) \times 2\% \times \$38{,}740$$

$$34 \times 0.02 \times \$38{,}740$$

$$= \$26{,}343.20 \text{ annual disability benefit}$$

B. Find the **monthly disability benefit.**

Annual disability ÷ 12

$26,343.20 ÷ 12 = $2195.266 = $2195.27 monthly disability benefit

21 + 65 − 52 = 34 × .02 × 38740 = 26343.2 ÷ 12 = 2195.266667

✓ SELF-CHECK Complete the problem, then check your answer in the back of the book.

1. Final average salary is $47,800, years worked are 15, retirement age is 60, your age is 50, and rate of benefits is 1.8%. Find the annual and monthly disability benefits. **$21,510; $1792.50**

CRITICAL THINKING

Refer to the first Example. Suppose that long-term disability costs an employer a typical rate of $0.12 per $10 of annual benefit received. What is the annual cost to the company of insuring Ms. Wormsley? ($316.12)

Warm-Up Exercises

1. 14 + (60 − 51) 23
2. 21 + (65 − 56) 30
3. 6 + (70 − 37) 39
4. 18 + (62 − 47) 33
5. \$47,500 × (23 × 2%) \$21,850
6. \$38,400 × (30 × 2.1%) \$24,192
7. \$26,500 × (28 × 1.8%) \$13,356

If an employee paid social security taxes for 5 of the 10 years before becoming disabled, the person would be eligible to receive disability benefits. Benefits are based on the table shown. To use the table, find the age and earnings closest to the age and earnings of the person.

F.Y.I.
The average monthly benefit for a disabled worker in 1991 is \$587, and the average payment to a disabled worker with a family is \$1022.

APPROXIMATE MONTHLY DISABILITY BENEFITS

(For Worker Becoming Disabled in 1992 With Steady Earnings)

Worker's Age	Worker's Family	Worker's Earnings In 1991					
		\$10,000	\$20,000	\$30,000	\$40,000	\$50,000	\$54,600 or More
25	Worker only	\$492	\$ 761	\$ 996	\$1122	\$1244	\$1250
	Worker, spouse, and child[2]	714	1142	1494	1684	1867	1875
35	Worker only	487	751	988	1112	1221	1224
	Worker, spouse, and child[2]	701	1126	1483	1686	1832	1837
45	Worker only	486	749	985	1082	1147	1149
	Worker, spouse, and child[2]	699	1124	1478	1622	1721	1723
55	Worker only	488	753	971	1037	1082	1083
	Worker, spouse, and child[2]	703	1129	1458	1556	1623	1624
64	Worker only	495	765	980	1037	1076	1077
	Worker, spouse, and child[2]	718	1148	1471	1556	1615	1616

[1] Use this column if you earn more than the maximum social security earnings base.
[2] Equals the maximum family benefit.

Note: The accuracy of these estimates depends on the pattern of the worker's actual past earnings.

EXAMPLE *Skill* 1 *Application* C *Term* Disability insurance

An employee, age 46, paid social security taxes for 18 years. The worker's last annual earnings were \$22,000. What will the worker and the worker's spouse and child receive for a monthly disability benefit?

SOLUTION Referring to the table, the worker's age is closest to 45 and the last annual earnings are closest to \$20,000. The approximate monthly disability benefit for the disabled worker, spouse, and child is \$1124.

✓ SELF-CHECK Complete the problem, then check your answer in the back of the book.

2. Worker's age is 57. Last annual earnings were \$29,900. Find the worker's approximate monthly disability benefit. **\$971**

ALTERNATIVE STRATEGIES: Enrichment

Have some students contact the Social Security Administration to obtain up-to-date information on disability benefits including the current approximate monthly dollar amounts.

PRACTICE AND APPLY

The following problems can be assigned for classwork and the answers checked in class to help students master the objective of the lesson.

- Guided Practice: 1–5, 14–18
- Independent Practice: 6–11, 19–23

PROBLEMS

In problems 3–8, find the annual and monthly disability benefits.

	(Years Worked	+	Expected Retirement Age	−	Present Age)	×	Rate of Benefits	×	Final Average Salary	=	Annual Disability Benefit	Monthly Disability Benefit
3.	(20	+	65	−	60)	×	2.0%	×	$40,000		**$20,000.00**	**$1666.67**
4.	(26	+	60	−	54)	×	2.1%	×	$56,000		**$37,632.00**	**$3136.00**
5.	(14	+	62	−	44)	×	1.8%	×	$35,700		**$20,563.20**	**$1713.60**
6.	(16	+	65	−	37)	×	2.0%	×	$38,450		**$33,836.00**	**$2819.67**
7.	(3	+	70	−	31)	×	1.75%	×	$29,840		**$21,932.40**	**$1827.70**
8.	(30	+	65	−	54)	×	2.1%	×	$71,910		**$61,914.51**	**$5159.54**

In problems 9–14, use the social security table to find the approximate monthly disability benefit.

	Worker's Age	Worker's Family	Earnings	Monthly Benefit
9.	64	Worker only	$48,950	**$1076**
10.	35	Worker, spouse, and child	$18,880	**$1126**
11.	54	Worker, spouse, and child	$40,000	**$1556**
12.	63	Worker only	$65,760	**$1077**
13.	36	Worker, spouse, and child	$26,970	**$1483**
14.	24	Worker only	$12,400	**$492**

15. Marci Unger, age 58.
Worked 28 years.
Expected retirement age: 67.
Rate of benefits: 2.0%.
Final average salary: $64,845.
What is the annual disability benefit? **$47,985.30**
What is the monthly disability benefit? **$3998.78**

16. Mark LaVine, age 58.
Worked 23 years.
Expected retirement age: 60.
Rate of benefits: 3.0%.
Final average salary: $47,147.
What is the annual disability benefit? **$35,360.25**
What is the monthly disability benefit? **$2946.69**

17. Joe Hogan, age 31.
Annual earnings of $33,960.
What will Joe, his wife, and child receive in social security disability benefits? **$1483**

18. Lisa Minatel, age 60.
Annual earnings of $37,000.
What will she receive in social security disability benefits? **$1037**

19. Paul Thornton had worked for the state for 18 years. He suffered a stroke and became disabled at age 49. His final average salary was $36,947.80. Normal retirement age is 65. The rate of benefits is 2.1%.
   a. What is Paul's annual disability benefit? **$26,380.73**
   b. What is Paul's monthly disability benefit? **$2198.39**

20. Theresa Gasiorowski had worked for Central State University for 13 years when she suffered a heart attack and became disabled. Theresa was 54 years of age and had planned to retire at the normal retirement age of 60. Her final average salary was $41,247.86. Central State's rate of benefits is 2%. What is Theresa's monthly disability benefit? **$1306.18**

21. Richard Wexler, age 58, had been an employee of Chambers, Inc., for 16 years when he became permanently disabled. He was covered under social security for all 16 years. His annual earnings last year were $41,870. Find the approximate monthly disability benefit for Richard, his wife, and child. **$1556**

F.Y.I.
In 1989, the average monthly benefit a disabled worker received from social security was $556.

22. Yolanda Winters, age 27, had been an employee of Etna-North Manufacturing for 7 years when she became permanently disabled. She was covered under social security for those 7 years. Her annual earnings last year were $14,560. Find the approximate monthly disability benefit she would receive. **$492**

23. George Hercule, age 23, had been an employee of Blendo Company for 5 years when he became permanently disabled. He was covered under social security for those 5 years. His annual earnings last year were $21,870. Find the approximate monthly disability benefit he would receive. **$761**

24. Henrietta Jordon, age 60, had been an employee of Giant Department Stores for 38 years when she became permanently disabled. She was covered under social security for all 38 years. Her annual earnings last year were $54,640. Find the approximate monthly disability benefit for Henrietta, her husband, and child. **$1616**

MAINTAINING YOUR SKILLS Look up the skills in parentheses if you need help or more practice.

Add. **(Skill 5)**

25. 18 + 17 **35**
26. 21 + 12 **33**
27. 7 + 32 **39**
28. 21 + 9 **30**

Solve. **(Skills 5, 6)**

29. 17 + (65 − 56) **26**
30. 9 + (60 − 31) **38**
31. 23 + (62 − 54) **31**
32. 16 + (70 − 41) **45**

Find the percentage. Round to the nearest cent. **(Skill 30)**

33. 2% of $47,965 **$959.30**
34. 2.1% of $36,741 **$771.56**
35. 1.8% of $29,176 **$525.17**
36. $51,416 × (20 × 2%) **$20,566.40**
37. $48,615 × (18 × 2.1%) **$18,376.47**

WRAP-UP

Go over Problem 19 as a class activity. Ask one person at a time to explain what calculations to do, then do them.

Assignment Guide

- Basic: 6–12, 19–24, 27–37
- Average: 12, 13, 24–26, 27–37 odd

ALTERNATIVE STRATEGIES: Reteaching

Emphasize that this formula is a little algebraic in that the computations within the parentheses must be performed first, then you find the percentage, and finally you multiply.

LESSON PLAN
12-5 Travel Expenses

FOCUS

The focus of this lesson is on computing the total business travel expenses for an employee traveling on business. Ask students why they think employees travel. (to sell products, attend conferences and meetings, and so on)

TEACH

Go over the four areas of travel expense. Point out that transportation expenses can include not only gas and tolls, but also airfare, taxi fare, or car rental costs. Point out also that tips are included in the meal expense. Additional expenses can include registration fees but will not include entertainment costs.

Warm-Up Exercises

1. 240 × $0.22 $52.80
2. 415 × $0.21 $87.15
3. $75.80 + $196.90 + $488 + $75 $835.70
4. (349 × $0.23) + $71.80 + $114.90 $266.97

12-5 Travel Expenses

OBJECTIVE
Compute the total business travel expense.

If you travel for your business, you will probably be reimbursed, or paid back, for all authorized expenses during your trip. **Travel expenses** usually include transportation, lodging, and meals.

Total Travel Expense = Cost of Transportation + Cost of Lodging + Cost of Meals + Additional Costs

EXAMPLE *Skills* 8, 5 *Application* A *Term* Travel expenses

The accounting department of the Diversified Sales Company will reimburse Roger Martin for attending a 3-day marketing conference. Roger drove to the conference, so Diversified will pay him $0.22 per mile. Roger's expenses included:

Tolls: $2.20
Hotel: $94.50 per night for 2 nights
Conference registration: $45
Meals (including tips):

Thursday		Friday		Saturday	
lunch	$ 9.47	breakfast	$ 8.50	breakfast	$ 7.89
dinner	$19.65	lunch	$15.00	lunch	$11.48
		dinner	$24.70		

Mileage: 240 miles round-trip

What is the total cost of sending Roger to the conference?

SOLUTION

A. Find the **cost of transportation.**
(240 × $0.22) + $2.20
$52.80 + $2.20 .... $ 55.00

B. Find the **cost of lodging.**
2 × $94.50 .... $189.00

C. Find the **cost of meals.**
$9.47 + $19.65 + $8.50 + $15.00 + $24.70 +
$7.89 + $11.48 .... $ 96.69

D. Find the **additional costs.**
Conference registration .... $ 45.00
Total Travel Expense $385.69

240 × .22 = 52.8 + 2.2 = 55 M+ 2 × 94.5 = 189 M+ 9.47 + 19.65 + 8.5 + 15 + 24.7 + 7.89 + 11.48 = 96.69 M+ 45 + RM = 385.69

COOPERATIVE LEARNING

If you can provide students with copies of the Internal Revenue Service guidelines, have them work in groups and look up the section on travel expenses to see what the IRS allows (mileage regulations, and so on). Students should then rework the Example using the IRS allowances. Point out that if a person is reimbursed by the company, he or she cannot claim the expense at tax time.

✔ SELF-CHECK Complete the problem, then check your answer in the back of the book.

1. Cost of transportation, $476; cost of lodging, $219; cost of meals, $147.80; additional costs, $89. Find the total travel expense. **$931.80**

PROBLEMS

Complete the table for these business trips.

	2.	3.	4.	5.	6.	7.
Name	T. Wills	B. Henry	M. Goss	V. Tarski	T. Lanza	B. Pappas
Miles Traveled	48	80	240	180	417	623
Cost at $0.21/mile	$10.08	**$16.80**	**$50.40**	**$37.80**	**$87.57**	**$130.83**
Meals	$27.80	$46.90	$ 71.95	$ 70.40	$238.51	$291.94
Hotel Room	$65.48	0	$189.40	$210.00	$314.90	$516.85
Total Expenses	**$103.36**	**$63.70**	**$311.75**	**$318.20**	**$640.98**	**$939.62**

8. Tomas Rogers. **$903.95**
Airfare: $488.
Hotel: 3 nights at $74.50 each
Meals: $117.45
Registration: $75
What is the total travel expense?

9. Carol Cipriani. **$524.70**
Train: $147.85.
Hotel: 2 nights at $110 each.
Meals: $71.85.
Conference registration: $85
What is the total travel expense?

10. Tim Kandis is a troubleshooter for dpa, Inc. This month his travel expenses included airplane fares of $217.60 and $147.80, 156 miles of driving at $0.23 per mile, taxicab fares of $7.25 and $4.50, and meals totaling $76.85. What was the total travel expense for the month? **$489.88**

11. Meg Larson, a sales representative for Curry Corporation, flew to New York to make a sales presentation. Airfare was $215. Meg rented a car for 3 days for $21.40 a day plus $0.32 a mile. She drove a total of 70 miles. Meg's hotel bill was $124.50 a night for 2 nights. Her meals cost $11.90, $24.85, $9.76, $14.91, $27.80, $9.80, and $14.90. What was Meg's total travel expense? **$664.52**

MAINTAINING YOUR SKILLS

Look up the skills in parentheses if you need help or more practice.

Multiply. **(Skill 8)**

12. 350 × $0.21 **$73.50**
13. 900 × $0.23 **$207**
14. 620 × $0.22 **$136.40**
15. 145 × $0.21 **$30.45**
16. 540 × $0.20 **$108**
17. 450 × $0.22 **$99**
18. 422 × $0.23 **$97.06**
19. 2 × $74.90 **$149.80**
20. 3 × $110.80 **$332.40**
21. 5 × $85.71 **$428.55**
22. 3 × $93.47 **$280.41**
23. 4 × $86.90 **$347.60**

Add. **(Skill 5)**

24. $55 + $190 + $100 **$345**
25. $74 + $87 + $219 **$380**
26. $74.85 + $217.47 + $117.95 **$410.27**
27. $184.73 + $67.52 + $347.85 **$600.10**

PRACTICE AND APPLY

The following problems can be assigned for classwork and the answers checked in class to help students master the objective of the lesson.

- Guided Practice: 1–4
- Independent Practice: 5–9

WRAP-UP

Have students compute the total travel expense: 200 miles at $0.22 per mile; 1 meal at $15; no hotel; registration fee of $75. ($134)

Assignment Guide

- Basic: 5–10, 12–27
- Average: 10, 11, 12–26 even

ALTERNATIVE STRATEGIES: Enrichment

Have students use a map to find the distance to the nearest large city. Then have them suggest typical hotel/motel costs for the area, and the cost of typical meals that they like. Use a conference registration fee of $50 and have students find the cost of going to this city and staying overnight for a conference.

LESSON PLAN
12-6 Employee Training

FOCUS
You can motivate this lesson by asking those students who have part-time jobs what kind of training they had, and how long the training lasted. The focus of this lesson is to compute the total employee training costs.

TEACH
Go over the three components of total training cost. Point out that there are usually initial training sessions for new employees to become familiar with both the company and their job. Some students may not be aware that there can be more training sessions later on. Point out that many companies periodically provide training sessions to improve efficiency, customer service, productivity, or to introduce updated equipment and techniques. College tuition may be reimbursed if a course is related directly to an employee's job.

Make sure students understand the concept of *release time*. Emphasize that this is a cost to the employer, who hopes to see the money come back in the form of improved performance by the employee.

12-6 Employee Training

OBJECTIVE
Compute the total employee training costs.

Your business may pay the expenses involved in training employees. Your company may send you to special job-related programs or may offer special training programs within the company. Expenses for training during regular work hours include the cost of **release time.** When you are granted release time, you are paid your regular wages or salary while you are away from your job.

$$\text{Total Training Costs} = \text{Cost of Release Time} + \text{Cost of Instruction} + \text{Additional Costs}$$

EXAMPLE *Skills* 8, 5 *Application* A *Term* Release time

The Acme Manufacturing Company chose 8 employees to attend a training program in the company. The employees were paid their regular wages while attending the two-day program. Their combined wages amounted to \$512 per day. The production control manager, who earns \$168 per day, was the course instructor. Refreshments were brought in twice a day at a cost of \$45.70 per day. Supplies and equipment for the program amounted to \$35 per person. What was the total cost for the training program?

SOLUTION

A. Find the **cost of release time.**
2 days × \$512.00 = \$1024.00 release time

B. Find the **cost of instruction.**
2 days × \$168.00 = \$336.00 instruction cost

C. Find the **additional costs.**
(2 × \$45.70) + (8 × \$35.00)
\$91.40 + \$280.00 = \$371.40 additional costs

D. Find the **total training cost.**
Cost of Release Time + Cost of Instruction + Additional Costs
\$1024.00 + \$336.00 + \$371.40 = \$1731.40 total training cost

2 [×] 512 [=] 1024 [M+] 2 [×] 168 [=] 336 [M+] 2 [×] 45.7 [=] 91.4 [M+] 8 [×] 35 [=] 280 [M+] [RM] 1731.40

✓ SELF-CHECK Complete the problem, then check your answer in the back of the book.

1. Cost of release time, \$1247; cost of instruction, \$250; additional costs, \$140. Find the total training cost. **\$1637**

CRITICAL THINKING
Ask students if they have thought about what they would like to do to earn a living. Do they have a specific job or career in mind? If so, do they know what kind of skills or education may be required to get and hold the job? Are they learning any of these skills in school? If not, why not?

PROBLEMS

	2.	3.	4.	5.	6.	7.
Number of Days	1	1	4	3	2	5
Daily Cost of Release Time	$417	$545	$345	$716	$ 96	$176
Daily Cost of Instruction	$ 75	$250	$175	$200	$150	$100
Daily Cost of Supplies	$ 50	$120	$100	$ 30	$ 25	$ 15
Total Training Cost	? **$542**	? **$915**	? **$2480**	? **$2838**	? **$542**	? **$1455**

8. 3 physicians. **$2454**
Attend 2-day seminar.
$1027 per day total for release time.
$175 per day total for instruction.
$25 per day total for supplies.
What is the total training cost?

9. 5 marketing representatives. **$4640**
Attend 3-day workshop.
$930 per day total for release time.
$85 per person for instruction.
$285 per person total travel expense.
What is the total training cost?

10. The payroll department is sending 8 payroll clerks to a 2-day seminar on the new tax law. Three of the clerks earn $52 per day, while the other 5 earn $60 per day. Registration costs $150 per person. Materials cost $25 per person. What will the 2-day seminar cost the payroll department? **$2312**

11. The Laser Research Company is sending their 5-person research team to a 2-day conference to learn a new automated lab procedure. One researcher earns $157.80 daily, two earn $136.80 daily, and two earn $119.20 daily. What will the 2-day conference cost the Laser Research Company in lost productivity for the 5 researchers? **$1339.60**

12. Six employees of the Solar Energy Company underwent a 5-day sales training program. Their wage rate averaged $9.65 per hour for an 8-hour day. The sales manager, at a daily salary of $146.53, conducted the 5-day session. Refreshments cost $37.50 per day. Sales kits were provided for each of the 6 employees at a cost of $36.45 each. What was the total cost of the sales training program? **$3454.85**

MAINTAINING YOUR SKILLS Look up the skills in parentheses if you need help or more practice.

Multiply. **(Skill 8)**

13. 4 × $145 **$580** **14.** 5 × $112 **$560** **15.** 8 × $74 **$592**

16. 3 × $716 **$2148** **17.** 2 × $217 **$434**

Divide. Round answers to the nearest cent. **(Skill 10)**

18. $47,650 ÷ 52 **$916.35** **19.** $34,840 ÷ 52 **$670** **20.** $28,645 ÷ 260 **$110.17**

Add. **(Skill 5)**

21. $1076 + $345 + $130 **$1551** **22.** $916 + $718 + $215 **$1849**

23. $1240.76 + $337.50 + $27.85 **$1606.11** **24.** $3167.25 + $175.86 + $47.95 **$3391.06**

Warm-Up Exercises

1. $219 + $74 + $388 $681
2. $747.50 + $87.60 + $219.20 $1054.30
3. ($417 × 2) + ($78 × 3) $1068
4. ($47.80 × 5) + ($74.65 × 4) $537.60
5. $27,820 ÷ 52 $535
6. $13,104 ÷ 260 $50.40

PRACTICE AND APPLY

The following problems can be assigned for classwork and the answers checked in class to help students master the objective of the lesson.

- Guided Practice: 1–4
- Independent Practice: 5–8, 10

WRAP-UP

Read Problem 10 to the class. Ask one student how he or she would find the cost of release time; then have that student do the computations. Do the same for the cost of instruction and for additional costs.

Assignment Guide

- Basic: 5–11, 13–24
- Average: 9, 11, 12, 14–24 even

ALTERNATIVE STRATEGIES: Reteaching

Use some Help Wanted ads from the newspaper and have students calculate the cost of training. Suppose a job pays $26,000 annually, and a person has to undergo six weeks of training before the individual is productive. How much does the training cost in just salary? ($26,000 × $\frac{6}{52}$ = $3000)

LESSON PLAN
Reviewing the Basics

The exercises on this page review skills, applications, and terms used in the unit. You can use the exercises to assess informally students' proficiency with this material.

The page can be used for guided practice and independent practice. You can work through a selection of the exercises together with students, and thus see immediately if they know how to do them, and you can then assign some of the exercises for independent practice. Be sure to go over the answers to all assigned exercises.

Reviewing the Basics

Skills

Solve. Round to the nearest cent.

(Skill 5)
1. $14,716 + $588.64 **$15,304.64**
2. $35,640 + $1603.80 + $712.80 **$37,956.60**

(Skill 8)
3. $643.75 × 2 **$1287.50**
4. 315 × $0.21 **$66.15**

(Skill 10)
5. $42,650 ÷ 52 **$820.19**
6. $26,790 ÷ 260 **$103.04**

(Skill 30)
7. 7.65% of $31,745 **$2428.49**
8. 4.6% of $17,895.75 **$823.20**

Round answers to the nearest tenth of a percent.

(Skill 31)
9. $14,716 is what percent of $45,617? **32.3%**
10. $4174.76 is what percent of $17,817.91? **23.4%**

Applications

Use the table to answer.

(Application C)
11. What is the base salary for a research assistant with 2 years of experience and a master's degree? **$30,300**

12. What is the base salary for a research assistant with 1 year of experience and a bachelor's degree? **$26,380**

RESEARCH ASSISTANT—BASE SALARY

Years Experience	Degree	
	Bachelor's	Master's
0	$25,125	$27,500
1	26,380	28,875
2	27,700	30,300
3	29,100	31,800

Terms

Use each term in one of the sentences.

c **13.** Employee benefits
d **14.** Release time
g **15.** Disability insurance
b **16.** Recruiting costs
a **17.** Salary scale
e **18.** Travel expenses

a. The personnel department may use a __?__ to compare the salaries of various jobs.

b. __?__ may include the cost of advertising in the newspaper.

c. Paid vacations and holidays are __?__ that most businesses give their employees.

d. When training sessions are held during work hours, the cost of __?__ is one of the expenses.

e. If a job requires travel, the employee may be reimbursed for __?__, which include transportation, lodging, and meals.

f. A __?__ is a raise in an employee's salary.

g. __?__ pays benefits to an individual who must miss work due to an illness or injury.

Refer to your reference files in the back of the book if you need help.

Unit Test

Lesson 12-1

1. The Hi-Tech Corporation hired Albert Jent for a new clerk-typist position at an annual salary of $14,500. The hiring costs included:

   Advertising: $315
   Interviews: Janet Gray, $75; Albert Jent, $85
   Office Agency fee: 20% of first year's salary

   What was the total cost of hiring Albert? **$3375**

Lesson 12-2

2. Tonia Irwin is a statistician, level 3, for the Heritage Insurance Corporation. Tonia receives a 4.1% cost-of-living increase and a 3.4% merit increase. What is her new salary? **$29,532.40**

Level	Statistician
1	$24,450
2	25,917
3	27,472

Lesson 12-3

3. Martha Hiller works for the General Gravel Corporation. She earns $14,820 per year. The corporation provides these benefits for Martha. To the nearest percent, what is the rate of benefits? **23%**

   Vacation: 2 weeks
   Holidays: 8 days
   Compensation insurance: 3.7%
   Unemployment insurance: 4.7%
   Social security: 6.2%
   Medicare: 1.45%

Lesson 12-4

4. Vic Iagulli had worked at Eastern State University for 16 years when he suffered a stroke and became permanently disabled. Vic was 49 years of age and had planned to retire at the normal retirement age at Eastern of 60. His final average salary was $32,417.80. Eastern's rate of benefits is 2.1%. What is Vic's monthly disability benefit? **$1531.74**

Lesson 12-5

5. Ursula Swede attended a professional conference last month. The payroll department reimbursed Ursula for these expenses:

   Airfare: $315.60
   Hotel: 2 nights at $87.95 each
   Meals: $117.80
   Registration: $55

   What was the total travel expense? **$664.30**

Lesson 12-6

6. The personnel department of the Krio Company is sending 3 of its employees to a local 2-day training program. The 3 employees' combined wages total $207.80 per day. The company paid the registration fee of $90 for each person. The company also paid for the employees' lunches for the 2 days at a total cost of $72.75. What was the total cost for the Krio Company to send their employees to the training program? **$758.35**

LESSON PLAN
Unit Test

Students should do the Unit Test on their own. Each problem on the test is keyed to a lesson in the unit. Students having difficulty with any particular problem should review the Example in the appropriate lesson and be assigned some of the Independent Practice problems for additional practice.

USING TECHNOLOGY

The cost of employee benefits is usually invisible to most employees, but it is not invisible to employers. Benefits are a large and significant cost to all businesses, big and small. Some corporations publish employee benefit booklets, which are individualized for each employee. They show the employee's benefits and the company's and employee's costs for each benefit.

Point out to students that one of the most important employee benefits is medical insurance. This insurance pays doctor bills and hospital bills when an employee is ill. Medical insurance is also one of the most costly benefits a business can provide. Generally, both employees and employers share the cost of this benefit.

A SPREADSHEET APPLICATION

Personnel

To complete this spreadsheet application, you will need the diskette *Spreadsheet Applications for Business Mathematics,* which accompanies this textbook.

Select option 12, Personnel, from the menu. Input the information in the following problems to find the individual benefits, total benefits, and rate of benefits.

Title	Annual Wage	Vacation	3.6% Comp. Ins.	8–day Sick Lv.	6.2% Soc. Sec.	1.45% Medicare	Total Benefits	Rate of Benefits
Dir.	$72,600	(3 wk) ?	?	?	?	?	?	? %
Asst. Dir.	48,900	(3 wk) ?	?	?	?	?	?	? %
Adm. Asst.	42,300	(2 wk) ?	?	?	?	?	?	? %
1st Clk.	27,650	(2 wk) ?	?	?	?	?	?	? %
2nd Clk.	18,540	(2 wk) ?	?	?	?	?	?	? %
Total	?	$?	?	?	?	?	?	? %

a. What are the total benefits for the director?
b. What is the rate of benefits for the assistant director?
c. What is the total compensation insurance collected for these 5 employees?
d. How much social security is withheld from the 1st clerk's pay? How much is contributed by the company?
e. How much more would it cost to give the administrative assistant a 3-week vacation?
f. How much would it cost to increase the number of sick days from 8 to 10 for each employee?

Title	Annual Wage	Vacation	6 Paid Holidays	4.6% Unemp. Ins.	6.2% Soc. Sec.	1.45% Medicare	Total Benefits	Rate of Benefits
Mgr.	$45,000	(4 wk) ?	?	?	?	?	?	? %
Asst. Mgr.	37,500	(4 wk) ?	?	?	?	?	?	? %
Ser. Tech.	28,500	(3 wk) ?	?	?	?	?	?	? %
Ser. Tech.	24,000	(3 wk) ?	?	?	?	?	?	? %
Janitor	18,000	(2 wk) ?	?	?	?	?	?	? %
Total	?	$?	?	?	?	?	?	? %

a. What is the rate of benefits for the manager?
b. What are the total benefits for the janitor?
c. What does vacation time cost for these 5 employees?
d. What is the total unemployment insurance collected for these 5 employees?
e. How much social security was withheld from the service technician earning $28,500 per year? How much was contributed by the employer?
f. How much could be saved by reducing the paid holidays to 5 days rather than 6?

A SIMULATION

Applying for a Job

Suppose you have decided to get a summer job to earn extra money. To decide what kind of work you might like to do, you talk to friends, family, teachers, and others. There are many factors to consider. Here are some of them.

Compensation	Transportation	Interest
hourly pay	location	interesting work
benefits	public transportation	work related to career goals
overtime pay	need for car	
tips	cost	
commission		

Hours	Conditions
regular hours	duties
night work	indoor work
weekend work	outdoor work
holiday work	clean, quiet environment
overtime work	

In comparing the wages of different jobs, consider costs too. For example, if you have to spend a lot of money for automobile expenses to drive to and from a job, you may end up with less money to spend than if you had a lower-paying job closer to home.

1. Look at the list of factors above. You may consider other factors important. List the ten factors that are most important to you, with the most important one first. (If you have never had a job, you may want to discuss the factors with someone who has.) **Answers will vary.**

Calculate the spendable income for each job.

	Hourly Rate	Hours per Week	Gross Pay	Deductions	Net Pay	Transportation Cost	Spendable Income
2.	$4.85	20	$97	$24.24	$72.76	$ 5	**$67.76**
3.	$5.00	15	**$75**	$18.24	**$56.76**	0	**$56.76**
4.	$6.25	40	**$250**	$69.13	**$180.87**	$15	**$165.87**
5.	$8.50	36	**$306**	$83.77	**$222.23**	$36	**$186.23**

LESSON PLAN
A Simulation

USING THE SIMULATION

Applying for a job is an individual experience and therefore students should work on this simulation individually. Each student should have copies of the job application form. You can begin by reading with students the introductory material at the top of the page.

If necessary, work with students as they complete the job application form. After students have finished the simulation, discuss it with them, particularly the questions on the last page about job interviews. There are many thought-provoking questions asked here that students should consider.

A SIMULATION (CONTINUED)

Finding a Job

There are many ways to find a job. Relatives and friends may know of job openings. Schools, churches, and other organizations may be able to help. Some cities have programs that find jobs for young people. State employment services list job openings. Newspaper help-wanted ads may be a good source. Some young people do odd jobs, such as gardening and painting, for their neighbors. If you want to do a particular kind of work, start early when writing or calling possible employers. You may also want to place a "Position Wanted" ad in a neighborhood newspaper.

HELP WANTED

CAR WASHER $4.90/hr. Tues–Sat, 9–6, 1 hr. lunch. Apply in person to Happy Time Auto Wash, 2167 Elm St. **$196**

CASHIER and sandwhich maker. 11–2 and 4–8, Mon. –Fri. $4.75/hr. Call 555-3597 for interview appointment. **$166.25**

CLERK TYPIST Must type at least 50 wpm and be willing to be trained as data input operator. 8–4:30, Mon.–Fri. $240/week. Assured Products, 621 East St., 555-3418 ex. 43. **$240**

COOK'S ASSISTANT for summer camp. No experience necessary, will train. 7hr./day, 7 days/wk. Meals, lodging, $200/wk. Green Mt. Camp, Woodstock, Vermont. **$200**

RECREATION AIDE for city parks. Supervise games, teach crafts. Tues–Sat, 10–6. $5/hr. Experience w/children required. Park Dept, 555-5000 ex. 104. **$200**

KENNEL HELP No experience necessary, love of animals needed. Clean, feed, exercise, bathe animals. 7–3:30 with 1/2 hr. lunch, Mon.–Fri. $4.90/hr. Wagon Wheel Kennels, 555-4225 before noon. **$196**

SWIMMING POOL SERVICE Cleaning & servicing pools, will train. $7.50/hr., 8–5 with 1 hr. lunch, Mon.–Fri. E-Z Pool Service, 555-8216 **$300**

6. Find the gross weekly pay for each of the jobs listed in the ads.

7. If you could have any of these jobs, which one would you choose? Why? Keep in mind the factors on page 359. If you have trouble choosing between two jobs, write down the reasons you like each one and compare them. This may help you narrow your choice to one job.
Answers will vary.

A SIMULATION (CONTINUED)

Job Application

When you apply for a job, you will be asked to fill out a job application form. You should be prepared to complete all items on the form. Be sure to receive permission to use the names of your personal character references.

Here is a sample of a job application form. Write down your answers as if you were filling out the form. **Answers will vary.**

EMPLOYMENT APPLICATION

Name ____________________

Address ____________________

Home Phone ____________________

Social Security Number ____________________

Date of Birth ____________________

In case of emergency, notify ____________________

Address ____________________

EDUCATIONAL BACKGROUND

Name and Address of School	Dates Attended		Date Graduated	Major Area of Study
	From	To		
Elementary School				
High School				
College				
Other				

EMPLOYMENT

Name and Address of Employer	Dates Employed		Position	Reason for Leaving
	From	To		

REFERENCES

Name __________ Address __________ Phone __________

Name __________ Address __________ Phone __________

Name __________ Address __________ Phone __________

MISCELLANEOUS

Hobbies ____________________

Hours Available for Interview ____________________

Number of Days per Week You Wish to Work: Minimum __________ Maximum __________

Are You Willing to Work Evenings? __________ Weekends? __________

A SIMULATION (CONTINUED)

Job Interviews

Usually an employer interviews many applicants for a job opening. It is important that you dress neatly, speak clearly, and are able to answer the interviewer's questions. Many people are nervous before and during an interview. It is helpful to know what type of questions an interviewer might ask. Here are some of the most common questions.

Write your answers to these questions. **Answers will vary.**

8. What kind of work are you most interested in? Why?
9. What are your future vocational or educational plans?
10. What do you know about our company? Why do you think you might like to work for us?
11. What jobs have you held? What did you learn from them?
12. What courses in school did you like best? least? Why?
13. What is your average grade in school? Do you think you have done the best academic work of which you are capable? If not, why not?
14. In what school activities have you participated? Which did you like most? Why? Have you been an officer in any activity?
15. How do you spend your spare time? What are your hobbies?
16. Do you prefer working with others or by yourself?
17. What is your major strength? What is your major weakness?
18. What have you done that shows initiative and willingness to work?
19. What qualifications do you have that make you feel you will be successful in your field?

An interview is an opportunity for you to ask questions about the job and the company. You may want to ask about pay, benefits, duties, tuition credit, opportunities for advancement, or other factors.

20. Write five specific questions you might ask if you were being interviewed for the job you chose on page 360.

If there is time, you and someone in your class might take turns interviewing each other.

U N I T

13

Production

The production department of your business makes, or *manufactures,* and packages the products you sell. The costs of manufacturing an item include both materials and labor. For you to break even, the income from your sales must cover your manufacturing costs. Sales beyond the *break-even point* result in profits. You may use *quality controls* to check for defective items during manufacturing. *Time studies* can tell you how much time is needed to make one item.

O U T L I N E

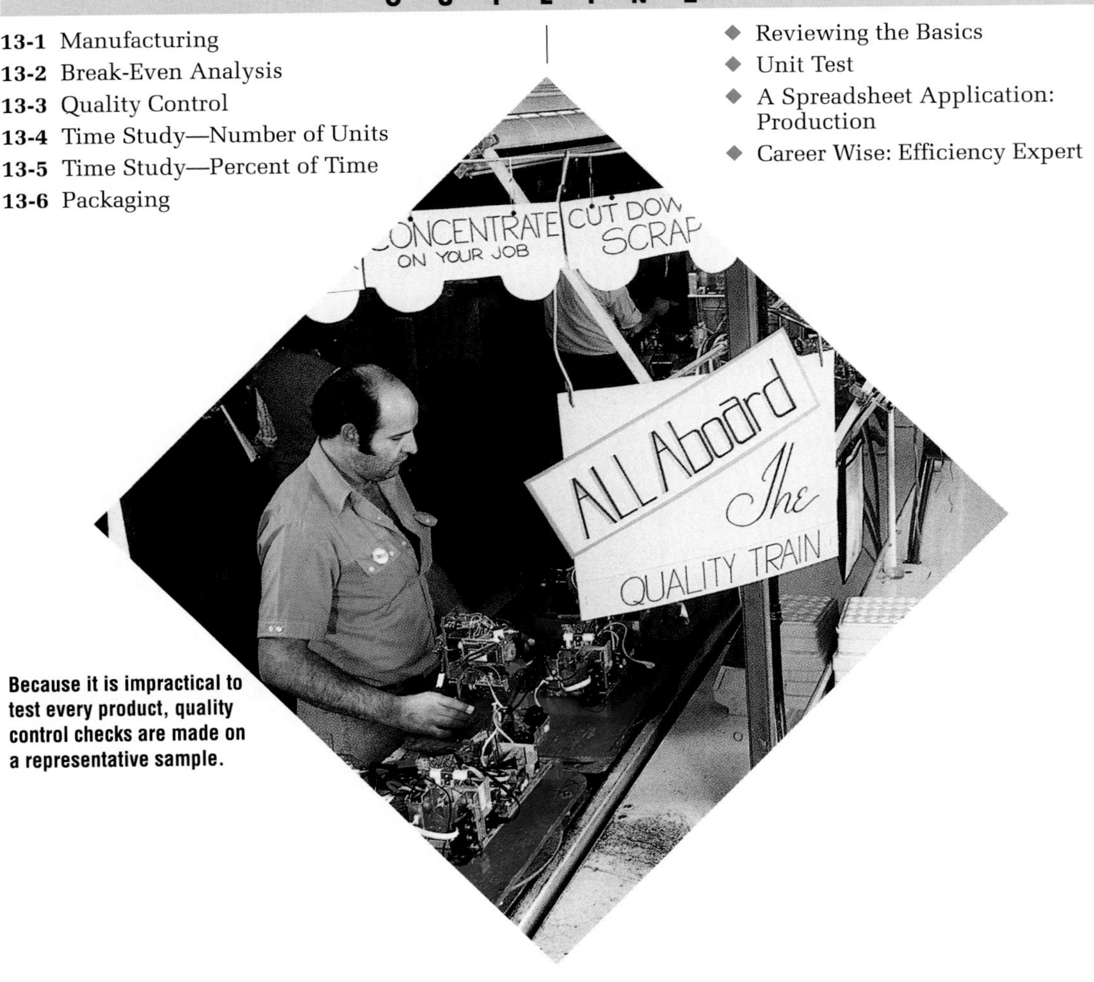

Because it is impractical to test every product, quality control checks are made on a representative sample.

UNIT 13 PRODUCTION

INTRODUCING THE UNIT

Introduce this unit by asking students to name well-known manufacturing companies. What are their main products? If appropriate, you may want to talk about local industry.

Ask students why some American companies manufacture their products in foreign countries. What does this practice do to the availability of jobs in the United States? Who are many Americans now competing with for manufacturing jobs? Are we winning or losing this competition?

LESSON PLAN
13-1 Manufacturing

FOCUS
Hold up a pen or pencil for the class to see and ask students to name some costs that might have gone into making it. When the costs of materials and labor are suggested, write these costs on the chalkboard and tell students that their sum is called the prime cost. The focus of this lesson is to compute the prime cost of manufacturing an item.

TEACH
Ask students to name some manufacturing companies in the United States and the products they make. Then have students name some companies in their state, county, or city. Ask students if they know how much money people are paid by the hour in these companies. Continue the discussion by considering some products made in foreign countries and ask students if they think the people who made these products earn more or less than United States' workers.

Warm-Up Exercises
1. $0.017 + $0.042 $0.059
2. $0.071 + $0.035 $0.106
3. $0.069 + $0.043 $0.112
4. $0.007 + $0.018 $0.025
5. ($0.98 ÷ 25) + ($9.75 ÷ 80) $0.161
6. ($1.19 ÷ 10) + ($14.70 ÷ 120) $0.242

13-1
Manufacturing

OBJECTIVE
Compute the prime cost of manufacturing an item.

The cost of manufacturing an item depends, in part, on the **direct material cost** and the **direct labor cost.** The direct material cost is the cost of the goods that you use to produce the item. The direct labor cost includes the wages paid to the employees who make the item. The **prime cost** is the total of the direct material cost and the direct labor cost. The prime cost is frequently expressed on a per-unit basis.

Prime Cost per Item = Direct Material Cost per Item + Direct Labor Cost per Item

EXAMPLE *Skills* 11, 12 *Application* A *Term* Prime cost

Electric Supply Inc. produces aluminum circuit housings. The machine operator stamps 20 housings from each strip of aluminum. Each strip costs $0.90. The operator can stamp 720 housings per hour. The direct labor charge is $12.50 per hour. To the nearest tenth of a cent, what is the prime cost of manufacturing a housing?

SOLUTION

A. Find the **direct material cost per item.**
$0.90 ÷ 20 = $0.045 direct material cost

B. Find the **direct labor cost per item.**
$12.50 ÷ 720 = $0.0173 = $0.017 direct labor cost

C. Find the **prime cost per item.**
Direct Material Cost + Direct Labor Cost
$0.045 + $0.017 = $0.062 or 6.2¢ cost per housing

✓ SELF-CHECK Complete the problems, then check your answers in the back of the book.

1. 40 brackets are made from a strip of metal costing $0.80. What is the direct material cost per item? **$0.02**
2. The direct material cost per item is $0.04. The direct labor cost per item is $0.18. Find the prime cost per item. **$0.22**

PROBLEMS

Round answers to the nearest tenth of a cent.

	Cost per Strip	Pieces per Strip	Direct Material Cost per Piece	Labor Cost per Hour	Pieces per Hour	Direct Labor Cost per Piece	Prime Cost
3.	$0.80	10	$0.08	$11.70	1000	$0.012	**$0.092**
4.	$0.75	25	**$0.03**	$13.50	1200	**$0.011**	**$0.041**
5.	$0.70	50	**$0.014**	$17.45	800	**$0.022**	**$0.036**

CULTURAL ANGLES
As a part of the homework assignment, ask each student to give one example of an item made by a country in Asia, Europe, Africa, Latin America, and South America. Let students work on this assignment for about one week.

Round answers to the nearest tenth of a cent.

6. 36 boxes per sheet of cardboard. Cost: $3.60 per sheet. Direct labor charge: $9.60 per hour. 200 boxes cut per hour. What is the prime cost per box? **$0.148**

7. 4200 washers per steel strip. Cost: $8.75 per strip. Direct labor charge: $12.65 per hour. 13,500 washers punched per hour. What is the prime cost per washer? **$0.003**

8. The Donnely Manufacturing Company manufactures plastic tablecloths. Each roll of printed plastic yields 120 tablecloths. Each roll costs $11.70. The direct labor charge is $10.20 per hour. The machine operator can cut and fold 90 tablecloths per hour. What is the prime cost of manufacturing each tablecloth? **$0.211**

9. The Northern Aluminum Company manufacturers aluminum products. One of their machines stamps housings for smoke alarms from strips of aluminum. Each strip of aluminum costs $0.90. The machine stamps 40 housings from each strip. The machine operator produces 10 housings per minute. The direct labor cost is $14.76 per hour. What is the prime cost of manufacturing each housing? **$0.048**

F.Y.I.
In fiscal 1991, the American Greeting Card Company earned $82.5 million on revenues of $1.43 billion.

10. Sue Clark is a machine operator at Modern Plastics, Inc. She molds 65 buckets from one container of molding plastic. Each container costs $2.98. Sue's machine molds one bucket every 4 seconds. The direct labor cost is $13.65 per hour. What is the prime cost of manufacturing one bucket? **$0.061**

11. A strip of heavy gauge tin makes 115 switch plates. The cost per strip is $0.98. The direct labor cost is $11.67 per hour. 1060 plates are stamped per hour. What is the prime cost of manufacturing each switch plate? **$0.02**

12. A strip of aluminum makes 20 fan blades. The aluminum costs $1.12 per strip. The direct labor cost is $15.65 per hour. One blade is stamped every 5 seconds. What is the prime cost of manufacturing each fan blade? **$0.078**

MAINTAINING YOUR SKILLS Look up the skills in parentheses if you need help or more practice.

Round answers to the nearest tenth of a cent. **(Skill 2)**

13. $0.21486 **21.5¢** 14. $0.05512 **5.5¢** 15. $0.00707 **0.7¢**

16. 34.19¢ **34.2¢** 17. 7.241¢ **7.2¢** 18. 1.987¢ **2.0¢**

Divide. Round answers to the nearest tenth of a cent. **(Skill 11)**

19. $0.90 ÷ 18 **$0.05** 20. $0.95 ÷ 5 **$0.19** 21. $0.92 ÷ 12 **$0.077** 22. $0.85 ÷ 12 **$0.071**

23. $1.47 ÷ 10 **$0.147** 24. $1.76 ÷ 5 **$0.352** 25. $17.85 ÷ 470 **$0.038** 26. $9.65 ÷ 330 **$0.029**

27. $17.90 ÷ 435 **$0.041** 28. $7.15 ÷ 120 **$0.06** 29. $11.72 ÷ 510 **$0.023** 30. $21.93 ÷ 1475 **$0.015**

PRACTICE AND APPLY

The following problems can be assigned for classwork and the answers checked in class to help students master the objective of the lesson.

- Guided Practice: 1–5
- Independent Practice: 6–8

WRAP-UP

Ask a student what he or she has learned in this lesson. Review the formula for computing the prime cost of an item.

Assignment Guide

- Basic: 6–11, 13–30
- Average: 9–12, 14–30 even

ALTERNATIVE STRATEGIES: Reteaching

Stress understanding of the terms: material cost and labor cost.

	Cost/ Strip	Pieces/ Strip	Direct Mat. Cost/Piece	Labor Cost per Hour	Pieces/ Hour	Dir. Labor Cost/Piece	Prime Cost
1)	$0.90	10	($0.09)	$14.00	100	($0.14)	($0.23)
2)	0.75	5	($0.15)	17.50	25	($0.70)	($0.85)

LESSON PLAN
13-2 Break-Even Analysis

FOCUS
In this lesson, students compute the break-even point in the number of manufactured units. Engage students in a discussion of the meaning of the term *break-even.* Ask them to give some examples from their own lives in which they have broken even.

TEACH
Discuss the concepts of fixed and variable costs. Point out that every family has fixed and variable costs and give some examples to illustrate these ideas. Then discuss the business examples of fixed and variable costs given in the text.

Write the formula on the chalkboard for reference as you work through the Example.

Using the Example, have students raise and lower the selling price per unit by increments of one cent to see how it affects the break-even point. They should use the range of values from $0.39 to $0.49 and arrange their results in a table.

Warm-up Exercises
1. $2.79 − $1.54 $1.25
2. $17.94 − $11.56 $6.38
3. $0.86 − $0.49 $0.37
4. $1.12 − $0.87 $0.25
5. $12.17 − $8.78 $3.39
6. $174.31 − $87.93 $86.38
7. $175,200 ÷ $7.40 23,676
8. $75,450 ÷ $0.50 150,900
9. $117,470 ÷ ($1.12 − $0.64) 244,729
10. $42,600 ÷ ($27.48 − $21.55) 7184

13-2 Break-Even Analysis

OBJECTIVE
Compute the break-even point in the number of manufactured units.

Your business can prepare a **break-even analysis** to determine how many units of a product must be made and sold to cover production expenses. The **break-even point** is the point at which income from sales equals the cost of production. Units sold after that point will result in a **profit** for your business.

To calculate the break-even point, you must know the total **fixed costs,** the **variable costs** per unit, and the selling price per unit. Fixed costs include rent, salaries, and other costs that are not changed by the number of units produced. Variable costs include the cost of raw materials, the cost of packaging, and any other costs that vary directly with the number of units produced.

$$\text{Break-Even Point in Units} = \frac{\text{Total Fixed Costs}}{\text{Selling Price per Unit} - \text{Variable Costs per Unit}}$$

EXAMPLE *Skills* 6, 11 *Application* A *Term* Break-even point

Token Metal Products, Inc., manufactures can openers. They plan to manufacture 750,000 hand-held can openers to be sold at $0.44 each. The fixed costs are estimated to be $142,570. Variable costs are $0.19 per unit. How many can openers must be sold for Token Metal Products, Inc., to break even?

SOLUTION Find the **break-even point in units.**

Total Fixed Costs ÷ (Selling Price per Unit − Variable Costs per Unit)

$142,570.00 ÷ ($0.44 − $0.19)
$142,570.00 ÷ $0.25 = 570,280 break-even point in units

.44 [−] .19 [=] 0.25 [M+] 142570 [÷] [RM] [=] 570280

✓ SELF-CHECK Complete the problems, then check your answers in the back of the book.

Find the break-even point in units.

1. Selling price is $0.79. Variable cost per unit is $0.29. Total fixed costs are $87,500. **175,000**
2. Selling price is $1.49. Variable cost per unit is $0.74. Total fixed costs are $124,500. **166,000**

CRITICAL THINKING
Write the formula to find the break-even point in units on the chalkboard. Ask students to suppose that a business wants to lower its break-even point for a product. What part of the formula can be manipulated by the business to do this? (selling price per unit) How? (increase the selling price per unit)

PROBLEMS

	Total Fixed Costs	÷	(Selling Price per Unit	−	Variable Costs per Unit)	=	Break-Even Point in Units
3.	$ 95,000	÷	($ 1.79	−	$ 0.79)	=	95,000 ?
4.	$ 148,000	÷	($ 0.59	−	$ 0.34)	=	592,000 ?
5.	$ 225,000	÷	($ 2.29	−	$ 1.79)	=	450,000 ?
6.	$ 478,400	÷	($ 3.49	−	$ 2.69)	=	598,000 ?
7.	$ 1,470,000	÷	($174.99	−	$ 98.75)	=	19,281 ?
8.	$12,476,500	÷	($417.79	−	$279.81)	=	90,423 ?

9. TenCo tennis rackets.
Fixed costs: $740,000.
Selling price per racket: $74.29.
Variable cost per racket: $48.76.
What is the break-even point in units? **28,986**

10. Farm Supply rural mailboxes.
Fixed costs: $192,800.
Selling price per mailbox: $39.99.
Variable cost per mailbox: $21.43.
What is the break-even point in units? **10,388**

11. Today's Music produces traditional CDs. The fixed costs total $2,417,950. The selling price per disc is $14.95. The variable cost per disc is $7.48. What is the break-even point in number of discs? **323,688**

12. Superior Custom Homes builds a limited number of luxury homes. The fixed costs of the operation total $2,157,750. The base selling price per custom home is $479,500. The base variable cost per custom home is $431,550. What is the break-even point in number of custom homes? **45**

13. True Bounce basketballs are manufactured by General Sports, Inc. They have total fixed costs of $3,110,400. The variable cost per basketball is $18.47. The selling price per basketball is $24.95. What is the break-even point in number of basketballs? **480,000**

14. The Alumni Association plans to produce and market monogrammed sweatshirts. They anticipate selling each sweatshirt for $24.95. Their variable costs per sweatshirt include $13.19 for supplies and $7.47 per shirt for direct labor. The fixed costs of the operation total $7450. How many monogrammed sweatshirts does the Alumni Association need to sell to break even? **1737**

MAINTAINING YOUR SKILLS Look up the skills in parentheses if you need help or more practice.

Subtract. **(Skill 6)**

15. $14.79 − $9.43 **$5.36**
16. $1.89 − $0.97 **$0.92**
17. $14.49 − $7.74 **$6.75**
18. $7.19 − $5.54 **$1.65**
19. $147.47 − $109.88 **$37.59**
20. $8129.00 − $6417.48 **$1711.52**

Divide. Round answers to the nearest whole number. **(Skill 11)**

21. $47,500 ÷ $0.25 **190,000**
22. $163,540 ÷ $0.20 **817,700**
23. $216,491 ÷ $0.79 **274,039**

PRACTICE AND APPLY

The following problems can be assigned for classwork and the answers checked in class to help students master the objective of the lesson.

- Guided Practice: 1–6
- Independent Practice: 7–11

WRAP-UP

Have students explain the meaning of each term in the formula used in this lesson.

Assignment Guide

- Basic: 7–13, 15–23
- Average: 12–14, 16–22 even

ALTERNATIVE STRATEGIES: Enrichment

There should be many examples around school similar to Problem 14. The Pep Club might be selling sweatshirts or the Band Boosters might be operating a concession stand at the games. In these examples, there are fixed and variable costs to consider when you are establishing the selling price and the break-even point. Have your students write proposals for these various activities identifying cost, selling price, and break-even point.

LESSON PLAN
13-3 Quality Control

FOCUS

Ask students if they have ever bought a product that was defective or know someone who did. Start a brief discussion as to why defective products reach the marketplace. The focus of this lesson is to compute the percent of defective goods and determine if the process is in or out of control.

TEACH

Explain to students that there is a strong emphasis in business today to produce high quality products that have no defects. Point out that many products made in the United States compete with foreign products, such as automobiles, and that consumers are demanding quality. Competition has forced business to strive to make the highest quality product possible and this benefits all consumers. Point out that quality control measures are much more demanding in most businesses today than they were a few years ago.

It is sometimes impractical for a business to test every item that is produced. In some cases, testing the item destroys it. For example, testing a fuse or the road life of a newly developed tire destroys the product. Most industries make decisions about the quality of their output based on a representative sample.

13-3 Quality Control

OBJECTIVE

Compute the percent of defective goods and determine if the process is in or out of control.

If your business deals with mass production, you need a **quality control** inspector to check the items that are manufactured. The inspector may examine a specified number of items. If the size of the item is incorrect or if the item is broken or damaged, it is classified as **defective.** The inspector computes the percent of the sample that is defective and plots the percent on a **quality control chart.** The quality control chart shows the percent of defective products that is allowable. If the actual percent of defective items is greater than the percent allowable, the process is said to be "out of control." Production must then be stopped and corrected.

$$\text{Percent Defective} = \frac{\text{Number Defective}}{\text{Total Number Checked}}$$

EXAMPLE *Skills* 1, 31 *Application* N *Term* Quality control

Bob Atkinson is a quality control inspector for Material Plating, Inc. At 9 a.m., he checked 25 stamped brackets produced during the previous hour and found 2 defective brackets. The process is in control if 5% or less of the sample is defective. Is the process in control or out of control? Plot the percent defective on a quality control chart.

SOLUTION

A. Find the **percent defective.**
Number Defective ÷ Total Number Checked
2 ÷ 25 = 0.08 = 8% defective
Process is out of control.

B. Plot the **percent defective** on a quality control chart.

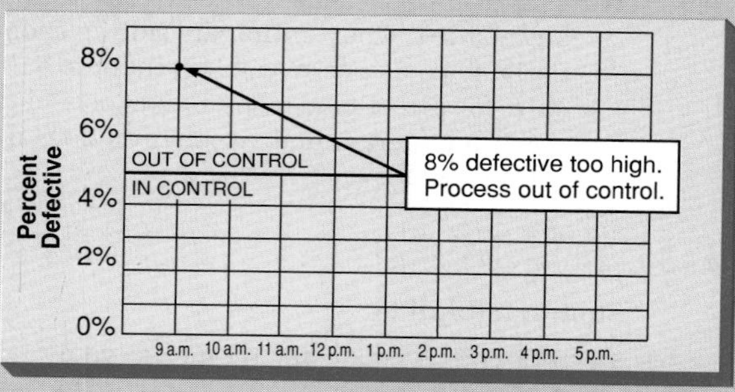

✓ SELF-CHECK Complete the problems, then check your answers in the back of the book.

1. Number defective is 5.
Number checked is 200.
What is the percent defective? **2.5%**

2. Number defective is 3.
Number checked is 50.
What is the percent defective? **6%**

MATHEMATICS NOTES

You may need to review graphing techniques with the students. Help them to control the size of their graphs by selecting appropriate intervals for the horizontal and vertical axes. Suggest the following intervals for the vertical axis for Problems 13 and 14.

Problem 13: 0%, 2%, 4%, 6%, 8%
Problem 14: 0%, 10%, 20%, 30%

PROBLEMS

If more than 5% of the sample is defective, the process is out of control.

	Number Defective	÷	Number Checked	=	Percent Defective	In or Out of Control?
3.	2	÷	50	=	4%	In
4.	1	÷	50	=	2%	In
5.	4	÷	50	=	8%	Out
6.	0	÷	50	=	0%	In
7.	3	÷	50	=	6%	Out
8.	5	÷	50	=	10%	Out

9. Sweaters.
100 in sample.
5 defective.
In control if 6% or less defective.
Is the process in or out of control?
In

10. Plastic flashlights.
25 in sample.
1 defective.
In control if 5% or less defective.
Is the process in or out of control?
In

11. Alice McHenry is a quality control inspector for Current Cassettes, Inc. The process is in control if 6% or less of each sample is defective. Alice checked a sample of 50 cassettes and found 4 defective cassettes. What percent of the sample is defective? Is the process in or out of control?
8%; Out

12. Tucker Jensen, a quality control inspector for Office Products Co., checked a sample of 250 ballpoint pens. Fourteen of the pens were defective. If more than 6% of the sample is defective, the process is out of control. What percent of the sample is defective? Is the process in or out of control? **5.6%; In**

For Exercises 13–17, draw a quality control chart, compute the percent defective in each sample, and plot the percents on the chart.

13. Every 2 hours, Bev Wallace checks samples from a punch press. Each sample contains 50 pieces. The operation is in control if 5% or less of each sample is defective. Bev found the following: **Check student's charts for problems 13–17.**

Time	9 a.m.	11 a.m.	1 p.m.	3 p.m.	5 p.m.
Number Defective	1	0	3	2	2
	2%	**0%**	**6%**	**4%**	**4%**

Out of control at 1 p.m.

14. Every hour, 10 compact disc players are taken off the production line at Modern Technologies and checked to see if they are defective. If more than 10% are defective, production is out of control. The number of defective players for the 9 a.m. to 5 p.m. checks are:

Time	9 a.m.	10 a.m.	11 a.m.	12 noon	1 p.m.	2 p.m.	3 p.m.	4 p.m.	5 p.m.
Number Defective	0	1	0	2	1	0	1	0	2

**14. 0%; 10%; 0%; 20%; 10%; 0%; 10%; 0%; 20%
Out of control at 12 noon and 5 p.m.**

Warm-up Exercises

1. 4 is what percent of 25?
   16%
2. 2 is what percent of 50?
   4%
3. 8 is what percent of 400?
   2%
4. 5 is what percent of 40?
   12.5%

PRACTICE AND APPLY

The following problems can be assigned for classwork and the answers checked in class to help students master the objective of the lesson.

- Guided Practice: 1–9
- Independent Practice: 10, 11, 13, 14

ALTERNATIVE STRATEGIES: Reteaching

As the FOCUS portion indicates, the objective of this lesson is to compute the percent of defective goods and determine if the process is in or out of control. Finding the percent defective is an application of Skill 31. Caution your students about the meaning of "5% or less", "not more than 5%", "up to 5%", etc...

WRAP-UP
Ask a few students to work Problem 16 and 17 at the chalkboard. Go over the results with the class.

Assignment Guide
- Basic: 10–15, 18–32
- Average: 12, 14–17, 18–32 even

15. Every 3 hours, Tom Smith pulls 25 bicycle rims off the assembly line and checks their diameter. If more than 6% of the rims are found to be defective, the assembly line is shut down. Determine whether the process is in or out of control at each time check. Draw a quality control chart.

Out of control at 12:30 p.m. and 3:30 p.m. Wed. and 3:30 p.m. Fri.

Time	9:30 a.m.	12:30 p.m.	3:30 p.m.	Day
Number Defective	0 0%	1 4%	1 4%	Monday
	1 4%	0 0%	0 0%	Tuesday
	1 4%	2 8%	2 8%	Wednesday
	0 0%	0 0%	1 4%	Thursday
	1 4%	1 4%	3 12%	Friday

16. Nora Stamos randomly checks 75 transistors 10 times a day. If more than 4% are defective, the production is stopped and corrections are made. Complete the table and draw a quality control chart.

Out of control at Checkpoint #8

Checkpoint No.	1	2	3	4	5	6	7	8	9	10
Number Defective	1	2	1	0	2	1	3	4	1	0
Percent Defective	?	?	?	?	?	?	?	?	?	?

1.3%; 2.7%; 1.3%; 0; 2.7%;1.3%; 4%; 5.3%; 1.3%; 0

17. Albert Burns checks circuit board production at different times during the day. He will check a different number each time. If more than 5% are defective, he will close down production and have corrections made. The circuit boards are checked electronically. Complete the table and draw a quality control chart.

In control at all times

Time	9:40 a.m.	10:15 a.m.	11:50 a.m.	1:30 p.m.	3:30 p.m.	4:30 p.m.
Number Checked	60	50	80	100	40	50
Number Defective	3	2	3	4	2	2
Percent Defective	?	?	?	?	?	?
	5%	4%	3.75%	4%	5%	4%

MAINTAINING YOUR SKILLS Look up the skills in parentheses if you need help or more practice.

Find the rate. Round answers to the nearest tenth of a percent. **(Skill 31)**

18. 2 is what percent of 50? **4%**
19. 6 is what percent of 100? **6%**
20. 4 is what percent of 25? **16%**
21. 2 is what percent of 60? **3.3%**
22. 3 is what percent of 40? **7.5%**
23. 4 is what percent of 150? **2.7%**
24. 4 is what percent of 130? **3.1%**
25. 2 is what percent of 200? **1%**

Which number is greater? **(Skill 1)**

26. 5.3% or 5% **5.3%**
27. 6% or 5.2% **6%**
28. 3% or 4% **4%**
29. 6% or 6.7% **6.7%**
30. 4.3% or 4% **4.3%**
31. 5% or 0% **5%**
32. 2% or 2.5% **2.5%**
33. 5% or 5% **Same**

13-4

Time Study—Number of Units

OBJECTIVE
Use time-study results to compute how many units can be produced.

Your business may conduct a **time study** to determine how long a particular job should take. A time study involves watching an employee complete a job, recording the time required for each task, and calculating the average time for each task. You can use the averages to determine how many units a worker can produce in a fixed period of time.

$$\text{Number of Units} = \frac{\text{Actual Time Worked}}{\text{Average Time Required per Unit}}$$

EXAMPLE *Skills* 5, 11 *Application* Q *Term* Time study

General Lamps, Inc., did a time study of Beth Peters's job as a carton packer. Beth's times were recorded. The averages were calculated.

$$\frac{\text{Sum of times}}{\text{Number of observations}}$$

Task	Observations in Seconds #1	#2	#3	#4	#5	Average Time
Pick up carton	3.9	5.4	3.9	4.4	3.9	4.3 sec.
Fill carton	12.0	13.0	14.5	12.5	11.0	12.6 sec.
Apply glue	14.5	12.5	13.0	13.5	13.5	13.4 sec.
Close carton	5.1	4.4	4.6	4.3	5.1	4.7 sec.
Remove filled carton	5.5	5.5	5.5	5.5	5.5	5.5 sec.

If Beth gets a 10-minute break each hour, how many cartons can she fill per hour?

SOLUTION

A. Find the **average time required per unit.**
4.3 + 12.6 + 13.4 + 4.7 + 5.5 = 40.5 seconds

B. Find the **actual time worked per hour.**
(60 − 10) minutes × 60 seconds = 3000 seconds

C. Find the **number of units per hour.**
Actual Time Worked ÷ Average Time Required per Unit
3000 ÷ 40.5 = 74.07 = 74 units per hour

4.3 [+] 12.6 [+] 13.4 [+] 4.7 [+] 5.5 [=] 40.5 [M+] 60 [−] 10 [=] 50 [×] 60 [=] 3000 [÷] [RM] [=] 74.074

✓ SELF-CHECK Complete the problems, then check your answers in the back of the book.

Find the number of units per hour.

1. Average time is 6 minutes per unit. 48 minutes worked per hour. **8**
2. Average time is $\frac{1}{4}$ hour per unit. $7\frac{1}{2}$ hours worked per day. **30**

LESSON PLAN 13-4 Time-Study Number of Units

FOCUS
The focus of this lesson is to use time-study results to compute the time spent on each task. Ask students why a business would want to know how long a particular job should take. List the reasons given on the chalkboard.

TEACH
Time-and-motion studies were first conducted by businesses in the early part of this century to get an idea of the average time within which a task should be completed. Point out that these time-studies can be useful to evaluate employees' performances and to improve work efficiency. For example, after conducting a time-study, an employer may simplify some tasks by changing the methods of performing them.

Warm-up Exercises
1. (7.8 + 7.4 + 7.7 + 7.7 + 7.6) ÷ 5 7.64
2. (31.6 + 32.3 + 32.0 + 31.7) ÷ 4 31.9
3. 3000 ÷ 17.5 171.4
4. 3300 ÷ 31.4 105
5. 50 ÷ 12.5 4
6. 55 ÷ 4.5 12.2
7. Convert 58 minutes to seconds. 3480
8. Convert 8 hours to seconds. 28,800

ALTERNATIVE STRATEGIES: Enrichment
Students may wish to research the history of time-and-motion studies. There are many names associated with time-and-motion studies and the usually associated incentive pay plans. They may wish to examine the contributions of Scanlon, Gilbreth, Taylor, Gantt, and so on. There are many abuses of time-and-motion studies and incentive pay plans by both labor and management. Your students may be able to identify some within your community and/or school.

PRACTICE AND APPLY

The following problems can be assigned for classwork and the answers checked in class to help students master the objective of the lesson.

- Guided Practice: 1–6
- Independent Practice: 7, 9

WRAP-UP

Ask a student to explain what he or she has learned in this lesson. Then use Problem 7 to summarize the objective of the lesson.

Assignment Guide

- Basic: 7–9, 11–18
- Average: 8, 10, 11–18 odd

PROBLEMS

	Actual Time Worked		÷	Average Time per Unit	=	Number of Units
3.	48 minutes	(× 60)	÷	43.5 seconds	=	66
4.	50 minutes	(× 60)	÷	39.7 seconds	=	76
5.	55 minutes	(× 60)	÷	65.6 seconds	=	50
6.	45 minutes	(× 60)	÷	5.5 seconds	=	491

7. Bagging and carrying out groceries.
Average time is 11.42 minutes per customer.
50 minutes worked per hour.
How many customers can be served per hour? **4**

8. Selling theater tickets.
Average time is 21.4 seconds per customer.
50 minutes worked per hour.
How many customers can be served per hour? **140**

9. Ben Krieger prepared a time study to determine the average time required to make a cash withdrawal at the 24-hour automatic teller machine.

Avg.	Task	Time in Seconds #1	#2	#3	#4	#5
1.7	Insert card in machine	1.0	2.0	1.5	2.5	1.5
14.9	Type in code	17.0	11.5	16.0	15.5	14.5
7.1	Specified amount wanted	7.4	6.9	7.1	7.0	7.1
1.5	Remove cash	1.5	1.0	2.0	1.5	1.5
1.2	Remove card	1.2	1.4	1.0	1.1	1.3

a. What is the average time required for each task?
b. What is the average time required to make a cash withdrawal? **26.4 sec.**
c. How many cash withdrawals can be made at a machine in 1 hour? **136**

10. The American Pump Company prepared a time study of Shirley Monroe's job as an assembler. How many units can Shirley complete in 1 hour if she takes one 10-minute break during the hour? **115**

Task	Time in Seconds #1	#2	#3	#4	#5
Move water pump to filter	3.7	4.2	3.9	3.8	3.7
Insert filter coil	5.1	5.2	4.9	5.2	5.1
Affix gasket	4.8	4.4	4.7	4.6	4.8
Hand thread bolts	6.1	6.3	6.3	6.2	6.4
Machine tighten bolts	6.1	6.0	6.2	6.3	6.2

MAINTAINING YOUR SKILLS Look up the skills in parentheses if you need help or more practice.

Add. **(Skill 5)**

11. 3.6 + 3.8 + 3.7 + 3.6 + 3.7 **18.4**
12. 14.2 + 14.1 + 13.8 + 14.0 + 13.9 **70**
13. 7.2 + 7.4 + 7.1 + 6.8 + 7.2 **35.7**
14. 1.1 + 0.9 + 0.8 + 1.2 + 1.0 **5**

Divide. Round answers to the nearest whole number. **(Skill 11)**

15. 3600 ÷ 4.5 **800**
16. 3000 ÷ 2.4 **1250**
17. 3300 ÷ 15.2 **217**
18. 8 ÷ 0.12 **67**

MATHEMATICAL NOTES

When going over the Example with the students, make sure they understand how to compute the average time for each task. Remind them that both the *average time* and the *actual time worked* should be converted to the same unit (hours, minutes, or seconds). For example, in Problem 7 the division will be 50 minutes ÷ 11.42 minutes. Instruct students to round any fractions in the final answers since only complete units are counted.

13-5

Time Study—Percent of Time

OBJECTIVE

Use time-study results to compute the percent of time spent on each task.

You can use a time study to determine what percent of an employee's time is spent on various activities during a workday.

$$\text{Percent of Time Spent on Activity} = \frac{\text{Time Spent on Activity}}{\text{Total Time}}$$

EXAMPLE *Skills* 5, 31 *Application* A *Term* Time study

Mike Reese works in the mail room of a large office. A time study showed that he spent his time on these activities. What percent of his time does Mike spend picking up and delivering mail?

Activity	Hours
Sorting mail	2.5
Picking up and delivering mail	3.0
Talking with employees	0.5
Taking coffee breaks	0.5
Making special deliveries	1.5
Total	8.0

SOLUTION Find the **percent of time spent on activity.**

Time Spent on Activity ÷ Total Time

3.0 ÷ 8.0 = 0.375 = 37.5% of time spent on activity

✓ SELF-CHECK Complete the problems, then check your answers in the back of the book.

1. In the example above, what percent of his time does Mike spend sorting mail? **31.25%**
2. In the example above, what percent of his time does Mike spend making special deliveries? **18.75%**

PROBLEMS

Round answers to the nearest hundredth of a percent.

	Activity	Hours	÷	Total Time	=	Percent of Time
3.	Loading furniture	3.00	÷	8	=	? **37.50%**
4.	Delivering furniture	4.50	÷	8	=	? **56.25%**
5.	Lunch	1.00	÷	8	=	? **12.5%**
6.	Returning truck	0.50	÷	8	=	? **6.25%**

7. Barb Lane, carton packer.
   Fills cartons 2.5 hours per day.
   Works 8 hours per day.
   What percent of her day is spent filling cartons? **31.25%**
8. Amy Narcussi, realtor.
   Answers phone 1 hour per day.
   Works 8 hours per day.
   What percent of her day is spent answering the phone? **12.5%**

LESSON PLAN 13-5 Time Study—Percent of Time

FOCUS

Introduce the lesson by asking students to compute the percent of time they spend watching TV each day. Then ask them to compute the percent of time spent in school each day. The focus of this lesson is to use time-study results to compute the percent of time spent on individual tasks.

TEACH

Point out that many companies compute the percent of time spent on a particular activity so that a certain limit is not exceeded. Limits may be established by labor-management contracts for a particularly dangerous or loud activity. Cost control measures might require a limit for a particularly costly activity. Ask students to suggest some activities in their own lives that can be looked at using the idea of percent of time. For example, what percent of the school day is spent in this class?

Warm-up Exercises

Find the percent (nearest hundredth).

1. 4.5 is what percent of 8? 56.25%
2. 2.4 is what percent of 9? 26.67%
3. 3.1 is what percent of 7.5? 41.33%
4. 1.7 is what percent of 10? 17%

ALTERNATIVE STRATEGIES: Reteaching

This lesson asks students to find what percent one part is of the total.

	Activity A	B	C	D	E	Total	What percent of the total is:			
1)	4.3	2.4	0.6	0.5	0.2	(8.0)	A?	(53.75%)	C?	(7.5%)
2)	1.2	0.7	0.5	0.8	1.8	(5.0)	B?	(14%)	D?	(16%)

PRACTICE AND APPLY

The following problems can be assigned for classwork and the answers checked in class to help students master the objective of the lesson.

- Guided Practice: 1–7
- Independent Practice: 8–11

WRAP-UP

Call upon a few students to name an activity they do each day and to compute the percent of time spent on that activity.

Assignment Guide

- Basic: 8–12, 14–21
- Average: 12, 13, 15–21 odd

9. Wilma Cole, receptionist.
Answers telephones 2.8 hours per day.
Works 7.5 hours per day.
What percent of her day is spent answering telephones? **37.33%**

10. Paul Hong, college professor.
Is in the classroom 3 hours per day.
Works 8.5 hours per day.
What percent of his day is spent in the classroom? **35.29%**

11. Ben Lucas is a maintenance programmer for the Ace Corporation. Ben works 7.5 hours per day. He spends about 2.25 hours each day testing programs on which he has made changes. What percent of Ben's day is spent testing these programs? **30%**

12. Ruth Loeb is a clerk at Violet's Dress Shoppe. She works 8 hours each day. A time study recorded her activities. What percent of the 8 hours does Ruth spend on each activity?

Activity	Hours	
Assisting customers	4.2	**52.5%**
Processing sales	1.1	**13.75%**
Restocking racks	0.9	**11.25%**
Pricing merchandise	0.8	**10%**
Break and miscellaneous	1.0	**12.5%**
Total	8.0	

13. Don Kurezlenski delivers bread for the City Bakery Company. He works 9 hours each day. A time study showed these figures. What percent of Don's workday is spent on each of the activities?

Activity	Hours	
Loading truck	1.3	**14.44%**
Driving truck	3.1	**34.44%**
Delivering bread	2.2	**24.44%**
Recording sale	0.5	**5.56%**
Checking out	0.3	**3.33%**
Break and miscellaneous	0.6	**6.67%**
Lunch	1.0	**11.11%**
Total	9.0	

MAINTAINING YOUR SKILLS Look up the skills in parentheses if you need help or more practice.

Add. **(Skill 5)**

14. 1.7 + 2.4 + 3.1 + 0.8 **8**

15. 2.1 + 1.8 + 3.3 + 0.9 + 1.9 **10**

16. 3.4 + 1.8 + 1.6 + 0.7 **7.5**

17. 3.2 + 2.1 + 1.6 + 1.2 + 0.9 **9**

Find the rate. Round answers to the nearest hundredth of a percent. **(Skill 31)**

18. 1.6 is what percent of 7.5? **21.33%**

19. 2.1 is what percent of 9? **23.33%**

20. 0.9 is what percent of 10? **9%**

21. 1.7 is what percent of 8? **21.25%**

13-6 Packaging

OBJECTIVE
Compute the dimensions of packaging cartons.

The production of your company's merchandise ends with **packaging**, placing the product in a container for shipment. The container eases handling and prevents breaking. The package may also identify the product and show it attractively. The size of the package depends on the size of the finished product.

Dimensions: Length, Width, Height

EXAMPLE *Skills* 16, 20 *Term* Packaging

The Cleaning Genie Company manufactures a cleaning solution that is sold in bottles with a diameter of 7 cm and a height of 20 cm. The company plans to package 12 bottles to a carton for shipping. The carton is made of 0.5-cm thick corrugated cardboard with 0.3-cm cardboard partitions. What are the dimensions of the package that the Cleaning Genie Company needs?

SOLUTION

A. Find the **length.**

4 bottles × 7 cm in diameter	28.0 cm
3 partitions × 0.3 cm wide	0.9 cm
2 ends × 0.5 cm wide	1.0 cm
Length	29.9 cm

B. Find the **width.**

3 bottles × 7 cm in diameter	21.0 cm
2 partitions × 0.3 cm	0.6 cm
2 ends × 0.5 cm	1.0 cm
Width	22.6 cm

C. Find the **height.**

1 bottle × 20 cm	20.0 cm
2 top flaps × 0.5 cm	1.0 cm
2 bottom flaps × 0.5 cm	1.0 cm
Height	22.0 cm

D. Find the **dimensions.**
Length: 29.9 cm; Width: 22.6 cm; Height: 22.0 cm

4 [×] 7 [=] 28 [M+] 3 [×] .3 [=] 0.9 [M+] 2 [×] .5 [=] 1 [M+] [RM] 29.9 [CM] 3 [×] 7 [=] 21 [M+] 2 [×] .3 [=] 0.6 [M+] 2 [×] .5 [=] 1 [M+] [RM] 22.6 [CM] 1 [×] 20 [=] 20 [M+] 2 [×] .5 [=] 1 [M+] 2 [×] .5 [=] 1 [M+] [RM] 22

LESSON PLAN
13-6 Packaging

FOCUS
Ask students to suggest some qualities that are important for good packaging, such as strength or attractiveness. In this lesson, students compute the dimensions of packaging cartons.

TEACH
Point out that before manufacturers ship such items as light bulbs, television sets, and cameras to their customers, they must carefully package these items to prevent breakage. In addition to protecting the product, the containers that are used for packaging are designed to make it easy to handle the product, to identify it, and to make it look attractive. The size, shape, and type of product will determine the container used for packaging.

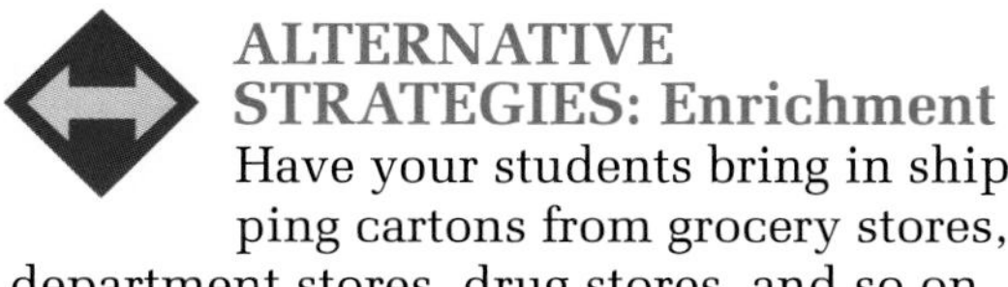

ALTERNATIVE STRATEGIES: Enrichment
Have your students bring in shipping cartons from grocery stores, department stores, drug stores, and so on. Cartons with partitions still in them would be a good way to demonstrate the physical reality of this lesson.

Warm-up Exercises

1. $4 \times 2\frac{1}{2}$ 10
2. 8×6.4 51.2
3. $8 + \frac{3}{16} + \frac{1}{4}$ $8\frac{7}{16}$
4. $10 + 1.4 + 3.2$ 14.6
5. $45 + \frac{5}{16} + \frac{1}{4}$ $45\frac{9}{16}$
6. $8 + 4.2 + 1.9$ 14.1
7. $7\frac{1}{2} + (8\frac{5}{16} \times 4)$ $40\frac{3}{4}$
8. $12.4 + (3 \times 1.4)$ 16.6

PRACTICE AND APPLY

The following problems can be assigned for classwork and the answers checked in class to help students master the objective of the lesson.

- Guided Practice: 1–14
- Independent Practice: 15, 17

COOPERATIVE LEARNING

Organize the class into small groups to work on Problems 15 and 17. Ask a representative from several groups to present their solutions to the class.

✓ SELF-CHECK Complete the problems, then check your answers in the back of the book.

1. 4 packages, each 5 inches wide, next to each other. Each end is $\frac{1}{4}$-inch cardboard. How wide is the carton? **1. $20\frac{1}{2}$ in**

2. 6 packages, each 12 cm long, next to each other. Each end is 0.5-cm cardboard. How long is the carton? **2. 73 cm**

PROBLEMS

Find the dimensions of the carton.

Height:

3. Top and bottom of 0.4 cm = ? **0.8 cm**
4. Height = ? **12.0 cm**
5. Total height = ? **12.8 cm**

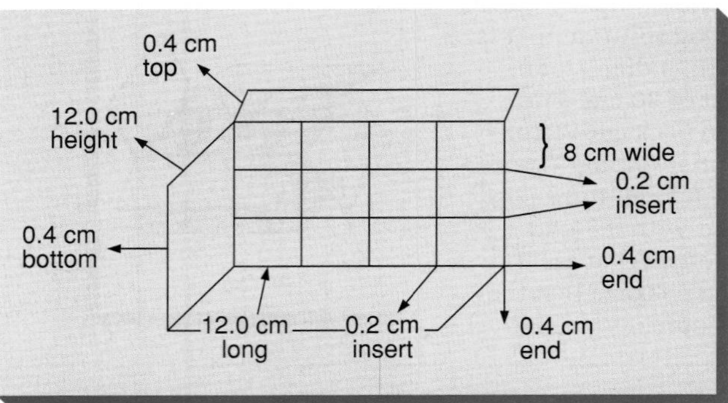

Width:

6. 3 packages, each 8 cm wide = ? **24.0 cm**
7. 2 inserts, each 0.2 cm = ? **0.4 cm**
8. 2 ends, each 0.4 cm = ? **0.8 cm**
9. Total width = ? **25.2 cm**

Length:

10. 4 packages, each 12 cm long = ? **48.0 cm**
11. 3 inserts, each 0.2 cm = ? **0.6 cm**
12. 2 ends, each 0.4 cm = ? **0.8 cm**
13. Total length = ? **49.4 cm**
14. Dimensions? **12.8 cm H x 25.2 cm W x 49.4 cm L**

Draw a sketch for each carton in Exercises 15–19.

15. The Clean Air Corporation packages 3 of these humidifiers per carton. The humidifiers are placed next to each other with a $\frac{3}{4}$-inch thick piece of cardboard between them. The carton is made of $\frac{5}{16}$-inch thick corrugated cardboard. What are the dimensions of the carton in which the humidifiers are packed?

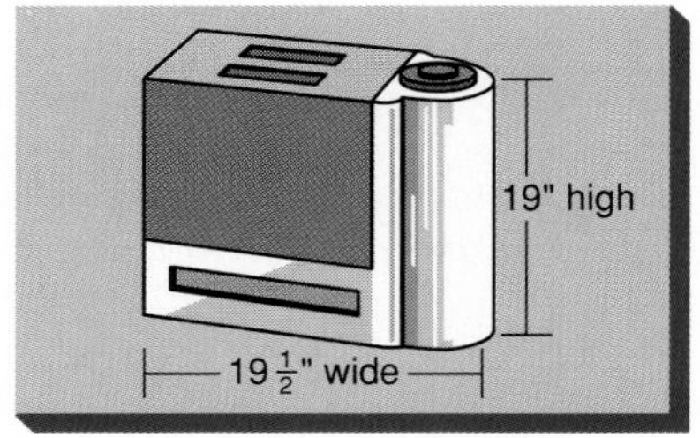

$19\frac{5}{8}$ in H x $20\frac{1}{8}$ in W x $24\frac{5}{8}$ in L

16. The Canned Food Company packages 12 cans of mixed vegetables to a package. Each can is 9 cm in diameter and 12 cm high. The carton is 0.5 cm thick and has 0.2-cm-thick partitions positioned between the cans.
    a. What are the dimensions of the carton if the cans are packaged in 1 row with 12 cans in the row? **13 cm H x 10 cm W x 111.2 cm L**
    b. What are the dimensions of the carton if the cans are arranged in 3 rows with 4 cans in each row? **13 cm H x 28.4 cm W x 37.6 cm L**

17. The Major Radio Corporation packages 6 of their weather-alert radios in 1 carton. The carton is made of $\frac{3}{8}$-inch-thick corrugated cardboard. Partitions between the radios are $\frac{1}{4}$-inch-thick.

$7\frac{1}{4}$ in H x $9\frac{1}{2}$ in W x $7\frac{1}{4}$ in L **a.** What are the dimensions of the carton if the weather-alert radios are arranged in 2 rows of 3, with each radio on its side?

$7\frac{1}{4}$ in H x 5 in W x 14 in L **b.** What are the dimensions of the carton if the weather-alert radios are arranged in 1 row of 6, with each radio on its side?

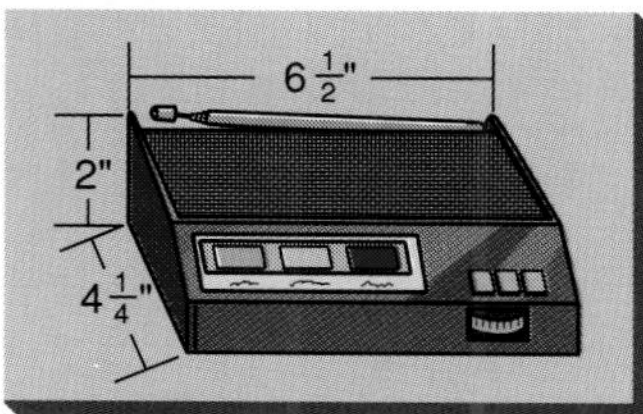

18. These 100-capacity tackle boxes are packaged 12 to a carton. The box is $10\frac{3}{4}$ in long, 6 in wide, and $6\frac{1}{2}$ in high. The carton is constructed of $\frac{1}{4}$-inch-thick cardboard. Cardboard spacers $\frac{1}{8}$-inch-thick are positioned between the boxes. What are the dimensions of the carton if the boxes are arranged in 2 layers, with each layer consisting of 2 rows of 3 boxes?

$13\frac{5}{8}$ in H x $18\frac{3}{4}$ in W x $22\frac{1}{8}$ in L

19. These boxed hand warmers are packaged 24 to a carton. The carton is made of 0.3 cm cardboard. There are no partitions.

a. What are the dimensions of the carton if the boxes are arranged standing up as shown, in 4 rows of 6 hand warmers?

14.6 cm H x 28.6 cm W x 9.6 cm L

b. What other arrangements could be used to package the hand warmers? Draw sketches and give the dimensions of the cartons.

Answers will vary.

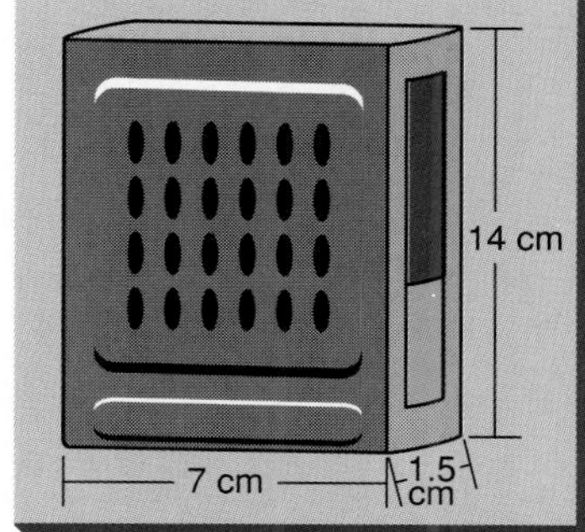

MAINTAINING YOUR SKILLS Look up the skills in parentheses if you need help or more practice.

Add. **(Skill 16)**

20. $\frac{1}{2} + \frac{1}{8}$ **$\frac{5}{8}$** **21.** $\frac{1}{8} + \frac{3}{16}$ **$\frac{5}{16}$** **22.** $\frac{1}{4} + \frac{3}{16}$ **$\frac{7}{16}$** **23.** $\frac{1}{8} + \frac{7}{16}$ **$\frac{9}{16}$**

24. $\frac{1}{2} + \frac{1}{4}$ **$\frac{3}{4}$** **25.** $\frac{3}{4} + \frac{9}{16}$ **$1\frac{5}{16}$** **26.** $\frac{3}{4} + \frac{3}{8} + \frac{1}{8}$ **$1\frac{1}{4}$**

27. $\frac{1}{8} + \frac{1}{16} + \frac{3}{16}$ **$\frac{3}{8}$** **28.** $\frac{3}{4} + \frac{7}{8} + \frac{3}{16}$ **$1\frac{13}{16}$** **29.** $\frac{1}{2} + \frac{3}{8} + \frac{7}{16}$ **$1\frac{5}{16}$**

Multiply. **(Skill 20)**

30. $6 \times \frac{1}{2}$ **3** **31.** $4 \times \frac{1}{8}$ **$\frac{1}{2}$** **32.** $5 \times \frac{1}{16}$ **$\frac{5}{16}$** **33.** $4 \times \frac{3}{8}$ **$1\frac{1}{2}$** **34.** $6 \times \frac{3}{4}$ **$4\frac{1}{2}$**

35. $6 \times \frac{5}{8}$ **$3\frac{3}{4}$** **36.** $4 \times \frac{1}{4}$ **1** **37.** $4 \times \frac{1}{2}$ **2** **38.** $6 \times \frac{5}{16}$ **$1\frac{7}{8}$** **39.** $8 \times \frac{1}{4}$ **2**

40. $6 \times \frac{3}{8}$ **$2\frac{1}{4}$** **41.** $4 \times \frac{9}{16}$ **$2\frac{1}{4}$** **42.** $2 \times \frac{5}{32}$ **$\frac{5}{16}$** **43.** $10 \times \frac{7}{8}$ **$8\frac{3}{4}$** **44.** $4 \times \frac{7}{32}$ **$\frac{7}{8}$**

WRAP-UP

Use Problem 17 to wrap up this lesson by working it with the students.

Assignment Guide

- Basic: 15–18, 20–42 even
- Average: 16, 18, 19, 21–43 odd

ALTERNATIVE ASSESSMENT

Engage students in a discussion of the production process. Have them relate the various topics studied in this unit so they can see the connections from manufacturing to packaging. Ask students how a company makes a profit? (Sales beyond the break-even point result in a profit.)

LESSON PLAN
Reviewing the Basics

The exercises on this page review skills, applications, and terms used in the unit. You can use the exercises to assess informally students' proficiency with this material.

The page can be used for guided practice and independent practice. You can work through a selection of the exercises together with students, and thus see immediately if they know how to do them, and you can then assign some of the exercises for independent practice. Be sure to go over the answers to all assigned exercises.

Reviewing the Basics

Skills

Round answers to the nearest cent.

(Skill 2)
1. \$0.1751 **\$0.18** **2.** \$4.0447 **\$4.04** **3.** \$147.7158 **\$147.72**

Solve.

(Skill 5)
4. 7.1 + 8.4 + 5.9 + 3.5 **24.9** **5.** 78.4 + 11.7 + 9.6 + 7.9 + 84.7 **192.3**

(Skill 6)
6. \$1.71 − \$1.43 **\$0.28** **7.** \$74.47 − \$56.92 **\$17.55** **8.** \$216.71 − \$149.84 **\$66.87**

(Skill 11)
9. \$146.70 ÷ 1600 **10.** \$74,750 ÷ \$0.56 **11.** \$1,714,700 ÷ \$2.10

9. \$0.09
10. 133,482
11. 816,524

(Skill 16)
12. $\frac{3}{4} + \frac{7}{16} + 5$ **$6\frac{3}{16}$** **13.** $7\frac{3}{4} + \frac{9}{16} + \frac{1}{8}$ **$8\frac{7}{16}$** **14.** $2\frac{3}{8} + \frac{1}{4} + 3$ **$5\frac{5}{8}$**

(Skill 20)
15. $6 \times \frac{3}{8}$ **$2\frac{1}{4}$** **16.** $4 \times \frac{7}{16}$ **$1\frac{3}{4}$** **17.** $6 \times 5\frac{3}{4}$ **$34\frac{1}{2}$**

(Skill 31)
18. 7 is what percent of 25? **28%** **19.** 4.5 is what percent of 14? **32.1%**

Applications

Use the graph to answer.

(Application N)
20. What was the income for March? **\$4500**

21. What was the income for June? **\$4000**

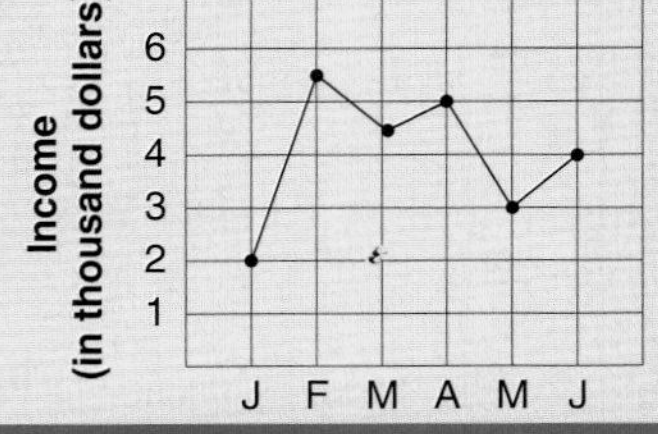

Find the mean.

(Application Q)
22. 7.2, 6.7, 7.1, 6.8 **6.95**

23. 17.5, 15.8, 16.5, 16.8, 16.1 **16.54**

Terms

Match each term with its definition on the right.

b **24.** Prime cost

d **25.** Break-even point

a **26.** Quality control

e **27.** Time study

f **28.** Packaging

a. a periodic inspection of items manufactured

b. the total of the direct labor cost and the direct material cost

c. the cost of wages paid to employees

d. the point at which income from sales equals the cost of production

e. determines how long a particular job should take

f. the final step in the production of a company's product

Refer to your reference files in the back of the book if you need help.

Unit Test

Lesson 13-1

1. Cosmo Steel produces cable from scrap steel that it buys for $248 per ton. The company produces 4 rolls of cable from each ton. The direct labor charge is $18.75 an hour. The employees produce 3 rolls of cable per hour. To the nearest tenth of a cent, what is the prime cost of manufacturing 1 roll of cable? **$68.25**

Lesson 13-2

2. Central Publishing Company plans to produce do-it-yourself books to be sold at $14.95 each. The fixed costs are estimated at $180,800. The variable costs are $10.43 per book. What is the estimated break-even point in units? **40,000**

Lesson 13-3

3. A quality control inspector for the Hu-Day Company found 3 out of 65 glass containers with cracks during the 5 p.m. inspection. The process is in control if the percent defective is 5% or less. Is the process in control or out of control? **In control**

Lesson 13-4

4. Family Food Stores made a time study of its stock operation. The store recorded the times that Kelly Booth spent shelving the contents of cartons.

	Observations in Seconds				
Task	#1	#2	#3	#4	#5
Pick up carton	5.0	5.5	4.0	5.5	4.5
Open carton	14.0	14.5	15.5	14.0	15.0
Place contents on shelf	243.0	254.0	252.5	243.5	254.5
Discard carton	5.0	5.5	5.0	5.0	5.5

If Kelly takes a 5-minute break each hour, how many cartons can she unpack in 1 hour? **12**

Lesson 13-5

5. Thomas Kuback is a medical office assistant for Family Practice, Inc. Thomas prepared a time study of his activities during the day. What percent of his time does Thomas spend retrieving files? **18.75%**

Activity	Hours
Interviewing patients	1.5
Processing records	2.5
Retrieving files	1.5
Lunch	1.0
Patient consultation	1.5

Lesson 13-6

6. Pro-Weight, Inc., ships its industrial weight containers in wooden boxes. The containers are arranged in 1 layer. Each box is made of $\frac{1}{4}$-inch-thick wood with $\frac{1}{8}$-inch-thick wood spacers. Each container is 5 inches high and $2\frac{1}{2}$ inches in diameter. What are the dimensions of each wooden box? **$5\frac{1}{2}$ in H x $8\frac{1}{4}$ in W x $8\frac{1}{4}$ in L**

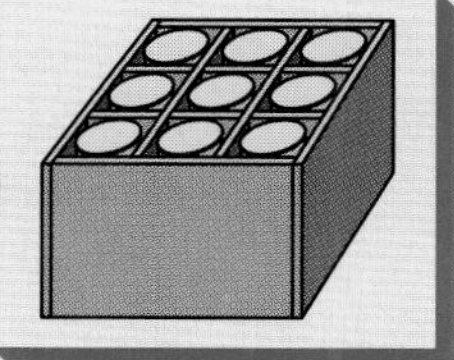

LESSON PLAN
Unit Test

Students should do the Unit Test on their own. Each problem on the test is keyed to a lesson in the unit. Students having difficulty with any particular problem should review the Example in the appropriate lesson and be assigned some of the Independent Practice problems for additional practice.

LESSON PLAN
A Spreadsheet Application

USING TECHNOLOGY

Students have learned in Unit 13 that quality control is an important part of the production process. This application reinforces the notion of a production process being in or out of control.

As a *critical thinking* question, ask students why a business would tolerate a certain percent of its products being defective. In other words, why isn't the percent of defective parts allowable 0%? (The answer to this question lies in the quality of the machines that make other things and in human error. It is very costly to make machines that manufacture other products and probably unrealistic to assume that machines made for most manufacturing processes could produce zero defects. When people are involved in the production process, a certain percentage of errors is always likely to occur.)

Point out that in certain industries, for example, the manufacturer of airplane engines, 100% of the engines must be perfect or many peoples' lives would be at risk.

A SPREADSHEET APPLICATION

Production

To complete this spreadsheet application, you will need the diskette *Spreadsheet Applications for Business Mathematics,* which accompanies this textbook.

Input the information in the following problems to find the percent defective and whether the process is in control or out of control.

	Number Tested	Number Defective	Percent Defective	Defective Parts Allowable (%)	Process In or Out of Control
1.	200	6	?	4%	?
2.	300	15	?	5%	?
3.	60	3	?	4%	?
4.	50	2	?	5%	?
5.	200	7	?	3%	?
6.	425	9	?	2%	?
7.	150	9	?	5%	?
8.	250	19	?	6%	?
9.	500	3	?	1%	?
10.	125	4	?	3%	?

11.

Day	Number Tested	Number Defective	Percent Defective	Defective Parts Allowable (%)	Process In or Out of Control
Mon.	180	9	?	4%	?
Tue.	180	7	?	4%	?
Wed.	190	7	?	4%	?
Thur.	180	6	?	4%	?
Fri.	180	8	?	4%	?

A SPREADSHEET APPLICATION

12. In #1, what percent defective parts were found?
13. In #1, is the process in or out of control?
14. In #2, what percent defective parts were found?
15. In #2, is the process in or out of control?
16. In #3, what percent defective parts were found?
17. In #3, is the process in or out of control?
18. What would happen in a situation such as that described in #3?
19. In #4, what percent defective parts were found?
20. In #4, is the process in or out of control?
21. What would happen in a situation such as that described in #4?
22. In #5, what percent defective parts were found?
23. In #5, is the process in or out of control?
24. In #6, is the process in or out of control?
25. In #7, is the process in or out of control?
26. In #8, is the process in or out of control?
27. In #9, is the process in or out of control?
28. In #10, is the process in or out of control?
29. Describe what happened with respect to being in or out of control in #11.
30. With respect to being in or out of control, why is what happened in #11 not uncommon?

LESSON PLAN
Career Wise

APPLICATIONS ON THE JOB

Discuss the meaning of the term *efficient* with students. Point out that people who work efficiently do not waste time or effort. Efficient procedures save a business money and in so doing can make it a stronger and more competitive business.

After students have answered the four questions, ask them to check the percents shown in the pie chart. They should divide the number of hours for each task by 1.58. Then ask them what the sum of the percents in the pie chart must be. (100%)

CAREER WISE

Efficiency Expert

Joshua Gomperts works as an efficiency expert for a mail-order sales company. Part of his job is to find ways to carry out the company's goals in an efficient manner. He also conducts time-management training sessions and discussion groups to get employee ideas and promote employee satisfaction.

To develop efficient procedures, Joshua gathers data regarding tasks and the time it takes to complete tasks. To evaluate the data, he uses statistical measures, such as averages. He also suggests how to arrange, label, and locate goods so that shippers can find them more easily and process orders more quickly.

Joshua prepared a survey for the 8 employees in the shipping department. Each employee was timed over a 5-week period and times needed to complete each task were averaged. The results are shown in the table at the left and the pie chart at the right.

TIME SPENT PROCESSING AN ORDER

Task	Time
Gather goods ordered	0.75 hours
Package the goods	0.25 hours
Prepare the invoice	0.33 hours
Ship the order	0.25 hours
Total	1.58 hours

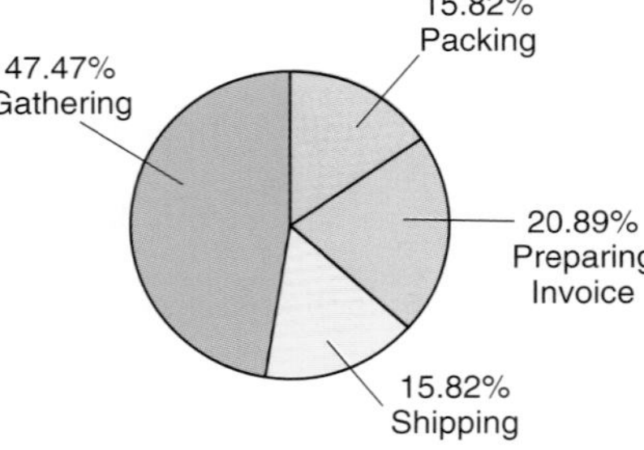

Check Your Understanding

1. Which tasks take the same amount of time to complete? What percent of 100%? **Packing and shipping; 15.82% each**
2. True or false: The time needed to gather the goods is about half of all the time it takes to process an order. **True**
3. Some employees recommended using small carts to get around the inventory area. If managers agree to this, do you think the 47.47% slice of the pie will get smaller? **Yes**
4. A new computer system will reduce the time needed to prepare an invoice to 0.1 hours. What percent of the new total time is this? Will the invoice prep slice of the pie be smaller? **About 7.4%; yes, a much smaller slice**

U N I T

14

Purchasing

The purchasing department orders merchandise for your business from suppliers, such as distributors, wholesalers, or manufacturers. Most suppliers sell goods to businesses at a *discount,* or a reduction in price. The lower your *net price* (the price you actually pay for an item), the higher your profit will be when you sell the item. Suppliers offer three kinds of discounts: *trade, chain,* and *cash.*

O U T L I N E

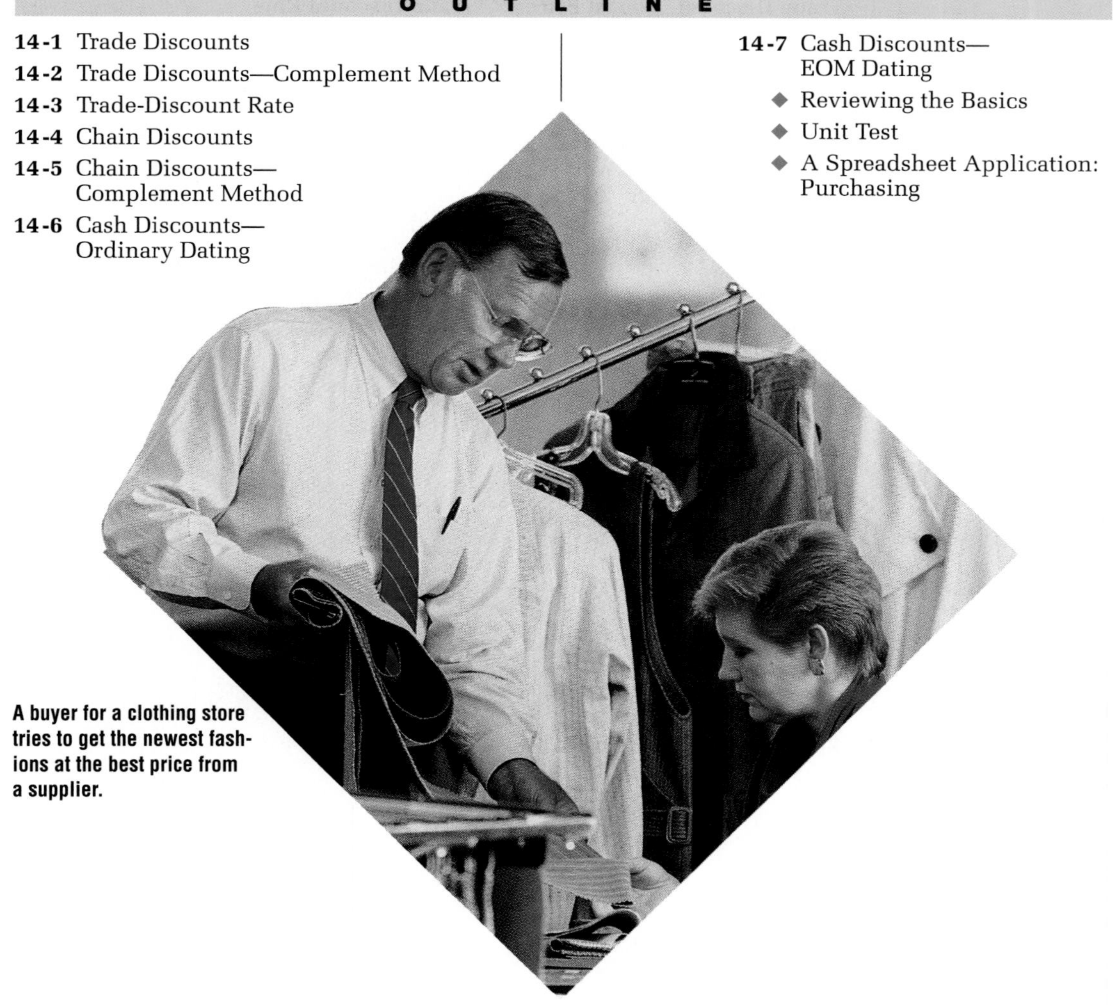

A buyer for a clothing store tries to get the newest fashions at the best price from a supplier.

UNIT 14
PURCHASING

INTRODUCING THE UNIT

Introduce this unit by inviting students to discuss their own buying habits. How do they select such items as stereo equipment, athletic shoes, clothing, and jewelry? Is price a major factor in the selection process? Have they ever considered that a retail store incurs costs to house the merchandise until it is sold?

383

FOCUS

Lead students in a discussion of the distribution channels of a product. Ask: How does a manufacturer of a bicycle actually get the product to the consumer. Show on the chalkboard one route that a product travels: Manufacturer to Wholesaler to Retailer to Consumer. Explain that through this route, the price of the bicycle can change because of discounts offered.

TEACH

Explain to students that wholesalers and retailers are called *middlemen.* Trade discounts offered to middlemen allow them to make a profit on the sale of an item. In most cases, a *merchant* wholesaler buys goods from a manufacturer, stores the goods, and then resells the goods to a retailer.

Stress that the list price, often referred to as the suggested retail price, is the retailer's *selling* price. The net price also called dealer's net or trade price, is the retailer's *buying* price.

Warm-Up Exercises

1. 30% of $74 $22.20
2. 25% of $76 $19
3. $228 − (20% of $228) $182.40
4. $476 − (15% of $476) $404.60
5. $774 − (10% of $774) $696.60

14-1 Trade Discounts

OBJECTIVE
Compute the trade discount and the net price.

Most suppliers provide a catalog from which your business can choose items to purchase. The catalog includes a description and the **list price,** or **catalog price,** of each item that the supplier sells. The list price is generally the price at which you may sell the item. You usually purchase the item at a **trade discount,** which is a discount from the list price. The trade discount is often expressed as a percent. The **net price** is the price you actually pay for the item.

Trade Discount = List Price × Trade-Discount Rate

Net Price = List Price − Trade Discount

EXAMPLE *Skills* 30, 6 *Application* A *Term* Trade discount

Johnson Bicycle Shop is purchasing 10-speed, 27-inch Trailblazer bicycles from a wholesaler. The list price in the wholesaler's catalog is $845. Johnson receives a 40% trade discount. What is the net price of each bicycle?

SOLUTION

A. Find the **trade discount.**
List Price × Trade-Discount Rate
$845.00 × 40% = $338.00 trade discount

B. Find the **net price.**
List Price − Trade Discount
$845.00 − $338.00 = $507.00 net price

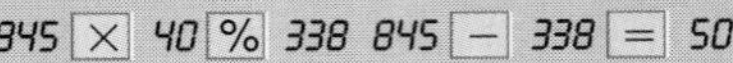

✔ SELF-CHECK Complete the problems, then check your answers in the back of the book.

1. List price: $380.
   Trade-discount rate: 30%.
   Find the trade discount.
   $114
2. List price: $380.
   Trade discount: $114.
   Find the net price.
   $266

PROBLEMS

Round answers to the nearest cent.

	3.	4.	5.	6.	7.	8.	9.
List Price	$140.00	$240.00	$400.00	$360.00	$174.00	$216.80	$94.75
Trade-Discount Rate	25%	10%	35%	15%	32%	27%	9%
Trade Discount	$35.00	$24	$140	$54	$55.68	$58.54	$8.53
Net Price	$105	$216	$260	$306	$118.32	$158.26	$86.22

CULTURAL ANGLES

Many Japanese automobiles are not only sold in the United States but are also manufactured in the U.S. Assign students the task of identifying one major Japanese auto maker that builds cars in the U.S. (Toyota, Honda) What is their channel of distribution? (Honda factory to Honda dealerships to consumers)

10. $47 list price for wok. 25% trade discount. What is the net price? **$35.25**

11. $219 list price for compact disc player. 42% trade discount. What is the net price? **$127.02**

12. An Easy-Wake clock radio carries a catalog price of $39.75. The Easy-Wake Company offers a 15% trade discount. What is the net price? **$33.79**

13. Giant discount stores purchases hand-held electronic calculators from Ace Wholesalers. Find the trade discount and net price for each calculator.

	Item	List	Trade-Discount Rate	Dis.	Net
a.	Basic	$19.95	20%	**$3.99**	**$15.96**
b.	Scientific	$24.49	23%	**$5.63**	**$18.86**
c.	Paper-tape	$34.95	28%	**$9.79**	**$25.16**

14. Hero, Inc., manufactures automobile parts and accessories. A portion of its "Parts and Accessories Price List" for auto dealers is shown. Hero offers the trade discounts listed to its dealers. Find the dealer price for each item. **A $51.97, HX $35.46, CQ $1.59, D $5.46**

Class	Part Number	Suggested List Price	Trade Discount
A	D2UZ 8200-B	$89.60	42%
HX	D27Z 82-A	$57.20	38%
CQ	COYT 8A178	$ 2.90	45%
D	C7ZZ 8B90	$ 7.80	30%

15. Commercial Paper Company offers a 55% discount to commercial users of their products. Pro Printers, Inc., placed this order. What is the net price?

Quantity	Units	Description	Price per Unit
25	SHEETS	H1163 32 × 40 14 PLY #33 MAT	$217.60
		Cut to 150 PCS. 11 × 14	
		125 PCS. 11 × 12	
		TOTAL	**$97.92** ?

16. Alvin Sales Agency receives a 32% trade discount on farm equipment from Land Equipment, Inc. What is the net price for each order?

Quantity	Stock No.	Description	Unit Price	
6	1-1	Heavy nuts	$0.45	**$ 1.84**
4	1-2	Starlock washer	$0.20	**$ 0.54**
2	61 HOB	Blades	$7.45	**$10.13**

MAINTAINING YOUR SKILLS Look up the skills in parentheses if you need help or more practice.

Find the percentage. Round answers to the nearest cent. **(Skill 30)**

17. $47.80 × 10% **$4.78**
18. $91.40 × 18% **$16.45**
19. $416.60 × 20% **$83.32**
20. $27.79 × 35% **$9.73**
21. $279.49 × 50% **$139.75**
22. $2178.99 × 42% **$915.18**

Subtract. **(Skill 6)**

23. $47.80 − $4.78 **$43.02**
24. $416.60 − $83.32 **$333.28**
25. $279.49 − $139.75 **$139.74**

PRACTICE AND APPLY

The following problems can be assigned for classwork and the answers checked in class to help students master the objective of the lesson.

- Guided Practice: 1–6
- Independent Practice: 7–9, 11–15 odd

WRAP-UP

Review the solution to Problem 15. Ask students which dollar amount is greater, the trade discount or the net price and why. (Trade discount is greater because the trade-discount rate is 55%.)

Assignment Guide

- Basic: 7–15, 17–25
- Average: 10–16 even, 17–25 odd

ALTERNATIVE STRATEGIES: Enrichment

Other channels of distribution for consumer goods are: (1) Manufacturer to Consumer and (2) Manufacturer to Retailer to Consumer. The first channel is used by craftpeople, who make such items as jewelry or leather goods and sell them at fairs. Some manufacturers also have their own retail stores, such as Firestone Tires, or have their own discount outlets where consumers can purchase the manufacturer's products at lower prices. The second type of channel often is used by makers of automobiles, furniture, and clothing.

LESSON PLAN 14-2 Trade Discounts—Complement Method

FOCUS

To motivate this lesson on computing the net price using the complement method, write *$60 is 100%* on the chalkboard, and explain that this is the list price of a jacket, or 100% of the price. Next, write *30% under 100%*, and explain that this is the trade-discount rate. Show the subtraction of 30% from 100% and ask students what the 70% represents. (the net price) Can the net price be found by using the 70%? Have a volunteer work the problem at the chalkboard using 70% of $60.

TEACH

After working through the Example of this lesson, rework the Example in Lesson 14-1 using the complement method.

Provide another problem for which the list price of an item is $14.25 and the trade-discount rate is 30%. Have two students find the net price, each using one of the two methods. After both students give their answers ($9.97 and $9.98), ask why there is a difference. (The difference is caused by rounding.)

14-2 Trade Discounts—Complement Method

OBJECTIVE
Compute the net price using the complement method.

The **complement method** is another method of finding the net price. When you receive a trade discount, you can subtract the discount rate from 100%. This gives you the complement of the trade-discount rate, or the percent you actually pay, sometimes called the net price rate. Multiply the complement by the list price to find the net price.

Net Price = List Price × Complement of Trade-Discount Rate

EXAMPLE *Skills* 30, 6 *Application* L *Term* Complement method

Johnson Bicycle Shop is also purchasing 3-speed, 26-inch Tour bicycles, which have a list price of $578. Johnson receives a trade discount of 40%. What is the net price of each bicycle?

SOLUTION

A. Find the **complement of trade-discount rate.**
100% − 40% = 60% complement of trade-discount rate

B. Find the **net price.**
List Price × Complement of Trade-Discount Rate
$578.00 × 60% = $346.80 net price

100 [−] 40 [=] 60 [M+] 578 [×] [RM] 60 [%] 346.8

✓ SELF-CHECK Complete the problems, then check your answers in the back of the book.

1. What is the complement of 30%? **70%**
2. List price: $380. Trade-discount rate: 30%. Find the net price. **$266**

PROBLEMS

Use the complement method to solve each problem. Round to nearest cent.

	3.	4.	5.	6.	7.	8.	9.
List Price	$120.00	$85.00	$27.00	$74.65	$317.48	$4.73	$417.89
Trade-Discount Rate	25%	20%	35%	15%	24%	18%	8%
Complement	75%	80%	65%	85%	76%	82%	92%
Net Price	$90	$68	$17.55	$63.45	$241.28	$3.88	$384.46

10. $48.77 list price for sprinkler. 32% trade discount. What is the net price? **$33.16**
11. $2178.90 list price for personal computer. 27% trade discount. What is the net price? **$1590.60**

CRITICAL THINKING

Among other functions, wholesalers provide a sales force for manufacturers. Ask students why this is advantageous to a manufacturer of pencils. (The cost of employing enough salespeople to sell its product to all the potential retailers would be enormous.)

12. Baxter Sales Agency receives a 27% trade discount on these items. What will Baxter pay for each order?

	Quantity	Stock No.	Description	Unit Price	Total List
$462.02	2	84-D	DAXCO cutter-complete	$316.45	$632.90
$26.83	5	84-06	Blades	$ 7.35	$ 36.75

13. Exact-Fit Automotive, Inc., receives a 41% trade discount on these items. What is the net price for each?

Number	Description	Quantity	List	Total	
ED-1	Water pump	2	$37.43	$ 74.86	**$44.17**
400	Headlight	4	$ 6.95	$ 27.80	**$16.40**
10S	Shock absorber	8	$17.75	$142.00	**$83.78**
PF-91	Sound module	3	$54.70	$164.10	**$96.82**

14. dba Charles receives a 54% trade discount from Commercial Clothes Company. Find the net price for each total listed.

	Order No.	Descript.	Style No.	No. of Units	Unit Price	Total
$73.09	11260467	Blazer	87024	2	$79.45	$158.90
$148.93	11260467	Jacket	87026	5	$64.75	$323.75
$123.33	11260467	Shirt	87233	9	$29.79	$268.11
$56.33	11260467	Shirt	87237	5	$24.49	$122.45
$55.17	11260467	Shirt	87533	6	$19.99	$119.94

15. The Star Sports Shop purchases fish finders from Icthos, Inc., for a list price of $179.79. Star Sports Shop receives a 28% trade discount.
   a. What net price does the Star Sports Shop pay for the fish finders?
   b. What is the dollar value of the discount? **a. $129.45 b. $50.34**

16. Do-It-All Lumber Co. receives a 19% discount from Wholesale Lumber Supply. Find the net price for each amount on the invoice.

Fixture Location	Description	Dept.	Qty.	Price	Amount	
NA3	Fastway	4	1	$8.25	$ 8.25	**$6.68**
EQ25	ITE	4	3	7.54	22.62	**$18.32**
BR15	Rite Corp.	4	6	1.94	11.64	**$9.43**
762OC	EFCO	4	2	2.42	4.84	**$3.92**

MAINTAINING YOUR SKILLS Look up the skills in parentheses if you need help or more practice.

Find the complement. **(Application L)**

17. 50% **50%**
18. 40% **60%**
19. 22% **78%**
20. 19% **81%**
21. 8% **92%**
22. 26% **74%**
23. 42% **58%**
24. 37% **63%**

Find the percentage. Round to nearest cent. **(Skill 30)**

25. $43.80 × 60% **$26.28**
26. $149.45 × 78% **$116.57**
27. $91.79 × 81% **$74.35**
28. $9.89 × 74% **$7.32**
29. $4171.84 × 63% **$2628.26**
30. $191.99 × 50% **$96.00**

Warm-Up Exercises

Find the complement.

1. 25% 75%
2. 32% 68%
3. 22% 78%

Calculate.

4. 75% of $96 $72
5. 68% of $524 $356.32
6. 78% of $124 $96.72

PRACTICE AND APPLY

The following problems can be assigned for classwork and the answers checked in class to help students master the objective of the lesson.

- Guided Practice: 1–6
- Independent Practice: 7–9, 11, 13

WRAP-UP

Ask individual students to give the definition of the terms *list price, trade discount, trade-discount rate,* and *net price.*

Assignment Guide

- Basic: 7–15, 17–30
- Average: 10–14 even, 15, 16, 18–30 even

ALTERNATIVE STRATEGIES: Reteaching

Mention that this is the same concept used in Lesson 5-7 on finding the sale price as a consumer.

Examples:
A. 25% off implies 75% is paid.
B. 18% off implies 82% is paid.

LESSON PLAN
14-3 Trade Discount Rate

FOCUS

Present the following situation to the class. In the most recent automobile parts catalogue, a headlight has a list price of $7. The manufacturer had previously offered Mercer Car Repair a trade-discount rate of 20%. An updated price sheet on headlights was sent to Mercer, which only gave the list price and the net price of items. On the sheet, the headlight has the same list price and a net price of $5.25. Does the new price reflect a better discount rate? (yes) In this lesson, students will learn how to compute the trade-discount rate when only the list price and net price are given.

TEACH

Stress that when finding the trade-discount rate, the list price is the divisor.

If students should check their answers by using the method of the previous lesson, (that is, multiplying the list price by the complement of the trade-discount rate to verify that the net price is the same as the one given in the problem), point out that rounding can cause a slight difference. Demonstrate the difference when checking the answer to Problem 4.

Warm-Up Exercises

1. $30 − $21 $9
2. $45 − $27 $18
3. $9 ÷ $30 0.3
4. $18 ÷ $45 0.4
5. ($145 − $116) ÷ $145 0.2
6. ($7.50 − $3.75) ÷ $7.50 0.5

14-3 Trade-Discount Rate

OBJECTIVE
Compute the trade-discount rate.

Some wholesalers and manufacturers publish only the list prices and net prices in their catalogs. If you know both of these prices, you can calculate the **trade-discount rate.**

Trade Discount = List Price − Net Price

$$\text{Trade-Discount Rate} = \frac{\text{Trade Discount}}{\text{List Price}}$$

EXAMPLE *Skills* 6, 31 *Application* A *Term* Trade-discount rate

The Auto Parts Manufacturing Company shows a $56 list price in its catalog for a water pump, #GF147. The net price to the retailer is $42. What is the trade-discount rate?

SOLUTION

A. Find the **trade discount.**
List Price − Net Price
$56.00 − $42.00 = $14.00 trade discount

B. Find the **trade-dicount rate.**
Trade Discount ÷ List Price
$14.00 ÷ $56.00 = 25% trade-discount rate

56 M+ − 42 = 14 ÷ RM 56 % 25

✓ SELF-CHECK Complete the problems, then check your answers in the back of the book.

1. List price: $140.
Net price: $84.
Find the trade discount. **$56**

2. Trade discount: $56.
List price: $140.
Find the trade-discount rate. **40%**

PROBLEMS

Round the trade-discount rate to the nearest tenth of a percent.

	3.	4.	5.	6.	7.	8.	9.
List Price	$5.00	$75.00	$60.00	$50.00	$112.60	$97.49	$5147.58
Net Price	$4.00	$50.00	$42.00	$27.50	$ 74.79	$68.78	$4258.69
Trade Discount	$1.00	? $25	? $18	? $22.50	? $37.81	? $28.71	? $888.89
Trade-Discount Rate	? 20.0%	? 33.3%	? 30.0%	? 45.0%	? 33.6%	? 29.4%	? 17.3%

COOPERATIVE LEARNING

Have small groups of students look through the yellow pages of the local telephone directory and make a list of the names of ten wholesalers and the major heading each is listed under. For example, one entry would be Brady Parts (company name), Air Conditioning Supplies & Parts—Wholesale (heading). Students should only list one company under each heading. Select a member of each group to compile the group's list, by headings, and present to the class the different types of wholesalers in the area.

10. List price is $198.49.
Net price is $149.27.
Estimate the trade discount and rate.
Compute the trade-discount rate. **$50; 25%; 24.8%**

11. List price is $102.79.
Net price is $78.56.
Estimate the trade discount and rate.
Compute the trade-discount rate. **$20; 20%; 23.6%**

12. Beauty Shop Wholesalers, Inc., offers discounts on most items they sell. What is the trade-discount rate for each of the items listed?

Item	List	Net	
Complexion brushes	$15.79	$11.68	**26.0%**
Lighted make-up mirror	$59.99	$41.45	**30.9%**
Cleansing brushes	$ 9.79	$ 6.11	**37.6%**
4-way make-up mirror	$47.49	$24.73	**47.9%**

13. Auto Paint Supply, Inc., varies the rate of discount with the item purchased. This invoice was received by Nu-Car Service. What is the trade-discount rate for each gross amount?

USE INVOICE NUMBER ON ALL CORRESPONDENCE — ALL ITEMS NET UNLESS OTHERWISE SPECIFIED

	Quantity Shipped	Quantity Ordered	Description	Unit Price	Gross Amount	Disc.	Net Amount
35%	4	4	0001-WH-TOUCH-UP	$3.98	$15.92	$5.57	$10.35
25%	2	2	0001-BL-TOUCH-UP	$3.98	$ 7.96	$1.99	$ 5.97
0%	2	2	0001-AVO-TOUCH-UP	$4.98	$ 9.96	0	$ 9.96
20%	4	4	0001-GLD-TOUCH-UP	$4.98	$19.92	$3.98	$15.94

14. Find the trade-discount rate for each gross amount listed on this invoice.

Quantity	Product Code	Price per Unit	Gross Amount	Net Amount	
140	00010	$ 7.99	$1118.60	$1006.74	**10%**
160	00015	$19.67	$3147.20	$2045.68	**35%**
110	00018	$24.84	$2732.40	$1366.20	**50%**
100	00022	$36.71	$3671.00	$2845.03	**22.5%**
200	00024	$14.35	$2870.00	$1937.25	**32.5%**
24	00500	$ 8.43	$ 202.32	$ 157.18	**22.3%**
48	00502	$ 6.20	$ 297.60	$ 201.40	**32.3%**

MAINTAINING YOUR

Subtract. **(Skill 6)**

15. $78.00 − $54.60 **$23.40**

16. $16.40 − $9.84 **$6.56**

17. $162.70 − $122.03 **$40.67**

18. $45.84 − $34.38 **$11.46**

19. $1075.79 − $788.32 **$287.47**

20. $2471.83 − $1971.92 **$499.91**

Find the rate. Round answers to the nearest tenth of a percent. **(Skill 31)**

21. $23.40 ÷ $78.00 **30%**

22. $6.56 ÷ $16.40 **40%**

23. $40.67 ÷ $162.70 **25.0%**

24. $214.91 ÷ $1073.78 **20.0%**

25. $1171.81 ÷ $11,721.49 **10.0%**

26. $1.78 ÷ $3.54 **50.3%**

PRACTICE AND APPLY

The following problems can be assigned for classwork and the answers checked in class to help students master the objective of the lesson.

- Guided Practice: 1–6
- Independent Practice: 7–9, 11, 12

WRAP-UP

Write the following on the chalkboard: _?_ − _?_ = _?_ ÷ _?_ = _?_.
Have individual students come to the chalkboard and write in the terms that correctly state how to compute the trade-discount rate.

Assignment Guide

- Basic: 7–13, 15–26
- Average: 10, 13, 14, 16–26 even

ALTERNATIVE STRATEGIES: Enrichment

Give the students this situation: Suppose a window manufacturer offers a 20% discount on a window with an $800 list price. Two weeks later the cost of production increases and the old list price becomes the new net price. What is the percent increase? ($800 × 80% = $640) and ($800 − $640) ÷ $640 = 25%

LESSON PLAN
14-4 Chain Discounts

FOCUS

Ask students what they would do if they were a wholesaler of stereo equipment and had 35 tape decks on hand that retailers were not buying. (increase the discount) Or, if there were three customers who always purchased large orders, how could you show them that you appreciated their business? (increase the discount) Explain that manufacturers and wholesalers sometimes give more than one discount to customers. These are called *chain discounts*.

TEACH

Be sure students understand that the discounts are deducted *successively* from the list price. You might want to show the computation of the Example vertically on the chalkboard:

List price $695.00
Less 1st discount
(20% × $695) .....−139.00
1st net price $556.00
Less 2nd discount
(10% × $556) .....− 55.60
Final net price $500.40

Warm-Up Exercises

1. 30% of $840 $252
2. 20% of $96 $19.20
3. $450 − (25% of $450) $337.50
4. $178 − (15% of $178) $151.30
5. $1478 − (35% of $1478) $960.70

14-4 Chain Discounts

OBJECTIVE
Compute the final net price after a chain of discounts.

A supplier may offer **chain discounts** to sell out a discontinued item or to encourage you to place a larger order. A chain discount is a series of trade discounts. For example:

35% less 20% less 15%

- This does not mean 35% + 20% + 15% off.
- 35% discount is deducted from list price to yield first net price.
- 20% of first net price is deducted from first net price to yield second net price.
- 15% of second net price is deducted from second net price to yield final net price.
- This is often written as 35/20/15.

EXAMPLE *Skills* 30, 2, 6 *Application* A *Term* Chain discount

The Green Lawn Corporation produces a $2\frac{1}{2}$- horsepower lawn mower, which is priced at $695. A new model is ready for production, so the $2\frac{1}{2}$-horsepower model is to be discontinued. Green Lawn is offering a chain discount of 20% less 10%. What is the final net price of the lawn mower?

SOLUTION

A. Find the **first discount.** ➡
List Price × Discount Rate
$695.00 × $20% = $139.00

B. Find the **first net price.**
List Price − Discount
$695.00 − $139.00 = $556.00

C. Find the **second discount.** ➡
$556.00 × 10% = $55.60

D. Find the **final net price.**
$556.00 − $55.60 = $500.40 final net price

695 × 20 % 139 695 − 139 = 556 × 10 % 55.6 556 − 55.6 = 500.4

✓ SELF-CHECK Complete the problems, then check your answers in the back of the book.

List price is $400. Chain discount is 30% less 25%.

1. Find the first net price. $280
2. Find the final net price. $210

PROBLEMS

	3.	4.	5.	6.	7.
List Price	$500	$780	$1240	$475	$2170
Chain Discount	30% less 20%	25% less 15%	20% less 15%	40% less 30%	40% less 25%
First Discount	$150	$195	? $248	? $190	? $868
First Net Price	$350	$585	? $992	? $285	? $1302
Second Discount	? $70	? $87.75	? $148.80	? $85.50	? $325.50
Final Net Price	? $280	? $497.25	? $843.20	? $199.50	? $976.50

USE OF CALCULATORS

Students who use a calculator will find their answer to the final net price for the raw-tip drill in Problem 11 as $41.45. The 1-cent difference is due to the calculator not rounding the second discount ($46.05 × 10% = $4.605; $46.05 − $4.605 = $41.445).

8. $74.60 list price for clock radio. 30/10 chain discount. What is the net price? **$47.00**

9. $125.75 list price for cordless phone. 40/20 chain discount. What is the net price? **$60.36**

10. Storm-Tite Inc. offers Mark Stores chain discounts of 30/15. What is the net price for each item?

Item	List Price	
Storm door	$147.80	**$87.94**
Storm window	$ 79.85	**$47.51**
Foil-backed insulation	$ 14.50/roll	**$8.63**

11. The General Saw Corporation offers trade discounts and additional discounts to encourage large orders. What is the net price per item if each invoice total is high enough to obtain the additional discount?

	Item	List Price	Trade Discount	Additional Discount
$46.93	Masonry drill	$98.80	50%	5% (invoice total over $250)
$41.44	Raw-tip drill	$76.75	40%	10% (invoice total over $500)
$35.02	Hole saw	$62.53	30%	20% (invoice total over $1000)

12. The Wholesale Supply Co. offers chain discounts of 35% less 20% less 15% to sell out a discontinued item. Find the final net price for an order of $1460. **$645.32**

13. What is the final net price for each of these overstocked items from the Everything Automotive Parts list?

Part Number	Suggested List Price	Chain Discount	
B7S	$914.80	40% less 25% less 15%	**$349.91**
C37X	$247.50	42% less 20% less 10%	**$103.36**
173A	$ 76.67	35% less 15% less 5%	**$40.24**
B62Y	$ 8.54	25% less 10% less 7%	**$5.36**

MAINTAINING YOUR SKILLS Look up the skills in parentheses if you need help or more practice.

Round to the place value indicated. **(Skill 2)**

14. $17.71155 (nearest cent) **$17.71**

15. $113.7051 (nearest dollar) **$114**

16. 17.98% (nearest tenth of a percent) **18.0%**

17. $8178.1449 (nearest cent) **$8178.14**

18. 71.47% (nearest percent) **71%**

19. $51.476 (nearest whole number) **$51**

Subtract. **(Skill 6)**

20. $17.60 − $5.28 **$12.32**

21. $94.35 − $37.74 **$56.61**

22. $214.80 − $21.48 **$193.32**

23. $567.88 − $141.97 **$425.91**

24. $1774.86 − $591.62 **$1183.24**

25. $24,599.99 − $11,069.55 **$13,530.44**

Find the percentage. Round to nearest cent. **(Skill 30)**

26. $217.80 × 35% **$76.23**

27. $670.45 × 25% **$167.61**

28. $814.86 × 15% **$122.23**

29. $3417.56 × 30% **$1025.27**

30. $7147.63 × 20% **$1429.53**

31. $56,147.71 × 10% **$5614.77**

PRACTICE AND APPLY

The following problems can be assigned for classwork and the answers checked in class to help students master the objective of the lesson.

- Guided Practice: 1–5
- Independent Practice: 6, 7, 9, 11

WRAP-UP

Write *15% less 10% less 5%* on the chalkboard. Call on three students to explain how each percent is used in calculating the final net price.

Assignment Guide

- Basic: 6–12, 14–31
- Average: 8–12, even, 13, 14–30 even

ALTERNATIVE STRATEGIES: Reteaching

Have students find the net price for an item with a list price of $100 and a chain discount of 30% less 20% less 10%. Write three choices for the answer on the chalkboard: (a) $40, (b) $63.40, and (c) $50.40. (c) Students who chose (a) added the discount rates (30% + 20% + 10%). Students who chose (b) did not subtract the discount amount from the list price ($100 × 30% = $30, $30 × 20% = $6, $6 × 10% = $0.60; $30 + $6 + $0.60 = $36.60; $100 − $36.60 = $63.40).

LESSON PLAN
14-5 Chain Discounts—Complement Method

FOCUS

Present a list of 10 automobile parts, each with a different list price. Tell students that the chain discounts are 30% less 20% less 10%. Ask them how many computational steps it would take to determine the net prices of the entire list. (30 multiplications and 30 subtractions) Explain that there is a more efficient way to finding the net prices using only 10 multiplications. In this lesson, students will learn how to compute the final net price using the complement method.

TEACH

After working the first two steps of the solution to the Example, explain to students that to find the net prices of the list discussed above, one simply multiplies each item's list price by 50.4%, the net-price rate.

Using the Example, ask students how they can find the discount without using the SED. (list price − net price = discount)

Warm-Up Exercises

Find the complements.

1. 40% 60%
2. 32% 68%

Find the product of the complements.

3. 20% less 10% 72%
4. 35% less 25% less 15% 41.4375%

14-5

Chain Discounts—Complement Method

OBJECTIVE

Compute the final net price using the complement method.

The complement method applied to trade discounts can also be applied to chain discounts. First subtract each discount from 100% to find the complements. Then multiply the complements to find the percent that you actually pay. This percent is called the **net-price rate.** To find the net price, multiply the net-price rate by the list price.

You can also use the complement method to find the discount. Find the complement of the net-price rate by subtracting it from 100%. That percent is called the **single equivalent discount (SED).** Multiply the SED by the list price to find the discount. You can check by subtracting the net price from the list price.

Net-Price Rate = Product of Complements of Chain-Discount Rates

Net Price = List Price × Net-Price Rate

SED = Complement of Net-Price Rate

Discount = List Price × SED

EXAMPLE *Skills* 30, 2, 6 *Application* A *Term* Chain discount

The Green Lawn Corporation manufactures an 8-horsepower riding mower, which is priced at $1450. A new model is ready for production, so the 8-horsepower model is to be discontinued. Green Lawn is offering a chain discount of 30% less 20% less 10%. What is the net price of the mower? What is the SED and the discount?

SOLUTION

A. Find the **net-price rate.**
Product of Complements of Chain-Discount Rates
70% × 80% × 90% = 50.4% net-price rate

B. Find the **net price.**
List Price × Net-Price Rate
$1450.00 × 50.4% = $730.80 net price

C. Find the **SED.**
Complement of the Net-Price Rate
100% − 50.4% = 49.6% SED

D. Find the **discount.**
List Price × SED
$1450.00 × 49.6% = $719.20 discount
Check: $1450.00 − $730.80 = $719.20

✓ SELF-CHECK Complete the problems, then check your answers in the back of the book.

List price is $560. Chain discount is 30% less 10%.

1. Find the net-price rate and the net price. **63%; $352.80**
2. Find the SED and the discount. **37%; $207.20**

MATHEMATICAL NOTES

Some students have difficulties when working with percents and decimals. Provide chalkboard practice in finding complements and multiplying complements using both percents and decimals. Point out that if a calculator is used to multiply complement rates, the percent key must be entered after the number. Tell them not to round on calculations until the final step in the problem.

PROBLEMS

	3.	4.	5.	6.	7.	8.
List Price	$620.00	$140.00	$436.00	$1237.00	$147.80	$96.46
Chain Discounts	30% less 20%	20% less 15%	40% less 20%	40% less 30%	30% less 20% less 10%	20% less 10% less 5%
Net Price	$347.20	$95.20	? **$209.28**	? **$519.54**	? **$74.49**	? **$65.98**
SED	? **44%**	? **32%**	? **52%**	? **58%**	? **49.6%**	? **31.6%**
Discount	? **$272.80**	? **$44.80**	? **$226.72**	? **$717.46**	? **$73.31**	? **$30.48**

9. $2150 list price for personal computer.
15/10 chain discount.
What is the net price? **$1644.75**
What is the discount? **$505.25**

10. $74.85 list price for calculator.
20/5 chain discount.
What is the net price? **$56.89**
What is the discount? **$17.96**

11. Outfitters, Inc., offers Clark's Clothes Co. chain discounts of 25/10. What is the net price for each item? What is the SED? **32.5%**

Item	List Price	
Storm boots	$149.79	**$101.11**
Storm coat	$219.49	**$148.16**
Storm gloves	$ 29.99	**$20.24**

12. The Globe Corporation offers trade discounts and additional discounts to encourage large orders. What is the net price per item and net discount if each invoice total is high enough to obtain the additional discount?

	Item Number	List Price	Trade Discount	Additional Discount
$64.40; $54.85	14ZB	$119.25	40%	10% (invoice total over $500)
$54.25; $30.51	11TC	$ 84.76	20%	20% (invoice total over $1000)
$34.03; $17.15	93MR	$ 51.18	30%	5% (invoice total over $250)

13. What is the net price and net discount for each of these overstocked items from the Car Catalog price list?

	Item	Suggested List Price	Chain Discount
$31.77; $46.68	Shock 147	$ 78.45	40% less 25% less 10%
$181.85; $159.97	Bumper 96	$341.82	30% less 20% less 5%
$35.35; $23.02	Rim 412	$ 58.37	25% less 15% less 5%
$101.14; $46.72	Quarter 753	$147.86	20% less 10% less 5%

MAINTAINING YOUR SKILLS Look up the skills in parentheses if you need help or more practice.

Round to the nearest cent. **(Skill 2)**

14. $19.7171 **$19.72**
15. $114.555 **$114.56**
16. $912.4142 **$912.41**
17. $2481.5764 **$2481.58**
18. $14.7210 **$14.72**

Find the complement. **(Application L)**

19. 30% **70%** **20.** 20% **80%** **21.** 25% **75%** **22.** 5% **95%** **23.** 10% **90%** **24.** 15% **85%** **25.** 7% **93%**

Find the percentage. **(Skill 31)**

26. $347.80 × 50.4% **$175.29**
27. $81.70 × 43.2% **$35.29**
28. $591.45 × 53.375% **$315.69**

PRACTICE AND APPLY

The following problems can be assigned for classwork and the answers checked in class to help students master the objective of the lesson.

- Guided Practice: 1–5
- Independent Practice: 6–8, 10

WRAP-UP

Ask students if they can find the net price using the SED. Ask a volunteer to find the net price of the mower in the Example using the SED. Ask which method is more efficient, using the SED or using the product of complements of the chain-discount rates. (the latter)

Assignment Guide

- Basic: 6–12, 14–28
- Average: 9, 11–13, 14–28 even

ALTERNATIVE STRATEGIES: Enrichment

Mention to students that the SED is used by buyers to select the supplier with the best deal. Have students solve the following problem. Three suppliers have the same list price. Chain discounts of 20% less 10% are offered by Supplier A and 10% less 10% less 10% by Supplier B. Supplier C offers a single discount of 27%. Ask students which supplier has the best price and why. (Supplier A; SED's are A: 28%, B: 27.1%, C: 27%.)

LESSON PLAN
14-6 Cash Discounts—Ordinary Dating

FOCUS

To introduce the Example, present the following. An order of $856.65 was sent to The Lighting Store, a retailer, last week by their supplier, Oak Hill Lighting, Inc. Ask students what the supplier needs to do to be paid for the lamp shipment. (bill or invoice the retailer) What would be an incentive to get The Lighting Store to pay the full amount of the invoice quickly? (offer a discount) Explain that discounts offered for timely payments are called *cash discounts*.

TEACH

Review the procedure for finding the number of days between two dates since the first step to solve the Example is to determine the last day a discount can be taken. Point out that the 3% discount can be taken up to and including 6/15.

Mention that using the complement method is a more direct way to find the cash price. (97% × $856.65 = $830.95)

Warm-Up Exercises

1. 2% of $140 $2.80
2. 3% of $350 $10.50
3. $90 − (4% of $90) $86.40
4. $436 − (2% of $436) $427.28
5. $148.50 − (1% of $148.50) $147.02

14-6

Cash Discounts—Ordinary Dating

OBJECTIVE

Compute the cash price when discount is based on ordinary dating.

You receive an **invoice** for each purchase you make from a supplier. The invoice lists the quantities and costs of the items purchased. To encourage prompt payment, the supplier may offer a **cash discount** if the bill is paid within a certain number of days. The exact terms of the discount are stated on the invoice. Many suppliers use **ordinary dating.** For example:

2/10, net 30 or **2/10, n/30**

2% cash discount if paid within 10 days of date of invoice.

Net price must be paid within 30 days of date of invoice.

Cash Discount = Net Price × Cash-Discount Rate

Cash Price = Net Price − Cash Discount

EXAMPLE *Skills* 30, 6 *Application* A *Term* Ordinary dating

This is part of the invoice that The Lighting Store received for a shipment of lamps. What is the cash price of the lamps if the bill is paid within 10 days?

Oak Hill Lighting, Inc. SHIP TO: THE LIGHTING STORE INVOICE NO: 4-3467

DATE	ORDER NO.		TERMS	ACCT. NO.
June 5, XX	LH 3019		2/10, NET 30	712E 4
STYLE	COLOR	QUANTITY	PRICE	AMOUNT
A9407	BGE	8	$37.25	$298.00
A9841	GRN	6	42.15	252.90
J2113	NVY	10	26.40	264.00
J2114	GRN	8	26.40	211.20
			TOTAL	$1026.10

SOLUTION

A. Find the **last day a discount can be taken.**
Date of invoice: June 5
Terms: 2/10, net 30
Discount can be taken until June 15, 10 days from date of invoice.

B. Find the **cash discount.**
Net Price × Cash-Discount Rate
$1026.10 × 2% = $20.522 = $20.52 cash discount

C. Find the **cash price.**
Net Price − Cash Discount
$1026.10 − $20.52 = $1005.58 cash price

1026.1 [×] 2 [%] 20.522 1026.1 [−] 20.52 [=] 1005.58

✓ SELF-CHECK Complete the problems, then check your answers in the back of the book.

Terms are 3/10, n/30. Net price is $640.

1. Find the cash discount. **$19.20** 2. Find the cash price. **$620.80**

COMMUNICATION SKILLS

Some suppliers offer terms such as 4/10, 2/20, net 30. Ask students what they think these terms mean. (4% discount if paid within 10 days of the invoice date, and 2% discount if paid within 20 days)

PROBLEMS

	Invoice Date	Terms	Date Paid	Net Price	Cash Discount	Cash Price
3.	3/19	2/10, n/30	3/28	$ 740.00	$14.80	? $725.20
4.	5/6	3/10, n/30	5/13	$ 516.40	? $15.49	? $500.91
5.	9/3	5/15, n/30	9/23	$ 348.64	? 0	? $348.64
6.	11/22	7/10, n/30	12/1	$4178.45	? $292.49	? $3885.96

F.Y.I.
The phrase "cash discount terms of 2/10, net 30, ROG" means that the discount period starts with the Receipt Of the Goods.

7. $6715.80 net price for hardware.
Terms are 4/10, n/30.
Date of invoice is October 7.
Date paid is October 14.
What is the cash price? **$6447.17**

8. $614.85 net price for auto parts.
Terms are 5/15, net 30.
Date of invoice is December 11.
Invoice is paid on December 23.
What is the cash price? **$584.11**

9. The net price of goods from Fashion Footwear, Inc., to the Mall Shoe Store was $916.78. The invoice was paid on February 28. According to the terms on the invoice, what was the cash price? **$898.44**

DATE SHIPPED	VIA	TERMS: 2%-20 DAYS	SALESPERSON	DATE OF INVOICE
2/24/-	OUR DELIVERY		65	2/23/-

10. The net price of goods from Women's Wear, Inc., to Violet's is $4217.86. If the invoice is paid on December 11, what is the cash price? **$3880.43**

CUSTOMER NUMBER 1238 SHIPPING DATE 11-22-XX INVOICE DATE 12-05-XX

CUST. ORDER NO.	CUST. ORDER DATE	TERMS	TYPE	SHPG. ORDER NO.
211	11-15-XX	8/10, n/30	02	847710

11. What is the cash price for these materials from Ace Tool & Die, Inc., to Summit Machine Shop? The invoice is paid on November 14. **$227.31**

CUSTOMER'S COPY
INVOICE NO. NO. 4559 DATE NOV. 7, 19XX TERMS 1%-10 days Net 30 days

YOUR ORDER NO.	PIECES	PATTERN NO.	WEIGHT	PRICE	AMOUNT
7552	240	641-7		$0.61	$146.40
1319	157	320-5		0.53	83.21
					$229.61

MAINTAINING YOUR SKILLS Look up the skills in parentheses if you need help or more practice.

Subtract. **(Skill 6)**

12. $516 − $10.32 **$505.68**
13. $916.70 − $27.50 **$889.20**
14. $3178.42 − $127.14 **$3051.28**
15. $119.20 − $11.92 **$107.28**
16. $9784.52 − $489.21 **$9295.31**
17. $96.20 − $0.48 **$95.72**

Find the percentage. Round to nearest cent. **(Skill 30)**

18. $415.00 × 2% **$8.30**
19. $147.84 × 3% **$4.44**
20. $916.74 × 8% **$73.34**
21. $714.50 × 1% **$7.15**

PRACTICE AND APPLY
The following problems can be assigned for classwork and the answers checked in class to help students master the objective of the lesson.

- Guided Practice: 1–5
- Independent Practice: 6–9

WRAP-UP
Review with students the ordinary dating method of cash discounts. Write *3/10 net 30* on the chalkboard and have individual students explain what each number represents.

Assignment Guide
- Basic: 6–10, 12–21
- Average: 10, 11, 13–21 odd

ALTERNATIVE STRATEGIES: Reteaching
Point out that to:

A. determine the last day for the discount, the number of days must be added to the date of the invoice. In the Example, June 5 + 10 days is June 15.

B. determine the cash price, multiply by the complement of the discount rate. In the Example, 2% implies 98% paid and 98% of $1026.10 = $1005.58.

LESSON PLAN
14-7 Cash Discounts—
EOM Dating

FOCUS

Ask students why ordinary dating could be inefficient for a retailer who receives many invoices during a month. (The person paying the invoices needs to keep track of every last day on each invoice to receive a cash discount.) Can they suggest a better way a supplier can offer cash discounts? Explain that many businesses prefer to pay their invoices after the end of the month. Thus, another method of granting cash discounts is based on end-of-month dating.

TEACH

Mention that the word *proximo*, or prox, which means "in the following month," is often used on invoices instead of EOM. The terms, such as 3/10, mean the same for both.

Point out that although the last day for payment of the net amount is not shown on an invoice (that is, net 30), it is implied that the retailer pay the full amount due by the end of the month. In the Example, Bob's Drug Store has 20 days after August 10, which is August 30, to pay $727.05.

Warm-Up Exercises

1. 2% of $1790 $35.80
2. 3% of $419.41 $12.58
3. $848 − (4% of $848) $814.08
4. 98% of $1790 $1754.20
5. $1790 − (2% of $1790) $1754.20

14-7

Cash Discounts—EOM Dating

OBJECTIVE
Compute the cash price when discount is based on end-of-month dating.

Many wholesalers use **end-of-month dating** when granting cash discounts. With this method of dating, your business receives a cash discount if you pay for your merchandise within a certain number of days after the end of the month. For example:

6/15 EOM ← A 6% cash discount if bill is paid within 15 days after the end of the month in which the invoice is issued.

Cash Discount = Net Price × Cash-Discount Rate

Cash Price = Net Price − Cash Discount

EXAMPLE *Skills* 30, 6 *Application* A *Term* End-of-month dating

Bob's Drugstore received this invoice from General Wholesalers. What is the last day that Bob's can take advantage of the 4% cash discount? What is the cash price?

DATE	INVOICE NO.	ACCT. NO.	STORE NO.	TERMS	VENDOR NO.
JUL 12	213788-6	208 712	34	4/10 EOM	12/7

SALESPERSON	VIA	ORDER REG. NO.
38L	OV DIRECT TRUCK	0547-6

CUSTOMER ORDER NO.	STYLE	QUANTITY	PRICE	AMOUNT
L3075	02673	10	$17.80	$178.00
L3076	02677	15	24.17	362.55
L3077	02681	25	7.46	186.50
			TOTAL	$727.05

SOLUTION

A. Find the **last day a discount can be taken.**
Date of invoice: July 12
Terms: 4/10 EOM
Discount can be taken until August 10, 10 days after the end of July.

B. Find the **cash discount.**
Net Price × Cash-Discount Rate
$727.05 × 4% = $29.082 = $29.08 cash discount

C. Find the **cash price.**
Net Price − Cash Discount
$727.05 − $29.08 = $697.97 cash price

CRITICAL THINKING

Have students state what the terms of an invoice mean in an if-then statement. For example, write Invoice date: July 10, 2/10 EOM on the chalkboard. Have one student begin the statement of the invoice, "If an invoice dated July 10 carries the terms 2/10 EOM, then . . ." Have another student continue, "a 2% discount can be taken up to and including August 10 and the net amount is due by August 30." Use terms of both EOM dating and ordinary dating.

✓ SELF-CHECK Complete the problems, then check your answers in the back of the book.

An invoice dated November 7 has terms of 2/10 EOM. Net price is $7100. The invoice is paid December 9.

1. Find the last day a discount can be taken. **December 10**
2. Find the cash price. **$6958**

PROBLEMS

	Invoice Date	Terms	Date Paid	Net Price	Cash Discount	Cash Price
3.	Jan. 21	2/10 EOM	Feb. 6	$ 615.00	$12.30	**$602.70**
4.	April 14	3/15 EOM	May 12	$ 1417.80	**$42.53**	**$1375.27**
5.	Sep. 2	4/10 EOM	Oct. 9	$ 7184.73	**$287.39**	**$6897.34**
6.	Dec. 19	5/10 EOM	Jan. 13	$16,517.84	**0**	**$16,517.84**

7. Net price is $916.40.
   Terms are 1/10 EOM.
   Invoice date is September 20.
   Invoice is paid on October 9.
   What is the cash price? **$907.24**

8. Net price is $7641.60.
   Terms are 6/10 EOM, n/30 EOM.
   Invoice date is May 20.
   Invoice is paid on June 2.
   What is the cash price? **$7183.10**

9. An invoice from Gentlemen's Clothiers, Inc., to Ducky's Tux Rentals shows a net price of $761.87. The terms are 5/15 EOM, n/30 EOM. The invoice is dated August 14. The invoice is paid September 15. What is the cash price? **$723.78**

10. An invoice from Krio Frozen Foods Corporation to Barr Wholesalers carries a net price of $26,715.81. The terms are 7/10 EOM. The invoice is dated Feb. 20. The invoice is paid March 7. What is the cash price? **$24,845.70**

11. An invoice from Classic Clothes to the Clothes Palace has a list price of $10,719.75. After a trade discount of 20%, the net price is $8575.80. A portion of the invoice is shown. What is the cash price if the invoice is paid on January 10? **$7889.74**

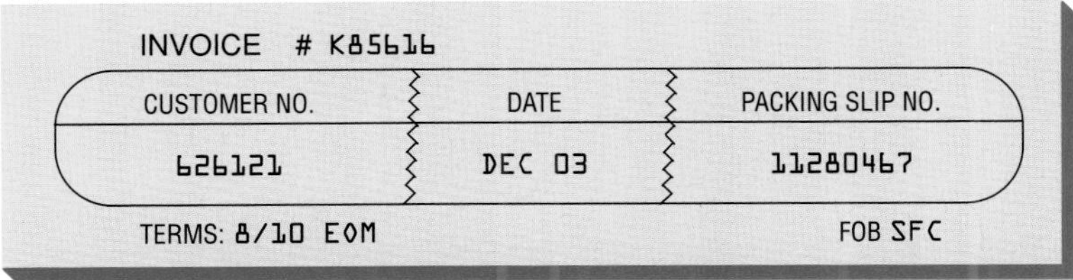

INVOICE # K85616

CUSTOMER NO.	DATE	PACKING SLIP NO.
626121	DEC 03	11280467

TERMS: 8/10 EOM FOB SFC

PRACTICE AND APPLY

The following problems can be assigned for classwork and the answers checked in class to help students master the objective of the lesson.

- Guided Practice: 1–4
- Independent Practice: 5–9

ALTERNATIVE STRATEGIES: Enrichment

Have students make conjectures about an invoice with these terms:

4/10, net 30 R.O.G.

R.O.G. stands for Receipt of Goods, thus a 4% discount can be taken for 10 days from the date the goods are received. R.O.G. is established by the retailer when the trucking company has the retailer sign the shipping papers. This is similar to a consumer signing for a UPS driver.

WRAP-UP

Go over the solution to Problem 7. Ask students which methods they used to find the cash price: (a) first finding the cash discount and then finding the cash price, or (b) using the complement method. Work the problem both ways at the chalkboard.

Assignment Guide

- Basic: 5–10, 12–20
- Average: 10, 11, 12–20 even

MAINTAINING YOUR SKILLS Look up the skills in parentheses if you need help or more practice.

Subtract. **(Skill 6)**

12. \$821.00 − \$16.42 **\$804.58**
13. \$1417.80 − \$14.18 **\$1403.62**
14. \$8614.50 − \$258.44 **\$8356.06**

Find the percentage. **(Skill 30)**

15. \$821.00 × 2% **\$16.42**
16. \$8614.50 × 3% **\$258.44**
17. \$12,643.17 × 8% **\$1011.45**
18. \$821.00 × 98% **\$804.58**
19. \$8614.50 × 97% **\$8356.07**
20. \$1417.80 × 99% **\$1403.62**

BUSINESS PROJECT

Have students in small groups call and visit some of the wholesalers on the list compiled from the telephone directory (see Lesson 14-3). Sample questions students can ask are: who are the manufacturers they buy products from; who are their customers; how do they give discounts, and so on. Each group should present its findings to the class.

Reviewing the Basics

Skills

Round to the nearest cent.

(Skill 1)

1. $9187.7649 — **$9187.76**
2. $514.4551 — **$514.46**
3. $71.8686 — **$71.87**
4. $217.6054 — **$217.61**

Solve. Round to the nearest cent.

(Skill 6)

5. $54.73 − $41.95 — **$12.78**
6. $4716.23 − $917.85 — **$3798.38**
7. $347.80 − $69.56 — **$278.24**
8. $7168.42 − $143.37 — **$7025.05**
9. $43,517.93 − $1305.54 — **$42,212.39**
10. $571.84 − $22.87 — **$548.97**

(Skill 30)

11. $179.47 × 3% — **$5.38**
12. $781.46 × 20% — **$156.29**
13. $297.97 × 2% — **$5.96**
14. $8654.50 × 8% — **$692.36**
15. $6247.75 × 90% — **$5622.98**
16. $787.14 × 50.4% — **$396.72**

Round to the nearest tenth of a percent.

(Skill 31)

17. $4.17 is what percent of $20.85? — **20.0%**
18. $72.73 is what percent of $289.99? — **25.1%**
19. $216.43 is what percent of $649.29? — **33.3%**
20. $1516.74 is what percent of $3794.49? — **40.0%**

Applications

(Application L)

Find the complement.

21. 30% — **70%**
22. 20% — **80%**
23. 68.4% — **31.6%**
24. 3% — **97%**
25. 50.4% — **49.6%**

Terms

Use each term in one of the sentences.

f **26.** Trade discount
c **27.** Chain discount
b **28.** Complement method
a **29.** Ordinary dating
d **30.** End-of-month dating
e **31.** Cash price

a. *3/5, net 30* is an example of ___?___.

b. The ___?___ is a way to find the net price when a trade-discount rate is given.

c. A series of discounts on an item is called a ___?___.

d. *9/5 EOM* is an example of ___?___.

e. Chain discounts are applied to the list price to get the net price. Cash discounts are applied to the net price to get the ___?___.

f. A discount from the catalog price is called a ___?___.

g. The catalog price is the ___?___.

Refer to your reference files in the back of the book if you need help.

LESSON PLAN
Reviewing the Basics

The exercises on this page review skills, applications, and terms used in the unit. You can use the exercises to assess informally students' proficiency with this material.

The page can be used for guided practice and independent practice. You can work through a selection of the exercises together with students, and thus see immediately if they know how to do them, and you can then assign some of the exercises for independent practice. Be sure to go over the answers to all assigned exercises.

LESSON PLAN
Unit Test

Students should do the Unit Test on their own. Each problem on the test is keyed to a lesson in the unit. Students having difficulty with any particular problem should review the Example in the appropriate lesson and be assigned some of the Independent Practice problems for additional practice.

Unit Test

Lesson 14-1

1. Pet Pride purchases its pet supplies from Wholesale Suppliers, Inc. The list price on one invoice was $714.86. Pet Pride receives a 20% trade discount. What is the net price for that invoice? **$571.89**

Lesson 14-2

2. Penn Skate purchases Ambassader roller skates that have a list price of $129.79 a pair. Penn Skate receives a trade discount of 25%. What is the net price that Penn Skate pays for each pair of skates? Use the complement method to solve. **$97.34**

Lesson 14-3

3. Heat Mfg., a manufacturer of small heaters, lists the price of its top-of-the-line heater as $179.99. The net price to the retailer is $116.99. What is the trade-discount rate? **35.0%**

Lesson 14-4

4. The Fitness Company manufactures an all-purpose exercise set that is priced at $947.50. When it introduced its new aerobic model, The Fitness Company offered a chain discount of 30% less 20% on the old model. What is the final net price of the all-purpose exercise set? **$530.60**

Lesson 14-5

5. Air-Hi produces a sports shoe that is priced at $114.49. When the new model was introduced, the $114.49 shoe was discounted at 15% less 5%. What is the net price of the old model using the complement method? **$92.45**

Lesson 14-6

6. The Capital Theatre received an invoice dated September 13 from Theatre Supply. The invoice carried terms of 2/10, net 30. The net price of the invoice was $561.87. What was the cash price of the invoice if it was paid on September 20? **$550.63**

Lesson 14-7

7. Custom Computer received this invoice from Hi-Tech, Inc. What is the last day that Custom Computer can make use of the 6% cash discount? What is the cash price? **August 10; $762.90**

DATE	INVOICE NO.	ACCT. NO.	STORE NO.	TERMS	VENDOR NO.
July 10	181818	75239	07	6/10 EOM	16

CUSTOMER ORDER NO.	TYPE	QUANTITY	PRICE	AMOUNT
F78906	08765	20	$10.56	$211.20
F90875	98765	35	6.98	244.30
L56329	09453	15	23.74	356.10
			TOTAL	$811.60

A SPREADSHEET APPLICATION

Purchasing

To complete this spreadsheet application, you will need the template diskette for *Mathematics with Business Applications.* Follow the directions in the *User's Guide* to complete this activity.

Input the information in the problems to find the net price and the cash price. Then answer the questions that follow.

	List Price	Chain Discounts	Net Price	Terms	Date of Invoice	Date Paid	Cash Price
1.	$ 500	30% less 20% less 10%	?	2/10, n/30	Jan. 23	Jan. 30	?
2.	$ 460	25% less 20%	?	net 30	Apr. 14	May 11	?
3.	$3820	15%	?	3/15, n/30	Nov. 11	Nov. 25	?
4.	$ 987	20% less 10% less 5%	?	2/10 EOM	Feb. 3	Mar. 9	?
5.	$ 987	25% less 10%	?	2/10, n/30	Feb. 3	Mar. 1	?
6.	$4750	30% less 10%	?	8/20, n/50	June 13	July 1	?
7.	$4750	38%	?	3/10, n/30	Mar. 23	Apr. 7	?
8.	$ 500	10% less 20% less 30%	?	3/15 EOM	May 21	June 3	?
9.	$8259	12.5% less 20% less 50%	?	2/30, n/60	Aug. 23	Sep. 23	?
10.	$8259	65%	?	2/30, n/60	Sep. 23	Oct. 23	?

11. What is the net price in #1?

12. What is the cash price in #1?

13. What is the cash price in #2?

14. What is the net price in #3?

15. What is the cash price in #3?

16. What is the net price in #4?

17. What is the net price in #5?

LESSON PLAN
A Spreadsheet Application

USING TECHNOLOGY

Invoices received by a business generally are sent to the accounts payable department to be paid. In a large business, many invoices have to be processed each day. Discuss the advantages that a person responsible for paying invoices has by using a computer. (keeping track of invoices; speed and accuracy of computations; data storage; ability to issue checks on the appropriate date)

As a *business note*, point out that **net 30** is the standard term for paying the net price. Some suppliers allow longer payment periods such as 45, 50, or 60 days. Point out that a payment made after the due date is considered delinquent.

You may wish to have students work with a partner to input the data and answer the questions for this application.

A SPREADSHEET APPLICATION (CONTINUED)

18. In #4 and #5, the list price is the same and the sum of the chain discounts is the same. Why isn't the net price the same? Can you make a general statement about which net price will be greater if the sum of the chain discounts is the same?

19. In #4, you get the cash discount by paying on March 9; in #5, you don't get the cash discount and you paid on March 1. Why?

20. By looking at the net prices for #6 and #7, compare chain discounts of 30% less 10% and a single trade discount of 38%.

21. When is the net price due in #7 if the cash discount is not taken?

22. What is the net price in #8?

23. Both #1 and #8 result in the same net price. How can this be?

24. What is the net price in #9?

25. What is the net price in #10?

26. How can the net prices in #9 and #10 be the same?

27. What is the cash price in #9?

28. What is the cash price in #10?

29. Both #9 and #10 have the same cash discount terms. Why do you get the cash discount in #10 but not in #9?

UNIT

15

Sales

After your business has purchased goods from suppliers, wholesalers, distributors, or manufacturers, you must determine the *retail price,* or selling price, of the product. You want to sell the product at a higher price than its *cost,* or the amount you paid for it. The difference between the cost and the selling price is your *markup,* or *gross profit.* When your markup is larger than your overhead or operating expenses, you make a *net profit.*

OUTLINE

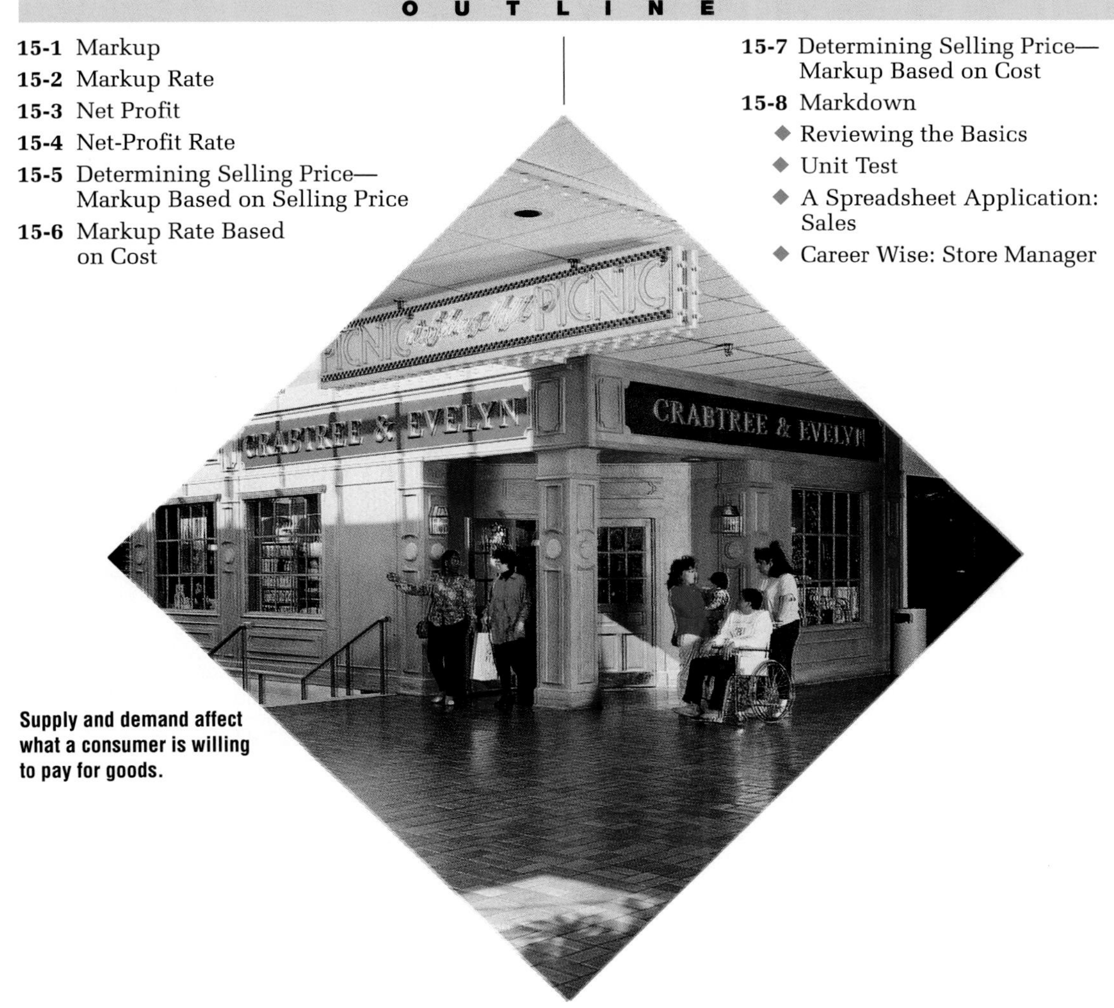

Supply and demand affect what a consumer is willing to pay for goods.

UNIT 15
SALES

INTRODUCING THE UNIT

Ask students if they wait for an item to "go on sale" before buying it. What do they think is a considerable markdown—10%? 20%? Perhaps they pay more attention to the actual dollar figure to be paid rather than percentages. Ask students why they think stores run sales (see Lesson 15-8). Compare markdown rates for different items.

LESSON PLAN
15-1 Markup

FOCUS

You can motivate this unit and lesson by asking students if they have ever bought anything on sale. Ask students why a store can sell something for less than the price marked on it. In this lesson, students will compute the markup in dollars of an item.

TEACH

Point out that the cost a business pays for an item is less than the retail price. The business must charge more than it paid for the item in order to make a profit. Go over the formula for finding the markup. Explain the other two terms, *gross profit* and *profit margin*. Explain that not all of the markup is profit, since it is out of this money that the business must pay for such costs as rent and employees' salaries.

Warm-Up Exercises

1. $9.79 – $7.83 $1.96
2. $34.49 – $25.87 $8.62
3. $197.97 – $138.58 $59.39
4. $419.19 – $251.51 $167.68
5. $2184.45 – $1201.45 $983.00
6. $9799.99 – $4899.99 $4900.00

15-1 Markup

OBJECTIVE
Compute the markup in dollars.

The **cost** of a product is the amount that your business pays for it. The cost includes expenses such as freight charges. Your business sells the product at a **selling price**, or **retail price**, which is higher than the cost. The **markup** is the difference between the selling price and the cost. The markup may also be called the **gross profit** or **profit margin**.

Markup = Selling Price – Cost

EXAMPLE *Skill* 6 *Application* A *Term* Markup

Discount Electronics, Inc., purchased cable-ready TV sets for $117.83 each. Discount Electronics sells the TVs for $199.99 each. What is the markup on each TV?

SOLUTION

Find the **markup.**

Selling Price – Cost
$199.99 – $117.83 = $82.16 markup

199.99 [–] 117.83 [=] 82.16

✓ SELF-CHECK Complete the problems, then check your answers in the back of the book.

Find the markup.

1. Selling price is $19.49. Cost is $7.63 **$11.86**
2. Selling price is $545.45. Cost is $272.72. **$272.73**

PROBLEMS

	Item	Selling Price	–	Cost	=	Markup
3.	Automobile	$16,497.00	–	$13,197.60	=	**$3299.40**
4.	Telephone	$ 49.79	–	$ 27.84	=	**$21.95**
5.	Watch	$ 97.87	–	$ 32.98	=	**$64.89**
6.	Book	$ 24.97	–	$ 18.43	=	**$6.54**
7.	Snowblower	$ 417.79	–	$ 235.84	=	**$181.95**
8.	Rosebush	$ 9.97	–	$ 6.68	=	**$3.29**
9.	Bicycle	$ 219.79	–	$ 148.93	=	**$70.86**
10.	Eraser	$ 0.35	–	$ 0.27	=	**$0.08**

BUSINESS NOTES

Sometimes a business may suffer a loss on an item, rather than make a profit on it. This happens when the cost of the item is greater than its selling price. These items are sometimes referred to as "loss leaders" in the retailing business.

11. Shoes.
Cost is $31.48.
Selling price is $79.59.
What is the markup? **$48.11**

12. Grape jam.
Cost is $0.56.
Selling price is $0.89.
What is the markup? **$0.33**

13. Basketball.
Cost is $12.78.
Selling price is $17.49.
What is the markup? **$4.71**

14. Cassette deck.
Cost is $17.97.
Selling price is $49.79.
What is the markup? **$31.82**

15. Mostly Kitchens, Inc., purchases paper towel holders at a cost of $7.14. Mostly Kitchens sells the paper towel holders for $10.99 each. What is the markup? **$3.85**

16. General Department Stores purchased bedroom curtains from The Linen Company for $6.19. General Department Stores sold the curtains for $9.79. What was the markup on the bedroom curtains? **$3.60**

17. Infinity Computers, Inc., advertised this monitor in yesterday's paper. The cost of the monitor to Infinity Computers is $297.88. What is the markup on the monitor? **$131.61**

18. Theresa Oakley operates a mini-engine repair shop. Recently she bought an older lawn mower for $20 at a local auction. She repaired the lawn mower for a cost of $9.17 in parts. She painted and polished the lawn mower for a cost of $6.43. When the lawn mower was finished, Theresa sold it for $75. What was the markup on the lawn mower? **$39.40**

19. Tasty Bakery Company sells chocolate eclairs for $15.48 a dozen. It costs Tasty Bakery $0.74 to produce each eclair.
   a. What is the markup on a dozen eclairs? **$6.60**
   b. What is the markup on each eclair? **$0.55**

20. Country Produce buys lettuce by the crate. Each crate contains 24 heads of lettuce. The cost of each crate to Country Produce is $11.76. Country Produce sells the lettuce for $0.64 per head.
   a. What is the markup per head of lettuce? **$0.15**
   b. What is the markup per crate of lettuce? **$3.60**

MAINTAINING YOUR SKILLS Look up the skill in parentheses if you need help or more practice.

Subtract. **(Skill 6)**

21. $54.45 − $43.56 **$10.89**
22. $1.74 − $1.57 **$0.17**
23. $216.96 − $130.18 **$86.78**
24. $491.79 − $418.02 **$73.77**
25. $24.49 − $23.27 **$1.22**
26. $643.89 − $321.94 **$321.95**
27. $14.79 − $11.54 **$3.25**
28. $4217.83 − $2319.81 **$1898.02**
29. $9179.84 − $4773.52 **$4406.32**

PRACTICE AND APPLY

The following problems can be assigned for classwork and the answers checked in class to help students master the objective of the lesson.

- Guided Practice: 1–6
- Independent Practice: 7–10, 12–16 even

WRAP-UP

Ask students for the new terms of the lesson. Ask different students what each term means. Ask how to find the dollar markup of an item.

Assignment Guide

- Basic: 7–18, 21–29
- Average: 11–17 even, 18–20, 22–26 even

ALTERNATIVE STRATEGIES: Reteaching

Stress that not all of the markup is profit. The cost of doing business (overhead), must come out of the markup. By doing these problems, the relationship between cost, markup, and selling price should become clearer.

	Cost	Markup	Selling Price
1)	$ 48.50	($36.49)	$ 84.99
2)	138.43	($61.36)	199.79
3)	($7.18)	$ 7.61	14.79

LESSON PLAN
15-2 Markup Rate

FOCUS

The focus of this lesson is to compute the markup as a percent of the selling price. Most businesses use percent rather than dollar amounts when talking about markups.

TEACH

Manufacturers and wholesalers usually express markup as a percent of their cost. Emphasize that retailers express markup as a percent of the selling price. Go over the formula. You can use this example to illustrate the difference in the two methods:

cost = $40
markup = $60
selling price = $100
markup for manufacturer or wholesaler = $\frac{60}{40}$ = 150%
markup for retailer = $\frac{60}{100}$ = 60%.

Ask students why they think retailers use the selling price in the formula. (Public sees 60% vs. 150%.)

Warm-Up Exercises

1. $179.79 − $107.87 $71.92
2. $44.56 − $22.28 $22.28
3. $71.92 is what percent of $179.79? 40%
4. $22.28 is what percent of $44.56? 50%
5. ($378.48 − $283.86) is what percent of $378.48? 25%

15-2 Markup Rate

OBJECTIVE
Compute the markup as a percent of the selling price.

The **markup rate** is the markup expressed as a percent. Businesses usually express the markup as a percent of the selling price.

$$\text{Markup Rate} = \frac{\text{Markup}}{\text{Selling Price}}$$

EXAMPLE *Skills* 6, 31 *Application* A *Term* Markup rate

Roy's Florist buys roses for $10.99 a dozen. It sells them for $18.95 a dozen. What is the markup rate based on the selling price?

SOLUTION

A. Find the **markup.**
Selling Price − Cost
$18.95 − $10.99 = $7.96 markup

B. Find the **markup rate** based on the selling price.
Markup ÷ Selling Price
$7.96 ÷ $18.95 = 0.420 = 42% markup rate

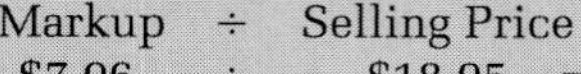
18.95 M+ − 10.99 = 7.96 ÷ RM 18.95 = 0.420

✓ SELF-CHECK Complete the problems, then check your answers in the back of the book.

Find the markup and the markup rate.

1. Selling price is $49.79. Cost is $34.85. **$14.94; 30%**
2. Selling price is $249.19. Cost is $161.97. **$87.22; 35%**

PROBLEMS

Round answers to the nearest tenth of a percent.

3. Video cassette recorder. Cost is $124.78. Selling price is $249.99 What is the markup rate based on the selling price? **50.1%**
4. PC desk. Cost is $74.63. Selling price is $119.79. What is the markup rate based on the selling price? **37.7%**
5. Cookware. Cost is $37.87. Selling price is $59.79. What is the markup rate based on the selling price? **36.7%**
6. Tricycle. Cost is $65.91 Selling price is $98.25. What is the markup rate based on the selling price? **32.9%**

PROBLEM SOLVING

When the cost of an item is doubled in order to get the selling price, it is called "keystoning." In Problem 8, what would the cost of the shirt have been if the cost had been keystoned to arrive at the selling price of $42.50? ($21.25)

7. Pete Kraemer is a buyer for The Sleep Store. He purchases Majestic twin-size box springs for $86.74 each. The Sleep Store sells them for $167.49 each. What is the markup for each box spring? What is the markup rate based on the selling price? **$80.75; 48.2%**

8. Rodolfo's purchases silk shirts for $18.43 each. The selling price of each shirt is $42.50. What is the markup rate based on the selling price of each shirt? **56.6%**

9. Automobile batteries cost Main Auto Supply $86.94. Main sells the batteries for $129.47. What is the markup rate based on the selling price? **32.8%**

10. Convenient Carryout purchases donuts for 90¢ a dozen. The selling price of the donuts is $1.99 per dozen. What is the markup rate based on the selling price? **54.8%**

11. Town Auto Sales pays $13,596, including freight, for the new model Tracker. Town Auto Sales sells each new Tracker for $16,995. What is the markup rate based on the selling price? **20.0%**

12. Video Rentals sells used cassettes for $24.95. The cost to Video Rentals is $17.86 per cassette. What is the markup rate based on the selling price? **28.4%**

13. A series L gas water heater sells for $219.29 at Carl's Appliances. Each water heater costs Carl's $152.90. What is the markup rate based on the selling price? **30.3%**

14. Dock 2 Imports held a special sale on wicker furniture. The selling price of each item was $99.95. What was the markup rate based on the selling price for each item? **34.5%; 41.5%; 25.0%**

ITEM NUMBER	DESCRIPTION	COST
WT145	Table	$65.49
WCR74	Chair	58.43
WCE04	Chaise	74.97

MAINTAINING YOUR SKILLS Look up the skills in parentheses if you need help or more practice.

Subtract. **(Skill 6)**

15. $25.45 − $19.32 **$6.13**
16. $79.49 − $63.59 **$15.90**
17. $119.29 − $83.50 **$35.79**
18. $272.21 − $163.33 **$108.88**
19. $2618.45 − $1309.23 **$1309.22**
20. $24,749 − $18,066.71 **$6682.29**

Find the rate. Round answers to the nearest tenth of a percent. **(Skill 31)**

21. $6.13 ÷ $25.45 **24.1%**
22. $35.79 ÷ $119.29 **30.0%**
23. $1309.22 ÷ $1309.23 **100.0%**
24. $216.45 ÷ $541.13 **40.0%**
25. $4176.48 ÷ $8352.96 **50%**
26. $10,874 ÷ $16,311 **66.7%**

PRACTICE AND APPLY

The following problems can be assigned for classwork and the answers checked in class to help students master the objective of the lesson.

- Guided Practice: 1–4
- Independent Practice: 5–8

WRAP-UP

Read Problem 6 to the class. Ask a student for the formula for the markup rate. Have one student find the markup and another find the markup rate.

Assignment Guide

- Basic: 5–13, 15–26
- Average: 9–14, 15–25 odd

ALTERNATIVE STRATEGIES: Enrichment

Generally, lower priced items with a high turnover rate have a smaller percent markup (smaller margin); higher priced items with a low turnover rate have a higher percent markup (greater margin). "Keystoning," which is a very common markup policy, represents a 50% markup based on the selling price (100% markup based on cost as we will see in Lesson 15-6)

LESSON PLAN
15-3 Net Profit

FOCUS
Ask students if they or their parents (or guardians) get to spend paychecks on whatever they want, or if they have certain obligations they must first meet with this money. Point out that businesses have similar obligations, which must be met from the markup dollars.

TEACH
Go over the terms *overhead* and *operating expenses.* Have students suggest other costs that could be included under operating expenses. The list may include maintenance, security, insurance, office or cleaning supplies, or running inventory. Mention that an overhead percent is usually used because some operating expenses are fixed, while others may vary from month to month. Emphasize that *net profit* is the amount of the *gross profit* that remains after the business's obligations have been met.

Warm-Up Exercises
1. $74.36 − $51.74 $22.62
2. $1.78 − $0.85 $0.93
3. 30% of $149.99 $45.00
4. 40% of $47.49 $19.00

15-3 Net Profit

OBJECTIVE
Compute the net profit in dollars.

The markup on the products that you sell must cover your **overhead** or **operating expenses.** These expenses include wages and salaries of employees, rent, utility charges, and taxes.

The exact overhead expense on each item sold is difficult to determine accurately. You may approximate the overhead expense of each item. For example, if your total overhead expenses are 40% of total sales, you may estimate the overhead expense of each item sold to be 40% of its selling price. When the markup of an item is greater than its overhead expense, you make a **net profit** on the item.

Overhead = Selling Price × Overhead Percent
Net Profit = Markup − Overhead

EXAMPLE *Skills* 30, 2 *Application* A *Term* Net profit

Jaworski Sport Shop purchases 15-foot rowboats for $44.98 each. Jaworski sells the boats for $89.99 each. Lou Jaworski estimates his overhead expenses to be 40% of the selling price of his merchandise. What is the net profit on each 15-foot rowboat?

SOLUTION

A. Find the **markup.**
Selling Price − Cost
$89.99 − $44.98 = $45.01 markup

B. Find the **overhead.**
Selling Price × Overhead Percent
$89.99 × 40% = $35.996 = $36.00 overhead

C. Find the **net profit.**
Markup − Overhead
$45.01 − $36.00 = $9.01 net profit

✔ SELF-CHECK Complete the problems, then check your answers in the back of the book.

Find the markup, overhead, and net profit.

1. Selling price is $140. Cost is $56. **$84; $70; $14** Overhead is 50% of selling price.
2. Selling price is $964. **$385.60;** Cost is $578.40. **$337.40; $48.20** Overhead is 35% of selling price.

PROBLEMS

	Markup	−	(Selling Price	×	Overhead Percent)	=	Net Profit
3.	$ 20	−	($ 40	×	30%)	=	**$8.00**
4.	$108	−	($180	×	40%)	=	**$36.00**
5.	$ 81	−	($270	×	20%)	=	**$27.00**

COMMUNICATION SKILLS
Some businesses are forced to go out of business when they can no longer make a net profit. Have students research and write a brief report on the bankruptcy process. If time permits, share these reports with the class.

Estimate the net profit based on the estimated overhead.

6. Power rake costs $49.74.
Selling price is $98.79.
Overhead is $19.64.
What is the net profit? **$29.41**

7. Light fixture costs $31.78.
Selling price is $89.79.
Overhead is $24.56.
What is the net profit? **$33.45**

8. Sweater costs $47.84.
Selling price is $119.49.
Overhead is 30% of selling price.
What is the net profit? **$35.80**

9. Pen/pencil set costs $41.89.
Selling price is $74.74.
Overhead is 35% of selling price.
What is the net profit? **$6.69**

10. Scroll Hardware Company purchases Allegré door hardware sets for $29.86 each. Scroll Hardware charges customers $49.99 to have the hardware installed. The labor and other overhead total $11.75 for each set installed. What is the net profit per set? **$8.38**

11. Storm King, Inc., purchases storm doors at a cost of $41.74 each. The selling price of an installed storm door is $119.50. The overhead for the storm door is estimated to be 40% of the selling price. What is the net profit? **$29.96**

12. Mary Callas is a buyer for Classic Shoes. She purchased some shoes for a cost of $13.89 a pair. Her store sells them for $29.99 a pair. Operating expenses are estimated to be 30% of the selling price. What is the net profit per pair of shoes? **$7.10**

13. Custom Comp, assemblers of specialized central processing units, delivered a system to Universal Delivery, Inc., for a selling price of $1,078,450. The cost of the unit to Custom Comp was $416,785. The system design, programming, and other overhead expenses were estimated to be 38.5% of the selling price. What net profit did Custom Comp make on the sale? **$246,461.75**

14. Thomas Gibson manages Craft Cave. He buys E-Z Trim knives at $11.28 a dozen. He sells them at $3.38 a pair. The overhead is estimated to be 25% of the selling price. What is the net profit per pair? **$0.65**

MAINTAINING YOUR SKILLS Look up the skills in parentheses if you need help or more practice.

Round answers to the nearest cent. **(Skill 2)**

15. $1.111 **$1.11**
16. $4.545 **$4.55**
17. $0.4581 **$0.46**
18. $9.8949 **$9.89**
19. $7.6661 **$7.67**
20. $96.4125 **$96.41**

Subtract. **(Skill 6)**

21. $71.86 − $45.47 **$26.39**
22. $9.78 − $9.43 **$0.35**
23. $47.81 − $21.93 **$25.88**

Find the percentage. Round answers to the nearest cent. **(Skill 30)**

24. $160 × 30% **$48**
25. $47.79 × 35% **$16.73**
26. $178.88 × 25% **$44.72**
27. $219.79 × 37.5% **$82.42**
28. $1789.49 × 40% **$715.80**
29. $18,749.45 × 31.4% **$5887.33**

PRACTICE AND APPLY

The following problems can be assigned for classwork and the answers checked in class to help students master the objective of the lesson.

- Guided Practice: 1–6
- Independent Practice: 7–11 odd

WRAP-UP

Have students explain the terms *overhead* and *net profit.* Write the formulas for finding each on the chalkboard. Read Problem 11 to the class. Have students substitute the numbers into the formulas.

Assignment Guide

- Basic: 7–12, 15–29
- Average: 8–12 even, 13, 14, 15–29 odd

ALTERNATIVE STRATEGIES: Reteaching

	Selling Price	Cost	Markup	Overhead Percent	Overhead	Net Profit
1)	$ 79.79	$38.56	($41.23)	25%	($19.95)	($21.28)
2)	119.95	68.47	($51.48)	30%	($35.99)	($15.49)
3)	2.49	1.12	($ 1.37)	18%	($ 0.45)	($ 0.92)

LESSON PLAN
15-4 Net Profit Rate

FOCUS

The focus of this lesson is to compute the net profit as a percent of the selling price. By using percents, a business can compare how profitable different items are that it sells.

TEACH

Review how to find the net profit of an item. Then go over the formula for finding the net-profit rate. Work through the Example. Remind students how to find the markup rate. Have them do so for the Example. ($26.21 ÷ $49.95 = 52.5%) Have students find the sum of the overhead rate and the net-profit rate. (30% + 22.5% = 52.5%) Point out that the gross profit was split into two parts—one part for overhead and one for net profit.

Warm-Up Exercises

1. $15.95 − $7.97 $7.98
2. 30% of $15.95 $4.79
3. $7.98 − $4.79 $3.19
4. $3.19 ÷ $15.95 0.20
5. ($260.99 − $140.20) − (25% of $260.99) $55.54
6. $55.54 ÷ $260.99 0.213

15-4 Net-Profit Rate

OBJECTIVE
Compute the net profit as a percent of the selling price.

You may want to know the **net-profit rate** of an item your business sells. The net-profit rate is net profit expressed as a percent of the selling price of the item.

$$\text{Net-Profit Rate} = \frac{\text{Net Profit}}{\text{Selling Price}}$$

EXAMPLE *Skills* 6, 30, 31 *Application* A *Term* Net-profit rate

East Imports sells a wok for $49.95. The cost of the wok to East Imports is $23.74. East Imports estimates the overhead expenses on the wok to be 30% of its selling price. What is the net-profit rate based on the selling price of the wok?

SOLUTION

A. Find the **net profit.**

Markup	−	Overhead		
($49.95 − $23.74)	−	($49.95 × 30%)		
$26.21	−	$14.99	=	$11.22 net profit

B. Find the **net-profit rate.**

Net Profit	÷	Selling Price				
$11.22	÷	$49.95	=	0.2246	=	22.5% net-profit rate

49.95 [−] 23.74 [M+] 49.95 [×] 30 [%] 14.985 [RM] 26.21 [−] 14.99 [=] 11.22 [÷] 49.95 [=] 0.2246

✓ SELF-CHECK Complete the problems, then check your answers in the back of the book.

Find the markup, overhead, net profit, and net-profit rate.

1. Selling price is $119.50. Cost is $44.78. **$74.72; $47.80** Overhead expense is 40% of selling price. **$26.92; 22.5%**
2. Selling price is $224.50. Cost is $102.60. **$121.90; $78.58;** Overhead expense is 35% of selling price. **$43.32; 19.3%**

PROBLEMS

Round answers to the nearest percent.

	Markup	−	(Selling Price	×	Overhead Percent)	=	Net Profit	Net-Profit Rate
3.	$ 40.00	−	($ 50.00	×	30%)	=	**$25**	**50%**
4.	$ 35.00	−	($ 70.00	×	25%)	=	**$17.50**	**25%**
5.	$ 81.47	−	($ 149.70	×	35%)	=	**$29.07**	**19%**
6.	$ 191.75	−	($ 449.49	×	20%)	=	**$101.85**	**23%**
7.	$7366.92	−	($8475.75	×	40%)	=	**$3976.62**	**47%**

CULTURAL ANGLES

Some of our cities were founded because of business. For example, St. Louis, MO was founded in 1764 by the French. They established this town as a post for fur trading. Ask students to name a nation whose people founded towns and cities in California. (Mexico)

Round answers to the nearest tenth of a percent.

8. Selling price is $22.49. **24.8%**
Markup is $11.45.
Overhead is $5.47.
What is the net-profit rate?

9. Selling price is $9.98. **22.0%**
Markup is $5.31.
Overhead is $2.47.
What is the net-profit rate?

10. Snow sled costs $21.74. **14.2%**
Selling price is $47.50.
Overhead is 40% of selling price.
What is the net-profit rate?

11. End table costs $55.67. **42.2%**
Selling price is $129.99.
Overhead is 15% of selling price.
What is the net-profit rate?

12. Court-Time, Inc., sells these tennis rackets. The cost to Court-Time is $37.48 per racket. Overhead is estimated to be 20% of the selling price. What is the net-profit rate? **22.3%**

Clean up with savings!
Tennis rackets $64.99
at Court-Time, Inc.

13. Value Discount Shops sell G.S. sweaters for $27.79 each. The markup on each sweater is $12.85. Overhead is estimated to be 30% of the selling price. What is the net-profit rate? **16.2%**

14. Dixie Deli purchases apple butter for $1.03 a jar. It sells the apple butter for $1.79 a jar. Dixie Deli estimates the overhead at 25% of the selling price. What is the net-profit rate on each jar? **17.3%**

15. Elaine Luttrell manufactures hunting decoys. She pays $4.79 for the parts for each decoy. Elaine assembles the decoys and sells them to sporting goods stores for $14.85. She estimates the overhead to be 37.5% of the selling price of the decoys. What is the net-profit rate per decoy? **30.2%**

16. The Sevas Die Shop purchases steel bars for $9.85 per bar. An automatic machine cuts and forms a bar into 30 steel pins. The 30 pins are packaged individually and marked to sell at 79¢ each. Overhead expenses are estimated to be 12.5% of the selling price per pin. What is the net-profit rate per pin? **45.9% or 45.6%**

17. Video Show, Inc., buys video cassettes for $52.56 a carton. A carton contains 24 cassettes. Video Show sells the cassettes for $4.95 each or $109.99 per carton. Overhead is estimated to be 32% of selling price.
   a. What is the net-profit rate on the sale of 1 cassette? **23.8%**
   b. What is the net-profit rate when a carton of cassettes is sold? **20.2%**

MAINTAINING YOUR SKILLS Look up the skills in parentheses if you need help or more practice.

Find the percentage. Round answers to the nearest cent. **(Skill 30)**

18. $49.49 × 30% **$14.85**
19. $9.78 × 20% **$1.96**
20. $134.49 × 25% **$33.62**
21. $749.79 × 15% **$112.47**
22. $1278.58 × 12.5% **$159.82**
23. $8645.15 × 37.5% **$3241.93**

Find the rate. Round answers to the nearest tenth of a percent. **(Skill 31)**

24. $5.86 ÷ $19.79 **29.6%**
25. $41.16 ÷ $99.49 **41.4%**
26. $67.50 ÷ $449.99 **15.0%**

PRACTICE AND APPLY

The following problems can be assigned for classwork and the answers checked in class to help students master the objective of the lesson.

- Guided Practice: 1–8
- Independent Practice: 9–13 odd

WRAP-UP

Ask students how to find the net-profit rate. Have students rework Problem 13 with an overhead rate of 25%. (21.2%)

Assignment Guide

- Basic: 9–15, 18–26
- Average: 10–14 even, 15–17, 18–26 even

ALTERNATIVE STRATEGIES: Reteaching

Finding the net profit rate is an application of Skill 31. Using the problems from the RETEACH in Lesson 15-3, find the net profit rate to the nearest tenth percent:

	Selling Price	Net Profit	Net Profit Rate
1)	$79.79	$21.28	(26.7%)
2)	119.95	15.49	(12.9%)
3)	2.49	0.92	(36.9%)

LESSON PLAN
15-5 Determining Selling Price—Markup Based on Selling Price

FOCUS
Many businesses use past sales histories to establish a markup rate for an item or for a department. The focus of this lesson is on computing the selling price of an item based on cost and markup rate.

TEACH
Emphasize the fact that the method of this lesson will work only after a pattern has been established for both sales and expenses. Once the patterns have been determined, the company can then set up a markup rate with which it is comfortable. Go over the method for finding the complement of the markup rate.

Warm-Up Exercises
Find the complements.

1. 30% 70%
2. 45% 55%
3. 50% 50%
4. $62\frac{1}{2}$% $37\frac{1}{2}$%

Calculate.

5. \$20 ÷ 40% \$50
6. \$66 ÷ 50% \$132
7. \$150 ÷ 30% \$500
8. \$35 is 35% of what amount? \$100

15-5

Determining Selling Price—Markup Based on Selling Price

OBJECTIVE
Compute the selling price of an item based on cost and markup rate.

You can use records of past sales and expenses to plan the markup rate needed to cover overhead expenses and yield a profit. You can use the cost of an item and the desired markup rate based on selling price to figure the best selling price.

$$\text{Selling Price} = \frac{\text{Cost}}{\text{Complement of Markup Rate}}$$

EXAMPLE *Skills* 28, 11 *Application* L *Term* Markup rate

VJ's Sport Store knows from past expense records that it must aim for a markup that is 40% of the selling price of its merchandise. The store received a shipment of running shoes at a cost of \$38.99 per pair. What is the minimum selling price that the store should charge?

		% of Total Sales
Sales for Month:	\$42,000	Sales
Cost of goods sold:	25,200	60%
Overhead expenses:	8,400	20%
Profit:	8,400	20%
	\$42,000	100%

SOLUTION

A. Find the **complement of markup rate.**
100% − 40% = 60% complement of markup rate

B. Find the **selling price.**
Cost ÷ Complement of Markup Rate
\$38.99 ÷ 60% = \$64.983 = \$64.98 selling price

✓ SELF-CHECK Complete the problems, then check your answers in the back of the book.

Find the selling price.

1. Cost is \$345. **\$460**
   Markup is 25% of selling price.
2. Cost is \$1.40. **\$2.00**
   Markup is 30% of selling price.

PROBLEMS

	3.	4.	5.	6.	7.	8.
Cost	\$6.50	\$120.00	\$86.74	\$420.00	\$47.84	\$212.60
Markup Rate (Selling Price Base)	35%	20%	50%	45%	37.5%	12.5%
Complement of Markup Rate	65%	? **80%**	? **50%**	? **55%**	? **62.5%**	? **87.5%**
Selling Price (Nearest Cent)	? **\$10**	? **\$150**	? **\$173.48**	? **\$763.64**	? **\$76.54**	? **\$242.97**

MATHEMATICAL NOTES
Remind students that Selling Price = Cost + Markup. For the Example, cost = 60%, markup = 40%, and selling price = 100%. Since the cost of \$38.99 is 60% of the selling price, the problem can be stated as \$38.99 is 60% of what number? Students can then see that in order to find the number, they need to divide \$38.99 by 60%.

9. Gold chain.
Cost is $7.48.
Markup is 60% of selling price.
What is the selling price? **$18.70**

10. Stemware.
Cost is $5.16.
Markup is 40% of selling price.
What is the selling price? **$8.60**

11. Paperback Paupers marks up paperbacks 35% of their selling price. A paperback costs Paupers $2.57. What is the selling price of the book? **$3.95**

F.Y.I.
A markup of 50% of cost is the same as a markup of $33\frac{1}{3}$% of selling price.

12. April's Shower purchases curtains from the manufacturer and marks them up 55% of the selling price. April's pays $8.10 per pair for a lined pair. What is the selling price for the curtain? **$18**

13. Wade Charles prices items at BarBells, Inc. BarBells has a storewide policy to mark up each item 53% of the selling price. What selling price does he place on these items?

ITEM	COST	
Adjustable bench	$23.50	**$50**
Weight rod	9.40	**$20**
Weight belt	11.75	**$25**

14. Hillary Lee manages the craft department of Super Craft, Inc. What selling price will she place on each of these items?

LINE	ITEM CODE	DESCRIPTION	UNIT	STORE COST	MARKUP RATE SELLING PRICE BASE	
14	063614	TAPE MASKING 300 IN ROLL	RL	0.545	55 %	**$1.21**
15	071358	PROTRACTOR W/RULER PKG	EA	0.105	50%	**$0.21**
16	031615	TWINE COTTON 4-PLY 430 FT	EA	0.302	57.5%	**$0.71**
17	062824	CHALK WHITE 12-STICK PKG	PK	0.149	60%	**$0.37**
18	120057	MODEL CEMENT PLASTIC 1 OZ	EA	0.218	48.2%	**$0.42**

Critical Thinking . . .

15. Suppose Hillary Lee in problem 14 decided to mark up all the items 55% of the selling price. If you purchase 1 of each item, how much more or less do you pay? **0**

MAINTAINING YOUR SKILLS Look up the skills in parentheses if you need help or more practice.

Find the complement. **(Application L)**

16. 35% **65%** **17.** 20% **80%** **18.** 50% **50%** **19.** 40% **60%** **20.** 37.5% **62.5%** **21.** 47.6% **52.4%**

Divide. Round answers to the nearest cent. **(Skill 11)**

22. $23.00 ÷ 65% **$35.38** **23.** $48.00 ÷ 80% **$60** **24.** $117.40 ÷ 50% **$234.80**

25. $471.83 ÷ 60% **$786.38** **26.** $978.56 ÷ 62.5% **$1565.70** **27.** $1748.72 ÷ 52.4% **$3337.25**

28. $7.53 ÷ 64.7% **$11.64** **29.** $71.46 ÷ 37.8% **$189.05** **30.** $12,317.87 ÷ 41.3% **$29,825.35**

PRACTICE AND APPLY

The following problems can be assigned for classwork and the answers checked in class to help students master the objective of the lesson.

- Guided Practice: 1–6
- Independent Practice: 7–9, 11

WRAP-UP

Have students give the formula for finding the selling price, using the complement of the markup rate. Ask a student to explain how to find the complement of a markup rate. Ask students to find the selling price if the cost of an item is $25 and the markup rate is 50%. ($50)

Assignment Guide

- Basic: 7–13, 15–30
- Average: 10, 12–14, 15–29 odd

ALTERNATIVE STRATEGIES: Reteaching

	Cost	Markup as a percent of Selling Price	Cost as a percent of Selling Price	Selling Price
1)	$ 48.22	35%	(65%)	($74.18)
2)	138.50	64%	(36%)	($384.72)
3)	3.50	50%	(50%)	($7.00)

LESSON PLAN
15-6 Markup Rate Based on Cost

FOCUS
In Lesson 15-2, students learned to find the markup rate based on the selling price. The focus of this lesson is on computing the markup rate based on cost.

TEACH
Point out that manufacturers know the cost of producing an item, including the cost of materials and labor. They also know the markup rate needed in order to make a certain level of gross profit. This is why manufacturers are more likely to base markup on the cost of an item.

Ask students why the markup rate on grocery items is so much lower than that on jewelry. (Answers may include the fact that groceries are a necessity for living; therefore, items are pretty well guaranteed to sell. Jewelry sales are less predictable, while certain expenses still must be covered.)

Warm-Up Exercises
1. $171.89 − $103.49 $68.40
2. $94.79 − $41.83 $52.96
3. $19.89 − $6.63 $13.26
4. $68.40 is what percent of $103.49? (nearest tenth) 66.1%
5. $13.26 is what percent of $6.63? 200%
6. ($94.79 − $41.83) is what percent of $41.83? (nearest tenth) 126.6%

15-6 Markup Rate Based on Cost

OBJECTIVE
Compute the markup rate based on cost.

Your business may use the cost of a product as the base for the markup rate. The markup rate in supermarkets is relatively low. For example, milk may be marked up 5% of the cost and other dairy products 20% to 30% of the cost. The markup of clothing may be from 80% to 140% of the cost. The markup of jewelry may be 100% or more of the cost.

$$\text{Markup Rate} = \frac{\text{Markup}}{\text{Cost}}$$

EXAMPLE *Skills* 6, 31 *Application* A *Term* Markup

Dinettes Inc. purchases a 5-piece dinette set for $180 from the manufacturer. It sells the 5-piece set for $288. What is the markup rate based on cost?

SOLUTION

A. Find the **markup.**
Selling Price − Cost
$288.00 − $180.00 = 108.00 markup

B. Find the **markup rate** based on cost.
Markup ÷ Cost
$108.00 ÷ $180.00 = 0.6 = 60% markup rate based on cost

288 [−] 180 [=] 108 [÷] 180 [=] 0.6

✓ SELF-CHECK Complete the problems, then check your answers in the back of the book.

Find the markup rate based on cost.

1. Selling price is $97.50. Cost is $58.50. **66.7%**
2. Selling price is $148. Cost is $74. **100%**

PROBLEMS

Round answers to the nearest tenth of a percent.

	(Selling Price −	Cost	= Markup) ÷	Cost	= Markup Rate
3.	($ 1.75	− $ 1.25	= **$0.50** ?)	÷ $ 1.25	= **40%** ?
4.	($ 50.00	− $ 20.00	= **$30** ?)	÷ $ 20.00	= **150%** ?
5.	($ 85.50	− $ 47.50	= **$38** ?)	÷ $ 47.50	= **80%** ?
6.	($791.00	− $316.40	= **$474.60** ?)	÷ $316.40	= **150%** ?
7.	($ 1.04	− $ 0.13	= **$0.91** ?)	÷ $ 0.13	= **700%** ?

CRITICAL THINKING
See Problem 13. The markup rate based on cost is more than 100%. If the markup rate is based on the selling price, why can the rate never be more than 100%? (The amount an item is marked up cannot be more than it sells for; the amount an item is marked up *can* be more than it cost the retailer.)

8. Selling price is $226.64.
Cost is $141.65.
What is the markup rate based on cost? **60%**

9. Selling price is $18.65 each.
Cost is $7.46 each.
What is the markup rate based on cost? **150%**

10. The Teen Jeans Co. buys 805 jeans at $11.99 per pair. They mark up each pair $7.99. What is the markup rate based on cost? **66.6%**

11. Quentin Clark, a salesperson at Doolittle's Department Store, sold a pair of golf shoes for $84.95. The shoes cost Doolittle's $27.84 a pair. What is the markup rate based on cost? **205.1%**

12. Trudi Alvarez is a buyer for the Small Tots Shoppe. She purchased an assortment of sunsuits for $2.74 each. They were marked to sell for $5.48 each. What is the markup rate based on cost? **100%**

13. Craig Nielson works at his parents' outdoor market. He sells tomatoes at $1.19 a pound. The tomatoes cost $0.39 a pound. What is the markup rate based on cost? **205.1%**

14. Martha Muntz is in charge of newspapers and magazines at Cramer's News. What is the markup rate based on cost for each item?

ITEM	COST	SELLING PRICE	
Local Sunday paper	$1.50	$2.50	**66.7%**
Map of city	0.50	1.25	**150%**
Out-of-state Sunday paper	2.50	6.50	**160%**
Antique magazine	1.25	2.75	**120%**

15. Discount Electronics, sells quality stereo components. The following list shows the cost and selling price of various components. Find the markup rate based on cost for each item.

ITEM	COST	SELLING PRICE	
P636 Receiver	$167.96	$419.89	**150%**
P552 CD player	224.69	374.49	**66.7%**
P9191 Cassette deck	199.99	399.99	**100%**
P88A Speakers (pair)	286.72	819.19	**185.7%**

MAINTAINING YOUR SKILLS Look up the skills in parentheses if you need help or more practice.

Subtract. **(Skill 6)**

16. $74.80 − $37.40 **$37.40**

17. $149.49 − $59.80 **$89.69**

18. $19.19 − $12.38 **$6.81**

19. $274.49 − $164.69 **$109.80**

20. $4738.50 − $3174.74 **$1563.76**

21. $25,789.49 − $20,631.59 **$5157.90**

Find the rate. Round answers to the nearest tenth of a percent. **(Skill 31)**

22. $89.69 ÷ $59.80 **150.0%**

23. $109.80 ÷ $164.69 **66.7%**

24. $5157.90 ÷ $20,631.59 **25.0%**

PRACTICE AND APPLY

The following problems can be assigned for classwork and the answers checked in class to help students master the objective of the lesson.

- Guided Practice: 1–8
- Independent Practice: 9–13 odd

WRAP-UP

Read Problem 11 to the class. Have one student explain how to find the markup rate based on cost; another to find the markup; and a third student to find the markup rate.

Assignment Guide

- Basic: 9–14, 16–24
- Average: 10–14 even, 15, 16–24 even

ALTERNATIVE STRATEGIES: Enrichment

The comments made under CRITICAL THINKING can be extended to further develop the relationship between the markup rate based on selling price and the markup rate based on cost.

Work with your students until they see that a 50% markup rate based on selling price (keystoning) is equivalent to a 100% markup rate based on cost. Use dollar figures to make this clear, e.g. $60 + $60 = $120 and $60 is 50% of $120 while $60 is 100% of $60.

LESSON PLAN 15-7 Determining Selling Price—Markup Based on Cost

FOCUS

Ask students where they may have heard of the term *retail selling price*. In this lesson, students learn to compute the selling price based on cost and markup rate.

TEACH

Emphasize the fact that there are two steps to finding the selling price: first to find the markup, then to add this to the cost. Remind students that markup usually is expressed as a percent of cost by manufacturers, and as a percent of selling price by retailers. Point out that markups of over 100% are not uncommon. These are encountered in the problems.

Warm-Up Exercises

1. 120% of $95 $114
2. 74% of $24.30 $17.98
3. $95 + $114 $209
4. $24.30 + $17.98 $42.28
5. $171.67 + (60% of $171.67) $274.67
6. $945.81 + (275% of $945.81) $3546.79

15-7 Determining Selling Price—Markup Based on Cost

OBJECTIVE
Compute the selling price based on cost and markup rate.

You can use the cost of an item and the desired markup rate based on cost to figure the selling price of an item.

Markup = Cost × Markup Rate

Selling Price = Cost + Markup

EXAMPLE *Skills* 30, 5 *Application* A *Term* Markup rate

Wholesale Jewelers, Inc., sells electronic digital watches to jewelry stores for $18.45 each. Wholesale Jewelers calculates the suggested retail price and attaches it to each watch. The retail price is computed by marking up the cost to the jewelry store by 160%. What is the suggested retail selling price?

SOLUTION

A. Find the **markup.**
Cost × Markup Rate
$18.45 × 160% = $29.52 markup

B. Find the **selling price.**
Cost + Markup
$18.45 + $29.52 = $47.97 selling price

18.45 [×] 160 [%] 25.92 18.45 [+] 25.92 [=] 47.97

✓ SELF-CHECK Complete the problems, then check your answers in the back of the book.

Find the markup and the selling price.

1. Cost is $50.
   Markup is 70% of cost.
   $35; $85

2. Cost is $140.
   Markup is 50% of cost.
   $70; $210

PROBLEMS

	3.	4.	5.	6.	7.	8.
Cost	$45.00	$96.49	$86.40	$16.40	$751.80	$14.24
Markup Rate	80%	100%	150%	225%	87.5%	400%
Markup	$36.00	$96.49	$129.60	$36.90	$657.83	$56.96
Selling Price	$81.00	$192.98	$216.00	$53.30	$1409.63	$71.20

ALTERNATIVE STRATEGIES: Reteaching

The selling price can be found directly by adding 100% to the markup rate based on the cost and multiplying that times the cost. It is time now to put all these together by completing the following table:

9. Handsaw.
Cost is $11.40.
Markup is 65% of cost.
What is the selling price? **$18.81**

10. Camera.
Cost is $37.48.
Markup is 180% of cost.
What is the selling price? **$104.94**

11. The Sports Wholesale Company purchases weight racks directly from the manufacturer for $10.80 each. The weight racks are marked up 210% of cost and sold to retail sporting good stores. What is the selling price of each weight rack? **$33.48**

12. The Haas Door Co. produces truck dock door seals. Their 9' × 10' seal costs $280 to produce. Markup is 75% of cost. What is the selling price? **$490**

13. XYZ Appliances buys Circle Clean washers from a distributor for $147.85 each. The markup is 100% of cost. What is the selling price? **$295.70**

14. Surface Combustion, Inc., calculates the cost of manufacturing a particular furnace to be $1214.78. Surface marks up each furnace 120% based on cost. What is the selling price? **$2672.52**

15. The Pet-Agree Co. buys dog collars from a manufacturer for $0.18 each. Pet-Agree marks up each collar 356% of cost. What is the selling price? **$0.82**

16. The Libbey Glass Company manufacturers stemware. Cost per gross (144 items) is $66.24. The stemware is sold at a markup of 115% based on cost.
   a. What is the selling price for a dozen stemware? **$11.87**
   b. What is the selling price for a single stemware? **$0.99**

17. Cook-n-Serve carries fancy oven mits that cost $0.86 a pair. Cook-n-Serve operates on a markup of 166% of cost.
   a. What is the selling price? **$2.29**
   b. What is the markup as a percent of the selling price? **62.4%**

18. Alvin Exporters buys engine gaskets from a U.S. manufacturer for $13.32 a dozen. Alvin marks up the gaskets 124% based on cost. The overhead is estimated to be 30% of the selling price.
   a. What is the selling price per gasket? **$2.49**
   b. What is the markup as a percent of the selling price? **55.4%**
   c. What is the net profit? **$0.63**
   d. What is the net profit as a percent of the selling price? **25.3%**

PRACTICE AND APPLY

The following problems can be assigned for classwork and the answers checked in class to help students master the objective of the lesson.

- Guided Practice: 1–9
- Independent Practice: 10–14 even

WRAP-UP

Ask students to explain the two steps in determining the selling price when markup is based on cost. Ask students what price they would sell an item for if the cost of making the item was $10 and the markup rate was 150% of cost. ($25)

Assignment Guide:

- Basic: 10–16, 19–30
- Average: 11–15 odd, 16–18, 19–29 odd

MAINTAINING YOUR SKILLS Look up the skills in parentheses if you need help or more practice.

Add. **(Skill 5)**

19. $85 + $44 **$129**
20. $144.47 + $185.52 **$329.99**
21. $1474.87 + $444.62 **$1919.49**

Find the percentage. Round answers to the nearest cent. **(Skill 30)**

22. $60.00 × 30% **$18.00**
23. $240.00 × 75% **$180.00**
24. $179.49 × 100% **$179.49**
25. $18.60 × 120% **$22.32**
26. $2417.60 × 150% **$3626.40**
27. $4.32 × 187% **$8.08**
28. $48.76 × 237.5%
29. $112.47 × 112.5%
30. $0.72 × 86.4%

28. **$115.81**
29. **$126.53**
30. **$0.62**

	Cost	Markup	Selling Price	Markup percent based on Selling Price	Cost
1)	$43.80	$19.10	($62.90)	(30.4%)	(43.6%)
2)	38.29	($41.50)	$79.79	(52.0%)	(108.4%)
3)	164.52	($109.68)	($274.20)	40.0%	(66.7%)
4)	7.59	($9.49)	($17.08)	(55.6%)	125.0%
5)	(629.99)	($370.00)	$999.99	37.0%	(58.7%)
6)	($1.11)	($2.38)	3.49	(68.2%)	214.4%

LESSON PLAN
15-8 Markdown

FOCUS

Ask students to tell about the last time they bought something that was advertised as "on sale." In this lesson, students will learn to compute the markdown in dollars and as a percent of the regular selling price.

TEACH

Have students discuss some of the reasons why businesses have sales. (Reasons may include clearing shelves of seasonal merchandise; getting rid of overstocked items; or moving outdated items to make room for newer models.)

Explain the terms *markdown* and *discount*. Do the same for *markdown rate*. Point out that markdown for a business is similar to that for a consumer (Lesson 5-6).

Warm-Up Exercises

Round percents to nearest tenth.

1. \$119.29 − \$35.79 \$83.50
2. \$83.50 ÷ \$119.29 70.0%
3. \$46.49 − \$27.89 \$18.60
4. \$18.60 ÷ \$46.49 40.0%
5. (\$179 − \$89) ÷ \$179 50.3%
6. \$49.79 − (20% of \$49.79) \$39.83

15-8 Markdown

OBJECTIVE

Compute the markdown in dollars and as a percent of the regular selling price.

Your business may sell some merchandise at sale prices to attract customers or to make room for new merchandise. The **markdown**, or **discount**, is the difference between the regular selling price of an item and its sale price. The **markdown rate** is the markdown expressed as a percent of the regular selling price of the item.

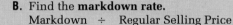

$$\text{Markdown} = \text{Regular Selling Price} - \text{Sale Price}$$

$$\text{Markdown Rate} = \frac{\text{Markdown}}{\text{Regular Selling Price}}$$

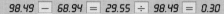

EXAMPLE *Skills* 31, 6 *Application* A *Term* Markdown rate

Ski's Sport Shop sells cross-country skis at a regular selling price of \$98.49. For one week only, Ski's has marked down the price to \$68.94. What is the markdown rate?

SOLUTION

A. Find the **markdown.**
Regular Selling Price − Sale Price
\$98.49 − \$68.94 = \$29.55 markdown

B. Find the **markdown rate.**
Markdown ÷ Regular Selling Price
\$29.55 ÷ \$98.49 = 0.3000 = 30% markdown rate

98.49 [−] 68.94 [=] 29.55 [÷] 98.49 [=] 0.30

✓ SELF-CHECK Complete the problems, then check your answers in the back of the book.

Find the markdown and markdown rate.

1. Regular selling price is \$80. Sale price is \$60. **\$20; 25%**
2. Regular selling price is \$174.79. Sale price is \$104.87. **\$69.92; 40%**

PROBLEMS

Round answers to the nearest tenth of a percent.

	(Regular Price	−	Sale Price	=	Markdown)	÷	Regular Price	=	Markdown Rate
3.	(\$ 25.00	−	\$ 20.00	=	\$5.00)	÷	\$ 25.00	=	20.0%
4.	(\$ 99.99	−	\$ 64.99	=	\$35.00)	÷	\$ 99.99	=	35.0%
5.	(\$ 247.87	−	\$ 149.99	=	\$97.88)	÷	\$ 247.87	=	39.5%
6.	(\$8674.50	−	\$4337.25	=	\$4337.25)	÷	\$8674.50	=	50.0%

COOPERATIVE LEARNING

Have students work through the following problem in small groups: Judy McGarry bought a stereo that originally cost \$350. The store offered it on sale for 10% off. Judy also used a gift certificate, worth \$25 off any purchase. She was given 5% off the original price for paying in cash. What was the final markdown rate that Judy got on the stereo? (22.1%)

Sale price is $1.29.
What is the markdown rate? **71.3%**

Sale price is $499.99. **11.5%**
What is the markdown rate?

9. Regular price is $750.
Sale price is $500. **33.3%**
What is the markdown rate?

10. Regular price is $79.29.
Sale price is $71.29. **10.1%**
What is the markdown rate?

11. A dozen eggs can be purchased for 59¢ with a coupon, 79¢ without. What is the markdown rate if the coupon is used? **25.3%**

12. Country Furniture has marked down this group of furniture.
a. What is the markdown rate on each item?
b. Based on markdown rate, which is the best buy? **Table: 54.8%**
c. Based on dollar markdown, which is the best buy? **Sofa: $490**

	Reg.	Sale
Sofa	$979 **50.1%**	$489
Love seat	$749 **52.1%**	$359
Chair	$499 **46.1%**	$269
Tables, ea.	$219 **54.8%**	$ 99

13. Free-form patio blocks are marked down.
a. Find the markdown rate for each.
b. Find the best buy, based on dollar markdown. **24 × 24: $2.30**
c. Find the best buy, based on percent markdown. **Outside: 36.0%**

FREE-FORM PATIO BLOCKS

	Reg.	Sale
Outside radius	$5.45 **36.0%**	$3.49
Border block	$4.45 **29.2%**	$3.15
Inside radius	$4.95 **31.9%**	$3.37
24" x 24" block	$6.45 **35.7%**	$4.15
16" x 16" block	$5.25 **34.5%**	$3.44

F.Y.I.
A markdown of 25% means you pay 75%.

Critical Thinking . . .

14. The Lion Store is participating in a national promotion of Port-a-Vac vacuum cleaners. The lowest-priced vacuum cleaner is marked down $5.25 to a sale price of $64.74. The middle-priced vacuum cleaner is marked down from $129.99 to $115.99. The top-priced vacuum cleaner is marked down $30.00 from the regular price of $149.99. What is the markdown rate for each? **Low 7.5%; Middle 10.8%; Top 20.0%**

MAINTAINING YOUR SKILLS Look up the skills in parentheses if you need help or more practice.

Subtract. **(Skill 6)**

15. $74.99 − $14.99 **$60.00**
16. $7.25 − $2.15 **$5.10**
17. $42.49 − $8.99 **$33.50**

Find the rate. (Round to nearest tenth percent.) **(Skill 31)**

18. $15.00 ÷ $50.00 **30.0%**
19. $4.99 ÷ $24.99 **20.0%**
20. $5.25 ÷ $25.99 **20.2%**
21. $71.79 ÷ $179.79 **39.9%**
22. $15.52 ÷ $44.49 **34.9%**
23. $9.00 ÷ $19.99 **45.0%**
24. $124.50 ÷ $829.29 **15.0%**
25. $269.50 ÷ $489.49 **55.1%**
26. $5.78 ÷ $17.35 **33.3%**
27. $16.50 ÷ $329.39 **5.0%**
28. $186.88 ÷ $747.50 **25.0%**
29. $8818.33 ÷ $26,455.00 **33.3%**

PRACTICE AND APPLY

The following problems can be assigned for classwork and the answers checked in class to help students master the objective of the lesson.

- Guided Practice: 1–7
- Independent Practice: 8–12 even

WRAP-UP

Have students explain the terms *markdown, discount,* and *markdown rate.* Have students explain how to find the markdown rate on an item that regularly sells for $150, but is on sale for $100.
($150 − $100 = $50
$50 ÷ $150 = $33\frac{1}{3}$%)

Assignment Guide

- Basic: 8–13, 15–29 odd
- Average: 9–13 odd, 14, 16–28 even

ALTERNATIVE STRATEGIES: Enrichment

Have your students find and bring to class newspaper ads that present markdowns in the various forms such as those given in Problem 14. Have students work in small groups to create their own problems from the information in the ads. For example, if they are given the regular price and the sale price, they would want to find the markdown and the markdown rate.

LESSON PLAN
Reviewing the Basics

The exercises on this page review skills, applications, and terms used in the unit. You can use the exercises to assess informally students' proficiency with this material.

The page can be used for guided practice and independent practice. You can work through a selection of the exercises together with students, and thus see immediately if they know how to do them, and you can then assign some of the exercises for independent practice. Be sure to go over the answers to all assigned exercises.

Reviewing the Basics

Skills

Round answers to the nearest tenth of a percent.

(Skill 2)

1. 10.15% **10.2%** **2.** 24.99% **25.0%** **3.** 147.864% **147.9%** **4.** 7.4996% **7.5%**

Solve.

(Skill 5)

5. $117.89 + $47.93 **$165.82** **6.** $523.47 + $67.84 **$591.31**

7. $916.47 + $56.38 **$972.85** **8.** $2.16 + $4.19 + $219.29 **$225.64**

(Skill 6)

9. $78.47 − $24.73 **$53.74** **10.** $217.25 − $31.45 **$185.80** **11.** $1417.45 − $221.84 **$1195.61**

Round answers to the nearest cent.

(Skill 11)

12. $24.50 ÷ 60% **$40.83** **13.** $419.71 ÷ 55% **$763.11** **14.** $947.84 ÷ 67.5% **$1404.21**

Write as a decimal.

(Skill 28)

15. 50% **0.5** **16.** 65% **0.65** **17.** $12\frac{1}{2}$% **0.125** **18.** 47.4% **0.474** **19.** $5\frac{1}{4}$% **0.0525**

Round answers to the nearest cent.

(Skill 30)

20. 64% of $245.00 **$156.80** **21.** 120% of $667.50 **$801.00** **22.** 15.6% of $14.78 **$2.31**

Round answers to the nearest tenth of a percent.

(Skill 31)

23. $7.96 is what percent of $39.79? **20.0%** **24.** $211.89 is what percent of $847.48 **25.0%**

Applications

Find the complement.

(Application L)

25. 40% **60%** **26.** 65% **35%** **27.** 2% **98%** **28.** 23.6% **76.4%**

Terms

Match each term with its definition on the right.

f **29.** Markup
a **30.** Markdown
c **31.** Net profit
b **32.** Net-profit rate
e **33.** Markdown rate

a. the difference between the regular selling price and the sale price of a product
b. the net profit expressed as a percent of the selling price of a product
c. the difference between the markup and the overhead when the markup is greater than the overhead
d. the amount that your business pays for a product
e. the discount expressed as a percent of the regular selling price of a product
f. the difference between the selling price and the cost of a product

Refer to the reference files in the back of the book if you need help.

Unit Test

Lesson 15-1

1. The Deckmasters Company buys an outdoor light fixture for $27.87. The company sells the fixture for $67.89. What is the markup? **$40.02**

Lesson 15-2

2. The Fruit Barn buys oranges for $0.33 a dozen. It sells them for $1.29 a dozen. What is the markup rate based on the selling price? **74.4%**

Lesson 15-3

3. Pipe Outfitters, Inc., an oil equipment wholesaler, purchases piping for $217.85 a length. It sells the lengths for $447.50. Pipe Outfitters estimates the overhead expenses for each length to be 30% of the selling price. What is the estimated net profit on each length? **$95.40**

Lesson 15-4

4. Secure-It, Inc., sells a dead bolt lock for $79.49. It costs Secure-It $41.67 to purchase the lock. The overhead is estimated to be 20% of the selling price of the lock. What is the estimated net-profit rate based on the selling price? **27.6%**

Lesson 15-5

5. Office Supplies, Inc., operates on a markup of 60% of the selling price of its merchandise. During December, the shop bought the supply of notebooks listed on the invoice. What selling price will Office Supply charge per notebook? **$3.70**

SHIP TO: OFFICE SUPPLIES, INC.	
ITEM DESCRIPTION	COST
007-31 Notebooks	$1.48 ea.

Lesson 15-6

6. Tia's, a shoe boutique, sells Italian-made shoes for $164.49. Tia's purchases the shoes for $97.88. To the nearest tenth of a percent, what is the markup rate based on cost? **68.1%**

Lesson 15-7

7. During the winter, Sun Grown buys oranges from overseas growers. Sun Grown determines the retail selling price of the oranges by marking up the cost by 125%. The fruit company bought its first shipment of winter oranges for $0.48 per dozen. What is the retail selling price per dozen of this shipment? **$1.08**

Lesson 15-8

8. Northside Auto Supply, an automobile supply shop, buys its products from Federal Auto Parts, Inc. Because of a downturn in sales, Federal has decided to reduce the price of its rebuilt generator. The regular selling price is $47.38. The generator's new selling price is $37.90. To the nearest tenth of a percent, what is the markdown rate? **20.0%**

LESSON PLAN
Unit Test

Students should do the Unit Test on their own. Each problem on the test is keyed to a lesson in the unit. Students having difficulty with any particular problem should review the Example in the appropriate lesson and be assigned some of the Independent Practice problems for additional practice.

LESSON PLAN
A Spreadsheet Application

USING TECHNOLOGY

After students have completed this application, take the time to go over the answers. In particular, questions 15, 16, 21, and 27 should be discussed. As you review the questions and answers, ask students to define key words such as *markup, markdown, selling price,* and *sale price.* Discuss also the meanings of basic concepts presented in the unit, such as percent markup based on cost or selling price.

A SPREADSHEET APPLICATION

Sales

To complete this spreadsheet application, you will need the diskette *Spreadsheet Applications for Business Mathematics,* which accompanies this textbook.

Select option 15, Sales, from the menu. Input the information in the following problems to find the cost (C), markup (MU), selling price (SP), percent markup based on cost (%M/C), percent markup based on selling price (%M/SP), markdown (MD), percent markdown (%MD), or sale price (SALE). Then answer the questions that follow.

	Cost	Mark-Up	Selling Price	%M/C	%M/S	%MD	MD	Sale
1.	$ 45.00	$?	$?	?	70%	20%	$?	$?
2.	124.50	?	?	?	50%	30%	?	?
3.	35.75	?	?	210%	?	25%	?	?
4.	124.50	?	?	100%	?	50%	?	?
5.	?	?	79.49	300%	?	25%	?	?
6.	?	?	145.78	85%	?	?	72.89	?
7.	?	?	12.96	?	25%	?	1.30	?
8.	?	?	4985.00	?	12%	?	997.00	?
9.	?	35.80	?	200%	?	?	?	42.96
10.	?	437.76	?	?	40%	?	?	711.36

11. What is the selling price in #1?

12. What is the sale price in #1?

13. What is the selling price in #2?

14. What is the selling price in #4?

A SPREADSHEET APPLICATION

(CONTINUED)

15. The cost is the same in #2 and #4, but the markup in #2 is 50% of the selling price and in #4 it is 100% of the cost; yet the selling price is the same in #2 and #4. Why?

16. How can the markup possibly be more than 100% in #3?

17. What is the markdown in #3?

18. What is the sale price in #3?

19. What is the cost in #5?

20. What is the percent markup based on selling price in #5?

21. In #5, the percent markup based on selling price and the percent markdown are the same (25%). The sale price equals the cost ($19.87). Will this always happen? Why or why not?

22. What is the markup in #6?

23. What is the percent markdown in #6?

24. What is the sale price in #6?

25. What is the cost in #7?

26. What is the sale price in #7?

27. Why are there percents over 100% for the percent markup based on cost, but none over 100% for the percent markup based on selling price?

28. What is the percent markup based on cost in #8?

29. What is the selling price in #9?

30. What is the cost in #10?

31. What is the percent markdown in #10?

LESSON PLAN
Career Wise

APPLICATIONS ON THE JOB

Students who have part-time jobs probably report to a manager. You might ask for volunteers who work to explain what their manager does. Ask students if they know of any franchises in the local area that are listed in the table. Can they name the franchises? Most students probably know the fast-food chains, but pursue other types of franchises with them also. Discuss how Jamal Mizrahi makes money for his fast-food franchise.

CAREER WISE

Store Manager

Jamal Mizrahi is the store manager at a fast-food franchise. As owner and manager, he purchases food supplies from the parent company. According to company policy and local conditions, he marks up the foods he sells. Beverages, sandwiches, side orders, and desserts each get different markups.

At the store, Mr. Mizrahi has given many local high school teenagers their first job. They learned how to "ring up a sale," as well as about paychecks and deductions.

The table and pie chart below show how the 50 largest franchises in the United States are divided. For example, there are 3 hotel-motel chains in the top 50 franchises. This is $\frac{3}{50} = 6\%$ of the top 50 franchises.

50 LARGEST FRANCHISES	
Fast foods	15
Auto rental, maintenance, supplies	6
Cleaning services	5
Printing/mailing	5
Personal fitness	4
Real estate	4
Hotels-motels	3
Convenience stores	2
Haircutters	1
Drug stores	1
Electronics	1
Hardware	1
Home decorating	1
Tax preparation	1

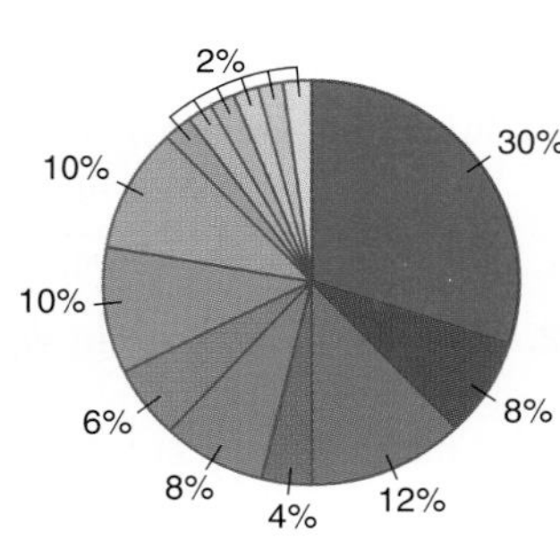

Check Your Understanding

1. What percent of the pie is represented by the fast-food industry? **30%**
2. The 10% slices correspond to what two industries?
   Cleaning; printing/mailing
3. The 2% slices of the pie together equal the slice of which one industry?
   Auto rental, maintenance, supplies
4. True or false: From the data, you can tell how many tax preparation franchises there are. Discuss your answer. **False.**
   You cannot tell whether there are 1 or 1000 tax preparation franchises. Nor can you conclude that there are as many tax preparation franchises as, say, hardware franchises.

U N I T

16

Marketing

The marketing department of your business estimates the number, the price, and the amount of sales you can expect to make on your product. This *market share* can be based on *opinion surveys* of people who buy your product and *sales projections,* or estimates of future sales. The marketing department also handles the advertising costs for your product.

OUTLINE

16-1 Opinion Surveys
16-2 Sales Potential
16-3 Market Share
16-4 Sales Projections
16-5 Sales Projections—Factor Method
16-6 Newspaper Advertising Costs
16-7 Television Advertising Costs
16-8 Pricing

- Reviewing the Basics
- Unit Test
- A Spreadsheet Application: Sales Analysis
- Cumulative Review (Skills/Applications) Units 12–16
- Cumulative Review Test Units 12–16

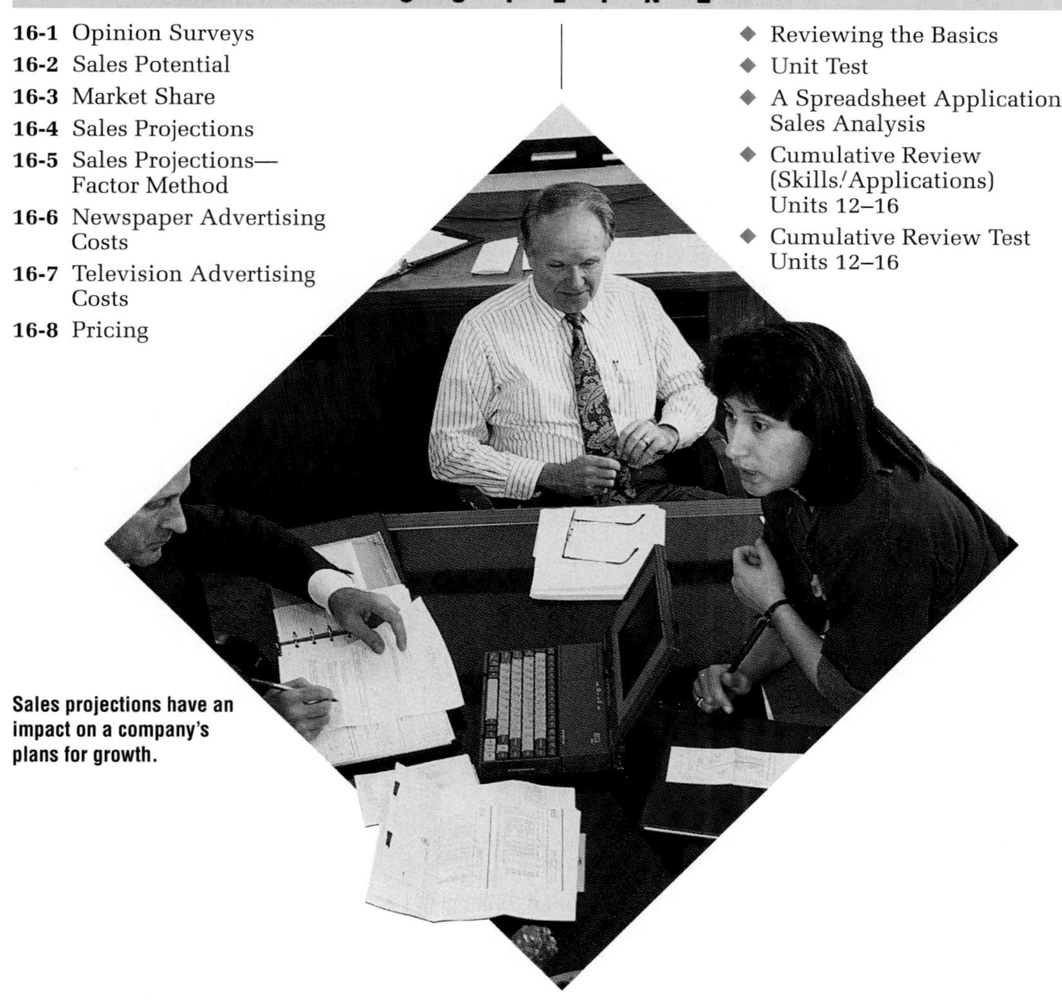

Sales projections have an impact on a company's plans for growth.

UNIT 16
MARKETING

INTRODUCING THE UNIT

Bring in ads for well-known products from magazines and newspapers. Open a discussion of why and how marketing strategies are so important to a company's success. Have students name some of their favorite *or* least favorite ad campaigns on television. Make sure students are aware of how much research has gone into a product before the public sees the first announcement of its being on the market.

425

LESSON PLAN 16-1 Opinion Surveys

FOCUS

Ask students if any of them have ever been a part of a survey that asks questions about a product. If any have, ask students what the product was, and any questions that they can recall being asked.

TEACH

Part of the market research involving a new product deals with opinion surveys. The panel of volunteers may range from several hundred to several thousand. Point out that an opinion research firm is hired so that the tests and surveys can be done objectively. The research firm selects individuals who are representative of the particular population (such as cat owners or couples with small children) the firm wishes to study.

As alternative questions to the Example, ask students to find the percent of responses that were "fair" from people age 18–40 (0.426%), and what percent of responses were from people under 18 (9.23%)

Warm-Up Exercises

1. 40 ÷ 50 0.8
2. 75 ÷ 400 0.1875
3. 1240 ÷ 2480 0.5
4. 165 ÷ 274 0.6022

Write as a percent.

5. 0.8 80%
6. 0.1875 18.75%
7. 0.5 50%
8. 0.6022 60.22%

16-1 Opinion Surveys

OBJECTIVE

Compute the rate of a particular response in an opinion survey.

When you develop a new product, you will want to know how well it is likely to sell. You may conduct a **product test** by asking a group of people to try the product. You may hire an **opinion research firm** that specializes in product testing to conduct the test. Panels of volunteers try the product and respond to questions in an **opinion survey.** The opinion research firm tabulates the answers to the questions and submits tables of results to you.

$$\text{Percent of Particular Response} = \frac{\text{Number of Times Particular Response Occurs}}{\text{Total Number of Responses}}$$

EXAMPLE *Skill* 31 *Application* C *Term* Opinion survey

Countryside Cereal Company conducted an opinion survey of 2600 people for its new Good Morning Cereal. This table shows the responses of the 2600 people in the survey.

		Age Group		
Opinion	**Under 18**	**18 to 40**	**40 or Over**	**Total**
Excellent	200	430	1450	2080
Good	25	108	215	348
Fair	10	12	99	121
Dislike	5	10	36	51
Total	240	560	1800	2600

What percent of the total responses were for "good"?

SOLUTION

Find the **percent of particular response.**

Number of Times Particular Response Occurs ÷ Total Number of Responses

348 ÷ 2600 = 0.133 = 13% percent of "good" responses

348 ÷ 2600 = 0.1338

✓ SELF-CHECK Complete the problems, then check your answers in the back of the book.

1. 1500 out of 2000 people surveyed own computers. What percent of the total surveyed own computers? **75%**
2. 460 out of 500 people surveyed like Mintie toothpaste. What percent of the total surveyed like it? **92%**

ALTERNATIVE ASSESSMENT

Have students write an opinion survey about some "business" at the school. Suggested topics may include cafeteria food, parking facilities, a piece of gym equipment, or a particular textbook. Have students choose a panel of volunteers and administer the survey. Students should then write a *constructive* paper on how the product or service can be improved.

PROBLEMS

	Product in Survey	Number of "Good" Responses	÷	Total Number of Responses	=	Percent of "Good" Responses
3.	Ice cream	320	÷	400	=	? **80%**
4.	Soup	75	÷	200	=	? **37.5%**
5.	Cleanser	600	÷	1000	=	? **60%**
6.	Movie	350	÷	1250	=	? **28%**
7.	Calculator	2047	÷	2300	=	? **89%**

8. Survey for orange juice.
375 responses.
180 "fair" responses.
What is the percent of "fair" responses? **48%**

9. Survey for silicone caulking.
525 responses.
50 negative responses.
What is the percent of negative responses? **9.5%**

10. Fisher Motors conducted a survey on its dealer service. The people surveyed were asked to choose one answer for this question: "If you DO NOT usually go to the dealer from whom you bought your car, why?"

The choices and number of responses received for each were:

37 Moved away from vicinity **12.3%**
14 Disliked quality of service **4.7%**
95 Service charges too high **31.7%**
110 Crowded service department **36.7%**
14 Location not convenient **4.7%**
30 Some other reason **10%**

What is the percent of each response?

11. Before beginning production of its new vacuum cleaner, the Devaney Company conducted an opinion survey of 1740 people. When asked if they would purchase the product, the people gave these responses:

	Age Group				
Response	**Under 25**	**25–34**	**35–49**	**50 or over**	**Total**
Definitely	53	151	126	75	**405** ?
Probably	75	184	203	140	**602** ?
Possibly	87	135	130	99	**451** ?
No	91	57	54	80	**282** ?
Total	306	527	513	394	1740

a. Find the total for each response. **Answers are listed in chart.**
b. Based on 1740 responses, what is the percent of each response?
c. What is the percent of "definitely" responses in each age group?
d. What age group is most likely to purchase the product?

b. 23.3%, 34.6%, 25.9%, 16.2%
c. 13.1%, 37.3%, 31.1%, 18.5%
d. 25-34

MAINTAINING YOUR SKILLS Look up the skill in parentheses if you need help or more practice.

Find the percentage. Round to the nearest tenth. **(Skill 30)**

12. 8% of 70 **5.6**
13. 70% of 80 **56**
14. 30% of 500 **150**
15. 25% of 1200 **300**

PRACTICE AND APPLY

The following problems can be assigned for classwork and the answers checked in class to help students master the objective of the lesson.

- Guided Practice: 1–5
- Independent Practice: 6–9

WRAP-UP

Ask students what is meant by a product test. Ask them to explain why an opinion research firm may be hired, and what types of panels of volunteers may be used.

Assignment Guide

- Basic: 6–10, 12–18
- Average: 10, 11, 12–18

ALTERNATIVE STRATEGIES: Enrichment

Many stores and restaurants have opinion survey cards that can be filled out. Ask students to bring in several from local businesses of which they are familiar. Have students complete the surveys, then form small groups to discuss how each business could use the cards to improve service to customers or the quality of an item.

LESSON PLAN
16-2 Sales Potential

FOCUS
The focus of this lesson is on computing the annual sales potential. Explain to students that before spending money on developing and introducing a new product, a manufacturer wants to know whether the new item will make a good profit or not.

TEACH
One of the first ways a manufacturer can determine the sales potential of a new product is to make a few items and test them on volunteers. This is a part of the Opinion Surveys of Lesson 16–1. The percent of the volunteers who would purchase the item can then be determined. From this point, the manufacturer must determine how often the item will be purchased and the total size of the market for the item. Government reports and statistical reports are often used to try to estimate the total market.

Warm-Up Exercises
1. 600 ÷ 2000 0.3
2. 450 ÷ 800 0.5625
3. 0.3 = ? % 30
4. 0.5625 = ? % 56.25
5. 40% of 6000 2400
6. 32% of 750 240
7. 15% of 100,000 × 3 45,000
8. 18% of 1,500,000 × 2 540,000

16-2 Sales Potential

OBJECTIVE
Compute the annual sales potential of a new product.

Before your business mass produces a new product, you may try to determine the product's **sales potential.** The sales potential is an estimate of the sales volume of a product during a specified period of time. You may manufacture a small number of the product for a selected group of people to try. This group is called a **sample.** The sales potential of your new product is based on: (1) the percent of the people in the sample who would purchase your product, (2) an estimate of the size of the market, and (3) the average number of times that an individual might purchase this type of product during a specified period of time. The market is the total number of people who might purchase the type of product that you make.

Annual Sales Potential = Estimated Market Size × Individual Rate of Purchase × Percent of Potential Purchasers

EXAMPLE *Skills* 31, 30 *Application* A *Term* Sales potential

The Ruston Corporation has developed a sun tan cream called Lite Stuff. Ruston chose a sample of teenagers to try the new product. Of the 3000 people in the sample, 1200 said they would purchase Lite Stuff. The Ruston Corporation estimates that there are 2,000,000 teenagers that buy sun tan creams. Ruston's surveys indicate that each teenager purchases about 3 tubes of sun tan cream per year. What is the sales potential for Lite Stuff for 1 year?

SOLUTION

A. Find the **percent of potential purchasers.**
1200 ÷ 3000 = 0.40 = 40% potential purchasers

B. Find the **annual sales potential.**
Estimated Market Size × Individual Rate of Purchase × Percent of Potential Purchasers
2,000,000 × 3 × 40% = 2,400,000 annual potential sales

1200 [÷] 3000 [=] 0.40 2000000 [×] 3 [×] 40 [%] 2400000

✓ SELF-CHECK Complete the problem, then check your answer in the back of the book.

1. Out of 2400 secretaries in a sample, 720 preferred a new computer printer ribbon. The estimated market size is 100,000. Each secretary uses about 30 ribbons a year. What is the annual sales potential? 900,000

COOPERATIVE LEARNING
Students should form groups of three and four. Have each group of students choose a potential new product such as a new toothpaste or a new baby toy. Have students research and report the potential market size. Sources of information can include *Vital Statistics*, the *Statistical Abstract of the United States*, or an almanac.

PROBLEMS

	2.	3.	4.	5.	6.
Number in Sample	400	500	1500	2000	2000
Number of Potential Purchasers	100	10	300	240	50
Percent of Potential Purchasers	25%	? **2%**	? **20%**	? **12%**	? **2.5%**
Estimated Market Size	800,000	1,800,000	5,700,000	25,000,000	32,600,000
Individual Rate of Purchase	2 cans	once	once	4 boxes	10 issues
Annual Sales Potential	? **400,000**	? **36,000**	? **1,140,000**	? **12,000,000**	? **8,150,000**

7. Child's car seat.
1200 people in sample.
360 would purchase seat.
Market size about 3,000,000.
Each family buys once.
What is the annual sales potential? **900,000**

8. Word-processing package.
8700 people in sample.
870 would purchase product.
Market size about 5,000,000.
Each person buys about 1 per year.
What is the annual sales potential? **500,000**

9. Glo dishwashing detergent is tested by 1000 people. Eighty-seven said they would buy it. The estimated market size is 4,500,000. The company estimates that each person would buy the detergent 12 times a year. What is the annual sales potential? **4,698,000**

10. The National Optometrics Company is marketing a new, softer, more pliable disposable contact lens. Out of a sample of 1200 users, 150 preferred the new lenses. There is an estimated total market of 1,500,000 contact lens users. The average contact lens wearer would purchase 12 lenses per year. What is the sales potential for the new lenses for 1 year? **2,250,000**

11. The Automotive Institute has developed a new spark plug for automobiles that will increase gas mileage. The marketing department had a sample of 2000 automobile owners use the new spark plug. Sixty owners responded favorably and stated that they would buy the spark plug. There are approximately 15,000,000 automobiles that could use this special spark plug. An average number of 4 spark plugs would be purchased per automobile per year. What is the sales potential for the new spark plug for 1 year? **1,800,000**

MAINTAINING YOUR SKILLS Look up the skill in parentheses if you need help or more practice.

Find the percentage. Round answers to the nearest hundredth. (Skill 30)

12. $4\frac{1}{4}$% of 96 **4.08**

13. $\frac{3}{8}$% of 120 **0.45**

14. 125% of \$4140 **\$5175**

15. 15.5% of 74.3 **11.52**

PRACTICE AND APPLY

The following problems can be assigned for classwork and the answers checked in class to help students master the objective of the lesson.

- Guided Practice: 1–4
- Independent Practice: 5–8

WRAP-UP

Write the formula for Annual Sales Potential on the chalkboard. Have students explain what each of the three factors means and how it might be estimated.

Assignment Guide

- Basic: 5–10, 12–15
- Average: 9–11, 12–15

ALTERNATIVE STRATEGIES: Reteaching

Point out that using a calculator makes working with the large numbers less time consuming. Using the memory key is also very helpful. Show how to solve these examples using a calculator.

A. 100 ÷ 400 = 0.25 times 1,500,000 = 375,000

B. 100 ÷ 300 = 0.333... times 1,500,000 = 500,000

C. 100 ÷ 300 = 0.333 times 1,500,000 = 495,500

Explain what effect rounding has on example C.

LESSON PLAN
16-3 Market Share

FOCUS
The focus of this lesson is on computing the market share of a new product. For a small business, this market may be a local area. For a big business, the market may be the entire United States, or expand to other countries in the world.

TEACH
Explain the term *market share* to students. Point out that a decreasing percent of the market share may be a signal that changes need to be made. These changes may be in the product itself, in management, or in sales and promotions.

Emphasize that in using the formula to compute market share, if the total product sales is expressed in units sold, then the total market sales must be expressed in units sold also. Likewise, if total product sales are expressed in dollar value of sales, then total market sales must be expressed in the same manner.

Warm-Up Exercises
Divide. Express as a percent and round to the nearest tenth of a percent.

1. 120,000 ÷ 2,000,000 6.0%
2. 900,000 ÷ 1,400,000 64.3%
3. 1,250,000 ÷ 15,000,000 8.3%
4. 1200 ÷ 6000 20.0%
5. n% of 90 = 27 30
6. n% of 80 = 160 200

16-3 Market Share

OBJECTIVE
Compute the market share of a new product.

To find out how well your product is selling in the marketplace, you may want to find your product's **market share.** Market share is that percent of the total market that purchases your product instead of a competitor's. You can calculate your percent of the total market sales by using either the number of units sold or the dollar value of sales.

$$\text{Market Share} = \frac{\text{Total Product Sales}}{\text{Total Market Sales}}$$

EXAMPLE *Skill* 31 *Application* A *Term* Market share

Amdex, an air conditioner manufacturer, sold 1,200,000 air conditioners during the year. During the same period, a total of 8,000,000 air conditioners were purchased in the entire U.S. market. What was Amdex's market share for the year?

SOLUTION Find the **market share.**

Total Product Sales ÷ Total Market Sales
1,200,000 ÷ 8,000,000 = 0.15 = 15% market share

✓ SELF-CHECK Complete the problems, then check your answers in the back of the book.

Find the market share.

1. Product sales total 2,000,000. Market sales total 20,000,000. 10%
2. Product sales total $4,000,000. Market sales total $180,000,000. 2.2%

PROBLEMS

	Company	Total Product Sales	÷	Total Market Sales	=	Market Share
3.	Diwan	$8,000,000	÷	$20,000,000	=	? 40%
4.	Morrill	$ 700,000	÷	$42,000,000	=	? 1.7%
5.	Hersh	$1,400,000	÷	$ 5,000,000	=	? 28%
6.	TBC	$9,000,000	÷	$45,000,000	=	? 20%

7. Typewriters.
   Product sales total 1,200,000 units.
   Market sales total 20,000,000 units.
   What is the market share? 6%
8. Lawn mowers.
   Product sales total 750,000 units.
   Market sales total 5,000,000 units.
   What is the market share? 15%

CRITICAL THINKING
Most of the problems in this lesson deal with a nationwide market sales total. In Problem 18, only the market sales total for Fresno County was considered. Ask students why Mountainside Nursery may not be interested in a larger market. (Most landscaping businesses are small and provide a local service only.)

9. Light bulbs.
Product sales total $8,000,000.
Market sales total $80,000,000.
What is the market share? **10%**

10. Carpets.
Product sales total $30,000,000.
Market sales total $900,000,000.
What is the market share? **3.3%**

11. Radial Tires, Inc., sells approximately 4,000,000 automobile tires per year. There are approximately 44,000,000 automobile tires sold per year in the entire market. What is Radial Tires' market share? **9.1%**

12. Tiger Paperbacks Inc. sells approximately 9,000,000 paperback books per year. Sales for the entire paperback book market total approximately 30,000,000 books per year. What is Tiger Paperbacks' market share of paperbacks? **30%**

13. The Mesquite Grill Company sells about $3,100,000 in outdoor grills annually. The total annual sales of outdoor grills in the entire market are about $8,500,000. What is Mesquite's market share for outdoor grills? **36.5%**

14. The Quality Linen Sheet Company has sales totaling approximately $1,400,000 in fitted sheets. Total sales of fitted sheets by all companies are approximately $5,350,000. What is Quality Linen's market share of fitted sheets? **26.2%**

15. Redbo Shoes had sales totaling approximately $975,000 in basketball shoes last year. Last year sales of all basketball shoes totaled approximately $12,450,000. What market share of basketball shoes did Redbo Shoes have? **7.8%**

16. Last year there were total sales of approximately $8,500,000 in dining room furniture. St. Helena Chair Company's sales of dining room furniture totaled approximately $900,000 last year. What was St. Helena Chair Company's market share of dining room furniture? **10.6%**

F.Y.I.
In 1991, AMD Corporation sold $200 million worth of its AM386 computer chips for a 14% share of the market (*Business Week*, December 9, 1991, p. 32).

17. Chateau Brothers, a company that manufactures games, sells an electronic video game called Match. Last year sales for Match totaled approximately $732,000. Total market sales for electronic video games were approximately $15,350,000. What market share did Chateau have? **4.8%**

18. Mountainside Nursery provides a landscaping service in Fresno County. Mountainside's landscaping business totaled approximately $789,400 for the year. The landscaping business for Fresno County totaled about $1,340,000 for the year. What market share did Mountainside have? **58.9%**

MAINTAINING YOUR SKILLS Look up the skills in parentheses if you need help or more practice.

Find the rate. Round answers to the nearest tenth of a percent. **(Skill 31)**

19. n% of 84 = 21 **25%**
20. n% of 50 = 35 **70%**
21. n% of 45 = 50 **111.1%**
22. n% of 36 = 90 **250%**

Divide. Round answers to the nearest tenth. **(Skill 10)**

23. 920 ÷ 40 **23**
24. 1145 ÷ 85 **13.5**
25. 868 ÷ 1224 **0.7**
26. 821 ÷ 15 **54.7**
27. 546 ÷ 42 **13**

PRACTICE AND APPLY

The following problems can be assigned for classwork and the answers checked in class to help students master the objective of the lesson.

- Guided Practice: 1–6
- Independent Practice: 7–11 odd

WRAP-UP

Ask students what is meant by *market share*. Read Problem 11 to the class. Ask a student to explain how to find the market share.

Assignment Guide

- Basic: 7–17, 19–26
- Average: 8–12 even, 13–18, 19–25 odd

ALTERNATIVE STRATEGIES: Enrichment

Have students look for articles in the *Wall Street Journal*, and various magazines, such as *Business Week*, *Standard & Poor Stock Analysis*, and *Forbes*, dealing with a firm's market share.

LESSON PLAN
16-4 Sales Projections

FOCUS

Many companies try to plan ahead to determine how many products to make, employees to hire, and so on. Ask students to suggest some reasons why this is necessary. (Projected sales indicate the future revenues for a company and give an indication of the need to hire more employees.) In this lesson, students learn to use a graph to compute projected sales.

TEACH

Many companies try to plan for several months or years ahead in terms of such things as sales, hiring needs, and raw materials. Emphasize that the starting point is based on past sales.

After a graph is constructed using the available data, it is used to predict future sales. Point out that the more consistent the shape of the graph, the more reliable predictions will be. However, no prediction can be 100% accurate due to unforeseen circumstances in the future.

16-4 Sales Projections

OBJECTIVE
Use a graph to compute projected sales.

A **sales projection** is an estimate of the dollar volume or unit sales that might occur during a future time period. A sales projection is usually based on past sales. You may use a sales projection to plan for production or purchasing. New products or changing economic conditions may result in sales figures that differ from the figures you projected.

You may use a graph to project a rough estimate of future sales:

1. Construct a graph of past sales.
2. Draw a straight line from the first year of data, approximately through the "middle" of the data, to the year for which the projection is being made.
3. Read the number or dollar value, the sales projection.

EXAMPLE *Application* N *Term* Sales projection

The marketing department of Stanley Stores, Inc., wishes to project sales for the year 2000. The sales history is:

Year	1986	1987	1988	1989	1990	1991	1992
Sales (in millions)	$2.5	$2.0	$3.0	$4.3	$3.3	$5.0	$4.5

Using a graph, what sales projection might the marketing department make for 2000?

SOLUTION

A. Graph the sales from 1986 through 1992.

B. Starting from sales in 1986, draw a straight line through the "middle" of the data to 2000.

C. Read the sales projection for 2000.

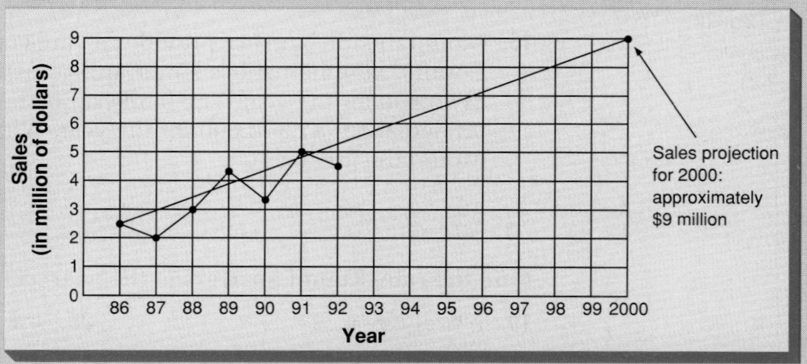

✓ SELF-CHECK Complete the problem, then check your answer in the back of the book.

1. Refer to the graph above. Read the approximate sales projections for 1993, 1995, and 1997. **$5.5 million; $6.5 million; $7.5 million**

CULTURAL ANGLES

Many major industries house their headquarters in Chicago, Illinois. Jean DuSable (1745? – 1818) was a businessman from Haiti who built a trading post in 1775 between the Chicago and Des Plaines rivers. The city of Chicago developed around his post. You might wish to have students research the kinds of businesses for which Chicago (or another city) is most noted.

PROBLEMS

Use the graph to estimate the sales projections for problems 2–7.

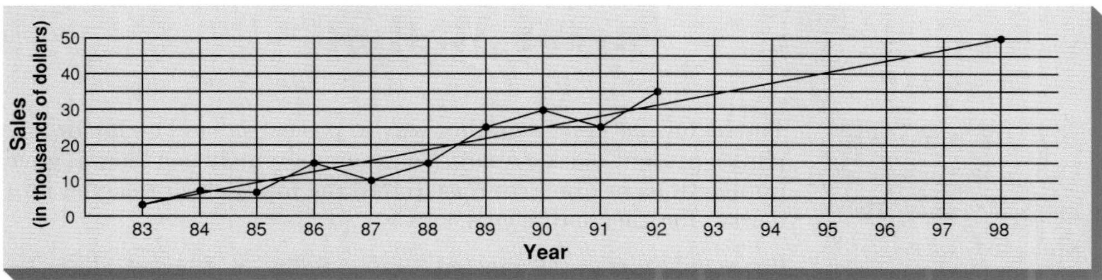

2. $35,000
3. $38,000
4. $41,000
5. $44,000
6. $47,000
7. $50,000

	2.	3.	4.	5.	6.	7.
Year	1993	1994	1995	1996	1997	1998
Sales Projection	?	?	?	?	?	?

Construct a graph for each problem and draw a straight line through the "middle" of the data to project sales.

8. Sales records for Target Sporting Goods show this data:

Year	1983	1984	1985	1986	1987	1988	1989	1990	1991	1992
Sales (in millions)	$4.4	$4.9	$5.0	$4.5	$5.1	$5.4	$5.0	$5.5	$5.7	$5.0

What sales might be projected for 1999? **$8.5**

9. The sales history of Bi-Low Markets shows:

Year	1965	1970	1975	1980	1985	1990
Sales (in millions)	$12	$14	$11	$12	$15	$14

Project sales for 1995, 2000, and 2005. **$14.4; $14.8; $15.2**

10. Digico manufactures computers. Production records for the first half of the year show this information for the Model XT360 computer:

Month	Jan.	Feb.	Mar.	Apr.	May	June
Production (in thousands of units)	14	13	12	13	14	13

Project production quantities for July, August, and September. **13; 13; 13**

MAINTAINING YOUR SKILLS Look up the skills in parentheses if you need help or more practice.

Multiply. Round answers to the nearest thousandth. **(Skill 8)**

11. $442.86 × 100 **$44,286.00**

12. $3240.03 × 1000 **$3,240,030.00**

13. 0.044 × 400 **17.6**

14. 0.00362 × 16.02 **0.058**

15. 0.725 × 143 **103.675**

Write each number in words. **(Skill 1)**

16. 407 **17.** 52 **18.** 0.6 **19.** 3104 **20.** 16.246

16. four hundred seven
17. fifty-two
18. six tenths
19. three thousand one hundred four
20. sixteen thousand two hundred forty-six

Warm-Up Exercises

1. Use the following information to construct a line graph.

Year:	Sales (in millions)
1989	$4
1990	$8
1991	$6
1992	$10
1993	$14

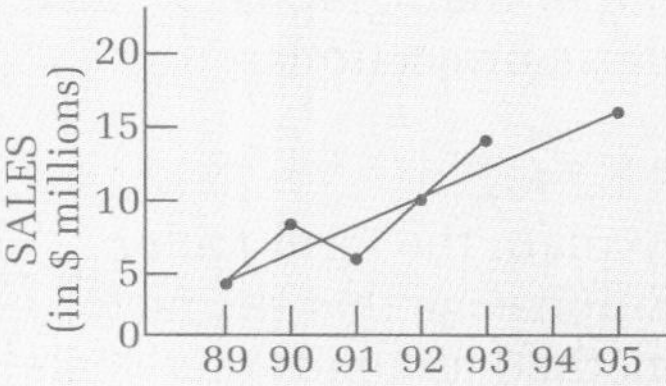

2. Project the sales to 1995. about $16 million

PRACTICE AND APPLY

The following problems can be assigned for classwork and the answers checked in class to help students master the objective of the lesson.

- Guided Practice: 1–7
- Independent Practice: 8

Note that answers for Problems 2–7 are approximations; thus, students' responses may be slightly different.

WRAP-UP

Ask students why a business might be concerned about future sales. Have them describe verbally how to use a graph of past sales to predict future sales.

Assignment Guide

- Basic: 8, 9, 11–20
- Average: 9, 10, 11–19 odd

ALTERNATIVE STRATEGIES: Reteaching

1. You may need to review with students how to read a line graph. Point out that it is used to show changes in the data over time and is used to show a trend.
2. For Problems 8, 9, and 10, you may want to provide students with graph paper. Emphasize the need for neatness and accurate placement of the points.

LESSON PLAN 16-5 Sales Projections—Factor Method

FOCUS

The focus of this lesson is on using the factor method to compute projected sales. This method uses percents, rather than graphs, to make these projections.

TEACH

Explain the term *factor*. Emphasize that the factor method assumes that years or months in the near future will be similar to what is happening in the market place at the current time. That is, if a company has a 20% market share now, and sales for next year for all items are $10,000,000, the company would expect to make 20% of the $10,000,000. Make sure that students understand that this is a projected figure, and not guaranteed sales for the company.

Warm-Up Exercises

1. 2% × $7,000,000 $140,000
2. 6% × $1,500,000,000 $90,000,000
3. 4.5% × $60,000,000 $2,700,000
4. 21% × $5,000,000 $1,050,000
5. $\frac{1}{2}$% × $30,000,000 $150,000
6. $\frac{3}{4}$% × $950,000 $7125

16-5 Sales Projections—Factor Method

OBJECTIVE
Use the factor method to compute projected sales.

The **factor method** is another way to project sales. The **factor** is your company's present market share. Your company may use federal government publications or other sources to find the total sales projected for the entire market for the coming year.

Projected Sales = Projected Market Sales × Market-Share Factor

EXAMPLE *Skill* 30 *Application* A *Term* Factor method

SafeAway Food Stores have a 4% share of the market for food sales in the Chicago Heights area. Food sales in the Chicago Heights area for next year are estimated to be $28,400,000. What is the projected sales figure for SafeAway Food Stores in the Chicago Heights area for next year?

SOLUTION

Find the **projected sales.**

Projected Market Sales × Market-Share Factor
$28,400,000 × 4% = $1,136,000 projected sales

28400000 [×] 4 [%] 1136000

✓ SELF-CHECK Complete the problems, then check your answers in the back of the book.

Find the projected sales.

1. Market-share factor: 7%. Projected market sales: $6,300,000. **$441,000**
2. Market-share factor: 14%. Projected market sales: $10,250,000. **$1,435,000**

PROBLEMS

	Company	Projected Market Sales	×	Market Share	=	Projected Sales
3.	ABS	$10,000,000	×	4%	=	**$400,000**
4.	Webb	$32,400,000	×	8%	=	**$2,592,000**
5.	Mairs	$54,000,000	×	7.5%	=	**$4,050,000**
6.	Beck	$70,000,000	×	3.2%	=	**$2,240,000**

7. Long-Life Lumber Distributors. 18% share of market. $1,800,000 estimated market total for next year. What is Long-Life's sales projection? **$324,000**
8. Quick Copy Machines. 15% share of market. $9,100,000 estimated market total for next year. What is Quick Copy's sales projection? **$1,365,000**

PROBLEM SOLVING

Have the students use Problem 14, **A Brief Case,** to answer the following.

1. Suppose Valley Dairy Outlets spend $200,000 to furnish the new store. How many years will it take to pay for the furnishings? (Need more information.)
2. Determine the average sale for Eagles three stores. ($146,000) Considering the two new stores, what are the projected average sales? ($146,000)

9. Bonaventura Bakery Inc. now has 13% of the market for Watson County. The total estimated sales of the county for next year are $3,000,000. What sales volume should Bonaventura Bakery project for next year? **$390,000**

10. Silver Oaks Bowling Lanes has traditionally had 5% of the bowling business in the East Delta area. The total estimated bowling business in this area for next year is $14,900,000. What bowling business can Silver Oaks project for next year? **$745,000**

11. Elliot's Auto Repair does 24% of the auto repair business in the town of Northport. The estimated auto repair business for next year is $542,600. What sales volume can Elliot's project for next year? **$130,224**

12. This year enrollment at Silverado Central University (SCU) accounted for 55% of all university students in the greater metropolitan area. Next year the total university enrollment in the greater metropolitan area is expected to be about 28,000 students. What enrollment figures can SCU project for next year? **15,400**

13. Based on past records, First National Savings Bank determined that it has a market share of 9.5% of all the savings accounts in Butler County. Estimated savings accounts in Butler County for next year are 470,000. Project the number of savings accounts that First National Savings Bank will have next year. **44,650**

14. Many retail businesses increase their market share by opening additional stores.

  a. With their 4 outlets, Valley Dairy Outlets now do 12% of the dairy business in Leland County. Estimated dairy products sales in Leland County for next year are $1,826,000. Valley is adding 1 more outlet next year and hopes to increase its market share to 18%. What is Valley's projected increase in sales for all 5 outlets? **$109,560**

  b. Eagle's 3 stores now do 15% of the bicycle sales business in North County. Estimated bicycle sales in the North County market area for next year are $2,920,000. Eagle is adding 2 stores and estimates the 2 stores will have total sales of $292,000 next year. What will be Eagle's projected market share next year? **25%**

MAINTAINING YOUR SKILLS Look up the skills in parentheses if you need help or more practice.

Find the percentage. Round answers to the nearest hundredth. **(Skill 30)**

15. 5.8% of 410 **23.78**	16. 9.81% of $32,500 **$3188.25**	17. 4.1% of $420,000 **$17,220**
18. 145% of 700 **1015**	19. $\frac{5}{8}$% of $800 **$5**	20. 0.8% of $60,000 **$480**
21. 2000% of $120 **$2400**	22. 1% of $85,000 **$850**	23. 15% of $120 **$18**

PRACTICE AND APPLY

The following problems can be assigned for classwork and the answers checked in class to help students master the objective of the lesson.

- Guided Practice: 1–7
- Independent Practice: 8–10

WRAP-UP

Have students give the new terms in this lesson, then explain each. Read Problem 9 to the class. Have a student explain how to find the projected sales volume.

Assignment Guide

- Basic: 8–13, 15–23
- Average: 11–14, 15–23 odd

ALTERNATIVE STRATEGIES: Enrichment

Have students use *Motor Vehicle Facts and Figures* or similar sources to find the present market share of the major automobile manufacturers in the United States. Using projections of the total sales of all automobiles for next year in the United States, have students find the potential sales of each major manufacturer.

LESSON PLAN
16-6 Newspaper Advertising Costs

FOCUS

In order to promote sales, many businesses advertise in newspapers. The focus of this lesson is on computing the cost of newspaper advertising.

TEACH

The cost of advertising depends on the following (1) whether the ad is in black and white or in color; (2) the daily circulation of the paper; (3) the number of column inches used; (4) whether the ad appears in a daily paper or the Sunday edition; and (5) whether or not the store has a contract with the newspaper. Discuss with students how each of the above affects the cost of the ad.

Mention that some newspapers may use another method of charging for advertisements. This method was based on the number of column inches and the rate per column inch.

Warm-Up Exercises

1. 900 × $0.75 $675
2. 400 × $1.05 $420
3. 800 × $0.705 $564
4. 3000 × $0.64 $1920

16-6 Newspaper Advertising Costs

OBJECTIVE
Compute the cost of advertising in a newspaper.

You may advertise your products or services in a newspaper to increase your sales. The cost of the advertisement depends on whether it is in color or black and white and on the amount of space it uses. The amount of space is figured in column inches. Generally, newspapers charge a certain rate for each column inch. Some newspapers offer a reduced rate if you contract weekly for a specified number of column inches.

Advertisement Cost = Number of Column Inches × Rate per Column Inch

EXAMPLE *Skills* 8, 2 *Application* A

Northshore Real Estate, Inc., contracted with *Hamilton News* for 126 inches of advertising each week. Northshore plans to advertise a new subdivision in the daily newspaper. The advertisement is the equivalent of 21 column inches. What is the cost of the advertisement each time it appears?

HAMILTON NEWS ADVERTISING RATES

	Per Column Inch Daily	Per Column Inch Sunday
Noncontract rates	$45.54	$55.28
Weekly Contract rates		
1 inch	35.57	43.52
2 inches	35.42	43.36
4 inches	35.28	43.36
8 inches	35.15	43.07
16 inches	34.90	42.80
31 inches	34.65	42.53
63 inches	34.40	42.26
94 inches	34.15	41.99
126 inches	33.90	41.72
252 inches	33.65	41.46

SOLUTION Find the **advertisement cost.**

Number of Column Inches × Rate per Column Inch
21 × $33.90 = $711.90 advertisement cost

✓ SELF-CHECK Complete the problems, then check your answers in the back of the book.

Use the table above to find the advertisement costs for Carpet Store. Carpet Store has contracted for 94 inches per week.

1. An advertisement in the Sunday paper is 12 column inches. **$503.88**
2. An advertisement in the daily paper is 10 column inches. **$341.50**

COMMUNICATION SKILLS

Have students work in cooperative groups to write an advertisement for selling a new product of their choice. One group should call the advertising department of the local newspaper for its ad rates. Groups should use the rates to calculate the cost of their ads and tell the other groups how they plan to promote their product and the estimated cost.

PROBLEMS

Use the table on page 436 to find the advertisement cost per run.

	Weekly Contract	Edition	Number of Column Inches	×	Rate per Column Inch	=	Advertisement Cost
3.	16 inches	Daily	15	×	$34.90	=	**$523.50**
4.	31 inches	Daily	9	×	**$34.65**	=	**$311.85**
5.	252 inches	Sunday	40	×	**$41.46**	=	**$1658.40**
6.	94 inches	Sunday	120	×	**$41.99**	=	**$5038.80**
7.	No Contract	Daily	5	×	**$45.54**	=	**$227.70**

8. MJ Sporting Goods.
Weekly contract for 63 inches.
26-inch ad on Sunday.
What is the cost of the advertisement? **$1098.76**

9. Maxson Auto Sales.
Weekly contract for 31 inches.
8-inch daily ad.
What is the cost of the advertisement? **$277.20**

10. SuperValue Supermarket has a weekly contract for 126 inches of advertising. In the Wednesday paper, SuperValue has an advertisement equivalent to 105 column inches. What is the cost of the advertisement? **$3559.50**

11. The Aspen Ski Lodge does not have a contract for advertising. Sunday's paper will carry an advertisement for the lodge of 12 column inches. How much does the advertisement cost? **$663.36**

12. Jo-Del Corporation placed this advertisement in the Monday *Hamilton News.* The advertisement is 2 columns wide and 2 inches deep.

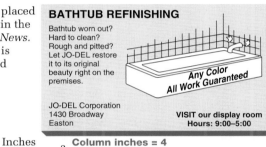

BATHTUB REFINISHING
Bathtub worn out? Hard to clean? Rough and pitted? Let JO-DEL restore it to its original beauty right on the premises.
Any Color
All Work Guaranteed
JO-DEL Corporation
1430 Broadway
Easton
VISIT our display room
Hours: 9:00–5:00

a. For how many column inches will Jo-Del be charged?
Columns Wide × Inches Deep = ? **Column inches = 4**

b. Jo-Del does not have a contract. What will the advertisement cost? **$182.16**

MAINTAINING YOUR SKILLS Look up the skill in parentheses if you need help or more practice.

Multiply. Round answers to the nearest thousandth. **(Skill 8)**

13. 4.9 × 1.8 **8.82**
14. 0.76 × 0.12 **0.091**
15. 0.003 × 10.6 **0.032**
16. 63.05 × 0.007 **0.441**
17. 0.982 × 0.15 **0.147**
18. 0.14 × 300 **42**
19. 66.00 × 0.25 **16.5**
20. 0.3498 × 25 **8.745**

PRACTICE AND APPLY

The following problems can be assigned for classwork and the answers checked in class to help students master the objective of the lesson.

- Guided Practice: 1–7
- Independent Practice: 8–10

WRAP-UP

Have students explain what is meant by a *column inch.* Have them find the advertising cost of 40 column inches at a rate of $0.50 per column inch. ($20)

Assignment Guide

- Basic: 8–11, 13–20
- Average: 11, 12, 14–20 even

ALTERNATIVE STRATEGIES: Reteaching

Relate this lesson to the real world by measuring some ads that appear in your local paper. Consider:

A. Car dealer ad: 3 columns wide × 10″ long = 30 column inches.

B. Sporting event: 2 columns wide × 4″ long = 8 column inches.

Use the rate schedule in the Example if you do not have local rates.

LESSON PLAN 16-7 Television Advertising Costs

FOCUS

Students probably are very familiar with the frequency and types of commercials that appear on television. In this lesson, students learn to compute the cost of television advertising.

TEACH

Advertising on local television is a way to reach thousands of consumers, while national television can reach millions. The cost of advertising also depends, to a certain extent, on whether the advertising is aimed at a local or a nation-wide audience. The most expensive time to advertise is during the evening hours of 5:00–11:00. Point out that no matter what time of the day, the 60-second advertisement costs twice the 30-second, which costs twice the 10-second one.

Warm-Up Exercises

1. $\frac{1}{2} \times 6000$ 3000
2. $\frac{1}{4} \times 2600$ 650
3. $\frac{3}{8} \times 2000$ 750
4. $\frac{3}{4} \times 9000$ 6750
5. $86.90 + 121.65 + 17.62$ = $226.17
6. $1240.50 + 982.80 + 3014.52$ = $5237.82
7. $12,940.70 + 1,290.35 + 967.58$ = $15,198.63

16-7 Television Advertising Costs

OBJECTIVE
Compute the cost of advertising on television.

You may advertise your products or services on television. The cost of a television advertisement depends on the time of day, program ratings, and the length of the advertisement. Television commercials are generally 10, 30, or 60 seconds long. The cost of a 10-second advertisement is usually one-half the cost of a 30-second advertisement. The cost of a 60-second advertisement is usually twice the cost of a 30-second advertisement.

$$\text{Cost of 10-Second Ad} = \frac{1}{2} \times \text{Cost of 30-Second Ad}$$

$$\text{Cost of 60-Second Ad} = 2 \times \text{Cost of 30-Second Ad}$$

EXAMPLE *Skill 20 Application A*

The marketing department of the Mcphee Company plans to advertise several new cosmetics on network television. The advertising campaign calls for these commercials:

Number	Length	Time
3	30-second	Daytime
2	10-second	Daytime
2	30-second	Prime time
4	60-second	Prime time

The rates are $8400 for a 30-second daytime commercial and $38,000 for a 30-second prime-time commercial. What is the total cost for this advertising campaign?

SOLUTION

A. Find the **cost of 30-second ads.**
Daytime: $8400 × 3 .... $ 25,200
Prime time: $38,000 × 2 .... $ 76,000

B. Find the **cost of 10-second ads.**
$\frac{1}{2}$ × Cost of 30-Second Ad
Daytime: ($\frac{1}{2}$ × $8400) × 2 .... $ 8400

C. Find the **cost of 60-second ads.**
2 × Cost of 30-Second Ad
Prime time: (2 × $38,000) × 4 .... $304,000

D. Find the **sum of the costs.** .... $ 413,600

8400 × 3 = 25200 M+ 38000 × 2 = 76000 M+ .5 × 8400 × 2 = 8400 M+ 2 × 38000 × 4 = 304000 M+ RM 413600

✔ SELF-CHECK

Complete the problems, then check your answers in the back of the book.

1. A 30-second ad costs $1200. Find the total cost for four 60-second ads. **$9600**
2. A 30-second ad costs $35,000. Find the total cost for four 30-second ads and five 10-second ads. **$227,500**

PROBLEMS

	Rate per 30-Second Ad	Number of 10-Second Ads	Number of 30-Second Ads	Number of 60-Second Ads	Total Cost
3.	$ 1,000	4	3	3	**$11,000**
4.	$ 500	3	4	0	**$2750**
5.	$30,000	5	5	2	**$345,000**
6.	$ 6,500	0	4	3	**$65,000**

F.Y.I.
Television network affiliated stations' average earnings amounted to $14.6 million in 1991.

7. Daytime local television. 60-second ad for dog food. 30-second ad costs $400. What is the cost of the 60-second ad? **$800**
8. Prime-time national television. 10-second ad for radio station. 30-second ad costs $21,400. What is the cost of the 10-second ad? **$10,700**
9. The manufacturers of Brighten mouthwash have planned an advertising campaign that includes these commercials:

Number	Length	Time
4	30-second	Daytime
3	10-second	Daytime
5	30-second	Prime time
6	60-second	Prime time

The rates are $2800 for a 30-second daytime commercial and $13,500 for a 30-second prime-time commercial. What is the cost of the advertising campaign? **$244,900**

F.Y.I.
A 60-second ad for a recent Super Bowl game cost approximately $1 million.

10. Pacafic New Car Sales is interested in sponsoring the sports coverage on the local television station. The rate per 30-second commercial is $475. Pacafic New Car Sales contracts for ten 30-second advertisements, five 10-second advertisements, and five 60-second advertisements. What is the total cost for these advertisements? **$10,687.50**
11. Alamo Amusement Park has planned a special television campaign. It will use a daytime show for twenty 10-second ads and five 30-second ads. It will use a prime-time show for ten 30-seconds ads and five 60-second ads. The rates are $250 per 30-second daytime ad and $1500 per 30-second prime-time ad. What is the cost of the television campaign? **$33,750**

MAINTAINING YOUR SKILLS

Look up the skill in parentheses if you need help or more practice.

Multiply. Express the answers in lowest terms. **(Skill 20)**

12. $\frac{1}{3} \times \frac{3}{5}$ **$\frac{1}{5}$**
13. $\frac{3}{12} \times \frac{4}{8}$ **$\frac{1}{8}$**
14. $6 \times 8\frac{1}{7}$ **$48\frac{6}{7}$**
15. $2\frac{3}{7} \times 1\frac{3}{8}$ **$3\frac{19}{56}$**
16. $4 \times 16\frac{1}{4}$ **65**

PRACTICE AND APPLY

The following problems can be assigned for classwork and the answers checked in class to help students master the objective of the lesson.

- Guided Practice: 1–6
- Independent Practice: 7–9

WRAP-UP

Ask students the three time lengths that television commercials are based on. Then have students explain the relationship in the cost of these three time lengths.

Assignment Guide

- Basic: 7–10, 12–16
- Average: 10, 11, 12–16

ALTERNATIVE STRATEGIES: Enrichment

Have students investigate and report on the cost of advertising in the following media.

A. radio
B. buses and trains
C direct mail
D. taxis
E. telephone directories
F. billboards
G. magazines
H. farm publications
I. flyers

16-8 Pricing

OBJECTIVE
Compute the selling price that will result in the highest possible net profit.

You must set the selling prices of your products high enough to cover expenses and to make a profit. You may determine the selling price of an item by estimating the net profit for each of several possible selling prices and choosing the selling price that will result in the highest profit. In manufacturing, the cost per item varies with the number manufactured and ultimately sold. Often, the higher the selling price of an item, the fewer the items that will be sold.

$$\text{Possible Net Profit} = \left(\text{Selling Price per Unit} - \text{Total Cost per unit}\right) \times \text{Estimated Unit Sales}$$

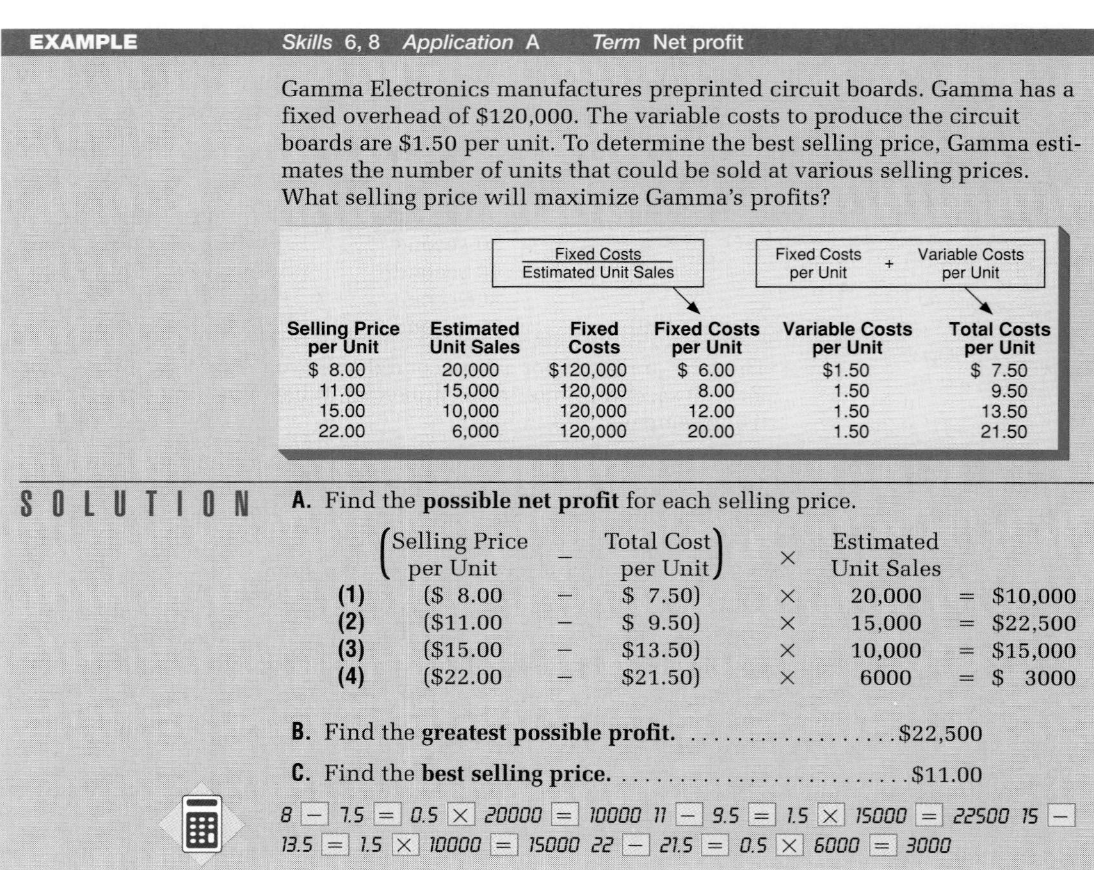

EXAMPLE *Skills* 6, 8 *Application* A *Term* Net profit

Gamma Electronics manufactures preprinted circuit boards. Gamma has a fixed overhead of $120,000. The variable costs to produce the circuit boards are $1.50 per unit. To determine the best selling price, Gamma estimates the number of units that could be sold at various selling prices. What selling price will maximize Gamma's profits?

Fixed Costs per Unit = Fixed Costs ÷ Estimated Unit Sales; Total Costs per Unit = Fixed Costs per Unit + Variable Costs per Unit

Selling Price per Unit	Estimated Unit Sales	Fixed Costs	Fixed Costs per Unit	Variable Costs per Unit	Total Costs per Unit
$ 8.00	20,000	$120,000	$ 6.00	$1.50	$ 7.50
11.00	15,000	120,000	8.00	1.50	9.50
15.00	10,000	120,000	12.00	1.50	13.50
22.00	6,000	120,000	20.00	1.50	21.50

SOLUTION

A. Find the **possible net profit** for each selling price.

	(Selling Price per Unit	−	Total Cost per Unit)	×	Estimated Unit Sales		
(1)	($ 8.00	−	$ 7.50)	×	20,000	=	$10,000
(2)	($11.00	−	$ 9.50)	×	15,000	=	$22,500
(3)	($15.00	−	$13.50)	×	10,000	=	$15,000
(4)	($22.00	−	$21.50)	×	6000	=	$ 3000

B. Find the **greatest possible profit.** $22,500

C. Find the **best selling price.** $11.00

8 − 7.5 = 0.5 × 20000 = 10000 11 − 9.5 = 1.5 × 15000 = 22500 15 − 13.5 = 1.5 × 10000 = 15000 22 − 21.5 = 0.5 × 6000 = 3000

LESSON PLAN
16-8 Pricing

FOCUS
Choose an item and ask students what they would be willing to pay for that item. Take the highest suggestion, add slightly more to the selling price, and ask if anyone would still be willing to buy the item. The focus of this lesson is on computing the selling price that will result in the highest possible net profit.

TEACH
Mention to students *the law of supply and demand*; the higher the price, the fewer people there are that will buy the item. Conversely, the lower the price, the more people there are that will buy it. Setting a selling price is important because both the business and the customer must be happy with it. The goal of the business is to obtain the highest possible net profit.

In the Example table, point out the rising selling price per unit versus the estimated unit sales. Have students list some costs that could be considered fixed or variable. Go over the steps involved in finding the possible net profit.

Warm-Up Exercises
1. 60,000 ÷ 20,000 3
2. 40,000 ÷ 10,000 4
3. 15,000 ÷ 5000 3
4. ($4 − $2) × 10,000 $20,000
5. ($2.50 − $1.10) × 8000 $11,200
6. ($12.00 − $7.00) × 5000 $25,000

BUSINESS NOTES
Point out that the price of an item is not the only factor people consider when deciding to buy it. Many people consider the reputations of the manufacturer and of the store, the store's location, service, and so on.

✔ SELF-CHECK Complete the problems, then check your answers in the back of the book.

Selling Price per Unit	Estimated Unit Sales	Fixed Costs	Fixed Costs per Unit	Variable Costs per Unit	Total Cost per Unit
$3.00	200,000	$200,000	$1.00	$0.05	$1.05
$4.50	150,000	$200,000	$1.33	$0.05	$1.38

1. Find the possible net profit for each selling price. **$390,000; $468,000**
2. Which selling price yields the greatest possible profit? **$4.50**

PROBLEMS

Round answers to the nearest cent.

	Selling Price per Unit	Estimated Unit Sales	Fixed Costs	Fixed Costs per Unit	Variable Costs per Unit	Total Cost per Unit	Possible Net Profit
3.	$45.00	10,000	$120,000	$12.00	$21.00	$33.00	**$120,000**
4.	$65.50	8000	$120,000	$15.00	$21.00	$36.00	**$236,000**
5.	$75.00	7000	$120,000	**$17.14**	$21.00	**$38.14**	**$258,020**
6.	$85.00	6000	$120,000	**$20.00**	$21.00	**$41.00**	**$264,000**

7. Complete the table for the Wood Specialties Company. Which selling price yields the greatest possible profit? How many units should be produced? **$3.75; 950,000**

Selling Price per Unit	Estimated Unit Sales	Fixed Costs	Fixed Costs per Unit	Variable Costs per Unit	Total Cost per Unit	Possible Net Profit
$ 3.75	950,000	$500,000	**$0.53**	$1.05	**$1.58**	**$2,061,500**
$ 4.50	700,000	$500,000	**$0.71**	$1.05	**$1.76**	**$1,918,000**
$ 8.50	300,000	$500,000	**$1.67**	$1.05	**$2.72**	**$1,734,000**
$10.25	250,000	$500,000	**$2.00**	$1.05	**$3.05**	**$1,800,000**

8. Software, Inc., developed these cost and price figures for a new video game. Which combination should the company choose to handle the smallest number of items and to exceed $500,000 in possible profit?

8. 30,000 items at $59.95

Selling Price per Unit	Estimated Unit Sales	Fixed Costs	Fixed Costs per Unit	Variable Costs per Unit	Total Cost per Unit	Possible Net Profit
$19.95	70,000	$500,000	**$7.14**	$11.75	**$18.89**	**$74,200**
$29.95	60,000	$500,000	**$8.33**	$11.75	**$20.08**	**$592,200**
$39.95	50,000	$500,000	**$10.00**	$11.75	**$21.75**	**$910,000**
$49.95	40,000	$500,000	**$12.50**	$11.75	**$24.25**	**$1,028,000**
$59.95	30,000	$500,000	**$16.67**	$11.75	**$28.42**	**$945,900**

MAINTAINING YOUR SKILLS Look up the skill in parentheses if you need help or more practice.

Subtract. **(Skill 6)**

9. 649.91 − 73.89 **576.02** **10.** 12.9 − 1.023 **11.877** **11.** 78 − 32.8 **45.2** **12.** 34.12 − 31 **3.12**

PRACTICE AND APPLY

The following problems can be assigned for classwork and the answers checked in class to help students master the objective of the lesson.

- Guided Practice: 1–4
- Independent Practice: 5, 6

WRAP-UP

Ask students to explain how to find the total cost per unit and the possible net profit. Ask students to describe, in general, what happens to the number of units sold as the price of an item increases.

Assignment Guide

- Basic: 5–7, 9–12
- Average: 7, 8, 9–12

ALTERNATIVE STRATEGIES: Reteaching

Be sure students understand that the fixed costs per unit is found by dividing the fixed costs by the estimated unit sales. Also point out that the total cost per unit is found by adding the fixed cost per unit plus the variable cost per unit. Carefully explain the computation in the example.

LESSON PLAN
Reviewing the Basics

The exercises on this page review skills, applications, and terms used in the unit. You can use the exercises to assess informally students' proficiency with this material.

The page can be used for guided practice and independent practice. You can work through a selection of the exercises together with students, and thus see immediately if they know how to do them, and you can then assign some of the exercises for independent practice. Be sure to go over the answers to all assigned exercises.

Reviewing the Basics

Skills

Solve.

(Skill 6)
1. $9.00 − $4.25 **$4.75**
2. $99.40 − $34.17 **$65.23**
3. $419.70 − $332 **$87.70**

(Skill 8)
4. 442 × $0.895 **$395.59**
5. $8.74 × 16,000 **$139,840**
6. 2.1 × 96.8 **203.28**

(Skill 6)
7. 275.75 − 45.4 **230.35**
8. 36.2 − 0.138 **36.062**
9. 2.5 − 1.95 **0.55**

(Skill 20)
10. $\frac{1}{2}$ × $471 **$235.50**
11. $\frac{1}{2}$ × $18,483 **$9241.50**
12. $\frac{1}{4}$ × $26,360 **$6590**

(Skill 30)
13. 35% of 19,840 **6944**
14. 21% of $16,820,000 **$3,532,200**

Applications

Round answers to the nearest tenth of a percent.

(Skill 31)
15. 1122 is what percent of 3670? **30.6%**
16. $27,320 is what percent of $83,580? **32.7%**

(Application C)
17. What is the total of the responses? "Very interested" is what percent of the total responses? **35%**

Opinion	City	Suburbs	Rural	Total
Very interested	210	1521	2040	**3771** ?
Interested	515	1235	1326	**3076** ?
Slightly interested	960	752	1158	**2870** ?
Not interested	340	430	281	**1051** ?
Total	?	?	?	?
	2025	**3938**	**4805**	**10,768**

(Application N)
18. What production level is projected for Lime Rock Mineral Company for 1996? **5.5**

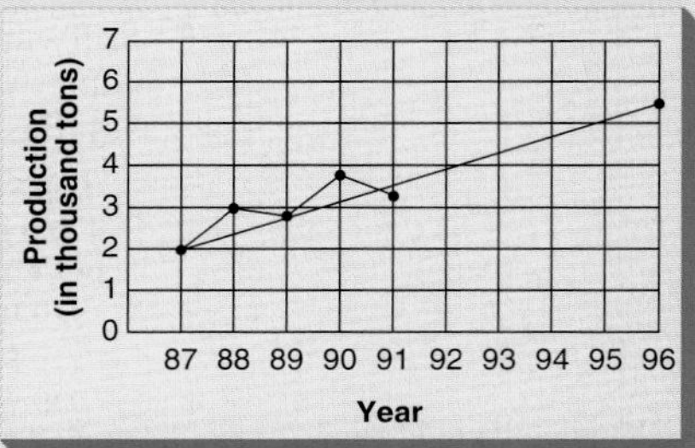

Terms

Write your own definition for each term. **Answers will vary.**

19. Opinion survey
20. Sales potential
21. Market share
22. Sales projection
23. Factor method
24. Net profit

Refer to your reference files in the back of the book if you need help.

Unit Test

Lesson 16-1

1. Willshire High School conducted an opinion survey of the student council's performance. Of the 500 students surveyed, 90 gave an "average" response. What percent of the total gave "average" responses? **18%**

Lesson 16-2

2. Of a sample of 800 swimming pool owners, 60 said they would buy Leisure Living's new chemical for swimming pools. There is a market of about 8,000,000 potential users. Each user would purchase the chemical twice per year. What is the sales potential for the chemical for 1 year? **1,200,000**

Lesson 16-3

3. Last year Madsen Wheel Works sold 736,000 wagons. Approximately 1,330,000 wagons were sold in the entire market last year. What was Madsen's market share? **55.3%**

Lesson 16-4

4. Use a graph to project Suntyme's sales for 1995 based on this sales history.

Approx. $6.5 million

Year	1989	1990	1991	1992
Sales (in millions)	$3.0	$4.0	$3.1	$4.9

Lesson 16-5

5. Willshire Clothing has 8.7% of the Willow Run metropolitan clothing market. Next year's clothing sales in the area are estimated to be $18,500,000. Use the factor method to project Willshire's sales for next year. **$1,609,500**

Lesson 16-6

6. Samson's Music Studio has a contract for 64 column inches of advertising in the *Herald* this week. What is the cost of a 40-column-inch advertisement placed on Sunday? **$1728**

	Per Column Inch	
Contract Rates	Daily	Sunday
16 inches	$35.80	$45.80
64 inches	35.00	43.20
126 inches	34.00	42.70

Lesson 16-7

7. Countryside Cereal plans to advertise on television. Costs for 30-second commercials are $1000 for daytime and $8000 for prime time. The cost of a 10-second commercial is $\frac{1}{2}$ the cost of a 30-second commercial. The cost of a 60-second commercial is 2 times the cost of a 30-second commercial. What is the total advertising cost? **$66,500**

Number	Length	Time
5	10-second	Daytime
4	30-second	Prime time
2	60-second	Prime time

Lesson 16-8

8. Complete the chart. Which selling price will maximize profits? **$25,000**

Selling Price per Unit	Estimated Unit Sales	Fixed Costs	Fixed Costs per Unit	Variable Costs per Unit	Total Costs per Unit	Possible Net Profit
$20,000	16	$200,000	$12,500	$2500	**$15,000**	**$80,000**
$25,000	14	$200,000	**$14,285.71**	$2500	**$16,785.71**	**$115,000.06**
$27,500	12	$200,000	**$16,666.67**	$2500	**$19,166.67**	**$99,999.96**

LESSON PLAN
Unit Test

Students should do the Unit Test on their own. Each problem on the test is keyed to a lesson in the unit. Students having difficulty with any particular problem should review the Example in the appropriate lesson and be assigned some of the Independent Practice problems for additional practice.

LESSON PLAN
A Spreadsheet Application

USING TECHNOLOGY

Inform students that sales analyses and other manipulations of financial data are done by computer programs called spreadsheets. During the past 10 years, the development of spreadsheet programs has resulted in the establishment of whole new companies. The growth of this new business has helped to create tens of thousands of new jobs in the United States.

A SPREADSHEET APPLICATION

Sales Analysis

To complete this spreadsheet application, you will need the diskette *Spreadsheet Applications for Business Mathematics,* which accompanies this textbook.

Input the information in the following problems to find the break-even point, gross sales, and gross profit.

1. Use the FIXED COSTS and VARIABLE COSTS given in the program. Change the SELLING PRICE and EXPECTED SALES to the amounts indicated.

	10—SELLING PRICE	11—EXPECTED SALES	BREAK-EVEN POINT	GROSS SALES	GROSS PROFIT
a.	$10.95	17,000	?	?	?
b.	11.45	15,500	?	?	?
c.	11.95	15,000	?	?	?
d.	12.45	14,000	?	?	?
e.	12.95	13,583	?	?	?
f.	13.45	11,500	?	?	?

2. Sobo Electronic Calculator Company developed these cost and price figures for their new Model 101 calculator.

FIXED COSTS:

1-RENT	2-TAXES	3-MARKETING	4-MANUFACTURING
10,200	62,400	75,000	81,500

VARIABLE COSTS:

5-SUPPLIES	6-LABOR	7-UTILITIES	8-PACKAGING	9-SHIPPING
1.50	3.75	0.85	0.25	0.25

Complete the table below to determine the best selling price and greatest profit.

	10—SELLING PRICE	11—EXPECTED SALES	BREAK-EVEN POINT	GROSS SALES	GROSS PROFIT
a.	$6.95	900,000	?	?	?
b.	7.25	800,000	?	?	?
c.	7.45	750,000	?	?	?
d.	8.45	400,000	?	?	?
e.	8.95	200,000	?	?	?
f.	9.45	50,000	?	?	?

Cumulative Review

Skills

Round answers to the nearest cent.

(Skill 2) **1.** \$7.848 **\$7.85** **2.** \$31.4545 **\$31.45** **3.** 79.9¢ **80¢** **4.** 12.3¢ **12¢**

Round answers to the nearest tenth of a percent.

5. 7.47% **7.5%** **6.** 34.66% **34.7%** **7.** 125.24% **125.2%** **8.** 212.625% **212.6%**

Solve. Round to nearest cent.

(Skill 5)
9. \$74.85 + \$125.14 **\$199.99**
10. \$517.87 + \$493.34 **\$1011.21**
11. \$19.47 + \$34.58 + \$7.86 **\$61.91**
12. \$4.98 + \$7.49 + \$2.19 **\$14.66**
13. \$1478.85 + \$2715.66 **\$4194.51**
14. \$916.43 + \$791.56 + \$288.72 **\$1996 .71**

(Skill 6)
15. \$74.49 − \$41.33 **\$33.16**
16. \$21.19 − \$7.79 **\$13.40**
17. \$46.22 − \$18.79 **\$27.43**
18. \$741.13 − \$564.45 **\$176.68**
19. \$9164.55 − \$7217.79 **\$1946 .76**
20. \$16,415.50 − \$3283.00 **\$13,132.50**
21. \$31,415.74 − \$23,567.79 **\$7847.95**

(Skill 8)
22. \$2.19 × 6 **\$13.14**
23. \$74.49 × 4 **\$297.96**
24. \$61.84 × 12 **\$742.08**
25. \$19.17 × 0.06 **\$1.15**
26. \$141.49 × 1.5 **\$212.24**
27. \$71.89 × 0.85 **\$61.11**

(Skill 11)
28. \$47.85 ÷ 12 **\$3.99**
29. \$617.81 ÷ 144 **\$4.29**
30. \$416.43 ÷ 0.6 **\$694.05**
31. \$19.48 ÷ 0.75 **\$25.97**
32. \$4217.55 ÷ 1.2 **\$3514.63**
33. \$27.45 ÷ 0.45 **\$61.00**

(Skill 16)
34. $\frac{1}{2} + \frac{1}{4}$ $\mathbf{\frac{3}{4}}$
35. $\frac{3}{8} + \frac{1}{4}$ $\mathbf{\frac{5}{8}}$
36. $\frac{3}{4} + \frac{7}{8}$ $\mathbf{1\frac{5}{8}}$
37. $1\frac{1}{2} + 2\frac{3}{8}$ $\mathbf{3\frac{7}{8}}$
38. $3\frac{1}{4} + 1\frac{7}{8}$ $\mathbf{5\frac{1}{8}}$
39. $\frac{1}{2} + 1\frac{5}{8} + 3\frac{1}{4}$ $\mathbf{5\frac{3}{8}}$

(Skill 20)
40. $\frac{1}{3} \times \frac{5}{8}$ $\mathbf{\frac{5}{24}}$
41. $\frac{3}{4} \times \frac{8}{9}$ $\mathbf{\frac{2}{3}}$
42. $\frac{1}{4} \times \frac{6}{7}$ $\mathbf{\frac{3}{14}}$
43. $\frac{1}{3} \times$ \$42 **\$14**
44. $\frac{1}{4} \times$ \$36 **\$9**
45. $\frac{1}{2} \times$ \$56 **\$28**

Write as a decimal.

(Skill 28)
46. 7% **0.07** **47.** 30% **0.3** **48.** 25% **0.25** **49.** 10% **0.1**
50. $1\frac{1}{2}$% **0.015** **51.** 5.5% **0.055** **52.** $4\frac{1}{4}$% **0.0425** **53.** 0.3% **0.003**

Solve. Round answers to the nearest cent or tenth of a percent.

(Skill 30)
54. 5% of \$174.50 **\$8.73**
55. 30% of \$417.80 **\$125.34**
56. $1\frac{1}{4}$% of \$241.70 **\$3.02**
57. 40% of \$617.84 **\$247.14**
58. 25% of \$1471.46 **\$367.87**
59. 6% of \$249.99 **\$15.00**

(Skill 31)
60. ?% of \$5000 = \$200 **4.0%**
61. ?% of 12 = 3 **25.0%**
62. ?% of \$19.79 = \$0.40 **2.0%**
63. ?% of \$416.78 = \$162.73 **39.0%**

LESSON PLAN
Cumulative Review (Skills/Applications) Units 12–16

The exercises on this page review skills, applications, and terms used in Units 12–16. You can use the exercises to assess informally students' proficiency with this material.

These pages can be used for guided practice and independent practice. You can work through a selection of the exercises together with students, and thus see immediately if they know how to do them. You can then assign some of the exercises for independent practice. Be sure to go over the answers to all assigned exercises.

LESSON PLAN
Cumulative Review
(Skills/Applications)
Units 12–16

Applications

Use the table to answer the following.

(Application C)

64. What is the regular price of a package of 100 mL beakers? **$9.75**

65. What is the sale price of a package of 250 mL beakers? **$14.24**

66. For 50 mL beakers, how much less is the sale price than the regular price? **$0.51**

PACKAGE OF 12 BEAKERS		
Size (mL)	Regular Price	Sale Price
50	$ 6.75	$ 6.24
100	9.75	8.75
250	15.75	14.24

67. For 100 mL beakers, the sale price is what percent of the regular price? **89.7%**

Find the complement.

(Application L)

68. 40% **60%** **69.** 25% **75%** **70.** 12.5% **87.5%** **71.** 5% **95%**

Use the graph to answer the following.

(Application N)

72. How many units were sold in 1988? **Approx. 3400**

73. In what year were the least units sold? **1986**

74. How many more units were sold in 1989 than in 1987? **Approx. 1500**

75. What is the sales projection for 1994? **Approx. 7000**

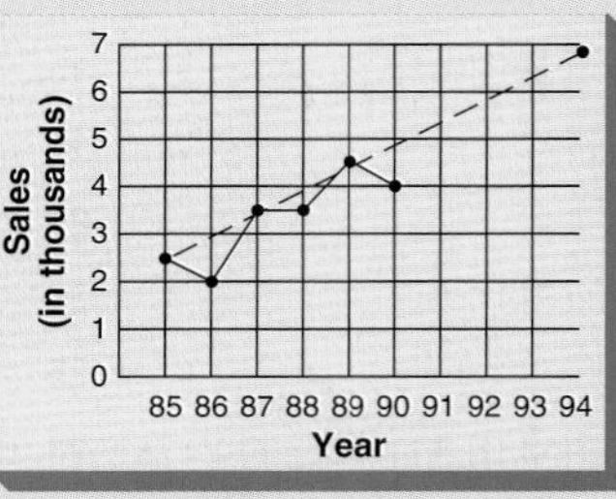

Find the mean.

(Application Q)

76. 10, 12, 17 **13** **77.** 18, 12, 11, 11 **13**

78. 174, 155, 163, 148 **160** **79.** 4.2, 1.6, 2.1, 4.0, 3.5 **3.08**

80. $14.51, $13.48, $14.76 **$14.25** **81.** $28,748, $30,615, $33,517 **$30,960**

Terms

Write your own definition for each term. **Answers will vary.**

82. Employee benefits **83.** Release time **84.** Salary scale

85. Travel expenses **86.** Disability insurance **87.** Prime cost

88. Break-even point **89.** Quality control **90.** Time study

91. Trade discount **92.** Chain discount **93.** Trade-discount rate

94. End-of-month dating **95.** Markup **96.** Markup rate

97. Net profit **98.** Markdown rate **99.** Opinion survey

100. Sales potential **101.** Market share **102.** Sales projection

Refer to your reference files in the back of the book if you need help.

Cumulative Review Test Units 12–16

Lesson 12-2

1. Sue Clark is a systems analyst for the Second National Bank. She presently earns $32,480 annually. Sue will receive a 4.8% cost-of-living increase and a 4.5% merit increase. What will Sue's new salary be? **$35,500.64**

Lesson 12-3

2. Tony Brown works for the Stark Corporation. He earns $21,840 per year. Stark provides these benefits for Tony. To the nearest percent, what is the rate of benefits? **23%**

Vacation: 2 weeks
Holidays: 8 days
Compensation insurance: 4.4%
Unemployment insurance: 3.6%
Social security: 6.2%
Medicare: 1.45%

Lesson 12-5

3. Meg Hart is a travel consultant. Her travel expenses this month include: airfare $1474, taxi fares $71.65, meals $416.50, hotels $821.37, and miscellaneous $61.40. What is Meg's total travel expense? **$2844.92**

Lesson 12-6

4. The personnel department of the Tru-Door Company is sending 5 of its employees to a local 2-day training program. The 5 employees' combined wages total $386 per day. The company paid the registration fee of $50 for each person. The company also paid for the employees' lunches for the 2 days at a total cost of $82.40. What was the total cost for the Tru-Door Company to send their employees to the training program? **$1104.40**

Lesson 13-1

5. Kasko Steel produces cable from scrap steel that it buys for $226 per ton. The company produces 4 rolls of cable from each ton. The direct labor charge is $38.78 an hour. The employees produce 2 rolls of cable per hour. To the nearest tenth of a cent, what is the prime cost of manufacturing 1 roll of cable? **$75.89**

Lesson 13-4

6. Giant Food Mart made a time study of its stock department. The store recorded the times that Joe Martin spent unpacking and shelving cartons of canned goods.

	Observations in Seconds				
Task	**#1**	**#2**	**#3**	**#4**	**#5**
Open carton	4.0	4.5	5.5	4.0	5.0
Remove cans	13.0	14.0	12.5	13.5	14.5
Place cans on shelf	12.0	11.6	12.8	12.2	11.4
Discard carton	5.0	5.5	5.0	5.0	5.5

If Joe takes a 10-minute break each hour, how many cartons can he unpack and shelve in 1 hour? **85**

LESSON PLAN Cumulative Review Test Units 12–16

Students should do the Cumulative Review Test on their own. Each problem on the test is keyed to a lesson in Units 12–16. Students having difficulty with any particular problem should review the Example in the appropriate lesson and be assigned some of the Independent Practice problems for additional practice.

Lesson 13-6

$8\frac{1}{4}$" H × $10\frac{1}{8}$" L × $10\frac{1}{8}$" W

7. Colonial Candle Company ships its top-of-the-line fancy candles in wooden boxes. The candles are arranged in 1 layer. Each box is made of $\frac{1}{8}$-inch thick wood with $\frac{1}{16}$-inch thick wood spacers. Each candle is 8 inches high and $3\frac{1}{4}$ inches in diameter. What are the dimensions of each wooden box?

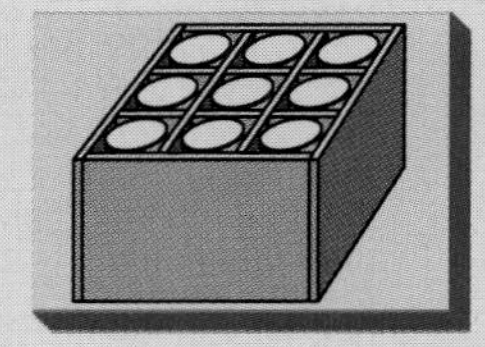

Lesson 14-3

8. Sun-Like, Inc. lists the price of its Sure-Tan Lamp as $47.49. The net price to the retailer is $37.99. What is the trade-discount rate? **20%**

Lesson 14-5

9. Gendron Wheel Co. produces a bicycle that is priced at $217.86. When a new model was introduced, the $217.86 bike was discounted at 30% less 15%. What is the net price of the old model using the complement method? **$129.63**

Lesson 14-6

10. DP Associates received an invoice from Commercial Computer Supply dated September 13. The invoice carried terms of 3/10, net 30. The net price of the invoice was $7419.80. What was the cash price of the invoice if it was paid on September 22? **$7197.21**

Lesson 15-3

11. Eagle-Hawk, Inc., a fireplace equipment wholesaler, purchases fireplace inserts for $147.85 each and sells them for $199.79. The company estimates the overhead expenses for each insert to be 20% of the selling price. What is the estimated net profit on each insert? **$11.98**

Lesson 15-7

12. Tree Grown determines the retail selling price of the oranges by marking up the cost by 140%. The fruit company bought its first shipment of winter oranges for $0.66 per dozen. What is the retail selling price per dozen of this shipment? **$1.58**

Lesson 16-2

13. Of a sample of 1200 swimming pool owners, 360 said they would buy Litehouse Living's new chemical for swimming pools. There is a market of about 4,000,000 potential users. Each user would purchase the chemical once per year. What is the sales potential for the chemical for 1 year? **1,200,000**

Lesson 16-6

14. Main Street Meat Market has a contract for 24 column inches of advertising at $35.15 per column inch in the *Herald* this year. What is the cost of an 8-column-inch advertisement? **$281.20**

Lesson 16-8

15. Complete the chart for Sailaway Corporation. Which selling price will maximize Sailaway's profits? **$45,000**

Selling Price per Unit	Estimated Unit Sales	Fixed Costs	Fixed Costs per Unit	Variable Costs per Unit	Total Costs per Unit	Possible Net Profit
$38,000	15	$200,000	$13,333.33	$4200	? **$17,533.33**	? **$307,000.05**
45,000	13	200,000	? **$15,384.62**	4200	? **$19,584.62**	? **$330,399.94**
53,000	9	200,000	? **$22,222.22**	4200	? **$26,422.22**	? **$239,200.02**

U N I T

17

Warehousing and Distribution

The warehousing and distribution department keeps a record of all merchandise, or *inventory,* stored in your warehouse. *Inventory cards* show the number of each item in stock, the number of incoming items (*receipts*), and the number of outgoing items (*issues*). Your inventory must be large enough to meet your sales needs. The size and value of your inventory will affect your insurance and taxes and the worth of your business. The warehousing and distribution department is also responsible for shipping your product.

O U T L I N E

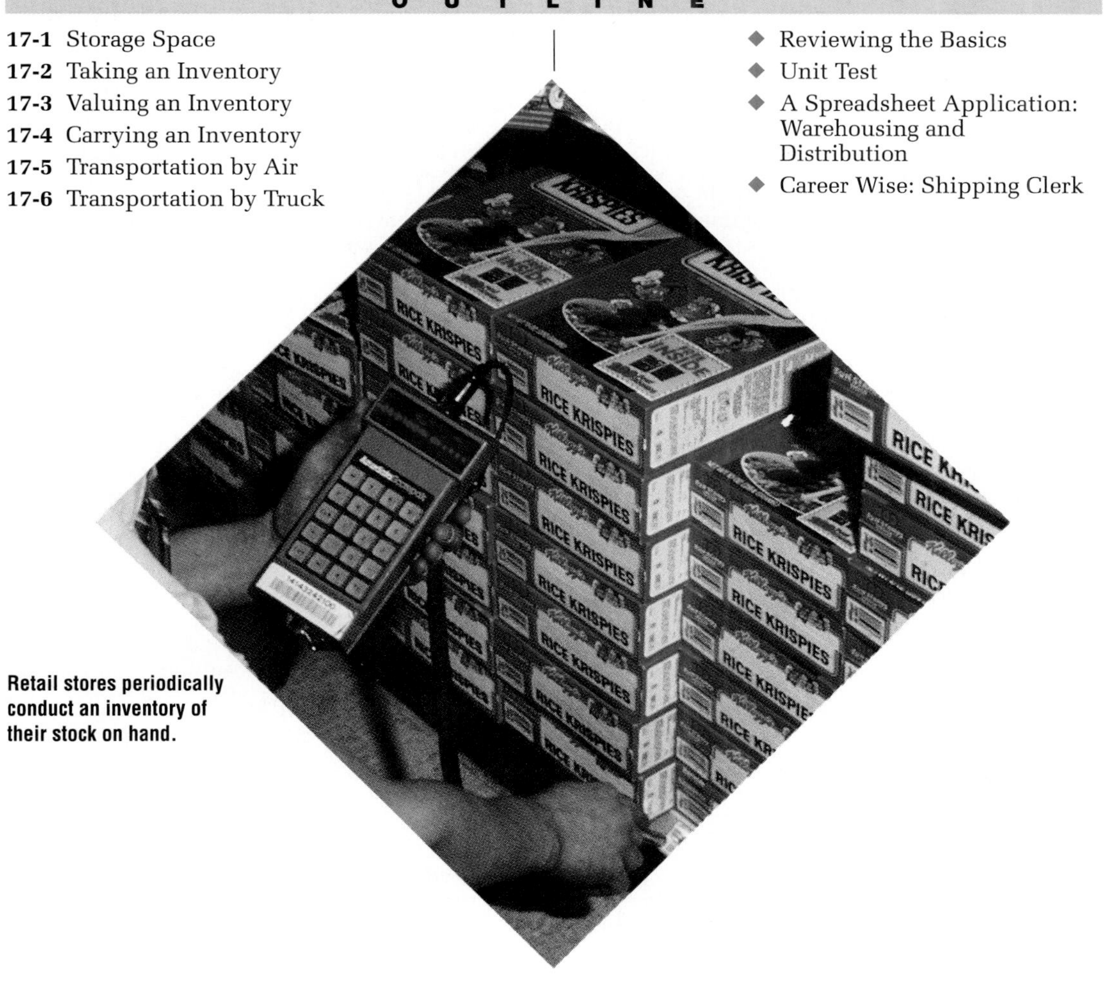

Retail stores periodically conduct an inventory of their stock on hand.

UNIT 17
WAREHOUSING AND DISTRIBUTION

INTRODUCING THE UNIT

Businesses keep a certain amount of merchandise on hand—this requires adequate storage space and keeping accurate records. There is a limit, however, to how much a business can keep on the premises. Some types of merchants, such as carpet and furniture dealers, constantly are getting new models to replace the old. Ask students how many have heard clearance sales advertised for local carpet or furniture dealers. The same principle applies to clothes retailers, who want to move last season's apparel out to make room for this season's garments.

LESSON PLAN
17-1 Storage Space

FOCUS

Students understand the idea of storage space. Ask them if they have any storage space at home or in school. (their lockers in school) Do they know where the school's supplies are stored? Are they aware of any warehouses in your town?

TEACH

Discuss the concept of volume. Remind students that volume is measured in cubic units. Use the example, 5 ft × 12 ft × 10 ft = 600 ft³, to illustrate how to write cubic units. In this lesson storage space is discussed in terms of the actual space that raw materials or products occupy. Mention that additional space may be required for aisle ways and pallets. Explain that bulk material (such as stone and sand) and odd-shaped items (such as rakes and shovels) may require extra storage space and special handling.

Warm-Up Exercises

1. 3 ft × 4 ft × 2 ft
   24 ft³
2. 2 m × 1.5 m × 1 m
   3.0 m³
3. 5 ft × 1.6 ft × 2 ft
   16 ft³
4. 4.0 m × 2.1 m × 1.8 m   15.12 m³
5. 24 ft³ × 300   7200 ft³
6. 15.12 m³ × 250
   3780 m³

17-1 Storage Space

OBJECTIVE
Compute the total storage space.

Your business needs **warehouse** or **storage space** in which to keep raw materials or products until you are ready to use them. You may need space in a large warehouse or only a small stockroom, depending on the size and quantity of the items you are storing.

Storage Space = Volume per Item × Number of Items

EXAMPLE *Skill* 8 *Application* Z *Term* Storage space

The PC-View Corporation manufactures personal computer monitors. Each monitor is packaged in a carton measuring 2 feet long, 1.5 feet wide, and 2 feet high. How many cubic feet of space does PC-View need to store 800 monitors?

SOLUTION

A. Find the **volume per item.**
2 ft × 1.5 ft × 2 ft = 6 ft³ volume per item

B. Find the **storage space.**
Volume per Item × Number of Items
6 ft³ × 800 items = 4800 ft³ storage space

2 [×] 1.5 [×] 2 [=] 6 [×] 800 [=] 4800

✓ SELF-CHECK Complete the problems, then check your answers in the back of the book.

Find the storage space needed.

1. Dimensions of item:
   2 ft × 3 ft × 3.5 ft.
   700 items. **14,700 ft³**
2. Dimensions of item:
   7 cm × 8 cm × 1 cm.
   1000 items. **56,000 cm³**

PROBLEMS

	Carton Dimensions (Length	×	Width	×	Height	=	Volume)	×	Number of Items	=	Storage Space
3.	(6.0 in	×	8.5 in	×	10.0 in	=	510 in³)	×	500	=	255,000 in³
4.	(2.5 ft	×	1.5 ft	×	1.0 ft	=	3.75 ft³)	×	200	=	750 ft³
5.	(1.2 m	×	0.6 m	×	0.4 m	=	0.288 m³)	×	500	=	144 m³
6.	(43 cm	×	24 cm	×	12 cm	=	12,384 cm³)	×	340	=	4,210,560 cm³
7.	(2.1 yd	×	1.8 yd	×	1.4 yd	=	5.292 yd³)	×	120	=	635.04 yd³
8.	(2.2 ft	×	1.5 ft	×	1.1 ft	=	3.63 ft³)	×	1200	=	4356 ft³

PROBLEM SOLVING

Have students determine the number of cubic inches of shelf space that is required to put 40 copies of their math textbook on a shelf piled 5 high in 8 rows. Is any space saved by stacking them 8 high in 5 rows? Have them experiment to find the most compact space required.

9. Washstand cupboard.
Carton is 3 ft by 1.8 ft by 2.5 ft.
100 cartons to be stored.
How much storage space is required? **1350 ft^3**

10. Table lamp.
Carton is 0.75 m by 0.6 m by 0.8 m.
700 cartons to be stored.
How much storage space is required? **252 m^3**

11. Green Shade study lamps are packed 2 in a carton. The dimensions of the carton are 2.4 feet by 1.2 feet by 1.8 feet. The School Supply Store wants to order 200 pairs of lamps. How much storage space will the School Supply Store need to store the lamps? **1036.8 ft^3**

12. Family Drugs has ordered 350 cartons of tape holders. Each carton of holders measures 1.6 feet by 0.8 feet by 0.5 feet. How much storage space is required? **224 ft^3**

13. Craft Electronics produces console television sets. The sets are packaged in cartons measuring 4 feet long, 2.5 feet wide, and 3 feet high. How much storage space is required for 1750 television sets? **52,500 ft^3**

14. The London Furniture Company packs its Queen Anne tables in cartons measuring 1.75 meters long, 0.8 meters wide, and 1.2 meters high. Empire Importing has ordered 500 tables. How much storage space is needed? **840 m^3**

15. The Great Western power lawn mower is housed in a carton measuring 2.5 feet by 1.3 feet by 0.8 feet. General Lawn Company ordered 875 Great Western lawn mowers. How much storage space is needed? **2275 ft^3**

16. Bananas are delivered to grocery stores in crates measuring 1 yard by 1.5 feet by 12 inches. How many cubic feet of storage space are required for 25 crates of bananas? **112.5 ft^3**

17. Fireplace hardware sets are delivered in boxes that are 0.8 meters by 25 centimeters by 15 centimeters. How many cubic meters of storage space are required to store 750 sets? **22.5 m^3**

18. Twenty-four clock radios are packed in a carton measuring 3 feet long, 2.4 feet wide, and 1.6 feet high. How much storage space is required for 200 cartons of 24 clock radios each? How much space is required for 3600 clock radios packaged in cartons of 24 clock radios each? **2304 ft^3; 1728 ft^3**

MAINTAINING YOUR SKILLS Look up the skill in parentheses if you need help or more practice.

Multiply. **(Skill 8)**

19. 2 ft × 3 ft × 5 ft **30 ft^3**

20. 1.4 m × 1.0 m × 0.8 m **1.12 m^3**

21. 8 in × 6 in × 4 in **192 in^3**

22. 18 mm × 15 mm × 9 mm **2430 mm^3**

23. 9.7 cm × 7.5 cm × 6.8 cm **494.7 cm^3**

24. 2.1 yd × 0.75 yd × 1.5 yd **2.3625 yd^3**

25. 45 ft^3 × 400 **18,000 ft^3**

26. 6.5 cubic m^3 × 620 **4030 m^3**

27. 112 in^3 × 250 **28,000 in^3**

PRACTICE AND APPLY

The following problems can be assigned for classwork and the answers checked in class to help students master the objective of the lesson.

- Guided Practice: 1–9
- Independent Practice: 10–14 even

WRAP-UP

Use Problem 12 to summarize the objective of the lesson.

Assignment Guide

- Basic: 10–16, 19–27
- Average: 11–15 odd, 16–18, 19–27 odd

ALTERNATIVE STRATEGIES: Reteaching

	Length	Width	Height	Volume per Item	Number of Items	Storage Space
1)	10 ft	8 ft	2 ft	(160 ft^3)	20	(3200 ft^3)
2)	9 in	7 in	3 in	(189 in^3)	50	(9450 in^3)
3)	12 cm	8 cm	6 cm	(576 cm^3)	75	(43,200 cm^3)

LESSON PLAN
17-2 Taking an Inventory

FOCUS
Ask students if they know how much clothing they own. How many pairs of pants, skirts, sweaters, and shoes does each student have? Call upon a student to tell the class how the exact number of each item of clothing he or she owns can be determined. (count each item)

TEACH
Although computers can keep track of inventories, many businesses still take a physical inventory at least once a year. This is done to verify that the items listed in the computer are actually in stock. Any items stolen, for example, would not be taken into account by the computer.

Some businesses also close several times a year to take inventory. Other businesses (such as department stores, drugstores, and supermarkets) constantly take inventory—called a perpetual inventory. A perpetual inventory is done either by hand or by using computerized cash registers, laser readers, or light-wand pens.

Discuss the idea that maintaining an inventory costs money. However, not having an item in inventory can result in the loss of a sale. Thus inventory control, which involves keeping the right level of items in stock, is an important function of a well-run business.

17-2 Taking an Inventory

OBJECTIVE
Compute total inventory.

An **inventory** shows the number of each item that your business has in stock. You may use an **inventory card** to keep a record of the items on hand. The number of incoming items, called **receipts,** are added to the previous inventory. The number of outgoing items, called **issues,** are subtracted from the previous inventory. This is frequently done on a computer.

Inventory = Previous Inventory + Receipts − Issues

EXAMPLE *Skills* 3, 4 *Application* A *Term* Inventory

The inventory card for Wholesale Paint Supply shows that 80 gallons of white latex paint, stock number WL-9, were in inventory on February 1. Each week, the inventory record is updated. How many gallons were on hand March 1?

SOLUTION

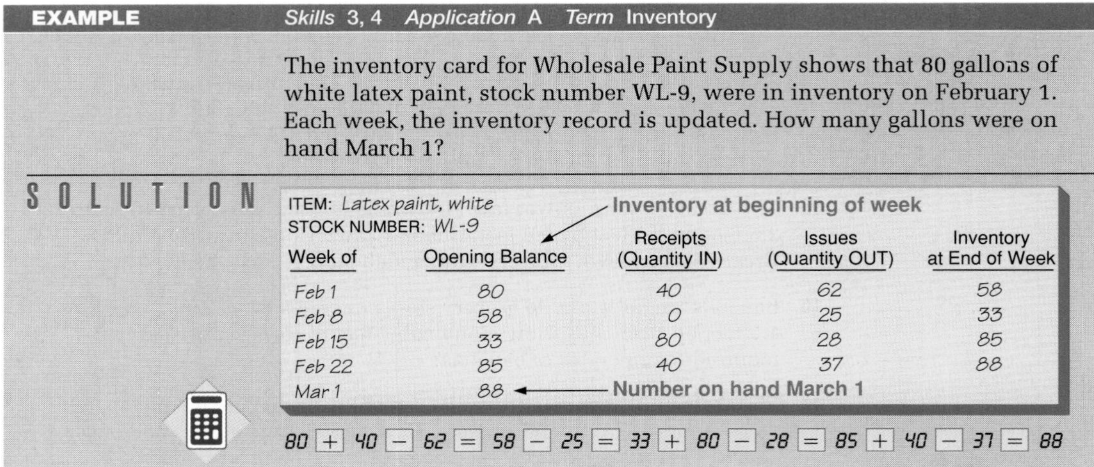

ITEM: *Latex paint, white*
STOCK NUMBER: *WL-9*

Week of	Opening Balance	Receipts (Quantity IN)	Issues (Quantity OUT)	Inventory at End of Week
Feb 1	80	40	62	58
Feb 8	58	0	25	33
Feb 15	33	80	28	85
Feb 22	85	40	37	88
Mar 1	88			

80 + 40 − 62 = 58 − 25 = 33 + 80 − 28 = 85 + 40 − 37 = 88

✓ SELF-CHECK Complete the problems, then check your answers in the back of the book.

Find the inventory.

1. Previous inventory, 400; receipts, 150; issues, 200. **350**
2. Previous inventory, 220, receipts, 70; issues 50. **240**

PROBLEMS

Complete the inventory record.

	Month	Opening Balance	+	Receipts (Quantity IN)	−	Issues (Quantity OUT)	=	Number on Hand
3.	January	300	+	120	−	80	=	? **340**
4.	February	340	+	80	−	100	=	? **320**
5.	March	? **320**	+	120	−	100	=	? **340**
6.	April	? **340**	+	0	−	80	=	? **260**

BUSINESS NOTES
Discuss the reasons why a business must know the number of each item that is in stock. If the stock is low, the manager might decide to reorder immediately. Emphasize that the number of items "in stock" is the same as the number of items "on hand." These terms refer to the total number of a particular item that a business has for sale, regardless of whether they are located in the stockroom or on the sales floor. The term "opening balance" is used interchangeably with "previous inventory."

7. Inventory of garden shovels.
80 on hand at beginning of week.
Received 40 during the week.
Issued 25 on Friday.
How many on hand at end of week? **95**

8. Inventory of 12-oz cans of baked beans.
Started the week with 148 cases.
Received 40 cases during the week.
Shipped out 65 cases during the week.
How many on hand at end of week? **123**

9. The Key Auto Supply Company takes an inventory on a monthly basis. What is the balance on hand for each item?

Item Number	Description	Beginning Balance	Quantity IN	Quantity OUT	Balance on Hand
X 23	10 W-30 Oil	147 qt	50 qt	78 qt	**119** ?
W 15	Washer solvent	91 gal	36 gal	25 gal	**102** ?
C 5	5-gallon can 30 W oil	17 cans	12 cans	3 cans	**26** ?

10. How many of each item does the central supply department have on hand at the end of this month?

STOCK NO.	DESCRIPTION	OPENING BALANCE	+ RECEIPTS	− ISSUES	= ON HAND
11398	STPLR-RS153	46	0	10	**36** ?
11402	BOX STPLS-RT11	117	0	50	**67** ?
11610	BOX PA CLPS-L3	18	144	21	**141** ?
11682	BOX PA CLPS-L7	7	36	9	**34** ?

11. How many dictionaries does the bookstore have in stock on each date?

LAST ACTION DATE	RECEIPTS	ISSUES	ON HAND
9/3	225		225
9/7		37	**188** ?
9/10		24	**164** ?
9/15		35	**129** ?
9/20	144		**273** ?
9/25		43	**230** ?

12. The Stock Status Summary for all of the stores in the Cushing Department Stores chain is printed monthly. How many of each item are in stock?

DATE 8-28		STOCK STATUS SUMMARY					
MATER'L CLASS	STOCK NUMBER	DESCRIPTION	OPENING BALANCE	TRANSACTIONS RECEIVED	ISSUED	ON HAND	
174	146301	MARKER-FP	117		26	?	**91**
174	146334	MARKER-W/2	34	12		?	**46**
175	121	NOTEBOOK	987	144	347	?	**784**
175	6841	NOTEBOOK	1413	72	465	?	**1020**

Warm-Up Exercises

1. 200 + 45 + 70 315
2. 310 − 100 − 50 160
3. 420 + 144 − 150 414
4. 1650 + 260 − 117 1793
5. 4163 + 225 − 168 4220
6. 7184 − 249 + 72 7007

PRACTICE AND APPLY

The following problems can be assigned for classwork and the answers checked in class to help students master the objective of the lesson.

- Guided Practice: 1–7
- Independent Practice: 8, 9, 11

ALTERNATIVE STRATEGIES: Enrichment

A perpetual inventory is maintained by use of a computerized cash register, laser readers, or light-wand pens as mentioned in the TEACH notes. Ask your students, in groups or individually, to investigate and report to the class on how these high tech items actually work. Either retailers or computer suppliers should be able to help. Time permitting, you may ask a computer systems person to talk with your class.

WRAP-UP

Ask various students to define the terms (1) *inventory*, (2) *receipts*, and (3) *issues*. Then work Problem 11.

Assignment Guide

- Basic: 8–16, 19–28
- Average: 10, 12–18, 20–28 even

13. Whitewater Supply Company carries this inventory card on inflatable vests. Find the balance for each date.

Part No. IV–17	Date	Receipts	Issues	Balance
Description:	6/1	76		76
Inflatable	6/12		36	**40** ?
vest	6/18	144		**184** ?
Location: 24	6/26		58	**126** ?

14. Complete this hardware supply inventory record for a deadbolt lock set.

Date	Receipts	Shipments	Balance
9/1			110
9/9	100		**210** ?
10/12		75	**135** ?
10/30		100	**35** ?
11/15	200		**235** ?
12/11		95	**140** ?

17.

Date	In	Out	Bal.
8/1			47
8/6		24	23
8/16	40		63
8/27		36	27
8/31	50		77

18.

Date	In	Out	Bal.
7/1			74
7/17		24	50
7/20	30		80
8/22		40	40
9/20	24		64
9/30		36	28

15. An inventory record for Sunrise Wholesale Pool Supply shows a receipt of 100 folding canvas lounge chairs on 4/1. Another shipment of 125 arrived on 4/15. One hundred ten chairs were shipped out on 5/10. Fifty arrived on 6/2 and 75 were shipped out on 6/17. Find the balance as of 6/17. **90**

16. On January 1, Sport Wholesalers started with an inventory of 175 basketballs. Shipments of 45, 76, and 25 were sent out on 1/13, 2/18, and 3/11, respectively. Receipts of 36, 144, and 72 were received on 1/20, 2/25, and 3/15, respectively. Find the balance as of 3/15. **281**

17. The Car Care Co. started on August 1 with a balance of 47 cases of car wax. On August 6, 24 cases were shipped out. On August 16, 40 cases were received. On August 27, 36 cases were shipped out, and on August 31, 50 cases were received. Prepare an inventory control card showing the balance after each transaction.

Critical Thinking . . .

18. **A Brief Case** The Wholesale Hardware Supply Company began July 1 with a balance of 74 gas chain saws. It received shipments of 30 and 24 chain saws on July 20 and September 20, respectively. It made shipments of 24, 40, and 36 chain saws on July 17, August 22, and September 30, respectively. Prepare an inventory control card showing the balance after each transaction.

MAINTAINING YOUR SKILLS Look up the skills in parentheses if you need help or more practice.

Add. **(Skill 3)**

19. 125 + 45 + 25 **195** **20.** 144 + 72 + 36 **252** **21.** 50 + 35 + 48 **133**

Subtract. **(Skill 4)**

22. 144 − 50 **94** **23.** 72 − 48 **24** **24.** 286 − 144 **142** **25.** 7146 − 5268 **1878**

26. 9178 − 493 **8685** **27.** 14,147 − 9749 **4398** **28.** 26,147 − 19,464 **6683**

BUSINESS PROJECT

Have some students contact a local retail business and investigate how that business maintains its inventory records. Ask these students to report their findings to the class.

17-3 Valuing an Inventory

OBJECTIVE
Use the average-cost method to compute the inventory value.

You must calculate the value of your inventory when you purchase insurance, pay taxes, or compute the worth of your business. The **average-cost method** is one way of calculating the value of an inventory. Because the cost of incoming items may change, you calculate the value of the inventory based on the average cost of the goods you received.

Inventory Value = Average Cost per Unit × Number on Hand

EXAMPLE *Skills* 8, 11 *Application* Q *Term* Average-cost method

Wholesale Paint Supply is valuing its inventory of white latex paint. On March 1, Wholesale Paint Supply had 88 gallons of white latex paint on hand. What is the value of the inventory on March 1?

ITEM: Latex paint, white
STOCK NUMBER: WL-9

Number on hand at beginning of week → Opening Balance receipts; Average cost in January → $2.15

Week of		Receipts	Unit Cost	Total Cost
Feb 1	Opening Balance:	80	$2.15	$172.00
		40	2.25	90.00
Feb 8		0	- -	- -
Feb 15		80	2.30	184.00
Feb 22		40	2.40	96.00
	Total	240		$542.00

SOLUTION

A. Find the **average cost per unit.**
Total Cost of Units ÷ Number Received
$542.00 ÷ 240 = $2.2583 = $2.26 average cost per unit

B. Find the **inventory value.**
Average Cost per Unit × Number on Hand
$2.26 × 88 = $198.88 inventory value

542 ÷ 240 = 2.2583333 2.26 × 88 = 198.88

✓ SELF-CHECK Complete the problems, then check your answers in the back of the book.

100 items purchased at $4.00 each. 60 items purchased at $3.80 each.

1. Find the average cost per unit. **$3.93**
2. Find the inventory value if 50 items were on hand. **$196.50**

LESSON PLAN 17-3 Valuing An Inventory

FOCUS

To motivate this lesson, ask students how a paint store could determine the total value of all the supplies it has in stock. Mention that paint bought from different suppliers could have different prices. Lead students to see that an average cost could be used to compute the value of the inventory.

TEACH

Use these figures to review the meaning of an average: $1.50, $2.25, $3.00, $1.75, $2.00, and $1.50. Ask students to find the average of the six figures. ($2.00)

Point out that a retailer may have identical items on the shelf that were received at different times and at different costs. When valuing an inventory, the average-cost method may be used to balance out the difference in cost of the various shipments received during a time period. You may want to refer to *average daily balance* from Lessons 6-5 and 6-6 and *average monthly expenditure* from Lesson 11-1 to help students understand the concept of average cost.

CRITICAL THINKING

Using Problems 11 and 12, ask students why a business may want a lower valued inventory or a higher valued inventory. Note that in Problem 13, the average cost does not always end up in between LIFO and FIFO. FISH (First In, Still Here) is what most businesses would like to avoid.

Warm-Up Exercises

1. 100 × $4.40 $440
2. 210 × $1.27 $266.70
3. 87 × $24.75 $2153.25
4. $740 ÷ 85 $8.71
5. $1651.80 ÷ 1230 $1.34
6. $4781.45 ÷ 240 $19.92

PRACTICE AND APPLY

The following problems can be assigned for classwork and the answers checked in class to help students master the objective of the lesson.

- Guided Practice: 1–3
- Independent Practice: 4–7

PROBLEMS

3. Top-Mart is valuing its inventory of popcorn poppers. On June 1, Top-Mart had 23 poppers on hand.
   a. What is the average cost per popper? **$2.15**
   b. What is the value of the inventory of poppers on June 1? **$49.45**

Week of	Receipts	Unit Cost	Total Cost
May 1	40	$2.10	$84.00
May 8	36	2.15	77.40
May 15	24	2.20	52.80
May 22	36	2.15	77.40
May 29	12	2.25	27.00
Total	148		$318.60

4. Sure Sound, Inc., is valuing its inventory of AM/FM cassette portable players. On April 1, Sure Sound had 53 players on hand.
   a. What is the average cost per player? **$6.05**
   b. What is the value of the inventory of players on April 1? **$320.65**

Date	Receipts	Unit Cost	Total Cost
January 12	50	$6.30	$315.00
February 11	65	6.10	396.50
March 21	80	5.85	468.00
Total	195		$1179.50

For problems 5 and 6, complete the tables and answer the questions.

5. a.

Date	Receipts	Unit Cost	Total Cost
3/10	1120	$0.25	**$280.00**
4/22	940	0.28	**263.20**
Total	**2060**		**$543.20**

b. Average cost per unit: **$0.26**. Number on hand on 4/30: 960.
c. Value of inventory: **$249.60**

6. a.

Date	Receipts	Unit Cost	Total Cost
12/18	650	$12.10	**$7865**
1/10	500	13.31	**6655**
2/6	450	13.98	**6291**
Total	**1600**		**$20,811**

b. Average cost per unit: **$13.01** Number on hand on 2/8: 674.
c. Value of inventory: **$8768.74**

7. Records for Kazrich Market show this opening balance and theses receipts for Hearthside soup in November. At the end of the month, 31 cans were on hand. What is the value of the inventory? **$9.61**

Date	Receipts	Unit Cost
11/1 Opening bal.	94	$0.32
11/10	72	0.30
11/17	36	0.32
11/24	48	0.31

8. TBA Inc's inventory records for packages of windshield wipers show this information. At the end of the month, 79 packages were on hand. Find the value of the inventory. **$280.45**

Date	Receipts	Unit Cost
5/1	93	$3.43
5/19	120	3.56
5/31	115	3.63

COMMUNICATION SKILLS

The acronyms FIFO and LIFO are used frequently in business. Encourage students to use them also in order to effectively communicate with one another when discussing methods of valuing inventory.

ALTERNATIVE STRATEGIES: Reteaching

Stress that with Average Cost you must round off the average cost per unit to the nearest cent before proceeding. Find the value of the inventory on the 30th using the Average Cost, FIFO, and LIFO Methods:

9. The Leisure Company, a swimming pool supply company, uses the FIFO (first in, first out) method of valuing inventory. The FIFO method assumes that the first items received are the first items sold.

Date	Receipts	Unit Cost	Total Cost
5/1	100	$4.65	$465.00
5/8	65	4.75	308.75
5/19	145	4.90	710.50
Total	310		$1484.25

a. Complete The Leisure Company's records for receipts of tubs of chlorine.

F.Y.I.
The industry average for small specialty retailers is an inventory turnover of 2.7 times a year.

As of May 20, The Leisure Company had sold 160 of the tubs of chlorine.

b. Use the FIFO method to calculate the cost of the 160 tubs sold. **$750**

c. Use the FIFO method to calculate the value of the 150 tubs in stock. **$734.25**

$465	100 at $4.65	= ?		
285	60 at $4.75	+ ?		
$750	160 Total	?		

Of the 65 received on 5/8

5 at $4.75	=	**$23.75**
145 at $4.90	+	**710.50**
150	Total	**$1484.25**

Critical Thinking . . .

10. Another method used to value an inventory is the LIFO (last in, first out) method. The LIFO method assumes that the last items received are the first items shipped.
a. Use the LIFO method to calculate the value of The Leisure Company's 150 tubs in stock in problem 9 above. **$702.50**
b. Use the average-cost method to calculate the value of The Leisure Company's 150 tubs in stock in problem 9 above. **$718.50**

11. In a rising economy (costs continue to rise), which of the three methods of valuing an inventory discussed in this lesson will result in the:
a. Highest valued inventory? Why? **FIFO Low cost items go first.**
b. Lowest valued inventory? Why? **LIFO High cost items go first.**

12. In a declining economy (costs continue to decline), which of the three methods of valuing an inventory will result in the:
a. Highest valued inventory? Why? **LIFO Low cost items go first.**
b. Lowest valued inventory? Why? **FIFO High cost items go first.**

13. Does the average-cost method of valuing an inventory always result in a value between the value determined using FIFO and LIFO?
No, only if costs consistently rise or decline.

MAINTAINING YOUR SKILLS Look up the skills in parentheses if you need help or more practice.

Multiply. **(Skill 8)**

14. 50 × $8.40 **$420**
15. 100 × $17.85 **$1785**
16. 72 × $36.90 **$2656.80**
17. 144 × $141.73 **$20,409.12**
18. 36 × $0.87 **$31.32**
19. 119 × $47.87 **$5696.53**
20. 216 × $247.81 **$53,526.96**
21. 483 × $79.86 **$38,572.38**

Divide. Round answers to the nearest cent. **(Skill 11)**

22. $1046.25 ÷ 125 **$8.37**
23. $438.70 ÷ 28 **$15.67**
24. $314.91 ÷ 92 **$3.42**
25. $30,613.74 ÷ 216 **$141.73**
26. $357.81 ÷ 410 **$0.87**
27. $9012.31 ÷ 187 **$48.19**
28. $13,111.54 ÷ 53 **$247.39**
29. $56,918.43 ÷ 719 **$79.16**

WRAP-UP

Ask a student to explain the meaning of the average-cost method. Then work Problem 7 to illustrate again how to find the value of an inventory.

Assignment Guide

Basic: 4–8, 14–29
Average: 8–13, 14–28 even

	Date	Unit Cost	In	Out	Bal
1)	1	$6.00	50		50
	8			30	(20)
	14	7.50	40		(60)
	22			25	(35)
	30	8.00	30		(65)

Avg Cost = $7 × 65 = $455
FIFO = ($8 × 30) + ($7.50 × 35) = *highest* $502.50
LIFO = ($6 × 50) + ($7.50 × 15) = *lowest* $412.50

LESSON PLAN
17-4 Carrying an Inventory

FOCUS

To motivate this lesson, ask students to assume that they are the owners of a large furniture store. Then ask them to think about how much inventory they would like to carry. Discuss the advantages or disadvantages of carrying different levels of inventory.

TEACH

Emphasize that inventory control is an essential aspect of a well-run business. If essential parts are not available when needed for production, for example, a manufacturing plant may have to shut down until they are available. An insufficient inventory may result in lost sales and customers. However, maintaining too much inventory costs extra money that could be used for other purposes. Through experience and the development of an historical record, most businesses learn how to maintain correct levels of inventories.

Warm-Up Exercises

1. 30% × $17,000
   $5100
2. 25% × $220,000
   $55,000
3. 20% × $55,500
   $11,100
4. 35% × $174,000
   $60,900

17-4 Carrying an Inventory

OBJECTIVE
Compute the annual cost of carrying an inventory.

Your business must keep a sufficient inventory of goods to meet production or sales needs. Based on past records, you may estimate the annual cost of maintaining, or carrying, the inventory at a certain level for the coming year. The annual cost of carrying the inventory includes taxes, insurance, storage fees, handling charges, and so on. The annual cost is often expressed as a percent of the value of the inventory.

Annual Cost of Carrying Inventory = Inventory Value × Percent

EXAMPLE *Skill 30 Application A*

Steel Facilitators, Inc., produces washers, nuts, bolts, and similar items. The company maintains an inventory of metal sheets and bars to be used in production. The company also maintains an inventory of finished products to fill customers' orders. Steel Facilitators estimates the annual cost of maintaining the inventory to be 25% of the value of the inventory. This year the company plans to maintain an inventory totaling $500,000 in value. What amount should be set as the estimated cost of carrying the inventory for the year?

SOLUTION Find the **annual cost of carrying inventory.**

Inventory Value × Percent
$500,000 × 25% = $125,000 annual cost of carrying inventory

500000 × 25 % 125000

✓ SELF-CHECK Complete the problems, then check your answers in the back of the book.

Find the annual cost of carrying an inventory.

1. Inventory value is $200,000. Estimated annual cost is 30% value. **$60,000**
2. Inventory value is $85,000. Estimated annual cost is 20% of value. **$17,000**

PROBLEMS

	3.	4.	5.	6.	7.	8.
Value of Inventory	$960.00	$14,000.00	$7800.00	$418.40	$1347.86	$5147.95
Percent	25%	35%	15%	22%	28.5%	32.5%
Annual Cost of Carrying Inventory	? **$240**	? **$4900**	? **$1170**	? **$92.05**	? **$384.14**	? **$1673.08**

MATHEMATICAL NOTES

Instruct students to use the following methods to reduce the amount of computations, as well as to double check for accuracy:

a. For Problem 9, change 25% to $\frac{1}{4}$, then divide $16,800 by 4.
b. For Problem 10, compute 32% of $82, then multiply the result by 1000. This method can also be applied to Problems 11, 12, and 13.
c. To check for accuracy in Problem 13a, add the expenses. The sum should be equal to 26% of $37,470.

9. Coral Fabric Company.
Maintains $16,800 inventory.
Costs about 25% of value to maintain.
What is the approximate annual cost of carrying the inventory? **$4200**

10. The Steel Assembly Company.
Maintains $82,000 inventory.
Costs about 32% of value to maintain.
What is the approximate annual cost of carrying the inventory? **$26,240**

11. Central Drug Stores estimates the cost of carrying its inventory of goods to be 18% of the value of the merchandise. About how much does it cost Central Drug Stores annually to carry a $21,560 inventory? **$3880.80**

12. Ace Iron and Steel, Inc., carries a $64,800 inventory. The company estimates the annual cost of carrying the inventory at 27% of the value of the inventory. What is the approximate annual cost of carrying the inventory? **$17,496**

13. Baker Wholesale Grocers, Inc., estimates the annual cost of carrying its inventory at 26% of the value of the inventory. The 26% is broken down as follows:

13a.	Type of Expense	Percent	14.
$2622.90	Spoilage and physical deterioration	7.0%	7000
4121.70	Interest	11.0%	11,000
1498.80	Handling	4.0%	4000
749.40	Storage facilities	2.0%	2000
374.70	Transportation	1.0%	1000
187.35	Taxes	0.5%	500
187.35	Insurance	0.5%	500
$9742.20	Total	26.0%	$26,000

a. Baker generally carries an inventory valued at $37,470. What is the approximate annual cost for each expense of carrying the inventory?

b. Baker was able to reduce its inventory to one half of its value and still meet customer orders. About how much did Baker save on the total cost of carrying its inventory? **$4871.10**

14. In problem 13, if Baker carried an inventory valued at $100,000, what would be the approximate annual cost for each expense of carrying the inventory? **See above.**

MAINTAINING YOUR SKILLS Look up the skill in parentheses if you need help or more practice.

Multiply. **(Skill 30)**

15. $120,000 × 20% **$24,000**
16. $96,400 × 25% **$24,100**
17. $146,500 × 30% **$43,950**
18. $196,750 × 15% **$29,512.50**
19. $71,630 × 35% **$25,070.50**
20. $347,840 × 28% **$97,395.20**
21. $516,780 × 37.5% **$193,792.50**
22. $7893 × 23.4% **$1846.96**
23. $11,746 × 12.5% **$1468.25**
24. $2,147,894 × 16% **$343,663.04**
25. $714,816 × 31.7% **$226,596.67**
26. $486.92 × 41.3% **$201.10**

PRACTICE AND APPLY

The following problems can be assigned for classwork and the answers checked in class to help students master the objective of the lesson.

- Guided Practice: 1–6
- Independent Practice: 7, 8, 10, 11

WRAP-UP

Call upon different students to explain how to compute the annual cost of carrying an inventory. Ask them to define the term *inventory value*.

Assignment Guide

- Basic: 7–13, 15–25 odd
- Average: 9, 12–14, 16–26 even

ALTERNATIVE STRATEGIES: Enrichment

The cost of carrying an inventory is given in the text as ranging from 18% (#11) to 32% (#10) of the value of the inventory. Ask your students to try and find out some actual figures in your area. Software suppliers with inventory control systems might have such information. Actual companies themselves would also have that information available. Twenty-five percent (25%) seems a good average.

LESSON PLAN 17-5 Transportation by Air

FOCUS

To introduce this lesson, ask students if they know the names of any air freight companies. Point out that the air freight industry has become one of the fastest growing and most competitive of all industries. TV commercials for Purolator, Federal Express, Emery, and UPS attest to this fact.

TEACH

Tell students that air freight is generally used when rapid delivery is desired. Air freight charges include pickup and delivery charges, the actual air freight charges, and a federal tax. Emphasize that the federal tax is usually applied to the air freight charges only, not to the pickup and delivery charges.

All of the companies mentioned in the Focus above publish their service rates. You might wish to call one of them and ask for a copy of their rates for use with students.

When discussing the Example, ask students why 700 and 650 are divided by 100 in Steps 1–3. (The air freight rate is based on 100-pound shipments.)

Warm-Up Exercises

1. 637 ÷ 100 6.37
2. 1480 ÷ 100 14.8
3. (164 ÷ 100) × \$61.70 \$101.19
4. \$58.75 × (820 ÷ 100) \$481.75
5. \$31.40 + \$481.75 + \$46.90 + \$24.09 \$584.14

17-5 Transportation by Air

OBJECTIVE
Compute the total shipping cost by air freight.

Your business may ship goods to its customers by **air freight.** Either you or the customer pays the shipping charges. The total cost of shipping goods by air freight includes pickup and delivery to the airport, air freight charges to the city of destination, local delivery from the airport to the final destination, and federal tax, where it applies.

Total Shipping Cost = Pickup Charge + Air Freight Charge + Delivery Charge + Federal Tax

EXAMPLE *Skills* 5, 8, 10, 30 *Application* A *Term* Air freight

The Innovative Auto Company is shipping automotive equipment by air freight. The packaged equipment weighs 650 pounds. Express Pickup Service charges \$6.85 for each 100 pounds or fractional part to transport the equipment to the airport. The air freight rate is \$67.50 per 100 pounds, charged on the actual number of pounds shipped. Speedy Delivery Service charges \$9.10 per 100 pounds, in increments of 50 pounds, to deliver the equipment. In addition, there is a 5% federal tax on the air freight charge because it involves interstate transportation. What is the total cost of shipping the equipment?

SOLUTION

A. Find the **pickup charge.**
\$6.85 × (700 ÷ 100) = \$ 47.95 pickup charge

B. Find the **air freight charge.**
\$67.50 × (650 ÷ 100) = \$438.75 air freight charge

C. Find the **delivery charge.**
\$9.10 × (650 ÷ 100) = \$ 59.15 delivery charge

D. Find the **federal tax** of 5%.
\$438.75 × 5% = \$21.9375 = \$ 21.94 federal tax
\$567.79 total shipping cost

700 ÷ 100 × 6.85 = 47.95 M+ 650 ÷ 100 × 67.5 = 438.75 M+ 650 ÷ 100 × 9.1 = 59.15 M+ 438.75 × 5 % 21.9375 M+ RM 567.7875

✔ SELF-CHECK Complete the problems, then check your answers in the back of the book.

1. Federal tax is 6% of the air freight charge. Air freight charge is \$246.50. Find the federal tax. **\$14.79**
2. Pickup charge is \$50.40. Air freight charge is \$378. Delivery charge is \$67.50. Federal tax is \$20.79. Find the total shipping cost. **\$516.69**

CULTURAL ANGLES

The use of air freight to ship goods to customers has brought foreign markets much closer to the United States. Ask one or two students to contact an air freight company in your area and ask for their service rates for worldwide shipments. Share this information with all members of the class.

PROBLEMS

	Goods	Cost of Pickup	+	Cost of Air Freight	+	Cost of Delivery	+	Federal Tax	=	Total Cost
3.	Auto parts	$102.00	+	$474.00	+	$63.00	+	$23.70	=	? $662.70
4.	Arc welders	$138.60	+	$828.00	+	$73.20	+	$45.54	=	? $1085.34
5.	Cameras	$ 25.00	+	$285.60	+	$37.50	+	$17.14	=	? $365.24
6.	Books	$ 76.50	+	$378.00	+	$67.50	+	$24.57	=	? $546.57

For problems 7 and 8, charges are per 100 pounds or fractional part.

7. 400 pounds of battery chargers.
$8.50 per 100 pounds for pickup.
$65.50 per 100 pounds for air freight
$7.50 per 100 pounds for delivery.
$5\frac{1}{2}$% federal tax on air freight charge. **$340.41**
What is the total shipping cost?

8. 540 pounds of 10-speed bicycles.
$5.60 per 100 pounds for pickup.
$62 per 100 pounds for air freight.
$6.40 per 100 pounds for delivery.
$6\frac{1}{2}$% federal tax on air freight charge. **$468.18**
What is the total shipping cost?

F.Y.I.
About 11.5% of airline revenues are derived from air freight and express package services.

9. A crate of musical instruments that weighs 470 pounds is shipped by air freight. The costs per 100 pounds or fractional part are: $5.00 for pickup, $78 for air freight, and $7.50 for delivery. A 5% federal tax on the air freight charge is added. What is the total shipping cost? **$472.00**

10. New Century is shipping 1460 pounds of prototype CAD-CAM equipment by air freight. Pickup is $6.50 per 100 pounds or fractional part. Delivery is $5.65 per 100 pounds or fractional part. The air freight is $64.50 per 100 pounds, charged on the actual number of pounds shipped. There is a 4% tax on the air freight and delivery charges. What is the total shipping cost? **$1165.01**

11. AB Consultants, Inc., is shipping 48 pounds of computer tapes by air freight from Detroit to Tampa. Reliable Service charges $6.85 per 100 pounds, in increments of 50 pounds, to transport the tapes to the airport. ATP Air Freight charges $0.395 per pound, actual weight, with a minimum charge of $30. Central Carrier charges $9.10 per 100 pounds or fractional part to deliver the tapes. There is a 6% federal tax on the air freight. What is the total shipping cost? **$44.33**

MAINTAINING YOUR SKILLS

Look up the skills in parentheses if you need help or more practice.

Find the percentage. Round answers to the nearest cent. **(Skill 30)**

12. $417.80 × 5% **$20.89**

13. $1216.90 × 6% **$73.01**

14. $641.85 × 5 % **$35.30**

15. $347.84 × 6 % **$22.61**

Add. **(Skill 5)**

16. $54.50 + $317.80 + $41.70 + $15.89 **$429.89**

17. $31.75 + $743.86 + $71.80 + $44.63 **$892.04**

PRACTICE AND APPLY

The following problems can be assigned for classwork and the answers checked in class to help students master the objective of the lesson.

- Guided Practice: 1–6
- Independent Practice: 7, 9

WRAP-UP

Tell students to close their books. Then call upon three students to come to the chalkboard and write the formula for finding the total shipping cost. Ask other students if all of the formulas are correct.

Assignment Guide

- Basic: 7–10, 12–17
- Average: 8, 10, 11, 13–17 odd

ALTERNATIVE STRATEGIES: Enrichment

Burlington, Federal Express, United Parcel Service, Emery, Purolator, and, to a lesser degree, the United States Postal Service are some air freight companies. Divide your students into groups and ask each group to investigate and report on one of the companies. You might wish to limit the assignment by asking them to report on one aspect of the service they provide. For example, ask about the overnight delivery of an envelope or a pouch; or the second day delivery of same.

LESSON PLAN 17-6 Transportation by Truck

FOCUS

To motivate this lesson, have students suggest some advantages of transportation by truck. List the advantages on the chalkboard. (convenience, flexible scheduling, and door-to-door service)

TEACH

Before working through the solution to the Example, practice reading the basic rates table with students. Then go over the three steps in the solution. Ask students to think about some of the factors that would need to be considered when choosing between shipping by air or by truck. Elicit factors such as size, weight, type of product, the distance involved, and cost. Ask students how they would ship flowers or pharmaceuticals. Which is more expensive, air freight or truck freight?

Warm-Up Exercises

1. 729 ÷ 100 7.29
2. 416 ÷ 100 4.16
3. 7.29 × $27.81 $202.73
4. 4.16 × $35.01 $145.64
5. (6150 ÷ 100) × $14.37 $883.76
6. (1547 ÷ 100) × $27.06 $418.62
7. (412 ÷ 100) × $21.99 $90.60

17-6 Transportation by Truck

OBJECTIVE
Compute the total shipping cost by truck.

Your business may ship its goods by truck. The **basic rate** for shipping by truck depends on the classification of the goods that are shipped. The classification is determined by the type and value of the goods. The basic rates differ according to the weight of the goods and the distance that they are being shipped, and they may also vary from carrier to carrier. The basic rates are often given per 100 pounds. Most companies charge only for the actual weight of the goods being shipped.

Total Shipping Cost = Weight × Basic Rate

EXAMPLE *Skills* 8, 10 *Application* C *Term* Basic rate

The King Spark Plug Co. is shipping 4470 pounds of spark plugs (class 100) via Motor Express. Motor Express will pick up the spark plugs at the manufacturing plant and deliver them to an automobile parts distributor. The distance is 325 miles. What is the total shipping cost?

MOTOR EXPRESS...BASIC RATES PER 100 POUNDS... CLASS 100 ITEMS

Weight Group (in pounds)	Distance (in miles) 1–100	101–200	201–300	301–400	401–500
0–500	$24.35	$28.17	$33.77	$36.67	$40.10
501–1000	19.81	22.54	26.51	29.28	32.08
1001–2000	14.37	16.63	19.85	21.61	23.66
2001–5000	11.94	13.81	16.55	17.95	19.65
5001–10,000	8.54	9.87	11.77	12.82	14.04

SOLUTION

A. Find the **basic rate.** (Refer to the table above.)
Weight group: 2001–5000 pounds
Distance: 301–400 miles........................$17.95 basic rate

B. Find the **weight** in hundreds of pounds.
4470 ÷ 100 = 44.7 weight

C. Find the **total shipping cost.**
Weight × Basic Rate
44.7 × $17.95 = $802.365 = $802.37 total shipping cost

4470 ÷ 100 = 44.7 × 17.95 = 802.365

✔ SELF-CHECK Complete the problems, then check your answers in the back of the book.

1. Using the table above, find the basic rate to ship 420 pounds 425 miles. $40.10
2. Find the total shipping cost. $168.42

ALTERNATIVE ASSESSMENT

Write the following on the chalkboard and have students complete each formula.

1. Storage Space = ?
2. Inventory = ?
3. Inventory Value = ?
4. Annual Cost of Carrying Inventory = ?
5. Total Shipping Cost (by air) = ?
6. Total Shipping Cost (by truck) = ?

PROBLEMS

Use the table on page 462 to find the basic rates.

	3.	4.	5.	6.	7.	8.
Weight (pounds)	9470	620	1476	3719	381	2147
Distance (miles)	125	251	18	376	391	474
Basic Rate	$9.87	$26.51	? $14.37	? $17.95	? $36.67	? $19.65
Total Shipping Cost	? $934.69	? $164.36	? $212.10	? $667.56	? $139.71	? $421.89

9. 2470 pounds of ski boots. Transported 315 miles. What is the total shipping cost? **$443.37**

10. 176 pounds of garden hose. Transported 437 miles. What is the total shipping cost? **$70.58**

11. A shipment of ski jackets weighing 384 pounds is transported by truck from the Out Country Clothing plant to its Waterside warehouse. The distance is 278 miles. What is the shipping cost? **$129.68**

12. A shipment of brooms and brushes is sent by truck from Green Bay, Wisconsin, to St. Louis, Missouri. The distance is 445 miles. The brooms and brushes weigh 2447 pounds. What is the shipping cost? **$480.84**

13. A 641-pound shipment of seat covers is transported from The Cover Factory, Inc., to the national distribution center in Chicago. The distance is 124 miles. What is the shipping charge? **$144.48**

14. A 500-pound shipment of darkroom equipment and chemicals is taken from Rochester, New York, to Cincinnati, Ohio. The distance is 491 miles. What is the shipping charge? **$200.50**

15. A shipment of drive shafts is being sent from Replacement Parts' central warehouse to a local distribution center in Vancouver, British Columbia. The distance is 454 miles. The drive shafts weigh a total of 6418 pounds. What is the shipping cost? **$901.09**

F.Y.I.
The vast network of highways, roads, and streets in the United States handles 135 million automobiles, 44 million trucks, and 1 million buses per day (Transport Topics, June 24, 1991).

MAINTAINING YOUR SKILLS

Look up the skill in parentheses if you need help or more practice.

Multiply. **(Skill 8)**

16. 4.70 × $32.13 **$151.01**
17. 11.60 × $20.40 **$236.64**
18. 41.73 × $15.30 **$638.47**
19. 14.84 × $31.02 **$460.34**
20. 35.72 × $12.84 **$458.64**
21. 17.13 × $12.66 **$216.87**
22. 7.56 × $27.81 **$210.24**
23. 20.72 × $17.64 **$365.50**
24. 81.48 × $10.25 **$835.17**
25. 9.12 × $33.00 **$300.96**
26. 7.73 × $25.41 **$196.42**
27. 1.43 × $19.35 **$27.67**
28. 19.11 × $22.62 **$432.27**
29. 97.87 × $12.51 **$1224.35**
30. 61.47 × $8.43 **$518.19**
31. 41.67 × $10.38 **$432.53**
32. 0.71 × $29.58 **$21.00**
33. 3.97 × $37.11 **$147.33**

PRACTICE AND APPLY

The following problems can be assigned for classwork and the answers checked in class to help students master the objective of the lesson.

- Guided Practice: 1–8
- Independent Practice: 9–11

WRAP-UP

Choose different combinations of weight groups and distances and ask students for the basic rates, using the table in the Example. For one choice of values, have students compute the total shipping cost.

Assignment Guide

- Basic: 9–14, 16–30 even
- Average: 12–15, 17–33 odd

ALTERNATIVE STRATEGIES: Reteaching

Review the use of the rate table on page 462 using these problems:

	Weight (Pounds)	Distance (Miles)	Basic Rate	Weight in Hundreds of Pounds	Total Shipping Cost
1)	480	98	($24.35)	(4.80)	($116.88)
2)	2470	234	($16.55)	(24.70)	($408.79)
3)	9426	489	($14.04)	(94.26)	($1323.41)

LESSON PLAN
Reviewing the Basics

The exercises on this page review skills, applications, and terms used in the unit. You can use the exercises to assess informally students' proficiency with this material.

The page can be used for guided practice and independent practice. You can work through a selection of the exercises together with students, and thus see immediately if they know how to do them, and you can then assign some of the exercises for independent practice. Be sure to go over the answers to all assigned exercises.

Reviewing the Basics

Skills

Round answers to the nearest hundred.

(Skill 2) **1.** 462 **500** **2.** 3475 **3500** **3.** 1147 **1100** **4.** 8169 **8200**

Solve.

(Skill 3) **5.** 47 + 64 **111** **6.** 473 + 349 **822** **7.** 719 + 582 **1301**

(Skill 4) **8.** 76 − 35 **41** **9.** 417 − 249 **168** **10.** 219 − 135 **84**

(Skill 8) **11.** $47.52 × 3.4 **$161.57** **12.** $71.83 × 12.44 **$893.57** **13.** $89.91 × 431.51 **$38,797.06**

Round answers to the nearest cent.

(Skill 10) **14.** $217.84 ÷ 19 **$11.47** **15.** $4175 ÷ 48 **$86.98** **16.** $1241.84 ÷ 36 **$34.50**

(Skill 30) **17.** 5.5% of $189.97 **$10.45** **18.** 35% of $256.73 **$89.86** **19.** 1.75% of $47.79 **$0.84**

Applications

Find the shipping charge.

(Application C)

20. Weight: 3478 pounds.
Distance: 1243 miles. **$1385.64**

21. Weight: 8217 pounds.
Distance: 715 miles. **$2177.51**

BASIC RATES PER 100 POUNDS

Weight Group (in pounds)	Distance (in miles)	
	500–1000	1001–1500
2001–5000	$31.12	$39.84
5001–10,000	26.50	35.25

Find the average cost to the nearest cent.

(Application Q)

22. 80 units at $3.50 each and 75 units at $3.65 each. **$3.57**

23. 215 units at $0.72 each and 100 units at $0.75 each. **$0.73**

Find the volume.

(Application Z)

24. Box is 6 ft by 3 ft by 8.5 ft. **153 ft^3**

25. Box is 15 m by 12 m by 7 m. **1260 m^3**

Terms

Match each term with its definition on the right.

b **26.** Inventory

a **27.** Average-cost method

d **28.** Air freight

c **29.** Basic rate

e **30.** Storage space

a. a way of calculating an inventory's value

b. the number of each item you have in stock

c. the charge per 100 pounds to ship goods by truck

d. a way of shipping goods to customers

e. a place in which to keep products until they are needed

f. includes taxes, storage fees, and handling costs

Refer to your reference files in the back of the book if you need help.

Unit Test

Lesson 17-1

1. The Data-Base Equipment Company manufactures laser printers. Each printer is stored in a box measuring 2 feet high, 1.5 feet wide, and 2.5 feet deep. How many cubic feet of space does Data-Base need to store 744 laser printers? **5580 ft^3**

Lesson 17-2

2. How many StraightAway models does Luxury Motor Coach have on its lot on October 1?

ITEM: StraightAway
STOCK NUMBER: JR 2201F

Month of	Opening Balance	Receipts	Issues	Inventory at End of Month
August	42	5	12	? **35**
September	? **35**	6	18	? **23**
October	? **23**			

Lesson 17-3

3. The Farm Store took inventory of its heavy-duty tarps. On March 1, the Farm Store had 40 heavy-duty tarps in stock. Using the average-cost method, what is the value of the inventory on March 1? **$401.20**

ITEM: Heavy Duty Tarps
STOCK NUMBER: 3251

Date		Receipts	Unit Cost	Total Cost
Feb 1	Opening Balance	50	$ 9.87	**$493.50** ?
Feb 18		25	10.12	**$253.00** ?
Feb 27		40	10.18	**$407.20** ?
	Total	? **115**		**$1153.70** ?

Lesson 17-4

4. City Cleaners maintains a soap inventory valued at $14,700. The cost of maintaining the inventory is approximately 20% of the value of the inventory. What is the approximate annual cost of carrying this inventory? **$2940**

Lesson 17-5

5. Cloud Aeronautics is shipping 186 pounds of flight instruments by air freight. The costs per 100 pounds or fractional part are: $6.76 for pick-up, $21.45 for air freight, and $5.75 for delivery. A 5% federal tax is added to the air freight charge. What is the total cost to deliver the instruments? **$70.07**

Lesson 17-6

6. The Barstow Company is shipping 846 pounds of automotive equipment. Commercial Trucking will handle the shipment. The distance is 180 miles. What is the total shipping cost? **$197.12**

COMMERCIAL TRUCKING CLASS 100 ITEMS BASIC RATES PER100 POUNDS

Weight Group (in pounds)	Distance (in miles)		
	1–100	101–200	201–300
0–500	$22.65	$25.60	$28.55
501–1000	$20.95	$23.30	$25.30

LESSON PLAN
Unit Test

Students should do the Unit Test on their own. Each problem on the test is keyed to a lesson in the unit. Students having difficulty with any particular problem should review the Example in the appropriate lesson and be assigned some of the Independent Practice problems for additional practice.

LESSON PLAN
A Spreadsheet Application

USING TECHNOLOGY
You may want to review the average-cost method for valuing an inventory before having students work on this application. (See Lesson 17-3.) Also, in Problem 13, ask students to write out or tell you the meanings of FIFO and LIFO.

Ask students why computers are used to keep track of inventories and their values. (accurate, efficient, cost effective) Also ask if a business would ever want to verify an inventory by making a physical count. (yes) Why? (Items may have been stolen or damaged.)

A SPREADSHEET APPLICATION

Warehousing and Distribution

To complete this spreadsheet application, you will need the template diskette for *Mathematics with Business Applications.* Follow the directions in the User's Guide to complete this activity.

Input the information in the following problems to find the balance on hand for each date, the average cost per unit, and the value of the inventory using the average-cost method. Then answer the questions that follow.

Date	Unit Cost	Receipts (Quantity IN)	Issues (Quantity OUT)	Balance on Hand
10/01	$17.45	100		?
10/05			75	?
10/09	17.20	80		?
10/12	17.00	250		?
10/19			98	?
10/26			123	?
10/29	16.50	100		?
10/30			54	?

What is the average cost per unit?

What is the inventory value using the average-cost method?

1. What is the balance on hand after 10/01?
2. What is the balance on hand after 10/05?

A SPREADSHEET APPLICATION (CONTINUED)

3. What is the value of the 80 items received on 10/09?

4. What is the balance on hand after 10/12?

5. What is the balance on hand after 10/26?

6. What is the ending balance on 10/30?

7. What is the average cost per item?

8. How did the computer calculate the average cost per unit?

9. What is the inventory value using the average-cost method?

10. How did the computer calculate the inventory value using the average cost per unit?

11. Suppose the unit cost for each receipt (10/01, 10/09, 10/12, and 10/29) equaled $17.00. What would the average cost per unit equal?

12. In the situation described in #11, what would be the inventory value using the average-cost method?

13. In the given situation, a declining economy, where unit costs are decreasing, would FIFO or LIFO result in the higher valued inventory? Why?

14. In the given situation, a declining economy, where unit costs are decreasing, would FIFO or LIFO result in the lower valued inventory? Why?

15. What is the value of this inventory using FIFO?

16. What is the value of this inventory using LIFO?

467

LESSON PLAN
Career Wise

APPLICATIONS ON THE JOB

Point out that a shipping clerk has an important responsibility in a mail-order business. If a package is not complete, an invoice is not correct, or the shipping schedule is not met, customers can become dissatisfied. This could hurt the profits of the business because sales may decline.

Ask students what it means to say that the shipping costs are a *function* of the amount of the purchase. (The costs can change when the amount of purchase changes.) Discuss the meanings of the open and closed dots on the step graph. Ask students if they think the invoices can be used to keep track of inventory. (yes) Why? (The invoices list what has been sold.)

CAREER WISE

Shipping Clerk

Mail-order shopping has become big business. Rosa DiPace knows this well, since she works for a mail-order house that boasts it can ship within 24 hours of the time the order is received. Hundreds of credit card orders are placed each day.

Rosa, along with others, is responsible for making sure that the package is complete, the invoice is correct, and that shipping schedules are met. Her company is liable for the goods until they are secured aboard the trucks that deliver them to customers.

The step graph shows the shipping cost as a function of the amount of the purchase. The graph shows, for example, that it costs $5.50 to ship an order valued at between $20.00 and $29.99.

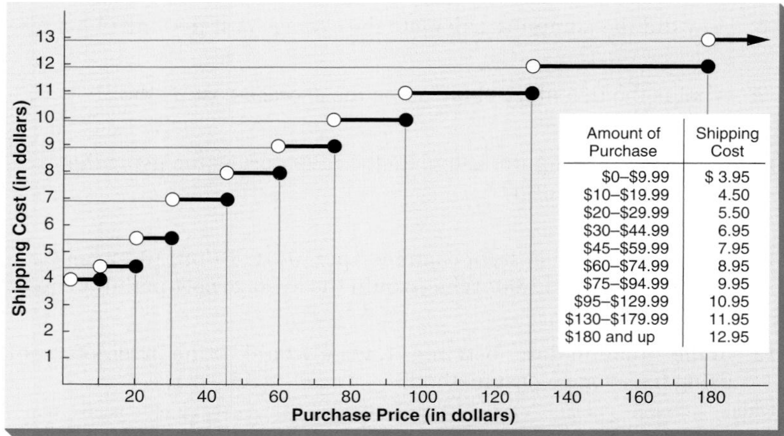

Amount of Purchase	Shipping Cost
$0–$9.99	$ 3.95
$10–$19.99	4.50
$20–$29.99	5.50
$30–$44.99	6.95
$45–$59.99	7.95
$60–$74.99	8.95
$75–$94.99	9.95
$95–$129.99	10.95
$130–$179.99	11.95
$180 and up	12.95

Check Your Understanding

1. True or false: It costs the same to ship an order valued at $110 as it does an order valued at $100. **True, both amounts are between $95.00 and $129.99.**
2. Find the difference between shipping an order valued at $48 and an order valued at $26. **$7.95 − $5.50 = $2.45**
3. True or false: The cost of shipping an order valued at $320 is twice that of an order valued at $160. **False, $12.95 is not twice $11.95.**
4. How much less does it cost to ship one order valued at $210 than an order valued at $100 and an order valued at $110? **$10.95 + $10.95 − $12.95 = $8.95**

U N I T

18

Services

Rent for your office space, maintenance costs of the building, and costs of utilities are all part of business *services*. Many businesses hire professional services for building maintenance and telephone installation and repair. In addition, your business may hire professional *consultants* to advise you on particular problems.

O U T L I N E

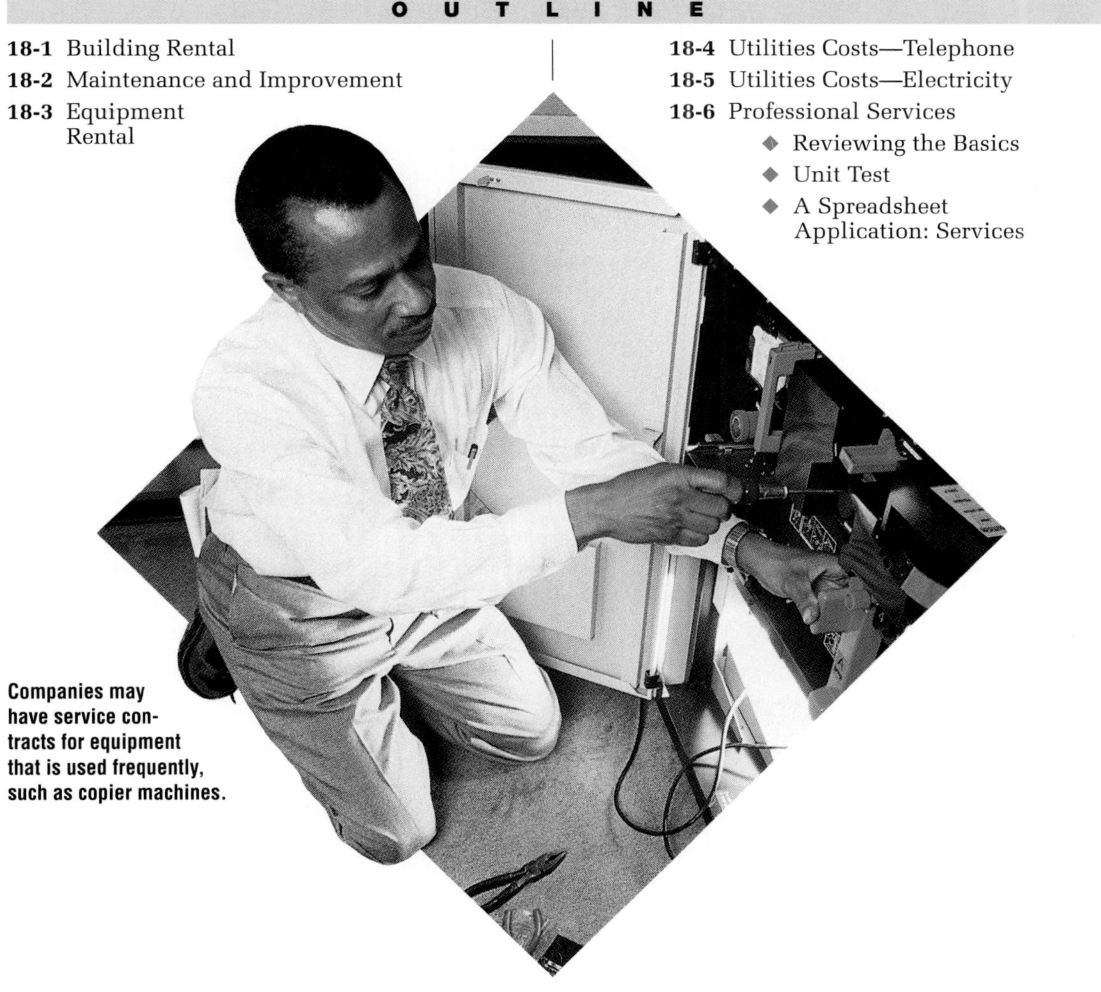

Companies may have service contracts for equipment that is used frequently, such as copier machines.

LESSON PLAN
UNIT 18
SERVICES

INTRODUCING THE UNIT

Businesses must pay many overhead expenses—utilities, rent, equipment, insurance—which are part of the cost of doing business. Make sure students understand that business managers, like themselves, try to get the most for their money. In this unit, students will explore some of the choices business managers face, and learn how to calculate these costs in order to make the best decision.

469

LESSON PLAN
18-1 Building Rental

FOCUS
Bring to class the commercial real estate section of your local newspaper. Pass it around the class and have individual students read various advertisements. Point out that rentals for most commercial properties are based on square footage. In this lesson, students will compute the monthly rental cost of a building or office.

TEACH
To develop students' sense of square footage and the cost in rental dollars, have two students measure your classroom. Use one of the square-foot prices from the newspaper and compute the rental cost.

Discuss with students the reasons for renting space rather than owning a building. (Retail stores have better exposure in a shopping center or mall; most small businesses cannot afford to buy or build a building; a company's headquarters could be in Chicago, but it needs office space for sales and service people in other areas.)

Warm-Up Exercises
1. 30 ft × 50 ft 1500 ft^2
2. 35 ft × 40 ft 1400 ft^2
3. 1500 × $5.50 $8250
4. 1400 × $9.60 $13,440
5. (6000 × $7.50) ÷ 12 $3750
6. (12,500 × $3.60) ÷ 12 $3750

18-1 Building Rental

OBJECTIVE
Compute the monthly rental charge.

Your business may rent or lease a building or a portion of a building, usually on an annual basis. The building owner may charge a certain rate per square foot per year. Your total monthly rental charge depends on the number of square feet that your business occupies.

$$\text{Monthly Rental Charge} = \frac{\text{Annual Rate} \times \text{Number of Square Feet}}{12}$$

EXAMPLE *Skills* 10, 2 *Application* X *Term* Rent

Ajax Assemblers, Inc., rents a portion of a building owned by The Gray Company. The floor space of Ajax's portion of the building measures 80 feet by 60 feet. The Gray Company charges an annual rate of $5.00 per square foot. To the nearest dollar, what is Ajax's monthly rental charge?

SOLUTION

A. Find the **number of square feet.**
Length × Width
80 ft × 60 ft = 4800 ft^2

B. Find the **monthly rental charge.**
(Annual Rate × Number of Square Feet) ÷ 12
($5.00 × 4800) ÷ 12
$24,000.00 ÷ 12 = $2000.00 monthly rental charge

✓ SELF-CHECK Complete the problems, then check your answers in the back of the book.

Find the number of square feet and the monthly rental charge.

1. 100 ft × 50 ft at $4.00 per square foot annually. **5000 ft^2; $1666.67**
2. 125 ft × 45 ft at $3.50 per square foot annually. **5625 ft^2; $1640.63**

PROBLEMS

Find the monthly rental charge to the nearest dollar.

	Dimensions	=	(Number of Square Feet	×	Annual Rate per Square Foot)	÷ 12	=	Monthly Rental Charge
3.	15 ft by 30 ft	=	(? 450 ft^2	×	$ 8.00)	÷ 12	=	$300
4.	40 ft by 20 ft	=	(? 800 ft^2	×	$ 7.50)	÷ 12	=	$500
5.	25 ft by 60 ft	=	(? 1500 ft^2	×	$10.00)	÷ 12	=	$1250

BUSINESS NOTES
Rental prices per square foot can vary considerably in an area. Ask students why. (The location can be in a more desirable and convenient area; for example, near a major highway.) Also, the **type** of space affects price; a warehouse, for instance, is less expensive because it doesn't require the amenities of an office,—carpeting, individual offices, furniture, and so on.

6. *Westside Herald* warehouse. Dimensions are 45 feet by 50 feet. Annual rental charge is $3.40 per square foot. What is the monthly rental charge? **$638**

7. Earring Tree mall store. Dimensions are 15 feet by 30 feet. Annual rental charge is $10.75 per square foot. What is the monthly rental charge? **$403**

8. The Miller Manufacturing Company is considering the rental of additional manufacturing space at $3.80 per square foot per year. The space Miller wants to rent measures 80 feet by 120 feet. What monthly rent will Miller pay for the additional space? **$3040**

9. The Cave has rented additional mall space to expand its arcade operation. The space measures 30 feet by 40 feet and rents for $15.60 per square foot per year. What monthly rent does the Cave pay for the additional space? **$1560**

10. The Flower Shoppe is opening a store in the warehouse district. The rent is $8.75 per square foot per year plus 5% of the store's gross sales. The area of the store is 2000 square feet. If The Flower Shoppe had $180,000 in gross sales the first year, what monthly rent will it pay? **$2208**

11. The Luncheonette rents a 30-foot by 50-foot area at the Nottingham Mall. The Luncheonette pays $9.20 per square foot per year plus $2\frac{1}{2}$% of gross sales. Last year The Luncheonette had $225,000 in gross sales. What was its monthly rent? **$1619**

12. The local office of Wall Medical Clinic is planning to open a branch office in the Heatherdowns area. The office in the advertisement suits its needs. What is the cost per square foot per year for this office? **$6.43**

HEATHERDOWNS AREA
OFFICE SPACE
1400 sq ft, carpeted and paneled, gas heat and air-conditioned. $750 per mo.
555-6619

13. Some businesses rent office space for 2 or 3 years at a time. Often, the lease for the office space includes a **rent escalation clause.** The rent escalation clause states that the rent may increase by a certain percent after a certain period of time.
    Robon Legal Clinic rents a suburban office space for $8.75 per square foot per year. The dimensions of the office are 35 feet by 40 feet. Robon signed a 2-year lease with a 6% rent escalation clause. The rent will increase by 6% for the second year of the lease. What is the monthly rental charge for each year of the lease? **$1021; $1082**

MAINTAINING YOUR SKILLS Look up the skill in parentheses if you need help or more practice.

Divide. Round answers to the nearest dollar. **(Skill 10)**

14. $18,900 ÷ 12 **$1575**
15. $25,800 ÷ 12 **$2150**
16. $14,949 ÷ 12 **$1246**
17. $9070 ÷ 12 **$756**
18. $31,645 ÷ 12 **$2637**
19. $216,847 ÷ 12 **$18,071**

PRACTICE AND APPLY

The following problems can be assigned for classwork and the answers checked in class to help students master the objective of the lesson.

- Guided Practice: 1–5
- Independent Practice: 6, 7, 9

WRAP-UP

Use one of the local advertisements for rental space, and as a class activity, calculate either the cost per year or the cost per square foot per month, depending on how the advertisement price is shown.

Assignment Guide

- Basic: 6–12, 14–19
- Average: 8, 10–13, 14–19

ALTERNATIVE STRATEGIES: Reteaching

For students who do not use a calculator, suggest that they use the short-cut method of multiplying and dividing by multiples of 10, 100, and 1000. For Problem 10, 5% of $180,000 can be computed by finding 5% of $180, then multiplying the result by 1000. For Problem 12, students can rewrite the annual cost ($9000) as $90 and the square footage (1400) as 14, and then divide. Point out that the result is the same as $9000 ÷ 1400.

LESSON PLAN 18-2 Maintenance and Improvement

FOCUS

Ask students if they have ever done any maintenance jobs, such as mowing a lawn or painting a home. If a student has done painting, ask how he or she charged for the service, by the hour or by the job. Who paid for the paint? Explain that if you own a home or building, the cost of maintenance is your responsibility.

TEACH

Students should be familiar with computing the labor charge, as they studied hourly pay in Lesson 1-1.

Discuss why businesses that own large buildings may prefer to have maintenance departments rather than hire an outside service firm. (Persons employed by the company are on site every day and can take care of most maintenance jobs, such as electrical or plumbing work. Cost may be less.)

Warm-Up Exercises

1. $18 × 6 $108
2. $8 × $1\frac{1}{2}$ × 20 $240
3. $6.40 × 1.5 × 7 $67.20
4. ($7.85 × 6 × 4) + (11 × $15.85) $362.75
5. $1216.43 + $17.89 $1234.32

18-2 Maintenance and Improvement

OBJECTIVE
Compute the total building maintenance charge.

If your business owns a building, you will need people to clean and maintain the building. You may have your own maintenance department, or you may hire a service firm. The total cost of a particular maintenance job generally includes a **labor charge** and a **materials charge.** The labor charge is the cost of paying the people who do the job and is calculated on an hourly basis for each service person.

Total Charge = Labor Charge + Materials Charge

EXAMPLE *Skill* 8 *Application* A *Term* Labor charge

Central Law Offices, Inc., hired Commercial Painting Service to paint its offices. Four painters worked 23 hours each to complete the job. The regular hourly rate for each painter is $19. All work was done on weekends, so the painters were paid time and a half. The painters used 42 gallons of paint, for which they charged $15.75 per gallon. What was the total charge for painting the offices?

SOLUTION

A. Find the **labor charge.**
($19.00 × 1.5) × 23 hrs. × 4 = $2622.00 labor charge

B. Find the **materials charge.**
42 × $15.75 = $661.50 materials charge

C. Find the **total charge.**
Labor Charge + Materials Charge
$2622.00 + $661.50 = $3283.50 total charge

19 [×] 1.5 [×] 23 [×] 4 [=] 2622 [M+] 42 [×] 15.75 [=] 661.5 [M+] [RM] 3283.5

✓ SELF-CHECK Complete the problems, then check your answers in the back of the book.

Find the total charge.

1. 4 people worked 10 hours each at $6.50 per hour. Materials cost $420. **$680**
2. 3 people worked 6 hours each at $8.00 per hour. **$277.65** 9 gallons of paint at $14.85/gal.

PROBLEMS

	(Time Required	×	Number of Employees	×	Hourly Rate	=	Labor Charge)	+	Materials Charge	=	Total
3.	(8 hours	×	2	×	$10.50	=	$168.00)	+	$425.00	=	$593.00
4.	(3 hours	×	12	×	$ 5.75	=	$207.00)	+	0	=	$207.00
5.	(4 $\frac{1}{2}$ hours	×	5	×	$ 9.80	=	$220.50)	+	$947.56	=	$1168.06

COOPERATIVE LEARNING

Have students work in small groups to solve Problem 12. After all groups have completed the problem, have one group's members show at the chalkboard how they solved the problem. Ask the other groups if they agree with the solution.

6. 2 people replace fluorescent lights. Hourly rate of $4.75 each. Job takes 4 hours. What is the labor charge? **$38.00**

7. 4 people clean office area. Hourly rate of $5.37 each. Job takes 12 hours. **$257.76** What is the labor charge?

8. Dr. Alice Desmond is moving from her present office to another office in the same building. It takes 4 people $3\frac{3}{4}$ hours to complete the move. The hourly rate per person is $12.80. What is the total charge? **$192.00**

9. City Cleaners hired 2 carpenters to remodel its store. The carpenters earned $14.95 each per hour. Each carpenter worked $21\frac{1}{2}$ hours. The materials charge was $1617.48. Find the total charge. **$2260.33**

10. The National Freight Company used 6 loads of crushed stone on a terminal lot at $92.75 a load. Ten people were each paid $5.15 an hour to spread the stone. It took 4 hours. What was the total charge? **$762.50**

11. Scott's Supermarket hired 5 people to refinish its floor. The regular hourly rate for each person was $6.15. Because the floor was refinished on Sunday, each worker was paid double time. The job took $5\frac{1}{2}$ hours to complete. The materials charge was $371.85. What was the total charge? **$710.10**

12. Miller's Department Store had a minor fire in its store. Miller's received these estimates for repair work.

Contractor	Hourly Rate	Time Required	Materials Charge
Carpenter	$17.15	$12\frac{1}{2}$ hours	$949.73
Electrician	$18.65	13 hours	$221.87
Painter	$ 9.85	$12\frac{1}{4}$ hours	$214.46

Miller's must decide if it should hire these 3 people or have its own employees repair the damage. Materials would cost Miller's a total of $1517.45. It would take 5 Miller's employees 13 hours each to do the job. Three of the employees each earn $9.40 an hour. The other 2 employees each earn $6.97 an hour. If the price is the only consideration, what should Miller's do? **Hire outside; save $101.72**

13. General Janitorial Service cleans the Second National Bank daily from Monday through Saturday. One General employee works at the bank for 3 hours each day. The employee is paid $4.60 per hour on weekdays and time and a half on Saturdays. When calculating Second National's total charge, General adds 75% of the cost of labor to cover its overhead. How much does General charge the Second National Bank per week? **$156.98**

MAINTAINING YOUR SKILLS Look up the skill in parentheses if you need help or more practice.

Multiply. **(Skill 8)**

14. $3.85 × 5 × 4 **$77.00**
15. $9.75 × 5 × 3 **$146.25**
16. $7.43 × 2.5 × 4 **$74.30**
17. $8.45 × $2\frac{1}{4}$ × 5 **$95.06**
18. $6.53 × $3\frac{3}{4}$ × 6 **$146.93**
19. $5.17 × $4\frac{1}{2}$ × 2 **$46.53**

PRACTICE AND APPLY

The following problems can be assigned for classwork and the answers checked in class to help students master the objective of the lesson.

- Guided Practice: 1–5
- Independent Practice: 6–12 even

WRAP-UP

Refer back to the Example and ask students if Commercial Painting Service didn't charge for time-and-a-half, how much money would Central Law Offices save? ($644)

Assignment Guide

- Basic: 6–12, 14–19
- Average: 7–13 even, 14–19

ALTERNATIVE STRATEGIES: Enrichment

Present the following problem for students to solve. Carson Furniture Company received estimates from two nurseries to landscape its grounds. **Nursery A:** 3 people to work 5 hours each at $8 per hour. Materials cost $1290. **Nursery B:** $1450 for the job (includes labor and materials). Which nursery had the better estimate? (Nursery A at $1410) Carson Furniture asked Nursery B for a breakdown of its costs, which were given as $1200 for materials and 2 people working 8 hours each. What hourly rate was Nursery B charging? ($15.63)

LESSON PLAN
18-3 Equipment Rental

FOCUS

Discuss with students the purpose of renting an item rather than buying it. Use an example of a movie video. If a person wants to see a movie, he or she can rent it, for about $2, rather than buy it, which might cost $40 or more. Explain that businesses also rent equipment and furniture when an item is needed only for a short period of time.

TEACH

When working through the Example, point out that the tax is applied to the rental charge for the length of time rented.

Instruct students to round the per-day or per-month cost to the nearest cent before multiplying by the number of days or months in order to avoid rounding differences.

You might mention that some rental contracts allow the rent to be applied toward the purchase of the item. This is also the case with some consumer goods, such as pianos or guitars.

Warm-Up Exercises

1. $180 + (6% of $180) $190.80
2. $71.86 + ($5\frac{1}{2}$% of $71.86) $75.81
3. 3 × (10% of $416.80) $125.04
4. 7 × (9% of $1417) $892.71
5. 12 × (11% of $71) $93.72

18-3 Equipment Rental

OBJECTIVE
Compute the total equipment rental cost.

Your business may find that it is more economical to rent than to buy certain equipment or furniture. Generally, the total cost of renting items is determined by the length of time for which you rent them. Some states charge a use tax on items that are rented.

Total Rental Cost = (Rental Charge × Time) + Tax

EXAMPLE *Skill* 30 *Application* A *Term* Rent

Tax-Aide, Inc., is renting new furniture for a small, temporary office. Office Rental Company charges 10% of the list price of new furniture per month. The list prices of the pieces that Tax-Aide is renting are:

Item	List Price
Desk, 60 in x 30 in	$259.95
File cabinet, 4-drawer	119.95
Swivel armchair	184.45
2 guest chairs ($97.45 ea)	194.90
Total	$759.25

In addition to the rental charge, there is a 5% tax. What is the total cost of renting the furniture for 4 months?

SOLUTION

A. Find the **rental charge per month.**
10% of list price total
$759.25 × 10% = $75.925 = $75.93 rental charge

B. Find the **tax.**
($75.93 × 4) × 5%
$303.72 × 5% = $15.186 = $15.19 tax

C. Find the **total rental cost.**
(Rental Charge × Time) + Tax
($75.93 × 4) + $15.19
$303.72 + $15.19 = $318.91 total rental cost

759.25 [×] 10 [%] 75.925 75.93 [×] 4 [=] 303.72 [M+] [×] 5 [%] 15.186 [+] [RM] 318.906

✓ SELF-CHECK Complete the problems, then check your answers in the back of the book.

Find the total rental cost.

1. Rental charge: 10% of $640. Tax rate: 6%. Time: 3 months. **$203.52**
2. Rental charge: 8% of $1470. Tax rate: 5.5%. Time: 5 months. **$620.34**

CRITICAL THINKING

To continue the discussion of Problem 8, ask students this question: If the rent of an item is 10% of the list price, should you rent it for 12 months? (No, rental cost is more than the cost to buy.) How long can you rent it before you match the list price? (10 months) Explain why. (10% × 10 months = 100% of list price) Using the video example discussed above, why would someone want to purchase a video? (They want to see it many times.)

PROBLEMS

	Item Price	List Charge	Monthly Charge	Monthly Time	Tax	Cost	Rental
3.	Word processor	$ 947.80	10% of list	4 months	4%	? $94.78	? $394.28
4.	File cabinet	$ 97.45	11% of list	12 months	6%	? $10.72	? $136.36
5.	Fax machine	$1141.80	9% of list	8 months	$6\frac{1}{4}$%	? $102.76	? $873.46

6. Backhoe rented for 3 weeks. Rental charge is $1475 per week. Tax is 5%. **$4646.25** What is the total rental cost?

7. Jackhammer rented for 7 days. Rental charge is $175.95 per day. No tax. **$1231.65** What is the total rental cost?

8. The list price of a personal computer is $2248. The monthly rental charge is $10\frac{1}{2}$% of the list price. There is a $5\frac{1}{4}$% tax per month. What is the total rental charge for 1 year? **$2981.19**

9. The Strong Cement Company plans to rent a bulldozer for 2 days. The rental charge is $212.45 per day plus a flat fee of $45 for delivery and pickup. There is no tax. What is the total rental cost? **$469.90**

10. The Legal Clinic is renting additional furniture for 6 months. The monthly rental charge is 9.5% of the list price. There is an 8% use tax. What is the total rental cost of the following: **$1474.26**

Item	List Price
3 desks with chairs	$329.95 each
10 guest chairs	69.25 each
1 typewriter	712.45 each

11. Harris and Sons, plumbers, rent a ditcher as they need it. The daily rental charge is $197.47. There is a $4\frac{1}{2}$% tax. What is the total cost of renting the ditcher for 10 days? **$2063.56**

12. Airline Travel needs an additional 4-drawer file cabinet for 6 months. Airline received cost information from 2 rental agencies and 1 used-equipment store. If price is the only consideration, which company should Airline contact? **CA Furniture Rental**

CA Furniture Rental	Direct Office Rental	Economy Office Furniture
Per month:	Per month:	• $119.49 purchase price
• $16.45 rental charge	• $17.49 rental charge	• 5% sales tax
• 5% insurance charge	• No insurance charge	• $12.50 delivery charge
• 5% use tax	• 5% use tax	

MAINTAINING YOUR SKILLS Look up the skill in parentheses if you need help or more practice.

Find the percentage. Round answers to the nearest cent. (Skill 30)

13. 5% of $516 **$25.80** **14.** 6% of $48.40 **$2.90** **15.** $5\frac{1}{2}$% of $146.30 **$8.05** **16.** 10% of $271.84 **$27.18**

PRACTICE AND APPLY

The following problems can be assigned for classwork and the answers checked in class to help students master the objective of the lesson.

- Guided Practice: 1–5
- Independent Practice: 6–8

WRAP-UP

Review the solution to Problem 8 with students. Ask: Is it more economical to rent the computer rather than buying it? (No, it cost $733.19 more to rent than to buy.)

Assignment Guide

- Basic: 6–11, 13–16
- Average: 9–12, 13–16

ALTERNATIVE STRATEGIES: Reteaching

Using the Example, have students calculate:

STEP 1: $759.25 × 10% = $75.925 = $75.93

STEP 2: $75.93 × 105% × 4 = $318.906

STEP 3: Show what happens if you don't round in Step 1. $759.25 × 10% × 105% × 4 = $318.885

LESSON PLAN
18-4 Utilities Costs—Telephone

FOCUS
Bring to class copies of an actual residence telephone bill, or prepare a facsimile of one. Distribute the copies to the class. Discuss with students the various charges shown on the bill, such as the basic monthly charge, long distance calls, and so on. Explain that businesses have similar bills, but that the method of billing can vary depending on the type of system, equipment, and service used.

TEACH
To continue the discussion of telephone costs, you might want to bring to class the local telephone directory. In most directories, monthly rates for both residence and business telephone services are given, such as call waiting and call forwarding. Generally, the monthly rate for a residence is lower than the rate for a business. For example, one telephone company charges $12.54 for a residence and $33.99 for a business.

Warm-Up Exercises
1. $26.15 + $22.15 + $1.45 $49.75
2. $26.15 + $0.80 + $3.25 + $2.50 + $0.98 $33.68
3. (114 − 73) × $0.08 $3.28
4. (214 − 165) × $0.09 $4.41
5. 3% of ($27 + $13) $1.20
6. 3% of ($26.15 + $34.70) $1.83

18-4 Utilities Costs—Telephone

OBJECTIVE
Compute the monthly telephone cost.

To operate your business, you will need several **utilities.** Utilities are public services, such as telephone, electricity, water, and gas. Each utility uses a different cost structure for charging its customers. For example, the **basic monthly charge** for your telephone service depends on the number of incoming lines, the type of equipment, and the type of service that you have. If you make more calls than the number included in your basic monthly charge, you must pay an additional amount. A federal excise tax is also added to your telephone charge each month.

Total Cost for Month = Basic Monthly Charge + Cost of Additional Calls + Federal Excise Tax

EXAMPLE *Skills* 5, 30 *Application* A *Term* Basic monthly charge

Andy's Laundry has 1 incoming telephone line. The basic monthly charge for this line is $26.15. The basic monthly charge includes 73 outgoing local calls per month. The cost of additional outgoing local calls is $0.08 each. Andy's made 92 local calls last month. A 3% federal excise tax is added to the bill. What is the total cost of Andy's telephone service for the month?

SOLUTION

A. Find the **cost of additional calls.**
(92 − 73) × $0.08
19 × $0.08 = $1.52 cost for additional calls

B. Find the **federal excise tax.**
($26.15 + $1.52) × 3%
$27.67 × 3% = $0.8301 = $0.83 excise tax

C. Find the **total cost for month.**
Basic Monthly Charge + Cost of Additional Calls + Federal Excise Tax
$26.15 + $1.52 + $0.83 = $28.50 total cost

92 [−] 73 [=] 19 [×] .08 [=] 1.52 [M+] 26.15 [+] 1.52 [=] 27.67 [×] 3 [%] .8301
26.15 [+] [RM] 1.52 [+] .83 [=] 28.5

✓ SELF-CHECK Complete the problems, then check your answers in the back of the book.

Find the total cost for the month.

1. Base charge of $26.15.
10 additional calls at $0.08 each.
3% federal tax. **$27.76**

2. Base charge of $28.45.
30 additional calls at $0.09 each.
3% federal tax. **$32.08**

ALTERNATIVE STRATEGIES: Enrichment
Point out to students that like a personal telephone expense, the monthly telephone cost for a business can vary. That is why it is considered a variable expense rather than a fixed expense, which does not vary from month to month. Finding the best telephone system and service is an important decision for a business. Businesses try to find ways of controlling this cost. One way to help reduce long distance calls is through the use of a facsimile (fax) machine. A fax machine can send a 1-page order or

PROBLEMS

	Basic Monthly Charge	Additional Local Calls	Cost per Additional Call	Cost of All Additional Calls	3% Federal Excise Tax	Total Cost for Month
3.	$26.15	15	$0.08	$1.20	$0.82	**$28.17**
4.	$28.75	8	$0.09	**$0.72**	**$0.88**	**$30.35**
5.	$42.35	21	$0.10	**$2.10**	**$1.33**	**$45.78**
6.	$15.25	0	$0.08	**0**	**$0.46**	**$15.71**

Add a 3% federal excise tax to find the total cost for the month.

7. Spotlight Discount has 1 telephone. Basic monthly charge is $26.15. Includes 73 outgoing local calls. Each additional call costs $0.08. Made 115 outgoing local calls. What is the total cost for the month? **$30.40**

8. Public Pharmacy has 1 telephone. Basic monthly charge is $27.50. Includes 80 outgoing local calls. Each additional call costs $0.09. Made 42 outgoing local calls. What is the total cost for the month? **$28.33**

9. Smith Medical Clinic has 2 telephones. The total basic monthly charge for both telephones is $31.65. The charge includes 160 outgoing local calls. The cost of additional local calls is $0.09 each. Smith Medical Clinic made 196 outgoing local calls this month. What is the total cost of telephone service, including tax, for the month? **$35.94**

For problems 10–11, refer to the chart at the right.

10. Charles Young, Inc., brokers for stocks and bonds, have 5 telephones for a total basic monthly charge of $88.60. The charge includes 500 local outgoing calls. The cost of additional local calls is $0.08 each. Last month, 863 outgoing local calls were made from the office. In addition, long-distance calls totaling $488.32 were made. Charles Young also has call forwarding and line-backer special services. The federal excise tax is applied to all charges. What is the total cost for the month? **$628.77**

11. Prime Realty Company has 3 telephone lines. The monthly charge of $47.85 includes 160 outgoing local calls. Each additional local call costs $0.11. Prime made 180 outgoing local calls this month. Prime also has call waiting and line-backer plus special services. What is the total cost of telephone service, including tax, for the month? **$64.01**

MONTHLY CHARGE PER SPECIAL SERVICE	
Call waiting	$9.60
Call forwarding	3.25
3-way calling	3.25
Line-backer	1.25
Line-backer plus	2.50

MAINTAINING YOUR SKILLS

Look up the skills in parentheses if you need help or more practice.

Add. **(Skills 5, 6, 8, and 30)**

12. $26.15 + $1.60 + $0.83 **$28.58**

13. $32.95 + $2.79 + $1.07 **$36.81**

14. $26.15 + [(103 − 73) × $0.08] **$28.55**

15. $37.45 + $2.45 + $3.25 + (3% of $43.15) **$44.44**

PRACTICE AND APPLY

The following problems can be assigned for classwork and the answers checked in class to help students master the objective of the lesson.

- Guided Practice: 1–4
- Independent Practice: 5–7, 9

WRAP-UP

As a class activity, rework the Example, but change the number of calls made to 105 and the tax to 5%.

Assignment Guide

- Basic: 5–10, 12–15
- Average: 8, 10, 11, 12–15

other document through the telephone to another party in 30 seconds, whereas it would probably take 5 minutes or more to speak directly to the other party. Have students research and report on the cost of using a fax machine.

LESSON PLAN
18-5 Utilities Costs—Electricity

FOCUS

To motivate this lesson on electricity costs, refer to the previous lesson's example. Ask students to name other utilities costs that Andy's Laundry will have in addition to the telephone expense. (electricity or gas, water) Explain that monthly electricity costs can vary for a business based on usage.

TEACH

As you work through the solution to the Example, review the meaning of each term as it is presented. Stress to students that organizing the data, as done in the solution, is an important problem-solving step.

You might want to mention that the cost of electricity for a business is not computed in the same way that it is for a home. Most homes do not have a demand charge. Homeowners usually pay just the energy charge and fuel adjustment charge.

Warm-Up Exercises

1. 100 × $6.54 $654
2. 1000 × $0.078 $78
3. (12,100 − 1000) × $0.065 $721.50
4. 12,100 × $0.016 $193.60
5. $654 + $78 + $721.50 + $193.60 $1647.10

18-5 Utilities Costs—Electricity

OBJECTIVE
Compute the monthly cost for electricity.

The monthly cost of electricity for your business depends on the **demand charge** and the **energy charge.** The demand charge is based on the **peak load** during the month. The peak load is the greatest number of **kilowatts** (kW) that your business uses at one time during the month. The energy charge is based on the total number of **kilowatt-hours** (kW·h) that your business uses during the month. Meters installed at your business record the kilowatt demand and the number of kilowatt-hours. The electric company may add a **fuel adjustment charge** to your monthly bill to help cover increases in the cost of fuel needed to produce your electricity.

Total Cost for Month = **Demand Charge** + **Energy Charge** + **Fuel Adjustment Charge**

EXAMPLE *Skill* 8 *Application* A *Term* Fuel adjustment charge

The Acme Manufacturing Company had a peak load of 100 kilowatts of electricity during April. The demand charge is $6.54 per kilowatt. Acme used a total of 30,000 kilowatt-hours of electricity during the month. The energy charge for the first 1000 kilowatt-hours is $0.076 per kilowatt-hour. The cost of the remaining kilowatt-hours is $0.058 per kilowatt-hour. The fuel adjustment charge for April is $0.0155 per kilowatt-hour. What is the total cost of the electricity that Acme used in April?

SOLUTION

A. Find the **demand charge.**
100 kW × $6.54 = $654.00 demand charge

B. Find the **energy charge.**
(1) 1000 kW·h × $0.076 = $76.00
(2) (30,000 − 1000) kW·h × $0.058 = $1682.00
$76.00 + $1682.00 = $1758.00 energy charge

C. Find the **fuel adjustment charge.**
30,000 kW·h × $0.0155 = $465.00 fuel adjustment charge

D. Find the **total cost for month.**
Demand Charge + Energy Charge + Fuel Adjustment Charge
$654.00 + $1758.00 + $465.00 = $2877.00 total cost

100 [×] 6.54 [M+] 654 1000 [×] .076 [=] 76 30000 [−] 1000 [=] 29000 [×] .058 [=] 1682 76 [+] 1682 [M+] 1758 30000 [×] .0155 [M+] 465 [RM] 2877

COMMUNICATION SKILLS

Have small groups of students contact their local utility companies for whatever information is available on how to conserve that particular company's resource (electricity, natural gas, water). Groups should write a brief report on their findings and present it to the class.

✓ **SELF-CHECK** Complete the problem, then check your answer in the back of the book.

1. Find the total cost for month.
   Demand Charge Energy Charge Fuel Adjustment Charge
   $(100 \times \$6.60) + [(1000 \times \$0.098) + (9000 \times \$0.067)] + (10,000 \times \$0.016)$
   $1521

PROBLEMS

2. Flagg Mattress Company.
   Used 17,000 kW·h in May.
   Peak load of 100 kW.
   Demand charge: $6.45 per kW.
   Energy charge: $0.07 per kW·h.
   Fuel adjustment: $0.018 per kW·h.
   What is the total cost for May?
   $2141

3. Town and Country Mall.
   Used 18,700 kW·h in April.
   Peak load of 90 kW.
   Demand charge: $5.90 per kW.
   Energy charge: $0.08 per kW·h.
   Fuel adjustment: $0.02 per kW·h.
   What is the total cost for April?
   $2401

F.Y.I.
Supplies of natural gas and heating oil are ample and shortages or sharp price rises are unlikely (*Kiplinger*, December 1991, p. 17).

4. City Deli used 8700 kilowatt-hours of electricity with a peak load of 100 kilowatts in June. The demand charge is $7.25 per kilowatt. The energy charge is $0.08 per kilowatt-hour for the first 1000 kilowatt-hours and $0.06 per kilowatt-hour for all kilowatt-hours over 1000. The fuel adjustment charge is $0.03 per kilowatt-hour. What is the total cost of electricity for City Deli for June? **$1528**

5. Pantry Supermarket used 21,400 kilowatt-hours of electricity last month. The peak load for the month was 120 kilowatts. The demand charge is $5.91 per kilowatt. The energy charge per kilowatt-hour is $0.0675 for the first 10,000 kilowatt-hours and $0.0455 for all kilowatt-hours over 10,000. The fuel adjustment charge is $0.015 per kilowatt-hour. What is the total cost of electricity for Pantry Supermarket for last month? **$2223.90**

6. *The City Journal* used 12,417 kilowatt-hours of electricity in August. The peak load during the month was 78 kilowatts. The demand charge is $6.96 per kilowatt. The energy charge per kilowatt-hour is $0.0872 for the first 10,000 kilowatt-hours and $0.0685 for all kilowatt-hours over 10,000. There is no fuel adjustment charge. What is the total cost of electricity for *The City Journal* in August? **$1580.44**

PRACTICE AND APPLY

The following problems can be assigned for classwork and the answers checked in class to help students master the objective of the lesson.

- Guided Practice: 1–3
- Independent Practice: 4, 5, 7

ALTERNATIVE STRATEGIES: Reteaching

Use the rates in the example for a total usage of 40,000 kilowatt hours. Follow this procedure.

STEP 1: Demand charge
$100 \times \$6.54 = \$\ 654.00$

STEP 2: Energy charge
$1000 \times \$0.076 = \$\ 76.00$
$(40,000 - 1000) \times \$0.058 = \$2262.00$

STEP 3: Fuel adjustment charge
$40,000 \times \$0.0155 = \$\ 620.00$

STEP 4: Find the total
$= \$3612.00$

WRAP-UP

Write on the chalkboard *Demand Charge, Energy Charge, Fuel Adjustment Charge*, and *Total Monthly Cost*. Using the data in Problem 4, have volunteers show how to compute each part.

Assignment Guide

- Basic: 4–8, 10–13
- Average: 6, 8, 9, 10–13

For problems 7–9, refer to the chart.

7. Riechle Manufacturing used 15,400 kilowatt-hours of electricity this month; the peak load was 80 kilowatts. What is the total cost of electricity for the month? **$1598.13**

8. Prompt Answering Service, Inc., used 7400 kilowatt-hours of electricity for July. Its peak load was 75 kilowatts. What is the total cost of electricity for the month? **$1060**

9. Combustible, Inc., used a total of 21,760 kilowatt-hours of electricity for the month. The peak load during the month was 170 kilowatts. What is the total cost of electricity for the month? **$2527.53**

NORTHWEST POWER	
Demand Charge	
First 50 kW	$6.54/kW
Over 50 kW	$5.91/kW
Energy Charge	
First 250 kW·h	$0.1055/kW·h
Next 750 kW·h	$0.0954/kW·h
Next 2000 kW·h	$0.0644/kW·h
Next 2000 kW·h	$0.0578/kW·h
Next 5000 kW·h	$0.0488/kW·h
Over 10,000 kW·h	$0.0455/kW·h
Fuel Adjustment	$0.017/kW·h

MAINTAINING YOUR SKILLS Look up the skill in parentheses if you need help or more practice.

Multiply. **(Skill 8)**

10. 100 × $6.47 **$647**

11. 1000 × $0.087 **$87**

12. 8700 × $0.063 **$548.10**

13. 9700 × $0.015 **$145.50**

CULTURAL ANGLES

Have students make a list of the major oil-producing countries in the world. Ask them to identify the countries in the Persian Gulf. Are there any major producers in South America? Is oil used to generate electricity? What sources of energy are used to generate electricity?

18-6 Professional Services

OBJECTIVE
Compute the total cost of professional services.

You may hire professional **consultants** to advise your business on a particular problem. The method of determining each **consultant's fee** may vary. Some consultants charge a flat fee, some charge a percent of the cost of the project, and some charge by the hour.

Total Cost = Sum of Consultants' Fees

EXAMPLE *Skills* 5, 8, 30 *Application* A *Term* Consultant's fee

Appleton Wholesale Grocers, Inc., plans to expand its use of computers to include billing, payroll, and inventory control. The company plans to construct a new building at a cost of $875,000 to house the new computer installation. Appleton hires an architect, a systems analyst, and a computer programmer. The architect charges 7% of the total cost of the building. The systems analyst charges a flat fee of $6000. The computer programmer charges $30 an hour and works 150 hours. What is the total cost of the professional services?

SOLUTION

Find the **sum of consultants' fees.**

Architect: $875,000 × 7% $61,250
Systems analyst: Flat fee 6000
Computer programmer: $30 × 150 4500
Total Cost $71,750

875000 [×] 7 [%] 61250 [M+] 6000 [M+] 30 [×] 150 [=] 4500 [M+] [RM] 71750

✓ SELF-CHECK Complete the problems, then check your answers in the back of the book.

Find the total cost.

1. Architect: 6% of $450,000. Lawyer: 20 hours at $75 per hour. Surveyor: $2500 flat fee. **$31,000**
2. Engineer: 25 hours at $42.50 per hour. Computer programmer: $1000 flat fee. Computer time: $450 flat fee. **$2512.50**

PROBLEMS

	Professional Service	Fee Structure	Project Information	Total Fee
3.	Patent attorney	$75.00 per hour	Worked 20 hours	**$1500** ?
4.	Industrial nurse	$10.25 per hour	Worked 40 hours	**$410** ?
5.	Productivity consultant	$5000 flat fee	Worked 25 hours	**$5000** ?
6.	Planning specialist	5% of project cost	$500,000 project	**$25,000** ?

ALTERNATIVE STRATEGIES: Enrichment
Have students research and report on the charges for consulting services in their area. Have them include lawyers, architects, real estate agents, engineers, computer services and so on.

LESSON PLAN 18-6 Professional Services

FOCUS
Ask students what they would do if they were in an automobile accident and the person in the other car decided to sue for damages. Would you go to court by yourself or would you hire a lawyer? Explain that lawyers are considered consultants. They are experts in their field of law, and they provide professional advice or services for a fee. Have students name other types of consultants. (realtors, stockbrokers, architects)

TEACH
You might want to have a student read the Example to the class. As each consultant and the method of charging for the service is mentioned, write the information on the chalkboard. Ask students to tell you what the total fee is for each consultant.

Discuss with students why they think consultants use different methods for charging for their services. For example, why would a lawyer decide to charge by the hour rather than by the job? (It is not always known how long a job will take.)

Warm-Up Exercises
1. $75 × 20 $1500
2. $135 × 6 $810
3. $2\frac{1}{2}$% of $3,500,000 $87,500
4. 6.5% of $650,000 $42,250
5. $2000 + (20 × $47.50) + (4.5% of $4,000,000) $182,950

PRACTICE AND APPLY

The following problems can be assigned for classwork and the answers checked in class to help students master the objective of the lesson.

- Guided Practice: 1–6
- Independent Practice: 7–10

WRAP-UP

After checking the solutions to the problems assigned in class, ask students what would be the hourly rate for the productivity consultant in Problem 5. ($200 per hour)

Assignment Guide

- Basic: 7–13, 15–20
- Average: 11–14, 15–20

7. Bond broker.
   Paid $1\frac{1}{2}$% of total bonds sold.
   $10,000,000 sold.
   What is the total fee? **$150,000**

8. Real estate broker.
   Paid $3\frac{1}{2}$% of total sale.
   Sale totals $615,500.
   What is the total fee? **$21,542.50**

9. The June Co. hired a staff development specialist to conduct a workshop for 83 of its employees. The specialist was paid $110 per person. What was the total cost of the specialist's services? **$9130**

10. Lakeside Hospital hired an industrial engineering firm to conduct a work sampling on the average nurse's day. Time Study, Inc., did the work sampling and charged $67.50 per hour. It took Time Study, Inc., 32 hours to complete the task. What did the work sampling cost Lakeside Hospital? **$2160**

11. The Central Farmers Co-Op plans to issue $20,000,000 in bonds to pay for an extensive expansion. One bond broker will sell the bonds for a fee of 3.25% of the $20,000,000 face value of the bonds. What will the broker's services cost Central? **$650,000**

12. Modern Service Stations plans to sell a large piece of commercial property. The selling price of the property is $975,000. Roula Mathis will handle the entire transaction for 6.5% of the selling price. What will it cost Modern Service Stations to have Roula handle the transaction? **$63,375**

13. Liberty Bank and Trust plans to build a new branch office. Liberty hired Ed Baker, an architectural consultant. Ed's fee is 7% of the cost of the project. Liberty hired Tina Pike as project engineer for the new branch. Tina's fee is 4.5% of the cost of the project. Liberty hired Sara Charles to take care of the legal aspects of the project. Sara charges $75 per hour. It took her 10.5 hours to prepare the legal documents. The cost of the branch office is estimated to be $4,575,000. What amount should Liberty plan to pay for the 3 professional services? **$526,912.50**

14. Super Food Mart has installed automated checkout equipment in its 6 stores. It hired Robert Case to consult on its installation. Case charged 6% of the installation cost. It also hired Marilyn Lee to instruct its employees on the use of the equipment. Marilyn conducted six 4-hour sessions at $125 per session. The cost of the installation totaled $68,970. What was the total cost for professional services? **$4888.20**

MAINTAINING YOUR SKILLS Look up the skills in parentheses if you need help or more practice.

Solve. **(Skills 5, 8, 30)**

15. $175.00 + $493.75 + $216.80 **$885.55**
16. $14,150.00 + $7185.90 + $417.95 **$21,753.85**
17. (5% of $1,750,000) + $14,750 **$102,250**
18. (7.5% of $863,900) + (25 × $9.50) **$65,030**
19. ($85 × 24.5) + $5000 + (2% of $2,500,000) **$57,082.50**
20. (6% of $3,500,000) + ($75 × 6.5) + $1475 **$211,962.50**

COOPERATIVE LEARNING

Have students work in cooperative groups to create a problem using the topic from one of the lessons in this unit. You might want to assign the groups a particular lesson. Then have each group give its problem to another group to solve. Encourage group participation in the discussion of the problems and their solutions.

Reviewing the Basics

Skills

Round answers to the nearest dollar.

(Skill 2)

1. $42.74 **$43**
2. $117.40 **$117**
3. $4171.60 **$4172**
4. $12,147.83 **$12,148**

Solve. Round answers to the nearest hundredth.

(Skill 3)

5. $179 + $464 + $958 **$1601**
6. $201 + $743 + $6198 **$7142**
7. $17,946 + $9217 **$27,163**

(Skill 4)

8. 743 − 295 **448**
9. 15,000 − 11,763 **3237**
10. 74,000 − 7193 **66,807**

(Skill 5)

11. 916.87 + 1714.31 **2631.18**
12. 716.47 + 176.178 **892.65**
13. 7.418 + 15.79 **23.21**

(Skill 6)

14. 8417.92 − 147.74 **8270.18**
15. 4684.75 − 619.87 **4064.88**
16. 219.47 − 156.26 **63.21**

(Skill 7)

17. 144 × 50 **7200**
18. 36 × 74 **2664**
19. 48 × 150 **7200**

(Skill 8)

20. 26.12 × 615 **16,063.8**
21. 4217.64 × 0.655 **2762.55**
22. 34,157 × 0.48 **16,395.36**

(Skill 10)

23. $4178 ÷ 12 **$348.17**
24. $22,196 ÷ 24 **$924.83**
25. $31,965 ÷ 36 **$887.92**

(Skill 30)

26. $417,800 × 35% **$146,230**
27. $125,750 × 6.5% **$8173.75**
28. $3,450,000 × 1.5% **$51,750**

Applications

Find the area of each rental space.

(Application X)

29. 30 ft by 40 ft **1200 ft²**
30. 25 ft by 45 ft **1125 ft²**
31. 120 ft by 150 ft **18,000 ft²**
32. 15 ft by 15 ft **225 ft²**
33. 10.5 ft by 48.5 ft **509.25 ft²**
34. 115.5 ft by 42.75 ft **4937.63 ft²**

Terms

Use each term in one of the sentences.

d 35. Rent

f 36. Labor charge

a 37. Basic monthly charge

c 38. Fuel adjustment charge

b 39. Consultant's fee

a. Your __?__ for telephone service may include a certain number of outgoing local calls.

b. You will pay a __?__ when you hire a professional person to advise your business on a particular project.

c. A __?__ may be added to your electricity bill.

d. Your business may __?__ rather than buy office space or equipment.

e. The __?__ is the greatest number of kilowatts that your business uses at one time during the month.

f. The cost of having a maintenance firm do a particular job may include a __?__ and a charge for materials.

Refer to your reference files in the back of the book if you need help.

LESSON PLAN
Reviewing the Basics

The exercises on this page review skills, applications, and terms used in the unit. You can use the exercises to assess informally students' proficiency with this material.

The page can be used for guided practice and independent practice. You can work through a selection of the exercises together with students, and thus see immediately if they know how to do them, and you can then assign some of the exercises for independent practice. Be sure to go over the answers to all assigned exercises.

LESSON PLAN
Unit Test

Students should do the Unit Test on their own. Each problem on the test is keyed to a lesson in the unit. Students having difficulty with any particular problem should review the Example in the appropriate lesson and be assigned some of the Independent Practice problems for additional practice.

Unit Test

Lesson 18-1

1. Moore's Nursery rents warehouse space at an annual rate of $4.50 per square foot. Moore's warehouse measures 60 feet by 150 feet. What is the monthly rental charge? **$3375**

Lesson 18-2

2. Town Mall hired 4 high school students to spring-clean the grounds. Each student worked for 2.5 hours at an hourly rate of $4.50. The materials charge was $46.75. What was the total charge for this service? **$91.75**

Lesson 18-3

3. Mary Taylor is opening a branch tax office for 3 months. She plans to rent this furniture at a monthly charge of 10% of the list price plus a 6% tax.

Item	List Price
2 desks	$279.79 each
2 desk chairs	97.50 each
4 guest chairs	74.75 each
1 file cabinet	112.45
1 bookcase	147.50

What is the total rental cost for the furniture? **$417.72**

Lesson 18-4

4. The Greco Import Store pays a basic monthly charge of $26.15 for telephone service. The charge includes 73 outgoing local calls. Each additional outgoing local call costs $0.08. Last month a total of 95 outgoing local calls were made from Greco's. A 3% federal tax was added to the bill for the month. What was the total cost of telephone service for the month? **$28.75**

Lesson 18-5

5. The Norris Company used 11,650 kilowatt-hours of electricity with a peak load of 120 kilowatts in April. The demand charge is $6.54 per kilowatt. The energy charge per kilowatt-hour is $0.076 for the first 10,000 kilowatt-hours and $0.058 for all kilowatt-hours over 10,000. The fuel adjustment charge is $0.0165 per kilowatt-hour. What is the total cost of electricity for The Norris Company for April? **$1832.73**

Lesson 18-6

6. Innovative Automotive, Inc., hired a consultant from Engineering Consortium to help develop a new engine. The consultant's fee was 7.3% of the cost of the project. The project cost $843,500. What was the total cost of the consultant's services? **$61,575.50**

A SPREADSHEET APPLICATION

Services

To complete this spreadsheet application, you will need the template diskette for *Mathematics with Business Applications*. Follow the directions in the *User's Guide* to complete this activity.

Input the information in the following problems to find the monthly rent for square footage, the monthly share of gross sales, and the total monthly rent.

	Store	Rental Space Dimensions	Rent per Sq Ft per Yr	Monthly Sq Ft Rent	Plus % of Gross Sales	Annual Gross Sales	Monthly Share Gross Sales	Total Monthly Rent
1.	Video	45' × 60'	$ 7.50	?	5%	$124,600	?	?
2.	Clothes	90' × 120'	2.75	?	4%	215,780	?	?
3.	Food	20' × 30'	10.25	?	0	–	0	?
4.	Arcade	60' × 75'	9.80	?	0	–	0	?
5.	Photo	50' × 50'	6.75	?	3%	156,860	?	?
6.	Shoe	40' × 65'	8.25	?	2.5%	453,780	?	?
7.	Lunch	35' × 50'	12.50	?	0	–	0	?
8.	Music	50' × 75'	9.50	?	4.5%	267,750	?	?
9.	Flower	25' × 30'	17.50	?	2%	98,574	?	?
10.	Deli	35' × 40'	13.60	?	0	–	0	?

LESSON PLAN
A Spreadsheet Application

USING TECHNOLOGY

After students have completed the application and you have checked their answers with them, ask who might use a computer to solve similar types of problems. (real estate agents, business owners) Ask also if they think the cost of rent is a significant cost for most businesses. (yes)

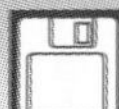

A SPREADSHEET APPLICATION

(CONTINUED)

11. In #1, what is the monthly rent for square footage?

12. In #1, what is the monthly share of gross sales?

13. In #1, what is the total monthly rent?

14. In #2, what is the monthly rent for square footage?

15. In #2, what is the monthly share of gross sales?

16. In #2, what is the total monthly rent?

17. In #1, the rent per square foot per year is almost $5 more than in #2, yet the total monthly rent is less in #1. State two reasons why this is the case.

18. In #3, what is the total monthly rent?

19. In #4, what is the total monthly rent?

20. In #5, what is the monthly rent for square footage?

21. In #5, what is the monthly share of gross sales?

22. In #5, what is the total monthly rent?

23. In #6, what is the *annual* rent for square footage?

24. In #6, what is the *annual* share of gross sales?

25. In #6, what is the total *annual* rent?

26. In #7, what is the total monthly rent?

27. In #8, what is the monthly rent for square footage?

28. In #8, what is the monthly share of gross sales?

29. In #8, what is the total monthly rent?

30. In #9, what is the total monthly rent?

31. In #10, what is the total *annual* rent?

32. In #3, #4, #7, and #10, why does the monthly rent for square footage equal the total monthly rent?

UNIT

19

Accounting

The accounting department prepares the company *payroll register,* which is a record of all your employees' income. The department also keeps records of all your business expenses and the value of your equipment. The value of your equipment goes down, or *depreciates,* yearly because of age or wear and tear. There are four ways to determine depreciation: *straight-line method, sum-of-the-years'-digits method, double-declining-balance method,* and *modified accelerated cost recovery system.*

OUTLINE

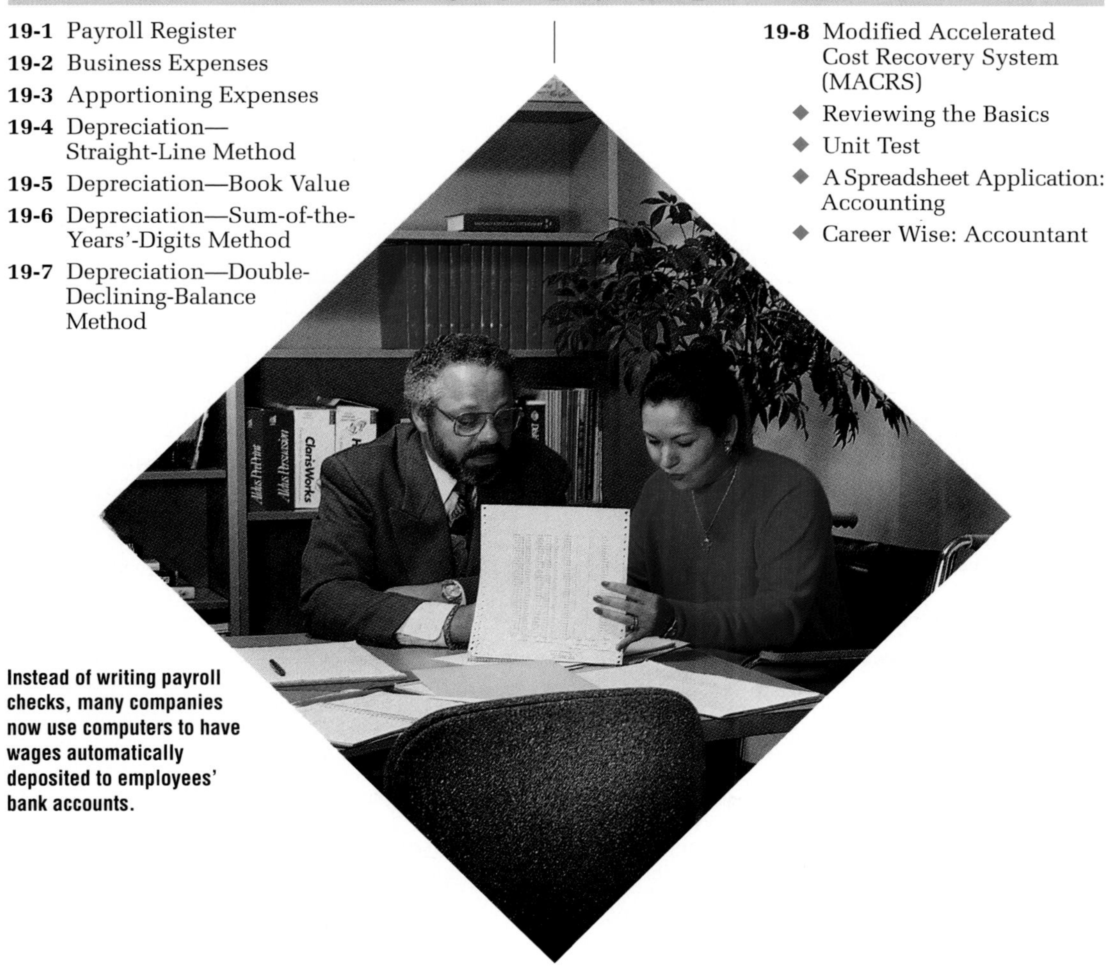

Instead of writing payroll checks, many companies now use computers to have wages automatically deposited to employees' bank accounts.

INTRODUCING THE UNIT

Point out that a major reason every business keeps accurate records of its expenses is for federal and state tax purposes. Taxes are not paid on business expenses, but these deductions must be substantiated by company records. Incomplete or inaccurate records can distort the financial results of a business and may lead to difficulties with the Internal Revenue Service.

LESSON PLAN
19-1 Payroll Register

FOCUS

Ask students who work to bring in a copy of their last pay stub. Make a list of the different types of deductions withheld. Point out that a company must keep a payroll register so that it can keep track of each employee's income and deductions.

TEACH

Companies must keep track of each employee's gross income, deductions, and net pay for auditing and tax purposes. This record may be completed weekly, biweekly, or monthly.

Remind students that computing commissions was covered in Lesson 1-6. Have students verify the gross pay for each employee listed, then subtract the deductions to verify the net pay for each.

19-1 Payroll Register

OBJECTIVE

Complete a payroll register.

A **payroll register** is a record of the gross income, deductions, and net income of your company's employees. You may use a computer to prepare your payroll register. If you prepare the register by hand, you will probably refer to tables to determine the amount of income tax to withhold from each employee's pay.

EXAMPLE *Skills* 5, 6, 8 *Application* C *Term* Payroll register

Natural Foods Center pays its employees weekly. Mary Clark prepares the payroll register for the center's 5 employees from the following information.

Name	Regular Pay	Overtime Pay	Income Tax Information	Health Insurance Coverage
T.L. Wright	$4.50/hr.	Time and a half	Married, 2 allow.	Family
J.A. Bruss	$6.45/hr.	Time and a half	Single, 1 allow.	—
H.T. Perkins	$325/week	—	Single, 2 allow.	Single
N.J. Nystrand	$435/week	—	Married, 3 allow.	Family
W.K. Fine	$280/week plus 5% commission	—	Married, 4 allow.	Family

SOLUTION Mary has prepared the payroll register for the week of March 2. What is the net pay for the week?

$280 plus 5% of $1860 in sales | From federal withholding tables | 6.2% of gross pay | 1.45% of gross pay

Payroll Register for Week of: *March 2,19–*

Name	Hours Worked Reg.	Hours Worked OT	Hourly Rate	Gross Pay	Deductions FIT	Soc Sec.	Medicare	Hosp. Ins.	Total	Net Pay
T.L. Wright	40	3	$4.50	$200.25	$ 8.00	$ 12.42	$ 2.90	$41.54	$64.86	$135.39
J.A. Bruss	40	—	6.45	258.00	28.00	16.00	3.74	20.77	68.51	189.49
H.T. Perkins	—	—	—	325.00	33.00	20.15	4.71	20.77	78.63	246.37
N.J. Nystrand	—	—	—	435.00	36.00	26.97	6.31	41.54	110.82	324.18
W.K. Fine	—	—	—	373.00	21.00	23.13	5.41	41.54	91.08	281.92
			Total	$1591.25	$126.00	$98.67	$23.07	$166.16	$413.90	$1177.35

8 [+] 12.42 [+] 2.90 [+] 41.54 [=] 64.86 [M+] 200.25 [−] [RM] 64.86 [=] 135.39

✔ SELF-CHECK Complete the problem, then check your answer in the back of the book.

1. Use this information to find the net pay: gross weekly pay, $396; FIT, $37; social security, $24.55; medicare, $5.74; health insurance, $42.75. **$285.96**

BUSINESS PROJECT

Have students visit local computer stores to see if any of them have payroll software. Have students note the computer models it is available for, and the cost. Students may ask a salesclerk for a sample printout of the payroll spreadsheet to bring to class.

PROBLEMS

	Name	Gross Pay	FIT	Soc. Sec.	Medicare	Total Ded.	Net Pay
2.	Abbot	$485.00	$ 56.00	$30.07	$7.03	$93.10	**$391.90**
3.	Eberly	$136.00	$ 11.00	$8.43	$1.97	$21.40	**$114.60**
4.	Jackson	$795.00	$107.00	$49.29	$11.53	**$167.82**	**$627.18**
5.	Lewton	$975.00	$157.00	$60.45	$14.14	**$231.59**	**$743.41**
6.	Timmons	$643.00	$100.00	$39.87	$9.32	**$149.19**	**$493.81**
7.	Young	$519.00	$ 63.00	$32.18	$7.53	**$102.71**	**$416.29**

8. Pre-Fab Manufacturing Company. Payroll for week. **$12,563.68**
Total gross pay: $16,478.43.
Total deductions: $3914.75.
What is the total net pay?

9. Lou's Variety Store. Payroll for week. **$318.92**
Total gross pay: $416.74.
Total deductions: $97.82.
What is the total net pay?

Use the tables on pages 642–643 for federal withholding tax (FIT). Use the social security tax rate of 6.2% and medicare tax rate of 1.45%.

10. The Goldstone Swimming Complex employs students in the summer. Goldstone pays a standard hourly rate of $5.00. The only deductions are federal withholding, social security, medicare, and city income tax (CIT). Complete the payroll register for the week.

Payroll Register for Goldstone Swimming Complex — Date *June 8, 19 –*

Employee	Income Tax Information	Hours Worked	Gross Pay	FIT	Soc. Sec.	Medicare	CIT	Total Ded.	Net Pay
Banks, C.	*Single, 1 allow.*	*24*	?	?	?	?	*$2.52*	?	?
Drake, H.	*Single, 0 allow.*	*34*	?	?	?	?	*3.57*	?	?
Faust, H.	*Single, 1 allow.*	*36*	?	?	?	?	*3.78*	?	?
Harakis, P.	*Single, 0 allow.*	*38*	?	?	?	?	*3.99*	?	?
Kendrick, P.	*Single, 0 allow.*	*36*	?	?	?	?	*3.78*	?	?
Reese, T.	*Single, 1 allow.*	*38*	?	?	?	?	*3.99*	?	?
Singleton, M.	*Single, 0 allow.*	*30*	?	?	?	?	*3.15*	?	?
Total			?	?	?	?	?	?	?

Warm-Up Exercises

1. $5.60 × 40 $224.00
2. $7.65 × 44 $336.60
3. $415 + (4% × $2619) $519.76
4. $7\frac{1}{2}$% of $6870.80 $515.31
5. $47.53 + $10.18 + $5.35 $63.06
6. $519.76 − $63.06 $456.70

CRITICAL THINKING

See Problem 12. Silver Wholesale wanted to give bonuses to each salesperson, ranging from $5 for the least productive person to $35 for the most productive person. Have students rank the employees from lowest to highest to determine the distribution of the bonuses. (Ranked by total sales divided by hours worked; ranking from lowest: Quinn, Xerxes, Grant, Iskersky, Miller, Talbot, Cramer)

PRACTICE AND APPLY

The following problems can be assigned for classwork and the answers checked in classs to help students master the objective of the lesson.

- Guided Practice: 1–7
- Independent Practice: 8–10

11. The Consortium, consulting engineers, pays its employees weekly salaries. Deductions include state and city income taxes. Complete the payroll register for the week.

Name	FIT Information	Weekly Salary	Deductions							Net Pay
			FIT	Soc. Sec. (6.2%)	Med.. (1.45%)	SIT	CIT	Hosp. Ins.	Total Ded.	
Cole	M, 3 allow.	$600.00	$62	$37.20	$8.70	$16.50	$10.50	$42.75	$177.65	$422.35
Dobbs	M, 2 allow.	585.00	$65	$36.27	$8.48	15.75	10.24	35.00	$170.74	$414.26
Haddad	S, 1 allow.	415.00	$52	$25.73	$6.02	11.41	7.26	17.50	$119.92	$295.08
Micelli	M, 1 allow.	564.80	$68	$35.02	$8.19	15.53	9.88	35.00	$171.62	$393.18
Nowak	S, 2 allow.	537.50	$69	$33.33	$7.79	14.78	9.41	17.50	$151.81	$385.69
Presser	M, 4 allow.	619.20	$57	$38.39	$8.98	17.03	10.84	42.75	$174.99	$444.21
Zatcoff	M, 0 allow.	612.44	$82	$37.97	$8.88	16.84	10.72	17.50	$173.91	$438.53
Total		$3933.94	$455	$243.91	$57.04	$107.84	$68.85	$208.00	$1140.64	$2793.30

12. The Silver Wholesale Clothes Co. hires extra sales personnel during its warehouse sale. The extra help earns $3.50 an hour plus 6% commission on all sales. Deductions include $4.00 per week for health insurance. Complete the payroll register.

Name	FIT Information	Hours Worked	Total Sales	Gross Pay	Deductions					Net Pay
					FIT	Soc. Sec. (6.2%)	Medicare (1.45%)	Insurance	Total	
Cramer	S, 1 allow.	30	$4200	$357.00	$43	$22.13	$5.18	$4.00	$74.31	$282.6
Grant	S, 0 allow.	30	2786	$272.16	$38	$16.87	$3.95	$4.00	$62.82	$209.3
Iskersky	M, 2 allow.	36	3865	$357.90	$31	$22.19	$5.19	$4.00	$62.38	$295.5
Miller	S, 1 allow.	40	4500	$410.00	$52	$25.42	$5.95	$4.00	$87.37	$322.6
Quinn	S, 0 allow.	25	1430	$173.30	$22	$10.74	$2.51	$4.00	$39.25	$134.0
Talbot	M, 3 allow.	40	5184	$451.04	$39	$27.96	$6.54	$4.00	$77.50	$373.5
Xerxes	S, 1 allow,	34	3146	$307.76	$36	$19.08	$4.46	$4.00	$63.54	$244.2
Total			$25,111	$2329.16	$261	$144.39	$33.78	$28.00	$467.17	$1861.9

ALTERNATIVE STRATEGIES: Reteaching

Point out that this lesson is a combination of Unit 1, Gross Income, and Unit 2, Net Income.

A. Review the following ways of calculating gross pay.

Hourly: (40 × $6.50) + (8 × 1.5 × $6.50) = $260 + $78 = $338.00

Commission: (5% of $100,000) + (7% of $100,000) + (8% of $50,000) = $5000 + $7000 + $4000 = $16,000

Hourly + Commission: ($4.25 × 36) + (3.5% × $3240) = $153 + $113.40 = $266.40

13. King Sporting Goods, Inc., pays its employees weekly. Deductions include federal withholding, state income tax, social security (6.2%), medicare (1.45%), and 1% of gross pay for unemployment compensation. Use this information to prepare the payroll for the week of November 30.

Name	FIT Information	Position	Pay Plan	Total Sales or Hours Worked
Adams	M, 2 allow.	Sales	$7\frac{1}{2}$% straight commission	$6740
Lightner	S, 1 allow.	Sales Manager	$7\frac{1}{2}$% straight commission	$7152
Rose	M, 3 allow.	Sales Manager	$420 + 6% commission	$3298
Ulrich	S, 0 allow.	Sales Trainee	5% straight commission	$4165
Goode	M, 1 allow,	Maintenance	$5.95 per hour	44 hours

KING SPORTING GOODS, INC.	Payroll Register				Week of: November 30, 19 –			
Name	Gross Pay	FIT	SIT	Soc. Sec.	Medicare	Tax for Unemploy. Comp.	Total Deductions	Net Pay
Adams	$505.50	$53	$20.22	$31.34	$7.33	$5.06	$116.95	$388.55
Lightner	$536.40	$81	21.46	$33.26	$7.78	$5.36	$148.86	$387.54
Rose	$617.88	$63	32.59	$38.31	$8.96	$6.18	$149.04	$468.84
Ulrich	$208.25	$27	8.33	$12.91	$3.02	$2.08	$53.34	$154.91
Goode	$261.80	$23	10.47	$16.23	$3.80	$2.62	$56.12	$205.68
Total	$2129.83	$247	$93.07	$132.05	$30.89	$21.30	$524.31	$1605.52

MAINTAINING YOUR SKILLS Look up the skills in parentheses if you need help or more practice.

Add. **(Skill 5)**

14. $17.96 + $8.44 **$26.40** **15.** $149.74 + $97.89 **$247.63** **16.** $212.47 + $197.83 **$410.30**

Subtract. **(Skill 6)**

17. $412.76 − $26.40 **$386.36** **18.** $676.83 − $212.91 **$463.92** **19.** $1864.75 − $417.93 **$1446.82**

WRAP-UP

Read Problem 10 to the class. Ask students how to find the gross pay, and how to find the net pay. Have one student look up the FIT. Ask a second student to compute FICA, and ask a third volunteer to compute the total deductions for Chuck Banks.

Assignment Guide

- Basic: 8–12, 14–19
- Average: 11–13, 15–19 odd

B. Review calculating deductions.

FIT on $338.00 for a single person claiming 1 exemption		$40.00
SIT if rate is 3.5%	$338.00 × 0.035	11.83
FICA if rate is 7.65%	$338.00 × 0.0765	25.86
CIT if rate is 2%	$338.00 × 0.02	6.76
Insurance deduction	40% × $60.00	24.00
	Total deductions	$108.45

C. Find the net pay. $338.00 − $108.45 = $229.55

LESSON PLAN 19-2 Business Expenses

FOCUS

Have students imagine that they own a lawn care business. Would they keep track of their expenses? (yes) How would you do this? (by listing types of expenses, such as lawn supplies, wages, gasoline, and so on) Explain that every company must keep an accurate record of its business expenses. In this lesson, students will compute what percent a particular business expense is of the total expenses.

TEACH

Discuss with students the types of business expenses listed in the Example. See if students can add any other categories to this list. Point out that the percent of Total (in the formula) can be used for tax purposes, to plan for future spending, to determine profitability, or to publish quarterly or annual reports. Try to bring to class a copy of a quarterly or annual report so that students can see how expenses are reported to the stockholders or to the public.

Warm-Up Exercises

1. $270 + $195 + $116 + $74 $655
2. $15,700 + $10,914 + $4560 $31,174
3. $145 is what % of $1160? 12.5%
4. $14,200 is what % of $42,600? 33.3%
5. $175,640 is what percent of $2,147,914? 8.2%

19-2 Business Expenses

OBJECTIVE

Compute the percent that a particular business expense is of the total expenses.

Your business must keep accurate records of all its expenses. You will use the information when you prepare income tax forms and when you calculate your company's profits. You may determine your total expenses monthly, quarterly, or annually. To plan for future spending, you may calculate the percent that each expense is of the total.

$$\text{Percent of Total} = \frac{\text{Particular Expense}}{\text{Total Expenses}}$$

EXAMPLE *Skills* 5, 31 *Application* A

Molded Plastic Products, Inc., manufactures plastic buckets, containers, and other products. Records of Molded Plastic's expenses for the first quarter of the year show:

Payroll	$42,171.84
Advertising	2500.00
Raw materials	12,417.83
Factory and showroom rental	4500.00
Office supplies	216.90
Insurance	517.85
Utilities	6417.93
Total	$68,742.35

What percent of total expenses did Molded Plastic spend on advertising during the quarter?

SOLUTION

Find the **percent of total.**

Particular Expense ÷ Total Expenses

$2500.00 ÷ $68,742.35 = 0.0363 = 3.6% of total was spent on advertising

2500 [÷] 68742.35 [=] .03636768

✓ SELF-CHECK Complete the problems, then check your answers in the back of the book.

1. Particular expenses: filing, $300; programming, $1200; payroll, $2500. **7.5%** Filing is what percent of the total?
2. Particular expenses: telephone, $175; electricity, $450; natural gas, $375; water, $200. **16.7%** Water is what percent of the total?

PROBLEMS

Find percents to the nearest tenth percent.

	Expenses for Supplies	÷	Total Expenses	=	Percent of Total
3.	$ 450	÷	$ 4500	=	? 10.0%
4.	$ 1400	÷	$ 28,000	=	? 5.0%
5.	$ 15,700	÷	$ 210,000	=	? 7.5%
6.	$132,685	÷	$5,415,750	=	? 2.4%

7. Trapper Financial Consultants. Expenses total $72,650 for quarter. Payroll for quarter is $48,435. Payroll expense is what percent of the total? **66.7%**

8. Moonlight Motel. Annual expenses total $214,500. Annual utilities cost $38,475. Utility cost is what percent of the total? **17.9%**

9. The Clear Pool Company had the following business expenses last year:

Payroll	$84,565
Advertising	2000
Equipment	21,495
Supplies	17,194
Insurance	2000
Utilities	1450
Rent	3000

What are the total business expenses? Payroll is what percent of the total?
$131,704 **64.2%**

10. Parson's Traditional Clothiers, Inc., had the following business expenses last month:

Payroll	$ 1645.80
Advertising	325.00
Cost of goods sold	18,175.75
Store rental	900.00
Supplies	45.00
Insurance	120.00
Utilities	217.89

Store rental is what percent of the total business expenses for the month? **4.2%**

MAINTAINING YOUR SKILLS Look up the skills in parentheses if you need help or more practice.

Add. **(Skill 5)**

11. $114,570 + 31,720 + 56,493 = **$202,783**

12. $2171.89 + 419.43 + 114.65 = **$2705.97**

13. $212.43 + 171.80 + 64.95 = **$449.18**

14. $714,789.43 + 24,561.78 + 7147.93 = **$746,499.14**

Find the rate. Round answers to the nearest tenth of a percent. **(Skill 31)**

15. $35,000 is what percent of $140,000? **25.0%**

16. $212 is what percent of $1060? **20.0%**

17. $78 is what percent of $471? **16.6%**

18. $7500 is what percent of $171,432? **4.4%**

19. $17,893.45 is what percent of $914,719.40? **2.0%**

PRACTICE AND APPLY

The following problems can be assigned for classwork and the answers checked in class to help students master the objective of the lesson.

- Guided Practice: 1–4
- Independent Practice: 5–8

WRAP-UP

Have students explain why companies need to keep track of the percent of total for their different types of expenses. Have students list what some of these expenses are.

Assignment Guide

- Basic: 3–8, 11–19
- Average: 9, 10, 12–18 even

ALTERNATIVE STRATEGIES: Enrichment

1. Have students use the Example to calculate what percent each expense is of the total.
2. Have students obtain a copy of the annual report of several companies. Have them offer an explanation of why certain businesses may have greater expenses in one category than another business.

LESSON PLAN 19-3 Apportioning Expenses

FOCUS

Each family has expenses. Ask students if they think that each person in a family contributes equally to the expenses, or whether the expenses vary from person to person. Point out that the expenses for a business can vary from department to department.

TEACH

Students may not be familiar with the term *apportion.* Make sure they understand that to apportion expenses means to distribute them by some method. One method is to use the square footage that each department occupies, as seen in the Example. Another method is to apportion expenses according to the gross sales of each department. Point out that apportioning expenses by square footage is quite common for places like shopping malls. Some common expenses, such as security, are apportioned by the size of each store in the mall.

Warm-Up Exercises

1. 45 ft × 65 ft 2925 ft²
2. 50 ft × 120 ft 6000 ft²
3. 2925 ÷ 45,000 0.065
4. 6000 ÷ 180,000 0.033
5. 0.065 × $15,000 $975
6. 0.033 × $240,000 $7920

19-3 Apportioning Expenses

OBJECTIVE

Compute a department's share of the total business expense.

Your business may **apportion,** or distribute, certain expenses among its departments. Each department is charged a certain amount of the particular expense. Often, the amount that each department is charged depends on the space that it occupies.

$$\text{Amount Paid} = \frac{\text{Square Feet Occupied}}{\text{Total Square Footage}} \times \text{Total Expense}$$

EXAMPLE *Skills* 10, 8 *Application* X *Term* Apportion

Mike's Discount Stores, Inc., occupies a building that contains 200,000 square feet. The cut-rate clothing department occupies an area that measures 100 feet by 150 feet. The annual cost of security for the entire building is $28,400. Mike's apportions the cost based on square feet occupied. What annual amount for security does Mike's charge to the cut-rate clothing department?

SOLUTION

A. Find the **square feet occupied.**
100 ft × 150 ft = 15,000 square feet occupied

B. Find the **amount paid.**
(Square Feet Occupied ÷ Total Square Footage) × Total Expense
15,000 ÷ 200,000 × $28,400
0.075 × $28,400 = $2130 amount paid

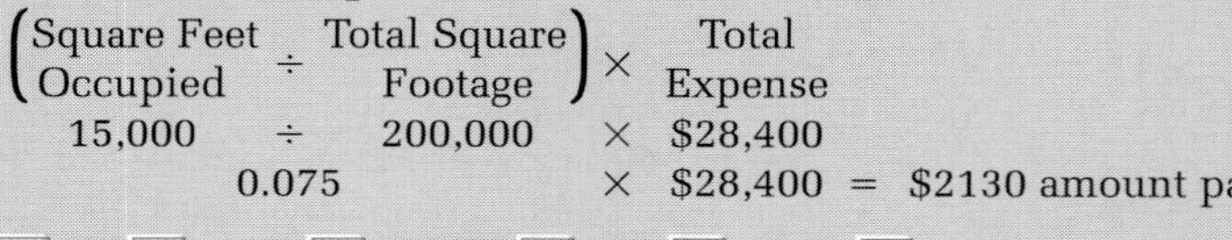

✓ SELF-CHECK Complete the problems, then check your answers in the back of the book.

1. The Law-for-All Clinic occupies 40 feet by 50 feet in a building with 50,000 square feet. Annual cost of janitorial service is $9600. What does The Law-for-All Clinic pay? **$384**

2. Family Shoes occupies 50 feet by 80 feet in a mall with 40,000 square feet. Monthly mall expenses are $12,500. What does Family Shoes pay? **$1250**

PROBLEMS

	(Square Feet Occupied	÷	Total Square Footage)	×	Total Expense	=	Amount Paid
3.	(3200	÷	160,000)	×	$62,500	=	**$1250**
4.	(15,000	÷	160,000)	×	$62,500	=	**$5859.38**
5.	(50,000	÷	160,000)	×	$62,500	=	**$19,531.25**

ALTERNATIVE ASSESSMENT

Bring in a newspaper that advertises office space for rent by the square foot. Have students compute the annual cost, then apportion the cost among three departments: two departments are equal in floor space, and the third department is twice as large as either of these two. (Apportioning would be done 50%—25%—25%.)

Round the intermediate answers to the nearest thousandth and the final answers to the nearest cent.

6. Building area: 980,000 sq ft.
Annual maintenance cost: $135,000.
Billing dept. is 40 ft by 60 ft.
What does the billing department pay annually for maintenance? **$270.00**

7. Building area: 450,000 sq ft.
Annual insurance charge: $5750.
Sales dept. is 30 ft by 40 ft.
What does the sales department pay annually for insurance? **$17.25**

8. The Dunberry Company apportions the annual cost of utilities among its departments. The total cost for the year was $186,470. The total area of the building is 750,000 square feet. The company's receiving department occupies an area that is 20 feet by 650 feet. What did the receiving department pay for utilities for the year? **$3169.99**

9. Tubular Assemblies, Inc., pays a total of $960,000 per year to rent its building. The total area of the building is 480,000 square feet. Tubular Assemblies apportions the rental charge among its departments. How much does each of these departments pay for rent?

a. Research and development, 45 ft by 60 ft **$5760.00**

b. Shipping, 50 ft by 120 ft **$12,480.00**

10. The Barteli Corporation owns and maintains a 5-store shopping mall. Barteli paid $12,000 for advertising last year. It apportions the cost based on square footage. How much does Barteli charge each store?

Shop	Dimensions	
Ken's Carry-Out	40 ft by 60 ft	**$4080.00**
The Jean Place	30 ft by 50 ft	**$2556.00**
Pan's Pizza	25 ft by 60 ft	**$2556.00**
Universal Travel	30 ft by 30 ft	**$1536.00**
The Jewel Store	25 ft by 30 ft	**$1272.00**

11. Some businesses apportion costs among their departments on the basis of gross sales. The gross sales for Tent Mart totaled $3,750,000 last year. It distributed these annual expenses:

Maintenance	Utilities	Security
$6000	$30,000	$4800

a. M: $1440.00
U: $7200.00
S: $1152.00
b. M: $1038.00
U: $5190.00
S: $ 830.40
c. M: $ 762.00
U: $3810.00
S: $ 609.60

a. The women's wear department had $900,000 in gross sales last year. How much did it pay for each of the annual expenses?

b. The shoe department had $650,000 in gross sales last year. How much did it pay for each of the annual expenses?

c. The toy department had $475,650 in gross sales last year. How much did it pay for each of the annual expenses?

MAINTAINING YOUR SKILLS Look up the skills in parentheses if you need help or more practice.

Divide. Round answers to the nearest thousandth. **(Skill 10)**

12. 900 ÷ 12,450 **0.072**

13. 2000 ÷ 96,500 **0.021**

14. 1200 ÷ 50,000 **0.024**

15. 2475 ÷ 475,000 **0.005**

PRACTICE AND APPLY

The following problems can be assigned for classwork and the answers checked in class to help students master the objective of the lesson.

- Guided Practice: 1–5
- Independent Practice: 6–8

WRAP-UP

Read Problem 7 to the class. Ask students how to compute the area of the Sales Department. Then ask students what it means to apportion expenses. Ask students to explain how they would find the Sales Department's share of the annual insurance charge.

Assignment Guide

- Basic: 6–10, 12–15
- Average: 9–11, 13, 15

ALTERNATIVE STRATEGIES: Reteaching

Students may use the short-cut methods of multiplying or dividing by multiples of 10, 100, 1000, and so on. For the Example 15,000 ÷ 200,000, students can find 15 ÷ 200 = 0.075.

LESSON PLAN
19-4 Depreciation—Straight-Line Method

FOCUS
The focus of this lesson is to use the straight-line method to compute the annual depreciation of an item. One item that students may already associate with depreciation is a car. Ask students if they think a new car has the same value after it is a year old.

TEACH
With many cars, the greatest depreciation usually occurs within the first year. Point out that the straight-line method is a way of averaging out the depreciation by assuming the same amount of depreciation each year.

Make sure students understand the terms *estimated life* and *resale value.* Point out that most mechanical items have some kind of resale value, even at the end of the item's expected life. The metal often has some value, and parts can often be sold for use in repairs of other items.

Point out that the Internal Revenue Service allows businesses to depreciate such items as trucks, machines, computers, and typewriters. A business can deduct the depreciation from its sales profits.

Warm-Up Exercises
1. $1850 − $150 $1700
2. $18,947 − $2550 $16,397
3. $1700 ÷ 5 $340
4. $16,397 ÷ 3 $5465.67
5. ($247.95 − $50) ÷ 4 $49.49
6. ($171,416 − $500) ÷ 10 $17,091.60

19-4

Depreciation—Straight-Line Method

OBJECTIVE
Use the straight-line method to compute the annual depreciation of an item.

For tax purposes, the Internal Revenue Service allows you to recognize the **depreciation** of many of the items that your business owns. Depreciation is a decrease in the value of an item because of its age or condition. The **straight-line method** is one way of determining the annual depreciation of an item. This method assumes that the depreciation is the same from year to year. To calculate the depreciation, you must know the original cost, the **estimated life,** and the **resale value** of the item. The estimated life of an item is the length of time, usually in years, that it is expected to last. The resale value is the estimated trade-in, salvage, or scrap value at the end of the item's expected life.

$$\text{Annual Depreciation} = \frac{\text{Original Cost} - \text{Resale Value}}{\text{Estimated Life}}$$

EXAMPLE *Skills* 4, 10 *Application* A *Term* Straight-line method

The law firm of Charles A. Adams, Inc., purchased a new copier that cost $1745. The life of the copier is estimated to be 5 years. The total resale value after 5 years of use is estimated to be $245. Using the straight-line method, find the annual depreciation of the copier.

SOLUTION
Find the **annual depreciation.**
(Original Cost − Resale Value) ÷ Estimated Life
($1745 − $245) ÷ 5
$1500 ÷ 5 = $300 annual depreciation

1745 [−] 245 [=] 1500 [÷] 5 [=] 300

✔ SELF-CHECK Complete the problems, then check your answers in the back of the book.

Find the annual depreciation.

1. $6000 original cost; 10 years estimated life; $1000 resale value. **$500**
2. $475 original cost; 3 years estimated life; $25 resale value. **$150**

PROBLEMS

	(Original Cost	−	Resale Value)	÷	Estimated Life	=	Annual Depreciation
3.	($ 1435	−	$ 235)	÷	3 years	=	**$400.00**
4.	($ 7186	−	$ 1000)	÷	5 years	=	**$1237.20**
5.	($ 14,750	−	$ 2000)	÷	10 years	=	**$1275.00**
6.	($115,476	−	$10,000)	÷	15 years	=	**$7031.73**

CULTURAL ANGLES
Currency can also go through depreciation. This happens when the demand of other countries to buy a nation's currency is less than the amount of currency the nation has in supply.

7. New radio tower. **$8500**
Cost is $175,000.
Estimated life of 20 years.
Salvage value estimated at $5000.
What is the annual depreciation?

8. New taxicab. **$5000**
Cost is $16,250.
Estimated life is 3 years.
Salvage value estimated at $1250.
What is the annual depreciation?

9. Central Dental Clinic recently purchased a new computer system for a total cost of $64,735. The estimated life of the system is 5 years. The trade-in value of the system after 5 years is estimated to be $10,000. What is the annual depreciation? **$10,947**

10. Tina Cole is a chartered financial advisor. She purchased equipment for her office for $7843. The trade-in value of the equipment is estimated to be $500 after 7 years of use. What is the annual depreciation? **$1049**

11. The Star Trucking Company purchased a new tractor unit for $125,000. After 4 years of useful life, the estimated salvage value is $10,000. What is the annual depreciation? **$28,750**

12. Tom Nichols is a representative for a landscape company. Tom uses his pickup truck entirely for business. He trades in his truck every 3 years. Tom paid $8500 for his present truck. He expects to receive $2000 for the truck when he trades it in after 3 years. What is the annual depreciation? **$2166.67**

13. Third National Bank recently purchased 3 new typewriters. Each typewriter cost $845. The estimated life of each typewriter is 6 years. The trade-in value of each typewriter is expected to be $125 at the end of the 6 years. What is the annual depreciation for all 3 typewriters? **$360**

14. Annie's Laundromat recently purchased 6 new washing machines at a cost of $325 each and 5 new clothes dryers at a cost of $268 each. The salvage value of the washers is expected to be $50 each after 4 years of use. The salvage value of the dryers is expected to be $75 each after 5 years of use. What is the total annual depreciation? **$605.50**

MAINTAINING YOUR SKILLS Look up the skills in parentheses if you need help or more practice.

Subtract. **(Skill 4)**

15. $1475 − 1189 = **$286**

16. $743 − 95 = **$648**

17. $725 − 56 = **$669**

18. $16,417 − 750 = **$15,667**

Divide. Round answers to the nearest cent. **(Skill 10)**

19. $14,718 ÷ 20 **$735.90**

20. $9171.45 ÷ 15 **$611.43**

21. $768.43 ÷ 5 **$153.69**

22. $4171.60 ÷ 3 **$1390.53**

Multiply. **(Skill 8)**

23. $1470.00 × 0.055 = **$80.85**

24. $963.49 × 0.05 = **$48.17**

25. $24.49 × 0.06 = **$1.47**

26. $11,419.90 × 0.075 = **$856.49**

PRACTICE AND APPLY

The following problems can be assigned for classwork and the answers checked in class to help students master the objective of the lesson.

- Guided Practice: 1–6
- Independent Practice: 7–9

WRAP-UP

Have students explain what is meant by *depreciation.* Ask them to explain the steps necessary to find the depreciation by the straight-line method. Have students compute the annual depreciation of an item that was worth $1000 new, and will be worth $100 at the end of its life expectancy in 5 years. ($180)

Assignment Guide

- Basic: 7–12, 15–26
- Average: 10–13, 15–25 odd

ALTERNATIVE STRATEGIES: Reteaching

1. Emphasize that students must subtract the resale value from the original cost before dividing by the estimated life.
2. In anticipation of the next lesson, be sure students understand that with the straight-line method, the depreciation is the same for each year.

LESSON PLAN 19-5 Depreciation— Book Value

FOCUS
Ask students if any of them have ever heard of the book value of a car. Point out that other items that depreciate in value have a book value. The focus of this lesson is to use the straight-line method to compute the book value of an item.

TEACH
Explain that the book value is used to determine the value of an item at any given time. This is needed if the business decides to sell an item before the end of its life expectancy. Make sure that students understand that the accumulated depreciation is the annual average depreciation times the age in the number of years. Accumulated depreciation can also be found by adding this year's depreciation to the accumulated depreciation for the previous year.

Warm-Up Exercises
1. $800 × 3 $2400
2. $12,400 × 5 $62,000
3. $147.80 × 2 $295.60
4. $4780.50 × 7 $33,463.50
5. $5000 − $2400 $2600
6. $44,987.45 − $33,463.50 $11,523.95

19-5 Depreciation— Book Value

OBJECTIVE
Use the straight-line method to compute the book value of an item.

Book value is the approximate value of an item after you have owned it and depreciated it for a period of time. The book value is the original cost minus the **accumulated depreciation.** The accumulated depreciation is the total depreciation to date. At the end of an item's life, its book value and resale value should be equal.

Book Value = Original Cost − Accumulated Depreciation

EXAMPLE *Skill* 7 *Application* A *Term* Book value

The law firm of Charles A. Adams, Inc., purchased a new copier for $1745. The total resale value after 5 years is estimated to be $245. Using the straight-line method, Charles A. Adams, Inc., determined that the copier will depreciate $300 per year. What will the book value be at the end of 4 years?

SOLUTION

A. Find the **accumulated depreciation** for fourth year.
$300 per year × 4 years = $1200 accumulated depreciation

B. Find the **book value.**
Original Cost − Accumulated Depreciation
$1745 − $1200 = $545 book value

✓ SELF-CHECK Complete the problems, then check your answers in the back of the book.

A computer system costs $96,000. Its estimated life is 4 years. The annual depreciation is $20,000. What will the book value be:

1. At the end of 3 years? **$36,000**   **2.** At the end of 4 years? **$16,000**

PROBLEMS

	Annual Depreciation	×	Number of Years	=	Accumlated Depreciation	Original Cost	−	Accumlated Depreciation	=	Book Value
3.	$ 400	×	6 years	=	**$2400**	$3500	−	**$2400**	=	**$1100**
4.	$16,000	×	3 years	=	**$48,000**	$86,500	−	**$48,000**	=	**$38,500**
5.	$ 50	×	2 years	=	**$100**	$280	−	**$100**	=	**$180**

Use the straight-line method of depreciation. Round answers to the nearest dollar.

6. $4800 for office furniture. Depreciates $430 per year. What is the book value after 3 years? **$3510**

7. $9165 for communications equipment. Depreciates $900 per year. What is the book value after 5 years? **$4665**

PROBLEM SOLVING
You might ask some students to show, by using the straight-line method, that the book value after the estimated life of an item equals its salvage or scrap value.

8. New facsimile transmitting equipment. Cost is $11,748. Estimated life is 5 years. Resale value of $3000. What is the book value after 3 years? **$6499.20**

9. New automobile. Cost is $18,417. Estimated life of 3 years. Resale value of $5000. What is the book value after 2 years? **$9472.34**

10. Tina Cole purchased new office equipment for $7843. The trade-in value of the equipment is estimated to be $500 after 7 years of use. Find the book value after each year of use.

Year	Annual Depreciation	Accumulated Depreciation	Book Value
1	$1049	$1049	$6749
2	1049	2098	**$5745**
3	1049	**$3147**	**$4696**
4	**$1049**	**$4196**	**$3647**
5	**$1049**	**$5245**	**$2598**
6	**$1049**	**$6294**	**$1549**
7	**$1049**	**$7343**	**$500**

11. Wayne Guest purchased a new over-the-road tractor for $114,760. The tractor is expected to have a trade-in value of $20,000 after 4 years. What is the book value at the end of each of the 4 years?

Year	Annual Depreciation	Accumulated Depreciation	Book Value
1	**$23,690**	**$23,690**	**$91,070**
2	**$23,690**	**$47,380**	**$67,380**
3	**$23,690**	**$71,070**	**$43,690**
4	**$23,690**	**$94,760**	**$20,000**

12. The units-of-production method is another method of computing depreciation. This method bases depreciation on the number of units produced rather than the years of estimated life.

$$\text{Depreciation} = \frac{\text{Number of Units Produced}}{\text{Estimated Units of Production}} \times \left(\text{Original Cost} - \text{Salvage Value}\right)$$

$167,375 a. Ross Manufacturing purchased a new automatic punchpress for $216,500. The press is expected to be worth $20,000 after 200,000 units are produced. Use the units-of-production method to calculate the book value of the press after 50,000 units are produced.

b. Sustom Plastics purchased a new folding machine for $51,615. The salvage value of the machine is estimated to be $5000 after 750,000 units are produced. Use the units-of-production method to calculate the book value after 150,000 units are produced. **$42,292**

MAINTAINING YOUR SKILLS Look up the skills in parentheses if you need help or more practice.

Multiply. **(Skills 7, 8)**

13. $550 × 3 **$1650**
14. $1240 × 5 **$6200**
15. $17.60 × 4 **$70.40**
16. $2500 × 7 **$17,500**
17. $11,400 × 9 **$102,600**
18. $8175 × 6 **$49,050**
19. $24,650 × 8 **$197,200**
20. $17.47 × 2 **$34.94**
21. $718.65 × 10 **$7186.50**

PRACTICE AND APPLY

The following problems can be assigned for classwork and the answers checked in class to help students master the objective of the lesson.

- Guided Practice: 1–5
- Independent Practice: 6–8

WRAP-UP

Have students list and define the new terms of the lesson. Have students find the book value of an item 4 years old that originally was worth $1000 and that depreciates $100 a year for a life expectancy of 8 years. ($600)

Assignment Guide

- Basic: 6–10, 12–20
- Average: 9–11, 12–20 even

ALTERNATIVE STRATEGIES: Enrichment

Have the students choose one item, such as a computer, with a cost of $3500, a salvage value of $500, and an estimated life of 5 years. Have them calculate the depreciation, accumulated depreciation, and the book value. Have them draw a line graph showing the accumulated depreciation for each year.

LESSON PLAN 19-6 Sum-of-the-Years'-Digits Method

FOCUS

The focus of this lesson is to use the sum-of-the-years'-digits method to compute the yearly depreciation. This method allows for greater depreciation in the first years of the life expectancy of an item.

TEACH

Go over the formula for depreciation. Point out that the original cost minus the resale value gives the total amount of depreciation. Have students verify that $\frac{4}{10} + \frac{3}{10} + \frac{2}{10} + \frac{1}{10} = 1$. This compares to the prior method of finding $\frac{1}{4} + \frac{1}{4} + \frac{1}{4} + \frac{1}{4} = 1$ for the straight-line method.

Make sure students understand that the numerator is the greatest number of years for the life expectancy, and that the numbers go in reverse order. That is, the first year's depreciation factor is not $\frac{1}{10}$, but $\frac{4}{10}$. The final year of 4 years is $\frac{1}{10}$, not $\frac{4}{10}$.

Warm-Up Exercises

1. $\frac{5}{15} \times \$87,600$ $29,200
2. $\frac{4}{10} \times \$17,100$ $6840
3. $\frac{3}{6} \times \$8470$ $4235
4. $\frac{9}{55} \times \$330,000$ $54,000
5. $\frac{6}{21} \times (\$758 - \$44)$ $204

19-6 Depreciation— Sum-of-the-Year's-Digits Method

OBJECTIVE
Use the sum-of-the-year's-digits method to compute the yearly depreciation.

The **sum-of-the-years'-digits method** is another way to determine the annual depreciation of an item. This method assigns a fraction of the total depreciation to each year of the item's life. More depreciation is allowed in the early years of the item's life than in the later years. For example, if the life of a machine is estimated to be 4 years, the years' digits are 4, 3, 2, and 1. The sum of the years' digits is 4 + 3 + 2 + 1 = 10. The fractions of the total depreciation are $\frac{4}{10}$ for the first year, $\frac{3}{10}$ for the second year, $\frac{2}{10}$ for the third year, and $\frac{1}{10}$ for the fourth year. To use the sum-of-the-years'-digits method, you must know the original cost, the estimated life, and the resale value.

$$\text{Depreciation for Year} = \text{Fraction of Total Depreciation} \times \left(\text{Original Cost} - \text{Resale Value}\right)$$

EXAMPLE *Skills* 20, 2 *Application* A *Term* Sum-of-the-years'-digits method

The General Assembly Company purchased a new numerically controlled riveter for $195,000. The expected life of the riveter is 5 years. At the end of the 5 years, the resale value of the machine is expected to be $15,000. General Assembly uses the sum-of-the-years'-digits method to calculate depreciation. To the nearest dollar, what is the depreciation for the first year of the life of the riveter?

SOLUTION

A. Find the **fraction of total depreciation.**

$$\frac{\text{Assigned Digit}}{\text{Sum of the Years' Digits}} = \frac{5}{5 + 4 + 3 + 2 + 1} = \frac{5}{15}$$

B. Find the **depreciation for year.**

$$\text{Fraction of Total Depreciation} \times \left(\text{Original Cost} - \text{Resale Value}\right)$$
$$\frac{5}{15} \times (\$195,000 - \$15,000)$$
$$\frac{5}{15} \times \$180,000 = \$60,000 \text{ depreciation for first year}$$

195000 [−] 15000 [=] 180000 [×] 5 [÷] 15 [=] 60000

✔ SELF-CHECK Complete the problems, then check your answers in the back of the book.

1. $\frac{4}{15} \times \$180,000 = ?$ depreciation for the second year. **$48,000**
2. $\frac{3}{15} \times \$180,000 = ?$ depreciation for the third year. **$36,000**

ALTERNATIVE STRATEGIES: Reteaching

1. Point out that the sum-of-the-years'-digits method can be found by the formula $\frac{n}{2}(n + 1)$
   For four years: $\frac{4}{2}(4 + 1) = 2(5) = 10$
   For seven years: $\frac{7}{2}(7 + 1) = 3.5(8) = 28$

PROBLEMS

Use the sum-of-the-years'-digits method to calculate depreciation. Round answers to the nearest dollar.

	Original Cost	Resale Value	Estimated Life	Depreciation 1st Year Fraction	Depreciation 1st Year Amount	Depreciation 2nd Year Fraction	Depreciation 2nd Year Amount
3.	$ 9000	$ 1200	3 years	$\frac{3}{6}$	$3900	$\frac{2}{6}$	**$2600**
4.	$ 14,700	$ 2100	6 years	$\frac{6}{21}$	$3600	$\frac{5}{21}$	**$3000**
5.	$125,000	$15,000	10 years	$\frac{10}{55}$	**$20,000**	$\frac{9}{55}$	**$18,000**
6.	$ 9500	$ 500	5 years	$\frac{5}{15}$	**$3000**	$\frac{4}{15}$	**$2400**

F.Y.I.
The Internal Revenue Service requires that the salvage value equal the book value at the end of the useful life.

7. Robotic delivery unit.
Estimated life of 2 years.
Original cost is $12,500.
Resale value is $3500.
What is the depreciation for the first year? **$6000**

8. Portable generator.
Estimated life of 3 years.
Original cost is $10,000.
Resale value is $400.
What is the depreciation for the first year? **$4800**

9. The original cost of a small delivery truck is $7800. After 4 years, the resale value is estimated to be $1000. What is the depreciation for each of the 4 years? **$2720; $2040; $1360; $680**

10. The Office Temp. Service recently purchased 2 new typewriters for a total cost of $2480. The typewriters have an estimated life of 5 years. After 5 years, the resale value of each typewriter is expected to be $250. What is the depreciation for each typewriter for each of the 5 years? **$330; $264; $198; $132; $66**

11. The Perfect Plastic Co. recently purchased a plastic-molding machine for $80,000. The estimated life of the machine is 4 years. The resale value after 4 years is expected to be $8000. What is the book value of the machine after each year of use?

Year	Depreciation for Year	Accumulated Depreciation	Book Value
First	**$28,800**	**$28,800**	**$51,200**
Second	**$21,600**	**$50,400**	**$29,600**
Third	**$14,400**	**$64,800**	**$15,200**
Fourth	**$7200**	**$72,000**	**$8000**

MAINTAINING YOUR SKILLS Look up the skills in parentheses if you need help or more practice.

Solve. Round answers to the nearest dollar. **(Skills 4, 20)**

12. $\frac{4}{15} \times$ $9000 **$2400** **13.** $\frac{2}{3} \times$ $12,630 **$8420** **14.** $\frac{3}{6} \times$ $4840 **$2420** **15.** $\frac{3}{10} \times$ $17,800 **$5340**

16. $\frac{6}{21} \times$ ($147,500 − $15,000) **$37,857**

17. $\frac{9}{55} \times$ ($7850 − $900) **$1137**

18. $\frac{7}{36} \times$ ($475 − $100) **$73**

2. Expand the example to show the depreciation for all 5 years.
$\frac{5}{15} \times$ $180,000 = $60,000
$\frac{4}{15} \times$ $180,000 = $48,000
$\frac{3}{15} \times$ $180,000 = $36,000
$\frac{2}{15} \times$ $180,000 = $24,000
$\frac{1}{15} \times$ $180,000 = $12,000

PRACTICE AND APPLY

The following problems can be assigned for classwork and the answers checked in class to help students master the objective of the lesson.

- Guided Practice: 1–6
- Independent Practice: 7–9

WRAP-UP

Have students find the sum-of-the-years'-digits for an item with a life expectancy of 8 years. (36) Ask them what fraction they would use to find the depreciation for the fifth year. ($\frac{4}{36}$)

Assignment Guide

- Basic: 7–10, 12–18
- Average: 10, 11, 12–18 even

LESSON PLAN 19-7 Depreciation—Double-Declining-Balance Method

FOCUS

In this lesson, students will learn to use the double-declining-balance method to compute the annual depreciation and book value. Some businesses prefer this method because the largest portion of depreciation of an item can be claimed in the first year of the item's life expectancy.

TEACH

Point out that this method of computing depreciation is different from the first two in that the book value does not need to be known in advance. It is computed each year, along with the depreciation for that year. Have students add the accumulated depreciation and the declining balance for each of the years in the Example. Students should find that the sum of the two figures is always $240,000, the original cost of the item. For Problem 5, suggest to students that they change $33\frac{1}{3}\%$ to the equivalent fraction $\frac{1}{3}$. They can then find $\frac{1}{3}$ of $42,000.

Warm-Up Exercises

1. 40% of $175,000 $70,000
2. 20% of $24,750 $4950
3. 50% of $470 $235
4. $66\frac{2}{3}\% \times \$12,900$ $8600
5. $2 \times (100\% \div 8)$ 25%
6. $2 \times (100\% \div 6)$ $33\frac{1}{3}\%$

19-7 Depreciation—Double-Declining-Balance Method

OBJECTIVE
Use the double-declining-balance method to compute the annual depreciation and book value.

Your business may use the **double-declining-balance method** to find the annual depreciation of an item. This method allows the highest possible depreciation in the first year of the item's life. The depreciation declines as the item approaches the end of its life. To use this method, you must know the item's original cost and estimated life.

$$\text{Annual Depreciation Rate} = 2 \times \frac{100\%}{\text{Estimated Life}}$$

$$\text{Depreciation for Year} = \text{Previous Declining Balance} \times \text{Annual Depreciation Rate}$$

EXAMPLE *Skill* 30 *Application* A *Term* Double-declining-balance method

Major Media Productions purchased production equipment for $240,000. Major Media estimates that the equipment will become obsolete in 5 years. What are the depreciation and book value for each of the 5 years?

SOLUTION

A. Find the **annual depreciation rate.**
$2 \times (100\% \div \text{Estimated Life})$
$2 \times (100\% \div 5) = 40\%$ annual rate

B. Find the **depreciation for year.**
Previous Declining Balance × Annual Depreciation Rate

40% of original cost

Previous Declining Balance − Depreciation for Year

Year	Depreciation for Year	Accumulated Depreciation	Declining Balance (Book Value)
First	$96,000.00	$96,000.00	$144,000.00
Second	57,600.00	153,600.00	86,400.00
Third	34,560.00	188,160.00	51,840.00
Fourth	20,736.00	208,896.00	31,104.00
Fifth	12,441.60	221,337.60	18,662.40.00 ← Resale value

40% of previous declining balance of $144,000

✓ SELF-CHECK Complete the problems, then check your answers in the back of the book.

A machine that cost $80,000 will be obsolete in 4 years.

1. Find the depreciation and book value for the first year. **$40,000; $40,000**
2. Find the depreciation and book value for the second year. **$20,000; $20,000**

COOPERATIVE LEARNING

Refer to the Self-Check problems. Have students complete the problem by finding the depreciation, accumulated depreciation, and declining balance for each of the four years. Have students use the resale value to then compute the depreciation by the straight-line method and the sum-of-the-years'-digits method. Have students compare methods, and describe which one they prefer and why.

PROBLEMS

	Estimated Life	Annual Depreciation Rate	×	Previous Declining Balance	=	Depreciation for Year
3.	10 years	20%	×	$ 16,450	=	$3290
4.	8 years	25%	×	$ 7400	=	$1850
5.	6 years	$33\frac{1}{3}$%	×	$ 42,000	=	$14,000
6.	3 years	$66\frac{2}{3}$%	×	$120,000	=	$80,000

7. New forklift cost $7200.
Estimated life of 6 years.
What is the depreciation for the first year? The book value?
$2400; $4800

8. New incinerator cost $135,000.
Estimated life of 4 years.
What is the depreciation for the first year? The book value?
$67,500; $67,500

9. The Packer Company purchased new inventory control equipment for $50,000. The life of the equipment is estimated to be 5 years. The resale value is expected to be $3888. Complete the depreciation table for each of the 5 years.

Year	Depreciation for Year	Accumulated Depreciation	New Declining Balance
First	$20,000	$20,000	$30,000
Second	12,000	$32,000	$18,000
Third	$7200	$39,200	$10,800
Fourth	$4320	$43,520	$6480
Fifth	$2592	$46,112	$3888

10. When the double-declining-balance method is used to calculate depreciation, the declining balance in the last year of an item's life might not equal the expected resale value. When this happens, the depreciation for the last year is adjusted so that the final declining balance and the resale value are the same.

Samsen's purchased office equipment for $10,000. The resale value is expected to be $500 after a 4-year life. Complete the depreciation table.

Year	Depreciation for Year	Accumulated Depreciation	New Declining Balance
1	$5000	$5000	$5000
2	$2500	$7500	$2500
3	$1250	$8750	$1250
4	$750	$9500	$500

This amount is computed to be $625. It must be increased so that the book value is $500.

MAINTAINING YOUR SKILLS Look up the skill in parentheses if you need help or more practice.

Find the percentage. Round answers to the nearest dollar. **(Skill 30)**

11. 40% of $1350 **$540**
12. 50% of $5000 **$2500**
13. 20% of $14,850 **$2970**
14. $66\frac{2}{3}$% of $9600 **$6400**
15. 25% of $415,600 **$103,900**
16. $33\frac{1}{3}$% of $1242 **$414**
17. 28.6% of $1490 **$426**
18. 22.2% of $12,560 **$2788**

PRACTICE AND APPLY
The following problems can be assigned for classwork and the answers checked in class to help students master the objective of the lesson.

- Guided Practice: 1–4
- Independent Practice: 5–8

WRAP-UP
Have a student give the formula for the annual depreciation rate to be used in computing depreciation by the double-declining-balance method. Have another student explain how to find the depreciation for the first year, and the declining balance. Have another student explain how to find each for the second year.

Assignment Guide
- Basic: 5–9, 11–18
- Average: 9, 10, 12–18 even

ALTERNATIVE STRATEGIES: Enrichment
Have the students choose one item, such as a computer, with a cost of $3500, a salvage value of $500, and an estimated life of 5 years. Have them calculate the depreciation, accumulated depreciation, and the book value. Have them draw a line graph showing the accumulated depreciation for each year. Compare this graph to the graph from Lesson 19-5.

LESSON PLAN 19-8 Modified Accelerated Cost Recovery Systems (MACRS)

FOCUS
One method of depreciation allows businesses to fully depreciate assets over a fixed number of years. In this lesson, students use the Modified Accelerated Cost Recovery System method to compute annual depreciation.

TEACH
This method of depreciation was first initiated in the Economy Recovery Act of 1981 and then modified by the Tax Reform Act of 1986 and the Tax Bill of 1989. It is used extensively for tax purposes. Items can be depreciated over less time than other methods, so businesses tend to replace certain items faster than before, thus stimulating the economy. You might mention that items having recovery periods of 4, 6, 8, . . . years under the modified system of 1986 had recovery periods of 3, 5, 7, . . . years under the Accelerated Cost Recovery System of 1981.

Warm-Up Exercises
Round to the nearest dollar.

1. 37% of $171,500 $63,455
2. 38% of $1280 $486.40
3. 8% of $4000 $320
4. 12% of $76,500 $9180
5. $1280 − $486 $794
6. $4000 − $320 $3680

19-8 Modified Accelerated Cost Recovery System (MACRS)

OBJECTIVE
Use the modified accelerated cost recovery system to compute annual depreciation and book value.

The **modified accelerated cost recovery system** (MACRS) is another method of computing depreciation. Introduced by the Tax Reform Act of 1986 and further modified by the Tax Bill of 1989, MACRS takes the place of the accelerated cost recovery system (ACRS) introduced by the Economic Recovery Tax Act of 1981. MACRS allows businesses to depreciate assets fully over a set period of time. This method encourages businesses to replace equipment earlier than they would if they used other depreciation methods. Under MACRS, assets can be depreciated fully over recovery periods of 4, 6, 8, 11, 16, or 21 years according to fixed percents.

Annual Depreciation = Original Cost × Fixed Percent

Book Value = Original Cost − Accumulated Depreciation

EXAMPLE *Skills* 30, 3 *Term* Modified accelerated cost recovery system

Prince Pizza purchased a new delivery van for $12,400 to use in delivering pizzas. MACRS allows delivery vans to be depreciated fully in 6 years according to 6 fixed percents: 20% the first year, 32% the second year, 19.2% the third year, 11.52% the fourth and fifth years, and 5.76% the sixth year. What are the annual depreciation and book value for each of the six years?

SOLUTION

A. Find the **annual depreciation.**

Original Cost	×	Fixed Percent		
$12,400.00	×	20.00%	=	$2480.00
$12,400.00	×	32.00%	=	$3968.00
$12,400.00	×	19.20%	=	$2380.80
$12,400.00	×	11.52%	=	$1428.48
$12,400.00	×	11.52%	=	$1428.48
$12,400.00	×	5.76%	=	$ 714.24

B. Find the **book value.**

Original Cost	−	Accumulated Depreciation		
$12,400.00	−	$ 2480.00	=	$9920.00
$12,400.00	−	$ 6448.00	=	$5952.00
$12,400.00	−	$ 8828.80	=	$3571.20
$12,400.00	−	$10,257.28	=	$2142.72
$12,400.00	−	$11,685.76	=	$ 714.24
$12,400.00	−	$12,400.00	=	0

$2480.00
+ $3968.00
$6448.00

$6448.00
+ $2380.80
$8828.80

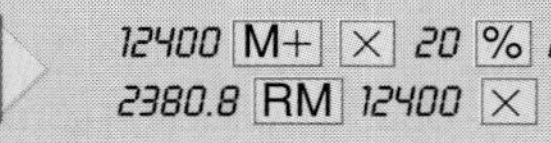

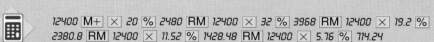

12400 M+ × 20 % 2480 RM 12400 × 32 % 3968 RM 12400 × 19.2 % 2380.8 RM 12400 × 11.52 % 1428.48 RM 12400 × 5.76 % 714.24

COMMUNICATION SKILLS
Have students write a brief report on the differences and similarities of the Accelerated Cost Recovery System (1981) and the Modified Accelerated Cost Recovery System (1986).

✔ SELF-CHECK Complete the problem, then check your answer in the back of the book.

$3400.00; $5440.00; $3264.00; $1958.40 $1958.40; $979.20

1. MACRS depreciates automobiles in 6 years according to the same 6 percents used for a delivery van in the example above. Find the annual depreciation for each of the 6 years for an automobile costing $17,000.

PROBLEMS

F.Y.I. Using the MACRS method, the salvage value is not used since 100% of the cost is depreciated.

Use the modified accelerated cost recovery system (MACRS) to find the annual depreciation. Round answers to the nearest cent.

		2.	3.	4.	5.	6.
	Cost	Taxi $14,600	Car $9240	Truck $24,700	Manufacturing Equipment $34,840	TeleComp $147,500
Year	Percent	Annual Dep.	Annual Dep.	Annual Dep.	Annual Dep.	Annual Dep.
1	20.00%	$2920.00	$1848.00	$4940.00	$6968.00	$29,500
2	32.00%	$4672.00	$2956.80	$7904.00	$11,148.80	$47,200
3	19.20%	$2803.20	$1774.08	$4742.40	$6689.28	$28,320
4	11.52%	$1681.92	$1064.45	$2845.44	$4013.57	$16,992
5	11.52%	$1681.92	$1064.45	$2845.44	$4013.57	$16,992
6	5.76%	$840.96	$532.22	$1422.72	$2006.78	$8496

7.

Yr.	Dep.
1	$ 6,797.28
2	$11,657.24
3	$ 8,325.24
4	$ 5,945.24
5	$ 4,250.68
6	$ 4,250.68
7	$ 4,250.68
8	$ 2,122.96
	$47,600.00

8.

Yr.	Dep.
1	$ 54,285
2	$ 74,025
3	$ 24,675
4	$ 11,515
	$164,500

7. Office computer system.
Original cost is $47,600.
Fully depreciated in 8 years.
Fixed percents are 14.28%, 24.49%, 17.49%, 12.49%, 8.93%, 8.93%, 8.93%, and 4.46%.
What is the depreciation for each of the 8 years?

8. Three-year-old racehorse.
Original cost is $164,500.
Fully depreciated in 4 years.
Fixed percents are 33%, 45%, 15%, and 7%.
What is the depreciation for each of the 4 years?

9. The Extended Care Center purchased a van for $48,260 to transport residents to and from a shopping center. The van is fully depreciated in 6 years. Fixed percents are 20%, 32%, 19.2%, 11.52%, 11.52%, and 5.76%. What are the depreciation and book value for each year?

Year	Percent	Cost	Depreciation	Accumulated Depreciation	Book Value
1	20.00%	$48,260.00	$ 9,652.00	$ 9,652.00	$38,608.00
2	32.00%	$48,260.00	$15,443.20	$25,095.20	$23,164.80
3	19.20%	$48,260.00	$ 9,265.92	$34,361.12	$13,898.88
4	11.52%	$48,260.00	$ 5,559.55	$39,920.67	$ 8,339.33
5	11.52%	$48,260.00	$ 5,559.55	$45,480.22	$ 2,779.78
6	5.76%	$48,260.00	$ 2,779.78	$48,260.00	($0.00)

PRACTICE AND APPLY

The following problems can be assigned for classwork and the answers checked in class to help students master the objective of the lesson.

- Guided Practice: 1–4
- Independent Practice: 5–7

ALTERNATIVE STRATEGIES: Reteaching

1. Emphasize that the depreciation must be for 4, 6, 8, 11, 16, or 21 years.
2. Point out that the cost of the item is used and that there is no salvage value involved.
3. Use the Example and show the calculations for all four methods.

WRAP-UP

Have students explain the differences between MACRS and the other methods for computing depreciation. Differences should include facts that the number of years allowed for depreciation is fixed, and items are allowed to fully depreciate.

Assignment Guide

- Basic: 7–9, 11–14
- Average: 8–10, 11–14

10. Wastewater treatment plants are depreciated fully in 16 years. Complete the depreciation table for a plant costing $1,400,000.

Year	Percent	Original Cost	Depre-ciation	Accumulated Depreciation	Book Value
1	5.00%	$1,400,000	$ 70,000	$ 70,000	$1,330,000
2	9.50%	$1,400,000	$133,000	$ 203,000	$1,197,000
3	8.55%	$1,400,000	$119,700	$ 322,700	$1,077,300
4	7.69%	$1,400,000	$107,660	$ 430,360	$ 969,640
5	6.93%	$1,400,000	$ 97,020	$ 527,380	$ 872,620
6	6.23%	$1,400,000	$ 87,220	$ 614,600	$ 785,400
7	5.90%	$1,400,000	$ 82,600	$ 697,200	$ 702,800
8	5.90%	$1,400,000	$ 82,600	$ 779,800	$ 620,200
9	5.90%	$1,400,000	$ 82,600	$ 862,400	$ 537,600
10	5.90%	$1,400,000	$ 82,600	$ 945,000	$ 455,000
11	5.90%	$1,400,000	$ 82,600	$1,027,600	$ 372,400
12	5.90%	$1,400,000	$ 82,600	$1,110,200	$ 289,800
13	5.90%	$1,400,000	$ 82,600	$1,192,800	$ 207,200
14	5.90%	$1,400,000	$ 82,600	$1,275,400	$ 124,600
15	5.90%	$1,400,000	$ 82,600	$1,358,000	$ 42,000
16	3.00%	$1,400,000	$ 42,000	$1,400,000	$ 0

MAINTAINING YOUR SKILLS Look up the skill in parentheses if you need help or more practice.

Find the percentage. Round answers to the nearest dollar. **(Skill 30)**

11. 38% of $74,500 **$28,310** **12.** 22% of $18,500 **$4070** **13.** 8% of $1475 **$118** **14.** 12% of $840 **$101**

Reviewing the Basics

Skills

Round answers to the nearest dollar.

(Skill 2)
1. $7151.444 **$7151** **2.** $71,417.20 **$71,417** **3.** $143.51 **$144**

Solve.

(Skill 4)
4. $24,600 − $6150 **$18,450** **5.** $214,650 − $17,172 **$197,478** **6.** $9170 − $917 **$8253**

(Skill 5)
7. $47.50 + $16.75 + $4.76 **$69.01** **8.** $4718.71 + $516.48 **$5235.19**
9. $2.49 + $1.79 **$4.28**

(Skill 6)
10. $5176.47 − $3247.98 **$1928.49** **11.** $22,417.87 − $5147.97 **$17,269.90**

(Skill 7)
12. $512 × 6 **$3072** **13.** $925 × 80 **$74,000** **14.** $4217 × 4 **$16,868**

(Skill 8)
15. 0.16 × $42,500 **$6800** **16.** 0.12 × $718 **$86.16** **17.** 0.37 × $3240 **$1198.80**

(Skill 20)
18. $\frac{4}{21}$ × $1344 **$256** **19.** $\frac{3}{10}$ × $185,750 **$55,725** **20.** $\frac{4}{15}$ × $23,480 **$6261.33**

Round answers to the nearest thousandth.

(Skill 10)
21. 2400 ÷ 37,000 **0.065** **22.** 2000 ÷ 180,000 **0.011** **23.** 1500 ÷ 50,000 **0.03**

Find the percentage. Round answers to the nearest cent.

(Skill 30)
24. 40% of $42,500 **$17,000.00** **25.** 12.5% of $8640 **$1080.00** **26.** 16% of $970 **$155.20**
27. 125% of $12,425 **$15,531.25** **28.** 10% of $115,475 **$11,547.50** **29.** 3% of $179.79 **$5.39**

Round answers to the nearest tenth of a percent.

(Skill 31)
30. $648 is what percent of $19,719? **3.3%**
31. $7845 is what percent of $49,417? **15.9%**
32. $13,415 is what percent of $871,980? **1.5%**
33. $20,450 is what percent of $147,850? **13.8%**

Applications

(Application X)
Find the area.
34. 120 ft by 50 ft **6000 ft^2** **35.** 40 ft by 60 ft **2400 ft^2**

Terms

Write your own definition for each term. **Answers will vary**

36. Payroll register **37.** Straight-line method
38. Apportion **39.** Sum-of-the-years'-digits method
40. Book value **41.** Double-declining-balance method
42. Modified accelerated cost recovery system

Refer to your reference files in the back of the book if you need help.

LESSON PLAN
Reviewing the Basics

The exercises on this page review skills, applications, and terms used in the unit. You can use the exercises to assess informally students' proficiency with this material.

The page can be used for guided practice and independent practice. You can work through a selection of the exercises together with students, and thus see immediately if they know how to do them, and you can then assign some of the exercises for independent practice. Be sure to go over the answers to all assigned exercises.

LESSON PLAN
Unit Test

Students should do the Unit Test on their own. Each problem on the test is keyed to a lesson in the unit. Students having difficulty with any particular problem should review the Example in the appropriate lesson and be assigned some of the Independent Practice problems for additional practice.

Unit Test

Yr.	Dep.	Dep.	Value
1	$36,000	$36,000	$18,000
2	$12,000	$48,000	$ 6,000

Lesson 19-1

1. Complete the payroll register. Use the social security tax rate of 6.2% and the medicare tax rate of 1.45%.

Name	Gross Income	FIT	Soc. Sec.	Medicare	SIT	Total Deductions	Net Pay
Cole	$415.00	$52	$25.73	$6.02	$8.30	$92.05	$322.95
Miller	442.00	57	$27.40	$6.41	8.84	$99.65	$342.35
Peters	455.00	46	$28.21	$6.60	9.10	$89.91	$365.09
Thomas	438.00	36	$27.16	$6.35	8.76	$78.27	$359.73
Total	$1750	$191	$108.50	$25.38	$35	$359.88	$1390.12

Lesson 19-2

2. Trace Realty had these expenses during the last quarter:

Payroll	$17,945
Advertising	1750
Office rental	2500
Office supplies	547
Utilities	723

What are the total expenses? To the nearest tenth of a percent, payroll is what percent of the total? **$23,465; 76.5%**

Lesson 19-3

3. The Mallard Aerodynamics Company pays $247,500 per year for security. The company apportions the cost among its departments based on space. The research department occupies an area that measures 40 feet by 60 feet. The building contains 1,500,000 square feet. How much does the research department pay for security? **$396**

Lesson 19-4

4. The Temp. Company purchased a new copy machine for $7860. The machine has an estimated life of 4 years. The resale value is expected to be $500. Use the straight-line method to find the annual depreciation. **$1840**

Lesson 19-5

5. Find the book value after each year of use for the copy machine in problem 4. **$6020; $4180; $2340; $500**

Lesson 19-6

6. The All-Purpose Card Co. purchased new store fixtures for a new branch store. The cost of the fixtures was $9860. The salvage value after 5 years is estimated to be $860. Use the sum-of-the-years'-digits method to find the depreciation for each of the 5 years. **$3000; $2400; $1800; $1200; $600**

Lesson 19-7

7. The original price of a proof and transit machine is $54,000. The machine is expected to have a life of 3 years. Use the double-declining-balance method to find the annual depreciation and the book value for the first 2 years. Round answers to the nearest dollar.

Lesson 19-8

8. Brown Construction purchased special tools for $8700. Brown will use the MACRS method to fully depreciate the tools over 4 years. The fixed percents for the 4 years are 33%, 45%, 15%, and 7%. Use the MACRS method to find the annual depreciation and book value for each of the 4 years.

Yr.	Annual Dep.	Accum. Dep.	Book Value
1	$2871	$2871	$5829
2	$3915	$6786	$1914
3	$1305	$8091	$609
4	$609	$8700	0

A SPREADSHEET APPLICATION

Accounting

To complete this spreadsheet application, you will need the template diskette for *Mathematics with Business Applications*. Follow the directions in the *User's Guide* to complete this activity.

Select option 19, Accounting, from the menu. Input the information in the following problems and the computer will automatically compute the depreciation schedule for these methods: straight-line, sum-of-the-year's-digits, double-declining-balance, and modified accelerated cost recovery system (MACRS). Then answer the questions that follow.

	Original Cost	Estimated Life	Resale Value
1.	$70,000	5 years	$20,000
2.	14,500	5 years	2500
3.	53,758	5 years	8758

4. In #1, using the straight-line method:
- **a.** What is the first year's depreciation?
- **b.** Is it the same each year?
- **c.** What is the book value after the fifth year?
- **d.** Will it always equal the resale value?

5. In #2, using the straight-line method:
- **a.** What is the annual depreciation?
- **b.** What is the book value after the fifth year?

6. In #3, using the straight-line method:
- **a.** What is the accumulated depreciation after the third year?
- **b.** What is the book value after the fourth year?

7. Using the sum-of-the-year's-digits method:
- **a.** What is the first year's depreciation in #1? in #2? in #3?
- **b.** Is it the same each year in each case?
- **c.** What is the second year's depreciation in #1? in #2? in #3?
- **d.** What is the book value after the fifth year in #1? in #2? into #3?
- **e.** Does the book value after the fifth year equal the resale value for each of the problems?

8. Using the double-declining-balance method:
- **a.** What is the first year's depreciation in #1? in #2? in #3?
- **b.** What is the second year's depreciation in #1? in #2? in #3?
- **c.** Does the book value after the fifth year automatically equal the resale value in any of the problems?
- **d.** What must be done to cause the book value after the fifth year to equal the resale value?

9. Using the modified accelerated cost recovery system method:
- **a.** What is the first year's depreciation in #1? in #2? in #3?
- **b.** How many years in the recovery period?

LESSON PLAN
A Spreadsheet Application

USING TECHNOLOGY

As students most likely realize from studying Unit 19, computing depreciations by using different methods can be a time-consuming task. In real-life situations, depreciation methods are programmed for use on a computer. Thus, once an accountant determines the appropriate method to use, the next task is to simply input the date and let the computer do the work.

The answers to the questions on this page should be provided to students so they can check their results. If time permits, you may want to go over the answers in class.

LESSON PLAN
Career Wise

APPLICATIONS ON THE JOB

Most students have probably heard about accountants and have some idea of what they do. Ask students how a person gets to be an accountant. (college education) Ask them also what accountants generally do for a business. (handle financial matters)

As you discuss the material on the page, ask students which of the two methods used yields the quickest depreciation for the new computer system on a year-to-year basis. (double-declining balance)

CAREER WISE

Accountant

Darryl Turner is an accountant for the Ace Technology Company. Prior to coming to Ace, he worked for an accounting firm that monitored expenses incurred by a small company that makes documentaries and music videos.

At Ace, Darryl's primary responsibility is handling the depreciation of the company's durable goods. This involves keeping purchase and use records. It also involves completing the income tax forms with regard to depreciation.

Recently, Darryl analyzed the straight-line (SL) and double-declining-balance (DDB) book values for a new $90,000 computer system with a useful life of 10 years and a resale value after that of $8500. His table shows the years of life, the SL book value, and the DDB book value. He also visualized the data in a graph.

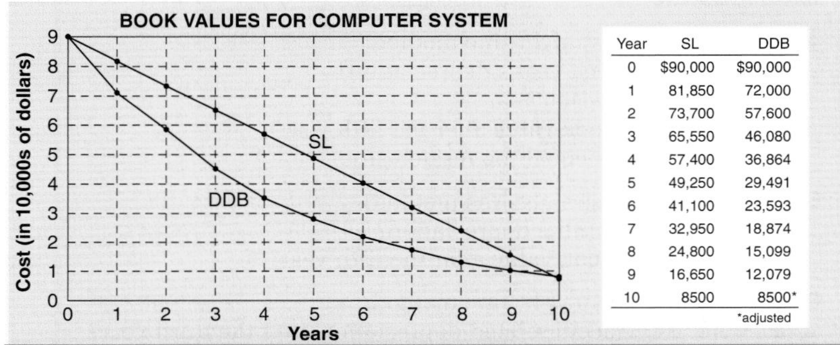

Year	SL	DDB
0	$90,000	$90,000
1	81,850	72,000
2	73,700	57,600
3	65,550	46,080
4	57,400	36,864
5	49,250	29,491
6	41,100	23,593
7	32,950	18,874
8	24,800	15,099
9	16,650	12,079
10	8500	8500*

*adjusted

Check Your Understanding

1. The table and graph show a value of $90,000 at year 0. Explain why this makes sense. **Both methods use the original cost at the start.**
2. Use the method in Lesson 19-4 to compute the book value after 2 years. **$73,700**
3. Use the method in Lesson 19-7 to compute the book value after 2 years. **$57,600**
4. Copy the table. Add a fourth column to show the difference between the SL book value and the DDB book value for each year. When is the difference the least? the greatest? Do the graphs confirm your answers? **Least: when the system is new and when the system is 10 years old; Greatest: midway through the life of the system. The graphs coincide at the start and end. The graphs are farthest apart vertically when the system is midway through its life.**

UNIT

20

Accounting Records

The accounting department also keeps other records that show the income, expenses, and value of your business. A *balance sheet* shows your cash, or *assets,* and the money you owe others, or your *liabilities.* An *income statement* details your income and operating expenses.

OUTLINE

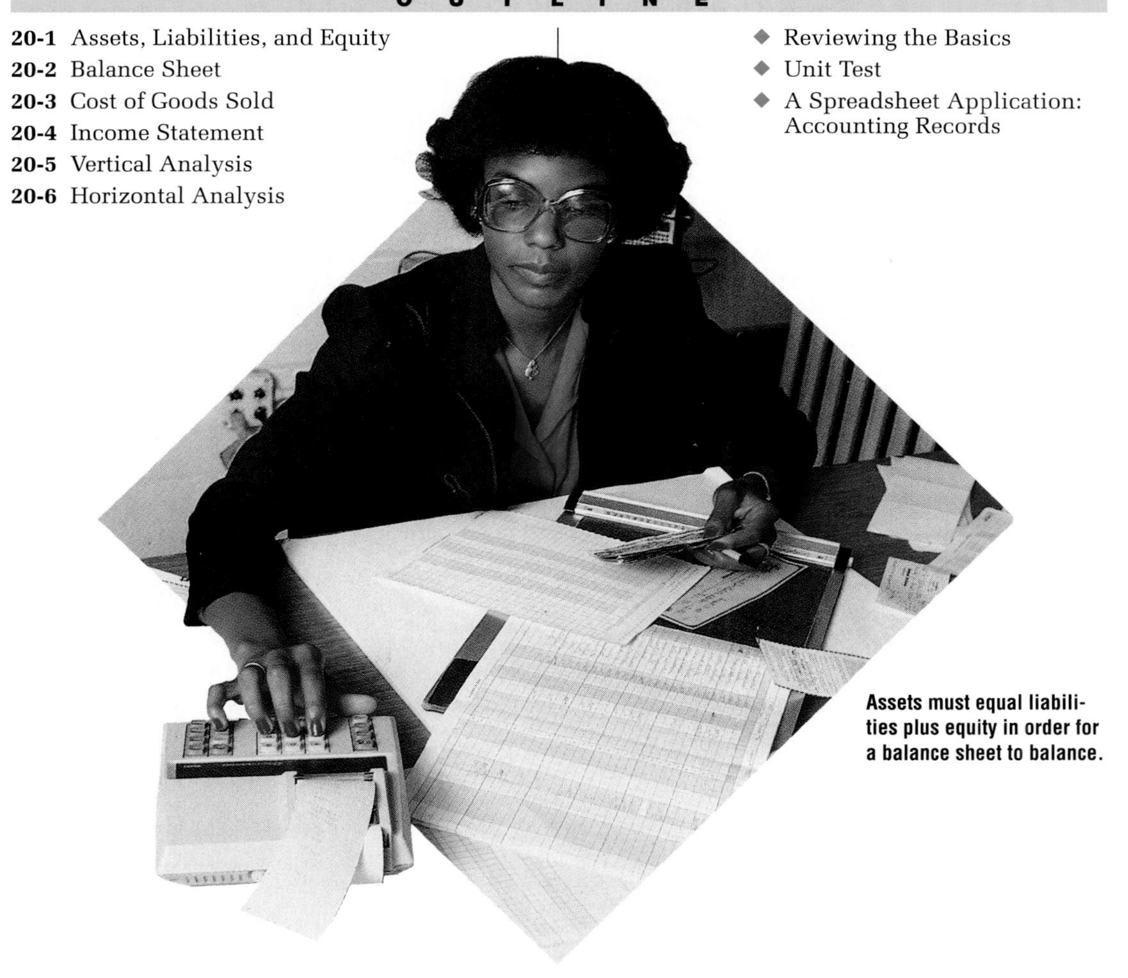

Assets must equal liabilities plus equity in order for a balance sheet to balance.

UNIT 20 ACCOUNTING RECORDS

INTRODUCING THE UNIT

You can help to strengthen students' reading and speaking skills by having them review a real balance sheet for a business. A brief discussion of such a balance sheet will be helpful to students in using the terms of this unit correctly.

LESSON PLAN
20-1 Assets, Liabilities, and Equity

FOCUS
Ask students if they have ever known anyone who started a business on his or her own. Discuss some of the steps involved, such as deciding on the type of business, the location, whether to rent or to buy the location, name of the business, and where to acquire merchandise.

TEACH
Students are probably not familiar with the terms *assets, liabilities,* and *equity.* Make sure students understand that assets can include anything of value owned by a business, including cash, supplies, and equipment. Point out that supplies and equipment can be owned even if they have not been paid for in full yet. A liability can be any money owed on any debt or loan. Owner's equity is the difference between the total assets and the total liabilities—it is what the business truly owns.

Use the following example. Have students place each item in either a column labeled *asset* or one labeled *liability*: $5200 in cash (a), $3650 in supplies (a), $12,720 in equipment (a), $5000 owed to bank (l), and $3650 in taxes (l).

Warm-Up Exercises
1. $5200 + $3650 + $12,720 $21,570
2. $41,850 + $27,400 + $1845 $71,095
3. $21,570 − $5850.75 $15,719.25
4. $71,095 − $24,214 $46,881

20-1 Assets, Liabilities, and Equity

OBJECTIVE
Compute the total assets, liabilities, and owner's equity.

When you start a business, you will need to buy either merchandise to sell or materials with which to make your products. You may need to purchase office supplies, equipment, buildings, or land. In addition, you must have cash to make change, pay bills, and meet other expenses. **Assets** are the total of your cash, the items that you have purchased, and any money that your customers owe you.

You may borrow money to start your business, or you may purchase merchandise on credit. **Liabilities** are the total amount of money that you owe to creditors. **Owner's equity, net worth,** or **capital** is the total value of assets that you own. Owner's equity plus liabilities equal assets.

Assets = Liabilities + Owner's Equity

Owner's Equity = Assets − Liabilities

Liabilities = Assets − Owner's Equity

EXAMPLE *Skills* 5, 6 *Application* A *Term* Owner's equity

Tina and John Agee recently opened The Clothing Store. They used $60,000 of their own money and a bank loan of $25,000. From the $85,000, Tina and John paid $15,000 for merchandise and $10,000 for supplies. This left a cash balance of $60,000. They received another shipment of $8000 worth of merchandise. They did not pay for this merchandise immediately. What was their owner's equity?

SOLUTION

A. Find the **assets.**

Cash: $85,000 − ($15,000 + $10,000)	$60,000
Merchandise: $15,000 + $8000	23,000
Supplies	10,000
Total Assets	$93,000

B. Find the **liabilities.**

Bank loan	$25,000
Unpaid merchandise	8,000
Total Liabilities	$33,000

C. Find the **owner's equity.**
Assets − Liabilities
$93,000 − $33,000 = $60,000 owner's equity

✓ SELF-CHECK Complete the problems, then check your answers in the back of the book.

1. Find the assets. Liabilities: $82,000. Owner's equity: $50,000. **$132,000**
2. Find the liabilities. Assets: $18,000 cash, $49,000 merchandise. Owner's equity: $25,000. **$42,000**

BUSINESS PROJECT
Have each student ask an adult to name some items that could be considered assets in their places of work. Do the same for liabilities. Have students compare their lists for similarities and differences. Is an item ever considered to be an asset by one company and a liability by another?

PROBLEMS

	Liabilities	+	Owner's Equity	=	Assets
3.	$27,000	+	$ 50,000	=	**$77,000**
4.	$14,750	+	$ 37,500	=	**$52,250**
5.	$38,750	+	**$36,250**	=	$ 75,000
6.	$54,690	+	**$88,960**	=	$143,650
7.	**$17,450**	+	$ 25,000	=	$ 42,450
8.	**$327,530**	+	$147,470	=	$475,000

9. $17,740 in cash.
$74,800 in merchandise.
$11,475 in supplies.
$22,480 owed to bank.
$917.80 owed in taxes.
What are the total assets?
What are the total liabilities?
What is the owner's equity?
$104,015; $23,397.80; $80,617.20

10. $11,420 in cash.
$17,590 in supplies.
$43,470 in equipment.
$7890 owed for supplies.
$12,740 owed to bank.
What are the total assets?
What are the total liabilities?
What is the owner's equity?
$72,480; $20,630; $51,850

11. Family Express Pharmacy has these assets and liabilities.

Cash: $4187	Supplies: $7185	Unpaid merchandise: $11,410
Inventory: $17,450	Building: $125,000	Taxes owed: $847
Equipment: $36,475	Land: $31,500	Real estate loan: $130,000

What are the total assets? What are the total liabilities? What is the owner's equity? **$221,797; $142,257; $79,540**

12. Howard's Jewelers has these assets and liabilities.

Cash on hand: $3417	Store fixtures: $1250	Unpaid merchandise: $6470
Customers owe: $6214	Building: $35,750	Taxes owed: $714.85
Inventory: $13,419	Land: $20,000	Wages owed: $274.35
Supplies: $417.50	Bank loan: $12,214	Mortgage loan: $31,340

What are the total assets? What are the total liabilities? What is the owner's equity? **$80,467.50; $51,013.20; $29,454.30**

MAINTAINING YOUR SKILLS Look up the skills in parentheses if you need help or more practice.

Add. **(Skill 5)**

13. $31,475.00 + 10,719.50 + 563.85 = **$42,758.35**

14. $9187.40 + 7341.85 + 1174.97 = **$17,704.22**

15. $12,471.80 + 7,518.75 + 431.83 = **$20,422.38**

16. $14,817.48 + 3,247.55 + 7,916.37 = **$25,981.40**

Subtract. **(Skill 6)**

17. $74,850.00 − 35,798.00 = **$39,052.00**

18. $147,875.00 − 74,917.00 = **$72,958.00**

19. $45,371.80 − 23,747.75 = **$21,624.05**

20. $12,147.85 − 7,591.97 = **$4,555.88**

PRACTICE AND APPLY

The following problems can be assigned for classwork and the answers checked in class to help students master the objective of the lesson.

- Guided Practice: 1–8
- Independent Practice: 9, 10

WRAP-UP

Have students list the new terms of this lesson and then tell what each term means. Have students give examples of assets and liabilities.

Assignment Guide:

- Basic: 9–11, 13–20
- Average: 11, 12, 14–20 even

ALTERNATIVE STRATEGIES: Reteaching

Using the comments in the TEACH section, review the definitions of **assets, liabilities,** and **owner's equity**. Stress the relationship in the accounting equation by completing this table with your students:

	Assets	Liabilities	Owner's Equity
1)	$357,000	$186,500	($170,500)
2)	$4.2 million	$2.7 million	($1.5 million)
3)	$12.6 billion	($7.8 billion)	$4.8 billion

LESSON PLAN
20-2 Balance Sheet

FOCUS
A balance sheet is used to keep a written record of a business's total assets, liabilities, and owner's equity. In this lesson, students will complete a balance sheet.

TEACH
Balance sheets may be completed over a monthly, quarterly, or annual basis. Balance sheets are usually a part of the financial report that is made available to stockholders.

Emphasize that the total assets must equal the total liabilities plus owner's equity. Point out that the double rule under the $93,000 is an accounting convention used to emphasize the equality, and to indicate that nothing can be added to or subtracted from these figures.

20-2

Balance Sheet

OBJECTIVE
Complete a balance sheet.

A **balance sheet** shows the financial position of your company on a certain date. You may prepare a balance sheet monthly, quarterly, or annually. The balance sheet shows your total assets, total liabilities, and owner's equity. The balance sheet is designed so that the assets appear on the left. The liabilities and owner's equity appear on the right. The sum of the assets must equal the sum of the liabilities and owner's equity.

Assets = Liabilities + Owner's Equity

EXAMPLE *Skill* 3 *Term* Balance sheet

The Agees used $60,000 of their own money plus a $25,000 loan to start their clothing store. They received two shipments of merchandise. They paid cash for the $15,000 shipment. They did not pay immediately for the $8000 shipment. They bought supplies for $10,000. The Agees have $60,000 left in cash. What does their balance sheet show?

SOLUTION

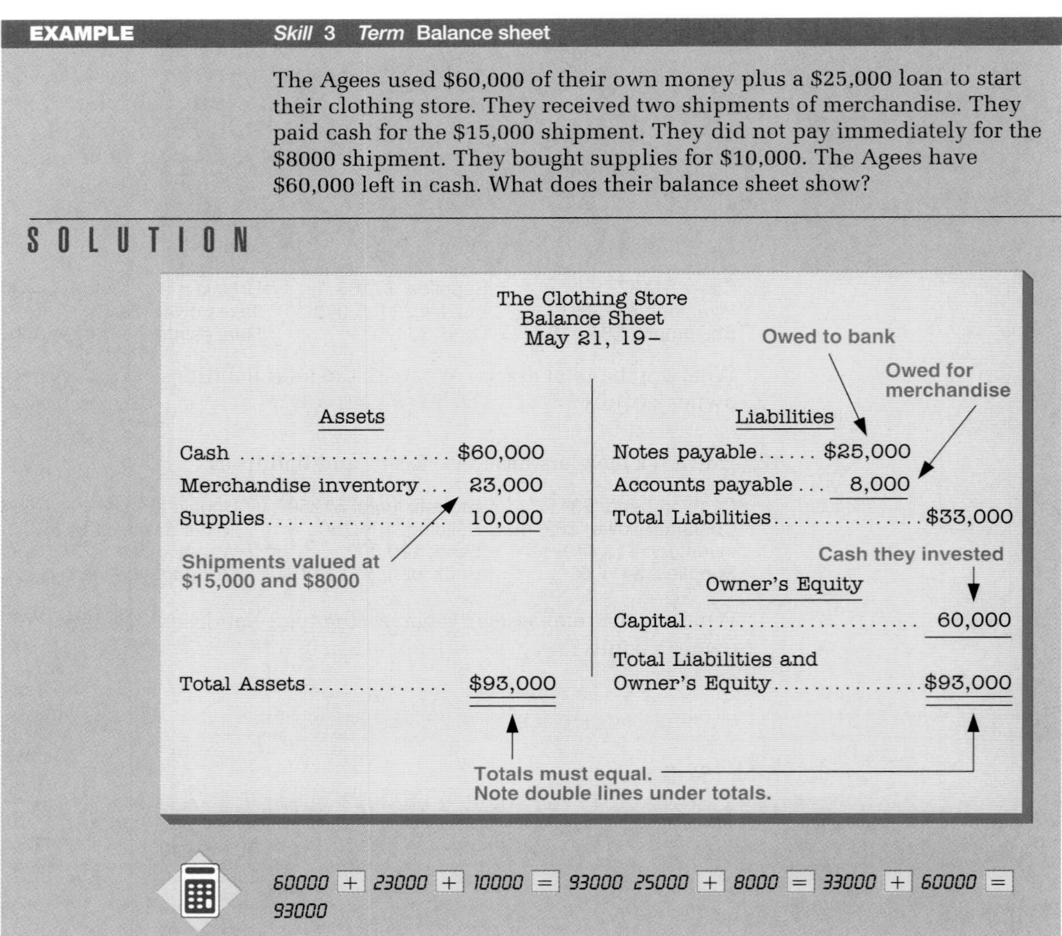

60000 + 23000 + 10000 = 93000 25000 + 8000 = 33000 + 60000 = 93000

✔ SELF-CHECK	Complete the problems, then check your answers in the back of the book.

Assets		Liabilities	
Cash	$25,000	Notes payable	$10,000
Inventory	30,000	Accounts payable	7,000
Supplies	12,000	Owner's equity	50,000

1. Total assets ..... ? $67,000

2. Total liabilities and owner's equity ...... ? $67,000

PROBLEMS

	3.	4.	5.	6.
Cash	$18,000	$ 45,000	$10,500	$125,400
Inventory	$11,500	$255,700	$41,200	$196,700
Supplies	$ 3,500	$ 11,400	$ 800	$ 12,800
Total Assets	**$33,000**	**$312,100**	**$52,500**	**$334,900**
Bank Loan	$10,500	$200,000	**$22,500**	$175,000
Taxes Owed	$ 1,800	$ 8,800	$ 5,000	**$ 9,900**
Total Liabilities	**$12,300**	**$208,800**	**$27,500**	**$184,900**
Owner's Equity	**$20,700**	**$103,300**	$25,000	$150,000

7. $25,000 in cash.
   $74,800 in merchandise.
   $65,000 owed to bank.
   What are the total assets?
   What are the total liabilities?
   What is the owner's equity?
   $99,800; $65,000; $34,800

8. $8450 in cash.
   $11,170 in equipment.
   $10,000 owed to bank.
   What are the total assets?
   What are the total liabilities?
   What is the owner's equity?
   $19,620; $10,000; $9620

9. Complete the balance sheet for Howard's Jewelers.

Howard's Jewelers
Balance Sheet
June 30, 19–

Assets		Liabilities	
Cash on hand	3417 00	Bank loan	12214 00
Accts. receivable	6214 00	Accts. payable	6470 00
Inventory	13419 00	Taxes owed	714 85
Supplies	417 50	Wages owed	274 35
Store fixtures	1250 00	Mortgage loan	31340 00
Building	35750 00	Total Liabilities	**51013 20**
Land	20000 00		
		Owner's Equity	?
		Capital	**29454 30**
		Total Liabilities	
Total Assets	**80467 50**	and Owner's Equity	**80467 50**

Warm-Up Exercises

1. $14,718.00 + $9543.20 + $816.00 $25,077.20
2. 2.1 + 3.7 + 1.7 + 8.5 16

Are the following totals equal?

3. $7210 + $847 + $9183 = $12,417 + $4823 Yes

PRACTICE AND APPLY

The following problems can be assigned for classwork and the answers checked in class to help students master the objective of the lesson.

- Guided Practice: 1–6
- Independent Practice: 7–9

ALTERNATIVE STRATEGIES: Enrichment

Annual reports of businesses will contain a balance sheet. Ask students to obtain annual reports from the companies themselves or from the library. Analyze them with respect to **assets, liabilities,** and **owner's equity**. Stress that a balance sheet shows the financial condition of a company on a given day (like a snapshot).

WRAP-UP

Have students explain what money amounts should be shown on each side of a balance sheet. Ask which two totals should be equal. (assets = liability + owner's equity)

Assignment Guide

- Basic: 7–11, 13–17
- Average: 10–12, 13–17 odd

10. Complete the balance sheet for Pet Supplies, Inc.

Pet Supplies, Inc.
Balance Sheet
Dec. 31, 19–

Assets		Liabilities	
Cash on hand	3,128.00	Bank loan	11,715.00
Accts. receivable	214.85	Accts. payable	474.84
Inventory	8,426.00	Taxes owed	461.72
Supplies	5,617.40	Wages owed	234.85
Store fixtures	2,405.75	Mortgage loan	54,175.00
Building	43,450.00	Total Liabilities	67,061.41
Land	24,500.00		
		Owner's Equity	
		Capital	20,680.59
Total Assets	87,742.00	Total Liabilities and Owner's Equity	87,742.00

Prepare a balance sheet for each business.

11. Wholesale Grocer Supply Co. had these assets and liabilities on December 31:

Assets		Liabilities	
Cash	$ 2,417,600	Accounts payable	$84,640,000
Accounts receivable	53,591,500	Notes payable	56,119,400
Inventory	48,478,600	Income taxes	975,450
Property	75,750,000	Other liabilities	864,560
Investments	475,000	**Total Liabilities**	**$142,599,410**
Other assets	18,791,500	**Owner's Equity, Capital**	**56,904,790**
Total Assets	**$199,504,200**	**Total Liabilities and Owner's Equity**	**$199,504,200**

12. Metal Abrasives, Inc., shows these assets and liabilities as of June 30:

Assets (in millions of dollars)		Liabilities (in millions of dollars)	
Cash	$ 1.1	Notes payable	$14.9
Accounts receivable	9.8	Accounts payable	4.3
Inventories	11.4	Income taxes	1.2
Property	19.7	Stock	10.3
Foreign investments	1.3	Other liabilities	4.1
Other assets	4.2		
		Total Liabilities	**$34.8**
Total Assets	**$47.5**	**Owner's Equity, Capital**	**$12.7**
		Total Liabilities and Owner's Equity	**$47.5**

MAINTAINING YOUR SKILLS Look up the skills in parentheses if you need help or more practice.

Add. **(Skills 3, 5)**

13. $14,780 + $13,190 **$27,970**

14. $147,560 + $93,480 **$241,040**

15. $4,175,000 + $897,400 **$5,072,400**

16. $416,750 + $318,430 + $84,970 **$820,150**

17. $1.7 + $2.3 + $4.5 + $1.4 **$9.9**

COOPERATIVE LEARNING

If you have a real balance sheet, you might want to organize the class into small groups to discuss it. Each student should have a copy of the balance sheet. Students can ask each other questions about the balance sheet.

20-3

Cost of Goods Sold

OBJECTIVE
Compute the cost of goods sold.

The balance sheet shows your total assets, total liabilities, and owner's equity at a given point in time. You also need to know whether the company is operating at a profit or loss. To determine if you are making money or losing money, you must know sales figures, expenses, and the **cost of goods sold.** The cost of goods sold is equal to the value of the beginning inventory plus the cost of any goods received (receipts) minus the value of the ending inventory.

Cost of Goods Sold = (Beginning Inventory + Receipts) − Ending Inventory

EXAMPLE *Skills* 5, 6, 8 *Application* A *Term* Cost of goods sold

The Clothing Store began the month with an inventory valued at $14,750. During the quarter, it received 100 belts that cost $6.49 each, 50 scarves at $8.24 each, 25 sweaters at $19.72 each, and 144 plastic raincoats at $2.50 each. The ending inventory was valued at $12,847. What was the cost of goods sold?

SOLUTION

A. Find the **receipts.**

100 belts	×	$ 6.49 each	=	$ 649.00
50 scarves	×	$ 8.24 each	=	412.00
25 sweaters	×	$19.72 each	=	493.00
144 raincoats	×	$ 2.50 each	=	360.00
		Total Receipts		$1914.00

B. Find the **cost of goods sold.**

(Beginning Inventory + Receipts) − Ending Inventory
($14,750.00 + $1914.00) − $12,847.00 = $3817.00 cost of goods sold

100 × 6.49 = 649 M+ 50 × 8.24 = 412 M+ 25 × 19.72 = 493 M+ 144 × 2.5 = 360 M+ RM 1914 + 14750 − 12847 = 3817

✔ SELF-CHECK Complete the problems, then check your answers in the back of the book.

Find the cost of goods sold.

	1.	2.
Beginning inventory	$156,470	$43,656
Receipts	21,960	11,712
Ending inventory	161,510	42,964
	$16,920	**$12,404**

PROBLEMS

3. Beginning inventory: $417,600.
Receipts: $75,800.
Ending inventory: $396,800.
Find the cost of goods sold. **$96,600**

4. Beginning inventory: $125,400.
Receipts: $31,200.
Ending inventory: $131,400.
Find the cost of goods sold. **$25,200**

ALTERNATIVE STRATEGIES: Reteaching

Complete this table with your students.

	Beginning Inventory	Receipts	Ending Inventory	Cost of Goods Sold
1)	$546,980	$52,850	$497,690	($102,140)
2)	$ 72,750	$14,759	$ 68,498	($ 19,011)
3)	$158,000	$61,900	($174,800)	$ 45,100

LESSON PLAN 20-3 Cost of Goods Sold

FOCUS

Students should be familiar with receipts that they receive when making purchases in stores. Mention that businesses also receive receipts when they buy goods. In this lesson, students will compute the cost of goods sold.

TEACH

Computing the cost of goods sold is a necessary step between completing the balance sheet and preparing the income statement (Lesson 20-4). The cost of goods sold can be computed using either the cost to the retailer or the selling price. Cost to the retailer is normally used.

Point out that the multiplications in the Example and in Problems 13 and 14 are referred to as *extensions* when they occur on an invoice. Make sure students understand that they must find the sum of all of the receipts, not just one of each category (belts, scarves, and so on).

Warm-Up Exercises

1. $16,850 + $2147 − $14,197 $4800
2. $96,471 + $11,516 − $98,583 $9404
3. $146,750 + [(3 × $179.80) + (6 × $114.39) + (144 × $46.90)] − $151,612 $3117.34
4. $63,419.85 + ($1216.48 + $9389.97 + $10,147.68) − $59,853.74 $24,320.24

PRACTICE AND APPLY

The following problems can be assigned for classwork and the answers checked in class to help students master the objective of the lesson.

- Guided Practice: 1–6
- Independent Practice: 7–9

WRAP-UP

Have a student give the formula for the cost of goods sold. Read Problem 8 to the class. Have a student tell which two numbers should be added and which one should be subtracted.

Assignment Guide

- Basic: 7–13, 15–24
- Average: 10–14, 15–23 odd

5. Beginning inventory: $75,470. Receipts: $14,650. Ending inventory: $72,170. Find the cost of goods sold. **$17,950**
6. Beginning inventory: $19,560. Receipts: $3780. Ending inventory: $20,450. Find the cost of goods sold. **$2890**
7. Beginning inventory: $112,475. Receipts: $42,164. Ending inventory: $96,512. Find the cost of goods sold. **$58,127**
8. Beginning inventory: $11,186. Receipts: $2871. Ending inventory: $12,074. Find the cost of goods sold. **$1983**
9. Miller Lamp Shop had a beginning inventory valued at $48,748.75. During the quarter, it had receipts of $9164.86. The value of the ending inventory was $50,041.93. Find the cost of goods sold. **$7871.68**
10. The Popcorn Castle had a beginning inventory valued at $2146.73. During the quarter, receipts totaled $516.65. The value of the ending inventory was $1989.83. Find the cost of goods sold. **$673.55**
11. Allison's Dress Shop began the quarter with an inventory valued at $21,647. During the quarter, 4 shipments were received valued at $2248.60, $1874.55, $2516.43, and $2050.74. The ending inventory for the quarter was valued at $20,416. What was the cost of goods sold? **$9921.32**
12. Central Auto Parts began the month with an inventory valued at $34,767.80. During the month, Central received 5 shipments valued at $1274.74, $4756.44, $983.45, $2465.39, and $416.93. Central's month-end inventory was valued at $36,193.48. Find the cost of goods sold. **$8471.27**
13. The Shoehorn Repair Shop started the quarter with an inventory valued at $2178.43. During the quarter, it received 5 boxes of replacement heels at $4.35 a box, 3 boxes of replacement soles at $6.15 a box, 4 cards of shoelaces at $7.65 a card, 6 spools of nylon thread at $12.13 a spool, and 2 large bottles of industrial strength adhesive at $9.87 a bottle. The ending inventory for the quarter was valued at $2241.83. Find the cost of goods sold. **$99.92**
14. Leather Limited started the month with an inventory valued at $18,719.45. During the month, it received 9 men's jackets at $47.53 each, 15 belts at $2.79 each, 4 ladies' long coats at $84.37 each, 5 men's hats at $8.71 each, and 5 ladies' jackets at $38.84 each. The month-end inventory was valued at $16,478.54. Find the cost of goods sold. **$3285.76**

MAINTAINING YOUR SKILLS Look up the skills in parentheses if you need help or more practice.

Add. **(Skill 5)**

15. $14,178.90 + $2417.36 **$16,596.26**
16. $8174.32 + $2461.89 **$10,636.21**
17. $56,147.83 + $9171.81 **$65,319.64**

Subtract. **(Skill 6)**

18. $16,596.80 − $13,743.34 **$2853.46**
19. $9771.83 − $5421.95 **$4349.88**
20. $147,178.66 − $93,147.02 **$54,031.64**

Multiply. **(Skill 8)**

21. 12 × $7.98 **$95.76**
22. 5 × $46.73 **$233.65**
23. 10 × $1.74 **$17.40**
24. 144 × $11.43 **$1645.92**

CULTURAL ANGLES

Tourism is a business in many countries. In 1987, the United States had tourist receipts of $14.7 billion. Have students use an almanac or other source (such as *The World Almanac and Book of Facts 1991*) to find the amount of tourist receipts for three other countries.

20-4 Income Statement

OBJECTIVE
Complete an income statement.

An **income statement,** or **profit-and-loss statement,** shows in detail your income and operating expenses. If your gross profit is greater than your total operating expenses, your income statement will show a **net income,** or **net profit.**

Gross Profit = Net Sales − Cost of Goods Sold

Net Income = Gross Profit − Total Operating Expenses

EXAMPLE *Skills* 3, 4 *Application* A *Term* Income statement

Three months after opening The Clothing Store, Tina and John Agee prepare an income statement. Sales for the first three months totaled $12,174. Merchandise totaling $173 was returned to them. The Agees' inventory records show that the goods they sold cost them $3817. Records show that their operating expenses totaled $8047. What is the net income?

SOLUTION

The Clothing Store
Income Statement
For the Quarter Ended June 30, 19–

Income:		
Sales	$12,174	
Less: Sales returns and allowances	173	
Net Sale		$12,001
Cost of goods sold		3,817
Gross profit on sales		$8,184
Operating Expenses:		
Salaries and wages	$6,400	
Delivery expenses	100	
Rent	900	
Advertising	75	
Utilities	120	
Supplies	50	
Depreciation	150	
Insurance	210	
Miscellaneous	42	
Total operating expenses		$8,047
Net Income		$137

Total Sales − Returns

Net Sales − Cost of Goods Sold

Gross Profit − Total Operating Expenses

12174 [−] 173 [=] 12001 [−] 3817 [=] 8184 6400 [+] 100 [+] 900 [+] 75 [+] 120 [+] 50 [+] 150 [+] 210 [+] 42 [=] 8047 [M+] 8184 [−] [RM] 8047 [=] 137

LESSON PLAN
20-4 Income Statement

FOCUS

An income statement will let a business know on a periodic basis whether it is making money or losing money. In this lesson, students will learn to complete an income statement.

TEACH

Explain that an income statement may be done annually, quarterly, or monthly. Unlike a balance sheet, an income statement must cover an entire accounting period.

Go over the different categories on the income statement in the Example. Point out that the gross profit on sales is the amount of markup on all items actually sold (not returned). Go over the operating expenses. Make sure students understand what a delivery expense is.

ALTERNATIVE STRATEGIES: Enrichment

Annual reports of businesses will contain income statements as well as balance sheets (see Lesson 20-2). Similar to 20-2 use annual reports from the companies themselves or from the library to find and analyze income statements. Stress that although a balance sheet is similar to a snapshot, an income statement is similar to a video in that it shows what has happened to a company over a period of time.

Warm-Up Exercises

1. Sales: $47,500; Returns: $7500; Net sales? $40,000
2. Gross profit: $21,420; Operating expenses: $9780; Net Income? $11,640
3. Net sales: $7490; Cost of goods sold: $3574; Gross profit? $3916

PRACTICE AND APPLY

The following problems can be assigned for classwork and the answers checked in class to help students master the objective of the lesson.

- Guided Practice: 1–5
- Independent Practice: 6, 7, 9

✓ SELF-CHECK Complete the problems, then check your answers in the back of the book.

1. Sales: $95,000.
   Returned merchandise: $3500.
   Cost of goods sold: $30,000.
   Total operating expenses: $34,700.
   Find the gross profit on sales.
   Find the net income.
   $61,500; $26,800
2. Sales: $475,000.
   Returned merchandise: $7500.
   Cost of goods sold: $178,500.
   Total operating expenses: $105,400.
   Find the gross profit on sales.
   Find the net income.
   $289,000; $183,600

PROBLEMS

	Total Sales	Returns	Net Sales	Cost of Goods Sold	Gross Profit	Operating Expenses	Net Income
3.	$ 14,700	$ 540	$14,160	$ 8,750	$5410	$ 725	**$ 4685**
4.	$ 38,900	$4120	**$ 34,780**	$ 11,175	**$ 23,605**	$18,900	**$ 4705**
5.	$ 3,750	$ 75	**$ 3675**	$ 1,740	**$ 1935**	$ 828	**$ 1107**
6.	$174,945	0	**$174,945**	$ 56,750	**$118,195**	$42,193	**$ 76,002**
7.	$674,916	$1274	**$673,642**	$417,916	**$255,726**	$96,419	**$159,307**

Complete the income statements.

8. Income:

Sales	$47,890	
Less: Sales returns and allowances	976	
Net sales	**$46,914**	?
Cost of goods sold		$21,742
Gross profit on sales	**$25,172**	?
Operating Expenses:		
Salaries and wages	$8,500	
Taxes	497	
Utilities	235	
Depreciation	975	
Total operating expenses	**$10,207**	?
Net Income	**$14,965**	?

9. Income:

Sales	$463,575	
Less: Sales returns and allowances	75,450	
Net sales	**$388,125**	?
Cost of goods sold		$231,000
Gross profit on sales	**$157,125**	?
Operating Expenses:		
Total operating expenses		$ 71,916
Net Income	**$85,209**	?

PROBLEM SOLVING

Have students unscramble these terms from the lesson:

1. TEN EOCINM (net income)
2. TEN SLASE (net sales)
3. GTRPOEAIN EPEXENSS (operating expenses)
4. IAREDPECITON (depreciation)

Prepare an income statement for each business in problems 10–13.

10. Last month Hurry-N-Go Service Station had total sales of $8961. Merchandise totaling $85 was returned. The goods that were sold cost Hurry-N-Go $5617. Operating expenses for the month were $718. **NS: $8876; GP: $3259; NI: $2541**

11. Advantage Housekeeping Service prepares an income statement monthly. For this past month, it collected $3150 from homeowners for services provided (sales). It had $70 in allowances for jobs that had to be redone. Salaries and wages totaled $1800, supplies cost $315, and other operating expenses totaled $94. **NS: $3080; GP: $2209; NI: $871**

12. During the past quarter, Henri's Clothes Company had total sales of $27,418 and returns of $220. The cost of goods sold amounted to $9193. Operating expenses for the quarter included: salaries and wages of $2000, real estate loan payment of $1210, advertising at $190, utilities and supplies of $195, bank loan payment of $350, and other operating expenses of $375. **NS: $27,198; GP: $18,005; OE: $4320; NI: $13,685**

13. Hi-Tech Aluminum Co. prepares an annual income statement for distribution to its stockholders. This past year, net sales totaled $121.4 million. Cost of goods sold totaled $63.1 million. Operating expenses included: wages and salaries of $23.4 million, depreciation and amortization of $3.3 million, general taxes of $1.1 million, interest paid totaling $1.0 million, income taxes of $8.7 million, and miscellaneous operating expenses of $1.2 million. **GP: $58.3; OE: $38.7; NI: $19.6**

14. If your total operating expenses are greater than your gross profit, your income statement will show a net loss.

Net Loss = Total Operating Expenses − Gross Profit

Find the net loss for this quarter.

Gross profit		1923 80
Operating Expenses:		
Wages	1675 65	
Supplies	483 00	
Miscellaneous	107 00	
Total		?
Net Loss		?

2265.65
341.85

MAINTAINING YOUR SKILLS Look up the skills in parentheses if you need help or more practice.

Add. **(Skill 5)**

15. $567.56 + 74.92 + 123.74 = **$766.22**

16. $18,516.50 + 8,743.91 + 191,434.84 = **$218,695.25**

17. $9871.47 + 219.50 + 4618.49 = **$14,709.46**

18. $231.42 + 71.85 + 63.79 = **$367.06**

Subtract. **(Skill 6)**

19. $174.93 − 84.79 = **$90.14**

20. $746.90 − 291.84 = **$455.06**

21. $7415 − 1700 = **$5715**

22. $135,463 − 96,394 = **$39,069**

WRAP-UP

Have students explain how to find net sales, gross profit, operating expenses, and net income. Also, have them explain the purpose of an income statement, and how the form differs from a balance sheet.

Assignment Guide

- Basic: 6–12, 15–22
- Average: 8, 10–14, 15–21 odd

LESSON PLAN
20-5 Vertical Analysis

FOCUS
Businesses often analyze their income statements to see how well they are doing. In this lesson, students learn three ways to analyze the information on these forms.

TEACH
When going over the formula for the percent of net sales, emphasize the fact that the "item" may be the sales returns, the costs of goods sold, or the gross profit on sales. Ask students which of the three a business would prefer to be as low as possible. (sales returns or cost of goods sold) Mention that "returns and allowances" are 1.4% of net sales.

Point out that for the current ratio, it is desirable for a business to have more assets than liabilities.

The numerator in the quick ratio, total assets − inventory, would include any cash or any items that can be sold for cash quickly. Point out that this ratio may be lower than the current ratio because the numerator of the current ratio includes items that may not be paid for yet (such as an unpaid shipment of goods).

20-5 Vertical Analysis

OBJECTIVE
Analyze balance sheets and income statements by finding what percent a part is of the whole and computing the current ratio and the quick ratio.

Your business may analyze its income statement by finding what percent any given item is of the net sales.

Your business may analyze its balance sheet by finding certain defined ratios. The **current ratio** is the ratio of total assets to total liabilities. The **quick ratio,** sometimes called the **acid-test ratio,** is the ratio of total assets minus inventory to total liabilities.

$$\text{Percent of Net Sales} = \frac{\text{Amount for Item}}{\text{Net Sales}}$$

$$\text{Current Ratio} = \frac{\text{Total Assets}}{\text{Total Liabilities}}$$

$$\text{Quick Ratio} = \frac{\text{Total Assets} - \text{Inventory}}{\text{Total Liabilities}}$$

EXAMPLE *Skill* 31 *Application* A *Term* Net sales

Tina and John Agee analyze their income statement for the quarter. The statement shows:

Income:		
Sales	$12,174	
Less: Sales returns and allowances	173	
Net sales		$12,001
Cost of goods sold		3,817
Gross profit on sales		$8,184

What is the gross profit as a percent of net sales?

SOLUTION

Find the **percent of net sales.**

Amount for Item ÷ Net Sales

$8184 ÷ $12,001 = 0.6819 = 68.2% of net sales

8184 ÷ 12001 = .681943

✓ SELF-CHECK Complete the problems, then check your answers in the back of the book.

1. Sales returns: $173.
   Net sales: $12,001.
   What are the sales returns as a percent of net sales? **1.4%**

2. Cost of goods sold: $3817.
   Net sales: $12,001.
   What is the cost of goods sold as a percent of net sales? **31.8%**

CRITICAL THINKING
In the Example, suppose that the Agees found that their "returns and allowances" were 40% of net sales. What conclusions could they reach about their merchandise? What course of action would you recommend?

(Possible answers: Merchandise is defective, or otherwise unsatisfactory; check merchandise before putting it out for sale; find another supplier.)

EXAMPLE *Skills* 22, 31 *Application* A *Terms* Current and quick ratios

The Agee's analyze their May 21 balance sheet. The balance sheet shows:

Assets		Liabilities	
Cash	$60,000	Notes payable	$25,000
Merchandise inventory	23,000	Accounts payable	8,000
Supplies	10,000	Total Liabilities	$33,000
Total Assets	$93,000	**Owner's Equity**	
		Capital	60,000
		Total Liabilities and Owner's Equity	$93,000

What are the current ratio and the quick ratio?

SOLUTION

A. Find the **current ratio.**

Total Assets ÷ Total Liabilities
$93,000 ÷ $33,000 = 2.81 *or* 2.8 to 1 *or* 2.8:1 current ratio

This means that the Agees' total assets are 2.8 times their total liabilities. This is a good fiscal position. A current ratio of at least 2 to 1 is considered good.

B. Find the **quick ratio.**

(Total Assets − Inventory) ÷ Total Liabilities
($93,000 − $23,000) ÷ $33,000 = 2.12 *or* 2.1 to 1 *or* 2.1:1 quick ratio

This means that the Agees' quick assets (liquid or quickly converted to cash) are 2.1 times their total liabilities. This is a very good fiscal position. A quick ratio of at least 1 to 1 is considered good.

✓ SELF-CHECK Complete the problems, then check your answers in the back of the book.

3. Total assets: $50,000.
Total liabilities: $30,000.
Find the current ratio. **1.7 to 1**

4. Total assets: $56,500.
Inventory: $16,500.
Total liabilities: $30,000.
Find the quick ratio. **1.3 to 1**

PROBLEMS

Round answers to the nearest tenth of a percent.

	Item	Amount for Item	÷	Net Sales	=	Percent of Net Sales
5.	Cost of goods sold	$20,000	÷	$40,000	=	**50.0**%
6.	Operating expenses	$12,500	÷	$40,000	=	**31.3**%
7.	Net income	$17,800	÷	$40,000	=	**44.5**%

Warm-Up Exercises

1. $94,000 − $13,800 $80,200
2. $516,920 − $21,471 $495,449
3. $13,800 is what % of $80,200? 17.2%
4. $21,471 is what percent of $495,449? 4.3%
5. $74,700 ÷ $36,000 = _?_:1 2.1
6. $37,500 ÷ $36,000 = _?_:1 1.0

PRACTICE AND APPLY

The following problems can be assigned for classwork and the answers checked in class to help students master the objective of the lesson.

- Guided Practice: 1–6, 8, 9
- Independent Practice: 7, 10, 11

ALTERNATIVE STRATEGIES: Reteaching

Using the balance sheet for Metal Abrasives, Inc., dated June 30, from Problem 12 on page 516 do these problems.

1) Cash is what percent of total assets?
($1.1 / $47.5 = 0.02315 = 2.3%)

2) What is the current ratio?
($47.5 / $34.8 = 1.3649, therefore 1.4:1)

3) Comment on the fiscal position based on the current ratio. (Not good; $1.40 in assets for every $1.00 in liabilities)

WRAP-UP

Have students give the formulas for the percent of net sales, the current ratio, and the quick ratio. Have them explain how each can be used to analyze a business's financial standing.

Assignment Guide

- Basic: 7, 10–13, 15–20
- Average: 12–14, 15–20

Find the ratios to the nearest tenth.

	Total Assets	Inventory	Total Liabilities	Current Ratio	Quick Ratio
8.	$ 78,500	$ 37,250	$ 40,000	**2.0:1**	**1.0:1**
9.	$325,800	$134,600	$163,200	**2.0:1**	**1.2:1**
10.	$ 4,897	$ 995	$ 2,750	**1.8:1**	**1.4:1**

11. Speedie-Quick's income statement for one month showed these figures.

a.

Net sales	$7690
Cost of goods sold	4137
Gross profit on sales	**$3553**

b. What is the gross profit as a percent of net sales? **46.2%**

12. The income statement for one quarter for Henri's Clothes Company showed these figures.

a.

Net sales	$27,198
Cost of goods sold	13,100
Gross profit on sales	**$14,098**

b.

Total operating expenses	$12,195
Net Income	**$1,903**

c. What is the cost of goods sold as a percent of net sales? **48.2%**
d. What is the gross profit on sales as a percent of net sales? **51.8%**
e. What are the operating expenses as a percent of net sales? **44.8%**
f. What is the net income as a percent of net sales? **7.0%**

13. Use the balance sheet for Howard's Jewelers, dated June 30, from problem 9 on page 515.
a. Find the current ratio. **1.6:1**
b. Find the quick ratio. **1.3:1**

Critical Thinking . . .

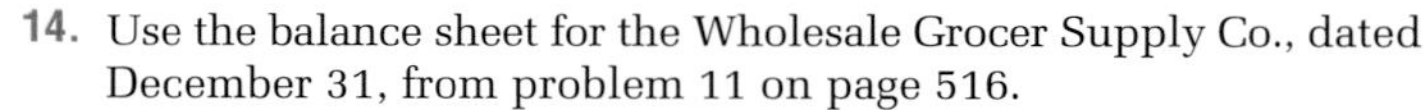

14. Use the balance sheet for the Wholesale Grocer Supply Co., dated December 31, from problem 11 on page 516.
a. Find the current ratio. **1.4:1**
b. Comment on the fiscal position based on the current ratio. **Not good**
c. Find the quick ratio. **1.1:1**
d. Comment on the fiscal position based on the quick ratio. **Good**

MAINTAINING YOUR SKILLS Look up the skills in parentheses if you need help or more practice.

Find the rate. Round answers to the nearest tenth of a percent. **(Skill 31)**

15. n% of 495 = 45 **9.1%**
16. n% of 120 = 30 **25%**
17. n% of $74 = $58 **78.4%**

Write as ratios with denominator of 1. Round answers to the nearest tenth. **(Skill 22)**

18. $174:$94 **1.9:1**
19. $74.90:$76.40 **1.0:1**
20. $74,910:$36,120 **2.1:1**

MATHEMATICAL NOTES

Remind students that in analyzing an income statement, the value for net sales is the base (denominator) for computing the percent and that the ratios are rounded to the nearest tenth compared to one.

BUSINESS NOTES

Commercial loan approval and credit ratings are frequently based partially on how close the current ratio is to 2:1 and the quick ratio is to 1:1. Loans would probably be denied to a business not in a healthy situation.

20-6 Horizontal Analysis

OBJECTIVE

Compare two income statements using horizontal analysis and compute the percentage of change.

Horizontal analysis is the comparison of two or more income statements for different periods. The comparison is done by computing **percent changes** from one income statement to another. When computing percent change, the dollar amount on the earlier statement is the **base figure.** The **amount of change** is the difference between the base figure and the corresponding figure on the current statement. If the amount for an item decreases from one income statement to the next, both the amount of change and the percent change are negative.

$$\text{Percent Change} = \frac{\text{Amount of Change}}{\text{Base Figure}}$$

EXAMPLE *Skill* 31 *Application* A *Term* Percent change

The Agees prepare a second income statement and compare it to the one for the previous quarter. To the nearest tenth of a percent, what is the percent change for each item?

SOLUTION

Current Figure − Base Figure → Amount of Change

	Last Quarter (Base)	This Quarter	Amount of Change	Percent Change
Net sales	$12,001	$13,174	$1,173	9.8%
Cost of goods sold	3817	4190	373	9.8%
Gross profit on sales	$ 8184	$ 8,984	$ 800	9.8%
Operating expenses	8047	7995	− 52	− 0.6%
Net income	$ 137	$ 989	$ 852	621.9%

Operating expenses decreased. Change is negative.

The Agee's net income increased by 621.9%

✔ SELF-CHECK Complete the problems, then check your answers in the back of the book.

Find the percent change from last quarter to this quarter.

1. Last quarter: $12,000. This quarter: $15,000. **25%**
2. Last quarter: $70,000. This quarter: $50,000. **−28.6%**

PROBLEMS

Round to the nearest tenth of a percent.

	Last Year (Base)	This Year	Amount of Change	Percent Change
3.	$720,000	$830,000	$110,000	**15.3%**
4.	$ 45,000	$ 36,000	− $ 9,000	**−20.0%**
5.	$114,750	$137,840	**$23,090** ?	**20.1%**

LESSON PLAN 20-6 Horizontal Analysis

FOCUS

Businesses can compare income statements from one accounting period to the next to see if their business is improving or getting worse. In this lesson, students learn to compare two income statements using horizontal analysis and compute the percent of change.

TEACH

The change in income statement amounts are of interest to managers, stockholders, and stock analysts. A horizontal analysis is one that looks at change over a period of time. The dollar amounts of like items are compared, and expressed as a percent of increase or decrease.

Make sure students use the earlier report as the base figure (denominator). The amount of change (numerator) will be positive for an increase and negative for a decrease. The corresponding percent will be positive or negative.

Have students list items for which a decrease from one report to the next is desirable. (operating expenses) Then ask students when an increase would be desirable. (gross profit on sales)

Warm-Up Exercises

1. $74,250 − $65,900 $8350
2. $657,918 − $686,578 −$28,660
3. $8350 is what percent of $65,900? 12.7%
4. −$28,660 is what percent of $686,578? −4.2%

ALTERNATIVE STRATEGIES: Reteaching

Review finding percent change by completing this table.

	Before	After	Amount of Change	Percent Change
1)	$37,500	$39,480	($1980)	(5.28%)
2)	275,684	312,739	(37,055)	(13.44%)
3)	$14.5 mil	$15.3 mil	($0.8 mil)	(5.52%)

PRACTICE AND APPLY

The following problems can be assigned for classwork and the answers checked in class to help students master the objective of the lesson.

- Guided Practice: 1–5
- Independent Practice: 6–8

WRAP-UP

Have students explain the purpose of doing a horizontal analysis. Read Problem 8 to the class. Ask students whether a business would rather see a positive or a negative percent change for the net sales.

Assignment Guide

- Basic: 6–11, 13–17
- Average: 9–12, 13–17 odd

6. Last year net sales were $150,000. This year net sales are $210,000. What is the amount of change? What is the percent change?
$60,000; 40.0%

7. Last week cost of goods sold was $4650. This week cost of goods sold is $3875. What is the amount of change? What is the percent change?
−$775; −16.7%

8. Last year net sales were $400,000. This year net sales are $600,000. What is the amount of change? What is the percent change?
$200,000; 50.0%

9. Last month net income was $42,476. This month net income is $51,419. What is the amount of change? What is the percent change?
$8943; 21.1%

10. Income statements for Hurry-N-Go Service Station showed these figures for March and April.

	March	April	Amount of Change	Percent Change
a. Net sales	$8876	$9172	? **$296**	? **3.3%**
b. Cost of goods sold	5617	5904	? **$287**	? **5.1%**
c. Gross profit on sales	3259	3268	? **$9**	? **0.3%**
d. Operating expenses	718	700	? **−$18**	? **−2.5%**
e. Net income	2541	2568	? **$27**	? **1.1%**

11. Income statements for Wholesale Grocer Supply Co. showed the following figures.

	Last Year (in thousands)	This Year (in thousands)	Amount of Change (in thousands)	Percent Change
a. Net sales	$117.4	$109.9	**−$7.5** ?	**−6.4%** ?
b. Cost of goods sold	56.9	54.7	**−$2.2** ?	**−3.9%** ?
c. Gross profit on sales	60.5	55.2	**−$5.3** ?	**−8.8%** ?
d. Operating expenses	35.7	36.2	**$0.5** ?	**1.4%** ?
e. Net income	24.8	19.0	**−$5.8** ?	**−23.4%** ?

12. Two annual income statements for Hi-Tech Aluminum Co. showed these figures.

	Last Year (in millions)	This Year (in millions)	Amount of Change (in millions)	Percent Change
a. Net sales	$121.4	$123.5	**$2.1**?	**1.7%**?
b. Cost of goods sold	63.1	64.6	**$1.5**?	**2.4%**?
c. Gross profit on sales	58.3	58.9	**$0.6**?	**1.0%**?
d. Operating expenses	38.7	36.8	**−$1.9**?	**−4.9%**?
e. Net income	19.6	22.1	**$2.5**?	**12.8%**?

MAINTAINING YOUR SKILLS Look up the skill in parentheses if you need help or more practice.

Find the rate. Round answers to the nearest tenth of a percent. **(Skill 31)**

13. $n\%$ of 95 = 19
20.0%

14. $n\%$ of 74 = 37
50.0%

15. $n\%$ of 1450 = 974
67.2%

16. $n\%$ of $14,176 = $4193
29.6%

17. $n\%$ of 748 = 1050
140.4%

ALTERNATIVE ASSESSMENT

Have students research business or financial magazines. Students should then compile a list of all the possible reasons they can find for why changes occur from one accounting period to another for each of these items:

1. net sales **2.** cost of goods sold
3. operating expenses **4.** net income

(Some areas to watch for are inflation, seasonal sales, foreign influence in the market, higher utility bills and wages.)

Reviewing the Basics

Skills

Solve.

(Skill 3)

1. $174,914 + $637,540 + $461,532 + $71,916 **$1,345,902**

2. $14,179,831 + $24,816,121 + $7,816,550 + $31,147,641 **$77,960,143**

(Skill 4)

3. $17,416 − $8154 **$9262**
4. $21,520 − $6473 **$15,047**
5. $54,318 − $18,433 **$35,885**
6. $41,117 − $24,823 **$16,294**

(Skill 5)

7. $4718.75 + $176.94 + $71.93 + $3491.89 **$8459.51**

8. $4214.91 + $15,714.70 + $871.85 + $15,716.95 **$36,518.41**

(Skill 6)

9. $8640.71 − $916.75 **$7723.96**
10. $41,917.62 − $2147.66 **$39,769.96**
11. $74,917.52 − $2147.73 **$72,769.79**
12. $516,531.78 − $391,478.91 **$125,052.87**

Round answers to the nearest tenth of a percent.

(Skill 31)

13. $17,800 is what percent of $91,400? **19.5%**
14. $7142 is what percent of $27,172? **26.3%**
15. n% of 91 = 27 **29.7%**
16. n% of 35 = 147 **420%**

Applications

Use the following formula to solve problems 17–20.

Gross Profit = Net Sales − Cost of Goods Sold

(Application A)

17. Net sales are $41,741. Cost of goods sold is $22,416. What is the gross profit? **$19,325**
18. Net sales are $28,419. Cost of goods sold is $17,816. What is the gross profit? **$10,603**
19. Net sales are $4,419,500. Cost of goods sold is $3,516,000. What is the gross profit? **$903,500**
20. Net sales are $8175. Cost of goods sold is $4395. What is the gross profit? **$3780**

Terms

Match each term with its definition on the right.

c 21. Owner's equity
e 22. Balance sheet
b 23. Income statement
d 24. Percent change

a. the total amount of money that you owe to creditors
b. shows your sales, operating expenses, and net profit or loss
c. the difference between your assets and liabilities
d. is negative if the amount decreases from one income statement to the next
e. shows your assets, liabilities, and owner's equity

Refer to your reference files in the back of the book if you need help.

LESSON PLAN
Reviewing the Basics

The exercises on this page review skills, applications, and terms used in the unit. You can use the exercises to assess informally students' proficiency with this material.

The page can be used for guided practice and independent practice. You can work through a selection of the exercises together with students, and thus see immediately if they know how to do them, and you can then assign some of the exercises for independent practice. Be sure to go over the answers to all assigned exercises.

LESSON PLAN
Unit Test

Students should do the Unit Test on their own. Each problem on the test is keyed to a lesson in the unit. Students having difficulty with any particular problem should review the Example in the appropriate lesson and be assigned some of the Independent Practice problems for additional practice.

Unit Test

Lesson 20-1

1. The Shoe Box had these assets and liabilities on June 30.

Cash: $4750	Unpaid merchandise: $6800
Inventory: $12,500	Taxes owed: $517
Supplies: $735	Bank loan: $5000
Store furnishings: $7215	

What are the total assets? What are the total liabilities? What is the owner's equity? **$25,200; $12,317; $12,883**

Lesson 20-2

2. Prepare a balance sheet for The Shoe Box. Use the information in problem 1.

Lesson 20-3

3. For one period, The Shoe Box had a beginning inventory of $11,450, receipts of $4360, and an ending inventory of $10,716. Find the cost of goods sold. **$5094**

Lesson 20-4

4. Global Manufacturing, Inc., prepares an annual income statement. This past year, Global had net sales of $789,640. The cost of goods sold totaled $341,620. Operating expenses included: wages and salaries totaling $107,119, depreciation and amortization totaling $5447, interest paid totaling $9179, product recall totaling $4172, and miscellaneous operating expenses of $3178. Prepare an income statement for Global Manufacturing. **Gross profit $448,020; Oper. exp. $129,095; Net income $318,925**

Lesson 20-5

5. Use the income statement for Global Manufacturing, Inc., in problem 4. To the nearest tenth of a percent, find the cost of goods sold as a percent of net sales. **43.3%**

Lesson 20-5

6. Use the balance sheet for The Shoe Box in problem 2.

a. Find the current ratio. **2.0:1**
b. Find the quick ratio. **1.0:1**

Lesson 20-6

7. Income statements for B.J. Benson's showed the following figures for two quarters. Round answers to the nearest tenth of a percent.

	First Quarter	Second Quarter	Amount of Change	Percent Change
a. Net sales	$8500	$9100	**$600**	**7.1%**
b. Cost of goods sold	4300	4500	**$200**	**4.7%**
c. Gross profit on sales	4200	4600	**$400**	**9.5%**
d. Operating expenses	2900	2800	**−$100**	**−3.4%**
e. Net income	1300	1800	**$500**	**38.5%**

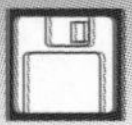

A SPREADSHEET APPLICATION

Accounting Records

To complete this spreadsheet application, you will need the template diskette for *Mathematics with Business Applications*. Follow the directions in the *User's Guide* to complete this activity.

Input the information in the problem to create a balance sheet and an income statement for The Home Furniture Mart. After you have finished entering the data, answer the questions that follow.

The Home Furniture Mart
Balance Sheet
March 31,19–

Assets (thousands)		Liabilities (thousands)	
Cash	$125.6	Accounts payable	$35.4
Accounts receivable	214.9	Notes payable	28.5
Inventory	56.3	Other liabilities	10.7
Supplies	8.6	Total Liabilities	?
Other Assets	24.6	Owner's Equity (thousands)	
Total Assets	?	Capital	?
		Total Liabilities and Owner's Equity	?

1. What are the total assets?
2. What are the total liabilities?
3. What is the owner's equity (capital)?
4. What are the total liabilities and owner's equity?
5. What is the current ratio?
6. What is the quick ratio?

LESSON PLAN
A Spreadsheet Application

USING TECHNOLOGY

After students have worked through the balance sheet and income statement for the Home Furniture Mart, and have answered the questions, you might want to discuss the advantages of using a computer for this purpose. Ask students if a person could have done the same job just as well as the computer. Lead students to understand that computers are really an indispensable tool for businesses today because of their ability to process vast amounts of information quickly.

A SPREADSHEET APPLICATION (CONTINUED)

The Home Furniture Mart
Income Statement
For the Quarter Ended March 31,19–

Income:		
Sales	$34,576	
Less: Returns	678	
Net sales		$?
Cost of goods sold		16,589
Gross profit on sales		$?
Operating Expenses:		
Salaries and wages	$ 9862	
Utilities	1423	
Supplies	856	
Total operating expenses		$?
Net Income		$?

7. What are the net sales?

8. What is the gross profit on sales?

9. What are the total operating expenses?

10. What is the net income?

11. What is the cost of goods sold as a percent of net sales?

12. What are the total operating expenses as a percent of net sales?

13. What is the net income as a percent of net sales?

14. Salaries and wages are what percent of the total operating expenses?

UNIT

21

Financial Management

Financial management of your business includes your investments, loans, and taxes. You may borrow money by taking a *commercial loan,* or business loan, from a bank. If you want to raise money, you can sell shares of your business, or stocks and bonds. Money that you don't need for day-to-day expenses can be invested to earn additional interest. Your business may invest its surplus cash in *United States Treasury Bills* or in other businesses with a high credit rating.

OUTLINE

21-1 Corporate Income Taxes
21-2 Issuing Stocks and Bonds
21-3 Borrowing
21-4 Investments
21-5 Investments—Net Interest
21-6 Growth Expenses

- Reviewing the Basics
- Unit Test
- A Spreadsheet Application: Financial Management
- Career Wise: Stockbroker
- A Simulation: Stock Market

Today, corporations own more than 90 percent of all stocks and bonds traded on a stock exchange.

UNIT 21
FINANCIAL MANAGEMENT

INTRODUCING THE UNIT

Discuss with students the fact that running a business successfully requires great care and organization in handling and recording all financial matters. In this unit, students will learn about ways to finance a business and how a business's finances are reported to the government.

LESSON PLAN 21-1 Corporate Income Taxes

FOCUS

To motivate this lesson, recall with students the purpose of federal income taxes. (to raise revenue to pay the costs of government and to provide public services) Explain that a large portion of money raised by income taxes is from corporations. Like individuals, corporations also have deductions, which are subtracted from the corporation's gross income to determine taxable income.

TEACH

Allow time for students to familiarize themselves with the tax table on this page before working the Example.

Point out that in a business that sells merchandise, gross income is usually the same as gross profit. Thus, the cost of goods sold is subtracted from net sales **(Net sales − cost of goods sold = Gross profit)** and is not reported as a business expense deduction.

Warm-Up Exercises

1. $74,916 + $21,493 + $8565 + $1314
   $106,288
2. $216,924 − ($56,173 + $31,250 + $12,196)
   $117,305
3. $42,473 × 15%
   $6370.95
4. $7500 + 25% of ($67,815 − $50,000)
   $11,953.75

21-1 Corporate Income Taxes

OBJECTIVE

Compute the taxable income and the corporate income tax.

Your business must pay federal income taxes. The tax rates vary depending on the size and type of your business. **Corporations,** businesses owned by stockholders, are subject to federal tax rates ranging from 15% to 39% of **taxable income.** Taxable income is the portion of your company's gross income that remains after normal business expenses are deducted. Normal business expenses include wages, rent, utilities, interest paid on loans, property taxes, depreciation, and so on. The structure for federal **corporate income taxes** is graduated.

If taxable income is:		Tax is:	
Over—	But not over—		Of the amount over—
0	50,000	15%	0
$ 50,000	$ 75,000	$ 7500 + 25%	$ 50,000
75,000	100,000	13,250 + 34%	75,000
100,000	—	21,750 + 39%	100,000

Taxable Income = Annual Gross Income − Deductions

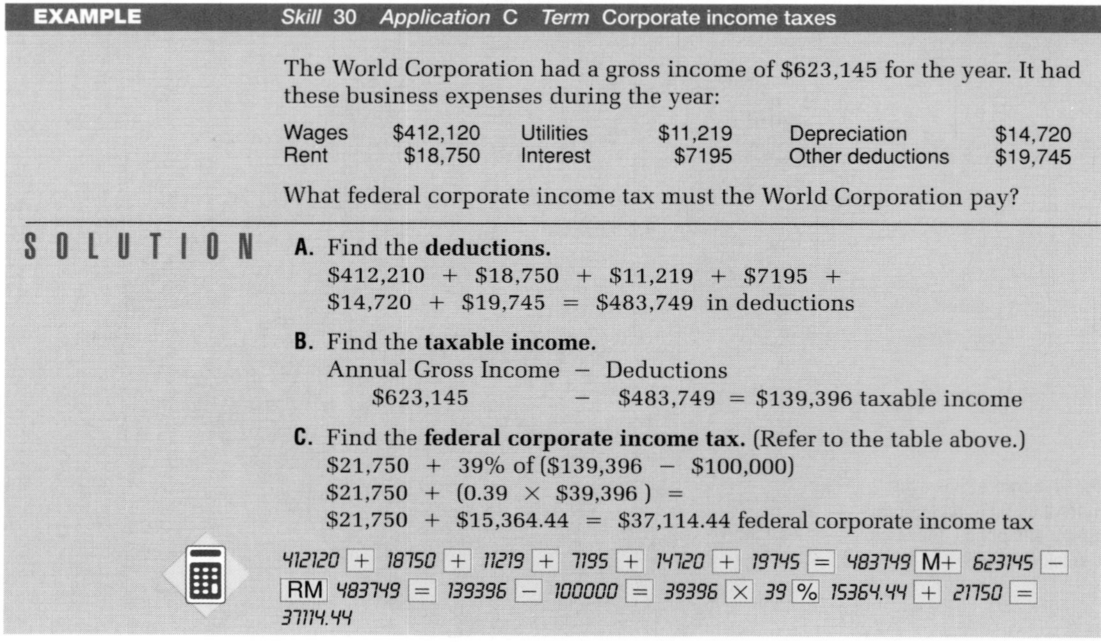

EXAMPLE *Skill* 30 *Application* C *Term* Corporate income taxes

The World Corporation had a gross income of $623,145 for the year. It had these business expenses during the year:

Wages	$412,120	Utilities	$11,219	Depreciation	$14,720
Rent	$18,750	Interest	$7195	Other deductions	$19,745

What federal corporate income tax must the World Corporation pay?

SOLUTION

A. Find the **deductions.**
$412,210 + $18,750 + $11,219 + $7195 + $14,720 + $19,745 = $483,749 in deductions

B. Find the **taxable income.**
Annual Gross Income − Deductions
$623,145 − $483,749 = $139,396 taxable income

C. Find the **federal corporate income tax.** (Refer to the table above.)
$21,750 + 39% of ($139,396 − $100,000)
$21,750 + (0.39 × $39,396) =
$21,750 + $15,364.44 = $37,114.44 federal corporate income tax

412120 [+] 18750 [+] 11219 [+] 7195 [+] 14720 [+] 19745 [=] 483749 [M+] 623145 [−] [RM] 483749 [=] 139396 [−] 100000 [=] 39396 [×] 39 [%] 15364.44 [+] 21750 [=] 37114.44

✓ SELF-CHECK Complete the problems, then check your answers in the back of the book.

Use the tax table above to find the federal corporate income tax.

1. Gross income: $145,000. Deductions: $60,000 **$16,650**
2. Gross income: $219,000. Deductions: $154,000. **$11,250**

CULTURAL ANGLES

Many large U.S. corporations operate businesses in other parts of the world. These businesses are also taxed in the country in which they operate. However, a tax credit, or deduction, is allowed for taxes paid to a foreign government. The tax structure of a country can affect a corporation's decision as to whether to operate there or not. If the people of a country are poor, the government relies heavily on business taxes. Have students investigate and report on the business tax structure of other countries.

PROBLEMS

Refer to the table on page 532 for federal corporate income taxes.

	Annual Gross Income	–	Deductions	=	Taxable Income	Total Tax
3.	$ 74,918	–	$ 38,172	=	$36,746	**$5511.90**
4.	$212,971	–	$125,539	=	**$87,432**	**$17,476.88**
5.	$279,434	–	$214,611	=	**$64,823**	**$11,205.75**
6.	$720,338	–	$621,913	=	**$98,425**	**$21,214.50**
7.	$916,418	–	$721,534	=	**$194,884**	**$58,754.76**

8. Johnson-Etna Mfg. Corp.
Annual gross income is $316,921.
Deductions total $234,847.
What is the federal corporate income tax? **$15,655.16**

9. The Towncraft Company.
Annual gross income is $721,173.
Deducticns total $670,224.
What is the federal corporate income tax? **$7737.25**

F.Y.I.
In 1989, the Internal Revenue Service collected $117 billion in corporate income taxes.

10. The Magno-Met Company had these business expenses for the year:

Wages	$816,147.80	Property taxes	$97,493.70
Rent	$60,000.00	Depreciation	$26,916.00
Utilities	$74,617.46	Other deductions	$34,124.76
Interest	$5119.54		

The Magno-Met Company had a gross income of $1,516,749 for the year.
a. What are the total business expenses? **$1,114,419.26**
b. What is the taxable income? **$402,329.74**
c. What is the federal corporate income tax for the year? **$139,658.60**

11. Gayle Sturgeon, D.V.M., formed a corporation. The corporation had these business expenses for the year:

Wages	$24,000.00	Property taxes	$7413.00
Utilities	$15,774.35	Depreciation	$3849.74
Interest	$1491.84	Other deductions	$4916.47

Dr. Sturgeon's gross income for the year was $134,176. What is Dr. Sturgeon's federal corporate income tax for the year? **$13,838.40**

MAINTAINING YOUR SKILLS Look up the skills in parentheses if you need help or more practice.

Find the percentage. Round answers to the nearest hundredth. **(Skill 30)**

12. 6% of 74 **4.44**
13. 7.65% of $80.77 **$6.18**
14. $\frac{7}{8}$% of 160 **1.4**
15. 145% of 350 **507.5**
16. 35% of 4700 **1645**
17. 34% of 816 **277.44**
18. $16\frac{1}{2}$% of 470 **77.55**
19. 2.64% of 1700 **44.88**

Subtract. **(Skill 6)**

20. $87,434 – $75,000 **$12,434**
21. $127,816 – $100,000 **$27,816**
22. 74.718 – 16.634 **58.084**

PRACTICE AND APPLY

The following problems can be assigned for classwork and the answers checked in class to help students master the objective of the lesson.

- Guided Practice: 1–5
- Independent Practice: 6–9

WRAP-UP

Have students name the various business expenses that can be deducted from gross income. Then ask a volunteer to explain how to find the tax for a corporation that has an annual taxable income of $76,500.

Assignment Guide

- Basic: 6–10, 12–22
- Average: 10, 11, 12–22 even

ALTERNATIVE STRATEGIES: Reteaching

Use the table on page 532 to complete these problems with your students:

	Gross Income	Deductions	Tax. Income	Income Tax
1)	$175,000	$ 87,600	($87,400)	($17,466.00)
2)	89,625	43,790	($45,835)	($ 6875.25)
3)	482,768	217,565	($265,203)	($86,179.17)
4)	263,900	198,735	($65,165)	($11,291.25)

LESSON PLAN
21-2 Issuing Stocks and Bonds

FOCUS
Lead students in a discussion about the cost of starting a business. What does a retail store need money for? (a store, merchandise to sell) A service company? (an office, equipment necessary to perform the service) Where does the money come from? Often the money comes from the savings of the person(s) starting the business. Explain that large businesses can get their money, or capital, by offering stock to the public. In this lesson, students will compute the selling expenses and the net proceeds from an issue of stocks and bonds.

TEACH
Explain to students that a business also may need long-term financing for such purposes as building a new plant or purchasing equipment, and that stocks and bonds are issued for these purposes.

Emphasize the difference between stocks and bonds. You may want to review with students the lessons on investing in stocks and bonds in Unit 10.

If available, bring some stock-offering announcements from newspapers to class for students to pass around. *The Wall Street Journal* is an excellent source for this purpose.

Warm-up Exercises
1. $31.50 × 400,000 $12,600,000
2. $24.75 × 1,500,000 $37,125,000
3. 6.5% × $12,600,000 $819.000
4. 4% × $37,125,000 $1,485,000

21-2 Issuing Stocks and Bonds

OBJECTIVE
Compute the selling expenses and the net proceeds from an issue of stocks and bonds.

Your business may raise money by issuing **stocks** or **bonds.** When you issue stocks, the buyer becomes a part owner of your business. When you issue bonds, the buyer is lending money to your business.

When you issue stocks or bonds, you must pay certain expenses. One expense is an **underwriting commission,** a commission to the investment banker who helps you distribute the stocks or bonds. Other expenses include accounting costs, legal fees, and printing costs. The amount that your business actually receives from the sale of the stocks or bonds after paying these expenses is the **net proceeds.**

Net Proceeds = Value of Issue − Total Selling Expenses

EXAMPLE *Skill* 30 *Application* A *Term* Net proceeds

The Landover Company is planning a major expansion program. To finance the program, Landover plans to sell an issue of 300,000 shares of stocks at $41.50 per share. The underwriting commission will be 6.5% of the value of the stocks. Accounting fees, legal fees, printing costs, and other expenses are estimated to be 0.9% of the value of the stocks. If all the shares of stock are sold, what net proceeds will Landover Company receive?

SOLUTION

A. Find the **value of issue.**
$41.50 × 300,000 = $12,450,000

B. Find the **total selling expenses.**
Underwriting commission:
$12,450,000 × 6.5% $809,250
Other expenses:
$12,450,000 × 0.9% $112,050
Total Selling Expenses $921,300

C. Find the **net proceeds.**
Value of Issue − Total Selling Expenses
$12,450,000 − $921,300 = $11,528,700 net proceeds

41.5 × 300000 = 12450000 M+ × 6.5 % 809250 RM 12450000 × .9 % 112050 + 809250 M− 921300 RM 11528700

✔ SELF-CHECK Complete the problems, then check your answers in the back of the book.

Find the net proceeds.

1. Value of stock: $8,000,000. Total selling expenses: 5.4% of value of stocks. **$7,568,000**
2. Value of stock: $37,450,000. Underwriting commission: $372,500. Total selling expenses: 0.3% of value of stock. **$36,965,150**

CRITICAL THINKING
There are pros and cons in issuing either stocks or bonds for long-term financing. Interest paid by a company on bonds is considered a debt and is deducted from gross profit, whereas stock dividends are issued after taxes have been paid. Ask students to suggest other reasons why a manager may prefer to raise money through bonds rather than through stocks. (Stockholders have some control over a company's management; a corporation shares its after-tax earnings with stockholders.)

PROBLEMS

	Underwriting Commission (Value of Issue × Percent = Dollar Amount)	+ Other Expenses	= Total Selling Expenses	Net Proceeds
3.	($6,800,000 × 6% = $408,000)	+ $47,500	$455,500	$6,344,500
4.	($950,000 × 9% = $85,500)	+ $26,500	$112,000	$838,000
5.	($21,000,000 × 5% = $1,050,000)	+ $54,650	$1,104,650	? $19,895,350

6. Value of stocks is $750,000. Underwriting commission is 10%. Other expenses total $21,640. What are the net proceeds? **$653,360**

7. Value of bonds is $7,435,750. Underwriting commission is 5.5%. Other expenses total $48,650. What are the net proceeds? **$6,978,133.75**

8. Universal Waste Disposal, Inc., sold 1,350,000 shares of stock at $24.625 per share. The investment banker's commission was 5% of the value of the stock. The other expenses were 1% of the value of the stock. What net proceeds did UWD, Inc., receive? **$31,249,125**

9. The Mercury Electric Company issued 20,000 shares of stocks at $62.50 per share. Find the net proceeds after these selling expenses are deducted. **$1,156,250**

Underwriting Expenses	
Commissions	$75,000
Legal fees	2500
Advertising	2500
Miscellaneous	1250

Other Expenses	
Printing costs	$4500
Legal fees	2500
Accounting fees	3500
Miscellaneous	2000

Use the table to solve problems 10 and 11.

10. The Hawthorne Company sold 25,000 bonds at $65 per bond. What are the net proceeds? **$1,519,375**

11. The Stock Paper Corporation sold 75,000 bonds for $35.75 per bond. What are the net proceeds? **$2,525,737.50**

Size of Issue (in millions)	Underwriting Commission	Other Expenses
Under $0.5	6.5%	1.5%
$0.5–$0.9	6%	1.3%
$1.0–$1.9	5.5%	1.0%
$2.0–$4.9	5%	0.8%

12. The Athens City Council needs to raise $2.8 million to continue required services for the remainder of the year. The council plans to sell bonds at $37.50 each to raise the money. Selling expenses are expected to be 5.8% of the value of the bonds. How many bonds must the council sell to obtain net proceeds of $2.8 million? **79,264**

MAINTAINING YOUR SKILLS Look up the skill in parentheses if you need help or more practice.

Find the percentage. Round answers to the nearest hundredth. **(Skill 30)**

13. 6.5% of $980,000 **$63,700**
14. 5% of $3,150,000 **$157,500**
15. 6% of $743,000 **$44,580**
16. 5.5% of $1.5 million **$82,500**

PRACTICE AND APPLY

The following problems can be assigned for classwork and the answers checked in class to help students master the objective of the lesson.

- Guided Practice: 1–5
- Independent Practice: 6, 7, 9

WRAP-UP

Go over the solutions to Problems 6 and 7. Ask one student to explain the relationship between the buyer and the corporation when stocks are purchased. Ask another student to explain the relationship when bonds are purchased.

Assignment Guide

- Basic: 6–11, 13–16
- Average: 8, 10–12, 13–16

ALTERNATIVE STRATEGIES: Enrichment

Problem 12 requires that students find how many bonds must be sold, which is a common situation in the business world. You may wish to have students work in cooperative groups to solve the problem. When all groups have completed the problem, have one group show its solution at the chalkboard.

LESSON PLAN
21-3 Borrowing

FOCUS

To motivate this lesson, present the following situation to the class. During October through December, a swimsuit manufacturer begins to encourage retailers to place early orders for swimsuits by offering trade discounts of 20%. In January, the company begins to manufacture the swimsuits and ships them to retailers. During April through June, most swimsuits are sold and the manufacturer receives payment from the retailers. Ask students how the company can afford to buy the materials needed for the manufacture of the swimsuits. (borrow money and pay back the loan after payments from retailers are received)

TEACH

Explain that banks offer business loans to companies that are similar to personal loans: they are short-term loans, usually for less than a year, and the maturity value of the loan is computed by adding the principal and the interest owed.

Mention that the prime rate is the interest rate banks set for their "best" corporate customers. Other firms are charged a higher interest rate depending on their financial rating; that is, the higher the risk for the bank, the higher the interest rate charged. You might ask a local bank representative to address the class about business loans and explain what businesses qualify as prime-rate customers and why.

21-3 Borrowing

OBJECTIVE

Compute the maturity value of a commerical loan.

Your business may borrow money to buy raw materials, products, or equipment. **Commercial loans,** or business loans, are similar to personal loans. The interest rates are generally one to two percentage points higher than the **prime rate.** The prime rate is the lowest rate of interest available to commercial customers at a given time.

The **maturity value** of your loan is the total amount you repay. The maturity value includes both the principal borrowed and the interest owed on the loan. Commercial loans usually charge ordinary interest at exact time. That is, the length of time of the loan is calculated by dividing the exact number of days of the loan by 360 days.

Interest = Principal × Rate × Time

Maturity Value = Principal + Interest Owed

EXAMPLE *Skills* 28, 10 *Application* A *Term* Maturity value

Harm's Drugstore borrowed $80,000 from First National Bank to pay for remodeling costs. The bank loaned the money at 10.5% ordinary interest for 60 days. What is the maturity value of the loan?

SOLUTION

A. Find the **interest owed.**

Principal × Rate × Time
$80,000 × 10.5% × $\frac{60}{360}$
($80,000 × 0.105 × 60) ÷ 360
$504,000 ÷ 360 = $1400 interest owed

B. Find the **maturity value.**

Principal + Interest Owed
$80,000 + $1400 = $81,400 maturity value

80000 M+ × 10.5 % 8400 × 60 ÷ 360 = 1400 + RM = 81400

✓ SELF-CHECK Complete the problems, then check your answers in the back of the book.

Find the maturity value of the loan.

1. $100,000 borrowed at 11.25% ordinary interest for 90 days. **$102,812.50**
2. $150,000 borrowed at 9.5% ordinary interest for 140 days. **$155,541.67**

MATHEMATICAL NOTES

For some problems in this lesson, the term (number of days divided by 360) of the loan can be written as an exact fraction. For example, 90 days ÷ 360 days = $\frac{1}{4}$. The fractions for 180 days, 120 days, 60 days, 135 days, and 270 days are $\frac{1}{2}$, $\frac{1}{3}$, $\frac{1}{6}$, $\frac{3}{8}$, and $\frac{3}{4}$, respectively. You may want to display these relationships on the chalkboard for students' reference.

PROBLEMS

Use ordinary interest at exact time to solve.

	Principal	×	Rate	×	Time	=	Interest	Maturity Value
3.	$70,000	×	10%	×	$\frac{90}{360}$	=	**$1750**	**$71,750**
4.	$95,000	×	11%	×	$\frac{100}{360}$	=	**$2902.78**	**$97,902.78**
5.	$37,500	×	10.5%	×	$\frac{120}{360}$	=	**$1312.50**	**$38,812.50**

6. $64,000 is borrowed for 60 days. Interest rate is 11.5%.
$1226.67 What is the interest owed?
$65,226.67 What is the maturity value?

7. $37,650 is borrowed for 180 days. Interest rate is 9.75%.
What is the interest owed? **$1835.44**
What is the maturity value? **$39,485.44**

8. Trust Bank, Inc., loaned $90,000 to Fernandez Home Builders, Inc. The term of the loan was 270 days. The interest rate was 11.75%. What was the maturity value of the loan? **$97,931.25**

9. To take advantage of a bicycle manufacturer's closeout special on touring bikes, Wheels, Inc., borrowed $50,000 from the Union Trust Company. Union Trust charged 10.75% interest on the loan. The term of the loan was 175 days. What was the maturity value of the loan? **$52,612.85**

10. The Gibraltar Construction Company was granted a 1-year construction loan of $650,000 to finance the construction of an apartment complex. Gibraltar borrowed the money from Citizens Trust Company at an interest rate of 10.25%. What was the maturity value of the loan? **$716,625**

11. The Solar Panel Manufacturing Company needs $1,450,000 for 270 days to help finance the production of an experimental solar hot water heater. The financial manager has arranged financing from 3 sources. Each loan charges ordinary interest at exact time.

Swancreek Trust Company
$500,000 loan for 270 days
Interest rate: 10.5%

Universal Investment Company
$450,000 loan for 270 days
Interest rate: 10.9%

Investment Bankers, Inc.
$500,000 loan for 270 days
Interest rate: 11.25% for first 90 days
11.00% for next 180 days

a. What is the total interest for the 3 loans? **$117,725**
b. What is the total maturity value? **$1,567,725**

MAINTAINING YOUR SKILLS

Look up the skill in parentheses if you need help or more practice.

Write the percents as decimals. **(Skill 28)**

12. 12.7% **0.127** **13.** 340% **3.4** **14.** 0.6% **0.006** **15.** $\frac{7}{10}$% **0.007** **16.** 0.05% **0.0005**

Warm-Up Exercises

Write as decimals.

1. 11.75% 0.1175
2. 9 1/2% 0.095
3. $42,100 × 0.1875
   $7893.75
4. $321,000 ÷ 360
   $891.67

PRACTICE AND APPLY

The following problems can be assigned for classwork and the answers checked in class to help students master the objective of the lesson.

- Guided Practice: 1–5
- Independent Practice: 6–8

WRAP-UP

At the end of December 1991, the prime rate was 6.5%. Rework the Example at the chalkboard using this rate, or the current rate. Have students explain the computations necessary for each step.

Assignment Guide

- Basic: 6–10, 12–16
- Average: 9–11, 12–16

ALTERNATIVE STRATEGIES: Reteaching

Assume ordinary interest and exact time as you complete this table.

	Principal	Rate	Time	Interest	Maturity Value
1)	$ 460,000	7%	120 days	($ 10,733.33)	($ 470,733.33)
2)	$2,850,000	8.25%	180 days	($117,562.50)	($2,967,562.50)
3)	$ 738,685	10.5%	300 days	($ 64,634.94)	($ 803,319.94)

LESSON PLAN
21-4 Investments

FOCUS

Ask students what they would do if, at the end of the month, their company had $20,000 in cash that was not needed for 6 months. (invest it) Point out that the investment should be *liquid*, or easily converted into cash in case the money is needed. One liquid investment is U.S. Treasury Bills, commonly referred to as T-Bills. In this lesson, students will compute the cost of a Treasury Bill.

TEACH

Explain to students that Treasury Bills are considered short-term investments for companies and individuals. Point out that the money can be reinvested at a T-Bill's maturity date; however, at that time, the interest rate could be higher or lower than the original rate.

After working the Example, have students determine the profit on the T-Bill at its maturity date. ($1516.67)

Warm-Up Exercises

1. ($80,000 × 9.4% × 120) ÷ 360 $2506.67
2. ($150,000 × 8.96% × 91) ÷ 360 $3397.33
3. $2506.67 + $35 $2541.67
4. $80,000 − $2541.67 $77,458.33

21-4 Investments

OBJECTIVE
Compute the cost of a Treasury Bill.

Your business may invest surplus cash that is not needed for day-to-day operations. One way to invest your money is to purchase **United States Treasury Bills.** You may purchase the bills through a bank. Some banks charge a service fee for the paperwork involved in obtaining the bills.

When you purchase a Treasury Bill, you are actually lending money to the government. In return, you receive interest at the rate that is in effect at the time you purchase the bill. The interest is ordinary interest at exact time. Treasury Bills are issued on a **discount** basis. That is, the interest is computed and then subtracted from the **face value** of the bill to determine the cost of the bill. The face value of the Treasury Bill is the amount of money you will receive on the **maturity date** of the bill. Maturity dates for Treasury Bills range from 91 days to a year.

Cost of Treasury Bill = (Face Value of Bill − Interest) + Service Fee

EXAMPLE *Skills* 8, 10 *Application* A *Term* United States Treasury Bill

The financial manager of the Osinski Manufacturing Company has decided to invest the company's surplus cash in a $50,000 United States Treasury Bill for 120 days. The interest rate is 9.25%. The bank charges a service fee of $25 to obtain the Treasury Bill. What is the cost of the Treasury Bill?

SOLUTION

A. Find the **interest.**

Principal × Rate × Time
$50,000.00 × 9.25% × $\frac{120}{360}$
($50,000.00 × 0.0925 × 120) ÷ 360
$555,000.00 ÷ 360 = $1541.666 = $1541.67

B. Find the **cost of Treasury Bill.**

(Face Value of Bill − Interest) + Service Fee
($50,000.00 − $1541.67) + $25.00
$48,458.33 + $25.00 = $48,483.33 cost of Treasury Bill

50000 M+ × 9.25 % 4625 × 120 ÷ 360 = 1541.666 M− RM 48458.333 + 25 = 48483.33

✓ SELF-CHECK Complete the problem, then check your answer in the back of the book.

1. $100,000 Treasury Bill purchased at 8.9% interest for 180 days. Service fee is $50. Find the cost of the Treasury Bill. **$95,600**

COMMUNICATION SKILLS

Provide the following information to students: A $10,000 T-Bill was purchased at 8% interest for 180 days. Write $10,000, $9600, and $400 on the chalkboard along with these terms: *face value, maturity date, interest, ordinary interest at exact time, cost,* and *discount basis.* Ask students what each figure represents using these terms. (Example: $10,000 is the face value of the T-Bill, the amount received on the maturity date. $400 is the interest earned based on ordinary interest at exact time. $9600 is the cost of the T-Bill on a discount basis.)

PROBLEMS

Use ordinary interest at exact time to solve.

	Face Value of Treasury Bill	Interest Rate	Time in Days	Interest	Bank Service Fee	Cost of Treasury Bill
2.	$ 70,000	9%	100	**$1750.00**	$25	**$68,275.00**
3.	$100,000	8.7%	120	**$2900.00**	$35	**$97,135.00**
4.	$150,000	9.35%	180	**$7012.50**	No fee	**$142,987.50**
5.	$ 85,000	10.213%	91	**$2194.38**	$30	**$82,835.62**

6. Face value of bill is $60,000.
Interest rate is 9.4%.
Bill matures in 120 days.
Bank service fee is $30.
What is the interest? **$1880**
What is the cost of the bill? **$58,150**

7. Face value of bill is $97,000.
Interest rate is 8.78%.
Bill matures in 100 days.
Bank service fee is $40.
What is the interest? **$2365.72**
What is the cost of the bill? **$94,674.28**

8. The J. P. Cotton Company purchased a $40,000 Treasury Bill at 10% ordinary interest for 120 days. The bank charged a service fee of $20. What was the interest? What was the cost of the Treasury Bill? **$1333.33** **$38,686.67**

9. The Pacer Investment Company purchased a $95,000 United States Treasury Bill at 9.24% interest. The Treasury Bill matures in 91 days. The bank service fee is $35. What is the cost of the Treasury Bill? **$92,816.12**

10. The Masters Printing Company had $115,000 in surplus cash. Sue Masters decided to invest in United States Treasury Bills at an interest rate of 10.45%. She purchased a 110-day bill with a face value of $115,000. The bank charged a $35 service fee. What was the cost of the Treasury Bill? **$111,362.99**

11. Robert Tompkins sold his apartment house for $240,000. After paying various expenses, he had $210,000 to invest. He purchased a $100,000 United States Treasury Bill for 120 days at an interest rate of 9.672%. He purchased a second Treasury Bill with a face value of $110,000. The second bill was for 91 days at an interest rate of 9.842%. Robert's bank does not charge a service fee on Treasury Bills of $100,000 or over. What was the total cost of the 2 bills? **$204,039.38**

MAINTAINING YOUR SKILLS

Look up the skills in parentheses if you need help or more practice.

Multiply. Round answers to the nearest thousandth. **(Skill 8)**

12. 17.49 × 8 **139.92** **13.** 7.475 × 8.1 **60.548** **14.** 0.57 × 0.41 **0.234** **15.** 21.43 × 0.214 **4.586**

Divide. Round answers to the nearest thousandth. **(Skill 10)**

16. 74 ÷ 6 **12.333** **17.** 7.47 ÷ 1000 **0.007** **18.** 971 ÷ 40 **24.275** **19.** 700 ÷ 32 **21.875**

PRACTICE AND APPLY

The following problems can be assigned for classwork and the answers checked in class to help students master the objective of the lesson.

- Guided Practice: 1–6
- Independent Practice: 6–8

WRAP-UP

Have individual students define the terms *discount basis, face value,* and *maturity date.* Ask students why businesses invest in T-Bills. (to have their extra cash earn interest)

Assignment Guide

- Basic: 6–10, 12–19
- Average: 9–11, 12–18 even

ALTERNATIVE STRATEGIES: Enrichment

There are treasury notes and treasury bills that earn interest at varying rates and are for varying durations of time. These rates and terms are normally published in the daily newspaper. Adjustable rate mortgage rates, as well as many other rates, are tied to treasury bill rates for various terms. To purchase a treasury note or bill you may have to find a federal reserve bank; where is the nearest such bank in your area and how does one go about purchasing a treasury note or bill?

LESSON PLAN
21-5 Investments—Net Interest

FOCUS
Within the context of this unit, discuss with students why financial managers need to have a strong background in mathematics, accounting, and business. (They have to obtain and use money to run an organization, which means analyzing when and where to borrow or invest money.) Point out that a financial manager can also make a short-term investment in commercial paper, which earns interest for the business.

TEACH
Emphasize the fact that commercial paper is a means by which companies can borrow money from other companies for a short period of time. Commercial paper is a promise by the borrowing company to pay back a set amount of money to the investing company at a specified date.

Point out that some features of commercial paper are similar to Treasury Bills; they can be purchased through banks and other financial institutions; the interest is ordinary interest at exact time, and so on. Mention that the interest rate on commercial paper usually is lower than the rate for a loan at commercial banks, which is why businesses issue commercial paper.

Warm-Up Exercises
1. ($70,000 × 10.4% × 120) ÷ 360 $2426.67
2. ($140,000 × 9.87% × 60) ÷ 360 $2303
3. $475 + $316.50 + $275 $1066.50
4. $4185.90 − $45 $4140.90

21-5 Investments—Net Interest

OBJECTIVE
Compute the net interest from commercial paper.

Your business may invest extra cash in **commercial paper.** Commercial paper is a 30-day to 270-day loan issued by a company with a very high credit rating. When your business invests in commercial paper, you are actually lending money to another company. Commercial paper earns ordinary interest at exact time. Usually, your business obtains commercial paper through a bank. The bank may charge a service fee. Your **net interest** is the total interest that you earn after any bank service fees have been deducted.

Net Interest = Total Interest Earned − Service Fee

EXAMPLE *Skills* 8, 10 *Application* A *Term* Commercial paper

Boat Works has a $75,000 cash surplus. The financial manager used the cash to make the following investments in commercial paper:

Hastings Corporation	$20,000 at 10.9% for 30 days
A.B.C. Financial Corporation	$40,000 at 10.5% for 60 days
Lake Erie Fish Company	$15,000 at 9.9% for 120 days

The bank charges a service fee of $35. What net interest will Boat Works earn?

SOLUTION

A. Find the **total interest earned.**

(1) Hastings Corporation
$20,000.00 × 10.9% × $\frac{30}{360}$ = $181.666 = $ 181.67

(2) A.B.C. Financial Corporation
$40,000.00 × 10.5% × $\frac{60}{360}$ = $ 700.00

(3) Lake Erie Fish Company
$15,000.00 × 9.9% × $\frac{120}{360}$ = $ 495.00

Total Interest Earned $1376.67

B. Find the **net interest.**
Total Interest Earned − Service Fee
$1376.67 − $35.00 = $1341.67 net interest

20000 [×] 10.9 [%] 2180 [×] 30 [÷] 360 [=] 181.666 [M+] 40000 [×] 10.5 [%] 4200 [×] 60 [÷] 360 [=] 700 [M+] 15000 [×] 9.9 [%] 1485 [×] 120 [÷] 360 [=] 495 [M+] [RM] 1376.666 [−] 35 [=] 1341.666

✓ SELF-CHECK Complete the problem, then check your answer in the back of the book.

1. $100,000 invested at 10.4% interest for 30 days; no service fee. Find the net interest. **$866.67**

ALTERNATIVE STRATEGIES: Reteaching
It seems appropriate that for one last time we find the number of days between two dates. We had to do this earlier for simple interest, finance charge using average daily balance, cash discounts, and we need to do it now for Problem 9. Using the table on page 645 complete this table:

PROBLEMS

Use ordinary interest at exact time to solve.

	Borrower	Value of Note	Interest Rate	Time in Days	Interest	Bank Service Fee	Net Interest
2.	Warehouse Corp.	$ 50,000	9.7%	60	$808.33	$20	$788.33
3.	Tel-Com Inc.	$120,000	10.3%	120	$4120	$25	$4095
4.	Furniture Factory	$250,000	10.7%	270	$20,062.50	No fee	? $20,062.50

5. $75,000 note for 90 days. Purchased from The Kord Corporation. Interest rate is 9.9%. Bank service fee is $20. What is the net interest? **$1836.25**

6. $50,000 note for 75 days. Purchased from Amos Packing Co. Interest rate is 10.2%. Bank service fee is $25. What is the net interest? **$1037.50**

7. The Carmine Coal Company has a $45,000 cash surplus. The financial manager made the following investments in commercial paper:

Rural Farm Electric	$25,000 at 10.5% for 270 days
Farmers' Grain Supply	$20,000 at 10.65% for 180 days

The bank charges a service fee of $15 for each note. What net interest will the Carmine Coal Company receive? **$3003.75**

8. Judy Myers, the financial manager of Myers Shoe Mfg. Co., made the following investments in commercial paper:

Piedmont Paper Co.	$75,000 at 10.75% for 270 days
Ace Development Co.	$80,000 at 10.5% for 180 days
Investment Properties Inc.	$95,000 at 11.5% for 30 days

The bank charges a service fee of $8 per note. What net interest will Myers Shoe Mfg. Co. receive? **$11,133.30**

9. **A Brief Case** On June 1, Adelphia Investments, Inc., received $272,500 for the sale of an industrial park. Harry Pappas, the financial manager for the company, invested the money in the following commercial paper:

Donovan Corporation	$75,000 at 10.4% until June 16
Commercial Telephone	$97,500 at 10.9% until September 25
Central Power Co.	$50,000 at 11.25% until October 20
General Land Development Co.	$50,000 at 11.4% until November 4

The bank service fee is $25 for each note under $51,000 and $15 for each note over $51,000. What net interest will Adelphia Investments, Inc., receive from these investments? **$8342.55**

MAINTAINING YOUR SKILLS Look up the skill in parentheses if you need help or more practice.

Multiply. Round answers to the nearest hundredth. **(Skill 8)**

10. 52.74 × 1000 **52,740** 11. 0.478 × 0.49 **0.23** 12. 4173 × 0.005 **20.87** 13. 7.165 × 0.047 **0.34**

PRACTICE AND APPLY

The following problems can be assigned for classwork and the answers checked in class to help students master the objective of the lesson.

- Guided Practice: 1–4
- Independent Practice: 5–7

WRAP-UP

As a class activity, work through the solution to Problem 7. Select two students to show the interest earned from each investment and another student to show how to find the net interest.

Assignment Guide

- Basic: 5–8, 10–13
- Average: 8, 9, 10–13

	Date of Origin	Due Date	Number of Days	Comment
1)	June 20	Dec. 20	(183)	(354 − 171)
2)	Mar. 27	Aug. 7	(133)	(219 − 86)
3)	Oct. 15	Feb. 20	(128)	Careful: [(365 + 51) − 288]
4)	Jan. 12, 1996	May 23, 1996	(132)	Careful: [(1 + 143) − 12]

Be careful in #3 because it goes over the end of the year (add 365) and in #4 because 1996 is a leap year (add 1).

LESSON PLAN
21-6 Growth Expenses

FOCUS
Ask students to imagine they are the president of their own business, which has sales annually of $5,000,000. Then ask them to suggest ways that they could expand their business to reach $10,000,000 in sales in five years. Write their ideas on the chalkboard. In this lesson, the focus is on computing the total cost of expanding a business.

TEACH
Point out that expanding a business costs money. Go over the various growth expenses listed in the Example for opening a new store. Then ask students to suggest some costs for the expansion ideas listed on the chalkboard. Try to focus on one particular business so the total costs for expansion can be computed.

21-6 Growth Expenses

OBJECTIVE
Compute the total cost of expanding a business.

You may expand your business in several different ways. You may purchase or build an addition to your present building. You may purchase another business to become part of your business. Your business may **merge,** or combine, with another business to form a new business. **Growth expenses** for your business may include construction fees, consultation fees, legal fees, and so on.

Total Cost of Expansion = Sum of Individual Costs

EXAMPLE *Skill* 5 *Application* A *Term* Growth Expenses

The owner of Posner's Deli plans to expand the business by opening a new store. Growth expenses include:

Marketing survey: $3500 plus $3.75 per person for 150 people interviewed
Land: $60,000
Building construction: $750,000
Architect's fee: 5% of the cost of construction
Surfacing parking lot: $0.90 per square foot for 6000 square feet
Legal fees: $975
Equipment and fixtures: $67,500
Additional stock: $13,000
Miscellaneous expenses: $4000

What is the total cost for the expansion of Posner's Deli?

SOLUTION

Find the **sum of individual costs.**

Marketing survey: $3500.00 + ($3.75 × 150)	$ 4062.50
Land	60,000.00
Building construction	750,000.00
Architect's fee: $750,000.00 × 5%	37,500.00
Surfacing parking lot: $0.90 × 6000	5400.00
Legal fees	975.00
Equipment and fixtures	67,500.00
Additional stock	13,000.00
Miscellaneous expenses	4000.00
Total Cost of Expansion	$942,437.50

3.75 [×] 150 [=] 562.5 [+] 3500 [=] 4062.5 [M+] 60000 [M+] 750000 [M+] [×] 5 [%] 37500 [M+] .9 [×] 6000 [=] 5400 [M+] 975 [M+] 67500 [M+] 13000 [M+] 4000 [M+] [RM] 942437.5

ALTERNATIVE ASSESSMENT
Call upon six volunteers to come to the chalkboard and write the objective of one lesson in the unit on the board. Then ask six other students to come to the board and write the formula for each objective below it. Call upon others to explain the meaning of each formula.

✔ SELF-CHECK Complete the problems, then check your answers in the back of the book.

Find the total cost of expansion.

1. Expansion costs for Wyatt Jewelry store: legal fees, $1500; display cases, $25,000; safe, $10,150; miscellaneous, $3200. **$39,850**

2. Expansion for Harriet's Fashions: new construction, $74,800; display racks, $9450; clothing stock, $57,500; advertisements, $500. **$142,250**

PROBLEMS

3. Expand housecleaning calls. Solicit new customers for $125. Purchase new van for $12,470. What is the total cost of expansion? **$12,595**

4. Expand lawn-care business. Purchase new truck for $10,640. Purchase supplies for $875. What is the total cost of expansion? **$11,515**

5. Grandview Farms is opening a new beef stick outlet in the Western Mall. Grandview pays rent in advance for 3 months. The rental charge is $1150 per month. Grandview makes a 20% down payment on refrigeration equipment that costs a total of $13,980. Grandview also purchased additional supplies for $5795. What is the total cost of expansion? **$12,041**

6. Outdoors, Inc., is adding a new department that will specialize in hunting equipment. Outdoors pays $775 for redecoration of an area of the store. Outdoors also makes a 30% down payment on new stock that costs a total of $17,785. What is the total of these growth expenses? **$6110.50**

7. The Antique Mart is expanding by adding 6 new stalls available for lease. The Mart is converting a storage area of 1800 square feet into these stalls. The costs of the expansions are:

Description	Cost
Construction permit	$120
Removal of two walls	$800
New lighting fixtures and installation	$2785
Carpet and installation	$18.90 per square yard
Down payment on fixtures	25% of $4260

What is the total cost for The Antique Mart to expand? **$8550**

Warm-Up Exercises

1. $4700 + ($17.50 × 200) $8200
2. $75,000 + $18,500 + $10,450 $103,950
3. 7.5% of $640,000 $48,000
4. 1.25% of $147,500 $1843.75

PRACTICE AND APPLY

The following problems can be assigned for classwork and the answers checked in class to help students master the objective of the lesson.

- Guided Practice: 1–4
- Independent Practice: 5, 6

ALTERNATIVE STRATEGIES: Enrichment

Have students investigate local costs for:

1) having a marketing survey done;
2) purchasing a plot of land in a commercial area;
3) having an architect draw the plans for a new building;
4) renting a new building (find the cost per square foot).

WRAP-UP

Engage students in a brief discussion of how to compute the total cost of expanding a business. Use a specific business as an example, such as a retail stereo store or an automobile repair shop.

Assignment Guide

- Basic: 5–8, 10–14
- Average: 7–9,10–14 even

8. Central National Bank plans to open a new branch office. Central purchased property for $34,500. Construction costs for a new building totaled $875,980. In addition, Central paid an architect's fee of 7.5% of the cost of construction. Legal fees for the expansion totaled $6000. New equipment and fixtures cost $24,675. Other expenses came to $7215. What was the total cost of Central's expansion? **$1,014,068.50**

9. JP Industries, Inc., is searching for a new business to buy. Finders, Inc., a company that specializes in locating firms for sale, has located a small machine plant. If JP Industries purchases the firm, it must pay Finders, Inc., a finder's fee of 4.5% of the total worth of the machine plant. To acquire the plant, JP Industries must pay the plant its total worth of $2.45 million. JP Industries must pay legal fees amounting to 1.25% of the total worth of the machine plant. In addition, JP Industries must pay the debts of the machine plant. The debts amount to $165,850 plus 12.4% interest for 1 year. What will be the total cost of the expansion? **$2,777,290.40**

MAINTAINING YOUR SKILLS Look up the skill in parentheses if you need help or more practice.

Add. **(Skill 5)**

10.	11.	12.	13.	14.
78.14	571.8	747.249	4.718	8456.87
+ 234.854	+ 28.912	+ 221.84	+ 0.643	+ 4318.93
312.994	**600.712**	**969.089**	**5.361**	**12,775.8**

BUSINESS PROJECT

Assign students to read the business section of their local newspaper in order to find a story about a business expansion and how it was done. Call upon a few students to give a brief report on what they have found in the newspapers.

Reviewing the Basics

Skills

Solve.

(Skill 5)

1. \$6718.75 + \$54,193.84 + \$715.66 **\$61,628.25**

2. \$76,000.00 + \$4218.74 + \$42,193.75 + \$21,424.95 **\$143,837.44**

(Skill 8)

3. \$86,000 × 0.095 × 120 **\$980,400**

4. \$35,000 × 0.105 × 180 **\$661,500**

5. \$17,500 × 0.0925 × 90 **\$145,687.50**

6. \$62,550 × 0.0975 × 270 **\$1,646,628.70**

Round answers to the nearest cent.

(Skill 10)

7. \$980,400 ÷ 360 **\$2723.33**

8. \$661,500 ÷ 360 **\$1837.50**

9. \$1,650,000 ÷ 360 **\$4583.33**

Write as a decimal.

(Skill 28)

10. 10.75% **0.1075** **11.** $9\frac{1}{2}$% **0.095** **12.** 9.475% **0.09475** **13.** 175% **1.75**

Find the percentage. Round answers to the nearest cent.

(Skill 30)

14. 11% of \$41,750 **\$4592.50**

15. 35% of \$24,760 **\$8666**

16. 2.1% of \$186 **\$3.91**

17. 9.875% of \$18,125 **\$1789.84**

18. $6\frac{1}{2}$% of \$3250 **\$211.25**

19. $\frac{1}{4}$% of \$86 **\$0.22**

Applications

Find the tax.

(Application C)

20. Taxable income is \$41,650. **\$6247.50**

21. Taxable income is \$93,460. **\$19,526.40**

If taxable income is:		Tax is:		
Over—	But not over—			Of the amount over—
0	50,000		15%	0
\$ 50,000	\$ 75,000	\$ 7,500 +	25%	\$ 50,000
75,000	100,000	13,250 +	34%	75,000
100,000	—	21,750 +	39%	100,000

Terms

Write your own definition for each term. **Answers will vary.**

22. Corporate income taxes

23. Net proceeds

24. Maturity value

25. United States Treasury Bill

26. Commercial paper

27. Growth expenses

Refer to your reference files in the back of the book if you need help.

LESSON PLAN
Reviewing the Basics

The exercises on this page review skills, applications, and terms used in the unit. You can use the exercises to assess informally students' proficiency with this material.

The page can be used for guided practice and independent practice. You can work through a selection of the exercises together with students, and thus see immediately if they know how to do them, and you can then assign some of the exercises for independent practice. Be sure to go over the answers to all assigned exercises.

LESSON PLAN
Unit Test

Students should do the Unit Test on their own. Each problem on the test is keyed to a lesson in the unit. Students having difficulty with any particular problem should review the Example in the appropriate lesson and be assigned some of the Independent Practice problems for additional practice.

Unit Test

Lesson 21-1

1. The Bellview Corporation had an annual gross income of $834,718 for the year. Business deductions totaled $697,497. What federal corporate income tax must Bellview pay for the year? **$36,266.19**

If taxable income is:		Tax is:	
Over—	But not over—		Of the amount over—
0	50,000	15%	0
$ 50,000	$ 75,000	$ 7500 + 25%	$ 50,000
75,000	100,000	13,250 + 34%	75,000
100,000	—	21,750 + 39%	100,000

Lesson 21-2

2. Core Control Inc. is issuing 500,000 shares of stock. Each share will be sold at $36.75. The underwriting commission is 3% of the value of the stock. The other expenses are expected to be 0.75% of the value of the stock. If all the shares of stock are sold, what net proceeds will Core Control receive? **$17,685,937.50**

Lesson 21-3

3. Mill Grove Fruit Farms borrowed $62,500 for 150 days at the prime rate of interest. When Mill Grove borrowed the money, the prime rate was 9.5%. The bank charged ordinary interest at exact time. What was the maturity value of the loan? **$64,973.96**

Lesson 21-4

4. Horton Manufacturing had $250,000 in surplus cash. The financial manager decided to invest in a United States Treasury Bill with a face value of $250,000. The bill matured in 126 days. The interest rate was 10.875% ordinary interest at exact time. The bank service fee was $35. What was the cost of the Treasury Bill? **$240,519.37**

Lesson 21-5

5. The Tom Brown Kitchen Supply Company sold its entire stock of kitchen cupboards for installation in an apartment complex. Tom Brown invested the $132,500 income from the sale in the following commercial paper:

Machinko Construction Co.	$90,000 at 9.95% for 30 days
Tobar Excavations, Inc.	$42,500 at 10.25% for 60 days

Both investments earn ordinary interest at exact time. The bank service fee is $45 for both. What net interest will the Tom Brown Kitchen Supply Company receive? **$1427.29**

Lesson 21-6

6. The Campus Bookstore opened a new branch store, which it rents for $1250 per month. The Campus Bookstore paid the rent in advance for 3 months. The Campus Bookstore purchased new furniture and store fixtures for $34,175. The Campus Bookstore also purchased additional merchandise for $7185. Miscellaneous expenses totaled $2465. What was the total cost of expansion? **$47,575**

A SPREADSHEET APPLICATION

Financial Management

To complete this spreadsheet application, you will need the template diskette for *Mathematics with Business Applications*. Follow the directions in the *User's Guide* to complete this activity.

Input information in the following problem to find the value of the issue, the total selling expenses, and the net proceeds. Then answer the questions that follow.

Stock Symbol	# of Shares	$ per Share	Value of Issue	Under. Comm. %	Selling Expense	Net Proceeds
AMR	500,000	$ 56.75	?	4.1%	?	?
CSX	250,000	36	?	6.8%	?	?
EDO	1,000,000	5.875	?	5%	?	?
GM	300,000	19.625	?	4.5%	?	?
IBM	200,000	127.875	?	3.75%	?	?
KFC	100,000	72.5	?	6%	?	?
MGM	750,000	11.375	?	5.5%	?	?
OEA	350,000	40.625	?	6.5%	?	?
QMS	600,000	19.25	?	5.875%	?	?
SBX	450,000	2.125	?	6.25%	?	?
URS	750,000	9.625	?	5.75%	?	?
WB	150,000	108.75	?	6.125%	?	?

1. What is the value of the issue for each of these stocks: AMR, EDO, IBM, MGM, QMS, and URS?
2. What is the selling expense for each of these stocks: CSX, GM, KFC, OEA, SBX, and WB?
3. What are the net proceeds for each of these stocks: AMR, CSX, EDO, GM, IBM, KFC, MGM, OEA, QMS, SBX, URS, and WB?
4. How many shares of stock must KFC sell at $72.50 per share to raise $10 million if the underwriter commission is 6%?
5. How many shares of stock must WB sell at $108.75 per share to raise $10 million if the underwriter commission is 6.125% and other expenses total $25,000?
6. How many shares of stock must EDO sell at $5.875 per share to raise $7.5 million if the underwriter commission is 5% and other expenses total 1.5% of the value of the stock?
7. If OEA could reduce the underwriter commission to 5.5%, how much more would the net proceeds be equal to?
8. If MGM's stock went up $0.50, how much more would the net proceeds be equal to?

LESSON PLAN
A Spreadsheet Application

USING TECHNOLOGY

The stock market is an excellent example of a financial institution that would not be able to function today were it not for the existence of powerful computers. You might want to suggest to students that they read the financial pages of the local newspaper to find the volume of trading on the New York Stock Exchange each day for a week. Point out that each transaction must be recorded and the proper documents sent to both the buyer and seller.

When students see stocks prices quoted in the newspaper, fractions are used; for example, $\$24\frac{3}{8}$ per share. Point out the fractions must be converted to decimals when data are used in computer calculations.

LESSON PLAN
Career Wise

APPLICATIONS ON THE JOB

As students work this page, ask them to explain what mergers and acquisitions are. They will probably need some help with these terms. The term *portfolio* also may need to be discussed. Point out to students that stockbrokers make money from commissions on each buy or sell order.

The line graph shows the daily fluctuations in the price of Xylux Company's stock. To buy a stock at a low price and sell it at a high price in a short period of time involves a great deal of risk. The reason is that one never knows for sure when a stock hits a low or a high. The price could always go lower or higher. Point out, however, that over long periods of time, such as five or ten years, the stock market has historically gone up. This fact makes stocks a good long-term investment.

CAREER WISE

Stockbroker

Vida Kasic is a stockbroker for a large investment firm. As a broker, she acts as an agent for someone who wants to buy or sell part ownership in a company.

Not only does Vida handle transactions for her clients, she also watches market trends. It is her hope that she can help people buy shares when the price is low and sell shares when the price increases.

Mergers and acquisitions are extremely important to Vida, since they can greatly affect the price of stock. When she finds that the health of the economy is changing, she might advise a client to alter his or her portfolio, the collection of investments the client has.

The line graph below shows the price of one share of Xylex Company stock for each business day in the month of July. Prices are closing prices.

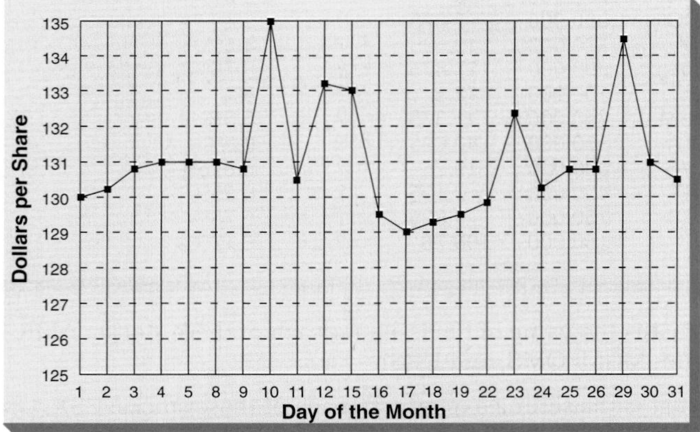

Check Your Understanding

1. Over which 3-day period did the price remain unchanged?
   July 4, 5, and 8
2. Over which two days was the greatest price increase?
   July 9 to July 10
3. Did the price rise or fall from July 23 to July 24? By how much?
   It fell about $2.25.
4. If 100 shares were bought on July 1 and sold on July 29 (at closing prices), was there a profit or loss? Of how much?
   Profit: 100 × ($134.5 − $130) = $450
5. What were the minimum and maximum closing prices during July?
   $129; $135.00
6. What was the average closing price for the stock in July?
   $2883 ÷ 23 = $131.00

A SIMULATION

Stock Market

Your employer has given you a $2000 bonus. You decide to invest the money in stocks. To help you decide which stocks to buy, you study the past performance of many stocks. You also read financial newspapers and magazines, and talk to people who are well-informed about the stock market.

52 Week High	52 Week Low	Stock	Dividend	Yield (%)	P.E. Ratio	Volume (000)	High	Low	Close	Net Change
26 1/2	23	Cablvsn	—	—	58	265	26 1/2	23 1/2	23	−3 1/2
22 3/8	20 1/2	LaZBoy	—	—	67	257	20	20 7/8	20	0
59	55 7/8	QuakrO	$0.25	6.8%	19	7446	57 3/8	55 1/8	56 3/8	1/2
8 1/2	5 3/8	TCBY	—	—	11	120	7 1/2	6	6 1/8	−1 1/2
44 3/8	34 1/2	Upjohn	1.16	7.4%	38	1456	44 1/2	41 7/8	42 1/4	3 3/8

1. You purchase 30 shares of LaZBoy stock at today's closing price. What is the cost of the stock? **$600.00**
2. Your stockbroker charges a 1.5% commission each time you buy or sell stock. What is the commission on this purchase? **$9.00**
3. What is the total amount you pay for the stock? **$609.00**
4. Of the $2000 that you want to invest in stock, how much is now available to be invested? **$1391.00**
5. LaZBoy declares a dividend of $1.25 per share. What is the annual yield? **6.25%**
6. What is the total dividend you receive on your 30 shares? **$37.50**
7. Including the dividend, how much money is now available to be invested in stocks? **$1428.50**

USING THE SIMULATION

Students can work the simulation individually, or you may wish to use it with *cooperative learning groups.* You can begin by reading with students the introductory material at the top of the page and discussing briefly the stock quotations. You also should discuss what a mutual fund is before students do the simulation. Show students a copy of the mutual fund quotations from the business section of a newspaper.

After students have finished the simulation, provide them with the answers so they can check their work.

A SIMULATION (CONTINUED)

Buying and Selling Stock

During the next year, you continue to study the stock market. Sometimes you buy stocks, starting with the money you have available after buying the LaZBoy stock. Sometimes you sell stock, either because the company is not doing well or because you want the money to buy stocks in a different company. You sell all your stock at the end of the year. While you owned some of the stocks, you received the dividends shown. You add the dividend to the money left to invest.

	Name of Stock	Number of Shares	Bought (B) Sold (S)	Price per Share	Amount Paid	Amount Received	Commission 1.5%	Total Cost	Dividend	Dividend Income	Total Income	Money Left to Invest
	LaZBoy	30	B	$20.000	$600.00	—	$9.00	$609.00	$1.25	$37.50	$37.50	$1428.50
8.	QuakrO	15	B	57.375	$860.63	—	$12.91	$873.54	0.25	$3.75	$3.75	$558.71
9.	TCBY	25	B	6.125	$153.13	—	$2.30	$155.43	—	—	—	$403.28
10.	Upjohn	8	B	43.000	$344.00	—	$5.16	$349.16	—	—	—	$54.12
11.	QuakrO	10	S	68.500	—	$685.00	$10.28	—	—	—	$674.72	$728.84
12.	TCBY	25	S	7.875	—	$196.88	$2.95	—	—	—	$193.93	$922.77
13.	Upjohn	16	B	40.125	$642.00	—	$9.63	$651.63	$1.16	$18.56	$18.56	$289.70
14.	CtyCm	60	B	3.500	$210.00	—	$3.15	$213.15	—	—	—	$76.55
15.	QuakrO	5	S	62.125	—	$310.63	$4.66	—	—	—	$305.97	$382.52
16.	LaZBoy	30	S	21.000	—	$630.00	$9.45	—	—	—	$620.55	$1003.07
17.	CtyCm	80	B	3.875	$310.00	—	$4.65	$314.65	—	—	—	$688.42
18.	Upjohn	24	S	44.875	—	$1077.00	$16.16	—	—	—	$1060.84	$1749.26
19.	CtyCm	140	S	3.125	—	$437.50	$6.56	—	—	—	$430.94	$2180.20

20. What is the total dividend you received? **$59.81**
21. Have you made a net profit or a net loss? How much? **Profit; $180.20**
22. What is the percent profit or loss on your $2000 investment? **9.01%**

A SIMULATION
(CONTINUED)

Comparing Mutual Funds

You realize at the end of the year that you do not have time to study the stock market. You decide to invest $2100 in three mutual finds. The following table shows the net asset value per share for each fund at the end of each month of the previous year.

23. On graph paper, set up a graph like the one at the bottom of the page. Make a line graph using the values in the table. Use a different color for each fund.

NET ASSET VALUE PER SHARE												
Month	Jan	Feb	Mar	Apr	May	June	July	Aug	Sep	Oct	Nov	Dec
South Bay Fund	$6.42	$6.55	$6.72	$6.89	$6.75	$6.81	$6.97	$7.14	$7.29	$7.50	$7.68	$7.90
Salinas Valley Fund	6.59	6.74	6.89	7.03	6.82	6.85	6.92	7.25	7.41	7.78	7.95	7.83
Central Section Fund	6.54	6.62	6.73	6.84	6.78	6.86	6.95	7.04	7.15	7.24	7.32	7.40

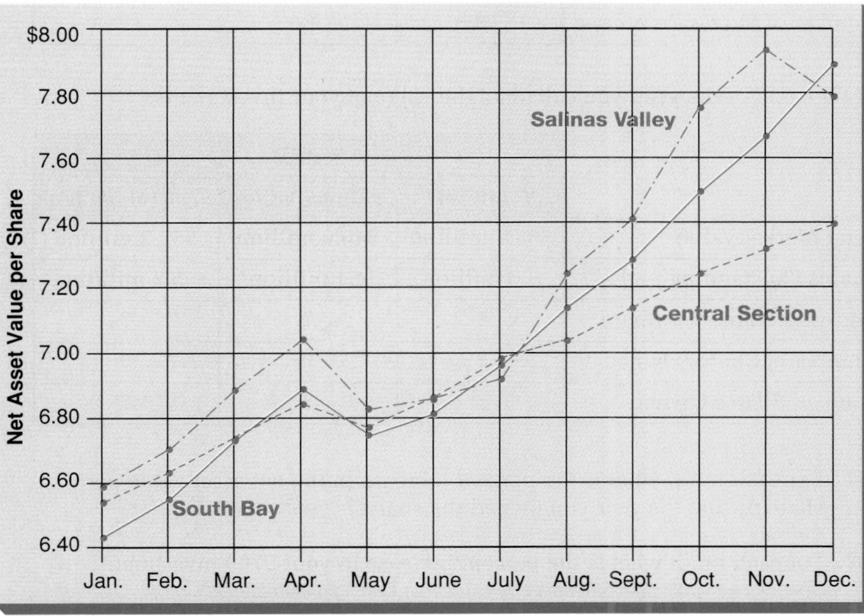

551

A SIMULATION (CONTINUED)

Investing in Mutual Funds

You decide to invest $700 in each of the three funds. South Bay and Salinas Valley have loading charges. Central Section is a no-load fund.

The table shows the total market value of each fund's portfolio and the number of shares outstanding on the day you make your investments. Calculate the number of shares of each fund you can purchase.

		FUND		
		South Bay	Salinas Valley	Central Section
	Total Market Value	$49.0 million	$30.6 million	$44.1 million
	Shares Outstanding	5.6 million	3.6 million	4.9 million
24.	Net Asset Value per Share	**$8.75**	**$8.50**	**$9.00**
	Amount Invested	$700	$700	$700
	Loading Rate	10%	8.5%	0
25.	Loading Charge	**$70**	**$59.50**	**0**
26.	Amount Invested Minus Loading Charge	**$630**	**$640.50**	**$700**
27.	Number of Shares Purchased	**72**	**75.35**	**77.78**

At the end of the year, you calculate the value of your investments.

		FUND		
		South Bay	Salinas Valley	Central Section
	Total Market Value	$63.4 million	$40.9 million	$52.2 million
	Shares Outstanding	6.0 million	4.1 million	5.2 million
28.	Net Asset Value per Share	**$10.57**	**$9.98**	**$10.04**
29.	Number of Shares Owned	**72**	**75.35**	**77.78**
30.	Value of Shares Owned	**$761.04**	**$751.99**	**$780.91**

31. For each fund, what is the percent increase in the net asset value per share during the year you owned the shares? **South Bay 17.22%; Salinas Valley 14.83%; Central Section 10.36%**
32. For each fund, what is the percent increase in your $700 investment? **South Bay 8.02%; Salinas Valley 6.91%; Central Section 10.36%**

UNIT 22

Corporate Planning

Corporate planning takes place for strategic short-term planning as well as extensive long-range planning. Economic considerations that play an important role in corporate planning include the *inflation rate* at the time, the size of the real and per capita *gross national product,* and the measure of the *consumer price index* for particular items. All these considerations, and more, go into planning and implementing a corporate *budget.*

OUTLINE

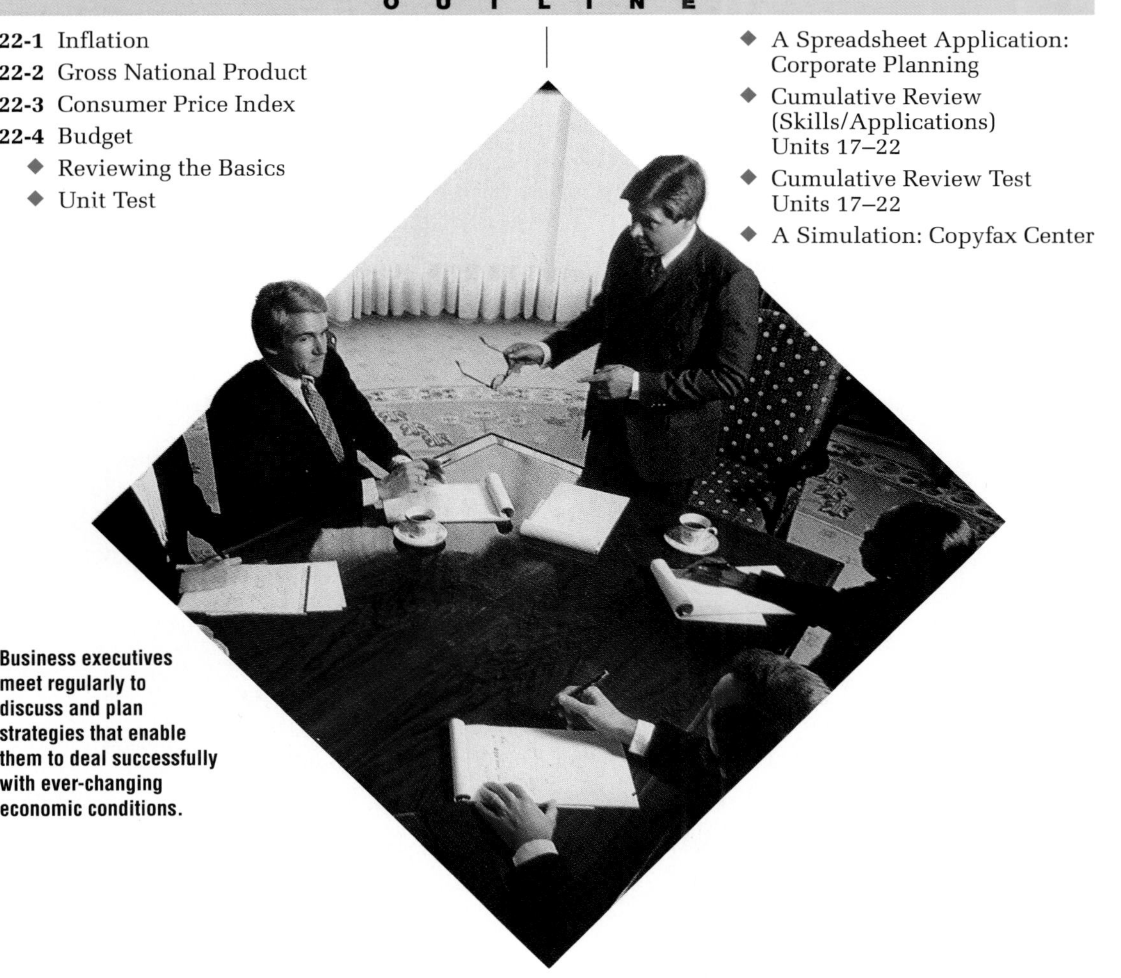

Business executives meet regularly to discuss and plan strategies that enable them to deal successfully with ever-changing economic conditions.

UNIT 22 CORPORATE PLANNING

INTRODUCING THE UNIT

Students will study the concepts of *inflation* and *GNP* in the first two lessons of this unit. As an introduction, bring to class newspaper articles that mention either or both of these terms and the effect they have on a particular business or country.

LESSON PLAN
22-1 Inflation

FOCUS

Draw a line graph on the chalkboard that shows the cost of a loaf of bread in 1967, 1981, and 1991 as \$0.35. \$0.75, and \$1.15, respectively. Ask students what the graph shows. (increasing price) Explain that this effect is called inflation. Have students suggest reasons as to why the loaf of bread has increased over the years. (higher cost of ingredients, wage increases, higher cost of utilities, transportation, and so on)

TEACH

Point out that inflation can have a serious effect on businesses, as well as on individuals. Discuss the effects listed in the introductory paragraph of this lesson. Mention that inflation has to be considered when a business tries to measure its profitability. It is also an important consideration when deciding which inventory valuation method to use.

Discuss the fact that some items, such as digital watches, television sets, and hand-held calculators, actually have decreased in price in recent years. This is the case in Problem 23. Ask students why this has happened. (Advances in technology have made it less expensive to produce these goods.)

22-1 Inflation

OBJECTIVE

Compute the inflation rate, current price, and original price.

Inflation is an economic condition during which there are price increases in the cost of goods and services. At the corporate level, inflation is observed as increases in (1) wholesale prices, (2) cost of utilities, (3) cost of production and shipping, and (4) demands for scarcer materials. Some of the causes of inflation are (1) heavy spending, resulting in high demand, (2) increased production costs while producers try to maintain profit levels, and (3) lack of competition.

The **inflation rate** is found by subtracting the original price from the current price and dividing the result by the original price. The inflation rate is expressed as a percent increase over a specified time period, usually to the nearest tenth of a percent.

The current price is found by adding the original price to the product of the original price times the inflation rate.

The original price is found by dividing the current price by 1 plus the inflation rate.

$$\text{Inflation Rate} = \frac{(\text{Current Price} - \text{Original Price})}{\text{Original Price}}$$

$$\text{Current Price} = \text{Original Price} + (\text{Original Price} \times \text{Inflation Rate})$$

$$\text{Original Price} = \frac{\text{Current Price}}{(1 + \text{Inflation Rate})}$$

EXAMPLE *Skills* 5, 10, 30, 31, 32 *Application* A *Term* Inflation rate

Liza Turner works in the summer as a junior staff reporter for the local newspaper. Liza was asked to write a story about the changes in the economy. One section was on inflation. She had the following data:

- New automobile costs \$15,000 today. Same model sold for \$12,500 a year ago.
- Home sold for \$130,000 2 years ago. Inflation rate over the 2-year period for homes is 10%.
- Current price for a lawn mower is \$395. Inflation rate over the last year for lawn mowers is 5%.

A. What is the inflation rate for the automobile?

B. What is the current price of the home?

C. What was the original price of the lawn mower last year?

MATHEMATICAL NOTES

Note that there are three formulas and that they can be derived from each other. Solve the first formula for the current price and you have the second formula. Solve the first formula for the original price and you have the third formula. Finding the inflation rate is an application of percent increase.

A. Find the **inflation rate** for the automobile.

(Current Price	−	Original Price)	÷	Original Price	
($15,000	−	$12,500)	÷	$12,500	
$2500			÷	$12,500	= 0.2 = 20% inflation rate

B. Find the **current price** of the home.

Original Price	+	(Original Price	×	Inflation Rate)	
$130,000	+	($130,000	×	10%)	
$130,000	+	$13,000			= $143,000 current price

C. Find the **original price** of the lawn mower.

Current Price	÷	(1 + Inflation Rate)	
$395.00	÷	(1 + 5%)	
$395.00	÷	1.05	= $376.1904 = $376.19 original price

15000 [−] 12500 [=] 2500 [÷] 12500 [=] 0.2

✔ SELF-CHECK Complete the problems, then check your answers in the back of the book.

1. Original price was $75. Current price is $80. Find the inflation rate. **6.7%**
2. Original price was $12,400. Inflation rate is 8.2%. Find the current price. **$13,416.80**
3. Current price is $28.05. Inflation rate is 10%. Find the original price. **$25.50**

PROBLEMS

	Inflation Rate	Current Price	Original Price
4.	? **6.4%**	$ 4670.00	$ 4389.80
5.	? **1.5%**	$ 79.99	$ 78.79
6.	5.0%	? **$154.35**	$ 147.00
7.	3.6%	? **$18,086.49**	$17,458.00
8.	0.7%	$ 1.29	? **$1.28**
9.	12.4%	$243,750.00	? **$216,859.43**

10. Original price: $16.95.
Current price: $18.49.
Find the inflation rate. **9.1%**

11. Original price: $3.50.
Inflation rate: 12.4%.
Find the current price. **$3.93**

12. Current price: $4370.
Inflation rate: 6.3%.
Find the original price. **$4111.01**

13. Original price: $0.988.
Current price: $1.249.
Find the inflation rate. **26.4%**

Warm-Up Exercises

Round to nearest tenth of a percent.

1. $1.82 is what percent of $45.97? 4.0%
2. $3001 is what percent of $9748? 30.8%
3. 6.5% of $17.89 $1.16
4. 12.4% of $23,750 $2945
5. $375.50 ÷ 104.2% $360.36
6. $1.97 ÷ 111.4% $1.77

PRACTICE AND APPLY

The following problems can be assigned for classwork and the answers checked in class to help students master the objective of the lesson.

- Guided Practice: 1–9
- Independent Practice 10–12, 14

ALTERNATIVE STRATEGIES: Reteaching

Point out the relationship of the three formulas using this data.

CP = $100 OP = $80 IR = 25%

1. IR = ($100 − $80) ÷ $80 = $20 ÷ $80 = 25%
2. CP = $80 + ($80 × 25%) = $80 + $20 = $100 or
   CP = (100% × $80) + (25% × $80) = 125% × $80 = $100
3. OP = $100 ÷ (1 + 25%) = $100 ÷ 1.25 = $80

WRAP-UP

Give students the following data: $2500, $2700, and 8%. Have individual students use the data to recreate the formulas of the lesson at the chalkboard. For example,

$$8\% = \frac{(\$2700 - \$2500)}{\$2500}.$$

Assignment Guide

- Basic: 10–18, 20–29
- Average: 13, 15–19, 21–29 odd

14. Original price: $134,980.
Inflation rate: 4.2%
Find the current price. **$140,649.16**

15. Current price: $27,495.
Inflation rate: 14.61%.
Find the original price. **$23,990.05**

16. Original price: $3680.
Current price: $11,850.
Find the inflation rate. **222.0%**

17. Original price: $0.49.
Inflation rate: 19.8%.
Find the current price. **$0.59**

F.Y.I.
Over the past 10 years, inflation has averaged 5% (Better Investing, December 1991, p. 36).

18. The inflation rate for grocery products is 5.7%. How much would a cart of groceries cost today if it cost $72.70 last year? **$76.84**

19. A month ago, a friend purchased a mini van for $16,865. The inflation rate for mini vans over the past month is 0.4%. What is the current price of a mini van? **$16,932.46**

20. At the grand opening of the Bayview Market in 1938, a 10-ounce box of cereal cost 10¢. Today the box costs $1.99. What is the inflation rate for cereal over that time period? **1890%**

21. At Bayview Market, a quart of milk could be purchased for $0.11 in 1938. The inflation rate for milk since that time is 600%. What does a quart of milk cost today? **$0.77**

22. At Bayview Market, 2 pounds of coffee cost $5.99 today. The rate of inflation for coffee since 1938 is 1715.2%. What did 2 pounds of coffee cost in 1938? **$0.33**

23. A hand-held calculator that had an original price of $79.95 currently sells for $19.95.

a. What is the inflation rate? **−75%**

b. What is true of the inflation rate when the current price is less than the original price? **It is negative.**

c. Instead of inflation, what might this be called? **Deflation**

MAINTAINING YOUR SKILLS Look up the skills in parentheses if you need help or more practice.

Find the percentage. Round answers to the nearest cent. **(Skill 30)**

24. 3.7% of $436.80 **$16.16**

25. 11.5% of $48.98 **$5.63**

26. 0.4% of $1.79 **$0.01**

Find the rate. Round answers to the nearest tenth of a percent. **(Skill 31)**

27. $12.85 is what percent of $257? **5.0%**

28. $3.79 is what percent of $199.99? **1.9%**

29. $2500 is what percent of $22,387.65? **11.2%**

BUSINESS NOTES

Inflation plays a very important role in business planning because the cost-of-living adjustments (COLA) to payroll are frequently dependent upon the inflation rate. Pension benefits, including social security, are also tied to the inflation rate. In times of high inflation, consumers tend to postpone purchases and this can result in a downturn in the economy. Inflation can cause savings to be worth less in terms of real buying power.

22-2

Gross National Product

OBJECTIVE
Understand the gross national product (GNP) and be able to compute the real GNP and the per capita GNP.

The **gross national product** is a measure of nation's economic performance. The GNP is the estimated total value of all goods and services produced by a nation during a year. Only goods, such as automobiles, machinery, food, and clothing, and services, such as haircuts, appliance repairs, and inventory costs, that add to the national income are included. The government uses the GNP to monitor the health of the country's economy.

Because of inflation, a nation's GNP can appear to be growing faster than it actually is. For this reason, the **real GNP**, or adjusted GNP, is corrected for inflation.

There **per capita GNP** provides an indication of the nation's standard of living. It is the GNP distributed over the population.

Real GNP = GNP − (GNP × Inflation Rate)

$$\text{Per Capita GNP} = \frac{\text{GNP}}{\text{Population}}$$

EXAMPLE *Skills* 1, 6, 11, 30 *Application* A *Term* Gross national product

The United States had a population of 248,700,000 in 1990, an inflation rate of 4.6%, and a GNP of $5200.8 billion. What is the real GNP? What is the per capita GNP?

SOLUTION

A. Find the **real GNP.** Rounded to the nearest tenth of a billion

GNP − (GNP × Inflation Rate)
$5200.8 billion − ($5200.8 billion × 4.6%)
$5200.8 billion − $239.2 billion = $4961.6 billion real GNP

B. Find the **per capita GNP.** 248,700,000 ÷ 1,000,000,000 = 0.2487 billion

GNP ÷ Population
$5200.8 billion ÷ $248,700,000, or 0.2487 billion
= $20,911.942 = $20,911.94 per capita GNP

5200.8 M+ × 4.6 % 239.2368 M− RM 4961.56

✔ SELF-CHECK Complete the problems, then check your answers in the back of the book.

1. GNP is $43,800,000. Inflation rate is 6.4%. Find the real GNP. **$40,996,800**
2. GNP is $92.8 million (or $92,800,000). Population is 78,650. Find the per capita GNP. **$1179.91**

Note that the very large numbers will cause some students to arrive at solutions with the decimal point misplaced. Have them drop the 6 zeros and then move the decimal point in the answer 6 places.

LESSON PLAN 22-2 Gross National Product

FOCUS

Ask students if they have heard or seen the term *gross national product* on television or in the newspaper. If so, can they define the term? Explain that GNP measures a country's economic growth. For example, the GNP is used to determine if the U.S. is producing more goods and services than in previous years. In this lesson, students will compute the real GNP and the per capita GNP.

TEACH

Have a student read the first paragraph of the introduction to the lesson. Mention that when goods, such as a pair of shoes, are included in the GNP, they are included only once. For example, a pair of shoes that sells for $75 has had different prices when sold by the manufacturer to the wholesaler or to the retail store. Only the retail price of $75 is used. In determining production, however, the year in which an item (such as an automobile) was manufactured is used, not the year it was sold.

Mention that the GNP is an important indicator to corporate planners because it helps to identify industries that are growing or not growing.

Point out that the per capita GNP is only a statistical average—it does not measure the real income of people.

ALTERNATIVE STRATEGIES: Reteaching

Spend ample time having students practice writing large numbers as decimals; for example, $45,800,000,000 can be written as $45.8 billion. Since most calculators cannot display some of the large numbers in this lesson, it will be necessary that students rewrite large numbers as decimals. Be sure students understand that the divisor and the dividend need to be expressed in the same way; that is, either in billions or millions.

Warm-Up Exercises

Round to the nearest tenth of a billion.

1. 6.2% of $45.8 billion $2.8 billion
2. 12.4% of $271.5 billion $33.7 billion
3. $45,800,000,000 − $2,839,600,000 $42,960,400,000
4. $52,845,000,000 ÷ 35,850,000 $1474.06
5. 4.82 trillion ÷ 274.1 million $17,584.82

PRACTICE AND APPLY

The following problems can be assigned for classwork and the answers checked in class to help students master the objective of the lesson.

- Guided Practice: 1–6
- Independent Practice: 7–11

WRAP-UP

Ask students what effect inflation has on the GNP. (Inflation reduces the GNP.) Also, ask students how to find the per capita GNP.

Assignment Guide

- Basic: 7–18, 21–26
- Average: 12–20, 21–25 odd

PROBLEMS

	GNP	Inflation Rate	Population	Real GNP	Per Capita GNP
3.	$123.6 million	4.3%	0.4 million	$118.3 mil	$309
4.	$57,459,650	0.4%	5,365,760	$57,229,811.40	$10.71
5.	$478.6 billion	6.2%	74.2 million	$448.9 bil	$6,450.13
6.	$756,500,000	2.1%	2,586,450	$740,613,500	$292.49
7.	$2,876.7 billion	7.8%	2,456,980,000	$2652.3 bil	$1,170.83
8.	$98,760,000,000	12.6%	31,478,000	$86,316.2 mil	$3,137.43

9. GNP is $768,000,000. Inflation rate is 4.6%. Find the real GNP. **$732,672,000**
10. GNP is $45,678,000,000. Population is 9,670,000. Find the per capita GNP. **$4723.68**
11. Australia has a population of 16,646,000 and a GNP of $220 billion. The inflation rate in Australia is 7.6%. Find the real GNP and the per capita GNP. **$203.28 billion; $13,216.39**
12. Greece has a GNP of $43,500,000,000. Its population is 10,066,000, and the inflation rate is 13.7%. Find the real GNP and the per capita GNP. **$37,540,500,000; $4321.48**

> **F.Y.I.**
> In 1992, the United States government stopped reporting the gross national product and instead reports the gross domestic product, which is a measure of products produced domestically.

13. Canada has a GNP of $486,000,000,000. It has a population of 26,527,000 and an inflation rate of 5.0%. Find the real GNP and the per capita GNP. **$461.7 billion; $18,320.96**
14. The GNP for Israel is $36 billion. Its population is 4.371 million. The rate of inflation in Israel is 20.2%. Find the real GNP and the per capita GNP. **$28.728 billion; $8236.10**
15. The GNP for Belgium is $153 billion. Its population is 9.898 million. The rate of inflation in Belgium is 3.1%. Find the real GNP and the per capita GNP. **$148.257 billion; $15,457.67**
16. The population of China is 1,130,065,000. China's GNP is $350,000,000,000, and the inflation rate is 16.3%. Find the real GNP and the per capita GNP. **$292.95 billion; $309.72**

17.

Nation	GNP	Population	Inflation Rate
France	$943 billion	56,184,000	3.5%
Italy	$825 billion	57,657,000	6.2%
U.K.	$758 billion	57,121,000	7.8%

Using the information above, determine:

a. Which nation has the highest real GNP. **France**
b. Which nation has the lowest real GNP. **U.K.**
c. Which nation has the highest per capita GNP. **France**
d. Which nation has the lowest per capita GNP. **U.K.**

18. Argentina has a GNP of $74.3 billion and a reported inflation rate of 3079%.
    a. What real GNP do you get using the formula? **−$2,213.397 billion**
    b. Why is this impossible? **There cannot be a negative GNP.**

CULTURAL ANGLES

Have students research the current GNP, population, and inflation rate of two countries of interest to them. Then have them compute the real GNP and per capita GNP.

CRITICAL THINKING

In Problem 20, the result in using the formula to find the real GNP is a negative number. (It is not possible to have a negative GNP.) Ask students why the formula does not work. (The inflation rate is over 100%.)

22-3

Consumer Price Index

OBJECTIVE
Compute the consumer price index (CPI), current cost, and cost in 1967 of any given commodity.

The **consumer price index** is a measure of the average change in prices of a certain number of goods and services. The year 1967 is used as the base year and the CPI for 1967 is set at 100. This means that a commodity that cost $100 in 1967 and that has a CPI of 159 today would cost $159 today. To find the CPI for any given commodity, divide the current cost by the cost in 1967 and then multiply by 100.

If you know the CPI for a given commodity and its cost in 1967, you can find the current cost by multiplying the cost in 1967 by the CPI and then dividing by 100.

If you know the CPI for a given commodity and its current cost, you can find the cost in 1967 by dividing the current cost by the CPI and then multiplying by 100.

$$\text{CPI} = \frac{\text{Current Cost}}{\text{Cost in 1967}} \times 100$$

$$\text{Current Cost} = \frac{(\text{Cost in 1967} \times \text{CPI})}{100}$$

$$\text{Cost in 1967} = \frac{\text{Current Cost}}{\text{CPI}} \times 100$$

EXAMPLE *Skills* 7, 8, 10 *Application* A *Term* Consumer price index

Liza Turner was asked to write a story about the consumer price index. She had the following data:

- The current cost of a briefcase is $48.50. The cost in 1967 was $19.95.
- The cost in 1967 of a good pair of sweat socks was $3.75. The CPI for sweat socks is 170.
- The current cost of a classical cassette is $8.00. The CPI for classical cassettes is 200.

A. What is the CPI for the briefcase?

B. What is the current cost of the sweat socks?

C. What was the cost of the cassette in 1967?

LESSON PLAN 22-3 Consumer Price Index

FOCUS

Ask students if they can name a term used in the news, other than GNP, to discuss the state of the economy. Mention the CPI and explain that it is one measure of the economy and inflation.

TEACH

Read to the class the definition found in the first sentence of the lesson. Point out that the CPI reflects month-to-month changes in the average costs of goods and services that a typical family would incur. Some of the categories used are housing, transportation, food and beverages, clothing, and medical care.

You might mention that cost-of-living adjustments in wages, discussed in Lesson 12-2, are frequently tied to the CPI, or the inflation rate.

As with the inflation rate, stress that not all commodities have a CPI greater than 100; that is, not all present costs are greater than costs in 1967. Problem 23 discusses some products that have decreased in cost. Relate it to Problem 23 from Lesson 22-1. Point out to students that although the inflation rate can be negative, the CPI cannot be. The CPI can be very small, but can never be negative.

ALTERNATIVE STRATEGIES: Enrichment

Have students investigate and report on the current CPI for various items that they or their family own. Have them compute the cost in 1967.

Warm-Up Exercises

Round to the nearest thousandth.

1. \$34.89 ÷ \$27.79
   1.255
2. \$49,625 ÷ \$29,999
   1.654
3. 1.654 × 100 165.4
4. \$19.49 × 262.5
   \$5116.13
5. \$5116.13 ÷ 100
   \$51.16
6. (\$362.75 ÷ 174.5) × 100 \$207.88

PRACTICE AND APPLY

The following problems can be assigned for classwork and the answers checked in class to help students master the objective of the lesson.

- Guided Practice: 1–9
- Independent Practice: 10–13

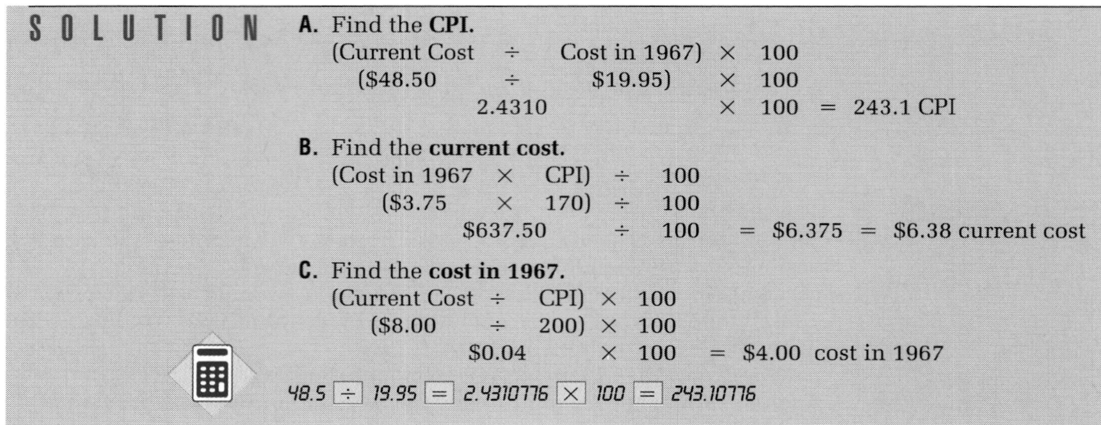

SOLUTION

A. Find the **CPI.**
(Current Cost ÷ Cost in 1967) × 100
(\$48.50 ÷ \$19.95) × 100
2.4310 × 100 = 243.1 CPI

B. Find the **current cost.**
(Cost in 1967 × CPI) ÷ 100
(\$3.75 × 170) ÷ 100
\$637.50 ÷ 100 = \$6.375 = \$6.38 current cost

C. Find the **cost in 1967.**
(Current Cost ÷ CPI) × 100
(\$8.00 ÷ 200) × 100
\$0.04 × 100 = \$4.00 cost in 1967

48.5 ÷ 19.95 = 2.4310776 × 100 = 243.10776

✓ SELF-CHECK Complete the problems, then check your answers in the back of the book.

1. The current cost is \$350. The cost in 1967 was \$140. What is the CPI? **250**
2. The CPI is 350. The cost in 1967 was \$765.95. What is the current cost? **\$2680.83**
3. The CPI is 140. The current cost is \$24.50. Find the cost in 1967. **\$17.50**

PROBLEMS

Round answers to the nearest tenth of a percent or to the nearest cent.

	Item	CPI	Current Cost	Cost in 1967
4.	Sport coat	**186.1**	\$139.50	\$ 74.95
5.	Dinner out for 2	**212.8**	\$ 45.75	\$ 21.50
6.	Apartment rent	110.6	**\$359.45**	\$325.00
7.	Lawn mower	156.7	**\$666.76**	\$425.50
8.	Bicycle	315.8	\$658.78	**\$208.61**
9.	Book	875.4	\$ 49.97	**\$5.71** ?

10. Current cost is \$48. Cost in 1967 was \$40. What is the CPI? **120**
11. Cost in 1967 was \$78. The CPI is 210. **\$163.80** What is the current cost?
12. The current cost is \$540. The CPI is 225. What was the cost in 1967? **\$240**
13. Current cost is \$125,500. Cost in 1967 was \$37,500. What is the CPI? **334.7**

PROBLEM SOLVING

Ask students what the following CPI numbers mean. Then have them find the cost of an item that sold for \$240 in 1967 using these numbers.

a. 50 (Cost now is half of what it was in 1967; \$120.)
b. 200 (Cost now is twice what it was in 1967; \$480.)
c. 100 (Cost now equals what it was in 1967; \$240.)
d. 0 (The item must be free now.)

Use the table for problems 14–20.

INDICATORS FOR JUNE	
Commodity	**CPI**
Fuel oil	504.8
Gasoline	364.6
Rent	238.9
Home	243.5
Groceries	315.6
Dining out	328.2
New car	312.7
Clothing	186.3

14. In 1967, Josh paid $4577.13 for a new car. What could he expect to pay today for a similar car? **$14,312.69**

15. In 1967, the Johnson family paid $48,650 for their home. What would a similar home sell for today? **$118,462.75**

F.Y.I.
Over the last 45 years, the consumer price index averaged 4.2% a year (Better Investing, December 1991, p. 36).

16. In 1967, the Johnson family paid $34.70 for their weekly groceries. How much could a similar family expect to pay for their weekly groceries today? **$109.51**

17. Last month's fuel oil bill for the Crowleys was $256.36. What would the Crowleys have paid for the same amount of fuel oil in 1967? **$50.78**

18. Bill and Mary Ellis spent $87.95 to have dinner out last Friday. How much would they have spent in 1967 to have a similar dinner out? **$26.80**

19. Alice McCarty pays $450 per month rent for her efficiency apartment. What would Alice have paid in 1967 for a similar apartment? **$188.36**

20. Tammy Spencer just purchased a new Treble four-door sedan for $18,457.50. What would a similar new car have cost Tammy in 1967? **$5902.62**

21. James Manos paid $11 for a haircut and styling that would have cost him $6.50 in 1967. What is the CPI? **169.2**

22. Tonia Curtis paid $6.75 for a movie ticket last weekend. In 1967, she would have paid $3.75 for the same ticket. What is the CPI? **180**

Critical Thinking . . .

23. Generally, in comparing prices from 1967 to the present:

a. If the price of a commodity was more in 1967 than it is now (for example, hand-held calculators, VCRs, kilowatts of electricity), would the CPI be more or less than 100 today? **Less than 100**

b. If something you purchased in 1967 was given away free today, what would the CPI equal? **0**

c. If something you would pay for today had been given away free in 1967, would it be possible to determine the CPI? **No**

MAINTAINING YOUR SKILLS Look up the skills in parentheses if you need help or more practice.

Divide. Round answers to the nearest tenth. **(Skill 11)**

24. $234.50 ÷ $78.25 **3.0**

25. $24.79 ÷ $18.45 **1.3**

26. $234,750.00 ÷ $489,980.00 **0.5**

27. $2.49 ÷ $5.79 **0.4**

Multiply. Round answers to the nearest cent. **(Skill 8)**

28. 234.3 × $45.75 **$10,719.23**

29. 156.8 × $378.89 **$59,409.95**

30. 112.4 × $7.89 **$886.84**

WRAP-UP

Call upon three students to each write one of the formulas of the lesson on the chalkboard. Then ask a volunteer to explain how current cost and cost in 1967 formulas can be derived from the CPI formula.

Assignment Guide

- Basic: 10–23, 26–36 even
- Average: 14–25, 27–37 odd

COOPERATIVE LEARNING

You might wish to form small groups to have students discuss the relationship between the GNP, inflation rate, CPI, and the economic health of a nation. Suggest that each group write a one-page statement that summarizes its discussion.

LESSON PLAN
22-4 Budget

FOCUS

Discuss the purpose of a budget with students and ask them if they have a budget for handling their spending money. Students having part-time jobs may allocate part of their earnings to a savings account, entertainment, clothing, hobbies, and so on.

TEACH

Stress the similarities between a personal budget as presented in Unit 11 and a corporate budget as presented here. How many times have we heard that something cannot be done because "it's not in the budget"? To paraphrase the old saying about "the hand that rocks the cradle rules the nation," you could add "the hand that makes the budget rules the company."

Work through the solution to the Example, having students check all the calculations.

22-4 Budget

OBJECTIVE

Allocate revenue and expenses and analyze a budget.

A **budget** is a financial plan that enables a business, a governmental agency, or an individual to identify what sources are expected to produce revenues (earn money) and what amounts are allocated to various departments or categories for expenses. Revenues and expenses are allocated as a percent of the total income. The main purpose of a budget is to monitor revenues and expenses. The actual amount spent must be compared with the budget allocation.

Budget Allocation = Total Income × Percent

Difference = Actual Amount − Budget Allocation

EXAMPLE *Skills* 6, 30 *Application* A *Term* Budget

The Kimberly Auto Company wanted to earn $2,457,890 in revenues for the year. The company expected to earn its revenues by the following sources: 75% from sales, 15% from services, and 10% from return on investments. At the end of the year, a budget analysis was sent to each manager showing the budgeted amounts and actual amounts of each area. Sales showed revenues of $1,755,965; services showed $369,240; and return on investments showed $245,790. Did the company reach its goal in revenues?

SOLUTION

A. Find the amount each source was budgeted, the **budget allocation.**

	Total Income	×	Percent		
Sales	$2,457,890	×	75%	=	$1,843,417.50
Services	$2,457,890	×	15%	=	$ 368,683.50
Return on investments	$2,457,890	×	10%	=	$ 245,789.00

B. Find the **difference** between the actual amount and budget allocation for each source.

	Actual Amount	−	Budget Allocation	=	Difference
Sales	$1,755,965.00	−	$1,843,417.50	=	−$87,452.50
Services	369,240.00	−	368,683.50	=	+556.50
Return on investments	245,790.00	−	245,789.00	=	+1.00
Total	$2,370,995.00	−	$2,457,890.00	=	−$86,895.00

The company was $86,895 short of its goal for the year.

2457890 M+ × 75 % 1843417.5 RM 2457890 × 15 % 368683.5 RM 2457890 × 10 % 245789

COMMUNICATION SKILLS

The skill of reading a budget can be enhanced by having students collect budgets from newspapers or magazines. There are federal, state, and municipal government budgets showing sources of revenue and categories of expenditures. There are budgets from school districts, county boards of mental health, and occasionally, some businesses. Have students collect as many budgets as possible and identify revenue sources and expenditures as both dollars and percents.

✓ SELF-CHECK Complete the problems, then check your answers in the back of the book.

Wagner Enterprises had a $400,000 budget. For each category, determine:

1. Amount budgeted **2.** Difference

	Expected Percent		Actual Amount	
Salaries	83%	**$332,000**	$360,000	**+$28,000**
Supplies	6%	**$24,000**	$ 20,000	**−$4000**
New equipment	5%	**$20,000**	$ 26,000	**+$6000**
Maintenance	4%	**$16,000**	$ 9,000	**−$7000**
Misc.	2%	**$8000**	$ 3,000	**−$5000**

PROBLEMS

	Total Revenue	Expected Percent	Actual Amount	Budget Allocation	Difference
3.	$ 80,000	25%	$ 15,000	**$20,000**	**−$5000**
4.	$ 120,000	60%	$ 74,000	**$72,000**	**+$2000**
5.	$ 860,000	7%	$ 64,000	**$60,200**	**+$3800**
6.	$ 987,500	80%	$ 760,000	**$790,000**	**−$30,000**
7.	$5,880,000	64%	$3,562,000	**$3,763,200**	**−$201,200**
8.	$7,644,000	8%	$ 756,253	**$611,520**	**+$144,733**

9. Allocate $150,000 as shown:
Sales: 70%. **$105,000**
Services: 20%. **$30,000**
Investments: 10%. **$15,000**

10. Allocate $600,000 as shown:
Interest: 15%. **$90,000**
Stocks: 65%. **$390,000**
Bonds: 20%. **$120,000**

11. Find the difference.

	Actual	Budget	
Sales	$460,000	$500,000	**−$40,000**
Services	150,000	180,000	**−$30,000**
Investments	60,000	50,000	**+$10,000**
Total	$670,000	$730,000	**−$60,000**

12. Find the difference.

	Actual	Budget	
Supplies	$155,400	$144,000	**+$11,400**
Equipment	450,600	390,000	**+$60,600**
Services	90,300	155,620	**−$65,320**
Total	$696,300	$689,620	**+$6,680**

13. The Downey Corporation is budgeting total revenues of $15,219,000 next year. Out of the total, 96% is expected to come from sales, 2% is expected to come from trading profits, and 2% is expected to come from other sources. How many dollars has Downey budgeted in expected revenues in each category? **$14,610,240; $304,380; $304,380**

Warm-Up Exercises

1. What is 34% of $375,680? $127,731.20
2. What is 24.6% of $375,680? $92,417.28
3. $127,731.20 is what percent of $375,680? 34%
4. $92,417.28 is what percent of $375,680? 24.6%
5. 34% + 24.6% + 18.6% + 22.8% 100%

PRACTICE AND APPLY

The following problems can be assigned for classwork and the answers checked in class to help students master the objective of the lesson.

- Guided Practice: 1–8
- Independent Practice: 9–13 odd

ALTERNATIVE STRATEGIES: Enrichment

Engage the class in a discussion of the topics presented in this unit. Focus on how the inflation rate, GNP, CPI, and budgets are used as tools by businesses to understand the state of the nation's economy and to make business decisions.

WRAP-UP

Have students suggest different numbers for the Example. Then have them solve the problem using the new numbers. They should follow the exact steps given in the book.

Assignment Guide

- Basic: 9–16, 18–23
- Average: 10–14 even, 15–17, 18–22 even

14. The Downey Corporation from problem 13 had actual revenues of $14,700,000 from sales; $240,000 from trading profits; and $120,000 from other sources. Did the company reach its revenue goals?
No, −$159,900

15. The Mason Plastics Company wanted to earn $1,500,000 in revenues for the year. The company expected to earn its revenues by the following sources: 85% of the total from sales, 10% from services, and 5% from return on investments. At the end of the year, budget analysis was sent to each manager showing the budgeted amounts and actual amounts of each area. Sales showed revenues of $1,590,000; services showed $120,000; and return on investments showed $34,000. Did the company reach its goal in revenues? **Yes, +$244,000**

16. The Posy Fruit Farm had planned earnings of $680,000 for the year. The farm expected its revenues to come from the following sources: 60% from apple sales, 20% from strawberry sales, 15% from blueberry sales, and 5% from cherry sales. At the end of the season, budget figures showed sales revenues of: apples, $420,000; strawberries, $91,320; blueberries, $91,340; and cherries, $39,900. Did the Posy Fruit Farm reach its goal in revenues? **No, −$37,440**

Critical Thinking . . .

17. Suppose you have a small lawn service and nursery business. Your expected revenues of $45,000 are to come from the following sources: 60% from lawn care, 20% from nursery stock sales, 10% from fall cleanup, 8% from landscaping supplies, and 2% from other sales. At the end of the year, your budget analysis indicated this income data: $24,500 from lawn care, $8300 from nursery stock sales, $4100 from fall cleanup, $2400 from landscaping supplies, and $1200 from other sales. Determine the amount budgeted for each source and find the difference between the actual amount and budget allocation.

Budget	Difference
$27,000	−$2500
$ 9,000	−$ 700
$ 4,500	−$ 400
$ 3,600	−$1200
$ 900	+$ 300
$45,000	−$4500

MAINTAINING YOUR SKILLS Look up the skill in parentheses if you need help or more practice.

Find the percentage. Round answers to the nearest cent. **(Skill 30)**

18. 78% of $348,970 **$272,196.60**

19. 43% of $7,385,450 **$3,175,743.50**

20. 4.6% of $23,684,780 **$1,089,499.88**

21. 14.7% of $836,712 **$122,996.66**

22. 83.7% of $3,769,628 **$3,155,178.64**

23. 1.6% of $598,684 **$9578.94**

Reviewing the Basics

Skills

Round to the nearest cent.

(Skill 2) **1.** $48.9858 **$48.99** **2.** $0.04179 **$0.04** **3.** $2841.9691 **$2841.97** **4.** $1.625 **$1.63**

Round to the nearest tenth percent.

5. 24.97% **25.0%** **6.** 8.7801% **8.8%** **7.** 17.405% **17.4%** **8.** 7.65% **7.7%**

Solve

(Skill 5) **9.** $3.95 + $2.49 **$6.44** **10.** $14.97 + $29.79 + $134.99 **$179.75**

(Skill 6) **11.** $212.95 − $168.75 **$44.20** **12.** $147,875.50 − $14,787.55 **$133,087.95**

(Skill 8) **13.** 1.124 × $475.75 **$534.74** **14.** 1.065 × $9568.85 **$10,190.83**

(Skill 11) **15.** $24.79 ÷ 1.045 **$23.72** **16.** $147,690 ÷ 1.105 **$133,656.11**

Find the percentage. Round answers to the nearest cent.

(Skill 30) **17.** 9.8% of $378.45 **$37.09** **18.** 15.48% of $37.79 **$5.85**

19. 109.6% of $24.95 **$27.35** **20.** 112.5% of $32,759.85 **$36,854.83**

Find the rate. Round answers to the nearest tenth percent.

(Skill 31) **21.** n% of $4.50 = $.13 **2.9%** **22.** n% of $379.89 = $402.64 **106.0%**

Find the base. Round answers to the nearest cent.

(Skill 32) **23.** 105.5% of n = $54.98 **$52.11** **24.** 5.3% of n = $12.85 **$242.45**

Applications

Find the amount withheld for social security.

(Application C)

25. $16,890 in 1986. **$1207.64**

26. $2186 in 1941. **$21.86**

27. $37,490 in 1990. **$2867.99**

28. $17,852 in 1981. **$1187.16**

Years	Tax Rate
1937–49	1.00%
1981	6.65%
1982–83	6.70%
1984	7.00%
1985	7.05%
1986–87	7.15%
1988–89	7.51%
1990–97	7.65%

Terms

Write your own definintion for each term. **Answers will vary.**

29. Inflation **30.** Inflation rate

31. Gross national product **32.** Real GNP

33. Per Capita GNP **34.** Consumer Price Index

35. Budget

Refer to your reference files in the back of the book if you need help.

LESSON PLAN
Reviewing the Basics

The exercises on this page review skills, applications, and terms used in the unit. You can use the exercises to assess informally students' proficiency with this material.

The page can be used for guided practice and independent practice. You can work through a selection of the exercises together with students, and thus see immediately if they know how to do them, and you can then assign some of the exercises for independent practice. Be sure to go over the answers to all assigned exercises.

LESSON PLAN
Unit Test

Students should do the Unit Test on their own. Each problem on the test is keyed to a lesson in the unit. Students having difficulty with any particular problem should review the Example in the appropriate lesson and be assigned some of the Independent Practice problems for additional practice.

Unit Test

Lesson 22-1

1. Last year a 2-liter bottle of grapefruit juice cost $0.99. Today it costs $1.09. What is the inflation rate? **10.1%**

Lesson 22-1

2. Ten years ago, a ballpoint pen cost $14.95. The inflation rate for the pen since that time is 16.8%. What does that pen cost today? **$17.46**

Lesson 22-1

3. The inflation rate for groceries over the past year has been 5.7%. What would you have paid a year ago for a bag of groceries that cost $34.80 today? **$32.92**

Lesson 22-2

4. The nation of Angola has a GNP of $4.7 billion and a population of 8,802,000. The inflation rate for Angola is 8.4%.
   a. What is the real GNP? **$4.3052 billion**
   b. What is the per capita GNP? **$533.97**

Lesson 22-3

5. The CPI for recreational services such as theater tickets is 180. Theater tickets for a Broadway show in New York City, which cost $52 today, would have cost how much in 1967? **$28.89**

Lesson 22-3

6. A new automobile purchased by the Jones family in 1967 for $8780 would cost $27,455 today. What is the consumer price index for automobiles? **312.7**

Lesson 22-3

7. The CPI for new homes is 243.5. A new home costing $54,800 in 1967 would cost how much today? **$133,438**

Lesson 22-4

8. The Able Manufacturing Company has budgeted total revenues of $3,800,000. Able allocates its annual budget expenditures according to the following percents:

		Allocated	Difference
Salaries	78.6%	$2,986,800	+$13,200
Research	16.4%	$623,200	+$19,800
Equipment	2.8%	$106,400	+$38,600
Supplies	2.2%	$83,600	−$8700

End-of-the-year data showed these actual expenditures: salaries, $3,000,000; research, $643,000; equipment, $145,000; and supplies, $74,900.
   a. How much is allocated to each of the categories?
   b. Find the difference between the actual amount and the budget allocation.

A SPREADSHEET APPLICATION

Corporate Planning

To complete this spreadsheet application, you will need the template diskette for *Mathematics with Business Applications.* Follow the directions in the *User's Guide* to complete this activity.

Input information in the following problems to find the original price, current price, inflation rate, real GNP, and CPI; and then, answer the questions that follow.

	Original Price	Current Price	Inflation Rate	Gross National Product (GNP)	Real GNP	Consumer Price Index (CPI)*
a.	$ 18.75	$ 21.45	?	$45,800,000,000	?	?
b.	$ 219.97	$299.99	?	$ 235.7 billion	?	?
c.	$ 47.79	?	6.2%	$ 562,000,000	?	?
d.	$8,750.00	?	10.4%	$ 73.9 million	?	?
e.	?	$ 4.75	7.2%	$ 7,650,000,000	?	?
f.	?	$847.50	1.6%	$1,750.6 million	?	?
g.	$ 649.98	?	?	$ 2,625.8 billion	?	215.6
h.	$ 98.98	?	?	$ 95,750,000	?	168.4
i.	?	$ 75.75	?	$ 1.4 billion	?	345.2
j.	?	$549.99	?	$97,748,251,000	?	116.4

***To work with the CPI, assume the original price occurred in 1967.**

1. What is the inflation rate in part **a**?

2. What is the real GNP in part **a**?

3. What is the CPI in part **a**?

4. What is the current price in part **c**?

5. What is the real GNP in part **c**?

6. What is the CPI in part **c**?

7. What is the original price in part **e**?

LESSON PLAN A Spreadsheet Application

USING TECHNOLOGY

Organize students into small groups to work together on this application. Point out to students that many large corporations employ economists and other financial experts in corporate planning departments to analyze the national business environment in order to make financial decisions. A key factor for every business is to price its products competitively but yet make a profit. Inflation rates need to be considered in pricing because the costs of materials or services have an important bearing on price.

Review the answers to all the questions. In particular, discuss Questions 18 through 21.

A SPREADSHEET APPLICATION
(CONTINUED)

8. What is the real GNP in part **e**?

9. What is the CPI in part **e**?

10. What is the current price in part **g**?

11. What is the inflation rate in part **g**?

12. What is the real GNP in part **g**?

13. What is the original price in part **i**?

14. What is the inflation rate in part **i**?

15. What is the real GNP in part **i**?

16. Explain the relationship between the inflation rate and the consumer price index. Knowing one, how can you find the other?

17. If the current price is less than the original price:
   a. What will be true about the inflation rate?
   b. What will be true about the CPI?
   c. Can the CPI ever be negative? Why or why not?
   d. How is the GNP related to the real GNP?

18. If the current price is equal to the original price:
   a. What is the inflation rate?
   b. What is the CPI?
   c. How is the GNP related to the real GNP?

19. If the inflation rate is greater than 100%:
   a. What happens to the real GNP? Is that possible?
   b. What happens to the CPI?

Cumulative Review

Skills

Round answers to the nearest dollar.

(Skill 2)

1. \$178.51 **\$179** 2. \$6.444 **\$6** 3. \$4218.17 **\$4218** 4. \$14.99 **\$15**

Solve. Round answers to the nearest cent.

(Skill 3)

5. 48 + 87 **135** 6. 724 + 57 **781** 7. \$694 + \$375 **\$1069**
8. \$2186 + \$3125 **\$5311** 9. \$85 + \$144 **\$229** 10. \$1596 + \$15,895 + \$48 **\$17,539**

(Skill 4)

11. \$85 − \$36 **\$49** 12. \$337 − \$98 **\$239** 13. \$400 − \$216 **\$184**
14. \$1345 − \$1155 **\$190** 15. \$8740 − \$984 **\$7756** 16. \$1946 − \$1473 **\$473**

(Skill 5)

17. \$14.95 + \$4.75 **\$19.70** 18. \$1.19 + \$0.45 **\$1.64** 19. \$22.25 + \$62.87 **\$85.12**
20. \$1500 + \$19.56 + \$3.52 **\$1523.08** 21. \$3187.48 + \$751.94 + \$100.53 **\$4039.95**

(Skill 6)

22. \$30.39 − \$17.91 **\$12.48** 23. \$500.99 − \$46.84 **\$454.15**
24. \$1543.60 − \$9.56 **\$1534.04** 25. \$531.25 − \$18.79 **\$512.46** 26. \$1327.81 − \$871.93 **\$455.88**

(Skill 7)

27. \$43 × 8 **\$344** 28. \$51 × 6 **\$306** 29. \$1500 × 5 **\$7500**
30. \$579 × 91 **\$52,689** 31. \$585 × 55 **\$32,175** 32. \$3591 × 72 **\$258,552**

(Skill 8)

33. \$140 × 0.06 **\$8.40** 34. \$2.85 × 0.12 **\$0.34** 35. \$26.24 × 0.86 **\$22.57**
36. \$793.02 × 4.70 **\$3727.19** 37. \$42.24 × 75 **\$3168** 38. \$1.59 × 57 **\$90.63**

(Skill 10)

39. \$75 ÷ 5 **\$15** 40. \$427 ÷ 24 **\$17.79** 41. \$5512 ÷ 52 **\$106**
42. \$329 ÷ 184 **\$1.79** 43. \$5275 ÷ 35 **\$150.71** 44. \$43,615 ÷ 360 **\$121.15**

(Skill 11)

45. \$30.65 ÷ 12 **\$2.55** 46. \$479.30 ÷ 24 **\$19.97** 47. \$5191.20 ÷ 365 **\$14.22**
48. \$8921.70 ÷ 4.12 **\$2165.46** 49. \$12,291.60 ÷ 21.7 **\$566.43**

(Skill 20)

50. \$85 × $\frac{4}{5}$ **\$68** 51. \$96 × $\frac{3}{8}$ **\$36** 52. \$32 × $\frac{3}{4}$ **\$24**
53. \$40,817 × $\frac{9}{10}$ **\$36,735.30** 54. $\frac{6}{15}$ × \$526 **\$210.40** 55. \$2274 × $\frac{2}{3}$ **\$1516**

Write as a decimal.

(Skill 28)

56. 7.65% **0.0765** 57. 8% **0.08** 58. 10.4% **0.104** 59. 74.35% **0.7435**

Solve. Round answers to the nearest cent.

(Skill 30)

60. 10% of \$475 **\$47.50** 61. 58% of \$856 **\$496.48** 62. 40% of \$8748 **\$3499.20**
63. $5\frac{1}{4}$% of \$506 **\$26.57** 64. $9\frac{1}{2}$% of \$14,612 **\$1388.14** 65. $3\frac{1}{2}$% of \$415,758 **\$14,551.53**

Solve. Round answers to the nearest tenth of a percent.

(Skill 31)

66. \$35 is what percent of \$84? **41.7%** 67. \$150 is what percent of \$315? **47.6%**
68. \$15,000 is what percent of \$70,000? **21.4%**
69. \$43,470 is what percent of \$210,000? **20.7%**

LESSON PLAN
Cumulative Review
(Skills/Applications)
Units 17–22

The exercises on this page review skills, applications, and terms used in Units 17–22. You can use the exercises to assess informally students' proficiency with this material.

These pages can be used for guided practice and independent practice. You can work through a selection of the exercises together with students, and thus see immediately if they know how to do them. You can then assign some of the exercises for independent practice. Be sure to go over the answers to all assigned exercises.

Applications

Find the tax bracket.

(Application C)

70. Taxable income is $45,690. **$0–$50,000**

71. Taxable income is $114,750. **Over $100,000**

72. Taxable income is $91,495. **$75,000–$100,000**

73. Taxable income is $18,710. **$0–$50,000**

Taxable Income		Your tax is:			
Over—	**But not over—**				**Of the amount over—**
0	$ 50,000			15%	0
$ 50,000	75,000	$ 7,500	+	25%	$ 50,000
75,000	100,000	13,250	+	34%	75,000
100,000	—	21,750	+	39%	100,000

Find the charge per 100 pounds.

$1002.23 **74.** Weight: 2490 pounds
Distance: 623 miles

$3466.20 **75.** Weight: 6540 pounds
Distance: 1300 miles

BASIC RATES PER 100 POUNDS		
Weight Group (in pounds)	**Distance (in miles)**	
	500-1000	**1001-1500**
2001-5000	$40.25	$66.20
5001-10,000	38.00	53.00

Find the average cost to the nearest cent.

(Application Q)

76. 2 units at $378 each and 3 units at $398 each. **$390**

77. 30 units at $2.35 each and 21 units at $2.25 each. **$2.31**

78. 75 units at $3.12 each and 14 units at $2.98 each. **$3.10**

79. 423 units at $0.56 each and 180 units at $0.60 each. **$0.57**

Find the area.

(Application X)

80. 29 m by 12 m **348 m²**

81. 15 ft by 12 ft **180 ft²**

82. 580 ft by 75 ft **43,500 ft²**

83. 727 in by 60 in **43,620 in²**

84. 7.5 ft by 14.5 ft **108.75 ft²**

85. 110 cm by 15 cm **1650 cm²**

Find the volume.

(Application Z)

86. 7 ft by 3 ft by 10 ft **210 ft³**

87. 2.7 m by 8 m by 11.4 m **246.24 m³**

88. 9 in by $13\frac{1}{2}$ in by 6 in **729 in³**

89. $1\frac{3}{4}$ ft by 4 ft by $5\frac{1}{2}$ ft **$38\frac{1}{2}$ ft³**

Terms

Write your own definition for each term. **Answers will vary.**

90. Inventory
91. U.S. Treasury Bill
92. Income statement
93. Inventory card
94. Apportion
95. Gross national product
96. Owner's equity
97. Book value
98. Net proceeds
99. Maturity value
100. Balance sheet
101. Consumer price index

Refer to your reference files in the back of the book if you need help.

Cumulative Review Test Units 17–22

Lesson 17-1

1. The Office Equipment Company manufactures some 3-drawer file cabinets. Each cabinet is stored in a box measuring 3.8 feet high, 1.5 feet wide, and 2.5 feet long. How many cubic feet of space does Office Equipment need to store 1200 file cabinets? **17,100 ft^3**

Lesson 17-2

2. How many RunAway models does Clarion Motor Coach have on its lot on October 1? **59**

Item: RunAway
Stock Number: KT 4403 G

Month of	Opening Balance	Receipts	Issues	Inventory at End of Month
August	75	12	8	? 79
September	79	15	35	? 59
October	?			

Lesson 17-3

3. In problem 2, the 75 in the opening balance were valued at $12,000 each, the 12 received in August were valued at $12,400 each, and the 15 received in September were valued at $12,600 each. Using the average-cost method, what is the value of the inventory? **$715,982.11**

Lesson 18-2

4. Motel Seven hired 4 high school students to clean its swimming pool. Each student worked for 2 hours at an hourly rate of $7.45. The materials charge was $43.70. What was the total charge for this service? **$103.30**

Lesson 18-5

5. The Huston Corporation used 15,400 kilowatt-hours of electricity with a peak load of 120 kilowatts in April. The demand charge is $6.40 per kilowatt. The energy charge per kilowatt-hour is $0.08 for the first 10,000 kilowatt-hours and $0.06 for all kilowatt-hours over 10,000. The fuel adjustment charge is $0.04 per kilowatt-hour. What is the total cost of electricity for the Huston Corporation for April? **$2508**

Lesson 19-1

6. Complete the payroll register. Use the social security tax rate of 6.2% and the medicare tax rate of 1.45%.

Name	Gross Income	FIT	Soc. Sec	Medicare	SIT	Total Ded.	Net Pay
Bisset	$300.00	$41.00	? $18.60	? $4.35	$4.50	? $68.45	? $231.55
Heller	435.70	64.00	? $27.01	? $6.32	6.54	? $103.87	? $331.83
Molnar	405.00	45.00	? $25.11	? $5.87	6.08	? $82.06	? $322.94
Stein	425.65	47.00	? $26.39	? $6.17	6.38	? $85.94	? $339.71
Total	? $1566.35	? $197	? $97.11	? $22.71	? $23.50	? $340.32	? $1226.03

LESSON PLAN
Cumulative Review Test Units 17–22

Students should do the Cumulative Review Test on their own. Each problem on the test is keyed to a lesson in Units 17–22. Students having difficulty with any particular problem should review the Example in the appropriate lesson and be assigned some of the Independent Practice problems for additional practice.

Lesson 19-3

7. The Foster Machine Company pays $125,000 per year for security. The company apportions the cost among its departments based on space. The research department occupies an area that measures 60 feet by 60 feet. The building contains 1,200,000 square feet. How much does the research department pay for security? **$375**

Lessons 19-4, 19-5

8. Central Copy purchased a new copy machine for $12,400. The machine has an estimated life of 4 years. The resale value is expected to be $1400. Use the straight-line method to find the annual depreciation. Find the book value after each year of use. **Depreciation: $2750 per year; Book value: $9650, $6900, $4150, $1400**

Lessons 20-1

9. The Running Shop had these assets and liabilities on May 31.

Cash: $4500	Unpaid merchandise: $13,470
Inventory: $28,000	Taxes owed: $975
Supplies: $356	Bank loan: $11,600
Store furnishings: $17,860	

What are the total assets? What are the total liabilities? What is the owner's equity? **$50,716; $26,045; $24,671**

Lesson 20-4

10. Globe Manufacturing, Inc., prepares an annual income statement. This past year, Globe had net sales of $800,000. The cost of goods sold totaled $375,750. Operating expenses included: wages and salaries totaling $165,970; depreciation and amortization totaling $19,590; interest paid totaling $15,540; product recall totaling $3596; and miscellaneous operating expenses of $6570. Prepare an income statement for Globe Manufacturing. **Gross profit: $424,250; Total operating expenses: $211,266; Net income: $212,984**

Lesson 21-2

11. Solar Energy, Inc., is issuing 500,000 shares of stock. Each share will be sold at $45. The underwriting commission is 3% of the value of the stock. The other expenses are expected to be 0.6% of the value of the stock. If all the shares of stock are sold, what net proceeds will Solar Energy receive? **$21,690,000**

Lesson 21-3

12. Hereford Ranch borrowed $48,500 for 120 days at the prime rate of interest. When Hereford Ranch borrowed the money, the prime rate was 10.8%. The bank charged ordinary interest at exact time. What was the maturity value of the loan? **$50,246**

Lesson 21-4

13. Carlton Manufacturing had $80,000 in surplus cash. The financial manager decided to invest in a United States Treasury Bill with a face value of $80,000. The bill matured in 126 days. The interest rate was 9.8% ordinary interest at exact time. The bank service fee was $25. What was the cost of the Treasury Bill? **$77,281**

Lesson 22-3

14. The CPI for clothing is 186.3. A sweater that cost $24.95 in 1967 would cost how much today? **$46.48**

A SIMULATION

Copyfax Center

You are the manager of Copyfax Center, a small business that does copying, printing, and word processing. Copyfax Center has 3 employees. Here is a chart of the hours they work each day, Monday through Friday.

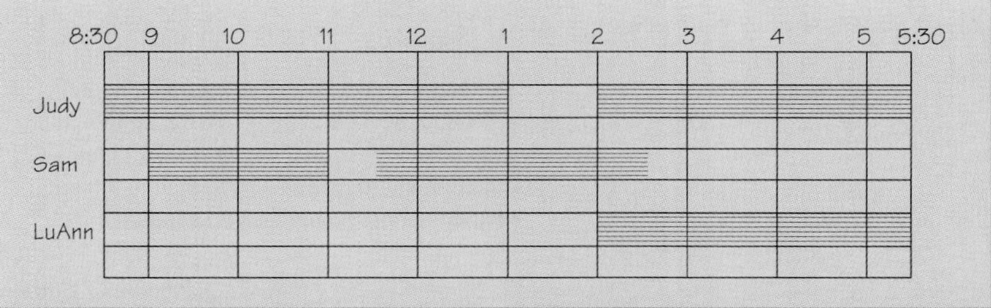

Each week you make up the weekly payroll. Each employee must pay 6.2% for social security, 1.45% for medicare, and 5% for state tax. You pay employees time and a half for overtime.

Use these work sheets to calculate this week's net pay for each employee. Use the tax tables on pages 642–643 to determine the federal income tax (FIT). This week Judy worked 3 hours overtime. Sam and LuAnn did not work overtime.

	Employee	Hourly Pay	Hours Worked		Gross Pay
			Regular	Overtime	
1.	Judy Olson	$7.60	40	3	$338.20
2.	Sam Borden	$6.50	25	0	$162.50
3.	LuAnn Tozzi	$4.50	17.5	0	$78.75

	Employee	Tax Status	FIT	Soc. Sec.	Medicare	State Tax	Health Ins.	Tot. Ded.	Net Pay
4.	J. Olson	Married 3 allow.	$21.00	$20.97	$4.90	$16.91	$8.50	$72.28	$265.92
5.	S. Borden	Single 2 allow.	$8.00	$10.08	$2.36	$8.13	$6.50	$35.07	$127.43
6.	L. Tozzi	Single 1 allow.	$2.00	$4.88	$1.14	$3.94	0	$11.96	$66.79

LESSON PLAN
A Simulation

USING THE SIMULATION

Students can work the simulation individually, or you may wish to use it with *cooperative learning groups.* Each student should have copies of the necessary forms. You can begin by reading with students the introductory material at the top of the page.

Check students work as they complete the forms and provide help as needed. After students have finished the simulation, provide them with completed forms so they can check their work.

A SIMULATION
(CONTINUED)

Pricing

The price that Copyfax Center charges for copies depends on the number of copies made. This chart shows the current prices.

Number of Copies	Price per Page*
1–24	$0.15
25–49	$0.135
50–99	$0.13
100–199	$0.125
200–499	$0.12
over 500	$0.115

*plus 6% sales tax

Regular customers get discounts. A few receive a trade discount of 20%. Others get a cash discount of 2% if they pay within 10 days. Both discounts are deducted before the sales tax is calculated.

Use these invoices to calculate the prices charged these customers.

7.

COPYFAX CENTER Invoice no. **2257** Date *June 15, 19—*

Customer *Christine's Hardware*

	Number of copies	800
a.	Price per page	$ 0.115
b.	Total price	$ 92.00
c.	Discount: *20* % ☒ trade ☐ cash	$ 18.40
d.	Net price	$ 73.60
e.	Sales tax	$ 4.42
f.	Invoice price	$ 78.02

8.

COPYFAX CENTER Invoice no. **2289** Date *June 17, 19—*

Customer *Mike's Travel Service*

	Number of copies	150
a.	Price per page	$ 0.125
b.	Total price	$ 18.75
c.	Discount: *2* % ☐ trade ☒ cash	$ 0.38
d.	Net price	$ 18.37
e.	Sales tax	$ 1.10
f.	Invoice price	$ 19.47

9. What is the price per page?
$0.097525

10. What is the price per page?
$0.1298

A SIMULATION
(CONTINUED)

Depreciation

Depreciation of Copyfax Center's equipment is considered a business expense.

Use this work sheet to calculate the depreciation of these items, using the straight-line method.

Item	11. High-Speed Copier	12. Word Processor	13. Computer Typesetter	14. Delivery Van
Year Purchased	1992	1989	1990	1992
Original Cost	$18,500	$4500	$9750	$14,248
Salvage Value	$ 1,000	$ 500	$ 750	$ 1,648
Total Depreciation	$17,500	$4000	$9000	$12,600
Estimated Life	5	5	6	6
Annual Depreciation	$3500	$800	$1500	$2100

As part of your long-range planning, you decide to calculate how much each item will be worth each year.

Use this work sheet to calculate the accumulated depreciation and the book value (B/V) of each item listed. A zero has been put in for any years that do not apply.

	15. High-Speed Copier		16. Word Processor		17. Computer Typesetter		18. Delivery Van	
Year	**A/D**	**B/V**	**A/D**	**B/V**	**A/D**	**B/V**	**A/D**	**B/V**
1989	0	0	$800	$3700	0	0	0	0
1990	0	0	$1600	$2900	$1500	$8250	0	0
1991	0	0	$2400	$2100	$3000	$6750	0	0
1992	$3500	$15,000	$3200	$1300	$4500	$5250	$2100	$12,148
1993	$7000	$11,500	$4000	$500	$6000	$3750	$4200	$10,048
1994	$10,500	$8000	0	0	$7500	$2250	$6300	$7948
1995	$14,000	$4500	0	0	$9000	$750	$8400	$5848
1996	$17,500	$1000	0	0	0	0	$10,500	$3748
1997	0	0	0	0	0	0	$12,600	$1648

A SIMULATION
(CONTINUED)

Balance Sheet

Each month you prepare a balance sheet. The balance sheet lists Copyfax Center's assets (what it owns), its liabilities (what it owes), and the owner's equity (assets minus liabilities).

You start by listing your assets and liabilities in a running account:

Copyfax Center	December 31, 19—
Cash on hand	$ 3000
Accounts receivable	$ 3500
Accounts payable	$10,500
Equipment (less accumulated depreciation)	$25,798
Inventory (paper, ink, etc.)	$ 2100
Prepaid insurance	$ 1800
Office supplies	$ 520
Taxes owed	$ 1500
Notes payable	$ 9500

19. Use this form to complete Copyfax Center's balance sheet. Enter the assets and liabilities from your running account.

COPYFAX CENTER
Balance Sheet Dec. 31, 19—

	Assets		Liabilities		
a.	Cash	$3000	Accounts payable	$ $10,500	h.
b.	Accounts receivable	$3500	Notes payable	$9500	i.
c.	Equipment	$25,798	Taxes owed	$1500	j.
d.	Inventory	$2100			k.
e.	Prepaid insurance	$1800	Total Liabilities	$ $21,500	l.
f.	Office supplies	$520	Owner's Equity	$ $15,218	m.
g.	Total Assets	$36,718	Total Liabilities and Equity	$ $36,718	n.

A SIMULATION (CONTINUED)

Income Statement

Each month you also prepare Copyfax Center's income statement. Here are the sales and expenses you have recorded during the month.

Copyfax Center			December 19—
Wages	$2510	Insurance	$ 600
Advertising	$ 135	Taxes	$ 500
Delivery	$ 160	Depreciation	$ 660
Postage	$ 300	Supplies	$ 170
Rent	$ 750	Gross sales	$8002
Utilities	$ 325	Sales discounts	$ 880

20. Use this form to prepare an income statement.

	Copyfax Center Income Statement for Month Ended December 31, 19—			
a.	**Income:**	Gross sales	$8002	
b.		Less sales discounts	$880	
c.		Net sales		$7122
d.	**Expenses:**	Wages	$2510	
e.		Advertising	$135	
f.		Delivery	$160	
g.		Postage	$300	
h.		Rent	$750	
i.		Utilities	$325	
j.		Insurance	$600	
k.		Taxes	$500	
l.		Depreciation	$660	
m.		Supplies	$170	
n.		**Total operating expenses**		$6110
o.		**Net Income**		$1012

A SIMULATION (CONTINUED)

Annual Report and Comparative Analysis

At the end of each year, you prepare an annual income statement. You can then compare Copyfax Center's finances for two years, in order to plan for next year.

This chart shows Copyfax Center's income and expenses for last year and this year. Fill in the missing numbers. Then, for each amount, calculate the percent increase from last year to this year.

21.

	Copyfax Center Annual Income Statement		Last Year	This Year	Percent Increase
a.	**Income:**	Gross sales	$89,560	$99,480	11.08%
b.		Less sales discounts	9985	10,750	7.66%
c.		Net sales	79,575	88,730	11.50%
d.	**Expenses:**	Wages	28,650	30,075	4.97%
e.		Advertising	1350	1550	14.81%
f.		Delivery	1520	1840	21.05%
g.		Postage	3260	3758	15.28%
h.		Rent	8400	8800	4.76%
i.		Utilities	3315	3900	17.65%
j.		Insurance	4100	4200	2.44%
k.		Taxes	4456	4826	8.30%
l.		Depreciation	2100	7900	276.19%
m.		Supplies	2136	2254	5.52%
n.		**Total operating expenses**	$59,287	$69,103	16.56%
o.		**Net Income**	$20,288	$19,627	−3.26%

Complete the following statements comparing last year's and this year's figures.

22. The expense with the largest percent increase was depreciation.

23. The next largest was delivery.

24. Net sales increased by 11.5 %.

25. Total operating expenses increased by 16.56 %.

26. Net income changed by −3.26 %.

The Bottom Line

What would you do differently next year to improve Copyfax Center's financial picture? Increase sales, decrease expenses

YOUR REFERENCE FILES

Skills

Thirty-two computational skills, from whole number and decimal operations through rates and percents, provide the basics you will need as a consumer and future business person.

Applications

Twenty-six mathematical applications, covering money, time, measurement, graphs, and more, provide the tools you will need for consumer and business use.

Terms

Key terms with their definitions from all lessons provide the background you will need in consumer and business situations.

579

SKILL

Numbers

Place Value

In 4532.869, give the place and value of each digit.

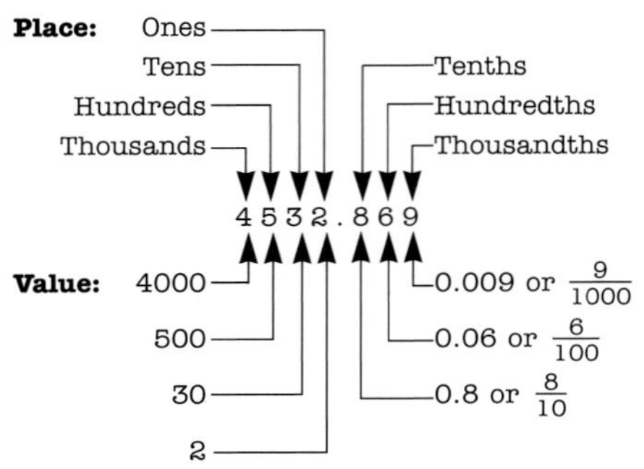

Give the place and value of the underlined digit. **See Selected Answers.**

1. $6\underline{5}$	**2.** $9\underline{6}$	**3.** $4\underline{7}2$
4. $\underline{2}3$	**5.** $10\underline{8}$	**6.** $\underline{5}36$
7. $15\underline{0}6$	**8.** $2\underline{8}21$	**9.** $7\underline{7}84$
10. $\underline{1}492$	**11.** $100\underline{9}$	**12.** $1\underline{9}26$
13. $0.\underline{3}7$	**14.** $1.6\underline{1}$	**15.** $2.73\underline{9}$
16. $\underline{6}.3$	**17.** $9.4\underline{2}$	**18.** $4.3\underline{7}$
19. $2\underline{4}.04$	**20.** $37.\underline{3}29$	**21.** $1.8\underline{2}4$
22. $\underline{4}93.89$	**23.** $90.25\underline{7}$	**24.** $23.\underline{5}72$
25. $12.76\underline{3}$	**26.** $0.\underline{0}78$	**27.** $5.46\underline{1}$

Numbers as Words

Write each number in words.

Number	**Words**
105	One hundred five
26	Twenty-six
17	Seventeen
\$98.09	Ninety-eight and $\frac{9}{100}$ dollars
\$33.13	Thirty-three and $\frac{13}{100}$ dollars

Write in words. **See Selected Answers.**

28. 18	**29.** 34	**30.** 159	**31.** 78
32. 103	**33.** 842	**34.** 207	**35.** 5012
36. 6005	**37.** 119	**38.** 72	**39.** 1240
40. 5102	**41.** 194	**42.** 6590	**43.** \$25.00
44. \$6.24	**45.** \$17.09	**46.** \$112.35	**47.** \$120.17
48. \$4.37	**49.** \$7749	**50.** \$65.90	**51.** \$0.50

Numbers as Decimals

Rewrite these large numbers as decimals.

\$14,700,000 → \$14.7 million

\$245,600 → \$245.6 thousand

Rewrite these decimals as numbers.

\$1.2 billion = \$1,200,000,000

\$43.6 thousand = \$43,600

Write in millions. **See Selected Answers.**

52. \$1,700,000 **53.** \$71,640,000 **54.** \$618,700,000

Write in thousands.

55. 18,400 **56.** \$9,640 **57.** 171,600

Write in billions.

58. \$16,450,000,000 **59.** 2,135,000,000

60. \$171,200,000,000

Write completely in numbers.

61. \$3.4 million **62.** 16.2 thousand **63.** \$11.2 billion

64. 17.2 million **65.** 7.3 billion **66.** \$74.21 thousand

67. 0.4 million **68.** \$0.5 thousand **69.** \$0.72 billion

Comparing Whole Numbers

Which number is greater: 5428 or 5431?

5428 = 5000 + 400 + 20 + 8

5431 = 5000 + 400 + 30 + 1

Same Same 30 is greater than 20.

So 5431 is greater than 5428.

Which number is greater?

70. 23 or 32

71. 54 or 45

72. 459 or 462

73. 741 or 835

74. 810 or 735

75. 125 or 211

76. 3450 or 6450

77. 5763 or 925

78. 1000 or 999

79. 444 or 4444

80. 3002 or 4000

81. 1236 or 820

82. 150 or 149

83. 493 or 650

84. $2000 or $1997

85. $101 or $99

86. $482 or $600

87. $39 or $93

88. $1686 or $1668

89. $568 or $742

90. $86,432 or $101,000

91. $791,000 or $768,000.

Comparing Decimals

Which number is greater: 24.93 or 24.86?

24.93 = 20 + 4 + 0.9 + 0.03

24.86 = 20 + 4 + 0.8 + 0.06

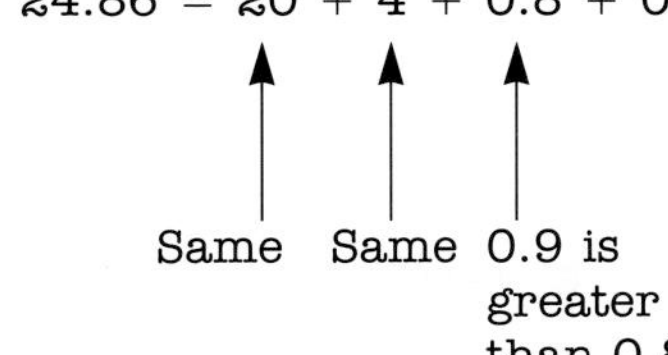
Same
Same
0.9 is greater than 0.8.

So 24.93 is greater than 24.86

Which number is greater?

92. 3.1 or 1.3

93. 1.2 or 2.0

94. 4.50 or 4.05

95. 25.1 or 20.8

96. 18.43 or 17.88

97. 56.84 or 58

98. 0.4 or 0.6

99. 0.01 or 0.1

100. 0.82 or 0.28

101. 0.5 or 0.06

102. 8.739 or 10

103. 0.002 or 0.020

104. $5.99 or $5.00

105. $10 or $8.50

106. $23.85 or $19.84

107. $11.19 or $19

108. $6.98 or $7.50

109. $83.59 or $600

110. $4327.75 or $6297.86

111. $8391.34 or $9521.39

112. $4640.66 or $4646.40

113. $2000.00 or $1997.98

SKILL 2

Round Numbers

Whole Numbers

Round 7863 to the nearest hundred.

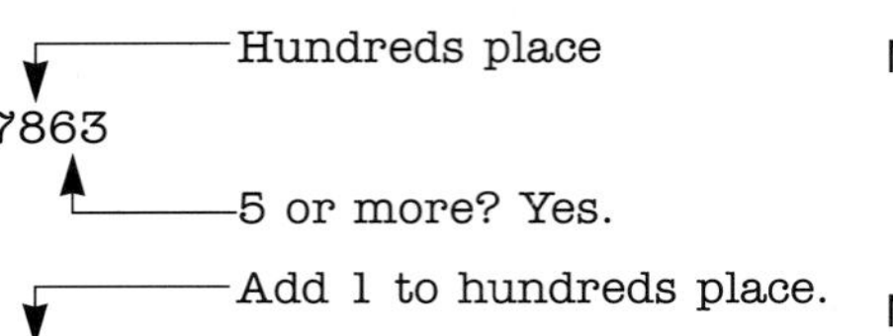

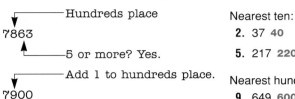

Add 1 to hundreds place.

7900

Change to zeros.

Round answers to the place value shown.

Nearest ten: **1.** 26 **30**

2. 37 **40** **3.** 68 **70** **4.** 195 **200**

5. 217 **220** **6.** 302 **300** **7.** 8099 **8100**

Nearest hundred: **8.** 119 **100**

9. 649 **600** **10.** 2175 **2200** **11.** 6042 **6000**

Nearest thousand: **12.** 7423 **7000**

13. 15,602 **16,000** **14.** 22,094 **22,000** **15.** 750 **1000**

Decimals

Round 0.6843 to the nearest thousandth.

Thousandths place

0.6843

5 or more? No.

Do not change.

0.684

Drop the final digit.

Round answers to the place value shown.

Nearest tenth: **16.** 0.63 **0.6**

17. 0.091 **0.1** **18.** 0.407 **0.4** **19.** 0.452 **0.5**

Nearest hundredth: **20.** 0.652 **0.65**

21. 0.474 **0.47** **22.** 0.168 **0.17** **23.** 0.355 **0.36**

Nearest thousandth: **24.** 0.4291 **0.429**

25. 0.6007 **0.601** **26.** 0.0097 **0.010** **27.** 0.2126 **0.213**

28. 6.3942 **6.394** **29.** 137.4920 **137.492** **30.** 9.9999 **10.000**

Mixed Practice

Round answers to the place value shown.

Nearest thousand:	**31.** 37,874 **38,000**	**32.** 19,266 **19,000**	**33.** 48,092 **48,000**
Nearest hundred:	**34.** 751 **800**	**35.** 919 **900**	**36.** 6771 **6800**
Nearest ten:	**37.** 26 **30**	**38.** 6533 **6530**	**39.** 575 **580**
Nearest one:	**40.** 6.2 **6**	**41.** 35.73 **36**	**42.** 17.392 **17**
Nearest tenth:	**43.** 189.673 **189.7**	**44.** 10.009 **10**	**45.** 0.07 **0.1**
Nearest hundredth:	**46.** 0.392 **0.39**	**47.** 152.430 **152.43**	**48.** 0.6974 **0.70**
Nearest thousandth:	**49.** 0.1791 **0.179**	**50.** 16.0005 **16.001**	**51.** 108.437 **108.437**

SKILL 3

Add Whole Numbers

Without Carrying

Add.

723	➧	723
154		154
+212		+212
		1089

1. 65 + 41 = 106
2. 76 + 32 = 108
3. 97 + 40 = 137
4. 32 + 25 = 57
5. 352 + 837 = 1189
6. 361 + 834 = 1195
7. 448 + 351 = 799
8. 125 + 604 = 729
9. 864 + 33 = 897
10. 721 + 77 = 798
11. 423 + 65 = 488
12. 108 + 91 = 199
13. 9037 + 1841 = 10,878
14. 9520 + 1379 = 10,899
15. 3924 + 5063 = 8987
16. 2840 + 1152 = 3992

With Carrying

Add.

		1 11
8679	➧	8 679
+9748		+9 748
		18,427

17. 32 + 39 = 71
18. 54 + 48 = 102
19. 187 + 23 = 210
20. 49 + 86 = 135
21. 728 + 169 = 897
22. 527 + 284 = 811
23. 845 + 697 = 1542
24. 697 + 546 = 1243
25. 3046 + 1592 = 4638
26. 7801 + 3564 = 11,365
27. 5246 + 6978 = 12,224
28. 8347 + 1528 = 9875
29. 8448 + 3753 = 12,201
30. 108 + 7665 = 7773
31. 9179 + 3608 = 12,787
32. 982 + 2165 = 3147

Mixed Practice

33. 1481 + 2317 = 3798
34. 8495 + 1417 = 9912
35. 5783 + 6535 = 12,318
36. 3950 + 1615 = 5565
37. 6259 + 1893 = 8152
38. 8347 + 1528 = 9875
39. 6845 + 2639 = 9484
40. 5692 + 1204 = 6896
41. 2642 + 4135 = 6777
42. 7921 + 2639 = 10,560
43. 7884 + 7069 = 14,953
44. 46,234 + 11,325 = 57,559
45. 17,694 + 15,893 = 33,587
46. 37,491 + 21,308 = 58,799
47. 59,641 + 27,840 = 87,481
48. 9100 + 536 + 2413 = 12,049
49. 7749 + 1240 + 6010 = 14,999
50. 6590 + 2408 + 5001 = 13,999
51. 5783 + 6535 + 2132 = 14,450
52. 6259 + 503 + 1893 = 8655

SKILL

Subtract Whole Numbers

Without Borrowing

Subtract.

$$\begin{array}{r} 9876 \\ -\,7545 \\ \hline \end{array} \quad \blacktriangleright \quad \begin{array}{r} 9876 \\ -\,7545 \\ \hline 2331 \end{array}$$

1. 784 − 453 **331**	**2.** 985 − 734 **251**	**3.** 693 − 542 **151**	**4.** 199 − 158 **41**
5. 7659 − 4217 **3442**	**6.** 8436 − 6223 **2213**	**7.** 5792 − 2481 **3311**	**8.** 4877 − 3614 **1263**
9. 6754 − 5643 **1111**	**10.** 1866 − 853 **1013**	**11.** 8191 − 171 **8020**	**12.** 1187 − 145 **1042**
13. 479 − 473 **6**	**14.** 3987 − 3085 **902**	**15.** 6358 − 127 **6231**	**16.** 1721 − 720 **1001**

With Borrowing

Subtract.

$$\begin{array}{r} 9672 \\ -\,4136 \\ \hline \end{array} \quad \blacktriangleright \quad \begin{array}{r} {\scriptstyle 6\ 12} \\ 96\not{7}\not{2} \\ -\,4136 \\ \hline 5536 \end{array}$$

$$\begin{array}{r} 8352 \\ -\,4584 \\ \hline \end{array} \quad \blacktriangleright \quad \begin{array}{r} {\scriptstyle 7\ 12\ 14\ 12} \\ \not{8}\not{3}\not{5}\not{2} \\ 4584 \\ \hline 3768 \end{array}$$

17. 100 − 36 **64**	**18.** 512 − 43 **469**	**19.** 602 − 503 **99**	**20.** 250 − 162 **88**
21. 6932 − 4674 **2258**	**22.** 8724 − 2932 **5792**	**23.** 4329 − 3163 **1166**	**24.** 9721 − 6842 **2879**
25. 6123 − 4214 **1909**	**26.** 9231 − 6453 **2778**	**27.** 7450 − 3783 **3667**	**28.** 7734 − 5935 **1799**
29. 8121 − 6846 **1275**	**30.** 9000 − 7997 **1003**	**31.** 9107 − 8248 **859**	**32.** 7734 − 5935 **1799**

Mixed Practice

33. 6140 − 3157 **2983**	**34.** 8005 − 6246 **1759**	**35.** 7000 − 5432 **1568**	**36.** 9297 − 9286 **11**	**37.** 9811 − 700 **9111**
38. 9148 − 954 **8194**	**39.** 2625 − 763 **1862**	**40.** 1850 − 975 **875**	**41.** 7469 − 5231 **2238**	**42.** 6342 − 5793 **549**
43. 10,743 − 7,842 **2901**	**44.** 16,947 − 14,523 **2424**	**45.** 22,493 − 5,967 **16,526**	**46.** 64,654 − 57,312 **7342**	**47.** 79,850 − 42,347 **37,503**
48. 172,493 − 67,254 **105,239**	**49.** 249,657 − 123,254 **126,403**	**50.** 300,692 − 147,593 **153,099**	**51.** 647,593 − 546,972 **100,621**	**52.** 800,000 − 627,351 **172,649**

SKILL 5

Add Decimals

Same Number of Places

Add.

178.79 596.24 +631.43	➧	178.79 596.24 +631.43 1406.46

1. 317.83 +161.16 **478.99**	**2.** 821.76 +178.23 **999.99**	**3.** 504.76 +296.36 **801.12**
4. 536.67 +197.47 **734.14**	**5.** 189.71 +601.48 **791.19**	**6.** 27.492 +31.608 **59.100**
7. 148.810 221.097 +173.206 **543.113**	**8.** 18.009 149.910 +251.295 **419.214**	**9.** 272.005 42.111 +502.543 **816.659**

Different Number of Places

Add.

0.91 6.647 5.3 +16	➧	0.910 6.647 5.300 +16.000 28.857

Placeholders

10. 10 + 6.3 **16.3**	**11.** 42.9 +21.64 **64.54**	**12.** 63.21 +42.327 **105.537**
13. 141.21 4.136 +32.003 **177.349**	**14.** 3.64 4.378 +0.213 **8.231**	**15.** 19.7 0.261 + 1.94 **21.901**
16. 18.09 49.91 +51.295 **119.295**	**17.** 0.161 18.24 + 6.059 **24.460**	**18.** 2.19 0.09 +0.7 **2.98**
19. 16.74 39.1 +26.492 **82.332**	**20.** 1.943 27.21 + 3.691 **32.844**	**21.** 2.5 1.692 +5.93 **10.122**

Mixed Practice

22. 60.148 +16.623 **76.771**	**23.** 271.195 +189.714 **460.909**	**24.** 163.2 + 37.915 **201.115**	**25.** 427.9 + 74.275 **502.175**	**26.** 693.642 +193.871 **887.513**
27. 973.573 81.91 + 9.20 **1064.683**	**28.** 69 27.198 +178.26 **274.458**	**29.** 0.64 0.378 +0.223 **1.241**	**30.** 0.007 0.986 +0.034 **1.027**	**31.** 19.7 0.261 + 1.94 **21.901**
32. 17.92 23.81 42.63 +15.27 **99.63**	**33.** 103.69 27.4 83.621 + 5.9 **220.611**	**34.** 327.219 27.543 111.621 +257.619 **724.002**	**35.** 471.6 2.937 57.11 +241.005 **772.652**	**36.** 327.915 1.002 40.07 +154.6 **523.587**

SKILL 6

Subtract Decimals

Same Number of Places

Subtract.

Example: 597.18 − 392.35 ➧ (regroup: 6 11) 597.18 − 392.35 = 204.83

1. 65.46 − 14.31 **51.15**	2. 48.58 − 15.47 **33.11**	3. 151.02 − 16.13 **134.89**
4. 36.25 − 13.67 **22.58**	5. 87.56 − 82.47 **5.09**	6. 51.634 − 27.849 **23.785**
7. 69.37 − 43.86 **25.51**	8. 89.63 − 7.99 **81.64**	9. 109.46 − 29.78 **79.68**
10. 521.52 − 38.56 **482.96**	11. 321.02 − 117.18 **203.84**	12. 572.24 − 283.35 **288.89**

Different Number of Places

Subtract.

Example: 86.9 − 3.84 ➧ (regroup: 8 10) 86.90 − 3.84 = 83.06

Placeholder

13. 79.6 − 8.75 **70.85**	14. 95.1 − 9.87 **85.23**	15. 100.1 − 15.78 **84.32**
16. 16.8 − 5.91 **10.89**	17. 36 − 16.4 **19.6**	18. 42 − 12.94 **29.06**
19. 17.9 − 9.83 **8.07**	20. 21 − 19.7 **1.3**	21. 67.2 − 9.76 **57.44**
22. 136.1 − 69.542 **66.558**	23. 771.9 − 394.27 **377.63**	24. 4578 − 878.127 **3699.873**

Mixed Practice

25. 87.56 − 82.47 **5.09**	26. 39.27 − 18.38 **20.89**	27. 36.1 − 16.117 **19.983**	28. 4.546 − 2.558 **1.988**	29. 653.05 − 327.19 **325.86**
30. 198.20 − 64.897 **133.303**	31. 854.01 − 649.656 **204.354**	32. 316.07 − 118.29 **197.78**	33. 800.04 − 242.17 **557.87**	34. 985.93 − 99.794 **886.136**
35. 6194.9 − 978.954 **5215.946**	36. 719.3 − 47.832 **671.468**	37. 5.9871 − 4.8693 **1.1178**	38. 17.9328 − 6.2973 **11.6355**	39. 843.002 − 64.973 **778.029**
40. 87.69 − 86.9975 **0.6925**	41. 4.97652 − 1.37846 **3.59806**	42. 3.29131 − 2.19378 **1.09753**	43. 6.962 − 4.21698 **2.74502**	44. 9.7 − 8.65947 **1.04053**

SKILL 7

Multiply Whole Numbers

Without Carrying

Multiply.

```
 442       442
×211  ➧   ×211
           442 ◄—— 1 × 442
         4 420 ◄——10 × 442
        88 400 ◄—200 × 442
        93,262 ◄—211 × 442
```

1. 73 × 21 = **1533**
2. 42 × 22 = **924**
3. 212 × 412 = **87,344**
4. 311 × 232 = **72,152**
5. 321 × 312 = **100,152**
6. 223 × 323 = **72,029**
7. 232 × 333 = **77,256**
8. 7143 × 102 = **728,586**
9. 8643 × 111 = **959,373**

With Carrying

Multiply.

```
 6524       6524
×273  ➧    ×273
           19 572 ◄—— 3 × 6524
          456 680 ◄——70 × 6524
        1 304 800 ◄—200 × 6524
        1,781,052 ◄—273 × 6524
```

10. 61 × 76 = **4636**
11. 78 × 36 = **2808**
12. 437 × 571 = **249,527**
13. 465 × 541 = **251,565**
14. 542 × 168 = **91,056**
15. 8023 × 532 = **4,268,236**
16. 64 × 27 = **1728**
17. 37 × 45 = **1665**
18. 68 × 71 = **4828**
19. 836 × 372 = **310,992**
20. 7501 × 447 = **3,352,947**
21. 5327 × 312 = **1,662,024**

Mixed Practice

22. 480 × 10 = **4800**
23. 230 × 300 = **69,000**
24. 641 × 237 = **151,917**
25. 231 × 122 = **28,182**
26. 122 × 40 = **4880**
27. 510 × 700 = **357,000**
28. 8233 × 2584 = **21,274,072**
29. 6010 × 6000 = **36,060,000**
30. 9000 × 7011 = **63,099,000**
31. 2793 × 1504 = **4,200,672**
32. 19,008 × 8000 = **152,064,000**
33. 8791 × 5000 = **43,955,000**
34. 6743 × 27 = **182,061**
35. 4231 × 253 = **1,070,443**
36. 8427 × 19 = **160,113**
37. 13,010 × 13 = **169,130**
38. 14,231 × 12 = **170,772**
39. 17,822 × 35 = **623,770**
40. 22,300 × 15 = **334,500**
41. 31,942 × 41 = **1,309,622**
42. 27,642 × 321 = **8,873,082**
43. 13,231 × 212 = **2,804,972**
44. 14,402 × 121 = **1,742,642**
45. 49,237 × 321 = **15,805,077**
46. 64,159 × 347 = **22,263,173**

SKILL 8

Multiply Decimals

Decimals Greater Than One

Multiply.

$$\begin{array}{r} 17.45 \\ \times 2.7 \\ \hline \end{array} \rightarrow \begin{array}{rl} 17.45 & \leftarrow 2 \text{ places} \\ \times 2.7 & \leftarrow +1 \text{ place} \\ \hline 47.115 & \leftarrow 3 \text{ places} \end{array}$$

1. 2.5 × 1.8 = 4.5
2. 8.36 × 1.5 = 12.54
3. 10.2 × 8.61 = 87.822
4. 15.36 × 5.3 = 81.408
5. 25.14 × 7.5 = 188.55
6. 19.36 × 7.12 = 137.8432
7. 27.06 × 8.53 = 230.8218
8. 4.367 × 8.5 = 37.1195
9. 5.564 × 7.9 = 43.9556
10. 32.63 × 9.2 = 300.196
11. 31.20 × 9.21 = 287.352
12. 6.715 × 9.03 = 60.63645

Decimals Less Than One

Multiply.

$$\begin{array}{r} 0.08 \\ \times 0.4 \\ \hline \end{array} \rightarrow \begin{array}{rl} 0.08 & 2 \text{ places} \\ \times 0.4 & +1 \text{ place} \\ \hline 0.032 & 3 \text{ places} \end{array}$$

13. 0.144 × 0.7 = 0.1008
14. 0.86 × 0.5 = 0.43
15. 0.96 × 0.1 = 0.096
16. 0.56 × 0.07 = 0.0392
17. 0.73 × 0.8 = 0.584
18. 0.05 × 0.9 = 0.045
19. 0.81 × 0.76 = 0.6156
20. 0.47 × 0.84 = 0.3948
21. 0.63 × 0.09 = 0.0567
22. 0.57 × 0.03 = 0.0171
23. 1.23 × 0.07 = 0.0861
24. 0.01 × 0.05 = 0.0005

Mixed Practice

25. 41.16 × 100 = 4116
26. 0.923 × 0.49 = 0.45227
27. 0.12 × 300 = 36
28. 67.32 × 10 = 673.2
29. 7.243 × 121 = 876.403
30. 557.4 × 100 = 55,740
31. 327.8 × 3.7 = 1212.86
32. 14.923 × 0.76 = 11.34148
33. 1.125 × 100 = 112.5
34. 0.009 × 1000 = 9
35. 2.014 × 40.7 = 81.9698
36. 2.854 × 0.04 = 0.11416
37. 6243.78 × 25.9 = 161,713.902
38. 5.9312 × 5.62 = 33.333344
39. 0.534 × 0.293 = 0.156462
40. 16.4591 × 51.23 = 843.199693
41. 96.00 × 0.875 = 84
42. 0.3172 × 0.2008 = 0.06369376
43. 0.1543 × 0.4931 = 0.07608533
44. 0.7984 × 0.0003 = 0.00023952

SKILL 9

Divide (Fractional Remainder)

Two-Digit Divisor

Divide. $46\overline{)703}$

$$\begin{array}{r} 1 \\ 46\overline{)703} \\ -46 \\ \hline 24 \end{array} \quad \Rightarrow \quad \begin{array}{r} 15 \\ 46\overline{)703} \\ -46\downarrow \\ \hline 243 \\ -230 \\ \hline 13 \end{array} \quad \Rightarrow \quad \begin{array}{r} 15\frac{13}{46} \\ 46\overline{)703} \\ -46\downarrow \\ \hline 243 \\ -230 \\ \hline 13 \end{array}$$

1. $27\overline{)63}$ $2\frac{9}{27}$
2. $38\overline{)89}$ $2\frac{13}{38}$
3. $41\overline{)84}$ $2\frac{2}{41}$
4. $46\overline{)613}$ $13\frac{15}{46}$
5. $53\overline{)754}$ $14\frac{12}{53}$
6. $61\overline{)685}$ $11\frac{14}{61}$
7. $21\overline{)1781}$ $84\frac{17}{21}$
8. $55\overline{)4273}$ $77\frac{38}{55}$
9. $43\overline{)2900}$ $67\frac{19}{43}$
10. $73\overline{)3956}$ $54\frac{14}{73}$
11. $23\overline{)1369}$ $59\frac{12}{23}$
12. $34\overline{)2167}$ $63\frac{25}{34}$
13. $81\overline{)5793}$ $71\frac{42}{81}$
14. $93\overline{)78}$ $\frac{78}{93}$

Three-Digit Divisor

Divide. $472\overline{)9463}$

$$\begin{array}{r} 2 \\ 472\overline{)9463} \\ -944 \\ \hline 2 \end{array} \quad \Rightarrow \quad \begin{array}{r} 20\frac{23}{472} \\ 472\overline{)9463} \\ -944\downarrow \\ \hline 23 \\ -0 \\ \hline 23 \end{array}$$

15. $114\overline{)1837}$ $16\frac{13}{114}$
16. $216\overline{)5417}$ $25\frac{17}{216}$
17. $321\overline{)4832}$ $15\frac{17}{321}$
18. $429\overline{)9038}$ $21\frac{29}{429}$
19. $892\overline{)7928}$ $8\frac{792}{892}$
20. $910\overline{)11,849}$ $13\frac{19}{910}$
21. $409\overline{)9429}$ $23\frac{22}{409}$
22. $900\overline{)7621}$ $8\frac{421}{900}$
23. $625\overline{)8742}$ $13\frac{617}{625}$
24. $710\overline{)15,694}$ $22\frac{74}{710}$
25. $843\overline{)17,691}$ $20\frac{831}{843}$
26. $937\overline{)22,474}$ $23\frac{923}{937}$

Mixed Practice

27. $19\overline{)342}$ 18
28. $46\overline{)782}$ 17
29. $71\overline{)1136}$ 16
30. $279\overline{)9778}$ $35\frac{13}{279}$
31. $35\overline{)2520}$ 72
32. $509\overline{)5089}$ $9\frac{508}{509}$
33. $621\overline{)8716}$ $14\frac{22}{621}$
34. $49\overline{)4959}$
35. $549\overline{)6937}$ $12\frac{349}{549}$
36. $953\overline{)9597}$ $10\frac{67}{953}$
37. $842\overline{)4975}$ $5\frac{765}{842}$
38. $87\overline{)2139}$
39. $473\overline{)85,642}$ $181\frac{29}{473}$
40. $192\overline{)91,845}$ $478\frac{69}{192}$
41. $622\overline{)54,641}$ $87\frac{527}{622}$
42. $812\overline{)55,692}$
43. $51\overline{)81,603}$ $1600\frac{3}{51}$
44. $23\overline{)44,364}$ $1928\frac{20}{23}$
45. $88\overline{)22,222}$ $252\frac{46}{88}$
46. $34\overline{)27,945}$
47. $417\overline{)86,902}$ $208\frac{166}{417}$
48. $71\overline{)103,205}$ $1453\frac{42}{71}$
49. $514\overline{)691,507}$ $1345\frac{177}{514}$
50. $64\overline{)237,605}$

34. $101\frac{10}{49}$ 38. $24\frac{51}{87}$ 42. $68\frac{476}{812}$ 46. $821\frac{31}{34}$ 50. $3712\frac{37}{64}$

S K I L L

Divide (Decimal Remainder)

Exact Quotient

Divide. 28)378

```
     13                 13.5
28)378      ➧    28)378.0
  -28              -28
   98               98
  -84              -84
   14               140
                   -140
```

1. 8)89	11.125	**2.** 10)123	12.3
3. 15)318	21.2	**4.** 16)264	16.5
5. 20)102	5.1	**6.** 50)530	10.6
7. 12)237	19.75	**8.** 25)1310	52.4
9. 32)1104	34.5	**10.** 96)2988	31.125

Rounded Quotient

Divide. 39)818

Round to the nearest hundredth.

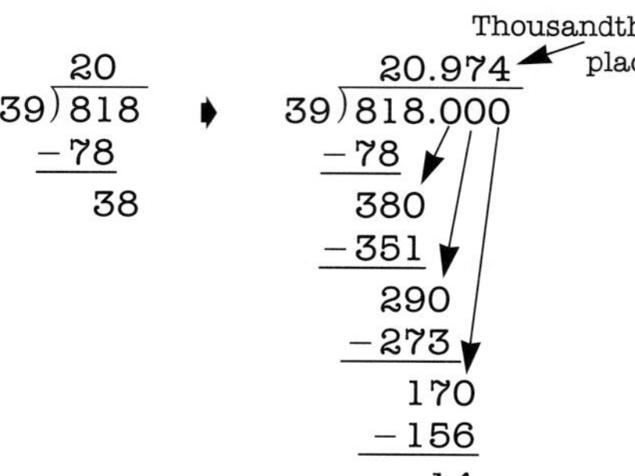

20.974 rounded to the nearest hundredth is 20.97 ← Skill 2

Round answers to the place value shown.

Nearest tenth:

		11. 14)319	22.8
12. 26)347	13.3	**13.** 23)371	16.1
14. 46)9415	204.7	**15.** 47)9719	206.8

Nearest hundredth:

		16. 19)427	22.47
17. 83)168	2.02	**18.** 47)432	9.19

Nearest thousandth:

		19. 37)402	10.865
20. 24)643	26.792	**21.** 21)452	21.524

Mixed Practice

Round answers to the nearest hundredth.

22. 28)1022	36.5	**23.** 24)209	8.71	**24.** 72)1665	23.13	**25.** 29)303	10.45
26. 85)1802	21.2	**27.** 24)747	31.13	**28.** 67)701	10.46	**29.** 71)400	5.63
30. 23)273	11.87	**31.** 36)630	17.5	**32.** 44)858	19.5	**33.** 59)852	14.44
34. 37)673	18.19	**35.** 41)9432	230.05	**36.** 73)1079	14.78	**37.** 24)994	41.42
38. 27)3365	124.63	**39.** 35)894	25.54	**40.** 42)5264	125.33	**41.** 110)4345	39.5

SKILL 11

Divide Decimals

Divisor Greater Than One

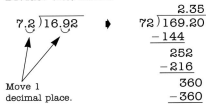

Divide. 7.2)16.92

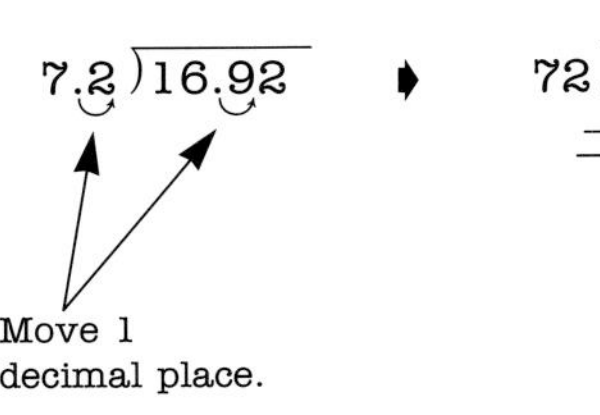


```
      2.35
72)169.20
  -144
    252
   -216
     360
    -360
```

Round answers to the nearest hundredth.

1. 2.5)11.5 **4.6**
2. 3.8)4.56 **1.2**
3. 3.2)18.272 **5.71**
4. 7.5)27.823 **3.71**
5. 3.15)53.55 **17**
6. 24.12)369.036 **15.3**
7. 4.08)26.52 **6.5**
8. 3.02)10.57 **3.5**
9. 4.23)181.5 **42.91**
10. 6.67)25.963 **3.89**

Divisor Less Than One

Divide. 0.032)14.400

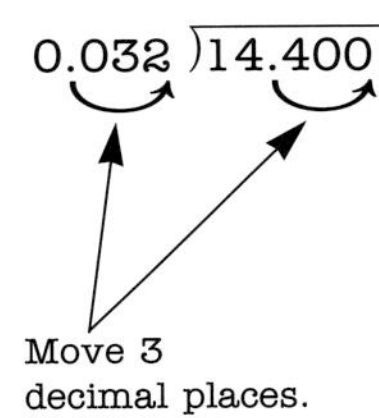


```
      450
32)14400
  -128
    160
   -160
     00
    -00
```

Round answers to the nearest hundredth.

11. 0.24)6.24 **26**
12. 0.372)6.324 **17**
13. 0.154)4.774 **31**
14. 0.48)2.938 **6.12**
15. 0.37)9.62 **26**
16. 0.21)1.374 **6.54**
17. 0.51)4.569 **8.96**
18. 0.67)2.693 **4.02**
19. 0.73)9.641 **13.21**
20. 0.81)11.632 **14.36**

Mixed Practice

Round answers to the nearest hundredth.

21. 3.12)4.386 **1.41**
22. 0.73)9.48 **12.99**
23. 0.136)33.32 **245**
24. 0.21)130.2 **620**
25. 6.94)8.378 **1.21**
26. 1.23)0.3813 **0.31**
27. 0.065)16.64 **256**
28. 8.34)7.416 **0.89**
29. 0.63)42.51 **67.48**
30. 2.91)5.932 **2.04**
31. 1.07)24.153 **22.57**
32. 1.1)29.9 **27.18**
33. 15.93)27.931 **1.75**
34. 12.12)36.422 **3.01**
35. 0.05)1.925 **38.5**
36. 2.03)21.249 **10.47**
37. 0.007)0.692 **98.86**
38. 1.05)25.421 **24.21**
39. 0.31)0.00354 **0.01**
40. 15.42)113.005 **7.33**
41. 5.02)86 **17.13**
42. 0.03)29 **966.67**
43. 5.4)0.062 **0.01**
44. 0.068)0.009 **0.13**

SKILL 12

Equivalent Fractions

Higher Terms

Complete the equivalent fraction.

$$\frac{3}{4} = \frac{?}{20}$$

$$\frac{3}{4} \overset{\times 5}{=} \frac{?}{20} \quad ➧ \quad \frac{3}{4} = \frac{15}{20}$$

Solve.

1. $\frac{5}{6} = \frac{?}{12}$ **10** **2.** $\frac{4}{9} = \frac{?}{27}$ **12** **3.** $\frac{7}{10} = \frac{35}{?}$ **50**

4. $\frac{11}{12} = \frac{22}{?}$ **24** **5.** $\frac{8}{17} = \frac{?}{51}$ **24** **6.** $\frac{15}{17} = \frac{90}{?}$ **102**

7. $\frac{11}{18} = \frac{?}{36}$ **22** **8.** $\frac{4}{19} = \frac{12}{?}$ **57** **9.** $\frac{16}{17} = \frac{48}{?}$ **51**

10. $\frac{11}{13} = \frac{?}{39}$ **33** **11.** $\frac{3}{4} = \frac{?}{32}$ **24** **12.** $\frac{15}{17} = \frac{60}{?}$ **68**

13. $\frac{7}{9} = \frac{?}{45}$ **35** **14.** $\frac{11}{12} = \frac{?}{72}$ **66** **15.** $\frac{5}{8} = \frac{45}{?}$ **72**

16. $\frac{11}{8} = \frac{33}{?}$ **24** **17.** $\frac{5}{24} = \frac{?}{96}$ **20** **18.** $\frac{23}{73} = \frac{?}{365}$ **115**

Lowest Terms

Reduce $\frac{12}{28}$ to lowest terms

$$\frac{12}{28} \overset{\div 4}{=} \frac{3}{7} \quad ➧ \quad \frac{12}{28} = \frac{3}{7}$$

Reduce to lowest terms.

19. $\frac{6}{9}$ $\frac{2}{3}$ **20.** $\frac{4}{16}$ $\frac{1}{4}$ **21.** $\frac{9}{18}$ $\frac{1}{2}$ **22.** $\frac{20}{22}$ $\frac{10}{11}$

23. $\frac{18}{27}$ $\frac{2}{3}$ **24.** $\frac{24}{32}$ $\frac{3}{4}$ **25.** $\frac{16}{48}$ $\frac{1}{3}$ **26.** $\frac{18}{42}$ $\frac{3}{7}$

27. $\frac{55}{66}$ $\frac{5}{6}$ **28.** $\frac{30}{50}$ $\frac{3}{5}$ **29.** $\frac{20}{54}$ $\frac{10}{27}$ **30.** $\frac{21}{27}$ $\frac{7}{9}$

31. $\frac{8}{14}$ $\frac{4}{7}$ **32.** $\frac{19}{57}$ $\frac{1}{3}$ **33.** $\frac{14}{28}$ $\frac{1}{2}$ **34.** $\frac{15}{25}$ $\frac{3}{5}$

35. $\frac{21}{28}$ $\frac{3}{4}$ **36.** $\frac{200}{365}$ $\frac{40}{73}$ **37.** $\frac{188}{366}$ $\frac{94}{183}$ **38.** $\frac{150}{365}$ $\frac{30}{73}$

39. $\frac{180}{360}$ $\frac{1}{2}$ **40.** $\frac{225}{365}$ $\frac{45}{73}$ **41.** $\frac{190}{360}$ $\frac{19}{36}$ **42.** $\frac{65}{75}$ $\frac{13}{15}$

43. $\frac{72}{468}$ $\frac{2}{13}$ **44.** $\frac{183}{366}$ $\frac{1}{2}$ **45.** $\frac{232}{1450}$ $\frac{4}{25}$ **46.** $\frac{1792}{5120}$ $\frac{7}{20}$

SKILL 13

Change Mixed Numbers/ Improper Fractions

Mixed Numbers to Improper Fractions

Write $3\frac{7}{8}$ as an improper fraction.

$$3\frac{7}{8} = \frac{(3 \times 8) + 7}{8}$$

$$3\frac{7}{8} = \frac{24 + 7}{8}$$

$$3\frac{7}{8} = \frac{31}{8}$$

Write as an improper fraction.

1. $6\frac{1}{4}$ $\frac{25}{4}$	**2.** $4\frac{3}{4}$ $\frac{19}{4}$	**3.** $7\frac{3}{8}$ $\frac{59}{8}$	**4.** $9\frac{1}{2}$ $\frac{19}{2}$
5. $6\frac{2}{3}$ $\frac{20}{3}$	**6.** $3\frac{5}{6}$ $\frac{23}{6}$	**7.** $7\frac{4}{5}$ $\frac{39}{5}$	**8.** $3\frac{1}{8}$ $\frac{25}{8}$
9. $4\frac{3}{10}$ $\frac{43}{10}$	**10.** $1\frac{1}{16}$ $\frac{17}{16}$	**11.** $4\frac{17}{32}$ $\frac{145}{32}$	**12.** $4\frac{1}{5}$ $\frac{21}{5}$
13. $5\frac{1}{3}$ $\frac{16}{3}$	**14.** $2\frac{1}{6}$ $\frac{13}{6}$	**15.** $2\frac{9}{10}$ $\frac{29}{10}$	**16.** $3\frac{1}{10}$ $\frac{31}{10}$
17. $3\frac{5}{16}$ $\frac{53}{16}$	**18.** $4\frac{9}{16}$ $\frac{73}{16}$	**19.** $5\frac{5}{32}$ $\frac{165}{32}$	**20.** $6\frac{3}{5}$ $\frac{33}{5}$
21. $7\frac{2}{11}$ $\frac{79}{11}$	**22.** $8\frac{5}{6}$ $\frac{53}{6}$	**23.** $11\frac{2}{5}$ $\frac{57}{5}$	**24.** $13\frac{1}{3}$ $\frac{40}{3}$

Improper Fractions to Mixed Numbers

Write $\frac{17}{3}$ as a mixed number.

$$\frac{17}{3} \;\rightarrow\; 3\overline{)17} = 5\frac{2}{3},\quad 17 - 15 = 2$$

← Skill 9

$$\frac{17}{3} = 5\frac{2}{3}$$

Write as a mixed number. Reduce any fractional parts to lowest terms.

25. $\frac{13}{2}$ $6\frac{1}{2}$	**26.** $\frac{18}{4}$ $4\frac{1}{2}$	**27.** $\frac{46}{8}$ $5\frac{3}{4}$	**28.** $\frac{19}{4}$ $4\frac{3}{4}$
29. $\frac{33}{6}$ $5\frac{1}{2}$	**30.** $\frac{30}{7}$ $4\frac{2}{7}$	**31.** $\frac{21}{9}$ $2\frac{1}{3}$	**32.** $\frac{28}{8}$ $3\frac{1}{2}$
33. $\frac{45}{18}$ $2\frac{1}{2}$	**34.** $\frac{49}{21}$ $2\frac{1}{3}$	**35.** $\frac{62}{21}$ $2\frac{20}{21}$	**36.** $\frac{23}{3}$ $7\frac{2}{3}$
37. $\frac{35}{6}$ $5\frac{5}{6}$	**38.** $\frac{47}{7}$ $6\frac{5}{7}$	**39.** $\frac{57}{18}$ $3\frac{1}{6}$	**40.** $\frac{37}{9}$ $4\frac{1}{9}$
41. $\frac{49}{8}$ $6\frac{1}{8}$	**42.** $\frac{63}{2}$ $31\frac{1}{2}$	**43.** $\frac{45}{7}$ $6\frac{3}{7}$	**44.** $\frac{71}{4}$ $17\frac{3}{4}$

SKILL

14 Change Fractions/Decimals

Fraction to Decimal

Write $\frac{5}{12}$ as a decimal. Round to the nearest hundredth.

$$\frac{5}{12} \rightarrow 12\overline{)5.000} = 0.416$$ ← Skill 10

0.416 rounded to the nearest hundredth is 0.42. ← Skill 2

$$\frac{5}{12} = 0.42$$

Write as a decimal. Round answers to the nearest hundredth.

1. $\frac{4}{5}$ **0.8**	**2.** $\frac{7}{20}$ **0.35**	**3.** $1\frac{1}{8}$ **1.13**	**4.** $\frac{3}{4}$ **0.75**
5. $2\frac{3}{7}$ **2.43**	**6.** $\frac{7}{12}$ **0.58**	**7.** $1\frac{11}{30}$ **1.37**	**8.** $\frac{2}{5}$ **0.4**
9. $\frac{3}{10}$ **0.3**	**10.** $3\frac{4}{25}$ **3.16**	**11.** $4\frac{3}{8}$ **4.38**	**12.** $\frac{7}{10}$ **0.7**
13. $3\frac{1}{12}$ **3.08**	**14.** $\frac{1}{15}$ **0.07**	**15.** $\frac{1}{30}$ **0.03**	**16.** $2\frac{9}{10}$ **2.9**
17. $5\frac{3}{20}$ **5.15**	**18.** $\frac{9}{20}$ **0.45**	**19.** $2\frac{3}{25}$ **2.12**	**20.** $\frac{1}{8}$ **0.13**
21. $7\frac{1}{7}$ **7.14**	**22.** $2\frac{5}{8}$ **2.63**	**23.** $\frac{4}{9}$ **0.44**	**24.** $1\frac{7}{8}$ **1.88**

Decimal to Fraction

Write 0.42 as a fraction in lowest terms.

$$0.42 = \frac{42}{100} = \frac{21}{50}$$ (Skill 1) ← Skill 12

$$0.42 = \frac{21}{50}$$

Write as a fraction in lowest terms.

25. 0.1 $\frac{1}{10}$	**26.** 3.7 $3\frac{7}{10}$	**27.** 0.30 $\frac{3}{10}$
28. 2.25 $2\frac{1}{4}$	**29.** 0.03 $\frac{3}{100}$	**30.** 0.09 $\frac{9}{100}$
31. 0.53 $\frac{53}{100}$	**32.** 1.75 $1\frac{3}{4}$	**33.** 0.003 $\frac{3}{1000}$
34. 0.010 $\frac{1}{100}$	**35.** 0.064 $\frac{8}{125}$	**36.** 4.206 $4\frac{103}{500}$
37. 4.444 $4\frac{111}{250}$	**38.** 0.732 $\frac{183}{250}$	**39.** 0.469 $\frac{469}{1000}$
40. 2.9 $2\frac{9}{10}$	**41.** 7.5 $7\frac{1}{2}$	**42.** 0.32 $\frac{8}{25}$
43. 1.08 $1\frac{2}{25}$	**44.** 0.06 $\frac{3}{50}$	**45.** 2.83 $2\frac{83}{100}$
46. 1.039 $1\frac{39}{1000}$	**47.** 0.105 $\frac{21}{200}$	**48.** 2.422 $2\frac{211}{500}$
49. 0.0005 $\frac{1}{2000}$	**50.** 0.2482 $\frac{1241}{5000}$	**51.** 1.6432 $1\frac{402}{625}$
52. 0.0058 $\frac{29}{5000}$	**53.** 0.0002 $\frac{1}{5000}$	**54.** 6.66 $6\frac{33}{50}$

SKILL 15

Add Fractions, Like Denominators

Fractions

Add.

Like denominators

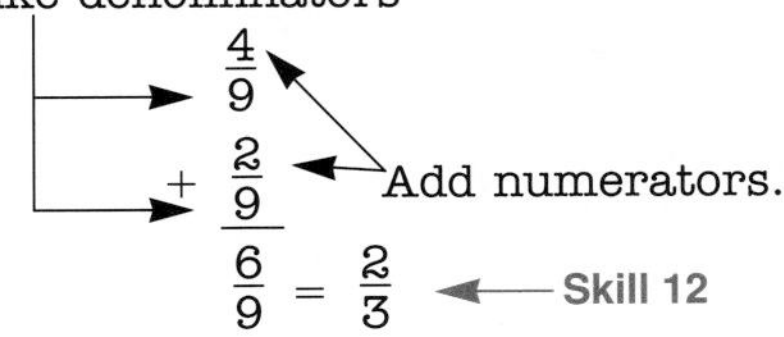

Express answers in lowest terms.

1. $\frac{4}{5} + \frac{3}{5}$ $1\frac{2}{5}$
2. $\frac{4}{7} + \frac{2}{7}$ $\frac{6}{7}$
3. $\frac{1}{8} + \frac{5}{8}$ $\frac{3}{4}$
4. $\frac{5}{9} + \frac{7}{9}$ $1\frac{1}{3}$
5. $\frac{5}{7} + \frac{6}{7}$ $1\frac{4}{7}$
6. $\frac{11}{12} + \frac{7}{12}$ $1\frac{1}{2}$
7. $\frac{13}{25} + \frac{16}{25}$ $1\frac{4}{25}$
8. $\frac{11}{30} + \frac{19}{30}$ 1
9. $\frac{15}{32} + \frac{27}{32}$ $1\frac{5}{16}$
10. $\frac{21}{40} + \frac{31}{40}$ $1\frac{3}{10}$
11. $\frac{22}{45} + \frac{31}{45}$ $1\frac{8}{45}$
12. $\frac{11}{50} + \frac{19}{50}$ $\frac{3}{5}$

Mixed Numbers

Add.

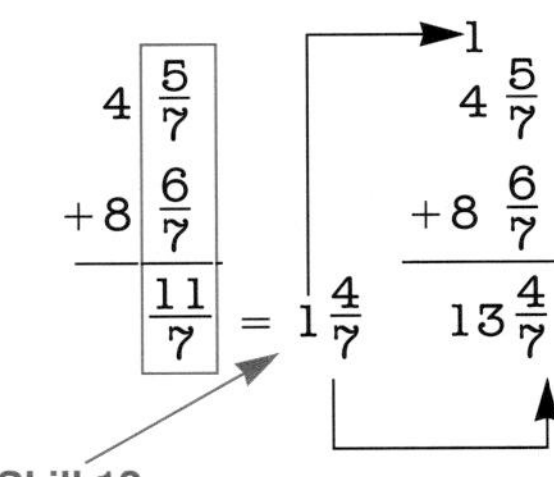

Express answers in lowest terms.

13. $4\frac{1}{3} + 7\frac{1}{3}$ $11\frac{2}{3}$
14. $13\frac{4}{5} + 8\frac{3}{5}$ $22\frac{2}{5}$
15. $5\frac{4}{7} + 8\frac{5}{7}$ $14\frac{2}{7}$
16. $12\frac{3}{8} + 14\frac{1}{8}$ $26\frac{1}{2}$
17. $6\frac{11}{12} + 5\frac{5}{12}$ $12\frac{1}{3}$
18. $2\frac{12}{13} + 7\frac{2}{13}$ $10\frac{1}{13}$
19. $8\frac{9}{16} + 8\frac{11}{16}$ $17\frac{1}{4}$
20. $14\frac{5}{24} + 15\frac{7}{24}$ $29\frac{1}{2}$
21. $4\frac{13}{32} + 5\frac{15}{32}$ $9\frac{7}{8}$
22. $15\frac{19}{45} + 6\frac{28}{45}$ $22\frac{2}{45}$
23. $7\frac{8}{35} + 4\frac{6}{35}$ $11\frac{2}{5}$
24. $9\frac{5}{32} + 14\frac{7}{32}$ $23\frac{3}{8}$

SKILL 16

Add Fractions, Unlike Denominators

Fractions

Add.

Unlike denominators → Like denominators

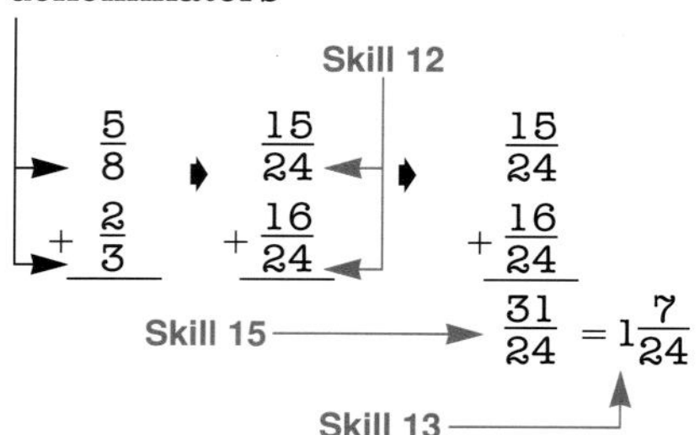

Express answers in lowest terms.

1. $\frac{1}{2} + \frac{3}{5}$ **$1\frac{1}{10}$**
2. $\frac{3}{4} + \frac{1}{6}$ **$\frac{11}{12}$**
3. $\frac{2}{7} + \frac{2}{3}$ **$\frac{20}{21}$**
4. $\frac{3}{8} + \frac{1}{5}$ **$\frac{23}{40}$**
5. $\frac{5}{6} + \frac{1}{3}$ **$1\frac{1}{6}$**
6. $\frac{7}{12} + \frac{3}{7}$ **$1\frac{1}{84}$**
7. $\frac{9}{11} + \frac{3}{10}$ **$1\frac{13}{110}$**
8. $\frac{13}{16} + \frac{9}{8}$ **$1\frac{15}{16}$**
9. $\frac{7}{20} + \frac{19}{30}$ **$\frac{59}{60}$**
10. $\frac{5}{18} + \frac{19}{24}$ **$1\frac{5}{72}$**
11. $\frac{1}{25} + \frac{13}{30}$ **$\frac{71}{150}$**
12. $\frac{2}{15} + \frac{29}{30}$ **$1\frac{1}{10}$**

Mixed Numbers

Add.

$$12\frac{2}{5} + 9\frac{3}{4} \rightarrow 12\frac{8}{20} + 9\frac{15}{20};\quad \frac{23}{20} = 1\frac{3}{20} \rightarrow 22\frac{3}{20}$$

Skill 12

Skill 15

Express answers in lowest terms.

13. $5\frac{1}{2} + 3\frac{2}{3}$ **$9\frac{1}{6}$**
14. $4\frac{3}{8} + 9\frac{3}{4}$ **$14\frac{1}{8}$**
15. $7\frac{5}{6} + 8\frac{1}{2}$ **$16\frac{1}{3}$**
16. $9\frac{7}{8} + 10\frac{7}{16}$ **$20\frac{5}{16}$**
17. $11\frac{2}{13} + 9\frac{15}{26}$ **$20\frac{19}{26}$**
18. $18\frac{2}{3} + 5\frac{7}{11}$ **$24\frac{10}{33}$**
19. $12\frac{4}{7} + 15\frac{7}{9}$ **$28\frac{22}{63}$**
20. $16\frac{5}{6} + 10\frac{6}{7}$ **$27\frac{29}{42}$**
21. $16\frac{5}{8} + 12\frac{1}{7}$ **$28\frac{43}{56}$**
22. $11\frac{4}{9} + 13\frac{3}{8}$ **$24\frac{59}{72}$**
23. $4\frac{2}{3} + 24\frac{7}{16}$ **$29\frac{5}{48}$**
24. $15\frac{5}{8} + 9\frac{4}{7}$ **$25\frac{11}{56}$**

S K I L L

Subtract Fractions, Like Denominators

Fractions

Subtract.

Like denominators

$$\begin{array}{r} \frac{11}{12} \\ -\frac{7}{12} \\ \hline \end{array} \quad \blacktriangleright \quad \begin{array}{r} \frac{11}{12} \\ -\frac{7}{12} \\ \hline \frac{4}{12} = \frac{1}{3} \end{array}$$

Subtract numerators.

Skill 13

Express answers in lowest terms.

1. $\frac{4}{9} - \frac{2}{9}$ $\frac{2}{9}$
2. $\frac{9}{8} - \frac{3}{8}$ $\frac{3}{4}$
3. $\frac{11}{12} - \frac{1}{12}$ $\frac{5}{6}$
4. $\frac{5}{6} - \frac{1}{6}$ $\frac{2}{3}$
5. $\frac{5}{7} - \frac{1}{7}$ $\frac{4}{7}$
6. $\frac{19}{27} - \frac{1}{27}$ $\frac{2}{3}$
7. $\frac{11}{16} - \frac{9}{16}$ $\frac{1}{8}$
8. $\frac{16}{25} - \frac{6}{25}$ $\frac{2}{5}$
9. $\frac{37}{40} - \frac{29}{40}$ $\frac{1}{5}$
10. $\frac{29}{36} - \frac{11}{36}$ $\frac{1}{2}$
11. $\frac{7}{10} - \frac{3}{10}$ $\frac{2}{5}$
12. $\frac{19}{36} - \frac{5}{36}$ $\frac{7}{18}$
13. $\frac{19}{24} - \frac{13}{24}$ $\frac{1}{4}$
14. $\frac{13}{28} - \frac{7}{28}$ $\frac{3}{14}$
15. $\frac{29}{60} - \frac{8}{60}$ $\frac{7}{20}$

Mixed Numbers

Subtract.

$$\begin{array}{r} 8\frac{11}{16} \\ -5\frac{9}{16} \\ \hline \end{array} \quad \blacktriangleright \quad \begin{array}{r} 8\boxed{\frac{11}{16}} \\ -5\boxed{\frac{9}{16}} \\ \hline \boxed{\frac{2}{16}} = \frac{1}{8} \end{array} \quad \blacktriangleright \quad \begin{array}{r} 8\frac{11}{16} \\ -5\frac{9}{16} \\ \hline 3\frac{1}{8} \end{array}$$

Skill 12

Express answers in lowest terms.

16. $7\frac{3}{4} - 6\frac{1}{4}$ $1\frac{1}{2}$
17. $8\frac{6}{7} - 7\frac{5}{7}$ $1\frac{1}{7}$
18. $10\frac{4}{5} - 6\frac{1}{5}$ $4\frac{3}{5}$
19. $12\frac{5}{8} - 9\frac{3}{8}$ $3\frac{1}{4}$
20. $2\frac{16}{21} - 1\frac{5}{21}$ $1\frac{11}{21}$
21. $7\frac{11}{24} - 5\frac{7}{24}$ $2\frac{1}{6}$
22. $14\frac{17}{20} - 8\frac{9}{20}$ $6\frac{2}{5}$
23. $15\frac{15}{32} - 10\frac{9}{32}$ $5\frac{3}{16}$
24. $14\frac{7}{8} - 9\frac{3}{8}$ $5\frac{1}{2}$
25. $12\frac{11}{13} - 10\frac{4}{13}$ $2\frac{7}{13}$
26. $18\frac{17}{20} - 15\frac{9}{20}$ $3\frac{2}{5}$
27. $35\frac{39}{40} - 27\frac{19}{40}$ $8\frac{1}{2}$
28. $74\frac{41}{75} - 29\frac{16}{75}$ $45\frac{1}{3}$
29. $103\frac{11}{35} - 78\frac{1}{35}$ $25\frac{2}{7}$

SKILL 18

Subtract Fractions, Unlike Denominators

Fractions

Subtract.

Unlike denominators — Like denominators

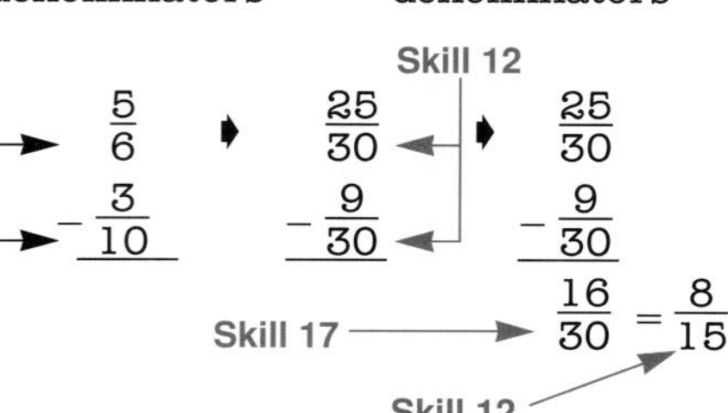

$$\frac{5}{6} - \frac{3}{10} \rightarrow \frac{25}{30} - \frac{9}{30} \rightarrow \frac{25}{30} - \frac{9}{30} = \frac{16}{30} = \frac{8}{15}$$

Express answers in lowest terms.

1. $\frac{3}{4} - \frac{3}{8}$ $\frac{3}{8}$
2. $\frac{1}{3} - \frac{2}{9}$ $\frac{1}{9}$
3. $\frac{9}{10} - \frac{3}{4}$ $\frac{3}{20}$
4. $\frac{4}{5} - \frac{2}{3}$ $\frac{2}{15}$
5. $\frac{5}{6} - \frac{7}{9}$ $\frac{1}{18}$
6. $\frac{5}{8} - \frac{5}{12}$ $\frac{5}{24}$
7. $\frac{3}{5} - \frac{8}{15}$ $\frac{1}{15}$
8. $\frac{11}{12} - \frac{3}{4}$ $\frac{1}{6}$
9. $\frac{4}{5} - \frac{13}{20}$ $\frac{3}{20}$
10. $\frac{6}{7} - \frac{10}{21}$ $\frac{8}{21}$
11. $\frac{1}{2} - \frac{2}{5}$ $\frac{1}{10}$
12. $\frac{3}{4} - \frac{11}{16}$ $\frac{1}{16}$

Mixed Numbers

Subtract.

$$9\frac{11}{12} - 7\frac{3}{8} \rightarrow 9\frac{22}{24} - 7\frac{9}{24} = 2\frac{13}{24}$$

Skill 12

Skill 17

Express answers in lowest terms.

13. $9\frac{3}{4} - 5\frac{1}{2}$ $4\frac{1}{4}$
14. $7\frac{5}{6} - 4\frac{2}{3}$ $3\frac{1}{6}$
15. $14\frac{4}{9} - 8\frac{1}{6}$ $6\frac{5}{18}$
16. $15\frac{3}{7} - 10\frac{1}{3}$ $5\frac{2}{21}$
17. $34\frac{11}{12} - 29\frac{4}{5}$ $5\frac{7}{60}$
18. $48\frac{4}{7} - 37\frac{5}{9}$ $11\frac{1}{63}$
19. $12\frac{5}{8} - 9\frac{1}{2}$ $3\frac{1}{8}$
20. $2\frac{16}{21} - 1\frac{2}{3}$ $1\frac{2}{21}$
21. $13\frac{3}{4} - 11\frac{3}{5}$ $2\frac{3}{20}$
22. $21\frac{5}{6} - 19\frac{5}{8}$ $2\frac{5}{24}$
23. $7\frac{11}{24} - 5\frac{1}{3}$ $2\frac{1}{8}$
24. $20\frac{15}{38} - 20\frac{1}{19}$ $\frac{13}{38}$

SKILL 19

Subtract Mixed Numbers, Borrowing

Whole Number and a Mixed Number

Subtract.

$$\begin{array}{r} 8 \\ -4\frac{3}{5} \\ \hline \end{array} \;\Rightarrow\; \begin{array}{r} 7\frac{5}{5} \\ -4\frac{3}{5} \\ \hline 3\frac{2}{5} \end{array}$$

Like denominators

Express answers in lowest terms.

1. $3-1\frac{1}{2}$ $1\frac{1}{2}$
2. $4-2\frac{3}{4}$ $1\frac{1}{4}$
3. $10-7\frac{9}{10}$ $2\frac{1}{10}$
4. $6-4\frac{3}{8}$ $1\frac{5}{8}$
5. $12-9\frac{5}{11}$ $2\frac{6}{11}$
6. $15-12\frac{7}{16}$ $2\frac{9}{16}$

Mixed Numbers with Unlike Denominators

Subtract.

$$\begin{array}{r} 6\frac{2}{3} \\ -2\frac{4}{5} \\ \hline \end{array} \;\Rightarrow\; \begin{array}{r} 6\frac{10}{15} \\ -2\frac{12}{15} \\ \hline \end{array} \;\Rightarrow\; \begin{array}{r} 5\frac{25}{15} \\ -2\frac{12}{15} \\ \hline 3\frac{13}{15} \end{array}$$

Skill 12

Express answers in lowest terms.

7. $5\frac{1}{2}-2\frac{4}{9}$ $3\frac{1}{18}$
8. $8\frac{1}{2}-3\frac{5}{8}$ $4\frac{7}{8}$
9. $18\frac{3}{4}-5\frac{5}{6}$ $12\frac{11}{12}$
10. $27\frac{2}{7}-13\frac{3}{4}$ $13\frac{15}{28}$
11. $28\frac{3}{7}-26\frac{9}{10}$ $1\frac{37}{70}$
12. $40\frac{19}{30}-39\frac{3}{4}$ $\frac{53}{60}$
13. $6\frac{3}{4}-3\frac{7}{8}$ $2\frac{7}{8}$
14. $16\frac{1}{12}-9\frac{3}{10}$ $6\frac{47}{60}$

Mixed Practice

15. $4-2\frac{5}{8}$ $1\frac{3}{8}$
16. $9-5\frac{11}{12}$ $3\frac{1}{12}$
17. $12-11\frac{7}{8}$ $\frac{1}{8}$
18. $16-7\frac{5}{11}$ $8\frac{6}{11}$
19. $8\frac{1}{3}-4\frac{5}{9}$ $3\frac{7}{9}$
20. $13\frac{1}{2}-7\frac{3}{5}$ $5\frac{9}{10}$
21. $14\frac{1}{5}-13\frac{3}{4}$ $\frac{9}{20}$
22. $13-6\frac{5}{8}$ $6\frac{3}{8}$
23. $27\frac{1}{4}-15\frac{15}{16}$ $11\frac{5}{16}$
24. $38\frac{3}{7}-29\frac{9}{10}$ $8\frac{37}{70}$
25. $18\frac{1}{12}-3\frac{11}{36}$ $14\frac{7}{9}$
26. $17\frac{8}{9}-5\frac{3}{8}$ $12\frac{37}{72}$
27. $19-5\frac{2}{7}$ $13\frac{5}{7}$
28. $16\frac{4}{5}-6\frac{3}{4}$ $10\frac{1}{20}$
29. $19\frac{11}{32}-7\frac{3}{16}$ $12\frac{5}{32}$
30. $25-8\frac{7}{10}$ $16\frac{3}{10}$

SKILL 20

Multiply Fractions/Mixed Numbers

Fractions

Multiply.

$\frac{5}{8} \times \frac{2}{3} = \frac{5 \times 2}{8 \times 3} = \frac{10}{24} = \frac{5}{12}$

Skill 12

Express answers in lowest terms.

1. $\frac{1}{2} \times \frac{2}{3}$ $\frac{1}{3}$ **2.** $\frac{1}{4} \times \frac{3}{5}$ $\frac{3}{20}$ **3.** $\frac{5}{6} \times \frac{3}{4}$ $\frac{5}{8}$

4. $\frac{8}{9} \times \frac{2}{7}$ $\frac{16}{63}$ **5.** $\frac{11}{12} \times \frac{5}{11}$ $\frac{5}{12}$ **6.** $\frac{4}{13} \times \frac{8}{9}$ $\frac{32}{117}$

7. $4 \times \frac{1}{2}$ 2 **8.** $17 \times \frac{2}{5}$ $6\frac{4}{5}$ **9.** $21 \times \frac{3}{7}$ 9

Mixed Numbers

Multiply.

$4\frac{1}{3} \times 2\frac{1}{4} = \frac{13}{3} \times \frac{9}{4} = \frac{117}{12} = 9\frac{3}{4}$

Skill 13

Express answers in lowest terms.

10. $1\frac{1}{2} \times 1\frac{1}{3}$ 2 **11.** $3\frac{2}{3} \times 4\frac{2}{5}$ $16\frac{2}{15}$

12. $\frac{1}{8} \times 4\frac{4}{5}$ $\frac{3}{5}$ **13.** $7 \times 8\frac{1}{3}$ $58\frac{1}{3}$

14. $12\frac{1}{2} \times 1\frac{1}{2}$ $18\frac{3}{4}$ **15.** $20\frac{1}{2} \times 2\frac{1}{4}$ $46\frac{1}{8}$

Mixed Practice

Express answers in lowest terms.

16. $\frac{2}{5} \times \frac{4}{7}$ $\frac{8}{35}$ **17.** $\frac{3}{8} \times \frac{4}{9}$ $\frac{1}{6}$ **18.** $\frac{2}{3} \times 1\frac{1}{8}$ $\frac{3}{4}$ **19.** $3\frac{2}{3} \times \frac{1}{4}$ $\frac{11}{12}$ **20.** $6 \times \frac{4}{5}$ $4\frac{4}{5}$

21. $\frac{3}{4} \times 8$ 6 **22.** $3\frac{1}{3} \times 4\frac{2}{5}$ $14\frac{2}{3}$ **23.** $2\frac{1}{4} \times 1\frac{2}{3}$ $3\frac{3}{4}$ **24.** $\frac{3}{4} \times 2\frac{5}{6}$ $2\frac{1}{8}$ **25.** $6 \times 3\frac{1}{4}$ $19\frac{1}{2}$

26. $18\frac{1}{2} \times 2\frac{2}{3}$ $49\frac{1}{3}$ **27.** $1\frac{3}{5} \times 5\frac{2}{6}$ $8\frac{8}{15}$ **28.** $5 \times 15\frac{1}{2}$ $77\frac{1}{2}$ **29.** $10 \times 6\frac{2}{5}$ 64 **30.** $2\frac{2}{7} \times 1\frac{1}{9}$ $2\frac{34}{63}$

31. $3\frac{4}{9} \times 7\frac{3}{8}$ $25\frac{29}{72}$ **32.** $\frac{7}{10} \times \frac{5}{9}$ $\frac{7}{18}$ **33.** $11\frac{3}{4} \times 8\frac{1}{3}$ $97\frac{11}{12}$ **34.** $\frac{11}{12} \times \frac{4}{33}$ $\frac{1}{9}$ **35.** $6\frac{7}{8} \times 10\frac{1}{3}$ $71\frac{1}{24}$

SKILL 21

Divide Fractions/Mixed Numbers

Fractions

Divide.

$$\frac{2}{3} \div \frac{3}{8} = \frac{2}{3} \times \frac{8}{3} = 1\frac{7}{9}$$

Skill 20

Express answers in lowest terms.

1. $\frac{1}{4} \div \frac{1}{8}$ **2**
2. $\frac{3}{8} \div \frac{1}{4}$ **$1\frac{1}{2}$**
3. $\frac{3}{4} \div \frac{2}{5}$ **$1\frac{7}{8}$**
4. $5 \div \frac{5}{6}$ **6**
5. $7 \div \frac{14}{15}$ **$7\frac{1}{2}$**
6. $8 \div \frac{4}{11}$ **22**
7. $\frac{3}{4} \div \frac{1}{8}$ **6**
8. $\frac{2}{5} \div \frac{5}{6}$ **$\frac{12}{25}$**
9. $\frac{4}{7} \div \frac{1}{28}$ **16**

Mixed Numbers

Divide.

Skill 13

$$5\frac{1}{6} \div 1\frac{1}{4} = \frac{31}{6} \div \frac{5}{4}$$
$$= \frac{31}{6} \times \frac{4}{5}$$
$$= \frac{124}{30}$$
$$= 4\frac{2}{15}$$

Skill 20

Express answers in lowest terms.

10. $3\frac{2}{3} \div 1\frac{1}{2}$ **$2\frac{4}{9}$**
11. $3\frac{5}{6} \div 1\frac{1}{5}$ **$3\frac{7}{36}$**
12. $2\frac{1}{6} \div 3\frac{1}{3}$ **$\frac{13}{20}$**
13. $4\frac{2}{3} \div 1\frac{3}{5}$ **$2\frac{11}{12}$**
14. $\frac{3}{4} \div 3\frac{1}{2}$ **$\frac{3}{14}$**
15. $\frac{5}{8} \div 2\frac{1}{2}$ **$\frac{1}{4}$**
16. $16 \div 1\frac{1}{8}$ **$14\frac{2}{9}$**
17. $11\frac{1}{3} \div 2\frac{1}{5}$ **$5\frac{5}{33}$**
18. $12 \div 2\frac{12}{17}$ **$4\frac{10}{23}$**
19. $13\frac{1}{7} \div 3\frac{1}{6}$ **$4\frac{20}{133}$**

Mixed Practice

Express answers in lowest terms.

20. $7\frac{1}{2} \div 1\frac{1}{4}$ **6**
21. $8\frac{1}{4} \div 5\frac{1}{2}$ **$1\frac{1}{2}$**
22. $\frac{5}{11} \div 2\frac{1}{5}$ **$\frac{25}{121}$**
23. $\frac{9}{10} \div \frac{3}{5}$ **$1\frac{1}{2}$**
24. $\frac{7}{12} \div \frac{1}{6}$ **$3\frac{1}{2}$**
25. $1\frac{3}{8} \div 2\frac{1}{16}$ **$\frac{2}{3}$**
26. $13\frac{1}{3} \div \frac{1}{10}$ **$133\frac{1}{3}$**
27. $8\frac{1}{3} \div \frac{5}{9}$ **15**
28. $\frac{19}{21} \div 3\frac{1}{2}$ **$\frac{38}{147}$**
29. $3\frac{5}{7} \div \frac{13}{21}$ **6**
30. $8 \div 1\frac{3}{5}$ **5**
31. $20 \div \frac{2}{3}$ **30**
32. $2\frac{1}{16} \div 11$ **$\frac{3}{16}$**
33. $5\frac{3}{8} \div 22$ **$\frac{43}{176}$**
34. $22 \div \frac{3}{5}$ **$36\frac{2}{3}$**
35. $\frac{11}{15} \div 1\frac{12}{13}$ **$\frac{143}{375}$**
36. $\frac{11}{12} \div \frac{11}{12}$ **1**
37. $2\frac{11}{14} \div \frac{3}{7}$ **$6\frac{1}{2}$**
38. $\frac{15}{16} \div \frac{5}{32}$ **6**
39. $19\frac{1}{2} \div 19\frac{1}{2}$ **1**

SKILL

22 Write Ratios

Compare Two Numbers

What is the ratio of desks to computers?

OFFICE INVENTORY		
Computers	Calculators	Desks
6	5	15

Ratio of desks to computers:

15 to 6 or 15:6 or $\frac{15}{6}$

Write the ratio.

DEPARTMENT SALES		
Meat	Produce	Dairy
$1500	$600	$1200

1. Ratio of meat to produce **1500:600**
2. Ratio of meat to dairy **1500:1200**
3. Ratio of produce to dairy **600:1200**
4. Ratio of dairy to produce **1200:600**
5. Ratio of produce to meat **600:1500**
6. Ratio of dairy to meat **1200:1500**

Ratios as Fractions

Write the ratio of tables to chairs as a fraction in lowest terms.

A restaurant has 40 chairs and 16 tables.

Ratio of tables to chairs is $\frac{16}{40}$.

$\frac{16}{40} = \frac{2}{5}$ ← Skill 12

Write the ratio as a fraction in lowest terms.

7. Nurse-to-patient ratio in a hospital is 6 nurses to 30 patients. $\frac{1}{5}$
8. Teacher-to-student ratio in a school is 8 teachers to 160 students. $\frac{1}{20}$
9. Door-to-window ratio in a house is 3 doors to 45 windows. $\frac{1}{15}$
10. Width-to-length ratio of a box is 22 cm to 30 cm. $\frac{11}{15}$
11. Room-to-desk ratio in a school is 40 rooms to 1000 desks. $\frac{1}{25}$
12. Land area-to-people ratio in a county is 6 km^2 to 138 people. $\frac{1}{23}$
13. Car-to-people ratio in a town is 4000 cars to 10,000 people. $\frac{2}{5}$
14. Hit-to-strikeout ratio of a batter is 15 hits to 25 strikeouts. $\frac{3}{5}$

SKILL 23

Proportions

Checking Proportions

Is this proportion true?

$\frac{5}{9} \stackrel{?}{=} \frac{35}{63}$

Cross multiply.

$\frac{5}{9} \stackrel{?}{=} \frac{35}{63}$ ➧ $5 \times 63 \stackrel{?}{=} 9 \times 35$

$315 = 315$

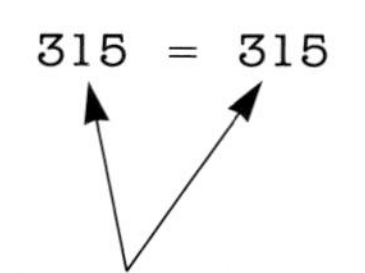

True. The products are equal.

Tell whether the proportions are true.

1. $\frac{1}{3} \stackrel{?}{=} \frac{4}{12}$ T
2. $\frac{2}{5} \stackrel{?}{=} \frac{4}{10}$ T
3. $\frac{4}{5} \stackrel{?}{=} \frac{16}{19}$ F
4. $\frac{5}{6} \stackrel{?}{=} \frac{25}{29}$ F
5. $\frac{3}{4} \stackrel{?}{=} \frac{6}{9}$ F
6. $\frac{1}{2} \stackrel{?}{=} \frac{11}{24}$ F
7. $\frac{3}{8} \stackrel{?}{=} \frac{9}{24}$ T
8. $\frac{2}{7} \stackrel{?}{=} \frac{14}{49}$ T
9. $\frac{16}{32} \stackrel{?}{=} \frac{1}{2}$ T
10. $\frac{13}{23} \stackrel{?}{=} \frac{1}{2}$ F
11. $\frac{10}{15} \stackrel{?}{=} \frac{5}{8}$ F
12. $\frac{22}{30} \stackrel{?}{=} \frac{11}{15}$ T

Solving Proportions

Solve for the number that makes the proportion true.

$\frac{7}{8} = \frac{28}{a}$

Cross multiply.

$\frac{7}{8} = \frac{28}{a}$ ➧ $7 \times a = 28 \times 8$

$7 \times a = 224$

$a = \frac{224}{7}$

$a = 32$

Solve.

13. $\frac{1}{2} = \frac{9}{h}$ 18
14. $\frac{2}{3} = \frac{12}{y}$ 18
15. $\frac{6}{a} = \frac{9}{11}$ $7\frac{1}{3}$
16. $\frac{8}{y} = \frac{12}{15}$ 10
17. $\frac{t}{6} = \frac{1}{11}$ $\frac{6}{11}$
18. $\frac{a}{7} = \frac{2}{19}$ $\frac{14}{19}$
19. $\frac{5}{9} = \frac{n}{10}$ $5\frac{5}{9}$
20. $\frac{4}{11} = \frac{x}{33}$ 12
21. $\frac{8}{13} = \frac{y}{20}$ $12\frac{4}{13}$
22. $\frac{42}{66} = \frac{y}{11}$ 7
23. $\frac{11}{33} = \frac{c}{60}$ 20
24. $\frac{8}{40} = \frac{5}{n}$ 25
25. $\frac{n}{7} = \frac{15}{21}$ 5
26. $\frac{2}{3} = \frac{18}{n}$ 27
27. $\frac{96}{16} = \frac{n}{4}$ 24
28. $\frac{13}{42} = \frac{65}{c}$ 210
29. $\frac{16}{3} = \frac{t}{12}$ 64
30. $\frac{h}{25} = \frac{15}{125}$ 3
31. $\frac{27}{y} = \frac{81}{45}$ 15
32. $\frac{a}{21} = \frac{9}{27}$ 7
33. $\frac{121}{11} = \frac{550}{h}$ 50

SKILL 24

Solve a Rate Problem

Equal Rates as a Proportion

Write the proportion.

300 words typed in 5 minutes. How many words in 3 minutes?

$$\frac{300 \text{ words}}{5 \text{ minutes}} = \frac{n \text{ words}}{3 \text{ minutes}}$$

$$\frac{300}{5} = \frac{n}{3}$$

Write the proportion.

1. 12 oranges cost 99¢. How much for 7 oranges? $\frac{99}{12} = \frac{x}{7}$
2. A car travels 84 km on 7 L of gas. How many kilometers on 13 L of gas? $\frac{84}{7} = \frac{x}{13}$
3. On a map, 5 cm represents 40 km. How many kilometers does 17 cm represent? $\frac{40}{5} = \frac{x}{17}$
4. A machine uses 50 kW•h in 6 hours. How many kilowatt-hours in 45 hours? $\frac{50}{6} = \frac{x}{45}$

Solve a Rate Problem

Find the number of boxes.

5000 envelopes in 10 boxes. 3000 envelopes in how many boxes?

$$\frac{5000 \text{ envelopes}}{10 \text{ boxes}} = \frac{3000 \text{ envelopes}}{n \text{ boxes}}$$

$$\frac{5000}{10} = \frac{3000}{n}$$

$n = 6$ ← Skill 23

Solve.

5. 5 apples cost 79¢. How much for 22 apples? **$3.48**
6. A machine produces 121 bolts in 3 hours. How many bolts in 15 hours? **605**
7. 36 pages typed in 4 hours. How many pages in 13 hours? **117 pp.**
8. A car travels 19 km on 2 L of gas. How many kilometers on 76 L of gas? **722 km**

Mixed Practice

Write the proportion and solve.

9. A machine produces 660 wheels in 4 hours. How many wheels are made in 9 hours? **1485 wheels**
10. Fuel costs 93¢ for 3 L. How much would it cost to fill a car with an 84-L tank? **$26.04**
11. The telephone rate to Europe is $6 for 3 minutes. How much for 14 minutes? **$28**
12. Sally can run 11 km in 60 minutes. How far can she run in 105 minutes? **19.25 km**
13. Juan can read 6 pages in 5 minutes. How long will it take him to read 40 pages? **$33\frac{1}{3}$ minutes**

SKILL 25

Compare Rates

Find the Unit Rate

Find the number of books.

4200 books in 200 boxes.
How many books in 1 box?

$$\frac{4200 \text{ books}}{200 \text{ boxes}} = \frac{n \text{ books}}{1 \text{ box}}$$

$n = 21$ ← Skill 24

The unit rate is 21 books per box.

Find the unit rate.

1. 3456 oranges in 12 crates. How many oranges in 1 crate? **288**
2. Ida prints 4950 words in 9 minutes. How many words in 1 minute? **550**
3. Lou drives 560 km in 7 hours. How many kilometers in 1 hour? **80**
4. $14.25 for 15 dozen eggs. How much for 1 dozen? **$.95**
5. Sam drives 384 km in 6 hours. How far does he drive in 1 hour? How far in 10 hours? **64; 640**
6. 2500 sheets of bond paper weigh 100 pounds. How many sheets weigh 1 pound? What will 10,000 sheets weigh? **25; 400 lb**

Compare Unit Rates

Which is faster?

Car A: 280 km in 4 hours
Car B: 455 km in 7 hours

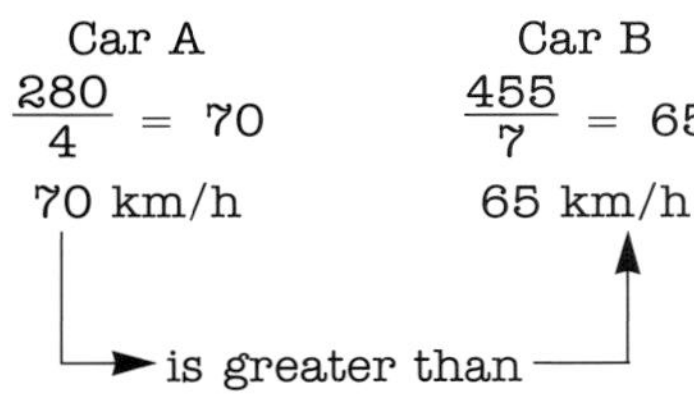

Car A is faster.

Compare the unit rates.

7. Which uses less fuel per kilometer? **A**
   Car A: 52 km on 6 L
   Car B: 59 km on 7 L
8. Which costs less per kilogram? **B**
   Carton A: $89.25 for 15 kg
   Carton B: $117.00 for 20 kg
9. Which shipment holds more per crate? **B**
   Shipment A: 1584 apples in 11 crates
   Shipment B: 1937 apples in 13 crates
10. Which makes more cogs per hour? **A**
    Machine A: 288 cogs in 4 hours
    Machine B: 414 cogs in 6 hours
11. Which gets more km per liter? **B**
    Car A: 162 km on 18 L of fuel
    Car B: 220 km on 20 L of fuel

SKILL 26

Write Decimals as Percents

Two or More Decimal Places

Write 0.037 as a percent.

0.037 = ?%

0.037 ➧ 0.037 (Move 2 places to the right.)

0.037 = 3.7%

Write as a percent.

1. 0.10 **10%**	**2.** 0.15 **15%**	**3.** 0.25 **25%**
4. 0.74 **74%**	**5.** 0.82 **82%**	**6.** 0.93 **93%**
7. 2.13 **213%**	**8.** 4.212 **421.2%**	**9.** 5.753 **575.3%**
10. 0.267 **26.7%**	**11.** 0.391 **39.1%**	**12.** 0.914 **91.4%**
13. 12.104 **1210.4%**	**14.** 0.625 **62.5%**	**15.** 10.82 **1082%**
16. 0.007 **0.7%**	**17.** 0.106 **10.6%**	**18.** 0.008 **0.8%**
19. 0.04 **4%**	**20.** 0.001 **0.1%**	**21.** 0.503 **50.3%**

Fewer Than Two Decimal Places

Write 0.5 as a percent.

0.5 = ?%

0.5 ➧ 0.50% (Use zero as placeholder. Move 2 places to the right.)

0.5 = 50%

Write as a percent.

22. 0.4 **40%**	**23.** 0.7 **70%**	**24.** 0.9 **90%**
25. 7.1 **710%**	**26.** 9.3 **930%**	**27.** 10.5 **1050%**
28. 11.0 **1100%**	**29.** 12.6 **1260%**	**30.** 17.1 **1710%**
31. 22.5 **2250%**	**32.** 29.0 **2900%**	**33.** 37.2 **3720%**
34. 0.1 **10%**	**35.** 7.5 **750%**	**36.** 1.8 **180%**
37. 16.3 **1630%**	**38.** 0.3 **30%**	**39.** 6.2 **620%**

Mixed Practice

Write as a percent.

40. 0.57 **57%**	**41.** 3.80 **380%**	**42.** 0.001 **0.1%**	**43.** 0.67 **67%**	**44.** 20.7 **2070%**
45. 8.8 **880%**	**46.** 0.2915 **29.15%**	**47.** 0.32 **32%**	**48.** 17.4 **1740%**	**49.** 3.003 **300.3%**
50. 139.25 **13,925%**	**51.** 0.2187 **21.87%**	**52.** 9.1 **910%**	**53.** 25.00 **2500%**	**54.** 12.004 **1200.4%**
55. 0.14 **14%**	**56.** 2.185 **218.5%**	**57.** 0.51 **51%**	**58.** 9.3 **930%**	**59.** 1.868 **186.8%**
60. 8.554 **855.4%**	**61.** 0.003 **0.3%**	**62.** 3.246 **324.6%**	**63.** 0.26 **26%**	**64.** 0.07 **7%**
65. 0.0032 **0.32%**	**66.** 3.642 **364.2%**	**67.** 0.29 **29%**	**68.** 0.052 **5.2%**	**69.** 3.42 **342%**
70. 0.30 **30%**	**71.** 2.1 **210%**	**72.** 0.0001 **0.01%**	**73.** 1.67 **167%**	**74.** 1.0 **100%**
75. 2.549 **254.9%**	**76.** 0.005 **0.5%**	**77.** 2.077 **207.7%**	**78.** 1.5 **150%**	**79.** 0.19 **19%**

SKILL 27

Write Fractions/ Mixed Numbers as Percents

Fractions

Write $\frac{3}{5}$ as a percent.

$\frac{3}{5} = ?\%$

$\frac{3}{5} = 0.6 = 60\%$

Skill 14 Skill 26

Write as a percent.

1. $\frac{1}{4}$ 25%
2. $\frac{2}{5}$ 40%
3. $\frac{3}{4}$ 75%
4. $\frac{7}{10}$ 70%
5. $\frac{1}{8}$ 12.5%
6. $\frac{3}{20}$ 15%
7. $\frac{7}{20}$ 35%
8. $\frac{12}{25}$ 48%
9. $\frac{7}{50}$ 14%
10. $\frac{7}{40}$ 17.5%
11. $\frac{5}{8}$ 62.5%
12. $\frac{11}{16}$ 68.75%

Mixed Numbers

Write $2\frac{3}{8}$ as a percent.

$2\frac{3}{8} = ?\%$

$2\frac{3}{8} = \frac{19}{8} = 2.375 = 237.5\%$

Skill 13 Skill 14 Skill 26

Write as a percent.

13. $2\frac{3}{4}$ 275%
14. $4\frac{5}{8}$ 462.5%
15. $6\frac{2}{3}$ 666.7%
16. $7\frac{1}{6}$ 716.7%
17. $8\frac{5}{9}$ 855.6%
18. $5\frac{4}{15}$ 526.7%
19. $8\frac{7}{40}$ 817.5%
20. $14\frac{3}{16}$ 1418.8%
21. $7\frac{11}{50}$ 722%

Mixed Practice

Write as a percent.

22. $\frac{3}{5}$ 60%
23. $5\frac{3}{8}$ 537.5%
24. $12\frac{5}{6}$ 1283.3%
25. $\frac{7}{12}$ 58.3%
26. $\frac{1}{9}$ 11.1%
27. $15\frac{3}{10}$ 1530%
28. $\frac{9}{40}$ 22.5%
29. $20\frac{4}{5}$ 2080%
30. $10\frac{6}{25}$ 1024%
31. $18\frac{5}{8}$ 1862.5%
32. $9\frac{3}{40}$ 907.5%
33. $\frac{15}{16}$ 93.8%
34. $8\frac{19}{50}$ 838%
35. $11\frac{5}{14}$ 1135.7%
36. $16\frac{9}{20}$ 1645%
37. $\frac{7}{9}$ 77.8%
38. $7\frac{3}{16}$ 718.8%
39. $5\frac{11}{15}$ 573.3%
40. $\frac{11}{30}$ 36.7%
41. $17\frac{7}{20}$ 1735%

SKILL 28

Write Percents as Decimals

Percent in Decimal Form

Write 8.75% as a decimal.

Drop % sign.

8.75% ➧ 0 08.75

Move 2 places to the left.

8.75% = 0.0875

Write as a decimal.

1.	10.5%	**0.105**	**2.**	15.7%	**0.157**	**3.**	40%	**0.40**
4.	85%	**0.85**	**5.**	120%	**1.20**	**6.**	137%	**1.37**
7.	6.7%	**0.067**	**8.**	7.1%	**0.071**	**9.**	8.9%	**0.089**
10.	95%	**0.95**	**11.**	119%	**1.19**	**12.**	7.9%	**0.079**
13.	17.2%	**0.172**	**14.**	85.6%	**0.856**	**15.**	100%	**1.00**
16.	1.35%	**0.0135**	**17.**	5.3%	**0.053**	**18.**	142%	**1.42**

Percent in Fractional Form

Write $\frac{3}{20}\%$ as a decimal.

$\frac{3}{20}\% = 0.15\%$ ➧ 0 00.15

Skill 14

$\frac{3}{20}\% = 0.0015$

Write as decimals.

19.	$\frac{7}{10}\%$ **0.007**	**20.**	$\frac{11}{20}\%$ **0.0055**	**21.**	$\frac{5}{8}\%$ **0.00625**
22.	$\frac{2}{25}\%$ **0.0008**	**23.**	$5\frac{3}{5}\%$ **0.056**	**24.**	$7\frac{1}{10}\%$ **0.071**
25.	$6\frac{4}{5}\%$ **0.068**	**26.**	$22\frac{3}{4}\%$ **0.2275**	**27.**	$30\frac{2}{5}\%$ **0.304**
28.	$89\frac{13}{20}\%$ **0.8965**	**29.**	$57\frac{21}{25}\%$ **0.5784**	**30.**	$12\frac{3}{4}\%$ **0.1275**

Mixed Practice

Write as a decimal.

31.	74.25% **0.7425**	**32.**	$6\frac{1}{4}\%$ **0.0625**	**33.**	$18\frac{4}{5}\%$ **0.1880**	**34.**	0.97% **0.0097**	**35.**	$14\frac{9}{20}\%$ **0.1445**
36.	$28\frac{3}{40}\%$ **0.28075**	**37.**	0.125% **0.00125**	**38.**	6.25% **0.0625**	**39.**	$\frac{7}{50}\%$ **0.0014**	**40.**	6.63% **0.0663**
41.	8.79% **0.0879**	**42.**	100% **1.00**	**43.**	$\frac{19}{20}\%$ **0.0095**	**44.**	127% **1.27**	**45.**	$\frac{1}{40}\%$ **0.00025**
46.	$20\frac{1}{4}\%$ **0.2025**	**47.**	$22\frac{1}{8}\%$ **0.22125**	**48.**	14.6% **0.146**	**49.**	$25\frac{1}{5}\%$ **0.252**	**50.**	18.9% **0.189**

SKILL 29

Write Percents as Fractions

Percent in Decimal Form

Write 37.5% as a fraction.

$37.5\% = 0.375 = \frac{3}{8}$

Skill 28 (0.375) Skill 14 ($\frac{3}{8}$)

Write as a fraction in lowest terms.

1–18 Refer to selected answers at the back of the text.

1. 45% **2.** 80% **3.** 175%

4. 200% **5.** 11.7% **6.** 0.1%

7. 0.15% **8.** 67.3% **9.** 10.6%

10. 78.55% **11.** 37.63% **12.** 51.42%

13. 50% **14.** 75% **15.** 10.1%

16. 0.12% **17.** 80.9% **18.** 42.1%

Percent in Fractional Form

Write $16\frac{2}{3}\%$ as a fraction.

$$16\frac{2}{3}\% = 16\frac{2}{3} \div 100$$

$$= \frac{50}{3} \div 100 \quad \text{Skill 13}$$

$$= \frac{50}{300} \quad \text{Skill 21}$$

$$= \frac{1}{6} \quad \text{Skill 12}$$

Write as a fraction in lowest terms.

19. $6\frac{1}{4}\%$ $\frac{1}{16}$ **20.** $37\frac{1}{2}\%$ $\frac{3}{8}$ **21.** $43\frac{3}{4}\%$ $\frac{7}{16}$

22. $3\frac{1}{2}\%$ $\frac{7}{200}$ **23.** $9\frac{1}{4}\%$ $\frac{37}{400}$ **24.** $10\frac{1}{3}\%$ $\frac{31}{300}$

25. $15\frac{1}{2}\%$ $\frac{31}{200}$ **26.** $20\frac{1}{2}\%$ $\frac{41}{200}$ **27.** $25\frac{1}{2}\%$ $\frac{51}{200}$

28. $15\frac{3}{4}\%$ $\frac{63}{400}$ **29.** $16\frac{2}{3}\%$ $\frac{1}{6}$ **30.** $33\frac{1}{3}\%$ $\frac{1}{3}$

Mixed Practice

Write as a fraction in lowest terms.

31. 30% $\frac{3}{10}$ **32.** 57.2% $\frac{143}{250}$ **33.** $31\frac{1}{4}\%$ $\frac{5}{16}$ **34.** 63.5% $\frac{127}{200}$ **35.** $12\frac{1}{2}\%$ $\frac{1}{8}$

36. $13\frac{1}{3}\%$ $\frac{2}{15}$ **37.** $26\frac{2}{3}\%$ $\frac{4}{15}$ **38.** 64.75% $\frac{259}{400}$ **39.** $11\frac{1}{9}\%$ $\frac{1}{9}$ **40.** 78.55% $\frac{1571}{2000}$

41. 50% $\frac{1}{2}$ **42.** 75% $\frac{3}{4}$ **43.** $1\frac{1}{4}\%$ $\frac{1}{80}$ **44.** 100% $\frac{1}{1}$ or 1 **45.** $12\frac{3}{4}\%$ $\frac{51}{400}$

46. $4\frac{3}{4}\%$ $\frac{19}{400}$ **47.** $5\frac{3}{8}\%$ $\frac{43}{800}$ **48.** 75.91% $\frac{7591}{10{,}000}$ **49.** $6\frac{2}{3}\%$ $\frac{1}{15}$ **50.** 121% $1\frac{21}{100}$

SKILL 30

Find the Percentage

Decimal Percents

Find 15.5% of 36.

15.5% of $36 = n$

Skill 28

$0.155 \times 36 = n$

$5.58 = n$

15.5% of $36 = 5.58$

Find the percentage.

1. 15% of 60 **9**
2. 30% of 72 **21.6**
3. 4% of 96 **3.84**
4. 9% of 122 **10.98**
5. 6.5% of 120 **7.8**
6. 8.3% of 150 **12.45**
7. 17.8% of 80 **14.24**
8. 31.2% of 140 **43.68**
9. 5.81% of 60 **3.486**
10. 7.32% of 45 **3.294**
11. 67.7% of 67 **45.359**
12. 8.92% of 35 **3.122**

Fractional Percents

Find $\frac{3}{4}\%$ of 1600.

$\frac{3}{4}\%$ of $1600 = n$

Skill 29

$\frac{3}{400} \times 1600 = n$

$12 = n$

$\frac{3}{4}\%$ of $1600 = 12$

Find the percentage.

13. $2\frac{1}{2}\%$ of 400 **10**
14. $4\frac{1}{2}\%$ of 200 **9**
15. $33\frac{1}{3}\%$ of 120 **40**
16. $3\frac{5}{6}\%$ of 600 **23**
17. $\frac{3}{4}\%$ of 800 **6**
18. $\frac{3}{5}\%$ of 50 **0.30**
19. $\frac{3}{8}\%$ of 600 **2.25**
20. $\frac{7}{8}\%$ of 80 **0.70**

Mixed Practice

Find the percentage.

21. 80% of 160 **128**
22. $8\frac{1}{3}\%$ of 72 **6**
23. 9.1% of 90 **8.19**
24. $15\frac{3}{5}\%$ of 90 **14.04**
25. $16\frac{2}{3}\%$ of 90 **15**
26. 45% of 72 **32.4**
27. $\frac{1}{4}\%$ of 800 **2**
28. 0.75% of 1000 **7.5**
29. 125% of 64 **80**
30. $12\frac{1}{4}\%$ of 65 **7.9625**
31. 7.2% of 127 **9.144**
32. $6\frac{1}{4}\%$ of 1600 **100**
33. $4\frac{1}{6}\%$ of 600 **25**
34. 16.5% of 84 **13.86**
35. $8\frac{1}{2}\%$ of 75 **6.375**
36. 24.7% of 80 **19.76**

SKILL 31

Find the Rate

Percentage Less Than Base

What percent of 75 is 60?

$n\%$ of 75 = 60

Write as a proportion and solve.

$$\frac{n}{100} = \frac{60}{75}$$

$75 \times n = 6000$

$n = 80$ ← Skill 23

80% of 75 = 60

Solve. Round answers to the nearest tenth of a percent.

1. $n\%$ of 40 = 20 **50%**
2. $n\%$ of 60 = 15 **25%**
3. $n\%$ of 90 = 18 **20%**
4. $n\%$ of 60 = 9 **15%**
5. $n\%$ of 70 = 25 **35.7%**
6. $n\%$ of 90 = 65 **72.2%**
7. $n\%$ of 65 = 58 **89.2%**
8. $n\%$ of 84 = 75 **89.3%**
9. $n\%$ of 113 = 79 **69.9%**
10. $n\%$ of 410 = 295.5 **72.1%**
11. $n\%$ of 94 = 82 **87.2%**
12. $n\%$ of 296 = 239.76 **81%**

Percentage Greater Than Base

What percent of 90 is 162?

$n\%$ of 90 = 162

Write as a proportion and solve.

$$\frac{n}{100} = \frac{162}{90}$$

$90 \times n = 16{,}200$

$n = 180$

180% of 90 = 162

Solve. Round answers to the nearest tenth of a percent.

13. $n\%$ of 20 = 30 **150%**
14. $n\%$ of 36 = 45 **125%**
15. $n\%$ of 50 = 60 **120%**
16. $n\%$ of 70 = 91 **130%**
17. $n\%$ of 60 = 80 **133.3%**
18. $n\%$ of 72 = 96 **133.3%**
19. $n\%$ of 24 = 31 **129.2%**
20. $n\%$ of 56 = 77 **137.5%**
21. $n\%$ of 37 = 148 **400%**
22. $n\%$ of 60 = 90 **150%**
23. $n\%$ of 81 = 415.75 **513.3%**
24. $n\%$ of 110 = 130 **118.2%**

Mixed Practice

Solve. Round answers to the nearest tenth of a percent.

25. $n\%$ of 125 = 100 **80%**
26. $n\%$ of 100 = 115 **115%**
27. $n\%$ of 130 = 60 **46.2%**
28. $n\%$ of 56 = 96 **171.4%**
29. $n\%$ of 89 = 49 **55.1%**
30. $n\%$ of 84 = 108 **128.6%**
31. $n\%$ of 115 = 92 **80%**
32. $n\%$ of 42 = 52.5 **125%**
33. $n\%$ of 64 = 76.8 **120%**
34. $n\%$ of 64 = 24 **37.5%**
35. $n\%$ of 204 = 51 **25%**
36. $n\%$ of 42 = 6.72 **16%**
37. $n\%$ of 36 = 50 **138.9%**
38. $n\%$ of 173 = 136.25 **78.8%**
39. $n\%$ of 18 = 36 **200%**
40. $n\%$ of 21 = 63 **300%**
41. $n\%$ of 120 = 80 **66.7%**
42. $n\%$ of 15.5 = 62 **400%**
43. $n\%$ of 90 = 4.5 **5%**
44. $n\%$ of 125 = 18.75 **15%**
45. $n\%$ of 75 = 110 **146.7%**

611

SKILL 32

Find the Base

Decimal Percents

42 is 37.5% of what number?

37.5% of $n = 42$

$0.375 \times n = 42$

Skill 28

$n = 42 \div 0.375$

$n = 112$

37.5% of $112 = 42$

Find the number.

1. 12.5% of $n = 9$ **72**
2. 62.5% of $n = 55$ **88**
3. 8.25% of $n = 664$ **8048.5**
4. 7% of $n = 3.5$ **50**
5. 5.75% of $n = 92$ **1600**
6. 9% of $n = 8.2$ **91.1**
7. 11% of $n = 37$ **336.4**
8. 3% of $n = 8.7$ **290**
9. 5% of $n = 3$ **60**
10. 12.5% of $n = 21$ **168**
11. 37.3% of $n = 50$ **134**
12. 81% of $n = 14$ **17.3**

Fractional Percents

3 is $6\frac{1}{4}\%$ of what number?

$6\frac{1}{4}\%$ of $n = 3$

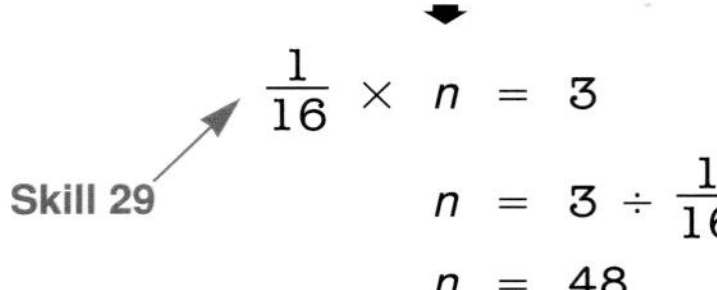

$\frac{1}{16} \times n = 3$

Skill 29

$n = 3 \div \frac{1}{16}$

$n = 48$

$6\frac{1}{4}\%$ of $48 = 3$

Find the number.

13. $2\frac{1}{2}\%$ of $n = 4$ **160**
14. $3\frac{1}{8}\%$ of $n = 2$ **64**
15. $33\frac{1}{3}\%$ of $n = 20$ **60**
16. $\frac{1}{4}\%$ of $n = 2$ **800**
17. $37\frac{1}{2}\%$ of $n = 34$ **$90\frac{2}{3}$**
18. $\frac{3}{5}\%$ of $n = 6$ **1000**
19. $45\frac{1}{2}\%$ of $n = 25$ **$54\frac{86}{91}$**
20. $\frac{3}{4}\%$ of $n = 7$ **$933\frac{1}{3}$**

Mixed Practice

Find the number.

21. 80% of $n = 60$ **75**
22. 12.5% of $n = 39$ **312**
23. 116% of $n = 2.9$ **2.5**
24. 10% of $n = 2.9$ **29**
25. 8% of $n = 4.2$ **52.5**
26. 40% of $n = 6.8$ **17**
27. $16\frac{2}{3}\%$ of $n = 13$ **78**
28. 145% of $n = 6.38$ **4.4**
29. $66\frac{2}{3}\%$ of $n = 120$ **180**
30. 25% of $n = 36$ **144**
31. $2\frac{2}{3}\%$ of $n = 6$ **225**
32. 90% of $n = 72$ **80**
33. 6% of $n = 3.3$ **55**
34. 7.8% of $n = 9$ **115.4**
35. $16\frac{3}{4}\%$ of $n = 33$ **$197\frac{1}{67}$**

Application A Formulas

Substituting in a Formula

To use a formula to solve a problem, substitute known values in the formula. Then solve. Remember to perform any computations within parentheses first.

▶ An airplane travels at a rate of 960 km/h. How far can the plane travel in 7 hours?

Formula: Distance = Rate × Time

Substitute: Distance = 960 × 7

Solve: Distance = 6720

The plane travels 6720 km in 7 hours.

Find the distance.

	Distance = Rate × Time		
	Rate	**Time**	**Distance**
1.	100 mi/h	6 h	600 mi
2.	55 km/h	4 h	220 km
3.	14 m/s	32 s	448 m
4.	30 mi/h	6 h	180 mi
5.	42 m/s	5 s	210 m

Find the area of the triangle.

	Area of Triangle = × Base × Height		
	Base	**Height**	**Area**
6.	1 m	0.5 m	0.25 m^2
7.	64 ft	38 ft	1216 ft^2
8.	18 in	5 in	45 in^2
9.	30 km	24 km	360 km^2

Find the percent of sales made.

	% of Sales = $\frac{\text{Sales Made}}{\text{Possible Sales}}$ × 100		
	Possible Sales	**Sales Made**	**Percent**
10.	$ 4,600	$ 1,900	41.3
11.	$19,100	$ 6,000	31.4
12.	$23,000	$13,000	56.5
13.	$64,000	$15,000	23.4
14.	$ 8,200	$ 3,000	36.6

Find the price of the stereo system.

	Price = Amp + CD Player + Speakers			
	Amp	**CD Player**	**Speakers**	**Price**
15.	$199.00	$ 99.95	$139.00	$ 437.95
16.	$395.00	$119.99	$259.00	$ 773.99
17.	$685.00	$249.00	$495.00	$1429.00
18.	$449.00	$188.00	$350.00	$ 987.00

Find the long-distance telephone charge.

	Charge = Cost per min × Minutes		
	Minutes	**Cost per min**	**Charge**
19.	15	$0.35	$ 5.25
20.	9	$1.12	$10.08
21.	31	$1.49	$46.19
22.	22	$0.54	$11.88
23.	38	$1.32	$50.16

Application B More Formulas

Multiplying or Dividing to Solve Formulas

To solve some formulas after substituting, you may have to multiply or divide both sides of the equation by the same number.

▶ An airplane cruising at a rate of 575 km/h has traveled 2645 km. How long did it take to make this trip?

Formula:	Distance = Rate × Time
Substitute:	2645 = 575 × Time
Divide:	$\frac{2645}{575} = \frac{575 \times \text{Time}}{575}$
Solve:	4.6 h = Time

Find the missing amount.

	Distance = Rate × Time		
	Rate	Time	Distance
1.	88 in/min	3.5 min	308 in
2.	60 km/h	3.25 h	195 km
3.	675 yd/h	13 h	8775 yd
4.	45 km/h	1.75 h	78.75 km
5.	24 km/h	8 h	192 km
6.	55 ft/h	$4\frac{1}{2}$ h	$247\frac{1}{2}$ ft
7.	3.2 m/s	30 s	96 m

Adding or Subtracting to Solve Formulas

To solve some formulas after substituting, you may have to add the same number to, or subtract it from, both sides of the equation.

▶ The temperature of a solution is 212°F (degrees Fahrenheit). What is this in degrees Celsius (°C)?

Formula:	°F = (1.8 × °C) + 32
Substitute:	212 = (1.8 × °C) + 32
Subtract:	− 32 − 32
	180 = (1.8 × °C)
Divide:	=
Solve:	100 = °C

Find the missing amount.

	°F = (1.8 × °C) + 32	
	Degrees Celsius	Degrees Fahrenheit
8.	50	122
9.	30	86
10.	0	32
11.	120	248
12.	35	95
13.	20	68
14.	144	291.2
15.	10	50
16.	−15	5

Application C Tables and Charts

Reading Tables and Charts

To read a table or chart, find the row containing one of the conditions of the information you seek. Run down the column containing the other condition until it crosses that row. Read the answer.

▶How many sales were made at the Lincoln office?

The table shows that there were 537 sales made at the Lincoln office.

IMAG CORP. SALES REPORT			
Office	Agents	Sales	Income
Bronx	252	640	$12,000,000
Calgary	204	307	1,800,000
Columbus	92	778	1,200,000
Dallas	50	761	1,300,000
Denver	76	398	1,500,000
Lincoln	35	537	1,500,000
San Diego	128	813	14,300,000
Toronto	710	390	7,900,000

Find the income for each of the following offices.

1. Columbus **$1,200,000**
2. Denver **$1,500,000**
3. Toronto **$7,900,000**
4. Bronx **$12,000,000**
5. Which offices made more than 500 sales? **Bronx, Columbus, Dallas, Lincoln, San Diego**
6. Which offices have less than 200 agents? **Columbus, Dallas, Denver, Lincoln, San Diego**

Using Tables and Charts

To classify an item, find the row or column that contains the given data. Then read the classification from the head of the row or column.

▶An order weighing 5 lb 5 oz cost $9.25 in shipping charges. To which zone was it delivered?

The table shows that the order was delivered within Zone 2.

DELIVERY CHART	Zone 1	Zone 2	Custom Delivery			
			Express		Express Plus	
			Ordered	Delivered	Ordered	Delivered
Find Your State at Right ▶ / Shipping Weight ▼	AL, AR, DC, DE, IA, IL, IN, KS, KY, MD, MI, MN, MO, MS, NC, NE, ND, OH, OK, PA, SC, SD, TN, VA, WI, WV	AK, AZ, CA, CO, CT, FL, GA, HI, ID, LA, MA, ME, MT, NH, NJ, NM, NV, NY, OR, RI, TX UT, VT, WA, WY	Mon. Tue. Wed. Thur. Fri. Sat. Sun.	Thur. Fri. Mon. Tue. Wed. Wed. Wed.	Mon. Tue. Wed. Thur. Fri. Sat. Sun.	Wed. Thur. Fri. Mon. Tue. Tue. Tue.
Minimum Charge	$ 4.00	$ 4.50	$ 9.00		$19.00	
2 lb 1 oz to 4 lb	6.50	6.75	11.00		21.00	
4 lb 1 oz to 8 lb	9.00	9.25	15.75		25.75	
8 lb 1 oz to 14 lb	11.25	11.75	22.25		32.25	
14 lb 1 oz to 20 lb	13.00	15.00	28.25		38.25	

Find the missing information.

7. An order weighing 16 lb cost $38.25 in shipping charges. How was the order sent? **Express plus**
8. An order weighing 9 lb 12 oz cost $11.25 in shipping charges. To what zone was it delivered? **Zone 1**
9. An order weighing 10 lb cost $22.25 in shipping charges.
   a. How was the order sent? **Express**
   b. When would the order be delivered if it was ordered on a Wednesday? **Mon.**

Application D Making Change

Making Change: Building Up

To make change, build up to the next nickel, dime, quarter, and so on, until the amount presented is reached.

▶ Compute the change from $5.00 for a purchase of $3.67.

CHANGE		
		$3.67
	3 pennies ⟶	+ .03
		3.70
	1 nickel ⟶	+ .05
		3.75
	1 quarter ⟶	+ .25
		4.00
	1 dollar ⟶	+1.00
	Amount presented ⟶	$5.00

Compute the change from $10. Use the fewest coins and bills.

1. $3.75 1 q., 1 do., 1 five do.
2. $7.78 2 p., 2 di., 2 do.
3. $1.99 1 p., 3 do., 1 five do.
4. $3.47 3 p., 2 q., 1 do. 1 five do.
5. $5.63 2 p., 1 di., 1 q., 4 do.
6. $9.19 1 p., 1 n., 3 q.

Compute the change. Use the fewest coins and bills.

7. Amount spent: $15.05 2 di., 3 q., 4 do.
   Amount presented: $20
8. Amount spent: $3.37 3p., 1 di., 2 q., 1 do.,
   Amount presented: $20 1 five do., 1 ten do.
9. Amount spent: $2.29 1p., 2 di., 2 q.
   Amount presented: $3

Making Change: Loose Coins

The amount presented may contain loose coins so that fewer coins and bills are returned in the change.

▶ Compute the change from $5.02 for a purchase of $3.67.

	Amount Presented	Purchase
a. Subtract the loose change.	$5.02	$3.67
	− .02	− .02
	$5.00	$3.65

b. Now compute the change from $5.00 for a purchase of $3.65.

CHANGE		
		$3.65
	1 dime ⟶	+ .10
		3.75
	1 quarter ⟶	+ .25
		4.00
	1 dollar ⟶	+1.00
		$5.00

Compute the change using the fewest coins and bills.

10. Amount spent: $5.26 3 q.
    Amount presented: $6.01
11. Amount spent: $7.79 1 q., 2 do.
    Amount presented: $10.04
12. Amount spent: $1.07 4 do.
    Amount presented: $5.07
13. Amount spent: $3.92 1 di., 1 do., 1 five do.
    Amount presented: $10.02
14. Amount spent: $15.21 1 n., 3 q., 4 do.
    Amount presented: $20.01
15. Amount spent: $7.81 1 q., 2 do.
    Amount presented: $10.06
16. Amount spent: $6.03 1 five do, 1 ten do.
    Amount presented: $21.03
17. Amount spent: $9.62 2 q., 1 do., 1 ten do.
    Amount presented: $21.12

Some students may suggest one $2 bill instead of 2 single dollars and one 50¢ coin instead of 2 quarters.

Application E Rounding Time: Nearest Quarter Hour

To round to the nearest quarter hour, find the quarter-hour interval that the given time is between. Subtract the earlier quarter from the given time. Subtract the given time from the later quarter. Round to whichever is closer.

▶ Round 6:19 to the nearest quarter hour.

6:19	6:30
− 6:15	− 6:19
4 minutes	11 minutes

4 is less than 11; round to 6:15

Round to the nearest quarter hour.

1. 8:39 — 8:45	**2.** 10:40 — 10:45	**3.** 6:55 — 7:00
4. 1:23 — 1:30	**5.** 11:17 — 11:15	**6.** 3:11 — 3:15
7. 12:09 — 12:15	**8.** 4:44 — 4:45	**9.** 3:32 — 3:30
10. 11:11 — 11:15	**11.** 8:51 — 8:45	**12.** 1:50 — 1:45
13. 7:35 — 7:30	**14.** 5:04 — 5:00	**15.** 9:40 — 9:45
16. 10:34 — 10:30	**17.** 12:39 — 12:45	**18.** 6:41 — 6:45
19. 6:01 — 6:00	**20.** 3:49 — 3:45	**21.** 11:36 — 11:30
22. 8:05 — 8:00	**23.** 10:10 — 10:15	**24.** 2:15 — 2:15

Application  Elapsed Time (Hours)

Finding Elapsed Time

To find the elapsed time, subtract the earlier time from the later time. If it is necessary to "borrow," rewrite the later time by subtracting 1 from the hours and adding 60 to the minutes.

▶ How much time has elapsed between 11:45 a.m. and 12:23 p.m.?

12:23	=	11:23 + :60	=	11:83
− 11:45	=	− 11:45	=	− 11:45
				38 min

Find the elapsed time.

1. From 6:05 p.m. to 8:51 p.m. — 2 h:46 min
2. From 4:18 a.m. to 6:52 a.m. — 2 h:34 min
3. From 8:30 a.m. to 11:45 a.m. — 3 h:15 min
4. From 6:45 p.m. to 11:59 p.m. — 5 h:14 min
5. From 10:45 a.m. to 11:36 a.m. — 51 min
6. From 3:45 p.m. to 5:30 p.m. — 1 h:45 min

Finding Elapsed Time Spanning 1 O'Clock

To find the elapsed time when the period spans 1 o'clock, add 12 hours to the later time before subtracting.

▶ How much time has elapsed between 7:45 p.m. and 2:52 a.m.?

2:52	=	2:52 + 12:00	=	14:52
− 7:45	=	− 7:45	=	− 7:45
				7h:7 min

Find the elapsed time.

7. From 8:15 a.m. to 1:30 p.m. — 5 h:15 min
8. From 4:55 a.m. to 1:00 p.m. — 8 h: 5 min
9. From 11:30 p.m. to 7:50 a.m. — 8 h:20 min
10. From 9:45 p.m. to 8:00 a.m. — 10 h:15 min
11. From 3:30 a.m. to 2:15 p.m. — 10 h:45 min

Application G Elapsed Time (Days)

To find the number of days **between** two dates, or **from** one date **to** the next, use the table on page 645. Find the position of each date in the year. When the dates run from one year to the next, add 365 to the later date. Then subtract the earlier date from the later.

Day No.	Jan.	Feb.	Mar.	Apr.	May	June	July	Aug.	Sept.
1	1	32	60	91	121	152	182	213	244
2	2	33	61	92	122	153	183	214	245
3	3	34	62	93	123	154	184	215	246
4	4	35	63	94	124	155	185	216	247
5	5	36	64	95	125	156	186	217	248
6	6	37	65	96	126	157	187	218	249

To find the number of days **from** one date **through** another, find the elapsed time between the dates and add 1 day.

▶ How many days have elapsed between April 4 and January 6?

Jan. 6:	(6 + 365)	=	371
Apr. 4:	− 94	=	− 94
			277 days

Find the elapsed time. Use the table on page 645.

1. Between August 10 and December 25
   137 days
2. From March 21 through September 30
   194 days
3. Between June 1 and December 30
   212 days
4. From May 21 to July 2
   42 days
5. From September 5 through May 20
   258 days

Application  Determining Leap Years

A year is a leap year if it is exactly divisible by 4. Exceptions are years that end in '00' (1900, 2000, and so on); they must be divisible by 400.

▶ Was 1848 a leap year?

1848 ÷ 4 = 462; 1848 was a leap year.

Find the leap years.

1. 1949	**2.** 1980 **Yes**	**3.** 2000 **Yes**
4. 1912 **Yes**	**5.** 2100	**6.** 1954
7. 1992 **Yes**	**8.** 1951	**9.** 2400 **Yes**
10. 1882	**11.** 1920 **Yes**	**12.** 2004 **Yes**

Application  Elapsed Time in a Leap Year

When the dates span February 29 in a leap year, add 1 day to the later date.

▶ How many days elapsed between January 5, 1980, and September 3, 1980?

Sept. 3, 1980:	(246 + 1)	=	247
Jan. 5, 1980:	− 5	=	− 5
			242 days

Find the elapsed time. Use the table on page 645.

1. January 12, 1978, to April 1, 1978
   79 days
2. July 1, 1979, through March 21, 1980
   265 days
3. February 5, 1984, to September 23, 1984
   231 days
4. January 9, 1988, through July 2, 1988
   176 days
5. February 21, 2000, to September 30, 2000
   222 days

618

Application J Fractional Parts of a Year

Changing Months Into a Part of a Year

To change months into a part of a year, write as a fraction with denominator 12. Express answers in lowest terms.

▶ 6 months is what part of a year?

$$6 \text{ months} = \frac{6}{12} \text{ year} = \frac{1}{2} \text{ year}$$

Change each term to a part of a year.

1. 9 months $\frac{3}{4}$ **yr.**
2. 7 months $\frac{7}{12}$ **yr.**
3. 12 months **1 yr.**
4. 20 months $1\frac{2}{3}$ **yr.**
5. 36 months **3 yr.**
6. 18 months $1\frac{1}{2}$ **yr.**
7. 21 months $1\frac{3}{4}$ **yr.**
8. 8 months $\frac{2}{3}$ **yr.**
9. 14 months $1\frac{1}{6}$ **yr.**

Changing Days Into a Part of a Year

To change days into a part of a year, write as a fraction with denominator 365 when using the exact year or 360 when using the ordinary year. Express answers in lowest terms.

▶ 240 days is what part of an exact year?

$$240 \text{ days} = \frac{240}{365} \text{ year} = \frac{48}{73} \text{ year}$$

Change each term to a part of a year.

10. 90 days as an exact year $\frac{18}{73}$ **yr.**
11. 300 days as an ordinary year $\frac{5}{6}$ **yr.**
12. 120 days as an exact year $\frac{24}{73}$ **yr.**
13. 175 days as an exact year $\frac{35}{73}$ **yr.**
14. 146 days as an ordinary year $\frac{73}{180}$ **yr.**
15. 250 days as an exact year $\frac{50}{73}$ **yr.**

Application  Chronological Expressions

COMMON CHRONOLOGICAL EXPRESSIONS (number of occurrences per year)		
	Weekly (52)	Biweekly (26)
Semimonthly (24)	Monthly (12)	Bimonthly (6)
	Quarterly (4)	
Semiannually (2)	Annually (1)	

▶ Monthly payments for 30 years. How many payments in all?

Monthly: 12 times a year
In 30 years: 12 × 30 = 360 payments

Find the number of occurrences. Use the table at the left.

1. Weekly for 1 year **52**
2. Biweekly for 2 years **52**
3. Quarterly for 4 years **16**
4. Semimonthly for 1 year **24**
5. Bimonthly for 2 years **12**
6. Annually for 25 years **25**
7. Monthly for year **6**
8. Weekly for 1.5 years **78**
9. Semimonthly for 2 years **60**

The Complement of a Number

To find the complement of a percent or decimal less than 1, subtract the percent from 100% or the decimal from 1.

▶ What is the complement of 65%?

The complement of 65% = 100% − 65% = 35%

▶ What is the complement of 0.25?

The complement of 0.25 = 1 − 0.25 = 0.75

Find the complement.

1. 0.15 **0.85**
2. 0.72 **0.28**
3. 0.35 **0.65**
4. 0.50 **0.50**
5. 0.80 **0.20**
6. 0.34 **0.66**
7. 91% **9%**
8. 85% **15%**
9. 47% **53%**
10. 66% **34%**
11. 7% **93%**
12. 45% **55%**
13. 3.5% **96.5%**
14. 56.6% **43.4%**
15. 75.9% **24.1%**

Reading Bar Graphs

To read a bar graph, find the bar that represents the information you seek. Trace an imaginary line from the top of the bar to the scale on the left. Read the value represented by the bar from the scale.

▶ In 1985, how much did it cost to paint the exterior of a nine-room house?

The bar graph shows that it cost about $800 to paint the exterior in 1985.

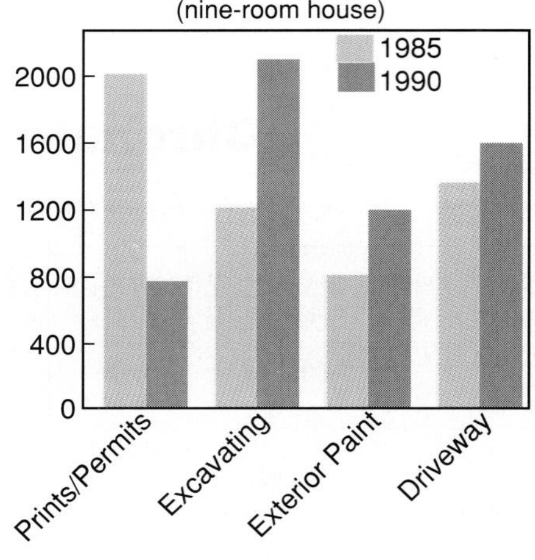

Answer the following. Use the bar graph shown.

1. In 1985, how much did it cost to construct the driveway? **$1300**
2. In 1990, how much did it cost to obtain the prints and permits? **$750**
3. In 1990, how much did it cost to excavate? **$2100**
4. What type of building costs decreased rather than increased? **Prints/Permits**
5. In 1990, how much more did it cost to excavate than in 1985? **$900**
6. In 1985, how much more did it cost to obtain prints and permits than in 1990? **$1250**
7. In 1990, how much more did it cost to paint the exterior of the house than in 1985? **$400**

Application N Reading Line Graphs

To read a line graph, find the point that represents the information you seek. Trace an imaginary line from the point to the scale on the left. Read the value represented by the point from the scale.

▶ What was the predicted and the actual amount of sales for the month of May?

The line graph shows that the predicted amount of sales was $10,000. The actual amount was $8000.

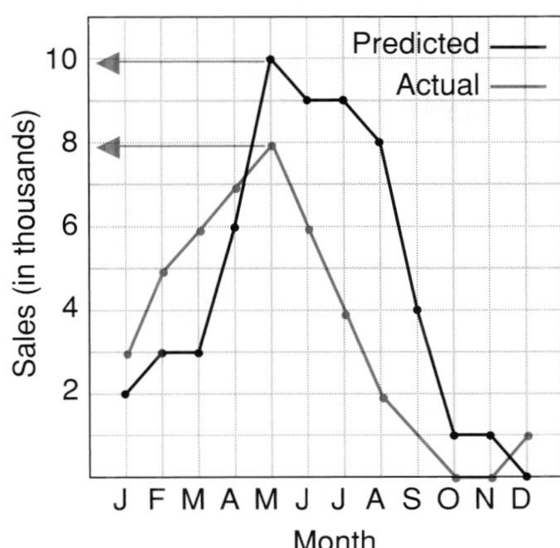

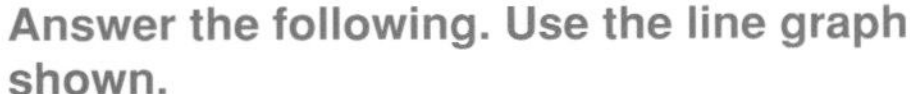

Answer the following. Use the line graph shown.

1. What was the predicted amount of sales for the month of August? **$8000**
2. Which month(s) actually had the least amount of sales? **October and November**
3. What was the actual amount of sales for the month of October? **$0**
4. In which month was the actual amount of sales equal to the amount predicted? **None**

Application O Reading Pictographs

To read a pictograph, find the scale to see how much each picture represents. Then multiply that amount times the number of pictures on each line to get the total amount.

▶ How many cars did Motor Mart sell?

Each picture represents 100 cars sold. The Motor Mart row has 7.5 pictures.

7.5 × 100 = 750 cars

Cars Sold

Mlra Motors

Motor Mart

Star Motors

= 100 cars

Answer the following. Use the pictograph shown.

1. How many cars did Mira Motors sell? **650**
2. How many cars did Star Motor sell? **1100**
3. How many more cars did Star Motors sell than Motor Mart? **350**

Application P Circle Graphs

Reading Circle Graphs

A circle graph is used to compare the parts to the whole. To find what part of the whole each section represents, multiply the amount, or percent, per section times the total amount.

▶ How much of the annual budget was spent on social services?

Social Services: 10¢ of the tax dollar

$$0.10 \times \$408{,}000 = \$40{,}800$$

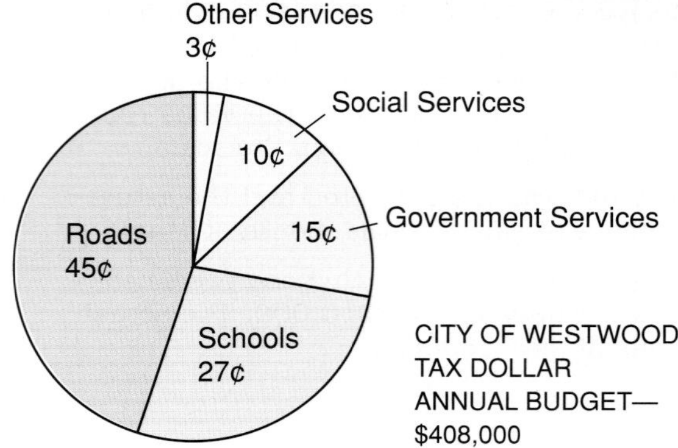

Answer the following. Use the circle graph shown.

1. How much of every tax dollar was spent to operate the schools? **$.27**
2. How much more of the annual budget was spent on city roads than on schools? **$73,440**
3. How much of the annual budget was spent to operate the schools? **$110,160**
4. Does the city spend more on the schools than on all other services combined? **No**

Constructing Circle Graphs

To construct a circle graph, find the percent of the total that each section represents. Multiply each percent times 360° to find the measure of the angle for each portion of the circle graph. Use a protractor to draw each portion.

▶ Construct that portion of a circle graph that represents the Bradford family spending 40¢ out of every dollar for food.

$$= 40\%;\ 0.40 \times 360° = 144°$$

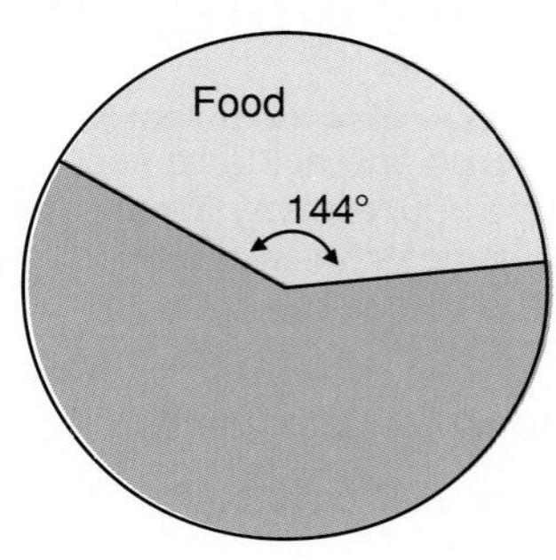

Check students' graphs.

5. Complete the circle graph shown. Use the following information.

   The Bradford family budgets the rest of each dollar as follows: rent, 30¢; books, 15¢; clothing and personal items, 10¢; and all other expenses, 5¢.
   108°; 54°; 36°; 18°

6. Draw a circle graph. Use the following information.

   A factory budgets each dollar spent as follows: salaries, 35¢; raw materials, 30¢; utilities, 10¢; plant maintenance, 15¢; and research and development, 10¢.
   126°; 108°; 36°; 54°; 36°

Application Q Mean

To find the mean (average) of a group of numbers, find the sum of the group and divide it by the number of items in the group.

▶ What is the mean of the following: 454, 376, 416, 472?

$$\text{Mean} = \frac{454 + 376 + 416 + 472}{4}$$

$$= 429.5$$

Find the mean for each group.

1. 43, 19, 61, 72, 81, 50 **$54\frac{1}{3}$**
2. 116, 147, 136, 151, 123, 117, 120 **130**
3. 4615, 5918, 7437, 8937 **6726.75**
4. 4.0, 3.5, 4.0, 3.0, 3.5, 3.0, 2.0, 0.5, 4.6, 5.0 **3.31**
5. $580,000; $625,000; $105,583; $733,358; $5,750,000; $255,000; $600,000 **$1,235,563**

Application  Median

To find the median of a group of numbers, arrange the items in order from smallest to largest. The median is the number in the middle. If there is an even number of items, find the mean of the two middle numbers.

▶ What is the median of the following: 454, 376, 416, 472?

Arrange in order: 376, <u>416</u>, <u>454</u>, 472

$$\text{Median} = \frac{416 + 454}{2} = 435$$

Find the median for each group.

1. 141, 136, 191, 187, 149, 148 **$148\frac{1}{2}$**
2. 17, 21, 30, 35, 27, 25, 15 **25**
3. 91, 92, 85, 98, 100, 76, 80, 75 **88**
4. 4.2, 3.7, 3.1, 4.8, 2.4, 3.0, 2.9 **3.1**
5. 0.07, 0.05, 0.10, 0.12, 0.09, 0.17, 0.01 **0.09**
6. $121,500; $49,750; $72,175; $65,449 **$68,812**
7. 2.06, 2.04, 2.00, 2.10, 2.08, 2.24, 1.55, 2.04, 2.13, 2.08, 2.09 **2.08**

Application  Mode

To find the mode of a group of numbers, look for the number that appears most often. A group may have no mode, or it may have more than one mode.

▶ What is the mode of the following: 92, 88, 76, 84, 92, 96, 1200?

The number that appears most often, the mode, is 92.

Find the mode for each group.

1. $51, $13, $24, $62, $55, $57, $24 **$24**
2. 800, 600, 800, 500, 600, 700 **600; 800**
3. 4.1, 4.7, 4.5, 4.3, 4.2, 4.4 **None**
4. 3, 3, 5, 3, 4, 2, 3, 2, 4, 4, 4 **3, 4**
5. 0.01, 0.1, 0.01, 1.0, 0.01, 0.11, 1.00 **0.01**
6. $45, $63, $27, $91, $65, $8, $43, $90 **None**

Application T Metric Length

The common units of length in the metric system are the millimeter, centimeter, meter, and the kilometer.

A millimeter(mm) is about the thickness of a U.S. dime.

A centimeter(cm) is about the thickness of a sugar cube.

A meter(m) is about the length of a baseball bat.

A kilometer(km) is 1000 m and is used to measure large distances.

Name the unit that is most commonly used to measure the object.

1. The length of a tennis court **m**
2. The distance from Detroit to Montreal **km**
3. The thickness of a magazine **mm**
4. The length of a standard paper clip **cm**
5. The length of your living room **m**
6. The thickness of a soda straw **mm**
7. The length of a postage stamp **mm**
8. The width of your foot **cm**

Application U Metric Mass and Volume

The common units of mass and volume in the metric system are the gram, kilogram, metric ton, liter, and kiloliter.

A gram(g) is about the mass of a paper clip.

A kilogram(kg) is about the mass of a hammer.

A metric ton(t) is 1000 kg and is used to measure very heavy objects.

A liter(L) is about the amount of liquid in a can of motor oil.

A kiloliter(kL) is 1000 L and is used to measure large volumes.

Name the unit that is most commonly used to measure the object.

1. The mass of your best friend **kg**
2. The amount of gas in a car's tank **L**
3. The mass of a postage stamp **g**
4. The volume of helium in a blimp **kL**
5. The mass of an adult blue whale **t**
6. The mass of a baseball **kg**
7. The volume of a large fuel tank **kL**
8. The mass of a dime **g**
9. The mass of a gumdrop **g**
10. The volume of air in a toy balloon **L**

Application V Perimeter

The distance around a shape is its perimeter. To find the perimeter of a shape, add the lengths of all its sides.

▶ What is the perimeter of the triangle at the right?

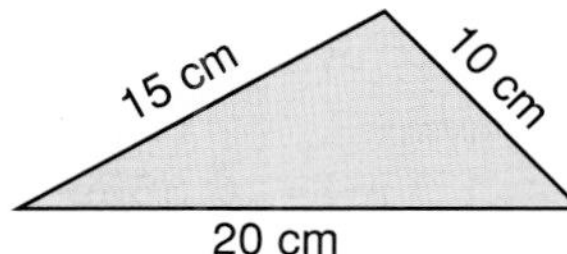

Perimeter:
10 cm + 15 cm + 20 cm = 45 cm

Find the perimeter.

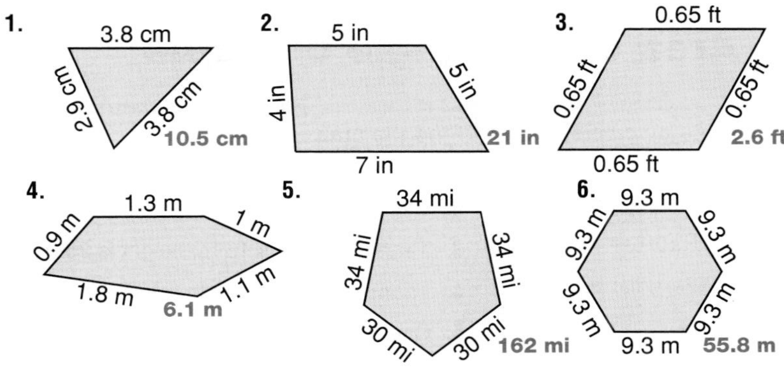

Application  W Circumference

The length of a diameter of a circle is twice the length of its radius. The circumference is the distance around a circle. The circumference is estimated by multiplying 3.14 times the length of a diameter.

▶ The diameter of a circle is 23 mm. What is its circumference?

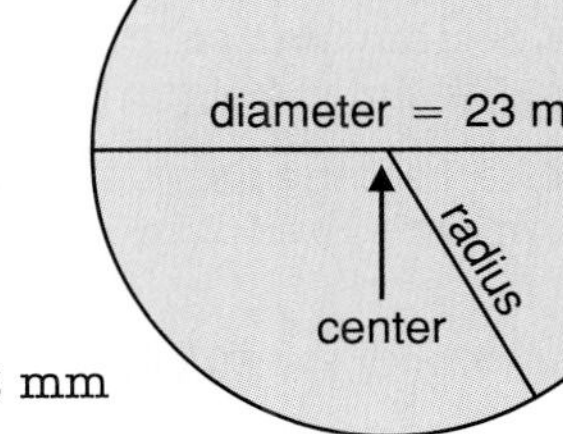

Circumference = 3.14 × Diameter
Circumference = 3.14 × 23 mm = 72.22 mm

Find the circumference. **Answers are approximations.**

1. Diameter: 4 mi **12.56 mi**
2. Diameter: 3.5 cm **10.99 cm**
3. Diameter: 0.8 m **2.512 m**
4. Radius: 35 mm **219.8 mm**
5. Radius: 0.75 ft **4.71 ft**
6. Radius: 13 in **81.64 in**
7. Diameter: 18 ft **56.52 ft**
8. Radius: 120 km **753.6 km**
9. Diameter: 12.5 mi **39.25 mi**

Application  Area of a Rectangle or Square

To find the area of a rectangle, multiply length times width. In a square, the length and width are equal.

▶ What is the area of a rectangular board that measures 3.5 m by 1.9 m?

Area = Length × Width
Area = 3.5 m × 1.9 m = 6.65 m^2

Find the area.

1. Rectangle: 25 in long, 6 in wide **150 in^2**
2. Square: 45.4 cm per side **2061.16 cm^2**
3. Rectangle: 0.9 m long, 0.75 m wide **0.675 m^2**
4. Square: $3\frac{1}{2}$ yd per side **$12\frac{1}{4}$ yd^2**
5. Rectangle: $4\frac{2}{3}$ ft long, $2\frac{1}{6}$ ft wide **$10\frac{1}{9}$ ft^2**

Application  Area of a Triangle or Circle

To find the area of a triangle or circle, use the formulas below.

Area of Triangle = $\frac{1}{2}$ × Base × Height
Area of Circle = 3.14 × $Radius^2$

▶ What is the area of a circle with a radius of 27 m?

Area of Circle = 3.14 × $Radius^2$
Area of Circle = 3.14 × 27^2
Area of Circle = 2289.06 m^2

Find the area.

1. Triangle: base is 3 in, height is 9 in **13.5 in^2**
2. Circle: radius is 5 cm **78.5 cm^2**
3. Triangle: base is 16 m, height is 25 m **200 m^2**
4. Triangle: base is 132 m, height is 9.6 m **633.6 m^2**
5. Circle: radius is $2\frac{1}{2}$ yd **19.625 yd^2**
6. Triangle: base is $9\frac{1}{3}$ ft, height is $3\frac{1}{4}$ ft **$15\frac{1}{6}$ ft^2**
7. Circle: radius is 86 mm **23,223.44 mm^2**

Application  Volume of a Rectangular Solid

To find the volume of a rectangular solid, multiply length times width times height.

▶ What is the volume of a rectangular box that is 6 m long, 4 m wide, and 2 m deep?

Volume = Length × Width × Height
Volume = 6 m × 4 m × 2 m
Volume = 48 m^3

Finds the volume.

1. 36 m long, 44 m wide, 328 m high **519,552 m^3**
2. 1.6 m long, 0.55 m wide, 0.8 m high **0.704 m^3**
3. 11 cm long, 12 cm wide, 3 cm high **396 cm^3**
4. 68 in long, 18 in wide, 18 in high **22,032 in^3**
5. $12\frac{1}{3}$ yd long, $4\frac{1}{4}$ yd wide, 3 yd high **157 yd^3**
6. $7\frac{1}{2}$ ft long, $3\frac{1}{2}$ ft wide, $5\frac{5}{8}$ ft high **$147\frac{21}{32}$ ft^3**

TERMS

accumulated depreciation (p. 498) The total depreciation to date.

acid-test ratio (p. 522) The ratio of total assets minus inventory to total liabilities.

age group (p. 246) A category to which an automobile is assigned by an insurance company depending on its age. Age group affects the base premium.

air freight (p. 460) A way of shipping goods to customers.

amount (p. 157) The balance in a savings account at the end of an interest period, computed by adding the principal and interest.

amount financed (p. 220) The portion of an item's purchase price that you still owe after making the down payment.

amount of change (p. 525) The difference between the base figure and the corresponding figure on the current income statement.

annual expenses (p. 318) Amounts for real estate taxes, insurance payments, contributions, and so on, that occur once a year.

annual interest rate (p. 155) The percent of the principal that you earn as interest based on one year.

annual percentage rate (APR) (p. 222) An index used to compare the relative cost of borrowing money.

annual premium (p. 246) The amount you pay each year for insurance coverage.

annual yield (p. 300) The total interest or return for one year expressed as a percent of the principal or amount invested.

apportion (p. 494) To distribute among several departments.

assessed value (p. 275) The value of property for tax purposes, found by multiplying the market value by the rate of assessment.

assets (p. 512) The total of your cash, the items that you have purchased, and any money that your customers owe you.

average-cost method (p. 455) A way of calculating the value of an inventory based on the average cost of the goods you received.

average daily balance (p. 206) The average of the balances in a charge account at the end of each day in the billing period.

balance (p. 130) The amount of money in a bank account.

balance sheet (p. 514) A statement that shows your total assets, total liabilities, and owner's equity.

bar graph (p. 38) A picture that graphically displays and compares numerical facts in the form of vertical or horizontal bars.

base figure (p. 525) The dollar amount on the earlier income statement when you are computing percent change.

TERMS

base premium (p. 246) The portion of your annual automobile insurance premium determined by the amount of coverage you want, the age group of your car, and the insurance-rating group.

base price (p. 240) The price of the engine, chassis, and any other piece of standard equipment for a new automobile.

basic monthly charge (p. 476) The regular monthly charge for a utility before any additions are made for extra services.

basic plan (p. 290) A health insurance program that includes only hospital and surgical-medical insurance. The basic plan does not include major medical insurance.

basic rate (p. 462) The charge, usually per 100 pounds, to ship goods by truck. It depends on the type of goods, the weight, and the distance to be shipped.

beneficiary (p. 294) The person named in a policy to receive the insurance benefits if the insured dies.

bodily injury insurance (p. 246) Insurance that protects you against financial loss if your automobile causes personal injury to someone.

bond (pp. 308, 534) A promise to pay a specified amount of money, with interest, on a certain date for lending money to a corporation or government.

book value (p. 498) The original cost of an item minus the accumulated depreciation.

break-even analysis (p. 366) An investigation of or by a manufacturing business to find out how many units of a product must be made and sold to cover production expenses.

break-even point (p. 366) The point at which income from sales equals the cost of production.

budget (p. 562) A financial plan that enables a business, a governmental agency, or an individual to identify what sources are expected to produce revenue (earn money) and what amounts are allocated to various departments or categories for expenses.

budget sheet (p. 318) An outline of your total monthly expenses, which can help you plan future expenditures and savings.

canceled check (p. 133) A check that the bank has paid by deducting money from your account.

capital (p. 512) The difference between your assets and liabilities.

cash discount (p. 394) A discount from a supplier to encourage prompt payment of a bill.

cash value (p. 296) The amount of money you will receive if you cancel a whole life insurance policy.

catalog price (p. 384) The price at which a business generally sells an item to a consumer.

certificate of deposit (p. 298) A written record of a special kind of savings account that earns higher interest than a regular account. You must leave the money on deposit for a specified time or be penalized for early withdrawal.

chain discount (p. 392) A series of trade discounts on an item.

check (p. 128) A written order directing the bank to deduct money from a checking account to make a payment.

check register (p. 130) A record of the deposits made into a checking account, electronic transfers, and of the checks written.

closed-end lease (p. 253) A type of automobile lease where you make a specified number of payments, return the car at the end of the lease period, and owe nothing unless you damaged the car or exceeded the mileage limit.

closing costs (p. 270) Fees paid at the time documents are signed transferring ownership of a home.

coinsurance clause (p. 292) A clause in your major medical policy stating the percent of the bill you must pay after subtracting the deductible.

collision insurance (p. 246) Insurance to pay for repairs to your automobile if it is involved in an accident.

collision waiver (p. 255) An agreement by which you can obtain complete insurance coverage on a rented automobile by paying an additional charge per day.

commercial loan (p. 536) A business loan.

commercial paper (p. 540) A 30- to 270- day loan issued by a company with a very high credit rating.

commission (p. 89) An amount of money that you are paid for selling a product or service, based on the dollar value or number of sales.

commission rate (p. 89) A specified amount of money that you are paid for each sale or a percent of the total value of your sales.

complement method (p. 386) A way to find the net price when a trade discount is given.

compound interest (p. 157) Interest earned not only on the original principal but also on the interest earned during previous interest periods.

comprehensive insurance (p. 246) Insurance that protects you from loss of or damage to your automobile caused by fire, vandalism, theft, and so on.

comprehensive plan (p. 290) A health insurance program that includes hospital, surgical-medical, and major medical insurance.

consultant (p. 481) A person who gives technical or expert advice.

consultant's fee (p. 481) The payment to a consultant for professional services.

consumer price index (CPI) (p. 559) A measure of the average change in prices of a certain number of goods and services.

corporate income taxes (p. 532) Federal and state taxes on the earnings of a corporation.

corporation (p. 532) A business made up of a number of owners, called stockholders.

cost (p. 404) The amount that a business pays for a product, including expenses such as freight charges.

cost of goods sold (p. 517) The value of the beginning inventory plus the cost of any goods received minus the value of the ending inventory.

cost-of-living adjustment (p. 344) A salary raise to help an employee keep up with inflation.

coupon (p. 177) A certificate or ticket that offers the customer a special discount on an item or service.

current ratio (p. 522) The ratio of total assets to total liabilities.

dealer's cost (p. 242) The price that the automobile dealer pays for a new car.

deductible clause (p. 246) A clause in an insurance policy stating that the insured must pay a portion of any repair bill, or a portion of the amount not covered by hospital or surgical-medical insurance.

defective (p. 368) Incorrect, broken, or damaged.

demand charge (p. 478) The charge on an electric bill based on the peak load during the month.

deposit (pp. 126, 146) An amount of money put into a bank account.

depreciation (pp. 250, 496) A decrease in the value of an item because of its age or condition.

destination charge (p. 240) The cost of shipping a new automobile from the factory to the dealer.

direct labor cost (p. 364) The wages paid to the employees who manufacture an item.

direct material cost (p. 364) The cost of the goods that you use to manufacture an item.

disability insurance (p. 348) Pays benefits to an individual who must miss work due to an illness or injury.

discount 1. (pp. 179, 418) The difference between the regular selling price of an item and its sale price. 2. (p. 538) The difference between the face value and maturity value of a bond or United States Treasury Bill.

discount rate (p. 179) The discount on an item expressed as a percent of its regular selling price.

dividends (p. 304) Return on an investment in the stock of a corporation.

double-declining-balance method (p. 502) A method of computing depreciation that allows the highest possible depreciation in the first year of an item's life.

TERMS

double time (p. 80) An overtime rate that is 2 times the regular hourly rate.

down payment (p. 220) The portion of an item's purchase price that you pay at the time of buying when you purchase the item with an installment loan.

driver-rating factor (p. 246) A number assigned by an insurance company to each driver of an automobile to reflect the driver's age, marital status, and so on. A higher driver-rating factor results in a higher annual premium.

emergency fund (p. 321) Money set aside for unpredictable expenses, such as medical bills or repair bills.

employee benefits (p. 346) Benefits an employee receives in addition to salary, such as health insurance, pension, or paid vacation.

end-of-month dating (p. 396) Terms such as "6/15 EOM" on an invoice. In this case, a 6% cash discount is offered if a bill is paid within 15 days after the end of the month in which the invoice is issued.

energy charge (p. 478) The charge on an electric bill based on the total number of kilowatt-hours used during the month.

escrow account (p. 280) An account often required by the lender of a mortgage loan to guarantee payment of the homeowner's real estate taxes and fire insurance premium.

estimated life (p. 496) The length of time, usually in years, that an item is expected to last.

exact interest (p. 218) Interest calculated by basing the length of time of a loan on a 365-day year.

expenditures (p. 316) Amounts of money paid out.

expense summary (p. 321) A totaling of actual monthly expenses used to compare the amounts spent to the amounts budgeted.

face value (pp. 294, 308, 538) The amount of money printed on an insurance policy, bond, or United States Treasury Bill.

factor (p. 434) A company's present market share.

factor method (p. 434) A way to project sales by using a company's present market share and the total estimated sales for the market in the future.

FIFO (first in, first out) method (p. 457) A method of valuing inventory which assumes that the first items received are the first items sold.

final payment (p. 228) The previous balance plus the current month's interest.

finance charge (p. 197) Interest that is charged to a customer for delaying payment of a bill.

fire protection class (p. 279) A number assigned to a home by an insurance company according to the quality of fire protection available in the area.

fixed costs 1. (p. 250) The costs of operating and maintaining an automobile that remain about the same regardless of the number of miles driven. 2. (p. 366) The costs of manufacturing that remain the same regardless of the number of units produced.

fixed expenses (p. 318) Amounts for rent or mortgage payments, installment payments, savings deposits, and so on, that do not vary from one month to the next.

fuel adjustment charge (p. 478) An additional charge on an electric bill to help cover increases in the cost of fuel needed to produce electricity.

graduated commission (p. 92) A method of payment in which the commission rate changes for each of several levels of sales, increasing as sales increase.

graduated income tax (p. 105) An income tax that involves a different tax rate for each of several levels of income.

gross national product (GNP) (p. 557) A measure of a nation's economic performance. It is the estimated total value of all goods and services produced by a nation during a year.

gross profit (p. 404) The difference between the selling price and the cost of a product when the selling price is greater than the cost.

group insurance (p. 109) Insurance offered to members of a group, such as the employees of a business, at a lower cost than individual insurance.

growth expenses (p. 542) Expenses from expanding a business, such as construction expenses, legal fees, or consultant's fees.

health insurance (p. 290) Insurance coverage against overwhelming medical expenses.

homeowner's insurance (p. 277) Insurance to protect yourself and your home from losses due to fire, theft of contents, or personal liability.

horizontal analysis (p. 525) The comparison of two or more income statements for different periods.

hospital insurance (p. 292) Insurance that pays most of the cost of hospitalization, including a semiprivate room and most laboratory tests.

hourly rate (p. 78) The amount of money earned per hour.

income statement (p. 519) A statement that shows in detail the sales, operating expenses, and net profit or loss of a business.

income tax (p. 100) A tax on the earnings, or income, of a person or business.

inflation (p. 554) An economic condition during which there are price increases in the cost of goods and services.

inflation rate (p. 554) Found by subtracting the original price from the current price and dividing the result by the original price.

installment loan (p. 220) A loan that is repaid in several equal payments over a specified period of time.

insurance-rating group (p. 246) A category to which an insurance company assigns an automobile depending on its size and value. The insurance-rating group affects the base premium.

Interest (pp. 155, 268) The amount of money paid for the use of money.

inventory (p. 452) The items that you have in stock.

inventory card (p. 452) A record of each item in stock.

invoice (p. 394) A written record of a purchase, listing the quantities and costs of the items purchased.

issues (p. 452) The number of outgoing items that are subtracted from the previous inventory.

kilowatt (kW) (p. 478) A unit of power equal to 1000 watts.

kilowatt-hour (kW•h) (p. 478) A unit to measure the consumption of electrical energy. It is equivalent to using one kilowatt of power constantly for one hour.

labor charge (p. 472) The cost of paying the people who do a job.

lease (p. 470) To agree to specific payments and other conditions in order to use property belonging to someone else for a period of time.

liabilities (p. 512) The total amount of money that you owe to creditors.

liability insurance (p. 246) Insurance that protects you against financial losses if your car is involved in an accident. Liability insurance includes bodily injury insurance and property damage insurance.

life insurance (p. 294) Financial protection that you purchase for your dependents in case of your death.

LIFO (last in, first out) method (p. 457) A method of valuing inventory which assumes that the last items received are the first items shipped.

limited-payment life insurance (p. 296) Lifetime insurance protection for which you pay premiums only for a specified number of years or until you reach a certain age.

line graph (p. 38) A picture used to compare facts over a period of time. An excellent way to show trends (increases or decreases).

list price (p. 384) The price at which a business generally sells an item to a consumer.

living expenses (p. 318) Amounts spent for food, clothing, utility bills, and so on, that vary from month to month.

loan value (p. 296) The amount of money the insurance company will lend you if you have a whole life insurance policy.

loss (p. 306) The difference between the cost and selling price of an item when the cost is greater than the selling price.

loss-of-use coverage (p. 277) Insurance coverage to pay for some of the expenses of living away from home while damage to your home is being repaired.

major medical insurance (p. 292) Insurance that pays for medical costs not covered by hospital or surgical-medical insurance.

markdown (pp. 179, 418) The difference between the regular selling price and the sale price of an item.

markdown rate (pp. 179, 418) The markdown on an item expressed as a percent of its regular selling price.

market (p. 428) The total number of people who might purchase the type of product being made.

market share (p. 430) The portion of the market that purchases a specific product.

market value (p. 275) The price at which a house or other article can be bought or sold.

markup (p. 404) The difference between the selling price and the cost of a product when the selling price is greater.

markup rate (p. 406) The markup expressed as a percent of the selling price or cost of an item.

materials charge (p. 472) The cost of the materials used to complete a job.

maturity date (p. 538) The date on which a note, bill, or bond is due and payable.

maturity value (pp. 218, 536) The total amount you repay on a loan, including both the principal borrowed and the interest owed.

medicare (p. 107) A federal health insurance program to provide hospitalization and medical services for people over 65. The money for this program is provided by a tax paid by employers and employees.

merge (p. 542) To combine with another business to form a new business.

merit increase (p. 344) A raise in your salary to reward you for the quality of your work.

mill (p. 275) A unit often used to express real estate tax rates, equal to $1.00 in tax for each $1000 of assessed value.

modified accelerated cost recovery system (MACRS) (p. 504) A method of computing depreciation that allows depreciation according to fixed percents over a set period of time.

mortgage loan (p. 266) A loan to finance the purchase of a home whereby the lender has the right to sell the property if the loan payments are not made.

net income (p. 519) The difference between the gross profit and the total operating expenses if the gross profit is greater than the operating expenses.

net interest (p. 540) The actual interest earned from investing in commercial paper after any bank service fees have been deducted.

net loss (p. 521) The difference between the markup, or gross profit, and the overhead, or operating expenses, when the overhead is greater than the markup.

net pay (p. 111) The amount of pay you have left after all tax withholdings and personal deductions have been subtracted.

net price (p. 384) The price a business actually pays for an item after discounts have been subtracted.

net-price rate (p. 392) The percent a business pays for an item, found by multiplying the complements of the chain discounts.

net proceeds (p. 534) The amount that a business actually receives from the sale of stocks or bonds after subtracting all selling expenses.

net profit (pp. 408, 519) The difference between the markup, or gross profit, and the overhead, or operating expenses, when the markup is greater than the overhead.

net-profit rate (p. 410) Net profit expressed as a percent of the selling price of an item.

net sales (p. 519) The total sales minus returns.

net worth (p. 512) The difference between your assets and liabilities.

nongroup insurance (p. 290) Insurance for people not enrolled in a group.

open-end lease (p. 253) A type of automobile lease where, at the end of the lease, you can buy the car for its expected value.

operating expenses (p. 408) Expenses of running a business, such as wages of employees, rent, utility charges, and taxes.

opinion-research firm (p. 426) A business that specializes in product testing and opinion surveys and the interpretation of the results.

opinion survey (p. 426) A series of questions answered by a group of volunteers. The results should approximate the opinions of a much larger group of people.

option (p. 240) An extra on a new automobile for convenience, safety, or appearance, such as a radio or air-conditioning.

ordinary dating (p. 394) Terms such as "2/10, net 30" on an invoice. In this case, a 2% cash discount is offered if the bill is paid within 10 days of date of invoice; the net price must be paid within 30 days.

ordinary interest (p. 218) Interest calculated by basing the length of time of a loan on a 360-day year.

outstanding check/deposit (p. 136) A check or deposit that appears in your check register but has not reached the bank in time to be listed on your statement.

overdraw (p. 128) To write checks in excess of the amount of money in a checking account.

overhead (p. 408) Expenses of running a

business, such as wages of employees, rent, utility charges, and taxes.

overtime pay (p. 80) The amount of money you earn when you work more than your regular hours. The overtime rate may be $1\frac{1}{2}$ or 2 times the regular hourly rate.

owner's equity (p. 512) The difference between your assets and liabilities.

packaging (p. 375) Placing a product in a container for shipment.

passbook (p. 150) A book given to you in which the bank records all the deposits and withdrawals for your account and any interest earned.

payroll register (p. 488) A record of the gross income, deductions and net income of a company's employees.

peak load (p. 478) The greatest number of kilowatts used by a consumer at one time during the month.

per capita GNP (p. 557) An indicator of the nation's standard of living. It is the GNP distributed over the population.

percent change (p. 525) The amount of change from an earlier income statement to the current one, divided by the dollar amount on the earlier statement.

periodic rate (p. 200) The monthly finance charge rate.

personal exemptions (p. 103) Subtractions made from your gross pay for yourself and every other person you support. They are used to determine your taxable wages.

personal liability and medical coverage (p. 277) Insurance coverage to protect you from financial losses if someone is injured on your property.

piecework (p. 85) Work for which you are paid according to the number of items you complete.

point (p. 271) One percent of the mortgage loan. Points are included in the closing costs.

premium 1. (p. 279) The amount paid for insurance coverage. 2. (p. 308) Bonds sold for more than face value.

previous balance (p. 200) The amount owed by a credit card user on the closing date of the last statement.

prime cost (p. 364) The total of the direct material cost and the direct labor cost.

prime rate (p. 536) The lowest rate of interest available on business loans at a given time.

principal (pp. 155, 218) The amount of money on which interest is paid.

product test (p. 426) An attempt to determine how well a product is going to sell by asking a group of people to try it.

profit (p. 366) The amount you receive after the costs of selling or production have been met.

profit-and-loss statement (p. 519) A statement

that shows in detail your sales, operating expenses, and net profit or loss.

profit margin (p. 404) The difference between the selling price and the cost of a product.

promissory note (p. 218) A type of single-payment loan. It is a written promise to pay a certain sum of money at a specific future date.

property damage insurance (p. 246) Insurance that protects you against financial loss if your automobile damages the property of other people.

quality control (p. 368) Periodic inspection of the products you manufacture.

quality control chart (p. 368) A chart to show the percent of defective products that is allowable in a manufacturing business.

quick ratio (p. 522) The ratio of total assets minus inventory to total liabilities.

quoted price (p. 308) The actual cost of a bond, usually a percent of the face value.

rate of assessment (p. 275) The percent used to calculate the assessed value of a home, given the market value.

real estate taxes (p. 275) Fees collected from the owners of property and used to maintain government operations.

real GNP (p. 557) The gross national product corrected for inflation.

rebate (p. 177) A special incentive to purchase a particular item where the consumer must mail to the manufacturer a rebate coupon along with the sales slip and the universal product code label from the item.

rebate schedule (p. 233) A table of percent refunds used to determine the part of the finance charge refunded when an installment loan is repaid before the final due date.

receipts (p. 452) The number of incoming items that are added to the previous inventory.

reconcile (p. 136) To obtain agreement between your check register and the bank statement.

recruiting costs (p. 342) Expenses of finding a new employee, such as advertising fees, interviewing costs, and hiring expenses.

refund (p. 177) A special incentive to purchase a particular item where the manufacturer or store coupon is redeemed at the time of purchase.

release time (p. 354) Time spent away from your job in training programs for which you are still paid your regular wages or salary.

rent (pp. 255, 470, 474) To pay a regular fee for the use of a car, equipment, or space in a building.

rent escalation clause (p. 471) A statement in a lease that the rent may increase by a certain percent after a certain period of time.

repayment schedule (p. 225) A schedule that

shows the distribution of interest and principal over the life of a loan.

replacement value (p. 277) The amount of money required to reconstruct a home if it is destroyed.

resale value (p. 496) The estimated trade-in or salvage value of an item at the end of its expected life.

residual value (p. 253) The expected value of an automobile at the end of the lease period.

retail price (p. 404) The price that a business expects a consumer to pay for a product.

salary (p. 87) A fixed amount of money that you earn on a regular basis.

salary scale (p. 344) A table of wages or salaries used to compare various jobs in a business.

sale price (p. 179) A purchase price that is lower than the regular selling price.

sales potential (p. 428) An estimate of the number of products that you might sell during a specified period of time.

sales projection (p. 432) An estimate of the dollar volume or unit sales expected to occur during a future time period.

sales quota (p. 93) The amount of sales expected of a sales representative for a given period of time.

sales receipt (pp. 170, 194) A sales slip or cash register tape that shows the selling price of the items purchased, any sales tax, and the total purchase price. If a credit card is used, the receipt also shows the name and signature of the purchaser and the account number.

sales tax (p. 168) A tax on the selling price of an item or service that you purchase.

sample (p. 428) A selected group of people, representative of a much greater number of people, who are asked to evaluate a new product.

selling price (p. 404) The price that a business expects a consumer to pay for a product.

service charge (p. 133) A charge by a bank for handling a checking account.

simple interest (p. 155) Interest paid only on the original principal.

simple interest installment loan (p. 222) A loan repaid with equal monthly payments where part of each payment is used to pay the interest on the unpaid balance of the loan and the remaining part is used to reduce the balance.

single equivalent discount (SED) (p. 392) One discount that is equal to a chain discount.

single-payment loan (p. 218) A loan that you repay with one payment after a specified period of time.

social security (p. 107) A federal law (FICA) to provide retirement income, survivor's benefits, and disability benefits. The money for this program is provided by a tax paid by employers and employees.

TERMS

statement (pp. 133, 152, 197) A record of an account sent by a bank or other business listing all transactions that have been recorded during a specified period of time.

sticker price (p. 240) The total of the base price, options, and destination charge for a new automobile.

stockbroker (p. 302) Someone who buys and sells stocks for other people.

stock certificate (p. 302) A printed form that proves part ownership of a corporation.

stocks (pp. 302, 534) Shares of ownership in the corporation issuing the stock.

storage space (p. 450) A place in which to keep products until they are needed.

straight commission (p. 89) A method of payment in which the only pay you receive is a commission on the sales you make.

straight-line method (p. 496) A method of computing depreciation which assumes that the depreciation is the same from year to year.

straight-time pay (p. 78) The total amount of money you earn for a pay period at your regular hourly rate.

sum-of-the-years'-digits method (p. 500) A method of computing depreciation that assigns a fraction of the total depreciation to each year of an item's life, with more depreciation allowed in the early years.

surgical-medical insurance (p. 292) Insurance that pays your doctor's fee, up to a certain amount, for surgery.

taxable income (p. 532) The portion of the gross income of a business that remains after normal business expenses are deducted.

tax rate 1. (p. 275) Expressed in mills and used to determine real estate taxes. 2. (p. 168) Expressed as a percent and used to determine sales taxes.

term (p. 218) The amount of time for which a loan is granted.

term life insurance (p. 294) A policy written for a specified amount of time, such as five years, or to a specified age. The policy expires at the end of the time unless renewed.

time and a half (p. 80) An overtime rate $1\frac{1}{2}$ times the regular hourly rate.

time study (p. 371) Determining how long a particular job should take.

trade discount (p. 384) A discount that a business receives when purchasing goods from a supplier.

trade-discount rate (p. 388) The trade discount on an item expressed as a percent of the list price.

travel expenses (p. 352) Money spent by an employee for expenses, such as transportation, lodging, and meals, while traveling on business.

underwriting commission (p. 534) The amount of money paid to the investment banker who helps a business distribute stocks or bonds.

unit (p. 499) One unit of insurance has a face value of $1000.

United States Treasury Bill (p. 538) A written promise by the federal government to repay the purchase price of the bill plus interest at a future date.

unit price (p. 173) The cost of an item expressed per unit of measure or count.

units-of-production method (p. 499) A method of computing depreciation based on the number of units produced.

universal life insurance (p. 296) Lifetime insurance protection, which is a combination of life insurance and a savings plan. You pay a minimum premium, with anything over the minimum going into a savings account.

unpaid balance (p. 203) The portion of the balance on your previous charge account or credit card statement that you have not yet paid.

used-car guide (p. 244) A monthly publication that gives price information on used cars purchased from dealers during the previous month.

utilities costs (p. 281, 476) Charges for public services, such as electricity or water.

variable costs 1. (p. 250) The costs of operating and maintaining an automobile, which increase as the number of miles driven increases. 2. (p. 366) Manufacturing costs that increase as the number of units produced increases.

wage scale (p. 344) A table of wages or salaries used to compare various jobs in a business.

warehouse (p. 450) A building or part of one used to store raw materials or products until they are needed.

weekly time card (p. 82) A record of the times you report for work and the times you leave.

whole life insurance (p. 296) Life insurance that offers financial protection for your entire life.

withdrawal (p. 148) An amount of money taken out of a savings account.

withholding allowance (p. 100) A limitation on the amount of federal income tax withheld from your pay. You may claim one allowance for yourself, one for your spouse if married, and one for every other person you support.

TABLE OF MEASURES

TIME

60 seconds (s) = 1 minute (min)
60 minutes = 1 hour (h)
24 hours = 1 day
7 days = 1 week

52 weeks = 1 year
12 months = 1 year
100 years = 1 century

METRIC SYSTEM

Length

10 millimeters (mm) = 1 centimeter (cm)
100 centimeters = 1 meter (m)
1000 meters = 1 kilometer (km)

Volume

1000 milliliters (mL) = 1 liter (L)
1000 cubic centimeters (cm^3) = 1 liter
10 milliliters = 1 centiliter (cL)
10 deciliters (dL) = 1 liter

Area

100 square millimeters (mm^2) = 1 square centimeter (cm^2)
10,000 square centimeters = 1 square meter (m^2)
10,000 square meters = 1 hectare (ha)

Mass

1000 milligrams (mg) = 1 gram (g)
1000 grams = 1 kilogram (kg)
1000 kilograms = 1 metric ton (t)

CUSTOMARY SYSTEM

Length

12 inches (in) = 1 foot (ft)
3 feet = 1 yard (yd)
5280 feet = 1 mile (mi)

Volume

8 fluid ounces (oz) = 1 cup (c)
2 cups = 1 pint (pt)
2 pints = 1 quart (qt)
4 quarts = 1 gallon (gal)

Area

144 square inches (in^2) = 1 square foot (ft^2)
9 square feet = 1 square yard (yd^2)
4837 square yards = 1 acre (A)

Weight

16 ounces = 1 pound (lb)
2000 pounds = 1 ton (t)

SINGLE Persons — WEEKLY Payroll Period

(For Wages Paid After December)

And the wages are—		And the number of withholding allowances claimed is—										
At least	But Less than	0	1	2	3	4	5	6	7	8	9	10
		The amount of income tax to be withheld shall be—										
$0	$25	$0	$0	$0	$0	$0	$0	$0	$0	$0	$0	$0
25	30	1	0	0	0	0	0	0	0	0	0	0
30	35	1	0	0	0	0	0	0	0	0	0	0
35	40	2	0	0	0	0	0	0	0	0	0	0
40	45	3	0	0	0	0	0	0	0	0	0	0
45	50	4	0	0	0	0	0	0	0	0	0	0
50	55	4	0	0	0	0	0	0	0	0	0	0
55	60	5	0	0	0	0	0	0	0	0	0	0
60	65	6	0	0	0	0	0	0	0	0	0	0
65	70	7	0	0	0	0	0	0	0	0	0	0
70	75	7	1	0	0	0	0	0	0	0	0	0
75	80	8	2	0	0	0	0	0	0	0	0	0
80	85	9	3	0	0	0	0	0	0	0	0	0
85	90	10	3	0	0	0	0	0	0	0	0	0
90	95	10	4	0	0	0	0	0	0	0	0	0
95	100	11	5	0	0	0	0	0	0	0	0	0
100	105	12	6	0	0	0	0	0	0	0	0	0
105	110	13	6	0	0	0	0	0	0	0	0	0
110	115	13	7	1	0	0	0	0	0	0	0	0
115	120	14	8	2	0	0	0	0	0	0	0	0
120	125	15	9	2	0	0	0	0	0	0	0	0
125	130	16	9	3	0	0	0	0	0	0	0	0
130	135	16	10	4	0	0	0	0	0	0	0	0
135	140	17	11	5	0	0	0	0	0	0	0	0
140	145	18	12	5	0	0	0	0	0	0	0	0
145	150	19	12	6	0	0	0	0	0	0	0	0
150	155	19	13	7	1	0	0	0	0	0	0	0
155	160	20	14	8	1	0	0	0	0	0	0	0
160	165	21	15	8	2	0	0	0	0	0	0	0
165	170	22	15	9	3	0	0	0	0	0	0	0
170	175	22	16	10	4	0	0	0	0	0	0	0
175	180	23	17	11	4	0	0	0	0	0	0	0
180	185	24	18	11	5	0	0	0	0	0	0	0
185	190	25	18	12	6	0	0	0	0	0	0	0
190	195	25	19	13	7	0	0	0	0	0	0	0
195	200	26	20	14	7	1	0	0	0	0	0	0
200	210	27	21	15	9	2	0	0	0	0	0	0
210	220	29	22	16	10	4	0	0	0	0	0	0
220	230	30	24	18	12	5	0	0	0	0	0	0
230	240	32	25	19	13	7	1	0	0	0	0	0
240	250	33	27	21	15	8	2	0	0	0	0	0
250	260	35	28	22	16	10	4	0	0	0	0	0
260	270	36	30	24	18	11	5	0	0	0	0	0
270	280	38	31	25	19	13	7	0	0	0	0	0
280	290	39	33	27	21	14	8	2	0	0	0	0
290	300	41	34	28	22	16	10	3	0	0	0	0
300	310	42	36	30	24	17	11	5	0	0	0	0
310	320	44	37	31	25	19	13	6	0	0	0	0
320	330	45	39	33	27	20	14	8	2	0	0	0
330	340	47	40	34	28	22	16	9	3	0	0	0
340	350	48	42	36	30	23	17	11	5	0	0	0
350	360	50	43	37	31	25	19	12	6	0	0	0
360	370	51	45	39	33	26	20	14	8	2	0	0
370	380	53	46	40	34	28	22	15	9	3	0	0
380	390	54	48	42	36	29	23	17	11	5	0	0
390	400	56	49	43	37	31	25	18	12	6	0	0
400	410	57	51	45	39	32	26	20	14	8	1	0
410	420	59	52	46	40	34	28	21	15	9	3	0
420	430	61	54	48	42	35	29	23	17	11	4	0
430	440	64	55	49	43	37	31	24	18	12	6	0
440	450	67	57	51	45	38	32	26	20	14	7	1
450	460	70	58	52	46	40	34	27	21	15	9	3
460	470	73	61	54	48	41	35	29	23	17	10	4
470	480	75	64	55	49	43	37	30	24	18	12	6
480	490	78	67	57	51	44	38	32	26	20	13	7
490	500	81	69	58	52	46	40	33	27	21	15	9
500	510	84	72	61	54	47	41	35	29	23	16	10
510	520	87	75	63	55	49	43	36	30	24	18	12
520	530	89	78	66	57	50	44	38	32	26	19	13
530	540	92	81	69	58	52	46	39	33	27	21	15

642

MARRIED Persons — **WEEKLY** Payroll Period

(For Wages Paid After December)

And the wages are–		And the number of withholding allowances claimed is—										
At least	But Less than	0	1	2	3	4	5	6	7	8	9	10
		The amount of income tax to be withheld shall be—										
$0	$70	$0	$0	$0	$0	$0	$0	$0	$0	$0	$0	$0
70	75	1	0	0	0	0	0	0	0	0	0	0
75	80	1	0	0	0	0	0	0	0	0	0	0
80	85	2	0	0	0	0	0	0	0	0	0	0
85	90	3	0	0	0	0	0	0	0	0	0	0
90	95	4	0	0	0	0	0	0	0	0	0	0
95	100	4	0	0	0	0	0	0	0	0	0	0
100	105	5	0	0	0	0	0	0	0	0	0	0
105	110	6	0	0	0	0	0	0	0	0	0	0
110	115	7	0	0	0	0	0	0	0	0	0	0
115	120	7	1	0	0	0	0	0	0	0	0	0
120	125	8	2	0	0	0	0	0	0	0	0	0
125	130	9	3	0	0	0	0	0	0	0	0	0
130	135	10	3	0	0	0	0	0	0	0	0	0
135	140	10	4	0	0	0	0	0	0	0	0	0
140	145	11	5	0	0	0	0	0	0	0	0	0
145	150	12	6	0	0	0	0	0	0	0	0	0
150	155	13	6	0	0	0	0	0	0	0	0	0
155	160	13	7	1	0	0	0	0	0	0	0	0
160	165	14	8	2	0	0	0	0	0	0	0	0
165	170	15	9	2	0	0	0	0	0	0	0	0
170	175	16	9	3	0	0	0	0	0	0	0	0
175	180	16	10	4	0	0	0	0	0	0	0	0
180	185	17	11	5	0	0	0	0	0	0	0	0
185	190	18	12	5	0	0	0	0	0	0	0	0
190	195	19	12	6	0	0	0	0	0	0	0	0
195	200	19	13	7	1	0	0	0	0	0	0	0
200	210	21	14	8	2	0	0	0	0	0	0	0
210	220	22	16	10	3	0	0	0	0	0	0	0
220	230	24	17	11	5	0	0	0	0	0	0	0
230	240	25	19	13	6	0	0	0	0	0	0	0
240	250	27	20	14	8	2	0	0	0	0	0	0
250	260	28	22	16	9	3	0	0	0	0	0	0
260	270	30	23	17	11	5	0	0	0	0	0	0
270	280	31	25	19	12	6	0	0	0	0	0	0
280	290	33	26	20	14	8	2	0	0	0	0	0
290	300	34	28	22	15	9	3	0	0	0	0	0
300	310	36	29	23	17	11	5	0	0	0	0	0
310	320	37	31	25	18	12	6	0	0	0	0	0
320	330	39	32	26	20	14	8	1	0	0	0	0
330	340	40	34	28	21	15	9	3	0	0	0	0
340	350	42	35	29	23	17	11	4	0	0	0	0
350	360	43	37	31	24	18	12	6	0	0	0	0
360	370	45	38	32	26	20	14	7	1	0	0	0
370	380	46	40	34	27	21	15	9	3	0	0	0
380	390	48	41	35	29	23	17	10	4	0	0	0
390	400	49	43	37	30	24	18	12	6	0	0	0
400	410	51	44	38	32	26	20	13	7	1	0	0
410	420	52	46	40	33	27	21	15	9	2	0	0
420	430	54	47	41	35	29	23	16	10	4	0	0
430	440	55	49	43	36	30	24	18	12	5	0	0
440	450	57	50	44	38	32	26	19	13	7	1	0
450	460	58	52	46	39	33	27	21	15	8	2	0
460	470	60	53	47	41	35	29	22	16	10	4	0
470	480	61	55	49	42	36	30	24	18	11	5	0
480	490	63	56	50	44	38	32	25	19	13	7	0
490	500	64	58	52	45	39	33	27	21	14	8	2
500	510	66	59	53	47	41	35	28	22	16	10	3
510	520	67	61	55	48	42	36	30	24	17	11	5
520	530	69	62	56	50	44	38	31	25	19	13	6
530	540	70	64	58	51	45	39	33	27	20	14	8
540	550	72	65	59	53	47	41	34	28	22	16	9
550	560	73	67	61	54	48	42	36	30	23	17	11
560	570	75	68	62	56	50	44	37	31	25	19	12
570	580	76	70	64	57	51	45	39	33	26	20	14
580	590	78	71	65	59	53	47	40	34	28	22	15
590	600	79	73	67	60	54	48	42	36	29	23	17
600	610	81	74	68	62	56	50	43	37	31	25	18
610	620	82	76	70	63	57	51	45	39	32	26	20
620	630	84	77	71	65	59	53	46	40	34	28	21

AMOUNT OF $1.00							
TOTAL INTEREST PERIOD	INTEREST RATE PER PERIOD						
	1.2500%	1.3750%	1.5000%	2.7500%	2.8750%	3.0000%	3.1250%
1	1.01250	1.01375	1.01500	1.02749	1.02875	1.03000	1.03125
2	1.02515	1.02768	1.03022	1.05575	1.05832	1.06090	1.06347
3	1.03797	1.04182	1.04567	1.08478	1.08875	1.09272	1.09671
4	1.05094	1.05614	1.06136	1.11462	1.12005	1.12550	1.13098
5	1.06408	1.07066	1.07728	1.14527	1.15225	1.15927	1.16632
6	1.07738	1.08538	1.09344	1.17676	1.18538	1.19405	1.20277
7	1.09085	1.10031	1.10984	1.20912	1.21946	1.22987	1.24036
8	1.10448	1.11544	1.12649	1.24237	1.25452	1.26677	1.27912
9	1.11829	1.13078	1.14339	1.27654	1.29059	1.30477	1.31909
10	1.13227	1.14632	1.16054	1.31165	1.32769	1.34391	1.36031
11	1.14642	1.16209	1.17795	1.34772	1.36586	1.38423	1.40282
12	1.16075	1.17806	1.19562	1.38478	1.40513	1.42576	1.44666
13	1.17526	1.19426	1.21355	1.42286	1.44553	1.46853	1.49187
14	1.18995	1.21068	1.23175	1.46199	1.48709	1.51258	1.53849
15	1.20482	1.22733	1.25023	1.50219	1.52984	1.55796	1.58657
16	1.21988	1.24421	1.26898	1.54350	1.57382	1.60470	1.63615
17	1.23513	1.26132	1.28802	1.58595	1.61907	1.65284	1.68728
18	1.25057	1.27866	1.30734	1.62956	1.66562	1.70243	1.74000
19	1.26620	1.29624	1.32695	1.67438	1.71351	1.75350	1.79438
20	1.28203	1.31406	1.34685	1.72042	1.76277	1.80611	1.85045
21	1.29806	1.33213	1.36706	1.76773	1.81345	1.86029	1.90828
22	1.31428	1.35045	1.38756	1.81635	1.86559	1.91610	1.96791
23	1.33071	1.36902	1.40838	1.86630	1.91922	1.97358	2.02941
24	1.34735	1.38784	1.42950	1.91762	1.97440	2.03279	2.09283

AMOUNT OF $1.00 AT 5.5%, COMPOUNDED DAILY, 365-DAY YEAR									
DAY	AMOUNT	DAY	AMOUNT	DAY	AMOUNT	DAY	AMOUNT	DAY	AMOUNT
1	1.00015	11	1.00165	21	1.00316	31	1.00468	50	1.00755
2	1.00030	12	1.00180	22	1.00331	32	1.00483	60	1.00907
3	1.00045	13	1.00196	23	1.00347	33	1.00498	70	1.01059
4	1.00060	14	1.00211	24	1.00362	34	1.00513	80	1.01212
5	1.00075	15	1.00226	25	1.00377	35	1.00528	90	1.01364
6	1.00090	16	1.00241	26	1.00392	36	1.00543	100	1.01517
7	1.00105	17	1.00256	27	1.00407	37	1.00558	110	1.01670
8	1.00120	18	1.00271	28	1.00422	38	1.00574	120	1.01823
9	1.00135	19	1.00286	29	1.00437	39	1.00589	130	1.01977
10	1.00150	20	1.00301	30	1.00452	40	1.00604	140	1.02131

644

ELAPSED TIME TABLE

						THE NUMBER OF EACH DAY OF THE YEAR						
Day No.	Jan.	Feb.	Mar.	Apr.	May	June	July	Aug.	Sept.	Oct.	Nov.	Dec.
1	1	32	60	91	121	152	182	213	244	274	305	335
2	2	33	61	92	122	153	183	214	245	275	306	336
3	3	34	62	93	123	154	184	215	246	276	307	337
4	4	35	63	94	124	155	185	216	247	277	308	338
5	5	36	64	95	125	156	186	217	248	278	309	339
6	6	37	65	96	126	157	187	218	249	279	310	340
7	7	38	66	97	127	158	188	219	250	280	311	341
8	8	39	67	98	128	159	189	220	251	281	312	342
9	9	40	68	99	129	160	190	221	252	282	313	343
10	10	41	69	100	130	161	191	222	253	283	314	344
11	11	42	70	101	131	162	192	223	254	284	315	345
12	12	43	71	102	132	163	193	224	255	285	316	346
13	13	44	72	103	133	164	194	225	256	286	317	347
14	14	45	73	104	134	165	195	226	257	287	318	348
15	15	46	74	105	135	166	196	227	258	288	319	349
16	16	47	75	106	136	167	197	228	259	289	320	350
17	17	48	76	107	137	168	198	229	260	290	321	351
18	18	49	77	108	138	169	199	230	261	291	322	352
19	19	50	78	109	139	170	200	231	262	292	323	353
20	20	51	79	110	140	171	201	232	263	293	324	354
21	21	52	80	111	141	172	202	233	264	294	325	355
22	22	53	81	112	142	173	203	234	265	295	326	356
23	23	54	82	113	143	174	204	235	266	296	327	357
24	24	55	83	114	144	175	205	236	267	297	328	358
25	25	56	84	115	145	176	206	237	268	298	329	359
26	26	57	85	116	146	177	207	238	269	299	330	360
27	27	58	86	117	147	178	208	239	270	300	331	361
28	28	59	87	118	148	179	209	240	271	301	332	362
29	29	...	88	119	149	180	210	241	272	302	333	363
30	30	...	89	120	150	181	211	242	273	303	334	364
31	31	...	90	...	151	...	212	243	...	304	...	365

Monthly Payments of $1,000 Loan

Rate	Number of Years of Loan							
(Percent)	5	10	15	20	25	30	35	40
7	$19.81	$11.62	$8.99	$7.76	$7.07	$6.66	$6.39	$6.22
7 1/2	20.04	11.88	9.28	8.06	7.39	7.00	6.75	6.59
8	20.28	12.14	9.56	8.37	7.72	7.34	7.11	6.96
8 1/2	20.52	12.40	9.85	8.68	8.06	7.69	7.47	7.34
9	20.76	12.67	10.15	9.00	8.40	8.05	7.84	7.72
9 1/2	21.01	12.94	10.45	9.33	8.74	8.41	8.22	8.11
10	21.25	13.22	10.75	9.66	9.09	8.78	8.60	8.50
10 1/2	21.50	13.50	11.06	9.99	9.45	9.15	8.99	8.89
11	21.75	13.78	11.37	10.33	9.81	9.53	9.37	9.29
11 1/2	22.00	14.06	11.69	10.67	10.17	9.91	9.77	9.69
12	22.25	14.35	12.01	11.02	10.54	10.29	10.16	10.09
12 1/2	22.50	14.64	12.33	11.37	10.91	10.68	10.56	10.49
13	22.76	14.94	12.66	11.72	11.28	11.07	10.96	10.90
13 1/2	23.01	15.23	12.99	12.08	11.66	11.46	11.36	11.31
14	23.27	15.53	13.32	12.44	12.04	11.85	11.76	11.72
14 1/2	23.53	15.83	13.66	12.80	12.43	12.25	12.17	12.13
15	23.79	16.14	14.00	13.17	12.81	12.65	12.57	12.54
15 1/2	24.06	16.45	14.34	13.54	13.20	13.05	12.98	12.95
16	24.32	16.76	14.69	13.92	13.59	13.45	13.39	13.36
17	24.86	17.38	15.40	14.67	14.38	14.26	14.21	14.19
18	25.40	18.02	16.11	15.44	15.18	15.08	15.03	15.02

ANNUAL PERCENTAGE RATE TABLE FOR MONTHLY PAYMENT PLANS

# of Pmts.	Annual Percentage Rate (Finance Charge per $100 of Amount Financed)									
▼	**2.00%**	**2.25%**	**2.50%**	**2.75%**	**3.00%**	**3.25%**	**3.50%**	**3.75%**	**4.00%**	**4.25%**
6	$0.58	$0.66	$0.73	$0.80	$0.88	$0.95	$1.02	$1.10	$1.17	$1.24
12	1.09	1.22	1.36	1.50	1.63	1.77	1.91	2.04	2.18	2.32
18	1.59	1.79	1.99	2.19	2.39	2.59	2.79	2.99	3.20	3.40
24	2.10	2.36	2.62	2.89	3.15	3.42	3.69	3.95	4.22	4.49
30	2.60	2.93	3.26	3.59	3.92	4.25	4.58	4.92	5.25	5.58
36	3.11	3.51	3.90	4.30	4.69	5.09	5.49	5.89	6.29	6.69
42	3.62	4.08	4.54	5.00	5.47	5.93	6.40	6.86	7.33	7.80
48	4.14	4.66	5.19	5.72	6.24	6.78	7.31	7.84	8.38	8.92
54	4.65	5.24	5.83	6.43	7.03	7.63	8.23	8.83	9.44	10.04
60	5.17	5.82	6.48	7.15	7.81	8.48	9.15	9.82	10.50	11.18
	4.50%	**4.75%**	**5.00%**	**5.25%**	**5.50%**	**5.75%**	**6.00%**	**6.25%**	**6.50%**	**6.75%**
6	$1.32	$1.39	$1.46	$1.54	$1.61	$1.68	$1.76	$1.83	$1.90	$1.98
12	2.45	2.59	2.73	2.87	3.00	3.14	3.28	3.42	3.56	3.69
18	3.60	3.80	4.00	4.21	4.41	4.61	4.82	5.02	5.22	5.43
24	4.75	5.02	5.29	5.56	5.83	6.10	6.37	6.64	6.91	7.18
30	5.92	6.25	6.59	6.92	7.26	7.60	7.94	8.28	8.61	8.96
36	7.09	7.49	7.90	8.30	8.71	9.11	9.52	9.93	10.34	10.75
42	8.27	8.74	9.91	9.69	10.16	10.64	11.12	11.60	12.08	12.56
48	9.46	10.00	10.54	11.09	11.63	12.18	12.73	13.28	13.83	14.39
54	10.65	11.26	11.88	12.49	13.11	13.73	14.36	14.98	15.61	16.23
60	11.86	12.54	13.23	13.92	14.61	15.30	16.00	16.70	17.40	18.10
	7.00%	**7.25%**	**7.50%**	**7.75%**	**8.00%**	**8.25%**	**8.50%**	**8.75%**	**9.00%**	**9.25%**
6	$2.05	$2.13	$2.20	$2.27	$2.35	$2.42	$2.49	$2.57	$2.64	$2.72
12	3.83	3.97	4.11	4.25	4.39	4.52	4.66	4.80	4.94	5.08
18	5.63	5.84	6.04	6.25	6.45	6.66	6.86	7.07	7.28	7.48
24	7.45	7.73	8.00	8.27	8.55	8.82	9.09	9.37	9.64	9.92
30	9.30	9.64	9.98	10.32	10.66	11.01	11.35	11.70	12.04	12.39
36	11.16	11.57	11.98	12.40	12.81	13.23	13.64	14.06	14.48	14.90
42	13.04	13.52	14.01	14.50	14.98	15.47	15.96	16.45	16.95	17.44
48	14.94	15.50	16.06	16.62	17.18	17.75	18.31	18.88	19.45	20.02
54	16.86	17.50	18.13	18.77	19.41	20.05	20.69	21.34	21.98	22.63
60	18.81	19.52	20.23	20.94	21.66	22.38	23.10	23.82	24.55	25.28
	9.50%	**9.75%**	**10.00%**	**10.25%**	**10.50%**	**10.75%**	**11.00%**	**11.25%**	**11.50%**	**11.75%**
6	$2.79	$2.86	$2.94	$3.01	$3.08	$3.16	$3.23	$3.31	$3.38	$3.45
12	5.22	5.36	5.50	5.64	5.78	5.92	6.06	6.20	6.34	6.48
18	7.69	7.90	8.10	8.31	8.52	8.73	8.93	9.14	9.35	9.56
24	10.19	10.47	10.75	11.02	11.30	11.58	11.86	12.14	12.42	12.70
30	12.74	13.09	13.43	13.78	14.13	14.48	14.83	15.19	15.54	15.89
36	15.32	15.74	16.16	16.58	17.01	17.43	17.86	18.29	18.71	19.14
42	17.94	18.43	18.93	19.43	19.93	20.43	20.93	21.44	21.94	22.45
48	20.59	21.16	21.74	22.32	22.90	23.48	24.06	24.64	25.23	25.81
54	23.28	23.94	24.59	25.25	25.91	26.57	27.23	27.90	28.56	29.23
60	26.01	26.75	27.48	28.22	28.96	29.71	30.45	31.20	31.96	32.71

ANNUAL PERCENTAGE RATE TABLE FOR MONTHLY PAYMENT PLANS

# of Pmts. **Annual Percentage Rate (Finance Charge per $100 of Amount Financed)**

▼	12.00%	12.25%	12.50%	12.75%	13.00%	13.25%	13.50%	13.75%	14.00%	14.25%
6	$3.53	$3.60	$3.68	$3.75	$3.83	$3.90	$3.97	$4.05	$4.12	$4.20
12	6.62	6.76	6.90	7.04	7.18	7.32	7.46	7.60	7.74	7.89
18	9.77	9.98	10.19	10.40	10.61	10.82	11.03	11.24	11.45	11.66
24	12.98	13.26	13.54	13.82	14.10	14.38	14.66	14.95	15.23	15.51
30	16.24	16.60	16.95	17.31	17.66	18.02	18.38	18.74	19.10	19.45
36	19.57	20.00	20.43	20.87	21.30	21.73	22.17	22.60	23.04	23.48
42	22.96	23.47	23.98	24.49	25.00	25.51	26.03	26.55	27.06	27.58
48	26.40	26.99	27.58	28.18	28.77	29.37	29.97	30.57	31.17	31.77
54	29.91	30.58	31.25	31.93	32.61	33.29	33.98	34.66	35.35	36.04
60	33.47	34.23	34.99	35.75	36.52	37.29	38.06	38.83	39.61	40.39

	14.50%	14.75%	15.00%	15.25%	15.50%	15.75%	16.00%	16.25%	16.50%	16.75%
6	$4.27	$4.35	$4.42	$4.49	$4.57	$4.64	$4.72	$4.79	$4.87	$4.94
12	8.03	8.17	8.31	8.45	8.59	8.74	8.88	9.02	9.16	9.30
18	11.87	12.08	12.29	12.50	12.72	12.93	13.14	13.35	13.57	13.78
24	15.80	16.08	16.37	16.65	16.94	17.22	17.51	17.80	18.09	18.37
30	19.81	20.17	20.54	20.90	21.26	21.62	21.99	22.35	22.72	23.08
36	23.92	24.35	24.80	25.24	25.68	26.12	26.57	27.01	27.46	27.90
42	28.10	28.62	29.15	29.67	30.19	30.72	31.25	31.78	32.31	32.84
48	32.37	32.98	33.59	34.20	34.81	35.42	36.03	36.65	37.27	37.88
54	36.73	37.42	38.12	38.82	39.52	40.22	40.92	41.63	42.33	43.04
60	41.17	41.95	42.74	43.53	44.32	45.11	45.91	46.71	47.51	48.31

	17.00%	17.25%	17.50%	17.75%	18.00%	18.25%	18.50%	18.75%	19.00%	19.25%	19.50%
6	$5.02	$5.09	$5.17	$5.24	$5.32	$5.39	$5.46	$5.54	$5.61	$5.69	$5.76
12	9.45	9.59	9.73	9.87	10.02	10.16	10.30	10.44	10.59	10.73	10.87
18	13.99	14.21	14.42	14.64	14.85	15.07	15.28	15.49	15.71	15.93	16.14
24	18.66	18.95	19.24	19.53	19.82	20.11	20.40	20.69	20.98	21.27	21.56
30	23.45	23.81	24.18	24.55	24.92	25.29	25.66	26.03	26.40	26.77	27.14
36	28.35	28.80	29.25	29.70	30.15	30.60	31.05	31.51	31.96	32.42	32.87
42	33.37	33.90	34.44	34.97	35.51	36.05	36.59	37.13	37.67	38.21	38.76
48	38.50	39.13	39.75	40.37	41.00	41.63	42.26	42.89	43.52	44.15	44.79
54	43.75	44.47	45.18	45.90	46.62	47.34	48.06	48.79	49.51	50.24	50.97
60	49.12	49.92	50.73	51.55	52.36	53.18	54.00	54.82	55.64	56.47	57.30

	19.75%	20.00%	20.25%	20.50%	20.75%	21.00%	21.25%	21.50%	21.75%	22.00%	22.25%
6	$5.84	$5.91	$5.99	$6.06	$6.14	$6.21	$6.29	$6.36	$6.44	$6.51	$6.59
12	11.02	11.16	11.31	11.45	11.59	11.74	11.88	12.02	12.17	12.31	12.46
18	16.36	16.57	16.79	17.01	17.22	17.44	17.66	17.88	18.09	18.31	18.53
24	21.86	22.15	22.44	22.74	23.03	23.33	23.62	23.92	24.21	24.51	24.80
30	27.52	27.89	28.26	28.64	29.01	29.39	29.77	30.14	30.52	30.90	31.28
36	33.33	33.79	34.25	34.71	35.17	35.63	36.09	36.56	37.02	37.49	37.95
42	39.30	39.85	40.40	40.95	41.50	42.05	42.60	43.15	43.71	44.26	44.82
48	45.43	46.07	46.71	47.35	47.99	48.64	49.28	49.93	50.58	51.23	51.88
54	51.70	52.44	53.17	53.91	54.65	55.39	56.14	56.88	57.63	58.38	59.13
60	58.13	58.96	59.80	60.64	61.48	62.32	63.17	64.01	64.86	65.71	66.57

647

AMOUNT PER $1.00 INVESTED, MONTHLY, QUARTERLY, AND DAILY COMPOUNDING						
Annual	Interest Period: 1 year			Interest Period: 4 year		
Rate	Monthly	Quarterly	Daily	Monthly	Quarterly	Daily
7.00%	1.072290	1.071859	1.072500	1.322053	1.319929	1.323094
7.25%	1.074958	1.074495	1.075185	1.335261	1.332961	1.336389
7.50%	1.077632	1.077135	1.077875	1.348599	1.346114	1.349817
7.75%	1.080312	1.079781	1.080573	1.362066	1.359388	1.363380
8.00%	1.082999	1.082432	1.083277	1.375666	1.372785	1.377079
8.25%	1.085692	1.085087	1.085988	1.389398	1.386306	1.390916
8.50%	1.088390	1.087747	1.088706	1.403264	1.399951	1.404891
8.75%	1.091095	1.090413	1.091430	1.417266	1.413723	1.419008
9.00%	1.093806	1.093083	1.094162	1.431405	1.427621	1.433265
9.25%	1.096524	1.095758	1.096900	1.445682	1.441647	1.447666
9.50%	1.099247	1.098438	1.099645	1.460098	1.455803	1.462212
9.75%	1.101977	1.101123	1.102397	1.474655	1.470088	1.476903
10.00%	1.104713	1.103812	1.105155	1.489354	1.484505	1.491742
10.25%	1.107455	1.106507	1.107921	1.504196	1.499055	1.506731
10.50%	1.110203	1.109207	1.110693	1.519183	1.513738	1.521869
10.75%	1.112958	1.111911	1.113473	1.534317	1.528555	1.537160
11.00%	1.115718	1.114621	1.116259	1.549598	1.543509	1.552604
11.25%	1.118485	1.117335	1.119052	1.565028	1.558600	1.568203
11.50%	1.121259	1.120055	1.121853	1.580608	1.573829	1.583959
11.75%	1.124039	1.122779	1.124660	1.596340	1.589197	1.599873
12.00%	1.126825	1.125508	1.127474	1.612226	1.604706	1.615946
12.25%	1.129617	1.128243	1.130295	1.628266	1.620357	1.632182
12.50%	1.132416	1.130982	1.133124	1.644462	1.636151	1.648580
12.75%	1.135221	1.133726	1.135959	1.660817	1.652089	1.665142
13.00%	1.138032	1,136475	1.138802	1.677330	1.668172	1.681871
13.25%	1.140850	1.139230	1.141651	1.694004	1.684402	1.698768
13.50%	1.143674	1.141989	1.144508	1.710841	1.700780	1.715835
13.75%	1.146505	1.144753	1.147371	1.727841	1.717308	1.733073
14.00%	1.149342	1.147523	1.150242	1.745006	1.733986	1.750484
14.25%	1.152185	1.150297	1.153121	1.762339	1.750815	1.768070
14.50%	1.155035	1.153076	1.156006	1.779840	1.767798	1.785832
14.75%	1.157891	1.155861	1.158898	1.797511	1.784935	1.803773
15.00%	1.160754	1.158650	1.161798	1.815354	1.802227	1.821894

INDEX

INDEX

SELECTED ANSWERS

Workshop 1 **Self-Check** **1.** two hundred forty-one **2.** seven and three hundred seventeen thousandths **3.** seven hundred sixty-one and thirteen one hundredths dollars or seven hundred sixty-one and $\frac{13}{100}$ dollars, or seven hundred sixty-one dollars and thirteen cents **Problems** **5.** $967.46 **7.** $647.56 **9.** $65.75 **11.** $95.07 **13.** fifteen and forty three one hundredths of a dollar or fifteen and $\frac{43}{100}$ dollars or fifteen dollars and 43 cents **15.** three hundred and five tenths **17.** 420,000 **19.** 960,000 **21.** 312,000 **23.** 868,000 **25.** 41,500 **27.** 17,800 **29.** 7290 **31.** 21,950 **33.** 3 **35.** 43 **37.** 21.2 **39.** 4718.2 **41.** 5.22 **43.** 31.80 **45.** 16,000,000 **47.** 15,749,000 **49.** 15,750,000 **51.** $14.72 **53.** $72.20 **55.** $119.90 **57.** $7.00 **59.** $152.00 **61.** $624.00 **63.** $3150 **65.** $520.00 **67.** $19,900 **69.** $4100 **71.** **a.** Seven and $\frac{17}{100}$ dollars **b.** Nine and $\frac{75}{100}$ dollars **c.** Forty-two and $\frac{91}{100}$ dollars **d.** Fifty-one and $\frac{80}{100}$ dollars **e.** Two hundred seventeen and $\frac{23}{100}$ dollars **f.** Five hundred thirty-one and $\frac{49}{100}$ dollars **g.** Nine thousand seven hundred fifty-one and $\frac{63}{100}$ dollars **h.** One thousand two hundred seventy and $\frac{82}{100}$ dollars **73.** **a.** $618,000 **b.** $719,000 **c.** $272,000 **d.** $915,000 **e.** $535,000 **f.** $420,000 **g.** $178,000 **h.** $572,000 **i.** $423,000

Workshop 2 **Self-Check** **1.** 8891 **2.** 759.80 **3.** 3079 **4.** 56.633 **5.** 284.4 **6.** 0.06 **7.** 62.313 **Problems** **9.** 554 **11.** 3943 **13.** 81.2 **15.** 0.4 **17.** 0.9 **19.** 2.244 **21.** $329.20 **23.** 29 miles to the gallon **25.** 1.36, 1.37, 1.39 **27.** 7.18, 7.38, 7.58 **29.** 15.231, 15.270, 15.321 **31.** 121.012, 121.021, 121.210 **33.** 0.1234, 0.1342, 0.1423 **35.** 10.100, 11.01, 11.11 **37.** 3.062, 3.09, 3.1 **39.** **a.** 513.12, 519.03, 532.626, 571.113, 587.41 **b.** 11.7, 32.615, 34.9, 67.192, 94.79 **c.** 15.04, 18.7, 26.311, 46.94, 71.21 **d.** 22.5, 38.9, 48.275, 67.21, 93.047 **41.** **a.** 873.15 and 877.142 **b.** 332.749 and 333.54 **43.** **a.** $2.45 **b.** $2.95 **c.** $2.45 **d.** $2.95 **e.** $2.95 **f.** $3.95 **g.** $4.45 **h.** $1.95 **i.** $2.95 **j.** $3.45 **45.** **a.** 2, 1, 3 **b.** 3, 1, 2 **c.** 1, 3, 2 **d.** 2, 1, 3 **e.** 3, 1, 2 **f.** 1, 3, 2

Workshop 3 **Self-Check** **1.** 41.76 **2.** 96.14 **3.** 174.70 **4.** 198.65 **5.** 198.208 **6.** 43.124 **7.** $44.06 **8.** $111.12 **Problems** **9.** 88.01 **11.** 1328.11 **13.** 86.007 **15.** 11.018 **17.** 205.2 **19.** 15.688 **21.** 463.491 **23.** 966.678 **25.** 645.356 **27.** 994.7351 **29.** $51.76 **31.** $130.40 **33.** 124.03 **35.** 303.1416 **37.** $123.96 **39.** $289.17 **41.** $148.87 **43.** $571.63 **45.** Subtotal $41.55, Total $44.87 **47.** **a.** 39.4 **b.** 48.8 **c.** 77.6 **d.** 75.5 **e.** 70.5

Workshop 4 **Self-Check** **1.** 53.61 **2.** 38.3 **3.** 610.08 **4.** 68.76 **5.** $16.78 **6.** $912.06 **7.** $332.15 **8.** $39.05 **Problems** **9.** 63.3 **11.** 11.33 **13.** 39.30 **15.** 4.639 **17.** 25.303 **19.** 5.9551 **21.** 55.617 **23.** 720.168 **25.** 407.22 **27.** 2.741 **29.** $67.69 **31.** $665.90 **33.** $264.03 **35.** $16,681.07 **37.** 117.5 **39.** 4.5 **41.** 22.9279 **43.** $28.91 **45.** $5.37 **47.** $863.15 **49.** **a.** $0.06 **b.** $0.88 **c.** $0.03 **d.** $2.00 **e.** $0.00 **f.** $1.89 **g.** $0.10 **h.** $2.58 **i.** $1.00 **j.** $0.16 **51.** **a.** Subtotal $132.55 **b.** Total deposit $112.55 **53.** **a.** Subtotal $96.59 **b.** Total deposit $56.59 **55.** **a.** Subtotal $577.13 **b.** Total deposit $502.13 **57.** **a.** $250.33 **b.** $102.60 **c.** $892.85 **d.** $378.72 **e.** $60.33 **f.** $1020.42 **g.** $500.60 **h.** $380.25 **i.** $210.80 **j.** $8.48

Workshop 5 **Self-Check** **1.** 8.151 **2.** 89.452 **3.** 0.0858 **4.** 0.0592 **5.** $35.28 **6.** 714 **7.** 4186.1 **8.** 3179.4 **Problems** **9.** 8.26 **11.** 21.2 **13.** 3.0601 **15.** 30.318 **17.** 0.0372 **19.** 0.0082 **21.** 0.0306 **23.** 0.003175 **25.** 0.16684 **27.** 2.18673 **29.** 0.000142 **31.** 1.866214 **33.** $45 **35.** $20.48 **37.** $43.10 **39.** 317 **41.** 65.17 **43.** 3285 **45.** 398.7 **47.** 72,716 **49.** 415 **51.** 69,178 **53.** 947,189 **55.** **a.** $3.87 **b.** $7.96 **c.** $2.98 **d.** $9.98 **e.** $8.76 **f.** $9.27 **g.** $7.49 **h.** $50.31 **57.** **a.** $314; $19.47; $4.55; $78.50; $14.13; $116.65; $197.35 **b.** $306; $18.97; $4.44; $76.50; $13.77; $113.68; $192.32 **c.** $350; $21.70; $5.08; $87.50; $15.75; $130.03; $219.97 **d.** $339.50; $21.05; $4.92; $84.88; $15.28; 126.13; $213.37 **e.** $388; $24.06; $5.63; $97; $17.46; $144.15; $243.85 **f.** $1697.50; $105.25; $24.62; $424.38; $76.39; $630.64; $1066.86

Workshop 6 **Self-Check** **1.** 6.2 **2.** 4.29 **3.** 29.4 **4.** 36 **5.** $1.36 **6.** 0.79 **7.** 1.389 **8.** 9.8628 **Problems** **9.** 1.8 **11.** 9.3 **13.** 7.4 **15.** 7.7 **17.** 8.5 **19.** 2.8 **21.** 73.2 **23.** 5.2 **25.** 1.83 **27.** 12.00 **29.** 53.39 **31.** 0.56 **33.** $2.66 **35.** $8.70 **37.** $4.31 **39.** $10.20 **41.** 0.93 **43.** 72.681 **45.** 4.298 **47.** 84.6265 **49.** 5.8969 **51.** 0.022098 **53.** 6.365218 **55.** 0.041549 **57.** 33 **59.** $4.09

61. **a.** 16.5 **b.** 18.7 **c.** 24.2 **d.** 29.6 **e.** 13.2 **f.** 6.6

Workshop 7 **Self-Check** **1.** 0.875 **2.** 0.6 **3.** 0.778 **4.** 0.667 **5.** $\frac{4}{10} = \frac{2}{5}$ **6.** $\frac{12}{100} = \frac{3}{25}$ **7.** $1\frac{125}{1000} = 1\frac{1}{8}$ **8.** $7\frac{82}{100} = 7\frac{41}{50}$ **Problems** **9.** 0.2 **11.** 0.714 **13.** 0.45 **15.** 0.444 **17.** 0.425 **19.** 0.7 **21.** 0.36 **23.** 0.417 **25.** 0.25 **27.** 0.13 **29.** 0.86 **31.** 0.3 **33.** 0.03 **35.** 0.93 **37.** 0.143 **39.** 0.583 **41.** 0.267 **43.** 0.567 **45.** 0.54 **47.** 0.982 **49.** $\frac{1}{4}$ **51.** $\frac{1}{2}$ **53.** $\frac{3}{10}$ **55.** $\frac{9}{20}$ **57.** $\frac{1}{8}$ **59.** $\frac{151}{200}$ **61.** $\frac{3}{16}$ **63.** $\frac{13}{40}$ **65.** $\frac{17}{50}$ **67.** $\frac{15}{16}$ **69.** $1\frac{1}{10}$ **71.** $7\frac{2}{25}$ **73.** $5\frac{117}{1000}$ **75.** $3\frac{7}{16}$ **77.** $21\frac{39}{40}$ **79.** $43\frac{11}{32}$ **81.**30.25 **83.** 0.25 **85.** $\frac{4}{5}$ **Applications** **87.** $52.50 **b.** $22.88 **c.** $12.63 **d.** $45.25 **e.** $8.38 **f.** $26.38 **g.** $39.50 **h.** $5.63 **89.** **a.** $222\frac{2}{25}$ **b.** $216\frac{8}{25}$ **c.** $216\frac{1}{100}$ **d.** $212\frac{29}{100}$ **e.** $210\frac{1}{20}$ **f.** $208\frac{21}{50}$ **c.** $208\frac{1}{4}$

Workshop 8 **Self-Check** **1.** 0.25 **2.** 0.375 **3.** 1.42 **4.** 0.09 **5.** 85% **6.** 7% **7.** 30% **8.** 155% **9.** 0.4% **Problems** **11.** 0.34 **13.** 0.60 **15.** 0.87 **17.** 0.99 **19.** 0.976 **21.** 0.818 **23.** 0.344 **25.** 0.736 **27.** 8.17 **29.** 3.76 **31.** 5.08 **33.** 17.83 **35.** 0.03 **37.** 0.09 **39.** 0.047 **41.** 0.072 **43.** 0.0803 **45.** 0.050921 **47.** 13% **49.** 35% **51.** 24% **53.** 97% **55.** 5% **57.** 1% **59.** 4.2% **61.** 9.65% **63.** 0.5% **65.** 0.61% **67.** 937.1% **69.** 720% **71.** 298% **73.** 700% **75.** 0.9% **77.** 4600% **79.** 10% **81.** 108,300% **83.** 1.123 **85.** 0.30; 0.50 **87.** **a.** 1.6% **b.** 0.2% **c.** 1.5% **d.** 4.5% **e.** 4.2% **f.** 1.3% **g.** 1.9% **89.** **a.** 0.017 **b.** 0.013 **c.** 0.011 **d.** 0.009 **e.** 0.007 **f.** 0.005

Workshop 9 **Self-Check** **1.** 28 **2.** 30 **3.** 1.5 **4.** $290 **Problems** **5.** 24.5 **7.** 10.8 **9.** 140.7 **11.** 100.8 **13.** 209.5 **15.** 3.8 **17.** 6.51 **19.** 82.08 **21.** 0.764 **23.** 780 **25.** 1677 **27.** 3170.5 **29.** 6835.5 **31.** 37,308.7 **33.** 6.1875 **35.** 7.48 **37.** 26.28 **39.** 192.075 **41.** 623 **43.** 741.258 **45.** $4.20 **47.** $27.30 **49.** $19.98 **51.** $1.62 **53.** $1.26 **55.** $0.90 **57.** $435 **59.** **a.** $9, $20.99 **b.** $10.20, $23.79 **c.** $12, $27.99 **61.** **a.** 54 **b.** 40 **c.** 33 **d.** 34 **e.** 108 **63.** **a.** $30.00 **b.** $51.00 **c.** $117.60 **d.** $117.00 **e.** $61.88 **f.** $149.36 **g.** $251.69 **h.** $175.00 **i.** $121.48 **j.** Negotiated

Workshop 10 **Self-Check** **1.** 5 **2.** 5.1 **3.** 130 **4.** $41.33 **Problems** **5.** 10 **7.** 80 **9.** 140.3 **11.** 743 **13.** 4.38 **15.** 3.05 **17.** $14.20 **19.** $77.33 **21.** 6.7 **23.** 33.9 **25.** 40.04 **27.** $408,000 **29.** 78.0 **31.** $150.00 **33.** 93.5 **35.** 94 **37.** **a.** 3 **b.** 51 **c.** 8 **d.** 5 **e.** 16 **f.** 17; 15; 21; 15; 15 **39.** $1.31 **41.** 10.25; 27; 3321

Workshop 11 **Self-Check** **1.** 7 h:15 min **2.** 5 h:15 min **3.** 5 h:55 min **4.** 2 h:57 min **Problems** **5.** 9 h:10 min **7.** 7 h:43 min **9.** 4 h:15 min **11.** 6 h:15 min **13.** 6 h:50 min **15.** 3 h:37 min **17.** 8 h:30 min **19.** 3 h:53 min **21.** 4 h:45 min **23.** 7 h:45 min **25.** 7 h:15 min **27.** 8 h:35 min **29.** 8 h **31.** 8 h:45 min **33.** 7 h:55 min **35.** 6 h:37 min **37.** 6 h:34 min **39.** 8 h:27 min **41.** **a.** 0 h:42 min **b.** 3 h:15 min **c.** 4 h:32 min **d.** 2 h:26 min **e.** 1 h:55 min **f.** 1 h:28 min **g.** 2 h:55 min

Workshop 12 **Self-Check** **1.** $1.25 **2.** $1.55 **3.** $2.34 **4.** $3.32 **5.** Size 38 or 39 **6.** Size 42 **Problems** **7.** $1.87 **9.** $2.46 **11.** $3.01 **13.** $1.79 **15.** $2.58 **17.** $1.40 **19.** **a.** 2 lb **b.** 11 lb **c.** 13 lb **d.** 10 lb **e.** 14 lb **f.** 8 lb **21.** 36 or 37 **23.** 33 **25.** 29 **27.** Bangor **29.** $54.00 **31.** $50.00 **33.** $56.00 **35.** $610; $620 **37.** $510; $520 **39.** $530; $540 **41.** $68 **43.** $70 **45.** $79

Workshop 13 **Self-Check**

1. **2.**

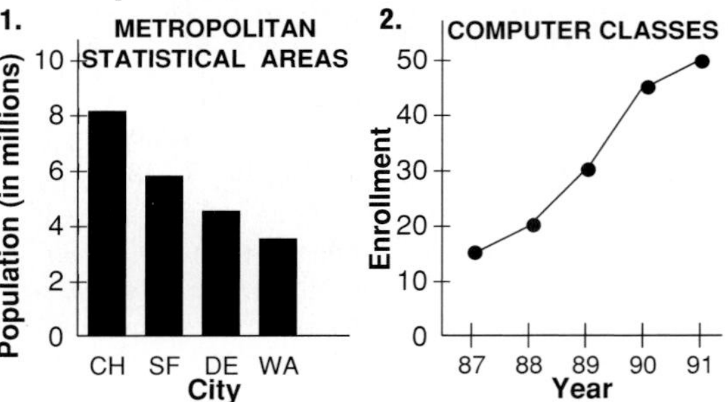

Problems **3.**

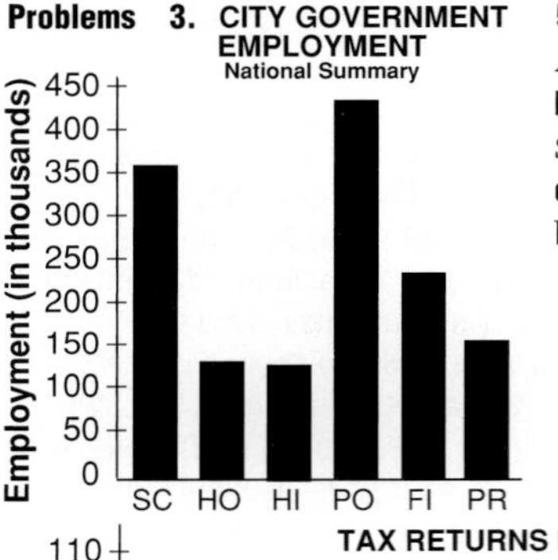

5. **a.** 1989; About $48 billion **b.** 1982; About $12 billion **c.** About $35 billion

7.

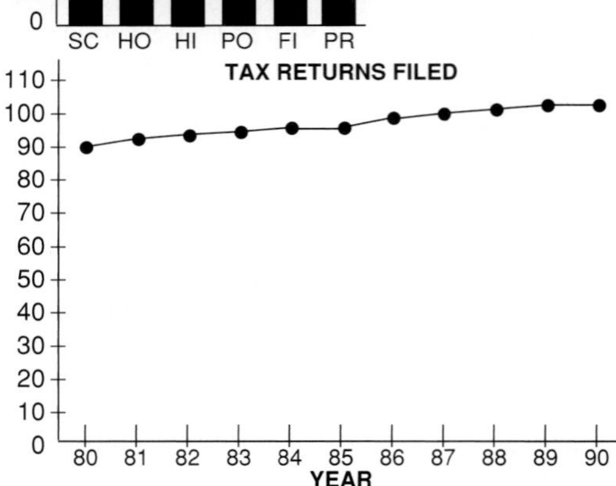

Workshop 14 **Self-Check** **1.** 108 in **2.** 150 mL **3.** 8 yd **4.** 3.5 m **Problems** **5.** 27 ft **7.** 112 oz **9.** 96 oz **11.** 3800 m **13.** 3200 g **15.** 2.25 gal **17.** 3.5 gal **19.** 2 kg **21.** 3.3 L **23.** 72 100 g **25.** 0.723 kg **27.** 180mm **29.** 40 in **31.** 11 pt **33.** 10 qt **35.** 29 pt **37.** 20 qt **39.** 8 c **41.** 240 in **43.** 3900 **45.** 2.13 L **47.** **a.** 36 ft **b.** 36 ft; 48 ft; 84 ft **c.** 21 ft; 18 ft 8 in; 39 ft 8 in **d.** 22 ft 6 in; 15 ft 4 in; 37 ft 10 in **e.** 22 ft 8 in; 26 ft 4 in; 49 ft **f.** 35 ft 4 in; 25 ft 6 in; 60 ft 10 in **g.** 16 ft 10 in; 15 ft 6 in; 32 ft 4 in; Total 381 ft 8 in

Workshop 15 **Self-Check** **1.** $40 - 30 = 10$; 16.495 **2.** $50 \div 7 = 7$; 6.32 **3.** $60 \times 10 = 600$; 602.6713 **4.** $5 \times 5 = 25$; $23\frac{11}{12}$ **Problems** **5.** 8000; 7789 **7.** 2; 2.14 **9.** 72; 73.9728 **11.** 7; 7.1 **13.** 6; $4\frac{1}{8}$ **15.** 3; $3\frac{1}{8}$ **17.** \$35; \$34.97 **19.** 80; 83.2 **21.** \$10; \$9.24 **23.** \$9; \$8.89 **25.** \$67; \$68.47 **27.** \$26

Workshop 16 **Self-Check:** **1.** 900; 1100; 1111 **2.** 15,000; 17,000; 16,784 **3.** \$13.00; \$15.00; \$15.01 **Problems:** **5.** 16,000; 16,846 **7.** 120; 120.4 **9.** 1200; 1195 **11.** \$73; \$74.96 **13.** \$400; \$462.81 **15.** \$23.00; \$23.39 **Applications:** **17.** \$12.00; \$11.79 **19.** \$70.00; \$71.20 **21.** 450; 430.28 **23.** **a.** 1200; Yes **b.** \$14; No; \$14.00 **c.** 48; No; 49.3 **d.** 360; Yes **e.** 130; No; 135.3 **f.** 45; Yes **g.** 4; No; 4.65 **h.** \$450; No; \$439.45

Workshop 17 **Self-Check:** **1.** $640 \div 80 = 8$; 8.07 **2.** $9000 \div 300 = 30$; 27.3 **3.** $\frac{1}{3} \times 900 = 300$; 324.1 **Problems:** **5.** 700; 697 **7.** 110; 107 **9.** 40; 42.72 **11.** \$900; \$895.50 **13.** 30; $31\frac{11}{14}$ **15.** \$50; \$48.75 **17.** \$4000; \$3958.33 **19.** \$80; \$81.82 **21.** \$33.00; \$32.06 **23.** \$2000; \$2050 **25.** \$70; \$17,500; \$18,200 **27.** \$160 to \$180; \$153.85 to \$173.08 **29.** \$18; \$17.88

Workshop 18 **Self-Check:** **1.** $4 \times 600 = 2400$; 2332 **2.** $4 \times \$30 = \$120$; \$119.09 **3.** $(3 \times \$2) + \$1 = \$6 + \$1 = \$7$; \$6.34 **Problems:** **5.** 2000; 2063 **7.** \$160; \$161.24 **9.** 36; 35.87 **11.** \$17; \$18.28 **13.** 3600; 3500.67 **15.** \$18; \$18.60 **17.** \$9; \$9.40 **19.** 2400; 2346 **21.** \$36; \$35.56 **23.** \$92; \$88.45

Workshop 19 **Self-Check:** **1.** \$3180 **Problems:** **3.** \$69.88 **5.** \$8784 **7.** \$444.50 **9.** Approximately 1495 rotations **11.** 224 miles **13.** 15 loaves

Workshop 20 **Self-Check:** **1.** Problem cannot be solved. Need number of payments. **Problems:** **3.** Cannot be solved. Need relationship between pints and pounds. **5.** \$24 **7.** Cannot be solved. Need weight of watermelons. **9.** 5' 11" **11.** Cannot be solved. Need cost of tennis balls. **13.** 19 **15.** Impossible to average 60 mph.

Workshop 21 **Self-Check:** **1.** \$14.38 **Problems:** **3.** ×, ×, +, \$216.60 **5.** ×; +; +; ×, compare, +; \$29.60 **7.** ×; ×; +; \$187.65 **9.** −; ×; +; 28.5 degrees C **11.** +; −; 223 **13.** ×; +; −; \$2250 **15.** ×; +; −; \$4.55

Workshop 22 **Self-Check:** **1.** No; he used 93 instead of \$0.93. Answer is \$9.30. **Problems:** **3.** No; Used 89¢ as \$89; estimate: \$4.50 **5.** Yes; No error **7.** No; Decimal point; estimate: \$50 **9.** No; Added instead of subtracted; estimate: \$30 **11.** About \$500 **13.** About \$1250 **15.** About \$200 **17.** About \$3 **19.** About 25

Workshop 23 **Self-Check:** **1.** 7 bicycles; 4 unicycles **Problems:** **3.** 6 bicycles; 3 tricycles **5.** 20 nickels; 20 quarters **7.** 25 **9.** 12 **11.** 4 pennies; 4 dimes; 1 quarter

Workshop 24 **Self-Check:** **1.** 81; 243; 729 **2.** 16; 22; 29 **Problems:** **3.** 34; 40; 46 **5.** 18; 15; 12 **7.** 25; 36; 49 **9.** 1093; 3280; 9841 **11.** 8 dimes; 12 quarters **13.** 33 **15.** 2:30 p.m. **17.** 4×4; 6×3 **19.** 11

Workshop 25 **Self-Check:** **1.** 7; 8; 9 **Problems:** **3.** 7 letters; 8 postcards **5.** All sums are 12. **7.** 2 **9.** Ford = 10; Chevy = 8; Dodge = 4 **11.** Sum is 27; Top: 8; 13; 6; Mid: 7; 9; 11; Bot: 12; 5; 10

Workshop 26 **Self-Check:** **1.** 29 days **Problems:** **3.** 5 **5.** \$15,000 **7.** 56 dozen **9.** 62 **11.** 6

Workshop 27 **Self-Check:** **1.** 11 **Problems:** **3.** Cork: 5¢; Bottle: \$1.05; **5.** 54 yards **7.** \$7962 **9.** Holstein **11.** 7; 9; 11 **13.** B = 3A − 2; 118 **15.** Harry = 197; Jerry = 210; Darrel = 192

Workshop 28 **Self-Check:** **1.** 3 **Problems:** **3.** 12 **5.** none **7.** **a.** 182 **b.** 198

Workshop 29 **Self-Check:** **1.** (18" + 7") − 10" **Problems:** **3.** 3 miles west and 6 miles north **5.** 8556.5 sq ft **7.** 15 **9.** (4 + 7) − 10 **11.** 342.26 sq ft

Workshop 30 **Self-Check:** **1.** $42{,}390 \div 5280 = 8$ **Problems:** **3.** **a.** 400 **b.** 225 **5.** 625 **7.** **a.** 352 ft **b.** 4400 ft **9.** Borrow 1 horse from neighbor; then Rick = 9; Mike = 6; and Peter = 2. **11.** There is no extra dollar.

Unit 1 Gross Income

1-1; pages 78–79 **Self-Check** **1.** $304 **2.** $296.25 **Problems** **3.** $288 **5.** $186 **7.** $453.13 **9.** $165 **11.** $812.50 **13.** $197.20 **15.** $444.06 **17.** $187 **19.** $215.63 **Maintaining Your Skills** **21.** 0.50 **23.** 0.10 **25.** $136.13 **27.** $471.25 **29.** 140 **31.** 100 **33.** 14,400 **35.** 1000

1-2; pages 80–81 **Self-Check** **1.** $441 **2.** $580.75 **Problems** **3.** $240; $72; $312 **5.** $192; $28.80; $220.80 **7.** $680; $442; $1122 **9.** $355.20; $115.20; $470.40 **11.** $329 **13.** $229.50 **15.** $570.24 **Maintaining Your Skills** **17.** $247.35 **19.** 1746.162 **21.** $441.60 **23.** $150 **25.** $125

1-3; pages 82–84 **Self-Check** **1.** $7\frac{1}{2}$ **2.** $8\frac{1}{4}$ **Problems** **3.** $7\frac{3}{4}$ **5.** $7\frac{3}{4}$ **7.** $8\frac{1}{2}$ **9.** $8\frac{1}{4}$ **11.** $8\frac{3}{4}$ **13.** 9 **15.** $49\frac{1}{2}$ **17.** $295.44 **19.** 8; $8\frac{1}{4}$; $7\frac{3}{4}$; 8; 8; 40; $300 **21.** S + S = 13 hrs; M − F = $33\frac{1}{2}$ hrs. **Maintaining Your Skills** **23.** $\frac{7}{8}$ **25.** $\frac{23}{24}$ **27.** $7\frac{3}{4}$ **29.** $3\frac{3}{4}$ **31.** 40,000 **33.** 68,000 **35.** 300

1-4; pages 85–86 **Self-Check** **1.** $448 **2.** $224.10 **Problems** **3.** $369 **5.** $269.70 **7.** $826.50 **9.** $263.86 **11.** $630 **13.** $401.25 **15.** $60.30 **17.** $69 **19.** $49 **21.** $1709.25 **Maintaining Your Skills** **23.** $264.00 **25.** $54.36 **27.** $128.00 **29.** $34.357 = $34.36 **31.** $1587.6 **33.** 456,000

1-5; pages 87-88 **Self-Check** **1.** $1650 **2.** $758.33 **Problems** **3.** $650 **5.** $1375 **7.** $1280.73 **9.** $240.38 **11.** $710 **13.** $2700; $623.08 **15.** $1450 **17.** $320.38 − $290.19 = $30.19 **19.** $3280 **Maintaining Your Skills** **21.** $285.80 **23.** 13.01 **25.** 559.38 **27.** 1329.73 **29.** 104 **31.** 104

1-6; pages 89–91 **Self-Check** **1.** $752 **2.** $7887 **3.** $160 **4.** $2125 **Problems** **5.** $7840 **7.** $1535.68 **9.** $139.35 **11.** $2394.60 **13.** $1495; $1600 **15.** $364; $850 **17.** $3521.10; $3521.10 **19.** $274.69 **21.** $181.50; $181.50 **23.** $8363.38 **25.** $188.21 **27.** $244.63 **Maintaining Your Skills** **29.** 2.13 **31.** 6.75 **33.** 9.25 **35.** $365.50 **37.** $331.20 **39.** 123.1 **41.** 521 **43.** 0.6 **45.** 0.8

1-7; pages 92–93 **Self-Check** **1.** $3050 **2.** $559.20 **Problems** **3.** $100; $360; $460 **5.** $90; $160; $40; $290 **7.** $482 **9.** $243.63 **11.** $142.50 **13.** $673.75 **Maintaining Your Skills** **15.** $1034.31 **17.** 1.960 **19.** 0.1025 **21.** 0.97 **23.** $105.00 **25.** $542.50

Reviewing the Basics, page 94 **Skills** **1.** $45.67 **3.** $0.05 **5.** $172.26 **7.** 20.534 **9.** $2.48 **11.** 0.00021 **13.** 4660 **15.** $19\frac{1}{2}$ **17.** $10\frac{3}{4}$ **19.** $341.00 **21.** 0.5 **23.** 0.75 **25.** 0.125 **27.** 0.0125 **29.** 3.85 **Applications** **31.** $8\frac{1}{2}$ hrs. **33.** 104 **35.** 16

Unit 2 Net Income

2-1; pages 100–102 **Self-Check** **1.** $26.00 **2.** $34.00 **Problems** **3.** $20 **5.** $33 **7.** $39 **9.** $4 **11.** $30 **13.** $45 **15.** $28 **17.** $46 **19.** $12 **Maintaining Your Skills** **21.** ones; 4 **23.** thousands; 6000 **25.** hundreds; 600 **27.** $27 **29.** $1.49 **31.** $276.10 **33.** $874.89

2-2; pages 103–104 **Self-Check** **1.** $668 **2.** $1445 **Problems** **3.** $280 **5.** $280 **7.** $3000 **9.** $2200; $377.50 **11.** $1662.30 **13.** $4220 **Maintaining Your Skills** **15.** 65.87 **17.** 31.903 **19.** 192 **21.** 12.5

2-3; page 105–106 **Self-Check** **1.** $33.75 **2.** $32.02 **Problems** **3.** $3200 **5.** $677 **7.** $56.29 **9.** $22.82 **Maintaining Your Skills** **11.** 0.547 **13.** $235.92 **15.** 16.54 **17.** 0.03 **19.** 12 **21.** 208

2-4; pages 107–108 **Self-Check** **1.** $148.80; $34.80 **2.** $21.70; $5.08 **Problems** **3.** $4.15; $0.97 **5.** 1.45%; $8.65; $2.02 **7.** 6.2%; 1.45%; $298.84; $69.89 **9.** $0; over $62,700; $141.38 **11.** $83.70; $19.58 **13.** $61; $29.26; $6.84 **15.** $60,500 YTD; $136.40; $79.75 **Maintaining Your Skills** **17.** $7.25 **19.** 0.15 **21.** 320 **23.** 23,980 **25.** 3800 **27.** 147,400

2-5; pages 109–110 **Self-Check** **1.** $40.00 **2.** $18.88 **Problems** **3.** $95 **5.** $44.75 **7.** $6.21 **9.** $35.00 **11.** $14.62 **Maintaining Your Skills** **13.** 13.05 **15.** 86.50 **17.** 7.68 **19.** 7.89

2-6; pages 111–114 **Self-Check** **1.** $215.32 **Problems** **3.** 17.00; 18.86; 4.41; 6.08; 4.56; 224.69 **5.** 47.00; 26.12; 6.11; 9.64; 7.37; 303.01 **7.** 54.00; 28.55; 6.68; 11.51; 8.06; 19.89; 331.81 **9.** $295.68 **Maintaining Your Skills** **11.** 67.316 **13.** 253.23 **15.** 482.337 **17.** 2171.16 **19.** 63.06 **21.** 65.38

Reviewing the Basics; page 115 **Skills** **1.** $239 **3.** $154.81 **5.** $40.10 **7.** $103.31 **9.** $24.01 **11.** $135.43 **13.** $19.00 **15.** $4.75 **17.** $4.76 **Applications** **19.** $63.68 **21.** 24 **23.** 72

Unit 3 Checking Account

3-1; pages 126–127 **Self-Check** **1.** $74.90 **2.** $105.00 **Problems** **3.** $41.80; $41.80 **5.** $178.20; $168.20 **7.** $1275.95; $1240.95 **9.** $435.44 **11.** $599.22 **13.** $151.45 **15.** $1238.03 **17.** $108.03 **Maintaining Your Skills** **19.** $396.05 **21.** $373.93 **23.** $507.00 **25.** $659.40 **27.** $653.32

3-2; pages 128–129 **Self-Check** **1.** $215.32 **2.** One hundred forty-three and $\frac{32}{100}$ **Problems** **3.** Forty and $\frac{40}{100}$ **5.** Sixty-three and $\frac{74}{100}$ **7.** Thirty-four and $\frac{6}{100}$ **9.** One thousand nine hundred seventeen and $\frac{00}{100}$ **11.** Two hundred one and $\frac{9}{100}$ **13.** Five thousand three hundred twenty-seven and $\frac{17}{100}$ **15.** No; Ninety-eight and $\frac{72}{100}$ **17.** Yes **19.** Yes **Maintaining Your Skills** **21.** $35.15 **23.** Nineteen and $\frac{25}{100}$ **25.** Four hundred thirty-five and $\frac{00}{100}$ **27.** Five thousand two hundred seventy-four and $\frac{19}{100}$ **29.** Six thousand one hundred ten and $\frac{50}{100}$

3-3; pages 130–132 **Self-Check** **1.** $1201.01 **2.** $1131.55 **Problems** **3.** $401.43 **5.** $366.72 **7.** $367.33 **9.** $211.86 **11.** $162.66; $109.63; $2.33

656

13. \$268.30; \$218.30; \$505.00; \$378.65; \$669.65 **15.** \$313.88; \$358.98; \$224.32; \$22.32; \$161.72; \$50.57; \$20.57; \$−28.35 overdrawn **Maintaining Your Skills** **17.** \$1245.96 **19.** \$466.21 **21.** \$816.55 **23.** \$247.97 **25.** \$77.09 **27.** \$356.91

3-4; pages 133–135 **Self-Check** **1.** \$258.70 **2.** \$63.90 **Problems** **3.** \$47.66 **5.** \$1026.12 **7.** \$15,533.36 **9.** \$709.66 **11.** \$401.24 **13.** 581.63; 2329.90; 1030.45; 7.17; 1888.25 **Maintaining Your Skills** **15.** \$1066.70 **17.** \$47,082.74 **19.** \$500.82 **21.** \$1362.97 **23.** 846.35 **25.** 865.464

3-5; pages 136–140 **Self-Check** **1.** \$275.49 **2.** \$916.33 **Problems** **3.** \$142.90; \$142.90 **5.** \$1446.89; \$1446.89 **7.** \$316.54; Yes **9.** **a.** \$44.89 **b.** \$191.37 **c.** \$794.83 **d.** \$794.83 **e.** Yes **Maintaining Your Skills** **11.** \$411.05 **13.** \$93.78 **15.** \$243.77 **17.** \$171.92 **19.** \$46.36

Reviewing the Basics; page 141 **Skills** **1.** Twenty-five and $\frac{79}{100}$ **3.** One thousand three hundred seventy-two and $\frac{35}{100}$ **5.** \$262.18 **7.** \$621.78 **9.** \$359.35 **11.** \$211.68 **Applications** **13.** \$825.29 **15.** c **17.** b **19.** d

Unit 4 Savings Accounts

4-1, pages 146–147 **Self-Check** **1.** \$163 **2.** \$101 **Problems** **3.** \$79.15; \$79.15 **5.** \$92.71; \$62.71 **7.** \$1032.58; \$1012.58 **9.** \$197.40 **11.** \$935.28 **13.** \$500.03 **15.** \$256.48 **17.** \$950.51 **Maintaining Your Skills** **19.** \$58.78 **21.** \$697.97 **23.** \$78.93 **25.** \$43.33 **27.** \$50.60

4-2, pages 148–149 **Self-Check** **1.** Forty-five and $\frac{76}{100}$ dollars **2.** \$291.42 **Problems** **3**. Seventeen and $\frac{35}{100}$ dollars **5.** Forty-four and $\frac{93}{100}$ dollars **7.** Four hundred six dollars **9.** Seven thousand eight hundred fifty-two and $\frac{3}{100}$ dollars **11.** **a.** 17594179 **b.** \$831.95 **c.** Eight hundred thirty-one and $\frac{95}{100}$ dollars **13.** **a.** 81-0-174927 **b.** \$318.29 **c.** Three hundred eighteen and $\frac{29}{100}$ dollars **15.** **a.** 06029175 **b.** \$76.60 **c.** Seventy-six and $\frac{60}{100}$ dollars **Maintaining Your Skills** **17.** Ninety-four and $\frac{78}{100}$ dollars **19.** One hundred sixty-two and $\frac{5}{100}$ dollars **21.** Four thousand two hundred eleven and $\frac{15}{100}$ dollars **23.** \$251.27 **25.** \$25,696.29

4-3, pages 150–151 **Self-Check** **1.** \$851.00 **2.** \$3191.30 **Problems** **3.** \$568.12 **5.** \$113.08 **7.** \$616.05 **9.** \$1357.10 **Maintaining Your Skills** **11.** \$438.00 **13.** \$1173.18 **15.** \$175 **17.** \$397.54 **19.** \$6075.25

4-4, pages 152–154 **Self-Check** **1.** \$869.50 **Problems** **3.** \$429.50 **5.** \$7529.03 **7.** \$1947.25 **9.** \$3832.44 **11.** \$579.86 **13.** \$18,219.75; \$18,339.93; \$17,939.93; \$18,559.39; \$18,159.39; \$18,059.39; \$18,179.24; \$18,179.24 **Maintaining Your Skills** **15.** \$1267.52 **17.** \$1047.62 **19.** \$4488.90 **21.** \$39.10 **23.** \$2000.90

4-5, pages 155–156 **Self-Check** **1.** \$6.50 **2.** \$56.25 **Problems** **3.** \$43.20; \$10.80 **5.** \$39.24; \$19.62 **7.** \$256.45; \$64.11 **9.** \$43.62 **11.** \$32.84 **13.** \$14.48 **15.** \$21.67 **Maintaining Your Skills** **17.** 0.5 **19.** 0.25 **21.** 6.25 **23.** 0.055 **25.** 0.095 **27.** 0.10625

4-6, pages 157–158 **Self-Check** **1.** \$2163.20 **Problems** **3.** \$2.01; \$404.01 **5.** \$342.38; \$18,602.38; \$348.79; \$18,951.17 **7.** \$824.18 **9.** \$907.95 **11.** \$9712.13 **13.** \$1965.65 **Maintaining Your Skills** **15.** 0.0525 **17.** 0.0575 **19.** 0.0725 **21.** \$49.40 **23.** \$1176.10 **25.** \$362 **27.** 0.33 **29.** 0.60

4-7, pages 159–161 **Self-Check** **1.** \$2230.88 **2.** \$4776.12 **Problems** **3.** \$1003.90; \$103.90 **5.** \$1555.41; \$215.41 **7.** \$4233.44; \$361.77 **9.** \$156.39 **11.** \$55.22 **13.** \$29.52 **15.** \$896.04; \$96.04 **Maintaining Your Skills** **17.** 2% **19.** 0.5% **21.** 4.25% **23.** 1.3125% **25.** 1.875% **27.** 1.625%

4-8, pages 162–163 **Self-Check** **1.** \$6022.62 **2.** \$22.62 **Problems** **3.** \$80301.60; \$301.60 **5.** \$6549.08; \$49.08 **7.** \$15600.30; \$279.30 **9.** \$31.41 **11.** \$384.84 **13.** \$1440.63 **Maintaining Your Skills** **15.** \$4085.24 **17.** \$554.99 **19.** \$953.72 **21.** \$7395.57 **23.** \$94.18 **25.** \$8057.49

Reviewing the Basics, page 164 **Skills** **1.** Four hundred seventy-nine dollars **3.** Three thousand ninety-one and $\frac{47}{100}$ dollars **5.** \$212.05 **7.** \$734.96 **9.** \$6723.06 **11.** 2.125**13.** 1.685 **15.** 8.25 **17.** 21.75 **19.** 0.0675 **21.** 0.1995 **Applications** **23.** 1.14339 **25.** 88 days **27.** $\frac{3}{12} = \frac{1}{4}$ **29.** $\frac{9}{12} = \frac{3}{4}$ **31.** 16 **33.–37.** Answers will vary.

Unit 5 Cash Purchase

5-1, pages 168–169 **Self-Check** **1.** \$21 **2.** \$513 **Problems** **3.** \$2.44 **5.** \$1.09 **7.** \$59.70 **9.** \$0.30 **11.** \$4.61 **13.** \$6.12 **15.** \$1.79 **17.** \$3.71 **Maintaining Your Skills** **19.** \$622.80 **21.** \$1.20 **23.** 4000 **25.** 2,437,000 **27.** 35,230 **29.** 1130 **31.** \$279.54 **33.** \$2.86 **35.** 358.0 **27.** 521.7

5-2, pages 170–172 **Self-Check** **1.** \$260.10 **Problems** **3.** \$1.22 **5.** \$12.47 **7.** \$64.15 **9.** \$63.62 **11.** \$305.83 **13.** \$132.23 **15.** \$31.15 **17.** \$139.92 **19.** \$29.98; \$119.95; \$23.96; \$26.94; \$7.72; \$208.55; \$12.51; \$221.06 **Maintaining Your Skills** **21.** 142.6152 **23.** 0.0321 **25.** 27.187 **27.** 776.156 **29.** 1382.65 **31.** 126.25 **33.** 1251.846

5-3, page 173–174 **Self-Check** **1.** 3.09¢ = 3.1¢ **2.** 12.375¢ = 12.38¢ **Problems** **3.** 8.3¢ **5.** 65.9¢ **7.** \$0.19 **9.** \$1.00 **11.** \$44.61 **13.** \$0.15 **15.** \$3.75 **17.** **a.** \$0.03 **b.** \$0.75 **c.** \$0.55 **d.** \$0.20 **Maintaining Your Skills** **19.** 4.62 **21.** 0.02 **23.** 20 **25.** 1350 **27.** 741.9 **29.** 1132.6 **31.** 87.16 **33.** 1000.00

5-4, pages 175–176 **Self-Check** **1.** 20 oz at 3.1¢ **2.** 100 at 2.39¢ each **Problems** **3.** 3.7¢; 4.0¢; small **5.** 66.1¢; 66.5¢; small **7.** b. **9.** 9 oz can **11.** 15-can box **13.** 40-oz jar **15.** 2.7¢; 2.6¢; 2.8¢; 3.1¢; Giant size 14-lb box **Maintaining Your Skills** **17.** 21.37 **19.** 145.05 **21.** $1586 **23.** 5.1

5-5, pages 177–178 **Self-Check** **1.** $1.09 **2.** $4.27 **Problems** **3.** $2.14 **5.** $2.39 **7.** $1.58 **9.** $1.44 **11.** $25.95 **13.** $57.67 **15.** $82.04 **Maintaining Your Skills** **17.** $1.30 19. $1.54 **21.** $0.86

5-6, pages 179–181 **Self-Check** **1.** $580 **2.** $35.80 **Problems** **3.** $70 **5.** $10.49 **7.** $2.99 **9.** $16.63 **11.** $2900 **13.** $0.57 **15.** $13.55 **17.** $87.48 **19.** $64 **21.** $7.00; $13.75; $14.75; $6.25; $19.50 **23.** White sale **Maintaining Your Skills** **25.** 69.776 **27.** 10.467 **29.** $112 **31.** 1265

5-7, pages 182–183 **Self-Check** **1.** $31.50 **2.** $175.20 **Problems** **3.** $48; $72 **5.** $2.75; $8.24 **7.** $292.50; $157.50 **9.** $61.08 **11.** $12.56 **13.** $27.88 – $19.79 = $8.09 = 29% of $27.88 **Maintaining Your Skills** **15.** $18.70 **17.** 25.76 **19.** 82.806 **21.** 0.0141 **23.** $4.49

Reviewing the Basics, page 184 **Skills** **1.** 14.4¢ **3.** 17.25¢ **5.** $1.72 **7.** 422¢ **9.** $279.20 **11.** $71.20 **13.** $37.52 **15.** $77.24 **17.** 7.08¢ **19.** $3.41 **21.** $31.69 **Applications** **23.** $279.98 **Terms** **25.** e **27.** a **29.** b

Cumulative Review: Units 1–5, pages 189-190 **Skills** **1.** Forty-seven and $\frac{59}{100}$ **3.** Two thousand thirty-five and $\frac{26}{100}$ **5.** $298 **7.** $33.18 **9.** $138.00 **11.** 15.4¢ **13.** 33.0¢ **15.** $22 **17.** $1809 **19.** $10,598 **21.** $1206 **23.** $664.83 **25.** $559.95 **27.** $112.91 **29.** $4135.35 **31.** $342 **33.** $0.20493 **35.** $0.222 **37.** $0.38 **39.** $134.44 **41.** $199.04 **43.** $325.15 **45.** $959.38 **47.** $422.29 **49.** 6 **51.** 0.5 **53.** 25.25 **55.** 0.0725 **57.** 0.125 **Applications** **9.** $9.99 **61.** $25 **63.** 1.12005 **65.** 1.21946 **67.** 2 hrs. 40 min. **69.** 9 hrs. 28 min. **71.** 19 **73.** 99 **75.** $\frac{1}{4}$ **77.** $\frac{5}{6}$ **79.** 156 **81.** 18 **83.–100.** Answers will vary.

Unit 6 Charge Accounts and Credit Cards

6-1; pages 194–196 **Self-Check** **1.** $6.96; $122.96 **2.** $5.00; $104.95 **Problems** **3.** $10.08; $154.08 **5.** $8.42; $137.97 **7.** $84.09 **9.** $28.26 **11.** $6.27 **13.** $62.74 **15.** $113.29 **17.** $44.67 **19.** $402.79 **21.** $491.77 **23.** $82.05 **Maintaining Your Skills** **25.** 0.08 **27.** 0.045 **29.** 0.092 **31.** $296.27 **33.** $87.27 **35.** 2.94 **37.** 7.052 **39.** $35.70

6-2; pages 197–199 **Self-Check** **1.** $597.50 **2.** $270.78 **Problems** **3.** $649.00 **5.** 337.66 **7.** 416.34 **9.** $375.66 **11.** $323.72 **13.** $796.35 **15.** $109.90; $188.73; $369.04 **Maintaining Your Skills** **17.** $405.04 **19.** $306.69 **21.** $309.30 **23.** $287.84 **25.** $3297.34

6-3; pages 200–202 **Self-Check** **1.** $2.70 **2.** $1.318 = $1.32 **Problems** **3.** $0.90 **5.** $2.37 **7.** $5.03 **9.** $2.55; $191.55 **11.** $7.46; $460.41 **13.** $19.23; $1076.52 **15.** $1.80; $123.42 **17.** $99.79; $1.77; $253.35 **Maintaining Your Skills** **19.** 43.16 **21.** 9.54 **23.** 3.00 **25.** $3.75 **27.** $0.44 **29.** 35.04 **31.** 0.528

6-4; pages 203–205 **Self-Check** **1.** $300; $4.50; $374.50 **2.** $70; $1.05; $166.05 **Problems** **3.** $400; $6; $486 **5.** $275; $4.13; $369.13 **7.** $416; $6.24; $644.74 **9.** $372.87; $5.59; $526.40 **11.** $1.32; $129.32 **13.** $1.95; $195.74 **15.** $374.29; $7.49; $461.09 **17.** $71.45; $307.68; $6.15; $130.48; $444.31 **Maintaining Your Skills** **19.** $497.58 **21.** $410.93 **23.** $369.10 **25.** $77.47 **27.** 8.8 **29.** 19.5 **31.** 3.5 **33.** 14.89

6-5; pages 206–210 **Self-Check** **1.** $5000 **2.** $400 **3.** $400 **4.** 19 **5.** $7600 **6.** 30 **7.** $13,000 **8.** $13,000 **9.** 30 **10.** $433.33 **11.** $2.48 **12.** $192.48 **Problems** **13.** $350.00; 30; $1500.00; $50 **15.** $243.55; $3.65 **17.** $4.85; $253.36 **19.** $82.73; $1.65; $152.95 **21.** $122.21; $2.44; $146.34 **Maintaining Your Skills** **23.** 2400 **25.** 825 **27.** 1578.5 **29.** 2163 **31.** 125 **33.** 36.45 **35.** 72 **37.** 283.4

6-6; pages 211–213 **Self-Check** **1.** $3500 **2.** $600 **3.** $600 **4.** $3000 **5.** $450 **6.** $450 **7.** $450 **8.** $7200 **9.** 30 **10.** $14,750; average daily balance = $491.67 **Problems** **11.** 12; $7200; 1; $740; 6; $4440; 1; $620; 10; $6200; 30; $19,200 $640.00 **13.** $147.76; $192.78; $3.86 **Maintaining Your Skills** **15.** 1800 **17.** 3081.68 **19.** 1097.58 **21.** 1783.04 **23.** 42.90 **25.** 256.89 **27.** 421.17 **29.** 574.84

Reviewing the Basics, page 214 **Skills** **1.** $173.14 **3.** $38.00 **5.** $537.38 **7.** $119.59 **9.** $32.11 **11.** $8320 **13.** $199.35 **15.** $1272.05 **17.** $68.71 **19.** $18.80 **21.** $735.89 **23.** 5.4321 **Applications** **25.** 26 days **27.** 19 days **29.** 61.83 **31.** $70.60 **Terms** **33.** c **35.** f **37.** a

Unit 7 Loans

7-1; pages 218–219 **Self-Check** **1.** $15 interest; $615 maturity value **2.** $19.73 interest; $819.73 maturity value **Problems** **3.** $913.50 **5.** $18.70; $868.70 **7.** $1349.21; $10,784.21 **9.** $403.20; $8803.20 **11.** $21,393.75 **13.** $48,670.63 **Maintaining Your Skills** **15.** 0.40 **17.** 0.07 **19.** 0.1564 **21.** $\frac{1}{2}$ **23.** $\frac{1}{3}$ **25.** 0.667 **27.** 0.35 **29.** 0.60

7-2; pages 220–221 **Self-Check** **1.** $272; $1088 **2.** $217.50; $507.50 **Problems** **3.** $512 **5.** $443.50; $1330.50 **7.** $1422; $8058 **9.** $1012 **11.** $7115.68 **13.** $3216 **15.** $790.36 **17.** 25% **Maintaining Your Skills** **19.** 0.32 **21.** 0.25 **23.** 0.20 **25.** 65 **27.** 142.50 **29.** $178.38 **31.** $17.63

7-3; pages 222–224 **Self-Check** **1.** $82.335 = 82.34; $1976.16; $326.16 **Problems** **3.** $111.07; $1999.26; $219.26 **5.** $155.25; $931.50; $31.50

7. $183.40; $200.80 **9.** $468.48 **11.** $134.76 **13.** 18%; $28.98 **15.** $687.68 **17.** 20%; $142.56 **19.** $614 **Maintaining Your Skills 21.** 20 **23.** 13.50 **25.** $53.13 **27.** 4158 **29.** 1209.19

7-4; pages 225–227 **Self-Check 1.** $18.87 **2.** $18.87; $294.33 **3.** $294.33; $1214.97 **Problems 5.** $86.40; $2313.60 **7.** $28.00; $57.51; $1622.49 **9.** $21.13; $44.88; $930.12 **11.** $30 **13.** $30; $49.35; $1450.65 **15.** $100; $205.40; $5794.60 **17.** $58.67; $154.77; $3045.23 **19.** $238.05; $112; $126.05; $8273.95 **21.** $200.75; $1035.78 **23.** $8.33; $204.79; $628.23 **25.** $4.21; $208.91; $212.48 **Maintaining Your Skills 27.** $600 **29.** $672 **31.** $610.71 **33.** $20.45 **35.** $864.29

7-5; pages 228–229 **Self-Check 1.** $800 + $8 = $808 **2.** $1280 + $16 = $1296 **Problems 3.** $4848 **5.** $12.17; $1472.97 **7.** $59.87; $3325.74 **9.** $103.23; $8361.23 **11.** $28.20; $1908.10 **13.** $2034.82 **15.** $724.47 **Maintaining Your Skills 17.** 600.38 **19.** 118.0248 **21.** 550.6677 **23.** 320.571

7-6; pages 230–232 **Self-Check 1.** $3.08; 10.50% **2.** $11.32; 10.50% **Problems 3.** 11.25% **5.** $14.05; 17.00% **7.** $14.23; 10.50% **9.** 13.00% **11.** 18.00% **13.** 10.00% **15.** 17.25% **17.** 14.50% **19.** Less than 10% **21.** 14.75%; 18.25% **23.** $128; 12.25%; $2115 **Maintaining Your Skills 25.** $19.44 **27.** $40.31 **29.** 30 **31.** 56.21 **33.** 70.16 **35.** 1.25 **37.** 421 **39.** 1.92

7-7; pages 233–234 **Self-Check 1.** 23.33% **2.** $58.97824 = $58.98 **Problems 3.** $18 **5.** 22.22%; $77.77 **7.** 14.29%; $40.24 **9.** $84.19 **11.** $84 **13.** $23.06 **15.** $781.66; $95.99 **Maintaining Your Skills 17.** 421.07 **19.** 340.00 **21.** 40.8 **23.** 6.895

Reviewing the Basics, page 235 **Skills 1.** $732.51 **3.** $84.60 **5.** $117.76 **7.** $3239.99 **9.** $162.24 **11.** 0.1263 **13.** 0.3862 **15.** $198 **17.** 0.75 **19.** 0.583 **21.** 0.2 **23.** 0.12 **25.** 0.153 **27.** 15% **29.** 13.7% **31.** 22.35% **33.** 45.4% **35.** 2051.2% **Applications 37.** $63.21 **39.** $\frac{5}{6}$ **41.** $\frac{7}{12}$ **43.** $\frac{185}{365}$ **45.** $\frac{90}{360}$ **47.–51.** Answers will vary.

Unit 8 Automobile

8-1; pages 240–241 **Self-Check 1.** $10,551 **2.** $12,497 **Problems 3.** $11,000 **5.** $13,145 **7.** $24,128 **9.** $18,335 **11.** $891.20 **Maintaining Your Skills 13.** 9064 **15.** 13,306 **17.** 9609.70

8-2; pages 242–243 **Self-Check 1.** $7200 + $2250 + $340 = $9790 **Problems 3.** $17,512 **5.** $10,018 **7.** $12,460.75 **9.** Sticker price: $10,760; est. dealer's cost: $9088 **11.** $13,643.80 **Maintaining Your Skills 13.** 56 **15.** 166.32 **17.** 4410 **19.** 6846.4 **21.** 6728.25 **23.** 5377.22

8-3; pages 244–245 **Self-Check 1.** $4445 **Problems 3.** $5350 **5.** $5000 **7.** $5100 **9.** $11,825 **Maintaining Your Skills 11.** 6045 **13.** 5125 **15.** 7975 **17.** 1980

8-4; pages 246–249 **Self-Check 1.** $233.20 + $62.00 + $204.00 = $499.20 base premium; $748.80 annual premium **Problems 3.** $576.00; $921.60 **5.** $708.40; $2762.76 **7.** $712.00; $961.20 **9.** $657.60; $2038.56 **11.** $517.20; $956.82 **13.** $524.40; $1494.54 **15.** $239.76 **17.** $2054.92; $160.72 **19.** $287.60 **Maintaining Your Skills 21.** 820.16 **23.** 1305.57 **25.** 185.96 **27.** 864 **29.** 2519.544 **31.** 7423.5 **33.** 1187.808

8-5; pages 250–252 **Self-Check 1.** $2600; $0.26/mile **2.** $3628.50; $0.30/mile **Problems 3.** $2250; $0.25 **5.** $5900; $0.40 **7.** $6634.40; $0.32 **9.** $6400; $0.26 **11.** $0.17 **13.** $0.31 **15.** $0.25 **17.** $0.46 **19.** $0.32; $1987.38 **21.** $3059.09; $0.36 **Maintaining Your Skills 23.** 21.75 **25.** 4.40 **27.** 0.04 **29.** 1.07 **31.** 3.21 **33.** 0.09 **35.** 0.28 **37.** 0.48

8-6; pages 253–254 **Self-Check 1.** $8101 **2.** $16,215 **Problems 3.** $5256; $5719 **5.** $5712; $7047 **7.** $22,440; $24,061 **9.** $7452 **11.** $15,560 **13.** $9852.00 + $851.50 = $10,703.50 **Maintaining Your Skills 15.** 169.5 **17.** 382.55

8-7; pages 255–257 **Self-Check 1.** $120.00 + 94.60 + 18.90 = $233.50 **2.** $233.50 ÷ 430 = $.543 **Problems 3.** $0.39 **5.** $224.88; $0.36 **7.** $76.89; $0.32 **9.** $183.02; $0.38 **11.** $274.34; $0.42 **13.** $621.95; $0.57 **15.** $147.87; $0.55 **17.** $135.49; $0.97 **19.** $0.67 **21.** $491; $0.33 **23.** $320.35; $0.76 **Maintaining Your Skills 25.** 508.35 **27.** 113.838 **29.** 23.82 **31.** 1.39 **33.** 1.47

Reviewing the Basics; page 258 **Skills 1.** $312.75 **3.** $223.11 **5.** $103.45 **7.** $69.54 **9.** $196.30 **11.** $131.04 **13.** $969.86 **15.** $117.90 **17.** $0.13 **19.** $0.11 **21.** $4195.80 **Applications 23.** $4200 **Terms 25.** e **27.** a **29.** b

Unit 9 Housing Costs

9-1; pages 266–267 **Self-Check 1.** $20,000; $60,000 **2.** $60,000; $140,000 **Problems 3.** $69,600 **5.** $24,700; $74,100 **7.** $9700; $38,800 **9.** $21,920 **11.** $195,415 **13.** $84,375 **15.** $70,000 **Maintaining Your Skills 17.** 21,000 **19.** 70,000 **21.** 21,275 **23.** 76,000 **25.** 86,250 **27.** 100,400

9-2; pages 268–269 **Self-Check 1.** $869.40; $208,656; $118,656 **Problems 3.** $737.80; $221,340; $151,340 **5.** $2637; $632,880; $407,880 **7.** $784.80; $235,440; $155,440 **9.** $209,640 **11.** $720.60; $129,708 **13.** $34,506 **Maintaining Your Skills 15.** 15,523.2 **17.** 5133 **19.** 80,041

9-3; pages 270–271 **Self-Check 1.** $45 + $1200 + $120 + $250 + $180 + $660 = $2455 **Problems 3.** $3645 **5.** $14,695 **7.** $2811.25 **9.** $26,407 **Maintaining Your Skills 11.** 3440 **13.** 1872 **15.** 360 **17.** 127.65 **19.** 815 **21.** 1728.13

9-4; pages 272–274 **Self-Check 1.** $799.77 **2.** $23.43 **3.** $79,953.37 **Problems 5.** $700; $37.80; $69,962.20

7. $1400; $92.80; $119,907.20 **9.** $460 **11.** $402.08; $36.80; $38,563.20 **13.** $600; $32.40; $59,967.60 **15.** $345; $39.12; $35,960.88 **17.** $833.94; $785.83; $48.11; $81,951.89 **19.** $5126.38; $195.45 **21. a.** $303.10; $310.00; $31,317.99. $300.13; $312.97; $31,005.02 **b.** $8.72; $604.38; $305.33. $308.26; $2.93; $305.33; 0 **Maintaining Your Skills** **23.** 91,760.45 **25.** 78,879.02 **27.** 14,807.3

9-5; pages 275–276 **Self-Check 1.** 0.0655 **2.** $28,000 **3.** $1834 **Problems 5.** $23,920 **7.** $1302.30 **9.** $5967.85 **11.** $14,742.51 **13.** $124,500 **Maintaining Your Skills 15.** 124.1556 **17.** 3.74928 **19.** 2289.544 **21.** 63,000 **23.** 302,496 **25.** 34,650

9-6; pages 277–278 **Self-Check 1.** $54,000 **2.** $21,600 **3.** $10,800 **Problems 5.** $95,000 **7.** $115,200; $23,040 **9.** $259,200; $129,600; $51,840; $25,920 **11.** $376,000; $37,600; $188,000; $75,200; $188,000; $3760 **Maintaining Your Skills** **13.** $24,000 **15.** 63,000 **17.** 489,600

9-7; pages 279–280 **Self-Check 1.** $481 **Problems** **3.** $303 **5.** $293 **7.** $882 **9.** $1836.08 **Maintaining Your Skills 11.** 112,000 **13.** 35,000 **15.** 10,710 **17.** 69,100

9-8; pages 281–284 **Self-Check 1.** No; $3100 × 35% = $1085 **2.** Yes; $3600 × 35% = $1260 **Problems 3.** $385 **5.** $308 **7.** $1904 **9.** $688.92; Yes; $693 **11.** $1584.80; No; $1505 **13.** $742.36; No; $696.50 **15.** No; $1540; Total = $1605.73 **17.** Mort. = $1234.80; Ins. = $40.08; Taxes = $357.85; Total = $1840.23; FHA; Yes; $1890 **Maintaining Your Skills 19.** 74,846.50 **21.** 826.70 **23.** 378.47 **25.** 515.13 **27.** 947.14 **29.** 1840 **31.** 1781.40 **33.** 8946 **35.** 98.1

Reviewing the Basics, page 285 **Skills 1.** $1615 **3.** $36,192 **5.** $13,600 **7.** $4993.81 **9.** $46.93 **11.** $373.72 **13.** 40,809.6 **15.** $1545.95 **17.** $2843.75 **Applications 19.** $909.60 **Terms 21.** b **23.** f **25.** e

Unit 10 Insurance and Investments

10-1; pages 290–291 **Self-Check 1.** $676 **2.** $56.33 **Problems 3.** $1064 **5.** $4200; $1050 **7.** $4500; $2700 **9.** $1570; $549.50; $22.90 **11.** $112.50 each **Maintaining Your Skills 13.** 5% **15.** 20% **17.** 201.85 **19.** 573.3

10-2; pages 292–293 **Self-Check 1.** $1250 **2.** $450 **Problems 3.** $580 **5.** $12,000; $2400; $2650 **7.** $3200; $890 **9.** $426.40 **Maintaining Your Skills** **11.** $498 **13.** 0.9 **15.** 10,679 **17.** 128 **19.** 703,560

10-3; pages 294–295 **Self-Check 1.** $55.50 **2.** $531.60 **Problems 3.** $19.10 **5.** 25; $2.99; $74.75 **7.** $240 **9.** $30.75 **11.** $325.55; $1360; $1627.75 **Maintaining Your Skills 13.** 544.714 **15.** 21.624 **17.** 18.4 **19.** 8.975 **21.** 3436.2 **23.** 0.12 **25.** 52 **27.** 73,200 **29.** 0.57

10-4; pages 296–297 **Self-Check 1.** $816.90 **2.** $2237.40 **Problems 3.** $728.50 **5.** $29.00 **7.** $1014.60; $119.40 **9.** $64.16; $5.04 **Maintaining Your Skills 11.** 3.05 **13.** 24.44 **15.** 275.85 **17.** 1461.54 **19.** 63.26 **21.** 371 **23.** 1749 **25.** 28,552.5

10-5; pages 298–299 **Self-Check 1.** $680.70 **2.** $18,169 **Problems 3.** $4823.37; $323.37 **5.** 1.375666; $12,380.99; $3380.99 **7.** $3271.24; $271.24 **9.** $548.45; $48.45 **11.** Granite Trust; $85.83 **Maintaining Your Skills 13.** 140 **15.** 340 **17.** 800 **19.** 0.067 **21.** 4.641 **23.** 2191.06 **25.** 7610

10-6; pages 300–301 **Self-Check 1.** 9.11% **2.** 9.31% **Problems 3.** $6446.97; $446.97; 7.45% **5.** 1.096524; $5482.62; $482.62; 9.65% **7.** 9.58% **9.** $859.88; 8.60% **11.** 9.00%; $372.33; 9.31% **13.** 8.75%; $827.45; 9.04% **Maintaining Your Skills 15.** 8.91% **17.** 2550.00% **19.** 30,000 **21.** 33,960

10-7; pages 302–303 **Self-Check 1.** $8775 **2.** $3286.50 **Problems 3.** $536.50 **5.** $36,625; $36,943 **7.** $12,150; $12,272 **9.** $12,225; $12,408.38 **11.** Ballon; $63.75 **13.** $24,118.17 **Maintaining Your Skills 15.** 3.125 **17.** 14.375 **19.** $18,200 **21.** $22,950

10-8; pages 304–305 **Self-Check 1.** $45; 5.32% **Problems 3.** 5.23% **5.** 7.63% **7.** 4.55% **9.** $807.50; 5.39% **11.** $740; 6.14% **13.** $0.32; $96; 0.85% **Maintaining Your Skills 15.** 0.12 **17.** 65.22 **19.** 25.83% **21.** 3.126% **23.** 13.145%

10-9; pages 306–307 **Self-Check 1.** $535.50 profit **2.** $2098 loss **Problems 3.** $4067.25 **5.** $5165.40; $759.60 loss **7.** $201.60 loss **9.** $11,500; $2,312.50 profit **11.** $923.44; $1857.56 loss **13.** $1774.10 profit **Maintaining Your Skills 15.** 381.6 **17.** $219.84 **19.** 14.92 **21.** 16.492 **23.** 108.4176

10-10, pages 308–310 **Self-Check 1.** $60; $805; 7.45% **2.** $725; $9200; 7.88% **Problems 3.** 8.42% **5.** $9237.50; $850; 9.20% **7.** $40; $875; 4.57% **9.** $375; $9437.50; 3.97% **11.** $5046.04; $652.50; 2009 **Maintaining Your Skills 13.** $13,987.50 **15.** 1080 **17.** 1034 **19.** 626.56 **21.** 31.34 **23.** 3.42

Reviewing the Basics, page 311 **Skills 1.** $74.14 **3.** $318.33 **5.** $42,337.50 **7.** $171,802.30 **9.** $2972.50 **11.** $227.04 **13.** $81.97 **15.** $10.80 **17.** 0.3125 **19.** 1.625 **21.** 1.25 **23.** 9% **25.** 18.8% **27.** 179% **Applications 29.** $576.40 **31–37.** Answers will vary.

Unit 11 Recordkeeping

11-1, pages 316–317 **Self-Check 1.** $774 **2.** $1533.55 **Problems 3.** $3440.00; $688 **5.** $8802.34; $1760.47 **7.** $545.45; $109.09 **9.** $131.23 **11.** $675.00 **13.** No; do not know their monthly net income. **Maintaining Your Skills 15.** $1537.42 **17.** $1213.50

11-2, pages 318–320 **Self-Check 1.** $1820 **2.** $1446.57 **Problems 3.** $751.50 **5.** $1742 **7.** $1796.67 **9.** Rent, life insurance, car insurance, car registration **11.** $1008.75 **13.** $226.79 **15.** No **Maintaining Your Skills 17.** $241.00 **19.** $25.74 **21.** $2014.61 **23.** $60.25 **25.** $6.44

11-3, pages 321–323 **Self-Check 1.** $8.90 less **2.** $24.42 more **Problems 3.** $15.05 less **5.** $4.79 more **7.** Telephone bill; Water bill **9.** $100.00; More; $16.70 **11.** Less; $12.86 **13.** No **15.** $1740.09; less; $56.57 **17.** Electric, heating fuel, water bill **19.** $162; less; $1.55 **21.** More; $31 **23.** No **Maintaining Your Skills 25.** $174.85 **27.** $2231.61 **29.** $6.84 **31.** $15.07 **33.** $352.19

Reviewing the Basics, page 324 **Skills 1.** $56.91 **3.** $111.11 **5.** $580.38 **7.** $75.05 **9.** $336.97 **11.** $104.96 **13.** 24.7% **15.** 9.7% **Applications 17.** $225.44 **19.** $399.34

Cumulative Review: Units 6–11, pages 329–330 **Skills 1.** 4.7 **3.** 96.201 **5.** $219.68 **7.** $75.00 **9.** $110 **11.** $1292 **13.** $35 **15.** $69 **17.** $1851 **19.** $104.21 **21.** $202.31 **23.** $1008.29 **25.** $56.12 **27.** $43.33 **29.** $557.30 **31.** $33,658.20 **33.** $17,080.80 **35.** $59.50 **37.** $4.34 **39.** $8.04 **41.** $13,502.67 **43.** 0.875 **45.** 0.65625 **47.** 74.8% **49.** 71.4% **51.** 0.45 **53.** 2.12 **55.** $381.12 **57.** $18.88 **59.** $5.04 **61.** $0.79 **Applications 63.** $60.05 **65.** 24 months **67.** 26 **69.** 48 **71.** 117 **73.** $\frac{1}{3}$ **75.** $\frac{1}{4}$ **77.** $\frac{182}{365}$ **79.** $8.54 **81.** $5.82 **83–105.** Answers will vary.

Unit 12 Personnel

12-1, pages 342–343 **Self-Check 1.** $6120 **2.** $23,795 **Problems 3.** $979.10 **5.** $2417.90 **7.** $12,880.70 **9.** $6128.65 **11.** $30,892.55 **Maintaining Your Skills 13.** $2325 **15.** $17,698 **17.** $5380.40 **19.** $13,791.58

12-2, pages 344–345 **Self-Check 1.** $21,035.70 **Problems 3.** $17,020 **5.** $30,299 **7.** $22,984 **9.** $40,457.60 **11.** $36,240.60 **13.** $14,872.00; $15,600.73; $7.50 **Maintaining Your Skills 15.** $22,975 **17.** $798.30 **19.** $468.10 **21.** $2220.95

12-3, pages 346–347 **Self-Check 1.** 30% **Problems 3.** 35% **5.** 32% **7.** $3807.11; 32.1% **9. a.** $1180, $1104.48, $1180, $1902.16, $444.86; $5811.50; $452, $423.07, $452, $728.62, $170.40; $2226.09; $572, $535.39, $572, $922.06, $215.64; $2817.09; $288, $269.57, $288, $464.26, $108.58; $1418.41 **b.** All 19% **Maintaining Your Skills 11.** $1854.97 **13.** $951.05 **15.** 40%

12-4, pages 348–351 **Self-Check 1.** $21,510; $1792.50; **2.** $971 **Problems 3.** $20,000.00, $1666.67; **5.** $20,563.20; $1713.60 **7.** $21,932.40; $1827.70 **9.** $1076 **11.** $1556 **13.** $1483 **15.** $47,985.30; $3998.78 **17.** $1483 **19. a.** $26,380.73; **b.** $2198.39 **21.** $1556 **23.** $761 **Maintaining Your Skills 25.** 35 **27.** 39 **29.** 26 **31.** 31 **33.** $959.30 **35.** $525.17 **37.** $18,376.47

12-5, pages 352–353 **Self-Check 1.** $931.80 **Problems 3.** $16.80; $63.70 **5.** $37.80; $318.20 **7.** $130.83; $939.62 **9.** $524.70 **11.** $664.52 **Maintaining Your Skills 13.** $207 **15.** $30.45 **17.** $99 **19.** $149.80 **21.** $428.55 **23.** $347.60 **25.** $380 **27.** $600.10

12-6, pages 354–355 **Self-Check 1.** $1637 **Problems 3.** $915 **5.** $2838 **7.** $1455 **9.** $4640 **11.** $1339.60 **Maintaining Your Skills 13.** $580 **15.** $592 **17.** $434 **19.** $670 **21.** $1551 **23.** $1606.11

Reviewing the Basics, page 356 **Skills 1.** $15,304.64 **3.** $1287.50 **5.** $820.19 **7.** $2428.49 **9.** 32.3% **Applications 11.** $30,300 **Terms 13.** c **15.** g **17.** a

Unit 13 Production

13-1, pages 364–365 **Self-Check 1.** $0.02 **2.** $0.22 **Problems 3.** $0.092 **5.** $0.014; $0.036 **7.** $0.003 **9.** $0.048 **11.** $0.02 **Maintaining Your Skills 13.** 21.5¢ **15.** 0.7¢ **17.** 7.2¢ **19.** $0.05 **21.** $0.077 **23.** $0.147 **25.** $0.038 **27.** $0.041 **29.** $0.023

13-2, pages 366–367 **Self-Check 1.** 175,000 **2.** 166,000 **Problems 3.** 95,000 **5.** 450,000 **7.** 19,281 **9.** 28,986 **11.** 323,688 **13.** 480,000 **Maintaining Your Skills 15.** $5.36 **17.** $6.75 **19.** $37.59 **21.** 190,000 **23.** 274,039

13-3, pages 368–370 **Self-Check 1.** 2.5% **2.** 6% **Problems 3.** 4%; In **5.** 8%; Out **7.** 6%; Out **9.** In **11.** 8%; Out **13.** 2%; 0%; 6%; 4%; 4%; Out of control at 1 p.m. **15.** 0%, 4%, 4%; 4%, 0%, 0%; 4%, 8%, 8%; 0%, 0%, 4%; 4%, 4%, 12% **17.** 5%; 4%; 3.75%; 4%; 5%; 4% **Maintaining Your Skills 19.** 6% **21.** 3.3% **23.** 2.7% **25.** 1% **27.** 6% **29.** 6.7% **31.** 5% **33.** Same

13-4, pages 371–372 **Self-Check 1.** 8 **2.** 30 **Problems 3.** 66 **5.** 50 **7.** 4 **9. a.** Avg. 1.7; 14.9; 7.1; 1.5; 1.2 **b.** 26.4 sec. **c.** 136 **Maintaining Your Skills 11.** 18.4 **13.** 35.7 **15.** 800 **17.** 217

13-5, pages 373–374 **Self-Check 1.** 31.25% **2.** 18.75% **Problems 3.** 37.50% **5.** 12.5% **7.** 31.25% **9.** 37.33% **11.** 30% **13.** 14.44%; 34.44%; 24.44%; 5.56%; 3.33%; 6.67%; 11.11% **Maintaining Your Skills 15.** 10 **17.** 9 **19.** 23.33% **21.** 21.25%

13-6, pages 375–377 **Self-Check 1.** $20\frac{1}{2}$ in **2.** 73 cm **Problems 3.** 0.8 cm **5.** 12.8 cm **7.** 0.4 cm **9.** 25.2 cm **11.** 0.6 cm **13.** 49.4 cm **15.** $19\frac{5}{8}$ in H × $20\frac{1}{8}$ in W × $24\frac{5}{8}$ in L **17. a.** $7\frac{1}{4}$ in H × $9\frac{1}{2}$ in W × $7\frac{1}{4}$ in L **b.** $7\frac{1}{4}$ in H × 5 in W × 14 in L **19. a.** 14.6 cm H × 28.6 cm W × 9.6 cm L **b.** Answers will vary. **Maintaining Your Skills 21.** $\frac{5}{16}$ **23.** $\frac{9}{16}$ **25.** $1\frac{5}{16}$ **27.** $\frac{3}{8}$ **29.** $1\frac{5}{16}$ **31.** $\frac{1}{2}$ **33.** $1\frac{1}{2}$ **35.** $3\frac{3}{4}$ **37.** 2 **39.** 2 **41.** $2\frac{1}{4}$ **43.** $8\frac{3}{4}$

Reviewing the Basics, page 378 **Skills 1.** $0.18 **3.** $147.72 **5.** 192.3 **7.** $17.55 **9.** $0.09 **11.** 816,524 **13.** $8\frac{7}{16}$ **15.** $2\frac{1}{4}$ **17.** $34\frac{1}{2}$ **19.** 32.1% **Applications 21.** $4000 **23.** 16.54 **Terms 25.** d **27.** e

Unit 14 Purchasing

14-1, pages 384–385 **Self-Check** **1.** $114 **2.** $266 **Problems** **3.** $105 **5.** $140; $260 **7.** $55.68; $118.32 **9.** $8.53; $86.22 **11.** $127.02 **13.** Dis.: **a.** $3.99 **b.** $5.63 **c.** $9.79; Net: **a.** $15.96 **b.** $18.86 **c.** $25.16 **15.** $97.92 **Maintaining Your Skills** **17.** $4.78 **19.** $83.32 **21.** $139.75 **23.** $43.02 **25.** $139.74

14-2, pages 386–387 **Self-Check** **1.** 70% **2.** $266 **Problems** **3.** $90 **5.** 65%; $17.55 **7.** 76%; $241.28 **9.** 92%; $384.46 **11.** $1590.60 **13.** $44.17; $16.40; $83.78; $96.82 **15. a.** $129.45 **b.** $50.34 **Maintaining Your Skills** **17.** 50% **19.** 78% **21.** 92% **23.** 58% **25.** $26.28 **27.** $74.35 **29.** $2628.26

14-3, pages 388–389 **Self-Check** **1.** $56 **2.** 40% **Problems** **3.** 20.0% **5.** $18; 30.0% **7.** $37.81; 33.6% **9.** $888.89; 17.3% **11.** $20; 20%; 23.6% **13.** 35%; 25%; 0%; 20% **Maintaining Your Skills** **15.** $23.40 **17.** $40.67 **19.** $287.47 **21.** 30% **23.** 25.0% **25.** 10.0%

14-4, pages 390–391 **Self-Check** **1.** $280 **2.** $210 **Problems** **3.** $70; $280 **5.** $248; $992; $148.80; $843.20 **7.** $868; $1302; $325.50; $976.50 **9.** $60.36 **11.** $46.93; $41.44; $35.02 **13.** $349.91; $103.36; $40.24; $5.36 **Maintaining Your Skills** **15.** $114 **17.** $8178.14 **19.** 51 **21.** $56.61 **23.** $425.91 **25.** $13,530.44 **27.** $167.61 **29.** $1025.27 **31.** $5614.77

14-5, pages 392–393 **Self-Check** **1.** 63%; $352.80 **2.** 37%; $207.20 **Problems** **3.** 44%; $272.80 **5.** $209.28; 52%; $226.72 **7.** $74.49; 49.6%; $73.31 **9.** $1644.75; $505.25 **11.** $101.11; $148.16; $20.24; 32.5% **13.** $31.77; $46.68; $181.85; $159.97; $35.35; $23.02; $101.14; $46.72 **Maintaining Your Skills** **15.** $114.56 **17.** $2481.58 **19.** 70% **21.** 75% **23.** 90% **25.** 93% **27.** $35.29

14-6, pages 394–395 **Self-Check** **1.** $19.20 **2.** $620.80 **Problems** **3.** $725.20 **5.** 0; $348.64 **7.** $6447.17 **9.** $898.44 **11.** $227.31 **Maintaining Your Skills** **13.** $889.20 **15.** $107.28 **17.** $95.72 **19.** $4.44 **21.** $7.15

14-7, pages 396–398 **Self-Check** **1.** December 10 **2.** $6958 **Problems** **3.** $602.70 **5.** $287.39; $6897.34 **7.** $907.24 **9.** $723.78 **11.** $7889.74 **Maintaining Your Skills** **13.** $1403.62 **15.** $16.42 **17.** $1011.45 **19.** $8356.07

Reviewing the Basics, page 399 **Skills** **1.** $9187.76 **3.** $71.87 **5.** $12.78 **7.** $278.24 **9.** $42,212.39 **11.** $5.38 **13.** 5.96 **15.** $5622.98 **17.** 20.0% **19.** 33.3% **Applications** **21.** 70% **23.** 31.6% **25.** 49.6% **Terms** **27.** c **29.** a **31.** e

Unit 15 Sales

15-1; pages 404–405 **Self-Check** **1.** $11.86 **2.** $272.73 **Problems** **3.** $3299.40 **5.** $64.89 **7.** $181.95 **9.** $70.86 **11.** $48.11 **13.** $4.71 **15.** $3.85 **17.** $131.61 **19.** $6.60; $0.55 **Maintaining Your Skills** **21.** $10.89 **23.** $86.78 **25.** $1.22 **27.** $3.25 **29.** $4406.32

15-2; pages 406–407 **Self-Check** **1.** $14.94; 30% **2.** $87.22; 35% **Problems** **3.** 50.1% **5.** 36.7% **7.** $80.75; 48.2% **9.** 32.8% **11.** 20.0% **13.** 30.3% **Maintaining Your Skills** **15.** $6.13 **17.** $35.79 **19.** $1309.22 **21.** 24.1% **23.** 100.0% **25.** 50%

15-3; pages 408–409 **Self-Check** **1.** $84; $70; $14 **2.** $385.60; $337.40; $48.20 **Problems** **3.** $8.00 **5.** $27.00 **7.** $33.45 **9.** $6.69 **11.** $29.96 **13.** $246,461.75 **Maintaining Your Skills** **15.** $1.11 **17.** $0.46 **19.** $7.67 **21.** $26.39 **23.** $25.88 **25.** $16.73 **27.** $82.42 **29.** $5887.33

15-4; pages 410–411 **Self-Check** **1.** $74.72; $47.80; $26.92; 22.5% **2.** $121.90; $78.58; $43.32; 19.3% **Problems** **3.** $25; 50% **5.** $29.07; 19% **7.** $3976.62; 47% **9.** 22.0% **11.** 42.2% **13.** 16.2% **15.** 30.2% **17. a.** 23.8% **b.** 20.2% **Maintaining Your Skills** **19.** $1.96 **21.** $112.47 **23.** $3241.93 **25.** 41.4%

15-5; pages 412–413 **Self-Check** **1.** $460 **2.** $2.00 **Problems** **3.** $10 **5.** 50%; $173.48 **7.** 62.5%; $76.54 **9.** $18.70 **11.** $3.95 **13.** $50; $20; $25 **15.** 0 **Maintaining Your Skills** **17.** 80% **19.** 60% **21.** 52.4% **23.** $60.00 **25.** $786.38 **27.** $3337.25 **29.** $189.05

15-6; pages 414–415 **Self-Check** **1.** 66.7% **2.** 100% **Problems** **3.** $0.50; 40% **5.** $38; 80% **7.** $0.91; 700% **9.** 150% **11.** 205.1% **13.** 205.1% **15.** 150%; 66.7%; 100%; 185.7% **Maintaining Your Skills** **17.** $89.69 **19.** $109.80 **21.** $5157.90 **23.** 66.7%

15-7; pages 416–417 **Self-Check** **1.** $35; $85 **2.** $70; $210 **Problems** **3.** $81.00 **5.** $129.60; $216.00 **7.** $657.83; $1409.63 **9.** $18.81 **11.** $33.48 **13.** $295.70 **15.** $0.82 **17.** $2.29; 62.4% **Maintaining Your Skills** **19.** $129 **21.** $1919.49 **23.** $180.00 **25.** $22.32 **27.** $8.08 **29.** $126.53

15-8; pages 418–419 **Self-Check** **1.** $20; 25% **2.** $69.92; 40% **Problems** **3.** $5.00; 20.0% **5.** $97.88; 39.5% **7.** 71.3% **9.** 33.3% **11.** 25.3% **13. a.** 36.0%; 29.2%; 31.9%; 35.7%; 34.5% **b.** 24 × 24; $2.30 **c.** Outside: 36.0% **Maintaining Your Skills** **15.** $60.00 **17.** $33.50 **19.** 20.0% **21.** 39.9% **23.** 45.0% **25.** 55.1% **27.** 5.0% **29.** 33.3%

Reviewing the Basics, page 420 **Skills** **1.** 10.2% **3.** 147.9% **5.** $165.82 **7.** $972.85 **9.** $53.74 **11.** $1195.61 **13.** $763.11 **15.** 0.5 **17.** 0.125 **19.** 0.0525 **21.** $801.00 **23.** 20.0% **Applications** **25.** 60% **27.** 98% **Terms** **29.** f **31.** c **33.** e

Unit 16 Marketing

16-1, pages 426–427 **Self-Check** **1.** 75% **2.** 92% **Problems** **3.** 80%8 **5.** 60% **7.** 89% **9.** 9.5% **11. a.** 405; 602; 451; 282 **b.** 23.3%; 34.6%

25.9%; 16.2% **c.** 13.1%; 37.3%; 31.1%; 18.5% **d.** 25–34 **Maintaining Your Skills** **13.** 56 **15.** 300

16-2, pages 428–429 **Self-Check** **1.** 900,000 **Problems** **3.** 2%; 36,000 **5.** 12%; 12,000,000 **7.** 900,000 **9.** 4,698,000 **11.** 1,800,000 **Maintaining Your Skills** **13.** 0.45 **15.** 11.52

16-3, pages 430–431 **Self-Check** **1.** 10% **2.** 2.2% **Problems** **3.** 40% **5.** 28% **7.** 6% **9.** 10% **11.** 9.1% **13.** 36.5% **15.** 7.8% **17.** 4.8% **Maintaining Your Skills** **19.** 25% **21.** 111.1% **23.** 23 **25.** 0.7 **27.** 13

16-4, pages 432–433 **Self-Check** **1.** $5.5 million; $6.5 million; $7.5 million **Problems** **3.** $38,000 **5.** $44,000 **7.** $50,000 **9.** $14.4; $14.8; $15.2 **Maintaining Your Skills** **11.** $44,286.00 **13.** 17.6 **15.** 103.675 **17.** Fifty-two **19.** Three thousand one hundred four

16-5, pages 434–435 **Self-Check** **1.** $441,000 **2.** $1,435,000 **Problems** **3.** $400.000 **5.** $4,050,000 **7.** $324,000 **9.** $390,000 **11.** $130,224 **13.** 44,650 **Maintaining Your Skills** **15.** 23.78 **17.** $17,220 **19.** $5 **21.** $2400 **23.** $18

16-6, pages 436–437 **Self-Check** **1.** $503.88 **2.** $341.50 **Problems** **3.** $523.50 **5.** $41.46; $1658.40 **7.** $45.54; $227.70 **9.** $277.20 **11.** $663.36 **Maintaining Your Skills** **13.** 8.82 **15.** 0.032 **17.** 0.147 **19.** 16.5

16-7, pages 438–439 **Self-Check** **1.** $9600 **2.** $227,500 **Problems** **3.** $11,000 **5.** $345,000 **7.** $800 **9.** $244,900 **11.** $33,750 **Maintaining Your Skills** **13.** $\frac{1}{8}$ **15.** $3\frac{19}{56}$

16-8, pages 440–441 **Self-Check** **1.** $390,000; $468,000 **2.** $4.50 **Problems** **3.** $120,000 **5.** $17.14; $38.14; $258,020 **7.** $0.53; $1.58; $2,061,500; $0.71; $1.76; $1,918,000; $1.67; $2.72; $1,734,000; $2.00; $3.05; $1,300,000; $3.75 selling price, 950,000 units **Maintaining Your Skills** **9.** 576.02 **11.** 45.2

Reviewing the Basics, page 442 **Skills** **1.** $4.75 **3.** $87.70 **5.** $139,840 **7.** 230.35 **9.** 0.55 **11.** $9241.50 **13.** 6944 **15.** 30.6% **Applications** **17.** 3771; 3076; 2870; 1051; 2025; 3938; 4805; 10,768

Cumulative Review: Units 12–16, pages 445–446 **Skills** **1.** $7.85 **3.** 80¢ **5.** 7.5% **7.** 125.2% **9.** $199.99 **11.** $61.91 **13.** $4194.51 **15.** $33.16 **17.** $27.43 **19.** $1946.76 **21.** $7847.95 **23.** $297.96 **25.** $1.15 **27.** $61.11 **29.** $4.29 **31.** $25.97 **33.** $61.00 **35.** $\frac{5}{8}$ **37.** $3\frac{7}{8}$ **39.** 5 **41.** $\frac{2}{3}$ **43.** $14 **45.** $28 **47.** 0.3 **49.** 0.1 **51.** 0.055 **53.** 0.003 **55.** $125.34 **57.** $247.14 **59.** $15.00 **61.** 25.0% **63.** 39.0% **Applications** **65.** $14.24 **67.** 89.7% **69.** 75% **71.** 95% **73.** 1986 **75.** Approx. 7000 **77.** 13 **79.** 3.08 **81.** $30,960 **83–101.** Answers will vary.

Unit 17 Warehousing and Distribution

17-1, pages 450–451 **Self-Check** **1.** 14,700 ft^3 **2.** 56,000 cm^3 **Problems** **3.** 510 in^3; 255,000 in^3 **5.** 0.288 m^3; 144 m^3 **7.** 5.292 yd^3; 635.04 yd^3 **9.** 1350 ft^3 **11.** 1036.8 ft^3 **13.** 52,500 ft^3 **15.** 2275 ft^3 **17.** 22.5 m^3 **Maintaining Your Skills** **19.** 30 ft^3 **21.** 192 in^3 **23.** 494.7 cm^3 **25.** 18,000 ft^3 **27.** 28,000 in^3

17-2, pages 452–454 **Self-Check** **1.** 350 **2.** 240 **Problems** **3.** 340 **5.** 320; 340 **7.** 95 **9.** 119; 102; 26 **11.** 188; 164; 129; 273; 230 **13.** 40; 184; 126 **15.** 90 **17.** Date 8/1, Bal. 47; Date 8/6, Out 24, Bal. 23; Date 8/16, In 40, Bal. 63; Date 8/27, Out 36, Bal. 27; Date 8/31, In 50, Bal. 77 **Maintaining Your Skills** **19.** 195 **21.** 133 **23.** 24 **25.** 1878 **27.** 4398

17-3, pages 455–457 **Self-Check** **1.** $3.93 **2.** $196.50 **Problems** **3.** **a.** $2.15 **b.** $49.45 **5.** **a.** $280.00; $263.20; 2060; $543.20 **b.** $0.26 **c.** $249.60 **7.** $9.61 **9.** **a.** $465.00; $306.75; $710.50; 310; $1484.25 **b.** $465; $285; $750 **c.** $23.75; $710.50; $734.25 **11.** **a.** FIFO; Low cost items go first. **b.** LIFO; High cost items go first. **13.** No, only if costs consistantly rise or decline. **Maintaining Your Skills** **15.** $1785 **17.** $20,409.12 **19.** $5696.53 **21.** $38,572.38 **23.** $15.67 **25.** $141.73 **27.** $48.19 **29.** $79.16

17-4, pages 458–459 **Self-Check** **1.** $60,000 **2.** $17,000 **Problems** **3.** $240 **5.** $1170 **7.** $384.14 **9.** $4200 **11.** $3880.80 **13.** **a.** $2622.90; $4121.70; $1498.80; $749.40; $374.70; $187.35; $187.35; $9742.20 **b.** $4871.10 **Maintaining Your Skills** **15.** $24,000 **17.** $43,950 **19.** $25,070.50 **21.** $193,792.50 **23.** $1468.25 **25.** $226,596.67

17-5, pages 460–461 **Self-Check** **1.** $14.79 **2.** $516.69 **Problems** **3.** $662.70 **5.** $365.24 **7.** $340.41 **9.** $472.00 **11.** $44.33 **Maintaining Your Skills** **13.** $73.01 **15.** $22.61 **17.** $892.04

17-6, pages 462–463 **Self-Check** **1.** $40.10 **2.** $168.42 **Problems** **3.** $934.69 **5.** $14.37; $212.10 **7.** $36.67; $139.71 **9.** $443.37 **11.** $129.68 **13.** $144.48 **15.** $901.09 **Maintaining Your Skills** **17.** $236.64 **19.** $460.34 **21.** $216.87 **23.** $365.50 **25.** $300.96 **27.** $27.67 **29.** $1224.35 **31.** $432.53 **33.** $147.33

Reviewing the Basics, page 464 **Skills** **1.** 500 **3.** 1100 **5.** 111 **7.** 1301 **9.** 168 **11.** $161.57 **13.** $38,797.06 **15.** $86.98 **17.** $10.45 **19.** $0.84 **Applications** **21.** $2177.51 **23.** $0.73 **25.** 1260 m^3 **Terms** **27.** a **29.** c

Unit 18 Services

18-1, pages 470–471 **Self-Check** **1.** 5000 ft^2; $1666.67 **2.** 5625 ft^2; $1640.63 **Problems** **3.** 450 ft^2; $300 **5.** 1500 ft^2; $1250 **7.** $403 **9.** $1560 **11.** $1619 **13.** $1021; $1082 **Maintaining Your Skills** **15.** $2150 **17.** $756 **19.** $18,071

18-2, pages 472–473 **Self-Check** **1.** $680 **2.** $277.65 **Problems** **3.** $168.00; $593.00 **5.** $220.50;

$1168.06 **7.** $257.76 **9.** $2260.33 **11.** $710.10 **13**. $156.98 **Maintaining Your Skills** **15.** $146.25 **17.** $95.06 **19.** $46.53

18-3, pages 474–475 **Self-Check** **1.** $203.52 **2.** $620.34 **Problems** **3.** $94.78; $394.28 **5.** $102.76; $873.46 **7.** $1231.65 **9.** $469.90 **11.** $2063.56 **Maintaining Your Skills** **13.** $25.80 **15.** $8.05

18-4, pages 476–477 **Self-Check** **1.** $27.76 **2.** $32.08 **Problems** **3.** $28.17 **5.** $2.10; $1.33; $45.78 **7.** $30.40 **9.** $35.94 **11.** $64.01 **Maintaining Your Skills** **13.** $36.81 **15.** $44.44

18-5, pages 478–480 **Self-Check** **1.** $1521 **Problems** **3.** $2401 **5.** $2223.90 **7.** $1598.13 **9.** $2527.53 **Maintaining Your Skills** **11.** $87 **13.** $145.50

18-6, pages 481–482 **Self-Check** **1.** $31,000 **2.** $2512.50 **Problems** **3.** $1500 **5.** $5000 **7.** $150,000 **9.** $9130 **11.** $650,000 **13.** $526,912.50 **Maintaining Your Skills** **15.** $885.55 **17.** $102,250 **19.** $57,082.50

Reviewing the Basics, page 483 **Skills** **1.** $43 **3.** $4172 **5.** $1601 **7.** $27,163 **9.** 3237 **11.** 2631.18 **13.** 23.21 **15.** 4064.88 **17.** 7200 **19.** 7200 **21.** 2762.55 **23.** $348.17 **25.** $887.92 **27.** $8173.75 **Applications** **29.** 1200 ft^2 **31.** 18,000 ft^2 **33.** 509.25 ft^2 **Terms** **35.** d **37.** a **39.** b

Unit 19 Accounting

19-1, pages 488–491 **Self-Check** **1.** $285.96 **Problems** **3.** $114.60 **5.** $231.59; $743.41 **7.** $102.71; $416.29 **9.** $318.92 **11.** $62, $37.20, $8.70, $177.65, $422.35; $65, $36.27, $8.48, $170.74, $414.26; $52, $25.73, $6.02, $119.92, $295.08; $68, $35.02, $8.19, $171.62, $393.18; $69, $33.33, $7.79, $151.81, $385.69; $57, $38.39, $8.98, $174.99, $444.21; $82, $37.97, $8.88, $173.91, $438.53; $3933.94, $455, $243.91, $57.04, $107.84, $68.85, $208.00, $1140.64, $2793.30 **13.** $505.50, $53, $31.34, $7.33, $5.06, $116.95, $388.55; $536.40, $81, $33.26, $7.78, $5.36, $148.86, $387.54; $617.88, $63, $38.31 $8.96, $6.18, $149.04, $468.84; $208.25, $27, $12.91, $3.02, $2.08, $53.34, $154.91; $261.80, $23, $16.23, $3.80, $2.62, $56.12, $205.68; $2129.83, $247, $93.07, $132.05, $30.89, $21.30, $524.31, $1605.52 **Maintaining Your Skills** **15.** $247.63 **17.** $386.36 **19.** $1446.82

19-2, pages 492–493 **Self-Check** **1.** 7.5% **2.** 16.7% **Problems 3.** 10.0% **5.** 7.5% **7.** 66.7% **9.** $131,704; 64.2% **Maintaining Your Skills** **11.** $202,783 **13.** $449.18 **15.** 25.0% **17.** 16.6% **19.** 2.0%

19-3, page 494–495 **Self-Check** **1.** $384 **2.** $1250 **Problems** **3.** $1250 **5.** $19,531.25 **7.** $17.25 **9.** $5760.00; $12,480.00 **11.** **a.** M: $1440.00; U: $7200.00; S: $1152.00 **b.** M: $1038.00; U: $5190.00; S: $830.40 **c.** M: $762.00; U: $3810.00; S: $609.60 **Maintaining Your Skills** **13.** 0.021 **15.** 0.005

19-4, page 496–497 **Self-Check** **1.** $500 **2.** $150 **Problems** **3.** $400.00 **5.** $1275.00 **7.** $8500 **9.** $10,947 **11.** $28,750 **13.** $360 **Maintaining Your Skills** **15.** $286 **17.** $669 **19.** $735.90 **21.** $153.69 **23.** $80.85 **25.** $1.47

19-5, page 498–499 **Self-Check** **1.** $36,000 **2.** $16,000 **Problems** **3.** $2400; $2400; $1100 **5.** $100; $100; $180 **7.** $4665 **9.** $9472.34 **11.** $23,690, $23,690, $91,070; $23,690, $47,380, $67,380; $23,690, $71,070, $43,690; $23,690, $94,760, $20,000 **Maintaining Your Skills** **13.** $1650 **15.** $70.40 **17.** $102,600 **19.** $197,200 **21.** $7186.50

19-6, page 500–501 **Self-Check** **1.** $48,000 **2.** $36,000 **Problems** **3.** $2600 **5.** $\frac{10}{55}$; $20,000; $\frac{9}{55}$; $18,000 **7.** $6000 **9.** $2720; $2040; $1360; $680 **11.** $28,800, $28,800, $51,200; $21,600, $50,400, $29,600; $14,400, $64,800, $15,200; $7200, $72,000, $8000 **Maintaining Your Skills** **13.** $8420 **15.** $5340 **17.** $1137

19-7, page 502–503 **Self-Check** **1.** $40,000; $40,000 **2.** $20,000; $20,000 **Problems** **3.** $3290 **5.** $33\frac{1}{3}$ %; $14,000 **7.** $2400; $4800 **9.** $20,000, $30,000; $32,000, $18,000; $7200, 39,200, $10,800; $4320, $43,520, $6480; $2592, $46,112, $3888 **Maintaining Your Skills** **11.** $540 **13.** $2970 **15.** $103,900 **17.** $426

19-8, page 504–505 **Self-Check** **1.** $3400.00; $5440.00; $3264.00; $1958.40; $1958.40; $979.20 **2.** $6958 **Problems** **3.** $1848.00; $2956.80; $1774.08; $1064.45; $1064.45; $532.22 **5.** $6968.00; $11,148.80; $6689.28; $4013.57; $4013.57; $2006.78 **7.** Yr. 1, Dep. $6797.28; Yr. 2, Dep. $11,657.24; Yr. 3, Dep. $8325.24; Yr. 4, Dep. $5945.24; Yr. 5, Dep. $4250.68; Yr. 6, Dep. $4250.68; Yr. 7, Dep. $4250.68; Yr. 8, Dep. $2122.96; $47,600.00 **9.** $48,260.00, $9,652.00, $9,652.00, $38,608.00; $48,260.00, $15,443.20, $25,095.20, $23,184.80; $48,260.00, $9,265.92, $34,361.12, $13,898.88; $48,260.00, $5559.55, $39,920.67; $8339.33; $48,260.00; $5559.55; $45,480.22, $2779.78; $48,260.00, $2779.78, $48,260.00, ($0.00) **Maintaining Your Skills** **11.** $28,310 **13.** $118

Reviewing the Basics, page 507 **Skills** **1.** $7151 **3.** $144 **5.** $197,478 **7.** $69.01 **9.** $4.28 **11.** $17,269.90 **13.** $74,000 **15.** $6800 **17.** $1198.80 **19.** $55,725 **21.** 0.065 **23.** 0.03 **25.** $1080 **27.** $15,531.25 **29.** $5.39 **31.** 15.9% **33.** 13.8% **Applications** **35.** 2400 ft^2

Unit 20 Accounting Records

20-1, pages 512–513 **Self-Check** **1.** $132,000 **2.** $42,000 **Problems** **3.** $77,000 **5.** $36,250 **7.** $17,450 **9.** $104,015; $23,397.80; $80,617.20 **11.** $221,797; $142,257; $79,540 **Maintaining Your Skills** **13.** $42,758.35 **15.** $20,422.38 **17.** $39,052.00 **19.** $21,624.05

20-2, pages 514–516 **Self-Check** **1.** $67,000 **2.** $67,000 **Problems** **3.** $33,000; $12,300; $20,700 **5.** $52,500; $22,500; $27,500 **7.** $99,800; $65,000; $34,800 **9.** 80,467.50; 51,013.20; 29,454.30; 80,467.50 **11.** Total Assets $199,504,200; Total Liabilities $142,599,410; Owner's Equity, Capital $56,904,790; Total Liabilities and Owner's Equity $199,504,200 **Maintaining Your Skills** **13.** $27,970 **15.** $5,072, 400 **17.** $9.9

20-3, pages 517–518 **Self-Check** **1.** $16,920 **2.** $12,404 **Problems** **3.** $96,600 **5.** $17,950 **7.** $58,127 **9.** $7871.68 **11.** $9921.32 **13.** $99.92 **Maintaining Your Skills** **15.** $16,596.26 **17.** $65,319.64 **19.** $4349.88 **21.** $95.76 **23.** $17.40

20-4, pages 519–521 **Self-Check** **1.** $61,500; $26,800 **2.** $289,000; $183,600 **Problems** **3.** $4685 **5.** $3675; $1935; $1107 **7.** $673,642; $255,726; $159,307 **9.** $388,125; $157,125; $85,209 loss **11.** NS: $3080; GP: $2209; NI: $871 **13.** GP: $58.3; OE: $38.7; NI: $19.6 **Maintaining Your Skills** **15.** $766.22 **17.** $14,709.46 **19.** $90.14 **21.** $5715

20-5, pages 522–524 **Self-Check** **1.** 1.4% **2.** 31.8% **3.** 1.7 to 1 **4.** 1.3 to 1 **Problems** **5.** 50.0% **7.** 44.5% **9.** 2.0:1; 1.2:1 **11.** $3553; 46.2% **13.** 1.6:1; 1.3:1 **Maintaining Your Skills** **15.** 9.1% **17.** 78.4% **19.** 1.0:1

20-6, pages 525–526 **Self-Check** **1.** 25% **2.** −28.6% **Problems** **3.** 15.3% **5.** $23,090; 20.1% **7.** −$775; −16.7% **9.** $8943; 21.1% **11.** **a.** −$7.5; −6.4% **b.** −$2.2; −3.9% **c.** −$5.3; −8.8% **d.** $0.5; 1.4% **e.** −$5.8; −23.4% **Maintaining Your Skills** **13.** 20.0% **15.** 67.2% **17.** 140.4%

Reviewing the Basics, page 527 **Skills** **1.** $1,345,902 **3.** $9262 **5.** $35,885 **7.** $8459.51 **9.** $7723.96 **11.** $72,769.79 **13.** 19.5% **15.** 29.7% **Applications** **17.** $19,325 **19.** $903,500 **Terms** **21.** c **23.** b

Unit 21 Financial Managment

21-1, pages 532–533 **Self-Check** **1.** $16,650 **2.** $11,250 **Problems** **3.** $5511.90 **5.** $64,823; $11,205.75 **7.** $194,884; $58,754.76 **9.** $7737.25 **11.** $13,838.40 **Maintaining Your Skills** **13.** $6.18 **15.** 507.5 **17.** 277.44 **19.** 44.88 **21.** $27,816

21-2, pages 534–535 **Self-Check** **1.** $7,568,000 **2.** $36,965,150 **Problems** **3.** $455,500; $6,344,500 **5.** $1,050,000; $1,104,650; $19,895,350 **7.** $6,978,133.75 **9.** $1,156,250 **11.** $2,525,737.50 **Maintaining Your Skills** **13.** $63,700 **15.** $44,580

21-3, pages 536–537 **Self-Check** **1.** $102,812.50 **2.** $155,541.67 **Problems** **3.** $1750; $71,750 **5.** $1312.50; $38,812.50 **7.** $1835.44; $39,485.44 **9.** $52,612.85 **11.** **a.** $117,725 **b.** $1,567,725 **Maintaining Your Skills** **13.** 3.4 **15.** 0.007

21-4, pages 538–539 **Self-Check** **1.** $95,600 **Problems** **3.** $2900; $97,135 **5.** $2194.38; $82,835.62 **7.** $2365.72; $94,674.28 **9.** $92,816.12 **11.** $204,039.38 **Maintaining Your Skills** **13.** 60.548 **15.** 4.586 **17.** 0.007 **19.** 21.875

21-5, pages 540–541 **Self-Check** **1.** $866.67 **Problems** **3.** $4120; $4095 **5.** $1836.25 **7.** $3003.75 **9.** $8342.55 **Maintaining Your Skills** **11.** 0.23 **13.** 0.34

21-6, pages 542–544 **Self-Check** **1.** $39,850 **2.** $142,250 **Problems** **3.** $12,595 **5.** $12,041 **7.** $8550 **9.** $2,777,290.40 **Maintaining Your Skills** **11.** 600.712 **13.** 5.361

Reviewing the Basics, page 545 **Skills** **1.** $61,628.25 **3.** $980,400 **5.** $145,687.50 **7.** $2723.33 **9.** $4583.33 **11.** 0.095 **13.** 1.75 **15.** $8666 **17.** $1789.84 **19.** $0.22 **Applications** **21.** $19,526.40 **Terms** **23–27.** Answers will vary.

Unit 22 Corporate Planning

22-1, pages 554–556 **Self-Check** **1.** 6.7% **2.** $13,416.80 **3.** $25.50 **Problems** **5.** 1.5% **7.** $18,086.49 **9.** $216,859.43 **11.** $3.93 **13.** 26.4% **15.** $23,990.05 **17.** $0.59 **19.** $16,932.46 **21.** $0.77 **23.** **a.** −75% **b.** It is negative **c.** Deflation **Maintaining Your Skills** **25.** $5.63 **27.** 5.0% **29.** 11.2%

22-2, pages 557–558 **Self-Check** **1.** $40,996,800 **2.** $1179.91 **Problems** **3.** $118.3 mil; $309 **5.** $448.9 bil; $6450.13 **7.** $2,652.3 bil; $1170.83 **9.** $732,672,000 **11.** $203.28 billion; $13,216.39 **13.** 461.7 billion; $18,320.96 **15.** $148.257 billion; $15,457.67 **17.** **a.** France **b.** U.K. **c.** France **d.** U.K.

22-3, pages 559–561 **Self-Check** **1.** 250.0 **2.** $2680.83 **3.** $17.50 **Problems** **5.** 212.8 **7.** $666.76 **9.** $5.71 **11.** $163.80 **13.** $334.7 **15.** $118,462.75 **17.** $50.78 **19.** $188.36 **21.** 169.2 **23.** **a.** Less than 100 **b.** 0 **c.** No **Maintaining Your Skills** **25.** 1.3 **27.** 0.4 **29.** $59,409.95

22-4, pages 562–564 **Self-Check** **1.** Budget; $332,000; $24,000; $20,000; $16,000; $8,000 **2.** Difference; +$28,000; −$4000; +$6000; −$7000; −$5000 **Problems** **3.** $20,000; −$5000 **5.** $60,200; +$3800 **7.** $3,763,200; −201,200 **9.** $105,000; $30,000; $15,000 **11.** −$40,000; −$30,000; +$10,000; −$60,000 **13.** $14,610,240; $304,380; $304,380 **15.** Yes; +$244,000 **17.** **Budget** $27,000; $9000; $4500; $3600; $900; total $45,000; **Difference** −$2500; −$700; −$400; −$1200; +$300; total −$4500 **Maintaining Your Skills** **19.** $3,175,743.50 **21.** $122,996.66 **23.** $9578.94

Reviewing the Basics, page 565 **Skills** **1.** $48.99 **3.** $2841.97 **5.** 25.0% **7.** 17.4% **9.** $6.44 **11.** $44.20 **13.** $534.74 **15.** $23.72 **17.** $37.09 **19.** $27.35 **21.** 2.9% **23.** $52.11 **Applications** **25.** $1207.64 **27.** $2867.99 **Terms** **29–35.** Answers will vary.

Cumulative Review: Units 17–22, pages 569–570 **Skills** **1.** \$179 **3.** \$4218 **5.** 135 **7.** \$1069 **9.** \$229 **11.** \$49 **13.** \$184 **15.** \$7756 **17.** \$19.70 **19.** \$85.12 **21.** \$4039.95 **23.** \$454.15 **25.** \$512.46 **27.** \$344 **29.** \$7500 **31.** \$32,175 **33.** \$8.40 **35.** \$22.57 **37.** \$3168 **39.** \$15 **41.** \$106 **43.** \$150.71 **45.** \$2.55 **47.** \$14.22 **49.** \$566.43 **51.** \$36 **53.** \$36,735.30 **55.** \$1516 **57.** 0.08 **59.** 0.7435 **61.** \$496.48 **63.** \$26.57 **65.** \$14,551.53 **67.** 47.6% **69.** 20.7% **Applications** **71.** over \$100,000 **73.** \$0–\$50,000 **75.** \$3466.20 **77.** \$2.31 **79.** \$0.57 **81.** 180 ft^2 **83.** 43,620 in^2 **85.** 1650 cm^2 **87.** 246.24 m^3 **89.** $38\frac{1}{2}$ ft^3

Skills File

1–pages 580–581 **1.** ones; 5 **3.** ten; 70 **5.** ones; 8 **7.** tens; 0 **9.** hundreds; 700 **11.** ones; 9 **13.** tenths; 0.3 **15.** thousandths; 0.009 **17.** hundredths; 0.02 **19.** ones; 4 **21.** hundredths; 0.02 **23.** thousandths; 0.007 **25.** thousandths; 0.003 **27.** thousandths; 0.001 **29.** Thirty-four **31.** Seventy-eight **33.** Eight hundred forty-two **35.** Five thousand twelve **37.** One hundred nineteen **39.** One thousand two hundred forty **41.** One hundred ninety-four **43.** Twenty-five dollars **45.** Seventeen and $\frac{9}{100}$ dollars **47.** One hundred twenty and $\frac{17}{100}$ dollars **49.** Seven thousand seven hundred forty-nine dollars **51.** Fifty cents or $\frac{50}{100}$ dollars **53.** \$71.64 million **55.** 18.4 thousand **57.** 171.6 thousand **59.** 2.135 billion **61.** \$3,400,000 **63.** \$11,200,000,000 **65.** 7,300,000,000 **67.** 400,000 **69.** \$720,000,000 **71.** 54 **73.** 835 **75.** 211 **77.** 5763 **79.** 4444 **81.** 1236 **83.** 650 **85.** \$101 **87.** \$93 **89.** \$742 **91.** \$791,000 **93.** 2.0 **95.** 25.1 **97.** 58 **99.** 0.1 **101.** 0.5 **103.** 0.020 **105.** \$10 **107.** \$19 **109.** \$600 **111.** \$9521.39 **113.** \$2000.00

2–page 582 **1.** 30 **3.** 70 **5.** 220 **7.** 8100 **9.** 600 **11.** 6000 **13.** 16,000 **15.** 1000 **17.** 0.1 **19.** 0.5 **21.** 0.47 **23.** 0.36 **25.** 0.601 **27.** 0.213 **29.** 137.492 **31.** 38,000 **33.** 48,000 **35.** 900 **37.** 30 **39.** 580 **41.** 36 **43.** 189.7 **45.** 0.1 **47.** 152.43 **49.** 0.179 **51.** 108.437

3–page 583 **1.** 106 **3.** 137 **5.** 1189 **7.** 799 **9.** 897 **11.** 488 **13.** 10,878 **15.** 8987 **17.** 71 **19.** 210 **21.** 897 **23.** 1542 **25.** 4638 **27.** 12,224 **29.** 12,201 **31.** 12,787 **33.** 3798 **35.** 12,318 **37.** 8152 **39.** 9484 **41.** 6777 **43.** 14,953 **45.** 33,587 **47.** 87,481 **49.** 14,999 **51.** 14,450

4–page 584 **1.** 331 **3.** 151 **5.** 3442 **7.** 3311 **9.** 1111 **11.** 8020 **13.** 6 **15.** 6231 **17.** 64 **19.** 99 **21.** 2258 **23.** 1166 **25.** 1909 **27.** 3667 **29.** 1275 **31.** 859 **33.** 2983 **35.** 1568 **37.** 9111 **39.** 1862 **41.** 2238 **43.** 2901 **45.** 16,526 **47.** 37,503 **49.** 126,403 **51.** 100,621

5–page 585 **1.** 478.99 **3.** 801.12 **5.** 791.19 **7.** 543.113 **9.** 816.659 **11.** 64.54 **13.** 177.349 **15.** 21.901 **17.** 24.460 **19.** 82.332 **21.** 10.122 **23.** 460.909 **25.** 502.175 **27.** 1064.683 **29.** 1.241 **31.** 21.901 **33.** 220.611 **35.** 772.652

6–page 586 **1.** 51.15 **3.** 134.89 **5.** 5.09 **7.** 25.51 **9.** 79.68 **11.** 203.84 **13.** 70.85 **15.** 84.32 **17.** 19.6 **19.** 8.07 **21.** 57.44 **23.** 377.63 **25.** 5.09 **27.** 19.983 **29.** 325.86 **31.** 204.354 **33.** 557.87 **35.** 5215.946 **37.** 1.1178 **39.** 778.029 **41.** 3.59806 **43.** 2.74502

7–page 587 **1.** 1533 **3.** 87,344 **5.** 100,152 **7.** 77,256 **9.** 959,373 **11.** 2808 **13.** 251,565 **15.** 4,268,236 **17.** 1665 **19.** 310,992 **21.** 1,662,024 **23.** 69,000 **25.** 28,182 **27.** 357,000 **29.** 36,060,000 **31.** 4,200,672 **33.** 43,955,000 **35.** 1,070,443 **37.** 169,130 **39.** 623,770 **41.** 1,309,622 **43.** 2,804,972 **45.** 15,805,077

8–page 588 **1.** 4.5 **3.** 87.822 **5.** 188.55 **7.** 230.8218 **9.** 43.9556 **11.** 287.352 **13.** 0.1008 **15.** 0.096 **17.** 0.584 **19.** 0.6156 **21.** 0.0567 **23.** 0.0861 **25.** 4116 **27.** 36 **29.** 876.403 **31.** 1212.86 **33.** 112.5 **35.** 81.9698 **37.** 161,713.902 **39.** 0.156462 **41.** 84 **43.** 0.07608533

9–page 589 **1.** $2\frac{9}{27}$ **3.** $2\frac{2}{41}$ **5.** $14\frac{12}{53}$ **7.** $84\frac{17}{21}$ **9.** $67\frac{19}{43}$ **11.** $59\frac{12}{23}$ **13.** $71\frac{42}{81}$ **15.** $16\frac{13}{114}$ **17.** $15\frac{17}{321}$ **19.** $8\frac{792}{892}$ **21.** $23\frac{22}{409}$ **23.** $13\frac{617}{625}$ **25.** $20\frac{831}{843}$ **27.** 18 **29.** 16 **31.** 72 **33.** $14\frac{22}{621}$ **35.** $12\frac{349}{549}$ **37.** $5\frac{765}{842}$ **39.** $181\frac{29}{473}$ **41.** $87\frac{527}{622}$ **43.** $1600\frac{3}{51}$ **45.** $252\frac{46}{88}$ **47.** $208\frac{166}{417}$ **49.** $1345\frac{177}{514}$

10–page 590 **1.** 11.125 **3.** 21.2 **5.** 5.1 **7.** 19.75 **9.** 34.5 **11.** 22.8 **13.** 16.1 **15.** 206.8 **17.** 2.02 **19.** 10.865 **21.** 21.524 **23.** 8.71 **25.** 10.45 **27.** 31.13 **29.** 5.63 **31.** 17.5 **33.** 14.44 **35.** 230.05 **37.** 41.42 **39.** 25.54 **41.** 39.5

11–page 591 **1.** 4.6 **3.** 5.71 **5.** 17 **7.** 6.5 **9.** 42.91 **11.** 26 **13.** 31 **15.** 26 **17.** 8.96 **19.** 13.21 **21.** 1.41 **23.** 245 **25.** 1.21 **27.** 256 **29.** 67.48 **31.** 22.57 **33.** 1.75 **35.** 38.5 **37.** 98.86 **39.** 0.01 **41.** 17.13 **43.** 0.01

12–page 592 **1.** 10 **3.** 50 **5.** 24 **7.** 22 **9.** 51 **11.** 24 **13.** 35 **15.** 72 **17.** 20 **19.** $\frac{2}{3}$ **21.** $\frac{1}{2}$ **23.** $\frac{2}{3}$ **25.** $\frac{1}{3}$ **27.** $\frac{5}{6}$ **29.** $\frac{10}{27}$ **31.** $\frac{4}{7}$ **33.** $\frac{1}{2}$ **35.** $\frac{3}{4}$ **37.** $\frac{94}{183}$ **39.** $\frac{1}{2}$ **41.** $\frac{19}{36}$ **43.** $\frac{2}{13}$ **45.** $\frac{4}{25}$

13–page 593 **1.** $\frac{25}{4}$ **3.** $\frac{59}{8}$ **5.** $\frac{20}{3}$ **7.** $\frac{39}{5}$ **9.** $\frac{43}{10}$ **11.** $\frac{145}{32}$ **13.** $\frac{16}{3}$ **15.** $\frac{29}{10}$ **17.** $\frac{53}{16}$ **19.** $\frac{165}{32}$ **21.** $\frac{79}{11}$ **23.** $\frac{57}{5}$ **25.** $6\frac{1}{2}$ **27.** $5\frac{3}{4}$ **29.** $5\frac{1}{2}$ **31.** $2\frac{1}{3}$ **33.** $2\frac{1}{2}$ **35.** $2\frac{20}{21}$ **37.** $5\frac{5}{6}$ **39.** $3\frac{1}{6}$ **41.** $6\frac{1}{8}$ **43.** $6\frac{3}{7}$

14–page 594 **1.** 0.8 **3.** 1.13 **5.** 2.43 **7.** 1.37 **9.** 0.3 **11.** 4.38 **13.** 3.08 **15.** 0.03 **17.** 5.15 **19.** 2.12 **21.** 7.14 **23.** 0.44 **25.** $\frac{1}{10}$ **27.** $\frac{3}{10}$ **29.** $\frac{3}{100}$ **31.** $\frac{53}{100}$ **33.** $\frac{3}{1000}$ **35.** $\frac{8}{125}$ **37.** $\frac{111}{250}$ **39.** $\frac{469}{1000}$ **41.** $7\frac{1}{2}$ **43.** $1\frac{2}{25}$ **45.** $2\frac{83}{100}$ **47.** $\frac{21}{200}$ **49.** $\frac{1}{2000}$ **51.** $1\frac{402}{625}$ **53.** $\frac{1}{5000}$

15–page 595 **1.** $1\frac{2}{5}$ **3.** $\frac{3}{4}$ **5.** $1\frac{4}{7}$ **7.** $1\frac{4}{25}$ **9.** $1\frac{5}{16}$ **11.** $1\frac{8}{45}$ **13.** $11\frac{2}{3}$ **15.** $14\frac{2}{7}$ **17.** $12\frac{1}{3}$ **19.** $17\frac{1}{4}$ **21.** $9\frac{7}{8}$ **23.** $11\frac{2}{5}$

16–page 596 **1.** $1\frac{1}{10}$ **3.** $\frac{20}{21}$ **5.** $1\frac{1}{6}$ **7.** $1\frac{13}{110}$ **9.** $\frac{59}{60}$ **11.** $\frac{71}{150}$ **13.** $9\frac{1}{6}$ **15.** $16\frac{1}{3}$ **17.** $20\frac{19}{26}$ **19.** $28\frac{22}{63}$ **21.** $28\frac{43}{56}$ **23.** $29\frac{5}{48}$

17–page 597 **1.** $\frac{2}{9}$ **3.** $\frac{5}{6}$ **5.** $\frac{4}{7}$ **7.** $\frac{1}{8}$ **9.** $\frac{1}{5}$ **11.** $\frac{2}{5}$ **13.** $\frac{1}{4}$ **15.** $\frac{7}{20}$ **17.** $1\frac{1}{7}$ **19.** $3\frac{1}{4}$ **21.** $2\frac{1}{6}$ **23.** $5\frac{3}{16}$ **25.** $2\frac{7}{13}$ **27.** $8\frac{1}{2}$ **29.** $25\frac{2}{7}$

18–page 598 **1.** $\frac{3}{8}$ **3.** $\frac{3}{20}$ **5.** $\frac{1}{18}$ **7.** $\frac{1}{15}$ **9.** $\frac{3}{20}$ **11.** $\frac{1}{10}$ **13.** $4\frac{1}{4}$ **15.** $6\frac{5}{18}$ **17.** $5\frac{7}{60}$ **19.** $3\frac{1}{8}$ **21.** $2\frac{3}{20}$ **23.** $2\frac{1}{8}$

19–page 599 **1.** $1\frac{1}{2}$ **3.** $2\frac{1}{10}$ **5.** $2\frac{6}{11}$ **7.** $3\frac{1}{18}$ **9.** $12\frac{11}{12}$ **11.** $1\frac{37}{70}$ **13.** $2\frac{7}{8}$ **15.** $1\frac{3}{8}$ **17.** $\frac{1}{8}$ **19.** $3\frac{7}{9}$ **21.** $\frac{9}{20}$ **23.** $11\frac{5}{16}$ **25.** $14\frac{7}{9}$ **27.** $13\frac{5}{7}$ **29.** $12\frac{5}{32}$

20–page 600 **1.** $\frac{1}{3}$ **3.** $\frac{5}{8}$ **5.** $\frac{5}{12}$ **7.** 2 **9.** 9 **11.** $16\frac{2}{15}$ **13.** $58\frac{1}{3}$ **15.** $46\frac{1}{8}$ **17.** $\frac{1}{6}$ **19.** $\frac{11}{12}$ **21.** 6 **23.** $3\frac{3}{4}$ **25.** $19\frac{1}{2}$ **27.** $8\frac{8}{15}$ **29.** 64 **31.** $25\frac{29}{72}$ **33.** $97\frac{11}{12}$ **35.** $71\frac{1}{24}$

21–page 601 **1.** 2 **3.** $1\frac{7}{8}$ **5.** $7\frac{1}{2}$ **7.** 6 **9.** 16 **11.** $3\frac{7}{36}$ **13.** $2\frac{11}{12}$ **15.** $\frac{1}{4}$ **17.** $5\frac{5}{33}$ **19.** $4\frac{20}{133}$ **21.** $1\frac{1}{2}$ **23.** $1\frac{1}{2}$ **25.** $\frac{2}{3}$ **27.** 15 **29.** 6 **31.** 30 **33.** $\frac{43}{176}$ **35.** $\frac{143}{375}$ **37.** $6\frac{1}{2}$ **39.** 1

22–page 602 **1.** 1500:600 **3.** 600:1200 **5.** 600:1500 **7.** $\frac{1}{5}$ **9.** $\frac{1}{15}$ **11.** $\frac{1}{25}$ **13.** $\frac{2}{5}$

23–page 603 **1.** T **3.** F **5.** F **7.** T **9.** T **11.** F **13.** 18 **15.** $7\frac{1}{3}$ **17.** $\frac{6}{11}$ **19.** $5\frac{5}{9}$ **21.** $12\frac{4}{13}$ **23.** 20 **25.** 5 **27.** 24 **29.** 64 **31.** 15 **33.** 50

24–page 604 **1.** $\frac{99}{12} = \frac{x}{7}$ **3.** $\frac{40}{5} = \frac{x}{17}$ **5.** $3.48 **7.** 117 pp. **9.** 1485 wheels **11.** $28 **13.** $33\frac{1}{3}$ minutes

25–page 605 **1.** 288 **3.** 80 **5.** 64; 640 **7.** A **9.** B **11.** B

26–page 606 **1.** 10% **3.** 25% **5.** 82% **7.** 213% **9.** 575.3% **11.** 39.1% **13.** 1210.4% **15.** 1082% **17.** 10.6% **19.** 4% **21.** 50.3% **23.** 70% **25.** 710% **27.** 1050% **29.** 1260% **31.** 2250% **33.** 3720% **35.** 750% **37.** 1630% **39.** 620% **41.** 380% **43.** 67% **45.** 880% **47.** 32% **49.** 300.3% **51.** 21.87% **53.** 2500% **55.** 14% **57.** 51% **59.** 186.8% **61.** 0.3% **63.** 26% **65.** 0.32% **67.** 29% **69.** 342% **71.** 210% **73.** 167% **75.** 254.9% **77.** 207.7% **79.** 19%

27–page 607 **1.** 25% **3.** 75% **5.** 12.5% **7.** 35% **9.** 14% **11.** 62.5% **13.** 275% **15.** 666.7% **17.** 855.6% **19.** 817.5% **21.** 722% **23.** 537.5% **25.** 58.3% **27.** 1530% **29.** 2080% **31.** 1862.5% **33.** 93.8% **35.** 1135.7% **37.** 77.8% **39.** 573.3% **41.** 1735%

28–page 608 **1.** 0.105 **3.** 0.40 **5.** 1.20 **7.** 0.067 **9.** 0.089 **11.** 1.19 **13.** 0.172 **15.** 1.00 **17.** 0.053 **19.** 0.007 **21.** 0.00625 **23.** 0.056 **25.** 0.068 **27.** 0.304 **29.** 0.5784 **31.** 0.7425 **33.** 0.1880 **35.** 0.1445 **37.** 0.00125 **39.** 0.0014 **41.** 0.0879 **43.** 0.0095 **45.** 0.00025 **47.** 0.22125 **49.** 0.252

29–page 609 **1.** $\frac{9}{20}$ **3.** $1\frac{3}{4}$ **5.** $\frac{117}{1000}$ **7.** $\frac{3}{2000}$ **9.** $\frac{53}{500}$ **11.** $\frac{3763}{10,000}$ **13.** $\frac{1}{2}$ **15.** $\frac{101}{1000}$ **17.** $\frac{809}{1000}$ **19.** $\frac{1}{16}$ **21.** $\frac{7}{16}$ **23.** $\frac{37}{400}$ **25.** $\frac{31}{200}$ **27.** $\frac{51}{200}$ **29.** $\frac{1}{6}$ **31.** $\frac{3}{10}$ **33.** $\frac{5}{16}$ **35.** $\frac{1}{8}$ **37.** $\frac{4}{15}$ **39.** $\frac{1}{9}$ **41.** $\frac{1}{2}$ **43.** $\frac{1}{80}$ **45.** $\frac{51}{400}$ **47.** $\frac{43}{800}$ **49.** $\frac{1}{15}$

30–page 610 **1.** 9 **3.** 3.84 **5.** 7.8 **7.** 14.24 **9.** 3.486 **11.** 45.359 **13.** 10 **15.** 40 **17.** 6 **19.** 2.25 **21.** 128 **23.** 8.19 **25.** 15 **27.** 2 **29.** 80 **31.** 9.144 **33.** 25 **35.** 6.375

31–page 611 **1.** 50% **3.** 20% **5.** 35.7% **7.** 89.2% **9.** 69.9% **11.** 87.2% **13.** 150% **15.** 120% **17.** 133.3% **19.** 129.2% **21.** 400% **23.** 513.3% **25.** 80% **27.** 46.2% **29.** 55.1% **31.** 80% **33.** 120% **35.** 25% **37.** 138.9% **39.** 200% **41.** 66.7% **43.** 5% **45.** 146.7%

32–page 612 **1.** 72 **3.** 8048.4 **5.** 1600 **7.** 336.4 **9.** 60 **11.** 134 **13.** 160 **15.** 60 **17.** $90\frac{2}{3}$ **19.** $54\frac{86}{91}$ **21.** 75 **23.** 2.5 **25.** 52.5 **27.** 78 **29.** 180 **31.** 225 **33.** 55 **35.** $197\frac{1}{67}$

Applications File

Page 613 **Application A** **1.** 600 mi **3.** 448 m **5.** 210 m **7.** 1216 ft^2 **9.** 360 km^2 **11.** 31.4

13. 23.4 **15.** $437.95 **17.** $1429.00 **19.** $5.25 **21.** $46.19 **23.** $50.16

Page 614 **Application B** **1.** 308 in **3.** 8775 yd **5.** 8 h **7.** 30 s **9.** 30 **11.** 248 **13.** 20 **15.** 50

Page 615 **Application C** **1.** $1,200,000 **3.** $7,900,000 **5.** Bronx, Columbus, Dallas, Lincoln, San Diego **7.** Express plus **9.** **a.** Express **b.** Mon.

Page 616 **Application D** **1.** 1 q., 1 do., 1 five do. **3.** 1 p., 3 do., 1 five do. **5.** 2 p., 1 di., 1 q., 4 do. **7.** 2 di., 3 q., 4 do., 4 do. **9.** 1 p., 2 di., 2 q. **11.** 1 q., 2 do. **13.** 1 di., 1 do., 1 five do. **15.** 1 q., 2 do. **17.** 2 q., 1 do., 1 ten do.

Page 617 **Application E** **1.** 8:45 **3.** 7:00 **5.** 11:15 **7.** 12:15 **9.** 3:30 **11.** 8:45 **13.** 7:30 **15.** 9:45 **17.** 12:45 **19.** 6:00 **21.** 11:30 **23.** 10:15 **Application F** **1.** 2 h:46 min **3.** 3 h:15 min **5.** 51 min **7.** 5 h:15 min **9.** 8 h:20 min **11.** 10 h:45 min

Page 618 **Application G** **1.** 137 days **3.** 212 days **5.** 258 days **Application H** **1.** No **3.** Yes **5.** No **7.** Yes **9.** Yes **11.** Yes **Application I** **1.** 79 days **3.** 231 days **5.** 222 days

Page 619 **Application J** **1.** $\frac{3}{4}$ yr. **3.** 1 yr. **5.** 3 yr. **7.** $1\frac{3}{4}$ yr. **9.** $1\frac{1}{6}$ yr. **11.** $\frac{5}{6}$ yr. **13.** $\frac{35}{73}$ yr. **15.** $\frac{50}{73}$ yr. **Application K** **1.** 52 **3.** 16 **5.** 12 **7.** 6 **9.** 60

Page 620 **Application L** **1.** 0.85 **3.** 0.65 **5.** 0.20 **7.** 9% **9.** 53% **11.** 93% **13.** 96.5% **15.** 24.1% **Application M** **1.** $1300 **3.** $2100 **5.** $900 **7.** $400

Page 621 **Application N** **1.** $8000 **3.** $0 **Application O** **1.** 650 **3.** 350

Page 622 **Application P** **1.** $0.27 **3.** $110,160 **5.** 108°; 54°; 36°; 18°

Page 623 **Application Q** **1.** $54\frac{1}{3}$ **3.** $6726\frac{3}{4}$ **5.** $1,235,563 **Application R** **1.** $148\frac{1}{2}$ **3.** 88 **5.** 0.09 **7.** 2.08 **Application S** **1.** $24 **3.** None **5.** 0.01

Page 624 **Application T** **1.** m **3.** mm **5.** m **7.** mm **Application U** **1.** kg **3.** g **5.** t **7.** kL **9.** g

Page 625 **Application V** **1.** 10.5 cm **3.** 2.6 ft **5.** 162 mi **Application W** **1.** 12.56 mi **3.** 2.512 m **5.** 4.71 ft **7.** 56.52 ft **9.** 39.25 mi

Page 626 **Application X** **1.** 150 in^2 **3.** 0.675 m^2 **5.** $10\frac{1}{9}$ ft^2 **Application Y** **1.** 13.5 in^2 **3.** 200 m^2 **5.** 19.625 yd^2 **7.** 23,223.44 mm^2 **Application Z** **1.** 519,552 m^3 **3.** 396 cm^3 **5.** $157\frac{1}{4}$ yd^3

CREDITS

Mechanical art by The Mazer Corporation.

Photographs:

iii: John Henley/Digital Stock Connection
iv: Blair Seitz/Digital Stock Connection
v: Don Mason/The Stock Market
vi: Jon Riley/Tony Stone
vii: Jim Erickson/The Stock Market
viii: Dan Bosler/Tony Stone
ix: Glencoe Stock
x: Bill Holden/Leo de Wys, Inc.
xi: Robert Daemmrich/Digital Stock Connection
xii: R. Ian Lloyd/Premium/Westlight
xiii: Glencoe Stock
1: Don Mason/The Stock Market
75: Dan Bosler/Tony Stone
77: Courtesy of International Business Machines Corporation
99: Robert Daemmrich
125: Stephen Marks/The Image Bank
145: Superstock
167: Superstock
193: Chuck Savage/The Stock Market
217: Michael Salas/The Image Bank
239: Gabe Palmer/The Stock Market
265: Henley & Savage/TSW-Click
289: Ron Chapple/FPG
315: Elliott Varner Smith
339: Blair Seitz/Digital Stock Connection
341: Jack Holtel
363: Brian Seed/TSW-Click
383: Gabe Palmer/The Stock Market
403: Jack Holtel
425: Robert Daemmrich
449: Bernsau/The Image Works
469: Robert Daemmrich
487: Jack Holtel
511: John Coletti/TSW-Click
531: New York Stock Exchange
553: Brett Froomer/The Image Bank

669